Biochemistry

Biochemistry

(with Medical Concepts, Clinical Correlates & Case Studies)

6e

Dr. U. Satyanarayana

M.Sc., Ph.D., F.I.C., F.A.C.B., F.A.C.B.I

Professor of Biochemistry & Director (Research)
Dr. Pinnamaneni Siddhartha Institute of Medical Sciences
(Dr. NTR University of Health Sciences)
Chinaoutpalli, Gannavaram (Mdl)
Krishna (Dist), A.P., India

Dr. U. Chakrapani

M.B.B.S., M.S., D.N.B.

Co-published with

ELSEVIER

Since 1960 BOOKS AND ALLIED

BOOKS & ALLIED Pvt. Ltd.

Biochemistry, 6e
Satyanarayana and Chakrapani

ELSEVIER

RELX India Pvt. Ltd.

Published by RELX India Pvt. Ltd., under special arrangement with Books & Allied (P) Ltd.

First Published: March 1999
Revised Reprint: August 2000
Second Revised Edition: June 2002
Revised Reprint: 2004, 2005
Third Revised Edition (multicolour): 2006
Revised Reprint: 2007, 2010
Fourth Revised Edition: 2013
Fifth Revised Edition: 2017
Revised and Updated Edition: 2020
Sixth Revised Edition: 2021

ISBN: 978-81-312-6435-5

e-Book ISBN: 978-81-312-6436-2

This edition of **Biochemistry, 6e** by Dr. U. Satyanarayana and Dr. U. Chakrapani is co-published by an arrangement between RELX India Private Limited and Books & Allied (P) Ltd.

RELX India Pvt. Ltd.

Registered Office: 818, 8th Floor Indraprakash Building, 21, Barakhamba Road, New Delhi-110001.

Corporate Office: 14th Floor, Building No. 10B, DLF Cyber City, Phase II, Gurgaon-122002, Haryana, India.

BOOKS & ALLIED (P) Ltd.

Registered Office: No. 1-E(1) 'Shubham Plaza', 83/1 Beliaghata Main Road, Kolkata-700010, West Bengal, India.

Cover Design : Depicts the universal energy currency of the living world—ATP, predominantly synthesized by the mitochondria of the cell (the functional unit of life), in comparison with the international currencies—$, £, €, ₹, ¥.

Authors sponsored and supported by ***Uppala Author-Publisher Interlinks***, Vijayawada-520008

Typeset at Books & Allied (P) Ltd., Kolkata-700010

Printed and bound in India at Sanat Printers, Kundli, Haryana

Special Acknowledgements to the Sixth Edition

We gratefully acknowledge the following faculty members who have critically reviewed every page of our book and provided inputs to update comprehensive knowledge on Medical/Clinical orientation of Biochemistry in the Sixth Edition.

Dr. (Ms) U. B. Vijaya Lakshmi, MD (Biochemistry)
Professor of Biochemistry
N.R.I. Medical College
Chinakakani, Guntur, A.P.

Dr. E. Suresh, MD (Biochemistry)
Assistant Professor of Biochemistry
Dr. Pinnamaneni Siddhartha Institute of Medical Sciences
Chinaoutpalli, A.P.

Dr. (Ms) P. Chudamani, MD (General Medicine)
Assistant Professor of General Medicine
N.R.I. Medical College
Chinakakani, Guntur, A.P.

The contributions of MD (Biochemistry) and MD (General Medicine) Professors are unique and significant, providing relevant and detailed information on 'Biochemical perspectives to Medicine', as required by MBBS students.

Dr. U. SATYANARAYANA
Dr. U. CHAKRAPANI

Preface to the Sixth Edition

We proudly present the Sixth Edition of 'Biochemistry', now entering 22nd year of its highly successful existence. This book has undergone metamorphosis and become more and more mature with successive editions.

Our book 'Biochemistry' is one the most preferred text books (in India and many other countries) by students as well as teachers of medical and other allied sciences. Most learners regard our book as 'Bible of Biochemistry'.

A unique, special and significant feature of Sixth Edition is the involvement of MD (Biochemistry) and MD (General Medicine) Professors who have critically reviewed every page of the book and provided inputs to update comprehensive knowledge on Medical/Clinical orientation of Biochemistry, with particular reference to health and disease.

Biochemistry is the running thread in the entire medical curriculum. The Sixth Edition gives more emphasis to the relevance and importance of Biochemistry to understand diseases and practice medicine. Further, wherever relevant, the subject of Biochemistry was integrated with other subjects of medical sciences.

Perhaps, this is the only book that gives specific and desired information to satisfy the MCI/NMC curriculum with competency map (competency code, with chapters and pages).

Each chapter has been thoroughly revised and updated with most recent, relevant and pertinent information. Medical concepts and clinical correlates have been expanded and updated, wherever relevant, biochemical aspects related to COVID-19 pandemic are incorporated.

The Sixth Edition of the book (with complimentary e-book) is more student friendly with a 3R motto-read, remember and reproduce. The colour figures are attractively done for better understanding of chemical structures and biochemical reactions. Based on their curriculum requirements, the students have a choice to learn the subject with or without structures.

I consider myself as an eternal reader and a regular student of Biochemistry. However, it is beyond my capability to keep track of the ever-growing advances in Biochemistry due to exponential growth of the subject. And, this makes me nervous, whenever I think of revising the book. I honestly and frankly admit that I have to depend on mature readers for subsequent editions of this book.

AN INVITATION TO READERS, WELL WISHERS AND SUBJECT EXPERTS

I have to admit that it is not all the time possible for me to meet the readers individually and get their feedback. I sincerely invite the readers, (particularly the students), my well wishers and experts in Biochemistry subject to feel free and write to me (Email ID : usatya52@gmail.com) expressing their frank opinions, critical comments and constructive suggestions. And this will help me to further improve the book in subsequent revisions.

Dr. U. SATYANARAYANA

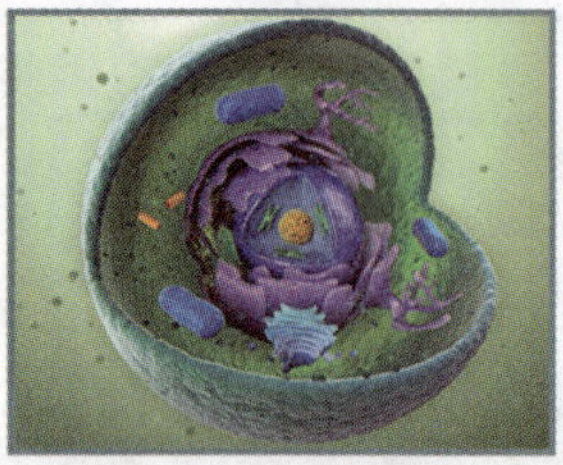

Preface to the First Edition

Biochemistry is perhaps the most fascinating subject as it deals with the chemical language of life, be it human, animal, plant or microorganism. No other science subject has as much application as biochemistry to the disciplines of medicine, health, veterinary, agriculture bioengineering and technology. This necessitates a totally different outlook for the books on biochemistry subject.

There are many biochemistry textbooks on the market. Some of them are purely basic while others are applied, and there are very few books which cover both these aspects together. For this reason, the students learning biochemistry in their undergraduate courses have to depend on multiple books to acquire a sound knowledge of the subject.

This book, 'Biochemistry' is unique with a simultaneous and equal emphasis on basic and applied aspects of biochemistry. This textbook primarily is an integration of medical and pure sciences, comprehensively written to meet the curriculum requirements of undergraduate courses in medical, dental, pharmacy, life-sciences and other categories (agriculture, veterinary, etc.) where students learn biochemistry as one of the subjects.

The tendency among the students (particularly medical) is to regard biochemistry as being mostly concerned with unimportant and complicated metabolic (chemical) pathways. This book gives a new orientation to the subject of biochemistry so that the students appreciate the great importance and significance of the application of biochemistry to medicine.

This book is designed to develop in students a sustained interest and enthusiasm to learn and develop the concepts in biochemistry in a logical and stepwise manner. It incorporates a variety of pedagogic aids, besides colour illustrations to help the students understand the subject quickly and to the maximum. The summary and biomedical/clinical concepts are intended for a rapid absorption and assimilation of the facts and concepts in biochemistry. The self-assessment exercises will stimulate the students to think rather than merely learn the subject. In addition, these exercises (essays, short notes, fill in the blanks, multiple choice questions) set at different difficulty levels, will cater to the needs of all the categories of learners.

It will not be out of place to mention here how-and when-the book was born. The entire book was written in the early morning hours (between 2 AM-6 AM; when the world around is fast asleep), during which period I carry out my intellectual activities. After a sound sleep, a fresh mind packed with creative ideas and innovative thoughts, has largely helped me to write this book. My wife pleaded with me that I should not write topics like diabetes, cancer, AIDS at home. In deference to her sentiment, I made a serious attempt to write those topics during my leisure time in the Department. But when I went through them in my serene mood of the early morning hours, I had to discard them in disappointment and rewrite them. Truly, each page of this book was conceived in darkness and born at daybreak!

This textbook is a distillation of my knowledge and teaching experience in biochemistry, acquired during the past 25 years. It contains predigested information on biochemistry for good understanding, assimilation and reproducibility. Each page is crafted with a fine eye. The ultimate purpose of this book is to equip the reader with comprehensive knowledge in biochemistry with reference to basic as well as applied aspects.

Although I have made every effort to make the book error free, I am under no illusion. I welcome comments, criticism and suggestions from the faculty, students and other readers, and this will help me make improvements in the next edition.

Dr. U. SATYANARAYANA

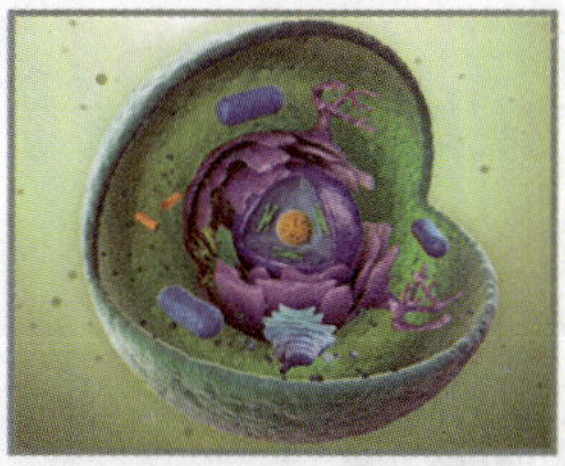

Acknowledgements to the Sixth Edition

I am highly grateful to the initiative, interest, enthusiasm, and dynamism of (late) Mr Arunabha Sen (Director, Books and Allied Pvt. Ltd., Kolkata) who was instrumental in publishing all my five books. We had 20 years of long and fruitful association. The sudden and premature demise of Mr Arunabha Sen (on 07-09-2018) remains a heavy professional and personal loss to me.

I thank Dr (Ms) U. B. Vijaya Lakshmi, MD, Professor of Biochemistry, NRI Medical College, Guntur, AP, who has thoroughly reviewed and significantly contributed to improve every page of the book with regard to the medical/clinically related Biochemistry, as per the MCI/NMC curriculum for MBBS students. I am grateful to Dr (Ms) E. Chudamani, MD, Assistant Professor of General Medicine, NRI Medical college, Guntur, AP. who reviewed the book and helped to improve the subject from clinicians' point of view. I acknowledge the contributions of Dr E. Suresh, MD, Assistant Professor of Biochemistry, Dr. PSIMS, Chinaoutpalli, in revising and proof-reading of the book.

I thank Ms Ruba Sen, Managing Director and Mr Depayan Sen, Director, Books and Allied Pvt. Ltd. for their support and encouragement to bring out the sixth edition of the book. I am indebted to Mr Shyamal Bhattacharya for his excellent page making and graphics work in the book. I am grateful to Mr Abhijit Ghosal for his help in cover design.

I express my gratitude to the officials of Elsevier (RELX India Pvt. Ltd.) for their constant support and encouragement – Mr Shanker Kaul, Managing Director; Mr Vidur Renjen, Director Education; Mr Sandeep Singh, Finance Manager; Ms Nimisha Goswami, Director Content Strategy – India/SEA; Ms Debyani Roy, Regional Manager IPR; Ms Shabina Nasim, Head of Content Project Management – APAC; Mr Arvind Kaul, Content Strategist; Mr Shubham Dixit, Content Project Manager; and Mr. Gopalakrishnan Venkatraman, Senior Graphic Artist.

I thank my wife Krishna Kumari, my younger son Amrut Pani and my daughters-in-law Hima Bindu and Oohasri for their constant encouragement. My special thanks to my grand daughter Maahe (10 years) and grand son Dev Sai (4 years) whose eversmiling faces, sweet words and deeds infuse energy into my academic activities. I am grateful to Uppala Author-Publisher Interlinks, Vijayawada for sponsoring and supporting me to bring out this edition.

Dr. U. SATYANARAYANA

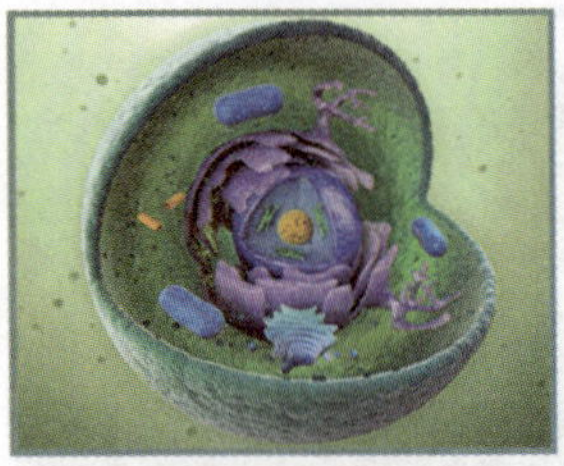

Acknowledgements to the First Edition

I owe a deep debt of gratitude to my parents, the late Sri U. Venkata Subbaiah, and Smt. Vajramma, for cultivating in me the habit of early rising. The writing of this book would never have been possible without this healthy habit. I am grateful to Dr. B. S. Narasinga Rao (former Director, National Institute of Nutrition, Hyderabad) for disciplining my professional life, and to my eldest brother Dr. U. Gudaru (former Professor of Power Systems, Walchand College of Engineering, Sangli) for disciplining my personal life.

My elder son, U. Chakrapani (MBBS) deserves a special place in this book. He made a significant contribution at every stage of its preparation—writing, verification, proof-reading and what not. I had the rare privilege of teaching my son as he happened to be a student of our college. And a major part of this book was written while he was learning biochemistry. Thus, he was the first person to learn the subject of biochemistry from my handwritten manuscript. The student-teacher relation (rather than the father-son) has helped me in receiving constant feedback from him and restructure the book in a way an undergraduate student would expect a biochemistry textbook to be.

Next, I thank Dr. G. Pitcheswara Rao (former Professor of Anatomy, SMC, Vijayawada) for his constructive criticism and advice, and Dr. B. Sivakumar (Director, National Institute of Nutrition, Hyderabad) for his helpful suggestions on the microfigures. I am grateful to my nephew, Mr. U. Srinivasa Rao, for helping me a drawing some figures. I acknowledge the help of my friend, Mr. P. Ramanujam (Lecturer in English, Andhra Loyola College, Viyayawada) for his encouragement and help at various states of manuscript preparation.

I am indebted to Dr. C. Nageswara Rao and Dr. C. Sudha (Pinnamaneni Polyclinic, Vijayawada) for their constant encouragement in my academic pursuits. I have been associated with them in research activities for over 15 years, the latest being as a Research Advisor to The KCP Neprho-Urological Centre and Research Foundation.

I express my gratitude to Mr. Arunabha Sen (Executive Director, Books & Allied (P) Ltd. and New Central Book Agency (P) Ltd., Calcutta) for readily conceding my request to publish the book in colour and taking all pains to bring it out to my satisfaction. I thank Mr. Abhijit Ghosal and M. Shyamal Bhattacharya for their excellent page-making and graphics-work in this book. I also thank Mr. Dinesh Bhunya and Mr. Subas Maitra for proof-reading and technical editing and to Mr. Sudip Mukherjee and Mr. Tarun Dey for data entry. I am also indebted to Mr. Bhaskar Das for cover design of this book.

Last but not least, I thank my wife Krishna Kumari and my younger son, Amrutpani, without whose cooperation and encouragement this book could never have been written. The manuscript was carefully nurtured like a new born baby and the book has now become a full-pledged member of our family.

Dr. U. SATYANARAYANA

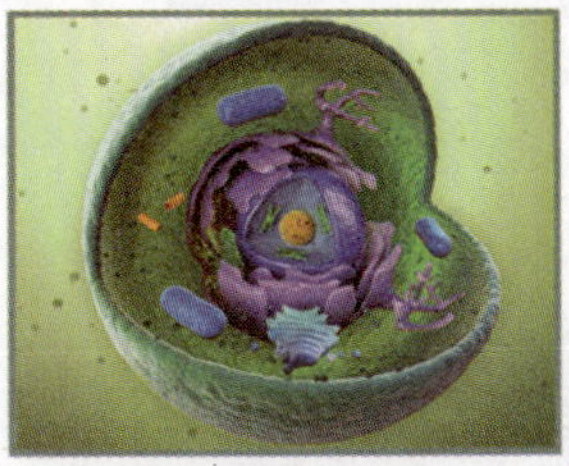

Scope of Biochemistry

The term Biochemistry was introduced by Carl Neuberg in 1903. Biochemistry broadly deals with the chemistry of life and living processes. There is no exaggeration in the statement, 'The scope of biochemistry is as vast as life itself!' Every aspect of life-birth, growth, reproduction, aging and death, involves biochemistry. For that matter, every movement of life is packed with hundreds of biochemical reactions. Biochemistry is the most rapidly developing and most innovative subject in medicine. This becomes evident from the fact that over the years, the major share of Nobel Prizes earmarked for Medicine and Physiology has gone to researchers engaged in biochemistry.

The discipline of biochemistry serves as a torch light to trace the intricate complexicities of biology, besides unravelling the chemical mysteries of life. Biochemical research has amply demonstrated that all living things are closely related at the molecular level. Thus biochemistry is the subject of unity in the diversified living kingdom.

Advances in biochemistry have tremendous impact on human welfare, and have largely benefited mankind and their living styles. These include the application of biochemistry in the laboratory for the diagnosis of diseases, the products (insulin, interferon, growth hormone etc.) obtained from genetic engineering, and the possible use of gene therapy in the near future.

ORGANIZATION OF THE BOOK

This textbook, comprising 43 chapters, is organized into seven sections in the heirarchical order of learning biochemistry.

- Section I deals with the chemical constituents of life—carbohydrates, lipids, proteins and amino acids, nucleic acids and enzymes.
- Section II physiological chemistry includes digestion and absorption, plasma proteins, hemoglobin and prophyrins, and biological oxidation.
- Section III incorporates all the metabolisms (carbohydrates, lipids, amino acids, nucleotides, minerals)
- Section IV covers hormones, organ function tests, water, electrolyte and acid-base balance, tissue proteins and body fluids, and nutrition.
- Section V is exclusively devoted to molecular biology and biotechnology (DNA-replication, recombination, and repair, transcription and translation, regulation of gene expression, recombinant DNA and biotechnology)
- Section VI gives relevant information on current topics such as human genome project, gene therapy, bioinformatics, prostaglandins, diabetes, cancer, AIDS etc.
- Section VII deals with the basic aspects for learning and understanding biochemistry (bioorganic chemistry, biophysical chemistry, tools of biochemistry, genetics, immunology).

Each chapter in this book is carefully crafted with colour illustrations, headings and subheadings to facilitate quick understanding. The important applications of biochemistry to human health and disease are put together as biomedical/clinical concepts. Icons are used at appropriate places to serve as 'landmarks'.

The origins of biochemical words, confusables in biochemistry, practical biochemistry and clinical biochemistry laboratory, case studies with biochemical correlations, given in the appendix are novel features.

The book is so organized as to equip the readers with a comprehensive knowledge of biochemistry, with special reference to biochemical perspectives to medicine.

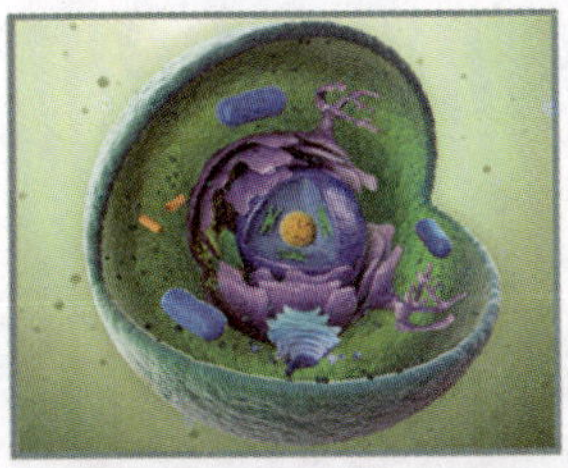

Contents

Competency map

BIOCHEMISTRY (CODE BI)

Topic: Basic Biochemistry

Competency Code	Competency	Core (Y/N)	Chapter Nos.	Page Nos.
BI1.1	Describe the molecular and functional organization of a cell and its sub-cellular components.	Y	1	3-8

Topic: Enzyme

BI2.1	Explain fundamental concepts of enzyme, isoenzyme, alloenzyme, coenzyme & co-factors. Enumerate the main classes of IUBMB nomenclature.	Y	6	87–117
BI2.2	Observe the estimation of SGOT & SGPT	Y	App-IV	732
BI2.3	Describe and explain the basic principles of enzyme activity	Y	6	90–94
BI2.4	Describe and discuss enzyme inhibitors as poisons and drugs and as therapeutic enzymes	Y	6	94–98
BI2.5	Describe and discuss the clinical utility of various serum enzymes as markers of pathological conditions.	Y	6	110–117
BI2.6	Discuss use of enzymes in laboratory investigations (Enzyme-based assays)	Y	6	110–117
BI2.7	Interpret laboratory results of enzyme activities & describe the clinical utility of various enzymes as markers of pathological conditions.	Y	6	110–117

Topic: Chemistry and Metabolism of Carbohydrates

Competency Code	Competency	Core (Y/N)	Chapter Nos.	Page Nos.
BI3.1	Discuss and differentiate monosaccharides, di-saccharides and polysaccharides giving examples of main carbohydrates as energy fuel, structural element and storage in the human body	Y	2	9–27
BI3.2	Describe the processes involved in digestion and assimilation of carbohydrates and storage.	Y	8	167–171
BI3.3	Describe and discuss the digestion and assimilation of carbohydrates from food.	Y	8	167–171
BI3.4	Define and differentiate the pathways of carbohydrate metabolism, (glycolysis, gluconeogenesis, glycogen metabolism, HMP shunt).	Y	13	242–243
BI3.5	Describe and discuss the regulation, functions and integration of carbohydrate along with associated diseases/disorders.	Y	13, 36	242–283, 651–656
BI3.6	Describe and discuss the concept of TCA cycle as a amphibolic pathway and its regulation.	Y	13	253–258
BI3.7	Describe the common poisons that inhibit crucial enzymes of carbohydrate metabolism (eg; fluoride, arsenate)	Y	13	256
BI3.8	Discuss and interpret laboratory results of analytes associated with metabolism of carbohydrates.	Y	13, 36	242–283, 646–656
BI3.9	Discuss the mechanism and significance of blood glucose regulation in health and disease.	Y	36	646–649
BI3.10	Interpret the results of blood glucose levels and other laboratory investigations related to disorders of carbohydrate metabolism.	Y	36	646–656

Topic: Chemistry and Metabolism of Lipids

BI4.1	Describe and discuss main classes of lipids (Essential/non-essential fatty acids, cholesterol and hormonal steroids, triglycerides, major phospholipids and sphingolipids) relevant to human system and their major functions.	Y	3	29–43

Competency Code	Competency	Core (Y/N)	Chapter Nos.	Page Nos.
BI4.2	Describe the processes involved in digestion and absorption of dietary lipids and also the key features of their metabolism	Y	8, 14	174-178, 285-329
BI4.3	Explain the regulation of lipoprotein metabolism & associated disorders.	Y	14	319–324
BI4.4	Describe the structure and functions of lipoproteins, their functions, interrelations & relations with atherosclerosis	Y	14	319–324
BI4.5	Interpret laboratory results of analytes associated with metabolism of lipids	Y	14	286, 316–318
BI4.6	Describe the therapeutic uses of prostaglandins and inhibitors of eicosanoid synthesis.	Y	32	618–622
BI4.7	Interpret laboratory results of analytes associated with metabolism of lipids.	Y	14	286, 316–318

Topic: Chemistry and Metabolism of Proteins

BI5.1	Describe and discuss structural organization of proteins.	Y	4	55–63
BI5.2	Describe and discuss functions of proteins and structure-function relationships in relevant areas eg, hemoglobin and selected hemoglobinopathies	Y	4, 10	67, 197–206
BI5.3	Describe the digestion and absorption of dietary proteins.	Y	8	171–174
BI5.4	Describe common disorders associated with protein metabolism.	Y	5	342–377
BI5.5	Interpret laboratory results of analytes associated with metabolism of proteins.	Y	4, 15	65, 377

Topic: Metabolism and homeostasis

BI6.1	Discuss the metabolic processes that take place in specific organs in the body in the fed and fasting states.	Y	16	382–388
BI6.2	Describe and discuss the metabolic processes in which nucleotides are involved.	Y	17	389–404
BI6.3	Describe the common disorders associated with nucleotide metabolism.	Y	17	398–399, 402
BI6.4	Discuss the laboratory results of analytes associated with gout & Lesch Nyhan syndrome.	Y	17	395–399

Competency Code	Competency	Core (Y/N)	Chapter Nos.	Page Nos.
BI6.5	Describe the biochemical role of vitamins in the body and explain the manifestations of their deficiency	Y	7	119–163
BI6.6	Describe the biochemical processes involved in generation of energy in cells.	Y	11	219–235
BI6.7	Describe the processes involved in maintenance of normal pH, water & electrolyte balance of body fluids and the derangements associated with these.	Y	21	469–485
BI6.8	Discuss and interpret results of Arterial Blood Gas (ABG) analysis in various disorders.	Y	21	481–485
BI6.9	Describe the functions of various minerals in the body, their metabolism and homeostasis.	Y	18	405–425
BI6.10	Enumerate and describe the disorders associated with mineral metabolism.	Y	18	407–425
BI6.11	Describe the functions of haem in the body and describe the processes involved in its metabolism and describe porphyrin metabolism.	Y	10	407–425
BI6.12	Describe the major types of haemoglobin and its derivatives found in the body and their physiological/ pathological relevance.	Y	10	196–217
BI6.13	Describe the functions of the kidney, liver, thyroid and adrenal glands.	Y	19, 20	196–217
BI6.14	Describe the tests that are commonly done in clinical practice to assess the functions of these organs (kidney, liver, thyroid and adrenal glands).	Y	19, 20	454–467, 443–447
BI6.15	Describe the abnormalities of kidney, liver, thyroid and adrenal glands.	Y	19, 20	461–464, 457–460, 441–443, 446, 447

Topic: Molecular Biology

Competency Code	Competency	Core (Y/N)	Chapter Nos.	Page Nos.
BI7.1	Describe the structure and functions of DNA and RNA and outline the cell cycle.	Y	5, 24	72–85 523–530
BI7.2	Describe the processes involved in replication & repair of DNA and the transcription & translation mechanisms.	Y	24, 25	523–538 539–559

Competency Code	Competency	Core (Y/N)	Chapter Nos.	Page Nos.
BI7.3	Describe gene mutations and basic mechanism of regulation of gene expression.	Y	24, 26	534–535 561–570
BI7.4	Describe applications of molecular technologies like recombinant DNA technology, PCR in the diagnosis and treatment of diseases with genetic basis.	Y	27	571–593
BI7.5	Describe the role of xenobiotics in disease	Y	31	613–617
BI7.6	Describe the anti-oxidant defence systems in the body.	Y	34	629–634
BI7.7	Describe the role of oxidative stress in the pathogenesis of conditions such as cancer, complications of diabetes mellitus and atherosclerosis.	Y	34	629–634

Topic: Nutrition

BI8.1	Discuss the importance of various dietary components and explain importance of dietary fibre.	Y	23	500–514
BI8.2	Describe the types and causes of protein energy malnutrition and its effects.	Y	23	514–515
BI8.3	Provide dietary advice for optimal health in childhood and adult, in disease conditions like diabetes mellitus, coronary artery disease and in pregnancy.	Y	23	513–514, 518
BI8.4	Describe the causes (including dietary habits), effects and health risks associated with being overweight/obesity.	Y	23	515–518
BI8.5	Summarize the nutritional importance of commonly used items of food including fruits and vegetables.(macro-molecules & its importance)	Y	23	500–513

Topic: Extracellular Matrix

BI9.1	List the functions and components of the extracellular matrix (ECM).	-	22	487–494
BI9.2	Discuss the involvement of ECM components in health and disease.	Y	22	487–494
BI9.3	Describe protein targeting & sorting along with its associated disorders.		25	558–559

Topic: Oncogenesis and Immunity

Competency Code	Competency	Core (Y/N)	Chapter Nos.	Page Nos.
BI10.1	Describe the cancer initiation, promotion oncogenes & oncogene activation. Also focus on p53 & apoptosis	Y	37	657–662
BI10.2	Describe various biochemical tumor markers and the biochemical basis of cancer therapy.	Y	37	662–665
BI10.3	Describe the cellular and humoral components of the immune system & describe the types and structure of antibody	Y	42, 9	697-701, 187-190
BI10.4	Describe & discuss innate and adaptive immune responses, self/non-self recognition and the central role of T-helper cells in immune responses.	Y	42	697–201
BI10.5	Describe antigens and concepts involved in vaccine development.	Y	42, 27	695, 588–589

Topic: Biochemical Laboratory Tests

BI11.2	Describe the preparation of buffers and estimation of pH.	Y	40	678
BI11.3	Describe the chemical components of normal urine.	Y	Inside back cover	–
BI11.4	Perform urine analysis to estimate and determine normal and abnormal constituents	Y	App-IV	729–730
BI11.5	Describe screening of urine for inborn errors & describe the use of paper chromatography	Y	41	688–689
BI11.6	Describe the principles of colorimetry	Y	41	691–693
BI11.7	Demonstrate the estimation of serum creatinine and creatinine clearance	Y	App–IV	731
BI11.8	Demonstrate estimation of serum proteins, albumin and A:G ratio	Y	App-IV	731
BI11.9	Demonstrate the estimation of serum total cholesterol and HDL- cholesterol	Y	App-IV	731
BI11.11	Demonstrate estimation of calcium and phosphorous	Y	App-IV	732

Competency Code	Competency	Core (Y/N)	Chapter Nos.	Page Nos.
BI11.12	Demonstrate the estimation of serum bilirubin	Y	App-IV	731
BI11.13	Demonstrate the estimation of SGOT/ SGPT	Y	App-IV	732
BI11.14	Demonstrate the estimation of alkaline phosphatase	Y	App-IV	732
BI11.15	Describe & discuss the composition of CSF	Y	22	496-497
BI11.16	Observe use of commonly used equipments/techniques in biochemistry laboratory including:			
	• pH meter	Y	41	687–696
	• Paper chromatography of amino acid	Y	41	688–689
	• Protein electrophoresis	Y	41, 9	690–691, 182–183
	• TLC, PAGE	Y	41	689–690
	• Electrolyte analysis by ISE	Y	–	–
	• ABG analyzer	Y	21	684–685
	• ELISA	Y	41	695–696
	• Immunodiffusion	Y	–	–
	• Autoanalyser	Y	App-V	737
	• Quality control	Y	App-V	736–737
	• DNA isolation from blood/ tissue	Y	–	–
BI11.17	Explain the basis and rationale of biochemical tests done in the following conditions:			
	• diabetes mellitus,	Y	36	651–656
	• dyslipidemia	Y	14	323–324
	• myocardial infarction,	Y	6	115
	• renal failure, gout,	Y	20, 17	461–464, 395–396
	• proteinuria,	Y	9, App-IV	185, 729
	• nephrotic syndrome,	Y	9, 14	183–185, 317
	• edema,	Y	9, 20	185, 515
	• jaundice,	Y	10, 20	215–216, 457–458
	• liver diseases, pancreatitis, disorders of acid-base balance,	–	8, 20, 21	180, 457–458, 481–483
	• thyroid disorders.	–	19	441–442

Competency Code	Competency	Core (Y/N)	Chapter Nos.	Page Nos.
BI11.18	Discuss the principles of spectrophotometry.	Y	41	691–693
BI11.19	Outline the basic principles involved in the functioning of instruments commonly used in a biochemistry laboratory and their applications.	Y	41	687–696
BI11.20	Identify abnormal constituents in urine, interpret the findings and correlate these with pathological states.	Y	App-IV	729–730
BI11.22	Calculate albumin: globulin (AG) ratio and creatinine clearance	Y	9, 20	183, 462–463
BI11.23	Calculate energy content of different food Items, identify food items with high and low glycemic index and explain the importance of these in the diet	Y	23	501, 506
BI11.24	Enumerate advantages and/or disadvantages of use of unsaturated, saturated and trans fats in food.	Y	23	508–509

CHEMICAL CONSTITUENTS OF LIFE

SECTION ONE

Biomolecules and the Cell

1

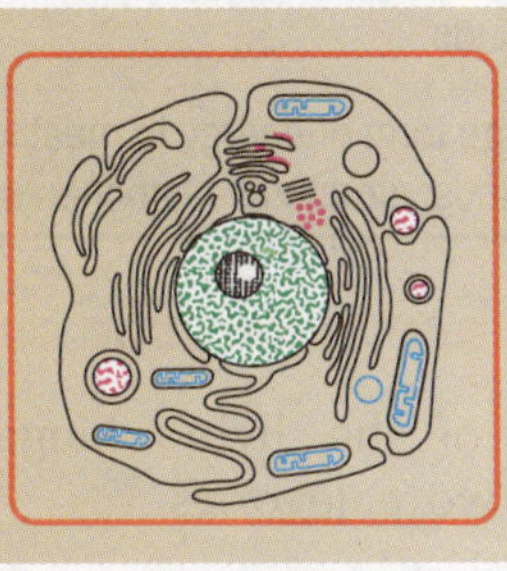

The cell speaks

"I am the unit of biological activity;
Organized into subcellular organelles;
Assigned to each are specific duties;
Thus, I truly represent life!"

The living matter is composed of mainly six elements—***carbon***, ***hydrogen***, ***oxygen***, ***nitrogen***, ***phosphorus*** and ***sulfur***. These elements together constitute about 90% of the dry weight of the human body. Several other functionally important elements are also found in the cells. These include Ca, K, Na, Cl, Mg, Fe, Cu, Co, I, Zn, F, Mo and Se.

Carbon—a unique element of life

Carbon is the most predominant and versatile element of life. It possesses a unique property to ***form infinite number of compounds***. This is attributed to the ability of carbon to form stable covalent bonds and C−C chains of unlimited length. It is estimated that about ***90% of compounds*** found in living system invariably ***contain carbon***.

Chemical molecules of life

Life is composed of lifeless chemical molecules. A single cell of the bacterium, *Escherichia coli* contains about 6,000 different organic compounds. It is believed that man may contain about 100,000 different types of molecules, although only a few of them have been characterized.

Complex biomolecules

The organic compounds such as ***amino acids***, ***nucleotides*** and ***monosaccharides*** serve as the ***monomeric units*** or building blocks of complex biomolecules—***proteins***, ***nucleic acids*** (DNA and RNA) and ***polysaccharides***, respectively. The important biomolecules (macromolecules) with their respective building blocks and major functions are given in ***Table 1.1***. As regards lipids, it may be noted that they are not biopolymers in a strict sense, but majority of them contain fatty acids.

Structural hierarchy of an organism

The ***macromolecules*** (proteins, lipids, nucleic acids and polysaccharides) form ***supramolecular assemblies*** (e.g. membranes) which in turn organize into organelles, cells, tissues, organs and finally the whole organism.

Table 1.1 The major complex biomolecules of cells

Biomolecule	Building block (repeating unit)	Major functions
1. Protein	Amino acids	Fundamental basis of structure and function of cell (static and dynamic functions).
2. Deoxyribonucleic acid (DNA)	Deoxyribonucleotides	Repository of hereditary information.
3. Ribonucleic acid (RNA)	Ribonucleotides	Essentially required for protein biosynthesis.
4. Polysaccharide (glycogen)	Monosaccharides (glucose)	Storage form of energy to meet short term demands.
5. Lipid	Fatty acids, glycerol	Storage form of energy to meet long term demands; structural components of membranes.

Chemical composition of man

The chemical composition of a normal man, weighing 65 kg, is given in ***Table 1.2***. Water is ***the solvent of life*** and contributes to more than 60% of the body weight. This is followed by protein (mostly in muscle) and lipid (mostly in adipose tissue). The carbohydrate content is rather low which is in the form of glycogen.

THE CELL

The cell is the structural and functional unit of life. It may be also regarded as the ***basic unit of biological activity.***

The concept of cell originated from the contributions of Schleiden and Schwann (1838). However, it was only after 1940, the complexities of cell structure were exposed.

Table 1.2 Chemical composition of a normal man (weight 65 kg)

Constituent	Percent (%)	Weight (kg)
Water	61.6	40
Protein	17.0	11
Lipid	13.8	9
Carbohydrate	1.5	1
Minerals	6.1	4

Prokaryotic and eukaryotic cells

The cells of the living kingdom may be divided into two categories

1. **Prokaryotes** (*Greek* : pro – before; karyon – nucleus) ***lack a well defined nucleus*** and possess relatively simple structure. These include the various bacteria.

2. **Eukaryotes** (*Greek* : eu – true; karyon – nucleus) possess a ***well defined nucleus*** and are more complex in their structure and function. The higher organisms (animals and plants) are composed of eukaryotic cells.

A comparison of the characteristics between prokaryotes and eukaryotes is listed in ***Table 1.3***.

EUKARYOTIC CELL

The human body is composed of about 10^{14} cells. There are about 250 types of specialized cells in the human body e.g. erythrocytes, nerve cells, muscle cells, β cells of pancreas. An eukaryotic cell is generally 10 to 100 μm in diameter. A diagrammatic representation of a typical rat liver cell is depicted in ***Fig.1.1***.

The plant cell differs from an animal cell by possessing a rigid cell wall (mostly composed of cellulose) and chloroplasts. The latter are the sites of photosynthesis.

Table 1.3 Comparison between prokaryotic and eukaryotic cells

Characteristic	Prokaryotic cell	Eukaryotic cell
1. Size	Small (generally 1-10 μm)	Large (generally 10-100 μm)
2. Cell membrane	Cell is enveloped by a rigid cell wall	Cell is enveloped by a flexible plasma membrane
3. Sub-cellular organelles	Absent	Distinct organelles are found (e.g. mitochondria, nucleus, lysosomes)
4. Nucleus	Not well defined; DNA is found as nucleoid, histones are absent	Nucleus is well defined, surrounded by a membrane; DNA is associated with histones
5. Energy metabolism	Mitochondria absent, enzymes of energy metabolism bound to membrane	Enzymes of energy metabolism are located in mitochondria
6. Cell division	Usually fission and no mitosis	Mitosis
7. Cytoplasm	Organelles and cytoskeleton absent	Contains organelles and cytoskeleton (a network of tubules and filaments)
8. Examples	Bacteria	Animals and plants

The cell consists of well defined subcellular organelles, enveloped by a plasma membrane. By differential centrifugation of tissue homogenate, it is possible to isolate each cellular organelle in a relatively pure form (***Refer Chapter 41***). The distribution of major enzymes and metabolic pathways in different cellular organelles is given in the chapter on enzymes (***Refer Table 6.7***). The subcellular organelles are briefly described in the following pages.

Nucleus

Nucleus is the ***largest cellular organelle***, surrounded by a double membrane nuclear envelope. The outer membrane is continuous with the membranes of endoplasmic reticulum. At certain intervals, the two nuclear membranes have nuclear pores with a diameter of about 90 nm. These pores permit the free passage of the products synthesized in the nucleus into the surrounding cytoplasm.

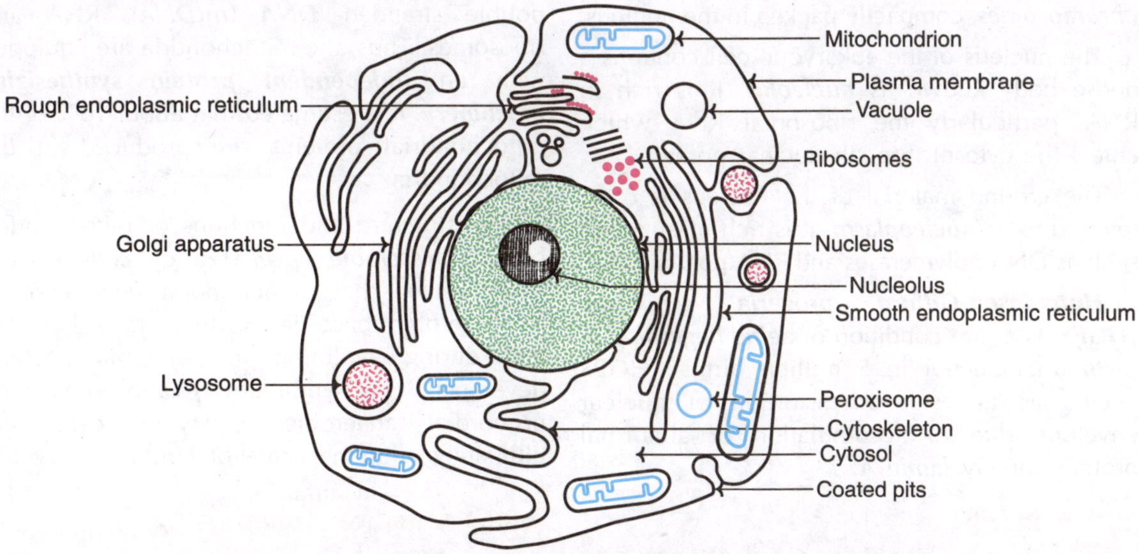

***Fig. 1.1 :** Diagrammatic representation of a rat liver cell.*

Medical Concepts/ Clinical Correlates

- Accumulation of lipofuscin, a pigment rich in lipids and proteins, in the cell has been implicated in ageing process.
- Leakage of lysosomal enzymes into the cell degrades several functional macromolecules and this may lead to certain disorders (e.g. arthritis).
- Zellweger syndrome is a rare disease, characterized by the absence of functional peroxisomes.
- Defect of a cytoskeleton protein, namely spectrin in RBC results in hereditary sperocytosis. This results in breakdown of RBC, leading to anemia.

Nucleus contains ***DNA***, the repository of genetic information. Eukaryotic DNA is associated with basic protein (histones) in the ratio of 1 : 1, to form ***nucleosomes***. An assembly of nucleosomes constitutes ***chromatin*** fibres of chromosomes (*Greek*: chroma – colour; soma – body). Thus, a single human chromosome is composed of about a million nucleosomes. The number of chromosomes is a characteristic feature of the species. ***Humans have 46 chromosomes***, compactly packed in the nucleus.

The nucleus of the eukaryotic cell contains a dense body known as ***nucleolus***. It is rich in RNA, particularly the ribosomal RNA which enters the cytosol through nuclear pores.

The ground material of the nucleus is often referred to as ***nucleoplasm***. It is rich in enzymes such as DNA polymerases and RNA polymerases.

Hutchinson-Gilford progeria syndrome (HGPS) is a rare condition of ***aging beginning at birth*** (incidence I in 5 million births). HGPS occurs as a result of distortion of nuclear envelope due to accumulation of abnormal protein namely ***lamin A***.

Mitochondria

The mitochondria (*Greek*: mitos – thread; chondros – granule) are the centres for the cellular respiration and energy metabolism. They are regarded as the ***power houses of the cell*** with variable size and shape. Mitochondria are rod-like or filamentous bodies, usually with dimensions of 1.0 × 3 µm. About 2,000 mitochondria, occupying about $^1/_5$th of the total cell volume, are present in a typical cell.

The mitochondria are composed of a double membrane system (***Refer Fig.11.5***). The outer membrane is smooth and completely envelops the organelle. The inner membrane is folded to form ***cristae*** (*Latin* – crests) which occupy a larger surface area. The internal chamber of mitochondria is referred to as ***matrix*** or ***mitosol***.

The components of ***electron transport chain*** and ***oxidative phosphorylation*** (flavoprotein, cytochromes b, c_1, c, a and a_3 and coupling factors) are buried in the inner mitochondrial membrane. The matrix contains several enzymes concerned with the energy metabolism of carbohydrates, lipids and amino acids (e.g., ***citric acid cycle, β-oxidation***). The matrix enzymes also participate in the synthesis of heme and urea. Mitochondria are the ***principal producers of ATP*** in the aerobic cells. ATP, the energy currency, generated in mitochondria is exported to all parts of the cell to provide energy for the cellular work.

The ***mitochondrial*** matrix contains a circular double stranded ***DNA*** (***mtDNA***), RNA and ribosomes. Thus, the mitochondria are equipped with an ***independent protein synthesizing machinery***. It is estimated that about 10% of the mitochondrial proteins are produced in the mitochondria.

The structure and functions of mitochondria closely ***resemble prokaryotic cells***. It is hypothesized that mitochondria have evolved from aerobic bacteria. Further, it is believed that during evolution, the aerobic bacteria developed a symbiotic relationship with primordial anaerobic eukaryotic cells that ultimately led to the arrival of aerobic eukaryotes.

Endoplasmic reticulum

The ***network of membrane enclosed spaces*** that extends throughout the cytoplasm constitutes endoplasmic reticulum (ER). Some of these thread-like structures extend from the nuclear pores to the plasma membrane.

A large portion of the ER is studded with ribosomes to give a granular appearance which is referred to as ***rough endoplasmic reticulum***. ***Ribosomes are the factories of protein biosynthesis***. During the process of cell fractionation, rough ER is disrupted to form small vesicles known as ***microsomes***. It may be noted that microsomes as such do not occur in the cell.

The smooth endoplasmic reticulum does not contain ribosomes. It is involved in the synthesis of lipids (triacylglycerols, phospholipids, sterols) and metabolism of drugs, besides supplying Ca^{2+} for the cellular functions.

Golgi apparatus

Eukaryotic cells contain a unique cluster of ***membrane vesicles*** known as ***dictyosomes*** which, in turn, constitute Golgi apparatus (or Golgi complex). The newly synthesized proteins are handed over to the Golgi apparatus which catalyse the addition of carbohydrates, lipids or sulfate moieties to the proteins. These chemical modifications are necessary for the transport of proteins across the plasma membrane.

Certain proteins and enzymes are enclosed in membrane vesicles of Golgi apparatus and secreted from the cell after the appropriate signals. The ***digestive enzymes of pancreas*** are produced in this fashion.

Golgi apparatus are also involved in the ***membrane synthesis***, particularly for the formation of intracellular organelles (e.g. peroxisomes, lysosomes).

Lysosomes

Lysosomes are spherical vesicles enveloped by a single membrane. Lysosomes are regarded as the ***digestive tract of the cell***, since they are actively involved in digestion of cellular substances—namely proteins, lipids, carbohydrates and nucleic acids. Lysosomal enzymes are categorized as ***hydrolases***. These include the enzymes (with substrate in brackets)—α-glucosidase (glycogen), cathepsins (proteins), lipases (lipids), ribonucleases (RNA).

The lysosomal enzymes are responsible for ***maintaining the cellular compounds in a dynamic state***, by their degradation and recycling. The degraded products leave the lysosomes, usually by diffusion, for reutilization by the cell. Sometimes, however, certain residual products, rich in lipids and proteins, collectively known as ***lipofuscin*** accumulate in the cell. Lipofuscin is the ***age pigment*** or wear and tear pigment which has been implicated in ageing process. As the cell dies, the lysosomes rupture and release hydrolytic enzymes that results in post-mortem autolysis.

The digestive enzymes of cellular compounds are confined to the lysosomes in the best interest of the cell. Escape of these enzymes into cytosol will destroy the functional macromolecules of the cell and result in many complications. The occurrence of several diseases (e.g. ***arthritis***, muscle diseases, allergic disorders) has been partly attributed to the ***release of lysosomal enzymes***.

Inclusion cell (I-cell) disease is a rare condition due to the absence of certain hydrolases in lysosomes. However, these enzyme are synthesized and found in the circulation. I-cell disease is due to ***a defect in protein targetting***, as the ***enzymes cannot reach lysosomes***.

Peroxisomes

Peroxisomes, also known as ***microbodies***, are single membrane cellular organelles. They are spherical or oval in shape and contain the enzyme ***catalase***. Catalase protects the cell from the toxic effects of H_2O_2 by converting it to H_2O and O_2. Peroxisomes are also involved in the oxidation of long chain fatty acids ($> C_{18}$), and synthesis of plasmalogens and glycolipids. Plants contain ***glyoxysomes***, a specialized type of peroxisomes, which are involved in the glyoxylate pathway.

Peroxisome biogenesis disorders (PBDs), are a group of rare diseases involving the enzyme activities of peroxisomes. The biochemical abnormalities associated with PBDs include increased levels of very long chain fatty acids (C_{24} and C_{26}) and decreased concentrations of plasmalogens. The most severe form of PBDs is ***Zellweger syndrome***, a condition characterized by the absence of functional peroxisomes. The victims of this disease may die within one year after birth.

Cytosol and cytoskeleton

The ***cellular matrix*** is collectively referred to as cytosol. Cytosol is basically a compartment containing several enzymes, metabolites and salts in an aqueous gel like medium. More recent studies however, indicate that the cytoplasm actually contains a complex network of protein filaments, spread throughout, that constitutes ***cytoskeleton***. The cytoplasmic filaments are of three types – ***microtubules***, actin filaments and intermediate filaments. The filaments which are polymers of proteins are responsible for the structure, shape and organization of the cell.

INTEGRATION OF CELLULAR FUNCTIONS

The eukaryotic cells perform a wide range of complex reactions/functions to maintain tissues, and for the ultimate well-being of the whole organism. For this purpose, the various intracellular processes and biochemical reactions are tightly controlled and integrated. Division of a cell into two daughter cells is good example of the orderly occurrence of an integrated series of cellular reactions.

Apoptosis is the ***programmed cell death*** or ***cell suicide***. This occurs when the cell has fulfilled its biological functions. Apoptosis may be regarded as a ***natural cell death*** and it differs from the cell death caused by injury due to radiation, anoxia etc. Programmed cell death is a highly regulated process.

SUMMARY

- *Life is composed of lifeless chemical molecules. The complex biomolecules, proteins, nucleic acids (DNA and RNA), polysaccharides and lipids are formed by the monomeric units amino acids, nucleotides, monosaccharides and fatty acids, respectively.*
- *The cell is the structural and functional unit of life. The eukaryotic cell consists of well defined subcellular organelles, enveloped in a plasma membrane.*
- *The nucleus contains DNA, the repository of genetic information. DNA, in association with proteins (histones), forms nucleosomes which, in turn, make up the chromosomes.*
- *The mitochondria are the centres for energy metabolism. They are the principal producers of ATP which is exported to all parts of the cell to provide energy for cellular work.*
- *Endoplasmic reticulum (ER) is the network of membrane enclosed spaces that extends throughout the cytoplasm. ER studded with ribosomes, the factories of protein biosynthesis, is referred to as rough ER. Golgi apparatus are a cluster of membrane vesicles to which the newly synthesized proteins are handed over for further processing and export.*
- *Lysosomes are the digestive bodies of the cell, actively involved in the degradation of cellular compounds. Peroxisomes contain the enzyme catalase that protects the cell from the toxic effects of* H_2O_2*. The cellular ground matrix is referred to as cytosol which, in fact, is composed of a network of protein filaments, the cytoskeleton.*
- *The eukaryotic cells perform a wide range of complex functions in a well coordinated and integrated fashion. Apoptosis is the process of programmed cell death or cell suicide.*

Carbohydrates

2

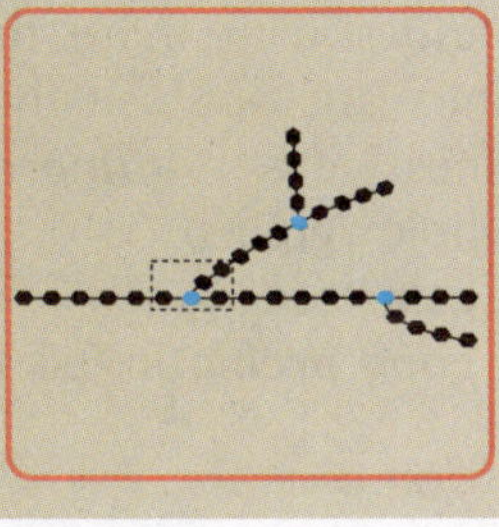

The carbohydrates speak

"We are polyhydroxyaldehydes or ketones;
Classified into mono-, oligo- and polysaccharides;
Held together by glycosidic bonds;
Supply energy and serve as structural constituents."

Carbohydrates are the most abundant organic molecules in nature. They are primarily composed of the elements ***carbon***, ***hydrogen* and *oxygen***. The name carbohydrate literally means '***hydrates of carbon***'.

Carbohydrates may be defined as polyhydroxyaldehydes or ketones or compounds which produce them on hydrolysis. The term 'sugar' is applied to carbohydrates soluble in water and sweet to taste.

The term ***glycome*** represents the entire ***spectrum of carbohydrates of an organism***, present in either free or complex form of molecules. ***Glycomics*** is the study of glycomes that includes physiological, pathological, genetic and various other aspects. Broadly, ***glycobiology*** is the comprehensive ***study of carbohydrates in health and disease***.

Functions of carbohydrates

Carbohydrates participate in a wide range of functions

1. They are the most abundant dietary ***source of energy*** (4 Cal/g) for all organisms.

2. Carbohydrates are ***precursors for*** many organic compounds (***fats, amino acids***).

3. Carbohydrates (as glycoproteins and glycolipids) participate in the structure of cell membrane (***glycocalyx***) and cellular functions such as cell growth, adhesion and fertilization.

4. They are structural components of many organisms. These include the fiber (cellulose) of plants, exoskeleton of some insects and the cell wall of microorganisms.

5. Carbohydrates also serve as the ***storage*** form of energy (***glycogen***) to meet the immediate energy demands of the body.

CLASSIFICATION OF CARBOHYDRATES

Carbohydrates are often referred to as ***saccharides*** (*Greek*: sakcharon–sugar). They are broadly classified into three major groups—***monosaccharides, oligosaccharides* and *polysaccharides***. This categorization is based on the number of sugar units. ***Mono- and oligosaccharides*** are sweet to taste, crystalline in

Table 2.1 Classification of monosaccharides with selected examples

Monosaccharides (empirical formula)	Aldose	Ketose
Trioses ($C_3H_6O_3$)	Glyceraldehyde	Dihydroxyacetone
Tetroses ($C_4H_8O_4$)	Erythrose	Erythrulose
Pentoses ($C_5H_{10}O_5$)	Ribose	Ribulose
Hexoses ($C_6H_{12}O_6$)	Glucose	Fructose
Heptoses ($C_7H_{14}O_7$)	Glucoheptose	Sedoheptulose

character and soluble in water, hence they are commonly known as ***sugars***.

Monosaccharides

Monosaccharides (*Greek* : ***mono-one***) are the simplest group of carbohydrates and are often referred to as ***simple sugars***. They have the general formula $C_n(H_2O)_n$, and they cannot be further hydrolysed. The monosaccharides are divided into different categories, based on the functional group and the number of carbon atoms

Aldoses : When the functional group in monosaccharides is an ***aldehyde*** $\left(-\overset{H}{\overset{|}{C}}=O\right)$, they are aldoses e.g. glyceraldehyde, glucose.

Ketoses : When the functional group is a ***keto*** $\left(-\overset{|}{C}=O\right)$ group, they are referred to as ketoses e.g. dihydroxyacetone, fructose.

Based on the number of carbon atoms, the monosaccharides are regarded as ***trioses (3C), tetroses (4C), pentoses (5C), hexoses (6C)*** *and* ***heptoses (7C)***. These terms along with functional groups are used while naming monosaccharides. For instance, ***glucose is an aldohexose while fructose is a ketohexose*** (***Table 2.1***).

The common monosaccharides and disaccharides are given in the ***Table 2.2***.

Oligosaccharides

Oligosaccharides (*Greek*: ***oligo-few***) contain ***2-10 monosaccharide*** molecules which are liberated on hydrolysis of carbohydrates. Based on the number of monosaccharide units present, the oligosaccharides are further subdivided to ***disaccharides***, ***trisaccharides*** etc.

Oligosaccharides with more than 3 monosaccharides units are not usually digested by human enzymes. e.g. ***raffinose*** (found in vegetables, whole grains) is a trisaccharide composed of galactose, glucose and fructose. ***Raffinose is a good prebiotic*** (bacterial food).

Polysaccharides

Polysaccharides (*Greek*: ***poly-many***) are polymers of monosaccharide units with high molecular weight (up to a million). They are usually tasteless (non-sugars) and form colloids with water. The polysaccharides are of two types – ***homopolysaccharides*** and ***heteropolysaccharides***.

MONOSACCHARIDES— STRUCTURAL ASPECTS

Stereoisomerism is an important character of monosaccharides. Stereoisomers are the compounds that have the same structural formulae but differ in their spatial configuration.

A carbon is said to be ***asymmetric when*** it is ***attached to four different atoms or groups***. The number of asymmetric carbon atoms (n) determines the ***isomers*** of a given compound which is equal to 2^n. ***Glucose*** contains 4 asymmetric carbons, and thus has ***16 isomers***.

Glyceraldehyde —the reference carbohydrate

Glyceraldehyde (triose) is the simplest monosaccharide with one asymmetric carbon atom. It exists as two stereoisomers and has been chosen as the ***reference carbohydrate*** to represent the structure of all other carbohydrates.

Table 2.2 Monosaccharides and disaccharides of biological importance

Monosaccharides	Occurrence	Biochemical importance
Trioses		
Glyceraldehyde	Found in cells as phosphate	Glyceraldehyde 3-phosphate is an intermediate in glycolysis
Dihydroxyacetone	Found in cells as phosphate	Its 1-phosphate is an intermediate in glycolysis
Tetroses		
D-Erythrose	Widespread	Its 4-phosphate is an intermediate in HMP pathway of carbohydrate metabolism
Pentoses		
D-Ribose	Widespread as a constituent of RNA and nucleotides	For the structure of RNA, and nucleotide coenzymes (ATP, NAD^+, $NADP^+$)
D-Deoxyribose	As a constituent of DNA	For the structure of DNA
D-Ribulose	Produced during metabolism	It is an important metabolite in hexose monophosphate shunt
D-Xylose	As a constituent of glycoproteins and gums	Involved in the function of glycoproteins
L-Xylulose	As an intermediate in uronic acid pathway	Excreted in urine in essential pentosuria
D-Lyxose	Heart muscle	As a constituent of lyxoflavin of heart muscle
Hexoses		
D-Glucose	As a constituent of polysaccharides (starch, glycogen, cellulose) and disaccharides (maltose, lactose, sucrose). Also found in fruits	The 'sugar fuel' of life; excreted in urine in diabetes. Structural unit of cellulose in plants
D-Galactose	As a constituent of lactose (milk sugar)	Converted to glucose, failure leads to galactosemia. Found in globosides
D-Mannose	Found in plant polysaccharides and animal glycoproteins	For the structure of polysaccharides
D-Fructose	Fruits and honey, as a constituent of sucrose and inulin	Its phosphates are intermediates of glycolysis
Heptoses		
D-Sedoheptulose	Found in plants	Its 7-phosphate is an intermediate in hexose monophosphate shunt, and in photosynthesis
Disaccharides	**Occurrence**	**Biochemical importance**
Sucrose	As a constituent of cane sugar and beet sugar, pineapple	Most commonly used table sugar supplying calories
Lactose	Milk sugar	Exclusive carbohydrate source to breast fed infants. Lactase deficiency (lactose intolerance) leads to diarrhea and flatulence
Maltose	Product of starch hydrolysis, occurs in germinating seeds	An important intermediate in the digestion of starch

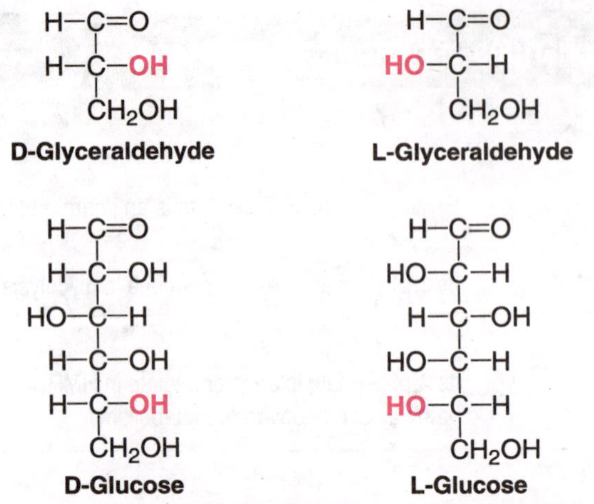

Fig. 2.1 : *D-and-L- forms of glucose compared with D- and L- glyceraldehydes (the reference carbohydrate).*

D- and L-isomers

The D- and L-isomers are ***mirror images of each other***. The spatial orientation of –H and –OH groups on the carbon atom (C_5 for glucose) that is adjacent to the terminal primary alcohol carbon determines whether the sugar is D- or L-isomer. If the ***–OH group*** is ***on the right*** side, the sugar is of ***D-series***, and if ***on the left*** side, it belongs to ***L-series***. The structures of D- and L-glucose based on the ***reference monosaccharide, D- and L-glyceraldehyde*** (glycerose) are depicted in ***Fig.2.1***.

It may be noted that the ***naturally occurring monosaccharides*** in the mammalian tissues are mostly of ***D-configuration***. The enzyme machinery of cells is specific to metabolise D-series of monosaccharides.

Optical activity of sugars

Optical activity is a characteristic feature of compounds with ***asymmetric carbon*** atom. When a beam of polarized light is passed through a solution of an optical isomer, it will be rotated either to the right or left. The term ***dextrorotatory* (d+) *and levorotatory* (l–)** are used to compounds that respectively rotate the plane of polarized light to the right or to the left.

An optical isomer may be designated as D(+), D(–), L(+) and L(–) based on its structural relation with glyceraldehyde. It may be noted that the D- and L-configurations of sugars are primarily based on the structure of glyceraldehyde, the optical activities however, may be different.

Racemic mixture : If ***d-*** and ***l-isomers*** are present in ***equal*** concentration, it is known as racemic mixture or ***dl mixture***. Racemic mixture ***does not exhibit any optical activity***, since the dextro- and levorotatory activities cancel each other.

In the medical practice, the term ***dextrose*** is used for ***glucose in solution***. This is because of the dextrorotatory nature of glucose.

Configuration of D-aldoses

The configuration of possible D-aldoses starting from D-glyceraldehyde is depicted in ***Fig.2.2***. This is a representation of ***Killiani-Fischer synthesis***, by increasing the chain length of an aldose, by one carbon at a time. Thus, starting with an ***aldotriose*** (***3C***), ***aldotetroses*** (***4C***), ***aldopentoses*** (***5C***) and ***aldohexoses*** (***6C***) are formed. Of the 8 aldohexoses, glucose, mannose and galactose are the most familiar. Among these, D-glucose is the only aldose monosaccharide that predominantly occurs in nature.

Configuration of D-ketoses

Starting from dihydroxyacetone (triose), there are five ketosugars which are physiologically important. Their structures are given in ***Fig.2.3***.

Epimers

If two monosaccharides ***differ*** from each other in their ***configuration around a single specific carbon*** (other than anomeric) atom, they are referred to as *epimers* to each other (***Fig.2.4***). For instance, ***glucose and galactose are epimers*** with regard to carbon 4 (***C_4-epimers***). That is, they differ in the arrangement of –OH group at C_4. Glucose and mannose are epimers with regard to carbon 2 (C_2-epimers).

The ***interconversion of epimers*** (e.g. glucose to galactose and vice versa) is known as epimerization, and a group of enzymes—namely—***epimerases*** catalyse this reaction.

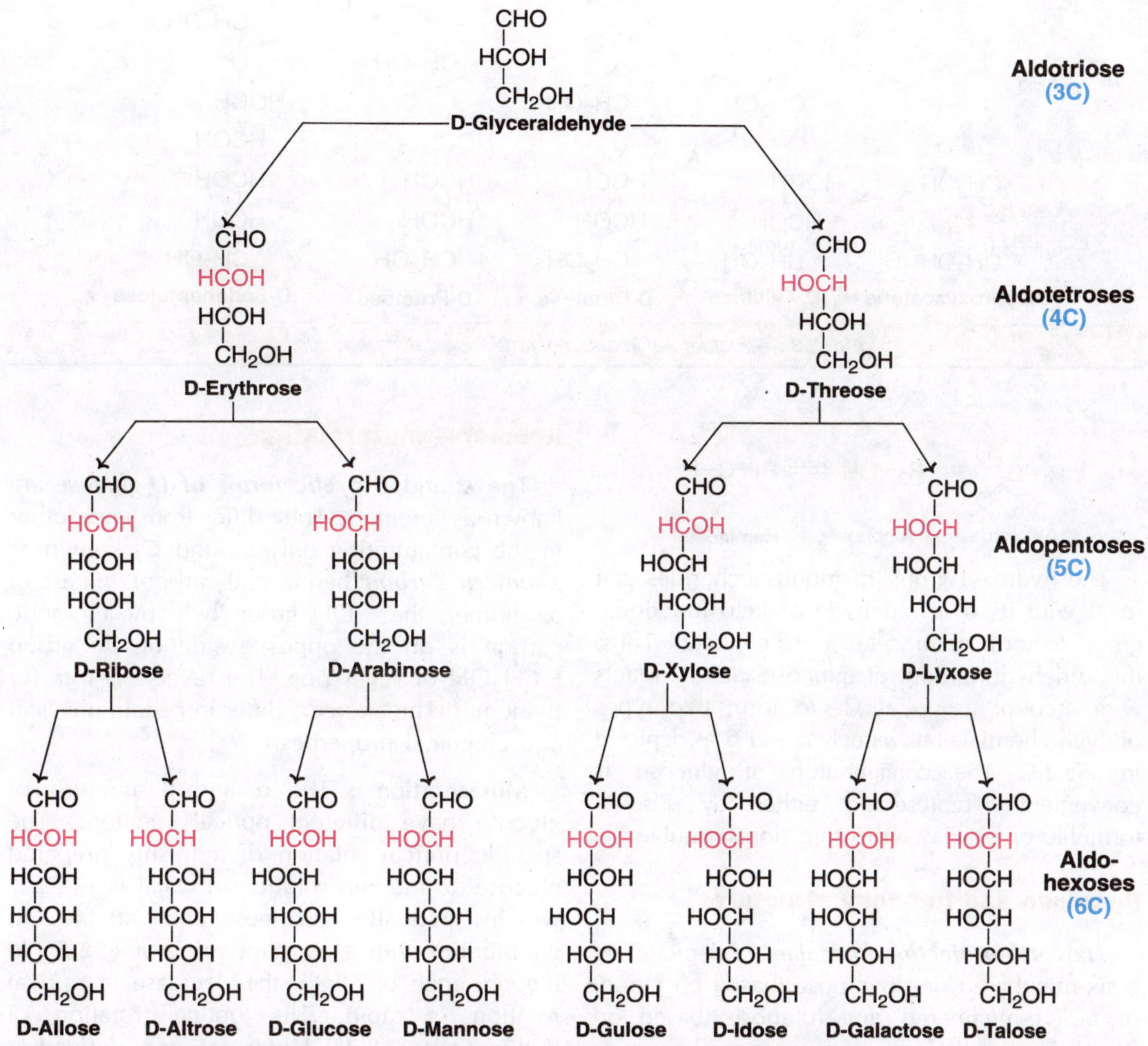

***Fig. 2.2 :** The structural relationship between D-aldoses shown in Fischer projection. (The configuration around C_2 (red) distinguishes the members of each pair).*

Enantiomers

Enantiomers are a special type of stereoisomers that are ***mirror images of each other***. The two members are designated as D- and L-sugars. Enantiomers of glucose are depicted in ***Fig.2.5***.

Majority of the sugars in the higher animals (including man) ***are of D-type*** (***Fig.2.5***).

The term ***diastereomers*** (diastereoisomers) is used to represent the ***stereoisomers*** that are ***not mirror images of one another***. When two diastereomers differ from each other at only one stereocenter (carbon), they are ***epimers***.

STRUCTURE OF GLUCOSE

For a better understanding of glucose structure, let us consider the formation of hemiacetals and hemiketals, respectively produced when an aldehyde or a ketone reacts with alcohol.

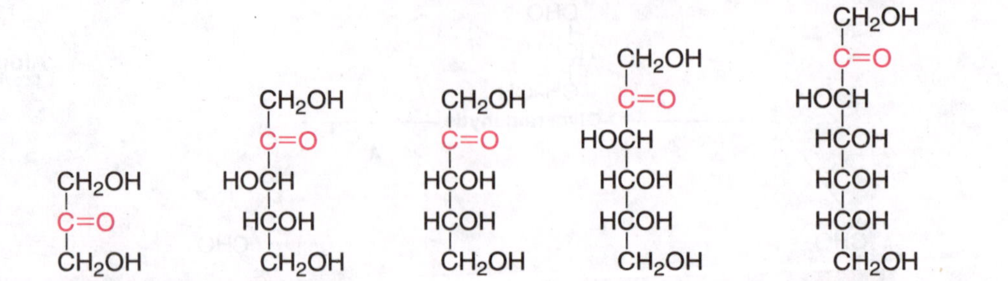

Dihydroxyacetone D-Xylulose D-Ribulose D-Fructose D-Sedoheptulose

Fig. 2.3 : *Structures of ketoses of physiological importance.*

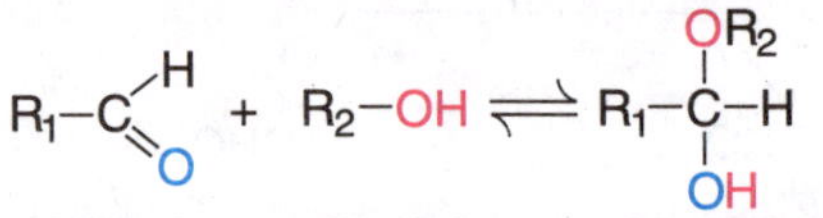

Aldehyde Alcohol Hemiacetal

The hydroxyl group of monosaccharides can react with its own aldehyde or keto functional group to form hemiacetal and hemiketal. Thus, the aldehyde group of glucose at C_1 reacts with alcohol group at C_5 to form two types of cyclic hemiacetals namely ***α*** and ***β***, as depicted in ***Fig.2.6***. The configuration of glucose is conveniently represented either by Fischer formulae or by Haworth projection formulae.

Pyranose and furanose structures

Haworth projection formulae are depicted by a six-membered ring pyranose (based on pyran) or a five-membered ring furanose (based on furan). The cyclic forms of glucose are known as α-D-glucopyranose and α-D-glucofuranose (***Fig.2.7***).

Anomers—mutarotation

The ***α*** and ***β cyclic forms of D-glucose*** are known as *anomers*. They differ from each other in the configuration only around ***C_1*** known as ***anomeric carbon*** (hemiacetal carbon). In case of α anomer, the −OH group held by anomeric carbon is on the opposite side of the group $-CH_2OH$ of sugar ring. The reverse is true for β-anomer. The anomers differ in certain physical and chemical properties.

Mutarotation : The α and β anomers of glucose have different optical rotations. The specific optical rotation of a freshly prepared glucose (α anomer) solution in water is +112.2° which gradually changes and attains an equilibrium with a constant value of +52.7°. In the presence of alkali, the decrease in optical rotation is rapid. The optical rotation of β-glucose is +18.7°. ***Mutarotation is defined as the change in the specific optical rotation representing the interconversion of*** α ***and*** β

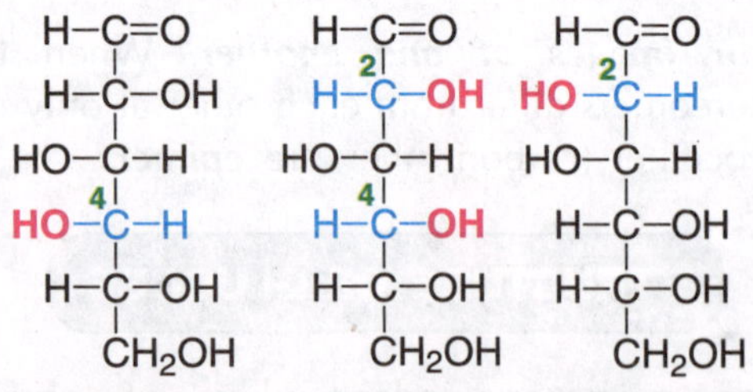

D-Galactose D-Glucose D-Mannose

Fig. 2.4 : *Structures of epimers (glucose and galactose are C_4-epimers while glucose and mannose are C_2-epimers).*

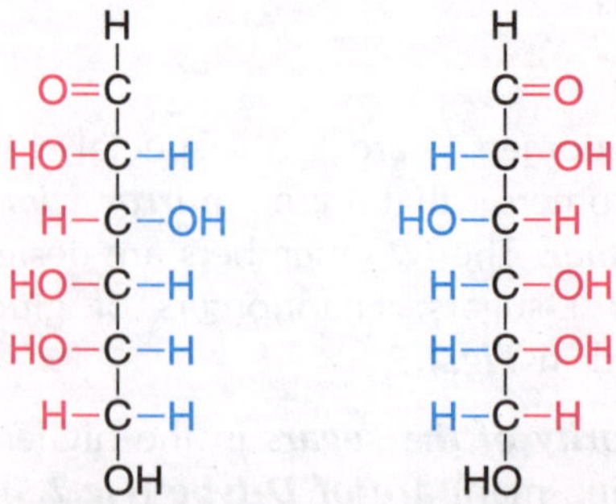

L-Glucose D-Glucose

Fig. 2.5 : *Enantiomers (mirror images) of glucose.*

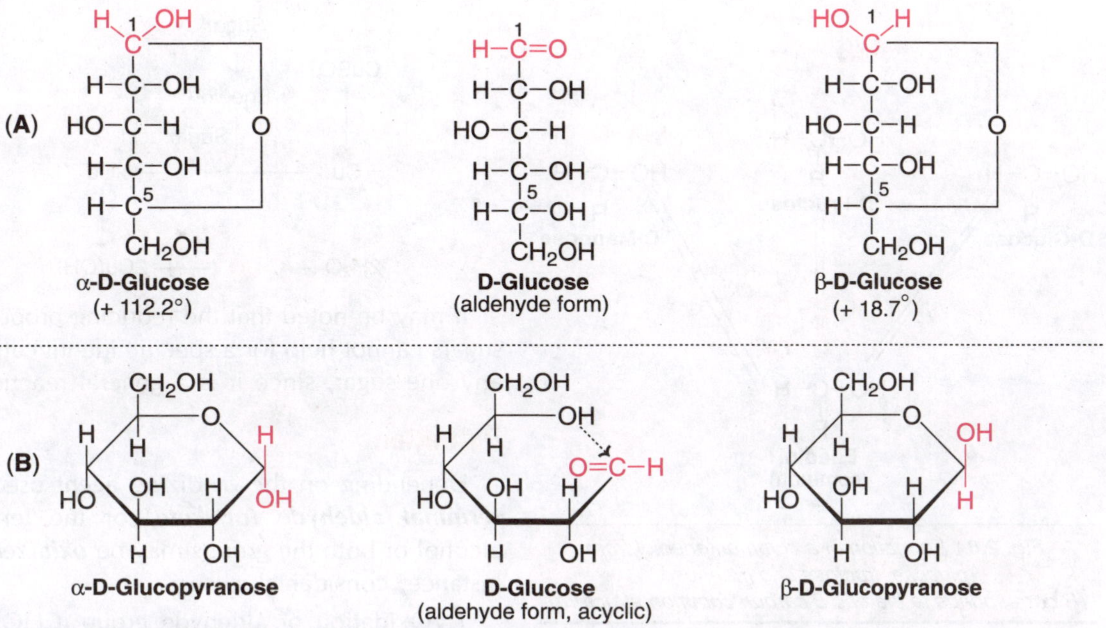

Fig. 2.6 : Mutarotation of glucose representing α and β anomers (A) Fischer projections (B) Haworth projections.

forms of D-glucose to an equilibrium mixture. Mutarotation depicted in ***Fig. 2.6***, is summarized below.

$$\underset{+\,112.2^\circ}{\alpha\text{-D-Glucose}} \rightleftharpoons \underset{+\,52.7^\circ}{\text{Equilibrium mixture}} \rightleftharpoons \underset{+\,18.7^\circ}{\beta\text{-D-Glucose}}$$

(Specific optical rotation $[\alpha]_D^{20}$)

The equilibrium mixture contains 63% β-anomer and 36% α-anomer of glucose with 1% open chain form. In aqueous solution, the β form is more predominant due to its stable conformation. The α and β forms of glucose are interconvertible which occurs through a linear form. The latter, as such, is present in an insignificant quantity.

Mutarotation of fructose : Fructose also exhibits mutarotation. In case of fructose, the pyranose ring (six-membered) is converted to ***furanose*** (five-membered) ring, till an equilibrium is attained. And fructose has a specific optical rotation of –92° at equilibrium.

The conversion of dextrorotatory (+) sucrose to levorotatory fructose is explained under inversion of sucrose (see later in this chapter).

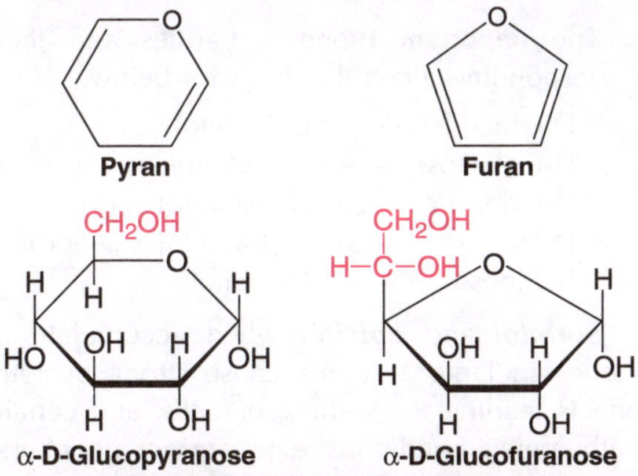

Fig. 2.7 : Structure of glucose-pyranose and furanose forms.

REACTIONS OF MONOSACCHARIDES

Tautomerization or enolization

The process of ***shifting a hydrogen atom from one carbon atom to another to produce enediols*** is known as ***tautomerization***. Sugars possessing anomeric carbon atom undergo tautomerization in alkaline solutions.

When glucose is kept in alkaline solution for several hours, it undergoes isomerization to

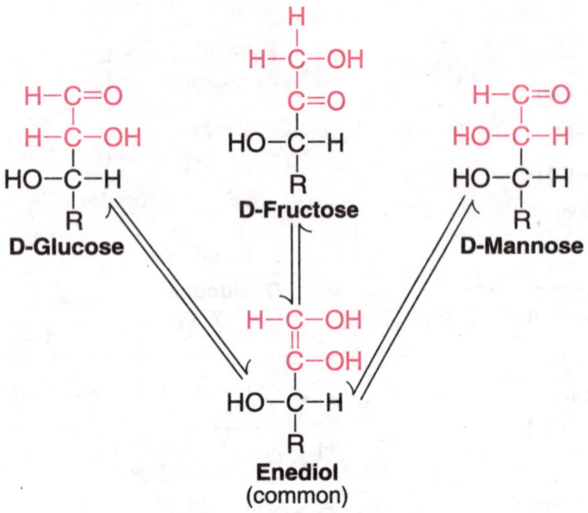

Fig. 2.8 : Formation of a common enediol from glucose, fructose and mannose (R corresponds to the end 3 carbon common structure).

form D-fructose and D-mannose. This reaction—known as the ***Lobry de Bruyn-von Ekenstein transformation***—results in the formation of a common intermediate—namely ***enediol***—for all the three sugars, as depicted in ***Fig.2.8***.

The ***enediols*** are highly reactive, hence sugars in alkaline solution are ***powerful reducing agents***.

Reducing properties

The sugars are classified as reducing or non-reducing. The reducing property is attributed to the ***free aldehyde*** or ***keto group*** of anomeric carbon.

In the laboratory, many tests are employed to identify the reducing action of sugars. These include ***Benedict's test, Fehling's test, Barfoed's test*** etc. The reduction is much ***more efficient in the alkaline medium*** than in the acid medium.

The ***enediol forms*** (explained above) or sugars reduce cupric ions (Cu^{2+}) of copper sulphate to cuprous ions (Cu^{+}), which form a yellow precipitate of cuprous hydroxide or a red precipitate of cuprous oxide as shown next.

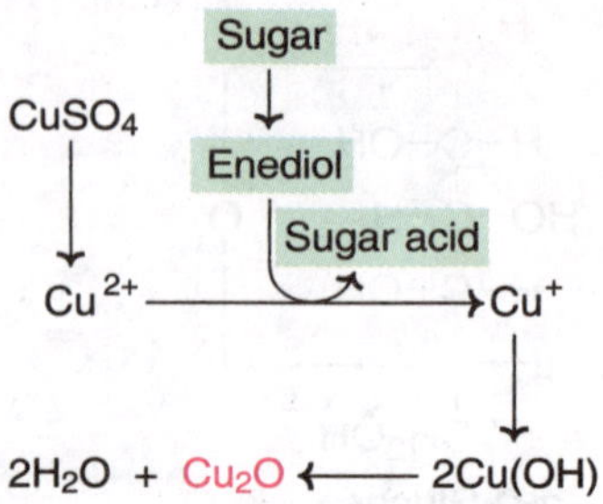

It may be noted that the reducing property of sugars cannot help for a specific identification of any one sugar, since it is a general reaction.

Oxidation

Depending on the oxidizing agent used, the ***terminal aldehyde (or keto)*** or the terminal alcohol or both the groups may be ***oxidized***. For instance, consider glucose :

1. Oxidation of aldehyde group (CHO $\longrightarrow$ COOH) results in the formation of ***gluconic acid***.

2. Oxidation of terminal alcohol group ($CH_2OH \longrightarrow COOH$) leads to the production of ***glucuronic acid***.

Reduction

When treated with reducing agents such as sodium amalgam, the ***aldehyde or keto group*** of monosaccharide is ***reduced*** to corresponding alcohol, as indicated by the general formula :

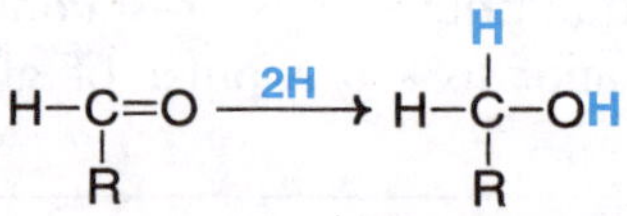

The important monosaccharides and their corresponding alcohols are given below.

D-Glucose $\longrightarrow$ D-Sorbitol
D-Galactose $\longrightarrow$ D-Dulcitol
D-Mannose $\longrightarrow$ D-Mannitol
D-Fructose $\longrightarrow$ D-Mannitol + D-Sorbitol
D-Ribose $\longrightarrow$ D-Ribitol

Sorbitol and ***dulcitol*** when accumulate in tissues in large amounts cause strong osmotic effects leading to swelling of ***cells***, and certain pathological conditions. e.g. ***cataract***, peripheral neuropathy, nephropathy. ***Mannitol is useful to reduce intracranial tension by forced diuresis*** (increased urine output).

Medical Concepts/ Clinical Correlates

- Glucose is the most important energy source of carbohydrates to the mammals (except ruminants). The bulk of dietary carbohydrate (starch) is digested and finally absorbed as glucose into the body.
- Dextrose (glucose in solution in dextrorotatory form) is frequently used in medical practice.
- Fructose is abundantly found in the semen which is utilized by the sperms for energy.
- Several diseases are associated with carbohydrates e.g., diabetes mellitus, glycogen storage diseases, galactosemia, lactose intolerance.

Dehydration

When treated with concentrated sulfuric acid, monosaccharides undergo dehydration with an elimination of 3 water molecules. Thus hexoses give hydroxymethyl furfural while pentoses give furfural on dehydration (***Fig.2.9***). These ***furfurals*** can ***condense with phenolic compounds*** (α-naphthol) to form ***coloured products***. This is the chemical basis of the popular ***Molisch test***.

In case of oligo- and polysaccharides, they are first hydrolysed to monosaccharides by acid, and this is followed by dehydration.

Bial's test : Pentoses react with strong HCl to form furfural derivatives which in turn react with ***orcinol to form green coloured complex***. Bial's test is useful for ***detection of xylose*** in urine in essential pentosuria.

Mucic acid test : ***Galactose*** when treated with nitric acid forms insoluble mucic acid crystals.

Osazone formation

Phenylhydrazine in acetic acid, when boiled with reducing sugars, forms osazones in a reaction summarized in ***Fig.2.10***.

As is evident from the reaction, the ***first two carbons*** (C_1 and C_2) are ***involved*** in osazone formation. The sugars that differ in their configuration on these two carbons give the same type of osazones, since the difference is masked by binding with phenylhydrazine. Thus ***glucose***, ***fructose*** and mannose give the same type (***needle-shaped***) osazones.

Reducing disaccharides also give osazones—***maltose sunflower-shaped***, and ***lactose powder-puff shaped***.

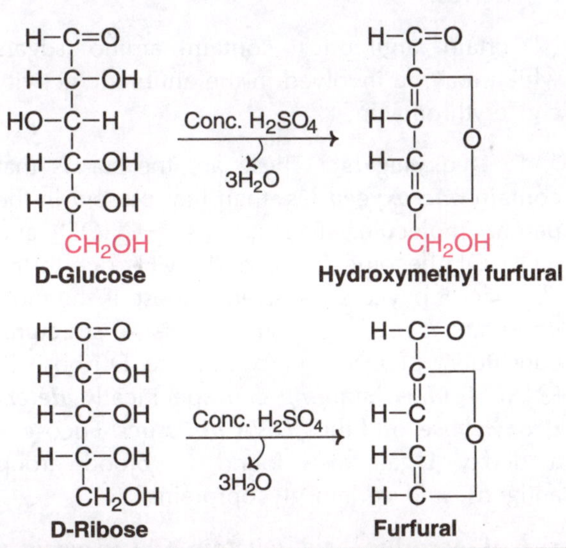

***Fig. 2.9 :** Dehydration of monosaccharides with concentrated H_2SO_4.*

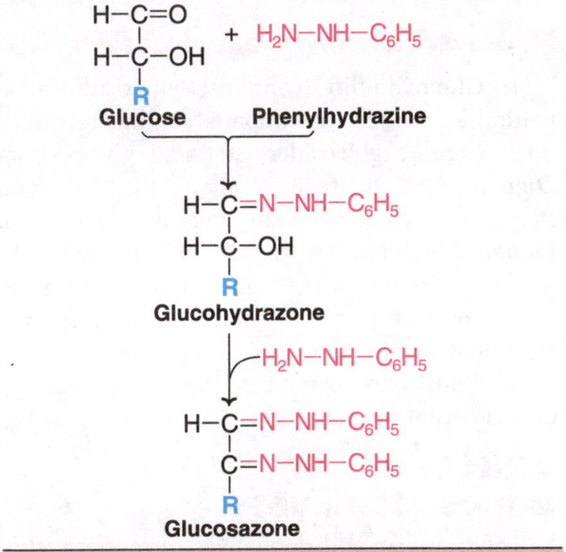

***Fig. 2.10 :** A summary of osazone formation (R represents C_3 to C_6 of glucose).*

Formation of esters

The alcoholic groups of monosaccharides may be esterified by non-enzymatic or enzymatic reactions. Esterification of carbohydrate with phosphoric acid is a common reaction in metabolism. Glucose 6-phosphate and glucose 1-phosphate are good examples. ATP donates the phosphate moiety in ester formation.

GLYCOSIDES

Glycosides are formed when the hemiacetal or hemiketal hydroxyl group (of anomeric carbon) of a carbohydrate reacts with a hydroxyl group of another carbohydrate or a non-carbohydrate (e.g. methyl alcohol, phenol, glycerol). The bond so formed is known as ***glycosidic bond*** and the ***non-carbohydrate*** moiety (when present) is referred to as ***aglycone***.

The monosaccharides are held together by glycosidic bonds to result in di-, oligo- or polysaccharides (see later for structures).

Naming of glycosidic bond : The nomenclature of glycosidic bonds is based on the ***linkages between the carbon atoms and the status of the anomeric carbon (α or β)***. For instance, lactose—which is formed by a bond between C_1 of β-galactose and C_4 of glucose—is named as $\beta(1 \rightarrow 4)$ glycosidic bond. The other glycosidic bonds are described in the structure of di- and polysaccharides.

Physiologically important glycosides

1. **Glucovanillin** (vanillin-D-glucoside) is a natural substance that imparts vanilla flavour.

2. **Cardiac glycosides** (steroidal glycosides) **:** ***Digoxin*** and digitoxin contain the aglycone steroid and they stimulate muscle contraction. ***Oubain*** which inhibits Na^+–K^+ ATPase and blocks active transport of Na^+ is also a cardiac glycoside.

3. **Streptomycin**, an antibiotic used in the treatment of tuberculosis is a glycoside.

4. **Phlorhizin** produces renal damage in experimental animals.

DERIVATIVES OF MONOSACCHARIDES

There are several derivatives of monosaccharides, some of which are physiologically important (***Fig.2.11***)

1. **Sugar acids :** Oxidation of aldehyde or primary alcohol group in monosaccharide results in sugar acids. Gluconic acid is produced from glucose by oxidation of aldehyde (C_1 group) whereas glucuronic acid is formed when primary alcohol group (C_6) is oxidized.

2. **Sugar alcohols** (***polyols***) **:** They are produced by reduction of aldoses or ketoses. For instance, sorbitol is formed from glucose, and mannitol from mannose.

3. **Alditols :** The monosaccharides, on reduction, yield polyhydroxy alcohols, known as alditols. Ribitol is a constituent of flavin coenzymes; glycerol and *myo*-inositol are components of lipids. Xylitol is a sweetener used in sugarless gums and candies.

4. **Amino sugars :** When one or more hydroxyl groups of the monosaccharides are replaced by amino groups, the products formed are amino sugars e.g. D-glucosamine, D-galactosamine. They are present as constituents of heteropolysaccharides.

N-Acetylneuraminic acid (NANA) is a derivative of N-acetylmannose and pyruvic acid. It is an important ***constituent of glycoproteins and glycolipids***. The term ***sialic acid*** is used to include NANA and its other derivatives.

Certain antibiotics contain amino sugars which may be involved in the antibiotic activity e.g. erythromycin.

5. **Deoxysugars :** These are the sugars that contain ***one oxygen less*** than that present in the parent molecule. The groups $-CHOH$ and $-CH_2OH$ become $-CH_2$ and $-CH_3$ due to the absence of oxygen. D-2-Deoxyribose is the most important deoxysugar since it is a structural constituent of DNA (in contrast to D-ribose in RNA). ***Feulgen staining*** can specifically ***detect deoxyribose***, and thus ***DNA in tissues***. Fucose is a deoxy L-galactose found in blood group antigens, and certain glycoproteins.

6. **L-Ascorbic acid** (vitamin C) **:** This is a ***water soluble vitamin***, the ***structure*** of which closely resembles that of a ***monosaccharide***.

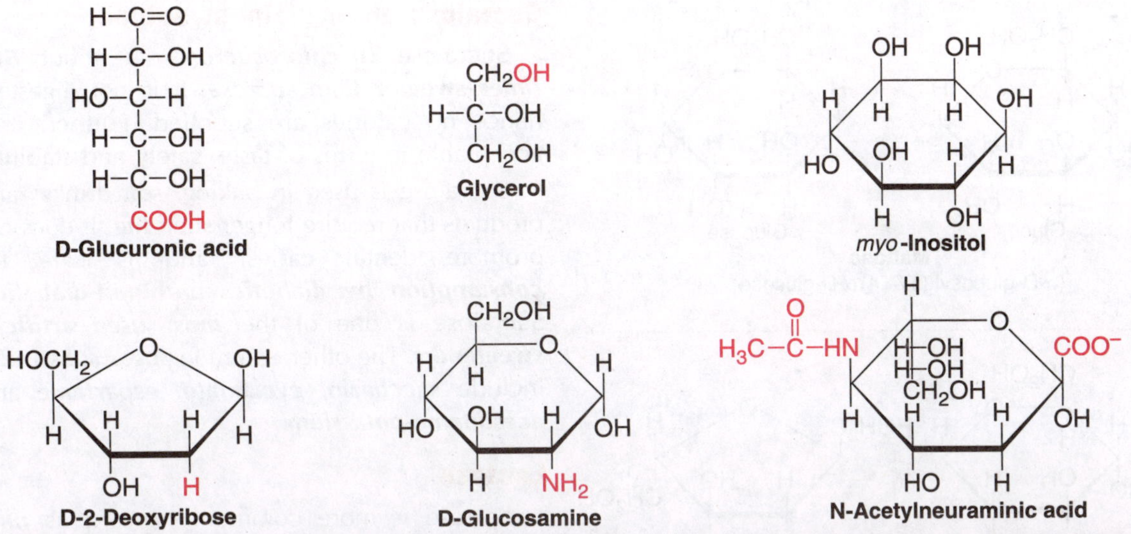

Fig. 2.11 : Structures of monosaccharide derivatives (selected examples).

DISACCHARIDES

Among the oligosaccharides, disaccharides are the most common (***Fig.2.12***). As is evident from the name, a disaccharide consists of two monosaccharide units (similar or dissimilar) held together by a ***glycosidic bond***. They are crystalline, water soluble and sweet to taste. The disaccharides are of two types

1. **Reducing** disaccharides with ***free aldehyde or keto*** group e.g. maltose, lactose.

2. **Non-reducing** disaccharides with ***no free aldehyde or keto*** group e.g. sucrose, trehalose.

Maltose

Maltose is composed of ***two α-D-glucose*** units held together by α (1 → 4) glycosidic bond. The free aldehyde group present on C_1 of second glucose answers the reducing reactions, besides the osazone formations (sunflower-shaped). Maltose can be hydrolysed by dilute acid or the enzyme maltase to liberate two molecules of α-D-glucose.

In ***isomaltose***, the glucose units are held together by α (1 → 6) glycosidic linkage.

Cellobiose is another disaccharide, identical in structure with maltose, except that the former has β (1 → 4) glycosidic linkage. Cellobiose is formed during the hydrolysis of cellulose.

Sucrose

Sucrose (***cane sugar***) is the sugar of commerce, mostly produced by sugar cane and sugar beets. Sucrose is made up of ***α-D-glucose and β-D-fructose***. The two monosaccharides are held together by a glycosidic bond ($\alpha_1 \rightarrow \beta_2$), between C_1 of α-glucose and C_2 of β-fructose. The reducing groups of glucose and fructose are involved in glycosidic bond, hence sucrose is a ***non-reducing sugar, and it cannot form osazones.***

Sucrose is an important source of dietary carbohydrate. It is sweeter than most other common sugars (except fructose) namely glucose, lactose and maltose. Sucrose is employed as a sweetening agent in food industry. The intestinal enzyme—sucrase—hydrolyses sucrose to glucose and fructose which are absorbed.

Inversion of sucrose

Sucrose, as such is dextrorotatory (+66.5°). But, when hydrolysed, sucrose becomes levorotatory (–28.2°). The process of ***change in optical rotation from dextrorotatory (+) to levorotatory (–)*** is referred to as inversion. The hydrolysed mixture of sucrose, containing

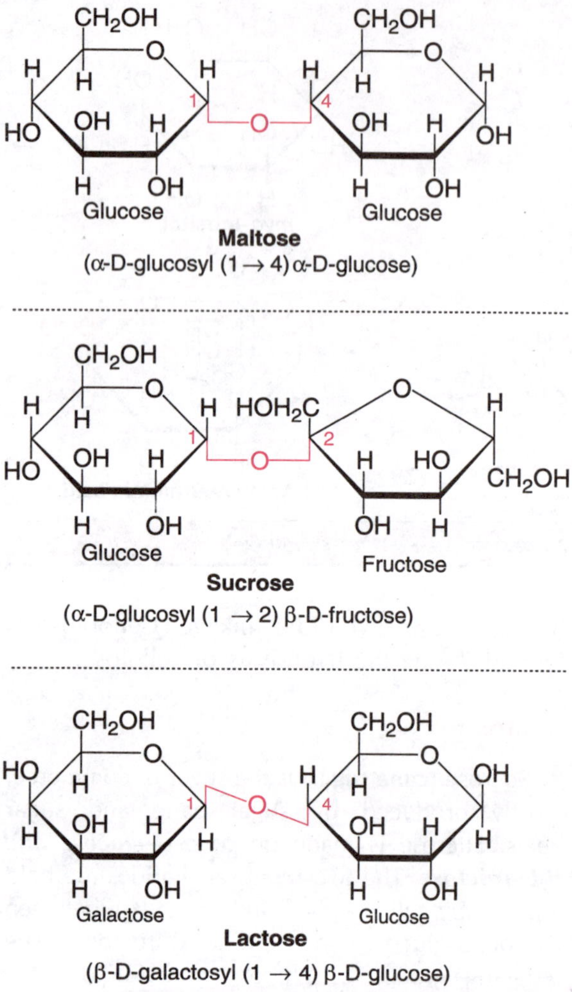

Fig. 2.12 : *Structures of disaccharides —maltose, sucrose and lactose.*

glucose and fructose, is known as ***invert sugar***. The process of inversion is explained below.

Hydrolysis of sucrose by the enzyme ***sucrase*** (***invertase***) or dilute acid liberates one molecule each of glucose and fructose. It is postulated that sucrose (dextro) is first split into α-D-glucopyranose (+52.5°) and β-D-fructofuranose, both being dextrorotatory. However, β-D-fructofuranose is less stable and immediately gets converted to β-D-fructopyranose which is strongly levorotatory (–92°). The overall effect is that dextro sucrose (+66.5°) on inversion is converted to levo form (–28.2°).

Sucralose–an artificial sweetener

Sucralose (***trichlorosucrose***) is about ***500 times sweeter than sucrose***. It is not digested, hence no calories are supplied. Further, it is acceptable in terms of taste, safety and stability.

Sucralose is used in baking, soft drinks and products that require longer shelf-life. It does not promote dental caries, and is ***safe for consumption by diabetics and non-diabetics***. Sucralose is one of the ***most used artificial sweeteners***. The other common ***sugar substitutes*** include ***saccharin***, ***cyclamate, aspartame*** and ***acesulfame-potassium***.

Lactose

Lactose is more commonly known as ***milk sugar*** since it is the disaccharide found in milk. Lactose is composed of ***β-D-galactose and β-D-glucose*** held together by β (1 → 4) glycosidic bond. The anomeric carbon of C_1 glucose is free, hence lactose exhibits reducing properties and forms osazones (powder-puff or hedgehog shape).

Lactose of milk is the most important carbohydrate in the nutrition of young mammals. It is hydrolysed by the intestinal enzyme lactase to glucose and galactose. Lactose is excreted in urine in people with lactose intolerance.

Lactulose

Lactulose is a ***synthetic disaccharide*** containing ***galactose*** and ***fructose***. It is neither digested nor absorbed in the intestine. Lactulose is useful for the treatment of ***hepatic encephalopathy***, a disorder characterized by elevated plasma ammonium levels. Lactulose converts ammonia (NH_3) in the lumen to ammonium ion (NH_4^+). This results in a reduction in the plasma NH_3, since NH_4^+ ions are not easily absorbed.

POLYSACCHARIDES

Polysaccharides (or simply ***glycans***) consist of repeat units of monosaccharides or their derivatives, held together by glycosidic bonds. They are primarily concerned with two important functions-structural, and storage of energy.

Polysaccharides are linear as well as branched polymers. This is in contrast to structure of proteins and nucleic acids which are

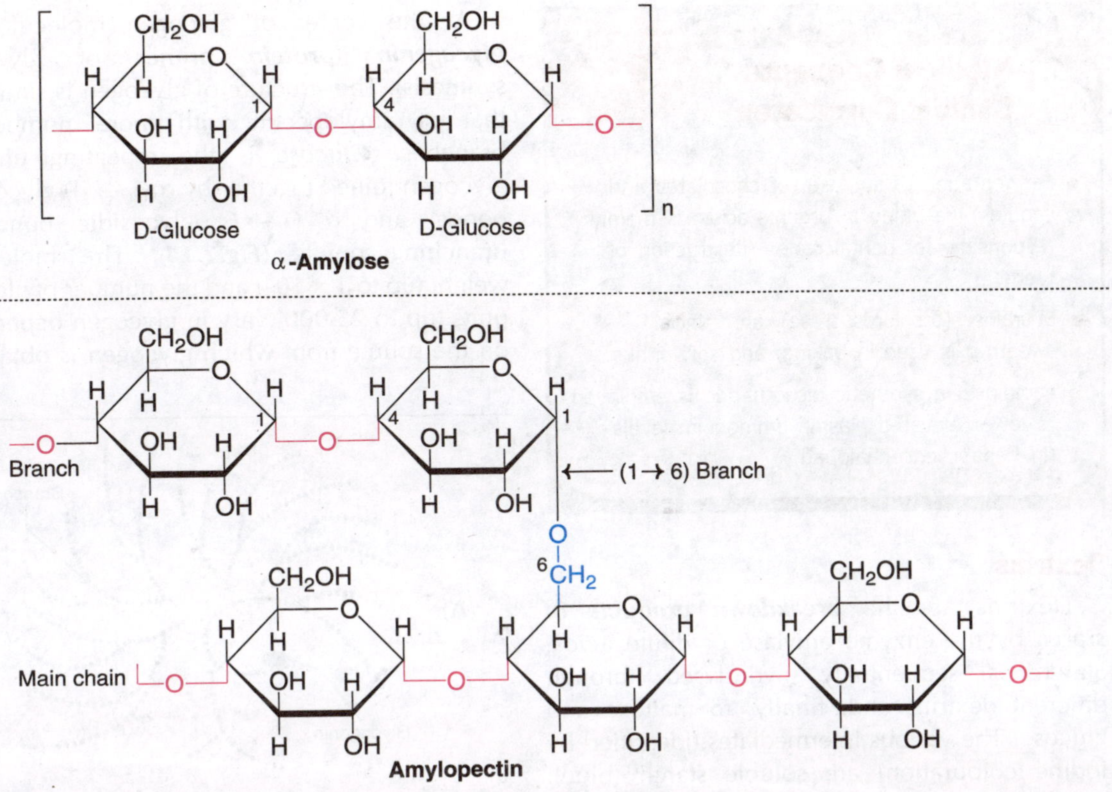

Fig. 2.13 : Structure of starch (α-amylose and amylopectin).

only linear polymers. The occurrence of branches in polysaccharides is due to the fact that glycosidic linkages can be formed at any one of the hydroxyl groups of a monosaccharide.

Polysaccharides are of two types

1. **Homopolysaccharides** on hydrolysis yield only a ***single type of monosaccharide***. They are named based on the nature of the monosaccharide. Thus, ***glucans*** are polymers of glucose whereas ***fructosans*** are polymers of fructose.

2. **Heteropolysaccharides** on hydrolysis yield a ***mixture*** of a few monosaccharides or their derivatives.

HOMOPOLYSACCHARIDES

Starch

Starch is the ***carbohydrate reserve of plants*** which is the most important dietary source for higher animals, including man. High content of starch is found in cereals, roots, tubers, vegetables etc. Starch is a homopolymer composed of ***D-glucose units*** held by α-glycosidic bonds. It is known as ***glucosan or glucan***.

Starch consists of two polysaccharide components-water soluble ***amylose*** (15-20%) and a water insoluble ***amylopectin*** (80-85%). Chemically, amylose is a long unbranched chain with 200–1,000 D-glucose units held by α (1 → 4) glycosidic linkages. Amylopectin, on the other hand, is a branched chain with α (1 → 6) glycosidic bonds at the branching points and α (1 → 4) linkages everywhere else (***Fig.2.13***). Amylopectin molecule containing a few thousand glucose units looks like a branched tree (20–30 glucose units per branch).

Starches are hydrolysed by ***amylase*** (pancreatic or salivary) to liberate dextrins, and finally maltose and glucose units. Amylase acts specifically on α (1 → 4) glycosidic bonds.

Medical Concepts/ Clinical Correlates

- Fermentation of sucrose (of chocolates and candies), resulting in organic acids, is mainly responsible for dental caries (destruction of teeth).
- Sucralose (trichlorosucrose), an artificial sweetner is used in baking, and soft drinks.
- Lactulose, a synthetic disaccharide is employed to lower elevated plasma ammonium levels (in hepatic encephalopathy).

Dextrins

Dextrins are the ***breakdown products of starch*** by the enzyme amylase or dilute acids. Starch is sequentially hydrolysed through different dextrins and, finally, to maltose and glucose. The various intermediates (identified by iodine colouration) are soluble starch (blue), ***amylodextrin*** (violet), ***erythrodextrin*** (red) and ***achrodextrin*** (no colour).

Dextrans

Dextrans are ***polymers of glucose***, produced by microorganisms. They are used as plasma volume expanders in transfusion, and ***chromatography*** (e.g. gel filtration).

Inulin

Inulin is a polymer of fructose i.e., ***fructosan.*** It occurs in dahlia bulbs, garlic, onion etc. It is a low molecular weight (around 5,000) polysaccharide easily soluble in water. Inulin is not utilized by the body. It is used for assessing kidney function through ***measurement of glomerular filtration rate (GFR).***

Glycogen

Glycogen is the ***carbohydrate reserve in animals,*** hence often referred to as ***animal starch.*** It is present in high concentration in liver, followed by muscle, brain etc. Glycogen is also found in plants that do not possess chlorophyll (e.g. yeast, fungi).

At the center of glycogen molecule lies ***glycogenin*** (***protein*** primer of glycogen synthesis). The structure of glycogen is similar to that of amylopectin with more number of branches. ***Glucose*** is the repeating unit in glycogen joined together by α $(1 \rightarrow 4)$ glycosidic bonds, and α $(1 \rightarrow 6)$ glycosidic bonds at branching points (***Fig.2.14***). The molecular weight (up to 1×10^8) and the number of glucose units (up to 25,000) vary in glycogen depending on the source from which glycogen is obtained.

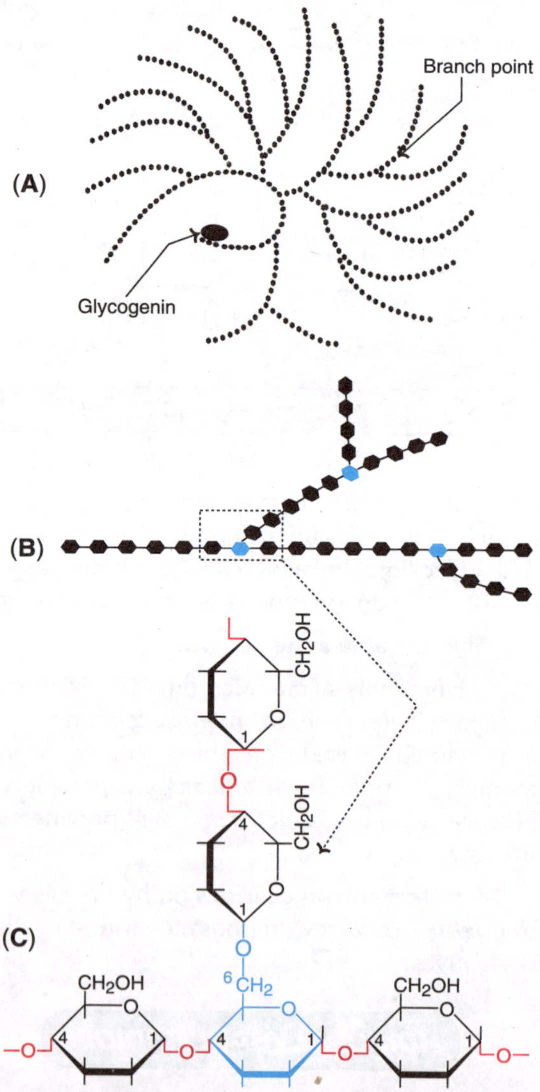

Fig. 2.14 : Structure of glycogen (A) General structure (B) and (C) Enlarged at a branch point.

Medical Concepts/ Clinical Correlates

- Accumulation of sorbitol and dulcitol in the tissues may cause certain pathological conditions e.g. cataract, nephropathy, peripheral neuropathy.
- Inulin, a polymer of fructose, is used to assess renal function by measuring glomerular filtration rate (GFR).
- Dextrans are used as plasma volume expanders in transfusion, and in gel-filtration chromatography.
- The non-digestible carbohydrates such as cellulose, play a significant role in human nutrition. These include decreasing the intestinal absorption of glucose and cholesterol, and increasing bulk of feces to avoid constipation.

Cellulose

Cellulose occurs ***exclusively in plants*** and it is the most abundant organic substance in plant kingdom. It is a predominant constituent of plant cell wall. Cellulose is totally absent in animal body.

Cellulose is composed of ***β-D-glucose*** units linked by **β** ***(1 → 4) glycosidic bonds*** (***Fig.2.15***). Cellulose ***cannot be digested by mammals***—including man—due to lack of the enzyme that cleaves β-glycosidic bonds (α amylase breaks α bonds only). Certain ruminants and herbivorous animals contain microorganisms in the gut which produce enzymes that can cleave β-glycosidic bonds. Hydrolysis of cellulose yields a disaccharide ***cellobiose,*** followed by ***β-D-glucose***.

Cellulose, though not digested, has great importance in human nutrition. It is a major constituent of ***fiber***, the ***non-digestable carbohydrate*** (others hemicellulose, lignin, pectins etc.). The functions of dietary fiber include decreasing the absorption of glucose and cholesterol from the intestine, besides increasing the bulk of feces. (***For details, Chapter 23***)

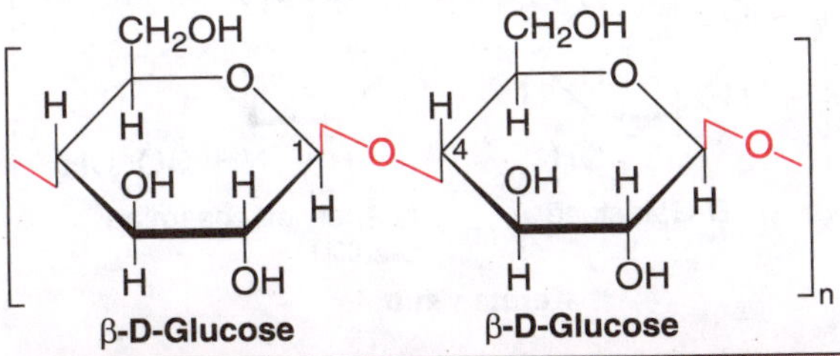

Fig. 2.15 : *Structure of cellulose (The repeating unit 'n' may be several thousands).*

Chitin

Chitin is composed of N-acetyl D-glucosamine units held together by β (1 → 4) glycosidic bonds. It is a structural polysaccharide found ***in the exoskeleton of some invertebrates*** e.g. insects, crustaceans.

HETEROPOLYSACCHARIDES

When the polysaccharides are ***composed of different types of sugars or their derivatives***, they are referred to as heteropolysaccharides or ***heteroglycans***.

MUCOPOLYSACCHARIDES

Mucopolysaccharides are heteroglycans made up of repeating units of sugar derivatives, namely ***amino sugars and uronic acids***. These are more commonly known as ***glycosaminoglycans (GAG)***. Acetylated amino groups, besides sulfate and carboxyl groups are generally present in GAG structure. The presence of sulfate and carboxyl groups contributes to acidity of the molecules, making them ***acid mucopolysaccharides***.

Some of the mucopolysaccharides are found in combination with proteins to form ***mucoproteins*** or ***mucoids*** or ***proteoglycans*** (***Fig.2.16***). Mucoproteins may contain up to 95% carbohydrate and 5% protein.

Mucopolysaccharides are essential components of tissue structure. The extracellular spaces of tissue (particularly connective tissue-cartilage, skin, blood vessels, tendons) consist of collagen and elastin fibers embedded in a matrix or ground substance. The ground substance is predominantly composed of GAG.

The important mucopolysaccharides include hyaluronic acid, chondroitin 4-sulfate, heparin, dermatan sulfate and keratan sulfate (***Fig.2.17***).

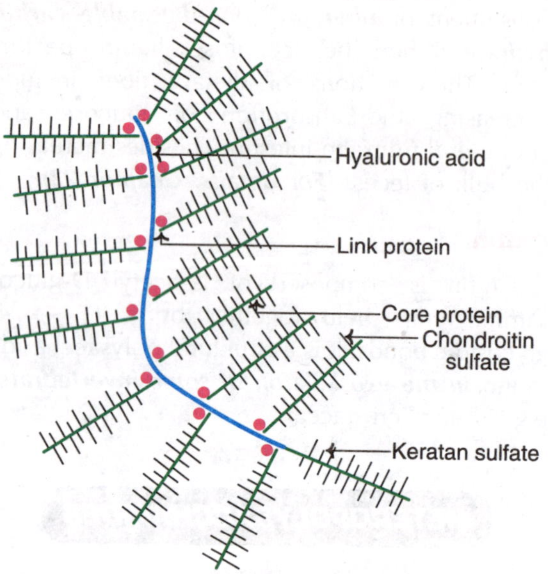

Fig. 2.16 : Diagrammatic representation of a proteoglycan complex.

Hyaluronic acid

Hyaluronic acid is an important GAG found in the ground substance of synovial fluid of joints and vitreous humor of eyes. It is also present as a ground substance in connective tissues, and forms a gel around the ovum. Hyaluronic acid serves as a ***lubricant and shock absorbant in joints.***

Hyaluronic acid is composed of alternate units of ***D-glucuronic acid*** and ***N-acetyl D-glucosamine***. These two molecules form disaccharide units held together by β (1 → 3) glycosidic bond (***Fig.2.17***). Hyaluronic acid contains about 250–25,000 disaccharide units (held by β 1 →4 bonds) with a molecular weight up to 4 million.

Hyaluronidase is an enzyme that breaks (β 1 →4 linkages) hyaluronic acid and other GAG. This enzyme is present in high concentration in testes, seminal fluid, and in certain snake and insect venoms. Hyaluronidase of ***semen*** is assigned an ***important role in fertilization*** as this enzyme clears the gel (hyaluronic acid) around the ovum allowing a better penetration of sperm into the ovum.

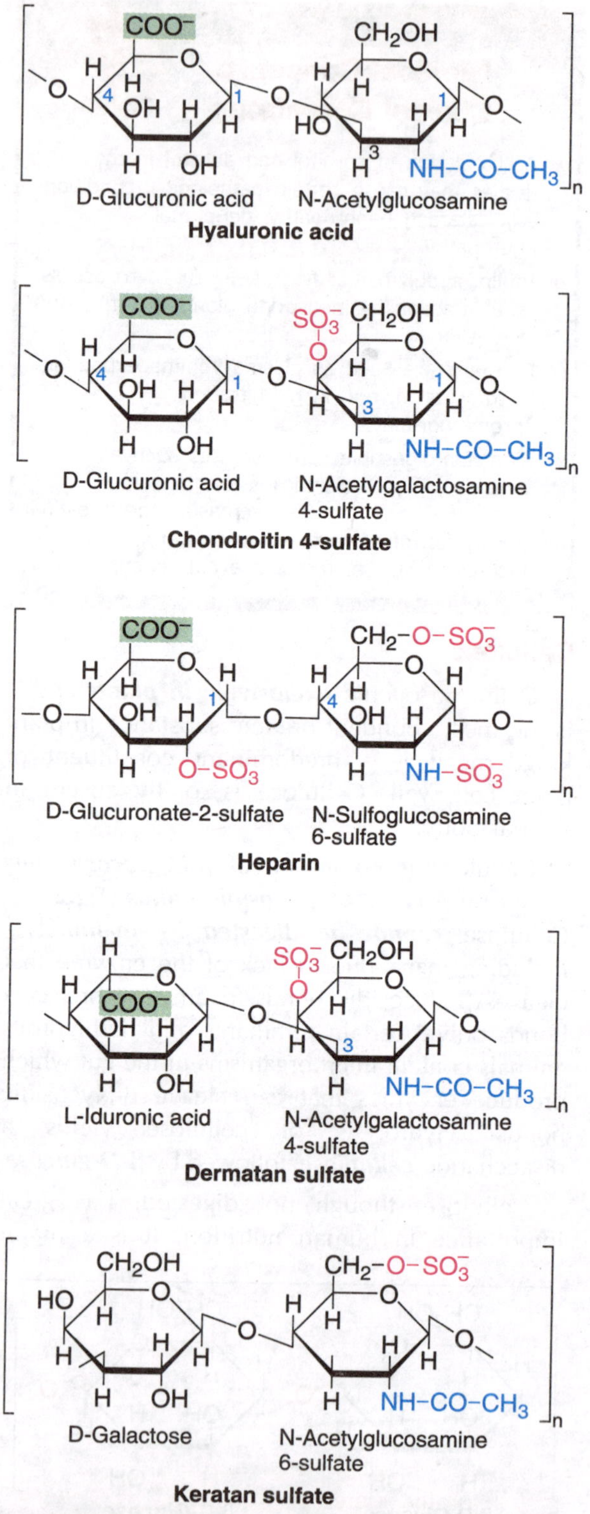

Fig. 2.17 : Structures of common glycosaminoglycans – the disaccharides as repeating units.

Table 2.3 A summary of glycosaminoglycans – composition, distribution and functions

Glycosaminoglycan	Composition	Tissue distribution	Function(s)
Hyaluronic acid	D-Glucuronic acid, N-acetylglucosamine	Connective tissue, synovial fluid, vitreous humor	Serves as a lubricant, and shock absorber. Promotes wound healing
Chondroitin sulfate	D-Glucuronic acid, N-acetylgalactosamine 4-sulfate	Cartilage, bone, skin, blood vessel walls	Helps to maintain the structure and shapes of tissues
Heparin	D-Glucuronate 2-sulfate, N-sulfoglucosamine 6-sulfate	Blood, lung, liver, kidney, spleen	Acts as an anticoagulant
Dermatan sulfate	L-Iduronic acid, N-acetyl-galactosamine 4-sulfate	Blood vessel valves, heart valves, skin	Maintains the shapes of tissues
Keratan sulfate	D-Galactose, N-acetyl-glucosamine 6-sulfate	Cartilage, cornea, connective tissues	Keeps cornea transparent

Hyaluronidase of bacteria helps their invasion into the animal tissues.

Chondroitin sulfates

Chondroitin 4-sulfate (*Greek*: chondros-cartilage) is a major constituent of various ***mammalian tissues*** (bone, cartilage, tendons, heart, valves, skin, cornea etc.). Structurally, it is comparable with hyaluronic acid. Chondroitin 4-sulfate consists of repeating disaccharide units composed of ***D-glucuronic acid and N-acetyl D-galactosamine 4-sulfate*** (***Fig.2.17***).

Heparin

Heparin is ***an anticoagulant*** (prevents blood clotting) that occurs ***in blood***, lung, liver, kidney, spleen etc. Heparin helps in the release of the enzyme lipoprotein lipase which helps in clearing the turbidity of lipemic plasma. Heparin is used in suspected cases of intramuscular coagulation.

Heparin is composed of alternating units of ***N-sulfo D-glucosamine 6-sulfate and glucuronate 2-sulfate*** (***Fig.2.17***).

Dermatan sulfate

Mostly found ***in skin***, dermatan sulfate is structurally related to chondroitin 4-sulfate. The only difference is that there is an inversion in the configuration around C_5 of D-glucuronic acid to form L-iduronic acid (***Fig.2.17***).

Keratan sulfate

It is a heterogeneous GAG with a variable sulfate content, besides small amounts of mannose, fructose, sialic acid etc. Keratan sulfate essentially consists of alternating units of D-galactosamine and N-acetylglucosamine 6-sulfate.

AGAR AND PECTINS

Agar, mostly found in sea weeds, is a polymer of galactose sulfate and glucose. Since agar is not digested, it serves as a dietary fiber (***Refer Chapter 23***). ***Agarose*** (with galactose and anhydrogalactose) is useful in the laboratory as a major component of microbial culture media, and in electrophoresis.

Pectins, found in apples and citrus fruits, contain galactouronate and rhamnose. Pectins, being non-digestible, are useful as ***dietary fiber***. They are also employed in the ***preparation of jellies***.

GLYCOPROTEINS

Several ***proteins*** are covalently ***bound to carbohydrates*** which are referred to as glycoproteins. The carbohydrate content of glycoprotein varies from 1% to 90% by weight.

Medical Concepts/ Clinical Correlates

- The mucopolysaccharide hyaluronic acid serves as a lubricant and shock absorbant in joints.
- Hyaluronic acid is used to enhance lubrication of eyes, and wound healing.
- The enzyme hyaluronidase of semen degrades the gel (contains hyaluronic acid) around the ovum. This allows effective penetration of sperm into the ovum.
- The mucopolysaccharide heparin is an anticoagulant (prevents blood clotting).
- The survival of Antarctic fish below –2°C is attributed to the antifreeze glycoproteins.
- Streptomycin is a glycoside employed in the treatment of tuberculosis.
- Mannitol is useful to reduce intracranial tension by forced diuresis.

Table 2.4 A selected list of glycoproteins and their major functions

Glycoprotein(s)	Major function(s)
Collagen	Structure
Hydrolases, proteases, glycosidases	Enzymes
Ceruloplasmin	Transport
Immunoglobulins	Defense against infection
Synovial glycoproteins	Lubrication
Thyrotropin, erythropoietin	Hormones
Blood group substances	Antigens
Fibronectin, laminin	Cell-cell recognition and adhesion
Intrinsic factor	Absorption of vitamin B_{12}
Fibrinogen	Blood clotting
Glycophorin	RBC structure

Sometimes the term ***mucoprotein*** is used for glycoprotein with ***carbohydrate concentration more than 4%***. Glycoproteins are very widely distributed in the cells and perform variety of functions. These include their role as enzymes, hormones, transport proteins, structural proteins and receptors. A selected list of glycoproteins and their major functions is given in ***Table 2.4***.

It is believed that ***alterations*** in the structures of certain ***glycoproteins*** of tumor cell membranes are responsible for spread of cancer (***metastasis***).

Glycophorin : It is a membrane protein of RBC. Glycophorin is heavily (60%) glycosylated, forming a hydrophilic charged coat, enabling RBC to circulate without adhering to other cells or vessel walls.

The carbohydrates found in glycoproteins include mannose, galactose, N-acetylglucosamine, N-acetylgalactosamine, xylose, L-fucose and N-acetylneuraminic acid (NANA). NANA is an important sialic acid (***See Fig.2.11***).

Lectins : These are a special category of glycoproteins involved in cell-cell recognition and adhesion besides degradation of blood cells and certain proteins. ***Selectins*** are a subgroup of lectins more specifically involved in cell-cell recognition and adhesion. Further, selectins also participate in the movement of immune cells (e.g. T-lymphocytes) to the sites of inflammation.

Antifreeze glycoproteins : The Antarctic fish live below –2°C, a temperature at which the blood would freeze. It is now known that these fish contain ***antifreeze glycoproteins*** which lower the freezing point of water and interfere with the crystal formation of ice.

Blood group substances

The blood group ***antigens*** (of erythrocyte membrane) contain carbohydrates as glycoproteins or glycolipids. N-Acetylgalactosamine, galactose, fucose, sialic acid etc. are found in the blood group substances.

AQUASOMES

Aquasomes (also called ***bodies of water***) are recently developed ***delivery systems*** for bioactive molecules — proteins, hormones, genes, antigens. They are spherical in shape containing central non-crystalline core, coated with ***carbohydrates*** (polyhydroxy oligomers). Aquasomes are successfully used as carriers of bioactive molecules to specific target sites.

SUMMARY

- *Carbohydrates are the polyhydroxyaldehydes or ketones, or compounds which produce them on hydrolysis. The term sugar is applied to carbohydrates soluble in water and sweet to taste. Carbohydrates are the major dietary energy sources.*
- *Carbohydrates are broadly classified into 3 groups—monosaccharides, oligosaccharides and polysaccharides. The monosaccharides are further divided into different categories based on the presence of functional groups (aldoses or ketoses).*
- *Glyceraldehyde (triose) is the simplest carbohydrate and is chosen as a reference to write the configuration of all other monosaccharides (D- and L- forms). If two monosaccharides differ in their structure around a single carbon atom, they are known as epimers. Glucose and galactose are C_4–epimers.*
- *D-Glucose is the most important naturally occurring aldose/monosaccharide. Glucose exists as α and β anomers with different optical rotations. The interconversion of α and β anomeric forms with change in the optical rotation is known as mutarotation.*
- *Monosaccharides participate in several reactions. These include oxidation, reduction, dehydration, osazone formation etc. Formation of esters and glycosides by monosaccharides is of special significance in biochemical reactions.*
- *Among the oligosaccharides, disaccharides are the most common. These include the reducing disaccharides namely lactose (milk sugar) and maltose (malt sugar) and the non-reducing sucrose (cane sugar).*
- *Polysaccharides are the polymers of monosaccharides or their derivatives, held together by glycosidic bonds. Homopolysaccharides are composed of a single monosaccharide (e.g., starch, glycogen, cellulose, inulin). Heteropolysaccharides contain a mixture of few monosaccharides or their derivatives (e.g., mucopolysaccharides).*
- *Starch and glycogen are the carbohydrate reserves of plants and animals respectively. Cellulose, exclusively found in plants, is the structural constituent. Inulin is utilized to assess kidney function by measuring glomerular filtration rate (GFR).*
- *Mucopolysaccharides (glycosaminoglycans) are the essential components of tissue structure. They provide the matrix or ground substance of extracellular tissue spaces in which collagen and elastin fibers are embedded. Hyaluronic acid, chondroitin 4-sulfate, heparin, are among the important glycosaminoglycans.*
- *Glycoproteins are a group of biochemically important compounds with a variable composition of carbohydrate (1-90%), covalently bound to protein. Several enzymes, hormones, structural proteins and cellular receptors are in fact glycoproteins.*

Self-Assessment Exercises

I. Essay questions

1. Define and classify carbohydrates with suitable examples. Add a note on the functions of carbohydrates.
2. Describe the structure and functions of mucopolysaccharides.
3. Give an account of the structural configuration of monosaccharides, with special reference to glucose.
4. Discuss the structure and functions of 3 biochemically important disaccharides.
5. Define polysaccharides and describe the structure of 3 homopolysaccharides.

II. Short notes

(a) Epimers, (b) Mutarotation, (c) Osazone formation, (d) Glycosidic bond, (e) Sugar derivatives, (f) Anomers, (g) Enediol, (h) Amino sugars, (i) Inversion of sucrose, (j) Deoxysugars.

III. Fill in the blanks

1. Name a non-reducing disaccharide ________________________.
2. The carbohydrate that is taken as a reference for writing the configuration of others ____________.
3. If two monosaccharides differ in configuration around a single carbon atom, they are known as ________________.
4. The α and β cyclic forms of D-glucose are referred to as ________________.
5. The non-carbohydrate moiety found in glycosides is known as ________________.
6. Give an example of a glycoside antibiotic ________________.
7. The glycosidic bonds at the branching points in the structure of starch are ____________.
8. The polysaccharide employed for the assessment of kidney function ____________.
9. The glycosaminoglycan that serves as a lubricant and shock absorbant of joints __________.
10. Name the sialic acid, mostly found in the structure of glycoproteins and glycolipids ________.

IV. Multiple choice questions

11. Ribose and deoxyribose differ in structure around a single carbon, namely
    (a) C_1 (b) C_2 (c) C_3 (d) C_4.
12. One of the following is not an aldose
    (a) Glucose (b) Galactose (c) Mannose (d) Fructose.
13. The glycosaminoglycan that serves as an anticoagulant
    (a) Heparin (b) Hyaluronic acid (c) Chondroitin sulfate (d) Dermatan sulfate.
14. The following polysaccharide is composed of β-glycosidic bonds
    (a) Starch (b) Glycogen (c) Dextrin (d) Cellulose.
15. The carbon atoms involved in the osazone formation
    (a) 1 and 2 (b) 2 and 3 (c) 3 and 4 (d) 5 and 6.

Lipids

3

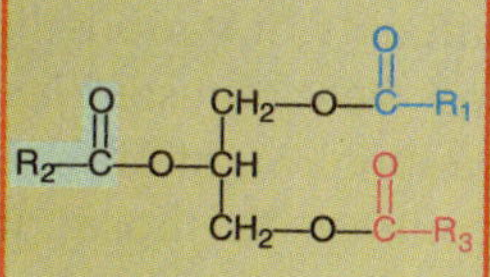

The fat speaks

"With water, I say, 'Touch me not';
To the tongue, I am tasteful;
Within limits, I am dutiful;
In excess, I am dangerous!"

Lipids (*Greek*: lipos–fat) are of great importance to the body as the chief concentrated storage form of energy, besides their role in cellular structure and various other biochemical functions. As such, lipids are a ***heterogeneous group*** of compounds and, therefore, it is difficult to define them precisely.

Lipids may be regarded as organic substances relatively insoluble in water, soluble in organic solvents (alcohol, ether etc.), actually or potentially related to fatty acids and utilized by the living cells.

Unlike the polysaccharides, proteins and nucleic acids, lipids are not polymers (do not have the same repeating units/monomers). Further, lipids are mostly small molecules.

Lipidomics is a recent term used for the comprehensive ***study of lipids*** that includes chemistry, functions, metabolism, pathology and various other aspects.

Classification of lipids

Lipids are broadly classified into simple, complex, derived and miscellaneous lipids, which are further subdivided into different groups

1. **Simple lipids : *Esters of fatty acids*** with alcohols. These are mainly of two types
   (a) **Fats and oils** (triacylglycerols) **:** These are esters of fatty acids with glycerol. The ***difference*** between fat and oil is only ***physical***. Thus, oil is a liquid while fat is a solid at room temperature.
   (b) **Waxes :** Esters of fatty acids (usually long chain) with alcohols other than glycerol. These alcohols may be aliphatic or alicyclic. Cetyl alcohol is most commonly found in waxes. Waxes are used in the preparation of candles, lubricants, cosmetics, ointments, polishes etc.

2. **Complex (or compound) lipids :** These are esters of fatty acids with alcohols containing ***additional groups*** such as phosphate, nitrogenous base, carbohydrate, protein etc. They are further divided as follows
   (a) **Phospholipids :** They contain phosphoric acid and frequently a nitrogenous base. This is in addition to alcohol and fatty acids.

(i) **Glycerophospholipids :** These phospholipids contain glycerol as the alcohol e.g., lecithin, cephalin.

(ii) **Sphingophospholipids :** Sphingosine is the alcohol in this group of phospholipids e.g., sphingomyelin.

(b) **Glycolipids :** These lipids contain a fatty acid, carbohydrate and nitrogenous base. The alcohol is sphingosine, hence they are also called as glycosphingolipids. Glycerol and phosphate are absent e.g., cerebrosides, gangliosides.

(c) **Lipoproteins :** Macromolecular complexes of lipids with proteins.

(d) **Other complex lipids :** Sulfolipids, aminolipids and lipopolysaccharides are among the other complex lipids.

3. **Derived lipids :** These are the derivatives obtained on the hydrolysis of group 1 and group 2 lipids which possess the characteristics of lipids. These include ***glycerol*** and other alcohols, ***fatty acids***, mono- and diacylglycerols, ***lipid (fat) soluble vitamins***, ***steroid hormones***, hydrocarbons and ketone bodies.

4. **Miscellaneous lipids :** These include a large number of compounds possessing the characteristics of lipids e.g., ***carotenoids***, squalene, ***hydrocarbons*** such as pentacosane (in bees wax), terpenes etc.

NEUTRAL LIPIDS : The lipids which are ***uncharged*** are referred to as neutral ***lipids.*** These are mono-, di-, and triacylglycerols, ***cholesterol*** and cholesteryl esters.

Functions of lipids

Lipids perform several important functions

1. They are the concentrated ***fuel reserve*** of the body (triacylglycerols).

2. Lipids are the ***constituents of membrane*** structure and regulate the membrane permeability (phospholipids and cholesterol).

3. They serve as a source of ***fat soluble vitamins*** (A, D, E and K).

4. Lipids are important as cellular metabolic regulators (steroid hormones and prostaglandins).

5. Lipids protect the internal organs, serve as ***insulating materials*** and give shape and smooth appearance to the body.

FATTY ACIDS

Fatty acids are ***carboxylic acids with hydrocarbon side chain***. They are the simplest form of lipids.

Occurrence

Fatty acids mainly occur in the esterified form as major constituents of various lipids. They are also present as free (unesterified) fatty acids. Fatty acids of animal orgin are much simpler in structure in contrast to those of plant origin which often contain groups such as epoxy, keto, hydroxy and cyclopentane rings.

Even and odd carbon fatty acids

Most of the fatty acids that occur in ***natural*** lipids are of ***even carbons*** (usually 14C – 20C). This is due to the fact that biosynthesis of fatty acids mainly occurs with the sequential addition of 2 carbon units. ***Palmitic acid (16C) and stearic acid (18C) are the most common***. Among the odd chain fatty acids, propionic acid (3C) and valeric acid (5C) are well known.

Saturated and unsaturated fatty acids

Saturated fatty acids ***do not contain double bonds***, while unsaturated fatty acids contain one or more double bonds. Both saturated and unsaturated fatty acids almost equally occur in the natural lipids. Fatty acids with one double bond are monounsaturated, and those with 2 or more double bonds are collectively known as ***polyunsaturated fatty acids (PUFA)***.

Nomenclature of fatty acids

The naming of a fatty acid (systematic name) is based on the hydrocarbon from which it is derived. The saturated fatty acids end with a suffix ***-anoic*** (e.g., octanoic acid) while the unsaturated fatty acids end with a suffix ***-enoic*** (e.g., octadecenoic acid). In addition to systematic names, fatty acids have common names which are more widely used (***Table 3.1***).

Numbering of carbon atoms : It ***starts from*** the ***carboxyl carbon*** which is taken as ***number 1***. The carbons adjacent to this (carboxyl C) are 2, 3, 4 and so on or alternately α, β, γ and so on. The terminal carbon containing methyl group is known omega (ω) carbon. Starting from the methyl end, the carbon atoms in a fatty acid are numbered as omega 1, 2, 3 etc. The numbering of carbon atoms in two different ways is given below

$$\underset{\omega_1}{\overset{7}{CH_3}}-\underset{\omega_2}{\overset{6}{CH_2}}-\underset{\omega_3}{\overset{5}{CH_2}}-\underset{\omega_4}{\overset{4}{CH_2}}-\underset{\omega_5}{\overset{3}{CH_2}}-\underset{\omega_6}{\overset{2}{CH_2}}-\overset{1}{COOH}$$

Length of hydrocarbon chain of fatty acids

Depending on the length of carbon chains, fatty acids are categorized into 3 groups—***short chain*** with ***less than 6 carbons***; ***medium chain*** with ***8 to 14 carbons*** and ***long chain*** with ***16 to 24 carbons***.

Shorthand representation of fatty acids

Instead of writing the full structures, biochemists employ shorthand notations (by numbers) to represent fatty acids. The general rule is that the total number of carbon atoms are written first, followed by the number of double bonds and finally the (first carbon) position of

TABLE 3.1 Selected examples of biochemically important fatty acids

Common Name	Systematic name	Abbreviation*	Structure
I. Saturated fatty acids			
Acetic acid	Ethanoic acid	2 : 0	CH_3COOH
Propionic acid	n-Propanoic acid	3 : 0	CH_3CH_2COOH
Butyric acid	n-Butanoic acid	4 : 0	$CH_3(CH_2)_2COOH$
Valeric acid	n-Pentanoic acid	5 : 0	$CH_3(CH_2)_3COOH$
Caproic acid	n-Hexanoic acid	6 : 0	$CH_3(CH_2)_4COOH$
Caprylic acid	n-Octanoic acid	8 : 0	$CH_3(CH_2)_6COOH$
Capric acid	n-Decanoic acid	10 : 0	$CH_3(CH_2)_8COOH$
Lauric acid	n-Dodecanoic acid	12 : 0	$CH_3(CH_2)_{10}COOH$
Myristic acid	n-Tetradecanoic acid	14 : 0	$CH_3(CH_2)_{12}COOH$
Palmitic acid	n-Hexadecanoic acid	16 : 0	$CH_3(CH_2)_{14}COOH$
Stearic acid	n-Octadecanoic acid	18 : 0	$CH_3(CH_2)_{16}COOH$
Arachidic acid	n-Eicosanoic acid	20 : 0	$CH_3(CH_2)_{18}COOH$
Behenic acid	n-Docosanoic acid	22 : 0	$CH_3(CH_2)_{20}COOH$
Lignoceric acid	n-Tetracosanoic acid	24 : 0	$CH_3(CH_2)_{22}COOH$
II. Unsaturated fatty acids			
Palmitoleic acid	*cis*-9-Hexadecenoic acid	16 : 1; 9	$CH_3(CH_2)_5CH = CH(CH_2)_7COOH$
Oleic acid	*cis*-9-Octadecenoic acid	18 : 1; 9	$CH_3(CH_2)_7CH = CH(CH_2)_7COOH$
Linoleic acid**	*cis, cis*-9,12–Octadecadienoic acid	18 : 2; 9, 12	$CH_3(CH_2)_4CH = CHCH_2CH = CH(CH_2)_7COOH$
Linolenic acid**	All *cis*–9,12,15–Octadecatrienoic acid	18 : 3; 9, 12, 15	$CH_3CH_2CH = CHCH_2CH = CHCH_2CH = CH(CH_2)_7COOH$
Arachidonic acid	All *cis*–5,8,11,14–Eicosatetraenoic acid	20 : 4; 5, 8, 11, 14	$CH_3(CH_2)_4CH = CHCH_2CH = CHCH_2CH = CHCH_2CH = CH(CH_2)_3COOH$

* *Total number of carbon atoms, followed by the number of double bonds and the first carbon position of the double bond(s).*

** *Essential fatty acids.*

double bonds, starting from the carboxyl end. Thus, saturated fatty acid, palmitic acid is written as 16 : 0, oleic acid as 18 : 1; 9, ***arachidonic acid*** as ***20 : 4; 5, 8, 11, 14***.

There are other conventions of representing the double bonds. Δ^9 indicates that the double bond is between 9 and 10 of the fatty acid. ω 9 represents the double bond position (9 and 10) from the ω end. Naturally occurring unsaturated fatty acids belong to ω 9, ω 6 and ω 3 series.

ω3 series	***Linolenic acid*** (18 : 3; 9, 12, 15)
ω6 series	***Linoleic acid*** (18 : 2; 9, 12) and ***arachidonic acid*** (20 : 4; 5, 8, 11, 14)
ω9 series	Oleic acid (18 : 1; 9)

The biochemically important saturated and unsaturated fatty acids are given in the ***Table 3.1***.

ESSENTIAL FATTY ACIDS

The fatty acids that cannot be synthesized by the body and, therefore, ***should be supplied in the diet*** are known as essential fatty acids (***EFA***). Chemically, they are ***polyunsaturated fatty acids***, namely ***linoleic acid*** (18 : 2; 9, 12) and ***linolenic acid*** (18 : 3; 9, 12, 15). ***Arachidonic acid*** (20 : 4; 5, 8, 11, 14) becomes essential, if its precursor linoleic acid is not provided in the diet in sufficient amounts. The structures of EFA are given in the ***Table 3.1***.

Biochemical basis for essentiality : Linoleic acid and linolenic acid are essential since ***humans lack the enzymes that can introduce double bonds beyond carbons 9 to 10.***

Functions of EFA : Essential fatty acids are required for the ***membrane structure*** and ***function, transport of cholesterol***, formation of lipoproteins, prevention of fatty liver etc. (***Chapter 23***). They are also needed for the synthesis of another important group of compounds, namely ***eicosanoids*** (***Chapter 32***).

Deficiency of EFA : The deficiency of EFA results in ***phrynoderma*** or ***toad skin***, characterized by the presence of horny eruptions on the posterior and lateral parts of limbs, on the back and buttocks, loss of hair and poor wound healing.

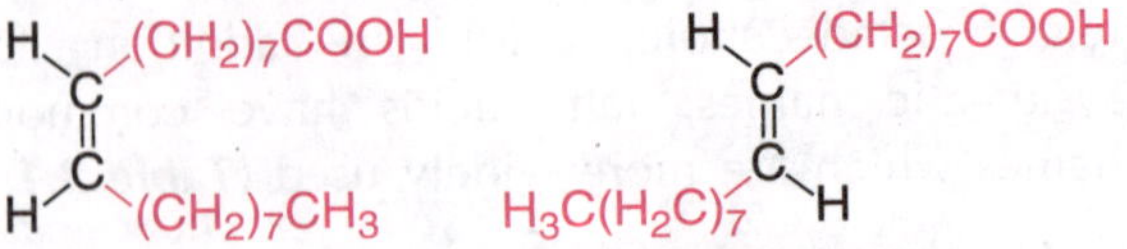

Fig. 3.1 : Cis-trans isomerism in unsaturated fatty acids.

Conditionally essential fatty acids

Certain ***fatty acids*** that become ***essential under some developmental*** or ***disease conditions*** are referred to as conditionally essential fatty acids. The most important among these is ***docosahexaenoic acid*** (***DHA***; ω-3 fatty acid; 22 : 6).

Humans can synthesize DHA, to a limited extent, from α-linoleic acid (ALA). However, the capacity to convert ALA to DHA is limited in preterm infant. Therefore, ***supplementation of DHA*** is required to meet the needs of ***developing brain.*** Many infant nutrient formulas have DHA added to them.

Fish oils are good sources of docosahexaenoic acid and eicosapentaenoic acid (EPA). Both of these are considered as heart disease fighting fatty acids.

Isomerism in unsaturated fatty acids

Unsaturated fatty acids exhibit ***geometric isomerism*** depending on the orientation of the groups around the double bond axis.

If the atoms or acyl groups are present on the same side of the double bond, it is a ***cis configuration***. On the other hand, if the groups occur on the opposite side, it is a ***trans configuration***. Thus oleic acid is a *cis* isomer while elaidic acid is a *trans* isomer, as depicted in ***Fig.3.1***. *Cis* isomers are less stable than *trans* isomers. Most of the naturally occurring unsaturated fatty acids exist as *cis* isomers.

In the *cis* isomeric form, there is a molecular binding at the double bond. Thus, oleic acid

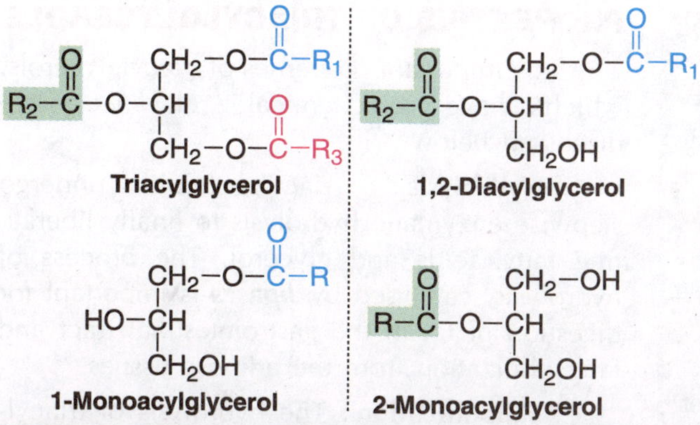

***Fig. 3.2 :** General structures of acylglycerols (For palmitoyl $R = C_{15}H_{31}$; for stearoyl $R = C_{17}H_{35}$; For linoleoyl $R = C_{17}H_{31}$)*

exists in an L-shape while elaidic acid is a straight chain. Increase in the number of double bonds will cause more bends (kinks) and arachidonic acid with 4 double bonds will have a U-shape. It is believed that *cis* isomers of fatty acids with their characteristic bonds will compactly pack the membrane structure.

Hydroxy fatty acids : Some of the fatty acids are hydroxylated. ***β-Hydroxybutyric acid***, one of the ketone bodies produced in metabolism, is a simple example of hydroxy fatty acids. Cerebronic acid and recinoleic acid are long chain hydroxy fatty acids.

Cyclic fatty acids : Fatty acids with cyclic structures are rather rare e.g., ***chaulmoogric acid*** found in chaulmoogra oil (used in leprosy treatment) contains cyclopentenyl ring.

Eicosanoids : These compounds are related to eicosapolyenoic fatty acids and include ***prostaglandins, prostacyclins, leukotrienes*** and thromboxanes. They are discussed together (***Chapter 32***).

TRANS FATTY ACIDS

Trans fatty acids (**TFA**) possess double bonds and are ***formed during hydrogenation*** (saturation) of natural oils. Further, small amounts of TFA are also formed in the rumen by the action of microorganisms on fat containing TFA. TFA are ***widely used in food industry*** (fast foods and ready to eat packaged foods) to ***increase the shelf life of various foods***.

TFA are known to be highly ***detrimental to health***, as their consumption is associated with ***increased risk*** of various diseases including diabetes, and ***cardiovascular diseases***. TFA have been implicated in systemic inflammation, insulin resistance, elevated LDL, reduced HDL etc. Therefore, TFA containing foods should be avoided in the diet.

Margarine is the most commonly used TFA containing fat in food industry. In recent years, with improved technology, soft margarine with low TFA content has been developed. However, the warning of health risk on TFA containing foods consumption continues.

TRIACYLGLYCEROLS

Triacylglycerols (***formerly triglycerides***) are the esters of glycerol with fatty acids. The fats and oils that are widely distributed in both plants and animals are chemically triacylglycerols. They are insoluble in water and non-polar in character and commonly known as ***neutral fats***.

Fats as stored fuel : Triacylglycerols are the most abundant group of lipids that primarily function as fuel reserves of animals. The fat reserve of normal humans (men 20%, women 25% by weight) is sufficient to meet the body's caloric requirements for 2-3 months.

Fats primarily occur in adipose tissue : Adipocytes of adipose tissue—predominantly found in the subcutaneous layer and in the abdominal cavity—are specialized for storage of triacylglycerols. The fat is stored in the form of globules dispersed in the entire cytoplasm. And surprisingly, triacylglycerols are not the structural components of biological membranes.

Structures of acylglycerols : Monoacylglycerols, diacylglycerols and triacylglycerols, respectively consisting of one, two and three molecules of fatty acids esterified to a molecule of glycerol, are known (***Fig.3.2***). Among these, triacylglycerols are the most important biochemically.

Simple triacylglycerols contain the same type of fatty acid residue at all the three carbons e.g., tristearoyl glycerol or tristearin.

Medical Concepts/ Clinical Correlates

- Lipids are important to the body as constituents of membranes, source of fat soluble (A, D, E and K) vitamins and metabolic regulators (steroid hormones and prostaglandins).
- Triacylglycerols (fats) primarily stored in the adipose tissue are concentrated fuel reserves of the body. Fats found in the subcutaneous tissue and around certain organs serve as thermal insulators.
- The unsaturated fatty acids–linoleic and linolenic acid—are essential to humans, the deficiency of which causes phrynoderma or toad skin.

Mixed triacylglycerols are more common. They contain ***2 or 3 different types of fatty acid*** residues. In general, fatty acid attached to C_1 is saturated, that attached to C_2 is unsaturated while that on C_3 can be either. Triacylglycerols are named according to placement of acyl radical on glycerol e.g., 1,3-palmitoyl 2-linoleoyl glycerol.

Triacylglycerols of plants, in general, have higher content of unsaturated fatty acids compared to that of animals.

Stereospecific numbering of glycerol

The structure of glycerol gives an impression that carbons 1 and 3 are identical. This is not true in a 3-dimensional structure. In order to represent the carbon atoms of glycerol in an unambiguous manner, biochemists adopt a ***s**tereospecific **n**umbering* (***sn***) and prefix glycerol with ***sn***.

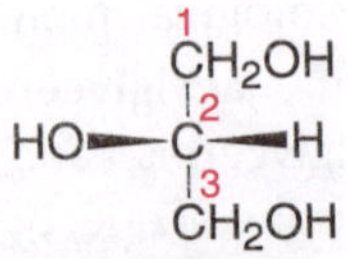

***sn*-Glycerol**

It should be noted that C_1 and C_3 are different. Cells possess enzymes that can distinguish these two carbons. Thus glycerokinase phosphorylates *sn*-3 (and not *sn*-1) glycerol to give *sn*-glycerol 3-phosphate.

PROPERTIES OF TRIACYLGLYCEROLS

A few important properties of triacylglycerols, which have biochemical relevance, are discussed below

1. **Hydrolysis :** Triacylglycerols undergo stepwise enzymatic hydrolysis to finally liberate free fatty acids and glycerol. The process of hydrolysis, catalysed by ***lipases*** is important for digestion of fat in the gastrointestinal tract and fat mobilization from the adipose tissues.

2. **Saponification :** The hydrolysis of triacylglycerols by alkali to produce glycerol and soaps is known as saponification.

$$\text{Triacylglycerol} + 3\ \text{NaOH} \longrightarrow \text{Glycerol} + 3\ \text{R-COONa (soaps)}$$

3. **Rancidity :** Rancidity is the term used to represent the ***deterioration of fats and oils*** resulting in an unpleasant taste. Fats containing unsaturated fatty acids are more susceptible to rancidity.

Rancidity occurs when fats and oils are exposed to air, moisture, light, bacteria etc. ***Hydrolytic rancidity*** occurs due to partial hydrolysis of triacylglycerols by bacterial enzymes. Oxidative rancidity is due to oxidation of unsaturated fatty acids. This results in the formation of unpleasant products such as dicarboxylic acids, aldehydes, ketones etc. Rancid fats and oils are unsuitable for human consumption.

Antioxidants : The ***substances which can prevent the occurrence of oxidative rancidity*** are known as antioxidants. Trace amounts of antioxidants such as tocopherols (vitamin E), hydroquinone, gallic acid and α-naphthol are added to the commercial preparations of fats and oils to prevent rancidity. Propyl gallate, butylated hydroxyanisole (BHA) and butylated hydroxytoluene (BHT) are the antioxidants used in food preservation.

4. **Lipid peroxidation *in vivo* :** In the living cells, lipids undergo oxidation to ***produce peroxides and free radicals which can damage the tissue***. The free radicals are believed to cause inflammatory diseases, ***ageing, cancer, atherosclerosis*** etc. It is fortunate that the cells possess antioxidants such as vitamin E, urate and

superoxide dismutase to prevent *in vivo* lipid peroxidation (***Chapter 34***).

Tests to check purity of fats and oils

Adulteration of fats and oils is increasing day by day. Several tests are employed in the laboratory to check the purity of fats and oils. Some of them are discussed hereunder

Iodine number : It is defined as the ***grams*** (number) ***of iodine absorbed by 100 g of fat*** or ***oil***. Iodine number is useful to know the relative unsaturation of fats, and is directly proportional to the content of unsaturated fatty acids. Thus lower is the iodine number, less is the degree of unsaturation. The iodine numbers of common oils/fats are given below.

Fat/oil	Iodine number
Coconut oil	7 — 10
Butter	25 — 28
Palm oil	45 — 55
Olive oil	80 — 85
Groundnut oil	85 — 100
Cottonseed oil	100 — 110
Sunflower oil	125 — 135
Linseed oil	175 — 200

Determination of iodine number will help to know the degree of adulteration of a given oil.

Saponification number : It is defined as the ***mg*** (number) ***of KOH required to hydrolyse (saponify) one gram of fat or oil***. Saponification number is a measure of the average molecular size of the fatty acids present. The value is higher for fats containing short chain fatty acids. The saponification numbers of a few fats and oils are given below

Human fat : 195–200
Butter : 230–240
Coconut oil : 250–260

Reichert-Meissl (RM) number : It is defined as the number of ml 0.1 N KOH required to completely neutralize the soluble volatile fatty acids distilled from 5 g fat. RM number is useful in testing the purity of butter since it contains a good concentration of volatile fatty acids (butyric acid, caproic acid and caprylic acid). This is in contrast to other fats and oils which have a negligible amount of volatile fatty acids. Butter has a RM number in the range 25-30, while it is less than 1 for most other edible oils. Thus any ***adulteration of butter can be easily tested*** by this sensitive RM number.

Acid number : It is defined as the number of mg of KOH required to completely neutralize free fatty acids present in one gram fat or oil. In normal circumstances, refined oils should be free from any free fatty acids. Oils, on decomposition—due to chemical or bacterial contamination—yield free fatty acids. Therefore, oils with increased acid number are unsafe for human consumption.

PHOSPHOLIPIDS

These are complex or ***compound lipids*** containing ***phosphoric acid***, in addition to fatty acids, nitrogenous base and alcohol (***Fig.3.3***). There are two classes of phospholipids

1. Glycerophospholipids (or phosphoglycerides) that contain glycerol as the alcohol.

2. Sphingophospholipids (or sphingomyelins) that contain sphingosine as the alcohol.

Glycerophospholipids

Glycerophospholipids are the major lipids that occur in biological membranes. They consist of glycerol 3-phosphate esterified at its C_1 and C_2 with fatty acids. Usually, C_1 contains a saturated fatty acid while C_2 contains an unsaturated fatty acid.

1. **Phosphatidic acid :** This is the simplest phospholipid. It does not occur in good concentration in the tissues. Basically, phosphatidic acid is an ***intermediate in the synthesis of triacylglycerols and phospholipids***.

The other glycerophospholipids containing different nitrogenous bases or other groups may be regarded as the derivatives of phosphatidic acid.

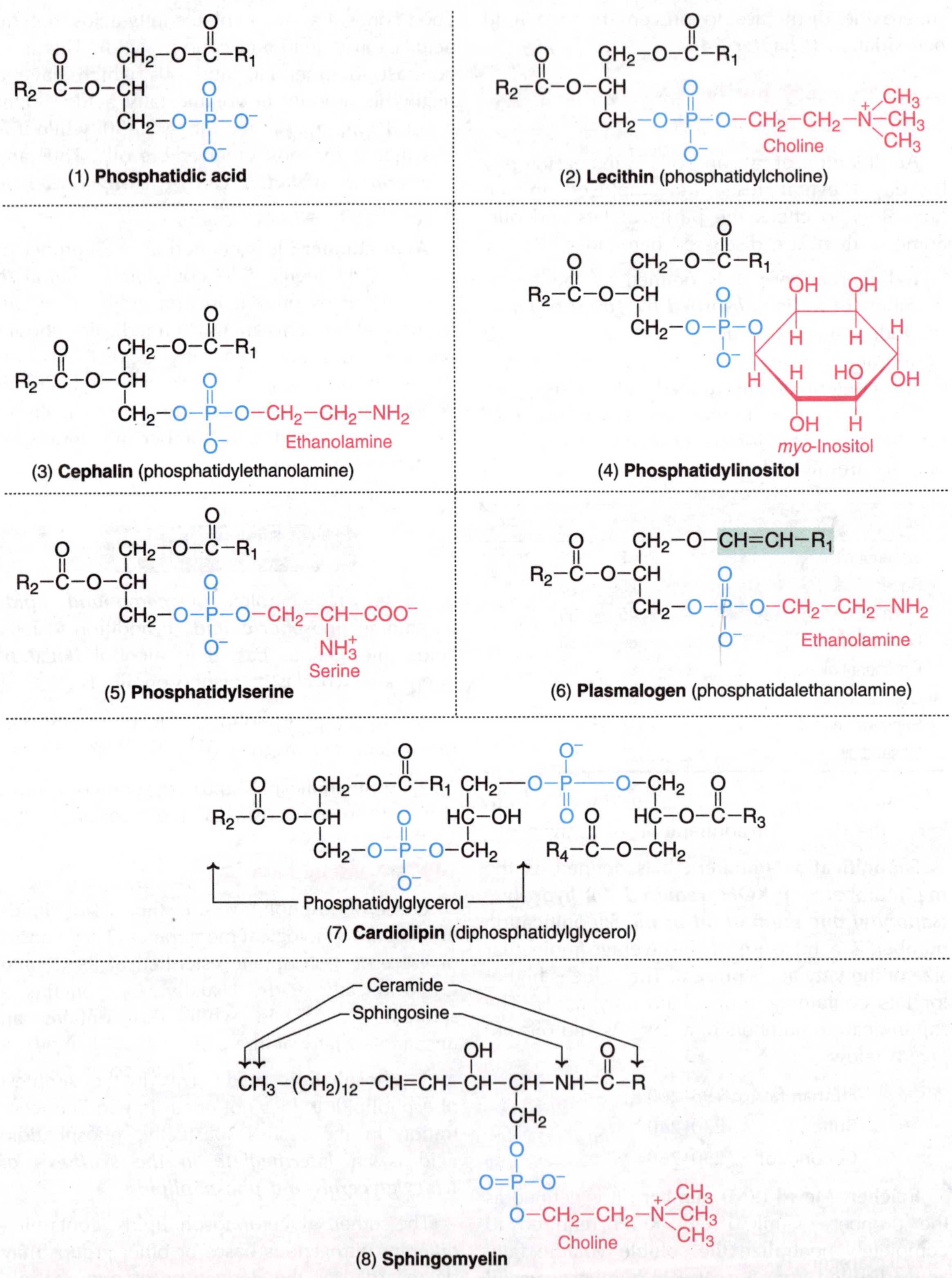

Fig. 3.3 : Structures of phospholipids.

> **Medical Concepts/ Clinical Correlates**
>
> - Trans fatty acids, widely used in food industry, are detrimental to health-linked with increased risk of diabetes, heart diseases etc.
> - Fats and oils on exposure to air, moisture, bacteria etc. undergo rancidity (deterioration). This can be prevented by the addition of certain antioxidants (vitamin E, hydroquinone, gallic acid).
> - In food preservation, antioxidants—namely propyl gallate, butylated hydroxyanisole and butylated hydroxytoluene—are commonly used.

2. **Lecithins (*phosphatidylcholine*):** These are the ***most abundant*** group of phospholipids in the cell membranes. Chemically, lecithin (*Greek* : lecithos—egg yolk) is a phosphatidic acid with ***choline as the base***. Phosphatidylcholines represent the ***storage form of body's choline***.

(a) **Dipalmitoyl lecithin** is an important phosphatidylcholine found in lungs that acts as a surfactant (details given later).

(b) **Lysolecithin** is formed by removal of the fatty acid either at C_1 or C_2 of lecithin.

3. **Cephalins (*phosphatidylethanolamine*) :** ***Ethanolamine*** is the ***nitrogenous base*** present in cephalins. Thus, lecithin and cephalin differ with regard to the base.

4. **Phosphatidylinositol :** The stereoisomer ***myo-inositol*** is attached to phosphatidic acid to give phosphatidylinositol (***PI***). This is an important component of cell membranes. The action of certain hormones (e.g. oxytocin, vasopressin) is mediated through PI.

5. **Phosphatidylserine :** The amino acid serine is present in this group of glycerophospholipids. ***Phosphatidylthreonine*** is also found in certain tissues and plays a role in ***apoptosis*** (programmed cell death).

6. **Plasmalogens :** When a fatty acid is attached by an ***ether*** linkage at C_1 of glycerol in the glycerophospholipids, the resultant compound is plasmalogen. Phosphati***dal***-ethanolamine is the most important which is similar in structure to phosphati***dyl***ethanolamine but for the ether linkage (in place of ester). An unsaturated fatty acid occurs at C_1. Choline, inositol and serine may substitute ethanolamine to give other plasmalogens.

7. **Cardiolipin :** It is so named as it was first isolated from heart muscle. Structurally, a cardiolipin consists of two molecules of phosphatidic acid held by an additional glycerol through phosphate groups. It is an important component of inner mitochondrial membrane and ***essential for mitochondrial function***. Decreased cardiolipin levels may result in mitochondrial dysfunction, aging, hypothyroidism, cardioskeletal myopathy (***Barth syndrome***). Cardiolipin is the only phosphoglyceride that possesses ***antigenic properties***.

In recent years, detection of ***anticardiolipin antibodies*** is used as a diagnostic tool in the evaluation of ***embryo transfers*** (in assisted reproductive technology) and ***organ transplantation***.

Sphingomyelins

Sphingosine is an amino alcohol present in sphingomyelins (sphingophospholipids). They do not contain glycerol at all. Sphingosine is attached by an amide linkage to a fatty acid to produce ***ceramide***. The alcohol group of sphingosine is bound to phosphorylcholine in sphingomyelin structure (***Fig.3.3***). Sphingomyelins are important constituents of myelin and are found in good quantity in brain and nervous tissues.

It is believed that sphingomyelins are involved in cell signaling, and in apoptosis.

Ceramide, acts as a second ***messenger*** (signaling molecule) by regulating programmed cell death (apoptosis), cell cycle and cell differentiation. A ceramide containing a 30-carbon fatty acid is a major component of skin, and it ***regulates skin's water permeability***.

Functions of phospholipids

Phospholipids constitute an important group of compound lipids that perform a wide variety of functions

1. In association with proteins, phospholipids form the structural ***components of membranes*** and regulate membrane permeability.

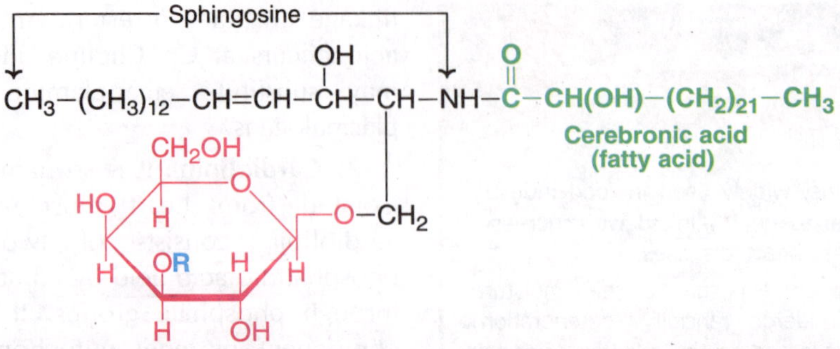

Fig. 3.4 : Structure of galactosylceramide (R = H). For sulfagalactosylceramide R is a sulfatide ($R = SO_4^{2-}$).

2. Phospholipids (lecithin, cephalin and cardiolipin) in the mitochondria maintain the ***conformation of electron transport chain components***, and thus cellular respiration.

3. Phospholipids participate in the ***absorption of fat*** from the intestine.

4. Phospholipids are essential for the synthesis of different lipoproteins, and thus participate in the ***transport of lipids***.

5. Accumulation of fat in liver (fatty liver) can be prevented by phospholipids, hence they are regarded as ***lipotropic factors***.

6. ***Arachidonic acid***, an unsaturated fatty acid liberated from phospholipids, serves as a precursor for the ***synthesis of eicosanoids*** (prostaglandins, prostacyclins, thromboxanes etc.).

7. Phospholipids participate in the ***reverse cholesterol transport*** and thus help in the removal of cholesterol from the body.

8. Phospholipids act as surfactants (agents lowering surface tension). For instance, ***dipalmitoyl phosphatidylcholine*** is an important lung surfactant.

9. ***Cephalins***, an important group of phospholipids ***participate in blood clotting***.

10. Phosphatidylinositol is the source of second messengers—***inositol triphosphate*** and ***diacylglyceol***, that are involved in the ***action of some hormones***.

Lung surfactant

Lung surfactant is mainly composed of phospholipid-dipalmitoyl lecithin, some proteins and carbohydrates. Dipalmitoyl lecithin is largely responsible for lung surfactant action (i.e., preventing adherence of inner surfaces of lungs), thereby reducing the workload of breathing. ***Infant respiratory distress syndrome*** (**IRDS**) in premature infants is ***due to the absence of lung surfactant***. This is caused by inability of immature lungs to synthesize dipalmitoyl lecithin.

Lecithin/sphingomyelin (**L/S**) ratio of amniotic fluid is an ***indicator of fetal maturity***. During the course of early fetal growth (up to 28th week), lung synthesizes more sphingomyelin, and later more lecithin. Thus, for a full term healthy infant, an ***L/S ratio above 2 indicates full lung maturity***. Low L/S ratio (<1) results in IRDS, a common cause of neonatal morbidity. IRDS is treated by administration of natural or artificial surfactants.

GLYCOLIPIDS

Glycolipids (***glycosphingolipids***) are important constituents of cell membrane and nervous tissues (particularly the brain). They occur as ***cell surface carbohydrates*** to form ***glycocalyx*** in plasma membranes.

Cerebrosides are the simplest form of glycolipids. They contain a ceramide (sphingosine attached to a fatty acid) and one or more sugars. Galactocerebroside (galactosylceramide) and glucocerebroside are the most important glycolipids. The structure of galactosylceramide is given in ***Fig.3.4***. It contains the fatty acid cerebronic acid. Sulfagalactosylceramide is the ***sulfatide*** derived from galactosylceramide.

Gangliosides : These are predominantly found in ganglions and are the most complex form of glycosphingolipids. They are the derivatives of

Medical Concepts/ Clinical Correlates

- The phospholipid – dipalmitoyl lecithin prevents the adherence of inner surface of the lungs, the absence of which is associated with respiratory distress syndrome in infants.
- Lecithin-sphingomyelin (L/S) ratio of amniotic fluid is an indicator of fetal maturity. L/S ratio above 2 indicates full lung maturity.
- Cephalins participate in blood clotting.

cerebrosides and contain one or more molecules of N-acetylneuraminic acid (NANA), the most important sialic acid. The structure of NANA is given in carbohydrate chemistry (***Refer Fig.2.11***).

The most important gangliosides present in the brain are GM_1, GM_2, GD, and GT, (***G*** represents ***g***anglioside while ***M, D*** and ***T*** indicate ***m***ono-, ***d***i- or ***t***ri- sialic acid residues, and the number denotes the carbohydrate sequence attached to the ceramide). The ganglioside, GM_2 that accumulates in Tay-Sachs disease is represented next (outline structure).

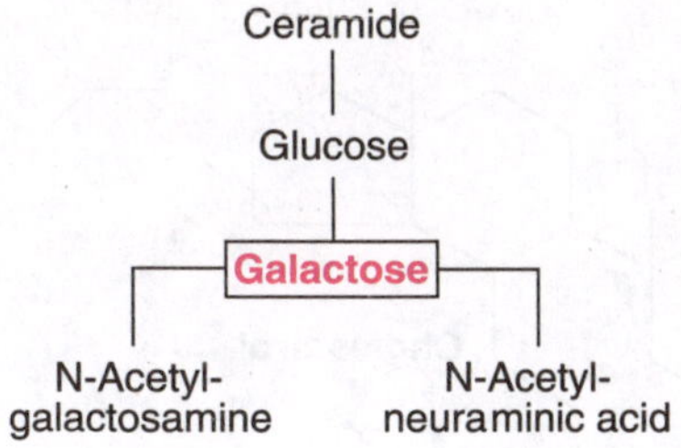

Globosides : A type of glycolipids with two or more sugars linked to ceramide the sugars usually present are glucose and galactose. Lactosylceramide (ceramide-galactose-glucose) is present in erythrocyte membrane.

LIPOPROTEINS

Lipoproteins are molecular complexes of lipids with proteins. They are the transport vehicles for lipids in the circulation. There are five types of lipoproteins, namely ***chylomicrons, very low density lipoproteins (VLDL), low density lipoproteins (LDL), high density lipoproteins (HDL) and free fatty acid-albumin complexes***. Their structure, separation, metabolism and diseases are discussed together (***Chapter 14***).

STEROIDS

Steroids are the compounds containing a cyclic steroid nucleus (or ring) namely ***cyclopentanoperhydrophenanthrene (CPPP)***. It consists of a phenanthrene nucleus (rings A, B and C) to which a cyclopentane ring (D) is attached.

The structure and numbering of CPPP are shown in ***Fig.3.5***. The steroid nucleus represents saturated carbons, unless specifically shown as double bonds. The methyl side chains (19 and 18) attached to carbons 10 and 13 are shown as single bonds. At carbon 17, steroids usually contain a side chain.

There are several steroids in the biological system. These include ***cholesterol, bile acids, vitamin D, sex hormones, adrenocortical hormones***, sitosterols, cardiac glycosides and alkaloids. If the steroid contains one or more hydroxyl groups it is commonly known as ***sterol*** (means solid alcohol).

CHOLESTEROL

Cholesterol, ***exclusively found in animals***, is the most abundant animal sterol. It is widely distributed in all cells and is a major component of cell membranes and lipoproteins. Cholesterol (*Greek* : chole–bile) was first isolated from bile. Cholesterol literally means 'solid alcohol from bile.'

Structure and occurrence

The structure of cholesterol ($C_{27}H_{46}O$) is depicted in ***Fig.3.5***. It has one hydroxyl group at C_3 and a double bond between C_5 and C_6. An 8 carbon aliphatic side chain is attached to C_{17}. Cholesterol contains a total of 5 methyl groups.

Due to the presence of an –OH group, cholesterol is weakly amphiphilic. As a structural component of plasma membranes, cholesterol is an important determinant of membrane permeability properties. The occurrence of cholesterol is much higher in the membranes of sub-cellular organelles.

Cholesterol is found in association with fatty acids to form cholesteryl esters (esterification occurs at the OH group of C_3).

Properties and reactions : Cholesterol is an yellowish crystalline solid. The crystals, under the microscope, show a notched (⏍) appearance. Cholesterol is insoluble in water and soluble in organic solvents such as chloroform, benzene, ether etc.

Several reactions given by cholesterol are useful for its qualitative identification and quantitative estimation. These include ***Salkowski test, Liebermann-Burchard reaction and Zak's test.***

Functions of cholesterol : Cholesterol is a poor conductor of heat and electricity, since it has a high dielectric constant. It is present in abundance in nervous tissues. It appears that cholesterol functions as an ***insulating cover*** for the transmission of electrical impulses in the nervous tissue. Cholesterol performs several other biochemical functions which include its role in ***membrane structure and function***, in the ***synthesis of bile acids, hormones*** (sex and cortical) and ***vitamin D*** (for details, ***Refer Chapters** 7* and ***19***).

Polyprenoids related to cholesterol

Isoprene units contain ***five carbons*** (–CH=$\overset{CH_3}{\overset{|}{C}}$–CH=C–). Some biological compounds, termed as polyprenoids, contain multiple isoprene units. Polyprenoids, although not steroids, are related to cholesterol since they are synthesized like cholesterol from isoprene units. Some examples are listed

- Ubiquinone (***coenzyme Q***) of electron transport chain.
- ***Dolichol*** involved in glycoprotein synthesis
- Plant-derived polyprenoids—***vitamins A, D, E and K, provitamin A*** (β-carotene) camphor, rubber.

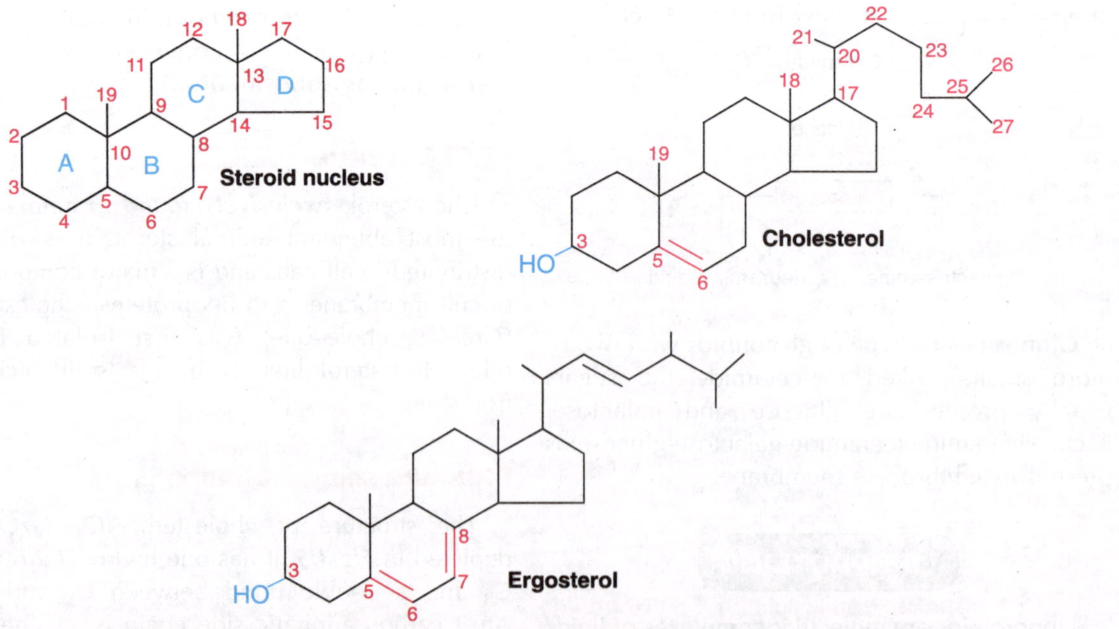

***Fig. 3.5 :** Structures of steroids (A, B, C—Perhydro-phenanthrene; D-Cyclopentane).*

Medical Concepts/ Clinical Correlates

- The action of certain hormones is mediated through phosphatidylinositol.
- Phospholipids are important for the synthesis and transport of lipoproteins and reverse transport of cholesterol.
- Patients with autoimmune diseases are susceptible to develop antiphospholipid (APL) antibodies. APL antibody syndrome is due to antibodies developed against phospholipid, namely cardiolipin.
- APL antibody syndrome has been implicated in pregnancy complications like miscarriage, eclampsia and preterm labour.

ERGOSTEROL

Ergosterol occurs in plants. It is also found as a structural constituent of membranes in yeast and fungi. Ergosterol (***Fig.3.5***) is an important precursor for vitamin D. When exposed to light, the ring B of ergosterol opens and it is converted to ergocalciferol, a compound containing vitamin D activity.

The other sterols present in plant cells include ***stigmasterol*** and ***β-sitosterol***.

AMPHIPATHIC LIPIDS

As per definition, lipids are insoluble (hydrophobic) in water. This is primarily due to the predominant presence of hydrocarbon groups. However, some of the lipids possess polar or hydrophilic groups which tend to be soluble in water. Molecules which ***contain both hydrophobic and hydrophilic groups*** are known as ***amphipathic*** (*Greek* : amphi-both, pathos—passion).

Examples of amphipathic lipids : Among the lipids, ***fatty acids, phospholipids***, sphingolipids, ***bile salts*** and ***cholesterol*** (to some extent) are amphipathic in nature.

Phospholipids have a hydrophilic head (phosphate group attached to choline, ethanolamine,

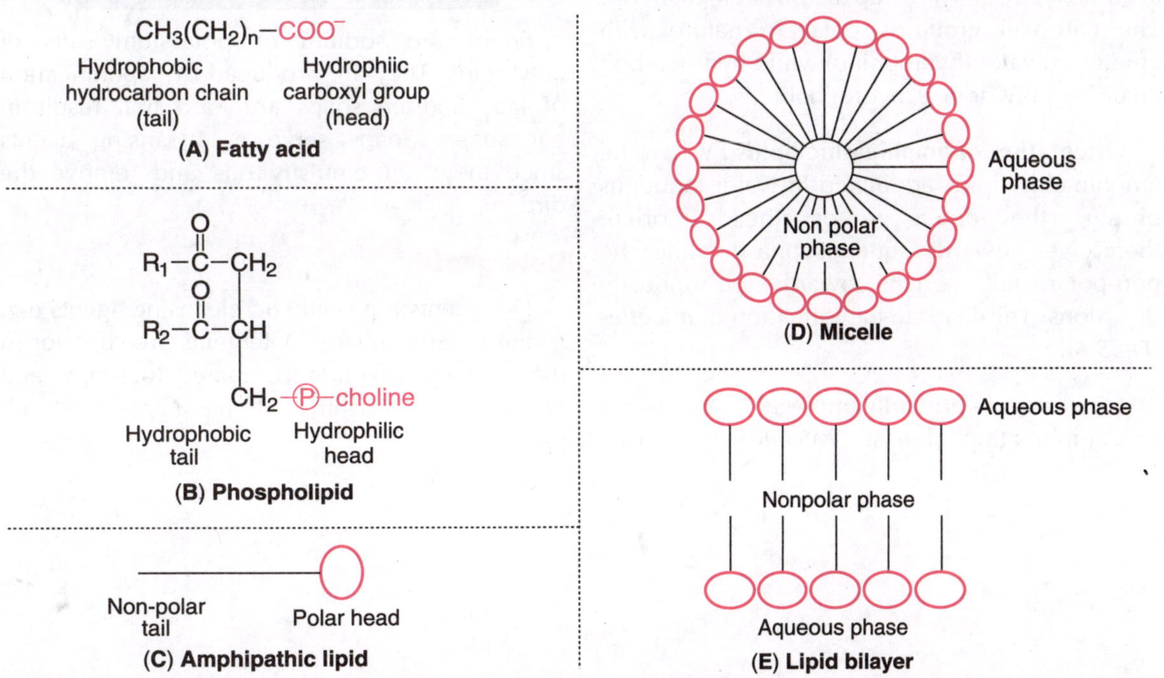

***Fig. 3.6 :** Summary of amphipathic lipids in the formation of micelle and lipid bilayer.*

Medical Concepts/ Clinical Correlates

- Cholesterol is essential for the synthesis of bile acids, hormones (sex and cortical) and vitamin D.
- Lipoproteins occur in the membrane structure, besides serving as a means of transport vehicles for lipids.
- Lipids are associated with certain disorders—obesity, atherosclerosis, and diabetes mellitus.
- Liposomes are used for administration of a variety of therapeutic substances (drugs, proteins, nucleic acids) in order to target specific organs or tissues.

inositol etc.) and a long hydrophobic tail. The general structure of an amphipathic lipid may be represented as a polar or hydrophilic head with a non-polar or hydrophobic tail (***Fig.3.6***).

Fatty acids contain a hydrocarbon chain with a carboxyl (COO^-) group at physiological p^H. The carboxyl group is polar in nature with affinity to water (hydrophilic) while hydrocarbon chain of fatty acid is hydrophobic.

Orientation of amphipathic lipids : When the amphipathic lipids are mixed in water (aqueous phase), the polar groups (heads) orient themselves towards aqueous phase while the non-polar (tails) orient towards the opposite directions. This leads to the formation of ***micelles*** (***Fig.3.6***).

Micelles are primarily molecular aggregates of amphipathic lipids. Micelle formation, facilitated by bile salts, is very important for lipid digestion and absorption (***Chapter 8***).

Membrane bilayers

In case of biological membranes, a bilayer of lipids is formed orienting the polar heads to the outer aqueous phase on either side and the nonpolar tails into the interior (***Fig.3.6***). The formation of a lipid bilayer is the basis of membrane structure.

Liposomes : They are produced when amphipathic lipids in aqueous medium are subjected to ***sonification***. They have intermittent aqueous phases in the lipid bilayer. Liposomes, in combination with tissue specific antigens, are used as carriers of drugs to target tissues.

Emulsions : These are produced when non-polar lipids (e.g. triacylglycerols) are mixed with water. The particles are larger in size and stabilized by emulsifying agents (usually amphipathic lipids), such as bile salts and phospholipids.

SOAPS AND DETERGENTS

Soaps are sodium or potassium salts of fatty acids. They are produced by saponification of fats. Sodium soaps are hard that result in bar soaps. Soaps serve as cleansing agents since they can emulsify oils and remove the dirt.

Detergents

Detergents are synthetic cleansing agents e.g. sodium lauryl sulfate. Detergents are superior in their cleansing action compared to soaps, and are used in washing clothes, and in tooth paste.

SUMMARY

- *Lipids are the organic substances relatively insoluble in water, soluble in organic solvents (alcohol, ether), actually or potentially related to fatty acids and are utilized by the body.*

- *Lipids are classified into simple (fats and oils), complex (phospholipids, glycolipids), derived (fatty acids, steroid hormones) and miscellaneous (carotenoids).*

- *Fatty acids are the major constituents of various lipids. Saturated and unsaturated fatty acids almost equally occur in natural lipids. The polyunsaturated fatty acids (PUFA) namely linoleic acid and linolenic acid are the essential fatty acids that need to be supplied in the diet.*

- *Triacylglycerols (simply fats) are the esters of glycerol with fatty acids. They are found in adipose tissue and primarily function as fuel reserve of animals. Several tests (iodine number, RM number) are employed in the laboratory to test the purity of fats and oils.*

- *Phospholipids are complex lipids containing phosphoric acid. Glycerophospholipids contain glycerol as the alcohol and these include lecithin, cephalin, phosphatidylinositol, plasmalogen and cardiolipin.*

- *Sphingophospholipids (sphingomyelins) contain sphingosine as the alcohol in place of glycerol (in glycerophospholipids). Phospholipids are the major constituents of plasma membranes.*

- *Cerebrosides are the simplest form of glycolipids which occur in the membranes of nervous tissue. Gangliosides are predominantly found in the ganglions. They contain one or more molecules of N-acetylneuraminic acid (NANA).*

- *Steroids contain the ring cyclopentanoperhydrophenanthrene. The steroids of biological importance include cholesterol, bile acids, vitamin D, sex hormones and cortical hormones. A steroid containing one or more hydroxyl groups is known as sterol.*

- *Cholesterol is the most abundant animal sterol. It contains one hydroxyl group (at C_3), a double bond (C_5–C_6) and an eight carbon side chain attached to C_{17}. Cholesterol is a constituent of membrane structure and is involved in the synthesis of bile acids, hormones (sex and cortical) and vitamin D.*

- *The lipids that possess both hydrophobic (non -polar) and hydrophilic (polar) groups are known as amphipathic. These include fatty acids, phospholipids, sphingolipids and bile salts. Amphipathic lipids are important constituents in the bilayers of the biological membranes.*

Self-Assessment Exercisesben

I. Essay questions

1. Write an account of classification of lipids with suitable examples.
2. Describe the structure and functions of phospholipids.
3. Discuss the saturated and unsaturated fatty acids of biological importance, along with their structures.
4. Describe the structure of steroids. Add a note on the functions of cholesterol.
5. Discuss the biological importance of amphipathic lipids.

II. Short notes

(a) Structure of triacylglycerols, (b) Glycolipids, (c) Essential fatty acids, (d) *Cis–trans* isomerism, (e) Rancidity, (f) Iodine number, (g) Phosphatidylinositol, (h) Sphingomyelins, (i) Steroid nucleus, (j) Micelles.

III. Fill in the blanks

1. The lipids that function as fuel reserve in animals ____________.
2. The isomerism associated with unsaturated fatty acids ____________.
3. The cyclic fatty acid employed in the treatment of leprosy ____________.
4. The lipids that are not the structural components of biological membranes ____________.
5. The prefix *sn* used to represent glycerol, *sn* stands for ____________.
6. The number of mg of KOH required to hydrolyse 1 g fat or oil is known as ____________.
7. The phospholipid that prevents the adherence of inner surfaces of lungs ____________.
8. The phospholipid that produces second messengers in hormonal action ____________.
9. Name the glycolipids containing N-acetylneuraminic acid ____________.
10. The steroids contain a cyclic ring known as ____________.

IV. Multiple choice questions

11. The nitrogenous base present in lecithin
 (a) Choline (b) Ethanolamine (c) Inositol (d) Serine.
12. The number of double bonds present in arachidonic acid
 (a) 1 (b) 2 (c) 3 (d) 4.
13. One of the following is an amphipathic lipid
 (a) Phospholipids (b) Fatty acid (c) Bile salts (d) All of the above.
14. Esterification of cholesterol occurs at carbon position
 (a) 1 (b) 2 (c) 3 (d) 4.
15. Name the test employed to check the purity of butter through the estimation of volatile fatty acids
 (a) Iodine number (b) Reichert-Meissl number (c) Saponification number (d) Acid number.

Proteins and Amino Acids 4

The proteins speak

"We are the basis of structure and function of life;
Composed of twenty amino acids, the building blocks;
Organized into primary, secondary, tertiary
and quaternary structure;
Classified as simple, conjugated and derived proteins."

Proteins are the ***most abundant organic molecules of the living system***. They occur in every part of the cell and constitute about 50% of the cellular dry weight. Proteins form the fundamental basis of structure and function of life.

Origin of the word 'protein'

The term **protein** is derived from a *Greek* word ***proteios***, meaning ***holding the first place***. Berzelius (Swedish chemist) suggested the name proteins to the group of organic compounds that are utmost important to life. Mulder (Dutch chemist) in 1838 used the term **proteins** for the high molecular weight nitrogen-rich and most abundant substances present in animals and plants.

The term ***proteomics*** is used for the entire ***study of galaxy of proteins*** that includes structure, function, genetics, pathology and various other aspects.

Functions of proteins

Proteins perform a great variety of specialized and essential functions in the living cells. These functions may be broadly grouped as ***static*** (***structural***) and ***dynamic***.

Structural functions : Certain proteins perform ***brick and mortar*** roles and are primarily responsible for structure and strength of body. These include ***collagen*** and ***elastin*** found in bone matrix, vascular system and other organs and ***α-keratin*** present in epidermal tissues.

Dynamic functions : The dynamic functions of proteins are more diversified in nature. These include proteins acting as ***enzymes, hormones, blood clotting factors, immunoglobulins***, membrane receptors, storage proteins, besides their function in genetic control, muscle contraction, respiration etc. Proteins performing dynamic functions are appropriately regarded as ***the working horses*** of cell.

Elemental composition of proteins

Proteins are predominantly constituted by five major elements in the following proportion.

Carbon	:	50 – 55%
Hydrogen	:	6 – 7.3%
Oxygen	:	19 – 24%
Nitrogen	:	13 – 19%
Sulfur	:	0 – 4%

Besides the above, proteins may also contain other elements such as P, Fe, Cu, I, Mg, Mn, Zn etc.

The content of ***nitrogen,*** an essential component of proteins, on an average is ***16%.*** Estimation of nitrogen in the laboratory (mostly by ***Kjeldahl's method***) is also used to find out the amount of protein in biological fluids and foods.

Proteins are polymers of amino acids

Proteins on complete hydrolysis (with concentrated HCl for several hours) yield L-α-amino acids. This is a common property of all the proteins. Therefore, ***proteins are the polymers of L-α-amino acids.***

Proteins are born at translation

A typical protein is born at translation (***protein biosynthesis***), and matures through post-translational process (Refer ***Chapter 25***). A protein performs specific and specialized functions, ages and dies (gets degraded) to release component amino acids.

STANDARD AMINO ACIDS

As many as 300 amino acids occur in nature—Of these, ***only 20***—known as standard amino acids are repeatedly ***found in the structure of proteins***, isolated from different forms of life—animal, plant and microbial. This is because of the universal nature of the genetic code available for the incorporation of only 20 amino acids when the proteins are synthesized in the cells. The process in turn is controlled by DNA, the genetic material of the cell. After the synthesis of proteins, some of the incorporated amino acids undergo modifications to form their derivatives.

AMINO ACIDS

Amino acids are a group of organic compounds containing two ***functional groups—amino*** and ***carboxyl***. The amino group (—NH_2) is basic while the carboxyl group (—COOH) is acidic in nature.

General structure of amino acids

The amino acids are termed as α-amino acids, if both the carboxyl and amino groups are attached to the same carbon atom, as depicted below

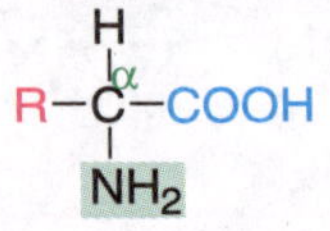

General structure

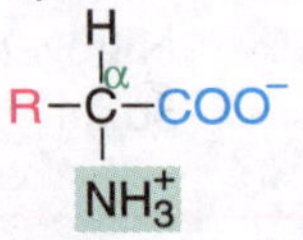

Exists as ion

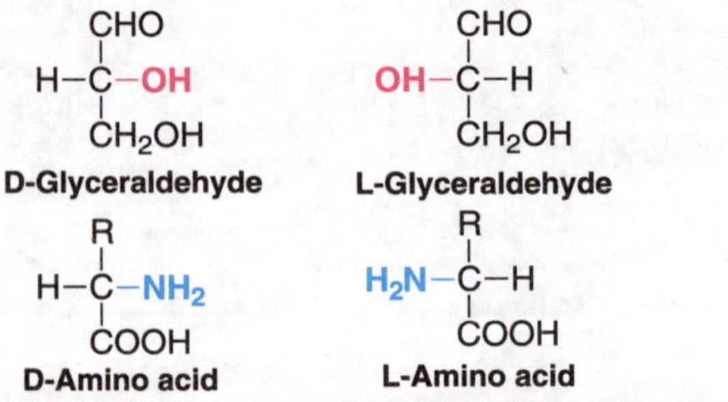

Fig. 4.1 : *D- and L-forms of amino acid based on the structure of glyceraldehyde.*

The α-carbon atom binds to a side chain represented by R which is different for each of the 20 amino acids found in proteins. The amino acids mostly exist in the ionized form in the biological system (shown above).

Optical isomers of amino acids

If a carbon atom is attached to four different groups, it is ***asymmetric*** and therefore exhibits ***optical isomerism***. The amino acids (except glycine) possess four distinct groups (R, H, COO^-, NH_3^+) held by α-carbon. Thus all the amino acids (except glycine where R = H) have optical isomers.

The structures of ***L- and D-amino acids*** are written based on the configuration of ***L- and D-glyceraldehyde*** as shown in ***Fig.4.1***. The proteins are composed of L-α-amino acids.

CLASSIFICATION OF AMINO ACIDS

There are different ways of classifying the amino acids based on the structure and chemical nature, nutritional requirement, metabolic fate etc.

A. Amino acid classification based on the structure

A comprehensive classification of amino acids is based on their structure and chemical nature. Each amino acid is assigned a 3 letter or 1 letter symbol. These symbols are commonly used to represent the amino acids in protein structure. The 20 amino acids found in proteins are divided into seven distinct groups.

In ***Table 4.1***, the different groups of amino acids, their symbols and structures are given. The salient features of different groups are described next

TABLE 4.1 Structural classification of L-α-amino acids found in proteins

Name	Symbol: 3 letters	Symbol: 1 letter	Structure	Special group present
I. Amino acids with aliphatic side chains				
1. Glycine	Gly	G	$H-CH(NH_3^+)-COO^-$	
2. Alanine	Ala	A	$CH_3-CH(NH_3^+)-COO^-$	
3. Valine	Val	V	$(H_3C)_2CH-CH(NH_3^+)-COO^-$	Branched chain
4. Leucine	Leu	L	$(H_3C)_2CH-CH_2-CH(NH_3^+)-COO^-$	Branched chain
5. Isoleucine	Ile	I	$CH_3-CH_2-CH(H_3C)-CH(NH_3^+)-COO^-$	Branched chain
II. Amino acids containing hydroxyl (—OH) groups				
6. Serine	Ser	S	$CH_2(OH)-CH(NH_3^+)-COO^-$	Hydroxyl
7. Threonine	Thr	T	$H_3C-CH(OH)-CH(NH_3^+)-COO^-$	Hydroxyl
Tyrosine	Tyr	Y	See under aromatic	Hydroxyl

Table 4.1 contd. next page

Name	Symbol: 3 letters	Symbol: 1 letter	Structure	Special group present
III. Sulfur containing amino acids				
8. Cysteine	Cys	C	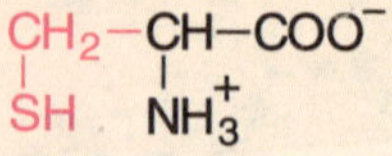	Sulfhydryl
Cystine	—	—	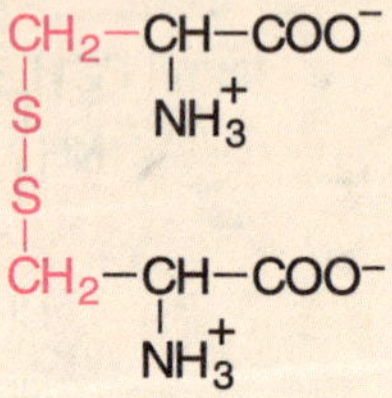	Disulfide
9. Methionine	Met	M	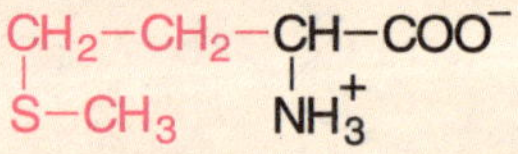	Thioether
IV. Acidic amino acids and their amides				
10. Aspartic acid	Asp	D	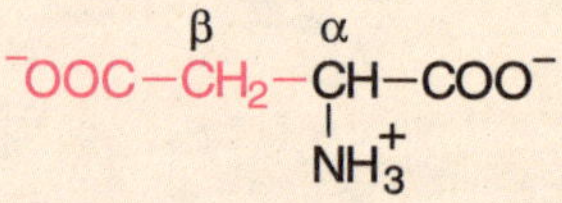	β-Carboxyl
11. Asparagine	Asn	N	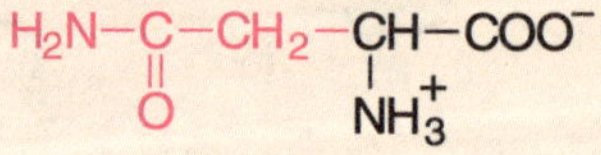	Amide
12. Glutamic acid	Glu	E	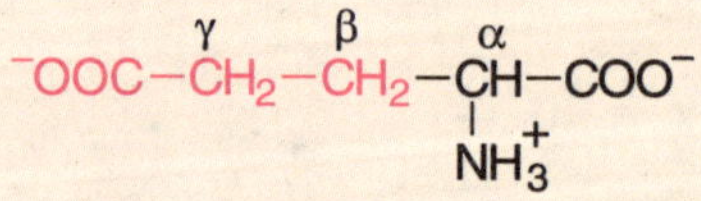	γ-Carboxyl
13. Glutamine	Gln	Q	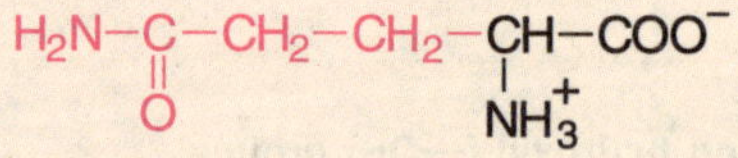	Amide
V. Basic amino acids				
14. Lysine	Lys	K	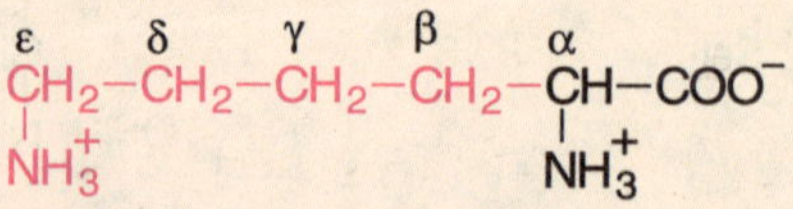	ε-Amino
15. Arginine	Arg	R	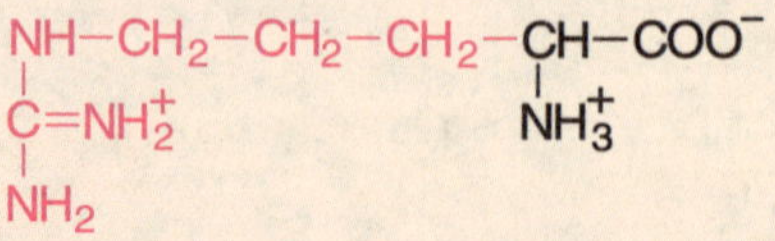	Guanidino
16. Histidine	His	H	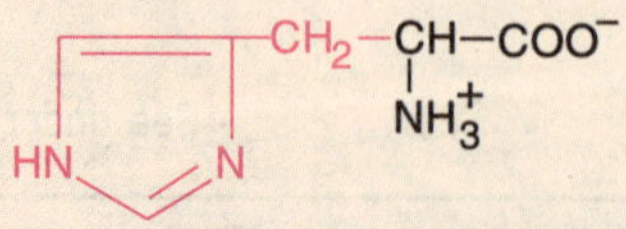	Imidazole

Table 4.1 contd. next page

Name	Symbol 3 letters	Symbol 1 letter	Structure	Special group present
VI. Aromatic amino acids				
17. Phenylalanine	Phe	F	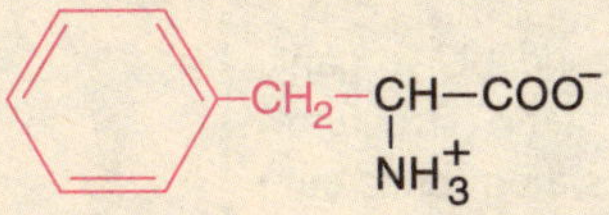	Benzene or phenyl
18. Tyrosine	Tyr	Y	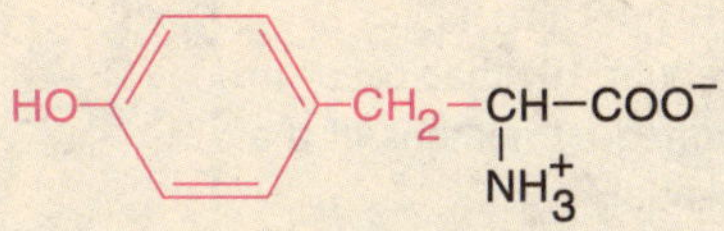	Phenol
19. Tryptophan	Trp	W	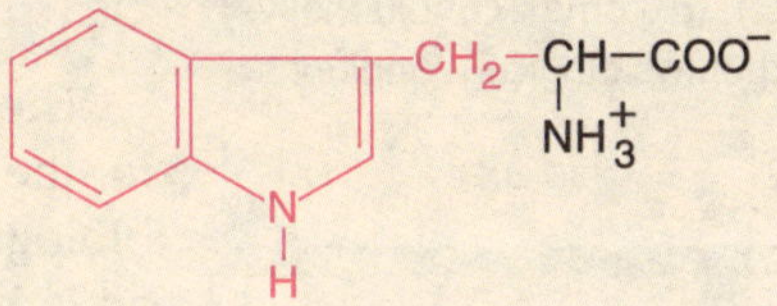	Indole
VII. Imino acid				
20. Proline	Pro	P	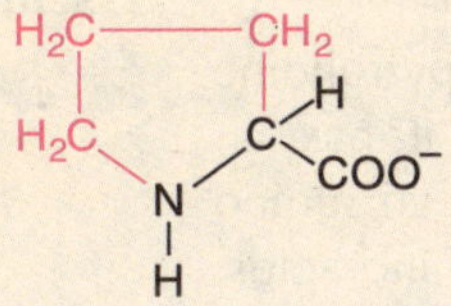 or 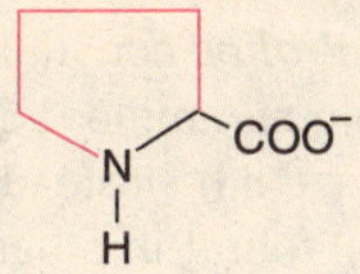	Pyrrolidine

***(Note** : R group is shown in red)*

1. **Amino acids with aliphatic side chains :** These are monoamino monocarboxylic acids. This group consists of the most simple amino acids—***glycine***, ***alanine***, ***valine***, ***leucine*** and ***isoleucine***. The last three amino acids (Leu, Ile, Val) contain branched aliphatic side chains, hence they are referred to as ***branched chain amino acids.***

2. **Hydroxyl group containing amino acids :** ***Serine***, ***threonine*** and ***tyrosine*** are hydroxyl group containing amino acids. Tyrosine—being aromatic in nature—is usually considered under aromatic amino acids.

3. **Sulfur containing amino acids :** ***Cysteine*** with sulfhydryl group and ***methionine*** with thioether group are the two amino acids incorporated during the course of protein synthesis. ***Cystine***, another important sulfur containing amino acid, is formed by condensation of two molecules of cysteine.

4. **Acidic amino acids and their amides :** ***Aspartic acid*** and ***glutamic acids*** are ***dicarboxylic monoamino acids*** while asparagine and glutamine are their respective amide derivatives. All these four amino acids possess distinct codons for their incorporation into proteins.

5. **Basic amino acids :** The three amino acids ***lysine***, ***arginine*** (with guanidino group) and ***histidine*** (with imidazole ring) are dibasic monocarboxylic acids. They are highly basic in character.

6. **Aromatic amino acids :** ***Phenylalanine***, ***tyrosine*** and ***tryptophan*** (with indole ring)

Medical Concepts/ Clinical Correlates

- Proteins are the most abundant organic molecules of life. They perform static (structural) and dynamic functions (enzymes, hormones etc.) in the living cells.
- Half of the amino acids (about 10) that occur in proteins have to be consumed by humans in the diet, hence they are essential.
- Some non-essential amino acids (glycine, arginine, tyrosine etc.) become conditionally essential in certain pathophysiological conditions (surgery, sepsis etc.).

are aromatic amino acids. Besides these, histidine may also be considered under this category.

7. **Imino acids :** ***Proline*** containing pyrrolidine ring is a unique amino acid. It has an imino group (=NH), instead of an amino group ($-NH_2$) found in other amino acids. Therefore, proline is an α-imino acid.

   Heterocyclic amino acids : Histidine, tryptophan and proline.

B. Classification of amino acids based on polarity

Amino acids are classified into 4 groups based on their polarity. Polarity is important for protein structure.

1. **Non-polar amino acids :** These amino acids are also referred to as hydrophobic (water hating). They have no charge on the 'R' group. The amino acids included in this group are — ***alanine***, ***leucine***, ***isoleucine***, valine, methionine, phenylalanine, tryptophan and proline.
2. **Polar amino acids with no charge on 'R' group :** These amino acids, as such, carry no charge on the 'R' group. They however possess groups such as hydroxyl, sulfhydryl and amide and participate in hydrogen bonding of protein structure. The simple amino acid glycine (where R = H) is also considered in this category. The amino acids in this group are— glycine, serine, threonine, cysteine, glutamine, asparagine and tyrosine.
3. **Polar amino acids with positive 'R' group :** The three amino acids lysine, arginine and histidine are included in this group.
4. **Polar amino acids with negative 'R' group :** The dicarboxylic monoamino acids— aspartic acid and glutamic acid are considered in this group.

C. Nutritional classification of amino acids

The 20 amino acids (***Table 4.1***) are required for the synthesis of variety proteins, besides other biological functions. However, all these 20 amino acids need not be taken in the diet. Based on the nutritional requirements, amino acids are grouped into three classes—***essential***, ***non-essential and conditionally essential***.

1. **Essential or indispensable amino acids :** The amino acids which ***cannot be synthesized by the body*** and, therefore, need to be supplied through the diet are called essential amino acids. They are required for proper growth and maintenance of the individual. The ten amino acids listed below are essential for humans (and also rats) :

   Arginine, **V**aline, **H**istidine, **I**soleucine, **L**eucine, **L**ysine, **M**ethionine, **Ph**enylalanine, **T**hreonine, **T**ryptophan.

   [The code ***A.V. HILL, MP., T. T.*** (first letter of each amino acid) may be memorized to recall essential amino acids. Other useful codes are H. VITTAL, LMP; PH. VILLMA, TT, PVT TIM HALL and MATTVILPhLy.]

   The two amino acids namely **a**rginine and **h**istidine can be synthesized by adults and not by growing children, hence these are considered as ***semi-essential amino acids*** (remember ***Ah***, to recall). Thus, 8 amino acids are absolutely essential while 2 are semi-essential.

2. **Non-essential or dispensable amino acids :** The body can synthesize about 10 amino acids to meet the biological needs, hence they need not be consumed in the diet. These are—glycine, alanine, serine, cysteine, aspartate, asparagine, glutamate, glutamine, tyrosine and proline.

3. **Conditionally essential or indispensable amino acids :** The synthesis of some non-essential amino acids is limited, although it is adequate to meet the body's requirement in a healthy state. In certain (pathophysiological) conditions like surgery, wound, sepsis, thermal injury etc., the ***non-essential amino acids become*** conditionally ***essential*** (indispensable). ***Arginine, cysteine, glycine, glutamine, proline*** and ***tyrosine*** are considered as conditionally essential amino acids, and ***need to be supplemented in the diet during pathophysiological conditions***.

D. Amino acid classification based on their metabolic fate

The carbon skeleton of amino acids can serve as a precursor for the synthesis of glucose (glycogenic) or fat (ketogenic) or both. From metabolic view point, amino acids are divided into three groups (for details, ***Refer Chapter 15***).

1. **Glycogenic amino acids :** These amino acids can serve as precursors for the formation of glucose or glycogen. e.g. ***alanine***, ***aspartate***, glycine, methionine etc.

2. **Ketogenic amino acids :** Fat can be synthesized from these amino acids. Two amino acids ***leucine*** and ***lysine*** are exclusively ketogenic.

3. **Glycogenic and ketogenic amino acids :** The four amino acids ***isoleucine***, phenylalanine, tryptophan, ***tyrosine*** are precursors for synthesis of glucose as well as fat.

Selenocysteine – the 21st amino acid

As already stated, 20 amino acids are commonly found in proteins. In recent years, a 21st amino acid namely selenocysteine has been added. It is found at the active sites of certain enzymes/proteins (***selenoproteins***). e.g. ***glutathione peroxidase***, glycine reductase, 5′-deiodinase, thioredoxin reductase. Selenocysteine is an unusual amino acid containing the trace element selenium in place of the sulfur atom of cysteine.

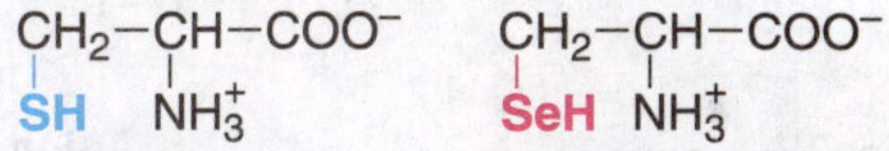

Cysteine ***Selenocysteine***

Incorporation of selenocysteine into the proteins during translation is carried out by the codon namely UGA. It is interesting to note that UGA is normally a stop codon that terminates protein biosynthesis. Another unique feature is that selenocysteine is enzymatically generated from serine directly on the tRNA (selenocysteine-tRNA), and then incorporated into proteins.

Pyrrolysine – the 22nd amino acid? : In the year 2002, some researchers have described yet another amino acid namely pyrrolysine as the 22nd amino acid present in protein. The stop codon UAG can code for pyrrolysine.

(**Note :** Selenocysteine and pyrrolysine are directly incorporated into proteins during translation)

Properties of amino acids

The amino acids differ in their physico–chemical properties which ultimately determine the characteristics of proteins.

A. Physical properties

1. **Solubility :** Most of the amino acids are ***usually soluble in water*** and insoluble in organic solvents.

2. **Melting points :** Amino acids generally melt at higher temperatures, often above 200°C.

3. **Taste :** Amino acids may be sweet (Gly, Ala, Val), tasteless (Leu) or bitter (Arg, Ile). ***Monosodium glutamate*** (MSG; ***ajinomoto***) is used as a flavoring agent in food industry, and Chinese foods to increase taste and flavor. In some individuals intolerant to MSG, ***Chinese restaurant syndrome*** (brief and reversible flu-like symptoms) is observed.

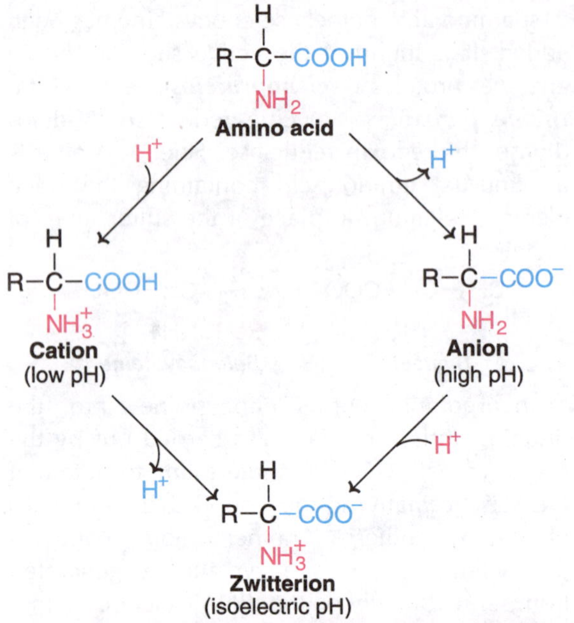

***Fig. 4.2 :** Existence of an amino acid as cation, anion and zwitterion.*

4. **Optical properties :** All the amino acids ***except glycine*** possess optical isomers due to the presence of asymmetric carbon atom. Some amino acids also have a second asymmetric carbon e.g. isoleucine, threonine. The structure of L- and D-amino acids in comparison with glyceraldehyde has been given (***See Fig.4.1***).

5. **Amino acids as ampholytes :** Amino acids contain both acidic (–COOH) and basic (–NH_2) groups. They ***can donate a proton or accept a proton***, hence amino acids are regarded as ampholytes.

Zwitterion or dipolar ion : The name *zwitter* is derived from the German word which means ***hybrid***. Zwitter ion (or dipolar ion) is a hybrid molecule containing ***positive and negative ionic groups***.

The amino acids rarely exist in a neutral form with free carboxylic (–COOH) and free amino (–NH_2) groups. In strongly ***acidic pH*** (low pH), the amino acid is positively charged (***cation***) while in strongly ***alkaline pH*** (high pH), it is negatively charged (***anion***). Each amino acid has a characteristic pH (e.g. leucine, pH 6.0) at which it carries both positive and negative charges and exists as zwitterion (***Fig.4.2***).

Isoelectric pH (symbol pI) is defined as the pH at which a ***molecule exists as a zwitterion*** or ***dipolar ion*** and carries no net charge. Thus, the molecule is electrically neutral.

The pI value can be calculated by taking the average pKa values corresponding to the ionizable groups. For instance, leucine has two ionizable groups, and its pI can be calculated as follows.

$$pI = \frac{pK_1(COO^-) + pK_2(NH_3^+)}{2}$$

$$pI = \frac{2.4 + 9.6}{2} = 6.0$$

Leucine exists as cation at pH below 6 and anion at pH above 6. At the isoelectric pH (pI = 6.0), leucine is found as zwitterion. Thus the pH of the medium determines the ionic nature of amino acids.

For the calculation of pI of amino acids with more than two ionizable groups, the pKas for all the groups have to be taken into account.

Titration of amino acids : The existence of different ionic forms of amino acids can be more easily understood by the titration curves. The graphic representation of leucine titration is depicted in ***Fig.4.3***. At low pH, leucine exists in

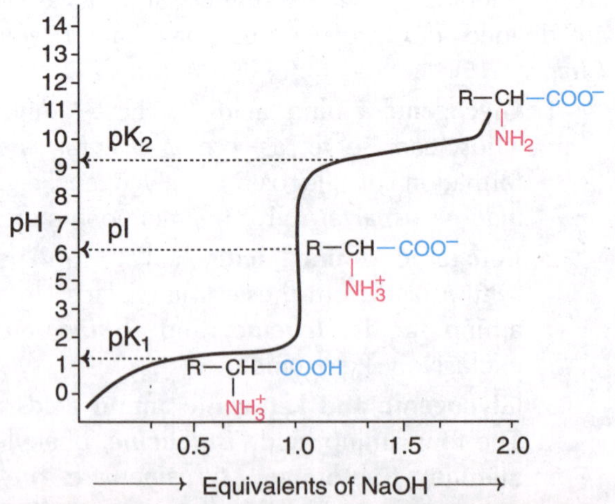

***Fig. 4.3 :** Titration curve of an amino acid-leucine ($R = (CH_3)_2$—CH—CH_2—; pK_1 = Dissociation constant for COOH; pI = Isoelectric pH; pK_2 = Dissociation constant for NH_3^+).*

Medical Concepts/ Clinical Correlates

- A protein is said to be complete (or first class) protein if all the essential amino acids are present in the required proportion by the human body e.g. egg albumin.
- Cooking results in protein denaturation exposing more peptide bonds for easy digestion.
- Monosodium glutamate (MSG; ajinomoto) is used as a flavoring agent in foods to increase taste and flavour. In some individuals intolerant to MSG, Chinese restaurant syndrome (brief and reversible flu-like symptoms) is observed.

a fully protonated form as cation. As the titration proceeds with NaOH, leucine loses its protons and at isoelectric pH (pI), it becomes a zwitterion. Further titration results in the formation of anionic form of leucine.

Some more details on isoelectric pH are discussed under the properties of proteins (p. 64).

B. Chemical properties

The general reactions of amino acids are mostly due to the presence of two functional groups namely carboxyl (−COOH) group and amino ($-NH_2$) group.

Reactions due to −COOH group

1. Amino acids form salts (−COONa) with bases and esters (−COOR′) with alcohols.
2. **Decarboxylation :** Amino acids undergo decarboxylation to produce corresponding amines.

$$R{-}\underset{\underset{NH_3^+}{|}}{CH}{-}COO^- \longrightarrow R{-}\underset{\underset{NH_3^+}{|}}{CH_2} + CO_2$$

   This reaction assumes significance in the living cells due to the formation of many ***biologically important amines***. These include ***histamine, tyramine and γ-amino butyric acid (GABA)*** from the amino acids histidine, tyrosine and glutamate, respectively.
3. **Reaction with ammonia :** The carboxyl group of dicarboxylic amino acids reacts with NH_3 to form amide

   Aspartic acid + NH_3 $\longrightarrow$ Asparagine

   Glutamic acid + NH_3 $\longrightarrow$ Glutamine

Reactions due to $-NH_2$ group

4. The amino groups behave as bases and combine with acids (e.g. HCl) to form salts ($-NH_3^+Cl^-$).
5. **Reaction with ninhydrin :** The α-amino acids react with ninhydrin to form a purple, blue or pink colour complex (***Ruhemann's purple***).

   Amino acid + Ninhydrin $\longrightarrow$ Keto acid + NH_3 + CO_2 + Hydrindantin

   Hydrindantin + NH_3 + Ninhydrin $\longrightarrow$ Ruhemann's purple

   Ninhydrin reaction is effectively used for the ***quantitative determination of amino acids and proteins.*** (***Note :*** Proline and hydroxyproline give yellow colour with ninhydrin).
6. **Colour reactions of amino acids :** Amino acids can be identified by specific colour reactions (***See Table 4.3***).
7. **Transamination :** Transfer of an amino group from an amino acid to a keto acid to form a new amino acid is a very important reaction in amino acid metabolism (details given in ***Chapter 15***).
8. **Oxidative deamination :** The amino acids undergo oxidative deamination to liberate free ammonia (***Refer Chapter 15***).

NON-STANDARD AMINO ACIDS

Besides the 20 standard amino acids (described above) present in the protein structure, there are several other amino acids which are biologically important. These include the ***amino acid derivatives*** found in proteins, ***non-protein amino acids*** performing specialized functions and the ***D-amino acids***.

Table 4.2 A selected list of important non-protein amino acids along with their functions

Amino acids	Function(s)
I. α-Amino acids	
Ornithine Citrulline Argininosuccinic acid	Intermediates in the biosynthesis of urea.
Thyroxine Triiodothyronine	Thyroid hormones derived from tyrosine.
S-Adenosylmethionine	Methyl donor in biological system.
Homocysteine	Intermediate in methionine metabolism. A risk factor for coronary heart diseases
Homoserine	Intermediate in threonine, aspartate and methionine metabolisms.
3, 4-Dihydroxy phenylalanine (DOPA)	A neurotransmitter, serves as a precursor for melanin pigment.
Creatinine	Derived from muscle and excreted in urine
Ovothiol	Sulfur containing amino acid found in fertilized eggs, and acts as an antioxidant
Azaserine	Anticancer drug
Cycloserine	Antituberculosis drug
II. Non-α-amino acids	
β-Alanine	Component of vitamin pantothenic acid and coenzyme A
β-Aminoisobutyric acid	End product of pyrimidine metabolism.
γ-Aminobutyric acid (GABA)	A neurotransmitter produced from glutamic acid
δ-Aminolevulinic acid (ALA)	Intermediate in the synthesis of porphyrin (finally heme)
Taurine	Found in association with bile acids.

A. Amino acid derivatives in proteins : The 20 standard amino acids can be incorporated into proteins due to the presence of universal genetic code. Some of these amino acids undergo specific modification after the protein synthesis (post-translational modifications) occurs.

- Collagen—the most abundant protein in mammals—contains ***4-hydroxyproline*** and ***5-hydroxylysine***.
- Histones—the proteins found in association with DNA—contain many methylated, phosphorylated or acetylated amino acids.
- ***γ-Carboxyglutamic acid*** is found in certain plasma proteins involved in blood clotting.

Peptidal amino acids and post-translational modifications

The ***modifications*** that occur in polypeptides ***during or after protein biosynthesis***, are referred to as post-translational modifications. Peptidal proline, peptidal lysine, and peptidal glutamate respectively get converted to hydroxyproline, hydroxylysine and γ-carboxyglutamate. The other post-translational modifications include methylation, acetylation, phosphorylation and formylation. These modifications are important for protein structure and function—solubility, stability, interaction with other proteins and biological activity.

B. Non-protein amino acids : These amino acids, although ***never found in proteins***, perform several biologically important functions. They may be either α-or non-α-amino acids. A selected list of these amino acids along with their functions is given in ***Table 4.2***.

C. D-Amino acids : The vast majority of amino acids isolated from animals and plants are of L-category. Certain D-amino acids are also

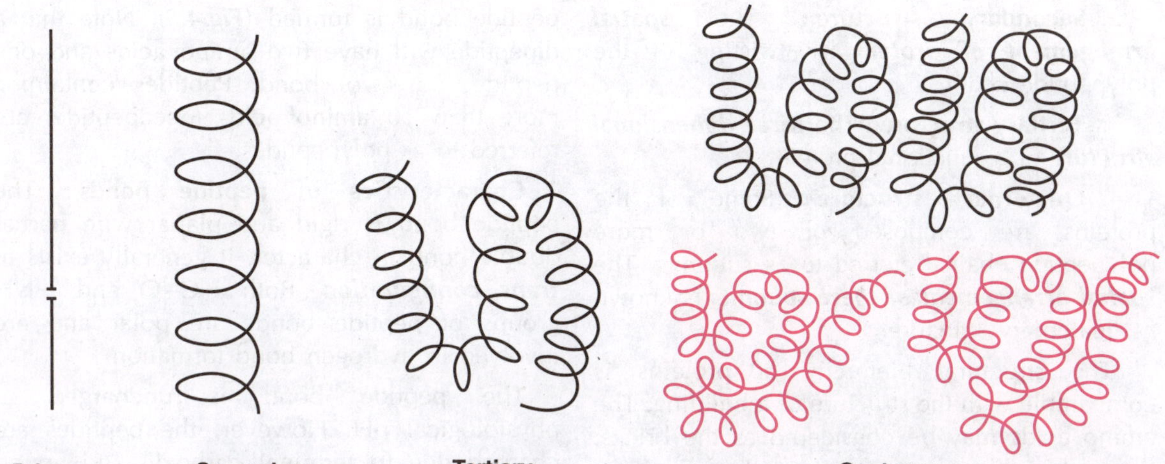

Fig. 4.4 : Diagrammatic representation of protein structure (Note : The four subunits of two types in quaternary structure).

found in the ***antibiotics*** (actinomycin-D, valinomycin, gramicidin-S). D-serine and D-aspartate are found in brain tissue. D-Glutamic acid and D-alanine are present in bacterial cell walls.

Amino acids useful as drugs

There a certain non-standard amino acids that are used as drugs.

- ***D-Penicillamine*** (D-dimethylglycine), a metabolite of penicillin, is employed in the chelation ***therapy of Wilson's disease***. This is possible since D-penicillamine can effectively chelate copper.
- ***N-Acetylcysteine*** is used in ***cystic fibrosis***, and chronic renal insufficiency, as it can function as an antioxidant.
- ***Gabapentin*** (γ-aminobutyrate linked to cyclohexane) is used as an ***anticonvulsant***.

Medical Concepts/ Clinical Correlates

- Certain non-standard amino acids are useful as drugs e.g. D-penicillamine in Wilson's disease, N-acetylcysteine in cystic fibrosis.
- Protein misfolding is associated with prion diseases, cystic fibrosis etc.

Potentially toxic amino acids

Certain non-protein amino acids, present in plants, have been identified to adversely affect human health.

- ***L-Homoarginine*** and β-N-oxalyldiaminopropionic acid (β-ODAP) of Lathyrus (Kesari dal) seeds results in ***lathyrism*** (an irreversible disease wherein individuals lose limbs control).
- The seeds of lathyrus also contain α,γ-diaminobutyric acid (an analogue of ornithine) that disrupts urea cycle, and causes ***ammonia toxicity***.
- L-β-Methylaminoalanine of cycad seeds is neurotoxic and has been implicated in neurodegenerative diseases e.g. ***Parkinson's dementia***, amyotrophic lateral sclerosis.

STRUCTURE OF PROTEINS

Proteins are the polymers of L-α-amino acids. The structure of proteins is rather complex which can be divided into 4 levels of organization (***Fig.4.4***) :

1. **Primary structure :** ***The linear sequence of amino acids*** forming the backbone of proteins (polypeptides).

2. **Secondary structure :** The ***spatial arrangement of protein*** by twisting of the polypeptide chain.

3. **Tertiary structure :** The ***three dimensional structure*** of a functional protein.

4. **Quaternary structure :** Some of the proteins are composed of two or more polypeptide chains referred to as subunits. The ***spatial arrangement of these subunits*** is known as quaternary structure.

[The structural hierarchy of proteins is comparable with the structure of a building. The amino acids may be considered as the bricks, the wall as the primary structure, the twists in a wall as the secondary structure, a full-fledged self-contained room as the tertiary structure. A building with similar and dissimilar rooms will be the quaternary structure].

The term ***protein*** is generally used for a polypeptide containing ***more than 50 amino acids.*** In recent years, however, some authors have been using '***polypeptide***' even if the number of amino acids is a few hundreds. They prefer to use protein to an assembly of polypeptide chains with quaternary structure.

PRIMARY STRUCTURE OF PROTEIN

Each protein has a unique sequence of amino acids which is determined by the genes contained in DNA. The primary structure of a protein is largely responsible for its function. A vast majority of genetic diseases are due to abnormalities in the amino acid sequences of proteins i.e. changes associated with primary structure of protein.

The amino acid composition of a protein determines its physical and chemical properties.

Peptide bond

The amino acids are held together in a protein by covalent peptide bonds or linkages. These bonds are rather strong and serve as the cementing material between the individual amino acids (considered as bricks).

Formation of a peptide bond : When the ***amino group*** of an amino acid combines with the ***carboxyl group*** of another amino acid, a peptide bond is formed (***Fig.4.5***). Note that a dipeptide will have two amino acids and one peptide (not two) bond. Peptides containing more than 10 amino acids (decapeptide) are referred to as polypeptides.

Characteristics of peptide bonds : The peptide bond is rigid and planar with partial double bond in character. It generally exists in ***trans configuration***. Both −C=O and −NH groups of peptide bonds are polar and are involved in hydrogen bond formation.

The peptide bond is uncharged at physiological pH. However, the peptides are charged due to terminal carboxyl and amino groups. Thus, ***peptides*** may be regarded as ***polyelectrolytes***.

Writing of peptide structures : Conventionally, the peptide chains are written with the free amino end (N-terminal residue) at the left, and the free carboxyl end (C-terminal residue) at the right. The amino acid sequence is read from N-terminal end to C-terminal end. Incidentally, the protein biosynthesis also starts from the N-terminal amino acid.

Shorthand to read peptides : The amino acids in a peptide or protein are represented by the 3-letter or one letter abbreviation. This is the ***chemical shorthand*** to write proteins.

Naming of peptides : For naming peptides, the amino acid suffixes ***-ine*** (glycine), ***-an*** (tryptophan), ***-ate*** (glutamate) are changed to ***-yl*** with the exception of C-terminal amino acid. Thus a tripeptide composed of an N-terminal

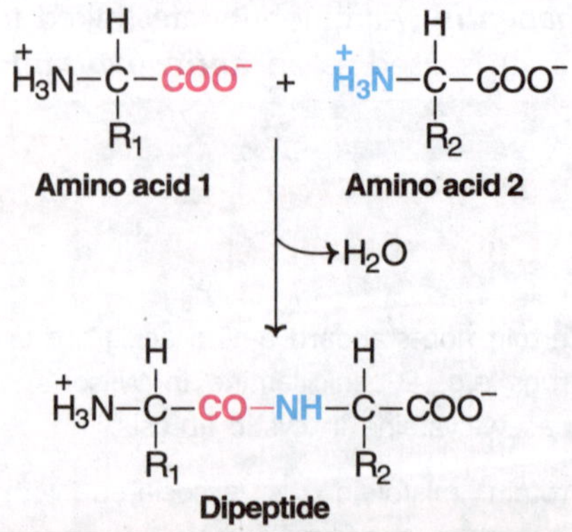

Fig. 4.5 : Formation of a peptide bond.

glutamate, a cysteine and a C-terminal glycine is called glutamyl-cysteinyl-glycine.

In the ***Fig.4.6***, the naming and representation of a tripeptide are shown.

$\overset{+}{H_3N}$—glutamate—cysteine—glycine—COO^-	Amino acids in a peptide
E — C — G	One letter symbols
Glu — Cys — Gly	Three letter symbols
Glutamyl — cysteinyl — glycine	Peptide name

***Fig. 4.6 :** Use of symbols in representing a peptide (**Note :** A tripeptide with 3 amino acids and two peptide bonds is shown; Free—NH_3^+ is on the left while free —COO^- is on the right).*

Dimensions of a peptide chain : The dimensions of a fully extended polypeptide chain are depicted in ***Fig.4.7***. The two adjacent α-carbon atoms are placed at a distance of 0.36 nm. The interatomic distances and bond angles are also shown in this figure.

Determination of primary structure

The primary structure comprises the identification of constituent amino acids with regard to their quality, quantity and sequence in a protein structure. ***A pure sample of a protein** or **a polypeptide is essential*** for the determination of primary structure which involves 3 stages :

1. Determination of amino acid composition.
2. Degradation of protein or polypeptide into smaller fragments.
3. Determination of the amino acid sequence.

1. Determination of amino acid composition in a protein : The protein or polypeptide is completely ***hydrolysed to liberate the amino acids*** which are quantitatively estimated. The hydrolysis may be carried out either by acid or alkali treatment or by enzyme hydrolysis. Treatment with enzymes, however results in smaller peptides rather than amino acids.

Pronase is a mixture of non-specific proteolytic enzymes that causes complete hydrolysis of proteins.

Separation and estimation of amino acids : The mixture of amino acids liberated by protein hydrolysis can be determined by ***chromatographic techniques***. The reader must refer ***Chapter 41*** for the separation and quantitative determination of amino acids. Knowledge on primary structure of proteins will be incomplete without a thorough understanding of chromatography.

2. Degradation of protein into smaller fragments : Protein is a large molecule which is sometimes composed of individual polypeptide chains. Separation of polypeptides is essential before degradation.

(a) **Liberation of polypeptides :** Treatment with urea or guanidine hydrochloride disrupts the non-covalent bonds and

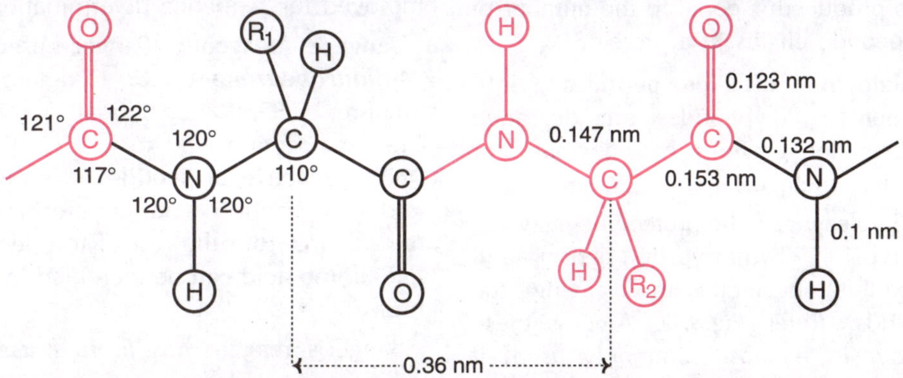

***Fig. 4.7 :** Dimensions of a fully extended polypeptide chain. (The distance between two adjacent α-carbon atoms is 0.36 nm).*

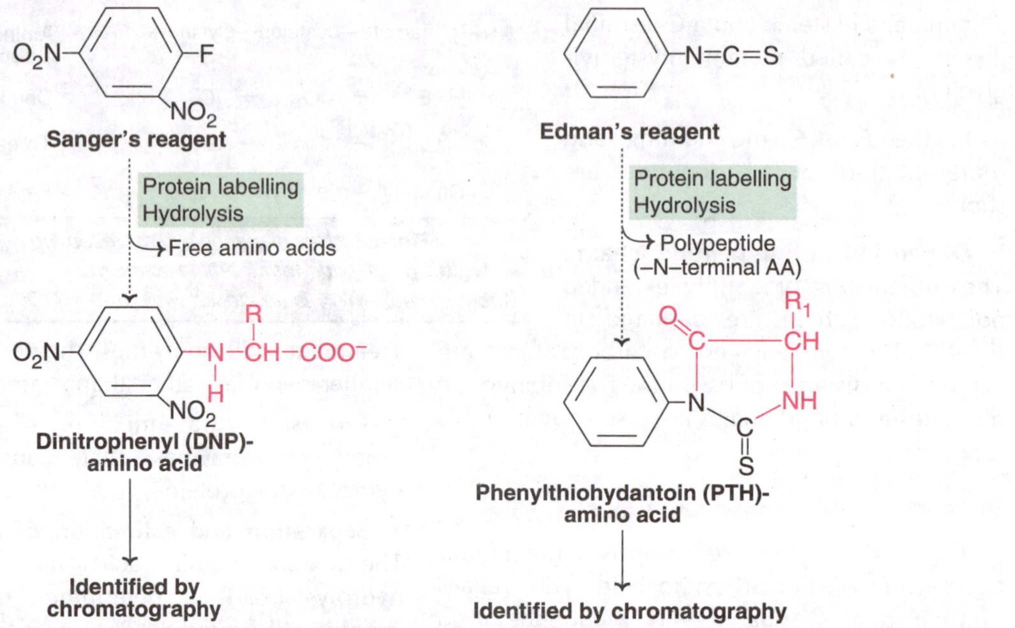

Fig. 4.8 : *Sanger's reagent (1-fluoro 2,4-dinitrobenzene) and Edman's reagent (Phenyl isothiocyanate) in the determination of amino acid sequence of a protein (AA-Amino acid).*

dissociates the protein into polypeptide units. For cleaving the disulfide linkages between the polypeptide units, treatment with performic acid is necessary.

(b) **Number of polypeptides :** The number of polypeptide chains can be identified by treatment of protein with ***dansyl chloride***. It specifically binds with N-terminal amino acids to form dansyl polypeptides which on hydrolysis yield N-terminal dansyl amino acid. The number of dansyl amino acids produced is equal to the number of polypeptide chains in a protein.

(c) **Breakdown of polypeptides into fragments :** Polypeptides are degraded into smaller peptides by enzymatic or chemical methods.

Enzymatic cleavage : The proteolytic enzymes such as trypsin, chymotrypsin, pepsin and elastase exhibit specificity in cleaving the peptide bonds (***Refer Fig.8.7***). Among these enzymes, trypsin is most commonly used. It hydrolyses the peptide bonds containing lysine or arginine on the carbonyl (–C=O) side of peptide linkage.

Chemical cleavage : ***Cyanogen bromide*** (CNBr) is commonly used to split polypeptides into smaller fragments. CNBr specifically splits peptide bonds, the carbonyl side of which is contributed by the amino acid methionine.

3. Determination of amino acid sequence : The polypeptides or their smaller fragments are conveniently utilized for the determination of sequence of amino acids. This is done in a step-wise manner to finally build up the order of amino acids in a protein. Certain reagents are employed for sequence determination (***Fig.4.8***).

Sanger's reagent : Sanger used ***1-fluoro 2,4-dinitrobenzene*** (FDNB) to determine insulin structure. ***FDNB*** specifically binds with N-terminal amino acid to form a dinitrophenyl (DNP) derivative of peptide. This on hydrolysis yields DNP-amino acid (N-terminal) and free amino acids from the rest of the peptide chain. DNP-amino acid can be identified by chromatography.

Sanger's reagent has ***limited use*** since the peptide chain is hydrolysed to amino acids.

Edman's reagent : ***Phenyl isothiocyanate*** is the Edman's reagent. It reacts with the

N-terminal amino acid of peptide to form a phenyl thiocarbamyl derivative. On treatment with mild acid, phenyl thiohydantoin (PTH)–amino acid, a cyclic compound is liberated. This can be identified by chromatography (***Fig.4.8***).

Edman's reagent has an advantage since a peptide can be sequentially degraded liberating N-terminal amino acids one after another which can be identified. This is due to the fact that the peptide as a whole is not hydrolysed but only releases PTH-amino acid.

Sequenator : This is an ***automatic machine*** to determine the amino acid sequence in a polypeptide (with around 100 residues). It is based on the principle of ***Edman's degradation*** (described above). Amino acids are determined sequentially from N-terminal end. The PTH-amino acid liberated is identified by high-performance liquid chromatography (HPLC). Sequenator takes about 2 hours to determine each amino acid.

Overlapping peptides

In the determination of primary structure of protein, several methods (enzymatic or chemical) are simultaneously employed. This results in the formation of overlapping peptides. This is due to the specific action of different agents on different sites in the polypeptide. Overlapping peptides are very useful in determining the amino acid sequence.

Reverse sequencing technique

It is the genetic material (chemically DNA) which ultimately determines the sequence of amino acids in a polypeptide chain. By analysing the nucleotide sequence of DNA that codes for protein, it is possible to translate the nucleotide sequence into amino acid sequence. This technique, however, fails to identify the disulfide bonds and changes that occur in the amino acids after the protein is synthesized (post-translational modifications).

SECONDARY STRUCTURE OF PROTEIN

The conformation of polypeptide chain by twisting or folding is referred to as secondary structure. The amino acids are located close to each other in their sequence. Two types of secondary structures, ***α-helix*** and ***β-sheet***, are mainly identified.

Indian scientist Ramachandran made a significant contribution in understanding the spatial arrangement of polypeptide chains.

α-Helix

α-Helix is the ***most common*** spiral structure of protein. It has a rigid arrangement of polypeptide chain. α-Helical structure was ***proposed by Pauling and Corey*** (1951) which is regarded as one of the milestones in the biochemistry research. The salient features of α-helix (***Fig.4.9***) are given below

1. The α-helix is a ***tightly packed*** coiled structure with amino acid side chains extending outward from the central axis.

2. The α-helix is ***stabilized by*** extensive ***hydrogen bonding***. It is formed between H atom attached to peptide N, and O atom attached to peptide C. The hydrogen bonds are individually weak but collectively, they are strong enough to stabilize the helix.

3. All the ***peptide bonds***, except the first and last in a polypeptide chain, participate in ***hydrogen bonding***.

4. ***Each turn*** of α-helix contains ***3.6 amino acids*** and travels a ***distance of 0.54 nm***. The spacing of each amino acid is 0.15 nm.

5. α-Helix is a ***stable conformation*** formed spontaneously with the lowest energy.

6. The ***right handed α-helix*** is more stable than left handed helix (a right handed helix turns in the direction that the fingers of right hand curl when its thumb points in the direction the helix rises).

7. Certain amino acids (particularly proline) disrupt the α-helix. Large number of acidic (Asp, Glu) or basic (Lys, Arg, His) amino acids also interfere with α-helix structure.

β-Pleated sheet

This is the second type of structure (hence β after α) proposed by Pauling and Corey. β-Pleated sheets (or simply β-sheets) are composed of two or more segments of ***fully extended peptide chains*** (***Fig.4.10***). In the

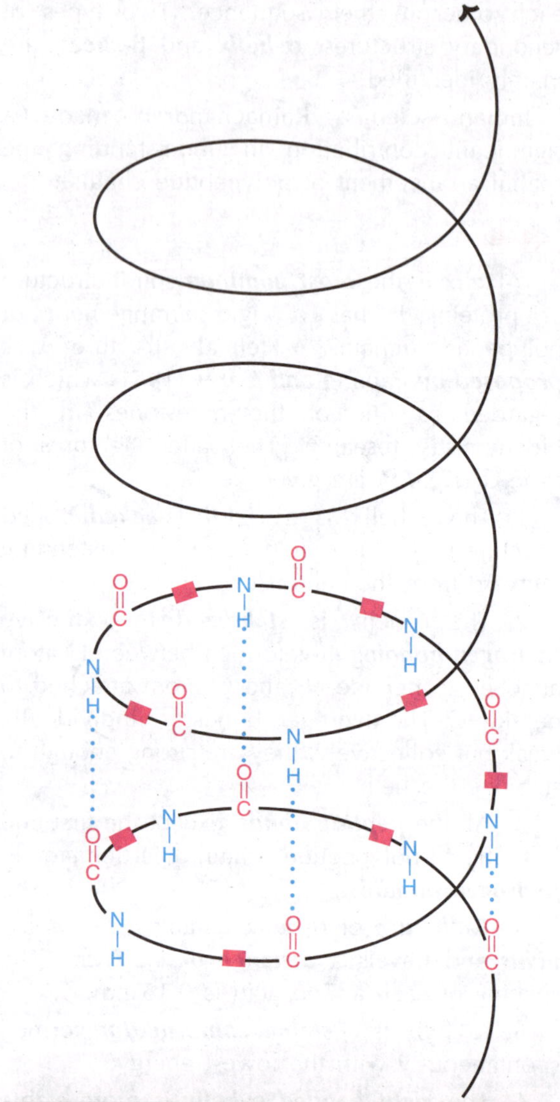

Fig. 4.9 : *Diagrammatic representation of secondary structure of protein—a right handed α-helix*

(▬ -Indicate $-\overset{H}{\overset{|}{C}}-R$ *groups of amino acids; dotted blue lines are hydrogen bonds; Note that only a few hydrogen bonds shown for clarity).*

β-sheets, the hydrogen bonds are formed between the neighbouring segments of polypeptide chain(s).

Parallel and anti-parallel β-sheets

The polypeptide chains in the β-sheets may be arranged either in parallel (the same direction) or anti-parallel (opposite direction). This is illustrated in ***Fig.4.10.***

β-Pleated sheet may be formed either by separate polypeptide chains (H-bonds are interchain) or a single polypeptide chain folding back on to itself (H-bonds are intrachain).

Occurrence of β-sheets : Many proteins contain β-pleated sheets. As such, the α-helix and β-sheet are commonly found in the same protein structure (***Fig.4.11***). In the globular proteins, β-sheets form the core structure.

Other types of secondary structures : Besides the α-and β-structures described above, the β-bends and nonrepetitive (less organised structures) secondary structures are also found in proteins.

SUPERSECONDARY STRUCTURE OF PROTEIN

The ***intermediate protein structures between secondary and tertiary structures*** are often referred to as supersecondary structures. These are combinations of α-helices and β-structures, connected through loops and bends that form specific patterns. Supersecondary structures are

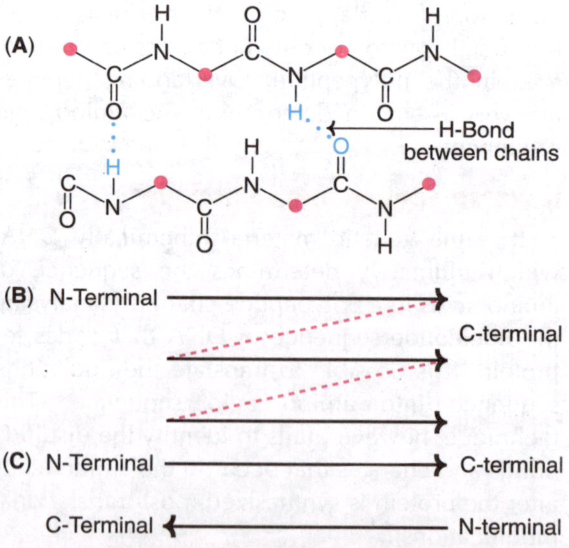

Fig. 4.10 : *Structure of β-pleated sheet* ***(A)*** *Hydrogen bonds between polypeptide chains* ***(B)*** *Parallel β-sheet* ***(C)*** *Antiparallel β-sheet.* ***(Note :*** *Red circles in A represent amino acid skeleton*$-\overset{H}{\overset{|}{C}}-R$ *).*

found in many proteins, and are stabilized by hydrogen bonding, salt bridges and hydrophobic interactions. The term ***motif*** is sometimes used to describe the supersecondary protein structures.

TERTIARY STRUCTURE OF PROTEIN

The ***three-dimensional arrangement of protein*** structure is referred to as tertiary structure. It is a compact structure with hydrophobic side chains held interior while the hydrophilic groups are on the surface of the protein molecule. This type of arrangement ensures stability of the molecule.

Bonds of tertiary structure : Besides the hydrogen bonds, ***disulfide bonds*** (–S–S), ***ionic interactions*** (electrostatic bonds), ***hydrophobic interactions*** and ***van der Waals forces*** also contribute to the tertiary structure of proteins.

Domains : The term domain is used to represent the ***basic units of protein structure*** (tertiary) ***and function***. A polypeptide with 200 amino acids normally consists of two or more domains.

QUATERNARY STRUCTURE OF PROTEIN

A great majority of the proteins are composed of single polypeptide chains. Some of the proteins, however, consist of two or more polypeptides which may be identical or unrelated. Such proteins are termed as ***oligomers*** and possess quaternary structure. The individual polypeptide chains are known as ***monomers, protomers*** or ***subunits***. A ***dimer*** consits of ***two*** polypeptides while a ***tetramer*** has four.

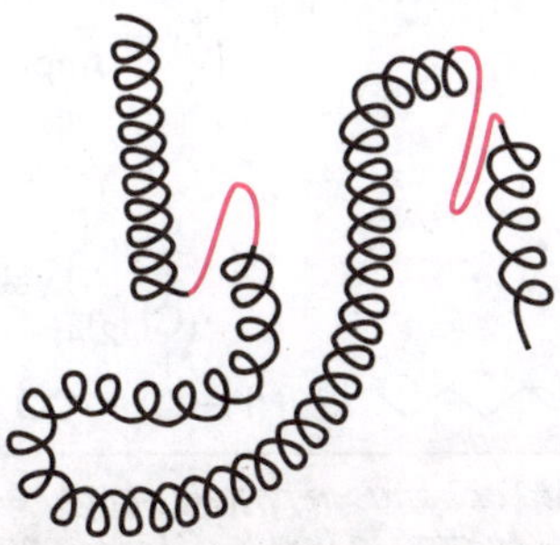

Fig. 4.11 : *Diagrammatic representation of a protein containing α-helix and β-pleated sheet (blue).*

Bonds in quaternary structure : The monomeric subunits are held together by non-convalent bonds namely ***hydrogen bonds, hydrophobic interactions and ionic bonds.***

Importance of oligomeric proteins : These proteins play a significant role in the regulation of metabolism and cellular function.

Examples of oligomeric proteins : ***Hemoglobin***, aspartate transcarbamoylase, ***lactate dehydrogenase.***

CIRCULAR PERMUTATION IN PROTEINS

Circular permutation refers ***to the changes in the order of amino acids in the peptide sequence of proteins.*** This may occur as post-translational modification, evolutionary change or artificially engineered mutation. e.g., concanavalin A and lectin are circularly permuted proteins (first discovered in 1970). Now more than 2000 such proteins are known.

Circular permutations are routinely engineered in the laboratory to improve catalytic activity or thermostability or to investigate properties of original proteins.

BONDS RESPONSIBLE FOR PROTEIN STRUCTURE

Protein structure is stabilized by two types of bonds—covalent and non-covalent.

1. **Covalent bonds :** The peptide and disulfide bonds are the strong bonds in protein structure. The formation of ***peptide bond*** and its chracteristics have been described.

Disulfide bonds : A disulfide bond (***–S–S***) is formed by the sulfhydryl groups (–SH) of two cysteine residues, to produce ***cystine*** (***Fig.4.12A***). The disulfide bonds may be formed in a single polypeptide chain or between different polypeptides. These bonds contribute to the structural conformation and stability of proteins.

2. **Non-covalent bonds :** There are, mainly, four types of non-covalent bonds.

(a) **Hydrogen bonds :** The hydrogen bonds are formed by sharing of hydrogen atoms between the nitrogen and carbonyl

Medical Concepts/ Clinical Correlates

- Collagen is the most abundant protein in mammals. It is rich in glycine, and contains hydroxyproline and hydroxylysine.
- Several biologically important peptides are known in the living organisms. These include glutathione for the maintenance of RBC structure and integrity; oxytocin that causes uterus contraction; vasopressin that stimulates retention of water by kidneys; enkephalins that inhibit the sense of pain in the brain.
- Antibiotics such as actinomycin, gramicidin, bacitracin and tyrocidin are peptide in nature.
- γ-Carboxyglutamic acid is an amino acid derivative found in certain plasma proteins involved in blood clotting.

oxygen of different peptide bonds (***Fig.4.12B***). Each hydrogen bond is weak but collectively they are strong. A large number of hydrogen bonds significantly contribute to the protein structure.

(b) **Hydrophobic bonds :** The non-polar side chains of neutral amino acids tend to be closely associated with each other in proteins (***Fig.4.12C***). As such, these are not true bonds. The occurrence of hydrophobic forces is observed in aqueous environment wherein the molecules are forced to stay together.

(c) **Electrostatic bonds :** These bonds are formed by interactions between negatively charged groups (e.g. COO^-) of acidic amino acids with positively charged groups (e.g. $-NH_3^+$) of basic amino acids (***Fig.4.12D***).

(d) **Van der Waals forces :** These are the non-covalent associations between electrically neutral molecules. They are formed by the electrostatic interactions due to permanent or induced dipoles.

EXAMPLES OF PROTEIN STRUCTURE

Structure of human insulin

Insulin consists of two polypeptide chains, A and B (***Fig.4.13***). The A chain has glycine at the N-terminal end and asparagine at the C-terminal end. The B chain has phenylalanine and alanine at the N- and C-terminal ends, respectively. Originally, insulin is synthesized as a single polypeptide ***preproinsulin*** which undergoes proteolytic processing to give ***proinsulin*** and ***finally insulin***.

Structure of collagen

Collagen is the ***most abundant protein*** in mammals, comprising approximately one-third

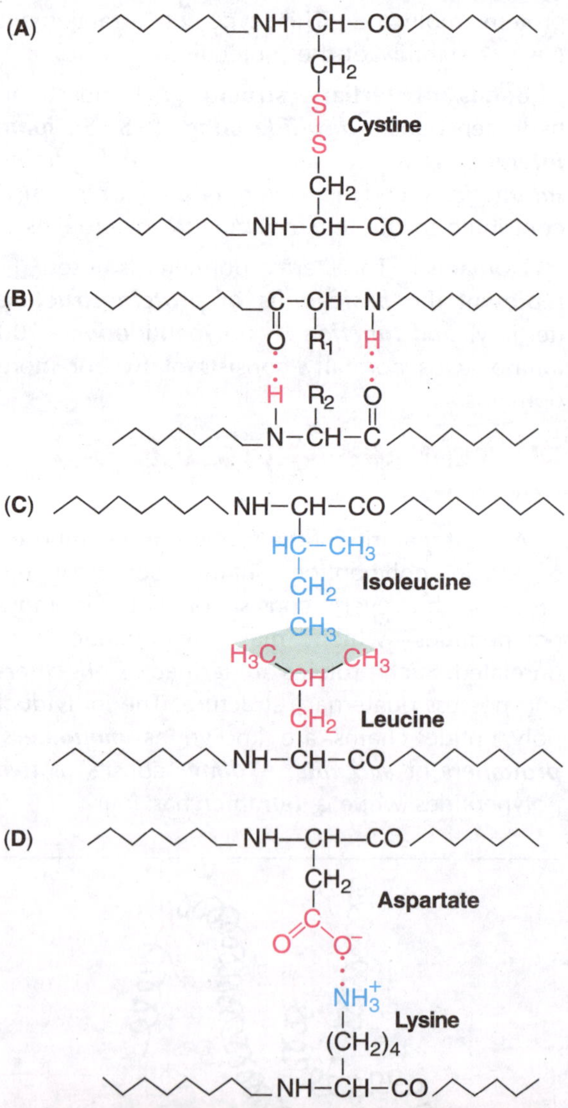

***Fig. 4.12 :** Major bonds in protein structure **(A)** Disulfide bond **(B)** Hydrogen bonds **(C)** Hydrophic bonds **(D)** Electrostatic bond.*
*(**Note : See Fig. 4.5** for peptide bond).*

of total body protein. Collagens are a group of structurally related and genetically distinct proteins. (at least 19 are known; they are numbered as I, II,.......XIX).

In principle, collagens are triple helical structures. The triple helix may occur throughout the molecule or only a part of the molecule. The structure of type I mature collagen (***Fig.22.1***) along with abnormalities associated with collagens are described in ***Chapter 22***.

In brief type I collagen has 3 polypeptide chains twisted around each other to form a rod like molecule of 1.4 nm diameter and 300 nm length. Collagen structures are unique with about $^1/_3$ rd of the amino acids contributed by glycine. Thus, the sequence of collagen is represented as $(Gly\text{-}x\text{-}y)_n$ where x and y represent other amino acids.

CIRCULAR PROTEINS

Circular proteins are ***formed by joining N and C terminal amino acids*** with a head-to-tail peptide bond. Circularization of proteins occurs by post-translational modifications. These proteins are less susceptible to proteolytic cleavage and exhibit enhanced stability and strong biological activity. Circular proteins are produced by bacteria and fungi e.g., gramicidin S, tyrocidin A.

Defensins are the circular peptides of animal origin that have been isolated from monkeys, baboons and humans. They are host defense peptides and are active against bacteria, fungi and viruses. Synthetic circular proteins have pharmaceutical applications.

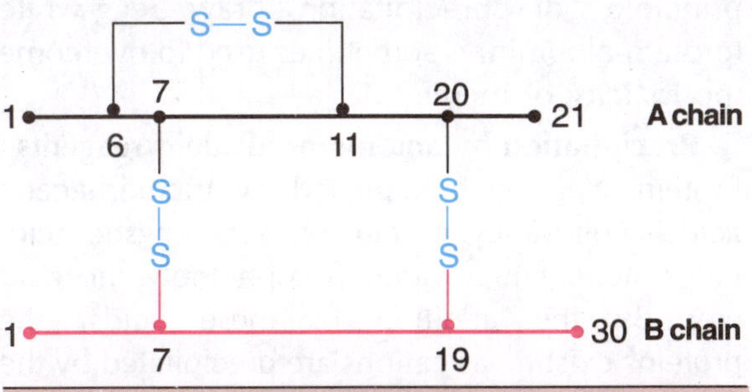

Fig. 4.13 : Diagrammatic representation of human insulin structure
(Note : A and B polypeptide chains are held by two disulfide linkages).

PROTEIN CONFORMATION— FOLDING AND MISFOLDING

The native conformation of proteins with specific folding and three dimensional structures is very essential for their biological activities. The process of protein folding is largely unknown,. However, a special group of proteins namely ***chaperones*** are known to facilitate protein folding.

Protein misfolding and diseases

Misfolded proteins may cause diseases e.g., ***cystic fibrosis, prion diseases*** (more details given ***Chapter 22***).

Methods to determine protein structure

For the determination of secondary and tertiary protein structures, ***X-ray crystallography*** is most commonly used. ***Nuclear magnetic resonance*** (NMR) spectra of proteins provides structural and functional information on the atoms and groups present in the proteins.

Methods for the isolation and purification of proteins

Proteins should be highly purified for detailed study of their physical, structural and functional properties.

Several methods are employed to isolate and purify proteins. Initially, proteins are fractionated by using different concentrations of ammonium sulfate or sodium sulfate. Protein fractionation may also be carried out by ultracentrifugation.

Protein separation is achieved by utilizing electrophoresis, isoelectric focussing, immuno-electrophoresis, ***ion-exchange chromatography, gel-filtration***, high performance liquid chromatography (HPLC) etc. The details of these techniques are described in ***Chapter 41***.

PROPERTIES OF PROTEINS

1. **Solubility :** Proteins form ***colloidal solutions*** instead of true solutions in water. This is due to huge size of protein molecules.

2. **Molecular weight :** The proteins vary in their molecular weights, which, in turn, is

dependent on the number of amino acid residues. ***Each amino acid*** on an average contributes to a ***molecular weight of about 110***. Majority of proteins/polypeptides may be composed of 40 to 4,000 amino acids with a molecular weight ranging from 4,000 to 440,000. A few proteins with their molecular weights are listed below :

Insulin-5,700; Myoglobin-17,000; Hemoglobin-64,450; Serum albumin-69,000.

3. **Shape :** There is a wide variation in the protein shape. It may be ***globular*** (insulin), oval (albumin) ***fibrous*** or elongated (fibrinogen).

4. **Isoelectric pH :** Isoelectric pH (pI) as a property of amino acids has been described. The nature of the amino acids (particularly their ionizable groups) determines the pI of a protein. The acidic amino acids (Asp, Glu) and basic amino acids (His, Lys, Arg) strongly influence the pI. At isoelectric pH, the proteins exist as ***zwitterions*** or ***dipolar ions***. They are electrically neutral (do not migrate in the electric field) with minimum solubility, maximum precipitability and least buffering capacity. The isoelectric pH(pI) for some proteins are given here

Pepsin-1.1; ***Casein-4.6***; Human albumin-4.7; Urease-5.0; Hemoglobin-6.7; Lysozyme-11.0.

p^I and protein staining : The staining properties of proteins are determined by isoelectric pH. ***Acidic dyes*** (e.g., eosin) dissociate into H^+ and dye which will bind to protein $-NH_3^+$ (protein cations). On the other hand, ***basic dyes*** (e.g., methylene blue hemotoxylin) dissociate into OH^- and dye which will then stain protein $-COO^-$ (protein anions).

5. **Acidic and basic proteins :** Proteins in which the ratio (ε Lys + ε Arg)/(ε Glu + ε Asp) is greater than 1 are referred to as basic proteins. For acidic proteins, the ratio is less than 1.

6. **Precipitation of proteins :** Proteins exist in colloidal solution due to hydration of polar groups ($-COO^-$, $-NH_3^+$, $-OH$). Proteins can be precipitated by dehydration or neutralization of polar groups.

Precipitation at pI : The proteins in general are least soluble at isoelectric pH. Certain proteins (e.g. casein) get easily precipitated when the pH is adjusted to pI (4.6 for casein). ***Formation of curd*** from milk is a marvellous ***example of slow precipitation of milk protein,*** casein at pI. This occurs due to the lactic acid produced by fermentation of bacteria which lowers the pH to the pI of casein.

Precipitation by salting out : The process of protein precipitation by the additional of neutral salts such as ***ammonium sulfate*** or ***sodium sulfate*** is known as salting out. This phenomenon is explained on the basis of ***dehydration of protein*** molecules by salts. This causes increased protein-protein interaction, resulting in molecular aggregation and precipitation.

The amount of salt required for protein precipitation depends on the size (molecular weight) of the protein molecule. In general, the higher is the protein molecular weight, the lower is the salt required for precipitation. Thus, serum ***globulins*** are ***precipitated by half saturation*** with ammonium sulfate while ***albumin*** is precipitated ***by full saturation***. ***Salting out*** procedure is conveniently used for separating serum albumins from globulins.

The addition of ***small quantities of neutral salts increases the solubility of proteins***. This process called as ***salting in*** is due to the diminished protein–protein interaction at low salt concentration.

Precipitation by salts of heavy metals : Heavy metal ions like Pb^{2+}, Hg^{2+}, Fe^{2+}, Zn^{2+}, Cd^{2+} cause precipitation of proteins. These metals being positively charged, when added to protein solution (negatively charged) in alkaline medium results in precipitate formation. Based on the principle of precipitation, raw egg-white (protein-albumin) is sometimes used to overcome the toxicity of mercury.

Precipitation by anionic or alkaloid reagents : Proteins can be precipitated by trichloroacetic acid, sulphosalicylic acid, phosphotungstic acid, picric acid, tannic acid, phosphomolybdic acid etc. By the addition of these acids, the proteins existing as cations are precipitated by the anionic form of acids to produce protein-sulphosalicylate, protein-tungstate, protein-picrate etc. Industrial ***tanning of leather*** is based on the principle of protein precipitation by tannic acid.

Medical Concepts/ Clinical Correlates

- Homocysteine has been implicated as a risk factor in the onset of coronary heart diseases.
- Several non-protein amino acids of biological importance are known. These include ornithine, citrulline and arginosuccinic acid (intermediates of urea synthesis), thyroxine and triiodothyronine (hormones), and β-alanine (of coenzyme A).
- The protein-free filtrate of blood, required for biochemical investigations (e.g. urea, sugar) can be obtained by using protein precipitating agents such as phosphotungstic acid and trichloroacetic acid.
- Heat coagulation test is most commonly employed to detect the presence of albumin in urine.

Precipitation by organic solvents : Organic solvents such as alcohol are good protein precipitating agents. They dehydrate the protein molecule by removing the water envelope and cause precipitation. The use of ***surgical spirit*** (about 20% alcohol) as a disinfectant is based on the ***precipitation of proteins*** and the death of bacteria.

7. **Colour reactions of proteins :** The proteins give several colour reactions which are often useful to identify the nature of the amino acids present in them (***Table 4.3***).

Biuret reaction : Biuret is a compound formed by heating urea to 180°C.

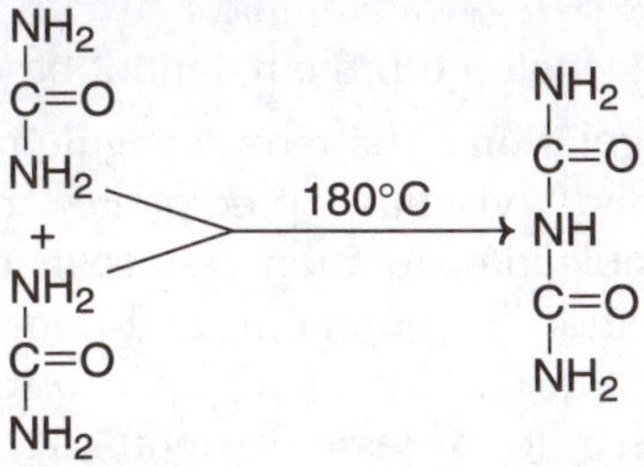

2 Moles of **urea** **Biuret**

When biuret is treated with dilute copper sulfate in alkaline medium, a purple colour is obtained. This is the basis of biuret test widely used for ***identification of proteins and peptides.***

Biuret test is answered by compounds containing two or more CO–NH groups i.e., peptide bonds. All proteins and peptides possessing at least two peptide linkages i.e., tripeptides (with 3 amino acids) give positive biuret test. Histidine is the only amino acid that answers biuret test. The principle of biuret test is conveniently used to detect the presence of proteins in biological fluids. The mechanism of biuret test is not clearly known. It is believed that the colour is due to the formation of a ***copper co-ordinated complex***, as shown below.

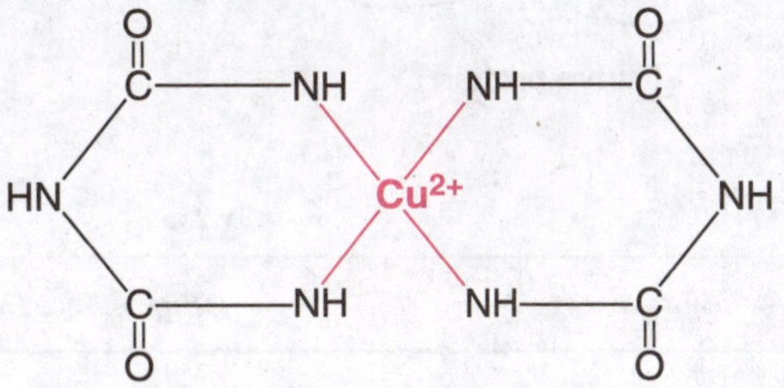

The presence of magnesium and ammonium ions interfere in the biuret test. This can be overcome by using excess alkali.

DENATURATION

The phenomenon of ***disorganization of native protein structure*** is known as denaturation. Denaturation results in the loss of secondary,

Table 4.3 Colour reactions of proteins/amino acids

Reaction	Specific group or amino acid
1. Biuret reaction	Two peptide linkages
2. Ninhydrin reaction	α-Amino acids
3. Xanthoproteic reaction	Benzene ring of aromatic amino acids (Phe, Tyr, Trp)
4. Milllons reaction	Phenolic group (Tyr)
5. Hopkins-Cole reaction	Indole ring (Trp)
6. Sakaguchi reaction	Guanidino group (Arg)
7. Nitroprusside reaction	Sulfhydryl groups (Cys)
8. Sulfur test	Sulfhydryl groups (Cys)
9. Pauly's test	Imidazole ring (His)
10. Folin–Coicalteau's test	Phenolic groups (Tyr)

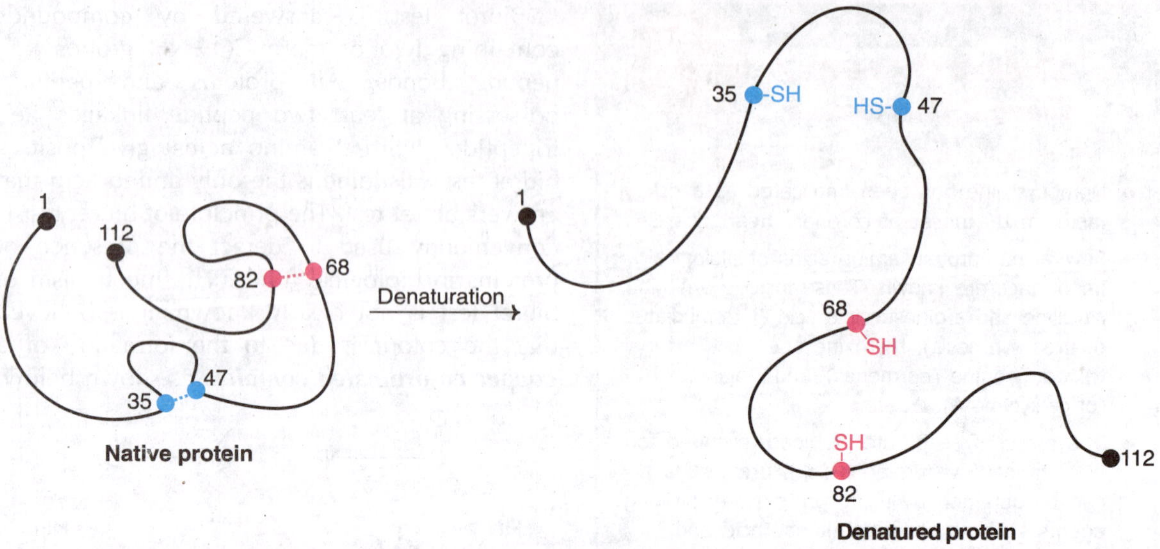

Fig. 4.14 : Denaturation of a protein.

tertiary and quaternary structure of proteins. This involves a change in physical, chemical and biological properties of protein molecules.

Agents of denaturation

Physical agents : ***Heat***, violent shaking, X-rays, UV radiation.

Chemical agents : ***Acids***, ***alkalies***, ***organic solvents*** (ether, alcohol), salts of heavy metals (Pb, Hg), urea, salicylate, detergents (e.g. sodium dodecyl sulfate).

Characteristics of denaturation

1. The ***native*** helical ***structure*** of protein is ***lost*** (***Fig.4.14***).

2. The ***primary structure*** of a protein with peptide linkages remains ***intact*** i.e., peptide bonds are not hydrolysed.

3. The protein ***loses*** its ***biological activity***.

4. Denatured protein becomes insoluble in the solvent in which it was originally soluble.

5 The viscosity of denatured protein (solution) increases while its surface tension decreases.

6. Denaturation is associated with increase in ionizable and sulfhydryl groups of protein. This is due to loss of hydrogen and disulfide bonds.

7. ***Denatured protein is more easily digested***. This is due to increased exposure of peptide bonds to enzymes. Cooking causes protein denaturation and, therefore, cooked food (protein) is more easily digested. Further, denaturation of dietary protein by gastric HCl enchances protein digestion by pepsin.

8. Denaturation is ***usually irreversible***. For instance, omelet can be prepared from an egg (protein-albumin) but the reversal is not possible.

9. Careful denaturation is sometimes reversible (known as ***renaturation***). Hemoglobin undergoes denaturation in the presence of salicylate. By removal of salicylate, hemoglobin is renatured.

10. Denatured protein cannot be crystallized.

Coagulation : The term 'coagulum' refers to a semi-solid viscous precipitate of protein. Irreversible denaturation results in coagulation. Coagulation is optimum and requires lowest temperature at isoelectric pH. Albumins and globulins (to a lesser extent) are coagulable proteins. ***Heat coagulation test is commonly used to detect the presence of albumin in urine***.

Flocculation : It is the process of protein precipitation at isoelectric pH. The precipitate is referred to as flocculum. Casein (milk protein) can be easily precipitated when adjusted to isoelectric pH (4.6) by dilute acetic acid. Flocculation is reversible. On application of heat, flocculum can be converted into an irreversible mass, coagulum.

NEPHELOMETRY AND TURBIDIMETRY

One of the properties of ***colloids*** (particle size 1–100 nm), based on the ***principle of scattering of light*** is measured by nephelometry or turbidimetry. Proteins are colloids, and these techniques are used to quantitatively measure them (particularly plasma proteins).

Nephelometry is based on the ***principle that dilute suspension of small particles (colloids) will scatter light passed through it***. The amount of emergent light (usually at an angle 30–90 degrees) is proportional to the turbidity and thus to the quantity of colloid. Nephelometry is usually employed to measure ***immunoglobulins***.

Turbidimetry is also based on the ***principle of scattering of light by colloidal particles***. In this technique, the emergent light is measured at 180°. In case of ***immunoturbidity***, the protein is precipitated by specific antibody and the turbidity of the ***antigen-antibody complex*** is measured.

CLASSIFICATION OF PROTEINS

Proteins are classified in several ways. Three major types of classifying proteins ***based on their function, chemical nature and solubility properties and nutritional importance*** are discussed here.

A. Functional classification of proteins

Based on the functions they perform, proteins are classified into the following groups (with examples)

1. **Structural proteins :** ***Keratin*** of hair and nails, collagen of bone.

2. **Enzymes or catalytic proteins :** Hexokinase, ***pepsin***.

3. **Transport proteins :** ***Hemoglobin***, serum albumin.

4. **Hormonal proteins :** ***Insulin***, growth hormone.

5. **Contractile proteins :** Actin, myosin.

6. **Storage proteins :** Ovalbumin, glutelin.

7. **Genetic proteins :** Nucleoproteins.

8. **Defense proteins :** Snake venoms, Immunoglobulins.

9. **Receptor proteins** for hormones, viruses.

B. Protein classification based on chemical nature and solubility

This is a more comprehensive and popular classification of proteins. It is based on the amino acid composition, structure, shape and solubility properties. Proteins are broadly classified into 3 major groups

1. **Simple proteins :** They are composed of ***only amino acid*** residues.

2. **Conjugated proteins :** Besides the amino acids, these proteins contain a non-protein moiety known as ***prosthetic group*** or conjugating group.

3. **Derived proteins :** These are the ***denatured*** or ***degraded products*** of simple and conjugated proteins.

The above three classes are further subdivided into different groups. The summary of protein classification is given in the ***Table 4.4***.

Among the ***simple proteins***, ***globular proteins*** are spherical in shape, soluble in water or other solvents and digestable e.g., albumin, globulin. ***Scleroproteins*** (fibrous proteins) are fiber like in shape, insoluble in water and resistant to digestion e.g. collagen, keratin.

The ***conjugated proteins*** may contain prosthetic groups such as nucleic acid, carbohydrate, lipid, metal etc. The ***Primary derived proteins*** are produced by agents such as heat, acids, alkalies etc., while the ***secondary derived proteins*** are hydrolytic products of proteins.

C. Nutritional classification of proteins

The nutritive value of proteins is determined by the composition of essential amino acids (described already). From the nutritional point of view, proteins are classified into 3 categories.

Table 4.4 Summary of classification of proteins

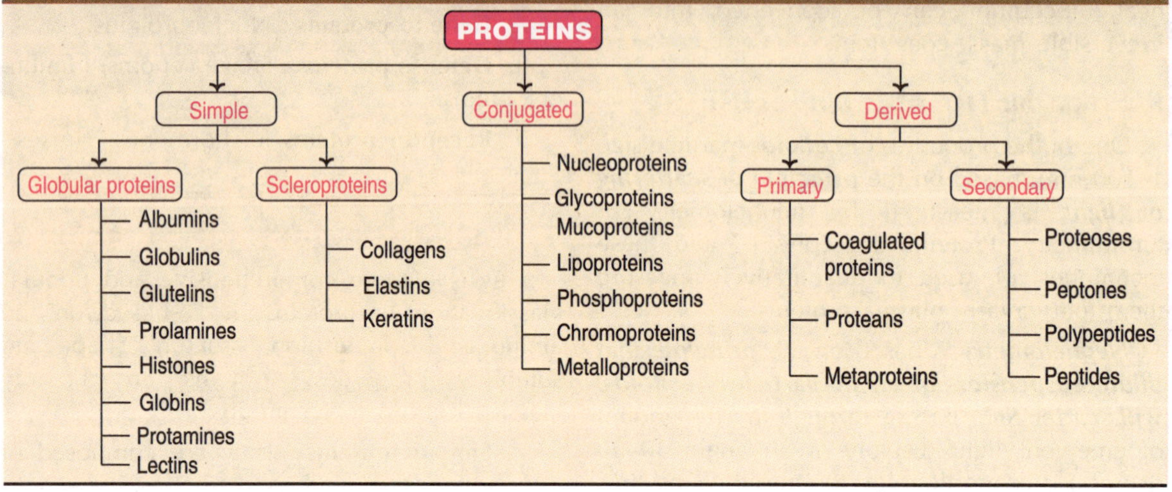

1. **Complete proteins :** These proteins have all the ten essential amino acids in the required proportion by the human body to promote good growth. e.g. ***egg albumin***, milk casein. They are also called as ***first class proteins*** or nutritionally rich proteins.

2. **Partially incomplete proteins :** These proteins ***partially lack one or more essential amino acids***, and can promote moderate growth. e.g. wheat and rice proteins (limiting Lys, Thr).

3. **Incomplete proteins :** These proteins ***completely lack one or more essential amino acids***. Hence they do not promote growth at all e.g. ***gelatin*** (lacks Trp), zein (lacks Trp, Lys).

BIOLOGICALLY IMPORTANT PEPTIDES

Several peptides occur in the living organisms that display a wide spectrum of biological functions. Generally, the term 'peptide' is applied when the number of constituent amino acids is less than 10. Some examples of biologically active peptides and their functions are described here.

1. Glutathione

It is a tripeptide composed of 3 amino acids. Chemically, glutathione is γ-glutamyl-cysteinyl-glycine. It is widely distributed in nature and exists in reduced or oxidized states.

$$\underset{\text{Reduced}}{2G{-}SH} \rightleftharpoons \underset{\text{Oxidized}}{G{=}S{-}S{-}G}$$

Functions : In a steady state, the cells generally maintain a ratio of about 100/1 of GSH to G−S−S−G. The ***reversible oxidation-reduction of glutathione is important*** for many of its biological functions.

- Glutathione serves as a coenzyme for certain enzymes e.g. prostaglandin PGE_2 synthetase, glyoxylase.
- It ***prevents the oxidation of sulfhydryl (−SH) groups*** of several proteins to disulfide (−S−S−) groups. This is essential for the protein function, including that of enzymes.
- It is believed that glutathione in association with glutathione reductase participates in the formation of correct disulfide bonds in several proteins.
- Glutathione (reduced) performs specialized functions in erythrocytes
  (i) It maintains ***RBC membrane structure*** and integrity.
  (ii) It ***protects hemoglobin*** from getting oxidized by agents such as H_2O_2.

Medical Concepts/ Clinical Correlates

- Nephelometry and turbidimetry are the techniques used to quantitatively measure proteins.
- Reduced glutathione (a tripeptide) is important to maintain the structure and functions of RBC.
- Pseudopeptides or duplicate peptides are designed by pharmaceutical industries to improve the efficacy of drugs.

- Glutathione is involved in the transport of amino acids in the intestine and kidney tubules via ***γ-glutamyl cycle*** or ***Meister cycle*** (***Refer Chapter 8***).
- Glutathione is involved in the ***detoxication*** process. The toxic substances (organophosphates, nitro compounds) are converted to mercapturic acids.
- Toxic amounts of peroxides and ***free radicals*** produced in the cells are ***scavanged by glutathione peroxidase*** (a selenium containing enzyme).

$$2\,GSH + H_2O_2 \xrightarrow{\text{Peroxidase}} G-S-S-G + 2\,H_2O$$

2. **Thyrotropin releasing hormone** (***TRH***) **:** It is a tripeptide secreted by hypothalamus. TRH stimulates pituitary gland to release thyrotropic hormone.

3. **Oxytocin :** It is a hormone secreted by posterior pituitary gland and contains 9 amino acids (nonapeptide). Oxytocin causes ***contraction of uterus***.

4. **Vasopressin** (***antidiuretic hormone, ADH***) **:** ADH is also a nonapeptide produced by posterior pituitary gland. It stimulates kidneys to retain water and thus ***increases the blood pressure***.

5. **Angiotensins :** Angiotensin I is a decapeptide (10 amino acids) which is converted to angiotensin II (8 amino acids). The later has more hypertensive effect. Angiotensin II also stimulates the release of aldosterone from adrenal gland.

6. **Methionine enkephalin :** It is a pentapeptide found in the brain and has opiate like function. It ***inhibits*** the sense of a ***pain***.

7. **Bradykinin and kallidin :** They are nona- and decapeptides, respectively. Both of them act as powerful vasodilators. They are produced from plasma proteins by snake venom enzymes.

8. **Peptide antibiotics :** Microbial antibiotics such as ***gramicidin***, ***bacitracin***, tyrocidin and actinomycin are peptide in nature.

9. **Aspartame :** It is a ***dipeptide*** (aspartyl-phenylalanine methyl ester), produced by a combination of aspartic acid and phenylalanine. Aspartame is about ***200 times sweeter than sucrose***, and is used as a low-calorie artificial sweetner in softdrink industry.

10. **Gastrointestinal hormones :** ***Gastrin***, ***secretin*** etc. are the gastrointestinal peptides which serve as hormones.

11. **Lethal peptides :** Bacterial peptides namely ***microcystin*** and ***nodularin*** promote hepatic tumors in small doses while in large doses, they are lethal.

PSEUDOPEPTIDES

Biological peptides are quickly degraded in the body, hence do not make good drugs. In pharmaceutical industry, duplicate peptides or pseudopeptides are designed. They have specific pharmacological activities with high bioavailability e.g., ***β-peptides***, ***γ-peptides***, ***oligoglycines***, ***peptide nucleic acid (PNA)***, DNA-cleaving pseudopeptides.

SUMMARY

- *Proteins are nitrogen containing, most abundant organic macromolecules widely distributed in animals and plants. They perform structural and dynamic functions in the organisms.*
- *Proteins are polymers composed of L-α-amino acids. They are 20 in number and classified into different groups based on their structure, chemical nature, nutritional requirement and metabolic fate. Selenocysteine has been recently identified as the 21st amino acid, and is found in certain proteins.*
- *Amino acids possess two functional groups namely carboxyl ($-COOH$) and amino ($-NH_2$). In the physiological system, they exist as dipolar ions commonly referred to as zwitterions.*
- *Besides the 20 standard amino acids present in proteins, there are several non-standard amino acids. These include the amino acid derivatives found in proteins (e.g. hydroxyproline, hydroxylysine) and, non-protein amino acids (e.g. ornithine, citrulline).*
- *The structure of protein is divided into four levels of organization. The primary structure represents the linear sequence of amino acids. The twisting and spatial arrangement of polypeptide chain is the secondary structure. Tertiary structure constitutes the three dimensional structure of a functional protein. The assembly of similar or dissimilar polypeptide subunits comprises quaternary structure.*
- *The determination of primary structure of a protein involves the knowledge of quality, quantity and the sequence of amino acids in the polypeptide. Chemical and enzymatic methods are employed for the determination of primary structure.*
- *The secondary structure of protein mainly consists of α-helix and/or β-sheet. α-Helix is stabilized by extensive hydrogen bonding. β-Pleated sheet is composed of two or more segments of fully extended polypeptide chains.*
- *The tertiary and quaternary structures of protein are stabilized by non-covalent bonds such as hydrogen bonds, hydrophobic interactions, ionic bonds etc.*
- *Proteins are classified into three major groups. Simple proteins contain only amino acid residues (e.g. albumin). Conjugated proteins contain a non-protein moiety known as prosthetic group, besides the amino acids (e.g. glycoproteins). Derived proteins are obtained by degradation of simple or conjugated proteins.*
- *In addition to proteins, several peptides perform biologically important functions. These include glutathione, oxytocin and vasopressin.*

Self-Assessment Exercises

I. Essay questions

1. Describe the classification of amino acids along with their structures.
2. Discuss the organization of protein structure. Give an account of the determination of primary structure of protein.
3. Describe the classification of proteins with suitable examples.
4. Write an account of non-standard amino acids.
5. Discuss the important biologically active peptides.

II. Short notes

(a) Essential amino acids, (b) Zwitterion, (c) Peptide bond, (d) Edman's reagent, (e) α-Helix, (f) β-Pleated sheet, (g) Denaturation, (h) Isoelectric point, (i) Glutathione, (j) Quaternary structure of protein.

III. Fill in the blanks

1. The average nitrogen content of proteins ________.
2. Proteins are the polymers of ________.
3. Name the sulfur containing essential amino acid ________.
4. The charged molecule which is electrically neutral is known as ________.
5. The non –α amino acid present in coenzyme A ________.
6. The bonds forming the backbone of protein structure ________.
7. The amino acid that is completely destroyed by acid hydrolysis of protein ________.
8. The number of peptide bonds present in a decapeptide ________.
9. The chemical name of Sanger's reagent ________.
10. The phenomenon of disorganization of native protein structure is known as ________.

IV. Multiple choice questions

11. The imino acid found in protein structure
    (a) Arginine (b) Proline (c) Histidine (d) Lysine.
12. The following is a non-protein amino acid
    (a) Ornithine (b) Homocysteine (c) Histamine (d) All of them.
13. The bonds in protein structure that are not broken on denaturation.
    (a) Hydrogen bonds (b) Peptide bonds (c) Ionic bond (d) Disulfide bonds.
14. Sequenator is an automatic machine to determine amino acid sequence in a polypeptide chain. The reagent used in sequenator is
    (a) Sanger's reagent (b) CNBr (c) Trypsin (d) Edman's reagent.
15. The reaction given by two or more peptide linkages is
    (a) Biuret test (b) Ninhydrin test (c) Xanthoproteic reaction (d) Pauley's test.

Nucleic Acids and Nucleotides

5

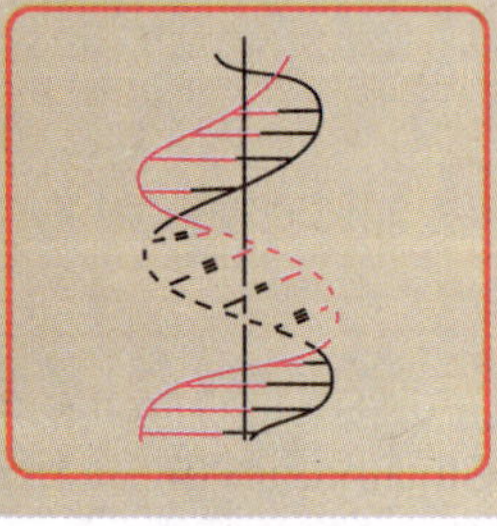

DNA, the bank of genetic information speaks

"I am the chemical basis of life and heredity!
Organized into genes that control every function;
Composed of repeating units of deoxyribonucleotides;
Arranged in a double helix, held by hydrogen bonds."

There are two types of nucleic acids, namely ***deoxyribonucleic acid*** (DNA) and ***ribonucleic acid*** (RNA). Primarily, nucleic acids serve as repositories and transmitters of genetic information.

Brief history

DNA was discovered in 1869 by Johann Friedrich Miescher, a Swiss researcher. The demonstration that DNA contained genetic information was first made in 1944, by Avery, Macleod and MacCary.

Functions of nucleic acids

DNA is the chemical basis of heredity and may be regarded as the ***reserve bank of genetic information***. DNA is exclusively responsible for maintaining the identity of different species of organisms over millions of years. Further, every aspect of cellular function is under the control of DNA. The ***DNA*** is organized into ***genes***, the fundamental units of ***genetic information***. The genes control the protein synthesis through the mediation of RNA, as shown below

DNA ⟶ RNA ⟶ Protein

The interrelationship of these three classes of biomolecules (DNA, RNA and proteins) constitutes the ***central dogma of molecular biology*** or more commonly the ***central dogma of life***.

Components of nucleic acids

Nucleic acids are the polymers of nucleotides (***polynucleotides***) held by 3′ and 5′ phosphate bridges. Thus, nucleic acids are built up by the monomeric units—nucleotides (In comparison, protein is a polymer of amino acids).

NUCLEOTIDES

Nucleotides are composed of a ***nitrogenous base***, a ***pentose sugar*** and a ***phosphate***. Nucleotides perform a wide variety of functions in the living cells, besides being the building blocks or

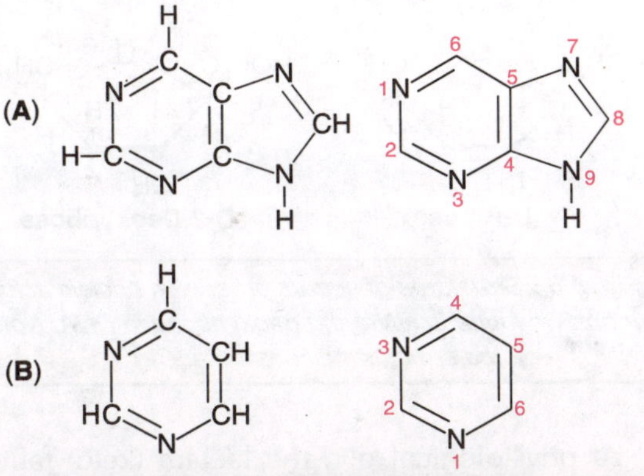

Fig. 5.1 : General structure of nitrogen bases (A) Purine (B) Pyrimidine (The positions are numbered according to the international system).

monomeric units in the nucleic acid (DNA and RNA) structure.

BIOLOGICALLY IMPORTANT NUCLEOTIDES AND THEIR FUNCTIONS

1. ***Purine and pyrimidine nucleotides***, besides being structural components of DNA and RNA, are involved in a wide variety of ***metabolic functions***—energy metabolism, protein synthesis, control of enzyme activities etc.

2. Nucleotides are structural components of some ***coenzymes of B-complex vitamins*** e.g. FAD, NAD^+.

3. Sugar derivative of nucleotides namely ***UDP-glucose*** participates in the synthesis of glycogen.

4. ***CDP-acylglycerol*** (a lipid derivative) is an intermediate in ***lipid biosynthesis***.

5. ***cAMP*** and ***cGMP*** are ***second messengers*** for some hormonal action.

6. ***Phosphoadenosine phosphosulfate (PAPS)*** and ***S-adenosylmethionine (SAM)*** respectively donate sulfate and methyl groups in biological reactions.

7. GTP and GDP are involved in a cascade of events of ***signal transduction*** pathways.

8. ***Synthetic analogs*** of nucleotides are employed in the ***treatment of cancer***, ***AIDS*** and suppression of immune response during organ transplantation (details given later).

STRUCTURE OF NUCLEOTIDES

As already stated, the nucleotide essentially consists of ***nucleobase***, ***sugar*** and ***phosphate***. The term ***nucleoside refers to base + sugar***. Thus, nucleotide is nucleoside + phosphate.

Purines and pyrimidines

The nitrogenous bases found in nucleotides (and, therefore, nucleic acids) are ***aromatic heterocyclic compounds***. The bases are of two types—purines and pyrimidines. Their general structures are depicted in ***Fig.5.1***. Purines are numbered in the anticlockwise direction while pyrimidines are numbered in the clockwise direction. And this is an internationally accepted system to represent the structure of bases.

Major bases in nucleic acids

The structures of major purines and pyrimidines found in nucleic acids are shown in ***Fig.5.2***. DNA and RNA contain the same purines

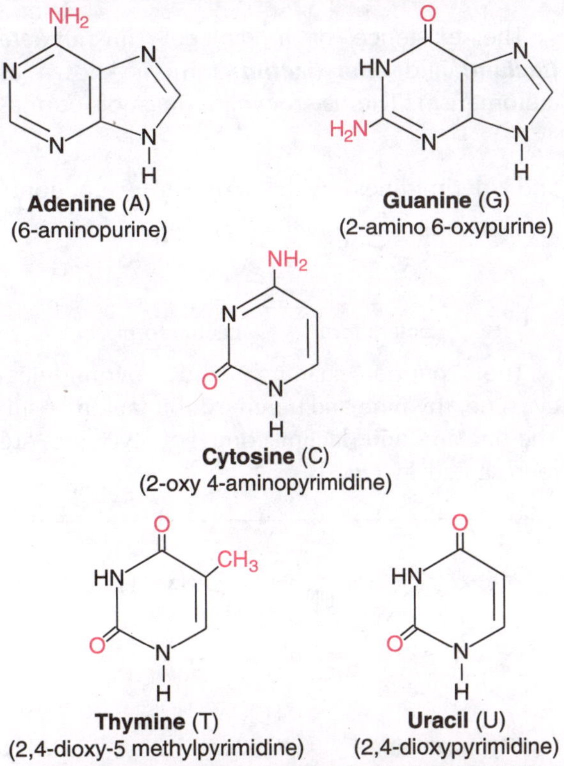

Fig. 5.2 : Structures of major purines (A, G) and pyrimidines (C, T, U) found in nucleic acids.

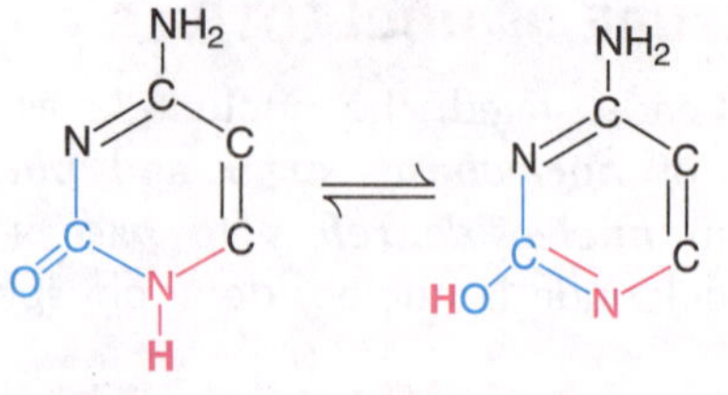

Lactam form **Lactim** form

Fig. 5.3 : *The tautomeric forms of cytosine.*

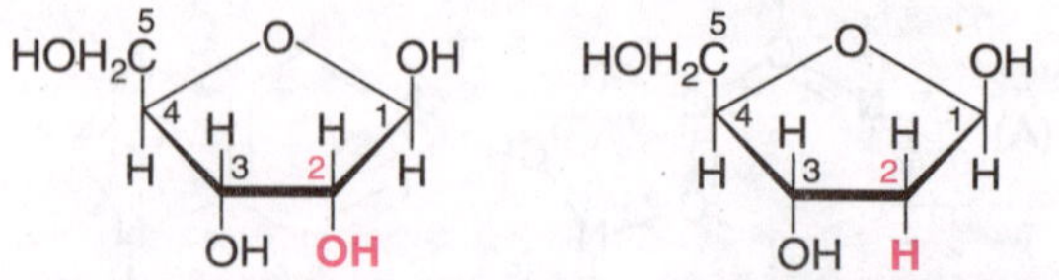

D-Ribose **D-2-Deoxyribose**

Fig. 5.5 : *Structures of sugars present in nucleic acids (ribose is found in RNA and deoxyribose in DNA; Note the structural difference at C_2).*

namely adenine (A) and guanine (G). Further, the pyrimidine cytosine (C) is found in both DNA and RNA. However, the nucleic acids differ with respect to the second pyrimidine base. ***DNA contains thymine (T) whereas RNA contains uracil (U)***. As is observed in the ***Fig.5.2***, thymine and uracil differ in structure by the presence (in T) or absence (in U) of a methyl group.

Tautomeric forms of purines and pyrimidines

The existence of a molecule in a ***keto (lactam)*** and ***enol (lactim)*** form is known as tautomerism. The heterocyclic rings of purines and pyrimidines with ***oxo*** $\left(\overset{O}{\overset{\|}{-C-}}\right)$ functional groups exhibit tautomerism as simplified below.

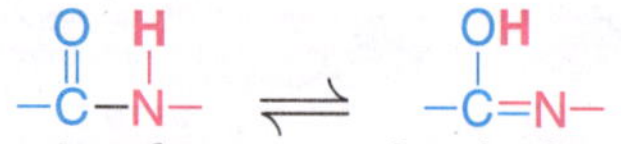

Lactam form **Lactim form**

The purine—guanine and pyrimidines-cytosine, thymine and uracil exhibit tautomerism. The lactam and lactim forms of cytosine are represented in ***Fig.5.3***.

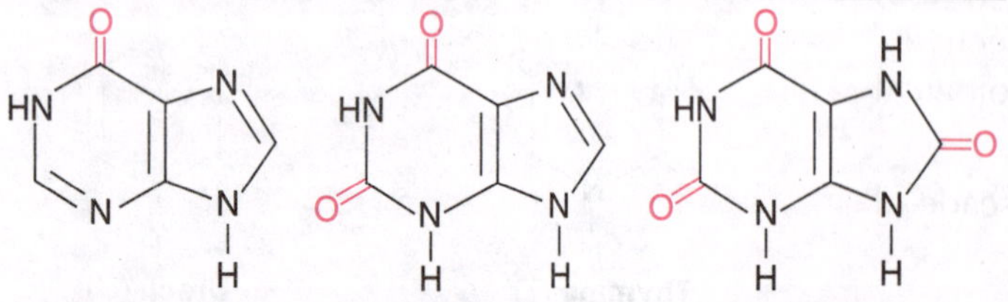

Hypoxanthine (6-oxypurine) **Xanthine** (2,6-dioxypurine) **Uric acid** (2,6,8-trioxypurine)

Fig. 5.4 : *Structures of some biologically important purines.*

At physiological pH, the lactam (keto) tautomeric forms are predominantly present.

Minor bases found in nucleic acids : Besides the bases described above, several minor and unusual bases are often found in DNA and RNA. These include ***5-methylcytosine***, ***N^4-acetylcytosine***, N^6-methyladenine, N^6, N^6-dimethyladenine, ***pseudouracil*** etc. It is believed that the unusual bases in nucleic acids will help in the recognition of specific enzymes.

Other biologically important bases : The bases such as ***hypoxanthine***, ***xanthine*** and ***uric acid*** (***Fig.5.4***) are present in the free state in the cells. The former two are the intermediates in purine synthesis while uric acid is the end product of purine degradation.

Purine bases of plants : Plants contain certain methylated purines which are of pharmacological interest. These include ***caffeine*** (of coffee), ***theophylline*** (of tea) and ***theobromine*** (of cocoa).

Sugars of nucleic acids

The five carbon monosaccharides (pentoses) are found in the nucleic acid structure. ***RNA*** contains ***D-ribose*** while ***DNA*** contains ***D-deoxyribose***. Ribose and deoxyribose differ in structure at C_2. Deoxyribose has one oxygen less at C_2 compared to ribose (***Fig.5.5***).

Nomenclature of nucleotides

The addition of a pentose sugar to base produces a nucleoside. If the sugar is ribose, ribonucleosides are formed. Adenosine, guanosine, cytidine

Table 5.1 Principal bases, nucleosides and nucleotides

Base	Ribonucleoside	Ribonucleotide (5′-monophosphate)	Abbreviation
Adenine (A)	Adenosine	Adenosine 5′-monophosphate or adenylate	AMP
Guanine (G)	Guanosine	Guanosine 5′-monophosphate or guanylate	GMP
Cytosine (C)	Cytidine	Cytidine 5′-monophosphate or cytidylate	CMP
Uracil (U)	Uridine	Uridine 5′-monophosphate or uridylate	UMP
Base	**Deoxyribonucleoside**	**Deoxyribonucleotide (5′-monophosphate)**	**Abbreviation**
Adenine (A)	Deoxyadenosine	Deoxyadenosine 5′-monophosphate or deoxyadenylate	dAMP
Guanine (G)	Deoxyguanosine	Deoxyguanosine 5′-monophosphate or deoxyguanylate	dGMP
Cytosine (C)	Deoxycytidine	Deoxycytidine 5′-monophosphate or deoxycytidylate	dCMP
Thymine (T)	Deoxythymidine	Deoxythymidine 5′-monophosphate or deoxythymidylate	dTMP

and uridine are the ribonucleosides of A, G, C and U respectively. If the sugar is a deoxyribose, ***deoxyribonucleosides*** are produced.

The term mononucleotide is used when a single phosphate moiety is added to a nucleoside. Thus adenosine monophosphate (AMP) contains adenine + ribose + phosphate.

The principal bases, their respective nucleosides and nucleotides found in the structure of nucleic acids are given in ***Table 5.1***. Note that the prefix 'd' is used to indicate if the sugar is deoxyribose (e.g. dAMP).

The binding of nucleotide components

The atoms in the purine ring are numbered as 1 to 9 and for pyrimidine as 1 to 6 (***See Fig.5.1***). The carbons of sugars are represented with an associated prime (′) for differentiation. Thus the pentose carbons are 1′ to 5′.

The pentoses are bound to nitrogenous bases by ***β-N-glycosidic bonds***. The N^9 of a purine ring binds with $C_{1(1')}$ of a pentose sugar to form a covalent bond in the purine nucleoside. In case of pyrimidine nucleosides, the glycosidic linkage is between N^1 of a pyrimidine and C'_1 of a pentose.

The hydroxyl groups of adenosine are esterified with phosphates to produce 5′- or 3′-monophosphates. 5′-Hydroxyl is the most commonly esterified, hence 5′ is usually omitted while writing nucleotide names. Thus AMP represents adenosine 5′-monophosphate. However, for adenosine 3′-monophosphate, the abbreviation 3′-AMP is used.

The structures of two selected nucleotides namely AMP and TMP are depicted in ***Fig.5.6***.

Nucleoside di- and triphosphates

Nucleoside monophosphates possess only one phosphate moiety (AMP, TMP). The addition of second or third phosphates to the nucleoside results in nucleoside diphosphate (e.g. ADP) or triphosphate (e.g. ATP), respectively.

The anionic properties of nucleotides and nucleic acids are due to the negative charges contributed by phosphate groups.

SYNTHETIC NUCLEOTIDE ANALOGS

It is possible to ***alter heterocyclic ring or sugar moiety***, and produce ***synthetic analogs of purines, pyrimidines***, nucleosides and nucleotides. Some of the synthetic analogs are highly useful in clinical medicine. The structures of selected purine and pyrimidine analogs are given in ***Fig.5.7***.

Medical Concepts/ Clinical Correlates

- DNA is the reserve bank of genetic information, ultimately responsible for the chemical basis of life and heredity.
- DNA is organized into genes, the fundamental units of genetic information. Genes control protein biosynthesis through the mediation of RNA.
- Nucleic acids are the polymers of nucleotides. Certain nucleotides serve as B-complex vitamin coenzymes (FAD, NAD^+, CoA), carriers of high energy intermediates (UDP-glucose, S-adenosylmethionine) and second messengers of hormonal action (cAMP, cGMP).

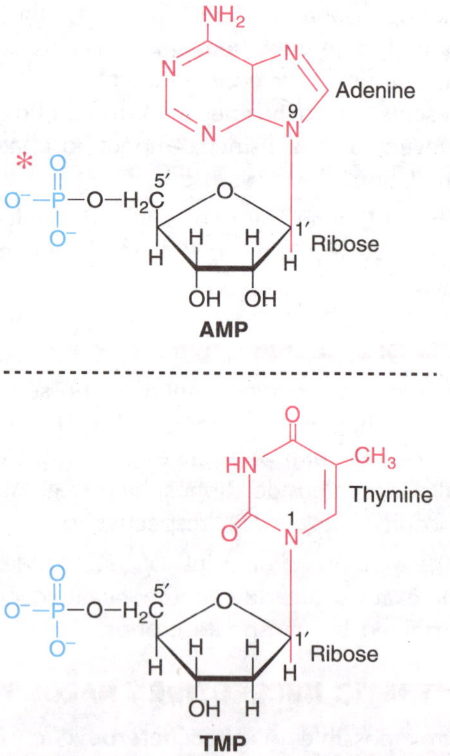

Fig. 5.6 : *The structures of adenosine 5′-monophosphate (AMP) and thymidine 5′-monophosphate (TMP) [* -Addition of second or third phosphate gives adenosine diphosphate (ADP) and adenosine triphosphate (ATP) respectively].*

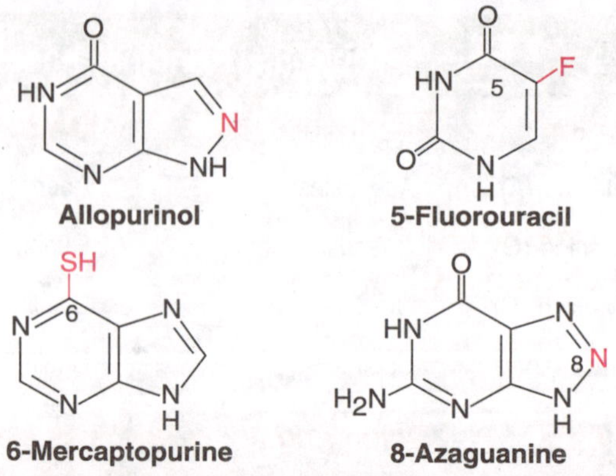

Fig. 5.7 : *Structures of selected purine and pyrimidine analogs.*

The pharmacological applications of certain analogs are listed below

1. ***Allopurinol*** is used in the treatment of hyperuricemia and ***gout*** (For details, ***Refer Chapter 17***).

2. ***5-Fluorouracil***, 6-mercaptopurine, 8-azaguanine, 3-deoxyuridine, 5- or 6-azauridine, 5- or 6-azacytidine and 5-iodouracil are employed in the ***treatment of cancers***. These compounds get incorporated into DNA and block cell proliferation.

3. ***Azathioprine*** (which gets degraded to 6-mercaptopurine) is used to suppress immunological rejection during ***transplantation***.

4. ***Arabinosyladenine*** is used for the treatment of ***neurological disease***, viral encephalitis.

5. ***Arabinosylcytosine*** is being used in cancer therapy as it interferes with DNA replication.

6. The drugs employed in the treatment of ***AIDS*** namely ***zidovudine*** or AZT (3-azido 2′,3′-dideoxythymidine) and didanosine (dideoxyinosine) are sugar modified ***synthetic nucleotide analogs*** (For their structure and more details ***Refer Chapter 38***).

STRUCTURE OF DNA

DNA is a ***polymer of deoxyribonucleotides*** (or simply deoxynucleotides). It is composed of monomeric units namely deoxyadenylate (dAMP), deoxyguanylate (dGMP), deoxycytidylate (dCMP) and deoxythymidylate (dTMP) (It may be noted here that some authors prefer to use TMP for deoxythymidylate, since it is found only in DNA).

Schematic representation of polynucleotides

The monomeric ***deoxynucleotides*** in DNA are ***held*** together by ***3′,5′-phosphodiester bridges*** (***Fig.5.8***). DNA (or RNA) structure is often represented in a short-hand form. The horizontal line indicates the carbon chain of sugar with base attached to $C_{1'}$. Near the middle of the horizontal line is $C_{3'}$ phosphate linkage while at the other end of the line is $C_{5'}$ phosphate linkage (***Fig.5.8***).

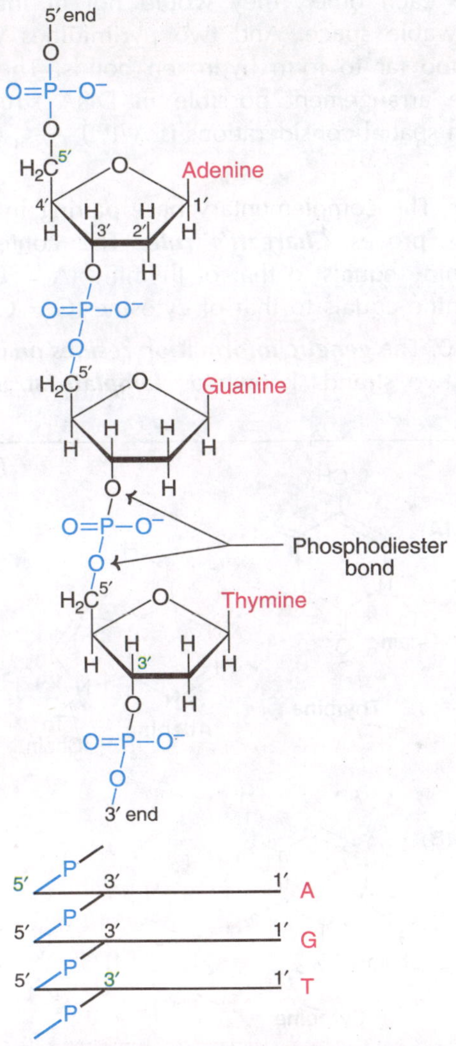

Fig. 5.8 : Structure of a polydeoxyribonucleotide segment held by phosphodiester bonds. On the lower part is the representation of short hand form of oligonucleotides.

Chargaff's rule of DNA composition

Erwin Chargaff in late 1940s quantitatively analysed the DNA hydrolysates from different species. He observed that in all the species he studied, DNA had equal numbers of adenine and-thymine residues (A = T) and equal numbers of guanine and cytosine residues (G = C). This is known as Chargaff's rule of ***molar equivalence between the purines and pyrimidines in DNA*** structure. The significance of Chargaff's rule was not immediately realised. The double helical structure of DNA derives its strength from Chargaff's rule (discussed later).

Single-stranded DNA, and RNAs which are usually single-stranded, do not obey Chargaff's rule. However, double-stranded RNA which is the genetic material in certain viruses satisfies Chargaff's rule.

DNA DOUBLE HELIX

The double helical structure of DNA was proposed by ***James Watson*** and ***Francis Crick*** in 1953 (Nobel Prize, 1962). The elucidation of DNA structure is considered as a ***milestone in*** the era of ***modern biology***. The structure of DNA double helix is comparable to a twisted ladder. The salient features of Watson-Crick model of DNA (now known as B-DNA) are described (***Fig.5.9***).

1. The DNA is a right handed double helix. It consists of ***two polydeoxyribonucleotide chains*** (strands) twisted around each other on a common axis.

2. The two strands are ***antiparallel***, i.e., one strand runs in the 5′ to 3′ direction while the other in 3′ to 5′ direction. This is comparable to two parallel adjacent roads carrying traffic in opposite direction.

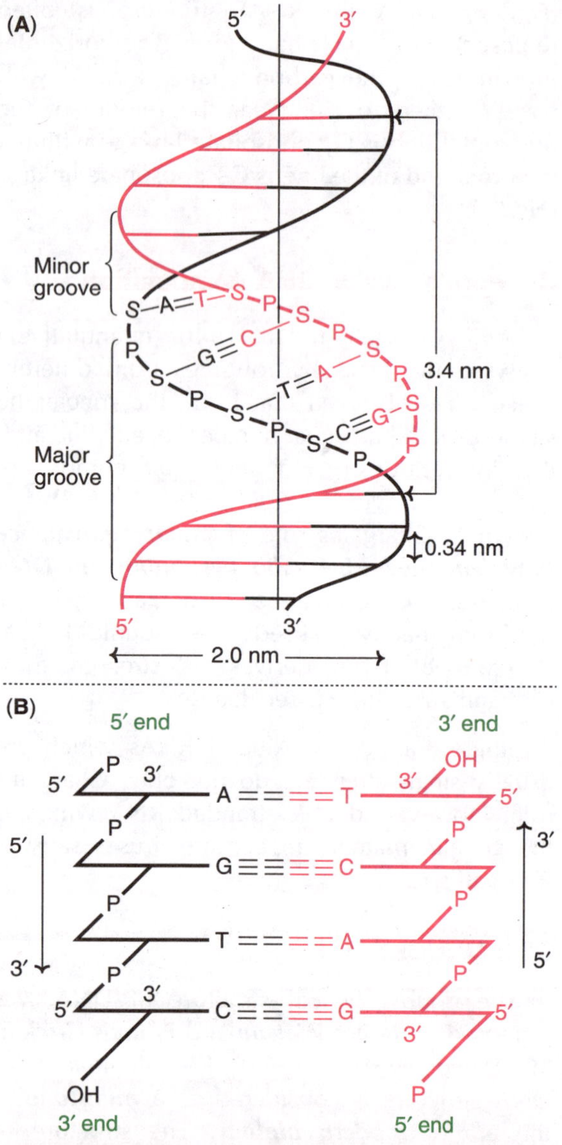

Fig. 5.9 : (A) Watson–Crick model of DNA helix (B) Complementary base pairing in DNA helix.

3. The width (or diameter) of a double helix is 20 A° (2 nm).

4. Each turn (pitch) of the helix is 34 A° (3.4 nm) with ***10 pairs of nucleotides***, each pair placed at a distance of about 3.4 A°.

5. Each strand of DNA has a hydrophilic ***deoxyribose phosphate backbone*** (3′-5′ phosphodiester bonds) on the outside (periphery) of the molecule while the hydrophobic bases are stacked inside (core).

6. The ***two polynucleotide*** chains are not identical but ***complementary*** to each other due to base pairing.

7. The two strands are ***held together by hydrogen bonds*** formed by complementary base pairs (***Fig.5.10***). The A-T pair has 2 hydrogen bonds while G-C pair has 3 hydrogen bonds. The G ≡ C is stronger by about 50% than A = T.

8. The ***hydrogen bonds*** are formed ***between a purine and a pyrimidine only***. If two purines face each other, they would not fit into the allowable space. And two pyrimidines would be too far to form hydrogen bonds. The only base arrangement possible in DNA structure, from spatial considerations is A-T, T-A, G-C and C-G.

9. The complementary base pairing in DNA helix proves ***Chargaff's rule***. The content of adenine equals to that of thymine (A = T) and guanine equals to that of cytosine (G = C).

10. The ***genetic information resides on*** one of the two strands known as ***template strand*** or

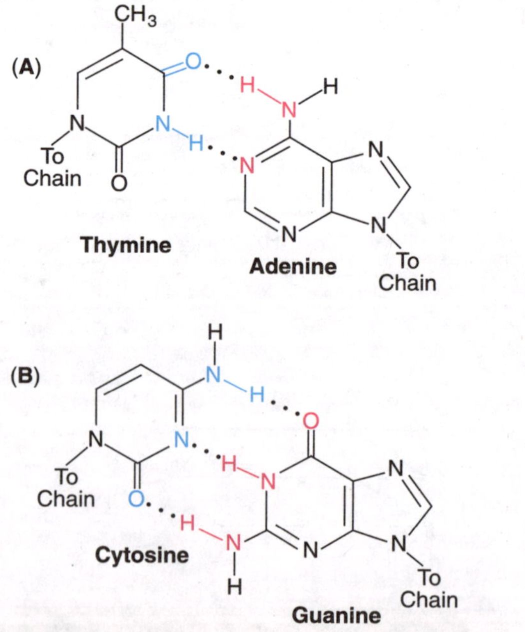

Fig. 5.10 : Complementary base pairing in DNA (A) Thymine pairs with adenine by 2 hydrogen bonds (B) Cytosine pairs with guanine by 3 hydrogen bonds.

Medical Concepts/ Clinical Correlates

- Uric acid is a purine, and the end product of purine metabolism, that has been implicated in the disorder gout.
- Certain purine bases from plants such as caffeine (of coffee), theophylline (of tea) and theobromine (of cocoa) are of pharmacological interest.
- Synthetic analogs of bases (5-fluorouracil, 6-mercaptopurine, 6-azauridine) are used to inhibit the growth of cancer cells.
- Certain antitumor drugs (e.g. cisplatin) can produce bent DNA structure and damage it.

sense strand. The opposite strand is antisense strand. The double helix has (wide) major grooves and (narrow) minor grooves along the phosphodiester backbone. Proteins interact with DNA at these grooves, without disrupting the base pairs and double helix.

Conformations of DNA double helix

Variation in the conformation of the nucleotides of DNA is associated with conformational variants of DNA. The double helical structure of DNA exists in at least 6 different forms-A to E and Z. Among these, B, A and Z forms are important (***Table 5.2***). The ***B-form of DNA*** double helix, described by Watson and Crick (discussed above), is the most predominant form ***under physiological conditions***. Each turn of the B-form has 10 base pairs spanning a distance of 3.4 nm. The width of the double helix is 2 nm.

The A-form is also a right-handed helix. It contains 11 base pairs per turn. There is a tilting of the base pairs by 20° away from the central axis.

The Z-form (Z-DNA) is a left-handed helix and contains 12 base pairs per turn. The polynucleotide strands of DNA move in a somewhat 'zig zag' fashion, hence the name Z-DNA.

It is believed that transition between different helical forms of DNA plays a significant role in regulating gene expression.

OTHER TYPES OF DNA STRUCTURE

It is now recognized that besides double helical structure, DNA also exists in certain unusual structures. It is believed that such structures are important for molecular recognition of DNA by proteins and enzymes. This is in fact needed for the DNA to discharge its functions in an appropriate manner. Some selected ***unusual structures of DNA*** are ***bent DNA***, ***triple-stranded DNA*** and ***four-standed DNA***.

THE SIZE OF DNA MOLECULE —UNITS OF LENGTH

DNA molecules are huge in size. On an average, a pair of B-DNA with a thickness of 0.34 nm has a molecular weight of 660 daltons.

For the measurement of lengths, DNA double-stranded structure is considered, and expressed in the form of ***base pairs*** (***bp***). A ***kilobase pair (kb)*** is 10^3 bp, and a ***megabase pair (Mb)*** is 10^6 bp and a gigabase pair (Gb) is 10^9 bp. The kb,

Table 5.2 Comparison of structural features of different conformations of DNA double helix

Feature	B-DNA	A-DNA	Z-DNA
Helix type	Right-handed	Right-handed	Left-handed
Helical diameter (nm)	2.37	2.55	1.84
Distance per each complete turn (nm)	3.4	3.2	4.5
Rise per base pair (nm)	0.34	0.29	0.37
Number of base pairs per complete turn	10	11	12
Base pair tilt	+19°	−1.2° (variable)	−9°
Helix axis rotation	Major groove	Through base pairs (variable)	Minor groove

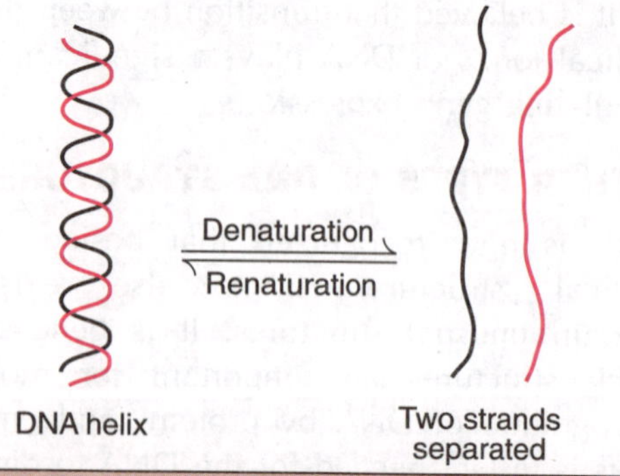

Fig. 5.11 : Diagrammatic representation of denaturation and renaturation of DNA.

Mb and Gb relations may be summarized as follows :

1 kb = 1000 bp

1 Mb = 1000 kb = 1,000,000 bp

1 Gb = 1000 Mb = 1,000,000,000 bp

It may be noted here that the lengths of RNA molecules (like DNA molecules) cannot be expressed in bp, since most of the RNAs are single-stranded.

The length of DNA varies from species to species, and is usually expressed in terms of base pair composition and ***contour length***. Contour length represents the ***total length of the genomic DNA in a cell***. Some examples of organisms with bp and contour lengths are listed.

- λ phage virus — 4.8×10^4 bp — contour length 16.5 μm.
- *E. coli* — 4.6×10^6 bp — contour length 1.5 μm.
- Diploid human cell (46 chromosomes) — 6.0×10^9 bp — contour length 2 meters.

It may be noted that the genomic DNA size is usually much larger than the size of the cell or nucleus containing it. For instance, in humans, a 2-meter long DNA is packed compactly in a nucleus of about 10μm diameter.

The genomic DNA may exist in linear or circular forms. Most DNAs in bacteria exist as closed circles. This includes the DNA of bacterial chromosomes and the extra-chromosomal DNA of plasmids. Mitochondria and chloroplasts of eukaryotic cells also contain circular DNA.

Chromosomal DNAs in higher organisms are mostly linear. Individual human chromosomes contain a single DNA molecule with variable sizes compactly packed. Thus the smallest chromosome contains 34 Mb while the largest one has 263 Mb.

DENATURATION OF DNA STRANDS

The two strands of DNA helix are held together by hydrogen bonds. Disruption of hydrogen bonds (by change in pH or increase in temperature) results in the separation of polynucleotide strands. This phenomenon of ***loss of helical structure of DNA*** is known as ***denaturation*** (***Fig.5.11***). The phosphodiester bonds are not broken by denaturation. Loss of helical structure can be measured by increase in absorbance at 260 nm (in a spectrophotometer). The phenomenon of increase in the absorbance of purines and pyrimidines, following denaturation is referred to as ***hyperchromicity***.

Melting temperature (***Tm***) is defined as the temperature at which half of the helical structure of DNA is lost. Since G-C base pairs are more stable (due to 3 hydrogen bonds) than A-T base pairs (2 hydrogen bonds), the Tm is greater for DNAs with higher G-C content. Thus, the Tm is 65°C for 35% G-C content while it is 70°C for 50% G-C content. Formamide destabilizes hydrogen bonds of base pairs and, therefore, lowers Tm. This chemical compound is effectively used in recombinant DNA experiments.

Renaturation (or ***reannealing***) is the process in which the ***separated complementary DNA strands can form a double helix.***

ORGANIZATION OF DNA IN THE CELL

As already stated, the double-stranded DNA helix in each chromosome has a length that is thousands times the diameter of the nucleus. For instance, in humans, a 2-meter long DNA is packed in a nucleus of about 10 μm diameter! This is made possible by a compact and marvellous packaging, and organization of DNA inside in cell.

Organization of prokaryotic DNA

In prokaryotic cells, the ***DNA is organized as a single chromosome*** in the form of a double-stranded circle. These bacterial chromosomes are packed in the form of nucleoids, by interaction with proteins and certain cations (polyamines).

Organization of eukaryotic DNA

In the eukaryotic cells, the DNA is associated with various ***proteins to form chromatin*** which then gets ***organized into*** compact structures namely ***chromosomes*** (***Fig.5.12***).

The DNA double helix is wrapped around the core proteins namely ***histones*** which are basic in nature. The core is composed of two molecules of histones (H2A, H2B, H3 and H4). Each core with two turns of DNA wrapped round it (approximately with 150 bp) is termed as a ***nucleosome***, the basic unit of chromatin. Nucleosomes are separated by spacer DNA to which histone H_1 is attached (***Fig.5.13***). This continuous string of nucleosomes, representing beads-on-a string form of chromatin is termed as 10 nm fiber. The length of the DNA is considerably reduced by the formation of 10 nm fiber. This 10-nm fiber is further coiled to produce 30-nm fiber which has a solenoid structure with six nucleosomes in every turn. These 30-nm fibers are further organized into loops by anchoring the fiber at A/T-rich regions namely scaffold-associated regions (SARS) to a protein scafold. During the course of mitosis, the loops are further coiled, the chromosomes condense and become visible.

STRUCTURE OF RNA

RNA is a ***polymer of ribonucleotides*** held together by ***3′,5′-phosphodiester bridges.*** Although RNA has certain similarities with DNA structure, they have specific differences

1. **Pentose :** The sugar in RNA is ribose in contrast to deoxyribose in DNA.

2. **Pyrimidine :** RNA contains the pyrimidine uracil in place of thymine (in DNA).

3. **Single strand :** RNA is usually a single-stranded polynucleotide. However, this strand may fold at certain places to give a double-stranded structure, if complementary base pairs are in close proximity.

4. **Chargaff's rule—not obeyed :** Due to the single-stranded nature, there is no specific relation between purine and pyrimidine contents. Thus the guanine content is not equal to cytosine (as is the case in DNA).

5. **Susceptibility to alkali hydrolysis :** Alkali can hydrolyse RNA to 2′,3′-cyclic diesters. This is possible due to the presence of a hydroxyl group at 2′ position. DNA cannot be subjected to alkali hydrolysis due to lack of this group.

6. **Orcinol colour reaction :** RNAs can be histologically identified by orcinol colour reaction due to the presence of ribose.

TYPES OF RNA

The three major types of RNAs with their respective cellular composition are given below

1. **Messenger RNA** (mRNA) : 5–10%
2. **Transfer RNA** (tRNA) : 10–20%
3. **Ribosomal RNA** (rRNA) : 50–80%

Besides the three RNAs referred above, other RNAs are also present in the cells. These include heterogeneous nuclear RNA (***hnRNA***), small nuclear RNA (***snRNA***), small nucleolar RNA (***snoRNA***) and small cytoplasmic RNA (***scRNA***). The major functions of these RNAs are given in ***Table 5.3***.

The RNAs are synthesized from DNA, and are primarily involved in the process of protein biosynthesis (***Chapter 25***). The RNAs vary in their structure and function. A brief description on the major RNAs is given.

Messenger RNA (mRNA)

The mRNA is synthesized in the nucleus (in eukaryotes) as ***heterogeneous nuclear RNA*** (***hnRNA***). hnRNA, on processing, liberates the functional mRNA which enters the cytoplasm to participate in ***protein synthesis***. mRNA has high molecular weight with a short half-life.

In general, mRNA of eukaryotes is more stable with longer half-life, compared to prokaryotic mRNA.

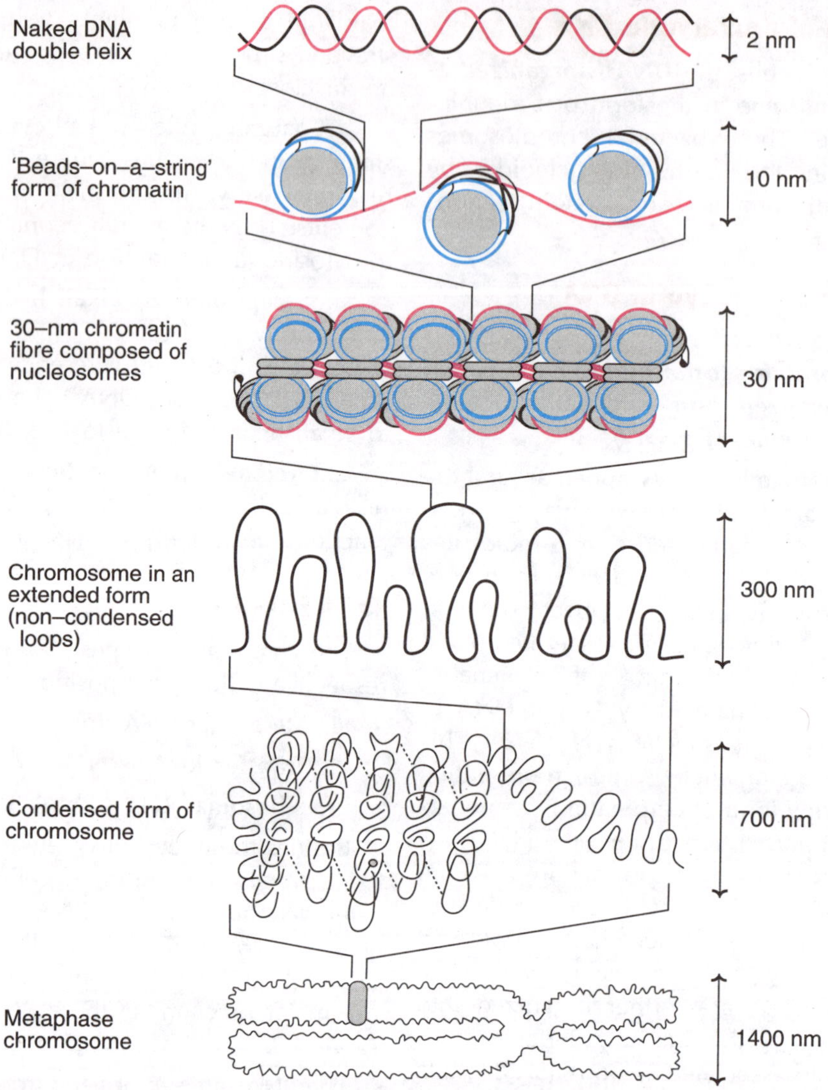

Fig. 5.12 : *Organization of eukaryotic DNA structure in the form of chromatin and chromosomes.*

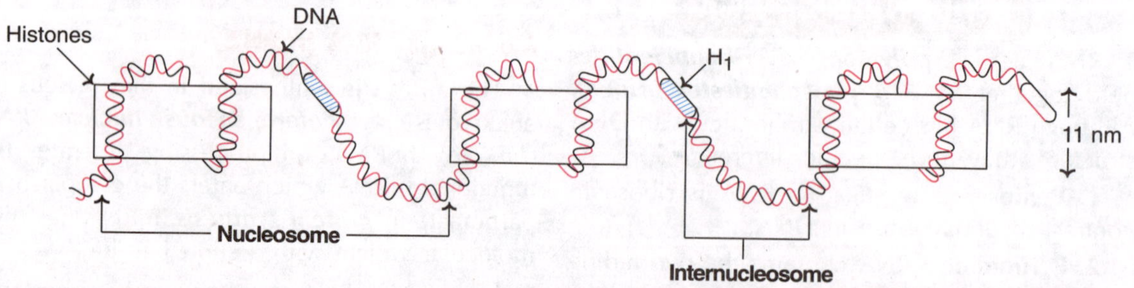

Fig. 5.13 : *Structure of nucleosomes.*

Medical Concepts/ Clinical Correlates

- MicroRNAs and silencing RNAs, capable of inhibiting gene expression, are potential agents for therapeutic purposes.
- It is possible to design recombinant ribozymes that can be used to cure diseases.
- Nucleic acid analogues (PNA, GNA) are important to investigate origins of life.

The eukaryotic mRNA is capped at the 5′-terminal end by 7-methylguanosine triphosphate. It is believed that this cap helps to prevent the hydrolysis of mRNA by 5′-exonucleases. Further, the cap may be also involved in the recognition of mRNA for protein synthesis.

The 3′-terminal end of mRNA contains a polymer of adenylate residues (20-250 nucleotides) which is known as ***poly (A) tail***. This tail may provide stability to mRNA, besides preventing it from the attack of 3′-exonucleases.

mRNA molecules often contain certain modified bases such as 6-methyladenylates in the internal structure.

Transfer RNA (tRNA)

Transfer RNA (***soluble RNA***) molecule contains 71-80 nucleotides (mostly 75) with a molecular weight of about 25,000. There are at least 20 species of tRNAs, corresponding to 20 amino acids present in protein structure. The structure of tRNA (for alanine) was first elucidated by Holley.

The structure of tRNA, depicted in ***Fig.5.14***, resembles that of a clover leaf. tRNA contains mainly four arms, each arm with a base paired stem.

1. **The acceptor arm :** This arm is capped with a sequence CCA (5′ to 3′). The amino acid is attached to the acceptor arm.

2. **The anticodon arm :** This arm, with the three specific nucleotide bases (anticodon), is responsible for the recognition of triplet codon of mRNA. The codon and anticodon are complementary to each other.

3. **The D arm :** It is so named due to the presence of dihydrouridine.

4. **The TΨC arm :** This arm contains a sequence of T, ***pseudouridine*** (represented by psi, Ψ) and C.

5. **The variable arm :** This arm is the most variable in tRNA. Based on this variability, tRNAs are classified into 2 categories :

(a) **Class I tRNAs :** The most predominant (about 75%) form with 3-5 base pairs length.

(b) **Class II tRNAs :** They contain 13-20 base pair long arm.

Base pairs in tRNA : The structure of tRNA is maintained due to the complementary base pairing in the arms. The four arms with their respective base pairs are given below

The acceptor arm	– 7 bp
The TΨC arm	– 5 bp
The anticodon arm	– 5 bp
The D arm	– 4 bp

Ribosomal RNA (rRNA)

The ribosomes are the factories of protein synthesis. The eukaryotic ribosomes are composed

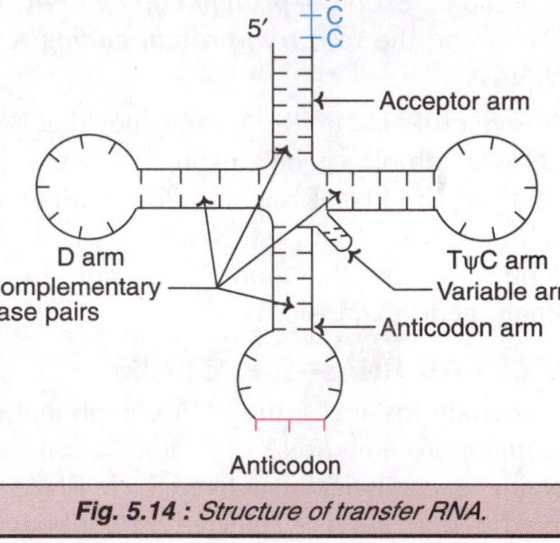

Fig. 5.14 : Structure of transfer RNA.

Table 5.3 Cellular RNAs and their function(s)

Type of RNA	Abbreviation	Function(s)
Messenger RNA	mRNA	Transfers genetic information from genes to ribosomes to synthesize proteins.
Heterogeneous nuclear RNA	hnRNA	Serves as a precursor for mRNA.
Transfer RNA	tRNA	Transfers amino acid to mRNA for protein biosynthesis.
Ribosomal RNA	rRNA	Provides structural framework for ribosomes.
Small nuclear RNA	snRNA	Involved in mRNA processing.
Long noncoding RNA	lncRNA	Regulates mRNA synthesis.
Small nucleolar RNA	snoRNA	Plays a key role in the processing of rRNA molecules.
Small cytoplasmic RNA	scRNA	Involved in the selection of proteins for export.
Transfer–messenger RNA	tmRNA	Mostly present in bacteria. Adds short peptide tags to proteins to facilitate the degradation of incorrectly synthesized proteins.
Micro-RNA and silencing RNA	miRNA and siRNA	Both inhibit gene expression.

of two major nucleoprotein complexes–60S subunit and 40S subunit. The 60S subunit contains 28S rRNA, 5S rRNA and 5.8S rRNA while the 40S subunit contains 18S rRNA. The function of rRNAs in ribosomes is not clearly known. It is believed that they play a significant role in the binding of mRNA to ribosomes and protein synthesis.

Other RNAs

The various RNAs and their respective functions are summarised in ***Table 5.3***.

The eukaryotic RNAs are considered under two broad categories—***protein coding RNAs*** (i.e. mRNAs) and the rest, ***non-protein coding RNAs*** (***ncRNAs***).

The microRNAs (miRNAs) and silencing RNAs (siRNAs) inhibit gene expression via the formation of RNA-RNA hybrids with target mRNA. miRNAs and siRNAs represent new potential agents for therapeutic purposes to regulate gene expression.

CATALYTIC RNAs—RIBOZYMES

In certain instances, the RNA component of a ribonucleoprotein (RNA in association with protein) is catalytically active. Such RNAs are termed as ribozymes. A selected list of ribozymes along with their biochemical functions is given in ***Table 5.4***.

Ribonuclease P (RNase P) is a ribozyme containing protein and RNA component. It cleaves tRNA precursors to generate mature tRNA molecules.

RNA molecules are known to adapt tertiary structure just like proteins (i.e. enzymes). The specific conformation of RNA may be responsible for its function as biocatalyst. It is believed that ***ribozymes*** (RNAs) were

Table 5.4 A selected list of ribozymes and the corresponding biochemical reactions

Ribozyme(s)	Biochemical reaction(s)
rRNA	Peptide bond formation in protein biosynthesis
RNase P	RNA cleavage and ligation
Self-splicing RNAs	DNA cleavage
RNAs of splicesome	RNA splicing
***In vitro* selected RNAs**	RNA polymerization, RNA phosphorylation RNA aminoacylation Glycoside bond formation Oxidation-reduction reactions Disulfide exchange

functioning as ***catalysts before the occurrence of protein enzymes***, during the course of evolution.

Recombinant ribozymes (rribozymes)

It is now possible to design recombinant ribozymes that will cleave any RNA. These ribozymes are now being considered as ***therapeutic agents to cure diseases***. Theoretically it is possible to selectively degrade faulty RNAs (mutated or inappropriately expressed RNAs in diseases) by rribozymes. This way specific RNAs can be eliminated from the cell that will help to inhibit the disease process.

NUCLEIC ACID ANALOGS

Nucleic acid analogs are structurally similar (analogous) to naturally occurring RNA and DNA. Examples of artificial nucleic acids include ***peptide nucleic acid (PNA), glucol nucleic acid (GNA)*** and ***threose nucleic acid (TNA)***. These artificial nucleic acids differ from natural DNA or RNA by changes in the backbone of the molecule.

Nucleic acid analogs are also called ***xeno nucleic acids (XNA)***. They are important to investigate possible scenarios of the origins of life, and to design new-to-nature forms of life.

SUMMARY

- *DNA is the chemical basis of heredity organized into genes, the basic units of genetic information.*
- *RNAs (mRNA, tRNA and rRNA) are produced by DNA which in turn carry out protein synthesis.*
- *Nucleic acids are the polymers of nucleotides (polynucleotides) held by 3′ and 5′ phosphodiester bridges. A nucleotide essentially consists of base + sugar (nucleoside) and phosphate.*
- *Besides being the constituents of nucleic acid structure, nucleotides perform a wide variety of cellular functions (e.g. energy carriers, metabolic regulators, second messengers etc.)*
- *Both DNA and RNA contain the purines-adenine (A) and guanine (G) and the pyrimidine-cytosine (C). The second pyrimidine is thymine (T) in DNA while it is uracil (U) in RNA. The pentose sugar, D-deoxyribose is found in DNA while it is D-ribose in RNA.*
- *The structure of DNA is a double helix (Watson-Crick model) composed of two antiparallel strands of polydeoxynucleotides twisted around each other. The strands are held together by 2 or 3 hydrogen bonds formed between the bases i.e. A = T; G ≡ C. DNA structure satisfies Chargaff's rule that the content of A is equal to T, and that of G equal to C.*
- *Besides the double helical structure, DNA also exists in certain unusual structures — bent DNA, triple-strand DNA, four-strand DNA.*
- *RNA is usually a single stranded polyribonucleotide. mRNA is capped at 5′ terminal end by 7-methylGTP while at the 3′-terminal end, it contains a poly A tail. mRNA specifies the sequence of amino acids in protein synthesis.*
- *The structure of tRNA resembles that of a clover leaf with four arms (acceptor, anticodon, D-, and TΨC) held by complementary base pairs. tRNA delivers amino acids for protein synthesis.*
- *Certain RNAs that can function as enzymes are termed as ribozymes. Ribozymes were probably functioning as catalysts before the occurrence of protein enzymes during evolution.*

Self-Assessment Exercises

I. Essay questions

1. Describe the structure of DNA.
2. Name different RNAs and discuss their structure.
3. Write an account of structure, function and nomenclature of nucleotides.
4. Describe the structure of nitrogenous bases present in nucleic acids. Add a note on tautomerism.
5. "The backbone of nucleic acid structure is 3′-5′ phosphodiester bridge."—justify.

II. Short notes

(a) Chargaff's rule, (b) Ribose and deoxyribose, (c) Hydrogen bonds in DNA, (d) Nucleoside, (e) Different forms of DNA, (f) Transfer RNA, (g) Purine bases of plants, (h) Complementary base pairs, (i) DNA denaturation, (j) hnRNA.

III. Fill in the blanks

1. The fundamental unit of genetic information is known as ____________.
2. DNA controls protein synthesis through the mediation of ____________.
3. Nucleic acids are the polymers of ____________.
4. The pyrimidine present in DNA but absent in RNA ____________.
5. Ribose and deoxyribose differ in their structure around carbon atom ____________.
6. Nucleotide is composed of ____________.
7. The scientist who observed that there exists a relationship between the contents of purines and pyrimidines in DNA structure (A = T; G = C) ____________.
8. The base pair G-C is more stable and stronger than A-T due to ____________.
9. Under physiological condition, the DNA structure is predominantly in the form ____________.
10. The acceptor arm of tRNA contains a capped nucleotide sequence ____________.

IV. Multiple choice questions

11. The nitrogenous base not present in DNA structure
    (a) Adenine (b) Guanine (c) Cytosine (d) Uracil.
12. The number of base pairs present in each turn (pitch) of B-form of DNA helix
    (a) 9 (b) 10 (c) 11 (d) 12.
13. The backbone of nucleic acid structure is constructed by
    (a) Peptide bonds (b) Glycosidic bonds (c) Phosphodiester bridges (d) All of them.
14. The following coenzyme is a nucleotide
    (a) FAD (b) NAD^+ (c) CoASH (d) All of them.
15. The nucleotide that serves as an intermediate for biosynthetic reaction
    (a) UDP-glucose (b) CDP-acylglycerol (c) S-Adenosylmethionine (d) All of them.

6

Enzymes

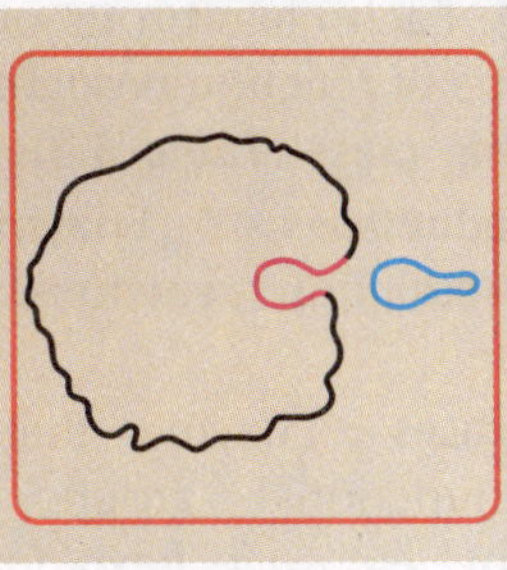

The enzymes speak

"We are the catalysts of the living world!
Protein in nature, and in action specific,
rapid and accurate;
Huge in size but with small active centres;
Highly exploited for disease diagnosis in lab centres."

Enzymes are biocatalysts – the catalysts of life. A ***catalyst*** is defined as a ***substance that increases the velocity*** or rate of a chemical reaction without itself undergoing any change in the overall process.

The student-teacher relationship may be a good example to understand how a catalyst works. The students often find it difficult to learn from a text-book on their own. The teacher explains the subject to the students and increases their understanding capability. Here, the teacher acts like a catalyst in enhancing the understanding ability of students. A good teacher is always a good catalyst in students' life!

Enzymes may be defined as ***biocatalysts synthesized by living cells. They are protein in nature (exception – RNA acting as ribozyme), colloidal and thermolabile in character, and specific in their action.***

In the laboratory, hydrolysis of proteins by a strong acid at 100°C takes at least a couple of days. The same protein is fully digested by the enzymes in gastrointestinal tract at body temperature (37°C) within a couple of hours. This remarkable difference in the chemical reactions taking place in the living system is exclusively due to enzymes. The very existence of life is unimaginable without the presence of enzymes.

HISTORICAL BACKGROUND

Berzelius in 1836 coined the term ***catalysis*** (*Greek* : to dissolve). In 1878, Kuhne used the word ***enzyme*** (*Greek* : in yeast) to indicate the catalysis taking place in the biological systems. Isolation of enzyme system from cell-free extract of yeast was achieved in 1883 by Buchner. He named the active principle as zymase (later found to contain a mixture of enzymes), which could convert sugar to alcohol. In 1926, James Sumner ***first*** achieved the ***isolation*** and crystallization of the enzyme ***urease*** from jack bean and identified it as a protein.

NOMENCLATURE AND CLASSIFICATION

In the early days, the enzymes were given names by their discoverers in an arbitrary manner. For example, the names pepsin, trypsin and chymotrypsin convey no information about the function of the enzyme or the nature of the substrate on which they act. Sometimes, the suffix-***ase*** was added to the substrate for naming the enzymes e.g. lipase acts on lipids; nuclease on nucleic acids; lactase on lactose. These are known as ***trivial names*** of the enzymes which, however, ***fail to give complete information of enzyme reaction*** (type of reaction, cofactor requirement etc.)

Enzymes are sometimes considered under two broad categories : (a) **Intracellular enzymes** – They are functional within cells where they are synthesized. (b) **Extracellular enzymes** – These enzymes are active outside the cell; all the digestive enzymes belong to this group.

The ***International Union of Biochemistry (IUB)*** appointed an ***Enzyme Commission in 1961***. This committee made a thorough study of the existing enzymes and devised some basic principles for the classification and nomenclature of enzymes. Since 1964, the ***IUB system of enzyme classification*** has been in force. Enzymes are divided into ***six major classes*** (in that order). Each class on its own represents the general type of reaction brought about by the enzymes of that class (***Table 6.1***).

1. **Oxidoreductases :** Enzymes involved in oxidation-reduction reactions e.g. ***alcohol dehydrogenase***.

2. **Transferases :** Enzymes that catalyse the transfer of functional groups (phosphoryl, glycosyl, methyl etc.) e.g. ***hexokinase***.

Table 6.1 Classification of enzymes

Enzyme class with examples*	Reaction catalysed
1. **Oxidoreductases** Alcohol dehydrogenase (alcohol : NAD^+ oxidoreductase E.C. 1.1.1.1.), cytochrome oxidase, L- and D-amino acid oxidases	Oxidation ⟶ Reduction $AH_2 + B \longrightarrow A + BH_2$
2. **Transferases** Hexokinase (ATP : D-hexose 6-phosphotransferase, E.C. 2.7.1.1.), transaminases, transmethylases, phosphorylase	Group transfer $A - X + B \longrightarrow A + B - X$
3. **Hydrolases** Lipase (triacylglycerol acyl hydrolase E.C. 3.1.1.3), choline esterase, acid and alkaline phosphatases, pepsin, urease	Hydrolysis $A - B + H_2O \longrightarrow AH + BOH$
4. **Lyases** Aldolase (fructose-bisphosphate aldolase, E.C. 4.1.2.13), fumarase, histidase	Addition ⟶ Elimination $A - B + X - Y \longrightarrow AX - BY$
5. **Isomerases** Triose phosphate isomerase (D-glyceraldehyde 3-phosphate ketoisomerase, E.C. 5.3.1.1), retinol isomerase, phosphohexose isomerase	Interconversion of isomers $A \longrightarrow A'$
6. **Ligases** Glutamine synthetase (L-glutamate ammonia ligase, E.C. 6.3.1.2), acetyl CoA carboxylase, succinate thiokinase	Condensation (usually dependent on ATP) $A + B \xrightarrow[\text{ATP} \;\to\; \text{ADP + Pi}]{} A - B$

**For one enzyme in each class, systematic name along with E.C. number is given in the brackets.*

3. **Hydrolases :** Enzymes that bring about hydrolysis of various compounds by addition of water e.g. ***lipase***.

4. **Lyases :** Enzymes specialised in the cleavage of bonds (C–C, C–O, C–N), resulting in double bonds e.g. ***aldolase***.

5. **Isomerases :** Enzymes involved in all the isomerization reactions e.g. ***retinol isomerase***.

6. **Ligases :** Enzymes catalysing the synthetic reactions (*Greek* : ligate—to bind) where two molecules are joined together and ATP is used e.g. ***succinate thiokinase***.

[The word ***OTHLIL*** (first letter in each class) may be memorised to remember the six classes of enzymes in the correct order].

Each class in turn is subdivided into many sub-classes which are further divided. A four digit ***E***nzyme ***C***ommission (***E.C.***) number is assigned to each enzyme representing the class (first digit), sub-class (second digit), sub-sub class (third digit) and the individual enzyme (fourth digit). Each enzyme is given a specific name indicating the substrate, coenzyme (if any) and the type of the reaction catalysed by the enzyme e.g. ***alcohol : NAD⁺ oxidoreductase E.C.1.1.1.1*** (***alcohol dehydrogenase***).

Translocases — a new class (*E.C.7*) of enzymes : The group of enzymes ***catalysing the transport of ions*** or molecules across biological membranes do not fit into any one of the 6 ***classes of enzymes***, described above. In 2018, the IUB classified these enzymes under a new enzyme class, namely translocases (E.C.7) e.g. ***ornithine translocase***.

CHEMICAL NATURE AND PROPERTIES OF ENZYMES

All the enzymes are invariably proteins. In recent years, however, a few RNA molecules have been shown to function as enzymes. Each enzyme has its own tertiary structure and specific conformation which is very essential for its catalytic activity. The functional unit of the enzyme is known as ***holoenzyme*** which is often made up of ***apoenzyme*** (the protein part) and a ***coenzyme*** (non-protein organic part).

The term ***prosthetic group*** is used when the non-protein moiety tightly (covalently) binds with the apoenzyme. Thus, prosthetic group is an integral part of enzyme structure. The coenzyme can be separated by dialysis from the enzyme while the prosthetic group cannot be.

The word ***monomeric enzyme*** is used if it is made up of a ***single polypeptide*** e.g. ribonuclease, trypsin. Some of the enzymes which possess more than one polypeptide (subunit) chain are known as ***oligomeric enzymes*** e.g. ***lactate dehydrogenase***, aspartate transcarbamoylase etc. There are certain ***multienzyme complexes*** possessing specific sites to catalyse different reactions in a sequence. Only the native intact multienzyme complex is functionally active and not the individual units, if they are separated e.g. ***pyruvate dehydrogenase***, fatty acid synthase, prostaglandin synthase etc. The enzymes exhibit all the general properties of proteins (***Chapter 4***).

Genetic engineering and modified enzymes

Recent advances in biotechnology have made it possible to ***modify the enzymes with desirable characters***-improved catalytic abilities, activities under unusual conditions. This approach is required since enzymes possess enormous potential for their use in medicine and industry.

Hybrid enzymes : It is possible to rearrange genes and produce ***fusion proteins***. e.g. a hybrid enzyme (of glucanase and cellulase) that can more efficiently hydrolyse barley β-glucans in beer manufacture.

Site-directed mutagenesis : This is a technique used to produce a ***specified mutation*** at a predetermined position in a DNA molecule. The result is ***incorporation of a desired amino acid*** (of one's choice) in place of the specified amino acid in the enzyme. By this approach, it is possible to produce an enzyme with desirable characteristics. e.g. ***tissue plasminogen activator*** (used to lyse blood clots in myocardial infarction) ***with increased half-life***. This is achieved by replacing asparagine (at position 120) by glutamine.

Medical Concepts/ Clinical Correlates

- The existence of life is unimaginable without the presence of enzymes—the biocatalysts.
- Majority of the coenzymes (TPP, NAD^+, FAD, CoA) are derived from B-complex vitamins in which form the latter exert their biochemical functions.

In recent years, it has also become possible to produce ***hybrid enzymes*** by rearrangement of genes. Another innovative approach is the production of ***abzymes*** or catalytic antibodies, the ***anti*b*ody* en*zymes***.

FACTORS AFFECTING ENZYME ACTIVITY

The contact between the enzyme and substrate is the most essential pre-requisite for enzyme activity. The important factors that influence the velocity of the enzyme reaction are discussed hereunder

1. Concentration of enzyme

As the concentration of the ***enzyme is increased, the velocity of the reaction*** proportionately ***increases*** (***Fig.6.1***). In fact, this property of enzyme is made use in determining the serum enzymes for the diagnosis of diseases. By using a known volume of serum, and keeping all the other factors (substrate, pH, temperature etc.) at the optimum level, the enzyme could be assayed in the laboratory.

2. Concentration of substrate

Increase in the substrate concentration gradually ***increases the velocity of enzyme reaction*** within the limited range of substrate levels. A ***rectangular hyperbola*** is obtained when velocity is plotted against the substrate concentration (***Fig.6.2***). Three distinct phases of the reaction are observed in the graph (A-linear; B-curve; C-almost unchanged).

Enzyme kinetics and K_m value

The enzyme (E) and substrate (S) combine with each other to form an unstable enzyme-substrate complex (ES) for the formation of product (P).

$$E + S \underset{k_2}{\overset{k_1}{\rightleftharpoons}} ES \xrightarrow{k_3} E + P$$

Here k_1, k_2 and k_3 represent the velocity constants for the respective reactions, as indicated by arrows.

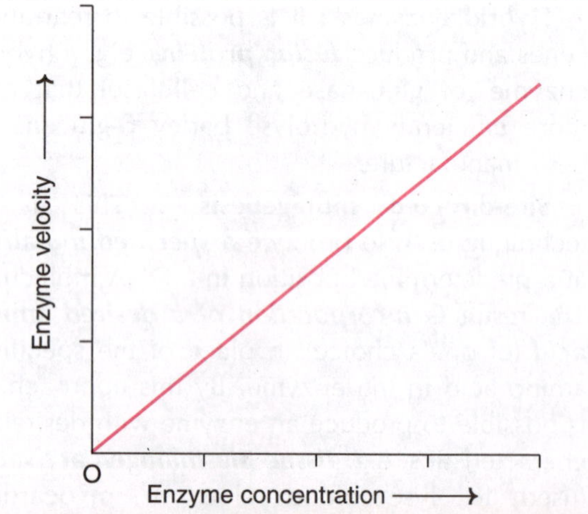

Fig. 6.1 : *Effect of enzyme concentration on enzyme velociy.*

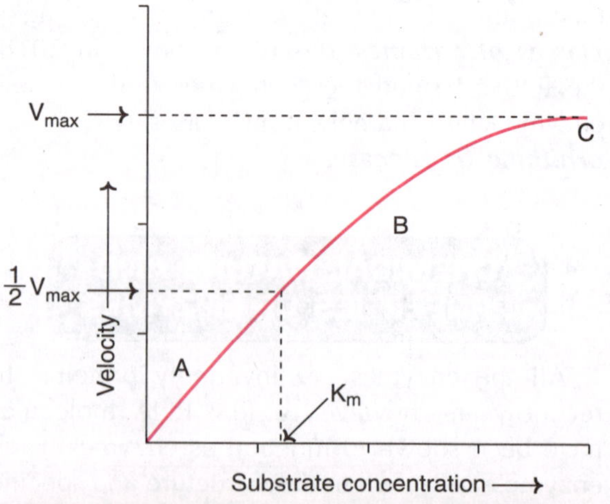

Fig. 6.2 : *Effect of substrate concentration on enzyme velocity (A-linear; B-curve; C-almost unchanged).*

K_m, the Michaelis-Menten constant (or ***Briggs and Haldane constant***), is given by the formula

$$K_m = \frac{k_2 + k_3}{k_1}$$

The following equation is obtained after suitable algebraic manipulation.

$$v = \frac{V_{max}[S]}{K_m + [S]} \quad \text{equation (1)}$$

where v = Measured velocity,

V_{max} = Maximum velocity,

S = Substrate concentration,

K_m = Michaelis – Menten constant.

Let us assume that the measured velocity (v) is equal to $\frac{1}{2}V_{max}$. Then the equation (1) may be substituted as follows

$$\frac{1}{2}V_{max} = \frac{V_{max}[S]}{K_m + [S]}$$

$$K_m + [S] = \frac{2V_{max}[S]}{V_{max}}$$

$$K_m + [S] = 2[S]$$

$$K_m = [S]$$

K stands for a constant and ***m*** stands for ***M***ichaelis (in K_m).

K_m or the ***Michaelis-Menten constant*** is defined as the ***substrate concentration*** (expressed in moles/l) ***to produce half-maximum velocity*** in an enzyme catalysed reaction. It indicates that half of the enzyme molecules (i.e. 50%) are bound with the substrate molecules when the substrate concentration equals the K_m value.

K_m value is a constant and a ***characteristic feature of a given enzyme*** (comparable to a thumb impression or signature). It is a representative for measuring the strength of ES complex. A ***low K_m value indicates a strong affinity between enzyme and substrate***, whereas a high K_m value reflects a weak affinity between them. For majority of enzymes, the K_m values are in the range of 10^{-5} to 10^{-2} moles. It may however, be noted that K_m is ***not dependent on the concentration of enzyme.***

Lineweaver-Burk double reciprocal plot : For the determination of K_m value, the substrate saturation curve (***Fig.6.2***) is not very accurate since V_{max} is approached asymptotically. By taking the reciprocals of the equation (1), a straight line graphic representation is obtained.

$$\frac{1}{v} = \frac{K_m + [S]}{V_{max}[S]}$$

$$\frac{1}{v} = \frac{K_m}{V_{max}} \times \frac{1}{[S]} + \frac{[S]}{V_{max}[S]}$$

$$\frac{1}{v} = \frac{K_m}{V_{max}} \times \frac{1}{[S]} + \frac{1}{V_{max}}$$

The above equation is similar to y = ax + b. Therefore, a plot of the reciprocal of the velocity $\left(\frac{1}{v}\right)$ vs. the reciprocal of the substrate concentration $\left(\frac{1}{[S]}\right)$ gives a straight line. Here the slope is K_m/V_{max} and whose y intercept is $1/V_{max}$.

The Lineweaver-Burk plot is shown in ***Fig.6.3***. It is much easier to calculate the K_m from the intercept on x-axis which is $-(1/K_m)$. Further, the double reciprocal plot is ***useful in understanding the effect of various inhibitions*** (discussed later).

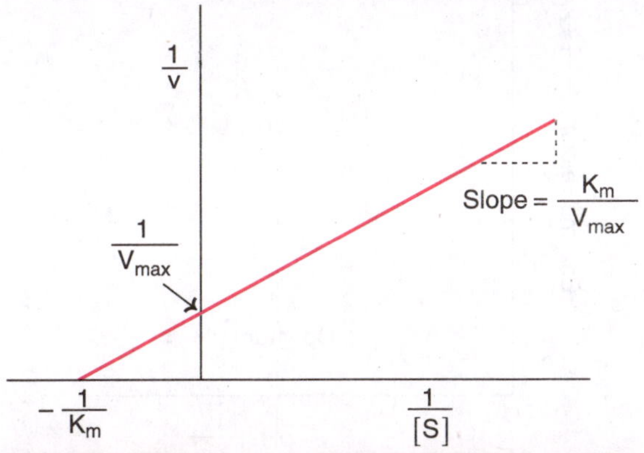

Fig. 6.3 : Lineweaver-Burk double reciprocal plot.

3. Effect of temperature

Velocity of an enzyme reaction ***increases with increase in temperature*** up to a maximum

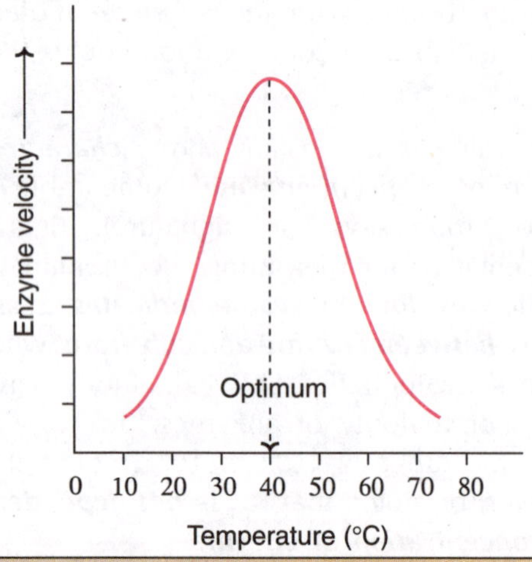

Fig. 6.4 : *Effect of temperature on enzyme velocity.*

and then declines. A ***bell-shaped curve*** is usually observed (***Fig.6.4***).

Temperature coefficient or Q_{10} is definec increase in enzyme velocity when t temperature is increased by 10°C. For a majority of enzymes, Q_{10} is 2 between 0°C and 40°C. Increase in temperature results in higher activation energy of the molecules and more molecular (enzyme and substrate) collision and interaction for the reaction to proceed faster.

The ***optimum temperature*** for most of the enzymes is between ***35°C–40°C***. However, a few enzymes (e.g. *Taq* DNA polymerase, muscle adenylate kinase) are active even at 100°C. Some plant enzymes like urease have optimum activity around 60°C. This may be due to very stable structure and conformation of these enzymes.

In general, when the enzymes are exposed to a temperature above 50°C, ***denaturation*** leading to derangement in the native (tertiary) structure of the protein and active site are seen. Majority of the enzymes become inactive at higher temperature (above 70°C).

Clinical significance : *Foods* can be preserved in ***refrigerators*** (at low temperatures) due to reduced bacterial enzyme activities. ***Certain surgeries*** are carried out by lowering the patient's body temperature (***induced hypothermia***), and thus the matabolic rate.

4. Effect of pH

Increase in the hydrogen ion concentration (pH) considerably influences the enzyme activity and a ***bell-shaped curve*** is normally obtained (***Fig.6.5***). Each enzyme has an optimum pH at which the velocity is maximum. Below and above this pH, the enzyme activity is much lower and at extreme pH, the enzyme becomes totally inactive.

Most of the ***enzymes*** of higher organisms show ***optimum activity around neutral pH (6-8)***. There are, however, many exceptions like pepsin (1-2), acid phosphatase (4-5) and alkaline phosphatase (10-11). Enzymes from fungi and plants are most active in acidic pH (4-6).

Hydrogen ions influence the enzyme activity by altering the ionic charges on the amino acids (particularly at the active site), substrate, ES complex etc.

5. Effect of product concentration

The accumulation of reaction products generally decreases the enzyme velocity. For certain enzymes, the products combine with the active site of enzyme and form a loose complex and, thus, inhibit the enzyme activity. In the living system, this type of inhibition is generally prevented by a quick removal of products formed. The end product inhibition by feedback mechanism is discussed later.

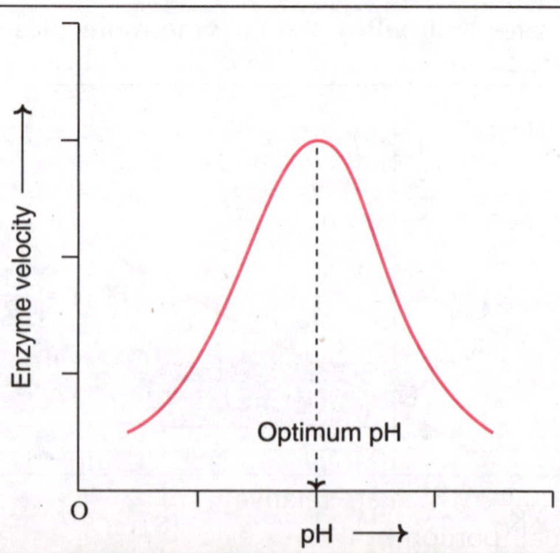

Fig. 6.5 : *Effect of pH on enzyme velocity.*

6. Effect of activators

Some of the enzymes require certain inorganic ***metallic cations*** like Mg^{2+}, Mn^{2+}, Zn^{2+}, Ca^{2+}, Co^{2+}, Cu^{2+}, Na^{+}, K^{+} etc. for their optimum activity. Rarely, anions are also needed for enzyme activity e.g. chloride ion (Cl^{-}) for amylase. Metals function as activators of enzyme velocity through various mechanisms—combining with the substrate, formation of ES-metal complex, direct participation in the reaction and bringing a conformational change in the enzyme.

Two categories of enzymes requiring metals for their activity are distinguished

- **Metal-activated enzymes :** The metal is not tightly held by the enzyme and can be exchanged easily with other ions

  e.g. ATPase (Mg^{2+} and Ca^{2+}), Enolase (Mg^{2+})
- **Metalloenzymes :** These enzymes hold the metals rather tightly which are not readily exchanged. e.g. ***alcohol dehydrogenase, carbonic anhydrase***, alkaline phosphatase, carboxypeptidase and aldolase contain ***zinc. Phenol oxidase (copper)***; Pyruvate oxidase (manganese); ***Xanthine oxidase (molybdenum)***; Cytochrome oxidase (iron and copper).

7. Effect of time

Under ideal and optimal conditions (like pH, temperature etc.), the time required for an enzyme reaction is less. Variations in the time of the reaction are generally related to the alterations in pH and temperature.

8. Effect of light and radiation

Exposure of enzymes to ultraviolet, beta, gamma and X-rays inactivates certain enzymes due to the formation of peroxides. e.g. UV rays inhibit salivary amylase activity.

ACTIVE SITE

Enzymes are big in size compared to substrates which are relatively smaller. Evidently, a small portion of the huge enzyme molecule is directly involved in the substrate binding and catalysis (***Fig.6.6***).

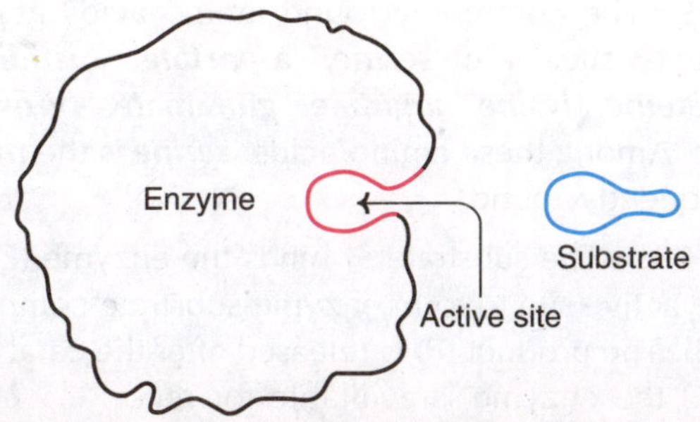

Fig. 6.6 : A diagrammatic representation of an enzyme with active site.

The active site (or active centre) of an enzyme represents as the small region at which the substrate(s) binds and participates in the catalysis.

Salient features of active site

1. The existence of active site is due to the tertiary structure of protein resulting in three-dimensional native conformation.

2. The active site is made up of amino acids (known as ***catalytic residues***) which are far from each other in the linear sequence of amino acids (primary structure of protein). For instance, the enzyme ***lysozyme*** has 129 amino acids. The ***active site*** is formed by the contribution of amino acid residues numbered ***35, 52, 62, 63 and 101***.

3. Active sites are regarded as ***clefts*** or ***crevices*** or pockets occupying a small region in a big enzyme molecule.

4. The active site is not rigid in structure and shape. It is rather ***flexible*** to promote the specific substrate binding.

5. Generally, the active site possesses a ***substrate binding site*** and a ***catalytic site.*** The latter is for the catalysis of the specific reaction.

6. The coenzymes or cofactors on which some enzymes depend are present as a part of the catalytic site.

7. The substrate(s) binds at the active site by weak noncovalent bonds.

8. Enzymes are specific in their function due to the existence of active sites.

9. The commonly found amino acids at the active sites are ***serine***, ***aspartate***, ***histidine***, ***cysteine***, ***lysine***, ***arginine***, ***glutamate***, ***tyrosine*** etc. Among these amino acids, ***serine*** is the most frequently found.

10. The substrate[S] binds the enzyme (E) at the active site to form enzyme-substrate complex (ES). The product (P) is released after the catalysis and the enzyme is available for reuse.

$$E + S \rightleftharpoons ES \rightarrow E + P$$

ENZYME INHIBITION

Enzyme ***inhibitor*** is defined as a substance which binds with the enzyme and brings about a ***decrease in catalytic activity*** of that enzyme. The inhibitor may be organic or inorganic in nature. There are three broad categories of enzyme inhibition

1. Reversible inhibition.
2. Irreversible inhibition.
3. Allosteric inhibition.

1. Reversible inhibition

The inhibitor binds non-covalently with enzyme and the enzyme inhibition can be reversed if the inhibitor is removed. The reversible inhibition is further sub-divided into

A. Competitive inhibition (***Fig.6.7A***)

B. Non-competitive inhibition (***Fig.6.7B***)

C. Uncompetitive inhibition.

A. Competitive inhibition

The inhibitor (I) which closely resembles the real substrate (S) is regarded as a ***substrate analogue***. The inhibitor competes with substrate and binds at the active site of the enzyme but does not undergo any catalysis. As long as the competitive inhibitor holds the active site, the enzyme is not available for the substrate to bind. During the reaction, ES and EI complexes are formed as shown below

$$E \xrightarrow{+S} ES \rightarrow E + P$$
$$E \xrightarrow{+I} EI$$

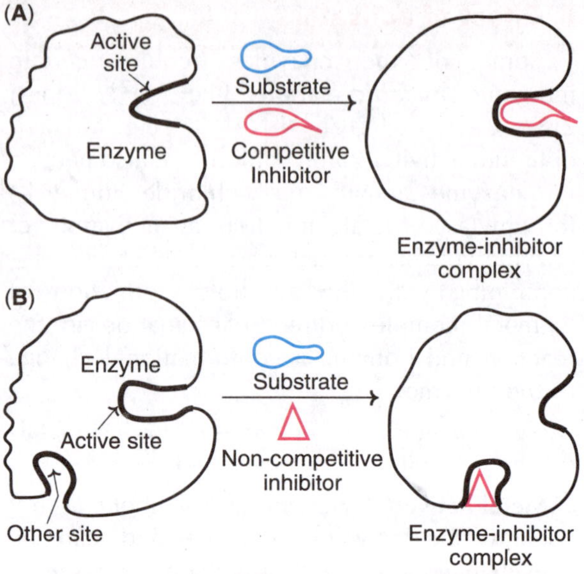

Fig. 6.7 : *A diagrammatic representation of (A) Competitive and (B) Non-competitive inhibition.*

The relative concentration of the substrate and inhibitor and their respective affinity with the enzyme determines the degree of competitive inhibition. The inhibition could be overcome by a high substrate concentration. In competitive inhibition, the K_m ***value increases*** whereas V_{max} remains ***unchanged*** (***Fig.6.8***).

The enzyme ***succinate dehydrogenase*** (SDH) is a classical example of competitive inhibition with succinic acid as its substrate. The compounds, namely, malonic acid, glutaric acid and oxalic acid, have ***structural similarity*** with succinic acid and compete with the substrate for binding at the active site of SDH.

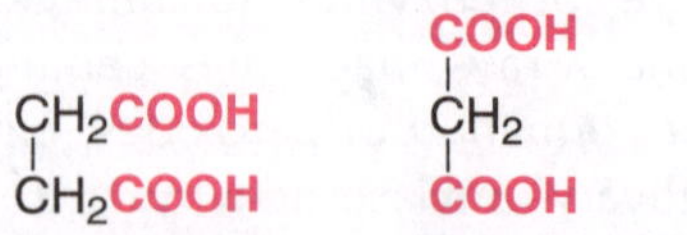

Succinic acid **Malonic acid**

Methanol is toxic to the body when it is converted to formaldehyde by the enzyme alcohol dehydrogenase (ADH). Ethanol can compete with methanol for ADH. Thus, ***ethanol can be used in the treatment of methanol poisoning***.

Table 6.2 Selected examples of enzymes with their respective substrates and competitive inhibitors

Enzyme	Substrate	Inhibitor(s)	Significance of inhibitor(s)
Xanthine oxidase	Hypoxanthine xanthine	Allopurinol	Used in the control of gout to reduce excess production of uric acid from hypoxanthine.
Monoamine oxidase	Catecholamines (epinephrine, norepinephrine)	Ephedrine, amphetamine	Useful for elevating catecholamine levels.
Dihydrofolate reductase	Dihydrofolic acid	Aminopterin, amethopterin, methotrexate	Employed in the treatment of leukemia and other cancers.
Acetylcholine esterase	Acetylcholine	Succinyl choline	Used in surgery for muscle relaxation, in anaesthetised patients.
Dihydropteroate synthase	Para aminobenzoic acid (PABA)	Sulfanilamide	Prevents bacterial synthesis of folic acid.
Vitamin K epoxide reductase	Vitamin K	Dicumarol	Acts as an anticoagulant.
HMG CoA reductase	HMG CoA	Lovastatin, pravastatin atorvastatin	Inhibit cholesterol biosynthesis

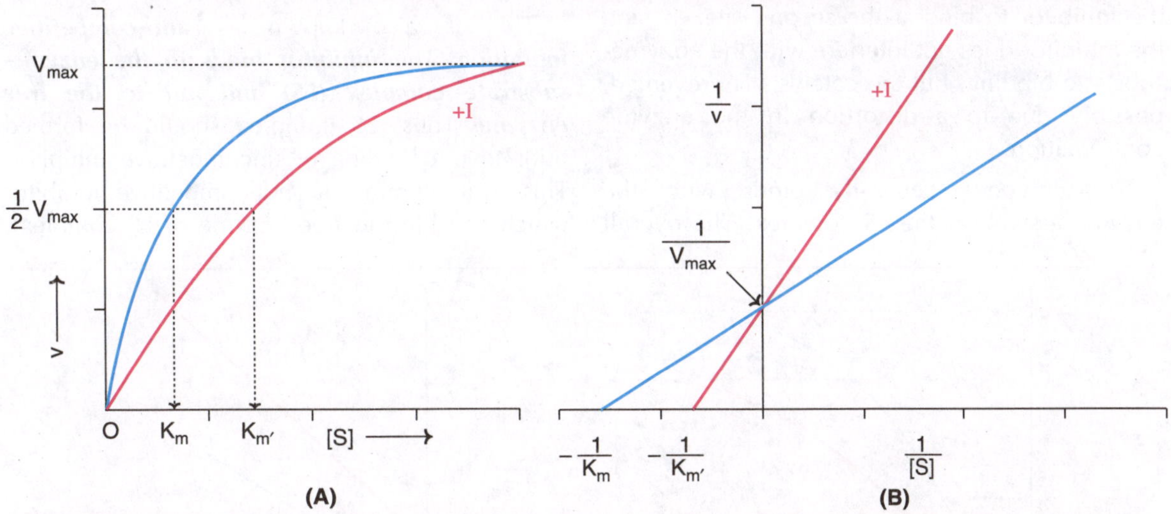

Fig. 6.8 : *Effect of competitive inhibitor (i) on enzyme velocity.* ***(A)*** *Velocity (v) versus substrate (S) plot.* ***(B)*** *Lineweaver-Burk plot (Red lines with inhibitor; competitive inhibitor increases* K_m*, unalters* V_{max}*).*

Some more examples of the enzymes with substrates and competitive inhibitors (of clinical and pharmacological significance) are given in ***Table 6.2***.

Antimetabolites : These are the chemical compounds that ***block the metabolic reactions*** by their inhibitory action on enzymes. Antimetabolites are usually structural analogues of substrates and thus are competitive inhibitors (***Table 6.2***). They are in use for cancer therapy, gout etc. The term ***antivitamins*** is used for the ***antimetabolites which block the biochemical actions of vitamins causing deficiencies, e.g. sulfanilamide, dicumarol***.

Medical Concepts/ Clinical Correlates

- Competitive inhibitors of certain enzymes are of great biological significance. Allopurinol, employed in the treatment of gout, inhibits xanthine oxidase to reduce the formation of uric acid. The other competitive inhibitors include aminopterin used in the treatment of cancers, sulfonilamide as antibacterial agent and dicumarol as an anticoagulant.
- Penicillin antibiotics irreversibly inhibit serine containing enzymes of bacterial cell wall synthesis.

B. Non-competitive inhibition

The inhibitor binds at a site other than the active site on the enzyme surface. This binding impairs the enzyme function. The inhibitor has ***no structural resemblance with the substrate***. However, there usually exists a strong affinity for the inhibitor to bind at the second site. In fact, the inhibitor does not interfere with the enzyme-substrate binding. But the catalysis is prevented, possibly due to a distortion in the enzyme conformation.

The inhibitor generally binds with the enzyme as well as the ES complex. The overall relation in non-competitive inhibition is represented below

$$\begin{array}{ccccc} E + S & \rightleftharpoons & ES & \longrightarrow & E + P \\ + & & + & & \\ I & & I & & \\ \upharpoonleft\downharpoonright & & \upharpoonleft\downharpoonright & & \\ EI + S & & EIS & & \end{array}$$

For non-competitive inhibition, the ***K_m value is unchanged*** while ***V_{max} is lowered*** (***Fig.6.9***).

Heavy metal ions (Ag^+, Pb^{2+}, Hg^{2+} etc.) can non-competitively ***inhibit*** the enzymes by binding with cysteinyl ***sulfhydryl groups***. The general reaction for Hg^{2+} is shown below

$$E{-}SH + Hg^{2+} \rightleftharpoons E{-}S\cdot\cdot Hg^{2+} + H^+$$

Heavy metals also lead to the formation of covalent bonds with carboxyl groups and histidine, often resulting in irreversible inhibition.

The most commonly used antidote for heavy metal poisoning is ***dimercaprol***, with –SH groups.

C. Uncompetitive inhibition

This is also known as ***anti-competitive inhibition***. The ***inhibitor binds to the enzyme-substrate complex (ES) and not to the free enzyme***. Thus, ES complex should be formed prior to the binding of uncompetitive inhibitor. This is in contrast to noncompetitive inhibitor which can bind to free enzyme or ES complex.

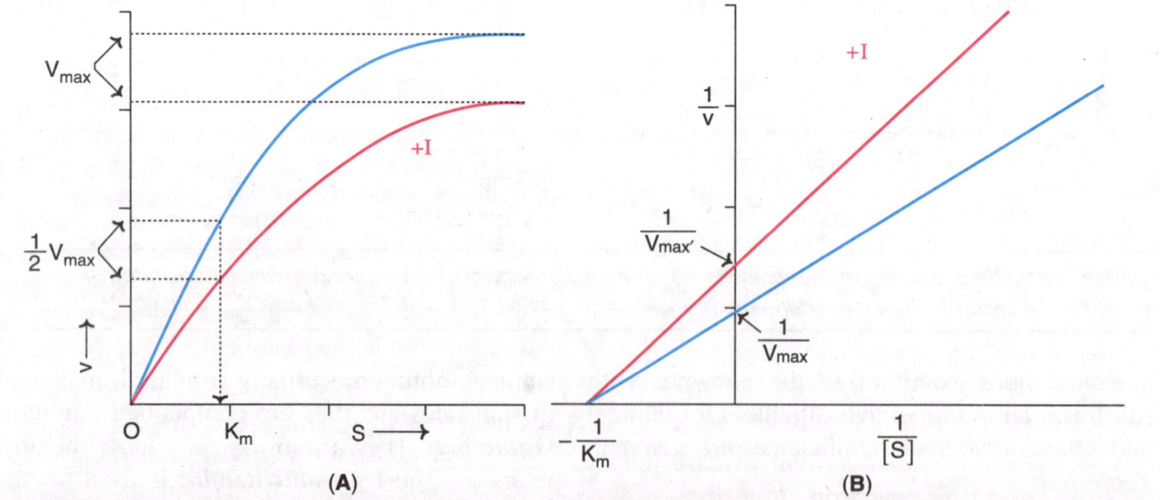

Fig. 6.9 : *Effect of non-competitive inhibitor (I) on enzyme velocity (A) Velocity (v) versus substrate (S) (B) Lineweaver-Burk plot (Red lines with inhibitor, non-competitive inhibitor does not change K_m but decreases V_{max}).*

An uncompetitive inhibitor need not have any structural resemblance with the substrate. Further, uncompetitive inhibition best works when the substrate concentration is high, and in this inhibition, both ***K_m and V_{max} are decreased*** e.g. inhibition of placental alkaline phosphatase by phenylalanine.

2. Irreversible inhibition

The inhibitors ***bind covalently with the enzymes and inactivate*** them, which is irreversible. These inhibitors are usually toxic substances that poison enzymes.

Iodoacetate is an irreversible inhibitor of the enzymes like papain and glyceraldehyde 3-phosphate dehydrogenase. Iodoacetate combines with sulfhydryl (–SH) groups at the active site of these enzymes and makes them inactive.

Diisopropyl fluorophosphate (DFP) is a ***nerve gas*** developed by the Germans during Second World War. DFP irreversibly binds with enzymes containing serine at the active site, e.g. ***serine proteases, acetylcholine esterase***.

Many ***organophosphorus*** insecticides like malathion are toxic to animals (including man) as they ***block the activity of acetylcholine esterase*** (essential for nerve conduction), resulting in paralysis of vital body functions.

Disulfiram (Antabuse®) is a drug used in the ***treatment of alcoholism***. It irreversibly inhibits the enzyme aldehyde dehydrogenase. Alcohol addicts, when treated with disulfiram become sick due to the accumulation of acetaldehyde, leading to alcohol avoidance. (***Note :*** Alcohol is metabolized by two enzymes. It is first acted upon by alcohol dehydrogenase to yield acetaldehyde. The enzyme aldehyde dehydrogenase converts acetaldehyde to acetic acid.)

The ***penicillin antibiotics*** act as irreversible inhibitors of serine – containing enzymes, and ***block the bacterial cell wall synthesis***.

Cyanide inhibits ***cytochrome oxidase*** (binds to iron atom) of electron transport chain. ***Fluoride inhibits enolase*** (by removing manganese), and thus ***glycolysis***.

Suicide inhibition

Suicide inhibition is a specialized form of irreversible inhibition. In this case, the original inhibitor (the structural analogue/competitive inhibitor) is converted to a more potent form by the same enzyme that ought to be inhibited. The so formed inhibitor binds irreversibly with the enzyme. This is in contrast to the original inhibitor which binds reversibly.

A good example of suicide inhibition is ***allopurinol*** (used in the treatment of gout, ***Refer Chapter 17***). Allopurinol, an inhibitor of xanthine oxidase, gets converted to ***alloxanthine, a more effective inhibitor*** of this enzyme.

The use of certain purine and pyrimidine analogues in cancer therapy is also explained on the basis suicide inhibition. For instance, ***5-fluorouracil*** gets converted to fluorodeoxy-uridylate which inhibits the enzyme thymidylate synthase, and thus nucleotide synthesis.

Some more examples of suicide inhibitors

- ***Aspirin*** inhibits ***cyclooxygenase of prosta-glandin synthesis.***
- ***Penicillin*** inhibits transpeptidase and blocks bacterial wall synthesis.
- ***Zidovudine*** (AZT) used in the treatment of ***AIDS*** inhibits the enzyme reverse trans-criptase.
- Sarin inhibits acetylcholinesterase.

3. Allosteric inhibition

The details of this type of inhibition are given under allosteric regulation as a part of the regulation of enzyme activity in the living system.

Enzyme inhibition by drugs

Enzymes are the natural targets for development of pharmacologic agents. Many of the drugs used in the treatment of diseases act as enzyme inhibitors. For example :

- Cholesterol lowering ***statin drugs*** (lovastatin) ***inhibit*** the enzyme ***HMG CoA reductase***.
- Drugs (***tenofovir***, ***emtricitabine***) employed to ***block HIV replication*** inhibit the enzyme viral reverse transcriptase.

- ***Hypertension*** is often treated by the drugs (***captopril***, enalapril) which inhibit angiotensin converting enzyme.
- Antibiotics-penicillin and amoxicillin inhibit enzymes of bacterial cell wall synthesis.

SUICIDE ENZYME

A suicidal ***enzyme*** possesses a rare and remarkable ***property of self-catalysed destruction.*** The synthesis of ***prostaglandins*** is switched off by this property of suicidal enzyme of ***cyclooxygenase*** (COX). Self destruction of COX is believed to be facilitated by the presence of another enzyme namely 15-hydroxy-prostaglandin dehydrogenase (***15 HPD***). This is evident from the fact that inhibiting the action of 15 HPD by indomethacin or sulfasalazine can increase prostaglandin levels.

ENZYME SPECIFICITY

Enzymes are highly specific in their action when compared with the chemical catalysts. The occurrence of thousands of enzymes in the biological system might be due to the specific nature of enzymes. ***Three types*** of enzyme specificity are well recognised–stereospecificity, reaction specificity, and substrate specificity.

Specificity is a characteristic property of the active site.

1. **Stereospecificity or optical specificity :** Stereoisomers are the compounds which have the same molecular formula, but differ in their structural configuration.

The ***enzymes act only on one isomer*** and, therefore, exhibit stereospecificity.

e.g. L-amino acid oxidase and D-amino acid oxidase act on L- and D-amino acids respectively; hexokinase acts on D-hexoses; glucokinase on D-glucose; amylase acts on α-glycosidic linkages; cellulase cleaves β-glycosidic bonds.

Stereospecificity is explained by considering three distinct regions of substrate molecule specifically binding with three complementary regions on the surface of the enzyme (***Fig.6.10***). The class of enzymes belonging to ***isomerases do not exhibit stereospecificity***, since they are specialized in the interconversion of isomers.

2. **Reaction specificity :** The ***same substrate*** can undergo ***different types of reactions***, each catalysed by a separate enzyme and this is referred to as reaction specificity. An amino acid can undergo transamination, oxidative deamination, decarboxylation, racemization etc. The enzymes however, are different for each of these reactions. (For details, ***refer Chapter 15***).

3. **Substrate specificity :** The substrate specificity varies from enzyme to enzyme. It may be either ***absolute***, ***relative*** or ***broad.***

- **Absolute substrate specificity :** Certain enzymes act ***only on one substrate*** e.g. ***glucokinase*** acts on glucose to give glucose 6-phosphate, urease cleaves urea to ammonia and carbon dioxide.
- **Relative substrate specificity :** Some enzymes ***act on structurally related substances.*** This, in turn, may be dependent on the specific group or a bond present. The action of ***trypsin*** is a good example for ***group specificity*** (***Refer Fig.8.6***). Trypsin hydrolyses peptide linkage involving arginine or lysine. Chymotrypsin cleaves peptide bonds attached to aromatic amino acids (phenylalanine, tyrosine and tryptophan). Examples of ***bond specificity***-glycosidases acting on glycosidic bonds of carbohydrates, lipases cleaving ester bonds of lipids etc.

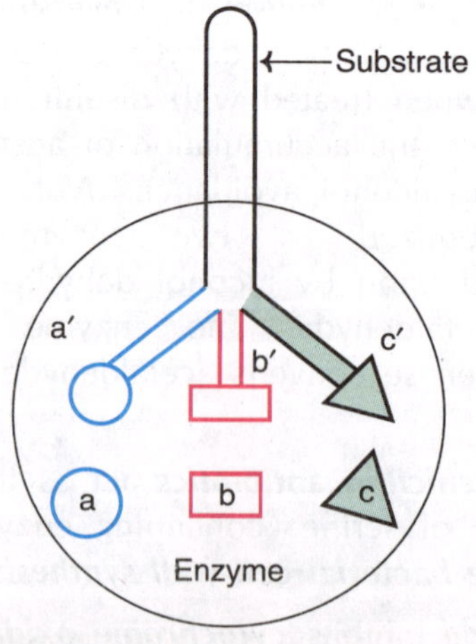

***Fig. 6.10 :** Diagrammatic representation of stereospecificity (a′, b′, c′)—three point attachment of substrate to the enzyme (a, b, c).*

- **Broad specificity :** Some enzymes ***act on closely related substrates*** which is commonly known as broad substrate specificity, e.g. ***hexokinase*** acts on glucose, fructose, mannose and glucosamine and not on galactose. It is possible that some structural similarity among the first four compounds makes them a common substrate for the enzyme hexokinase.

COENZYMES

The protein part of the enzyme, on its own, is not always adequate to bring about the catalytic activity. Many enzymes require certain ***non-protein small additional factors***, collectively referred to as cofactors for catalysis. The cofactors may be organic or inorganic in nature.

The non-protein, organic, low molecular weight and dialysable substance associated with enzyme function is known as coenzyme.

The functional enzyme is referred to as ***holoenzyme*** which is made up of a protein part ***(apoenzyme)*** and a non-protein part ***(coenzyme)***. The term ***prosthetic group*** is used when a non-protein moiety is ***tightly bound*** to the enzyme which is not easily separable by dialysis. The term ***activator*** refers to the ***inorganic cofactor*** (like Ca^{2+}, Mg^{2+}, Mn^{2+} etc.) necessary to enhance enzyme activity. It may, however, be noted that some authors make no distinction between the terms cofactor, coenzyme and prosthetic group and use them interchangeably.

Coenzymes are second substrates : Coenzymes are often regarded as the second substrates or ***co-substrates***, since they have affinity with the enzyme comparable with that of the substrates. Coenzymes ***undergo alterations*** during the enzymatic reactions, which are ***later regenerated***. This is in contrast to the substrate which is converted to the product.

Coenzymes participate in various reactions involving ***transfer of atoms or groups*** like hydrogen, aldehyde, keto, amino, acyl, methyl, carbon dioxide etc. Coenzymes play a decisive role in enzyme function.

Coenzymes act as substrate shuttles : Certain coenzymes ***transport substrate from one point of cell to another***, and thus to serve as recyclable shuttles. The coenzyme UDP combines with substrate glucose (UDP-glucose), coenzyme A binds to acetyl group (acetyl CoA), and tetrahydrofolate combines with methyl group (methyl THF). In all these cases, the coenzymes facilitate the recognition of substrates by their target enzymes.

Coenzymes from B-complex vitamins : Most of the coenzymes are the derivatives of water soluble B-complex vitamins. In fact, the ***biochemical functions of B-complex vitamins*** are exerted through their respective ***coenzymes***. The chapter on vitamins gives the details of structure and function of the coenzymes (***Chapter 7***). In ***Table 6.3***, a summary of the vitamin related coenzymes along with their functions is given.

Non-vitamin coenzymes : Not all coenzymes are vitamin derivatives. There are some other organic substances, which have no relation with vitamins but function as coenzymes. They may be considered as non-vitamin coenzymes e.g. ***ATP, CDP, UDP*** etc. The important non-vitamin coenzymes along with their functions are given in ***Table 6.4***.

Nucleotide coenzymes : Some of the coenzymes possess nitrogenous base, sugar and phosphate. Such coenzymes are, therefore, regarded as nucleotides e.g. NAD^+, $NADP^+$, FMN, FAD, coenzyme A etc.

Protein coenzymes : ***Thioredoxin*** is a protein that serves as a coenzyme for the enzyme ribonucleotide reductase (***Chapter 17***).

Coenzymes do not decide enzyme specificity : A particular coenzyme may participate in catalytic reactions along with different enzymes. For instance, NAD^+ acts as a coenzyme for lactate dehydrogenase and alcohol dehydrogenase. In both the enzymatic reactions, NAD^+ is involved in hydrogen transfer. The ***specificity of the enzyme is mostly dependent on the apoenzyme and not on the coenzyme***.

Table 6.3 Coenzymes of B-complex vitamins*

Coenzyme (abbreviation)	Derived from vitamin	Atom or group transferred	Dependent enzyme (example)
Thiamine pyrophosphate (TPP)	Thiamine	Aldehyde or keto	Transketolase
Flavin mononucleotide (FMN)	Riboflavin	Hydrogen and electron	L - Amino acid oxidase
Flavin adenine dinucleotide (FAD)	Riboflavin	"	D - Amino acid oxidase
Nicotinamide adenine dinucleotide(NAD^+)	Niacin	"	Lactate dehydrogenase
Nicotinamide adenine dinucleotide phosphate ($NADP^+$)	"	"	Glucose 6-phosphate dehydrogenase
Lipoic acid	Lipoic acid	"	Pyruvate dehydrogenase complex
Pyridoxal phosphate (PLP)	Pyridoxine	Amino or keto	Alanine transaminase
Coenzyme A (CoA)	Pantothenic acid	Acyl	Thiokinase
Tetrahydrofolate (FH_4)	Folic acid	One carbon (formyl, methenyl etc.)	Formyl transferase
Biocytin	Biotin	CO_2	Pyruvate carboxylase
Methylcobalamin; Deoxyadenosyl cobalamin	Cobalamin	Methyl/isomerisation	Methylmalonyl CoA mutase

***** *Details for each coenzyme are given in Chapter 7 on vitamins*

MECHANISM OF ENZYME ACTION

Catalysis is the prime function of enzymes. Enzymes are powerful catalysts. The nature of catalysis taking place in the biological system is similar to that of non-biological catalysis. For any chemical reaction to occur, the reactants have to be in an activated state or transition state.

Enzymes lower activation energy : The energy required by the reactants to undergo the reaction is known as ***activation energy.*** The reactants when heated attain the activation energy. The catalyst (or the enzyme in the biological system) reduces the activation energy and this causes the reaction to proceed at a lower temperature. Enzymes do not alter the equilibrium constants, they only enhance the velocity of the reaction.

The role of catalyst or enzyme is comparable with a tunnel made in a mountain to reduce the barrier as illustrated in ***Fig.6.11***. The enzyme lowers energy barrier of reactants, thereby making the reaction go faster. The ***enzymes reduce the activation energy of the reactants*** in such a way that all the biological systems occur at body temperature (below 40°C).

Table 6.4 Coenzymes not related to B-complex vitamins

Coenzyme	Abbreviation	Biochemical functions
Adenosine triphosphate	ATP	Donates phosphate, adenosine and adenosine monophosphate (AMP) moieties.
Cytidine diphosphate	CDP	Required in phospholipid synthesis as carrier of choline and ethanolamine.
Uridine diphosphate	UDP	Carrier of monosaccharides (glucose, galactose), required for glycogen synthesis.
S - Adenosylmethionine (active methionine)	SAM	Donates methyl group in biosynthetic reactions.
Phosphoadenosine phosphosulfate (active sulfate)	PAPS	Donates sulfate for the synthesis of mucopolysaccharides.

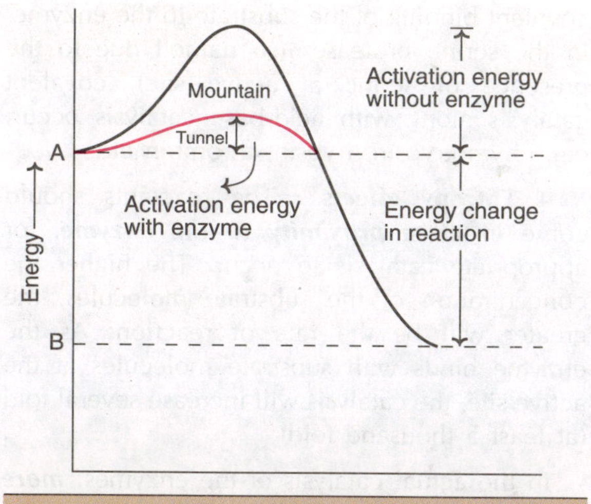

Fig. 6.11 : Effect of enzyme on activation energy of a reaction (A is the substrate and B is the product. Enzyme decreases activation energy).

Enzyme-substrate (ES) complex formation

The prime requisite for enzyme catalysis is that the substrate (S) must combine with the enzyme (E) at the active site to form enzyme-substrate complex (ES) which ultimately results in the product formation (P).

$$E + S \rightleftharpoons ES \longrightarrow E + P$$

A few theories have been put forth to explain mechanism of enzyme-substrate complex formation.

Lock and key model or Fischer's template theory

This theory was proposed by a German biochemist, Emil Fischer. This is in fact the very first model proposed to explain an enzyme catalysed reaction.

According to this model, the structure or conformation of the ***enzyme is rigid***. The substrate fits to the binding site (now active site) just as a key (substrate) fits into the key hole (active site) of a lock (enzyme). Thus the active site of an enzyme is a rigid and pre-shaped template where only a specific substrate can bind. This model does not give any scope for the flexible nature of enzymes, hence the model totally ***fails to explain*** many facts of ***enzymatic reactions***, the most important being the effect of allosteric modulators.

Induced fit theory or Koshland's model

Koshland, in 1958, proposed a more acceptable and realistic model for enzyme-substrate complex formation. As per this model, the ***active site is not rigid and pre-shaped***. The essential features of the substrate binding site are present at the nascent active site. The ***interaction of the substrate with the enzyme induces*** a fit or ***a conformation change in the enzyme***, resulting in the formation of a strong substrate binding site. Further, due to induced fit, the appropriate amino acids of the enzyme are repositioned to form the active site and bring about the catalysis (***Fig.6.12***).

Induced fit model has ***sufficient experimental evidence*** from the X-ray diffraction studies. Koshland's model also explains the action of allosteric modulators and competitive inhibition on enzymes.

Substrate strain theory

In this model, the substrate is strained due to the induced conformation change in the enzyme. It is also possible that when a substrate binds to the preformed active site, ***the enzyme induces a***

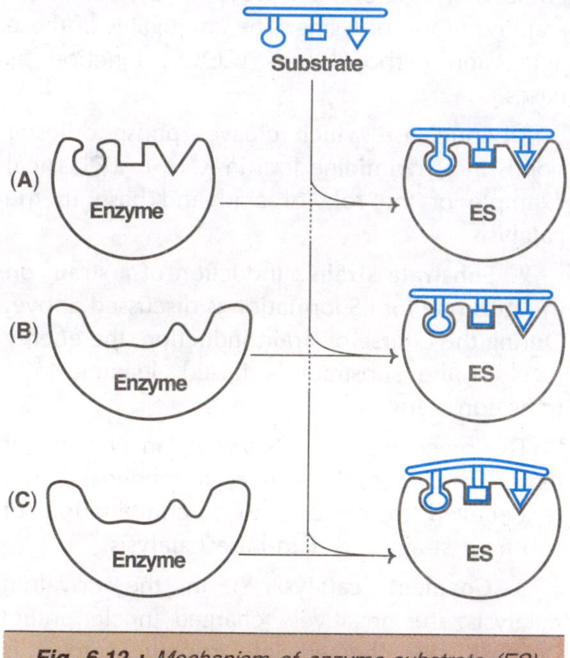

Fig. 6.12 : Mechanism of enzyme-substrate (ES) complex formation (A) Lock and key model (B) Induced fit theory (C) Substrate strain theory.

strain to the substrate. The strained substrate leads to the formation of product.

In fact, a combination of the induced fit model with the substrate strain is considered to be operative in the enzymatic action.

MECHANISM OF ENZYME CATALYSIS

The formation of an enzyme-substrate complex (ES) is very crucial for the catalysis to occur, and for the product formation. It is estimated that an enzyme catalysed reaction proceeds 10^6 to 10^{12} times faster than a non-catalysed reaction. The enhancement in the rate of the reaction is mainly due to four processes :

1. Acid-base catalysis;
2. Substrate strain;
3. Covalent catalysis;
4. Entropy effects.

1. **Acid-base catalysis :** Role of acids and bases is quite important in enzymology. At the physiological pH, ***histidine*** is the most important amino acid, the ***protonated*** form of which functions as an ***acid*** and its corresponding conjugate as a base. The other acids are –OH group of tyrosine, –SH group of cysteine, and ε-amino group of lysine. The conjugates of these acids and carboxyl ions (COO^-) function as bases.

Ribonuclease which cleaves phosphodiester bonds in a pyrimidine loci in RNA is a classical example of the role of acid and base in the catalysis.

2. **Substrate strain :** Induction of a strain on the substrate for ES formation is discussed above. During the course of strain induction, the energy level of the substrate is raised, leading to a transition state.

The mechanism of lysozyme (an enzyme of tears, that cleaves β-1,4 glycosidic bonds) action is believed to be due to a combination of substrate strain and acid-base catalysis.

3. **Covalent catalysis :** In the covalent catalysis, the negatively charged (nucleophilic) or positively charged (electrophilic) group is present at the active site of the enzyme. This group attacks the substrate that results in the covalent binding of the substrate to the enzyme. In the serine proteases (so named due to the presence of serine at active site), covalent catalysis along with acid-base catalysis occur, e.g. chymotrypsin, trypsin, thrombin etc.

4. **Entropy effects :** The reactants should come in ***close proximity to the enzyme***, for appropriate catalysis to occur. The higher the concentration of the substrate molecules, the greater will be the rate of reaction. As the enzyme binds with substrate molecules at the active site, the catalysis will increase several fold (at least a thousand fold)

In the actual catalysis of the enzymes, ***more*** than one of the ***processes*** – acid-base catalysis, substrate strain, covalent catalysis and proximity catalysis ***are simultaneously operative***. This will help the substrate(s) to attain a transition state leading to the formation of products.

THERMODYNAMICS OF ENZYMATIC REACTIONS

The enzyme catalysed reactions may be broadly grouped into three types based on thermodynamic (energy) considerations.

1. **Isothermic reactions :** The energy exchange between reactants and products is negligible e.g. glycogen phosphorylase

$$\text{Glycogen} + \text{Pi} \longrightarrow \text{Glucose 1-phosphate}$$

2. **Exothermic (exergonic) reactions :** Energy is liberated in these reactions e.g. urease

$$\text{Urea} \longrightarrow NH_3 + CO_2 + \text{energy}$$

3. **Endothermic (endergonic) reactions :** Energy is consumed in these reactions e.g. glucokinase

$$\text{Glucose} + \text{ATP} \longrightarrow \text{Glucose 6-phosphate} + \text{ADP}$$

SINGLE MOLECULE ENZYMOLOGY

Traditionally, an enzyme is assayed, for its activity, using a large ensembles of molecules. This results in the data that will give an average activity of the said enzyme, across multiple cycles of catalysis. This type of studies do not reveal the specific properties of individual enzyme molecules.

Medical Concepts/ Clinical Correlates

- In the living system, the regulation of enzyme activities occurs through allosteric inhibition, activation of latent enzymes, compartmentation of metabolic pathways, control of enzyme synthesis and degradation.
- Feedback (or end product) inhibition is a specialized form of allosteric inhibition that controls several metabolic pathways e.g. CTP inhibits aspartate transcarbamoylase; Cholesterol inhibits HMG CoA reductase. The end product inhibition is utmost important to cellular economy since a compound is synthesized only when required.

Recent developments in ***nanotechnology*** have made it possible to observe the catalytic events of individual enzymes and substrate molecules. ***Single molecule enzymology*** is the process of ***single molecule studies of enzyme catalysis*** to ***measure the catalytic events and individual steps***. Further, single enzyme technology is useful to provide insight into the underlying mechanisms in the enzyme catalysed reactions.

Single molecule enzymes are usually isolated by using ***microwell array technology***. The catalytic events involving individual enzyme and substrate molecules are observed by fluorescence microscopy.

REGULATION OF ENZYME ACTIVITY IN THE LIVING SYSTEM

In biological system, regulation of enzyme activities occurs at different stages in one or more of the following ways to achieve cellular economy.

1. Allosteric regulation
2. Activation of latent enzymes
3. Compartmentation of metabolic pathways
4. Control of enzyme synthesis
5. Enzyme degradation
6. Isoenzymes

1. Allosteric regulation and allosteric inhibition

Some of the enzymes possess additional sites, known as allosteric sites (*Greek* : allo–other), besides the active site. Such enzymes are known as allosteric enzymes. The ***allosteric sites*** are unique places on the enzyme molecule.

Allosteric effectors : Certain substances referred to as allosteric modulators (effectors or modifiers) ***bind at the allosteric site and regulate the enzyme activity***. The enzyme activity is increased when a positive (+) allosteric effector binds at the allosteric site known as activator site. On the other hand, a negative (–) allosteric effector binds at the allosteric site called inhibitor site and inhibits the enzyme activity.

Classes of allosteric enzymes : Enzymes that are regulated by allosteric mechanism are referred to as allosteric enzymes. They are divided into two classes based on the influence of allosteric effector on K_m and V_{max}.

- **K-class of allosteric enzymes**, the effector ***changes the*** K_m and not the V_{max}. Double reciprocal plots, similar to competitive inhibition are obtained e.g. phosphofructokinase.
- **V-class of allosteric enzymes**, the effector ***alters the*** V_{max} and not the K_m. Double reciprocal plots resemble that of non-competitive inhibition e.g. acetyl CoA carboxylase.

Conformational changes in allosteric enzymes : Most of the allosteric enzymes are oligomeric in nature. The subunits may be identical or different. The non-covalent reversible binding of the effector molecule at the allosteric site brings about a conformational change in the active site of the enzyme, leading to the inhibition or activation of the catalytic activity (***Fig.6.13***). In the concerted model, allosteric enzymes exist in two conformational

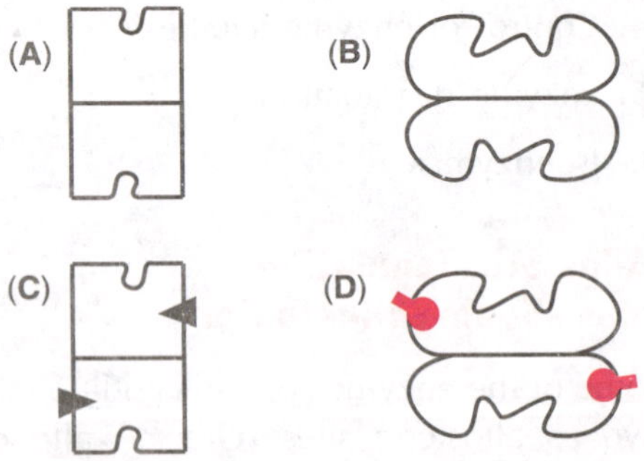

***Fig. 6.13 :** Diagrammatic representation of an allosteric enzyme **(A)** T-form; **(B)** R-form; **(C)** Effect of allosteric inhibitor; **(D)** Effect of allosteric activator.*

states – the T (tense or taut) and the R (relaxed). The T and R states are in equilibrium.

$$T \underset{\text{Allosteric inhibitor}}{\overset{\text{Allosteric activator (or) substrate}}{\rightleftarrows}} R$$

Allosteric inhibitors favour T state whereas activators and substrates favour R state. The substrate can bind only with the R form of the enzyme. The concentration of enzyme molecule in the R state increases as more substrate is added, therefore the binding of the substrate to the allosteric enzyme is said to be cooperative. Selected examples of allosteric enzymes responsible for rapid control of biochemical pathways are given in ***Table 6.5***.

Feedback regulation

The process of ***inhibiting the first step by the final product***, in a series of enzyme catalysed reactions of a metabolic pathway is referred to as feedback regulation. Look at the series of reactions given below

$$A \xrightarrow{e_1} B \xrightarrow{e_2} C \xrightarrow{e_3} D \xrightarrow{e_4} E$$

A is the initial substrate, B, C, and D are the intermediates and E is the end product, in a pathway catalysed by four different enzymes (e_1, e_2, e_3, e_4). The very first step ($A \rightarrow B$ by the enzyme e_1) is the most effective for regulating the pathway, by the final end product E. This type of control is often called ***negative feedback regulation*** since increased levels of end product will result in its (e_1) decreased synthesis. This is a real ***cellular economy to save the cell from the wasteful expenditure*** of synthesizing a compound which is already available within the cell.

Feedback inhibition or ***end product inhibition*** is a specialised type of allosteric inhibition necessary to control metabolic pathways for efficient cellular function.

Aspartate transcarbamoylase (ATCase) is a good example of an allosteric enzyme inhibited by a feedback mechanism. ATCase catalyses the very first reaction in pyrimidine biosynthesis.

Table 6.5 Some enzymes with allosteric effectors

Enzyme	Metabolic pathway	Allosteric Inhibitor	Allosteric Activator
Hexokinase	Glycolysis	Glucose 6-phosphate	—
Phosphofructokinase	Glycolysis	ATP	AMP, ADP
Isocitrate dehydrogenase	Krebs cycle	ATP	ADP, NAD^+
Pyruvate carboxylase	Gluconeogenesis	—	Acetyl CoA
Fructose 1,6 - bisphosphatase	Gluconeogenesis	AMP	—
Carbamoyl phosphate synthetase I	Urea cycle	—	N - Acetylglutamate
Tryptophan oxygenase	Tryptophan metabolism	—	L - Tryptophan
Acetyl CoA carboxylase	Fatty acid synthesis	Palmitate	Citrate

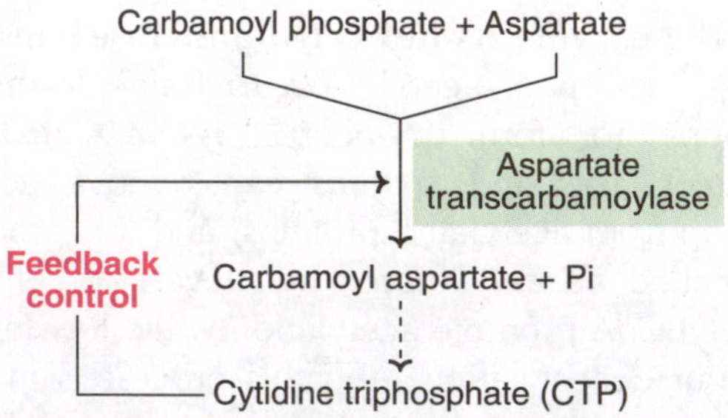

Carbamoyl phosphate undergoes a sequence of reactions for synthesis of the end product, CTP. When CTP accumulates, it allosterically inhibits the enzyme aspartate transcarbamoylase by a feedback mechanism.

Feedback regulation or feedback inhibition? Sometimes a distinction is made between these two usages. Feedback regulation represents a phenomenon while feedback inhibition involves the mechanism of regulation. Thus, in a true sense, they are not synonymous. For instance, dietary cholesterol decreases hepatic cholesterol biosynthesis through feedback regulation. This does not involve feedback inhibition, since dietary cholesterol does not directly inhibit the regulatory enzyme HMG CoA reductase. However, the activity of gene encoding this enzyme is reduced (repression) by cholesterol.

2. Active and inactive enzymes

Latent enzymes, as such, are inactive. Some enzymes are synthesized as ***proenzymes*** or ***zymogens*** which undergo irreversible covalent activation by the breakdown of one or more peptide bonds. For instance, proenzymes –namely chymotrypsinogen, pepsinogen and plasminogen, are respectively – converted to the active enzymes chymotrypsin, pepsin and plasmin.

Certain enzymes exist in the active and inactive forms which are interconvertible, depending on the needs of the body. The interconversion is brought about by the reversible covalent modifications, namely ***phosphorylation*** and ***dephosphorylation***, and oxidation and reduction of disulfide bonds.

Protein phosphorylation : It is highly versatile and selective process. Protein phosphorylation, that occurs ***at hydroxyl groups*** of certain amino acids (***tyrosine, threonine*** etc.) may increase or decrease enzyme activity. Protein phosphorylation and dephosphorylation are respectively catalysed by protein kinases and protein phosphatases. The interplay between these two groups of enzymes is involved in cellular coordinated functioning, integration of biochemical reactions and overall regulation of cellular activities.

Selected examples of phosphorylation and dephosphorylation along with enzyme active states are given in ***Table 6.6***, and one of them is discussed below.

Glycogen phosphorylase is a muscle enzyme that breaks down glycogen to provide energy. This enzyme is a homodimer (two identical

Table 6.6 Selected examples of phosphorylated enzymes (phospho-E) and dephosphorylated enzymes (dephospho-E), along with their activity states

Enzyme (pathway)	Activity state	
	High	Low
HMG CoA reductase kinase (cholesterol synthesis)	Phospho-E	Dephospho-E
Citrate lyase (fatty acid synthesis)	Phospho-E	Dephospho-E
Fructose 2,6-bisphosphatase (glycolysis)	Phospho-E	Dephospho-E
Glycogen phosphorylase (glycogen breakdown)	Phospho-E	Dephospho-E
Glycogen synthase (glycogen synthesis)	Dephospho-E	Phospho-E
Pyruvate dehydrogenase (link between glycolysis and Krebs cycle)	Dephospho-E	Phospho-E
Acetyl CoA carboxylase (fatty acid synthesis)	Dephospho-E	Phospho-E
HMG CoA reductase (cholesterol synthesis)	Dephospho-E	Phospho-E

subunits) and exists in two interconvertible forms. Phosphorylase b (dephospho enzyme) is inactive which is converted by phosphorylation of serine residues to active form phosphorylase a. The inactive enzyme phosphorylase b is produced on dephosphorylation as illustrated below.

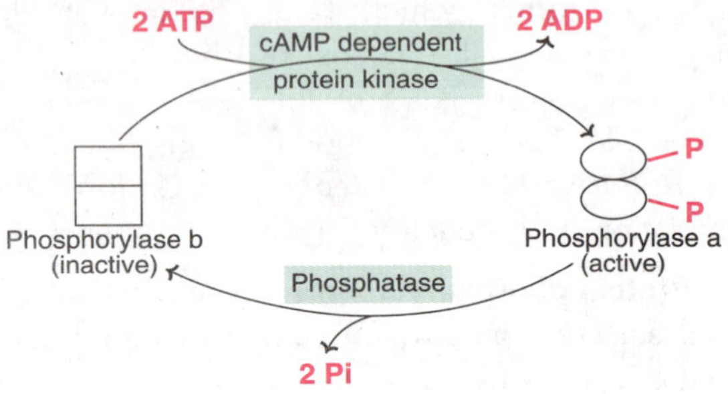

Active enzymes with sulfhydryl (–SH) groups **:** A few enzymes are active only with sulfhydryl (–**SH**) groups, e.g. succinate dehydrogenase, urease. Substances like glutathione bring about the stability of these enzymes.

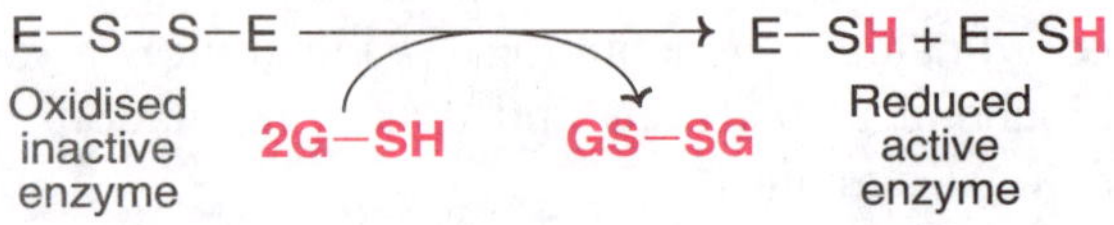

Protein acetylation : The process of acetylation and deacetylation of proteins, particularly certain enzymes, is closely linked with metabolic regulation e.g. acetyl CoA synthetase, glutamate dehydrogenase, carbamoyl phosphate synthetase I.

The enzyme ***lysine acetyltransferase*** transfers acetyl group of acetyl CoA to lysine (ε-amino group) to form ***N-acetyl lysine***. ***Histone deacetylases*** and ***sirtuins*** can remove acetyl group from acetylated proteins, and this process is called ***deacetylation***. High levels of cellular acetyl CoA promote acetylation while low levels favour deacetylation. And this process is in the best interest of cellular metabolic regulation.

3. Compartmentation

There are certain substances in the body (e.g. fatty acids, glycogen) which are synthesized and also degraded. There is no point for simultaneous occurrence of both the pathways. Generally, the ***synthetic*** (anabolic) and ***breakdown*** (catabolic) ***pathways are operative in different cellular organelles*** to achieve maximum economy. For instance, enzymes for fatty acid synthesis are found in the cytosol whereas enzymes for fatty acid oxidation are present in the mitochondria.

The intracellular location of certain enzymes and metabolic pathways is given in ***Table 6.7***.

4. Control of enzyme synthesis

Most of the enzymes, particularly the rate limiting ones, are present in very low concentration. Nevertheless, the amount of the enzyme directly controls the velocity of the reaction, catalysed by that enzyme. ***Many rate***

Table 6.7 Distribution of certain enzymes and metabolic pathways in cellular organelles

Organelle	Enzyme/metabolic pathway
Cytoplasm	Aminotransferases; peptidases; glycolysis; hexose monophosphate shunt; fatty acid synthesis; purine and pyrimidine catabolism.
Mitochondria	Fatty acid oxidation; amino acid oxidation; Krebs cycle; urea synthesis; electron transport chain and oxidative phosphorylation.
Nucleus	Biosynthesis of DNA and RNA.
Endoplasmic reticulum (microsomes)	Protein biosynthesis; triacylglycerol and phospholipid synthesis; steroid synthesis and reduction; cytochrome P_{450}; esterase.
Lysosomes	Lysozyme; phosphatases; phospholipases; hydrolases; proteases; lipases; nucleases.
Golgi apparatus	Glucose 6-phosphatase; 5′-nucleotidase; glucosyl- and galactosyl-transferases.
Peroxisomes	Catalase; urate oxidase; D-amino acid oxidase; long chain fatty acid oxidation.

limiting enzymes have short half-lives. This helps in the efficient regulation of the enzyme levels.

There are two types of enzymes—(a) ***Constitutive enzymes*** (house-keeping enzymes)—the levels of which are not controlled and remain fairly constant. (b) ***Adaptive enzymes***—their concentrations increase or decrease as per body needs and are well-regulated. The synthesis of enzymes (proteins) is regulated by the genes (***Refer Chapter 26***).

Induction and repression : The term induction is used to represent increased synthesis of enzyme while repression indicates its decreased synthesis. Induction or repression, ultimately determines the enzyme concentration at the gene level through the mediation of hormones or other substances.

Examples of enzyme induction : The hormone ***insulin induces*** the synthesis of ***glycogen synthetase***, glucokinase, phosphofructokinase and pyruvate kinase. All these enzymes are involved in the utilization of glucose. The hormone cortisol induces the synthesis of many enzymes e.g. pyruvate carboxylase, tryptophan oxygenase and tyrosine aminotransferase.

Examples of repression : In many instances, substrate can repress the synthesis of enzyme. Pyruvate carboxylase is a key enzyme in the synthesis of glucose from non-carbohydrate sources like pyruvate and amino acids. If there is sufficient glucose available, there is no necessity for its synthesis. This is achieved through repression of ***pyruvate carboxylase by glucose***.

5. Enzyme degradation

Enzymes are not immortal, since it will create a series of problems. There is a lot of variability in the half-lives of individual enzymes. For some, it is in days while for others in hours or in minutes, e.g. LDH_4 — 5 to 6 days; LDH_1 — 8 to 12 hours; amylase — 3 to 5 hours.

In general, the key and ***regulatory enzymes are most rapidly degraded***. If not needed, they immediately disappear and, as and when required, they are quickly synthesized. Though not always true, an enzyme with long half-life is usually sluggish in its catalytic activity.

6. Isoenzymes

Multiple forms of the same ***enzyme*** will also help in the regulation of enzyme activity, Many of the isoenzymes are tissue-specific. Although isoenzymes of a given enzyme catalyse the same reaction, they differ in K_m, V_{max} or both. e.g. isoenzymes of LDH and CPK.

UNITS OF ENZYME ACTIVITY

Enzymes are never expressed in terms of their concentration (as mg or µg etc.), but are expressed only as activities. Various methods have been introduced for the estimation of enzyme activities (particularly for the plasma enzymes). In fact, the activities have been expressed in many ways, like ***King-Armstrong units, Somogyi units, Reitman-Frankel units***, spectrophotometric units etc.

Katal

In order to ***maintain uniformity*** in the expression of enzyme activities (as units) worldover, the ***Enzyme Commission of IUB*** has suggested radical changes. A new unit – namely katal (abbreviated as kat) – was introduced. ***One kat denotes the conversion of one mole substrate per second*** (mol/sec). Activity may also be expressed as millikatals (mkat), microkatals (µkat) and so on. However, enzymes activities are less commonly expressed in the form of katals.

International Units (IU)

Some workers prefer to use standard units or ***SI units*** (System International) or simply ***enzyme unit*** (U). One SI unit or International Unit (IU) is defined as the amount of enzyme activity that catalyses the ***conversion of one micromol of substrate per minute***. SI units and katal are interconvertible.

$$1 \text{ IU} = 16.67 \text{ nkat}$$

Laboratory use of enzyme units

In the clinical laboratories, however, the units— namely katal or SI units—are yet to find a place. Many investigators ***still use the old units***

Medical Concepts/ Clinical Correlates

- A few RNA molecules (ribozymes) function as non-protein enzymes. It is believed that ribozymes were functioning as biocatalysts before the occurrence of protein enzymes during evolution.
- Certain enzymes are utilized as therapeutic agents. Streptokinase is used to dissolve blood clots in circulation while asparaginase is employed in the treatment of leukemias.

like King-Armstrong units, Somogyi units etc. while expressing the enzyme activities. It is therefore, essential that the units of enzyme activity, along with the normal values, be invariably stated while expressing the enzymes for comparison.

In this book, enzyme units (expressed as U/l) are more frequently used. This is done in accordance with the units expressed in standard medicine text books.

NON-PROTEIN ENZYMES

Ribozymes

Ribozymes are a group of ***ribonucleic acids*** that function as biological ***catalysts***, and they are regarded as non-protein enzymes.

Altman and his coworkers, in 1983, found that ***ribonuclease P*** – an enzyme till then known to cleave precursors of tRNAs to give tRNAs – was functional ***due to RNA*** component present in the enzyme and not the protein part of the enzyme (Refer ***Chapter 4***).

The RNA part isolated from ribonuclease P exhibited a true enzyme activity and also obeyed Michaelis-Menten kinetics. Later studies have proved that RNA, in fact, can function as an enzyme and bring about the catalysis.

RNA molecules are known to adapt a tertiary structure just as in the case of proteins (i.e. enzymes). The specific conformation of RNA may be responsible for its function as biocatalyst. It is believed that ***ribozymes (RNAs) were functioning as catalysts*** before the occurrence of protein enzymes ***during evolution***.

APPLICATIONS OF ENZYMES

Certain enzymes are useful as therapeutic agents, analytical reagents, in genetic manipulations and for industrial applications (***Table 6.8***).

Enzymes as therapeutic agents

1. **Streptokinase** prepared from streptococcus is useful ***for clearing the blood clots***. Streptokinase activates plasma plasminogen to plasmin

Table 6.8 A selected list of applications of enzymes

Enzyme	Application
Therapeutic applications	
Streptokinase/urokinase	To remove blood clots
Asparaginase	In cancer therapy
Papain	Anti-inflammatory
α_1-Antitrypsin	To treat emphysema
Pancreatic enzymes (trypsin, lipase)	For digestion (in pancreatic diseases)
Analytical application reagents (for estimation)	
Glucose oxidase and peroxidase	Glucose
Urease	Urea
Cholesterol oxidase	Cholesterol
Uricase	Uric acid
Lipase	Triacylglycerols
Luciferase	To detect bacterial contamination of foods
Alkaline phosphatase/ horse radish peroxidase	In the analytical technique ELISA
Applications in genetic engineering	
Restriction endonucleases	Gene transfer, DNA finger printing
Taq DNA polymerase	Polymerase chain reaction
Industrial applications	
Rennin	Cheese preparation
Glucose isomerase	Production of high fructose syrup
α-Amylase	In food industry
Proteases	Washing powder

which, in turn, attacks fibrin to convert into soluble products.

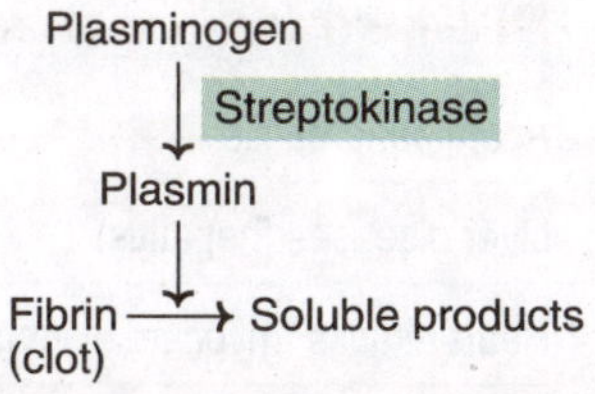

Streptokinase and tissue plasminogen activator (tPA) are used in the treatment of acute myocardial infarction.

2. The enzyme ***asparaginase*** is used in the ***treatment of leukemias***. Tumor cells are dependent on asparagine of the host's plasma for their multiplication. By administering asparaginase, the host's plasma levels of asparagine are drastically reduced. This leads to depression in the viability of tumor cells.

3. Certain recombinant enzymes have been approved for the treatment of some lysosomal storage diseases e.g. Gaucher disease (β-glucosidase), Pompe disease (α-glucosidase), Fabry disease (α-galactosidase A).

Enzymes as analytical reagents

Some enzymes are ***useful in the clinical laboratory for the measurement of substrates***, drugs, and even the activities of other enzymes. The biochemical compounds (e.g. glucose, urea, uric acid, cholesterol) can be more accurately and specifically estimated by enzymatic procedures compared to the conventional chemical methods. A good example is the ***estimation of plasma glucose by glucose oxidase and peroxidase method***.

Immobilized enzymes

Enzymes can be used as catalytic agents in industrial and medical applications. Some of these enzymes are immobilized by binding them to a solid, insoluble matrix which will not affect the ***enzyme stability or its catalytic activity***. Beaded gels and cyanogen bromide activated sepharose are commonly used for immobilization of enzymes. The bound enzymes can be preserved for long periods without loss of activity.

Glucose oxidase and peroxidase, immobilized and coated on a strip of paper, are used in the clinical laboratory for the detection of glucose in urine.

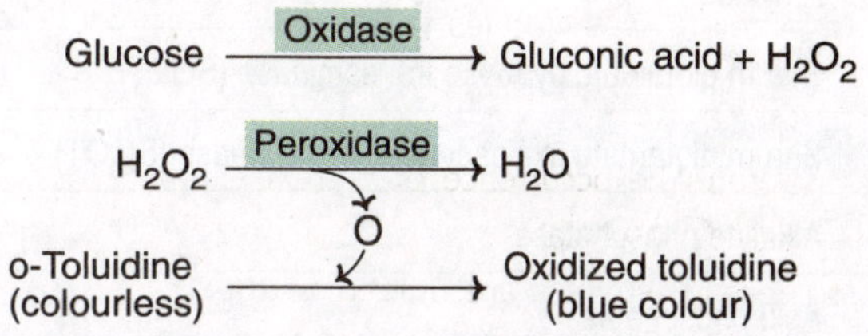

The intensity of the blue colour depends on the concentration of glucose. Hence, the strip method is useful for semi-quantitative estimation of glucose in urine.

PLASMA/SERUM ENZYMES

Enzymes in the circulation are divided into two groups – plasma functional and plasma non-functional.

1. Plasma specific or plasma functional enzymes

Certain enzymes are normally present in the plasma and they have specific functions to perform. Generally, these enzyme activities are higher in plasma than in the tissues. They are mostly synthesized in the liver and enter the circulation e.g. lipoprotein lipase, plasmin, thrombin, choline esterase, ceruloplasmin etc.

2. Non-plasma specific or plasma non-functional enzymes

These enzymes are either totally absent or present at a low concentration in plasma compared to their levels found in the tissues. The digestive enzymes of the gastrointestinal tract (e.g. amylase, pepsin, trypsin, lipase etc.) present in the plasma are known as ***secretory enzymes.*** All the other plasma enzymes associated with metabolism of the cell are collectively referred to as ***constitutive enzymes*** (e.g. lactate dehydrogenase, transaminases, acid and alkaline phosphatases, creatine phosphokinase).

Table 6.9 Important enzymes in the diagnosis of diseases

Serum enzyme (elevated)	Disease (most important)
Amylase	Acute pancreatitis
Serum glutamate pyruvate transaminase (SGPT)	Liver diseases (hepatitis)
Serum glutamate oxaloacetate transaminase (SGOT)	Heart attacks (myocardial infarction)
Alkaline phosphatase	Rickets, obstructive jaundice
Acid phosphatase	Cancer of prostate gland
Lactate dehydrogenase (LDH)	Heart attacks, liver diseases
Creatine phosphokinase (CPK)	Myocardial infarction (early marker)
Aldolase	Muscular dystrophy
5′-Nucleotidase	Hepatitis
γ-Glutamyl transpeptidase (GGT)	Alcoholism

DIAGNOSTIC IMPORTANCE OF ENZYMES

Estimation of enzyme activities in biological fluids (particularly plasma/serum) is of great clinical importance. Particularly non-plasma specific enzymes are very important for the diagnosis and prognosis of several diseases.

The normal serum level of an enzyme indicates the balance between its synthesis and release in the routine cell turnover. The raised enzyme levels could be due to cellular damage, increased rate of cell turnover, proliferation of cells, increased synthesis of enzymes etc. Serum enzymes are conveniently used as ***markers*** to detect the cellular damage which ultimately helps ***in the diagnosis of diseases***.

(**Note :** Ther term ***biomarker*** refers to any ***laboratory analyte*** (***enzyme, protein, antigen,*** antibody, metabolite etc.) ***that is useful for the diagnosis/prognosis of any disease***. Biomarker is a vague term, and less frequently used by biochemists.)

A summary of the important enzymes useful for the diagnosis of specific diseases is given in ***Table 6.9***. Detailed information on the diagnostic enzymes ***including reference values*** is provided in ***Table 6.10***. A brief account of selected diagnostic enzymes is discussed

Amylase **:** The activity of serum amylase is increased in ***acute pancreatitis*** (reference 30-150 U/l). The peak value is observed within 8-12 hours after the onset of disease which returns to normal by 3rd or 4th day. Elevated activity of amylase is also found in urine of the patients of acute pancreatitis. Serum amylase is also important for the diagnosis of chronic pancreatitis, acute parotitis (mumps) and obstruction of pancreatic duct.

Alanine transaminase (ALT/SGPT) **:** SGPT is elevated in ***acute hepatitis*** of viral or toxic origin, jaundice and cirrhosis of liver (reference 10-45 IU/l).

Aspartate transaminase (AST/SGOT) **:** SGOT activity in serum is increased in ***myocardial infarction*** and also in liver diseases (reference 10-50 U/l).

It may be noted that ***SGPT is more specific for the diagnosis of liver diseases while SGOT is for heart diseases***. This is mainly because of their cellular distribution – SGPT is a cytosomal enzyme while SGOT is found in cytosol and mitochondria.

Table 6.10 Increase in plasma (serum) enzymes in the diagnosis of diseases

Enzymes	Reference value	Disease(s) in which increased
I. **Digestive enzymes**		
Amylase	30–150 U/l	Acute pancreatitis, mumps (acute parotitis), obstruction in pancreatic duct, severe diabetic ketoacidosis.
Lipase	30–150 U/l	Acute pancreatitis, moderate elevation in carcinoma of pancreas.
II. **Transaminases**		
Alanine transaminase (ALT) or serum glutamate pyruvate transaminase (SGPT)	10–45 U/l	Acute hepatitis (viral or toxic), jaundice, cirrhosis of liver.
Aspartate transaminase (AST) or serum glutamate oxaloacetate transaminase (SGOT)	10–50 U/l	Myocardial infarction, liver diseases, liver cancer, cirrhosis of liver.
III. **Phosphatases**		
Alkaline phosphatase (ALP) (pH optimum 9–10)	50–100 U/l	Bone diseases (related to higher osteoblastic activity)-rickets, Pagets' disease, hyperparathyroidism, carcinoma of bone. Liver diseases-obstructive jaundice (cholestasis), infective hepatitis, cirrhosis of liver.
Acid phosphatase (ACP) (pH optimum 4–6)	0.5–5.0 U/l	Prostatic carcinoma i.e. cancer of prostate gland (tartarate liabile ACP serves as a marker for diagnosis and follow up), Pagets' disease.
IV. **Enzymes of carbohydrate metabolism**		
Aldolase	1–8 U/l	Muscular dystrophy, liver diseases, myocardial infarction, myasthenia gravis, leukemias
Lactate dehydrogenase (LDH)	200–450 U/l	Myocardial infarction, acute infective hepatitis, muscular dystrophy, leukemia, pernicious anaemia.
V. **Miscellaneous enzymes**		
Creatine kinase (CK) or creatine phosphokinase (CPK)	Males : 50–200 U/l Females : 30–150 U/l	Myocardial infarction (CK useful for early detection), muscular dystrophy, hypothyroidism, alcoholism.
Cholinesterase (ChEI)	5–10 U/l	Nephrotic syndrome, myocardial infarction
5′-Nucleotidase or nucleotide phosphatase (NTP)	2–10 U/l	Hepatitis, obstructive jaundice, tumors
γ-Glutamyl transpeptidase (GGT)	5–50 U/l	Alcoholism, infective hepatitis, obstructive jaundice.
Ceruloplasmin (ferrooxidase)	20–60 mg/dl	Bacterial infections, collagen diseases, cirrhosis, pregnancy.

Table 6.11 Decrease in plasma (serum) enzymes in certain diseases		
Enzyme	**Reference values**	**Disease(s) in which decreased**
Amylase	30–150 U/l	Liver diseases
Ceruloplasmin	20–60 mg/dl	Wilson's disease (hepatolenticular degeneration)
Glucose 6-phosphate dehydrogenase (G6PD) in RBC	7–12 U/g Hb	Congenital deficiency with hemolytic anemia

Alkaline phosphatase (ALP) **:** It is elevated in certain bone and liver diseases (reference 50-100 U/l). ALP is useful for the diagnosis of ***rickets***, hyperparathyroidism, carcinoma of bone, and ***obstructive jaundice.***

Acid phosphatase (ACP) **:** It is increased in the ***cancer of prostate gland*** (reference 0.5-5.0 U/l). The tartarate labile ACP (reference <1 KA units/dl) is useful for the diagnosis and prognosis of prostate cancers i.e. ACP is a good tumor marker.

Lactate dehydrogenase (LDH) **:** LDH is useful for the diagnosis of ***myocardial infarction, infective hepatitis***, leukemia and muscular dystrophy (reference 200-450 U/l). LDH has five isoenzymes, the details of which are described later.

Creatine kinase (CK) **:** It is elevated in ***myocardial infarction*** (early detection) and muscular dystrophy (reference males : 50–200 U/l; females : 30–150 U/l). CK has three isoenzymes (described later).

γ-Glutamyl transpeptidase (GGT) **:** It is a sensitive diagnostic marker for the detection of ***alcoholism***. GGT is also increased in infective hepatitis and obstructive jaundice.

Decreased plasma enzyme activities

Sometimes, the plasma activities of the enzymes may be lower than normal which could be due to decreased enzyme synthesis or congenital deficiency. In ***Table 6.11***, the decreased plasma enzymes in certain disorders are given.

ISOENZYMES

The ***multiple forms of an enzyme*** catalysing the same reaction are ***isoenzymes*** or ***isozymes.*** They, however, ***differ in their physical*** and ***chemical properties*** which include the structure, electrophoretic and immunological properties, K_m and V_{max} values, pH optimum, relative susceptibility to inhibitors and degree of denaturation.

Explanation for the existence of isoenzymes

Many possible reasons are offered to explain the presence of isoenzymes in the living systems.

1. Isoenzymes synthesized from ***different genes*** e.g. malate dehydrogenase of cytosol is different from that found in mitochondria.

2. Oligomeric enzymes consisting of ***more*** than one type of ***subunits*** e.g. lactate dehydrogenase and creatine phosphokinase.

3. An enzyme may be active as monomer or oligomer e.g. glutamate dehydrogenase.

4. In glycoprotein enzymes, differences in carbohydrate content may be responsible for isoenzymes e.g. alkaline phosphatase.

Isoenzymes of lactate dehydrogenase (LDH)

Among the isoenzymes, LDH has been the most thoroughly investigated.

LDH whose systematic name is L-lactate-NAD^+ oxidoreductase (E.C. 1.1.1.27) catalyses the interconversion of lactate and pyruvate as shown below

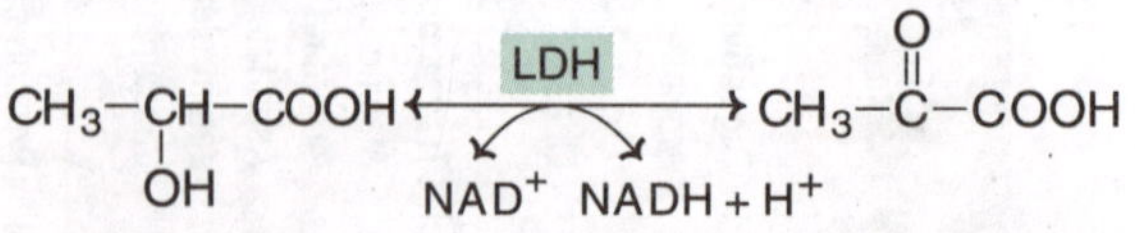

Lactic acid **Pyruvic acid**

Table 6.12 Lactate dehydrogenase (LDH) isoenzymes and their characteristics

Isoenzyme	Subunit constitution	Principal tissue of origin	Electrophoretic mobility	Whether destroyed by heat (at 60 °C)	Percentage in serum of healthy people
LDH_1	H_4	Heart and RBC	Fastest	No	25%
LDH_2	H_3M	Heart and RBC	Faster	No	35%
LDH_3	H_2M_2	Brain and kidney	Fast	Partially	27%
LDH_4	HM_3	Liver and skeletal muscle	Slow	Yes	8%
LDH_5	M_4	Skeletal muscle and liver	Slowest	Yes	5%

LDH has five distinct isoenzymes LDH_1, LDH_2, LDH_3, LDH_4 and LDH_5. They can be separated by electrophoresis (cellulose or starch gel or agarose gel). LDH_1 has more positive charge and fastest in electrophoretic mobility while LDH_5 is the slowest.

Structure of LDH isoenzymes : LDH is an oligomeric (tetrameric) enzyme made up of four polypeptide subunits. Two types of subunits namely M (for muscle) and H (for heart) are produced by different genes. M–subunit is basic while H subunit is acidic. The isoenzymes contain either one or both the subunits giving LDH_1 to LDH_5. The characteristic features of LDH isoenzymes are given in ***Table 6.12***.

Significance of differential catalytic activity : ***LDH_1 (H_4)*** is predominantly found in ***heart muscle*** and is inhibited by pyruvate – the substrate. Hence, pyruvate is not converted to lactate in cardiac muscle but is converted to acetyl CoA which enters citric acid cycle. LDH_5 (M_4) is mostly present in skeletal muscle and the inhibition of this enzyme by pyruvate is minimal, hence pyruvate is converted to lactate. Further, LDH_5 has low K_m (high affinity) while ***LDH_1*** has ***high K_m*** (low affinity) ***for pyruvate***. The differential catalytic activities of LDH_1 and LDH_5 in heart and skeletal muscle, respectively, are well suited for the aerobic (presence of oxygen) and anaerobic (absence of oxygen) conditions, prevailing in these tissues.

Medical Concepts/ Clinical Correlates

- Elevation in serum enzyme activities is of great importance for the diagnosis of several diseases. e.g. amylase–acute pancreatitis; ALT–hepatitis; alkaline phosphatase–obstructive jaundice; CPK-MI;
- Lowered body temperature (hypothermia) is accompanied by a decrease in enzyme activities. This principle is exploited to reduce metabolic demand during open heart surgery or transportation of organs for transplantation surgery.

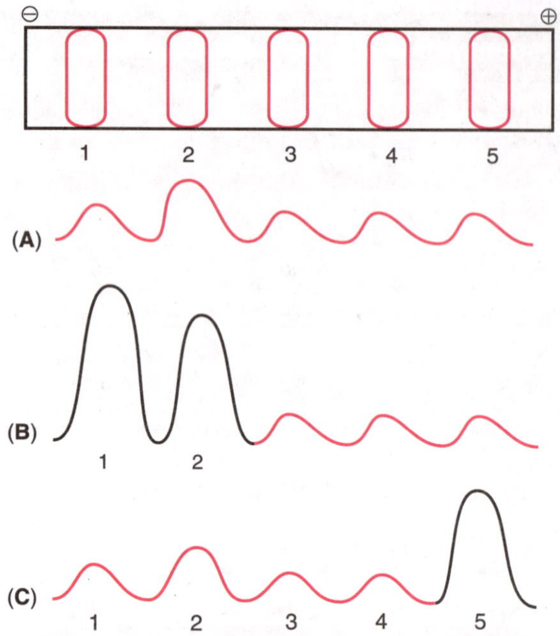

Fig. 6.14 : Electrophoresis of lactate dehydrogenase with relative proportions of isoenzymes ***(A)*** *Normal serum* ***(B)*** *Serum from a patient of myocardial infarction (LDH_1 and $LDH_2\uparrow$)* ***(C)*** *Serum from a patient of liver disease ($LDH_5\uparrow$).*

Diagnostic importance of LDH : Isoenzymes of LDH have immense value in the diagnosis of heart and liver related disorders (***Fig.6.14***). In healthy individuals, the activity of LDH_2 is higher than that of LDH_1 in serum. In the case of ***myocardial infarction***, ***LDH_1*** is much greater than LDH_2 and this happens within 12 to 24 hours after infarction. Increased activity of ***LDH_5*** in serum is an indicator of ***liver diseases***. LDH activity in the RBC is 80–100 times more than that in the serum. Hence for estimation of LDH or its isoenzymes, serum should be totally free from hemolysis or else false positive results will be obtained.

Isoenzymes of creatine phosphokinase

Creatine kinase (CK) or creatine phosphokinase (CPK) catalyses the interconversion of phosphocreatine (or creatine phosphate) to creatine.

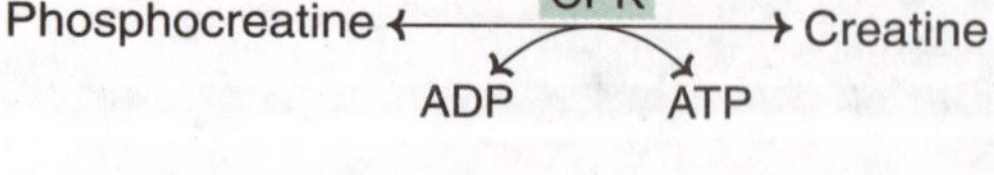

CPK exists as three isoenzymes. Each isoenzyme is a dimer composed of two subunits—M (muscle) or B (brain) or both.

Isoenzyme	Subunit	Tissue of origin
CPK_1	BB	Brain
CPK_2	MB	Heart
CPK_3	MM	Skeletal muscle

In healthy individuals, the isoenzyme CPK_2 (MB) is almost undetectable in serum with less than 2% of total CPK. After the myocardial infarction (MI), within the first 6-18 hours, CPK_2 increases in the serum to as high as 20% (against 2% normal). CPK_2 isoenzyme is not elevated in skeletal muscle disorders. Therefore, estimation of the enzyme ***CPK_2 (MB) is the earliest reliable indication of myocardial infarction***.

(**Note :** Recently, ***two mitochondrial isoenzymes of CK*** (mtCK) have been indentified. The mtCK are described in biological oxidation).

Isoenzymes of alkaline phosphatase

As many as six isoenzymes of alkaline phosphatase (ALP) have been identified. ALP is a monomer, the isoenzymes are due to the ***difference in the carbohydrate content*** (sialic acid residues). The most important ALP isoenzymes are α_1-ALP, α_2-heat labile ALP, α_2-heat stable ALP, pre-β ALP, γ-ALP etc.

Increase in α_2-heat labile ALP suggests hepatitis whereas pre β-ALP indicates bone diseases.

Isoenzymes of alcohol dehydrogenase

Alcohol dehydrogenase (ADH) has two heterodimer isoenzymes. Among the white Americans and Europeans, $\alpha\beta_1$ isoenzyme is predominant whereas in Japanese and Chinese (Orientals) $\alpha\beta_2$ is mostly present. The isomer $\alpha\beta_2$ more rapidly converts alcohol to acetaldehyde.

Accumulation of acetaldehyde is associated with tachycardia (increase in heart rate) and facial flushing among Orientals which is not commonly seen in whites. It is believed that ***Japanese and Chinese have increased sensitivity to alcohol due to the presence of $\alpha\beta_2$–isoenzyme of ADH.***

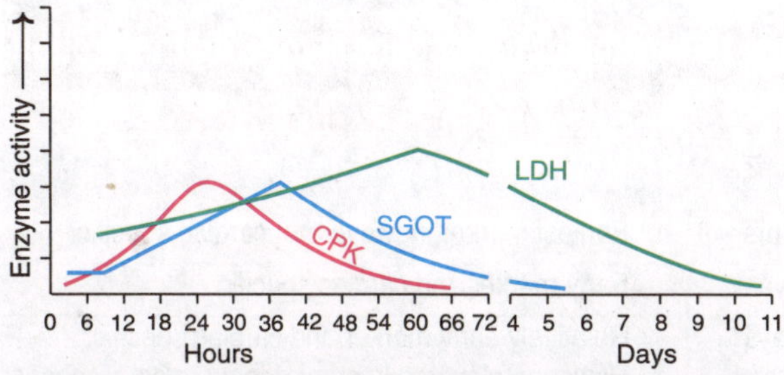

Fig. 6.15 : Enzyme pattern in myocardial infarction (CPK-Creatine phosphokinase; SGOT-Serum glutamate oxaloacetate transaminase; LDH-Lactate dehydrogenase).

ENZYME PATTERN IN DISEASES

For the right diagnosis of a particular disease, it is always better to estimate a few (three or more) serum enzymes, instead of a single enzyme. Examples of enzyme patterns in important diseases are given here.

Enzymes in myocardial infarction (MI)

The enzymes – namely creatine phosphokinase (***CPK***), aspartate transaminase (***AST***) and lactate dehydrogenase (***LDH***)—are important for the diagnosis of myocardial infarction (***MI***). The elevation of these enzymes in serum in relation to hours/days of MI is given in the ***Fig.6.15***.

Creatine phosphokinase (precisely isoenzyme ***MB***) is the first enzyme to be released into circulation within 6-18 hours after the infarction. Therefore, CPK estimation is highly useful for the early diagnosis of MI. This enzyme reaches a peak value within 24-30 hours, and returns to normal level by the 2nd or 3rd day.

Aspartate transaminase (AST or SGOT) rises sharply after CPK, and reaches a peak within 48 hours of the myocardial infarction. AST takes 4-5 days to return to normal level.

Lactate dehydrogenase (LDH_1) generally rises from the second day after infarction, attains a peak by the 3rd or 4th day and takes about 10-15 days to reach normal level. Thus, LDH is the last enzyme to rise and also the last enzyme to return to normal level in MI.

Non-enzymatic biomarkers in MI

Besides the enzymes, certain non-enzymatic biomarkers are equally ***important for the diagnosis*** of myocardial infraction.

Cardiac troponins (CT) : Although not enzymes, the proteins cardiac troponins are highly useful for the early diagnosis of MI. Among these, ***troponin I*** (***i***nhibitory element of actomysin ATPase) and ***troponin T*** (***t***ropomyosin binding element) are important. **C**ardiac troponin I (CTI) is released into circulation within four hours after the onset of MI, reaches a peak value by 12–24 hours, and remains elevated for about a week. In recent years, ***high sensitivity cardiac troponins*** (hsCT) are being used for early and more reliable diagnosis of MI.

The protein ***myoglobin*** is also an early marker for the diagnosis of MI. However, it is not specific to cardiac diseases. High serum concentrations of ***brain natriuretic peptide (BNP)*** and ***atrial natriuretic peptide (ANP)*** are marker for congestive cardiac failure. In the ***Table 6.13***, a summary of the diagnostic markers used in MI is given.

Enzymes in liver diseases

The following enzymes—when elevated in serum – are useful for the diagnosis of liver dysfunction due to ***viral hepatitis*** (jaundice), ***toxic hepatitis***, ***cirrhosis*** and ***hepatic necrosis***

1. ***Alanine transaminase***
2. Aspartate transaminase
3. Lactate dehydrogenase.

The enzymes that markedly increase in intrahepatic and extrahepatic cholestasis are :

(1) Alkaline phosphatase, (2) 5′-Nucleotidase

Serum-***γ-glutamyl transpeptidase*** is useful in the diagnosis of ***alcoholic*** liver diseases.

Enzymes in muscle diseases

In the ***muscular dystrophies***, serum levels of certain muscle enzymes are increased. These include creatine phosphokinase, aldolase and aspartate transaminase.

Enzymes in cancers

Increase in the serum ***acid phosphatase*** (tartarate labile) is specific for the detection of ***prostatic carcinoma***.

Table 6.13 Summary of diagnostic markers used for the evaluation of acute myocardial infarction

Diagnostic marker	Time of peak elevation	Time of return to normal level	Diagnostic importance
Myoglobin	4-6 hrs	20–25 hrs	Earliest marker, however not cardiac specific.
Cardiac troponin I	12-24 hrs	5-9 days	Early marker and cardiac specific.
Cardiac troponin T	18-36 hrs	5-14 days	Relatively early marker and cardiac specific. However, elevated in other degenerative diseases.
Creatine phosphokinase (MB)	20-30 hrs	24-48 hrs	Cardiac specific and early marker.
Lactate dehydrogenase (LDH_1)	48-72 hrs	10-15 days	Relatively late marker and cardiac specific.
Aspartate transaminase	30-48 hrs	4-6 days	Not cardiac specific.

[**Note : *Prostate specific antigen*** (PSA; mol wt. 32 KD), though not an enzyme, is a more reliable marker for the detection of prostate cancer. Normal serum concentration of PSA is 1-4 ng/ml].

Neuron–specific enolase serves as a marker for lung cancer, neuroblastoma, pheochromocytoma etc.

DIAGNOSTIC IMPORTANCE OF ENZYMES IN OTHER BODY FLUIDS AND TISSUES

Besides serum/plasma enzymes, enzyme estimations in other body fluids, and tissues also have some diagnostic importance. A few examples are lised.

Table 6.14 Serum enzyme profiles (patterns) in diseases

Disease/enzyme(s)	Significance
I **Myocardial infarction** (Refer **Table 6.13** and **Fig. 6.15**)	
II **Hepatic disease**	
Alanine transaminase (ALT)	Markedly elevated in viral hepatitis
Aspartate transaminse (AST)	Increased in liver diseases. Significantly elevated in obstructive jaundice (gall stones).
γ-Glutamyl transpeptidase (GGT)	Markedly increased in alcoholic liver diseases.
5′-Nucleotidase	Elevated in hepatic cholestasis.
III **Muscle disease**	
Creatine kinase (CK)	Markedly increased in muscle disease (CK-MM more sensitive).
Aldolase (ALD)	Early marker (not specific)
Aspartate transaminase (AST)	Significantly increased, although not specific.
IV **Bone disease**	
Alkaline phosphatase (ALP)	Increased in rickets and Paget's disease.
V **Pancreatic disease**	
Amylase	Significantly elevated in acute pancreatitis.
Lipase	Markedly increased in acute pancreatitis.
VI **Prostate cancer**	
Acid phosphatase (ACP)	Marker enzyme for prostate cancer.
Prostate specific antigen (PSA)	Significantly elvated in prostate cancer (not an enzyme).

Urine : Urinary ***amylase*** is increased in acute pancreatitis. β-N-Acetylgalactosidase in urine is elevated in renal graft dysfunction. ***β-Glucuronidase*** is increased in the cancers of urinary bladder, pancreas etc.

Cerebrospinal fluid : ***Lactate dehydrogenase*** is increased in CSF in meningitis.

Gastric juice : ***β-Glucuronidase*** activity is increased in gastric carcinoma.

Feces : Fecal ***trypsin*** levels are decreased in cystic fibrosis.

Liver : ***Glucose 6-phosphatase*** in liver is significantly lower in type I glycogen storage disease.

Muscle : ***Phosphorylase*** activity in muscle is decreased in McArdle's disease.

Erythrocytes : ***Glucose 6-phosphate dehydrogenase*** deficiency in RBC causes hemolytic anemia. Decreased ***transketolase*** activity in erythrocytes is used for diagnosis of ***thiamine deficiency***.

In addition to the above, ***cultured fibroblasts***, and ***amniotic cells*** are frequently used for the ***diagnosis of inborn errors of metabolism*** e.g. phenylalanine hydroxylase deficiency in phenylketonuria (in cultured amniotic cells).

SUMMARY

- *Enzymes are the protein biocatalysts synthesized by the living cells. They are classified into six major classes—oxidoreductases, transferases, hydrolases, lyases, isomerases and ligases.*
- *An enzyme is specific in its action, possessing active site, where the substrate binds to form enzyme-substrate complex, before the product is formed.*
- *Factors like concentration of enzyme, substrate, temperature, pH etc. influence enzyme activity. The substrate concentration to produce half-maximal velocity is known as Michaelis constant (K_m value).*
- *Enzyme activities are inhibited by reversible (competitive, and non-competitive), irreversible and allosteric manner.*
- *Many enzymes require certain non-protein substances called cofactors (coenzymes) for their action. Most of the coenzymes are derivatives of B-complex vitamins (e.g. NAD^+, FAD, TPP etc.)*
- *The mechanism of enzyme action is explained by lock and key model (of Fischer), more recently induced fit model (of Koshland) and substrate strain theory.*
- *In the living system, there is a constant regulation of enzyme levels, brought about by allosteric mechanism, activation of proenzymes, synthesis, degradation of enzymes etc.*
- *Estimation of serum enzymes is of great help in the diagnosis of several diseases. Serum amylase and lipase are increased in acute pancreatitis; alanine transaminase in hepatitis; aspartate transaminase, lactate dehydrogenase (LDH) and creatine phosphokinase (CPK) in myocardial infarction; alkaline phosphatase in rickets.*
- *Isoenzymes are the multiple forms of an enzyme catalysing the same reaction which however, differ in their physical and chemical properties. LDH has five isoenzymes while CPK has three. LDH_1 and CPK_2 are very important in the diagnosis of MI.*

THE READER MUST REFER APPENDIX VI (Page Nos : 738 - 756) FOR RELATED CLINICAL CASE STUDIES (Nos : 36, 37, 38, 40, 41, 42, 43, 44)

Self-Assessment Exercises

I. Essay questions

1. What are enzymes? Describe their classification and nomenclature.
2. Write an account of the various factors affecting enzyme activity.
3. Describe the mechanism of enzyme action.
4. What are coenzymes? Write briefly on the role of coenzymes in enzyme action.
5. Write an account of the importance of serum enzymes in the diagnosis of diseases.

II. Short notes

(a) Enzyme specificity, (b) Competitive inhibition, (c) Coenzymes, (d) Allosteric enzymes, (e) Isoenzymes, (f) K_m value, (g) Serum enzymes in myocardial infarction, (h) Lactate dehydrogenase, (i) Role of metals in enzyme action, (j) Active site.

III. Fill in the blanks

1. The literal meaning of enzyme is __________.
2. The class of enzymes involved in synthetic reactions are __________.
3. The non-protein part of holoenzyme __________.
4. Enzymes lose the catalytic activity at temperature above 70°C due to __________.
5. Examples of two enzymes containing zinc are __________ and __________.
6. The place at which substrate binds with the enzyme __________.
7. The enzyme glucose 6-phosphate dehydrogenase requires the coenzyme __________.
8. The E.C. number for alcohol dehydrogenase is __________.
9. Phsophofructokinase is allosterically activated by __________.
10. The very first enzyme elevated in serum in myocardial infarction __________.

IV. Multiple choice questions

11. Pepsin is an example for the class of enzymes namely
    (a) Oxidoreductases (b) Transferases (c) Hydrolases (d) Ligases.
12. The coenzyme not involved in hydrogen transfer
    (a) FMN (b) FAD (c) $NADP^+$ (d) FH_4.
13. In the feedback regulation, the end product binds at
    (a) Active site (b) Allosteric site (c) E-S complex (d) None of these.
14. γ-Glutamyl transpeptidase activity in serum is elevated in
    (a) Pancreatitis (b) Muscular dystrophy (c) Myocardial infarction (d) Alcoholism.
15. In recent years, a non-protein compound has been identified to bring about catalysis in biological system. The name of the compound is
    (a) DNA (b) RNA (c) Lipids (d) Carbohydrates.

7 Vitamins

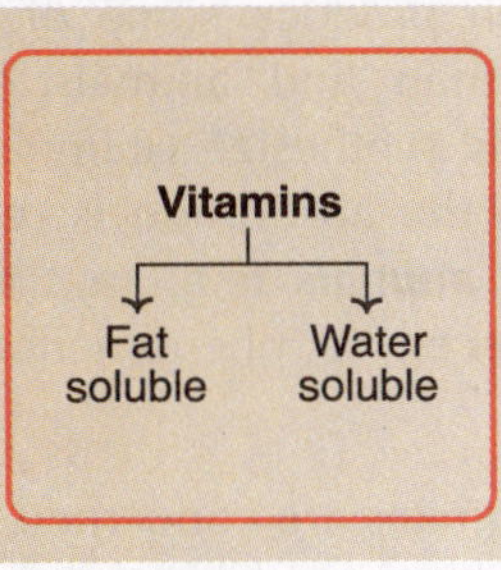

The vitamins speaks

"We are for growth, health and welfare of organisms;
Discharge our duties directly or through coenzymes;
Deficiency symptoms are our alert signals;
Satisfied we shall be, with additional supplements."

It is difficult to define vitamins precisely. ***Vitamins may be regarded as organic compounds required in the diet in small amounts to perform specific biological functions for normal maintenance of optimum growth and health of the organism***. Generally, vitamins are not synthesized by the body, and therefore need to be supplied through the diet. The bacterium ***E.coli*** does not require any vitamin, as it ***can synthesize all of them***. It is believed that during the course of evolution, the ability to synthesize vitamins was lost. Hence, the higher organisms have to obtain them from diet.

History and nomenclature

In the beginning of 20th century, it was clearly understood that the diets containing purified carbohydrate, protein, fat and minerals were not adequate to maintain the growth and health of experimental rats, which the natural foods (such as milk) could do.

Hopkins coined the term ***accessory factors*** to the unknown and essential nutrients present in the natural foods. Funk (1913) isolated an active principle (an amine) from rice polishings and, later in yeast, which could cure beri-beri in pigeons. He coined the term ***vitamine*** (*Greek* : vita-life) to the accessory factors with a belief that all of them were ***amines***. It was later realised that only few of them are amines. The term *vitamin*, however, is continued without the final letter 'e'.

The usage of A, B and C to vitamins was introduced in 1915 by McCollum and Davis. They first felt there were only two vitamins—***fat soluble A*** and ***water soluble B*** (anti-beriberi factor). Soon another water soluble anti-scurvy factor named vitamin C was described. Vitamin A was later found to possess two components-one that prevents night blindness (vitamin A) and another anti-ricket factor named as vitamin D. A fat soluble factor called vitamin E, in the absence of which rats failed to reproduce properly, was discovered. Yet another fat soluble vitamin concerned with coagulation was discovered in mid 1930s. It was named as vitamin K. In the sequence of alphabets it should have been F, but ***K*** was preferred to reflect its function (***k***oagulation).

As regards the water soluble factors, vitamin C was identified as a pure substance and named as ascorbic acid. Vitamin B was found to be a complex mixture and nomenclature also became complex. B_1 was clearly identified as anti-beri-beri factor. Many investigators carried out intensive research between 1920 and 1930 and went on naming them as the water soluble vitamins B_2, B_3, B_4, B_5, B_6, B_7, B_8, B_9, B_{10}, B_{11} and B_{12}. Some of them were found to be mixtures of already known vitamins. And for this reason, a few members (numbers!) of the B-complex series disappeared from the scene. Except for B_1, B_2, B_6 and B_{12}, names are more commonly used for other B-complex vitamins.

Classification of vitamins

There are about 15 vitamins, essential for humans. They are classified as ***fat soluble*** (A, D, E and K) and ***water soluble*** (C and B-group) vitamins as shown in the ***Table 7.1***. The B-complex vitamins may be sub-divided into ***energy-releasing*** (B_1, B_2, B_6, biotin etc.), ***hematopoietic*** (folic acid and B_{12}), ***neurotropic*** (B_1, B_6, B_{12}), and ***sulfur-containing*** (B_1, B_5, B_7). Most of the ***water soluble vitamins exert the functions through their respective coenzymes*** while only one fat soluble vitamin (K) has been identified to function as a coenzyme.

Synthesis of vitamins by intestinal bacteria

Vitamins, as per the definition, are not synthesized in the body. However, the bacteria of the gut can produce some of the vitamins, required by man and animals. The bacteria mainly live and synthesize vitamins in the colon region, where the absorption is relatively poor. ***Some of the animals*** (e.g. rat, deer etc.) ***eat their own feces***, a phenomenon known as ***coprophagy***.

As far as humans are concerned, it is believed that the normal intestinal ***bacterial synthesis***, and absorption of ***vitamin K*** and ***biotin may be sufficient*** to meet the body requirements. For other B-complex vitamins, the synthesis and absorption are relatively less. Administration of anitibiotics often kills the vitamin synthesizing bacteria present in the gut, hence additional consumption of vitamins is recommended.

Table 7.1 Classification of vitamins

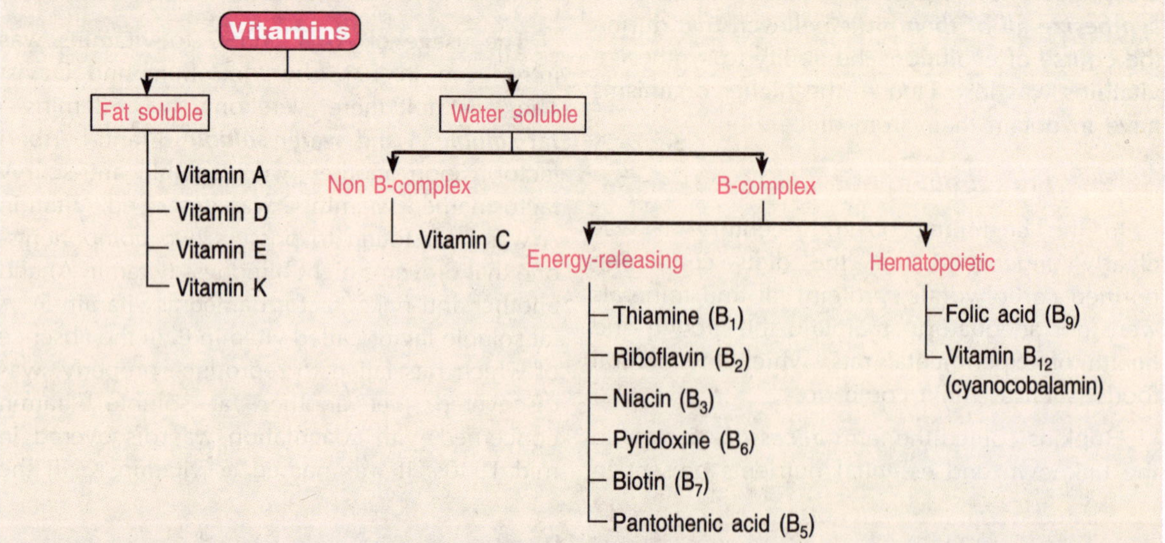

Medical Concepts/ Clinical Correlates

- It is believed that during the course of evolution, the ability to synthesize vitamins was lost by the higher organisms, hence they should be supplied through the diet.
- For humans, the normal intestinal bacterial synthesis of vitamin K and biotin is almost sufficient to meet the body's requirements.
- Administration of antibiotics often destroys the vitamin synthesizing bacteria in the gut, hence additional supplementation of vitamins is recommended during antibiotic therapy.

Fat soluble vitamins—general

The four vitamins, namely vitamin A, D, E, and K are known as fat or ***lipid soluble***. Their availability in the diet, absorption and transport are ***associated with fat***. They are soluble in fats and oils and also the fat solvents (alcohol, acetone etc.). Fat soluble vitamins can be stored in liver and adipose tissue. They are not readily excreted in urine. Excess consumption of these vitamins (particularly A and D) leads to their accumulation and toxic effects.

All the fat soluble vitamins are ***isoprenoid compounds***, since they are made up of one or more of five carbon units namely ***isoprene units*** ($-CH=C.CH_3-CH=CH-$). Fat soluble vitamins perform diverse functions. Vitamin K has a specific coenzyme function.

Water soluble vitamins—general

The water soluble vitamins are a ***heterogenous group*** of compounds since they differ chemically from each other. The only common character shared by them is their solubility in water. Most of these vitamins are readily excreted in urine and they are ***not toxic to the body***. Water soluble vitamins are not stored in the body in large quantities (except B_{12}). For this reason, they must be continuously supplied in the diet. ***Generally, vitamin deficiencies are multiple rather than individual with overlapping symptoms***. It is often difficult to pinpoint the exact biochemical basis for the symptoms.

The dietary deficiency of a single B-complex vitamin is rather rare, since nutritionally poor diets are often with ***multiple nutrient or vitamin deficiencies***.

The water soluble vitamins form ***coenzymes*** (***Refer Table 6.3***) that participate in a variety of biochemical reactions, related to either energy generation or hematopoiesis. It may be due to this reason that the deficiency of vitamins results in a number of overlapping symptoms. The common symptoms of the deficiency of one or more vitamins involved in energy metabolism include dermatitis, glossitis (red and swollen tongue), cheilitis (rupture at the corners of lips), diarrhea, mental confusion, depression and malaise.

Deficiency of vitamins B_1, B_6 and B_{12} is more closely associated with ***neurological*** manifestations.

Vitamers

The term vitamers represents the chemically similar substances that possess ***qualitatively similar vitamin activity***. Some examples of vitamers are given below

- Retinol, retinal and retinoic acid are vitamers of vitamin A.
- Pyridoxine, pyridoxal and pyridoxamine are vitamers of vitamin B_6.

Provitamins are the precursors or inactive forms of vitamins e.g. β-carotene for vitamin A, cholecalciferol for vitamin D.

Recommended dietary allowance (RDA) of vitamins

RDA in general, represents the ***quantities of nutrients to be provided in the diet daily***. From the functional point of view, RDA of a given nutrient is adequate for normal health and metabolic integrity. The RDAs of vitamins are given in this chapter.

(**Note** : Some other terms–***reference daily intake*** or ***recommended daily intake*** (***RDI***), ***daily reference values*** (***DRV***), ***daily value*** (***DV***) ***of foods*** are also used in place of RDA)

INDIVIDUAL VITAMINS

In the following pages, the individual members of the fat soluble and water soluble vitamins are discussed with regard to the chemistry, biochemical functions, recommended dietary/daily allowances (RDA), dietary sources, deficiency manifestations etc.

VITAMIN A

The fat soluble vitamin A, as such is present only in foods of animal origin. However, its ***provitamins, carotenes*** are found ***in plants***.

It is recorded in the history that Hippocrates (about 500 B.C.) cured night blindness. He prescribed to the patients ox liver (in honey), which is now known to contain high quantity of vitamin A.

Chemistry

In the recent years, the term vitamin A is collectively used to represent many structurally related and biologically active molecules (***Fig.7.1***). The term ***retinoids*** is often used to include the natural and synthetic forms of vitamin A. ***Retinol, retinal*** and ***retinoic acid*** are regarded as ***vitamers*** of vitamin A.

1. **Retinol (vitamin A alcohol) :** It is a primary alcohol containing ***β-ionone ring***. The side chain has ***two isoprenoid units***, four double bonds and one hydroxyl group. Retinol is present in animal tissues as retinyl ester with long chain fatty acids.

2. **Retinal (vitamin A aldehyde) :** This is an aldehyde form obtained by the ***oxidation*** of retinol. Retinal and retinol are interconvertible. ***Previously***, the name ***retinine*** was used for retinal.

3. **Retinoic acid (vitamin A acid) :** This is produced by the ***oxidation of retinal***. However, retinoic acid cannot give rise to the formation of retinal or retinol.

4. **β-Carotene (provitamin A) :** This is found in ***plant foods***. It is cleaved in the intestine to produce two moles of retinal. ***In humans, this conversion is inefficient***, hence β-carotene possesses about one-sixth vitamin A activity compared to that of retinol.

Absorption, transport and mobilization

Dietary retinyl esters are hydrolysed by pancreatic or intestinal brush border hydrolases in the intestine, releasing retinol and free fatty acids. Carotenes are hydrolysed by ***β-carotene 15-15′-dioxygenase*** of intestinal cells to release 2 moles of retinal which is reduced to retinol. In the intestinal mucosal cells, retinol is reesterified to long chain fatty acids, incorporated into chylomicrons and transferred to the lymph. The retinol esters of chylomicrons are taken up by the liver and stored (***Fig.7.2***).

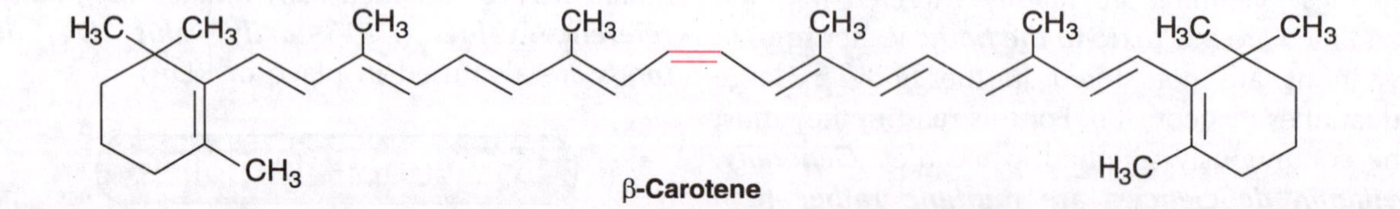

Fig. 7.1 : Structures of vitamin A and related compounds (Red colour represents the substituent groups in the respective compounds).

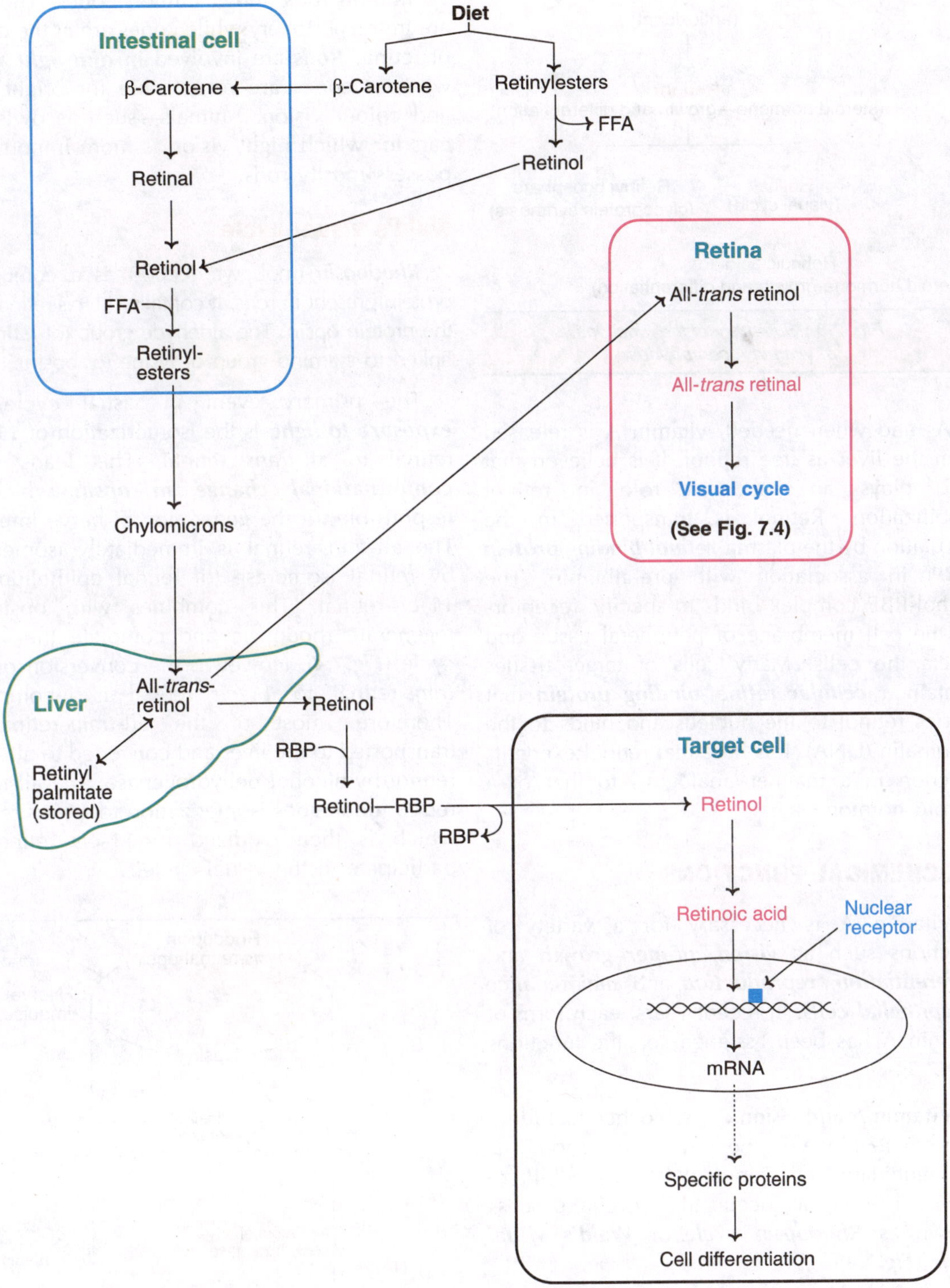

Fig. 7.2 : *Summary of vitamin A absorption, transport and biochemical functions (FFA-Free fatty acid; RBP-Retinol binding protein).*

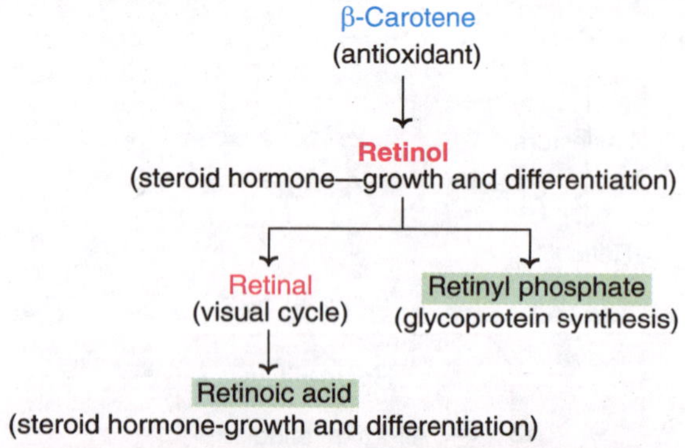

***Fig. 7.3 :** Summary of the functions of vitamin A compounds.*

As and when needed, vitamin A is released from the liver as free retinol. It is believed that zinc plays an important role in retinol mobilization. Retinol is transported in the circulation by the plasma ***retinol binding protein*** (***RBP***) in association with pre-albumin. The retinol-RBP complex binds to specific receptors on the cell membrane of peripheral tissue and enters the cells. Many cells of target tissues contain a ***cellular retinol-binding protein*** that carries retinol to the nucleus and binds to the chromatin (DNA). It is here that retinol exerts its function in a manner analogous to that of a steroid hormone.

BIOCHEMICAL FUNCTIONS

Vitamin A is necessary for a variety of functions such as ***vision***, ***proper growth*** and ***differentiation***, ***reproduction*** and ***maintenance of epithelial cells***. In recent years, each form of vitamin A has been assigned specific functions (***Fig.7.3***).

Vitamin A and vision : The biochemical function of vitamin A in the process of vision was first elucidated by George Wald (Nobel Prize 1968). The events occur in a cyclic process known as ***Rhodopsin cycle*** or ***Wald's visual cycle*** (***Fig.7.4***).

Rods and cones

The retina of the eye possesses two types of cells—rods and cones. The human eye has about 10 million rods and 5 million cones. The rods are in the periphery while cones are at the centre of retina. ***Rods*** are involved in ***dim light vision*** whereas cones are responsible for bright light and colour vision. Animals—such as owls and cats for which night vision is more important—possess mostly rods.

Wald's visual cycle

Rhodopsin (mol. wt. 35,000) is a conjugated protein present in rods. It contains 11-*cis* retinal and the protein opsin. The aldehyde group (of retinal) is linked to ε-amino group of lysine (of opsin).

The primary event in visual cycle, on ***exposure to light***, is the isomerization of 11-*cis*-retinal to all-*trans* retinal. This leads to a ***conformational change in opsin*** which is responsible for the generation of nerve impulse. The all-*trans*-retinal is immediately isomerized by retinal isomerase (of retinal epithelium) to 11-*cis*-retinal. This combines with opsin to regenerate rhodopsin and complete the visual cycle (***Fig.7.4***). However, the conversion of all *trans*-retinal to 11-*cis* retinal is incomplete. Therefore, most of the all-*trans*-retinal is transported to the liver and converted to all-*trans* retinol by alcohol dehydrogenase. The all-*trans*-retinol undergoes isomerization to 11-*cis* retinol which is then oxidized to 11-*cis* retinal to participate in the visual cycle.

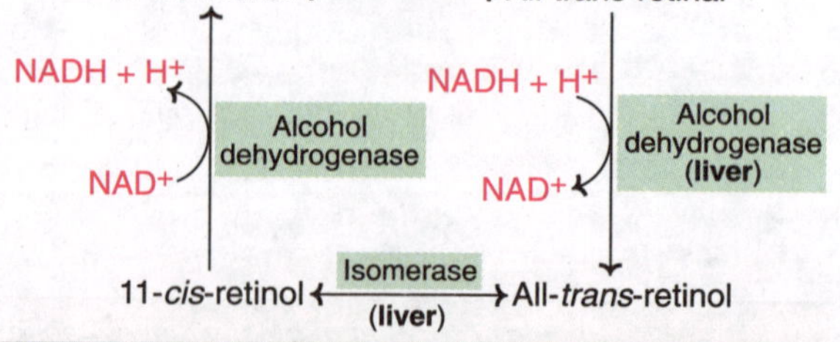

***Fig. 7.4 :** Wald's visual cycle.*

Dark adaptation time : When a person shifts from a bright light to a dim light (e.g. entry into a dim cine theatre), rhodopsin stores are depleted and vision is impaired. This results in a ***temporary blindness***. However, within a few minutes, known as dark adaptation time, rhodopsin is resynthesized and vision is improved. Dark adaptation time is ***increased in vitamin A deficient individuals***.

Bleaching of rhodopsin : When exposed to light, the colour of rhodopsin changes from red to yellow, by a process known as bleaching. Bleaching occurs in a few milliseconds and many unstable intermediates are formed during this process.

Rhodopsin ⟶ Prelumirhodopsin ⟶ Lumirhodopsin
↓
All-trans-retinal + Opsin ⟵ Metarhodopsin II ⟵ Metarhodopsin I

Visual cascade and cGMP : When light strikes the retina, a number of biochemical changes leading to membrane hyperpolarization occur resulting in the genesis of nerve impulse. The ***hyperpolarization*** of the membrane is brought about by a visual cascade involving cyclic GMP. When a photon (from light) is absorbed by rhodopsin, metarhodopsin II is produced. The protein ***transducin*** is activated by metarhodopsin II. This involves an exchange of GTP for GDP on inactive transducin. The activated transducin activates cyclic GMP phosphodiesterase. This enzyme degrades cyclic GMP in the rod cells (***Fig.7.5***). A rapid decrease in cyclic GMP closes the Na^+ channels in the membranes of the rod cells. This results in hyperpolarization which is an excitatory response transmitted through the neuron network to the visual cortex of the brain.

Colour vision

Cones are specialized in bright and colour vision. Visual cycle comparable to that present in rods is also seen in cones. The colour vision is governed by colour sensitive pigments–***porphyropsin (red), iodopsin (green) and cyanopsin (blue)***. All these pigments are retinal-opsin complexes. When bright light strikes the retina, one or more of these pigments are bleached, depending on the particular colour of light. The pigments dissociate to all-*trans*-retinal and opsin, as in the case of rhodopsin. And this reaction passes on a nerve impulse to brain as a specific colour—red when porphyropsin splits, green when iodopsin splits or blue for cyanopsin. Splitting of these three pigments in different proportions results in the perception of different colours by the brain.

Other biochemical functions of vitamin A

1. ***Retinoic acid*** functions almost like ***steroid hormones***. It regulates gene expression, and thus is involved in the cell growth and differentiation. Retinoic acid binds to nuclear receptors that in turn binds to DNA and regulates transcription of specific genes (***Fig.7.2***).

2. Vitamin A is essential to ***maintain healthy epithelial tissue***. This is due to the fact that retinol and retinoic acid are required to prevent keratin synthesis (responsible for horny surface). ***Retinol creams*** are used to treat ***acne***, prevent wrinkles and signs of ***aging***.

3. Retinyl phosphate synthesized from retinol is necessary for the ***synthesis of*** certain ***glycoproteins***, and mucopolysaccharides which are required for ***growth and mucus secretion***.

4. Retinol is necessary for normal ***reproduction***. It acts like a hormone and regulates gene expression.

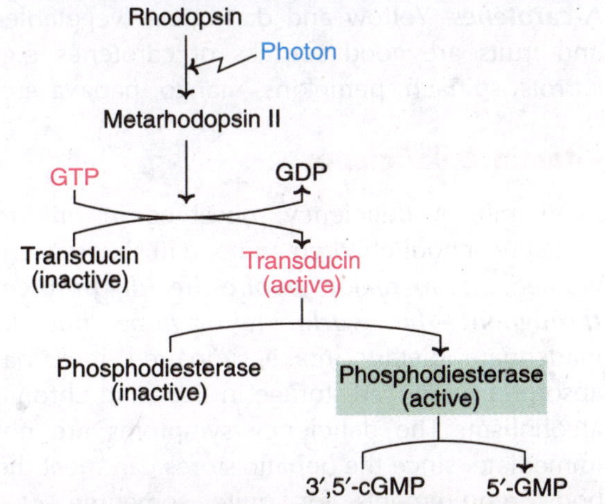

Fig. 7.5 : *The visual cascade involving cyclic guanosine monophosphate (3′,5′-cGMP).*

5. Vitamin A is considered to be essential for the maintenance of ***proper immune system*** to fight against various infections.

6. Cholesterol synthesis requires vitamin A. Mevalonate, an intermediate in the cholesterol biosynthesis, is diverted for the synthesis of coenzyme Q in vitamin A deficiency. It is pertinent to note that the discovery of coenzyme Q was made in vitamin A deficient animals.

7. ***Carotenoids*** (most important β-carotene) function as ***antioxidants*** and reduce the risk of cancers initiated by free radicals and strong oxidants. β-Carotene is found to be beneficial to prevent heart attacks. This is also attributed to the antioxidant property.

Recommended dietary allowance (RDA)

The requirement of vitamin A for adults in the form of retinol (or retinol equivalents) is around 600 μg/day (or 4800 μg of β-carotene). It may be noted that one International Unit (IU) equals to 0.3 mg of retinol. The requirement increases in pregnant women, and lactating mothers.

Dietary sources

Animal sources contain (preformed) vitamin A. The best sources are ***liver***, ***kidney***, egg yolk, milk, cheese, butter. ***Fish*** (cod or shark) liver ***oils*** are very rich in vitamin A.

Vegetable sources contain the provitamin A-***carotenes***. Yellow and dark green vegetables and fruits are good sources of carotenes e.g. carrots, spinach, pumpkins, mango, papaya etc.

Vitamin A deficiency

Vitamin A deficiency, observed in one in three pre-school children and one in six pregnant women, is a ***major public health problem throughout the world***. It may be due to inadequate dietary intake, impaired intestinal absorption, reduced storage in liver and chronic alcoholism. The deficiency symptoms are not immediate, since the hepatic stores can meet the body requirements for quite sometime (2-4 months). The deficiency manifestations are related to the eyes, skin and growth.

Deficiency manifestations of the eyes : ***Night blindness*** (***nyctalopia***) is one of the earliest symptoms of vitamin A deficiency. The individuals have difficulty to see in dim light since the dark adaptation time is increased.

Severe deficiency of vitamin A leads to ***xerophthalmia***. This is characterized by dryness in conjunctiva and cornea, and keratinization of epithelial cells. In certain areas of conjunctiva, white triangular plaques known as ***Bitot's spots*** are seen. Nyctalopia and Bitot's spots are reversible with vitamin A supplementation.

If xerophthalmia persists for a long time, corneal ulceration and degeneration occur. This results in the destruction of cornea, a condition referred to as ***keratomalacia***, causing total blindness. Therefore, adequate intake of vitamin A is necessary for the prevention of blindness.

Other deficiency manifestations

Effect on growth : Vitamin A deficiency results in growth retardation due to impairment in skeletal formation.

Effect on reproduction : The reproductive system is adversely affected in vitamin A deficiency. Degeneration of germinal epithelium leads to sterility in males.

Effect on skin and epithelial cells : The skin becomes rough and dry. ***Keratinization*** of epithelial cells of gastrointestinal tract, urinary tract and respiratory tract is noticed. This leads to increased bacterial infection. Vitamin A deficiency is associated with formation of ***urinary stones***.

The plasma level of ***retinol binding protein is decreased*** in vitamin A deficiency.

Hypervitaminosis A

Excessive consumption of vitamin A leads to toxicity. The symptoms of hypervitaminosis A include dermatitis raised intracranial tension, enlargement of liver, skeletal decalcification, tenderness of long bones, loss of weight, irritability, loss of hair, joint pains etc. ***Elderly people*** are ***more susceptible*** to vitamin A toxicity, hence overdoses should be avoided.

Total serum vitamin A level (normal 20–50 μg/dl) is elevated in hypervitaminosis A. Free retinol or retinol bound to plasma lipoproteins is actually harmful to the body. It is now believed that the vitamin A toxicosis symptoms appear only after retinol binding capacity of retinol binding protein exceeds.

Higher concentration of retinol increases the synthesis of lysosomal hydrolases. The manifestations of ***hypervitaminosis A*** are attributed to the ***destructive action of hydrolases***, particularly on the cell membranes.

BENEFICIAL EFFECTS OF β-CAROTENE

Increased consumption of β-carotene is associated with ***decreased*** incidence of ***heart attacks***, skin and lung cancers. This is attributed to the ***antioxidant role of β-carotene*** which is independent of its role as a precursor of vitamin A. Ingestion of high doses of β-carotene for long periods are not toxic like vitamin A.

VITAMIN D

Vitamin D is a fat soluble vitamin (however, now considered as a hormone). It resembles ***sterols in structure*** and functions like a ***hormone***.

Chemistry

Ergocalciferol (vitamin D_2) is formed from ergosterol and is present in plants (***Fig.7.6***). Cholecalciferol (vitamin D_3) is found in animals. Both the sterols are similar in structure except that ergocalciferol has an additional methyl group and a double bond. ***Ergocalciferol*** and ***cholecalciferol*** are sources for vitamin D activity and are referred to as ***provitamins***.

During the course of cholesterol biosynthesis (***Chapter 14***), ***7-dehydrocholesterol*** is formed as an intermediate. On exposure to sunlight, 7-dehydrocholesterol is converted to cholecalciferol in the skin (***Fig.7.7***). Vitamin D is regarded as ***sun-shine vitamin***.

The synthesis of vitamin D_3 in the skin is proportional to the exposure to sunlight. Dark skin pigment (melanin) adversly influences the synthesis of cholecalciferol. (**Note :** The term vitamin D_1 is no more in use. It was originally used for fat soluble crystalline material, which later turned out to be a mixture).

Absorption, transport and storage

Vitamin D is absorbed in the small intestine for which bile is essential. Through lymph, vitamin D enters the circulation bound to plasma α_2-globulin and is distributed throughout the body. Liver and other tissues store small amounts of vitamin D.

METABOLISM AND BIOCHEMICAL FUNCTIONS

Vitamins D_2 and D_3, as such, are not biologically active. They are metabolized identically in the body and converted to active forms. The metabolism and biochemical functions of vitamin D are depicted in ***Fig.7.8***.

Synthesis of 1,25-DHCC

Cholecalciferol is first hydroxylated at 25th position to ***25-hydroxycholecalciferol (25-OHD₃)*** by a specific hydroxylase present in liver. 25-OH D_3 is the major storage and circulatory form of vitamin D. Kidney possesses a specific enzyme, ***25-hydroxycholecalciferol (calcidiol) 1-hydroxylase*** which hydroxylates 25-hydroxy-cholecalciferol at position 1 to produce ***1,25-dihydroxycholecalciferol (1,25-DHCC)***. 1,25 DHCC contains 3 hydroxyl groups (1,3 and 25 carbon) hence referred to as

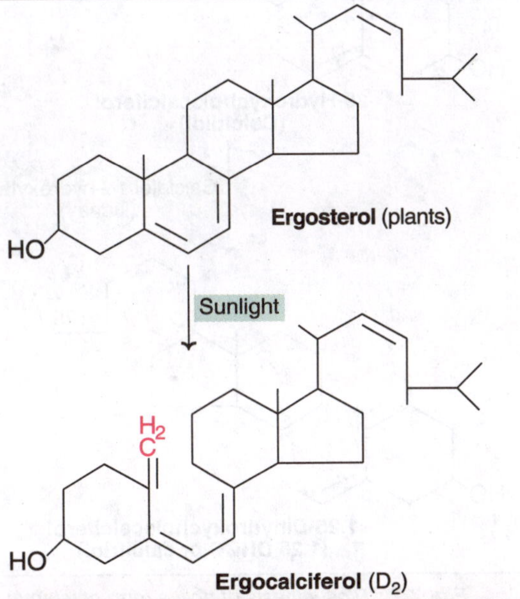

Fig. 7.6 : Formation of ergocalciferol from ergosterol.

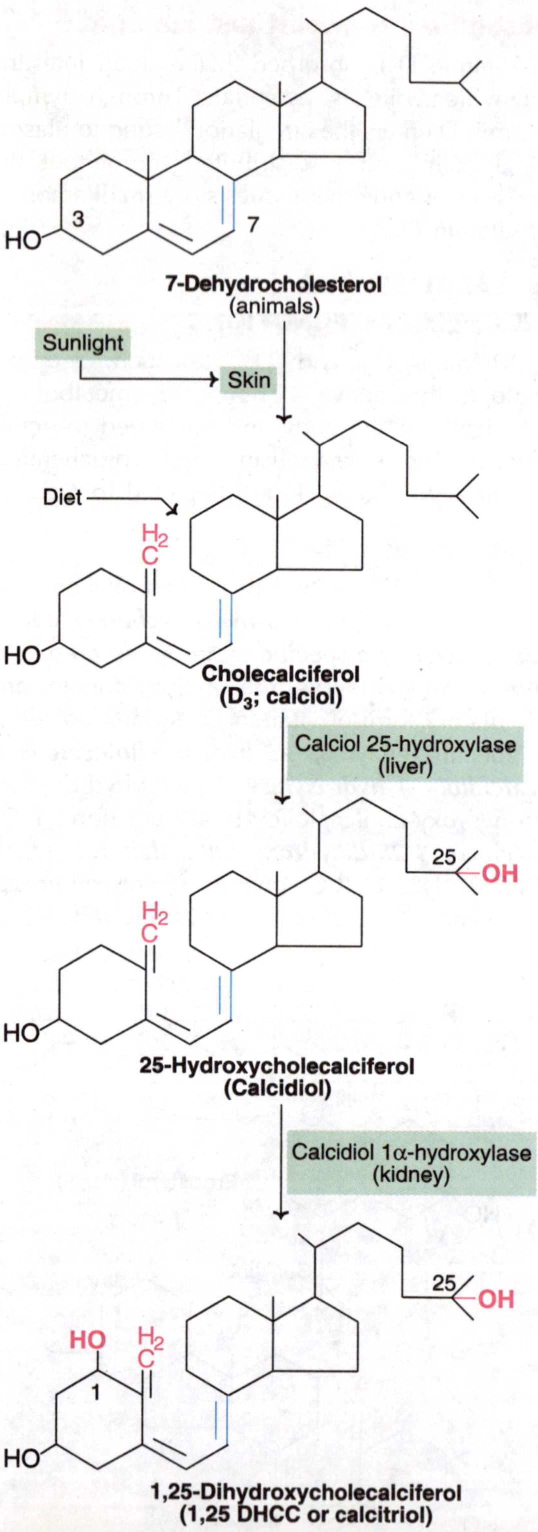

***Fig. 7.7 :** Biosynthesis of active form of vitamin D-calcitriol (1,25 DHCC).*

calcitriol. Both the hydroxylase enzymes (of liver and kidney) require cytochrome P_{450}, NADPH and O_2 for the hydroxylation process.

Regulation of the synthesis of 1,25-DHCC

The concentration of 1,25-DHCC is regulated by plasma levels of ***calcium*** and ***phosphate.*** They control hydroxylation reaction at position 1. Low plasma phosphate increases the activity of 25-hydroxycholecalciferol 1-hydroxylase. Low plasma calcium enhances the production of ***parathyroid hormone*** which in turn activates 1-hydroxylase. Thus the action of phosphate is direct while that of calcium is indirect on kidney 1-hydroxylase.

Biochemical functions

Calcitriol (***1,25-DHCC***) is the biologically ***active*** form of vitamin D, with a main function to maintain calcium homeostasis. It ***regulates the plasma levels of calcium and phosphate.*** Calcitriol acts at 3 different levels (intestine, kidney and bone) to maintain plasma calcium (normal 9–11 mg/dl).

1. **Action of calcitriol on the intestine :** Calcitriol ***increases the intestinal absorption of calcium*** and phosphate. In the intestinal cells, calcitriol binds with a cytosolic receptor to form a calcitriol-receptor complex. This complex then approaches the nucleus and interacts with a specific DNA leading to the synthesis of a specific ***calcium binding protein*** (***calbindin***). This protein increases the calcium uptake by the intestine. The mechanism of action of calcitriol on the target tissue (intestine) is similar to the action of a ***steroid hormone.***

2. **Action of calcitriol on the bone :** In the osteoblasts of bone, calcitriol stimulates calcium uptake for deposition as calcium phosphate. Thus calcitriol is essential for ***bone formation.*** The bone is an important reservoir of calcium and phosphate. Calcitriol along with parathyroid hormone increases the mobilization of calcium and phosphate from the bone. This causes elevation in the plasma calcium and phosphate levels.

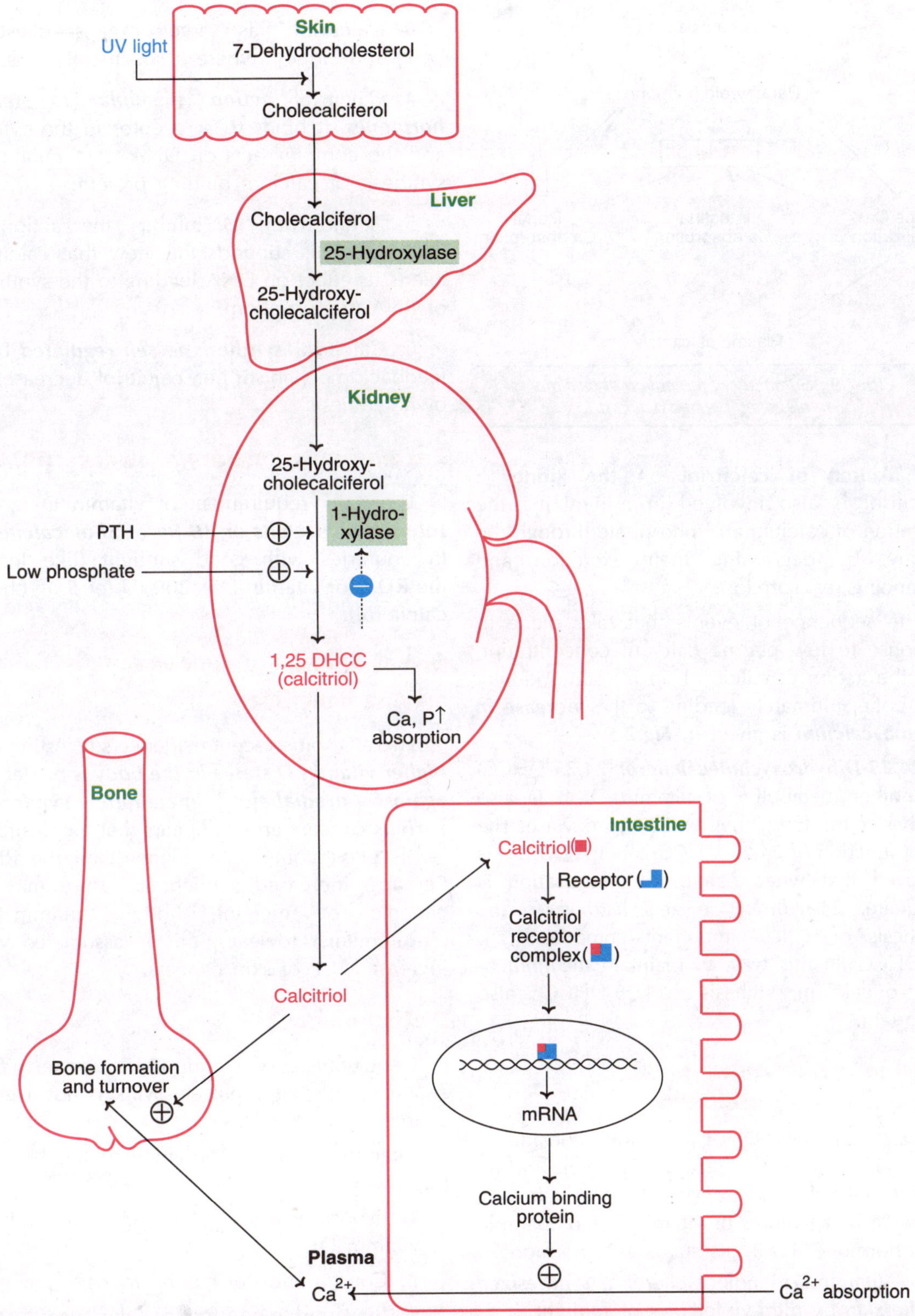

***Fig. 7.8 :** Metabolism and biochemical functions of vitamin D (1, 25 DHCC-1, 25-Dihydroxycholecalciferol, also called as calcitriol is the active form of vitamin D; PTH-Parathyroid hormone).*

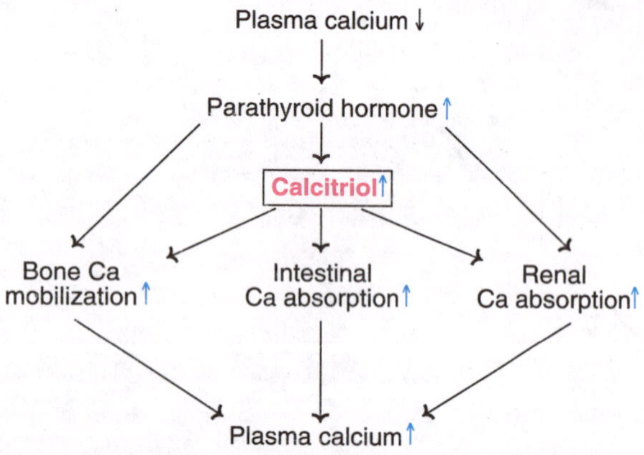

Fig. 7.9 : *Summary of the action of calcitriol in elevating plasma calcium.*

3. **Action of calcitriol on the kidney :** Calcitriol is also involved in minimizing the excretion of calcium and phosphate through the kidney, by decreasing their excretion and enhancing reabsorption.

The sequence of events that take place in response to low plasma calcium concentration and the action of calcitriol on intestine, kidney and bone, ultimately leading to the ***increase in plasma calcium*** is given in ***Fig.7.9***.

24,25-Dihydroxycholecalciferol (24,25-DHCC) is another metabolite of vitamin D. It is also synthesized in the kidney by 24-hydroxylase. The exact function of 24,25-DHCC is not known. It is believed that when calcitriol concentration is adequate, 24-hydroxylase acts leading to the synthesis of a less important compound 24, 25-DHCC. In this way, to maintain the homeostasis of calcium, synthesis of 24,25-DHCC is also important.

Vitamin D is a hormone and not a vitamin—justification

Calcitriol (1,25-DHCC) is now considered as an important ***calciotropic hormone***, while cholecalciferol is the prohormone. The following characteristic features of vitamin D (comparable with hormone) justify its status as a hormone.

1. Vitamin D_3 (cholecalciferol) is ***synthesized in the skin*** by ultra-violet rays of sunlight.

2. The biologically active form of vitamin D, calcitriol is produced in the kidney.

3. Calcitriol has target organs—intestine, bone and kidney, where it specifically acts.

4. Calcitriol ***action*** is ***similar to steroid hormones***. It binds to a receptor in the cytosol and the complex acts on DNA to stimulate the synthesis of calcium binding protein.

5. Actinomycin D inhibits the action of calcitriol. This supports the view that calcitriol exerts its effect on DNA leading to the synthesis of RNA (transcription).

6. Calcitriol ***synthesis*** is ***self-regulated*** by a feedback mechanism i.e., calcitriol decreases its own synthesis.

Recommended dietary allowance (RDA)

The daily requirement of vitamin D is ***400 International Units*** or ***10 μg of cholecalciferol***. In countries with good sunlight (like India), the RDA for vitamin D is 200 IU (or 5 μg cholecalciferol).

Higher consumption of vitamin D may be beneficial

There is some recent evidence suggesting that ***higher vitamin D status in the body is protective against prediabetes, metabolic syndrome, various cancers etc.*** It appears that the desirable levels of vitamin D are higher than the RDA. Perhaps, increased sunlight exposure may be capable of meeting body's vitamin D requirement. However, this is associated with increased risk of skin cancers.

Dietary sources

Good sources of vitamin D include ***fatty fish, fish liver oils***, egg yolk etc. Milk is not a good source of vitamin D.

Vitamin D can be provided to the body in three ways

1. Exposure of skin to sunlight for synthesis of vitamin D;

2. Consumption of natural foods;

3. By irradiating foods (like yeast) that contain precursors of vitamin D and fortification of foods (milk, butter etc.).

Deficiency symptoms

Vitamin D deficiency is relatively less common, since this vitamin can be synthesized in the body. However, ***insufficient exposure to sunlight*** and consumption of diet lacking vitamin D ***results*** in its ***deficiency***.

Vitamin D deficiency occurs in strict vegetarians, chronic alcoholics, individuals with liver and kidney diseases or fat malabsorption syndromes. In some people, who ***cover the entire body*** (***burqa***) for religious customs, ***vitamin D deficiency*** is also ***observed***, if the requirement is not met through diet.

Deficiency of vitamin D causes ***rickets in children and osteomalacia in adults***. Rickets is derived from an old *English* word 'wrickken', meaning to twist. Osteomalacia is derived from Greek (osteon-bone; malakia-softness). Vitamin D is often called as antirachitic vitamin.

Rickets in children is characterized by bone deformities due to incomplete mineralization, resulting in soft and pliable bones and delay in teeth formation. The weight-bearing bones are bent to form ***bow-legs***. In rickets, the plasma level of calcitriol is decreased and ***alkaline phosphatase activity is elevated***. Alkaline phosphatase is concerned with the process of bone formation. There is an overproduction of alkaline phosphatase related to more cellular activity of the bone. It is believed to be due to a vain attempt to result in bone formation.

In case of ***osteomalacia (adult rickets)*** demineralization of the bones occurs (bones become softer), increasing their susceptibility to fractures.

Diagnosis of vitamin D deficiency : Estimation of plasma levels of 25-hydroxycholecalciferol (Reference range 30–40 ng/ml) is normally employed to evaluate vitamin D deficiency.

Renal rickets (renal osteodystrophy)

This is seen in patients with chronic renal failure. Renal rickets is mainly due to ***decreased synthesis of calcitriol*** in kidney. It can be treated by administration of calcitriol.

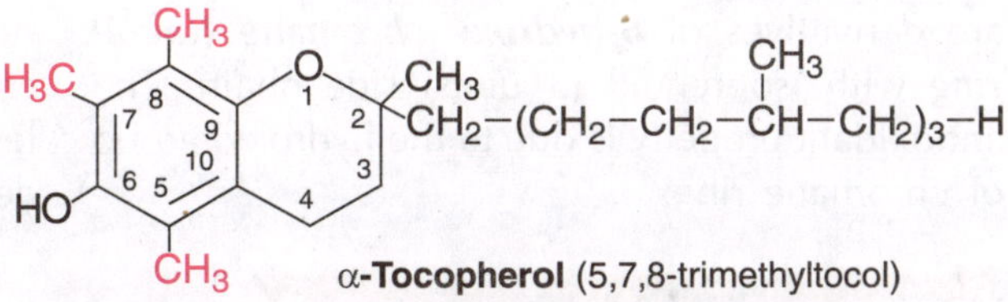

α-Tocopherol (5,7,8-trimethyltocol)
β-Tocopherol (5,8-dimethyltocol)
γ-Tocopherol (7,8-dimethyltocol)

Fig. 7.10 : *Structure of α-tocopherol (**Note :** The tocopherols differ in the substitution of methyl groups, represented in red).*

Hypervitaminosis D

Vitamin D is stored mostly in liver and slowly metabolised. Among the vitamins, vitamin D is the ***most toxic in overdoses*** (10-100 times RDA). Toxic effects of hypervitaminosis D include demineralization of bone (resorption) and increased calcium absorption from the intestine, leading to elevated calcium in plasma (hypercalcemia). Prolonged hypercalcemia is associated with deposition of calcium in many soft tissues such as kidney and arteries.

Although very high dietary consumption of vitamin D is toxic, excessive exposure to sunlight does not result in vitamin D poisoning. This is due to the fact that there is a limited capacity to form vitamin in the skin.

VITAMIN E

Vitamin E (***tocopherol***) is a naturally occurring antioxidant. It is essential for normal reproduction in many animals, hence known as ***anti-sterility vitamin***. Vitamin E is described as a '***vitamin in search of a disease***.' This is due to the lack of any specific vitamin E deficiency disease in humans.

Evans and his associates (1936) isolated the compounds of vitamin E activity and named them as tocopherols (*Greek* : tokos-child birth; pheros-to bear; ol-alcohol).

Chemistry

Vitamin E is the name given to a group of tocopherols and tocotrienols. About eight ***tocopherols (vitamin E vitamers)*** have been identified–***α, β, γ, δ*** etc. Among these, ***α-tocopherol*** is the most active. The tocopherols are derivatives of ***6-hydroxy chromane (tocol) ring*** with isoprenoid (3 units) side chain. The antioxidant property is due to the hydroxyl group of chromane ring.

Absorption, transport and storage

Vitamin E is absorbed along with fat in the small intestine. Bile salts are necessary for the absorption. In the liver, it is incorporated into lipoproteins (VLDL and LDL) and transported. Vitamin E is stored in adipose tissue, liver and muscle. The normal plasma level of tocopherol is less than 1 mg/dl.

Biochemical functions

Most of the functions of vitamin E are related to its ***antioxidant*** property. It prevents the non-enzymatic oxidations of various cell components (e.g. unsaturated fatty acids) by molecular oxygen and free radicals such as superoxide (O_2^-) and hydrogen peroxide (H_2O_2). The element ***selenium helps*** in these functions.

Vitamin E is lipophilic in character and is found in association with lipoproteins, fat deposits and cellular membranes. ***It protects the polyunsaturated fatty acids (PUFA) from peroxidation reactions***. Vitamin E acts as a scavenger and gets itself oxidized (to quinone form) by free radicals (R) and spares PUFA, as shown below

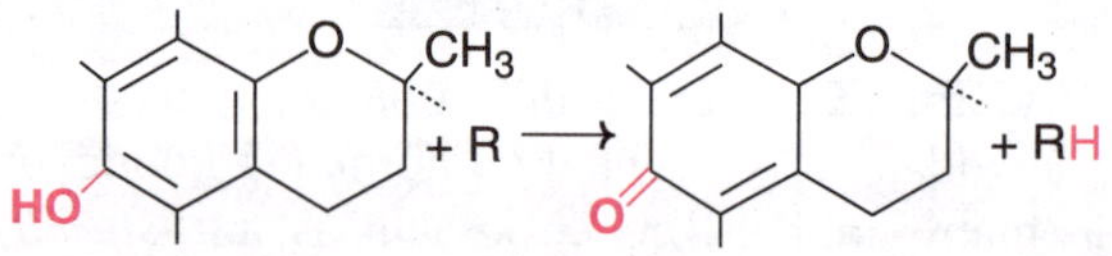

The biochemical functions of vitamin E, related either directly or indirectly to its antioxidant property, are given hereunder

1. Vitamin E is essential for the ***membrane structure and integrity*** of the cell, hence it is regarded as a ***membrane antioxidant***.

2. It prevents the peroxidation of poly-unsaturated fatty acids in various tissues and membranes. It ***protects RBC from hemolysis*** by oxidizing agents (e.g. H_2O_2).

3. It is closely associated with ***reproductive functions*** and prevents sterility. Vitamin E preserves and maintains germinal epithelium of gonads for proper reproductive function.

4. It ***increases the synthesis of heme*** by enhancing the activity of enzymes δ-amino-levulinic acid (ALA) synthase and ALA dehydratase.

5. It is required for ***cellular respiration***–through electron transport chain (believed to stabilize coenzyme Q).

6. Vitamin E prevents the oxidation of vitamin A and carotenes.

7. It is required for proper storage of creatine in skeletal muscle.

8. Vitamin E is needed for optimal absorption of amino acids from the intestine.

9. It is involved in proper synthesis of nucleic acids.

10. Vitamin E ***protects liver*** from being damaged by toxic compounds such as carbon tetrachloride.

11. It works in association with vitamins A, C, and β-carotene, to delay the onset of cataract.

12. Vitamin E has been recommended for the ***prevention of chronic diseases*** such as ***cancer and heart diseases***. Clinical trials in this regard are rather disappointing, hence it is no more recommended. However, some clinicians continue to use it particularly in subjects susceptible to heart attacks. It is believed that vitamin E prevents the oxidation of LDL. (**Note :** The oxidized LDL have been implicated to promote heart diseases.)

Vitamin E and selenium

The element selenium is found in the enzyme ***glutathione peroxidase*** that destroys free radicals. Thus, Se is also involved in antioxidant functions like vitamin E, and both of them ***act synergistically***. To a certain extent, Se can spare the dietary requirement of vitamin E, and vice versa.

Recommended dietary allowance (RDA)

Intake of vitamin E is directly ***related to the consumption of polyunsaturated fatty acids*** (PUFA) i.e., requirement increases with increased intake of PUFA. A daily consumption of about

Medical Concepts/ Clinical Correlates

- Vitamin A deficiency causes night blindness; vitamin D deficiency rickets (in children) or osteomalacia (in adults); vitamin E deficiency minor neurological symptoms; vitamin K deficiency bleeding.
- Fat soluble vitamins are not readily excreted in urine, hence excess consumption leads to their accumulation and toxic effects.

10 mg (15 IU) of α-tocopherol for man and ***8 mg*** (12 IU) for woman is recommended. One mg of α-tocopherol is equal to 1.5 IU. Vitamin E supplemented diet is advised for pregnant and lactating women.

Dietary sources

Many vegetable oils are rich sources of vitamin E. ***Wheat germ oil***, cotton seed oil, peanut oil, corn oil and sunflower oil are the good sources of this vitamin. It is also present in meat, milk, butter and eggs.

Deficiency symptoms

The symptoms of vitamin E deficiency vary from one animal species to another. In many ***animals***, the deficiency is associated with ***sterility***, degenerative changes in muscle, megaloblastic anaemia and changes in central nervous system. Severe symptoms of vitamin E deficiency are not seen in humans except increased fragility of erythrocytes and minor neurological symptoms.

Toxicity of vitamin E

Among the fat soluble vitamins (A, D, E, K), vitamin E is the ***least toxic***.

VITAMIN K

Vitamin K is the only fat soluble vitamin with a specific coenzyme function. It is required for the production of blood clotting factors, essential for coagulation (in *German–**K**oagulation*; hence the name K for this vitamin).

Chemistry

Vitamin K exists in different forms (***Fig.7.11***). Vitamin K_1 (phylloquinone) is present in plants. Vitamin K_2 (menaquinone) is produced by the intestinal bacteria and also found in animals. Vitamin K_3 (menadione) is a synthetic form.

All the three vitamins (K_1, K_2, K_3) are ***naphthoquinone derivatives***. Isoprenoid side chain is present in vitamins K_1 and K_2. The three vitamins are stable to heat. Their activity is, however, lost by oxidizing agents, irradiation, strong acids and alkalies.

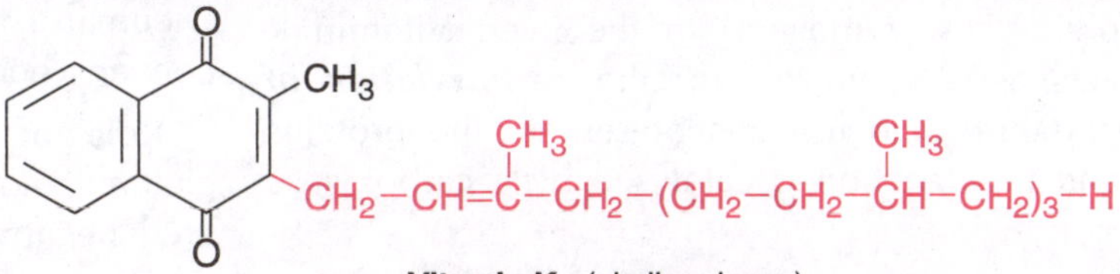

Vitamin K_1 (phylloquinone)

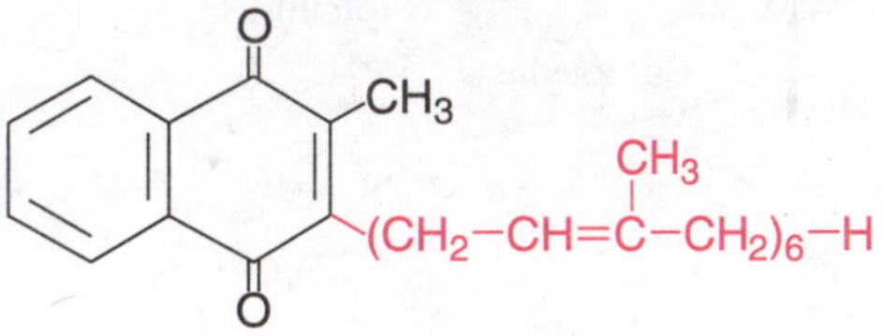

Vitamin K_2 (menaquinone)

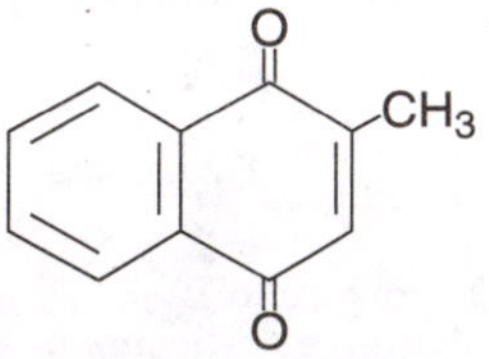

Vitamin K_3 (menadione)

Fig. 7.11 : *Structures of vitamin K.*

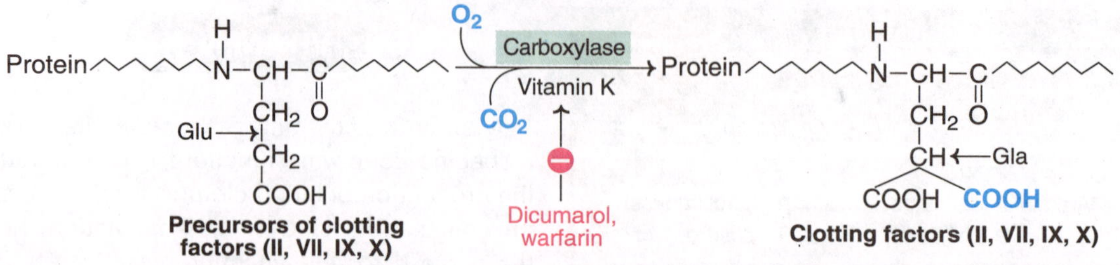

Fig. 7.12 : Vitamin K dependent carboxylation of the precursors of clotting factors.

Absorption, transport and storage

Vitamin K is taken in the diet or synthesized by the intestinal bacteria. Its absorption takes place along with fat and is dependent on bile salts. Vitamin K is transported along with LDL and is stored mainly in liver.

Biochemical functions

The functions of vitamin K are concerned with blood clotting process. It brings about the ***post-translational*** (after protein biosynthesis in the cell) ***modification of certain blood clotting factors***. The clotting factors ***II*** (***prothrombin***), ***VII***, ***IX*** and ***X*** are synthesized as inactive precursors (zymogens) in the liver. Vitamin K acts as a coenzyme for the ***carboxylation of glutamic acid*** residues present in the proteins and this reaction is catalysed by a carboxylase (microsomal). It involves the conversion of glutamate (Glu) to γ-carboxyglutamate (Gla) and requires vitamin K, O_2 and CO_2 (***Fig.7.12***). The formation of ***γ-carboxyglutamate*** is inhibited by ***dicumarol***, an anticoagulant found in spoilt sweet clover. ***Warfarin*** is a synthetic analogue that can inhibit vitamin K action (***Fig.7.13***).

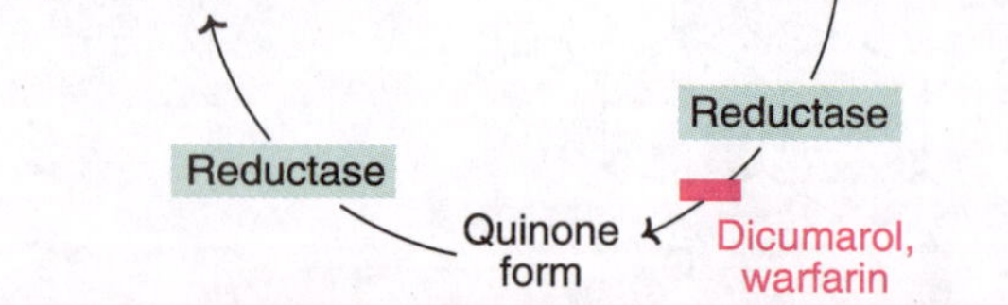

Fig. 7.13 : Summary of vitamin K cycle in carboxylation reaction.

Vitamin K is also required for the carboxylation of glutamic acid residues of ***osteocalcin***, and the ***matrix Gla protein in bone***, and ***nephrocalcin*** in kidney.

The mechanism of carboxylation is not fully understood. It is known that a 2,3-epoxide derivative of vitamin K is formed as an intermediate during the course of the reaction. Dicumarol inhibits the enzyme (reductase) that converts epoxide to active vitamin K.

Role of Gla in clotting : The γ-carboxyglutamic acid (Gla) residues of clotting factors are negatively charged (COO^-) and they combine with positively charged calcium ions (Ca^{2+}) to form a complex. The mechanism of action has been studied for prothrombin. The prothrombin –Ca complex binds to the phospholipids on the membrane surface of the platelets (***Fig.7.14***). This leads to the increased conversion of prothrombin to thrombin.

Recommended dietary allowance (RDA)

Strictly speaking, there is no RDA for vitamin K, since it can be ***adequately synthesized in the gut***. It is however, recommended that half of the body requirement is provided in the diet, while the other half is met from the bacterial synthesis. Accordingly, the ***suggested RDA*** for an adult is ***70-100 μg/day***.

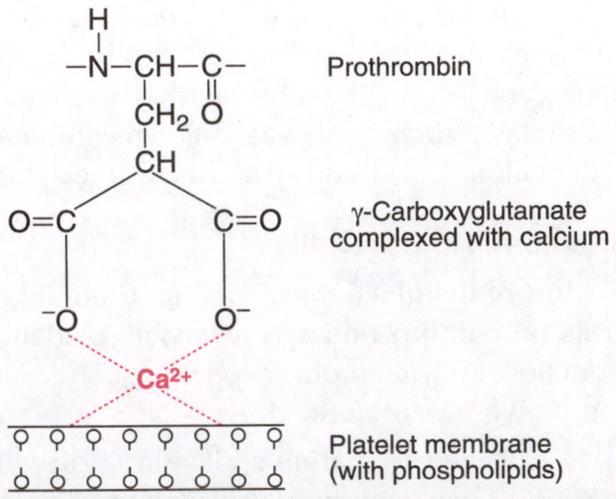

Fig. 7.14 : Mechanism of action of γ-carboxyglutamate of prothrombin in blood clotting.

Dietary sources

Cabbage, ***cauliflower***, tomatoes, alfa alfa, spinach and other green vegetables are good sources. It is also present in egg yolk, meat, liver, cheese and dairy products.

Deficiency symptoms

The deficiency of vitamin K is ***uncommon***, since it is present in the diet in sufficient quantity and/or is adequately synthesized by the intestinal bacteria. However, vitamin K deficiency may occur due to its faulty absorption (lack of bile salts), loss of vitamin into feces (diarrheal diseases) and administration of antibiotics (killing of intestinal flora).

Deficiency of vitamin K leads to the lack of active prothrombin in the circulation. The result is that blood coagulation is adversely affected. The individual bleeds profusely even for minor injuries. ***The blood clotting time is increased***.

Hypervitaminosis K

Administration of large doses of vitamin K produces hemolytic anaemia and jaundice, particularly in infants. The toxic effect is due to increased breakdown of RBC.

Antagonists of vitamin K

The compounds—namely ***heparin***, ***bishydroxy-coumarin***—act as anticoagulants and are antagonists to vitamin K. The salicylates and dicumarol are also antagonists to vitamin K. Dicumarol is structurally related to vitamin K and acts as a ***competitive inhibitor*** in the synthesis of active prothrombin.

VITAMIN C (ASCORBIC ACID)

Vitamin C is a water soluble ***versatile vitamin***. It plays an important role in human health and disease.

Scurvy has been known to man for centuries. It was the first disease found to be associated with diet. In the sixteenth century, about 10,000 mariners died of a miraculous disease (scurvy) due to lack of fresh vegetables in their diet. James Lind, a surgeon of the English Navy, in 1753 published 'Treatise on Scurvy'. Based on Lind's observations, the Royal Navy since 1795 used to supply lime or lemon juice to all the crews. The English Navy used to carry crates of lemons, hence they were popularly known as ***Limeys***.

Chemistry

Ascorbic acid is a hexose (6 carbon) derivative and closely ***resembles mono-saccharides*** in structure (***Fig.7.15***). The acidic property of vitamin C is due to the enolic hydroxyl groups. It is a strong reducing agent. L-Ascorbic acid undergoes oxidation

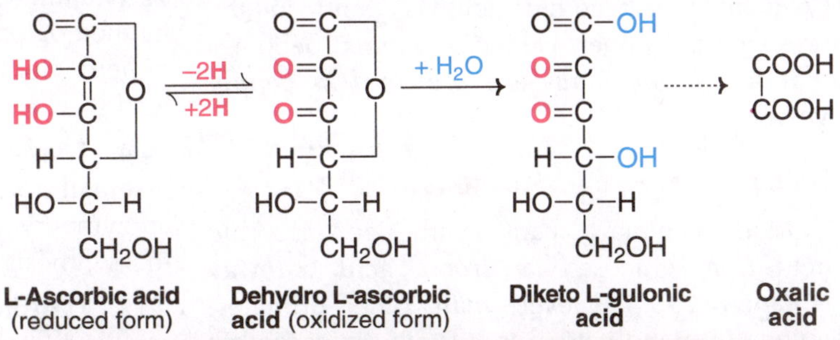

Fig. 7.15 : Structures of vitamin C (ascorbic acid) and its related compounds.

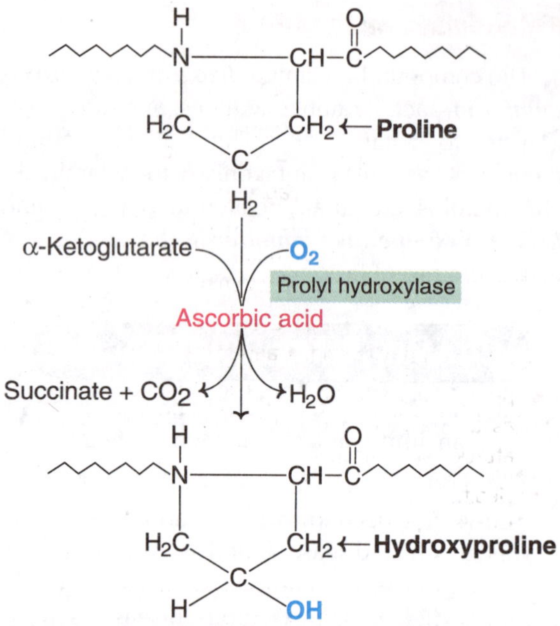

Fig. 7.16 : *Ascorbic acid dependent hydroxylation of proline of protocollagen.*

to form dehydroascorbic acid and this reaction is reversible. Both ***ascorbic acid and dehydroascorbic acid are biologically active***. However, D-ascorbic acid is inactive. The plasma and tissues predominantly contain ascorbic acid in the reduced form. The ratio of ***ascorbic acid to dehydroascorbic acid*** in many tissues is ***15 : 1***. On hydration, dehydroascorbic acid is irreversibly converted to 2,3-diketogulonic acid which is inactive. Hydration reaction is almost spontaneous, in alkaline or neutral solution. It is for this reason that oxidation of vitamin C is regarded as biological inactivation (formation of diketogulonic acid). Oxidation of ascorbic acid is rapid in the presence of copper. Hence ***vitamin C*** becomes ***inactive, if the foods are prepared in copper vessels***.

Biosynthesis and metabolism

Many animals can synthesize ascorbic acid from glucose via ***uronic acid pathway*** (***Chapter 13***). However, ***man,*** other primates, guinea pigs and bats cannot synthesize ascorbic acid due to the ***deficiency of*** a single enzyme namely ***L-gulonolactone oxidase***.

Vitamin C is rapidly absorbed from the intestine. It is not stored in the body to a significant extent. Ascorbic acid is excreted in urine as such, or as its metabolites—diketogulonic acid and oxalic acid (***Fig.7.15***).

Biochemical functions

Most of the functions of vitamin C are related to its property to undergo reversible oxidation–reduction i.e., interconversion of ascorbic acid and dehydroascorbic acid.

1. **Collagen formation :** Vitamin C plays the ***role of a coenzyme in hydroxylation of proline and lysine*** while protocollagen is converted to collagen (i.e. ***post-translational modification***). The hydroxylation reaction is catalysed by ***lysyl hydroxylase*** (for lysine) and ***prolyl hydroxylase*** (for proline). This reaction is dependent on vitamin C, molecular oxygen and α-ketoglutarate (***Fig.7.16***).

Hydroxyproline and hydroxylysine are essential for the collagen cross-linking and the strength of the fiber. In this way, vitamin C is necessary for ***maintenance of normal connective tissue and wound healing process***.

2. **Bone formation :** Bone tissues possess an ***organic matrix, collagen*** and the inorganic calcium, phosphate etc. Vitamin C is required for bone formation.

3. **Iron and hemoglobin metabolism :** Ascorbic acid ***enhances iron absorption*** by keeping it in the ferrous form. This is due to the reducing property of vitamin C. It helps in the formation of ferritin (storage form of iron) and mobilization of iron from ferritin.

Vitamin C is useful in the reconversion of methemoglobin to hemoglobin. The degradation of hemoglobin to bile pigments requires ascorbic acid.

4. **Tryptophan metabolism :** Vitamin C is essential for the ***hydroxylation of tryptophan*** (enzyme-hydroxylase) to hydroxytryptophan in the synthesis of serotonin.

5. **Tyrosine metabolism :** Ascorbic acid is required for the oxidation of p-hydroxy phenylpyruvate to homogentisic acid in tyrosine metabolism. Further, ascorbic acid is

also needed by dopamine β-hydroxylase, involved in the ***synthesis of catecholamines*** (norepinephrine and epinephrine).

6. **Folic acid metabolism :** The active form of the vitamin folic acid is tetrahydrofolate (FH_4). Vitamin C is ***needed for the formation of FH_4*** (enzyme-folic acid reductase). Further, in association with FH_4, ascorbic acid is involved in the maturation of erythrocytes.

7. **Synthesis of corticosteroid hormones :** Adrenal gland possesses high levels of ascorbic acid, particularly in periods of stress. It is believed that vitamin C is necessary for the ***hydroxylation reactions*** in the synthesis of corticosteroid hormones.

8. **Sparing action of other vitamins :** Ascorbic acid is a ***strong antioxidant***. It spares vitamin A, vitamin E, and some B-complex vitamins from oxidation.

9. **Immunological function :** Vitamin C enhances the ***synthesis of immunoglobulins*** (antibodies) and increases the phagocytic action of leukocytes.

10. **Preventive action on cataract :** Vitamin C reduces the risk of cataract formation.

11. **Preventive action on chronic diseases :** As an antioxidant, vitamin C reduces the risk of cancer, cataract, and coronary heart diseases.

Recommended dietary allowance (RDA)

About ***40-50 mg*** vitamin C intake per day will meet the adult requirement. Additional intakes (20-40% increase) are recommended for women during pregnancy and lactation.

Dietary sources

Citrus fruits, gooseberry (***amla***), guava, green vegetables (cabbage, spinach), tomatoes, potatoes (particularly skin) are rich in ascorbic acid. High content of vitamin C is found in adrenal gland and gonads. ***Milk*** is a ***poor source*** of ascorbic acid.

Deficiency symptoms

The deficiency of ascorbic acid (mostly found in children and elderly people) results in ***scurvy***. This disease is characterized by spongy and ***sore gums, loose teeth***, anemia, swollen joints, fragile blood vessels, decreased immunocompetence, delayed wound healing, sluggish hormonal function of adrenal cortex and gonads, haemorrhage, osteoporosis etc. Most of these symptoms are related to impairment in the synthesis of collagen and/or the antioxidant property of vitamin C.

Infantile scurvy (***Barlow's disease***) was reported as early as 1934. It occurs ***in infants*** aged 6-12 months due to ***vitamin C deficiency***.

Medical Concepts/ Clinical Correlates

- Vitamin C deficiency causes scurvy. The manifestations of scurvy are related to the impairment in the synthesis of collagen and/or antioxidant property of vitamin C.
- Megadoses of vitamin C are used in common cold, wound healing, trauma etc.
- β-Carotene, vitamin E and ascorbic acid serve as antioxidants and reduce the risk of heart attacks and cancers.
- Health experts recommend that all newborns (mostly deficient in vitamin K) get a dose of vitamin K at birth for proper blood clotting.

Megadoses of vitamin C and its controversy

Linus Pauling (1970) first advocated the consumption of megadoses of ascorbic acid (even up to 18 g/day, 300 times the daily requirement!) ***to prevent and cure common cold***. He is remembered as a scientist who suggested 'keep vitamin C in gunny bags and eat in grams.' This generated a lot of controversy worldover. It is now clear that megadose of vitamin C does not prevent common cold. But the duration of cold and the severity of symptoms are reduced. It is believed that ascorbic acid promotes leukocyte function.

Megadoses (1-4 g/day) of vitamin C are still continued in common cold, wound healing,

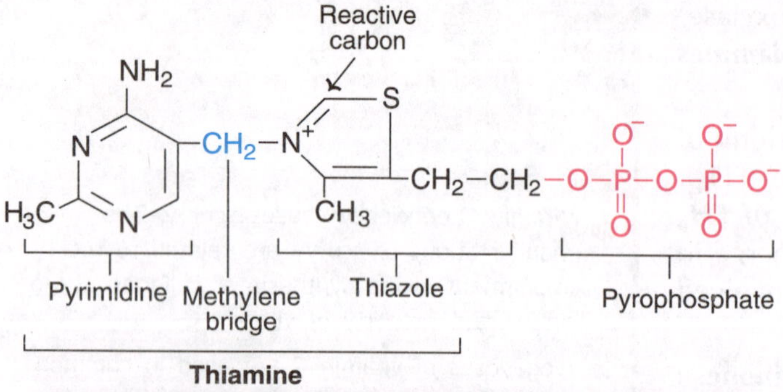

Fig. 7.17 : Structures of thiamine and thiamine pyrophosphate (TPP).

trauma etc. As an antioxidant, ascorbic acid certainly provides some health benefits.

Ascorbic acid, as such, has not been found to be toxic. But, dehydroascorbic acid (oxidized form of ascorbic acid) is toxic. Further, oxalate is a major metabolite of vitamin C. ***Oxalate*** has been implicated in the ***formation of kidney stones.*** However, there are controversial reports on the megadoses of vitamin C leading to urinary stones.

THIAMINE (VITAMIN B_1)

Thiamine (***anti-beri-beri*** or antineuritic ***vitamin***) is water soluble. It has a specific coenzyme, ***thiamine pyrophosphate (TPP)*** which is mostly associated with carbohydrate metabolism.

Chemistry

Thiamine contains a pyrimidine ring and a thiazole ring held by a methylene bridge (***Fig.7.17***). Thiamine is the only natural compound with thiazole ring.

The alcohol (OH) group of thiamine is esterfied with phosphate (2 moles) to form the ***coenzyme***, ***thiamine pyrophosphate*** (TPP or ***cocarboxylase***). The pyrophosphate moiety is donated by ATP and the reaction is catalysed by the enzyme thiamine pyrophosphate transferase.

Biochemical functions

The coenzyme, thiamine pyrophosphate or cocarboxylase is intimately connected with the energy releasing reactions in the carbohydrate metabolism (***Fig.7.18***).

1. The enzyme ***pyruvate dehydrogenase*** catalyses (oxidative decarboxylation) the irreversible conversion of pyruvate to acetyl CoA. This reaction is dependent on TPP, besides the other coenzymes (details given in carbohydrate metabolism, ***Chapter 13***).

2. **α-Ketoglutarate dehydrogenase** is an enzyme ***of the citric acid cycle***. This enzyme is comparable with pyruvate dehydrogenase and requires TPP.

3. **Transketolase** is dependent on TPP. This is an enzyme of the hexose monophosphate shunt (***HMP shunt***).

4. **The branched chain α-keto acid dehydrogenase (decarboxylase)** catalyses the oxidative decarboxylation of branched chain amino acids (valine, leucine and isoleucine) to the respective keto acids. This enzyme also requires TPP.

5. TPP plays an important role in the ***transmission of nerve impulse***. It is believed that TPP is required for acetylcholine synthesis and the ion translocation of neural tissue.

Recommended dietary allowance (RDA)

The daily requirement of thiamine depends on the intake of carbohydrate. A dietary supply of ***1-1.5 mg/day*** is recommended for adults (about 0.5 mg/1,000 Cals of energy). For children, RDA is 0.7-1.2 mg/day. The requirement marginally ***increases in pregnancy and lactation*** (2 mg/day), ***old age and alcoholism***.

Dietary sources

Cereals, pulses, oil seeds, nuts and yeast are good sources. Thiamine is mostly concentrated

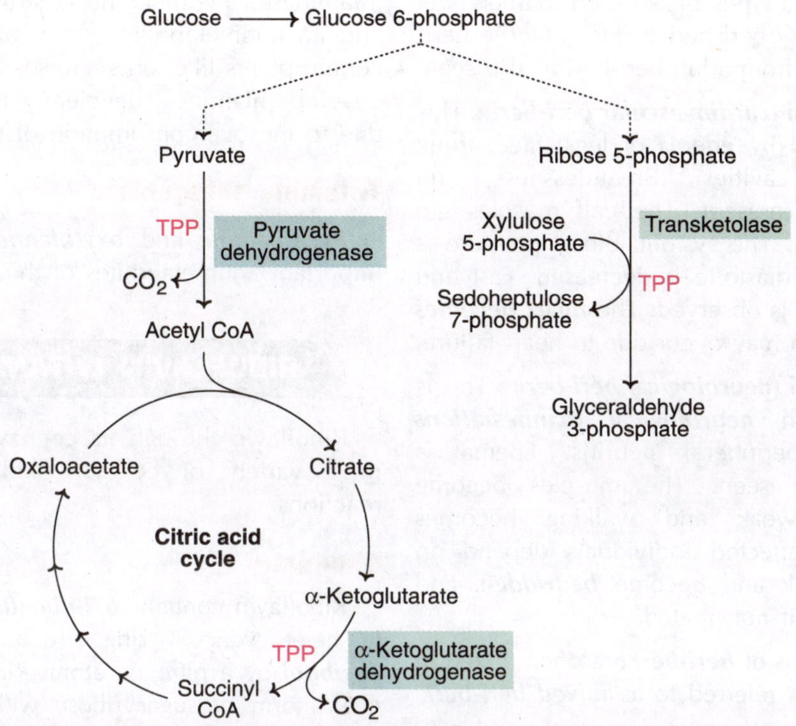

Fig. 7.18 : Summary of the reactions dependent on thiamine pyrophosphate (TPP).

in the ***outer layer*** (bran) of cereals. Polishing of rice removes about 80% of thiamine. Vitamin B_1 is also present in animal foods like pork, liver, heart, kidney, milk etc. In the ***parboiled*** (boiling of paddy with husk) and milled ***rice***, thiamine is not lost in polishing. Since thiamine is a water soluble vitamin, it is extracted into the water during cooking process. Such water should not be discarded.

Deficiency symptoms

The deficiency of vitamin B_1 results in a condition called ***beri-beri*** [*Sinhalese* : I cannot (said twice)]. Beri-beri is mostly seen in populations ***consuming exclusively polished rice*** as staple food. The early symptoms of thiamine deficiency are loss of appetite (anorexia), weakness, constipation, nausea, depression, ***peripheral neuropathy***, irritability etc. Numbness in the legs complaints of 'pins and needles sensations' are reported.

Biochemical changes in B_1 deficiency

1. Carbohydrate metabolism is impaired. Accumulation of ***pyruvate*** occurs in the tissues which is harmful. Pyruvate concentration in plasma is elevated and it is also excreted ***in urine***.

2. Normally, pyruvate does not cross the blood-brain barrier and enter the brain. However, in thiamine deficiency, an ***alteration occurs in the blood-brain barrier permitting the pyruvate to enter the brain directly***. It is believed that ***pyruvate accumulation in brain*** results in disturbed metabolism that may be responsible for ***polyneuritis***.

3. Thiamine deficiency leads to ***impairment in nerve impulse transmission*** due to lack of TPP.

4. The transketolase activity in erythrocytes is decreased. Measurement of ***RBC transketolase*** activity is a reliable ***diagnostic test*** to assess ***thiamine deficiency***.

In adults, two types of beri-beri, namely wet beri-beri and dry beri-beri occur. Infantile beri-beri that differs from adult beri-beri is also seen.

Wet beri-beri (***cardiovascular beri-beri***) **:** This is characterized by ***edema*** of legs, face, trunk and serous cavities. Breathlessness and palpitation are present. The calf muscles are slightly swollen. The systolic blood pressure is elevated while diastolic is decreased. Fast and bouncing pulse is observed. The ***heart becomes weak*** and death may occur due to heart failure.

Dry beri-beri (***neurological beri-beri***) **:** This is associated with ***neurological manifestations*** resulting in peripheral neuritis. Edema is not commonly seen. The muscles become progressively weak and walking becomes difficult. The affected individuals depend on support to walk and become ***bedridden***, and may even ***die***, if not treated.

The symptoms of ***beri-beri*** are often mixed in which case it is referred to as ***mixed beri-beri***.

Infantile beri-beri : This is seen in ***infants born*** to mothers suffering from ***thiamine deficiency***. The breast milk of these mothers contains low thiamine content. Infantile beri-beri is characterized by sleeplessness, restlessness, vomiting, convulsions and bouts of screaming due to cardiac dilatation.

Wernicke-Korsakoff syndrome

This disorder also known as ***cerebral beri-beri***, is mostly ***seen in chronic alcoholics***. The body demands of thiamine increase in alcoholism. Insufficient intake or impaired intestinal absorption of thiamine will lead to this syndrome. It is characterized by loss of memory, apathy and a rhythmical to and fro motion of the eye balls.

Thiamine deficiency due to thiaminase and pyrithiamine

The enzyme thiaminase is present in certain seafoods. Their inclusion in the diet will destroy thiamine by a cleavage action (pyrimidine and thiazole rings split) and lead to deficiency. Incidence of beri-beri in some parts of Japan is attributed to the consumption of fish (rich in thiaminase). Pyrithiamine, a structural analogue and an antimetabolite of thiamine is found in certain plants like ferns. Horses and cattle often develop thiamine deficiency (fern poisoning) due to the overconsumption of the plant fern.

Thiamine antagonists

Pyrithiamine and ***oxythiamine*** are the two important antimetabolites of thiamine.

RIBOFLAVIN (VITAMIN B_2)

Riboflavin through its coenzymes takes part in a variety of cellular oxidation–reduction reactions.

Chemistry

Riboflavin contains ***6,7-dimethyl isoalloxazine*** (a heterocyclic 3 ring structure) attached to ***D-ribitol*** by a nitrogen atom. Ribitol is an open chain form of sugar ribose with the aldehyde group (CHO) reduced to alcohol (CH_2OH).

Riboflavin is stable to heat but sensitive to light. When exposed to ultra-violet rays of sunlight, it is converted to ***lumiflavin*** which exhibits yellow fluorescence. The substances namely ***lactoflavin*** (from milk), ***hepatoflavin*** (from liver) and ***ovoflavin*** (from eggs) which were originally thought to be different are structurally identical to riboflavin.

Coenzymes of riboflavin

Flavin mononucleotide (FMN) and flavin adenine dinucleotide (FAD) are the two coenzyme forms of riboflavin. The ribitol (5 carbon) is linked to a phosphate in FMN. FAD is formed from FMN by the transfer of an AMP moiety from ATP (***Fig.7.19***).

Biochemical functions

The flavin coenzymes (mostly FAD and to a lesser extent FMN) participate in many ***redox reactions*** responsible for energy production. The functional unit of both the coenzymes is isoalloxazine ring which serves as an acceptor of two hydrogen atoms (with electrons). FMN or FAD undergo identical reversible reactions accepting two hydrogen atoms forming $FMNH_2$ or $FADH_2$ (***Fig.7.20***).

Enzymes that use flavin coenzymes (FMN or FAD) are called ***flavoproteins***. The coenzymes (prosthetic groups) often bind rather tightly, to the protein (apoenzyme) either by non-covalent bonds (mostly) or covalent bonds in the holoenzyme. Many flavoproteins contain metal atoms (iron, molybdenum etc.) which are known as ***metalloflavoproteins***.

The coenzymes, FAD and FMN are associated with certain enzymes involved in carbohydrate, lipid, protein and purine metabolisms, besides the electron transport chain. A few examples are listed in ***Table 7.2***. Further details are given in the respective chapters.

Recommended dietary allowance (RDA)

The daily requirement of riboflavin for an adult is ***1.2-1.7 mg***. Higher intakes (by 0.2-0.5 mg/day) are advised for pregnant and lactating women.

Dietary sources

Milk and milk products, ***meat***, eggs, liver, kidney are rich sources. Cereals, fruits, vegetables and fish are moderate sources.

Deficiency symptoms

Riboflavin deficiency symptoms include ***cheilosis*** (fissures at the corners of the mouth), ***glossitis*** (tongue smooth and purplish) and ***dermatitis***. Riboflavin deficiency as such is uncommon. It is mostly seen along with other vitamin deficiencies. Chronic alcoholics are susceptible to B_2 deficiency. Assay

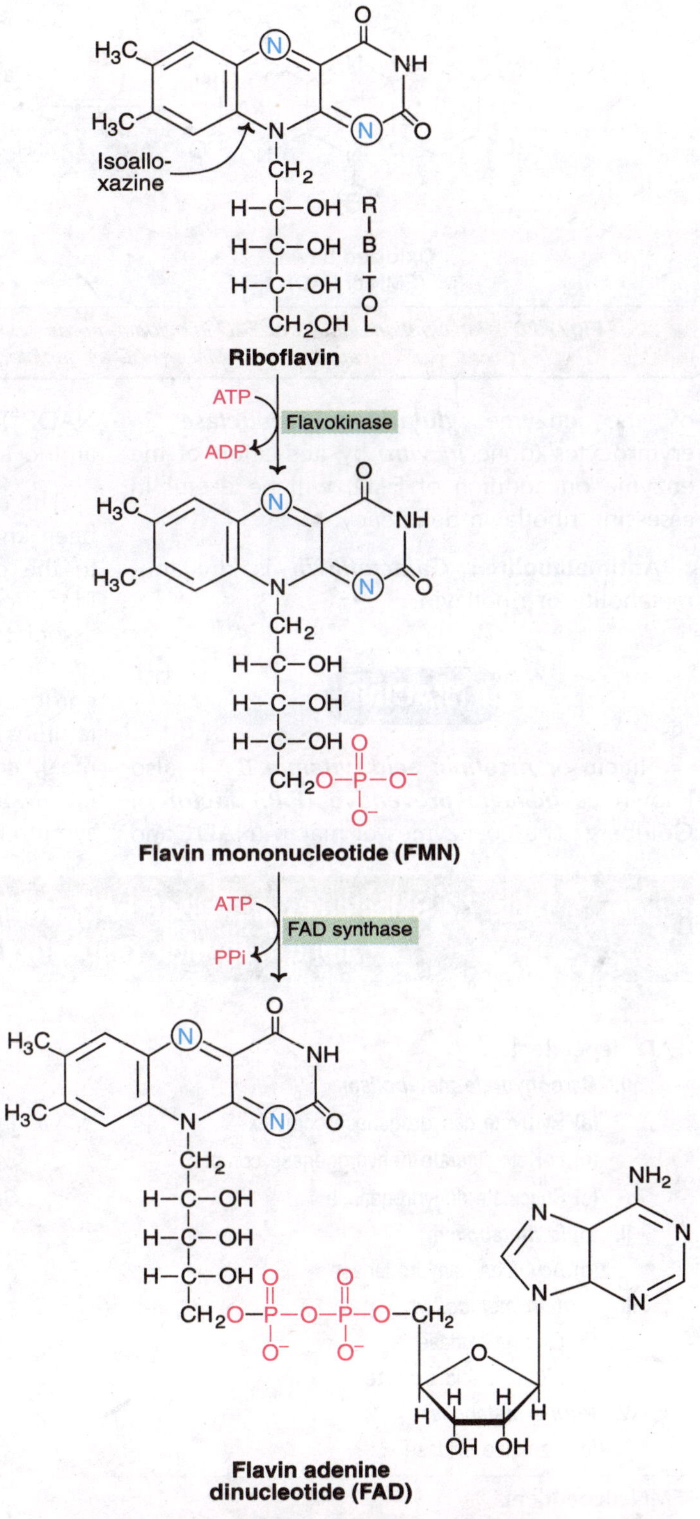

Fig. 7.19 : Structures and biosynthesis of flavin mononucleotide (FMN) and flavin adenine dinucleotide (FAD).

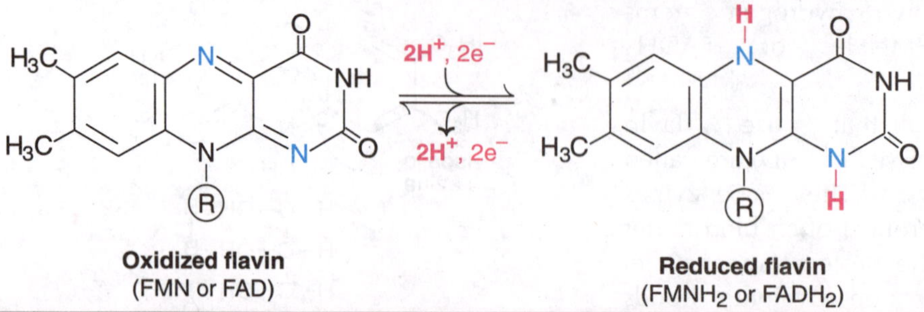

Fig. 7.20 : *Participation of FMN or FAD in oxidation-reduction reactions (R-represents the rest of the structure of FMN or FAD as depicted in* **Fig. 7.19***).*

of the enzyme ***glutathione reductase*** in erythrocytes (done *in vitro* by activation of the enzyme on addition of FAD) will be useful in assessing riboflavin deficiency.

Antimetabolite : ***Galactoflavin*** is an anti-metabolite of riboflavin.

NIACIN

Niacin or ***nicotinic acid*** (***vitamin*** B_3) is also known as ***pellagra preventive (P.P.) factor*** of Goldberg. The coenzymes of niacin (NAD^+ and $NADP^+$) can be synthesized by the essential amino acid, ***tryptophan***.

The disease pellagra (*Italian* : rough skin) has been known for centuries. However, its relation to the deficiency of a dietary factor was first identified by Goldberger. Goldberger and his associates conducted an interesting experiment for this purpose. Twelve convicts were promised pardon if they consumed diet of pellagrous families for one year. The diet consisted of corn meal, corn starch, rice, sweet potato and pork fat. More than half of the subjects showed symptoms of pellagra in less than a year, while

Table 7.2 Selected examples of FAD and FMN dependent enzymes along with their respective reactions

Enzyme	Reaction
FAD dependent	
I. *Carbohydrate metabolism*	
(a) Pyruvate dehydrogenase complex*	Pyruvate ⟶ Acetyl CoA
(b) α-Ketoglutarate dehydrogenase complex*	α-Ketoglutarate ⟶ Succinyl CoA
(c) Succinate dehydrogenase	Succinate ⟶ Fumarate
II. *Lipid metabolism*	
(d) Acyl CoA dehydrogenase	Acyl CoA ⟶ α, β-Unsaturated acyl CoA
III. *Protein metabolism*	
(e) Glycine oxidase	Glycine ⟶ Glyoxylate + NH_3
(f) D-Amino acid oxidase	D-Amino acid ⟶ α-Keto acid + NH_3
IV. *Purine metabolism*	
(g) Xanthine oxidase	Xanthine ⟶ Uric acid
FMN dependent	
L-Amino acid oxidase	L-Amino acid ⟶ α-Keto acid + NH_3

* *Dihydrolipoyl dehydrogenase component requires FAD*

no such symptoms were observed in other prisoners on a regular diet. Administration of dried meat or liver to the patients cured pellagra (Goldberger, 1928).

Much before it was recognized as a vitamin, nicotinic acid was well known as a chemical compound, produced by the oxidation of ***nicotine*** (present in tobacco leaves). The term '***niacin***' was coined and ***more commonly used for nicotinic acid***. This was done to emphasize the role of niacin as a vitamin and avoid the impression that nicotinic acid is an oxidized form of nicotine. However, most of the authors use niacin and nicotinic acid synonymously.

Chemistry and synthesis of coenzymes

Niacin is a ***pyridine derivative***. Structurally, it is ***pyridine 3-carboxylic acid***. The amide form of niacin is known as niacinamide or nicotinamide.

Dietary nicotinamide, niacin and tryptophan (an essential amino acid) contribute to the synthesis of the coenzymes–***nicotinamide adenine dinucleotide (NAD^+) and nicotinamide adenine dinucleotide phosphate ($NADP^+$)***. The pathway for the biosynthesis of NAD^+ and $NADP^+$ is depicted in ***Fig.7.21***. Nicotinamide is deaminated in the body to niacin. Niacin then undergoes a series of reactions to produce NAD^+ and $NADP^+$. Tryptophan produces quinolinate which then forms nicotinate mononucleotide and, ultimately, NAD^+ and $NADP^+$. ***Sixty milligrams of tryptophan is equivalent to 1 mg of niacin for the synthesis of niacin coenzymes***. Phosphoribosyl pyrophosphate and ATP, respectively, provide ribose phosphate and AMP moieties for the synthesis of NAD^+. Glutamine donates amide group. In the structure of the coenzymes, nitrogen atom of nicotinamide carries a positive charge (due to the formation of an extra bond, N is in quaternary state), hence the coenzymes are NAD^+ and $NADP^+$. Nicotinamide, liberated on the degradation of NAD^+ and $NADP^+$ is mostly ***excreted in urine as N-methylnicotinamide***.

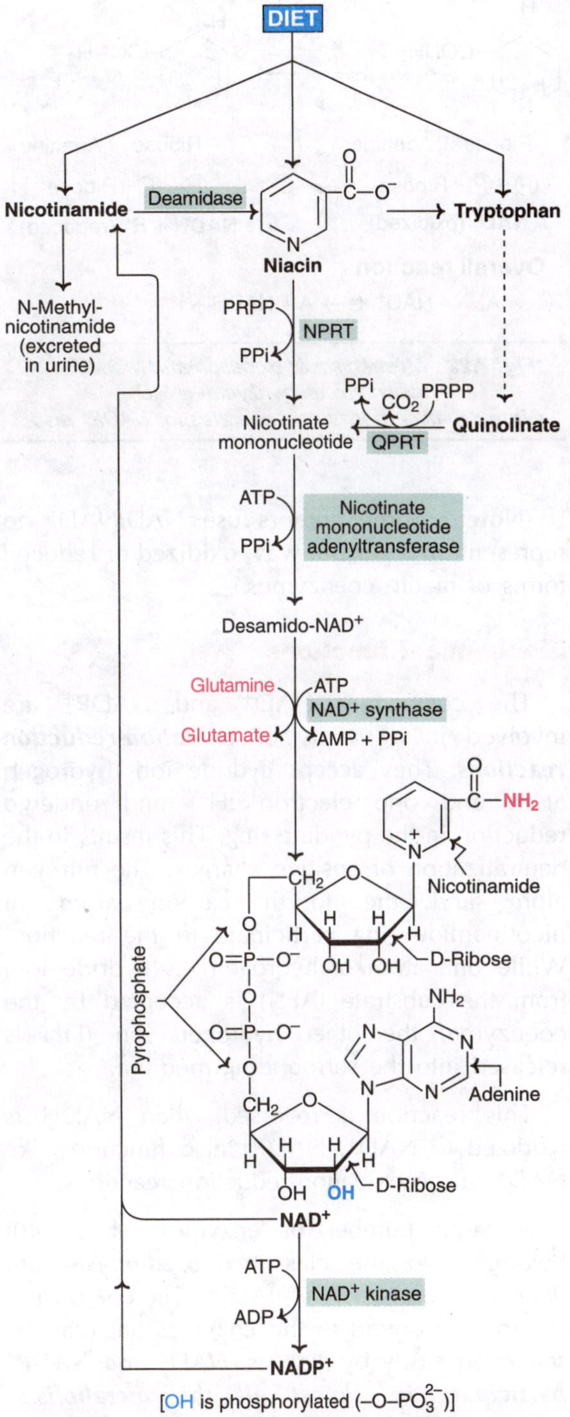

Fig. 7.21 : Outline of the biosynthesis of nicotinamide nucleotides, NAD^+ and $NADP^+$ (NPRT-Nicotinate phosphoribosyltransferase; QPRT-Quinolinate phosphoribosyltransferase; PPi-Pyrophosphate; PRPP-Phosphoribosyl pyrophosphate).

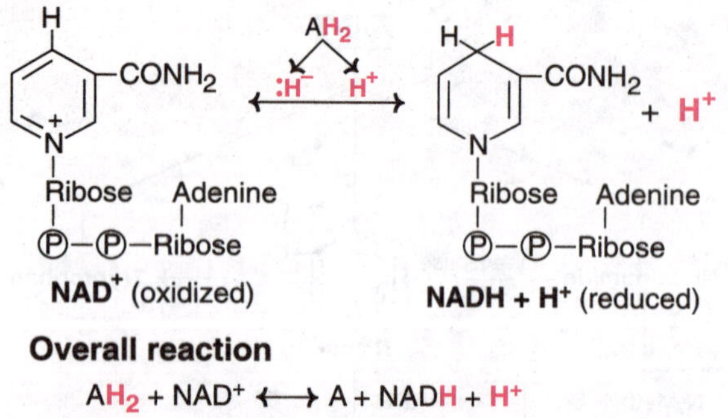

Fig. 7.22 : *Mechanism of oxidation and reduction of nicotinamide coenzyme—NAD^+*
*(**Note :** Similar mechanism operates for $NADP^+$ also).*

(**Note :** Some authors use NAD/NADP to represent, in a general way, oxidized or reduced forms of niacin coenzymes.)

Biochemical functions

The coenzymes NAD^+ and $NADP^+$ are involved in a variety of ***oxidation-reduction reactions.*** They accept hydride ion (hydrogen atom and one electron $:H^-$) and undergo reduction in the pyridine ring. This results in the neutralization of positive charges. The nitrogen atom and the fourth carbon atom of nicotinamide ring participate in the reaction. While one atom of hydrogen (as hydride ion) from the substrate (AH_2) is accepted by the coenzyme, the other hydrogen ion (H^+) is released into the surrounding medium.

This reaction is reversed when NADH is oxidized to NAD^+. $NADP^+$ also functions like NAD^+ in the oxidation-reduction reactions.

A large number of enzymes (about 40) belonging to the class ***oxidoreductases*** are dependent on NAD^+ or $NADP^+$. The coenzymes are loosely bound to the enzymes and can be separated easily by dialysis. ***NAD^+ and $NADP^+$ participate in almost all the metabolisms*** (carbohydrate, lipid, protein etc.). Some enzymes are exclusively dependent on NAD^+ whereas some require only $NADP^+$. A few enzymes can use either NAD^+ or $NADP^+$. Selected examples of enzymes and the reactions they catalyse are given in ***Table 7.3.***

NADH produced is oxidized in the electron transport chain ***to generate ATP***. ***NADPH*** is also important for many ***biosynthetic reactions*** as it donates reducing equivalents.

Recommended dietary allowance (RDA)

The daily requirement of niacin for an adult is ***15-20 mg*** and for children, around 10-15 mg. Very often, the term ***niacin equivalents (NE)*** is used while expressing its RDA. ***One NE = 1 mg niacin or 60 mg of tryptophan***. Instead of mg, the daily requirements are known as niacin equivalents. Pregnancy and lactation in women impose an additional metabolic burden and increase the niacin requirement.

Dietary sources

The rich natural sources of niacin include ***liver***, ***yeast***, whole grains, cereals, pulses like beans and peanuts. Milk, fish, eggs and vegetables are moderate sources. The essential amino acid tryptophan can serve as a precursor for the synthesis of nicotinamide coenzymes. On an average, 1 g of a good quality protein containing about 60 mg of tryptophan is equivalent to 1 mg of niacin ***(conversion ratio 60 : 1)*** for the synthesis of nicotinamide coenzymes. Tryptophan has many other essential and important functions in the body, hence dietary ***tryptophan cannot totally replace niacin.*** Increased conversion of tryptophan to niacin has been reported in low protein diet and starvation. Tryptophan can replace niacin to an extent of 10% for the synthesis of coenzymes. Therefore, both niacin and tryptophan have to be invariably provided in the diet.

Is niacin a vitamin?

The amino acid tryptophan can form niacin coenzymes right in the body. Therefore, ***strictly speaking, niacin is not a vitamin.***

Deficiency symptoms

Niacin deficiency results in a condition called ***pellagra*** (*Italian*: rough skin). This disease involves skin, gastrointestinal tract and central

Table 7.3 Selected examples of NAD^+ and $NADP^+$ dependent enzymes along with their respective reactions

Enzyme	Reaction
NAD^+ dependent	
I. ***Carbohydrate metabolism***	
(a) Glyceraldehyde 3-phosphate dehydrogenase	Glyceraldehyde 3-phosphate ⟶ 1, 3-Bisphosphoglycerate
(b) Lactate dehydrogenase	Pyruvate ⟶ Lactate
(c) Pyruvate dehydrogenase complex	Pyruvate ⟶ Acetyl CoA
(d) α-Ketoglutarate dehydrogenase complex	α-Ketoglutarate ⟶ Succinyl CoA
II. ***Lipid metabolism***	
(e) β-Hydroxy acyl CoA dehydrogenase	β-Hydroxy acyl CoA ⟶ β-Keto acyl CoA
(f) β-Hydroxybutyrate dehydrogenase	β-Hydroxybutyrate ⟶ Acetoacetate
(g) Alcohol dehydrogenase	Alcohol ⟶ Acetaldehyde
III. ***Protein metabolism***	
(h) Branched chain α-keto acid dehydrogenase	α-Keto acids of branched chain amino acids (Leu, Ile, Val) ⟶ Corresponding acyl CoA thioesters
(i) Tyramine dehydrogenase	Tyramine ⟶ p-Hydroxyphenyl acetate
NAD^+ or $NADP^+$ dependent	
(a) Glutamate dehydrogenase	Glutamate ⟶ α-Ketoglutarate + NH_3
(b) Isocitrate dehydrogenase	Isocitrate ⟶ Oxalosuccinate
$NADP^+$ dependent	
(a) Glucose 6-phosphate dehydrogenase	Glucose 6-phosphate ⟶ 6-Phosphogluconolactone
(b) Malic enzyme	Malate ⟶ Pyruvate
NADPH dependent	
(a) 3-Ketoacyl reductase	3-Ketoacyl enzyme ⟶ 3-Hydroxy acyl enzyme
(b) HMG CoA reductase	HMG CoA ⟶ Mevalonate
(c) Squalene epoxidase	Squalene ⟶ Squalene oxide
(d) Cholesterol 7α-hydroxylase	Cholesterol ⟶ 7α-Hydroxy cholesterol
(e) Phenylalanine hydroxylase	Phenylalanine ⟶ Tyrosine
(f) Dihydrofolate reductase	Folic acid ⟶ Tetrahydrofolic acid.

nervous system. The symptoms of pellagra are commonly referred to as ***three Ds***. The disease also progresses in that order ***dermatitis, diarrhea, dementia***, and if not treated may rarely lead to ***death*** (4th D). Pellagra is frequently observed in ***Hartnup's disease*** (**Refer Chapter 8**) and in ***carcinoid syndrome*** (**Refer Chapter 15**).

Dermatitis (inflammation of skin) is usually found in the areas of the skin exposed to sunlight (neck, dorsal part of feet, ankle and parts of face). Diarrhea may be in the form of loose stools, often with blood and mucus. Prolonged diarrhea leads to weight loss. Dementia is associated with degeneration of nervous tissue. The symptoms of dementia include anxiety, irritability, poor memory, insomnia (sleeplessness) etc.

Pellagra is mostly seen among people whose staple diet is corn or maize. Niacin present in maize is unavailable to the body as it is in bound form, and tryptophan content is low in maize.

Multiple causes of pellagra

It is now accepted that pellagra is caused by ***deficiency of niacin and tryptophan***. In addition, deficiencies of ***riboflavin*** or ***pyridoxine*** (required for the synthesis of nicotinamides from tryptophan) are also important. Further, in the outbreaks of ***pellagra, women*** are ***more*** (more than twice) ***affected than men***. This is attributed to ***inhibition of tryptophan metabolism by estrogen metabolites***.

Therapeutic uses of niacin

Administration of niacin (not niacinamide) in pharmacological doses (2-4 g/day, 200 times the RDA) results in a number of biochemical effects in the body, not related to its function as a vitamin. Most of the effects are believed to be due to the influence of niacin on cyclic AMP levels.

1. Niacin ***inhibits lipolysis*** in the adipose tissue and decreases the circulatory fatty acids.

2. Triacylglycerol synthesis in the liver is decreased.

3. The serum levels of low density lipoproteins (LDL), very low density lipoproteins (VLDL), ***triacylglycerol*** and ***cholesterol*** are ***lowered***. Hence niacin is used in the treatment of ***hyperlipoproteinemia type II b*** (elevation of LDL and VLDL). Niacin can be taken along with statins.

Toxicity of niacin

Although megadoses of niacin are ***useful for the treatment of hyperlipidemia***, there are certain ***harmful side effects*** also.

1. Glycogen and fat reserves of skeletal and cardiac muscle are depleted.

2. There is a tendency for the increased levels of glucose and ***uric acid*** in the circulation.

3. Prolonged use of niacin results in elevated serum levels of certain enzymes, suggesting liver damage.

PYRIDOXINE (VITAMIN B_6)

Vitamin B_6 is used to collectively represent the three compounds namely ***pyridoxine, pyridoxal and pyridoxamine*** (the ***vitamers of B_6***).

Chemistry

Vitamin B_6 compounds are ***pyridine derivatives***. They differ from each other in the structure of a functional group attached to 4th carbon in the pyridine ring. Pyridoxine is a primary alcohol, pyridoxal is an aldehyde form while pyridoxamine is an amine (***Fig.7.23***). Pyridoxamine is mostly present in plants while pyridoxal and pyridoxamine are found in animal foods. Pyridoxine can be converted to pyridoxal and pyridoxamine, but the latter two cannot form pyridoxine.

Synthesis of coenzyme

The active form of vitamin B_6 is the coenzyme ***pyridoxal phosphate (PLP)***. PLP can be synthesized from the three compounds pyridoxine, pyridoxal and pyridoxamine. B_6 is excreted in urine as 4-pyridoxic acid. The different forms of B_6 and their inter-relationship are depicted in ***Fig.7.23***.

Biochemical functions

Pyridoxal phosphate (PLP), the coenzyme of vitamin B_6 is found attached to the ε-amino group of lysine in the enzyme. PLP is closely associated with the metabolism of amino acids. The synthesis of certain specialized products such as serotonin, histamine, niacin coenzymes from the amino acids is dependent on pyridoxine. Pyridoxal phosphate participates in reactions like ***transamination, decarboxylation, deamination, transsulfuration, condensation*** etc.

1. **Transamination :** Pyridoxal phosphate is involved in the transamination reaction (by transaminase) converteing ***amino acids to keto acids***. The keto acids enter the citric acid cycle and get oxidized to generate energy. Thus B_6 is

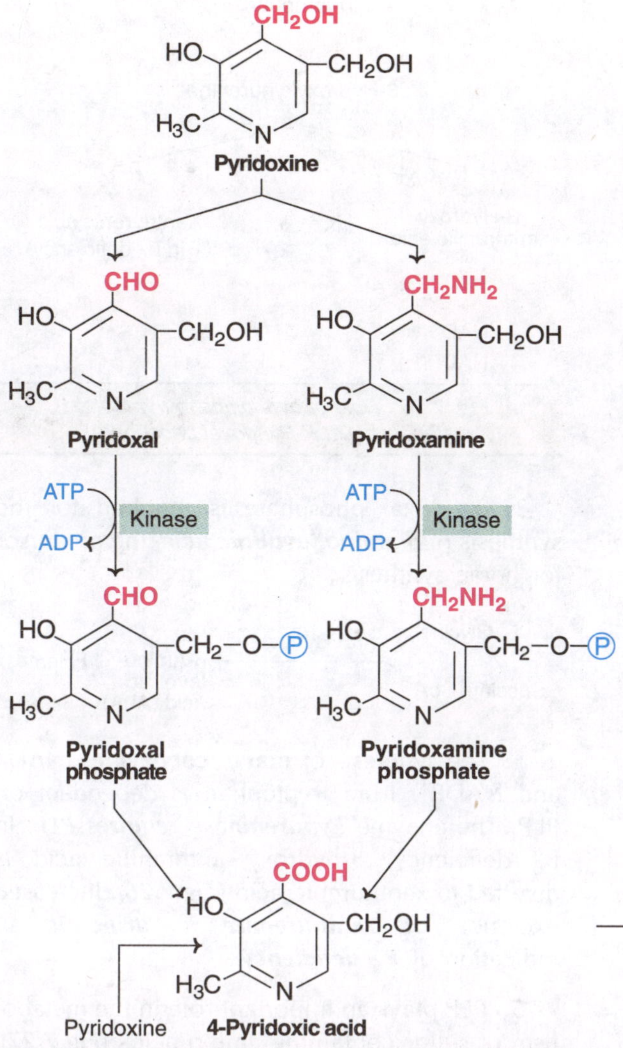

Fig. 7.23 : Pyridoxine, its derivatives and coenzyme.

an energy releasing vitamin. It integrates carbohydrate and amino acid metabolisms (***Fig.7.24***).

During the course of transamination, PLP interacts with amino acid to form a ***Schiff base (Fig.7.25)***. The amino group is handed over to PLP to form pyridoxamine phosphate and the keto acid is liberated.

2. **Decarboxylation :** Some of the α-amino acids undergo decarboxylation to form the respective amines. This is carried out by a group of enzymes called ***decarboxylases*** which are dependent on PLP. Many ***biogenic amines*** with important functions are synthesized by PLP decarboxylation.

(a) **Serotonin** (5-hydroxytryptamine, 5 HT), produced from tryptophan is important in nerve impulse transmission (neurotransmitter). It regulates sleep, behaviour, blood pressure etc.

$$\text{Tryptophan} \longrightarrow \text{5-Hydroxytryptophan} \xrightarrow[CO_2]{PLP} \text{5-Hydroxytryptamine}$$

(b) **Histamine** is a vasodilator and lowers blood pressure. It stimulates gastric HCl secretion and is involved in inflammation and allergic reactions.

$$\text{Histidine} \xrightarrow[CO_2]{PLP} \text{Histamine}$$

(c) Glutamate on decarboxylation gives ***γ-amino butyric acid*** (GABA). GABA inhibits the transmission of nerve impulses, hence it is an inhibitory neurotransmitter.

$$\text{Glutamate} \xrightarrow[CO_2]{PLP} \text{GABA}$$

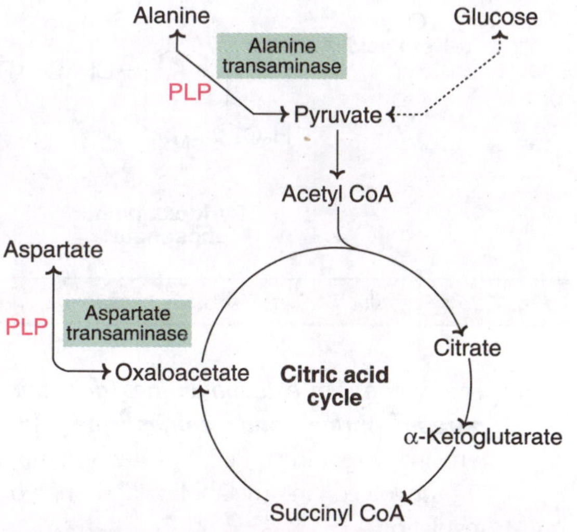

Fig. 7.24 : *Pyridoxal phosphate (PLP) integrates amino acid and carbohydrate metabolisms (Alanine and aspartate are the amino acids respectively converted to pyruvate and oxaloacetate, the keto acids).*

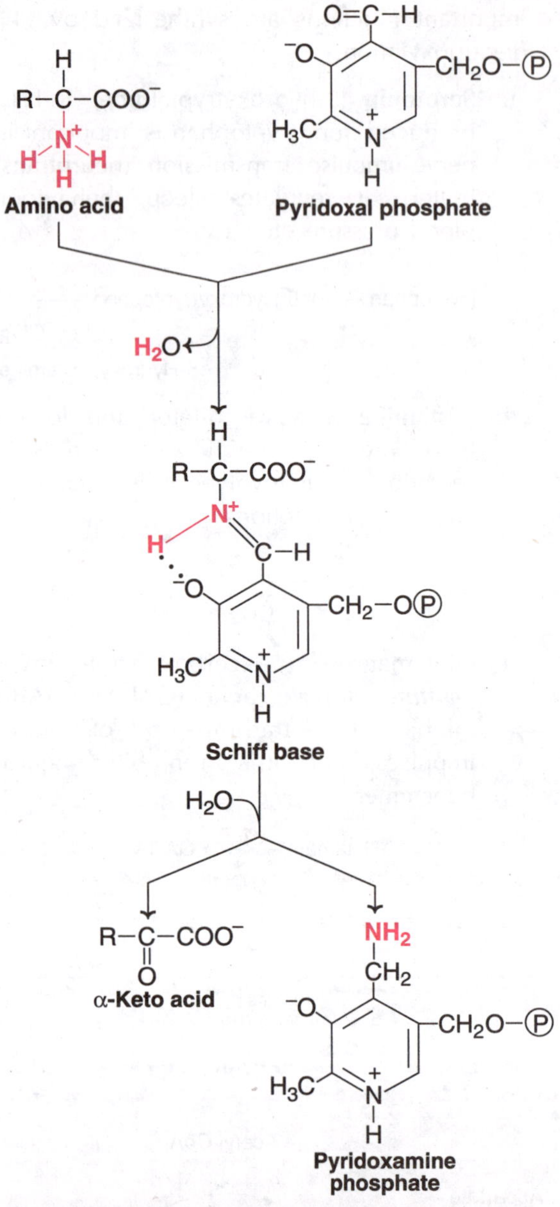

Fig. 7.25 : Formation of Schiff base in transamination.

(d) The synthesis of ***catecholamines*** (***dopamine, norepinephrine and epinephrine***) from tyrosine require PLP. Catecholamines are involved in metabolic and nervous regulation.

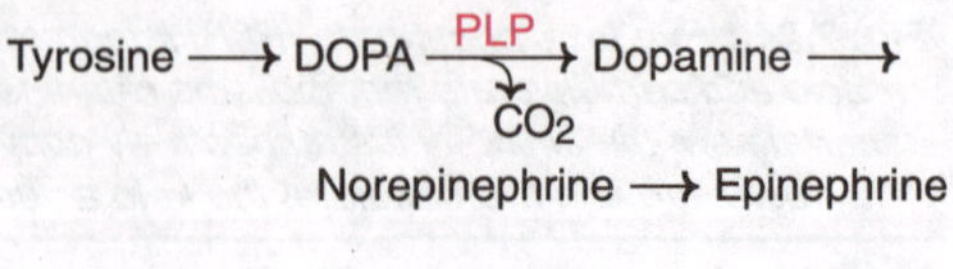

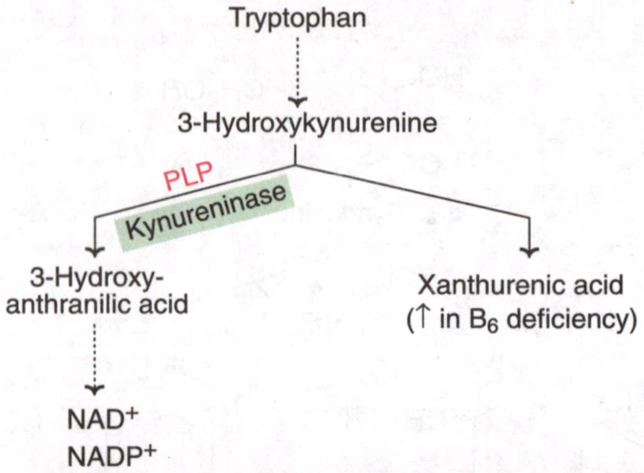

Fig. 7.26 : Role of pyridoxine in tryptophan metabolism (PLP-Pyridoxal phosphate).

3. Pyridoxal phosphate is required for the synthesis of ***δ-amino levulinic acid***, the precursor for heme synthesis.

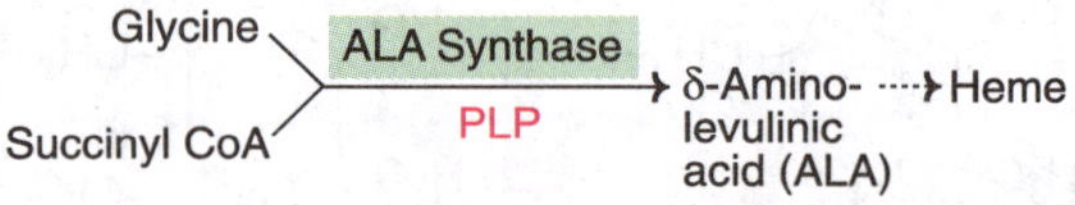

4. The synthesis of ***niacin coenzymes*** (NAD^+ and $NADP^+$) from tryptophan is dependent on PLP. The enzyme ***kynureninase requires PLP***. In B_6 deficiency, 3-hydroxy anthranilic acid is diverted to xanthurenic acid (***Fig.7.26***). Increased excretion of ***xanthurenate in urine*** is an indication of ***B_6 deficiency***.

5. PLP plays an important role in the metabolism of sulfur containing amino acids (***Fig.7.27***).

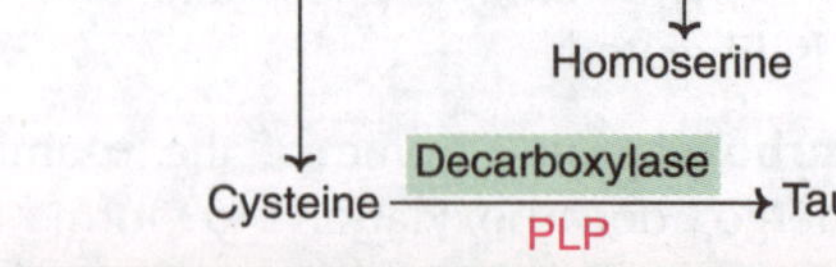

Fig. 7.27 : Role of pyridoxine in the metabolism of sulfur amino acids (PLP-Pyridoxal phosphate).

Medical Concepts/ Clinical Correlates

- Megadoses of niacin are useful in the treatment of hyperlipidemia.
- Long term use of isoniazid for the treatment of tuberculous causes vitamin B_6 deficiency.
- Folic acid supplementation lowers elevated plasma homocysteine levels, and reduces the risk associated with atherosclerosis and thrombosis.

Transsulfuration (transfer of sulfur) from homocysteine to serine occurs in the synthesis of cysteine. This is carried out by a PLP dependent cystathionine synthase. Taurine, a decarboxylated (PLP dependent) product of cysteine, is involved in the conjugation of bile acids.

6. **Deamination** of hydroxyl group containing amino acids requires PLP.

$$\text{Serine} \xrightarrow[\text{PLP}]{\text{Dehydratase}} \text{Pyruvate} + NH_3$$

$$\text{Threonine} \xrightarrow[\text{PLP}]{\text{Dehydratase}} \alpha\text{-Ketobutyrate} + NH_3$$

7. **Serine** is synthesized from glycine by a PLP dependent enzyme hydroxymethyltransferase.

8. The enzyme ***glycogen phosphorylase*** (that cleaves glycogen to glucose 1-phosphate) contains PLP, covalently bound to lysine residue. Phosphorylase structure is stabilized by PLP, for effective enzymatic function.

9. PLP is needed for the ***absorption of amino acids*** from the intestine.

10. Adequate intake of B_6 is useful to ***prevent hyperoxaluria*** and urinary stone formation.

11. Vitamin B_6 is involved in the action of steroid hormones. PLP removes hormone receptor complex from DNA binding, and thus terminates hormonal action.

Recommended dietary allowance (RDA)

The requirement of pyridoxine for an adult is ***1.5-2.0 mg/day***. During pregnancy, lactation and old age, an intake of 2.5 mg/day is recommended. As is observed from the coenzyme function, pyridoxine is closely associated with protein (amino acid) metabolism. The daily requirements of B_6 are calculated on the assumption that the intake of protein is <100 g/day.

Dietary sources

Animal sources such as ***egg yolk***, ***fish***, milk, meat are rich in B_6. Wheat, corn, cabbage, roots and tubers are good vegetable sources.

Deficiency symptoms

Pyridoxine deficiency is associated with ***neurological symptoms*** such as depression, irritability, nervousness and mental confusion. Convulsions and peripheral neuropathy are observed in severe deficiency. These symptoms are related to the decreased synthesis of biogenic amines (serotonin, GABA, norepinephrine and epinephrine). In children, B_6 deficiency with a drastically reduced GABA production results in convulsions (epilepsy).

Decrease in hemoglobin levels, associated with hypochromic microcytic anaemia, is seen in B_6 deficiency. This is due to a ***reduction in heme production.***

The synthesis of niacin coenzymes (NAD^+ and $NADP^+$) from tryptophan is impaired. ***Xanthurenic acid***, produced in high quantities is ***excreted in urine***, which serves as a reliable index (particularly after ***tryptophan load test***) for ***B_6 deficiency***. Vitamin B_6 status can be assessed by a more sensitive assay of erythrocyte transaminases.

Dietary deficiency of pyridoxine is rather rare and is mostly observed in women taking oral contraceptives, alcoholics and infants.

Drug induced B_6 deficiency

Isoniazid (isonicotinic acid hydrazide, INH) is a drug frequently used for the treatment of

tuberculosis. It combines with pyridoxal phosphate to form inactive hydrazone derivatives which inhibit PLP dependent enzymes. Tuberculous patients, on long term use of isoniazid, develop peripheral neuropathy which responds to B_6 therapy.

The drug ***penicillamine*** (β-dimethyl cysteine) is used in the treatment of patients with rheumatoid arthritis, Wilson's disease and cystinuria. This drug also reacts with PLP to form inactive thiazolidine derivative.

Administration of drugs namely isoniazid and penicillamine should be accompanied by pyridoxine supplementation to avoid B_6 deficiency.

Pyridoxine antagonists

Isoniazid, ***deoxypyridoxine*** and methoxy pyridoxine are the antagonists of vitamin B_6.

Toxic effects of overdose vitamin B_6

Excess use of vitamin B_6 (2.5 g/day) in the women of premenstrual syndrome is associated with sensory neuropathy. Some workers have suggested that vitamin B_6 more than 200 mg/day may cause neurological damage.

BIOTIN

Biotin (formerly known as ***anti-egg white injury factor, vitamin B_7*** or ***vitamin H***) is a sulfur containing B-complex vitamin. It directly participates as a coenzyme in the carboxylation reactions.

Boas (1927) observed that rats fed huge quantity of raw egg white developed dermatitis and nervous manifestations, besides retardation in growth. She however, found that feeding cooked egg did not produce any of these symptoms. It was later shown that the egg white injury in rats and chicks was due to the presence of an anti-vitamin in egg white. The ***egg-white injury factor*** was identified as a glycoprotein–***avidin*** and ***biotin*** was called as ***anti-egg white injury factor***.

Chemistry

Biotin is a heterocyclic sulfur containing monocarboxylic acid. The structure is formed by fusion of ***imidazole*** and ***thiophene rings*** with a valeric acid side chain (***Fig.7.28***). Biotin is covalently bound to ε-amino group of lysine to form biocytin in the enzymes. ***Biocytin*** may be regarded as the ***coenzyme*** of biotin.

Biochemical functions

Biotin serves as a carrier of CO_2 in ***carboxylation reactions***. The reaction catalysed by ***pyruvate carboxylase***, converting pyruvate to oxaloacetate has been investigated in detail. This enzyme has biotin bound to the apoenzyme linked to the ε-amino group of lysine, forming the active enzyme (holoenzyme). Biotin–enzyme reacts with CO_2 in presence of ATP (provides energy) to form a carboxybiotin–enzyme complex. This high energy complex hands over the CO_2 to pyruvate (carboxylation reaction) to produce oxaloacetate (***Fig.7.29***).

As a coenzyme, biotin is involved in various metabolic reactions.

1. **Gluconeogenesis and citric acid cycle :** The ***conversion of pyruvate to oxaloacetate*** by biotin dependent ***pyruvate carboxylase*** (described above) is essential for the synthesis of glucose from many non-carbohydrate sources. Oxaloacetate so formed is also required for the continuous operation of citric acid cycle.

2. **Fatty acid synthesis :** Acetyl CoA is the starting material for the synthesis of fatty acids. The very first step in fatty acid synthesis is a carboxylation reaction.

$$\text{Acetyl CoA} \xrightarrow[\text{Acetyl CoA carboxylase}]{\text{Biotin}} \text{Malonyl CoA}$$

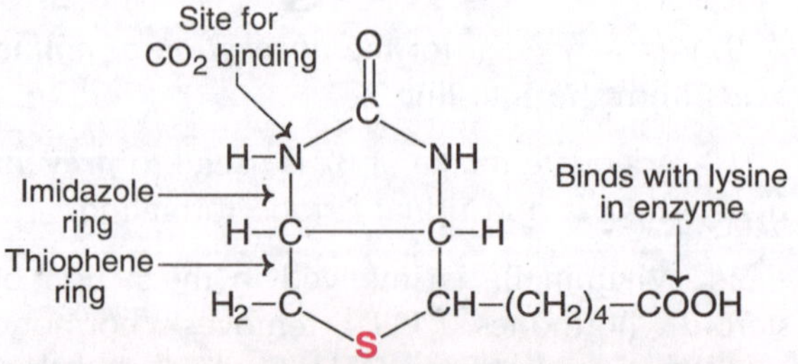

Fig. 7.28 : Structure of biotin with binding sites.

3. **Propionyl CoA** is produced in the metabolism of certain amino acids (valine, isoleucine, threonine etc.) and degradation of odd chain fatty acids. Its further metabolism is dependent on biotin.

Propionyl CoA —(Biotin; Propionyl CoA carboxylase)→ Methylmalonyl CoA

4. In the metabolism of ***leucine***, the following reaction is dependent on biotin.

β-Methylcrotonyl CoA —(Biotin; β-Methylcrotonyl CoA carboxylase)→ β-Methyl glutaconyl CoA

[**Note :** It was once believed that all the carboxylation reactions in the biological system are dependent on biotin. This was later proved to be wrong. There are a few carboxylation reactions which do not require biotin e.g. formation of carbamoyl phosphate in urea cycle, incorporation of CO_2 in purine synthesis.]

5. **Regulation of cell cycle :** Biotin is involved in cell cycle regulation through its action to biotinylate key nuclear proteins.

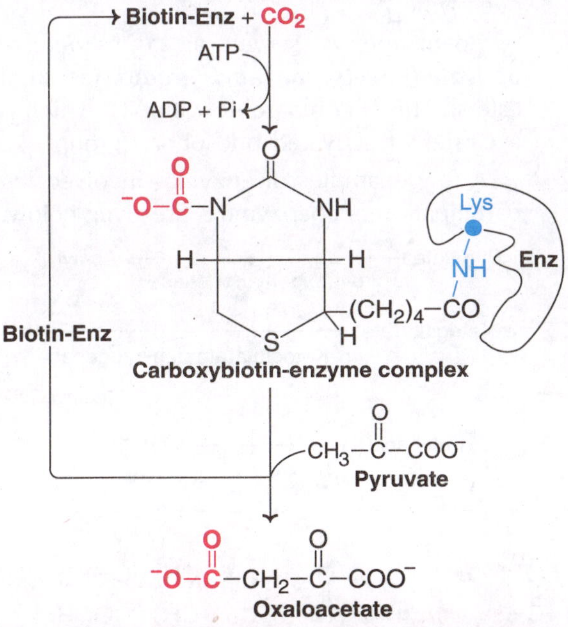

***Fig. 7.29 :** Role of biotin in the carboxylation reaction, catalysed by the enzyme pyruvate carboxylase (Enz-Enzyme).*

6. **Regulation of gene expression :** Biotinylated histones play a key role in gene expression and DNA repair.

Recommended dietary allowance (RDA)

A daily intake of about ***20-30 µg*** is recommended for adults. In fact, biotin is normally synthesized by the intestinal bacteria. However, to what extent the synthesized biotin contributes to the body requirements is not clearly known.

Dietary sources

Biotin is widely distributed in both animal and plant foods. The rich sources are ***liver, kidney***, egg yolk, milk, tomatoes, grains etc.

Deficiency symptoms

The symptoms of biotin deficiency include ***anemia, loss of appetite, nausea, dermatitis,*** glossitis etc. Biotin deficiency may also result in depression, hallucinations, and dermatitis.

Biotin ***deficiency is uncommon***, since it is well distributed in foods and also supplied by the intestinal bacteria. The deficiency may be associated with the following two causes.

1. Destruction of intestinal flora due to prolonged use of drugs such as ***sulfonamides***.

2. High consumption of raw eggs. The raw egg white contains a glycoprotein–***avidin***, which tightly binds with biotin and blocks its absorption from the intestine. An intake of about 20 raw eggs per day is needed to produce biotin deficiency in humans. Consumption of an occasional raw egg will not result in deficiency.

Antagonists

Desthiobiotin, biotin sulphonic acid are biotin antagonists.

PANTOTHENIC ACID

Pantothenic acid (*Greek* : pantos–everywhere), or ***vitamin B_5***, formerly known as chick anti-dermatitis factor (or filtrate factor) is widely distributed in nature. It's metabolic role as ***coenzyme A*** is also widespread.

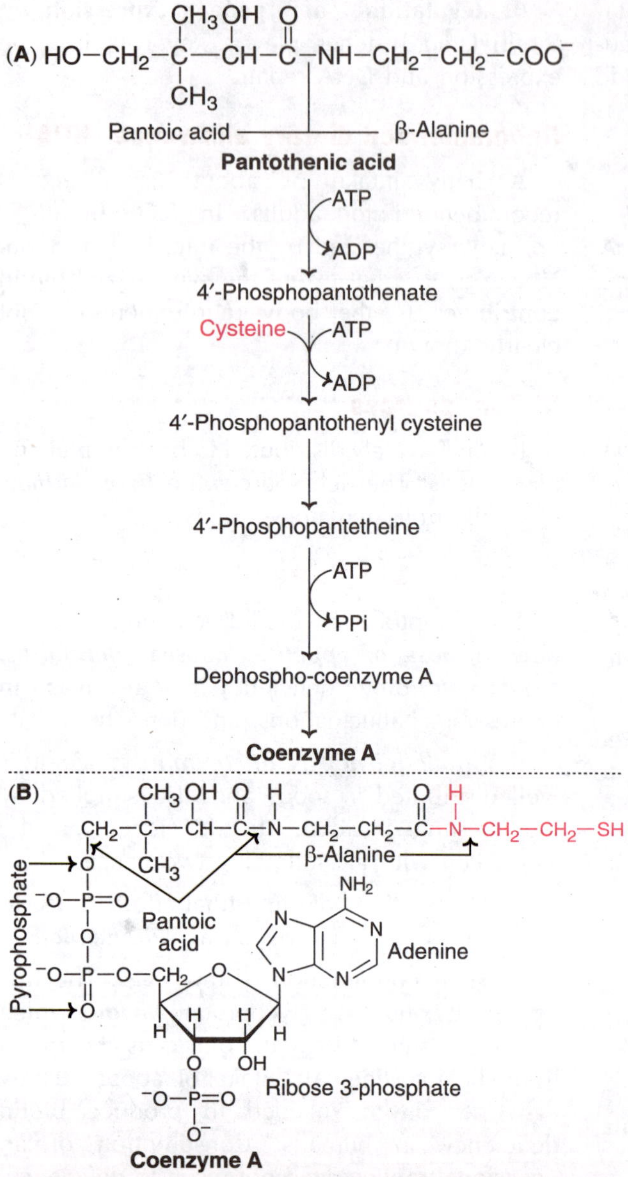

Fig. 7.30 : (A) Summary of the synthesis of coenzyme A from pantothenic acid (B) Structure of coenzyme A.

Chemistry and synthesis of coenzyme A

Pantothenic acid consists of two components, ***pantoic acid and β-alanine***, held together by a peptide linkage. Synthesis of coenzyme A from pantothenate occurs in a series of reactions (***Fig.7.30***). Pantothenate is first phosphorylated to which cysteine is added. Decarboxylation, followed by addition of AMP moiety and a phosphate (each from ATP) results in coenzyme A. The structure of coenzyme A consists of pantothenic acid joined to β-mercaptoethanolamine (thioethanolamine) at one end. On the other side, pantothenic acid is held by a phosphate bridge to adenylic acid. The adenylic acid is made up of adenine, and a phosphate linked to carbon-3 of ribose.

Biochemical functions

The functions of pantothenic acid are exerted through ***coenzyme A or CoA (A for acetylation)***. Coenzyme A is a central molecule involved in all the metabolisms (carbohydrate, lipid and protein). It plays a unique role in integrating various metabolic pathways. More than ***70 enzymes that depend on coenzyme A*** are known.

Coenzyme A has a terminal thiol or sulfhydryl group (–SH) which is the reactive site, hence CoA-SH is also used. Acyl groups (free fatty acids) are linked to coenzyme A by a thioester bond, to give acyl CoA. When bound to acetyl unit, it is called acetyl CoA. With succinate, succinyl CoA is formed. There are many other compounds bound to coenzyme A.

Coenzyme A serves as a ***carrier of activated acetyl or acyl groups*** (as thiol esters). This is comparable with ATP which is a carrier of activated phosphoryl groups.

A few examples of enzymes involved the participation of coenzyme A are given below.

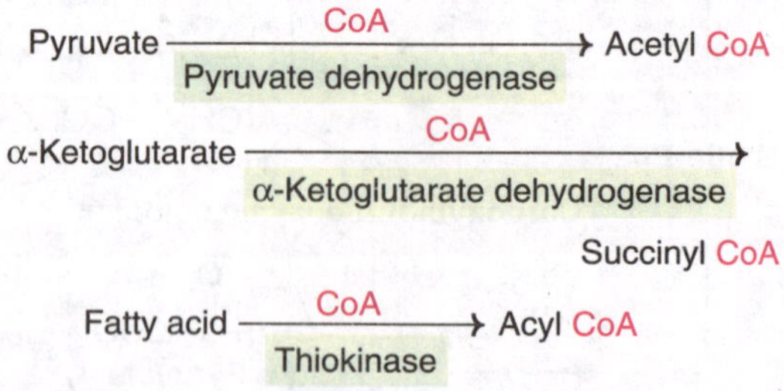

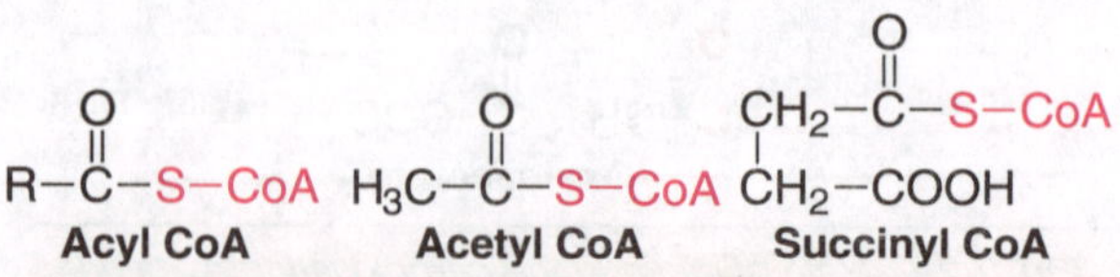

Fig. 7.31 : Selected examples of compounds bound to coenzyme A.

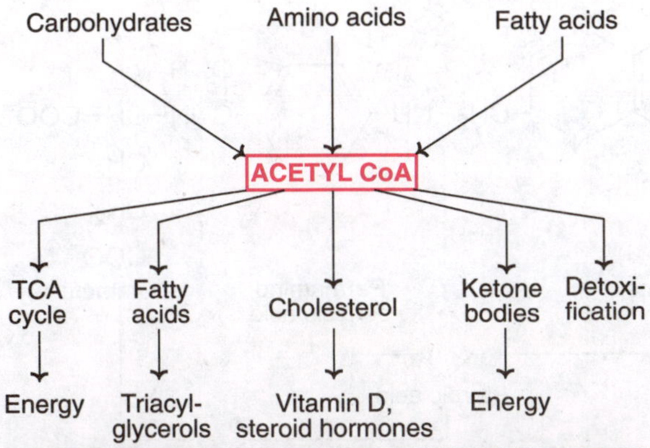

Fig. 7.32 : An overview of formation and utilization of acetyl CoA.

In some of the metabolic reactions, group transfer is important which occurs in a coenzyme A bound form.

Acetyl CoA + Choline ⟶ Acetylcholine + CoA

Acetyl CoA + Oxaloacetate ⟶ Citrate + CoA

Succinyl CoA + Acetoacetate ⟶ Acetoacetyl CoA + Succinate

Coenzyme A may be regarded as a ***coenzyme of metabolic integration***, since acetyl CoA is a central molecule for a wide variety of biochemical reactions, as illustrated in ***Fig.7.32***.

Succinyl CoA is also involved in many reactions, including the synthesis of porphyrins of heme.

Besides the various functions through coenzyme A, ***pantothenic acid*** itself is a ***component of fatty acid synthase complex*** and is involved in the formation of fatty acids.

Recommended dietary allowance (RDA)

The requirement of pantothenic acid for humans is not clearly known. A daily intake of about ***5-10 mg*** is advised for adults.

Dietary sources

Pantothenic acid is one of the ***most widely distributed*** vitamins found in plants and animals. The rich sources are egg, liver, meat, yeast, milk etc.

Deficiency symptoms

It is a surprise to biochemists that despite the involvement of pantothenic acid (as coenzyme A) in a great number of metabolic reactions, its deficiency manifestations have not been reported in humans. This may be due to the widespread distribution of this vitamin or the symptoms of pantothenic acid may be similar to other vitamin deficiencies. Dr. Gopalan, a world renowned nutritionist from India, linked the ***burning feet syndrome*** (pain and numbness in the toes, sleeplessness, fatigue etc.) with pantothenic acid deficiency.

Pantothenic acid deficiency in experimental animals results in anemia, fatty liver, decreased steroid synthesis etc.

FOLIC ACID

Folic acid or ***folacin*** (*Latin* : folium-leaf) or ***vitamin*** $\boldsymbol{B_9}$ is abundantly found in green leafy vegetables. It is important for ***one carbon metabolism*** and is required for the synthesis of certain amino acids, purines and the pyrimidine-thymine.

Chemistry

Folic acid consists of three components–***pteridine ring, p-amino benzoic acid (PABA)*** and ***glutamic acid***. There are multiple forms of folic acid with variable number of glutamic acid (1-7) residues. Folic acid mostly has one glutamic acid residue and is known as ***pteroyl-glutamic acid (PGA)***.

The active form of folic acid is ***tetrahydrofolate (THF or*** $\boldsymbol{FH_4}$***)***. It is synthesized from folic acid by the enzyme ***dihydrofolate reductase***. The reducing equivalents are provided by 2 moles of NADPH. The hydrogen atoms are present at positions 5, 6, 7 and 8 of THF (***Fig.7.33***).

Absorption, transport and storage

Most of the dietary folic acid found as polyglutamate with 3-7 glutamate residues (held by peptide bonds) is not absorbed in the intestine. The enzyme ***folate conjugase*** present in duodenum and jejunum splits the glutamate residues. Only the monoglutamate of folic acid is absorbed from the intestine. However, inside the cells, tetrahydrofolates are found as polyglutamates (with 5-6 amino acid residues) derivatives, which appear to be biologically most potent. As ***polyglutamate***, folic acid is stored to some extent in the liver. The body can store 10-12 mg of folic acid that will usually last for 2-3 months. In the circulation, N^5-methyl tetrahydrofolate is abundantly present.

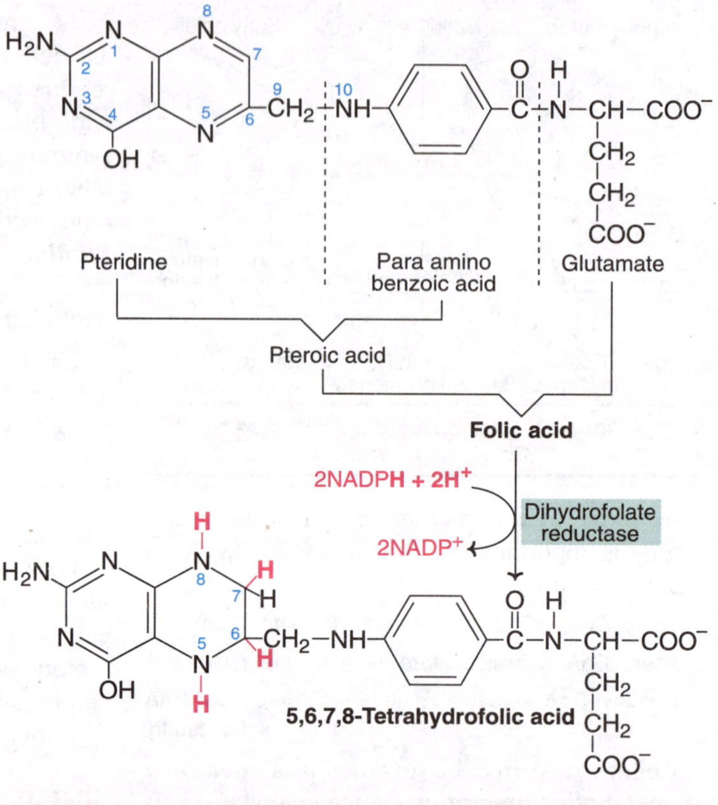

Fig. 7.33 : Conversion of folic acid to tetrahydrofolic acid (THF).

Biochemical functions

Tetrahydrofolate (THF or FH_4), the coenzyme of folic acid, is actively involved in the ***one carbon metabolism***. THF serves as an acceptor or donor of one carbon units (formyl, methyl etc.) in a variety of reactions involving amino acid and nucleotide metabolism.

The one carbon units bind with THF at position N^5 or N^{10} or on both N^5 and N^{10} of pteroyl structure. The attachment of formyl (—CHO) at position 5 of THF gives N^5-formyl tetrahydrofolate which is commonly known as ***folinic acid or citrovorum factor***. The other commonly found one carbon moieties and their binding with THF are given below.

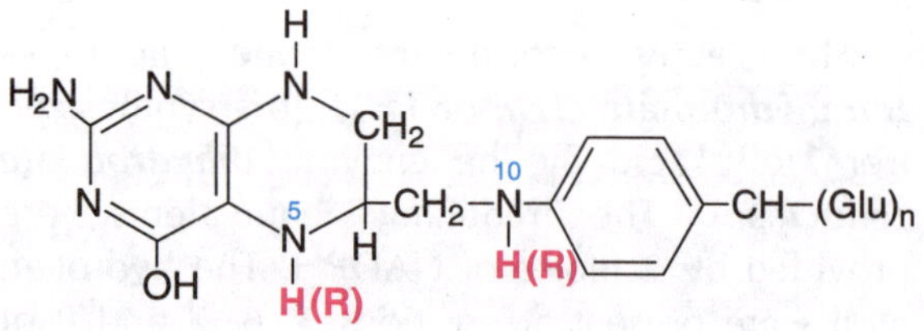

THF-1 carbon derivative	R group (one carbon unit)
N^5-Formyl THF	—CHO
N^{10}-Formyl THF	—CHO
N^5-Formimino THF	—CH=NH
N^5, N^{10}-Methenyl THF	≡CH
N^5, N^{10}-Methylene THF	$=CH_2$
N^5 – Methyl THF	$—CH_3$

The essential functions of THF in one carbon metabolism are summarized in ***Fig.7.34***.

The interrelationship between the various 1-carbon THF derivatives along with their involvement in the synthesis of different compounds is given in ***Fig.15.32*** (***Chapter 15***). Many important compounds are synthesized in one carbon metabolism.

1. ***Purines*** (carbon 2, 8) which are incorporated into DNA and RNA.

2. ***Pyrimidine*** nucleotide–deoxythymidylic acid (dTMP), involved in the synthesis of DNA.

3. ***Glycine***, serine, ethanolamine and choline are produced.

4. ***N-Formylmethionine***, the initiator of protein biosynthesis (in prokaryotes) is formed.

Tetrahydrofolate is mostly trapped as N^5-methyl THF in which form it is present in the circulation. Vitamin B_{12} is needed for the conversion of N^5-methyl THF to THF, in a reaction wherein homocysteine is converted to methionine. This step is essential for the

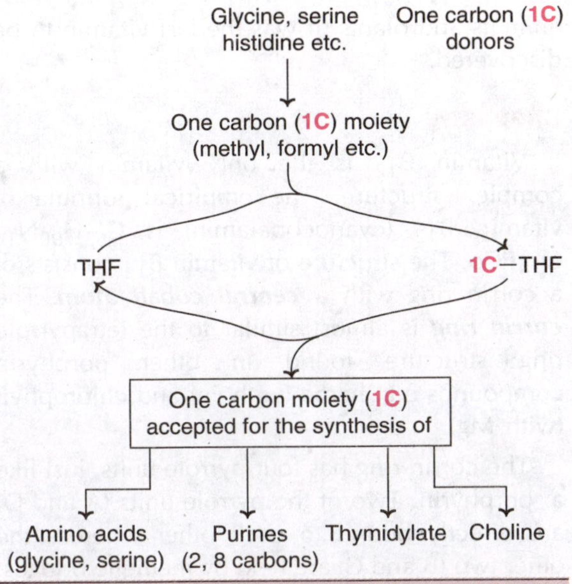

Fig. 7.34 : *An overview of one carbon metabolism (THF-Tetrahydrofolate).*

liberation of free THF and for its repeated use in one carbon metabolism. In B_{12} deficiency, conversion of N^5-methyl THF to THF is blocked (more details given under vitamin B_{12}).

Methylene, methenyl, and formyl-THF are interconvertible. When one carbon folates are in excess, formyltetrahydrofolates get oxidized to yield CO_2. By this way, a pool of free folate is maintained.

Recommended dietary allowance (RDA)

The daily requirement of folic acid is around ***200 μg***. In the women, higher intakes are recommended during pregnancy (400 μg/day) and lactation (300 μg/day).

Dietary sources

Folic acid is widely distributed in nature. The rich sources are ***green leafy vegetables***, whole grains, cereals, liver, kidney, yeast and eggs. Milk is rather a poor source of folic acid.

Deficiency symptoms

Folic acid deficiency is probably the ***most common vitamin deficiency***, observed ***primarily in the pregnant women***, throughout the world. The pregnant women, lactating women, women on oral contraceptives, and alcoholics are also susceptible to folate deficiency. The folic acid deficiency may be due to (one or more causes) ***inadequate dietary intake, defective absorption, use of anti-convulsant drugs*** (phenobarbitone, dilantin, phenytoin), ***and increased demand.***

In folic acid deficiency, decreased production of purines and dTMP is observed which impairs DNA synthesis. Due to a block in DNA synthesis, the maturation of erythrocytes is slowed down leading to macrocytic RBC. The rapidly dividing cells of bone marrow are seriously affected. The ***macrocytic anemia*** (abnormally large RBC) associated with megaloblastic changes in bone marrow is a characteristic feature of folate deficiency.

Folic acid deficiency in pregnant women may cause ***neural tube defects in the fetus***. Hence high doses of folic acid are recommended in pregnancy to prevent birth defects.

Folic acid is associated with the metabolism of histidine. ***Formiminoglutamate (FIGLU)***, formed in histidine metabolism transfers the one carbon fragment, formimino group (–CH=NH) to tetrahydrofolate to produce N^5-formimino THF. In case of folic acid deficiency, ***FIGLU accumulates and is excreted in urine. Histidine load test*** utilizing the excretion of FIGLU in urine is used to assess folic acid deficiency.

Medical Concepts/ Clinical Correlates

- Distinct deficiency conditions of certain B-complex vitamins are known

Thiamine	—	Beri-beri
Niacin	—	Pellagra
Folic acid	—	Macrocytic anemia
Riboflavin	—	Cheilosis, glossitis
Pyridoxine	—	Peripheral neuropathy
Cobalamin	—	Pernicious anemia

- B-complex vitamin deficiencies are usually multiple rather than individual with overlapping symptoms.
- A combined therapy of vitamin B_{12} and folic acid is commonly employed to treat the patients of megaloblastic anemia.
- Decreased folic acid and increased homocysteine are associated with higher incidence of neural tube defects in the fetus.

Folic acid and hyperhomocysteinemia

Elevated plasma levels of homocysteine are associated with increased risk of atherosclerosis, thrombosis and coronary heart disease. Hyperhomocysteinemia is mostly due to functional folate deficiency caused by impairment to form methyltetrahydrofolate by the enzyme methylene tetrahydrofolate reductase (***See Fig.7.39***). This results in a failure to convert homocysteine to methionine. Folic acid supplementation reduces hyperhomocysteinemia, and thereby the risk for various health complications.

Folic acid antagonists

Aminopterin and amethopterin (also called as methotrexate) are structural analogues of folic acid. They competitively ***inhibit dihydrofolate reductase*** and block the formation of THF. The biosynthesis of purines, thymine nucleotides, and hence DNA is impaired. This results in the blockage of cell proliferation. ***Aminopterin and methotrexate are successfully used in the treatment of many cancers, including leukemia***.

Trimethoprim (a component of the drug septran or bactrim) and ***pyrimethamine*** (***antimalarial drug***) are structurally related to folic acid. They inhibit dihydrofolate reductase, and the formation of THF. Trimethoprim is used to treat bacterial infections of sore throat, urinary tract, gastrointestinal tract etc.

Sulfonamides are ***structural analogues of PABA***. They competitively inhibit the enzyme (dihydropteroate synthase) responsible for the incorporation of PABA into pteridine to produce folic acid. For this reason, sulfonamides are used as antibacterial drugs. Sulfonamides, have no effect on human body, since folic acid is not synthesized and supplied through the diet.

It is evident from the above that ***inhibitors*** of folate metabolism are ***useful for treatment of cancer, bacterial infections, malaria***.

COBALAMIN (VITAMIN B_{12})

Vitamin B_{12} is also known as ***anti-pernicious anemia vitamin***. It is a unique vitamin, synthesized by only microorganisms and not by animals and plants. It was the last vitamin to be discovered.

Chemistry

Vitamin B_{12} is the only vitamin with a complex structure. The empirical formula of vitamin B_{12} (cyanocobalamin) is $C_{63}H_{90}N_{14}O_{14}PCo$. The structure of vitamin B_{12} consists of a corrin ring with a ***central cobalt atom***. The ***corrin ring*** is almost similar to the tetrapyrrole ring structure found in other porphyrin compounds e.g. heme (with Fe) and chlorophyll (with Mg).

The corrin ring has four pyrrole units, just like a porphyrin. Two of the pyrrole units (A and D) are directly bound to each other whereas the other two (B and C) are held by methene bridges. The groups namely methyl, acetamide and propionamide are the substituents on the pyrrole rings. Vitamin B_{12} has cobalt atom in a coordination state of six. Cobalt present at the centre of the corrin ring is bonded to the four pyrrole nitrogens. Cobalt also holds (below the corrin plane) dimethylbenzimidazole (DMB) containing ribose 5-phosphate and amino-isopropanol. A nitrogen atom of dimethylbenzimidazole is linked to cobalt. The amide group of aminoisopropanol binds with D ring of corrin. The cobalt atom also possesses a sixth substituent group located above the plane of corrin ring (***Fig.7.35***). The substituent group may be one of the following

1. Cyanide (predominant) in ***cyanocobalamin*** (B_{12a})
2. Hydroxyl in ***hydroxycobalamin*** (B_{12b})
3. Nitrite in ***nitrocobalamin*** (B_{12c}).

There are two coenzyme forms of vitamin B_{12} (***Fig.7.36***).

(a) **5′-Deoxyadenosyl cobalamin**, cyanide is replaced by 5′ deoxyadenosine forming an unusal carbon cobalt bond.

(b) **Methylcobalamin** in which cyanide is replaced by methyl group.

Absorption, transport and storage

The vitamin B_{12} is present in the diet in a bound form to proteins. B_{12} is liberated by the

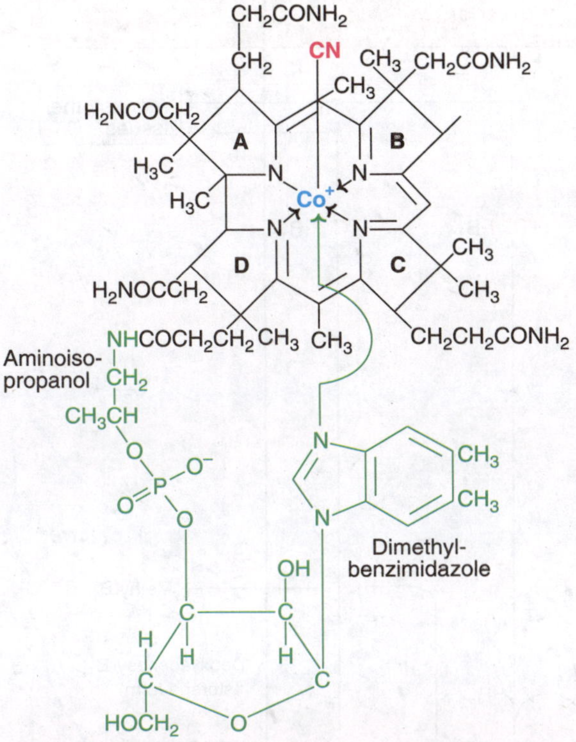

Fig. 7.35 : Structure of vitamin B_{12} (cyanocobalamin).

enzymes (acid hydrolases) in the stomach and binds to ***cobalophilin*** (a protein, secreted in the saliva). In the duodenum, cobalophilin-B_{12} complex is hydrolysed to release vitamin B_{12}.

The dietary source of B_{12} is known as ***extrinsic factor of Castle***. The stomach secretes a special protein called ***intrinsic factor (IF)***. It is a glycoprotein (8-15% carbohydrate) with a molecular weight around 50,000. Intrinsic factor is resistant to proteolytic digestive enzymes.

The cobalamin–IF complex travels through the gut. The complex binds to specific receptors on the surface of the mucosal cells of the ileum. The binding of the complex and entry of B_{12} into the mucosal cells is mediated by Ca^{2+} ions. In the mucosal cells, B_{12} is converted to methylcobalamin (***Fig.7.37***). It is then transported in the circulation in a bound form to proteins namely ***transcobalamins*** (TC-I, TC-II). Methylcobalamin is mostly bound to TC-I (90%) and to a lesser degree to TC-II (10%). It is believed that TC-I acts as a repository of B_{12}, while TC-II mediates the tissue uptake of B_{12}. Methylcobalamin which is in excess is taken up by the liver, converted to deoxyadenosyl B_{12} and stored in this form. It is believed that liver can store about 4-5 mg, an amount sufficient to meet the body requirements of B_{12} for 4-6 years.

Biochemical functions

About ten enzymes requiring vitamin B_{12} have been identified. Most of them are found in bacteria (glutamate mutase, ribonucleotide reductase etc.). There are only ***two reactions in mammals*** that are dependent on vitamin B_{12}.

1. **Synthesis of methionine from homocysteine :** Vitamin B_{12}, as methylcobalamin is used in this reaction. This is an important

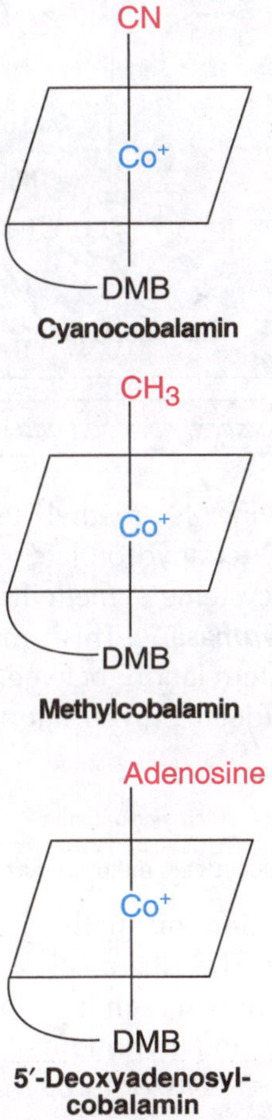

Fig. 7.36 : *Coenzyme derivatives of vitamin B_{12}* ***(Note :*** *Corrin ring represented diagrammatically is identical in all; DMB-Dimethylbenzimidazole).*

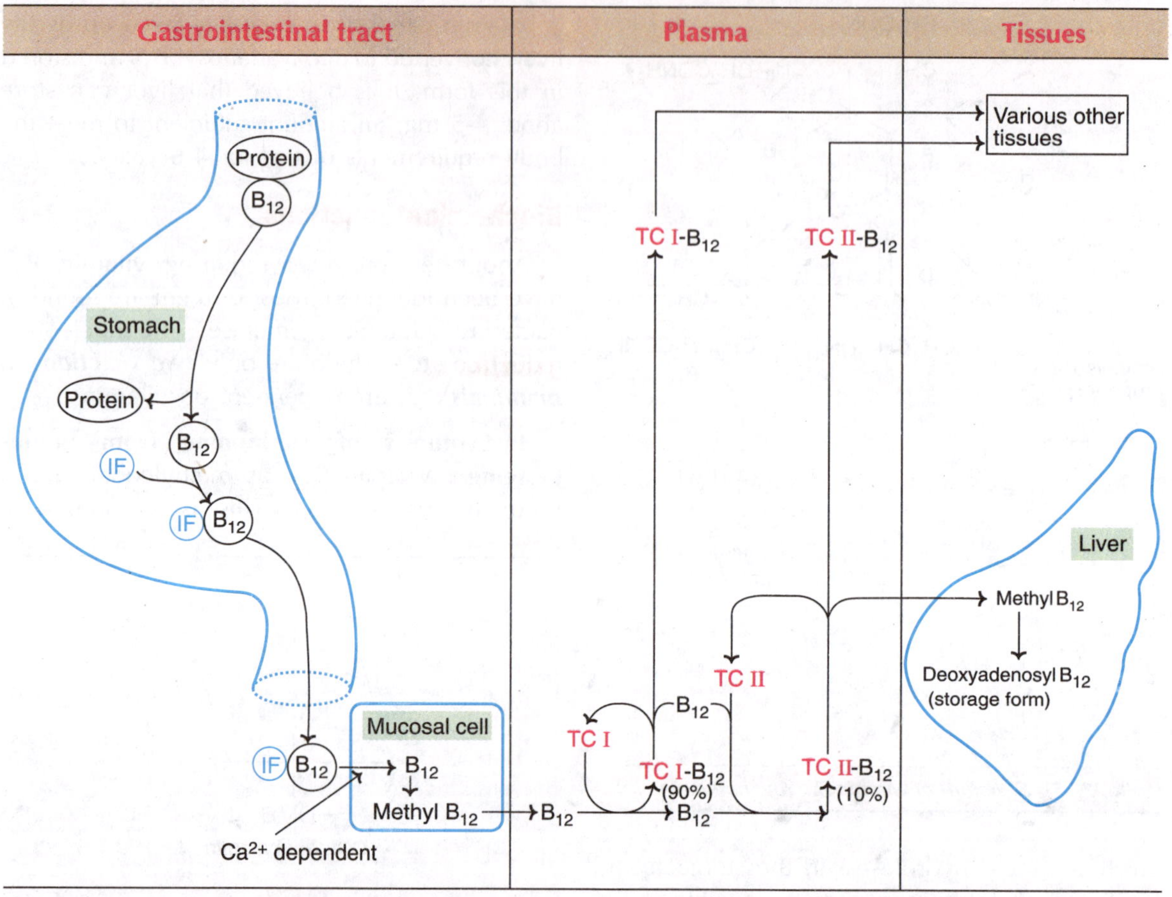

Fig. 7.37 : Absorption, transport and storage of vitamin B_{12} (IF-Intrinsic factor; TC-Transcobalamins (TC–I, TC–II).

reaction involving N^5-methyl tetrahydrofolate from which tetrahydrofolate is liberated (enzyme-homocysteine methyltransferase or ***methionine synthase***). This metabolic step signifies the interrelation between vitamin B_{12} and folic acid (details given later)

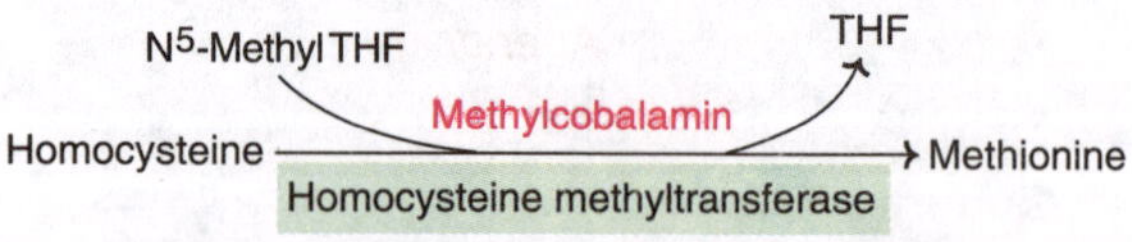

2. **Isomerization of methymalonyl CoA to succinyl CoA :** The degradation of odd chain fatty acids, certain amino acids (valine, isoleucine etc.) and pyrimidines (thymine and uracil) produce directly or through the mediation of propionyl CoA, an important compound methylmalonyl CoA. This is converted by the enzyme ***methylmalonyl CoA mutase*** to succinyl CoA in the presence of B_{12} coenzyme, deoxyadenosyl cobalamin (***Fig.7.38***). This reaction involves hydrogen transfer and intramolecular rearrangement. In B_{12} deficiency, methylmalonyl CoA accumulates and is excreted in urine as methylmalonic acid.

Recommended dietary allowance (RDA)

A daily intake ***1-2 µg*** of vitamin B_{12} is adequate to meet the ***adult*** requirements. For children, ***0.5-1.5 µg/day*** is recommended. During ***pregnancy and lactation***, the requirement is ***3 µg/day***.

Dietary sources

Foods of ***animal origin*** are the only sources for vitamin B_{12}. The rich sources are ***liver, kidney, milk, curd, eggs,*** fish, pork and chicken. Curd is a better source than milk, due to the synthesis of B_{12} by *Lactobacillus*.

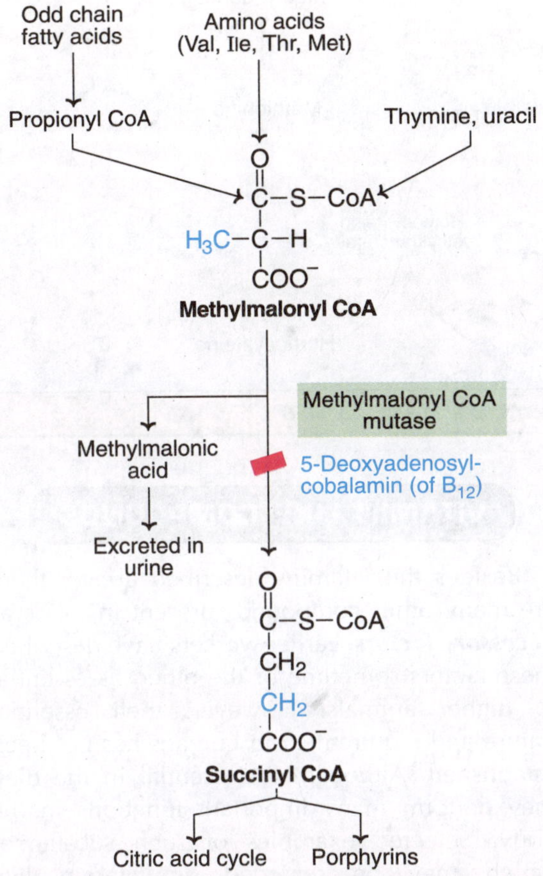

Fig. 7.38 : *Role of vitamin B_{12} in isomerization of methylmalonyl CoA to succinyl CoA (▬-Blockade in B_{12} deficiency).*

Vitamin B_{12} is synthesized only by micro-organisms (anaerobic bacteria). Plants cannot synthesize, hence B_{12} is never found in plant foods. Animals obtain B_{12} either by eating foods, derived from other animals or from the intestinal bacterial synthesis. Small quantities of vitamin ***B_{12}, synthesized by bacteria on the surface of fruits***, may partially meet the requirements.

Deficiency symptoms

The most important disease associated with vitamin B_{12} deficiency is ***pernicious anemia***. It is characterized by low hemoglobin levels, decreased number of erythrocytes and neurological manifestations. One or more of the following causes are attributed to the occurrence of pernicious anemia.

1. ***Autoimmune destruction*** of gastric parietal cells that secrete intrinsic factor. In the absence of IF, vitamin B_{12} cannot be absorbed.

2. ***Hereditary malabsorption*** of vitamin B_{12}.

3. Partial or total ***gastrectomy*** – these individuals become intrinsic factor deficient.

4. Insufficient production of IF and/or gastric HCl, occasionally seen in older people.

5. Dietary deficiency of B_{12} is seen among ***strict vegetarians*** of low socioeconomic group in the developing countries (India, Srilanka).

From the foregoing discussion, it is clear that pernicious anemia is more a disease of the stomach than due to the deficiency of vitamin B_{12}.

B_{12} deficiency is also associated with ***neuronal degeneration and demyelination of nervous system***. The symptoms include paresthesia (numbness and tingling) of fingers and toes. In advanced stages, confusion, loss of memory and even psychosis may be observed. The neurological symptoms of pernicious anemia are believed to be due to the accumulation of methylmalonyl CoA that interferes in myelin sheath formation in two possible ways.

1. The biosynthesis of fatty acids, required for myelin formation, is imparied. This is because, ***methylmalonyl CoA acts as a competitive inhibitor of malonyl CoA in fatty acid synthesis.***

2. Methylmalonyl CoA can substitute malonyl CoA in fatty acid synthesis, resulting in a new type of branched chain fatty acids, that will disrupt the normal membrane structure.

The excretion of ***methylmalonic acid*** (elevated) ***in urine*** and estimation of serum B_{12} level are used to assess B_{12} deficiency. In recent years, ***Schilling test*** that determines the absorption of vitamin B_{12} from intestinal tract is employed to detect B_{12} deficiency.

Treatment

Vitamin B_{12} is administered in therapeutic doses (100-1000 μg) intramuscularly. Folic acid administration can also reverse hematological abnormalities observed in B_{12} deficiency. However, the neurological symptoms persist. Therefore, a combined supplementation of B_{12} and folate is employed to treat the patients with megaloblastic anemias.

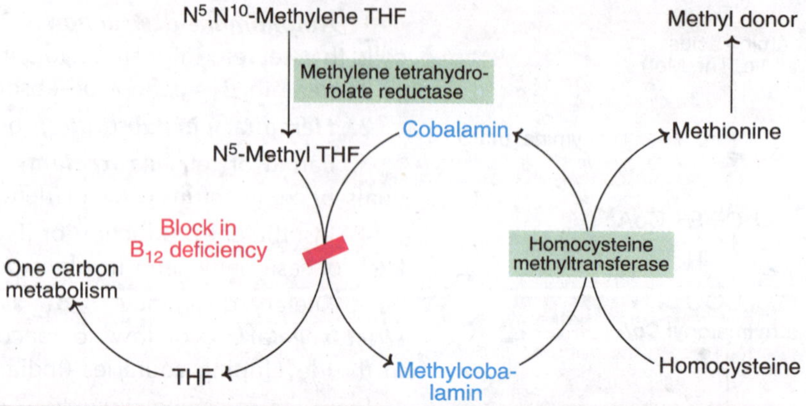

Fig. 7.39 : *Interrelationship between folic acid and vitamin B_{12}*

INTERRELATION BETWEEN FOLIC ACID AND VITAMIN B_{12}—FOLATE TRAP OR METHYL TRAP HYPOTHESIS

The deficiency of either folate or vitamin B_{12} results in a similar type of anemia. This suggests a probable biochemical interrelation between these two vitamins. There is only one metabolic reaction known, common to folate and vitamin B_{12} (***Fig.7.39***).

In vitamin B_{12} deficiency, increased folate levels are observed in plasma. The activity of the enzyme ***homocysteine methyltransferase*** (methionine synthase) ***is low in B_{12} deficiency***. As a result, the only major pathway for the conversion of N^5-methyl THF to tetrahydrofolate is blocked and body THF pool is reduced. Essentially, almost the entire body folate becomes trapped as N^5-methyl THF. This is known as ***folate trap or methyl trap***. In this manner, B_{12} deficiency results in decreased folate coenzymes that leads to reduced nucleotide and DNA synthesis.

Although the tissue folate levels are adequate or high, there is a ***functional folate deficiency*** due to the lack of THF pool. The outcome is the development of ***megaloblastic anemia***. Administration of the amino acid methionine has been shown to partially correct the symptoms of B_{12} deficiency. This is due to the fact that the formation of N^5-methyl THF is inhibited by S-adenosylmethionine. A combined therapy of vitamin B_{12} and folic acid is generally employed to treat the patients with megaloblastic anemia.

VITAMIN LIKE COMPOUNDS

Besides the vitamins described above, there are many other compounds present in foods as ***accessory factors***. Earlier workers have described these factors sometime or the other, as essential to higher animals. However, their essential nature and requirement in humans has not been established. Although not essential in the diet, they perform many important functions in the body. Selected examples of such substances which may be regarded as vitamin like compounds are described here.

CHOLINE

Choline is ***trimethylhydroxy ethanolamine***.

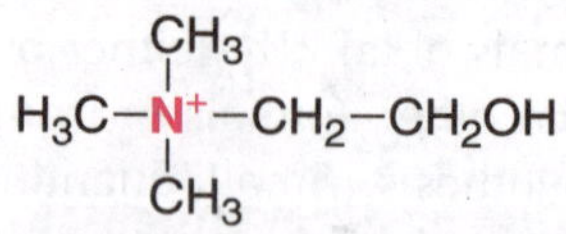

It can be synthesized in the body (from serine). It is also available from many dietary sources (e.g. milk, eggs, liver, cereals etc.). ***Radiolabelled choline*** (^{11}C-choline) is used in ***medical imaging***.

Biochemical functions

1. Choline, as a component of ***phospholipids*** (lecithins), is involved in membrane structure and lipid transport.

Medical Concepts/ Clinical Correlates

- Sulfonamides serve as antibacterial drugs by inhibiting the incorporation of PABA to produce folic acid.
- Drugs that inhibit folate metabolism are used in the treatment of cancers (e.g. methotrexate) in control of bacterial (e.g. trimethoprim) and malarial (e.g. pyrimethamine) infections.
- Alcoholics are more prone for B-complex vitamin deficiencies. Thiamine deficiency is observed in about 80%, pyridoxine in over 50% and folic acid deficiency in about 30% of alcoholics.
- Lipoic acid is therapeutically useful as an antioxidant to prevent stroke, myocardial infarction etc.

2. Choline prevents the accumulation of fat in liver (as ***lipotropic factor***). It promotes the synthesis of phospholipids and lipoproteins and the disposal of ***triacylglycerols*** from liver.

3. Due to the presence of three methyl groups (one carbon fragments), choline is actively involved in ***one carbon metabolism***.

4. Choline is a precursor for the synthesis of acetylcholine which is required for transmission of nerve impulse.

Choline—an essential nutrient?

As such, choline can be synthesized and reutilized in humans. This may however, be insufficient to meet the body needs. Some workers label choline as an essential dietary nutrient with RDA in the range of 400–500 mg/day.

INOSITOL

Inositol is ***hexahydroxy-cyclohexane***. It is also known as *myo*-inositol or *meso*-inositol.

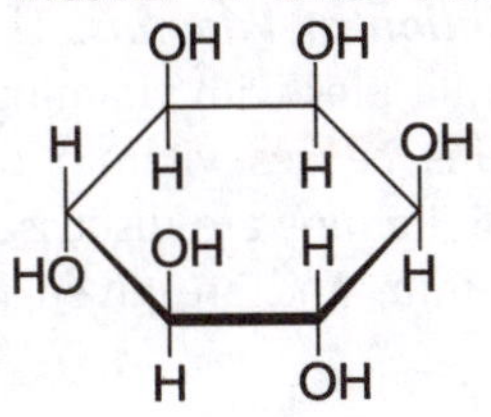

Biochemical functions

1. Inositol is required for the synthesis of ***phosphatidylinositol*** (lipositol) which is a constituent of cell membrane.

2. It acts as a ***lipotropic factor*** (along with choline) and prevents the accumulation of fat in liver.

3. For some hormones, inositol acts as a ***second messenger*** at the membrane level for the release of Ca^{2+} ions.

4. Phytin is hexaphosphate of inositol found in plants. It prevents the absorption of iron and calcium from the intestine.

5. Inositol is used in the ***treatment of polycystic ovary syndrome*** (***PCOS***). It helps by increasing insulin sensitivity which improves ovarian function.

LIPOIC ACID

Lipoic acid (thioctic acid) is a sulfur containing fatty acid (6,8-dithiooctanoic acid). It exists in an oxidized and reduced form. Lipoic acid is fat as well as water soluble.

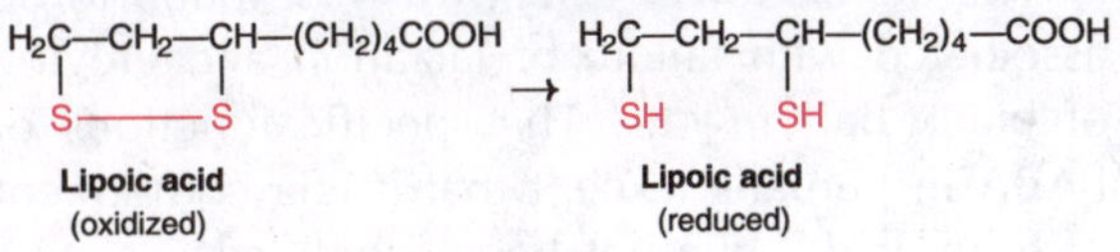

Biochemical functions

Lipoic acid is involved in the decarboxylation reactions along with other vitamins (thiamine, niacin, riboflavin and pantothenic acid). The conversion of pyruvate to acetyl CoA (by pyruvate dehydrogenase) and α-ketoglutarate to succinyl CoA (by α-ketoglutarate dehydrogenase) require lipoic acid.

Therapeutic uses of lipoic acid

In recent years, administration of high doses (100–600 mg/day) of lipoic acid (or dihydrolipoic acid) is gaining importance. Being fat and water soluble, it can comfortably reach various tissues. The therapeutic applications of lipoic acid are related to its antioxidant property (regarded as universal antioxidant), some of them are listed

- Reduces the free radicals in brain that otherwise contribute to Alzheimer's disease and multiple sclerosis.

- Lipoic acid stimulates production of glutathione (GSH), besides helping in the recyling of vitamins E and C.
- Reduces insulin resistance, and brings down plasma low density lipoproteins.
- May be useful in the prevention of stroke and myocardial infarction.

PARA AMINOBENZOIC ACID

Para aminobenzoic acid (PABA) is a structural constituent of folic acid. PABA may be regarded as a ***vitamin in another vitamin*** (folic acid)

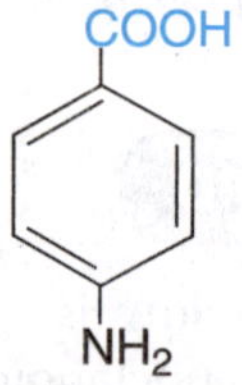

PABA

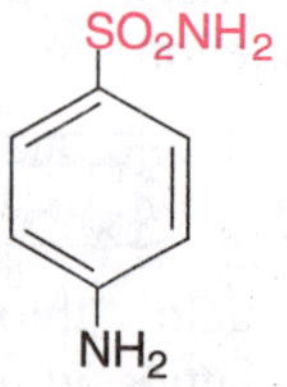

Sulfanilamide

The deficiency of PABA was first found to be associated with failure of lactation and greying of black hair in rats. The specific functions of PABA in humans, except that it is a component of folic acid, have not been identified.

PABA is synthesized by the bacteria and is essential for their growth. The sulfa drug sulfanilamide (p-amino benzene sulfanilamide) is a structural analogue of PABA. Sulfanilamide competes with PABA and acts as a bacteriostatic agent. Ingestion of large doses of PABA will compete with the action of drugs and therefore should be avoided during sulfanilamide therapy (***trade name—sulfonamides***).

BIOFLAVONOIDS

Szent-Gyorgi and his associates (1936) observed that flavonoids, isolated from lemon peel (known as citrin) were responsible for maintenance of normal capillary permeability. The term ***vitamin P*** (P for ***permeability***) was used to this group of substances. However, they are commonly known as bioflavonoids.

Bioflavonoids act as antioxidants and protect ascorbic acid from being destroyed. It is suggested that this antioxidant property may be responsible for maintenance of capillary permeability. Bioflavonoids have been used to correct the vascular abnormality in humans.

Bioflavonoids are found in peel and pulp of citrus fruits, tobacco leaves and many vegetables. The requirement of these compounds in humans has not been established.

TABLE 7.4 Selected list of antivitamins/ vitamin antagonists along with corresponding vitamins

Antivitamin/ vitamin antgonists	Vitamin
Dicumarol Warfarin	Vitamin K
Thiaminase Pyrithiamine Oxythiamine	Thiamine
Galactoflavin	Riboflavin
Isoniazid Deoxypyridoxine	Pyridoxine
Avidin Desthiobiotin	Biotin
Aminopterin Methotrexate Trimethoprim	Folic acid
Sulfonilamide	Para-aminobenzoic acid

ANTIVITAMINS

Antivitamins are ***antagonistic to*** (oppose and block) the ***action of vitamins***. They usually have structural similarities with vitamins. Administration of antivitamins causes vitamin deficiencies. The common antivitamins are discussed as antagonists for each vitamin, and are given in ***Table 7.4***.

SUMMARY

- *Vitamins are accessory food factors required in the diet. They are classified as fat soluble (A, D, E and K) and water soluble (B-complex and C).*
- *Vitamin A is involved in vision, proper growth, differentiation and maintenance of epithelial cells. Its deficiency results in night blindness.*
- *The active form of vitamin D is calcitriol which functions like a steroid hormone and regulates plasma levels of calcium and phosphate. Vitamin D deficiency leads to rickets in children and osteomalacia in adults.*
- *Vitamin E is a natural antioxidant necessary for normal reproduction in many animals.*
- *Vitamin K has a specific coenzyme function. It catalyses the carboxylation of glutamic acid residues in blood clotting factors (II, VII, IX and X) and converts them to active form.*
- *Vitamin C is involved in the hydroxylation of proline and lysine in the formation of collagen. Scurvy is caused by ascorbic acid deficiency. Therapeutic use of megadoses of vitamin C, to cure everything from common cold to cancer, has become controversial.*
- *Thiamine (B_1), as a cocarboxylase (TPP) is involved in energy releasing reactions. Its deficiency leads to beri-beri.*
- *The coenzymes of riboflavin (FAD and FMN) and niacin (NAD^+ and $NADP^+$) take part in a variety of oxidation-reduction reactions connected with energy generation. Riboflavin deficiency results in cheilosis and glossitis, while niacin deficiency pellagra.*
- *Pyridoxal phosphate (PLP), the coenzyme of vitamin B_6, is mostly associated with amino acid metabolism. PLP participates in transamination, decarboxylation, deamination and condensation reactions.*
- *Biotin (anti-egg white injury factor) participates as a coenzyme in carboxylation reactions of gluconeogenesis, fatty acid synthesis etc.*
- *Coenzyme A (of pantothenic acid) is involved in the metabolism of carbohydrates, lipids and amino acids, and their integration.*
- *Tetrahydrofolate (THF), the coenzyme of folic acid participates in the transfer of one carbon units (formyl, methyl etc.) in amino acid and nucleotide metabolism. Megaloblastic anemia is caused by folic acid deficiency.*
- *Vitamin B_{12} has two coenzymes, deoxyadenosylcobalamin and methylcobalamin. B_{12} deficiency results in pernicious anemia.*
- *Certain vitamin like compounds (choline, inositol, PABA, lipoic acid) participate in many biochemical reactions.*

THE READER MUST REFER APPENDIX VI (Page Nos : 738 – 756) FOR RELATED CLINICAL CASE STUDIES (Nos : 3, 4, 5, 6, 7, 8, 9, 35)

Self-Assessment Exercises

I. Essay questions

1. Classify vitamins and briefly discuss their functions and deficiency disorders.
2. Describe the chemistry, biochemical functions, daily requirements, sources and deficiency manifestations of vitamin A.
3. Write an account of folic acid involvement in one carbon metabolism.
4. Discuss the biochemical functions of vitamin C. Add a note on the therapeutic use of megadoses of this vitamin.
5. Write briefly about the coenzymes involved in oxidation-reduction reactions.

II. Short notes

(a) Vitamin D is a hormone-justify, (b) Thiamine pyrophosphate, (c) Coenzymes of niacin, (d) Pyridoxal phosphate in transamination, (e) Folate trap, (f) Tocopherol, (g) Vitamin K in carboxylation, (h) Biocytin, (i) Choline, (j) Pernicious anemia.

III. Fill in the blanks

1. The A in coenzyme A stands for____________________.
2. The vitamin containing isoalloxazine ring____________________.
3. The vitamin that is regarded as a vitamin in search of a disease____________________.
4. Anti-tuberculosis drug, isonicotinic acid hydrazide (INH) leads to the deficiency of vitamin____________________.
5. The egg injury factor present in raw egg white____________________.
6. The 'burning feet syndrome' in man is associated with the deficiency of____________________.
7. The vitamin that is synthesized by only microorganisms____________________.
8. The three Ds in pellagra stand for, ____________, ____________ and ____________.
9. The fat soluble vitamin required for carboxylation reaction____________________.
10. FIGLU (formimino glutamic acid) is excreted in urine in the deficiency of vitamin____________________.

IV. Multiple choice questions

11. Which one of the vitamin A functions as a steroid hormone
    (a) Retinal (b) Retinol (c) Provitamin A (d) β-Carotene.
12. The functionally active form of vitamin D is
    (a) Cholecalciferol (b) Ergocalciferol (c) Dehydrocholesterol (d) Calcitriol.
13. The metabolite excreted in urine in thiamine deficiency
    (a) Pyruvate (b) Glucose (c) Xanthurenic acid (d) FIGLU.
14. The coenzyme directly concerned with the synthesis of biogenic amines
    (a) TPP (b) $NADP^+$ (c) Biotin (d) Pyridoxal phosphate.
15. Folic acid antagonist(s) used in the treatment of cancer
    (a) Methotrexate (b) Trimethoprim (c) Sulfonamide (d) All the three.

PHYSIOLOGICAL BIOCHEMISTRY

SECTION TWO

Digestion and Absorption

8

The natural foodstuffs speak

"Complex is the ingested food,
But digested to simpler products,
Absorbed by intestinal mucosal cells,
Assimilated and utilized by all cells."

Food is the basic and essential requirement of man for his very existence. The food we eat consists of carbohydrates, proteins, lipids, vitamins and minerals. The bulk of the food ingested is mostly in a complex form which cannot, as such, be absorbed by the body.

Digestion is a process involving the hydrolysis of large and complex organic molecules of foodstuffs into smaller and preferably water-soluble molecules which can be easily absorbed by the gastrointestinal tract for utilization by the organism. Digestion of macromolecules also promotes the absorption of fat soluble vitamins and certain minerals.

Cooking of the food, and mastication (in the mouth) significantly ***improve the digestibility*** of foodstuffs by the enzymes.

Gastrointestinal tract

Digestion as well as absorption are complicated processes that occur in the gastrointestinal tract (***GIT***) involving many organs (***Fig.8.1***). The essential organs with their respective major functions are given in ***Table 8.1***. The digestive organs possess a large reserve capacity. For instance, pancreas secretes enzymes 5-10 fold higher than required for digestion of foods normally ingested.

Table 8.1 Organs of gastrointestinal tract with their major functions in digestion and absorption

Organ	Major function(s)
Mouth	Production of saliva containing α-amylase; partial digestion of polysaccharides
Stomach	Elaboration of gastric juice with HCl and proteases; partial digestion of proteins
Pancreas	Release of $NaHCO_3$ and many enzymes required for intestinal digestion
Liver	Synthesis of bile acids
Gall bladder	Storage of bile
Small intestine	Final digestion of foodstuffs; absorption of digested products
Large intestine	Mostly absorption of electrolytes; bacterial utilization of certain non-digested and/or unabsorbed foods

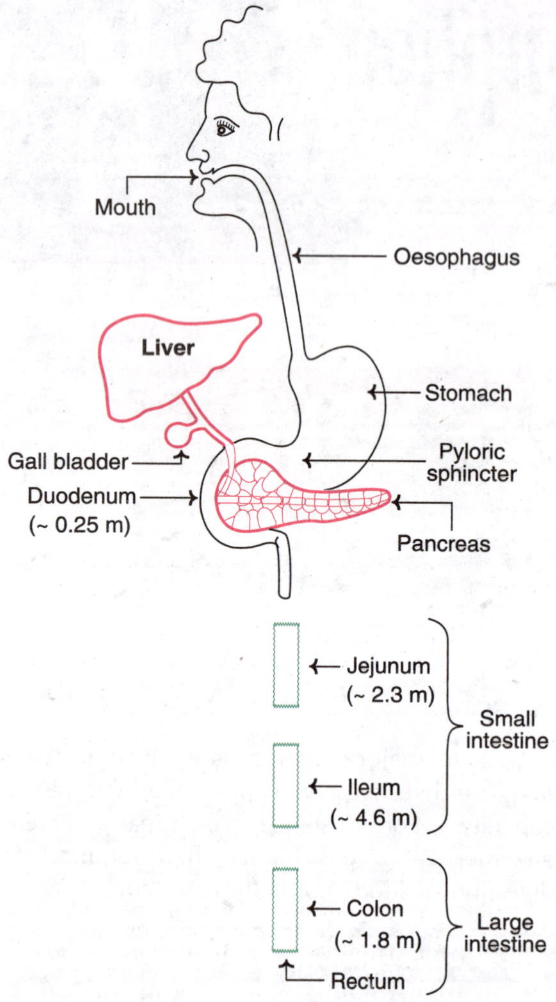

Fig. 8.1 : Diagrammatic representation of gastrointestinal tract.

The digestion and absorption of individual foods, namely carbohydrates, proteins, lipids and nucleic acids, is described here. The ***gastrointestinal hormones*** are discussed under hormones (***Chapter 19***).

CARBOHYDRATES

The principal dietary carbohydrates are ***polysaccharides (starch, glycogen), disaccharides (lactose, sucrose)*** and, to a minor extent, monosaccharides (glucose, fructose). The structures of carbohydrates are described in ***Chapter 2***.

Digestion

The digestion of carbohydrates occurs briefly in mouth and ***largely in the intestine***. The polysaccharides get hydrated during heating which is essential for their efficient digestion. The hydrolysis of glycosidic bonds is carried out by a group of enzymes called ***glycosidases*** (***Fig.8.2***).

Digestion in the mouth : Carbohydrates are the only nutrients for which the digestion begins in the mouth to a significant extent. During the process of mastication, ***salivary α-amylase (ptyalin)*** acts on starch randomly and cleaves α-1,4-glycosidic bonds. The products formed include α-limit dextrins, maltotriose and maltose.

Carbohydrates not digested in the stomach : The enzyme salivary amylase is inactivated by ***high acidity (low pH)*** in the stomach. Consequently, the ongoing degradation of starch is stopped.

Digestion in the small intestine : The acidic dietary contents of the stomach, on reaching small intestine, are neutralized by bicarbonate produced by pancreas. The ***pancreatic α-amylase*** acts on starch and specifically acts on α-1,4-glycosidic bonds and not on α-1,6-bonds. The resultant products are disaccharides (maltose, isomaltose) and oligosaccharides (***Fig.8.3***).

The final digestion of di- and oligosaccharides to monosaccharides (***Fig.8.4***) primarily occurs at the mucosal lining of the upper jejunum. This is carried out by ***oligosaccharidases*** (e.g. glucoamylase acting on amylose) ***and disaccharidases*** (e.g. maltase, sucrase, lactase). The

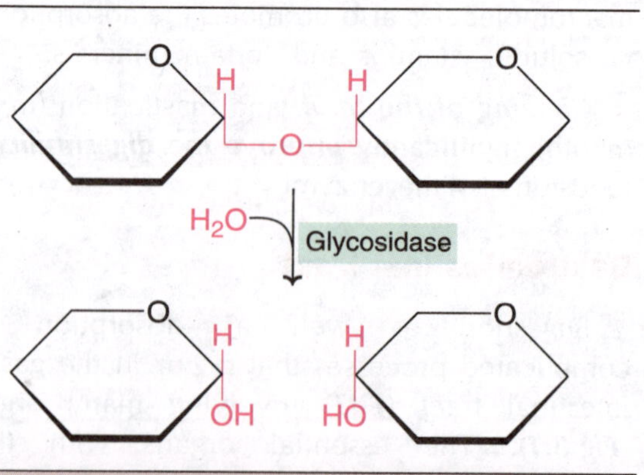

Fig. 8.2 : Hydrolysis of a glycosidic bond.

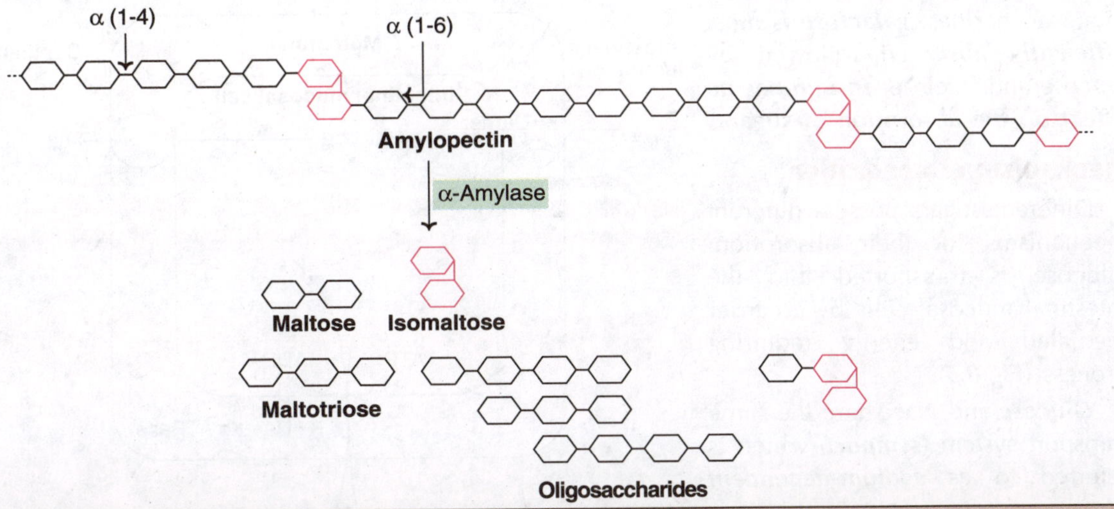

Fig. 8.3 : Degradation of amylopectin by salivary or pancreatic α-amylase.

enzyme ***sucrase*** is capable of hydrolysing a large quantity of table sugar (sucrose). In contrast, ***lactase (β-galactosidase)*** is the rate-limiting, and, consequently, the utilization of milk sugar (lactose) is limited in humans.

Absorption of monosaccharides

The principal monosaccharides produced by the digestion of carbohydrates are glucose, fructose and galactose. Of these, glucose accounts for nearly 80% of the total monosaccharides. The absorption of sugars mostly takes place in the ***duodenum*** and ***upper jejunum*** of small intestine.

There exists considerable variation in the absorption of different monosaccharides. The relative rates of absorption of important monosaccharides in comparison with glucose are given below

Glucose	—	***100***
Galactose	—	110
Fructose	—	43
Mannose	—	20
Xylose	—	15
Arabinose	—	9

It is observed that hexoses are more rapidly absorbed than pentoses. Further, among the

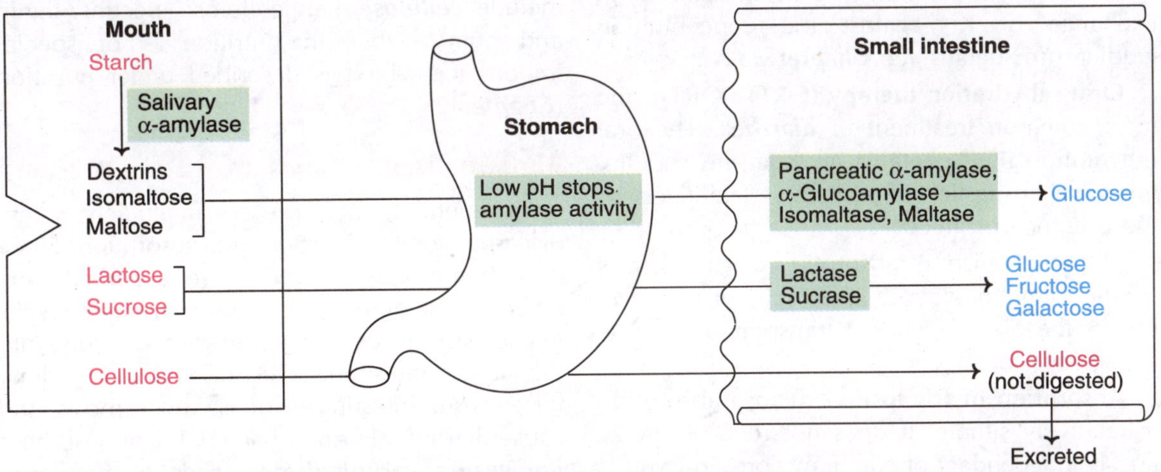

Fig. 8.4 : Overview of digestion of carbohydrates.

monosaccharides, ***galactose is most efficiently absorbed*** followed by glucose and fructose. ***Insulin has no effect on the absorption of sugars.***

Mechanism of absorption

Different sugars possess different mechanisms for their absorption. Glucose is transported into the intestinal mucosal cells by a carrier mediated and energy requiring process (***Fig.8.5***).

Glucose and ***Na***$^+$ share the same transport system (***symport***) which is referred to as ***sodium-dependent glucose transporter***. This is mediated by **S**odium **GL**ucose **cotransporter-I** (***SGLT-I***). The concentration of Na^+ is higher in the intestinal lumen compared to mucosal cells. Na^+, therefore, moves into the cells along its concentration gradient and simultaneously glucose is transported into the intestinal cells. This is mediated by the same carrier system. Thus, Na^+ diffuses into the cell and it drags glucose along with it. The intestinal Na^+ gradient is the immediate energy source for glucose transport. This energy is indirectly supplied by ATP since the reentry of Na^+ (against the concentration gradient) into the intestinal lumen is an energy-requiring active process. The enzyme ***Na***$^+$***-K***$^+$ ***ATPase*** is involved in the transport of Na^+ in exchange of K^+ against the concentration gradient (for details ***see Chapter 33***).

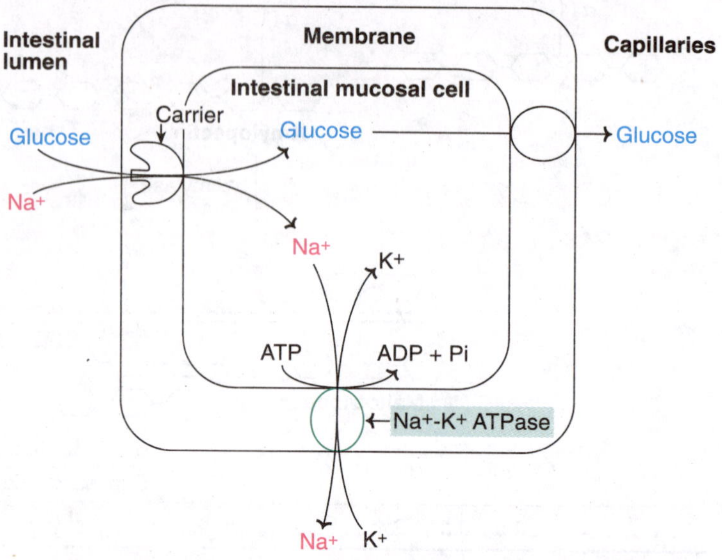

Fig. 8.5 : *Transport of glucose across intestinal epithelium (**Note :** Transport of amino acids also occurs by a similar mechanism; replace glucose in figure by amino acid).*

Oral rehydration therapy (ORT) : ORT is the most common treatment of ***diarrhea***. The oral rehydration fluid contains glucose and sodium. Intestinal absorption of sodium is facilitated by the presence of glucose.

The mechanism of absorption of galactose is similar to that of glucose. The inhibitor ***phlorizin*** blocks the Na^+ dependent transport of glucose and galactose.

Absorption of fructose : Fructose absorption is relatively simple. It does not require energy and is independent of Na^+ transport. Fructose is transported by ***facilitated diffusion mediated by a carrier***. Inside the epithelial cell, most of the fructose is converted to glucose. The latter then enters the circulation.

Pentoses are absorbed by a process of simple diffusion.

Non-digestible carbohydrates

The plant foods are rich in fibrous material which cannot be digested either by the human enzymes or intestinal bacteria. The ***fibers*** are chemically ***complex carbohydrates*** which include ***cellulose***, hemicellulose, pectins, lignin and gums. Fiber in nutrition is of special importance which is described under nutrition (***Chapter 23***).

Abnormalities of carbohydrate digestion

In general, humans possess an efficient system of carbohydrate digestion and absorption. Since only the monosaccharides are absorbed, any defect in the activities of ***disaccharidases*** results in the passage of undigested disaccharides into the large intestine. The disaccharides draw water from intestinal mucosa by osmosis and cause diarrhea. Further, bacterial action of these ***undigested carbohydrates leads to flatulence*** (increased intestinal motility, cramps, and irritation).

Medical Concepts/ Clinical Correlates

- Cooking of food significantly improves the digestibility by enzymes.
- Lactose intolerance due to a defect in the enzyme lactase (β-galactosidase) is very common. The treatment advocated is severe restriction of lactose (milk and milk products) in the diet.
- Flatulence, occurring after ingestion of certain non-digestible oligosaccharides, is characterized by increased intestinal motility, cramps and irritation.

Lactose intolerance

Lactose intolerance is the ***most common disorder of carbohydrate digestion*** in humans. This is due to a defect in the enzyme ***lactase*** (β-galactosidase). It is estimated that more than half of the world's adult population is affected by lactose intolerance. It is more commonly found in Africans (blacks) and Asians (upto 90% of adults) compared to Europeans.

Continued consumption of lactose by lactose intolerant individuals causes typical symptoms of ***flatulence*** (described later).

Lactose intolerance may be ***primary (congenital)*** or ***secondary (acquired)***. Acquired lactose intolerance may occur due to a sudden and high intake of milk-based diets. Lactase is an inducible enzyme. Therefore in acquired intolerance, if milk is withdrawn temporarily, diarrhea will be limited. For lactose intolerant people, consumption of ***curd*** is benefical, since lactobacilli present in curd contain the enzyme lactase. Further, ***yeast rich*** in lactase, can also be used for treatment of lactose intolerance.

The best treatment for lactose intolerance is ***elimination of lactose from the diet*** (i.e. severe restriction of milk and dairy products).

The problem of flatulence

Flatulence is characterized by ***increased intestinal motility, cramps and irritation***. This occurs after ingestion of certain carbohydrates and is explained as follows.

Certain disaccharides (e.g. lactose in lactose intolerance) and oligosaccharides (e.g. raffinose) are not digested by humans. They can be degraded by the bacteria present in ileum (lower part of small intestine) to liberate monosaccharides. The latter can be metabolized by the bacteria.

As the monosaccharides are utilized by the intestinal bacteria, gases such as ***hydrogen, methane and carbon dioxide***—besides lactate and short chain fatty acids—are released. These compounds cause flatulence.

The occurrence of flatulence after the ingestion of ***leguminous seeds*** (bengal gram, redgram, beans, peas, soya bean) is very common. They contain several ***nondigestible oligosaccharides*** by human intestinal enzymes. These compounds are degraded and utilised by intestinal bacteria causing flatulence. ***Raffinose*** containing galactose, glucose and fructose is a predominant oligosaccharide found in leguminous seeds.

PROTEINS

The proteins subjected to digestion and absorption are obtained from two sources—dietary and endogenous.

The digestion and absorption of proteins is very efficient in healthy humans, hence very little protein (about 5-10 g/day) is lost through feces. Dietary proteins are ***denatured on cooking*** and therefore, ***easily digested***.

Proteins are degraded by a class of enzymes—namely ***hydrolases***—which specifically cleave the peptide bonds, hence known as ***peptidases***. They are divided into two groups

1. **Endopeptidases (proteases)** which attack the internal peptide bonds and release peptide fragments, e.g. ***pepsin***, ***trypsin***.

2. **Exopeptidases** which act on the peptide bonds of terminal amino acids. Exopeptidases are subdivided into ***carboxypeptidases*** (act on C-terminal amino acid) and ***aminopeptidases*** (act on N-terminal amino acid).

I. Digestion of proteins by gastric secretion

Protein digestion begins in the stomach. ***Gastric juice*** produced by stomach contains hydrochloric acid and a protease proenzyme (zymogen) namely pepsinogen.

Hydrochloric acid : The ***pH*** of the stomach is **< *2*** due to the presence of HCl, ***secreted by parietal (oxyntic) cells*** of gastric gland. This acid performs two important functions-***denaturation*** of proteins and killing of certain microorganisms. The denatured proteins are more susceptible to proteases for digestion.

Pepsin : Pepsin (*Greek* : pepsis—digestion) is produced by the serous cells of the stomach as pepsinogen, the inactive zymogen or proenzyme. Pepsinogen is converted to active pepsin either by autocatalysis, brought about by other pepsin molecules or by gastric HCl (pH < 2).

Pepsin A is the most predominant gastric protease which preferentially ***cleaves peptide bonds*** formed by amino groups of ***phenylalanine or tyrosine or leucine***.

Pepsin digestion of proteins results in peptides and a few amino acids which act as stimulants for the release of the hormone cholecystokinin from the duodenum.

Rennin : This enzyme, also called chymosin, is found in the ***stomach of infants and children***. Rennin is involved in the curdling of milk. It converts milk protein casein to calcium paracaseinate which can be effectively digested by pepsin. Rennin is ***absent in adults***.

II. Digestion of proteins by pancreatic proteases

The proteases of pancreatic juice are secreted as zymogens (proenzymes) and then converted to active forms. These processes are initiated by the release of two polypeptide hormones, namely ***cholecystokinin*** and ***secretin*** from the intestine.

Specificity and action of pancreatic proteases : Trypsin, chymotrypsin and elastase are endopeptidases active at neutral pH. Gastric HCl is neutralized by pancreatic $NaHCO_3$ in the intestine and this creates favourable pH for the action of proteases.

The substrate specificity of pancreatic proteases is depicted in ***Fig.8.6***. For instance, trypsin cleaves the peptide bonds, the carbonyl (−CO−) group of which is contributed by arginine or lysine.

Action of carboxypeptidases : The pancreatic carboxypeptidases (A and B) are metalloenzymes that are dependent on Zn^{2+} for their catalytic activity, hence they are sometimes called ***Zn-proteases***. They also possess certain degree of substrate specificity in their action. For example, carboxypeptidase B acts on peptide bonds of COOH-terminal amino acid, the amino group of which is contributed by arginine or lysine (***Fig.8.6***).

The combined action of pancreatic proteases results in the formation of free amino acids and small peptides (2-8 amino acids).

III. Digestion of proteins by small intestinal enzymes

The luminal surface of intestinal epithelial cells contains ***aminopeptidases*** and ***dipeptidases***. Aminopeptidase is a non-specific exopeptidase which repeatedly cleaves N-terminal amino

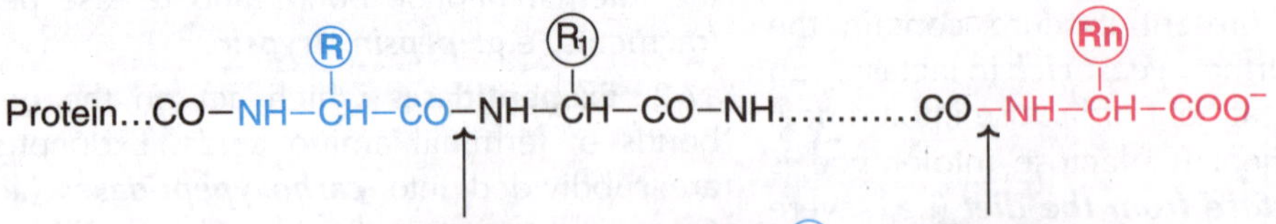

Enzyme	Nature of (R)	Enzyme	Nature of (Rn)
Pepsin A	Tyr, Phe, Leu	Carboxypeptidase A	Ala, Ile, Leu, Val
Trypsin	Arg, Lys	Carboxypeptidase B	Arg, Lys
Chymotrypsin	Trp, Tyr, Phe, Leu, Met		
Elastase	Ala, Gly, Ser		

***Fig. 8.6 :** Digestion of proteins—Specificity of enzyme cleavage of peptide bonds. (R_1 can be from any amino acid).*

acids one by one to produce free amino acids and smaller peptides. The dipeptidases act on different dipeptides to liberate amino acids (***Fig.8.7***).

Absorption of amino acids and dipeptides

The free amino acids, dipeptides and to some extent tripeptides are absorbed by intestinal epithelial cells.

The ***small intestine*** possesses an efficient system to absorb free amino acids. L-Amino acids are more rapidly absorbed than D-amino acids. The transport of L-amino acids occurs by an active process (against a concentration gradient), in contrast to D-amino acids which takes place by a simple diffusion.

Mechanism of amino acid absorption

Amino acids are primarily absorbed by a ***similar*** mechanism, as described for the transport ***of D-glucose***. It is basically a ***Na^+-dependent active process*** linked with the transport of Na^+. As the Na^+ diffuses along the concentration gradient, the amino acid also enters the intestinal cell. Both Na^+ and amino acids share a common carrier and are transported together. The energy is supplied indirectly by ATP (for details, ***see absorption of monosaccharides*** and ***Fig.8.5***).

Another transport system to explain the mechanism of amino acid transfer across membrane in the intestine and kidney has been put forth. This is known as ***γ-glutamyl cycle or Meister cycle*** and involves a tripeptide namely ***glutathione*** (γ-glutamylcysteinylglycine). Three ATP are utilized for the transport of a single amino acid by this cycle. For this reason, Meister cycle is not a common transport system for amino acid. However, this cycle is operative for rapid transport of cysteine and glutamine.

The γ-glutamyl cycle appears to be important for the metabolism of glutathione, since this tripeptide undergoes rapid turnover in the cells. There may be more physiological significance of γ-glutamyl cycle.

Absorption of intact proteins and polypeptides

For a short period, immediately after birth, the small intestine of infants can absorb intact proteins and polypeptides. The uptake of proteins occurs by a process known as ***endocytosis*** or ***pinocytosis***. The macromolecules are ingested by formation of small vesicles of plasma membrane followed by their internalization. The direct absorption of intact proteins is very important for the transfer of maternal immunoglobulins (γ-globulins) to the offspring.

The intact proteins and polypeptides are not absorbed by the adult intestine. However, the macromolecular absorption in certain individuals appears to be responsible for antibody formation that often causes ***food allergy***.

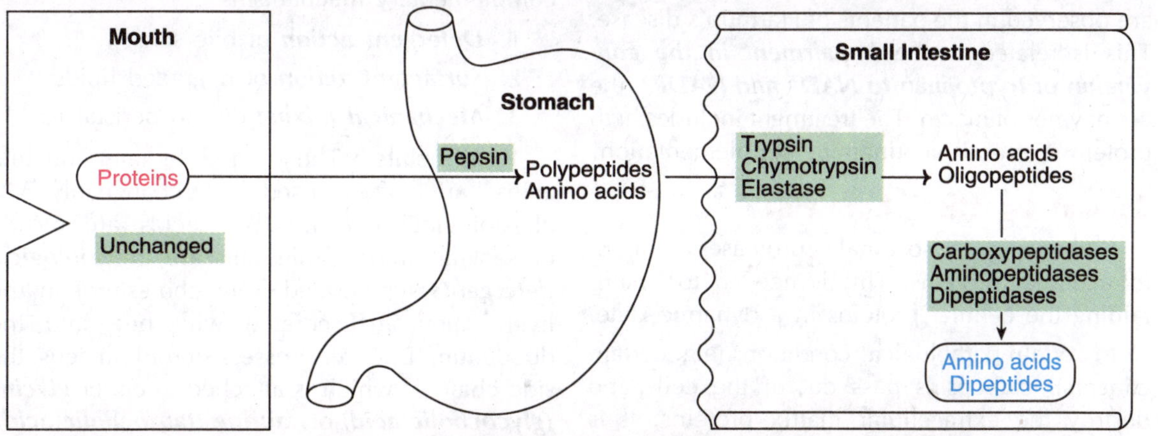

Fig. 8.7 : Overview of digestion of proteins.

Medical Concepts/ Clinical Correlates

- Direct intestinal absorption of proteins and polypeptides is observed in the infants, immediately after birth. This is important for the transfer of maternal immunoglobulins (via breast-feeding) to the offspring.
- In some adults, macromolecular (protein) absorption by intestine is responsible for antibody formation, often causing food allergy. Celiac disease is an allergic reaction caused by wheat protein - gluten.

Abnormalities of protein digestion and amino acid absorption

Any defect in the pancreatic secretion impairs protein and fat digestion. This causes the loss of undigested protein in the feces along with the abnormal appearance of lipids. Deficiency of pancreatic secretion may be due to pancreatitis, cystic fibrosis or surgical removal of pancreas.

Hartnup's disease (neutral amino aciduria)

Hartnup is the name of the family in whom this disease was first discovered. It is characterized by the ***inability*** of intestinal and renal epithelial cells ***to absorb neutral amino acids*** (tryptophan, alanine, serine, threonine, valine). ***Tryptophan*** absorption is most severely affected with a result that typical symptoms of ***pellagra*** are observed in the patients of Hartnup's disease. This is related to the ***impairment in the conversion of tryptophan to NAD^+ and $NADP^+$***, the coenzymes of niacin. The treatment includes high protein diet and nicotinamide supplementation.

Cellular digestion of proteins

Cathepsins, lysosomal proteases, digest intracellular proteins. This is needed for maintaining the cellular proteins in a dynamic state.

In certain pathological conditions (e.g. certain cancers), cathepsins pass out of the cells and destroy the extracellular matrix proteins. It is believed that ***damage to extracellular matrix may result in cancer cells to grow freely***.

LIPIDS

Lipids are insoluble or sparingly soluble in aqueous solution. The digestive enzymes, however, are present in aqueous medium. This poses certain problems for the digestion and absorption of lipids. Fortunately, the digestive tract possesses specialized machinery to

1. Increase the surface area of lipids for digestion;
2. Solubilize the digested products for absorption.

Minor digestion of lipids in the stomach

In case of infants, the milk fat (with short chain fatty acids) can be hydrolysed by ***gastric lipase*** to some extent. This is because the stomach pH of infants is close to neutrality, ideal for gastric lipase action.

Emulsification of lipids in the small intestine

Emulsification is the phenomenon of ***dispersion of lipids into smaller droplets due*** to reduction in the surface tension. This is accompanied by increase in the surface area of lipid droplets. Emulsification is ***essential for*** effective ***digestion*** of lipids, since the enzymes can act only on the surface of lipid droplets. More correctly, lipases act at the interfacial area between the aqueous and lipid phase.

The process of emulsification occurs by three complementary mechanisms

1. ***Detergent action*** of bile salts;
2. ***Surfactant action*** of degraded lipids;
3. ***Mechanical mixing*** due to peristalsis.

1. **Bile salts :** The terms bile salts and bile acids are often used interchangeably. At physiological pH, the bile acids are mostly present as anions. Bile salts are the ***biological detergents*** synthesized from cholesterol in the liver. They are secreted with bile into the duodenum. Bile salts possess steroid nucleus, the side chain of which is attached to either ***glycine (glycocholic acid)*** or ***taurine (taurocholic acid)***. For the synthesis and other details on bile acids, refer cholesterol metabolism (***Chapter 14***). Bile

salts are the most effective biological emulsifying agents. They interact with lipid particles and the aqueous duodenal contents and convert them into smaller particles (emulsified droplets). Further, bile salts stabilize the smaller particles by preventing them from coalescing.

2. **Surfactant action of degraded lipids :** The initial digestive products of lipids (catalysed by lipase) namely free fatty acids, mono-acylglycerols promote emulsification. These compounds along with phospholipids are known as surfactants.

3. Besides the action of bile salts and surfactants, the ***mechanical mixing due to peristalsis*** also helps in the emulsification of lipids. The smaller lipid emulsion droplets are good substrates for digestion.

Digestion of lipids by pancreatic enzymes

The pancreatic enzymes are primarily responsible for the degradation of dietary triacylglycerols, cholesteryl esters and phospholipids.

Degradation of triacylglycerols (fat)

Pancreatic lipase is the major enzyme that digests dietary fats. This enzyme preferentially cleaves fatty acids (particularly long chain, above 10 carbons) at position 1 and 3 of triacylglycerols. The products are ***2-monoacylglycerol*** (formerly 2-monoglyceride) and ***free fatty acids*** (***Fig.8.8***). The activity of pancreatic lipase is inhibited by bile acids which are present along with the enzyme in the small intestine. This problem is overcome by a small protein, ***colipase*** (mol. wt. 12,000). It is also secreted by pancreas as procolipase and converted to active form by trypsin. Colipase binds at the lipid-aqueous interface and helps to anchor and stabilize lipase.

Lipid esterase is a less specific enzyme present in pancreatic juice. It acts on monoacylglycerols, cholesteryl esters, vitamin esters etc. to liberate free fatty acids. The presence of bile acids is essential for the activity of lipid esterase.

Degradation of phospholipids

Phospholipases are enzymes responsible for the hydrolysis of phospholipids. Pancreatic juice is rich in phospholipase A_2 which cleaves the fatty acid at the 2nd position of phospholipids. The products are a free fatty acid and a lysophospholipid. Phospholipase A_2 is secreted as a zymogen which is activated in the intestine by the action of trypsin.

An overview of the digestion of lipids is given in ***Fig.8.9***.

Absorption of lipids

The former and present theories to explain the absorption of lipids are briefly described hereunder

1. **Lipolytic theory put forth by Verzar :** According to this, ***fats*** are ***completely hydrolysed to glycerol and free fatty acids***. The latter are absorbed either as soaps or in association with bile salts.

2. **Partition theory proposed by Frazer :** This theory states that the ***digestion of triacylglycerols is partial*** and not complete. The partially digested triacylglycerols, in association with bile salts, form emulsions. The lipids are taken up by the intestinal mucosal cells. As per this theory, resynthesis of lipids is not necessary for their entry into the circulation.

3. **Bergstrom theory :** This is a more recent and comprehensive theory to explain lipid absorption. It has almost replaced the earlier theories, and is briefly described hereunder

The primary products obtained from the lipid digestion are ***2-monoacylglycerol, free fatty acids and free cholesterol***.

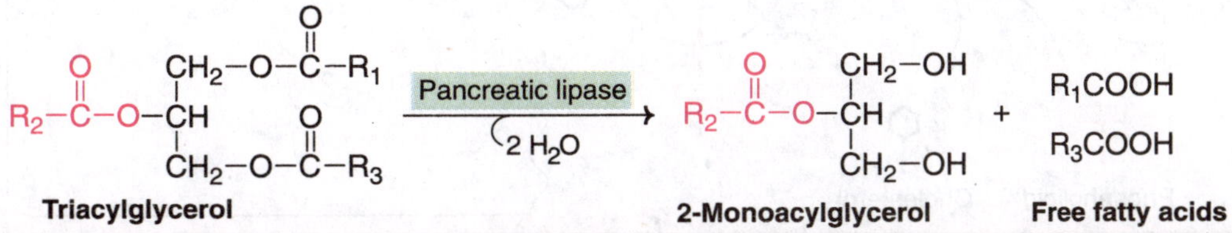

Fig. 8.8 : *Enzymatic cleavage of dietary fat.*

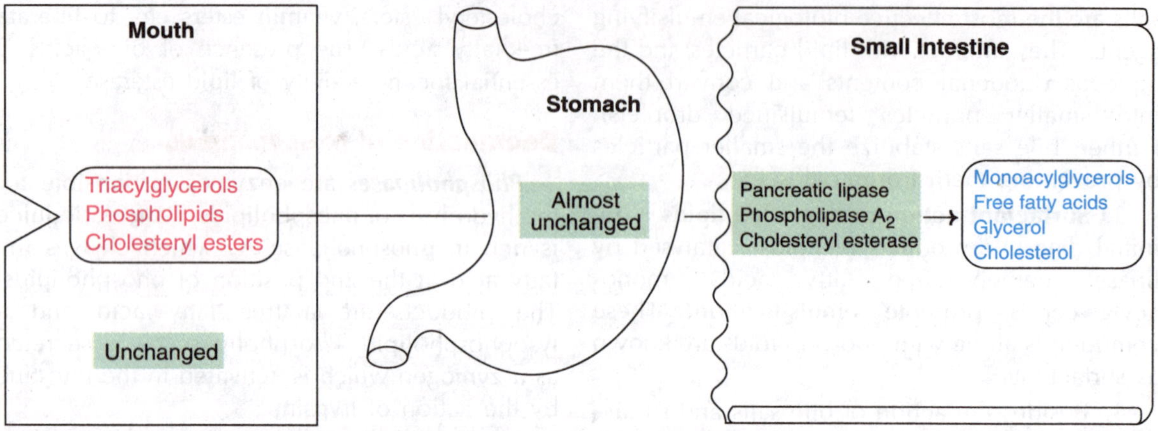

Fig. 8.9 : Overview of digestion of lipids.

Role of bile salts in lipid absorption

Besides their participation in digestion, bile salts are ***essential for absorption of lipids***. Bile salts form ***mixed micelles*** with lipids. These micelles are smaller in size than the lipid emulsion droplets (utilized for digestion, described above). The micelles have a disk like shape with lipids (monoacylglycerol, fatty acids, cholesterol and phospholipids) at the interior and bile salts at the periphery.

Mechanism of lipid absorption

The ***mixed micelles*** serve as the major vehicles for the ***transport of lipids*** from the intestinal lumen to the membrane of the intestinal mucosal cells, the site of lipid absorption. The lipid components pass through the unstirred fluid layer and are absorbed through the plasma membrane by diffusion (***Fig.8.10***). Absorption is almost complete for monoacylglycerols and free fatty acids which are slightly water soluble. However, for water insoluble lipids, the absorption is incomplete. For instance, less than 40% of the dietary cholesterol is absorbed.

The micelle formation is also essential for the absorption of fat soluble vitamins, particularly vitamins A and K.

The ***efficiency of lipid absorption*** is dependent on the quantity of ***bile salts*** to solubilize digested lipids in the mixed micelles.

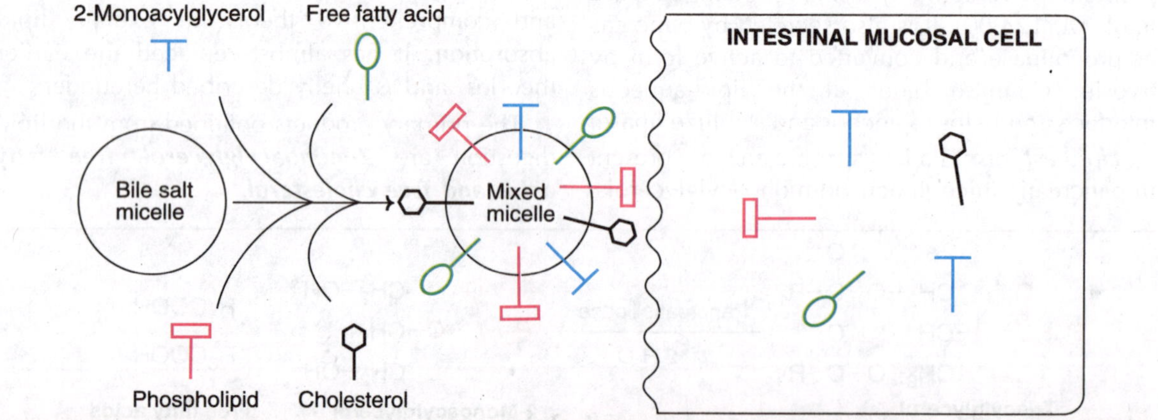

Fig. 8.10 : Absorption of lipids by intestinal mucosal cell.

Medical Concepts/ Clinical Correlates

- Emulsification of lipids is essential for their effective digestion, since lipases can act only on the surface of lipid droplets. Bile salts are the most efficient biological emulsifying agents.
- Pharmacological interventions (e.g. orlistat, olestra) to block fat digestion and/or absorption so as to prevent obesity are in recent use.
- Intestinal absorption of cholesterol can be inhibited by certain drugs e.g. ezetimibe.
- Steatorrhea, characterized by the loss of lipids in feces, is commonly associated with impaired pancreatic function and biliary obstruction.

Synthesis of lipids in the intestinal mucosal cells

The ***fatty acids of short and medium chain*** length (< 10 carbons), after their absorption into the intestinal cells, ***do not undergo any modification***. They enter the portal circulation and are transported to the liver in a bound form to albumin.

The long chain fatty acids are activated by ***thiokinase*** (fatty acyl CoA synthetase) in the intestinal cells. The acyl CoA derivatives so formed combine with 2-monoacylglycerols to produce ***triacylglycerols***. These reactions are catalysed by a group of enzymes, namely acyltransferases (***Fig.8.11***). Further, within the intestinal cells, cholesterol is converted to cholesterylester, and phospholipids are regenerated from the absorbed lysophospholipids. The ***newly synthesized lipids are usually different from those consumed in the diet***.

Secretion of lipids from the intestinal mucosal cells

The lipids that are resynthesized (described above) in the intestinal cells are hydrophobic in nature. They are put together as lipid droplets and surrounded by a thin layer consisting of mostly apolipoproteins (A_I and B-48) and phospholipids. This package of lipids enveloped in the layer stabilizes the droplets and increases their solubility. These particles are known as ***chylomicrons***.

Chylomicrons migrate to the plasma membrane of intestinal mucosal cells. They are released into the lymphatic vessels by exocytosis. The presence of chylomicrons (*Greek:* chylos–juice) gives the lymph a milky appearance, which is observed after a lipid-rich meal. Chylomicrons enter the large body veins via the thoracic duct. Blood from here flows to the heart and then to the peripheral tissues (muscle, adipose tissue) and, finally, to the liver. Adipose tissue and muscle take up a large proportion of dietary lipids from chylomicrons for storage and transport. It is believed that this bypass arrangement (passage of chylomicrons through peripheral tissues) protects the liver from a lipid overload after a meal.

Abnormalities of lipid digestion and absorption

The gastrointestinal tract possesses an efficient system for digestion and absorption of lipids. It can comfortably handle as much as 4 times the normal daily intake of lipids.

Steatorrhea : It is a condition characterized by the ***loss of lipids in the feces***. Steatorrhea may be due to

1. A defect in the secretion of bile or pancreatic juice into the intestine;

2. Impairment in the lipid absorption by the intestinal cells.

Steatorrhea is commonly seen in disorders associated with pancreas, biliary obstruction, severe liver dysfunction etc.

Cholesterol stones (gall stone)

Cholesterol stone formation in gall-bladder (***gall stones***) is a frequent health complication. It is found ***more*** commonly ***in females*** than in males often in association with obesity. Cholesterol gall stones are formed when liver secretes bile (containing phospholipids, bile acids etc.), supersaturated with respect to cholesterol.

Gall stones, when small, sometimes, can freely pass through from the common bile duct into intestine, and therefore cause no complications. However, when the gall stone size is larger, it can block the opening of the gall bladder or the common bile duct. Consequently, bile salts entry into intestine may be blocked, resulting in impaired fat digestion and absorption.

Small gall stones, sometimes, can be dissolved by administration of chenodeoxy cholic acid. Surgical intervention may be required to remove gall stones or inflammed gall blader.

OBESITY AND FAT ABSORPTION

Obesity is a major problem in many parts of the world as the availability of food is generally abundant and overeating is common. Intake of lipids largely contributes to obesity. In recent years, pharmacological interventions to prevent fat digestion, absorption, and thus obesity are in use. Two approaches are given below

1. Pancreatic lipase degrades dietary triacylglycerol to fatty acids and glycerol which are absorbed. ***Orlistat*** is a non-hydrolysable analog of triacylglycerol, and is a powerful ***inhibitor of pancreatic lipase***, hence prevents fat digestion, and absorption.

2. ***Olestra*** is a synthetic lipid, produced by esterification of natural fatty acids with sucrose (instead of glycerol). Olestra tastes like a natural lipid. However, it cannot be hydrolysed and therefore, gets excreted.

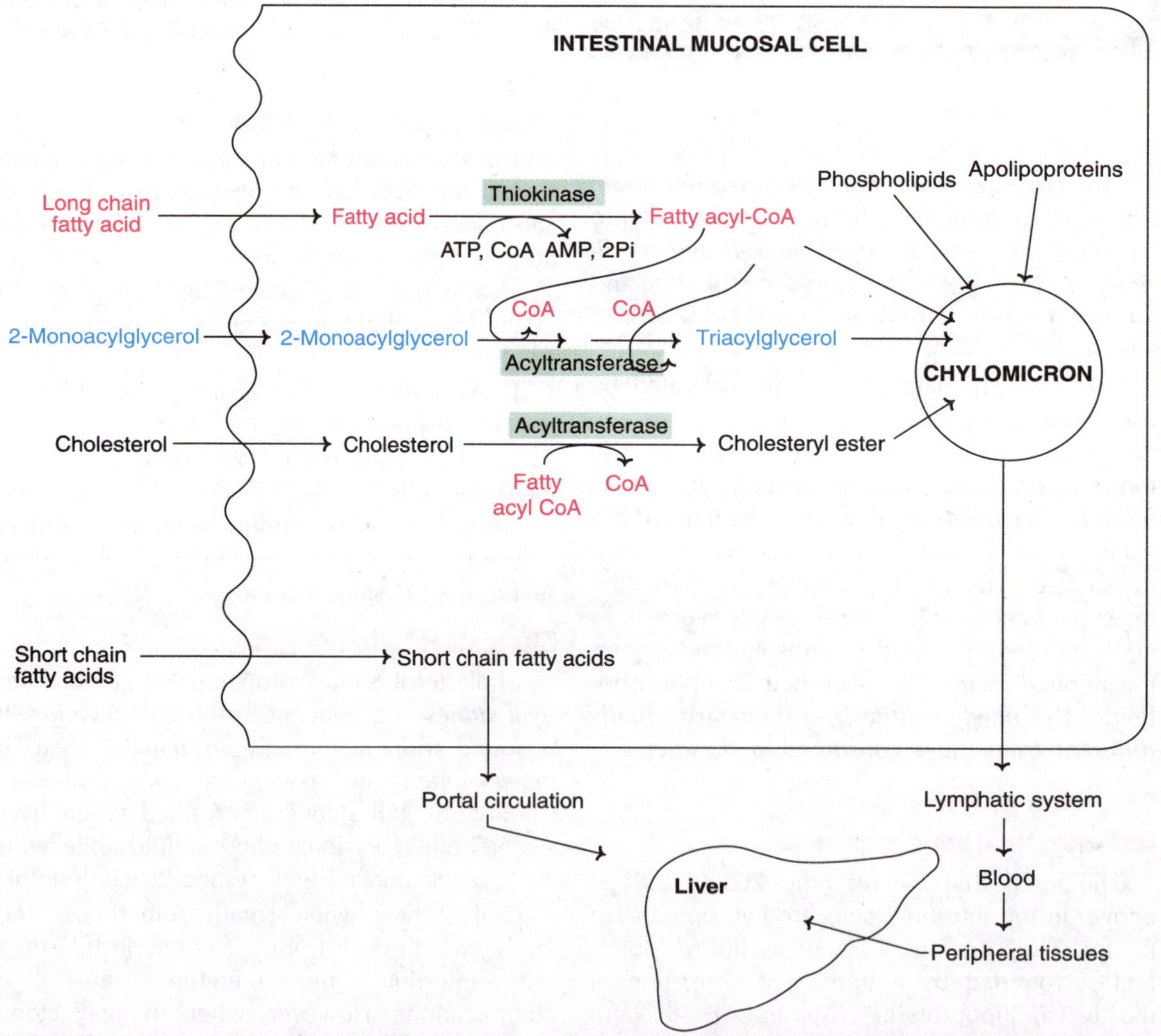

Fig. 8.11 : Formation and secretion of chylomicrons by intestinal mucosal cells.

Medical Concepts/ Clinical Correlates

- Excess bilirubin (e.g. in hemolytic anemia) in bile precipitates with calcium to form insoluble, Ca-bilirubinate that leads to formation of gall stones.
- Gastric ulcers are mainly caused by the bacterium *H. pylori*. The antibiotics that eliminate this bacterium are effective in the treatment.
- Pancreatitis is caused by autodigestion of pancreas, obstruction by gallstones, and excessive consumption of alcohol.

NUCLEIC ACIDS

Nucleic acids (DNA and RNA), and their bases purines and pyrimidines can be synthesized in the body, and thus they are dietarily non-essential.

The digestion of dietary nucleic acids is carried out in the small intestine, primarily by the enzymes of pancreatic juice. ***Ribonucleases*** and ***deoxyribonucleases***, respectively, hydrolyse RNA and DNA to oligonucleotides (***Fig.8.12***). The latter are degraded by phosphodiesterases to form mononucleotides. Nucleotidases act on nucleotides to liberate phosphate and nucleosides. The nucleosides may be either directly absorbed or degraded to free bases before absorption. Some of the unabsorbed purines are metabolized by intestinal bacteria.

The ***dietary purines and pyrimidines*** are ***not*** of much ***utility for the synthesis of tissue nucleic acids***. Further, the purines after their absorption are mostly converted to uric acid by the intestinal mucosal cells and excreted in the urine.

ABNORMALITIES RELATED TO DIGESTION AND ABSORPTION

The following are the major abnormalities (of interest to biochemists) concerned with digestion and absorption of food in the gastrointestinal tract.

Lactose intolerance, ***deficiency of sucrase***, ***Hartnup's disease*** and ***steatorrhea*** have already been described. ***Peptic ulcer***, ***pancreatitis*** and ***celiac disease*** are other important disorders associated with digestive system.

Peptic ulcers

Gastric and duodenal ulcers are collectively known as peptic ulcers. Ulceration occurs due to the autodigestion of mucosa by the gastric secretions (pepsin and HCl). In the patients of peptic ulcer, gastric HCl is always present in the pyloric regions of stomach and the duodenum. Gastic ulcers are mainly caused by the bacterium ***Helicobacter pylori*** which lives in the nutrient-rich gastric mucosa. *H. pylori* induces chronic inflammation in the stomach tissues, which gets exposed to acid damage. For this reason, the best mode of ***treatment for gastric ulcers is the use of antibiotics that eliminate H. pylori.***

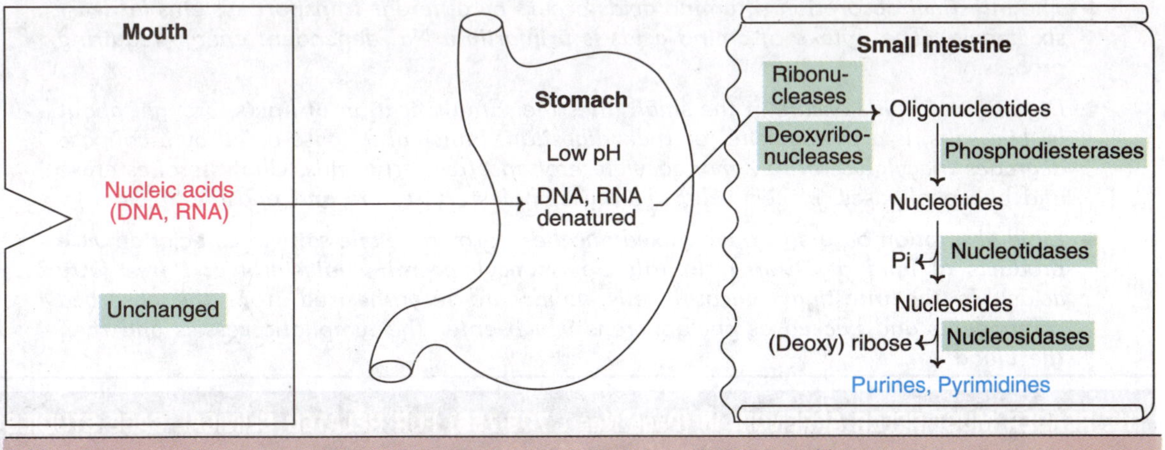

Fig. 8.12 : Overview of digestion of nucleic acids.

Achlorhydria is a less serious disorder involving the failure to secrete gastric HCl.

Pancreatitis

Inflammation of the pancreas is known as pancreatitis. Acute pancreatitis is caused by the autodigestion of pancreas due to the unusual conversion of zymogens into the active enzymes by trypsin. In normal circumstances, this is prevented by trypsin inhibitor.

Acute pancreatitis is a life-threatening disorder. Measurement of serum ***amylase*** (highly elevated) is used ***in the diagnosis of pancreatitis***. Excessive ***consumption of alcohol*** over a long period is blamed as the prime cause of chronic pancreatitis.

Celiac disease (celiac sprue)

Celiac disease is a disease of ***malabsorption caused by immunomediated damage*** to small intestine. This occurs in some individuals as a result of ingestion of gluten (or gliadin derived from gluten), a protein found in wheat, barley and rye.

SUMMARY

- *Digestion is a process that converts complex foodstuffs into simpler ones which can be readily absorbed by the gastrointestinal tract.*
- *Stomach, duodenum and upper part of small intestine are the major sites of digestion. The small intestine is the prime site for the absorption of digested foods.*
- *The digestion of carbohydrates is initiated in the mouth by salivary α-amylase and is completed in the small intestine by pancreatic α-amylase, oligosaccharidases and disaccharidases.*
- *Monosaccharides are the final absorbable products of carbohydrate digestion. Glucose is transported into the intestinal mucosal cells by a carrier mediated,* Na^+*-dependent energy requiring process.*
- *Lactose intolerance due to a defect in the enzyme lactase (β-galactosidase) resulting in the inability to hydrolyse lactose (milk sugar) is the common abnormality of carbohydrate digestion.*
- *Protein digestion begins in the stomach by pepsin, which is aided by gastric HCl. Pancreatic proteases (trypsin, chymotrypsin and elastase) and intestinal aminopeptidases and dipeptidases complete the degradation of proteins to amino acids and some dipeptides.*
- *The intestinal absorption of amino acids occurs by different transport systems (at least six known). The uptake of amino acids is primarily a* Na^+*-dependent energy requiring process.*
- *Digestion of lipids occurs in the small intestine. Emulsification of lipids, brought about by bile salts, is a prerequisite for their digestion. Pancreatic lipase aided by a colipase degrades triacylglycerol to 2-monoacylglycerol and free fatty acids. Cholesterol esterase and phospholipases, respectively, hydrolyse cholesteryl esters and phospholipids.*
- *Lipid absorption occurs through mixed micelles, formed by bile salts in association with products of lipid digestion (primarily 2-monoacylglycerol, cholesterol and free fatty acids). In the intestinal mucosal cells, lipids are resynthesized from the absorbed components and packed as chylomicrons which enter the lymphatic vessels and then the blood.*

THE READER MUST REFER APPENDIX VI (Page Nos : 738-756) FOR RELATED CLINICAL CASE STUDIES (Nos : 11, 12)

Self-Assessment Exercises

I. Essay questions

1. Write an account of the digestion and absorption of lipids.
2. Describe briefly the digestion of carbohydrates and proteins.
3. Give an account of the Na^+ dependent intestinal transport of glucose and amino acids.
4. Describe the role of intestine in the digestion of foodstuffs.
5. Write briefly on the enzymes of gastrointestinal tract involved in the digestion of foodstuffs.

II. Short notes

(a) Mixed micelles, (b) Lactose intolerance, (c) Salivary amylase, (d) Disaccharidases, (e) γ-Glutamyl cycle, (f) Zymogens, (g) Specificity of proteases, (h) Bile salts, (i) Synthesis of chylomicrons in the intestinal mucosal cells, (j) Pancreatic juice.

III. Fill in the blanks

1. Cellulose is not digested in humans due to lack of the enzyme that hydrolyses _______________ bonds.
2. The most important carbohydrate associated with flatulence caused by ingestion of leguminous seeds _______________.
3. Lactose intolerance is caused by the deficiency of the enzyme _______________.
4. The non-digested carbohydrates are collectively called _______________.
5. Gastric HCl is secreted by _______________.
6. Name of the peptide believed to be involved in the transport of amino acids _______________.
7. The disease characterized by impairment in the absorption of neutral amino acids _______________.
8. Trypsin hydrolyses peptide bonds, the carbonyl group of which is contributed by the amino acids _______________ or _______________.
9. The inhibition of the enzyme pancreatic lipase by bile salts is overcome by a protein, namely _______________.
10. The vehicles for the transport of lipids from the intestinal lumen to the membrane of mucosal cells _______________.

IV. Multiple choice questions

11. Transport of glucose from the lumen to the intestinal mucosal cells is coupled with diffusion of
    (a) Na^+ (b) K^+ (c) Cl^- (d) HCO_3^-.
12. The key enzyme that converts trypsinogen to trypsin is
    (a) Secretin (b) Chymotrypsin (c) Elastase (d) Enteropeptidase.
13. The products obtained by the action of pancreatic lipase on triacylglycerols are
    (a) Glycerol and free fatty acids (b) 1-Acylglycerol and free fatty acids (c) 2-Acylglycerol and free fatty acids (d) 3-Acylglycerol and free fatty acids.
14. The lipoproteins synthesized in the intestinal mucosal cells from the absorbed lipids are
    (a) High density lipoproteins (b) Chylomicrons (c) Low density lipoproteins (d) Very low density lipoproteins.
15. Salivary α-amylase becomes inactive in the stomach primarily due to
    (a) Inactivation by low pH (b) Degradation by gastric pepsin (c) Inhibition by Cl^- (d) Inhibition by peptides.

9 Plasma Proteins

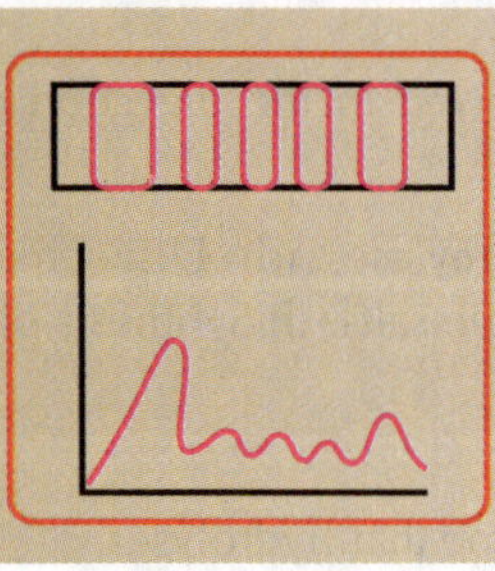

The official 'spokesperson' of plasma proteins, albumin, speaks

"I am the most abundant plasma protein;
Produced exclusively by the liver;
Perform osmotic, transport and nutritive functions;
Estimated in lab to assess liver function."

The ***plasma*** is the ***liquid medium of blood*** (55-60%), in which the cell components—namely erythrocytes, leukocytes, platelets—are suspended. If blood containing anticoagulants (e.g. heparin, potassium oxalate) is centrifuged, the plasma separates out as a supernatant while the cells remain at the bottom. The packed cell ***volume*** or ***hematocrit*** is about ***45%***.

The term serum is applied to the liquid medium which separates out after the blood clots (coagulates). Serum does not contain fibrinogen and other clotting factors. Thus, the main ***difference between plasma and serum is the presence or absence of fibrinogen***. (**Note :** Plasma is a physiological fluid while serum is prepared in laboratory)

Importance of blood

The total volume of blood in an adult is around 4.5 to 5 liters. Blood performs several diversified functions. These include respiration, excretion, acid-base maintenance, water balance, transport of metabolites, hormones and drugs, body defense and coagulation.

Separation of plasma proteins

The total concentration of ***plasma proteins*** is about ***6-8 g/dl***. The plasma is a complex mixture of proteins, and several techniques are employed to separate them. In ***salting out process*** (use of varying concentrations of ammonium sulfate or sodium sulfate), the plasma proteins can be separated into three groups—namely albumin, globulins and fibrinogen. Total serum proteins can be estimated by ***Biuret method*** while serum albumin by ***Bromocresol green (BCG) method***.

Electrophoresis : The ***most commonly employed analytical technique*** for the separation of plasma (serum) proteins. The principles of electrophoresis are described in ***Chapter 41***. Paper or agar gel electrophoresis with veronal buffer (pH-8.6) separates plasma proteins into 5 distinct bands namely ***albumin, α_1, α_2, β and γ globulins*** (***Fig.9.1***). The concentration of each one of these fractions can be estimated by a ***densitometer***.

Abnormal electrophoretic pattern

Electrophoresis of serum proteins is conveniently used for the diagnosis of certain diseases

1. **Multiple myeloma :** A sharp and distinct ***M band*** appears in the γ-globulin fraction.

2. **Acute infections :** α_1- and α_2-***globulins*** are increased.

3. **Nephrotic syndrome :** Decreased albumin with sharp and ***prominent*** α_2-***globulin***.

4. **Primary immune deficiency :** Diminished γ-globulin band.

5. α_1-**Antitrypsin deficiency :** Diminished α_1-globulin band.

Albumin/globulin (A/G) ratio : ***The albumin*** concentration of plasma is ***3.5 to 5.0 g/dl*** while that of total globulins is 2.5 to 3.5 g/dl. The ***normal A/G ratio is 1.2 to 1.5 : 1***. The A/G ratio is lowered either due to decrease in albumin or increase in globulins, as found in the following conditions

1. Decreased synthesis of albumin by liver—usually found in liver diseases and severe protein malnutrition.

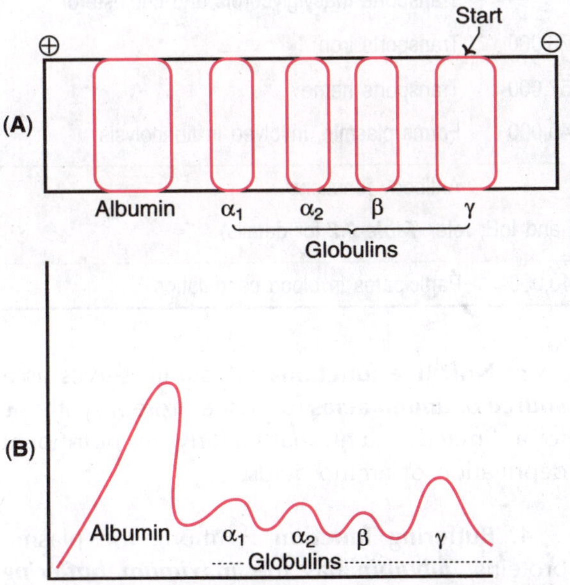

Fig. 9.1 : Electrophoresis of serum proteins—(A) Separated bands, (B) Densitometer scannning.

2. Excretion of albumin into urine in kidney damage.

3. Increased production of globulins associated with chronic infections, multiple myelomas etc.

Components of plasma proteins

The important plasma proteins along with their characteristics (based on electrophoretic pattern) and major functions are given in ***Table 9.1***. Some selected plasma proteins are discussed hereunder.

ALBUMIN

Albumin is the ***major constituent*** (60%) of plasma proteins with a concentration of ***3.5–5.0 g/dl***. Human albumin has a molecular weight of 69,000, and consists of a single polypeptide chain of 585 amino acids with 17 disulfide bonds.

Synthesis of albumin

Albumin is exclusively synthesized by the ***liver***. For this reason, measurement of serum albumin concentration is conveniently ***used to assess liver function*** (synthesis decreased in liver diseases). Liver produces about 12 g albumin per day which represents 25% of the total hepatic protein synthesis. Albumin has an ***half-life of 20 days***.

Functions of albumin

Plasma albumin performs ***osmotic***, ***transport*** and ***nutritive functions***

1. **Osmotic function :** Due to its high concentration and low molecular weight, albumin contributes to ***75–80% of the total*** plasma ***osmotic pressure*** (25 mm Hg). Thus, albumin plays a predominant role in maintaining blood volume and body fluid distribution. Decrease in plasma albumin level results in a fall in osmotic pressure, leading to enhanced fluid retention in tissue spaces, causing edema. The ***edema*** observed in ***kwashiorkor***, a disorder of protein-energy malnutrition, is attributed to a drastic ***reduction in plasma albumin*** level.

Table 9.1 Important plasma proteins, characteristics and major functions

Protein	Plasma concentration	Molecular weight	Major function(s)
Albumin	3.5–5.0 g/dl	69,000	Osmotic, transport, nutritive and buffering
Prealbumin	25–30 mg/dl	61,000	Transports thyroxine to some extent
α_1-Globulins	0.3–0.5 g/dl	—	—
α_1-Antitrypsin	< 0.2 g/dl	54,000	Inhibitor of trypsin
α_1-Lipoproteins (HDL)	0.2–0.3 g/dl	—	Transports cholesterol and phospholipids
Orosomucoid	< 0.1 g/dl	44,000	Binds with progesterone
Retinol binding protein (RBP)	3–6 mg/dl	21,000	Transports vitamin A
Thyroxine binding globulin (TBG)	1–2 mg/dl	58,000	Transports thyroid hormones
Transcortin or cortisol binding protein (CBG)	3–4 mg/dl	52,000	Major transporter of steroid hormones (e.g. cortisol, corticosterone)
α_2-Globulins	0.4–0.8 g/dl	—	—
α_2-Macroglobulin	0.2–0.3 g/dl	800,000	Antitrypsin and antiplasmin activity
Haptoglobins (Hp 1-1; Hp 2-1 and Hp 2-2)	< 0.3 g/dl	90,000	Binds with plasma free hemoglobin and prevents its excretion
Prothrombin	< 0.02 g/dl	63,000	Participates in blood coagulation
Ceruloplasmin	< 0.03 g/dl	150,000	Transport of copper; oxidation of Fe^{2+} to Fe^{3+}.
β-Globulins	0.6–1.1 g/dl	—	—
β-Lipoproteins (LDL)	0.2–0.5 g/dl	—	Transports triacylglycerols and cholesterol
Transferrin	0.2–0.3 g/dl	76,000	Transports iron
Hemopexin	< 0.1 g/dl	57,000	Transports heme
Plasminogen	< 0.05 g/dl	140,000	Forms plasmin, involved in fibrinolysis
γ-Globulins	0.8–1.8 mg/dl	—	Antibody functions
(Immunoglobulins—IgG, IgA, IgM, IgD and IgE; refer ***Table 9.2*** for details)			
Fibrinogen	0.2–0.4 g/dl	340,000	Participates in blood coagulation

2. **Transport functions :** Plasma albumin binds to several biochemically important compounds and transports them in the circulation. These include ***free fatty acids***, ***bilirubin***, steroid hormones, calcium and copper.

[**Note :** Besides albumin, there are several other ***plasma transport proteins*** e.g. prealbumin, retinol binding protein, thyroxine binding protein, transcortin etc., as given in ***Table 9.1***].

3. **Nutritive functions :** Albumin serves as a ***source of amino acids*** for tissue protein synthesis to a limited extent, particularly in nutritional deprivation of amino acids.

4. **Buffering function :** Among the plasma proteins, ***albumin*** has the ***maximum buffering capacity***. However, the buffering action of albumin in plasma is not significant compared to bicarbonate buffer system.

Clinical significance of albumin

1. Albumin, binding to certain compounds in the plasma, ***prevents them from crossing the blood-brain barrier*** e.g. albumin-bilirubin complex, albumin-free fatty acid complex.

2. ***Hypoalbuminemia*** (lowered plasma albumin), observed in ***malnutrition, protein losing enteropathy, nephrotic syndrome*** and ***cirrhosis of liver*** is associated with edema.

The osmotic pressure of albumm significantly contributes to maintaining plasma volume, and fluid volume of interstitial fluids. In hypo-albuminemia (albumin <2g/dl), plasma osmotic pressure is decreased, leading to flow of water from plasma to interstitial compartment that results in ***edema of legs and other body parts.***

3. Albumin is excreted into urine (***albuminuria***) in nephrotic syndrome and in certain inflammatory conditions of urinary tract. ***Microalbuminuria*** (30-300 mg/day) is clinically important for predicting the future risk of renal diseases (***Refer Chapter 36***).

4. Albumin is therapeutically useful for the ***treatment of burns, and hemorrhage***.

GLOBULINS

Globulins constitute several proteins that are separated into four distinct bands (***α_1, α_2, β and γ-globulins***) on electrophoresis (***See Fig.9.1***). Globulins, in general, are bigger in size than albumin. In ***Table 9.1***, the important globulins are given, some of them are discussed hereunder.

α_1-Antitrypsin

α_1-Antitrypsin, more recently called as ***α-anti-proteinase***, is a glycoprotein with 394 amino acids and a molecular weight of 54,000. It is a major constituent of α_1-globulin fraction of plasma proteins with a normal concentration of about 200 mg/dl. α_1-Antitrypsin is a serine protease inhibitor. It combines with trypsin, elastase and other protease enzymes and inhibits their activity.

Clinical significance of α_1-antitrypsin

α_1-Antitrypsin (α_1-AT) deficiency has been implicated in two diseases, namely, ***emphysema*** and ***α_1-AT deficiency liver disease.***

Emphysema (*Greek* : emphusan—to inflate) is a term used to represent the ***abnormal distension of lungs*** by air. At least 5% of emphysema cases are due to the deficiency of α_1-AT. This is associated with ***lung infections*** (e.g. pneumonia) and increase in the activity of macrophages to release elastase that damages lung tissues. In the normal circumstances, elastase activity is inhibited by α_1-AT.

Effect of smoking on α_1-AT : The amino acid methionine at position 358 of α_1-AT is involved in binding with proteases. Smoking causes oxidation of this methionine to methionine sulfoxide. As a result, α_1-AT with ***methionine sulfoxide*** cannot bind and inactivate proteases. Emphysema is more commonly associated with heavy smoking and the situation becomes worse in persons with α_1-AT deficiency.

α_1-Antitrypsin deficiency and liver disease : This is due to the accumulation of a mutant α_1-AT which aggregates to form polymers. These polymers, in turn cause liver damage (hepatitis) followed by accumulation of collagen resulting in fibrosis (cirrhosis).

α_2-Macroglobulin

α_2-Macroglobulin concentration in plasma is elevated in ***nephrotic syndrome***. This is due to the fact that majority of the low molecular weight proteins are lost in urine (***proteinuria***) in this disorder.

HAPTOGLOBIN

Haptoglobin (***Hp***), a glycoprotein, is an ***acute phase protein***. Its plasma concentration is increased in several inflammatory conditions.

Functions of haptoglobin

Haptoglobin binds with the free hemoglobin (known as extra-corpuscular hemoglobin) that spills into the plasma due to hemolysis. The ***haptoglobin-hemoglobin (Hp-Hb) complex***

Medical Concepts/ Clinical Correlates

- Albumin, the most abundant plasma protein, is involved in osmotic function, transport of several compounds (fatty acids, steroid hormones), besides the buffering action.
- Hypoalbuminemia and albuminuria are observed in nephrotic syndrome.
- α_1-Antitrypsin deficiency has been implicated in emphysema (abnormal distension of lungs by air) which is more commonly associated with heavy smoking.
- Haptoglobin prevents the possible loss of free hemoglobin from the plasma through the kidneys by forming haptoglobin-hemoglobin complex.

(mol. wt. 155,000) cannot pass through glomeruli of kidney while free Hb (mol. wt. 65,000) can. ***Haptoglobin***, therefore, ***prevents the loss of free Hb into urine***.

Clinical significance of Hp : Plasma concentration of Hp is ***decreased in hemolytic anemia***. This is explained as follows. The half-life of Hp is about 5 days while that of Hp-Hb complex is 90 min. In hemolytic anemia, free Hb in plasma is elevated leading to increased formation of Hp-Hb complex. This complex in turn, is rapidly cleared from the plasma resulting in decreased Hp levels.

CERULOPLASMIN

Ceruloplasmin is a blue coloured, ***copper—containing α_2-globulin*** with a molecular weight of 150,000. Its plasma concentration is about 30 mg/dl. Ceruloplasmin binds with almost 90% of plasma copper (6 atoms of Cu bind to a molecule). This binding is rather tight and, as a result, copper from ceruloplasmin is not readily released to the tissues. Albumin carrying only 10% of plasma copper is the major supplier of copper to the tissues. Ceruloplasmin possesses oxidase activity, and it is associated with ***Wilson's disease*** which is discussed under copper metabolism (***Chapter 18***).

TRANSFERRIN

Transferrin (Tf) is a glycoprotein. It is a ***transporter of iron in the circulation***. Plasma transferrin levels are increased in iron deficiency anemia, and decreased in nephrotic syndrome.

ACUTE PHASE PROTEINS

Acute phase response refers to a non-specific response to the stimulus of infection, injury, various ***inflammatory conditions*** (affecting tissue/ organs), cancer etc. This phase is associated with a characteristic pattern of changes in certain plasma proteins, collectively referred to as acute phase proteins e.g. ***α_1-antitrypsin, α_2-macroglobulin, ceruloplasmin***, complement proteins, C-reactive protein. During the acute phase, synthesis of certain plasma proteins decreases, and they are regarded as ***negative acute phase proteins*** e.g. albumin, transferrin.

C-reactive protein (CRP)

CRP is a ***major component of acute phase proteins***. It is produced in the liver and is present in the circulation in minute concentration (< 1 mg/dl). C-reactive protein (***C*** stands for ***c***arbohydrate) is involved in the promotion of immune system through the activation of complement cascade.

Estimation of CRP in serum is important for the evaluation of acute phase response. The response of CRP to surgery is depicted in ***Fig.9.2***. In a normal surgery, serum CRP increases and returns

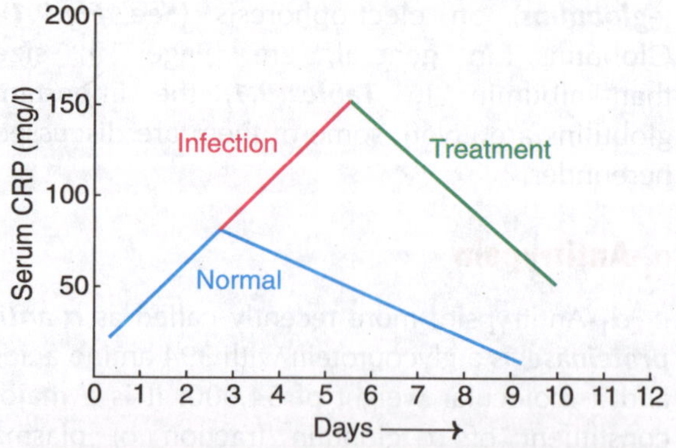

***Fig. 9.2 :** The response of C-reactive protein (CRP) in response to surgery (The normal acute phase is depicted by blue line, the development of infection by red line and the response after treatment by green line).*

to normal level within 7-10 days. If the recovery is complicated by any infection, it will be reflected by the continuous elevation of CRP which requires further treatment.

Increased levels of ***high sensitivity CRP*** (***hs-CRP***) in the circulation (reference range ***1-3 mg/dl***) are useful for predicting the ***risk of coronary heart disease***.

IMMUNOGLOBULINS

The higher vertebrates, including man, have evolved a defense system to protect themselves against the invasion of foreign substances—a virus, a bacterium or a protein. The defense strategies of the body are collectively referred to as ***immunity***, and are briefly described under immunology (***Chapter 42***). Immunoglobulins (or antibodies) are described here.

Immunoglobulins—basic concepts

Immunoglobulins, a specialised group of proteins, are mostly associated with ***γ-globulin fraction*** (on electrophoresis) of plasma proteins. Some immunoglobulins however, separate along with β and α-globulins. Therefore, it should be noted that γ-globulin and immunoglobulin are not synonymous. ***Immunoglobulin is a functional term*** while ***γ-globulin is a physical term***.

Structure of immunoglobulins

All the immunoglobulin (***Ig***) molecules basically consist of two identical ***heavy (H) chains*** (mol. wt. 53,000 to 75,000 each) and two identical ***light (L) chains*** (mol. wt. 23,000 each) held together by disulfide linkages and non-covalent interactions (***Fig.9.3***). Thus, immunoglobulin is a Y-shaped tetramer (H_2L_2). Each heavy chain contains approximately 450 amino

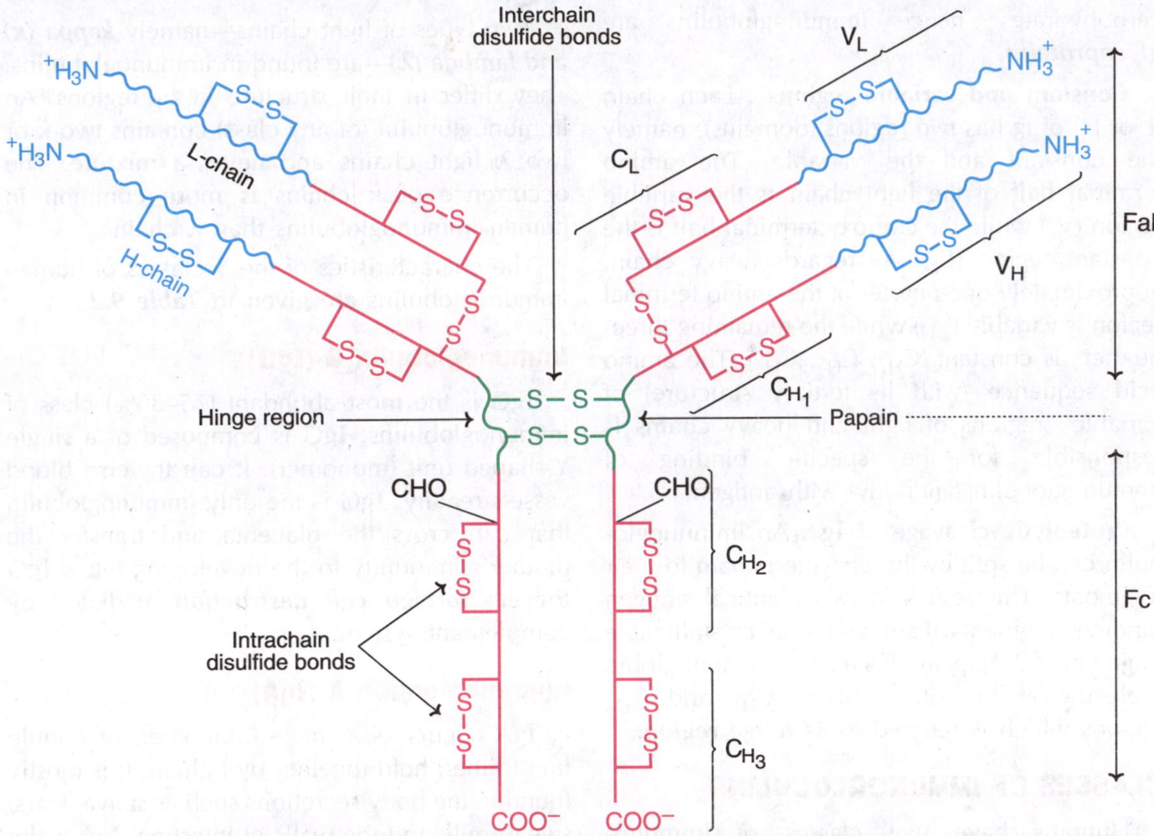

Fig. 9.3 : *Diagrammatic representation of human IgG molecule (V—variable region, C—constant region; CHO—carbohydrate; Each heavy chain is composed of four units—V_H, C_{H_1}, C_{H_2}, C_{H_3} while light chain consists of two units—V_L, C_L).*

Table 9.2 Characteristics of human immunoglobulins

Type	H-Chain	L-Chains	Molecular formula	Molecular weight	Percentage carbohydrate	Serum conc. mg/dl	Major function(s)
IgG	γ	κ or λ	$\gamma_2\kappa_2$ or $\gamma_2\lambda_2$	~150,000	3	800–1,500	Mostly responsible for humoral immunity
IgA	α	κ or λ	$(\alpha_2\kappa_2)_{1-3}$ or $(\alpha_2\lambda_2)_{1-3}$	$\sim(160,000)_{1-3}$	8	150–400	Protects the body surfaces
IgM	μ	κ or λ	$(\mu_2\kappa_2)_5$ or $(\mu_2\lambda_2)_5$	~900,000	12	50–200	Humoral immunity, serves as first line of defense
IgD	δ	κ or λ	$(\delta_2\kappa_2$ or $\delta_2\lambda_2)$	~180,000	13	1–10	B-cell receptor ?
IgE	ε	κ or λ	$\varepsilon_2\kappa_2$ or $\varepsilon_2\lambda_2$	~190,000	12	0.02–0.05	Humoral sensitivity and histamine release.

acids while each light chain has 212 amino acids. The heavy chains of Ig are linked to carbohydrates, hence immunoglobulins are ***glycoproteins***.

Constant and variable regions : Each chain (L or H) of Ig has two regions (domains), namely the constant and the variable. The amino terminal half of the light chain is the variable region (V_L) while the carboxy terminal half is the constant region (C_L). As regards heavy chain, approximately one-quarter of the amino terminal region is variable (V_H) while the remaining three-quarters is constant (C_{H_1}, C_{H_2}, C_{H_3}). The amino acid sequence (with its tertiary structure) of variable regions of light and heavy chains is responsible for the specific binding of immunoglobulin (antibody) with antigen.

Proteolytic cleavage of Ig : An immunoglobulin can be split by the enzyme ***papain*** to their fragments. This results in two identical antigen binding fragments (Fab) and one crystallizable fragment (Fc). Papain cleaves the immunoglobin molecule at the site between C_{H1} and C_{H2} regions which is referred to as ***hinge region.***

CLASSES OF IMMUNOGLOBULINS

Humans have five classes of immunoglobulins—namely ***IgG, IgA, IgM, IgD*** and ***IgE***—containing the heavy chains γ, α, μ, δ and ε, respectively. The type of heavy chain ultimately determines the class and the function of a given Ig.

Two types of light chains—namely ***kappa (κ) and lambda (λ)***—are found in immunoglobulins. They differ in their structure in C_L regions. An immunoglobulin (of any class) contains two κ or two λ light chains and never a mixture. The occurrence of κ chains is more common in human immunoglobulins than λ chains.

The characteristics of the 5 classes of human immunoglobulins are given in ***Table 9.2***.

Immunoglobulin G (IgG)

IgG is the most abundant (75–80%) class of immunoglobulins. IgG is composed of a single Y-shaped unit (monomer). It can traverse blood vessels readily. IgG is the only immunoglobulin that can cross the placenta and transfer the mother's immunity to the developing fetus. IgG triggers ***foreign cell destruction*** mediated by complement system.

Immunoglobulin A (IgA)

IgA occurs as a single (monomer) or double unit (dimer) held together by J chain. It is mostly found in the body secretions such as saliva, tears, sweat, milk and the walls of intestine. IgA is the most predominant antibody in the colostrum, the initial secretion from the mother's breast after a baby is born. The IgA molecules bind with

Medical Concepts/ Clinical Correlates

- Immunoglobulins (antibodies), a specialized group of plasma globular proteins, are actively involved in immunity. IgG and IgM are primarily concerned with humoral immunity while IgE is associated with allergic reactions.
- Multiple myeloma, a plasma cell cancer disease of bone marrow, is characterized by overproduction of abnormal immunoglobulins (mostly IgG). Laboratory diagnosis of multiple myeloma can be made by the presence of a distinct M band on plasma/serum electrophoresis.

bacterial antigens present on the body (outer epithelial) surfaces and remove them. In this way, IgA ***prevents the foreign substances*** from entering the body cells.

Immunoglobulin M (IgM)

IgM is the largest immunoglobulin composed of 5 Y-shaped units (IgG type) held together by a J polypeptide chain. Thus IgM is a pentamer. Due to its large size, IgM cannot traverse blood vessels, hence it is restricted to the blood stream. IgM is the first antibody to be produced in response to an antigen and is the most ***effective against invading microorganisms***. It may be noted that IgM can simultaneously combine with 5 antigenic sites due to its pentameric structure.

Immunoglobulin D (IgD)

IgD is composed of a single Y-shaped unit and is present in a low concentration in the circulation. IgD molecules are present on the surface of B cells. Their function, however, is not known for certain. Some workers believe that IgD may function as ***B-cell receptor***.

Immunoglobulin E (IgE)

IgE is a single Y-shaped monomer. It is normally present in minute concentration in blood. IgE levels are elevated in individuals with ***allergies*** as it is associated with the body's allergic responses. The IgE molecules tightly bind with mast cells which release histamine and cause allergy.

Production of immunoglobulins by multiple genes

As already discussed, immunoglobulins are composed of light and heavy chains. Each light chain is produced by 3 separate genes, namely a variable region (V_L) gene, a constant region (C_L) gene and a joining region (J) gene. Each heavy chain is produced by at least 4 different genes—a variable region (V_H) gene, a constant region (C_H) gene, a joining region (J) gene and diversity region (D) gene. Thus multiple genes are responsible for the synthesis of any one of the immunoglobulins.

Antibody diversity : A person is capable of generating antibodies to almost an unlimited range of antigens (more than one billion!). It should, however, be remembered that humans do not contain millions of genes to separately code for individual immunoglobulin molecules. The antibody diversity is achieved by two special processes, namely ***combination of various structural genes*** and ***somatic mutations***.

MULTIPLE MYELOMA

Multiple myeloma, a ***plasma cell cancer***, constitutes about 1% of all cancers affecting the population. Females are more susceptible than males for this disorder and it usually occurs in the age group 45-60 years.

Abnormal Ig production : Multiple myeloma is due to the ***malignancy of a single clone of plasma cells in the bone marrow***. This results in the overproduction of abnormal immunoglobulins, mostly (75%) ***IgG*** and in some cases (25%) IgA or IgM. IgD type multiple myeloma found in younger adults is less common (<2%) but more severe. In patients of multiple myeloma, the synthesis of normal immunoglobulins is diminished causing depressed immunity. Hence recurrent infections are common in these patients.

Electrophoretic pattern : The plasma of multiple myeloma patients shows a characteristic pattern of electrophoresis. There is a sharp and distinct band (M band, for myeloma globulin) between β-and γ-globulins. Further, this ***M band*** almost replaces the γ-globulin band due to the diminished synthesis of normal γ-globulins.

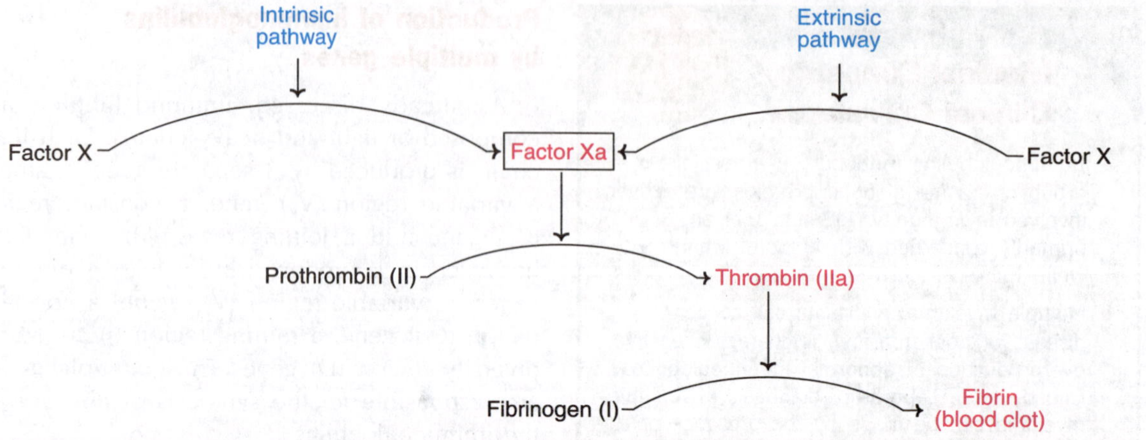

Fig. 9.4 : Overview of blood clotting with the final common pathway.

Bence Jones proteins : Henry Bence Jones first described them in 1847. These are the ***light chains*** (κ or λ) of ***immunoglobulins*** that are ***synthesized in excess***. Bence Jones proteins have a molecular weight of 20,000 or 40,000 (for dimer). In about 20% of the patients of multiple myeloma, Bence Jones proteins are excreted in urine which often damages the renal tubules.

Amyloidosis is characterized by the deposits of light chain fragments in the tissue (liver, kidney, intestine) of multiple myeloma patients.

The presence of Bence Jones proteins in urine can be detected by specific tests.

1. **Electrophoresis** of a concentrated urine is the best test to detect Bence Jones proteins in urine.

2. **The classical heat test** involves the ***precipitation*** of Bence Jones proteins when slightly acidified urine is ***heated to 40-50°C***. This precipitate ***redissolves*** on further heating of urine to boiling point. It reappears again on cooling urine to ***about 70°C***.

3. **Bradshaw's test** involves layering of urine on concentrated HCl that forms a ***white ring*** of ***precipitate***, if Bence Jones proteins are present.

BLOOD CLOTTING

The term ***hemostasis*** is applied to the sequence of ***physiological responses to stop bleeding*** (loss of blood after an injury). This is carried out by blood clotting.

Blood clotting or coagulation is the body's major ***defense mechanism against blood loss***. A blood clot is formed as a result of a series of reactions involving nearly 20 different substances, most of them being ***glycoproteins***, ***synthesized by the liver***.

Blood clotting process involves two independent pathways

1. **The extrinsic pathway** is the initial process in clotting and involves the ***factors that are not present in the blood*** (hence the name).

2. **The intrinsic pathway** involves a series of reactions participated by the ***factors present in the blood***.

Strictly speaking, the extrinsic and intrinsic pathways are not independent, since they are coupled together. Further, the final reactions are identical for both pathways that ultimately lead to the activation of prothrombin to thrombin and the conversion of fibrinogen to fibrin clot (***Fig.9.4***).

The blood ***coagulation factors*** in human plasma along with their common names and molecular weights are listed in ***Table 9.3***. All but two of these factors are designated by a Roman numeral. It should, however, be noted that the ***numbers*** represent the ***order of their discovery*** and not the order of their action. The cascade of blood clotting process is depicted in ***Fig.9.5*** and the salient features are discussed below. The active form of a factor is designated by a subscript ***a.*** The active clotting factors (with exception of fibrin) are serine proteases.

Conversion of fibrinogen to fibrin

Fibrinogen (factor I) is a soluble glycoprotein that constitutes 2-3% of plasma proteins (plasma concentration 0.3 g/dl). Fibrinogen consists of 6 polypeptide chains-two A α, two B β and two γ making the structure $(A\ \alpha)_2\ (B\ \beta)_2\ \gamma_2$.

Fibrinogen undergoes proteolytic cleavage catalysed by thrombin to release small ***fibrinopeptides (A and B)***. This results in the formation of fibrin monomers which can stick together to form hard clots (***Fig.9.6***). Clot formation is further stabilized by covalent cross-linking between glutamine and lysine residues. This reaction cross-links fibrin clots and is catalysed by fibrin stabilizing factor (XIII). The red colour of the clot is due to the presence of red cells entangled in the fibrin cross-links.

Conversion of prothrombin to thrombin

Prothrombin (II) is the inactive zymogen form of thrombin (IIa). The activation of prothrombin occurs on the platelets and requires the presence of factors Va and Xa, besides phospholipids and Ca^{2+}.

The extrinsic pathway

The extrinsic pathway is very rapid and occurs in response to ***tissue injury***. This pathway essentially involves the conversion of proconvertin (VII) to its active form (VIIa) and the generation factor Xa. The tissue factor (III), found to be necessary to accelerate the action VIIa on a factor X, is present in lung and brain.

The intrinsic pathway

The intrinsic pathway is rather slow. It involves the participation of a contact system (wounded surface) and a series of factors to generate factor Xa.

Table 9.3 Blood coagulation factors in humans

Factor number	Common name(s)	Subunit molecular weight
I	Fibrinogen	340,000
II	Prothrombin	720,000
III	Tissue factor, thromboplastin	370,000
IV	Calcium (Ca^{2+})	—
V	Proaccelerin, labile factor	330,000
VII	Proconvertin, serum prothrombin conversion accelerator (SPCA)	50,000
VIII	Antihemophilic factor A, antihemophilic globulin (AHG)	330,000
IX	Christmas factor, antihemophilic factor B,	56,000
	Plasma thromboplastin component (PTC)	
X	Stuart-Prower factor	56,000
XI	Plasma thromboplastin antecedent (PTA)	160,000
XII	Hageman factor	80,000
XIII	Fibrin-stabilizing factor (FSF), fibrinoligase, Liki Lorand factor	320,000
—	Prekallikrein	88,000
—	High molecular weight kininogen (HMK)	150,000

Note : *The numbers represent the order of their discovery and not the order of their action. Factor Va was once referred to as factor VI, hence there is no factor VI.*

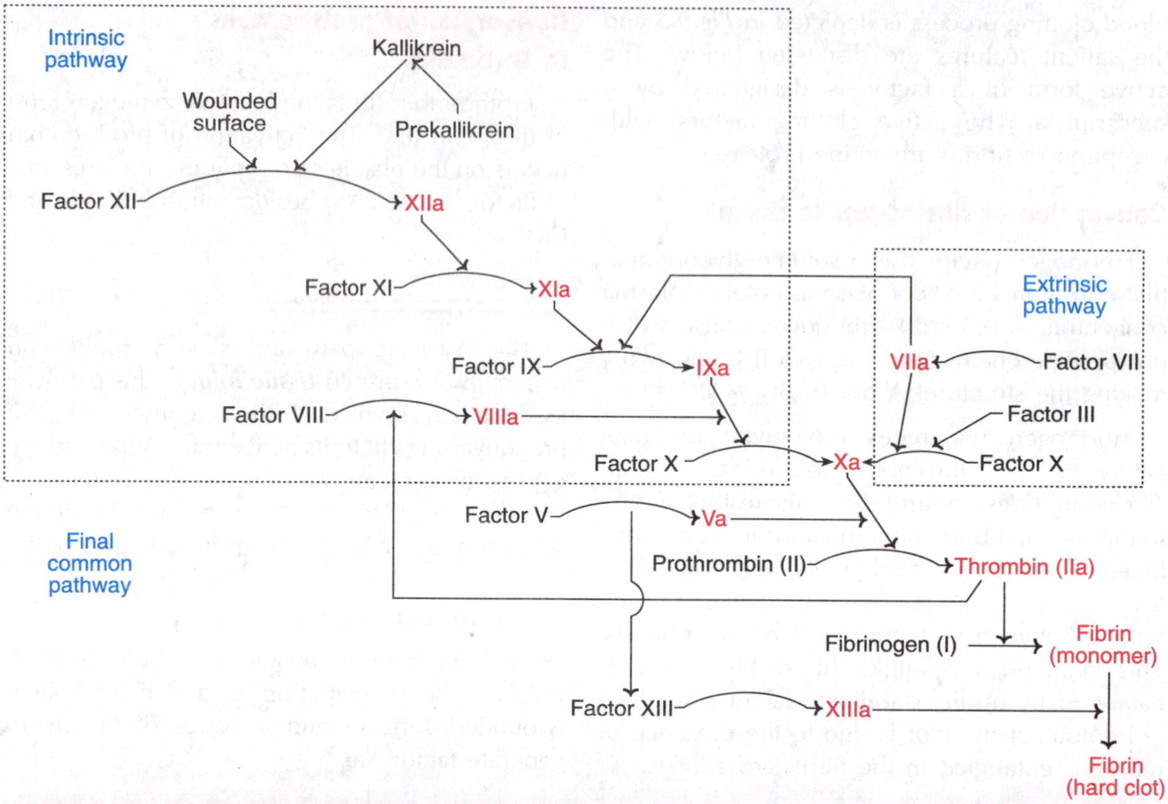

***Fig. 9.5 :** The blood clotting cascade in humans (the active forms of the factors are represented in red with subscript 'a').*

The Hageman factor (XII) is activated (XIIa) on exposure to activating wound surface containing collagen or platelet membranes. The formation of XIIa is accelerated by kallikrein and HMK. The activated Hageman factor (XIIa) activates factor XI. The XIa activate the Christmas factor (IX). The Christmas factor is also activated by active proconvertin (VIIa).

In the next step, the Stuart factor (X) is activated by Christmas factor (IXa) and this reaction requires the presence of antihemophilic factor (VIIIa), Ca^{2+} and phospholipids.

The extrinsic and intrinsic pathways lead to the formation of factor Xa which then participates in the final common pathway to ultimately result in the formation of fibrin clot.

Anticoagulants

Several substances, known as anticoagulants, are in use to ***inhibit the blood clotting***. Calcium is required for certain reactions of blood coagulation. The substances which bind with

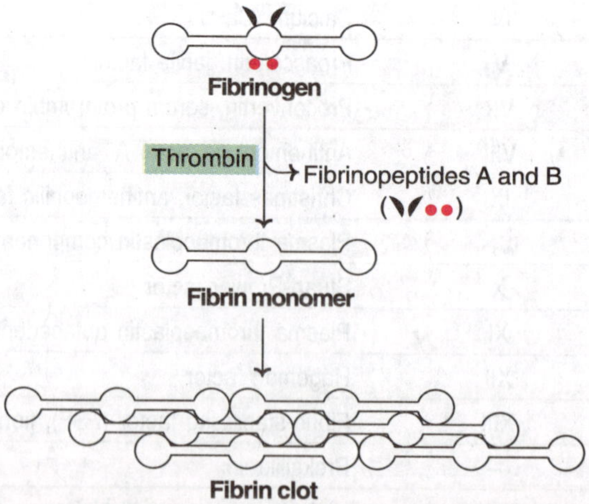

***Fig. 9.6 :** Diagrammatic representation of fibrin clot formation from fibrinogen.*

Medical Concepts/ Clinical Correlates

- Blood clotting or coagulation is the body's major defense mechanism against blood loss. Defects in clotting factors cause coagulation abnormalities such as hemophilia A (deficiency of factor VIII) and Christmas disease (deficiency of factor IX).
- Anticoagulants inhibit blood clotting. These include heparin, oxalate, fluoride, EDTA and citrate.
- Certain blood tests in the laboratory require unclotted blood. Anticoagulants are employed for collecting such specimens.

Ca^{2+} are very effective as anticoagulants. These include ***oxalate***, ***fluoride***, ***EDTA*** and ***citrate***.

Heparin is an anticoagulant used ***to maintain normal hemostasis***. It is a heteropolysaccharide found in many tissues including mast cells in the endothelium of blood vessels. Heparin combines with antithrombin III which in turn, inhibits the clotting factors II, IX, X, XI, XII and kallikrein. Heparin can be administered to patients during and after surgery to retard blood clotting.

The blood contains another anticoagulant—namely ***protein C***—which is activated by thrombin. Active protein C hydrolyses and inactivates clotting factors V and VIII.

Low molecular weight heparins (LMWH) are a group of anticoagulant medications. They are used in the prevention of blood clots and in the treatment of myocardial infarction, deep vein thrombosis and pulmonary embolism. LMWH are obtained by fractionation and depolymerisation of heparin.

Warfarin, a vitamin K antagonist may be considered as an ***oral anticoagulant***. This acts by reducing the synthesis of certain clotting factors (II, VII, IX and X).

Fibrinolysis

The term fibrinolysis refers to the dissolution or lysis of blood clots. ***Plasmin*** is mostly responsible for the dissolution of fibrin clots. Plasminogen, synthesized in the kidney, is the inactive precursor of plasmin. Tissue plasminogen activator (TPA) and urokinase convert plasminogen to plasmin.

Streptokinase is a therapeutic fibrinolytic agent which activates plasminogen. It is administered immediately after symptoms of myocardial infarction, pulmonary embolism, and deep vein thrombosis in legs.

Abnormalities in blood clotting

Several abnormalities associated with blood clotting are known. These are due to defects in ***clotting factors*** which may be inherited or acquired. Hemophilia, Von Willebrand's disease etc., are examples of inherited disorder while afibrinogenemia is an acquired disease.

Hemophilia A (classical hemophilia) : This is a ***sex-linked*** disorder transmitted by females affecting males. Hemophilia A is the most common clotting abnormality and is due to the ***deficiency of antihemophilic factor (VIII)***. The affected individuals have prolonged clotting time and suffer from internal bleeding (particularly in joints and gastrointestinal tract). Hemophilia A has gained importance, since the Royal families of Britain are among the affected individuals.

Hemophilia B (Christmas disease) : This is due to the deficiency of ***Christmas factor (IX)***. The clinical symptoms are almost similar to that found in hemophilia A.

Von Willebrand's disease : It is characterized by failure of platelets to aggregate and is due to a defect in the platelet adherence factor.

SUMMARY

- *The total concentration of plasma proteins is about 6-8 g/dl. Electrophoresis separates plasma proteins into 5 distinct bands, namely albumin, α_1, α_2, β and γ globulins.*
- *Albumin is the major constituent (60%) of plasma proteins with a concentration 3.5 to 5.0 g/dl. It is exclusively synthesized by the liver. Albumin performs osmotic, transport and nutritive functions.*
- *α_1-Antitrypsin is a major constituent of α_1 globulin fraction. α_1-Antitrypsin deficiency has been implicated in emphysema and a specific liver disease.*
- *Haptoglobin (Hp) binds with free hemoglobin (Hb) that spills into the plasma due to hemolysis. The Hp-Hb complex cannot pass through the glomeruli, hence haptoglobin prevents the loss of free hemoglobin into urine.*
- *Alterations in the acute phase proteins (e.g. α_1-antitrypsin, ceruloplasmin, C-reactive protein) are observed as a result of non-specific response to the stimulus of infection, injury, inflammation etc. Estimation of serum C-reactive protein is used for the evaluation of acute phase response.*
- *Immunoglobulins are specialized proteins to defend the body against the foreign substances. They are mostly associated with γ-globulin fraction of plasma proteins. The immunoglobulins essentially consist of two identical heavy chains and two identical light chains, held together by disulfide linkages.*
- *Five classes of immunoglobulins—namely IgG, IgA, IgM, IgD and IgE—are found in humans. IgG is most abundant and is mainly responsible for humoral immunity. IgA protects body surfaces. IgM serves as a first line of defense for humoral immunity while IgE is associated with allergic reactions.*
- *Multiple myeloma is due to the malignancy of a single clone of plasma cells in the bone marrow. This causes the overproduction of abnormal IgG. The plasma of multiple myeloma patients on electrophoresis shows a distinct M-band.*
- *Blood clotting is the body's major defense mechanism against blood loss. The extrinsic and intrinsic pathways lead to the formation of factor Xa which then participates in the final common pathway to activate prothrombin to thrombin. Fibrinogen is then converted to fibrin clot.*
- *Plasmin is mostly responsible for the dissolution of fibrin clots. Plasminogen, synthesized by the kidney, is the inactive precursor of plasmin. Tissue plasminogen activator (TPA) and urokinase convert plasminogen to plasmin.*

THE READER MUST REFER APPENDIX VI (Page Nos : 738 - 756) FOR RELATED CLINICAL CASE STUDIES (Nos : 32, 57, 67)

Self-Assessment Exercises

I. Essay questions

1. Describe the characteristics and major functions of plasma proteins.
2. Give an account of different types of immunoglobulins along with their functions.
3. Discuss the cascade of blood clotting process.
4. Describe the structure of different immunoglobulins.
5. Discuss the role of acute phase proteins in health and disease.

II. Short notes

(a) Electrophoresis of plasma proteins, (b) Functions of albumin, (c) α_1-Antitrypsin, (d) Haptoglobin, (e) Immunoglobulin G, (f) Multiple myeloma, (g) Bence-Jones proteins, (h) Fibrinogen, (i) Anticoagulants, (j) Hemophilia.

III. Fill in the blanks

1. The difference between plasma and serum is the presence or absence of __________.
2. The most commonly employed technique for separation of plasma proteins __________.
3. Haptoglobin binds and prevents the excretion of the compound __________.
4. The cells responsible for the production of immunoglobulins __________.
5. The immunoglobulin that can cross the placenta and transfer the mother's immunity to the developing fetus __________.
6. The immunoglobulins that can bind with mast cells and release histamine __________.
7. Bence-Jones proteins are precipitated when urine is heated to __________.
8. The major component of acute phase proteins used for the evaluation of acute phase response __________.
9. The extrinsic and intrinsic pathways result in the formation of a common activated factor __________.
10. The factor mostly responsible for the lysis of blood clot __________.

IV. Multiple choice questions

11. Hemophilia A is due to the deficiency of clotting factor
    (a) X (b) V (c) VIII (d) II.
12. Plasma albumin performs the following functions
    (a) Osmotic (b) Transport (c) Nutritive (d) All of them.
13. The immunoglobulin present in most abundant quantity
    (a) IgG (b) IgA (c) IgM (d) IgE.
14. Name the immunoglobulin involved in body allergic reactions
    (a) IgA (b) IgE (c) IgD (d) IgM.
15. The following anticoagulant binds with Ca^{2+} and prevents blood clotting
    (a) Heparin (b) Oxalate (c) Protein C (d) All of them.

Hemoglobin and Porphyrins

10

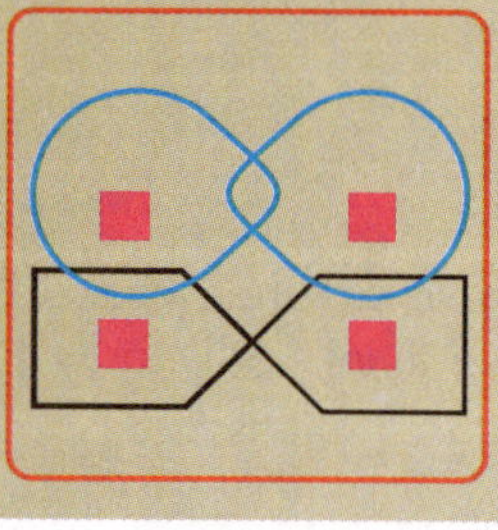

The hemoglobin speaks

"I am the red of blood, responsible for respiration;
Deliver O_2 to tissues and return CO_2 to lungs;
Influenced by factors pH, BPG and Cl^- in my functions;
Disturbed in my duties by structural abnormalities."

The structure, functions and abnormalities of hemoglobin, the synthesis and degradation of heme, the porphyrin containing compounds are discussed in this chapter.

HEMOGLOBIN

Hemoglobin (Hb) is the red blood pigment, exclusively found in ***erythrocytes*** (*Greek*: erythrose—red; kytos—a hollow vessel). The normal concentration of Hb in blood in males is 14–16 g/dl, and in females 13–15 g/dl. Hemoglobin performs two important biological functions concerned with respiration

1. Delivery of O_2 from the lungs to the tissues.

2. Transport of CO_2 and protons from tissues to lungs for excretion.

Structure of hemoglobin

Hemoglobin (mol. wt. 64,450) is a conjugated protein, containing ***globin***—the apoprotein part—and the ***heme***—the non-protein part (prosthetic group). Hemoglobin is a ***tetrameric allosteric protein*** (***Fig.10.1***).

Structure of globin : Globin consists of four polypeptide chains of two different primary structures (monomeric units). The common form of adult hemoglobin (HbA_1) is made up of two α-chains and two β-chains ($\alpha_2\beta_2$). Some authors consider hemoglobin consisting of two identical dimers—$(\alpha\beta)_1$ and $(\alpha\beta)_2$. Each α-chain contains 141 amino acids while β-chain contains 146 amino acids. Thus HbA_1 has a total of 574

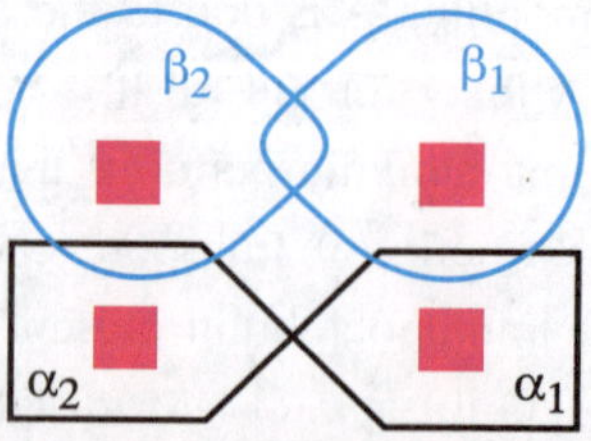

***Fig. 10.1 :** Diagrammatic representation of hemoglobin with 2α and 2β chains (Red blocks–Heme).*

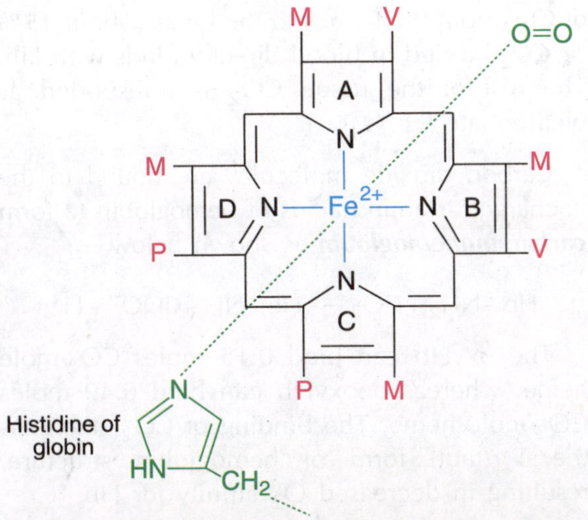

Fig. 10.2 : Structure of heme [M-Methyl (– CH_3); V-Vinyl (– CH_2– CH_2–); P-Propionyl (– CH_2– CH_2– CH_2).

amino acid residues. The four subunits of hemoglobin are held together by non-covalent interactions—hydrophobic, ionic and hydrogen bonds. Each subunit contains a heme group.

Structure of heme : The characteristic red colour of hemoglobin (ultimately blood) is due to heme. Heme contains a porphyrin molecule namely ***protoporphyrin IX***, with iron at its center. Protoporphyrin IX consists of four ***pyrrole*** rings to which four methyl, two propionyl and two vinyl groups are attached (***Fig.10.2***).

Heme is a common prosthetic group present in cytochromes, in certain enzymes e.g. catalase, tryptophan pyrolase, and chlorophyll (Mg^{2+}).

Other forms of hemoglobin

Besides the adult hemoglobin (HbA_1) described above, other minor hemoglobins are also found in humans (***Table 10.1***). In adults a small fraction (< 5%) of hemoglobin, known as HbA_2 is present. HbA_2 is composed of two α and two δ (delta) chains. ***Fetal hemoglobin (HbF)*** is synthesized during the fetal development and a little of it may be present even in adults. ***Hereditary persistence of fetal hemoglobin (HPFH)*** is a condition in which fetal hemoglobin synthesis is not terminated at birth but continues into adulthood. ***Glycosylated hemoglobin (HbA_{1c})***, formed by covalent binding of glucose is also found in low concentration. It is increased in diabetes mellitus which is utilized for the prognosis of these patients (Refer ***Chapter 36***).

Myoglobin

Myoglobin (Mb) is monomeric oxygen binding ***hemoprotein*** found in ***heart*** and ***skeletal muscle***. It has a single polypeptide (153 amino acids) chain with heme moiety. Myoglobin (mol. wt. 17,000) structurally resembles the individual subunits of hemoglobin molecule. For this reason, the more complex properties of hemoglobin have been conveniently elucidated through the study of myoglobin.

Myoglobin functions as a ***reservoir for oxygen.*** It further serves as oxygen carrier that promotes the transport of oxygen to the rapidly respiring muscle cells.

As a result of muscle injury, myoglobin is quickly released into circulation. Although not specific, myoglobin is an early ***marker*** for the diagnosis of ***myocardial infarction. Myoglobinuria*** in adults is due to trauma or muscle necrosis while in children it is often linked with viral infections.

Functions of hemoglobin

Hemoglobin is largely responsible for the ***transport of O_2*** from lungs to tissues. It also helps to ***transport CO_2*** from the tissues to the lungs.

Binding of O_2 to hemoglobin

One molecule of hemoglobin (with four hemes) can bind with four molecules of O_2. This is in contrast to myoglobin (with one heme) which can bind with only one molecule of oxygen. In other words, each heme moiety can bind with one O_2.

Table 10.1 Normal major types of hemoglobins

Type	Composition and symbol	Percentage of total hemoglobin
HbA_1	$\alpha_2\beta_2$	90%*
HbA_2	$\alpha_2\delta_2$	< 5%
HbF	$\alpha_2\gamma_2$	< 2%
HbA_{1c}	$\alpha_2\beta_2$-glucose	< 5%*

(*Adult Hb is around 95% that includes HbA, and HbA_{1C})

Oxygen dissociation curve : The ***binding*** ability of ***hemoglobin with O_2*** at different partial pressures of oxygen (***pO_2***) can be measured by a graphic representation known as O_2 dissociation curve. The curves obtained for hemoglobin and myoglobin are depicted in ***Fig.10.3***.

It is evident from the graph that myoglobin has much higher affinity for O_2 than hemoglobin. Hence O_2 is bound more tightly with myoglobin than with hemoglobin. Further, pO_2 needed for half saturation (50% binding) of myoglobin is about 1 mm Hg compared to about 26 mm Hg for hemoglobin.

Transport of O_2 to the tissues

In the lungs, where the concentration of O_2 is high (hence high pO_2), the hemoglobin gets fully saturated (loaded) with O_2. Conversely, at the tissue level, where the O_2 concentration is low (hence low pO_2), the oxyhemoglobin releases (unloads) its O_2 for cellular respiration. This is often mediated (in muscle) by binding O_2 to myoglobin which serves as the immediate reservoir and supplier of O_2 to the tissues (***Fig.10.4***).

Transport of CO_2 by hemoglobin

In aerobic metabolism, for every molecule of O_2 utilized, one molecule of CO_2 is liberated. Hemoglobin actively participates in the transport

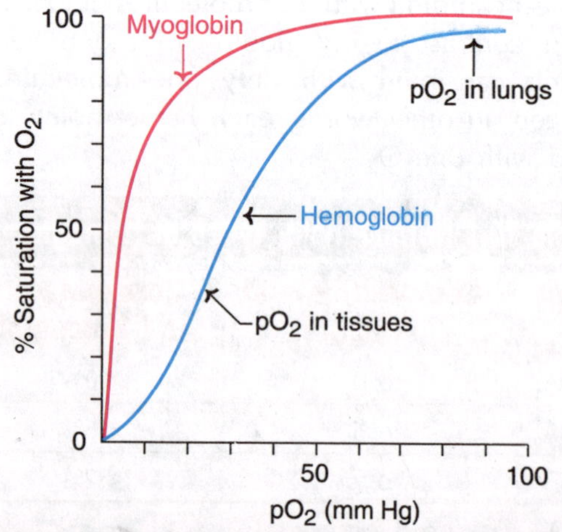

Fig. 10.3 : Oxygen dissociation curves of hemoglobin and myoglobin (pO_2– Partial pressure of oxygen).

of CO_2 from the tissues to the lungs. About ***15% of CO_2*** carried in blood directly binds with Hb. The rest of the tissue CO_2 is transported as bicarbonate (HCO_3^-).

Carbon dioxide molecules are bound to the uncharged α-amino acids of hemoglobin to form ***carbaminohemoglobin*** as shown below

$$Hb-NH_2+CO_2 \rightleftharpoons Hb-NH-COO^- + H^+$$

The oxyHb can bind 0.15 moles CO_2/mole heme, whereas deoxyHb can bind 0.40 moles CO_2/mole heme. The binding of CO_2 stabilizes the T (taut) form of hemoglobin structure, resulting in decreased O_2 affinity for Hb.

Hemoglobin also helps in the transport of CO_2 as bicarbonate, as explained below (***Fig.10.5***).

As the CO_2 enters the blood from tissues, the enzyme ***carbonic anhydrase*** present in ***erythrocytes*** catalyses the formation of carbonic acid ***(H_2CO_3)***. Bicarbonate (HCO_3^-) and proton (H^+) are released on dissociation of carbonic acid. Hemoglobin acts as a buffer and immediately binds with protons. It is estimated that for every 2 protons bound to Hb, 4 oxygen molecules are released to the tissues. In the lungs, binding O_2 to Hb results in the release of protons. The bicarbonate and protons combine to form carbonic acid. The latter is acted upon by carbonic anhydrase to release CO_2, which is exhaled.

BOHR EFFECT

The binding of oxygen to hemoglobin decreases with increasing H^+ concentration (lower pH) or when the hemoglobin is exposed to increased partial pressure of CO_2 (pCO_2). This phenomenon is known as Bohr effect. It is due to a change in the binding affinity of oxygen to hemoglobin. Bohr effect causes a ***shift in the oxygen dissociation curve to the right*** (***Fig.10.6***).

Bohr effect is primarily responsible for the release of O_2 from the oxyhemoglobin to the tissue. This is because of increased pCO_2 and decreased pH in the actively metabolizing cells.

EFFECT OF 2,3-BISPHOSPHOGLYCERATE ON O_2 AFFINITY OF Hb

2,3-Bisphosphoglycerate (***2,3-BPG***; formerly, 2,3-diphosphoglycerate) is the most abundant organic phosphate in the erythrocytes. Its molar concentration is approximately equivalent to that of hemoglobin. 2,3-BPG is produced in the erythrocytes from an intermediate (1,3-bisphosphoglycerate) of glycolysis. This short pathway, referred to as ***Rapoport-Leubering shunt***, is described in carbohydrate metabolism (***Chapter 13***).

Binding of 2,3-BPG to deoxyhemoglobin

2,3-BPG regulates the binding of O_2 to hemoglobin. It specifically binds to deoxyhemoglobin (and not to oxyhemoglobin) and decreases the O_2 affinity to Hb. The effect of 2,3-BPG on Hb may be summarized as follows

$$\underset{\text{OxyHb}}{HbO_2} + \text{2,3-BPG} \longrightarrow \underset{\text{DeoxyHb bound to 2,3-BPG}}{\text{Hb-2,3-BPG}} + O_2$$

The reduced affinity of O_2 to Hb facilitates the ***release of O_2*** at the partial pressure found in the tissues. Thus, 2,3-BPG shifts the oxygen dissociation curve to the right (***Fig.10.7***).

Mechanism of action of 2,3-BPG

One molecule of 2,3-BPG binds with one molecule (tetramer) of deoxyhemoglobin in the central cavity of the four subunits. This central pocket has positively charged (e.g. histidine, lysine) two β-globin chains. Ionic bonds (salt bridges) are formed between the positively charged amino acids (of β globins) with the negatively charged phosphate groups of 2,3-BPG (***Fig.10.8***). The binding of 2,3-BPG ***stabilizes the deoxygenated***

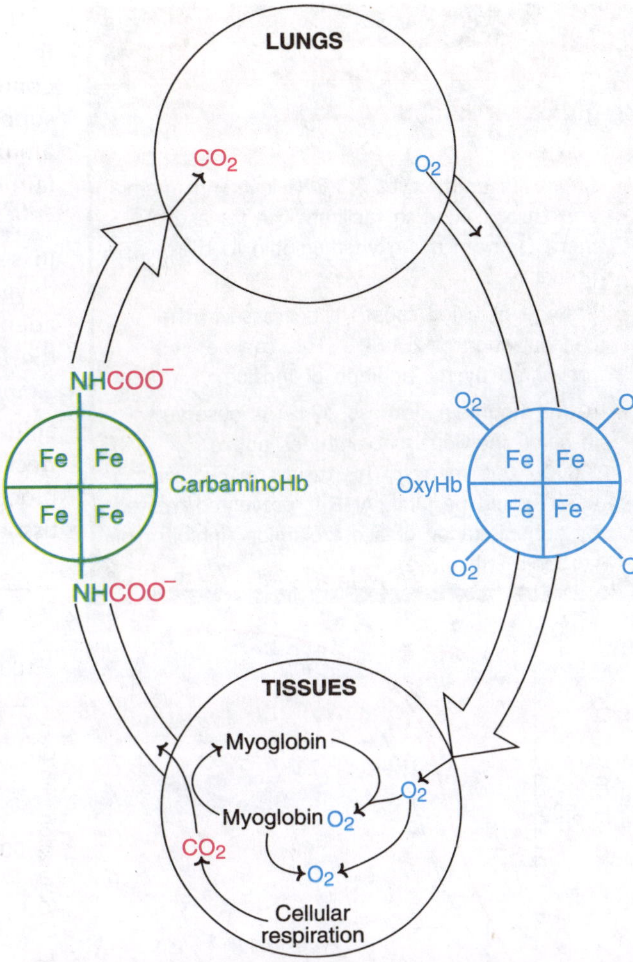

Fig. 10.4 : *Diagrammatic representation of transport of O_2 and CO_2 by hemoglobin.*

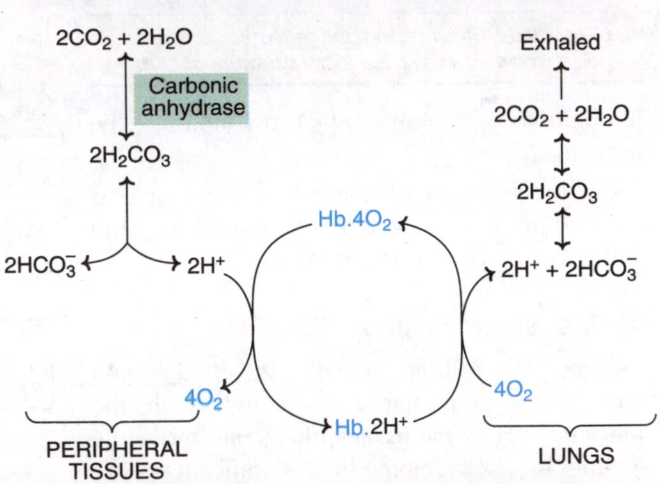

Fig. 10.5 : *Transport of CO_2 through the mediation of hemoglobin.*

Medical Concepts/ Clinical Correlates

- Increased erythrocyte 2,3-BPG levels in anemia and chronic hypoxia facilitate the release of more O_2 from the oxyhemoglobin to the tissues.
- Storage of blood causes a decrease in the concentration of 2,3-BPG. This can be prevented by the addition of inosine.
- Acute mountain sickness (AMS) is observed in some individuals at altitudes above 3,000-4,000 metres. This results in hypoxia which could be fatal. AMS is prevented/treated by administration of acetazolamide (inhibitor of carbonic anhydrase).

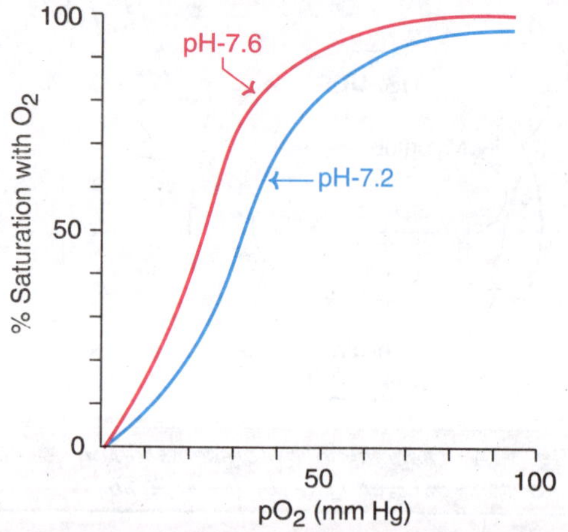

Fig. 10.6 : Effect of pH (Bohr effect) on oxygen dissociation curve (pO_2-Partial pressure of O_2).

hemoglobin (T-form) by cross-linking the β-chains.

On oxygenation of hemoglobin, 2,3-BPG is expelled from the pocket and the oxyhemoglobin attains the R-form of structure.

Clinical significance of 2,3-BPG

Since the binding of 2,3-BPG with hemoglobin is primarily associated with the release of O_2 to the tissues, this small molecule assumes a lot of biomedical significance. The erythrocyte levels of 2,3-BPG are related to tissue demands of oxygen supply.

1. **In hypoxia :** The concentration of ***2,3-BPG*** in erythrocytes is ***elevated*** in chronic hypoxic conditions associated with difficulty in O_2 supply. These include adaptation to high altitude, obstructive pulmonary emphysema (airflow in the bronchioles blocked) etc.

2. **In anemia :** 2,3-BPG levels are ***increased*** in severe anemia in order to cope up with the oxygen demands of the body. This is an adaptation to supply as much O_2 as possible to the tissue, despite the low hemoglobin levels.

3. **In blood transfusion :** Storage of blood in acid citrate-dextrose medium results in the ***decreased concentration of 2,3-BPG***. Such blood when transfused fails to supply O_2 to the tissues immediately.

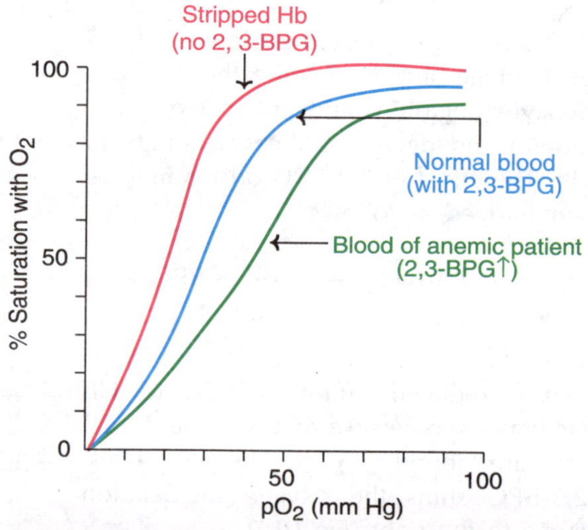

Fig. 10.7 : Effect of pH (Bohr effect) on oxygen dissociation curve (pO_2-Partial pressure of O_2).

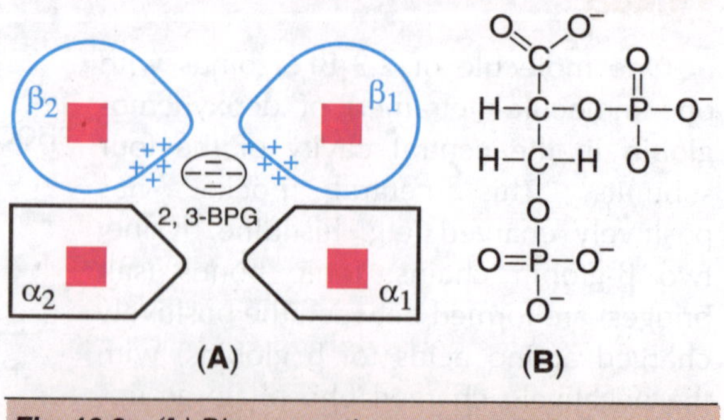

*Fig. 10.8 : **(A)** Diagrammatic representation of binding of 2,3-BPG to deoxyhemoglobin; **(B)** Structure of 2,3-BPG.*

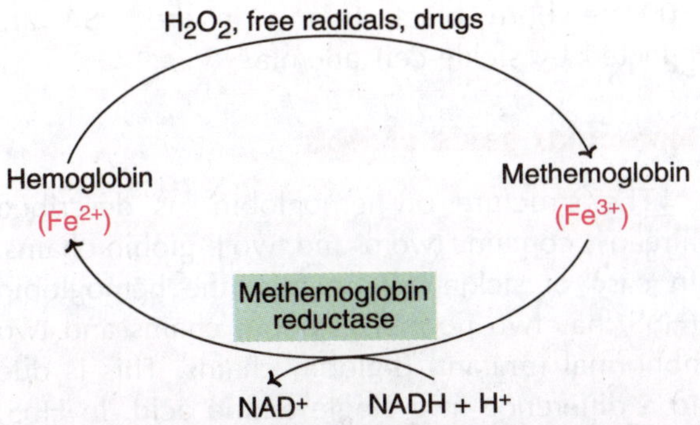

Fig. 10.9 : *Conversion of hemoglobin to methemoglobin and vice versa.*

Addition of ***inosine*** (hypoxanthine-ribose) to the stored blood prevents the decrease of 2,3-BPG. The ribose moiety of inosine gets phosphorylated and enters the hexose monophosphate pathway and finally gets converted to 2,3-BPG.

4. **Fetal hemoglobin (HbF) :** The binding of 2,3-BPG to fetal hemoglobin is very weak. Therefore, HbF has ***higher affinity for O_2*** compared to adult hemoglobin (HbA). This may be needed for the transfer of oxygen from the maternal blood to the fetus.

HEMOGLOBIN DERIVATIVES

Hemoglobin (specifically heme) combines with different ligands and forms hemoglobin derivatives. The normal blood contains oxyHb and deoxyHb. Besides these, ***methemoglobin*** (metHb) and ***carboxyhemoglobin*** are the other important Hb derivatives. The Hb derivatives have characteristic colour and they can be detected by absorption spectra.

Methemoglobin

For the biological function of hemoglobin—to carry oxygen—the iron should remain in the ferrous (Fe^{2+}) state. Hemoglobin (Fe^{2+}) can be oxidized to methemoglobin (Fe^{3+}). In normal circumstances, however, molecular oxygen does not oxidize Hb, it only loosely binds to form oxyhemoglobin.

The oxidation of hemoglobin to methemoglobin (metHb) may be caused in the living system by ***H_2O_2, free radicals and drugs***. The methemoglobin (with Fe^{3+}) is unable to bind to O_2. Instead, a water molecule occupies the oxygen site in the heme of metHb.

In normal circumstances, the occasional oxidation of hemoglobin is corrected by the enzyme ***methemoglobin reductase*** present in erythrocytes (***Fig.10.9***).

Carboxyhemoglobin (COHb)

Carbon monoxide (CO) is a toxic compound (an industrial pollutant) that can bind with Hb in the same manner as O_2 binds. However, CO has about 200 times more affinity than O_2 for binding with Hb.

Clinical manifestations of CO toxicity are observed when the COHb concentration exceeds 20%. The symptoms include headache, nausea, breathlessness, vomiting and irritability. Administration of O_2 through oxygen masks will help to reverse the manifestations of CO toxicity.

ABNORMAL HEMOGLOBINS

Abnormal hemoglobins are the ***result of mutations in the genes that code for α or β chains of globin***. As many as 400 mutant hemoglobins are known. About 95% of them are due to alteration in a single amino acid of globin.

Hemoglobinopathies

It is a term used to describe the disorders caused by the synthesis of abnormal hemoglobin molecule or the production of insufficient quantities of normal hemoglobin or rarely both.

Sickle-cell anemia (HbS) and ***hemoglobin C disease (HbC)*** are the classical examples of abnormal hemoglobins. ***Thalassemias***, on the other hand, are caused by decreased synthesis of normal hemoglobin.

SICKLE-CELL ANEMIA OR SICKLE-CELL HEMOGLOBIN

Sickle-cell anemia (HbS) is the ***most common*** form of ***abnormal hemoglobins***. It is so named

Medical Concepts/ Clinical Correlates

- Hemoglobin (Fe^{2+}) on oxidation by H_2O_2, free radicals or drugs, forms methemoglobin (Fe^{3+}) which cannot transport O_2.
- Carboxyhemoglobin is produced when carbon monoxide, an industrial pollutant, binds to hemoglobin. The clinical manifestations of CO toxicity (> 20% COHb) include headache, nausea, breathlessness and vomiting.

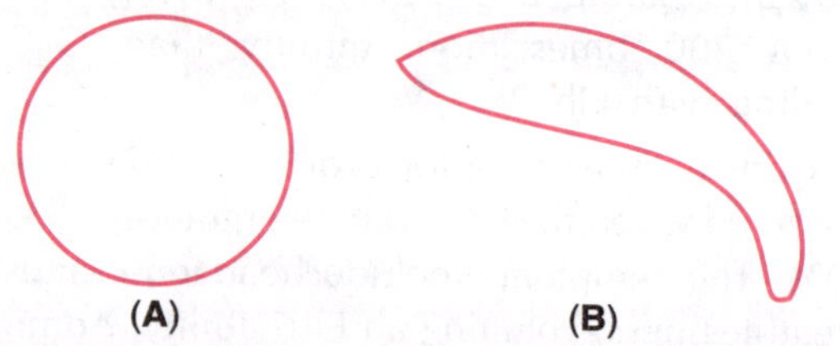

Fig. 10.10 : *Erythrocytes : (A) From a normal person; (B) From a patient of sickel-cell anemia.*

because the erythrocytes of these patients adopt a sickle shape (***crescent like***) at low oxygen concentration (***Fig.10.10***).

Occurrence of the disease

Sickle-cell anemia is largely confined to tropical areas of the world. It ***primarily occurs in the black population***. It is estimated that 1 in 500 newborn black infants in the USA are affected by sickle-cell anemia.

Molecular basis of HbS

The structure of hemoglobin (as described already) contains two α-and two β-globin chains. In case of sickle-cell anemia, the hemoglobin (HbS) has two normal α-globin chains and two abnormal (mutant) β-globin chains. This is due to a difference in a single amino acid. In HbS, ***glutamate at sixth position of β-chain is replaced by valine*** (Glu $\beta_6 \rightarrow$ Val).

Sickle-cell anemia is due to a change (***missense mutation***) in the single nucleotide (thymine → adenine) of β-globin gene. This error causes the formation of altered codon (GUG in place of GAG) which leads to the incorporation of valine instead of glutamate at the sixth position in β-chain (***Fig.10.11***).

Homozygous and heterozygous HbS : Sickle-cell anemia is said to be homozygous, if caused by inheritance of two mutant genes (one from each parent) that code for β-chains. In case of heterozygous HbS, only one gene (of β-chain) is affected while the other is normal. The erythrocytes of heterozygotes contain both HbS and HbA and the disease is referred to as ***sickle-cell trait*** which is ***more common in blacks*** (almost 1 in 10 are affected). The individuals of sickle-cell trait lead a ***normal life***, and do not usually show clinical symptoms. This is in contrast to homozygous sickle-cell anemia.

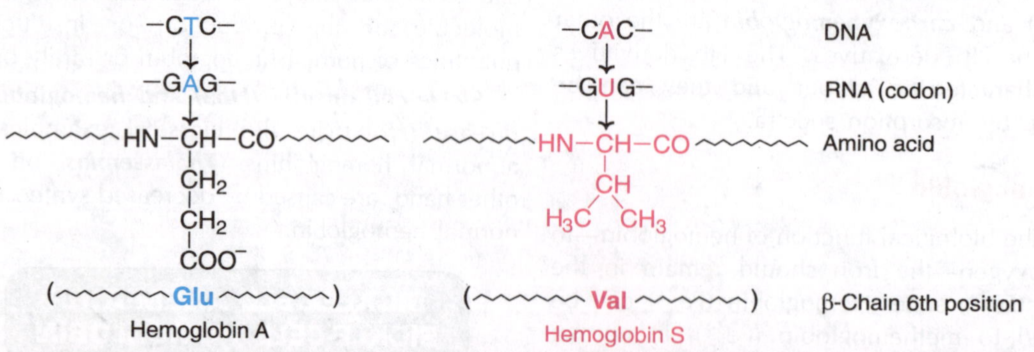

Fig. 10.11 : *Formation of β-chain of hemoglobin in normal and sickle cell anemia (**Note :** Single base mutation in DNA (T → A) causes replacement of glutamate by valine at 6th position of β-chain).*

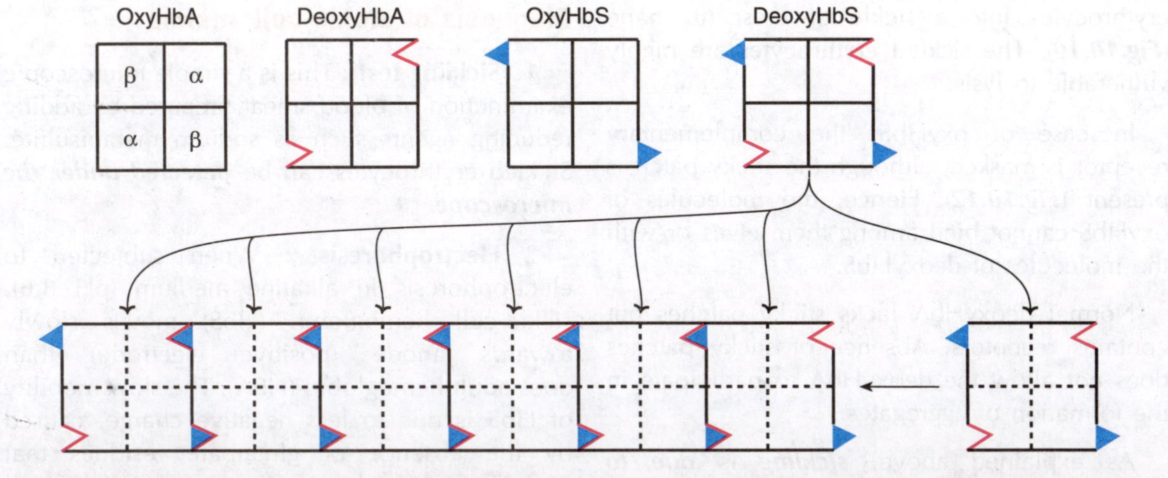

Fig. 10.12 : Diagrammatic representation of sticky patch (Blue) and sticky patch receptor (>) in the formation of long aggregates of deoxyhemoglobins.

Abnormalities associated with HbS

Sickle-cell anemia is characterized by the following abnormalities

1. **Life-long hemolytic anemia :** The sickled ***erythrocytes*** are fragile and their ***continuous breakdown*** leads to life-long anemia.

2. **Tissue damage and pain :** The sickled cells block the capillaries resulting in poor blood supply to tissues. This leads to extensive damage and ***inflammation*** of certain tissues causing pain.

3. **Increased susceptibility to infection :** Hemolysis and tissue damage are accompanied by increased susceptibility to infection and diseases.

4. **Premature death :** Homozygous individuals of sickle-cell anemia die before they reach adulthood (< 20 years).

Mechanism of sickling in sickle-cell anemia

Glutamate is a polar amino acid and it is replaced by a non-polar valine in sickle-cell hemoglobin. This causes a marked ***decrease in the solubility of HbS*** in deoxygenated form (T-form). However, solubility of oxygenated HbS is unaffected.

Sticky patches and formation of deoxyhemoglobin fibres

The substitution of valine for glutamate results in a sticky patch on the outer surface of β-chains. It is present on oxy- and deoxyhemoglobin S but absent on HbA. There is a site or receptor complementary to sticky patch on deoxyHbS.

The sticky patch of one deoxyHbS binds with the receptor of another deoxyHbS and this process continuous resulting in the formation of long aggregate molecules of deoxyHbS (***Fig.10.12***). Thus, the polymerization of deoxy-HbS molecules leads to long fibrous precipitates (***Fig.10.13***). These stiff fibres distort the

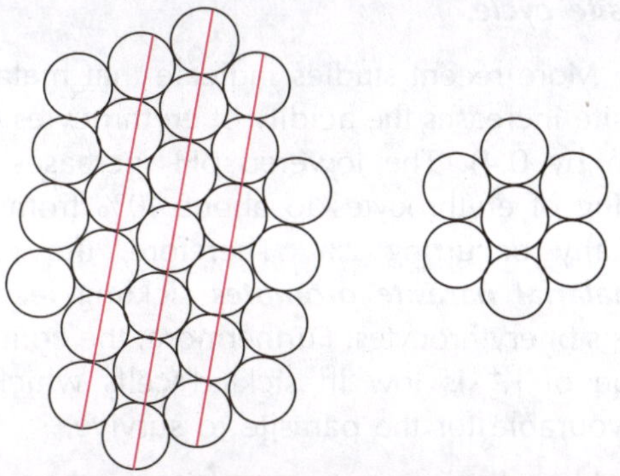

Fig. 10.13 : Diagrammatic representation a fibre of aggregated deoxyhemoglobin.

erythrocytes into a sickle or crescent shape (***Fig.10.10***). The sickled erythrocytes are highly vulnerable to lysis.

In case of oxyHbS, the complementary receptor is masked, although the sticky patch is present (***Fig.10.12***). Hence, the molecules of oxyHbS cannot bind among themselves or with the molecules of deoxyHbS.

Normal deoxyHbA lacks sticky patches but contains receptors. Absence of sticky patches does not allow the deoxyHbA to participate in the formation of aggregates.

As explained above, ***sickling*** is ***due to polymerization of deoxyHbS***. Therefore, if HbS is maintained in the oxygenated form (or with minimum deoxyHbS), sickling can be prevented.

Sickle-cell trait provides resistance to malaria

The incidence of sickle-cell disease coincides with the high incidence of malaria in tropical areas of the world (particularly among the black Africans).

Sickle-cell trait (heterozygous state with about 40% HbS) provides resistance to malaria which is a major cause of death in tropical areas. This is explained as follows

1. Malaria is a parasitic disease caused by *Plasmodium falciparum* in Africa. The malarial parasite spends a part of its life cycle in erythrocytes. ***Increased lysis of sickled cells*** (shorter life span of erythrocytes) ***interrupts the parasite cycle.***

2. More recent studies indicate that malarial parasite increases the acidity of erythrocytes (pH down by 0.4). The lowered pH increases the sickling of erythrocytes to about 40% from the normally occurring 2%. Therefore, the entry of ***malarial parasite promotes*** sickling leading to lysis of erythrocytes. Furthermore, the concentration of K^+ is low in sickled cells which is unfavourable for the parasite to survive.

Sickle-cell trait appears to be an ***adaptation for the survival*** of the individuals ***in malaria-infested regions.***

Diagnosis of sickle-cell anemia

1. **Sickling test :** This is a simple microscopic examination of blood smear prepared by adding reducing agents such as sodium metabisulfite. Sickled erythrocytes can be ***detected under the microscope.***

2. **Electrophoresis :** When subjected to electrophoresis in alkaline medium (pH 8.6), sickle-cell hemoglobin (HbS) moves slowly towards anode (positive electrode) than does adult hemoglobin (HbA). The slow mobility of HbS is due to less negative charge, caused by the absence of glutamate residues that carry negative charge. In case of sickle-cell trait, the fast moving HbA and slow moving HbS are observed. The electrophoresis of hemoglobin obtained from lysed erythrocytes can be routinely used for the diagnosis of sickle-cell anemia and sickle-cell trait (***Fig.10.14***).

3. Tests to ***detect genes*** of sickle-cell disease.

Management of sickle-cell disease

Administration of ***sodium cyanate inhibits sickling of erythrocytes,*** Cyanate increases the affinity of O_2 to HbS and lowers the formation of deoxyHbS. However, it causes certain side-effects like peripheral nerve damage.

In patients with severe anemia, repeated ***blood transfusion is required.*** This ***may result*** in iron overload and ***cirrhosis of liver.***

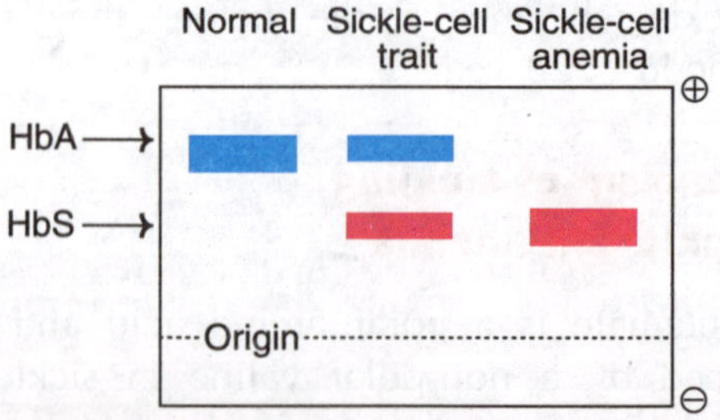

Fig. 10.14 : *Electrophoresis of hemoglobins at pH 8.6 (HbA–Normal adult hemoglobin; HbS–Sickle cell hemoglobin).*

Replacement of HbS with other forms of hemoglobins has been tried. Fetal hemoglobin (HbF) reduces sickling. Sickle-cell disease awaits gene-replacement therapy!

Hemoglobin C disease

Cooley's hemoglobinemia (HbC) is characterized by substitution of ***glutamate by lysine in the sixth position of β-chain***. Due to the presence of lysine, HbC moves more slowly on electrophoresis compared to HbA and HbS. HbC disease occurs only in blacks. Both homozygous and heterozygous individuals of HbC disease are known. This disease is characterized by mild hemolytic anemia. No specific therapy is recommended.

Hemoglobin D

This is caused by the ***substitution of glutamine in place of glutamate in the 121st position of β-chain***. Several variants of HbD are identified from different places indicated by the suffix. For instance, HbD (Punjab), HbD (Los Angeles). HbD, on electrophoresis moves along with HbS.

Hemoglobin E

This is the most common abnormal hemoglobin after HbS. It is estimated that about 10% of the population in South-East Asia (Bangladesh, Thailand, Myanmar) suffer from HbE disease. In India, it is prevalent in West Bengal. HbE is characterized by ***replacement of glutamate by lysine at 26th position of β-chain***. The individuals of HbE (either homozygous or heterozygous) have no clinical manifestations.

THALASSEMIAS

Thalassemias are a group of hereditary hemolytic disorders characterized by ***impairment/imbalance in the synthesis of globin chains of Hb***.

Thalassemias (*Greek*: thalassa-sea) mostly occur in the regions surrounding the Mediterranean sea, hence the name. These diseases, however, are also prevalent in Central Africa, India and the Far East.

Molecular basis of thalassemias

Hemoglobin contains 2α and 2β globin chains. The synthesis of individual chains is so coordinated that each α-chain has a β-chain partner and they combine to finally give hemoglobin ($\alpha_2\beta_2$). Thalassemias are characterized by a ***defect in the production of α-or β-globin chain***. There is however, no abnormality in the amino acids of the individual chains.

Thalassemias occur due to a variety of molecular defects

1. Gene deletion or substitution,
2. Underproduction or instability of mRNA,
3. Defect in the initiation of chain synthesis,
4. Premature chain termination.

α-Thalassemias

α-Thalassemias are caused by a ***decreased*** synthesis ***or*** total ***absence of α-globin chain of Hb***. There are four copies of α-globin gene, two on each one of the chromosome 16. Four types of α-thalassemias occur which depend on the number of missing α-globin genes. The salient features of different α-thalassemias are given in ***Table 10.2***.

1. **Silent carrier** state is due to loss of one of the four α-globin genes with no physical manifestations.

2. **α-Thalassemia *trait*** caused by loss of two genes (both from the same gene pair or one from each gene pair). Minor anemia is observed.

3. **Hemoglobin H disease**, due to missing of three genes, is associated with moderate anemia.

4. **Hydrops fetalis** is the most severe form of α-thalassemias due to lack of all the four genes. The fetus usually survives until birth and then dies.

β-Thalassemias

Decreased synthesis or total ***lack*** of the formation of ***β-globin chain*** causes β-thalassemias. The production of α-globin chain

Table 10.2 Summary of different types of α-thalassemias

Type of thalassemia	Number of missing genes	Schematic representation of genes on chromosome 16	Clinical symptoms
Normal	Nil	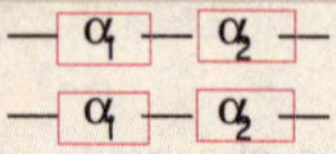	Nil
Silent carrier	1	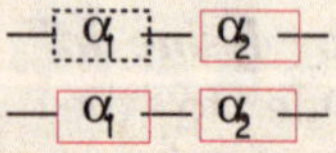	No symptoms
α-Thalassemia trait (heterozygous form)	2	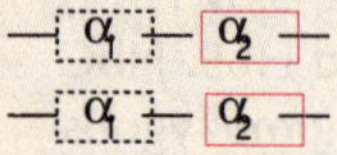	Minor anemia
Hemoglobin H disease	3	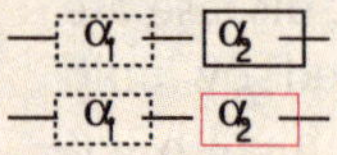	Mild to moderate anemia may lead normal life.
Hydrops fetalis	4	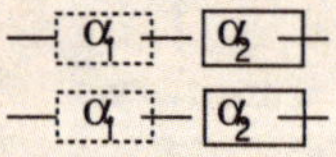	Fetal death usually occurs at birth.

(▭ : represent functional genes; ⬚ : represent missing genes)

continues to be normal, leading to the formation of a globin tetramer (α_4) that precipitate. This causes premature death of erythrocytes. There are mainly two types of β-thalassemias (***Fig.10.15***)

1. **β-Thalassemia minor :** This is an heterozygous state with a ***defect in only one of the two β-globin gene pairs on chromosome 11***. This disorder, also known as β-***thalassemia trait***, is usually asymptomatic, since the individuals can make some amount of β-globin from the affected gene.

2. **β-Thalassemia major :** This is a homozygous state with a ***defect in both the genes responsible for β-globin synthesis***. The infants born with β-thalassemia major are healthy at birth since β-globin is not synthesized during the fetal development. They become severely anemic and die within 1-2 years. Frequent blood transfusion is required for these children. This is associated with iron overload which in turn may lead to death within 15-20 years.

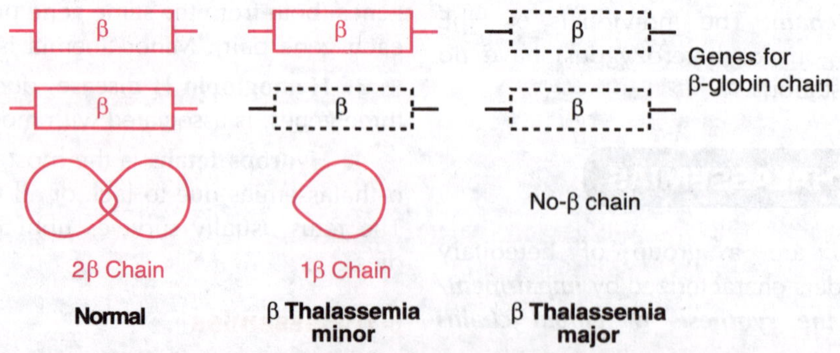

***Fig. 10.15:** Diagrammatic representation of gene deletions in β-thalassemias (each one of the chromosome pair of 11 has one gene for β-globin).*

Medical Concepts/ Clinical Correlates

- Sickle cell hemoglobin (HbS) causes hemolytic anemia, increased susceptibility to infection and premature death. However, HbS offers protection against malaria.
- Thalassemias are hemolytic disorders caused by impairment/imbalance in the synthesis of globin chains of Hb. These include α-thalassemia trait, hydrops fetalis and β-thalassemias.

PORPHYRINS

Porphyrins are cyclic compounds composed of ***4 pyrrole rings*** held together by ***methenyl*** (=CH−) ***bridges*** (***Fig.10.16***). Metal ions can bind with nitrogen atoms of pyrrole rings to form complexes. Heme is an iron-containing porphyrin (***See Fig.10.2***) while chlorophyll is a magnesium-containing porphyrin. Thus heme and chlorophyll are the classical examples of metalloporphyrins.

Presentation and nomenclature of porphyrins

The structure of porphyrins ($C_{20}H_{14}N_4$) has four ***pyrrole rings*** namely I, II, III and IV. Naturally occurring porphyrins contain substituent groups replacing the 8 hydrogen atoms of the porphyrin nucleus.

Hans Fischer, the father of porphyrin chemistry, proposed a shorthand model for presentation of porphyrin structures. Accordingly, each pyrrole ring is represented as a bracket. Thus porphyrin has 4 closed brackets with the 8 substituent positions numbered as shown in ***Fig.10.16***.

Type I porphyrins : When the substituent groups on the 8 positions are symmetrically arranged they are known as type I porphyrins, e.g. uroporphyrin I.

Type III porphyrins : They contain asymmetric groups at the 8 positions and are more common in the biological system. Originally, Fischer placed them as IX series hence they are more popularly known as type IX porphyrins. It may be observed that the structure of uroporphyrin is asymmetric since on ring IV, the order of substituent groups is reversed (P, A instead of A, P).

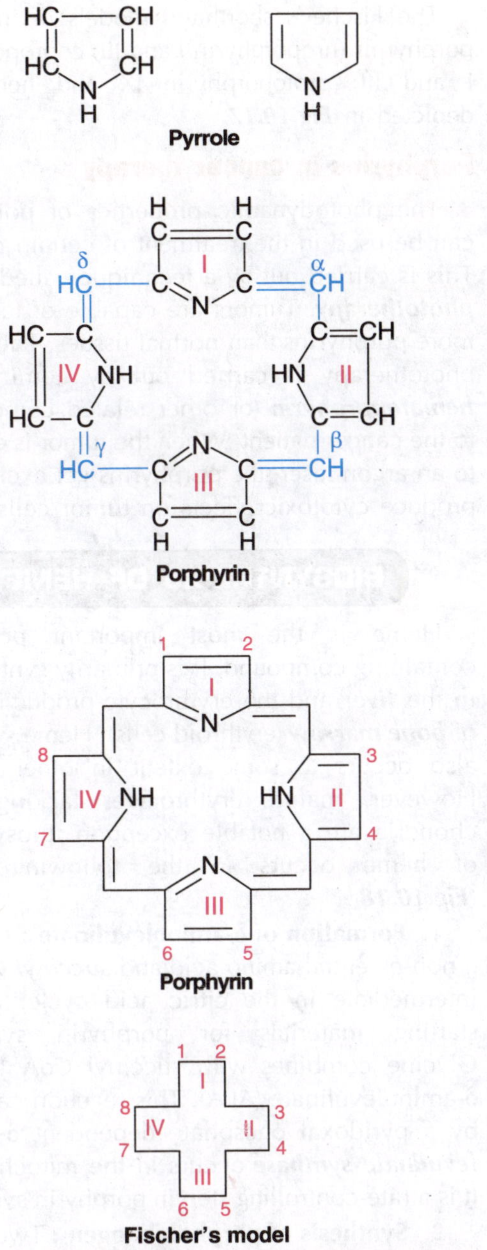

Fig. 10.16 : Structures of pyrrole and porphyrin [I-IV are pyrrole rings; 1-8 are substituent positions; α, β, γ, δ are methylene (− CH=) bridges.]

The Fischer's shorthand models of important porphyrins (uroporphyrin I and III; coproporphyrin I and III; protoporphyrin IX and heme) are depicted in ***Fig.10.17***.

Porphyrins in cancer therapy

The photodynamic properties of porphyrins can be used in the treatment of certain cancers. This is carried out by a technique called ***cancer phototherapy***. Tumors are capable of taking up more porphyrins than normal tissues. The cancer phototherapy is carried out by administering ***hematoporphyrin*** (or other related compounds) to the cancer patient. When the tumor is exposed to an argon laser, the porphyrins get excited and produce cytotoxic effects on tumor cells.

BIOSYNTHESIS OF HEME

Heme is the most important porphyrin containing compound. It is primarily synthesized in the ***liver*** and the erythrocyte-producing cells of ***bone marrow*** (erythroid cells). Heme synthesis also occurs to some extent in other tissues. However, mature erythrocytes lacking mitochondria are a notable exception. Biosynthesis of heme occurs in the following stages (***Fig.10.18***).

1. **Formation of δ-aminolevulinate : *Glycine***, a non-essential amino acid and ***succinyl CoA***, an intermediate in the citric acid cycle, are the starting materials for porphyrin synthesis. Glycine combines with succinyl CoA to form δ-aminolevulinate (ALA). This reaction catalysed by a pyridoxal phosphate dependent ***δ-aminolevulinate synthase*** occurs in the mitochondria. It is a rate-controlling step in porphyrin synthesis.

2. **Synthesis of porphobilinogen :** Two molecules of δ-aminolevulinate condense to form porphobilinogen (PBG) in the cytosol. This reaction is catalysed by a Zn-containing enzyme ***ALA dehydratase***. It is sensitive to inhibition by heavy metals such as lead.

3. **Formation of porphyrin ring :** Porphyrin synthesis occurs by condensation of four molecules of porphobilinogen. The four pyrrole rings in porphyrin are interconnected by methylene ($-CH_2$) bridges derived from α-carbon of glycine.

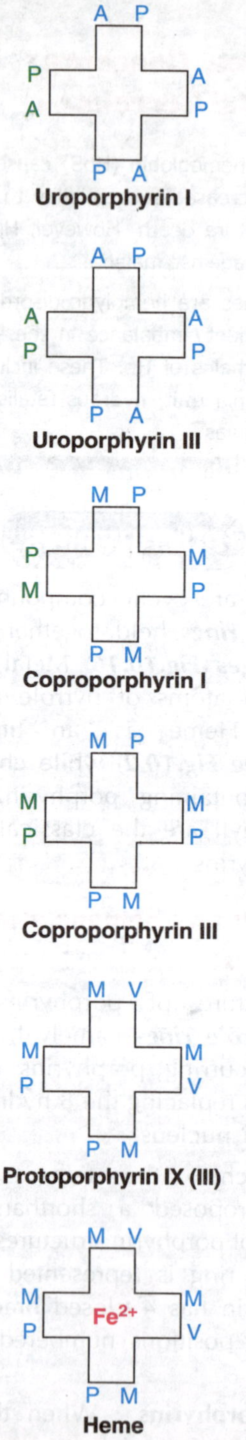

Fig. 10.17 : *Fischer's shorthand models of physiologically important porphyrins [A–Acetate ($-CH_3COO^-$); P–Propionyl ($-CH_2CH_2COO^-$); M–Methyl ($-CH_3$); V–Vinyl ($-CH=CH_2$)].*

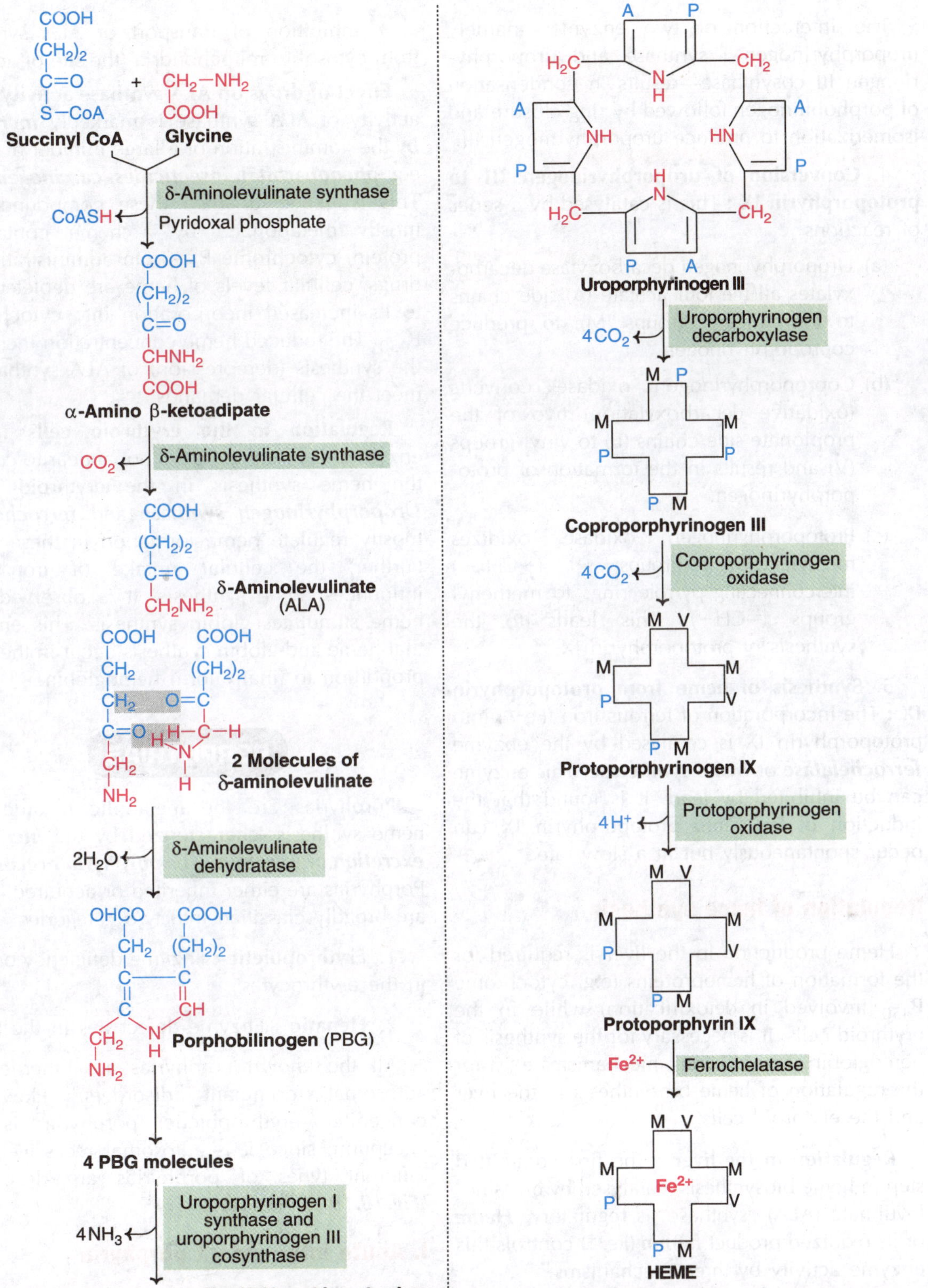

Fig. 10.18 contd. next column

***Fig. 10.18 :** Biosynthesis of heme [A-Acetyl (— CH_2— COO^-); P-Propionyl (— CH_2— CH_2— COO^-); M-Methyl (— CH_3); V-Vinyl (— CH= CH_2)].*

The interaction of two enzymes—namely uroporphyrinogen I synthase and uroporphyrinogen III cosynthase—results in condensation of porphobilinogen followed by ring closure and isomerization to produce uroporphyrinogen III.

4. **Conversion of uroporphyrinogen III to protoporphyrin IX :** This is catalysed by a series of reactions

(a) Uroporphyrinogen decarboxylase decarboxylates all the four acetate (A) side chains to form methyl groups (M), to produce coproporphyrinogen.

(b) Coproporphyrinogen oxidase converts (oxidative decarboxylation) two of the propionate side chains (P) to vinyl groups (V) and results in the formation of protoporphyrinogen.

(c) Protoporphyrinogen oxidase oxidizes methylene groups ($-CH_2-$) interconnecting pyrrole rings to methenyl groups ($=CH-$). This leads to the synthesis of protoporphyrin IX.

5. **Synthesis of heme from protoporphyrin IX :** The incorporation of ferrous iron (Fe^{2+}) into protoporphyrin IX is catalysed by the enzyme ***ferrochelatase*** or ***heme synthetase***. This enzyme can be inhibited by lead. It is found that the induction of Fe^{2+} into protoporphyrin IX can occur spontaneously but at a slow rate.

Regulation of heme synthesis

Heme production in the liver is required for the formation of hemoproteins (e.g. cytochrome P_{450} involved in detoxification) while in the erythroid cells, it is necessary for the synthesis of hemoglobin. Two different mechanisms exist for the regulation of heme biosynthesis in the liver and the erythroid cells.

Regulation in the liver : The first committed step in heme biosynthesis, catalysed by δ-aminolevulinate (ALA) synthase, is regulatory. ***Heme*** or its oxidized product ***hemin*** (Fe^{3+}) controls this enzyme activity by three mechanisms

1. Feedback regulation
2. Repression of ALA synthatase
3. Inhibition of transport of ALA synthase from cytosol to mitochondria (the site of action).

Effect of drugs on ALA synthase activity : The activity of ALA synthase is markedly ***increased by*** the administration of a large number of drugs e.g. ***phenobarbital, insecticides, carcinogens*** etc. This is expected since these compounds are mostly metabolized by a heme containing protein, cytochrome P_{450}. On administration of drugs, cellular levels of heme are depleted due to its increased incorporation into cytochrome P_{450}. The reduced heme concentration increases the synthesis (derepression) of ALA synthase to meet the cellular demands.

Regulation in the erythroid cells : The enzyme ALA synthase does not appear to control the heme synthesis in the erythroid cells. ***Uroporphyrinogen synthase*** and ***ferrochelatse*** mostly regulate heme formation in these cells. Further, the cellular uptake of iron also influences heme synthesis. It is observed that heme stimulates globin synthesis. This ensures that heme and globin synthesis occur in the right proportion to finally form hemoglobin.

PORPHYRIAS

Porphyrias are the metabolic disorders of heme synthesis, characterized by the ***increased excretion of porphyrins or porphyrin precursors***. Porphyrias are either inherited or acquired. They are broadly classified into two categories

1. **Erythropoietic :** Enzyme deficiency occurs in the erythrocytes.

2. **Hepatic :** Enzyme defect lies in the liver.

All the known porphyrias are inherited as autosomal dominant disorders. However, congenital erythropoietic porphyria is an exception, since it is autosomal recessive. The different types of porphyrias are described (***Fig.10.19***, ***Table 10.3***)

I. Acute intermittent porphyria

This disorder occurs due to the deficiency of the enzyme ***uroporphyrinogen I synthase.*** Acute intermittent porphyria is characterized by increased excretion of porphobilinogen and δ-aminolevulinate. The urine gets darkened on exposure to air due to the conversion of porphobilinogen to porphobilin and porphyrin. The other characteristic features of acute intermittent porphyria are as follows

- It is usually expressed after puberty in humans.
- The symptoms include abdominal pain, vomiting and cardiovascular abnormalities. The neuropsychiatric distrubances observed in these patients are believed to be due to reduced activity of tryptophan pyrrolase (caused by depleted heme levels), resulting in the accumulation of tryptophan and 5-hydroxytryptamine.

Table 10.3 A general summary of porphyrias

Type of porphyria	Enzyme defect	Characteristics
Hepatic		
Acute intermittent porphyria	Uroporphyrinogen I synthase	Abdominal pain, neuropsychiatric symptoms
Porphyria cutanea tarda	Uroporphyrinogen decarboxylase	Photosensitivity
Hereditary coproporphyria	Corpoporphyrinogen oxidase	Abdominal pain, photosensitivity, neuropsychiatric symptoms
Variegate porphyria	Protoporphyrinogen oxidase	Abdominal pain, photosensitivity, neuropsychiatric symptoms
Erythropoietic		
Congenital erythropoietic porphyria	Uroporphyrinogen III cosynthase	Photosensitivity, increased hemolysis
Protoporphyria	Ferrochelatase	Photosensitivity

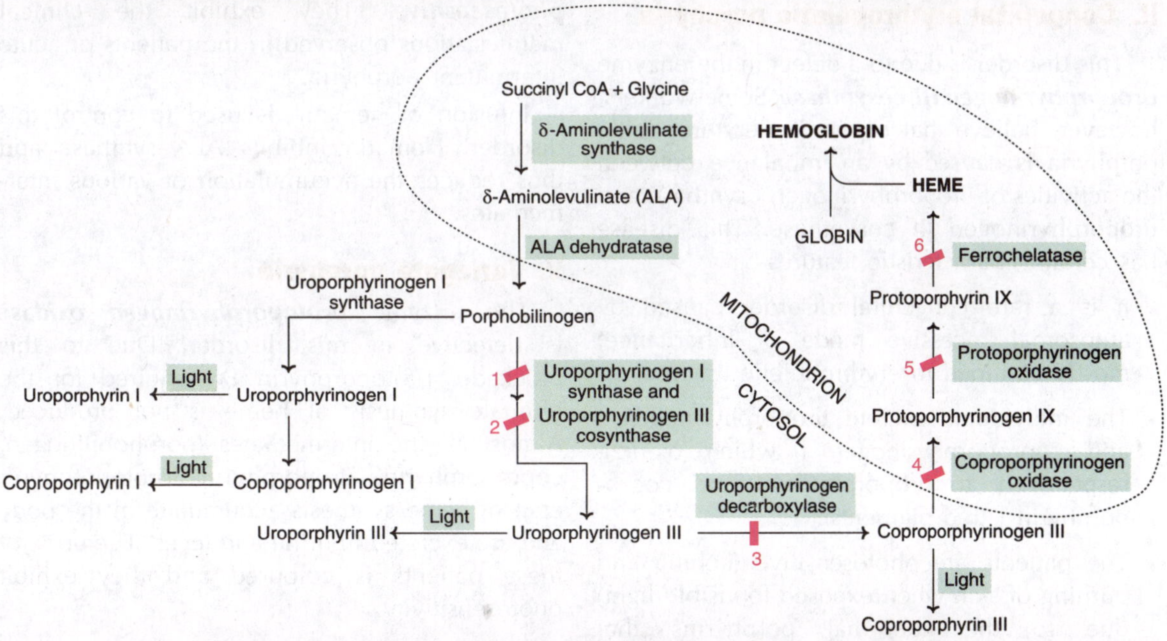

***Fig. 10.19 :** Summary of heme synthesis along with porphyrias (1–Acute intermittent porphyria; 2–Congenital erythropoietic porphyria; 3–Porphyria cutanea tarda; 4–Hereditary coproporphyria; 5–Variegate porphyria; 6–Protoporphyria).*

- The symptoms are more severe after administration of drugs (e.g. barbiturates) that induce the synthesis of cytochrome P_{450}. This is due to the increased activity of ALA synthase causing accumulation of PBG and ALA.
- These patients are not photosensitive since the enzyme defect occurs prior to the formation of uroporphyrinogen.

Acute intermittent porphyria is treated by administration of hematin which inhibits the enzyme ALA synthase and the accumulation of porphobilinogen.

[The disease—acute intermittent porphyria—has historical importance. King George III (1760-1820) ruled England during the period of American revolution. He was a victim of this disease and possessed the characteristic manifestations (such as red colour urine) and was considered mad. The decisions taken by the deranged King due to acute intermittent porphyria had led to a war followed by American Independence. It is widely believed that American history would have been different, had George III not inherited this metabolic disorder!]

II. Congenital erythropoietic porphyria

This disorder is due to a defect in the enzyme ***uroporphyrinogen III cosynthase***. Some workers, however, believe that congenital erythropoeitic porphyria is caused by an imbalance between the activities of uroporphyrinogen I synthase and uroporphyrinogen III cosynthase. This disease has certain characteristic features

- It is a rare congenital disorder caused by autosomal recessive mode of inheritance, mostly confined to erythropoietic tissues.
- The individuals excrete uroporphyrinogen I and coproporphyrinogen I which oxidize respectively to uroporphyrin I and coproporphyrin I (red pigments).
- The patients are photosensitive (itching and burning of skin when exposed to visible light) due to the abnormal porphyrins that accumulate.
- Increased hemolysis is also observed in the individuals affected by this disorder.

III. Porphyria cutanea tarda

This is also known as ***cutaneous hepatic porphyria*** and is the most common porphyria, usually associated with liver damage caused by alcohol overconsumption or iron overload. The partial deficiency of the enzyme ***uroporphyrinogen decarboxylase*** appears to be responsible for the occurrence of porphyria cutanea tarda. The other characteristic features include

- Increased excretion of uroporphyrins (I and III) and rarely porphobilinogen.
- ***Cutaneous photosensitivity*** is the most important clinical manifestation of these patients.
- Liver exhibits fluorescence due to high concentration of accumulated porphyrins.

IV. Hereditary coproporphyria

This disorder is due to a defect in the enzyme ***coproporphyrinogen oxidase***. As a result of this, coproporphyrinogen III and other intermediates (ALA and PBG) of heme synthesis prior to the blockade are excreted in urine and feces. The ***victims of*** hereditary coproporphyria are ***photosensitive***. They exhibit the clinical manifestations observed in the patients of acute intermittent porphyria.

Infusion of hematin is used to control this disorder. Hematin inhibits ALA synthase and thus reduces the accumulation of various intermediates.

V. Variegate porphyria

The enzyme ***protoporphyrinogen oxidase*** is defective in this disorder. Due to this blockade, protoporphyrin IX required for the ultimate synthesis of heme is not produced. Almost all the intermediates (porphobilinogen, coproporphyrin, uroporphyrin, protoporphyrin etc.) of heme synthesis accumulate in the body and are excreted in urine and feces. The urine of these patients is coloured and they exhibit photosensitivity.

VI. Protoporphyria

This disorder, also known as ***erythropoietic protoporphyria***, is caused by a deficiency of the

Medical Concepts/ Clinical Correlates

- Administration of porphyrins can be used in the treatment of certain cancers by phototherapy.
- Abnormalities in heme synthesis cause porphyrias which may be erythropoietic (enzyme defect in RBC) or hepatic (enzyme defect in liver). Porphyrias are associated with elevated excretion of porphyrins, neuropsychiatric disturbances and cardiovascular abnormalities.

enzyme ***ferrochelatase***. Protoporphyrin IX accumulates in the tissues and is excreted into urine and feces. Reticulocytes (young RBC) and skin biopsy exhibit red flourescence.

Acquired (toxic) porphyrias

The porphyrias, though not inherited, may be acquired due to the toxicity of several compounds. Exposure of the body to heavy metals (e.g. lead), toxic compounds (e.g. hexachlorobenzene) and drugs (e.g. griseofulvin) inhibits many enzymes in heme synthesis. These include ALA dehydratase, uroporphyrin I synthase and ferrochelatase.

DEGRADATION OF HEME TO BILE PIGMENTS

Erythrocytes have a life span of 120 days. At the end of this period, they are removed from the circulation. Erythrocytes are taken up and ***degraded by the macrophages*** of the reticuloendothelial (RE) system in the spleen and liver. The hemoglobin is cleaved to the protein part globin and non-protein heme. About 6 g of hemoglobin per day is broken down, and resynthesized in an adult man (70 kg).

Fate of globin : The globin may be reutilized as such for the formation of hemoglobin or degraded to the individual amino acids. The latter undergo their own metabolism, including participation in fresh globin synthesis.

Sources of heme : It is estimated that about 80% of the heme that is subjected for degradation comes from the erythrocytes and the rest (20%) comes from immature RBC, myoglobin and cytochromes.

Heme oxygenase : A complex microsomal enzyme namely heme oxygenase utilizes NADPH and O_2 and cleaves the methenyl bridges between the two pyrrole rings (A and B) to form ***biliverdin***. Simultaneously, ferrous iron (Fe^{2+}) is oxidized to ferric form (Fe^{3+}) and released. The products of heme oxygenase reaction are biliverdin (a green pigment), Fe^{3+} and carbon monoxide (CO). Heme promotes the activity of this enzyme.

Biliverdin is excreted in birds and amphibia while in mammals it is further degraded.

Biliverdin reductase : Biliverdin's methenyl bridges (between the pyrrole rings C and D) are reduced to methylene group to form ***bilirubin*** (yellow pigment). This reaction is catalysed by an NADPH dependent soluble enzyme, biliverdin reductase (***Fig.10.20***). ***One gram of hemoglobin on degradation finally yields about 35 mg bilirubin***. Approximately 250-350 mg of bilirubin is daily produced in human adults. The term bile pigments is used to collectively represent bilirubin and its derivatives.

Transport of bilirubin to liver : Bilirubin is lipophilic and therefore insoluble in aqueous solution. Bilirubin is transported in the plasma in a bound (non-covalently) form to albumin. Albumin has two binding sites for bilirubin—a high affinity site and a low affinity site. Approximately 25 mg of bilirubin can bind tightly to albumin (at high affinity sites) per 100 ml of plasma. The rest of the bilirubin binds loosely (at the low affinity sites) which can be easily detached from albumin to enter the tissues. Certain drugs and antibiotics (e.g sulfonamides, salicylates) can displace bilirubin from albumin. Due to this, bilirubin can enter the central nervous system and cause damage to neurons.

As the ***albumin-bilirubin complex*** enters the liver, bilirubin dissociates and is taken up by sinusoidal surface of the hepatocytes by a carrier mediated active transport. The transport system

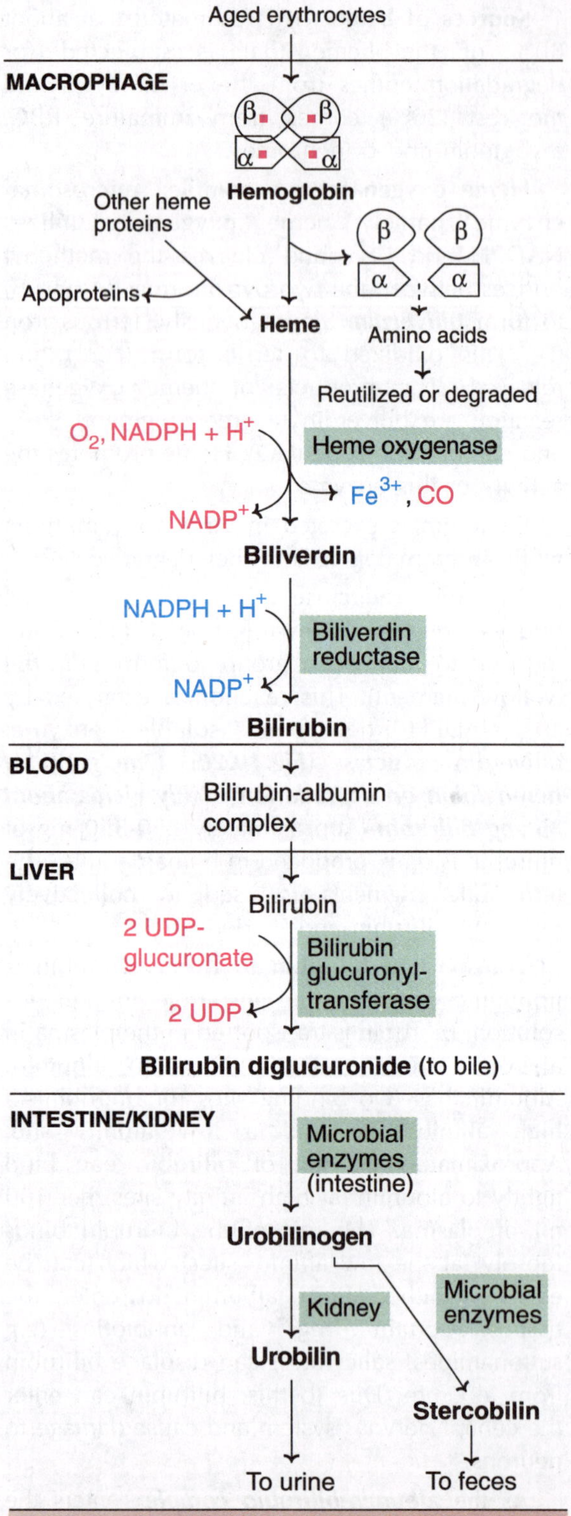

Fig. 10.20 : Degradation of heme to bile pigments (Note : Colours used in structures represent change in the specific reaction only).

has a very high capacity and therefore is not a limitation for further metabolism of bilirubin. Inside the hepatocytes, bilirubin binds to a specific intracellular protein namely ***ligandin***.

Conjugation of bilirubin

In the liver, bilirubin is conjugated with two molecules of glucuronate supplied by UDP-glucuronate. This reaction, catalysed by ***bilirubin glucuronyltransferase*** (of smooth endoplasmic reticulum) results in the formation of a water soluble bilirubin diglucuronide (***Figs.10.20 and 10.21***). When bilirubin is in excess, bilirubin monoglucuronides also accumulate in the body. The enzyme bilirubin glucuronyl-transferase can be induced by a number of drugs (e.g. phenobarbital),

Excretion of bilirubin into bile

Conjugated bilirubin is excreted into the bile canaliculi against a concentration gradient which then enters the bile. The transport of ***bilirubin diglucuronide*** is an active, energy-dependent

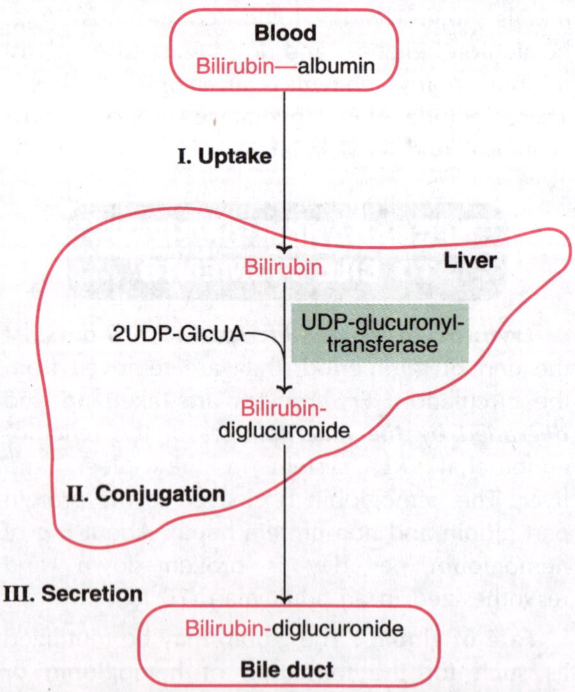

Fig. 10.21 : Summary of bilirubin metabolism (UDP-GlcUA—UDP-glucuronic acid).

and rate limiting process. This step is easily susceptible to any impairment in liver function. Normally, there is a good coordination between the bilirubin conjugation and its excretion into bile. Thus almost all the bilirubin (> 98%) that enters bile is in the conjugated form.

Fate of bilirubin

Bilirubin glucuronides are hydrolysed (deconjugated) in the intestine by specific bacterial enzymes namely ***β-glucuronidases*** to liberate bilirubin. The latter is then converted to urobilinogen (colourless compound), a small part of which may be reabsorbed into the circulation. Urobilinogen can be converted to urobilin (an yellow colour compound) in the kidney and excreted. The characteristic ***yellow colour of urine is due to urobilin.***

A major part of urobilinogen is converted by bacteria to stercobilin which is excreted along with feces. The characteristic ***brown colour of feces is due to stercobilin.***

JAUNDICE

The normal ***serum*** total ***bilirubin*** concentration is in the range of ***0.2 to 1.0 mg/dl.*** Of this, about 0.2-0.6 mg/dl is unconjugated while 0.2 to 0.4 mg/dl is conjugated bilirubin.

Jaundice (*French* : *Jaune*-yellow) is a clinical condition characterized by yellow colour of the white of the eyes (sclerae), skin and nail bed. It is caused by the deposition of bilirubin due to its elevated levels in the serum. The term ***hyperbilirubinemia*** is often used to represent the increased concentration of serum bilirubin. (***Note :*** For more details on jaundice, refer ***Chapter 20***)

Classification of jaundice

Jaundice (also known as ***icterus***) may be more appropriately considered as a symptom rather than a disease. It is rather difficult to classify jaundice, since it is frequently caused due to multiple factors. ***For the sake of convenience to understand***, jaundice is classified into three major types—***hemolytic, hepatic and obstructive.***

1. **Hemolytic jaundice :** This condition is associated with ***increased hemolysis of erythrocytes*** (e.g. incompatible blood transfusion, malaria, sickle-cell anemia). This results in the overproduction of bilirubin beyond the ability of the liver to conjugate and excrete the same. It should, however be noted that liver possesses a large capacity to conjugate about 3.0 g of bilirubin per day against the normal bilirubin production of 0.3 g/day.

In hemolytic jaundice, more bilirubin is excreted into the bile leading to the increased formation of urobilinogen and stercobilinogen. Hemolytic jaundice is characterized by

- ***Elevation*** in the serum ***unconjugated bilirubin.***
- Increased excretion of urobilinogen in urine.
- Dark brown colour of feces due to high content of stercobilinogen.

2. **Hepatic (hepatocellular) jaundice :** This type of jaundice is caused by ***dysfunction of the liver*** due to damage to the parenchymal cells. This may be attributed to ***viral infection*** (viral hepatitis), ***poisons*** and ***toxins*** (chloroform, carbon tetrachloride, phosphorus etc.) cirrhosis of liver, cardiac failure etc. Among these, viral hepatitis is the most common.

Damage to the liver adversely affects the bilirubin uptake and its conjugation by liver cells. Hepatic jaundice is characterized by

- Increased levels of ***conjugated and unconjugated bilirubin in the serum.***
- Dark coloured urine due to the excessive excretion of bilirubin and urobilinogen.
- Increased activities of ***alanine transaminase*** (SGPT) and ***aspartate transaminase*** (SGOT) released into circulation due to damage to hepatocytes.
- The patients pass pale coloured stools due to the absence of stercobilinogen.
- The affected individuals experience nausea and anorexia (loss of appetite).

3. **Obstructive (regurgitation) jaundice :** This is due to an obstruction in the bile duct that prevents the passage of bile into the intestine. The obstruction may be caused by gall stones,

> **Medical Concepts/ Clinical Correlates**
>
> - Jaundice is caused by elevated serum bilirubin (normal < 1.0 mg/dl) levels and is characterized by yellow coloration of white of the eyes, and skin.
> - Bilirubin is known to function as an antioxidant.
> - Phototherapy (by exposure to blue light) is used to control severe cases of neonatal physiologic jaundice.

tumors etc. Obstructive jaundice is characterized by

- ***Increased*** concentration of ***conjugated bilirubin in serum***.
- Serum ***alkaline phosphatase*** is elevated as it is released from the cells of the damaged bile duct.
- Dark coloured urine due to elevated excretion of bilirubin and clay coloured feces due to absence of stercobilinogen.
- The patients experience nausea and gastrointestinal pain.

JAUNDICE DUE TO GENETIC DEFECTS

There are certain hereditary abnormalities that cause jaundice.

Neonatal-physiologic jaundice

Physiological jaundice is ***not truly a genetic defect***. It is caused by ***increased hemolysis*** coupled with immature hepatic system for the uptake, conjugation and secretion of bilirubin. The activity of the enzyme ***UDP-glucuronyltransferase is low*** in the newborn. Further, there is a limitation in the availability of the substrate UDP-glucuronic acid for conjugation. The net effect is that in some infants the serum uncojugated bilirubin is highly elevated (may go beyond 25mg/dl), which can cross the blood-brain barrier. This results in ***hyperbilirubinemic toxic encephalopathy*** or ***kernicterus*** that causes mental retardation. The drug phenobarbital is used in the treatment of neonatal jaundice, as it can induce bilirubin metabolising enzymes in liver. In some neonates, blood transfusion may be necessary to prevent brain damage.

Phototherapy : Bilirubin can absorb blue light (420–470 nm) maximally. Phototherapy deals with the exposure of the jaundiced neonates to blue light. By a process called ***photoisomerization***, the toxic native unconjugated bilirubin gets converted into a non-toxic isomer namely ***lumirubin***. Lumirubin can be easily excreted by the kidneys in the unconjugated form (in contrast to bilirubin which cannot be excreted). Serum bilirubin is monitored every 12–24 hours, and phototherapy is continuously carried out till the serum bilirubin becomes normal (< 1 mg/dl).

Crigler-Najjar syndrome type I

This is also known as congenital non-hemolytic jaundice. It is a rare disorder and is due to a defect in the hepatic enzyme ***UDP-glucuronyltransferase***. Generally, the children die within first two years of life.

Crigler-Najjar syndrome type II

This is again a rare hereditary disorder and is due to a less severe defect in the bilirubin conjugation. It is believed that hepatic UDP-glucuronyltransferase that catalyses the addition of second glucuronyl group is defective. The serum bilirubin concentration is usually less than 20 mg/dl and this is less dangerous than type I.

Gilbert's disease : This is not a single disease but a combination of disorders. These include

1. A defect in the uptake of bilirubin by liver cells.

2. An impairment in conjugation due to reduced activity of UDP-glucuronyltransferase.

3. Decreased hepatic clearance of bilirubin.

SUMMARY

- *Hemoglobin (HbA_1, mol. wt. 64,450) is a conjugated protein containing globin, the apoprotein and the heme, the nonprotein moiety (prosthetic group). It is a tetrameric, allosteric protein with 2α and 2β polypeptide chains held by non-covalent interactions. Each subunit contains a heme with iron in the ferrous state.*
- *Hemoglobin is responsible for the transport of O_2 from lungs to the tissues. Each heme (of Hb) can bind with one molecule of O_2 and this is facilitated by cooperative heme-heme interaction.*
- *Hemoglobin actively participates in the transport of CO_2 from tissues to lungs. Increased partial pressure of CO_2 (pCO_2) accompanied by elevated H^+ decreases the binding of O_2 to Hb, a phenomenon known as Bohr effect.*
- *The four compounds namely 2,3-bisphosphoglycerate, CO_2, H^+ and Cl^- are collectively known as allosteric effectors. They interact with hemoglobin and facilitate the release of O_2 from oxyHb.*
- *Sickle-cell anemia (HbS) is a classical example of abnormal hemoglobins. It is caused when glutamate at 6th position of β-chain is replaced by valine. HbS is characterized by hemolytic anemia, tissue damage, increased susceptibility to infection and premature death. Sickle-cell anemia, however offers resistance to malaria.*
- *Thalassemias are a group of hereditary hemolytic disorders characterized by impairment/imbalance in the synthesis of globin (α or β) chain of Hb. Hydrops fetalis, the most severe form of α-thalassemia is characterized by the death of infant at birth. β-Thalassemia major is another serious disorder with severe anemia and death of child within 1-2 years.*
- *Heme is the most important porphyrin compound, primarily synthesized in the liver from the precursors-glycine and succinyl CoA. Heme productioin is regulated by δ-aminolevulinate synthase.*
- *Porphyrias are the metabolic disorders of heme synthesis, characterized by the increased excretion of porphyrins or their precursors. Acute intermittent porphyria occurs due to the deficiency of the enzyme uroporphyrinogen I synthase and is characterized by increased excretion of porphobilinogen and δ-aminolevulinate. The clinical symptoms include neuropsychiatric disturbances and cardiovascular abnormalities.*
- *Heme is degraded mainly to bilirubin, an yellow colour bile pigment. In the liver, it is conjugated to bilirubin diglucuronide, a more easily excretable form into bile.*
- *Jaundice is a clinical condition caused by elevated serum bilirubin concentration (normal <1.0 mg/dl). Jaundice is of three types-hemolytic (due to increased hemolysis), hepatic (due to impaired conjugation) and obstructive (due to obstruction in the bile duct).*

THE READER MUST REFER APPENDIX VI (Page Nos : 738 – 756) FOR RELATED CLINICAL CASE STUDIES (Nos : 10, 16, 31, 34, 36, 37, 38, 39, 70)

Self-Assessment Exercises

I. Essay questions

1. Describe the structure of hemoglobin and discuss oxygen transport.
2. Write an account of hemoglobinopathies with special reference to sickle-cell anemia.
3. Discuss the biosynthesis of heme. Add a note on the regulation of heme synthesis.
4. What are porphyrias? Describe any three porphyrias in detail.
5. Write an account of the degradation of heme to bile pigments. Add a note on jaundice.

II. Short notes

(a) Methemoglobin, (b) Heme—heme interaction, (c) Bohr effect, (d) 2,3—BPG, (e) Sickle cell anemia and malaria, (f) Thalassemias, (g) Acute intermittent porphyria, (h) Heme oxygenase, (i) Bilirubin diglucuronide, (j) Carboxyhemoglobin.

III. Fill in the blanks

1. The total number of amino acids present in adult hemoglobin ______________.
2. The oxidation of ferrous (Fe^{2+}) iron to ferric (Fe^{3+}) iron in hemoglobin results in the formation of a compound namely ______________.
3. The enzyme that catalyses the formation of carbonic acid ______________.
4. Name the compound that is increased in RBC of anemic patients to facilitate the supply of O_2 to the tissues ______________.
5. Sickling of RBC in sickle-cell anemia is due to polymerization of ______________.
6. The disorders characterized by decreased synthesis or total absence of globin chains of hemoglobin are collectively known as ______________.
7. The intermediate of citric acid cycle that is involved in heme synthesis ______________.
8. The enzyme defect in acute intermittent porphyria ______________.
9. The enzyme that is regulated by feedback inhibition in heme synthesis is ______________.
10. The product formed when heme oxygenase cleaves heme ______________.

IV. Multiple choice questions

11. The characteristic red colour of hemoglobin is due to
    (a) Heme (b) α-Globin (c) β-Globin (d) All of them.
12. The number of heme groups present in myoglobin
    (a) 1 (b) 2 (c) 3 (d) 4.
13. The patients of sickle-cell anemia are resistant to
    (a) Filaria (b) Malaria (c) Diabetes (d) Trypanosomiasis.
14. The compound that facilitates the release of O_2 from oxyhemoglobin
    (a) 2, 3-BPG (b) H^+ (c) $C1^-$ (d) All of them.
15. Name the amino acid that directly participates in the synthesis of heme
    (a) Methionine (b) Aspartate (c) Glycine (d) Tryptophan.

Biological Oxidation

11

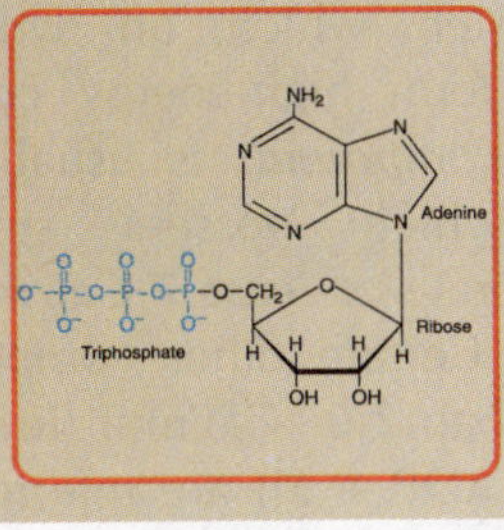

The high energy compound, ATP speaks

"I am the energy currency of the cell!
Continuous consumption and regeneration is my thrill;
Without me, all biochemical functions come to a standstill;
Existence of life is unimaginable without my will."

For a better understanding of biological oxidation, it is worthwhile to have a basic knowledge of bioenergetics and the role of high-energy compounds in biological processes.

BIOENERGETICS

Bioenergetics or ***biochemical thermodynamics*** deals with the study of energy changes (transfer and utilization) in biochemical reactions. The reactions are broadly classified as ***exergonic*** (energy releasing) and ***endergonic*** (energy consuming). Bioenergetics is concerned with the initial and final states of energy component of the reactants and not the mechanism of chemical reactions.

Free energy

The energy actually ***available to do work*** (utilizable) is known as free energy. Changes in the free energy (ΔG) are valuable in predicting the feasibility of chemical reactions. The reactions can occur spontaneously if they are accompanied by decrease in free energy.

During a chemical reaction, heat may be released or absorbed. ***Enthalpy (ΔH)*** is a measure of the change in heat content of the reactants, compared to products.

Entropy (ΔS) represents a change in the randomness or ***disorder of reactants and products***. Entropy attains a maximum as the reaction approaches equilibrium. The reactions of biological systems involve a temporary decrease in entropy.

The relation between the changes of ***free energy (ΔG), enthalpy (ΔH)*** and ***entropy (ΔS)*** is expressed as

$$\Delta G = \Delta H - T\Delta S$$

T represents the absolute temperature in Kelvin (K = 273 + °C).

The term ***standard free energy*** represented by $\Delta G°$ (note the superscript°) is often used. It indicates the ***free energy change*** when the reactants or products are at a concentration of 1 mol/l at pH 7.0.

Negative and positive ΔG

If free energy change (ΔG) is represented by a negative sign, there is a loss of free energy. The reaction is said to be ***exergonic***, and ***proceeds spontaneously***. On the other hand, a positive ΔG indicates that energy must be supplied to the reactants. The reaction ***cannot proceed spontaneously*** and is ***endergonic*** in character.

The ***hydrolysis of ATP*** is a classical example of ***exergonic*** reaction

$$ATP + H_2O \longrightarrow ADP + Pi \quad (\Delta G° = -7.3 \text{ Cal/mol})$$

The reversal of the reaction (ADP + Pi → ATP) is endergonic and occurs only when there is a supply of energy of at least 7.3 Cal/mol (ΔG° is positive).

The ***free energy*** change becomes ***zero*** (ΔG = 0) when a reaction is ***at equilibrium***.

At a constant temperature and pressure, ΔG is dependent on the actual concentration of reactants and products. For the conversion of reactant A to product B (A → B), the following mathematical relation can be derived

$$\Delta G = \Delta G° + RT \ln \frac{[B]}{[A]}$$

where ΔG° = Standard free energy change

R = Gas constant (1.987 Cal/mol)

T = Absolute temperature (273 + °C)

ln = Natural logarithm

[B] = Concentration of product

[A] = Concentration of reactant.

ΔG° is related to equilibrium constant (Keq)

When a reaction $A \rightleftharpoons B$ is at equilibrium (eq), the free energy change is zero. The above equation may be written as

$$\Delta G = 0 = \Delta G° + RT \ln \frac{[B] \text{ eq.}}{[A] \text{ eq.}}$$

Hence ΔG° = – RT ln Keq.

ΔG is an additive value for pathways

Biochemical pathways often involve a ***series of reactions***. For such reactions, ***free energy change is an additive value***. The sum of ΔG is crucial in determining whether a particular pathway will proceed or not. As long as the sum of ΔGs of individual reactions is negative, the pathway can operate. This happens despite the fact that some of the individual reactions may have positive ΔG.

HIGH-ENERGY COMPOUNDS

Certain compounds are encountered in the biological system which, on hydrolysis, yield energy. The term high-energy compounds or ***energy rich compounds*** is usually applied to substances which possess sufficient free energy to liberate ***at least 7 Cal/mol*** at pH 7.0 (***Table 11.1***). Certain other compounds which liberate less than 7.0 Cal/mol (lower than ATP hydrolysis to ADP + Pi) are referred to as low-energy compounds.

Table 11.1 Standard free energy of hydrolysis of some important compounds

Compounds	ΔG° (Cal/mol)
High-energy phosphates	
Phosphoenol pyruvate	–14.8
Carbamoyl phosphate	–12.3
Cyclic AMP	–12.0
1,3-Bisphosphoglycerate	–11.8
Phosphocreatine	–10.3
Acetyl phosphate	–10.3
S-Adenosylmethionine*	–10.0
Pyrophosphate	–8.0
Acetyl CoA**	–7.7
ATP → ADP + Pi	–7.3
Low-energy phosphates	
ADP → AMP + Pi	–6.6
Glucose 1-phosphate	–5.0
Fructose 6-phosphate	–3.8
Glucose 6-phosphate	–3.3
Glycerol 3-phosphate	–2.2

* *Sulfonium compound*

** *Thioester*

Table 11.2 High-energy compounds

Class	Bond	Example(s)
Pyrophosphates	—C—Ⓟ—Ⓟ	ATP, pyrophosphate
Acyl phosphates	O ‖ —C—O~Ⓟ	1,3-Bisphosphoglycerate, carbamoyl phosphate, acetyl phosphate
Enol phosphates	—CH ‖ —C—O~Ⓟ	Phosphoenol pyruvate
Thiol esters (thioesters)	C ‖ —C—O~S—	Acetyl CoA, acyl CoA
Guanidio phosphates (Phosphagens)	| —N~P	Phosphocreatine, phosphoarginine

All the high-energy compounds—when hydrolysed—liberate more energy than that of ATP. These include ***phosphoenol pyruvate, 1,3-bisphosphoglycerate, phosphocreatine*** etc. Most of the high-energy compounds contain phosphate group (exception acetyl CoA) hence they are called high-energy phosphate compounds.

Classification of high-energy compounds

There are at least 5 groups of high-energy compounds (***Table 11.2***).

1. ***Pyrophosphates*** e.g. ATP.
2. ***Acyl phosphates*** e.g. 1,3-bisphosphoglycerate.
3. ***Enol phosphates*** e.g. phosphoenolpyruvate.
4. ***Thioesters*** e.g. acetyl CoA.
5. ***Phosphagens*** e.g. phosphocreatine.

Phosphagens : These are high-energy ***phosphate storage forms.*** Phosphocreatine is found in vertebrate skeletal muscle, heart, brain and spermatozoa while ***arginine phosphate*** occurs in ***invertebrate muscle***. Phosphagens serve as reserviors and help to maintain continuous supply of energy for biological functions (as described in ATP-ADP cycle).

High-energy bonds : The high-energy compounds possess ***acid anhydride bonds*** (mostly phosphoanhydride bonds) which are formed by the condensation of two acidic groups or related compounds. These bonds are referred to as high-energy bonds, since the free energy is liberated when these bonds are hydrolysed. Lipmann suggested use of the ***symbol ~ to represent high-energy bond.*** For instance, ATP is written as AMP~P~P. (the short wavy twist or ***line*** ~ is called ***squiggle***)

ATP – the most important high-energy compound

Adenosine triphosphate (ATP) is a unique and the most important high-energy molecule in the living cells. It consists of an adenine, a ribose and a triphosphate moiety (***Fig.11.1***). ATP is a high-energy compound due to the presence of two phosphoanhydride bonds in the triphosphate unit. ATP serves as the ***energy currency of the cell***, as is evident from the ATP-ADP cycle.

ATP-ADP Cycle

The hydrolysis of ATP is associated with the release of large amount of energy.

$$ATP + H_2O \longrightarrow ADP + Pi + 7.3\ Cal.$$

The energy liberated is utilized for various processes like muscle contraction, active transport etc. ***ATP can also act as a donor of high-energy*** phosphate to low-energy compounds, to make them energy rich. On the other hand, ***ADP can accept high-energy phosphate*** from the compounds such as creatine phosphate, possessing higher free energy content to form ATP.

ATP serves as an immediately available energy currency of the cell which is constantly

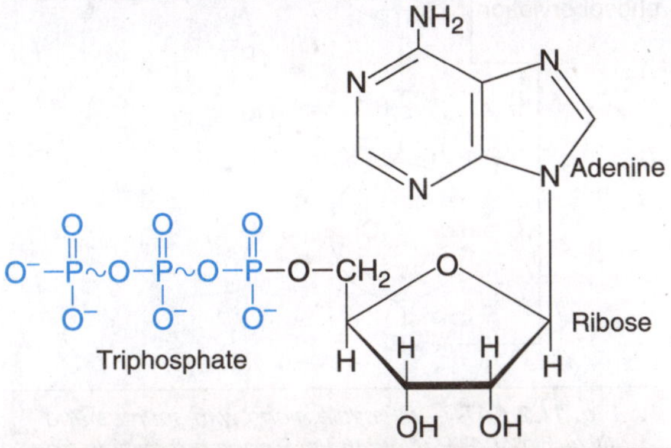

Fig. 11.1 : Structure of ATP.

being utilized and regenerated. This is represented by ATP-ADP cycle, the fundamental basis of energy exchange reactions in living system (***Fig.11.2***). The ***turnover of ATP*** is very ***high***.

ATP acts as an ***energy link between the catabolism*** (degradation of molecules) and ***anabolism*** (synthesis) in the biological system.

Synthesis of ATP

ATP can be synthesized in two ways

1. **Oxidative phosphorylation :** This is the major source of ATP in aerobic organisms. It is ***linked with the mitochondrial electron transport chain*** (details described later).

2. **Substrate level phosphorylation :** ATP may be directly synthesized during ***substrate oxidation in the metabolism***. The high-energy compounds such as phosphoenolpyruvate and 1,3-bisphosphoglycerate (intermediates of glycolysis) and succinyl CoA (of citric acid cycle) can transfer high-energy phosphate to ultimately produce ATP.

Storage forms of high-energy phosphates

Phosphocreatine (creatine phosphate) stored in vertebrate muscle and brain is an energy-rich compound. In ***invertebrates, phosphoarginine*** (arginine phosphate) replaces phosphocreatine.

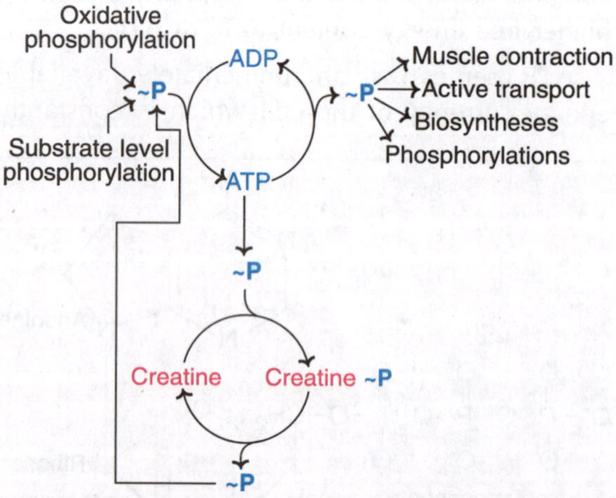

Fig. 11.2 : ATP-ADP cycle along with sources and utilization of ATP (Note that ~P does not exist in free form, but is only transferred).

BIOLOGICAL OXIDATION

Oxidation is defined as the ***loss of electrons*** and ***reduction as the gain of electrons***. This may be illustrated by the interconversion of ferrous ion (Fe^{2+}) to ferric ion (Fe^{3+}).

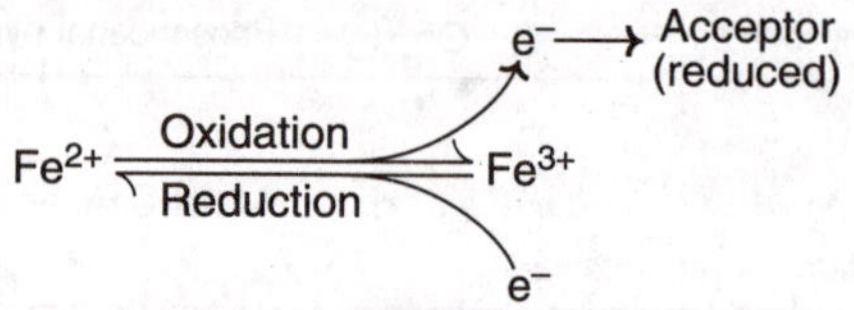

The electron lost in the oxidation is accepted by an acceptor which is said to be reduced. Thus the ***oxidation-reduction*** is a tightly ***coupled process***.

The general principle of oxidation-reduction is applicable to biological systems also. The oxidation of NADH to NAD^+ coupled with the reduction of FMN to $FMNH_2$ is illustrated

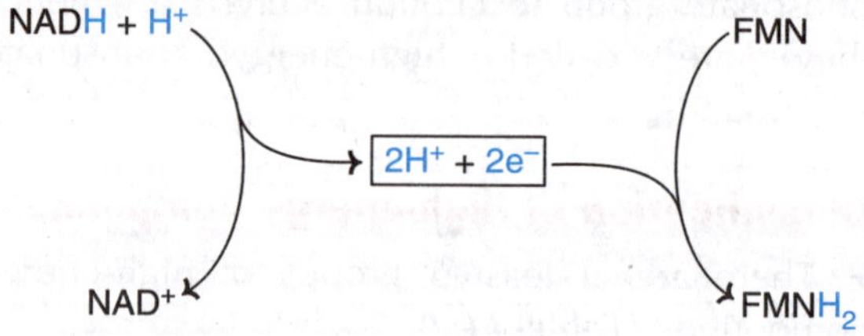

In the above illustration, there are two redox pairs NADH/NAD^+ and FMN/$FMNH_2$. The redox pairs differ in their tendency to lose or gain electrons.

Redox potential (E_0)

The ***oxidation-reduction*** potential or, simply, redox potential, is a ***quantitative measure of the tendency of a redox pair to lose or gain electrons***. The redox pairs are assigned specific ***standard redox potential*** (E_0 volts) at pH 7.0 and 25°C.

The redox potentials of some biologically important redox systems are given in ***Table 11.3***. The more negative redox potential represents a greater tendency (of reductant) to lose electrons.

Table 11.3 Standard redox potential (E_0) of some oxidation-reduction systems

Redox pair	E_0 Volts
Succinate/α-ketoglutarate	– 0.67
$2H^+/H_2$	– 0.42
$NAD^+/NADH$	– 0.32
$NADP^+/NADPH$	– 0.32
$FMN/FMNH_2$ (enzyme bound)	– 0.30
Lipoate (ox/red)	– 0.29
$FAD/FADH_2$	– 0.22
Pyruvate/lactate	– 0.19
Fumarate/succinate	+ 0.03
Cytochrome b (Fe^{3+}/Fe^{2+})	+ 0.07
Coenzyme Q (ox/red)	+ 0.10
Cytochrome c_1 (Fe^{3+}/Fe^{2+})	+ 0.23
Cytochrome c (Fe^{3+}/Fe^{2+})	+ 0.25
Cytochrome a (Fe^{3+}/Fe^{2+})	+ 0.29
$\frac{1}{2}O_2/H_2O$	+ 0.82

On the other hand, a more positive redox potential indicates a greater tendency (of oxidant) to accept electrons. The ***electrons flow from a redox pair with more negative E_0 to another redox pair with more positive E_0***.

The redox potential (E_0) is directly related to the change in the free energy (ΔG°).

ELECTRON TRANSPORT CHAIN

The energy-rich ***carbohydrates*** (particularly glucose), ***fatty acids*** and ***amino acids*** undergo a series of metabolic reactions and, finally, get ***oxidized to CO_2 and H_2O***. The reducing equivalents from various metabolic intermediates are transferred to coenzymes NAD^+ and FAD to produce, respectively, ***NADH and $FADH_2$***. The latter two reduced coenzymes pass through the electron transport chain (***ETC***) or ***respiratory chain*** and, finally, reduce oxygen to water. The passage of electrons through the ETC is associated with the loss of free energy. A part of this free energy is utilized to generate ATP from ADP and Pi (***Fig.11.3***). It must be remembered that the life of higher animals is absolutely dependent on supply of O_2 for ETC and generation of ATP.

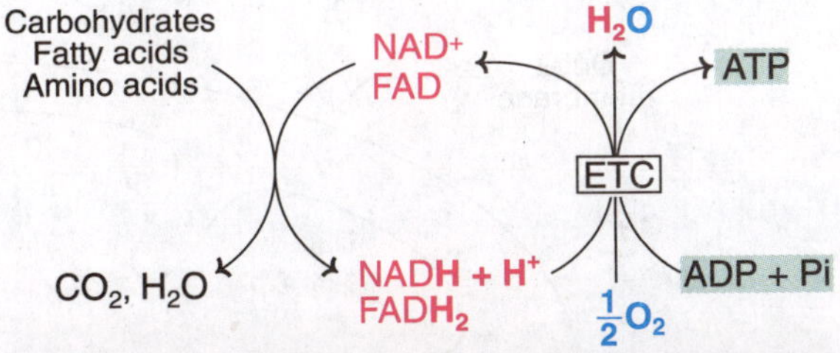

***Fig. 11.3 :** Overview of biological oxidation (ETC-Electron transport chain).*

An overview of the ETC is depicted in ***Fig.11.4***.

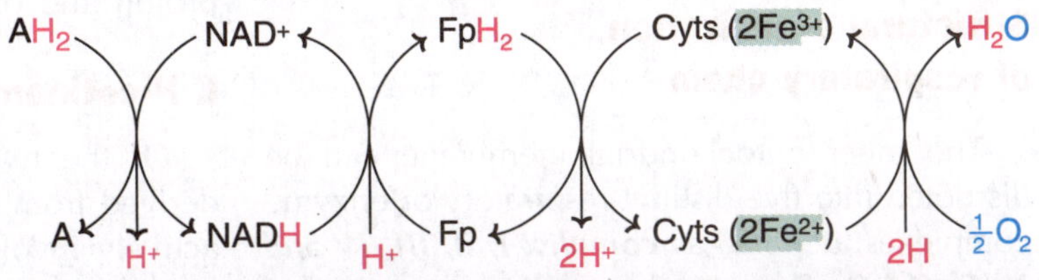

***Fig. 11.4 :** Overview of electron transport chain (A–Substrate; F_p–Flavoprotein; Cyts–Cytochromes).*

Mitochondria – the power houses of cell

The mitochondria are the ***centres for metabolic oxidative reactions*** to generate reduced coenzymes (NADH and $FADH_2$) which, in turn, are utilized in ETC to liberate energy in the form of ATP. For this reason, mitochondrion is appropriately regarded as the ***power house of the cell***.

Mitochondrial organization

The mitochondrion consists of ***five distinct parts***. These are the outer membrane, the inner membrane, the intermembrane space, the cristae and the matrix (***Fig.11.5***).

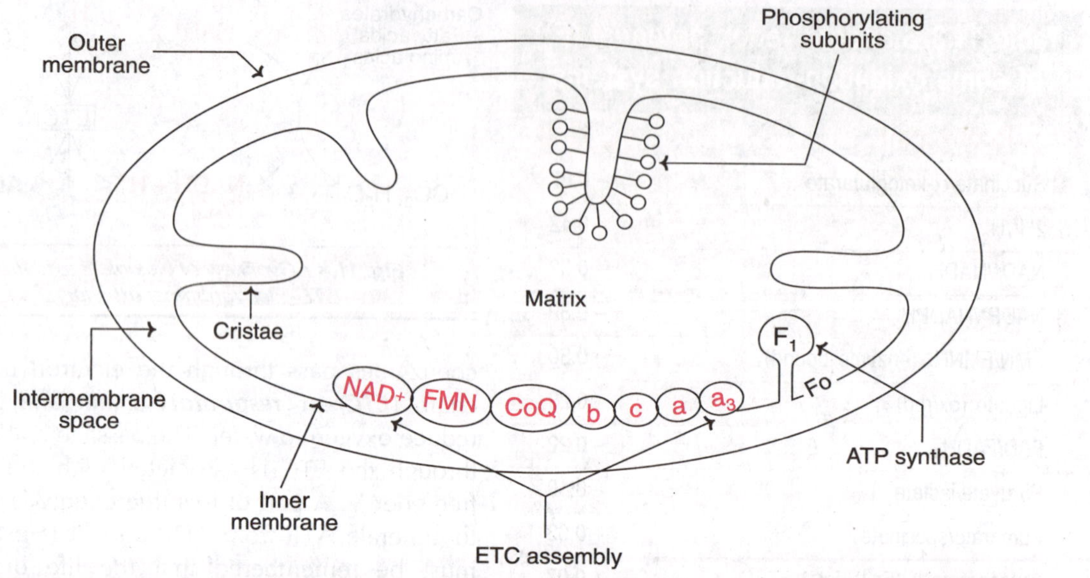

Fig. 11.5 : Structure of mitochondrion depicting electron transport chain (ETC) (F_o, F_1—Protein subunits).

Inner mitochondrial membrane : The ***electron transport chain and ATP synthesizing system*** are located on the ***inner mitochondrial membrane*** which is a specialized structure, rich in proteins. It is impermeable to ions (H^+, K^+, Na^+) and small molecules (ADP, ATP). This membrane is highly folded to form ***cristae***. The surface area of inner mitochondrial membrane is greatly increased due to cristae. The inner surface of the inner mitochondrial membrane possesses specialized particles (that look like lollipops), the ***phosphorylating subunits*** which are the centres for ATP production.

Mitochondrial matrix : The interior ground substance forms the matrix of mitochondria. It is rich in the ***enzymes*** responsible for the ***citric acid cycle, β-oxidation of fatty acids*** and ***oxidation of amino acids***.

Structural organization of respiratory chain

The inner mitochondrial membrane can be disrupted into five distinct respiratory or enzyme complexes, denoted as ***complex I, II, III, IV and V*** (***Fig.11.6***). The complexes I-IV are carriers of electrons while ***complex V*** is responsible for ATP ***synthesis***. Besides these enzyme complexes, there are certain mobile electron carriers in the respiratory chain. These include NADH, coenzyme Q, cytochrome C and oxygen.

The enzyme complexes (I-IV) and the mobile carriers are collectively involved in the transport of electrons which, ultimately, combine with oxygen to produce water. The largest proportion of the oxygen supplied to the body is utilized by the mitochondria for the operation of electron transport chain.

Components and reactions of the electron transport chain

There are five distinct carriers that participate in the electron transport chain (ETC). These carriers are sequentially arranged (***Fig.11.7***) and are responsible for the transfer of electrons from a given substrate to ultimately combine with proton and oxygen to form water.

I. Nicotinamide nucleotides

Of the two coenzymes NAD^+ and $NADP^+$ derived from the vitamin niacin, NAD^+ is more actively involved in the ETC. NAD^+ is reduced to ***NADH + H^+*** by dehydrogenases with the removal of two hydrogen atoms from the substrate (AH_2). The substrates include

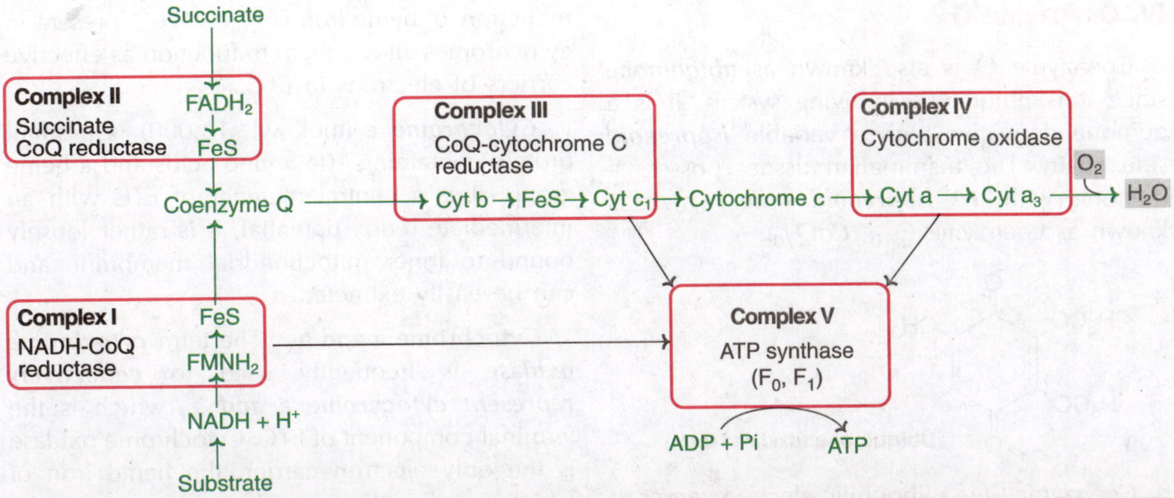

Fig. 11.6 : Multiprotein complexes in electron transport chain.

glyceraldehyde-3 phosphate, pyruvate, isocitrate, α-ketoglutarate and malate.

$$AH_2 + NAD^+ \rightleftharpoons A + NADH + H^+$$

NADPH + H⁺ produced by $NADP^+$-dependent dehydrogenase is ***not usually a substrate for ETC***. NADPH is more effectively utilized for anabolic reactions (e.g. fatty acid synthesis, cholesterol synthesis).

II. Flavoproteins

The enzyme NADH dehydrogenase (NADH-coenzyme Q reductase) is a flavoprotein with FMN as the prosthetic group. The coenzyme FMN accepts two electrons and a proton to form $FMNH_2$. ***NADH dehydrogenase*** is a complex enzyme closely associated with non-heme iron proteins (NHI) or iron-sulfur proteins (FeS).

$$NADH + H^+ + FMN \longrightarrow NAD^+ + FMNH_2$$

Succinate dehydrogenase (succinate-coenzyme Q reductase) is an enzyme found in the inner mitochondrial membrane. It is also a flavoprotein with FAD as the coenzyme.This can accept two hydrogen atoms ($2H^+ + 2e^-$) from succinate.

$$\text{Succinate} + FAD \longrightarrow \text{Fumarate} + FADH_2$$

III. Iron-sulfur proteins

The iron-sulfur (FeS) proteins exist in the oxidized (Fe^{3+}) or reduced (Fe^{2+}) state. About half a dozen FeS proteins connected with respiratory chain have been identified. However, the mechanism of action of iron-sulfur proteins in the ETC is not clearly understood.

One FeS participates in the transfer of electrons from FMN to coenzyme Q. Other FeS proteins associated with cytochrome b and cytochrome c_1 participate in the transport of electrons.

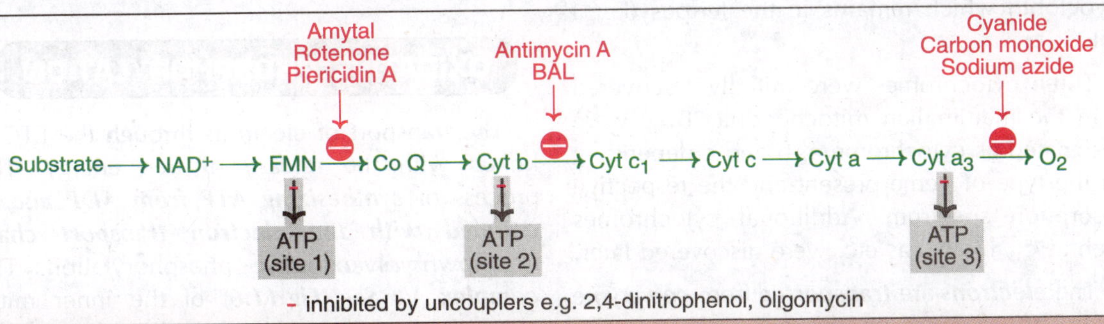

Fig. 11.7 : Electron transport chain with sites of ATP synthesis and inhibitors (BAL–British antilewisite).

IV. Coenzyme Q

Coenzyme Q is also known as ***ubiquinone*** since it is ubiquitous in living system. It is a quinone derivative with a variable ***isoprenoid*** side chain. The mammalian tissues possess a quinone with ***10 isoprenoid units*** which is known as coenzyme Q_{10} (***CoQ_{10}***).

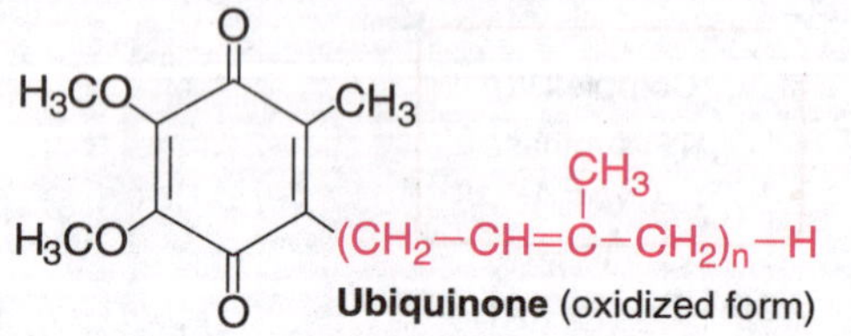

Ubiquinone (oxidized form)

Coenzyme Q is a lipophilic electron carrier. It can accept electrons from $FMNH_2$ produced in the ETC by NADH dehydrogenase or $FADH_2$ produced outside ETC (e.g. succinate dehydrogenase, acyl CoA dehydrogenase).

Coenzyme Q is not found in mycobacteria. Vitamin K performs similar function as coenzyme Q in these organisms. Coenzyme Q has no known vitamin precursor in animals. It is directly ***synthesized in the body***. (Refer cholesterol biosynthesis, ***Chapter 14***)

V. Cytochromes

The cytochromes are ***conjugated proteins*** containing ***heme group***. The latter consists of a porphyrin ring with iron atom. The heme group of cytochromes differ from that found in the structure of hemoglobin and myoglobin. The iron of heme in cytochromes is alternately oxidized (Fe^{3+}) and reduced (Fe^{2+}), which is essential for the transport of electrons in the ETC. This is in contrast to the heme iron of hemoglobin and myoglobin which remains in the ferrous (Fe^{2+}) state.

Three cytochromes were initially discovered from the mammalian mitochondria. They were designated as cytochrome a, b and c depending on the type of heme present and the respective absorption spectrum. Additional cytochromes such as c_1, b_1, b_2, a_3 etc. were discovered later.

The ***electrons*** are ***transported*** from ***coenzyme Q to cytochromes*** (in the order) ***b, c_1, c, a and a_3***. The property of reversible oxidation-reduction of heme iron $Fe^{2+} \rightleftharpoons Fe^{3+}$ present in cytochromes allows them to function as effective carriers of electrons in ETC.

Cytochrome c (mol. wt. 13,000) is a small protein containing 104 amino acids and a heme group. It is a central member of ETC with an intermediate redox potential. It is rather loosely bound to inner mitochondrial membrane and can be easily extracted.

Cytochrome a and a_3 : The term ***cytochrome oxidase*** is frequently used to ***collectively represent cytochrome a and a_3*** which is the terminal component of ETC. Cytochrome oxidase is the only electron carrier, the heme iron of which can directly react with molecular oxygen. Besides heme (with iron), this oxidase also contains copper that undergoes oxidation-reduction ($Cu^{2+} \rightleftharpoons Cu^{+}$) during the transport of electrons.

In the final stage of ETC, the transported electrons, the free protons and the molecular oxygen combine to produce water.

Inherited disorders of ETC

Mitochondrial encephalopathy, lactic acidosis, and stroke (MELAS), are inherited disorders due to the deficiency of NADH-Q oxidoreductase (complex I) or cytochrome oxidase (complex IV). It is caused by a mutation in mitochondrial DNA. MELAS are believed to be involved in ***Alzheimer's disease*** and diabetes mellitus.

Fatal infantile mitochondrial myopathy is caused by reduced or absence of most oxido-reductases in ETC.

OXIDATIVE PHOSPHORYLATION

The transport of electrons through the ETC is linked with the release of free energy. The process of ***synthesizing ATP from ADP and Pi coupled with the electron transport chain*** is known as oxidative phosphorylation. The ***complex V*** (***See Fig.11.6***) of the inner mito-chondrial membrane is the ***site of oxidative phosphorylation (OXPHOS)***.

Medical Concepts/ Clinical Correlates

- The most important function of food is to supply energy to the living cells. This is finally achieved through biological oxidation.
- The supply of O_2 is very essential for the survival of life (exception—anaerobic bacteria).
- ATP, the energy currency of the cell, acts as a link between the catabolism and anabolism in the living system. The major production of body's ATP occurs in the mitochondria through oxidative phosphorylation coupled with respiration.
- Respiratory chain or electron transport chain (ETC) is blocked by site specific inhibitors such as rotenone, amytal, antimycin A, BAL, carbon monoxide and cyanide.

P : O Ratio

The P : O ratio refers to the ***number of inorganic phosphate*** molecules utilized for ATP generation for every ***atom of oxygen*** consumed.

(**Note :** ***Earlier, a P : O ratio of 3*** was assigned to ***NADH*** while it was ***2*** for ***$FADH_2$***. This was based on the belief that two protons were pumped at each complex (I, III and IV) resulting in the generation of one ATP, and thus a total of 3 for NADH and 2 for $FADH_2$. Studies in recent years suggest that this is not correct).

As per the recent evidence, the flow of 2 electrons (from NADH + H^+) through complexes I, III and IV results in the translocation of 10 protons from mitochondrial matrix. The synthesis of one ATP requires the involvement of 4 protons. Thus a total of 2.5 ATP (***P : O ratio 2.5***) are generated from ***NADH*** and ***1.5 ATP*** (***P : O ratio 1.5***) from ***$FADH_2$***.

It may be summarized that the oxidation of one ***NADH*** through ETC coupled with oxidative phosphorylation generates ***2.5 ATP*** while oxidation of one ***$FADH_2$*** yields ***1.5 ATP***. This difference is due to the fact that the coupling sites for NADH oxidation are 3 with 10 protons whereas for $FADH_2$ there are only two sites with 6 protons.

Sites of oxidative phosphorylation in ETC

There are ***three sites*** in the ETC that are exergonic to result in the synthesis of 3 ATP molecules (***See Fig.11.7***).

1. ***Oxidation of $FMNH_2$*** by coenzyme Q.
2. ***Oxidation of cytochrome b*** by cytochrome c_1.
3. ***Cytochrome oxidase*** reaction.

Each one of the above reactions represents a ***coupling site*** for ATP production. There are only two coupling sites for the oxidation of $FADH_2$ (P : O ratio 1.5), since the first site is bypassed.

Energetics of oxidative phosphorylation

The transport of electrons from redox pair $NAD^+/NADH$ (E_0 = – 0.32) to finally the redox pair $\frac{1}{2}O_2/H_2O$ (E_0 = + 0.82) may be simplified and represented in the following equation

$$\frac{1}{2}O_2 + NADH + H^+ \longrightarrow H_2O + NAD^+$$

The redox potential difference between these two redox pairs is 1.14 V, which is equivalent to an energy 52 Cal/mol.

2.5 ATP are synthesized in the ETC when NADH is oxidized which equals to 18.25 Cal (each ATP = 7.3 Cal).

The efficiency of energy conservation is calculated as

$$\frac{18.25 \times 100}{52} = 35.10\%.$$

Therefore, when NADH is oxidized, about ***35% of energy is trapped in the form of 2.5 ATP*** and the ***remaining is lost as heat***. The heat liberation is not a wasteful process, since it allows ETC to go on continuously to generate ATP. Further, ***this heat is necessary to maintain body temperature***.

MECHANISM OF OXIDATIVE PHOSPHORYLATION

Several hypotheses have been put forth to explain the process of oxidative phosphorylation. The most important among them—namely, chemical coupling, and chemiosmotic—are discussed in the next page.

Chemical coupling hypothesis

This hypothesis was put forth by Edward Slater (1953). According to chemical coupling hypothesis, during the course of electron transfer in respiratory chain, a series of ***phosphorylated high-energy intermediates*** are first produced which are utilized for the synthesis of ATP. These reactions are believed to be analogous to the substrate level phosphorylation that occurs in glycolysis or citric acid cycle. However, this hypothesis lacks experimental evidence, since all attempts, so far, to isolate any one of the high-energy intermediates have not been successful.

Chemiosmotic hypothesis

This mechanism, originally proposed by Peter Mitchell (1961), is now ***widely accepted***. It explains how the transport of electrons through the respiratory chain is effectively utilized to produce ATP from ADP + Pi. The concept of chemiosmotic hypothesis is comparable with energy stored in a battery separated by positive and negative charges.

Proton gradient : The inner mitochondrial membrane, as such, is impermeable to protons (H^+) and hydroxyl ions (OH^-). The transport of electrons through ETC is coupled with the ***translocation of protons (H^+)*** across the inner mitochondrial membrane (coupling membrane) from the matrix to the intermembrane space. The pumping of protons results in an ***electrochemical or proton gradient***. This is due to the accumulation of more H^+ ions (low pH) on the outer side of the inner mitochondrial membrane than the inner side (***Fig.11.8***). The ***proton gradient*** developed due to the electron flow in the respiratory chain is ***sufficient to result in the synthesis of ATP from ADP and Pi***.

Enzyme system for ATP synthesis : ATP synthase, present in the complex V, utilizes the proton gradient for the synthesis of ATP. This enzyme is also known as ***ATPase***, since it can hydrolyse ATP to ADP and Pi. ***ATP synthase***, is a complex enzyme and consists of two functional subunits, namely F_1 and F_0 (***Fig.11.9***). Its ***structure is comparable with 'lollipops'.***

The protons that accumulate on the intermembrane space re-enter the mitochondrial matrix leading to the synthesis of ATP.

Rotary motor model for ATP generation

Paul Boyer in 1964 proposed (Nobel Prize, 1997) that a conformational change in the mitochondrial membrane proteins leads to the synthesis of ATP. The original Boyer hypothesis, now considered as ***rotary motor/engine driving model*** or ***binding change model***, is widely accepted for the generation ATP.

The enzyme ATP synthase is F_0F_1 complex (of complex V). The F_0 subcomplex is composed of channel protein 'C' subunits to which F_1-ATP synthase is attached (***Fig.11.10***). F_1-ATP synthase consists of a central γ subunit surrounded by alternating α and β subunits (α_3 and β_3).

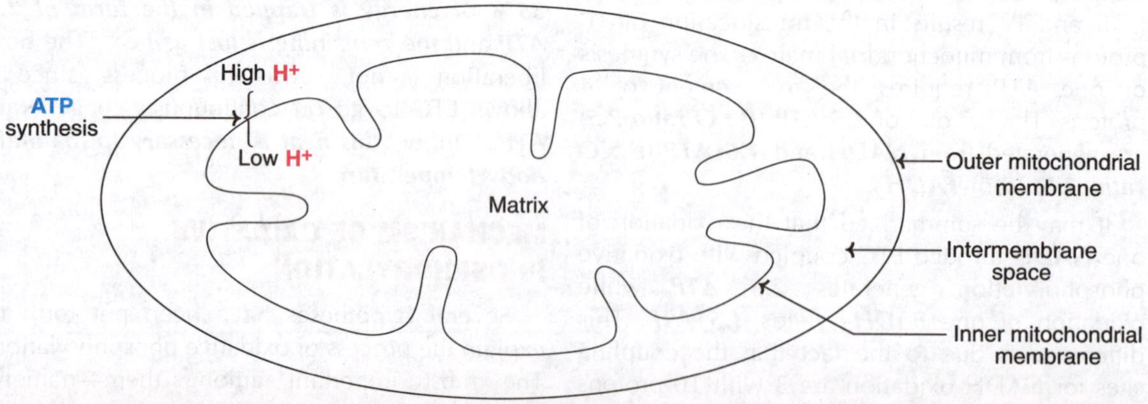

Fig. 11.8 : Outline of chemiosmotic hypothesis for oxidative phosphorylation.

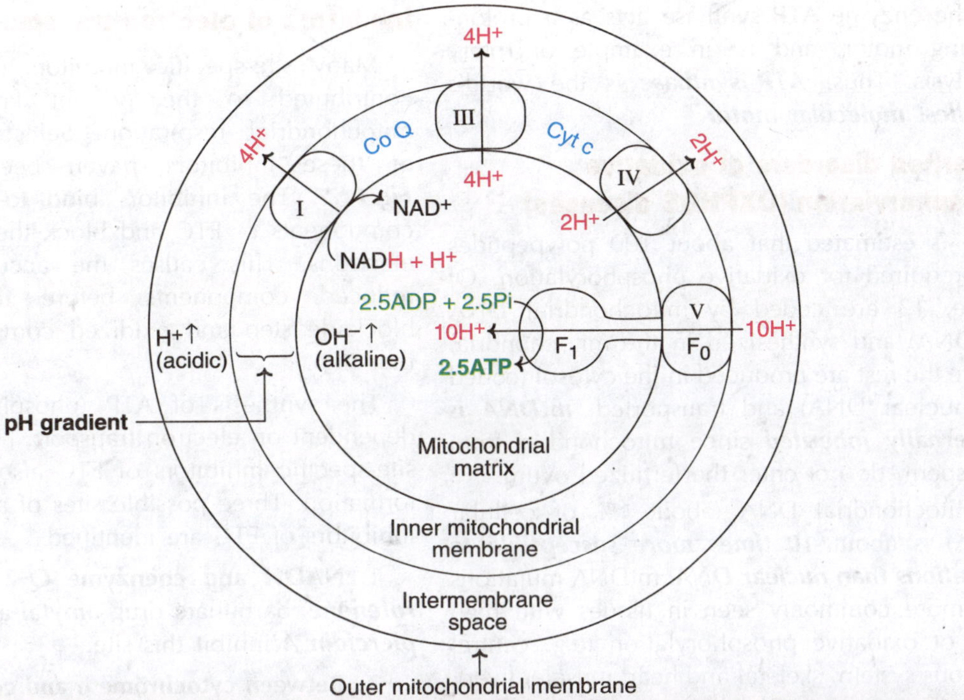

Fig. 11.9 : Diagrammatic representation of chemiosmotic hypothesis for oxidative phosphorylation (I, III, IV and V–Respiratory chain complexes; F_o, F_1–Protein subunits for phosphorylation).

In response to the proton flux, the γ subunit physically rotates. This induces ***conformational changes in the β_3 subunits that finally lead to the release of ATP.***

According to the binding change mechanism, the three β subunits of F_1-ATP synthase adopt different conformations. One subunit has ***open*** (***O***) conformation, the second has ***loose*** (***L***) conformation while the third one has ***tight*** (***T***) ***conformation*** (***Fig.11.11***).

By an unknown mechanism, protons induce the rotation of γ subunit, which in turn induces conformation changes in β subunits. The substrates ADP and Pi bind to β subunit in L-conformation. The L site changes to T conformation, and this leads to the synthesis of ATP. The O site changes to L conformation which binds to ADP and Pi. The T site changes to O conformation, and releases ATP. This cycle of conformation changes of β subunits is repeated. And 2.5 ATP are generated for each revolution (***Fig.11.11***).

It may be noted that the ***ATP release*** in O conformation is ***energy dependent (and not ATP synthesis)*** and very crucial in rotary motor model for ATP generation.

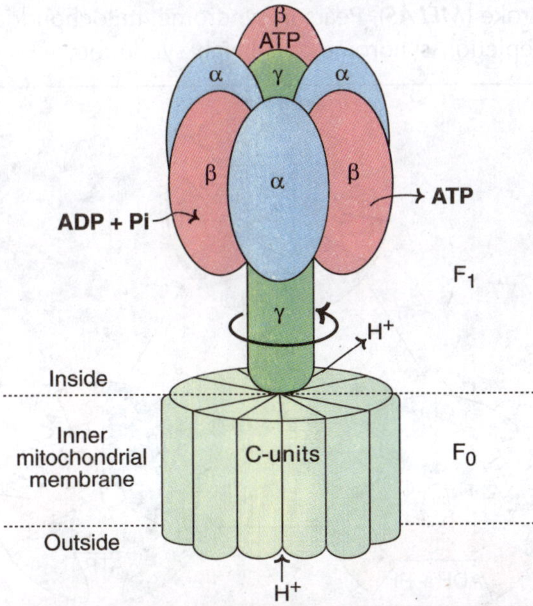

Fig. 11.10 : Structure of mitochondrial ATP synthase (F_oF_1) complex (C units-channel protein subunits; α, β, and γ are the subunits of F_1-ATP synthase).

The enzyme ATP synthase acts as a proton-driving motor, and is an example of rotary catalysis. Thus, ***ATP synthase*** is the world's ***smallest molecular motor***.

Inherited disorders of oxidative phosphorylation (OXPHOS diseases)

It is estimated that about 100 polypeptides are required for oxidative phosphorylation. Of these, 13 are coded by mitochondrial DNA (mtDNA) and synthesized in the mitochondria, while the rest are produced in the cytosol (coded by nuclear DNA) and transported. ***mtDNA is maternally inherited*** since mitochondria from the sperm do not enter the fertilized ovum.

Mitochondrial DNA (about 1% of cellular DNA) is about ***10 times more susceptible to mutations than nuclear DNA***. mtDNA mutations are more commonly seen in tissues with high rate of oxidative phosphorylation (e.g. central nervous system, skeletal and heart muscle, liver).

Leber's hereditary optic neuropathy is an example for mutations in mtDNA. This disorder is characterized by loss of bilateral vision due to ***neuroretinal degeneration***.

The other OXPHOS diseases include **m**itochondrial **e**ncephalopathy, **l**actic acidosis, and **s**troke (***MELAS***), Pearson syndrome, mitochondrial depletion syndrome and Leigh syndrome.

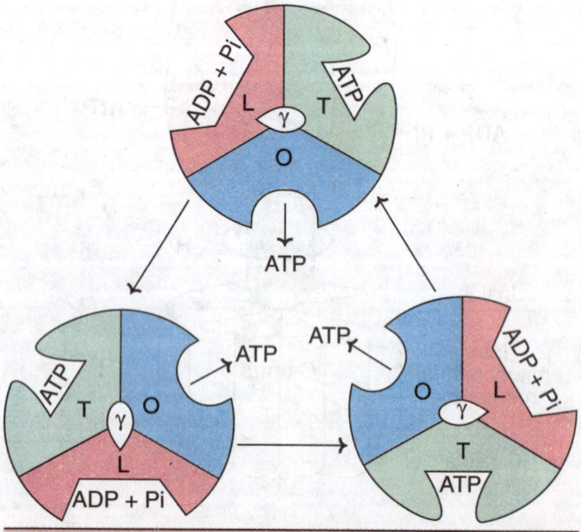

Fig. 11.11 : *The binding change model (rotary motor/engine driving model) for ATP synthesis by F_1-ATP synthase.*

Inhibitors of electron transport chain

Many site-specific inhibitors of ETC have contributed to the present knowledge of mitochondrial respiration. Selected examples of these inhibitors haven been given in ***Fig.11.7***. The inhibitors bind to one of the components of ETC and block the transport of electrons. This causes the accumulation of reduced components before the inhibitor blockade step and oxidized components after that step.

The synthesis of ATP (phosphorylation) is dependent on electron transport. Hence, all the site-specific inhibitors of ETC also inhibit ATP formation. Three possible sites of action for the inhibitors of ETC are identified

1. **NADH and coenzyme Q :** Fish poison ***rotenone***, barbituate drug ***amytal*** and antibiotic ***piercidin A*** inhibit this site.

2. **Between cytochrome b and c_1 :** ***Antimycin A***— an antibiotic, ***British antilewisite (BAL)***— an antidote used against war-gas—are the two important inhibitors of the site between cytochrome b and c_1.

3. **Inhibitors of cytochrome oxidase :** Carbon monoxide, cyanide, hydrogen sulphide and azide effectively inhibit cytochrome oxidase. ***Carbon monoxide*** reacts with reduced form of the cytochrome while ***cyanide*** and ***azide*** react with oxidized form.

Cyanide poisoning

Cyanide is probably the most potent inhibitor of ETC. It binds to Fe^{3+} of cytochrome oxidase blocking mitochondrial respiration leading to cell death. Cyanide poisoning causes death due to tissue asphyxia (mostly of central nervous system). Cyanide is a popular suicidal agent commonly ***used by militants, terrorists*** and warriors.

Cyanide containing compounds include ***hydrogen cyanide*** gas and certain cyanide salts e.g. ***potassium cyanide***. If cyanide is inhaled, it causes coma, seizures and cardiac arrest, and finally death in a few seconds. At lower doses, cyanide causes headache, weakness, giddiness, confusion and difficulty in breathing.

For the treatment of cyanide poisoning, specially designed ***antidote kits*** are available. The kits usually contain ***amyl nitrate*** and ***sodium nitrite***.

The nitrites convert hemoglobin into methemoglobin. Cyanide binds to methemoglobin forming cyanomethemoglobin. This results in the release of cyanide from cytochrome oxidase.

In recent years, ***hydroxycobalamin*** (a form of ***vitamin*** B_{12}) is ***used as an antidote*** for cyanide poisoning. The reaction involves the ***conversion of cyanide to a harmless cyanocobalamin***.

INHIBITORS OF OXIDATIVE PHOSPHORYLATION

Uncouplers

The mitochondrial transport of electrons is tightly coupled with oxidative phosphorylation (ATP synthesis). In other words, ***oxidation and phosphorylation proceed simultaneously***. There are certain ***compounds that can uncouple*** (or delink) ***the electron transport from oxidative phosphorylation***. Such compounds, known as uncouplers, increase the permeability of inner mitochondrial membrane to protons (H^+). The result is that ATP synthesis does not occur. The energy linked with the transport of electrons is dissipated as heat. ***The uncouplers allow*** (often at accelerated rate) ***oxidation of substrates*** (via NADH or $FADH_2$) ***without ATP formation.***

The uncoupler, ***2,4-dinitrophenol (DNP)***, has been extensively studied. It is a small lipophilic molecule. DNP is a proton-carrier and can easily diffuse through the inner mitochondrial membrane. In the people seeking to lose weight, DNP was used as a drug. However, this is now discontinued, as it produces hyperthermia and other side effects. In fact, Food and Drug Administration (USA) has banned the use of DNP.

The other uncouplers include ***dinitrocresol***, ***pentachlorophenol*** and ***trifluorocarbonylcyanide phenylhydrazone*** (FCCP). The last compound (FCCP) is said to be 100 times more effective as an uncoupler than dinitrophenol. When administered ***in high doses***, the drug ***aspirin*** acts as an uncoupler.

Physiological uncouplers : Certain physiological substances which act as uncouplers at higher concentration have been identified. These include ***thermogenin***, ***thyroxine*** and ***long chain free fatty acids***. The unconjugated bilirubin is also believed to act as an uncoupler. This is, however, yet to be proved beyond doubt.

Significance of uncoupling

Uncoupling of respiration from oxidative phosphorylation under natural conditions assumes biological significance. The ***maintenance of body temperature*** is particularly important in hairless animals, hibernating animals and the animals adapted to cold. These animals possess a specialized tissue called ***brown adipose tissue*** in the upper back and neck portions. The mitochondria of brown adipose tissue are rich in electron carriers and are specialized to carry out an ***oxidation uncoupled from phosphorylation***. This causes liberation of heat when fat is oxidized in the brown adipose tissue. Brown adipose tissue may be considered as a site of ***non-shivering thermogenesis***. The presence of active brown adipose tissue in certain individuals is believed to protect them from becoming obese. The excess calories consumed by these people are burnt and ***liberated as heat***, instead of being stored as fat.

Thermogenin (or ***uncoupling protein***, UCPI) is a physiological uncoupler, located in the inner mitochondrial membrane of brown adipose tissue. It ***blocks the formation of ATP, and generates heat***. This assumes significant in the newborn, and during hibernation in animals.

Ionophores : The term 'ionophores' is used to collectively represent the lipophilic substances that ***promote the transport of ions across biological membranes***.

All the uncouplers (described above) are, in fact, proton ionophores.

The antibiotics ***valinomycin***, ***gramicidin A*** and ***nigercin*** act as ***ionophores*** for K^+ ions. Both these compounds are also capable of dissipating proton gradient across the inner mitochondrial membrane and inhibit oxidative phosphorylation.

Other inhibitors of oxidative phosphorylation

Oligomycin : This antibiotic prevents the mitochondrial oxidation as well as phosphorylation. It ***binds with the enzyme ATP synthase*** and blocks the proton (H^+) channels. It thus prevents the translocation (re-entry) of protons into the mitochondrial matrix. Due to this, protons get accumulated at higher concentration in the intermembrane space. Electron transport (respiration) ultimately stops, since protons cannot be pumped out against steep proton gradients.

Atractyloside : This is a plant toxin and inhibits oxidative phosphorylation by an indirect mechanism. Adenine nucleotide carrier system facilitates the transport of ATP and ADP. Atractyloside inhibits adenine nucleotide carrier and, thus, ***blocks the adequate supply of ADP,*** thereby preventing phosphorylation.

TRANSPORT OF REDUCING EQUIVALENTS—SHUTTLE PATHWAYS

The inner mitochondrial membrane is impermeable to NADH. Therefore, the NADH produced in the cytosol cannot directly enter the mitochondria. Two pathways—namely ***glycerol-phosphate shuttle*** and ***malate-aspartate shuttle***—are operative to do this job. They ***transport*** the ***reducing equivalents from cytosol to mitochondria and not vice versa.***

I. Glycerol-phosphate shuttle

Cytosolic glycerol 3-phosphate dehydrogenase oxidizes NADH to NAD^+. The reducing equivalents are transported through glycerol 3-phosphate into the mitochondria. Glycerol 3-phosphate dehydrogenase—present on outer surface of inner mitochondrial membrane—reduces FAD to $FADH_2$. Dihydroxyacetone phosphate escapes into the cytosol and the shuttling continues as depicted in ***Fig.11.12.*** $FADH_2$ gets oxidized via ETC to generate ***1.5 ATP.***

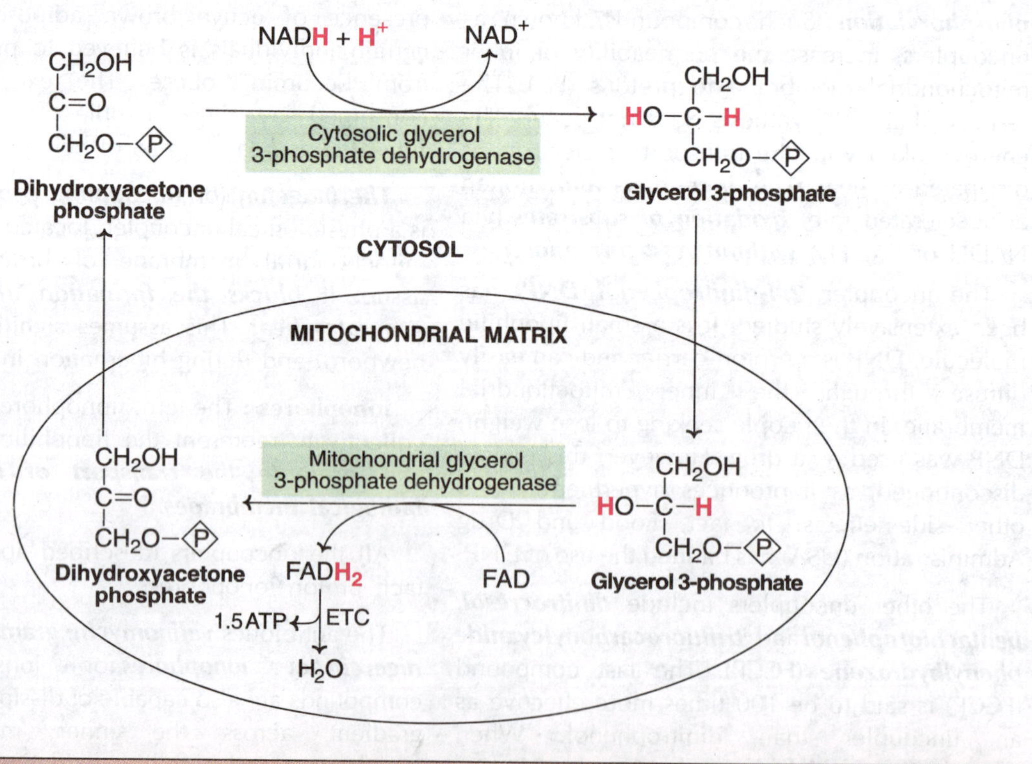

Fig. 11.12 : Glycerol-phosphate shuttle (reducing equivalents transported are shown in blue).

Medical Concepts/ Clinical Correlates

- Cyanide is a popular suicidal agent that causes death by blocking ETC. Recently, hydroxycobalamin (binds with cyanide to form harmless cynocobalamin) is used as an antidote for cyanide poisoning.
- Uncoupling of respiration from oxidative phosphorylation under natural conditions assumes biological significance. The brown adipose tissue, rich in electron carriers, brings about oxidation uncoupled from phosphorylation. The presence of active brown adipose tissue in some individuals is believed to protect them from becoming obese. This is because the excess calories consumed by these people are burnt and liberated as heat instead of being stored as fat.
- Inherited disorders of oxidative phosphorylation caused by the mutations in mitochondrial DNA have been identified e.g. Leber's hereditary optic neuropathy.

II. Malate-aspartate shuttle

In the cytosol, oxaloacetate accepts the reducing equivalents (NADH) and becomes malate. Malate then enters mitochondria where it is oxidized by mitochondrial malate dehydrogenase. In this reaction, NADH and oxaloacetate are regenerated. NADH gets oxidized via electron transport chain and ***2.5 ATP*** are produced. This is in contrast to glycerolphosphate shuttle where only 1.5 ATP are produced.

In the mitochondria, oxaloacetate participates in transamination reaction with glutamate to produce aspartate and α-ketoglutarate. The aspartate enters the cytosol and transaminates with α-ketoglutarate to give oxaloacetate and glutamate. The malate-aspartate shuttle is shown in ***Fig.11.13***.

Creatine phosphate shuttle

Creatine phosphate is a high energy compound, closely linked with energy transfer reactions (see ATP-ADP cycle, ***Fig.11.2***). Creatine phosphate shuttle involves ***transfer of high***

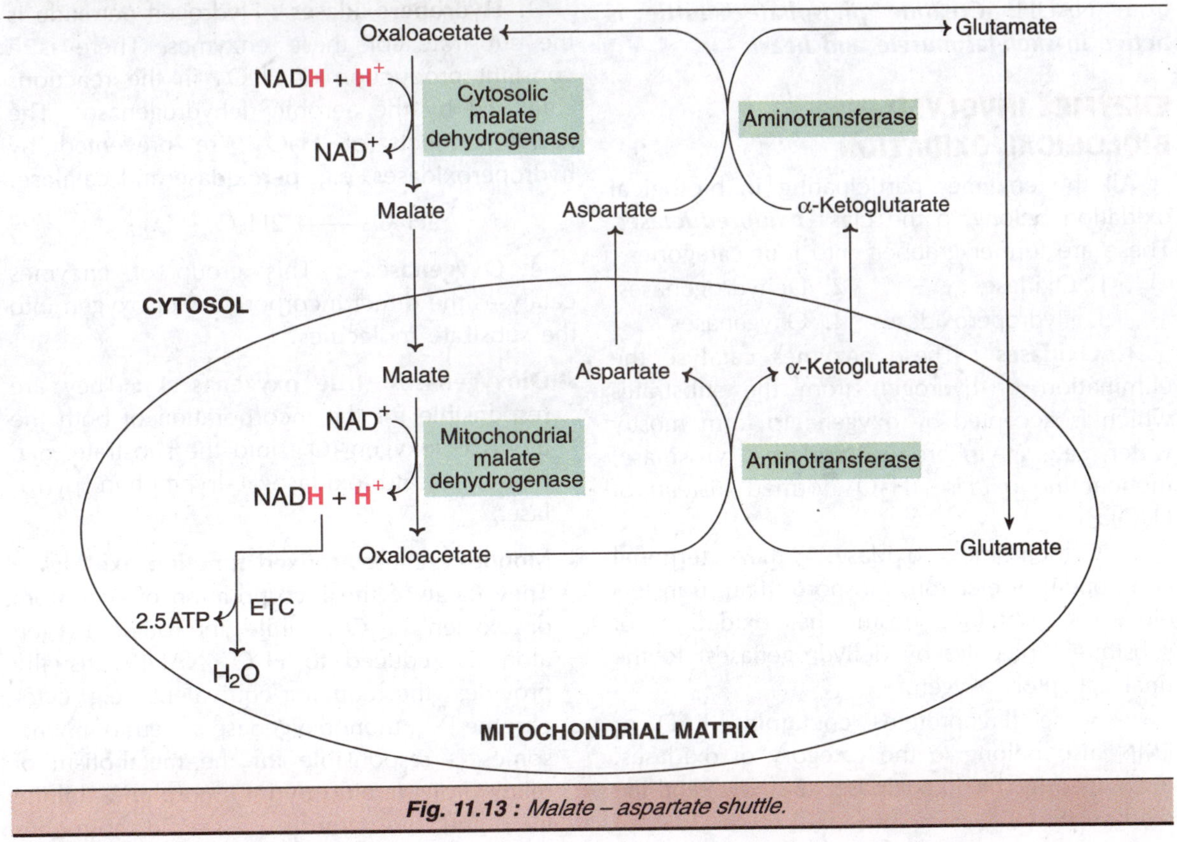

Fig. 11.13 : *Malate – aspartate shuttle.*

energy phosphate from mitochondria (creatine combines with ATP to form creatine phosphate) ***to the cytosol*** in the active tissues such as skeletal muscle and heart. In this way, creatine phosphate shuttle functions as an energy buffer for transfer of high energy phosphate and generation of ATP in the extra—mitochondrial system.

Mitochondrial isoenzymes of creatine kinase (mtCK)

In addition to the three cytosolic isoenzymes of CK (BB, MB, MM described in ***Chapter 6***), two mitochondrial CK (mtCK) have been identified—***ubiquitous*** and ***sarcomeric*** (umtCK and smtCK). Both mtCK are octamers consisting of four dimers each. mtCK are directly involved in the formation of phosphocreatine from mitochondrial ATP and ***participate in creatine phosphate shuttle.***

Shuttle pathways and tissues

Liver and ***heart*** utilize ***malate-aspartate shuttle***, and yield 2.5 ATP per mole of NADH. Most of the other tissues, however, employ glycerol-phosphate shuttle and liberate 1.5 ATP from NADH. ***Creatine phosphate shuttle is active in skeletal muscle and heart.***

ENZYMES INVOLVED IN BIOLOGICAL OXIDATION

All the enzymes participating in biological oxidation belong to the class ***oxidoreductases.*** These are further grouped into four categories

1. Oxidases 2. Dehydrogenases
3. Hydroperoxidases 4. Oxygenases.

1. **Oxidases :** These enzymes catalyse the elimination of hydrogen from the substrates which is accepted by oxygen to form mostly water, e.g. cytochrome oxidase, tyrosinase, monoamine oxidase (H_2O_2 formed instead of H_2O).

Cytochrome oxidase, the terminal component of electron transport chain, transfers electrons (obtained from the oxidation of substrate molecules by dehydrogenases) to the final acceptor, oxygen.

Some flavoproteins containing FAD or FMN also belong to the category of oxidases. e.g., L-amino acid oxidase (FMN), xanthine oxidase (FAD).

2. **Dehydrogenases :** As the name indicates, these enzymes cannot utilize oxygen as hydrogen acceptor. They ***catalyse the reversible transfer of hydrogen from one substrate to another*** and, thus, bring about oxidation-reduction reactions. There are a large number of enzymes belonging to this group

- NAD^+ dependent dehydrogenases, e.g. alcohol dehydrogenase, glycerol 3-phosphate dehydrogenase.
- $NADP^+$ dependent dehydrogenases, e.g. HMG CoA reductase, enoyl reductase.
- FMN dependent dehydrogenases, e.g. NADH dehydrogenase.
- FAD dependent dehydrogenases, e.g. succinate dehydrogenase, acyl CoA dehydrogenase.
- The cytochromes : All the cytochromes of electron transport chain (b, c_1 and c) except the terminal cytochrome oxidase ($a+a_3$) belong to this group.

3. **Hydroperoxidases :** Hydrogen peroxide is the substrate for these enzymes. There is a constant production of H_2O_2 in the reactions catalysed by the aerobic dehydrogenases. The harmful effects of H_2O_2 are prevented by hydroperoxidases, e.g. peroxidase and catalase.

$$2H_2O_2 \longrightarrow 2H_2O + O_2$$

4. **Oxygenases :** This group of enzymes catalyses the direct incorporation of oxygen into the substrate molecules.

- **Dioxygenases (true oxygenases) :** They are responsible for the incorporation of both the atoms of oxygen (O_2) into the substrate, e.g. homogentisate oxidase, L-tryptophan pyrrolase.
- **Monooxygenases (mixed function oxidases) :** They catalyse the incorporation of one atom of oxygen ($\frac{1}{2}O_2$) while the other oxygen atom is reduced to H_2O. NADPH usually provides the reducing equivalents, e.g. cytochrome P_{450} monooxygenase system of microsomes is responsible for the metabolism of many drugs (amino pyrine, morphine, aniline etc.) and biosynthesis of steroid hormones

(from cholesterol). The action of Cyt P_{450} is depicted here.

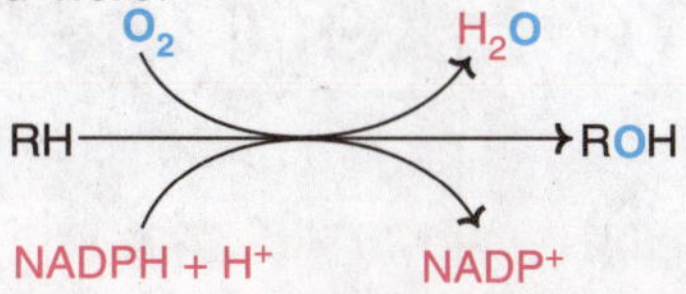

Electron transport in prokaryotes

In contrast to eukaryotes, the prokaryotes ***lack mitochondria***. However, prokaryotes possess a separate system for biological oxidation. A set of electron carriers (different from that found in mitochondria) and ***enzymes*** of oxidative phosphorylation are ***bound to the inner cell membrane in prokaryotes***. This arrangement of oxidative machinery is one of the reasons to believe that mitochondria of higher organisms have descended from prokaryotic cells.

SUMMARY

- *Bioenergetics deals with the study of energy changes in biochemical reactions. Change in free energy (ΔG) is valuable in predicting the feasibility of a reaction. A negative and a positive ΔG, respectively, represent an exergonic (energy-releasing) and endergonic (energy-consuming) reactions.*
- *High-energy compounds ($\Delta G > -7.0$ Cal/mol) play a crucial role in the energy transfer of biochemical reactions (e.g. ATP, phosphocreatine, phosphoenolpyruvate).*
- *ATP is the energy currency of the cell. ATP-ADP cycle acts as a connecting energy link between catabolic and anabolic reactions.*
- *Respiratory chain or electron transport chain (ETC) located in the inner mitochondrial membrane represents the final stage of oxidizing the reducing equivalents (NADH and $FADH_2$) derived from the metabolic intermediates to water.*
- *ETC is organized into five distinct complexes. The complexes I to IV are electron carriers while complex V is responsible for ATP production. The components of ETC are arranged in the sequence*
  $NAD^+ \longrightarrow FMN \longrightarrow CoQ \longrightarrow Cyt\ b \longrightarrow Cyt\ c_1 \longrightarrow Cyt\ c \longrightarrow Cyt\ a + a_3 \longrightarrow O_2$
- *The process of synthesizing ATP from ADP and Pi coupled with ETC is known as oxidative phosphorylation. NADH oxidation with a P : O ratio 2.5 indicates that 2.5 ATP are synthesized while $FADH_2$ oxidation (P : O ratio 1.5) results in the production of 1.5 ATP.*
- *Among the hypotheses put forth to explain the mechanism of oxidative phosphorylation, the chemiosmotic hypothesis (of Mitchell) is widely accepted. The rotary motor model (of Boyer) involving the conformation changes in the β-subunits of ATP synthase explains the ATP generation.*
- *NADH produced in the cytosol cannot directly enter mitochondria. Glycerol-phosphate shuttle (generates 1.5 ATP) and malate-asparate shuttle (generates 2.5 ATP) operate to overcome the difficulty.*
- *There are many inhibitors of electron transport chain (rotenone, amytal, antimycin, CO,-CN, H_2S etc.) and oxidative phosphorylation (oligomycin, atractyloside). Uncouplers (e.g. dinitrophenol) are the substances that delink ETC from oxidative phosphorylation.*
- *The enzymes participating in biological oxidation belong to the class oxidoreductases. There are five groups, namely oxidases, aerobic dehydrogenases, anaerobic dehydrogenases, hydroperoxidases and oxygenases.*

THE READER MUST REFER CHAPTER 34 FOR DETAILS ON FREE RADICALS AND ANTIOXIDANTS.

Self-Assessment Exercises

I. Essay questions

1. Write an account of the high-energy compounds in metabolism
2. Describe the components of electron transport chain and discuss the oxidation of NADH.
3. Define oxidative phosphorylation. Discuss chemiosmotic hypothesis in detail.
4. Give an account of the enzymes involved in biological oxidation.
5. Discuss about the inhibitors of ETC and oxidative phosphorylation.

II. Short notes

(a) High-energy bonds, (b) Uncouplers, (c) P : O ratio, (d) Redox loops, (e) ATP synthase, (f) Cytochromes, (g) Sites of oxidative phosphorylation, (h) Coenzyme Q, (I) Redox potential, (j) ATP as energy currency.

III. Fill in the blanks

1. The relation between the change of free energy (ΔG), enthalpy (ΔH) and entropy (ΔS) is expressed by the equation _____________.
2. A negative sign of free energy indicates that the reaction is _____________.
3. The bonds responsible for a majority of high-energy compounds are _____________.
4. The storage form of high-energy compound in invertebrates is _____________.
5. A more negative redox potential represents a greater tendency to lose _____________.
6. The electron transport chain is located in _____________.
7. The prosthetic group present in cytochromes _____________.
8. The component of electron transport chain which can directly react with molecular oxygen _____________.
9. The site of ETC inhibited by cyanide _____________.
10. Superoxide is converted to H_2O_2 by the enzyme _____________.

IV. Multiple choice questions

11. Name the compound with the greatest standard free energy.
    (a) ATP (b) Phosphocreatine (c) Cyclic AMP (d) Phosphoenolpyruvate.
12. One of the following components of ETC possesses isoprenoid units
    (a) Coenzyme Q (b) Cytochrome (c) Cytochrome b (d) Non-heme iron.
13. The P : O ratio for the oxidation of $FADH_2$ is
    (a) 1 (b) 1.5 (c) 2.5 (d) 4.
14. Inner mitochondrial membrane is impermeable to
    (a) H^+ (b) K^+ (c) OH^- (d) All of them.
15. ATP synthase activity is associated with the mitochondrial enzyme complex
    (a) V (b) III (c) IV (d) I.

METABOLISMS

SECTION THREE

12 Introduction to Metabolism

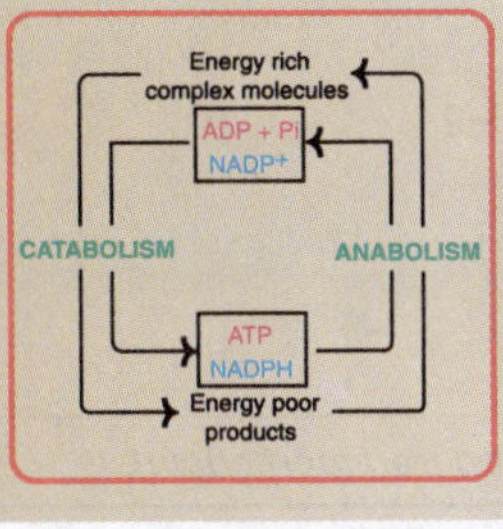

The metabolism introduces itself

"I represent the chemical reactions of life;
Composed of catabolism and anabolism;
Catabolism is degradative to generate energy;
Anabolism is synthetic that consumes energy."

Hundreds of reactions simultaneously take place in a living cell, in a well-organized and integrated manner. The entire spectrum of ***chemical reactions, occurring in the living system,*** are collectively referred to as ***metabolism.*** Knowledge of metabolism is necessary to properly understand the abnormalities associated with diseases.

A ***metabolic pathway*** (or metabolic map) constitutes a series of enzymatic reactions to produce specific products. The term ***metabolite*** is applied to a substrate or an intermediate or a product in the metabolic reactions.

Metabolism is broadly divided into two categories (***Fig.12.1***).

1. **Catabolism** : The ***degradative processes*** concerned with the breakdown of complex molecules to simpler ones, with a concomitant release of energy.

2. **Anabolism** : The ***biosynthetic reactions*** involving the formation of complex molecules from simple precursors.

A clear demarcation between catabolism and anabolism is rather difficult, since there are several intermediates common to both the processes. The term ***amphibolism*** is also in use for reactions which are both catabolic and anabolic in nature.

Amphibolic pathways (or cycles), regarded as ***crossroads*** of ***metabolism***, serve as links between anabolic and catabolic pathways e.g. Krebs cycle.

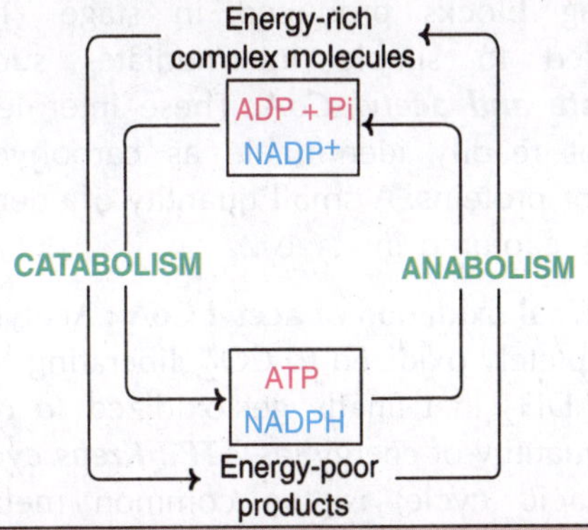

Fig. 12.1 : An outline of catabolism and anabolism.

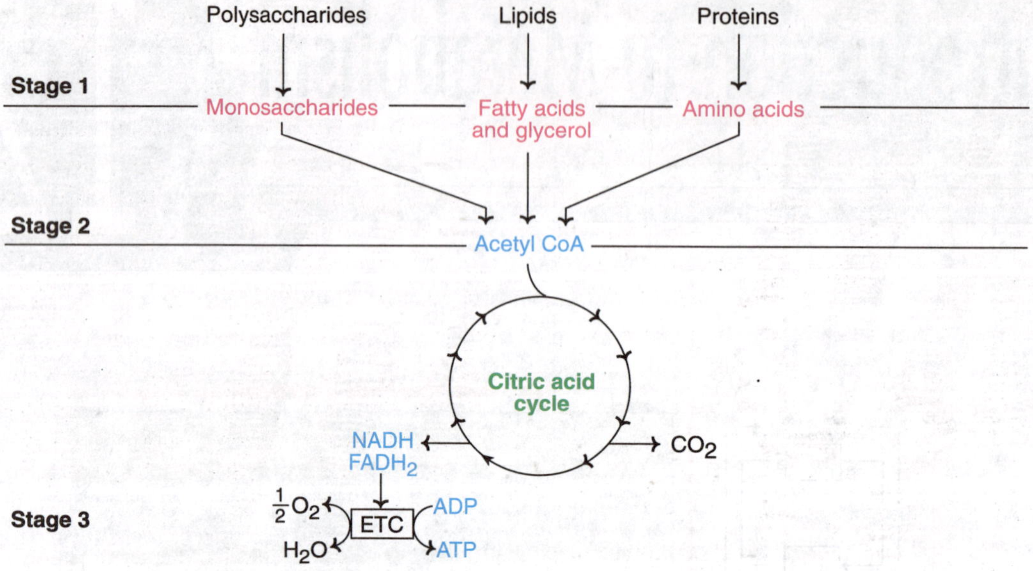

***Fig. 12.2 :** The three stages of catabolism (ETC–Electron transport chain).*

Catabolism

The very purpose of catabolism is to trap the energy of the biomolecules in the form of ATP and to generate the substances (precursors) required for the synthesis of complex molecules. Catabolism occurs in three stages (***Fig.12.2***).

1. **Conversion of complex molecules into their building blocks :** Polysaccharides are broken down to monosaccharides, lipids to free fatty acids and glycerol, proteins to amino acids.

2. **Formation of simple intermediates :** The building blocks produced in stage (1) are degraded to simple intermediates such as ***pyruvate and acetyl CoA***. These intermediates are not readily identifiable as carbohydrates, lipids or proteins. A small quantity of energy (as ATP) is captured in stage 2.

3. **Final oxidation of acetyl CoA :** Acetyl CoA is completely oxidized to CO_2, liberating NADH and $FADH_2$ that finally get oxidized ***to release*** large quantity of energy (as ***ATP***). ***Krebs cycle*** (or citric acid cycle) is the common metabolic pathway involved in the final oxidation of all energy-rich molecules. This pathway accepts the carbon compounds (pyruvate, succinate etc.) derived from carbohydrates, lipids or proteins.

Anabolism

For the synthesis of a large variety of complex molecules, the starting materials are relatively few. These include pyruvate, acetyl CoA and the intermediates of citric acid cycle. Besides the availability of precursors, the anabolic reactions are dependent on the ***supply of energy*** (as ATP or GTP) and reducing equivalents (as NADPH + H^+).

The anabolic and catabolic pathways are not reversible and operate independently. As such, the metabolic pathways occur in specific cellular locations (mitochondria, microsomes etc.) and are controlled by different regulatory signals.

The terms—intermediary metabolism and energy metabolism—are also in use. ***Intermediary metabolism*** refers to the entire range of catabolic and anabolic reactions, not involving nucleic acids. ***Energy metabolism*** deals with the metabolic pathways concerned with the storage and liberation of energy.

Types of metabolic reactions

The biochemical reactions are mainly of four types

1. Oxidation-reduction. 2. Group transfer.
3. Rearrangement and isomerization.
4. Make and break of carbon-carbon bonds.

These reactions are catalysed by specific enzymes—more than 2,000 known so far.

Methods employed to study metabolism

The metabolic reactions do not occur in isolation. They are interdependent and integrated into specific series that constitute ***metabolic pathways.*** It is, therefore, not an easy task to study metabolisms. Fortunately, the ***basic metabolic pathways in most organisms are essentially identical.*** For this reason, many organisms can be used to understand metabolisms.

Several methods are employed to elucidate biochemical reactions and the metabolic pathways. These experimental approaches may be broadly divided into 3 categories

1. Use of whole organisms or its components.
2. Utility of metabolic probes.
3. Application of isotopes.

The actual methods employed may be either ***in vivo*** (in the living system) or ***in vitro*** (in the test tube) or, more frequently, both.

1. **Use of whole organism or its components :**

(a) **Whole organisms :** The ultimate aim of a biochemist is to know the metabolism in the organism as a whole. Glucose tolerance test (GTT), employed to measure the response of man (or other animals) towards carbohydrate metabolism is a good example of the use of whole organism.

(b) **Isolated organs, tissue slices, whole cells**, subcellular organelles, cell-free systems and recently purified components are frequently used to elucidate biochemical reactions and metabolic pathways.

2. **Utility of metabolic probes :** Two types of metabolic probes are commonly used to trace out biochemical pathways. These are metabolic ***inhibitors*** and ***mutations.*** In both the cases, there is a specific blockade in a metabolic reaction which helps to understand the pathway. Inhibitors of electron transport chain have been largely responsible to elucidate the sequence of electron carriers (***Chapter 11***). The ***inborn errors of metabolism*** in higher organisms and the genetic manipulations in the microorganisms have also contributed a lot to the understanding of metabolisms.

3. **Application of isotopes :** Isotopes are the atoms with the same number of protons but different neutrons. By use of isotopes, the molecules of the living system can be labelled without altering their chemical properties. Application of isotopes in biochemistry has revolutionized ***the study of metabolisms.*** More details on the utility of isotopes in biochemistry are given elsewhere (***Chapter 41***).

SUMMARY

- *The wide range of chemical reactions occurring in the living system are collectively known as metabolism. Catabolism is concerned with the degradation of complex molecules to simpler ones coupled with the liberation of energy (ATP). On the other hand, anabolism deals with the synthetic reactions converting simple precursors to complex molecules, coupled with the consumption of energy (ATP). A metabolic pathway constitutes a series of enzymatic reactions to produce specific products.*
- *Several methods are employed to study metabolism. These include the use of the whole organism or its components (organ, tissue, cells, organelles etc.), utility of metabolic probes (inhibitors and mutations) and application of isotopes.*

Metabolism of Carbohydrates 13

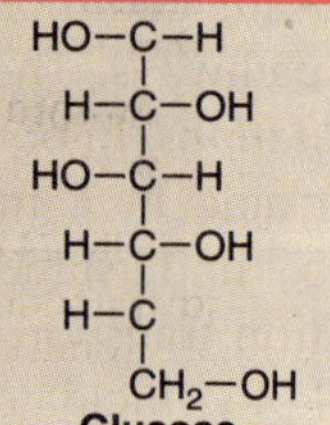

Glucose

The official spokesperson of carbohydrate metabolism, 'glucose', speaks

"I burn myself to provide fuel to life!
Generated through gluconeogenesis by my friends;
Engaged in the synthesis of lipids, amino acids;
Deranged in my duties due to diabetes mellitus."

THE READER MUST REFER CHAPTER 36 FOR BLOOD GLUCOSE REGULATION, DIABETES MELLITUS ETC.

Carbohydrates are the ***major source of energy*** for the living cells. As such, carbohydrates are the first cellular constituents, synthesized by green plants during photosynthesis from carbon dioxide and water, on absorption of light. Thus, ***light is the ultimate source of energy for all biological processes.***

The monosaccharide ***glucose is the central molecule in carbohydrate metabolism*** since all the major pathways of carbohydrate metabolism are connected with it (***Fig.13.1***). Glucose is utilized as a source of energy, it is synthesized from non-carbohydrate precursors and stored as glycogen to release glucose as and when the need arises. The other monosaccharides important in carbohydrate metabolism are fructose, galactose and mannose.

The ***fasting blood glucose*** level in normal individuals is ***70-100 mg/dl*** (4.5-5.5 mmol/l) and it is very efficiently maintained at this level (for details refer ***Chapter 36***). Liver plays a key role in monitoring and stabilizing blood glucose levels. Thus ***liver*** may be appropriately considered as ***glucostat monitor***.

Major pathways of carbohydrate metabolism

The important pathways of carbohydrate metabolism are listed

1. **Glycolysis** (Embden-Meyerhof Parnas pathway) : The oxidation of ***glucose to pyruvate and lactate***.

2. **Citric acid cycle** (Krebs cycle or tricarboxylic acid cycle) : The ***oxidation of acetyl***

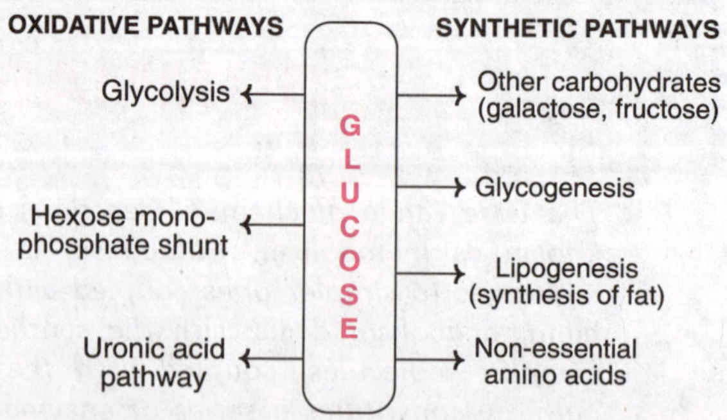

*Fig. 13.1 : Overview of glucose metabolism. (Note : For **majority of the pathways**, glucose participates as glucose 6-phosphate).*

CoA to CO_2. Krebs cycle is the final common oxidative pathway for carbohydrates, fats or amino acids, through acetyl CoA.

3. **Gluconeogenesis** : The ***synthesis of glucose*** from non-carbohydrate precursors (e.g. amino acids, glycerol etc.).

4. **Glycogenesis :** The ***formation of glycogen*** from glucose.

5. **Glycogenolysis** : The ***breakdown of glycogen*** to glucose.

6. **Hexose monophosphate shunt** (pentose phosphate pathway or direct oxidative pathway) **:** This pathway is an alternative to glycolysis and TCA cycle for the ***oxidation of glucose*** (directly to carbon dioxide and water).

7. **Uronic acid pathway** : Glucose is converted to ***glucuronic acid***, pentoses and, in some animals, to ascorbic acid (not in man). This pathway is also an alternative oxidative pathway for glucose.

8. **Galactose metabolism :** The pathways concerned with the conversion of galactose to glucose and the synthesis of lactose.

9. **Fructose metabolism :** The oxidation of fructose to pyruvate and the relation between fructose and glucose metabolism.

10. **Amino sugar and mucopolysaccharide metabolism :** The synthesis of amino sugars and other sugars for the formation of mucopolysaccharides and glycoproteins.

An overview of carbohydrate metabolism showing major pathways and important products is shown in ***Fig. 13.2***.

Entry of glucose into cells

Glucose concentration is very low in the cells compared to plasma (for humans < 100 mg/dl). However, glucose does not enter the cells by simple diffusion. Two specific transport systems are recognized for the entry of glucose into the cells

1. **Insulin-independent transport system of glucose :** This is a carrier mediated uptake of glucose which is not dependent on the hormone insulin. This is operative in ***hepatocytes, erythrocytes and brain.***

2. **Insulin-dependent transport system :** This occurs in ***muscle and adipose tissue.***

Glucose transporters (GLUT) : In recent years, at least six glucose transporters (GLUT-1 to GLUT-5 and GLUT-7) in the cell membranes have been identified. They exhibit tissue specificity. For instance, GLUT-1 is abundant in

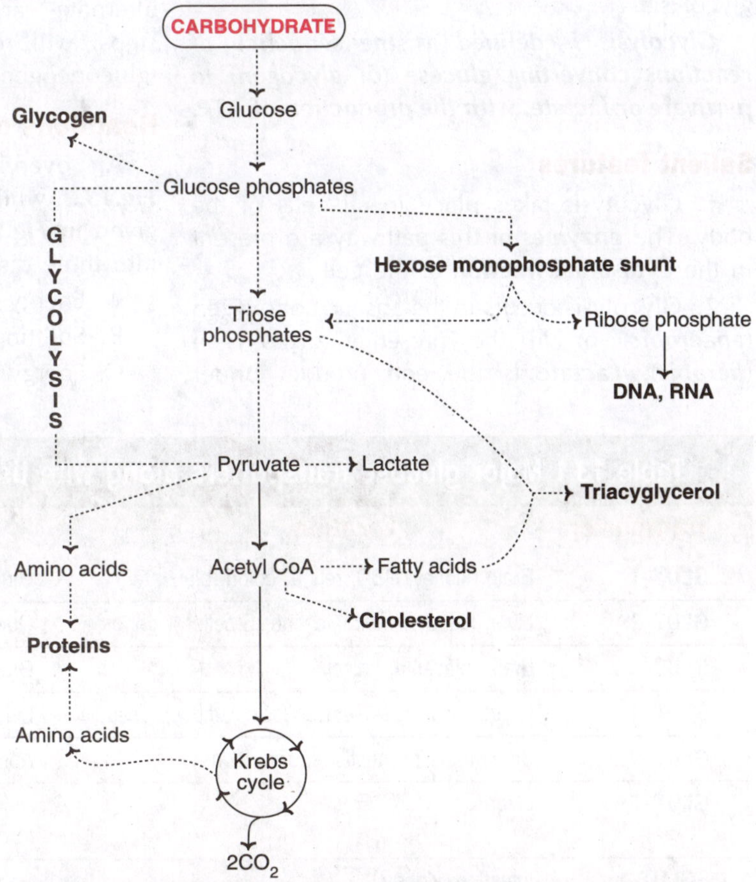

Fig. 13.2 : *An outline of carbohydrate metabolism depicting major pathways and products.*

erythrocytes whereas GLUT-4 is abundant in skeletal muscle and adipose tissue (***Table 13.1***).

Insulin increases the number and promotes the activity of ***GLUT-4*** in skeletal muscle and adipose tissue. In ***type 2 diabetes*** mellitus, ***insulin resistance*** is observed in these tissues. This is due to the reduction in the quantity of GLUT-4 in insulin deficiency.

GLYCOLYSIS

Glycolysis is derived from the *Greek* words (glycose—sweet or sugar; lysis—dissolution). It is a universal pathway in the living cells. The complete pathway of glycolysis was elucidated in 1940. This pathway is often referred to as ***Embden-Meyerhof-Parnas pathway (EMP pathway)*** in honour of the biochemists who made a major contribution to the knowledge of glycolysis.

Glycolysis is defined as the sequence of reactions converting glucose (or glycogen) to pyruvate or lactate, with the production of ATP.

Salient features

1. Glycolysis takes place in all cells of the body. The ***enzymes*** of this pathway are present in the ***cytosomal fraction*** of the cell.

2. Glycolysis occurs in the absence of oxygen (***anaerobic***) or in the presence of oxygen (***aerobic***). ***Lactate*** is the end product under anaerobic condition. In the aerobic condition, ***pyruvate*** is formed, which is then oxidized to CO_2 and H_2O.

3. Glycolysis is a major pathway for ***ATP synthesis in*** tissues lacking mitochondria, e.g. ***erythrocytes, cornea, lens*** etc.

4. Glycolysis is very ***essential for brain*** which is dependent on glucose for energy. The glucose in brain has to undergo glycolysis before it is oxidized to CO_2 and H_2O.

5. Glycolysis (anaerobic) may be summarized by the net reaction

Glucose + 2ADP + 2Pi $\longrightarrow$ 2Lactate + 2ATP

6. Glycolysis is a central metabolic pathway with many of its intermediates providing branch points to other pathways. Thus, the intermediates of glycolysis are useful for the synthesis of amino acids and fat.

7. Reversal of glycolysis along with the alternate arrangements at the irreversible steps, will result in the synthesis of glucose (gluconeogenesis).

Reactions of glycolysis

An overview of glycolysis is depicted in ***Fig.13.3***, while detailed sequence of reactions is given in ***Fig.13.3***. The pathway can be divided into three distinct phases

A. Energy investment phase or priming stage

B. Splitting phase

C. Energy generation phase.

Table 13.1 Major glucose transporters along with tissue location and functions

Transporter	Location	Major function(s)
GLUT–1	Brain, kidney, RBC, retina, colon, placenta	Glucose uptake
GLUT–2	Liver, kidney, small intestine, β-cells of pancreas	Quick uptake and release of glucose
GLUT–3	Brain, placenta, kidney	Glucose uptake
GLUT–4	Heart, muscle, skeletal muscle, adipose tissue	Insulin mediated glucose uptake
GLUT–5	Small intestine, testis, sperm, kidney	Absorption of fructose
GLUT–7	Liver	Release of glucose from endoplasmic reticulum to cytosol
SGLUT–1	Intestine, kidney	Sodium dependent active uptake of glucose against concentration gradient (cotransport of Na and glucose)

The sequence of reactions are discussed below.

A. **Energy investment phase (preparatory phase)**

1. Glucose is phosphorylated to glucose 6-phosphate by ***hexokinase*** or ***glucokinase*** (both are ***isoenzymes***). This is an irreversible reaction, dependent on ATP and Mg^{2+}. The enzyme ***hexokinase*** is present in almost ***all the tissues***. It catalyses the phosphorylation of various ***hexoses*** (fructose, mannose etc.), has ***low*** K_m for substrates (about 0.1 mM) and is ***inhibited by glucose 6-phosphate***.

   Glucokinase present ***in liver***, catalyses the phosphorylation of ***only glucose***, ***has high*** K_m ***for glucose*** (10 mM) and is ***not inhibited by glucose 6-phosphate***.

   Due to high affinity (low K_m), glucose is utilized by hexokinase even at low concentration, whereas glucokinase acts only at higher levels of glucose i.e., after a meal when blood glucose concentration is above 100 mg/dl.

   Glucose 6-phosphate is impermeable to the cell membrane. It is a ***central molecule with a variety of metabolic fates***—glycolysis, glycogenesis, gluconeogenesis and pentose phosphate pathway.

2. Glucose 6-phosphate undergoes isomerization to give fructose 6-phosphate in the presence of the enzyme phosphohexose isomerase and Mg^{2+}.

3. Fructose 6-phosphate is phosphorylated to fructose 1,6-bisphosphate by phosphofructokinase (PFK). This is an ***irreversible committed*** and a ***regulatory step*** in glycolysis.

B. **Splitting phase**

4. The six carbon fructose 1,6-bisphosphate is split (hence the name glycolysis) to two three-carbon compounds, glyceraldehyde 3-phosphate and dihydroxyacetone phosphate by the enzyme aldolase (fructose 1,6-bisphosphate aldolase).

5. The enzyme phosphotriose isomerase catalyses the reversible interconversion of glyceraldehyde 3-phosphate and dihydroxyacetone phosphate. Thus, two molecules of glyceraldehyde 3-phosphate are obtained from one molecule of glucose.

C. **Energy generation phase (pay off phase)**

6. Glyceraldehyde 3-phosphate dehydrogenase converts glyceraldehyde 3-phosphate to 1,3-bisphosphoglycerate. This step is important as it is involved in the formation of NADH + H^+ and a high energy compound 1,3-bisphosphoglycerate. ***Iodoacetate*** and ***arsenate inhibit*** the enzyme glyceraldehyde 3-phosphate dehydrogenase. In aerobic condition, NADH passes through the electron transport chain and 5 ATP (2 × 2.5 ATP) are synthesized by oxidative phosphorylation.

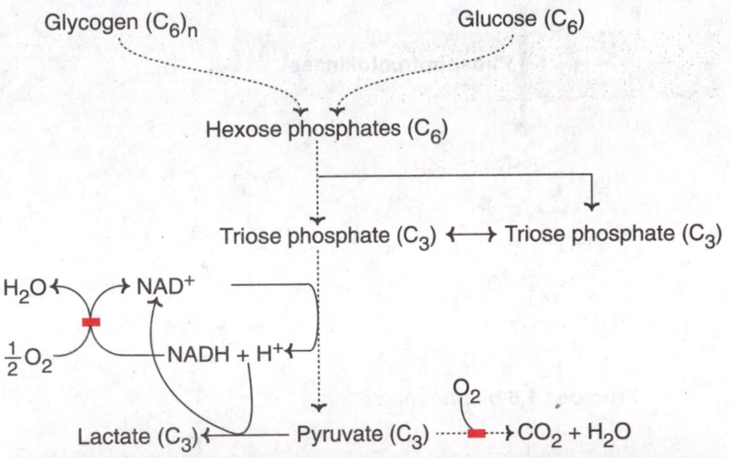

Fig. 13.3 : *An overview of glycolysis (▬ Reactions blocked under anaerobic conditions; In RBC, these two reactions do not occur due to lack of mitochondria)*

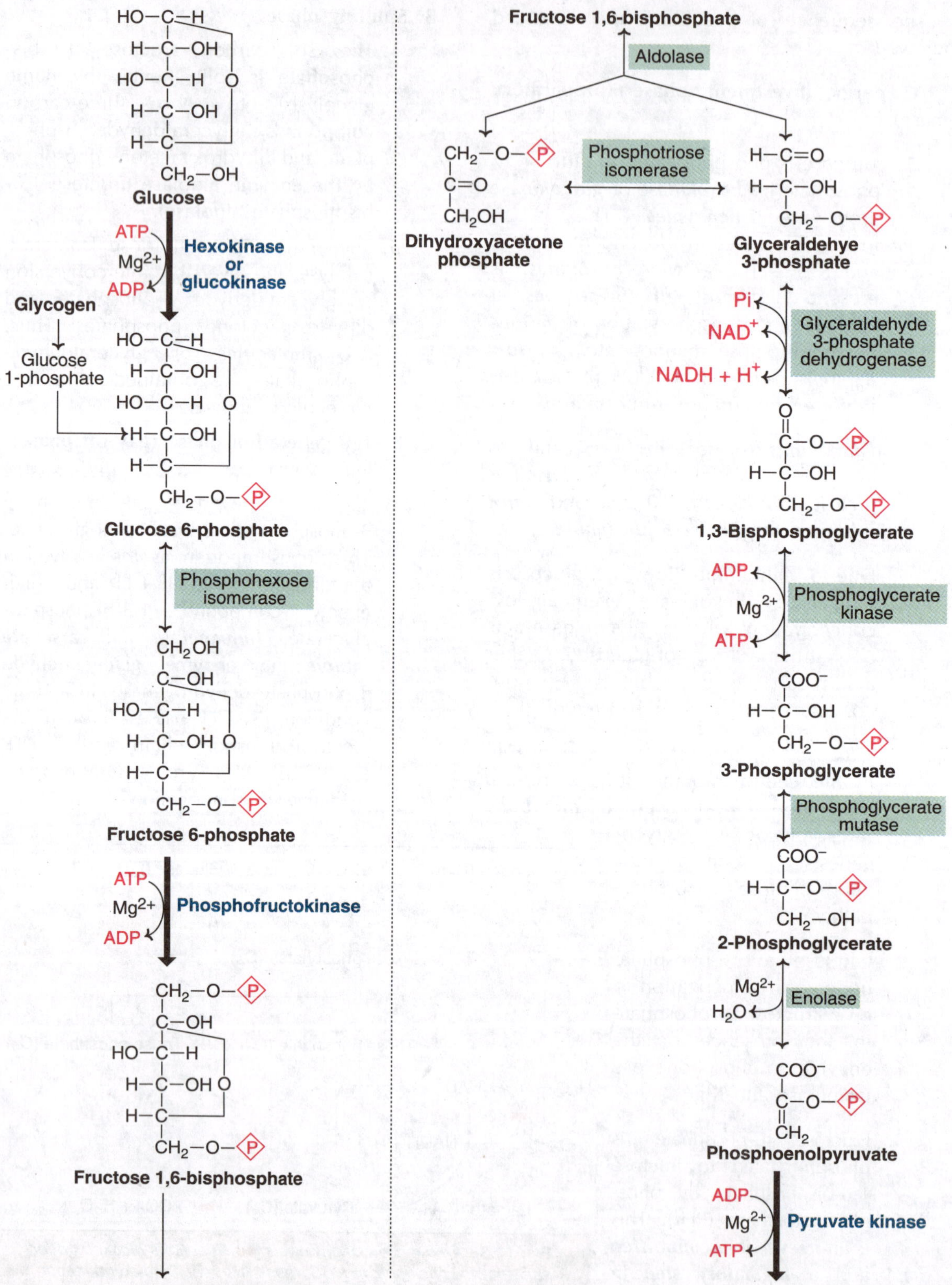

Fig 13.4 contd. next column

Fig 13.4 contd. next page

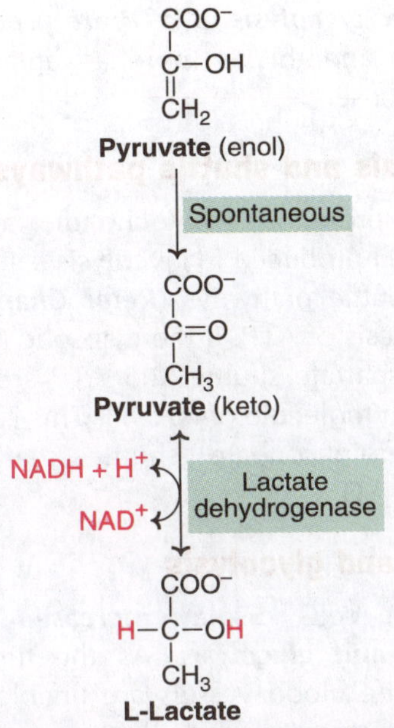

***Fig. 13.4 :** The reactions in the pathway of glycolysis (The three steps catalysed by hexokinase, phosphofructokinase and pyruvate kinase, shown in thick lines are irreversible).*

7. The enzyme phosphoglycerate kinase acts on 1,3-bisphosphoglycerate resulting in the synthesis of ATP and formation of 3-phosphoglycerate. This step is a good example of ***substrate level phosphorylation***, since ATP is synthesized from the substrate without the involvement of electron transport chain. ***Phosphoglycerate kinase reaction is reversible, a rare example among the kinase reactions.***

8. 3-Phosphoglycerate is converted to 2-phosphoglycerate by phosphoglycerate mutase. This is an isomerization reaction.

9. The high energy compound phosphoenol pyruvate is generated from 2-phosphoglycerate by the enzyme ***enolase***. This enzyme requires Mg^{2+} or Mn^{2+} and is ***inhibited*** by ***fluoride.*** For blood ***glucose estimation*** in the laboratory, fluoride is added to the blood to ***prevent glycolysis by the cells***, so that blood glucose is correctly estimated. (Fluoride combines with Mg^{2+} and phosphate to form a complex that binds with active site of enolase and blocks access of substrate. Thus, fluoride is an ***unusual competitive inhibitor***).

10. The enzyme pyruvate kinase catalyses the transfer of high energy phosphate from phosphoenol pyruvate to ADP, leading to the formation of ATP. This step also is a ***substrate level phosphorylation.*** This reaction is irreversible.

Conversion of pyruvate to lactate—significance

Under anaerobic conditions (lack of O_2), ***pyruvate is reduced by NADH to lactate*** in presence of the enzyme ***lactate dehydrogenase*** (competitive inhibitor—oxamate). The NADH utilized in this step is obtained from the reaction catalysed by glyceraldehyde 3-phosphate dehydrogenase. The ***formation of lactate allows the regeneration of NAD^+*** which can be reused by glyceraldehyde 3-phosphate dehydrogenase so that glycolysis proceeds even in the absence of oxygen to supply ATP.

The occurrence of uninterrupted glycolysis is very essential in skeletal muscle during strenous exercise where oxygen supply is very limited. ***Glycolysis in the erythrocytes leads to lactate production***, since mitochondria—the centres for aerobic oxidation—are absent. ***Brain***, ***retina***, skin, renal medulla and gastrointestinal tract ***derive most of their energy from glycolysis***.

Lactic acidosis

Lactic acid is a three carbon hydroxy acid. Elevation of lactic acid in the circulation (normal plasma 4–15 mg/dl) may occur due to its increased production or decreased utilization. ***Mild forms*** of lactic acidosis (not life-threatening) are associated with ***strenuous exercise***, ***shock***, respiratory diseases, cancers, thiamine deficiency, alcoholism, von Gierke's disease etc.

Severe forms of lactic acidosis are observed due to impairment/collapse of circulatory system which is often encountered in ***myocardial infarction***, ***pulmonary embolism***, ***uncontrolled hemorrhage and severe shock***. This type of lactic acidosis is due to inadequate supply of O_2 to the tissues with a drastic reduction in ATP synthesis (since the cells have to survive in anaerobic conditions) which may even lead to death. The term ***oxygen debt*** refers to the excess amount of O_2 required to recover. In clinical practice, measurement of plasma lactic acid is useful to know about the oxygen debt, and monitor the patient's recovery, and save the patient from morbidity and mortality.

Production of ATP in glycolysis

The details of ATP generation in glycolysis (from glucose) are given in ***Table 13.2***. Under ***anaerobic conditions, 2 ATP*** are synthesized while, under ***aerobic conditions, 7 ATP*** are synthesized, if malate the shuttle pathway operates.

When the glycolysis occurs from glycogen, one more ATP is generated. This is because no ATP is consumed for the activation of glucose (glycogen directly produces glucose 1-phosphate which forms glucose 6-phosphate). Thus, in ***anaerobic glycolysis, 3 ATP are produced from glycogen*** and this is more advantageous than from glucose.

Glycolysis and shuttle pathways

In the presence of mitochondria and oxygen, the NADH produced in glycolysis can participate in the shuttle pathways (***Refer Chapter 11***) for the synthesis of ATP. If the cytosolic NADH uses malate-aspartate shuttle, 2.5 ATP are generated from each molecule of NADH. This is in contrast to glycerolphosphate shuttle that produces only 1.5 ATP.

Cancer and glycolysis

Cancer cells display ***increased uptake of glucose, and glycolysis***. As the tumors grow rapidly, the blood vessels are unable to supply adequate oxygen, and thus a condition of hypoxia exists. Due to this, anaerobic glycolysis predominantly occurs to supply energy. The cancer cells get adapted to hypoxic glycolysis through the involvement of a transcription factor

Table 13.2 Generation of ATP in glucose metabolism

Pathway	Enzyme (method of ATP synthesis)	Number of ATP synthesized
Glycolysis	Glyceraldehyde 3-phosphate dehydrogenase (2 NADH, ETC, oxidative phosphorylation)	5
	Phosphoglycerate kinase (substrate level phosphorylation)	2
	Pyruvate kinase (substrate level phosphorylation)	2
	Two ATP are consumed in the reactions catalysed by hexokinase and phosphofructokinase	–2
	Net ATP synthesis in glycolysis in aerobic condition	***7***
	Pyruvate dehydrogenase (2 NADH, ETC, oxidative phosphorylation)	5
Citric acid cycle	Isocitrate dehydrogenase (2 NADH, ETC, oxidative phosphorylation)	5
	α-Ketoglutarate dehydrogenase (2 NADH, ETC, oxidative phosphorylation)	5
	Succinate thiokinase (substrate level phosphorylation)	2
	Succinate dehydrogenase (2 $FADH_2$, ETC, oxidative phosphorylation)	3
	Malate dehydrogenase (2 NADH, ETC, oxidative phosphorylation)	5
	Total ATP per mole of glucose under aerobic condition	***32***
	Total ATP per mole of glucose under anaerobic condition	2

Note : *Values represent ATP synthesized as per the new P:O ratios of 2.5 for NADH and 1.5 for* $FADH_2$.

Medical Concepts/ Clinical Correlates

- Glycolysis is an important source of energy supply to brain, retina, skin and renal medulla.
- The crucial significance of glycolysis is its ability to generate ATP in the absence of oxygen.
- Skeletal muscle, during strenuous exercise, requires the occurrence of uninterrupted glycolysis. This is due to the limited supply of oxygen.
- The cardiac muscle cannot survive for long in the absence of oxygen since it is not well adapted for glycolysis under anaerobic conditions.
- The occurrence of glycolysis is very much elevated in rapidly growing cancer cells.
- Glycolysis in erythrocytes is associated with 2,3-bisphosphoglycerate (2,3-BPG) production. In the presence of 2,3-BPG, oxyhemoglobin unloads more oxygen to the tissues.

namely ***hypoxia-inducible transcription factor*** (***HIF***). HIF increases the synthesis of glycolytic enzymes and the glucose transporters. However, the cancer cells cannot grow and survive without proper vascularization. One of the modalities of ***cancer treatment is to use drugs that can inhibit vascularization of tumors***.

Irreversible steps in glycolysis

Most of the reactions of glycolysis are reversible. However, the three steps catalysed by the enzymes ***hexokinase*** (or glucokinase), ***phosphofructokinase*** and ***pyruvate kinase,*** are irreversible. These three stages mainly regulate glycolysis. The ***reversal of glycolysis***, with ***alternate arrangements made at the three irreversible stages***, leads to the synthesis of glucose from pyruvate (***gluconeogenesis***).

Regulation of glycolysis

The three enzymes namely hexokinase (and glucokinase), phosphofructokinase and pyruvate kinase, catalysing the ***irreversible reactions regulate glycolysis***.

Hexokinase is inhibited by glucose 6-phosphate. This enzyme prevents the accumulation of glucose 6-phosphate due to product inhibition. Glucokinase, which specifically phosphorylates glucose, is an ***inducible enzyme***. The substrate glucose, probably through the involvement of insulin, induces glucokinase.

Phosphofructokinase (PFK) is the most important regulatory enzyme in glycolysis. This enzyme catalyses the ***rate limiting committed step.*** PFK is an allosteric enzyme regulated by allosteric effectors. ATP, citrate and H^+ ions (low pH) are the most important allosteric inhibitors, whereas, fructose 2,6-bisphosphate, ADP, AMP and Pi are the allosteric activators.

Role of fructose 2,6-bisphosphate in glycolysis

Fructose 2,6-bisphosphate (F2,6-BP) is considered to be the most important ***regulatory factor*** (activator) for controlling PFK and, ultimately, ***glycolysis*** in the liver. F2,6-BP is synthesized from fructose 6-phosphate by the enzyme phosphofructokinase called PFK-2 (PFK-1 is the glycolytic enzyme). F2,6-BP is hydrolysed by fructose 2,6-bisphosphatase. The function of synthesis and degradation of F2,6-BP is brought

Table 13.3 Regulation of glycolysis

Enzyme	Activation	Inhibition
Hexokinase	—	Glucose 6-phosphate
Glucokinase	Insulin	Glucagon
Phosphofructokinase	Insulin, fructose 2,6-bisphosphate, ADP, AMP, Pi	Glucagon, ATP, citrate
Pyruvate kinase	Insulin, fructose 1,6-bisphosphate	Glucagon, ATP, cyclic AMP

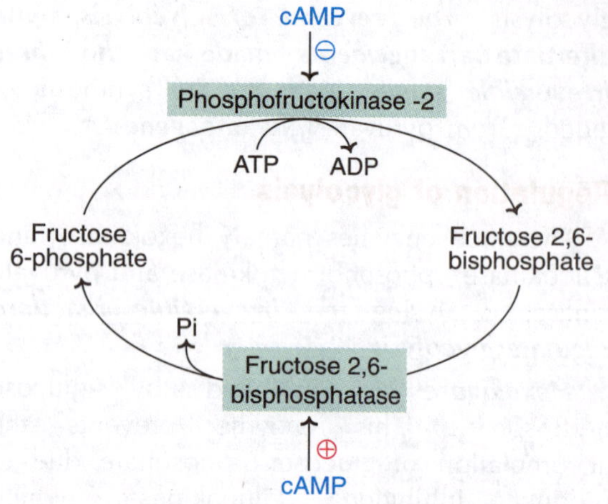

Fig. 13.5 : *Regulation of fructose 2,6-bisphosphatase.*

out by a ***single enzyme*** (same polypeptide with two active sites) which is referred to as ***bifunctional enzyme (Fig.13.5)***. In fact, the combined name of phosphofructokinase-2/fructose 2,6-bisphosphatase is used to refer to the enzyme that synthesizes and degrades F2,6-BP.

The activity of PFK-2 and fructose 2,6-bisphosphatase is controlled by covalent modification which, in turn, is regulated by cyclic AMP (cAMP is the second messenger for certain hormones). Cyclic AMP brings about dephosphorylation of the bifunctional enzyme, resulting in inactivation of active site responsible for the synthesis of F2,6-BP but activation of the active site responsible for the hydrolysis of F2,6-BP.

Pyruvate kinase also regulates glycolysis. This enzyme is inhibited by ATP and activated by F1,6-BP. Pyruvate kinase is active (a) in dephosphorylated state and inactive (b) in phosphorylated state. Inactivation of pyruvate kinase by phosphorylation is brought about by cAMP-dependent protein kinase. The hormone—***glucagon inhibits*** hepatic glycolysis by this mechanism (***Fig.13.6***).

Pasteur effect

The ***inhibition of glycolysis by oxygen*** (aerobic condition) is known as Pasteur effect. It was discovered by Louis Pasteur, more than a century ago, while studying fermentation by yeast. He observed that when anaerobic yeast cultures were exposed to air, the utiliziation of glucose decreased by nearly seven fold.

In the aerobic condition, the levels of glycolytic intermediates from fructose 1,6-bisphosphate onwards decrease while the earlier intermediates accumulate. This clearly indicates that Pasteur effect is due to the inhibition of the enzyme phosphofructokinase. The inhibitory effect of citrate and ATP (produced in the presence of oxygen) on phosphofructokinase explains the Pasteur effect.

Crabtree effect

The phenomenon of inhibition of oxygen consumption by the addition of glucose to tissues having high aerobic glycolysis is known as Crabtree effect. Basically, this is ***opposite*** to that of ***Pasteur effect.*** Crabtree effect is due to increased competition of glycolysis for inorganic phosphate (Pi) and NAD^+ which limits their availability for phosphorylation and oxidation.

Glycolysis and dental caries

Dental caries refers to the ***destruction or decalcification of hard teeth due to organic acids*** released by bacterial infections. The anaerobic bacteria (e.g. ***Streptococcus mutans, Lactobacillus*** sps) that colonize the oral cavity contribute to the development of dental caries. These bacteria grow optimally on refined and fermentable sugars (e.g. sucrose of chocolates,

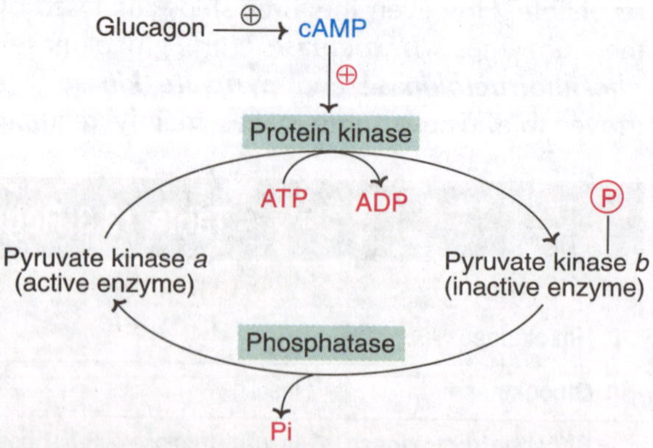

Fig. 13.6 : *Regulation of pyruvate kinase.*

candies) by utilizing anaerobic glycolysis. ***Lactic acid and other acids*** (produced by bacteria) ***erode tooth enamel and dentin***, and dissolve hydroxyapatite matrix of teeth that ***results in cavity formation***. Low levels of ***fluoride***, from tooth pastes or when applied topically can inhibit the enzyme enolase and ***reduce glycolysis and thus tooth decay***. Further, fluoride integrates into tooth surface to form fluoroapatite which offers resistance to demineralization.

RAPOPORT-LEUBERING CYCLE

This is a ***supplementary pathway to glycolysis*** which is ***operative in the mature erythrocytes*** of man and other mammals. Rapoport-Leubering cycle is mainly concerned with the synthesis of ***2,3-bisphosphoglycerate (2,3-BPG)*** in the RBC. 1,3-Bisphosphoglycerate (1,3-BPG) produced in glycolysis is converted to 2,3-BPG by the enzyme 2,3-bisphosphoglycerate mutase (***Fig.13.7***). 2,3-BPG is hydrolysed to 3-phosphoglycerate by bisphosphoglycerate phosphatase. It is now believed that bisphosphoglycerate mutase is a ***bifunctional enzyme*** with mutase and phosphatase activities catalysed by two different sites present on the same enzyme.

About 15-25% of the glucose that gets converted to lactate in erythrocytes goes via 2,3-BPG synthesis.

Significance of 2,3-BPG

1. Production of 2,3-BPG allows the glycolysis to proceed without the synthesis of ATP. Rapoport-Leubering cycle, therefore is a ***shunt pathway of glycolysis*** to dissipate or waste the energy not needed by erythrocytes.

2. 2,3-BPG, is not a waste molecule in RBC. It combines with hemoglobin (Hb) and reduces Hb affinity with oxygen. Therefore, in the presence of 2,3-BPG, ***oxyhemoglobin unloads more oxygen to the tissues.***

Increase in erythrocyte 2,3-BPG is observed in hypoxic condition, high altitude, fetal tissues, anemic conditions etc. In all these cases, 2,3-BPG will enhance supply of oxygen to tissues.

3. Glycolysis in the erythrocytes is linked with 2,3-BPG production and oxygen transport. In the deficiency of the enzyme hexokinase, glucose is not phosphorylated, hence the synthesis and concentration of 2,3-BPG are low in RBC. The hemoglobin exhibits high oxygen affinity in hexokinase-defective patients. On the other hand, in the patients with pyruvate kinase deficiency, the level of 2,3-BPG in erythrocytes is high, resulting in low oxygen affinity. (For a more detailed discussion on the functions of 2,3-BPG, refer ***Chapter 10***).

CONVERSION OF PYRUVATE TO ACETYL CoA

Pyruvate, occupying a central place in metabolic pathways, is converted to acetyl CoA

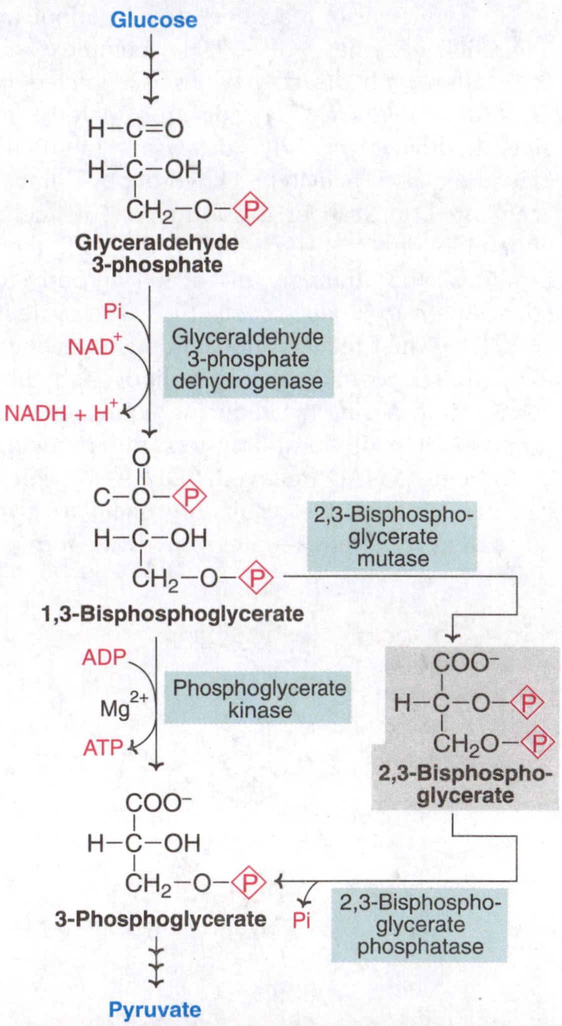

***Fig. 13.7 :** Rapoport-Leubering cycle for the synthesis of 2,3-bisphosphoglycerate (2,3-BPG).*

by ***oxidative decarboxylation***. This is an irreversible reaction, catalysed by a multienzyme complex, known as ***pyruvate dehydrogenase complex*** (***PDH***), which is found only in the mitochondria. High activities of PDH are found in cardiac muscle and kidney. The enzyme PDH requires five cofactors (coenzymes), namely—***TPP, lipoamide, FAD, coenzyme A*** and ***NAD***$^+$ (lipoamide contains lipoic acid linked to ε-amino group of lysine). The overall reaction of PDH is

$$\text{Pyruvate} + NAD^+ + \text{CoA} \xrightarrow{\text{PDH}} \text{Acetyl CoA} + CO_2 + NADH + H^+$$

Reactions of PDH complex

The sequence of reactions brought about by different enzymes of PDH complex in association with the coenzymes is depicted in ***Fig.13.8***. Pyruvate is decarboxylated to give hydroxyethyl TPP, catalysed by PDH (decarboxylase activity). Dihydrolipoyl transacetylase brings about the formation of acetyl lipoamide (from hydroxethyl-TPP) and then catalyses the transfer of acetyl group to coenzyme A to produce acetyl CoA. The cycle is complete when reduced lipoamide is converted to oxidized lipoamide by dihydrolipoyl dehydrogenase, transferring the reducing equivalents to FAD. $FADH_2$, in turn, transfers the reducing equivalents to NAD^+ to give $NADH + H^+$, which can pass through the respiratory chain to give 2.5 ATP (5 ATP from 2 moles of pyruvate formed from glucose) by oxidative phosphorylation.

The intermediates of PDH catalysed reaction are not free but bound with enzyme complex. ***In mammals, the PDH complex has an approximate molecular weight of 9 × 10^6. It contains 60 molecules of dihydrolipoyl-transacetylase and about 20–30 molecules each of the other two enzymes (pyruvate dehydrogenase and dihydrolipoyl dehydrogenase).***

A comparable enzyme with PDH is ***α-ketoglutarate dehydrogenase complex*** of citric acid cycle which catalyses oxidative decarboxylation of α-ketoglutarate to succinyl CoA.

Arsenic poisoning : The enzymes PDH and α-ketoglutarate dehydrogenase are ***inhibited by arsenite***. Arsenite ***binds*** to thiol (–SH) groups of ***lipoic acid*** and makes it unavailable to serve as a cofactor.

Regulation of PDH

Pyruvate dehydrogenase is a good example for ***end product*** (acetyl CoA, NADH) ***inhibition***. Besides this, PDH is also regulated by phosphorylation and dephosphorylation (***Fig.13.9***) PDH is active as a dephosphoenzyme while it is inactive as a phosphoenzyme. PDH phosphatase activity is promoted by Ca^{2+}, Mg^+ and insulin (in adipose tissue). It is of interest to note that calcium released during muscle contraction stimulates PDH (by increasing phosphatase activity) for energy production.

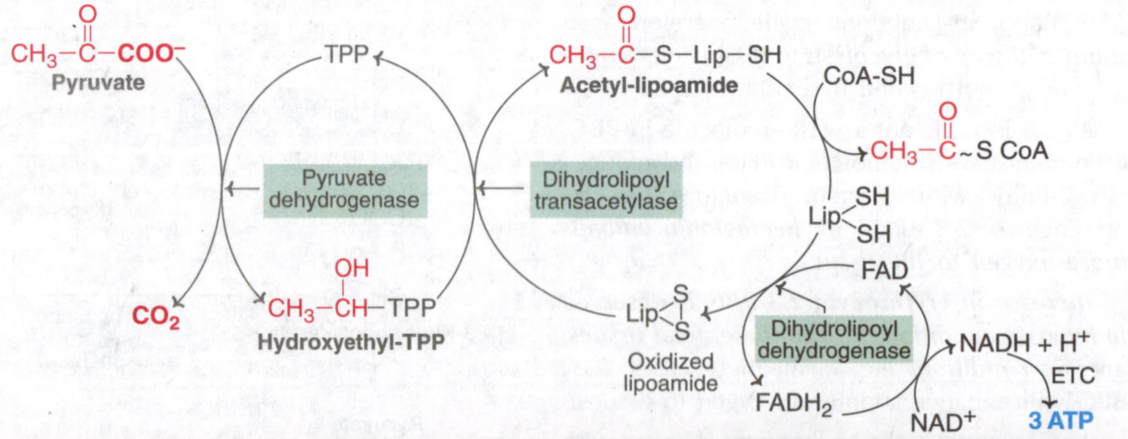

Fig. 13.8 : The mechanism of action of pyruvate dehydrogenase complex. (**Note :** *The reaction involving the conversion of pyruvate to acetyl CoA requires five coenzymes—TPP, lipoamide, CoASH, FAD and NAD*$^+$).

PDH kinase (responsible to form inactive PDH) is promoted by ATP, NADH and acetyl CoA, while it is inhibited by NAD^+, CoA and pyruvate. The net result is that in the presence of high energy signals (ATP, NADH), the PDH is turned off.

During fasting, fatty acid oxidation is enhanced, resulting in acetyl CoA and ATP that inhibit PDH activity and also glycolysis. Thus, ***carbohydrate metabolism is spared in a fasting state***.

Biochemical importance of PDH

1. ***Lack of TPP*** (due to deficiency of thiamine) inhibits PDH activity resulting in the ***accumulation of pyruvate***.

2. In the ***thiamine deficient alcoholics***, pyruvate is rapidly converted to lactate, resulting in ***lactic acidosis***.

3. In patients with inherited deficiency of PDH, lactic acidosis is observed.

4. PDH activity is inhibited by arsenic and mercuric ions. This is brought about by binding of these ions with –SH groups of lipoic acid.

Metabolic importance of pyruvate

Pyruvate is an important metabolite, as it occupies a ***junction point between various metabolic pathways***. Besides its conversion to acetyl CoA (utilized in a wide range of metabolic reactions-citric acid cycle, fatty acid synthesis etc.), ***pyruvate is a good substrate for gluconeogenesis***.

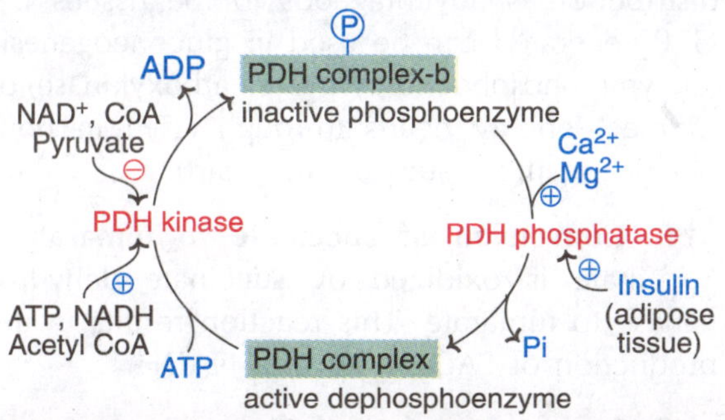

Fig. 13.9 : Regulation of pyruvate dehydrogenase (PDH) complex.

CITRIC ACID CYCLE

The citric acid cycle (***Krebs cycle*** or tricarboxylic acid—***TCA cycle***) is the most important metabolic pathway for the energy supply to the body. About 65-70% of the ATP is synthesized in Krebs cycle. ***Citric acid cycle essentially involves the oxidation of acetyl CoA to CO_2 and H_2O.*** This cycle utilizes about two-thirds of total oxygen consumed by the body. The name TCA cycle is used, since, at the outset of the cycle, tricarboxylic acids (citrate, cis-aconitate and isocitrate) participate.

TCA cycle—the central metabolic pathway

The citric acid cycle is the ***final common oxidative pathway for carbohydrates, fats and amino acids***. This cycle not only supplies energy but also provides many intermediates required for the synthesis of amino acids, glucose, heme etc. Krebs cycle is the most important central pathway ***connecting almost all the individual metabolic pathways*** (either directly or indirectly).

Brief history

The citric acid cycle was ***proposed by Hans Adolf Krebs in 1937***, based on the studies of oxygen consumption in pigeon breast muscle. The cycle is named in his honour (Nobel Prize for Physiology and Medicine in 1953.)

[**Note :** It is of interest to note that the original manuscript on TCA cycle submitted by Krebs to the journal 'Nature' was not accepted. He published it in another journal Enzymoligia. Krebs used to carry the rejection letter (of Nature) with him, and advise the researches never to be discouraged by research paper rejection].

Location of TCA cycle

The enzymes of TCA cycle are located in ***mitochondrial matrix***, in close proximity to the electron transport chain. This enables the synthesis of ATP by oxidative phosphorylation without any hindrance.

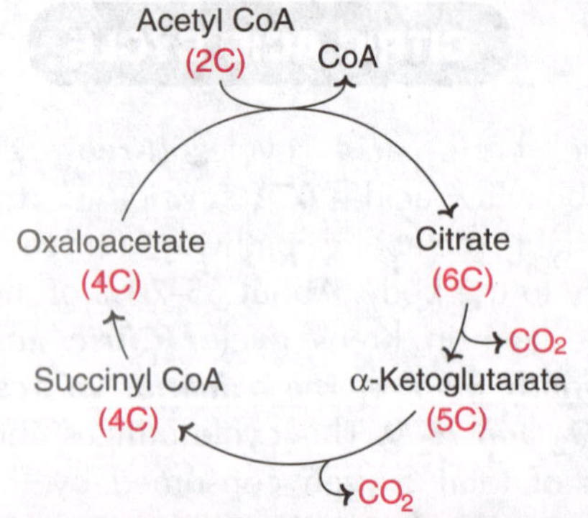

Fig. 13.10 : An overview of Krebs cycle.

TCA cycle—an overview

Krebs cycle basically involves the combination of a two carbon acetyl CoA with a four carbon oxaloacetate to produce a six carbon tricarboxylic acid, citrate. In the reactions that follow, the two carbons are oxidized to CO_2 and oxaloacetate is regenerated and recycled. ***Oxaloacetate is considered to play a catalytic role in citric acid cycle.*** An overview of Krebs cycle is depicted in ***Fig.13.10***.

TCA cycle—an open cycle

Krebs cycle is a cyclic process. However, it should not be viewed as a closed circle, since ***many compounds enter the cycle and leave***. TCA cycle is comparable to a heavy traffic circle in a national highway with many connecting roads. Each intermediate of the cycle connecting another pathway is a road!

Reactions of citric acid cycle

Oxidative decarboxylation of pyruvate to acetyl CoA by pyruvate dehydrogenase complex is discussed above. This step is a connecting link between glycolysis and TCA cycle. A few authors, however, describe the conversion of pyruvate to acetyl CoA along with citric acid cycle. The events of TCA cycle are described hereunder (***Fig.13.11***).

1. **Formation of citrate :** Krebs cycle proper starts with the condensation of acetyl CoA and oxaloacetate, catalysed by the enzyme citrate synthase.

2. and 3. **Citrate is isomerized to isocitrate** by the enzyme aconitase. This is achieved in a two stage reaction of dehydration followed by hydration through the formation of an intermediate—*cis*-aconitate.

4. and 5. **Formation of α-ketoglutarate :** The enzyme isocitrate dehydrogenase (ICD) catalyses the conversion (oxidative decarboxylation) of isocitrate to oxalosuccinate and then to α-ketoglutarate. The formation of NADH and the liberation of CO_2 occur at this stage.

6. **Conversion of α-ketoglutarate to succinyl CoA** occurs through ***oxidative decarboxylation***, catalysed by α-ketoglutarate dehydrogenase complex. This enzyme is dependent on five cofactors—***TPP, lipoamide, NAD⁺, FAD and CoA***. The mechanism of the reaction is analogous to the conversion of pyruvate to acetyl CoA (***See Fig.13.8***).

7. **Formation of succinate :** Succinyl CoA is converted to succinate by succinate thiokinase. This reaction is coupled with the phosphorylation of GDP to GTP or ADP to ATP. This is a ***substrate level phosphorylation***.

Two ***isoenzymes*** of ***succinate thiokinase*** has been identified. One ***dependent on GDP*** is present in tissues like liver and kidney where gluconeogenesis occurs. The second succinate thiokinase isoenzyme is ***dependent on ADP*** and present in non-gluconeogenic tissues (***brain, heart***).

This type of succinate thiokinase isoenzyme distribution is advantageous to the tissues i.e. GTP generated can be used in gluconeogenesis (enzyme-phosphoenolpyruvate carboxykinase) or to meet energy needs (through ATP) in non-gluconeogenic tissues (brain, heart).

8. **Conversion of succinate to fumarate :** Succinate is oxidized by succinate dehydrogenase to fumarate. This reaction results in the production of $FADH_2$ and not NADH.

9. **Formation of malate :** The enzyme fumarase catalyses the conversion of fumarate to malate with the addition of H_2O.

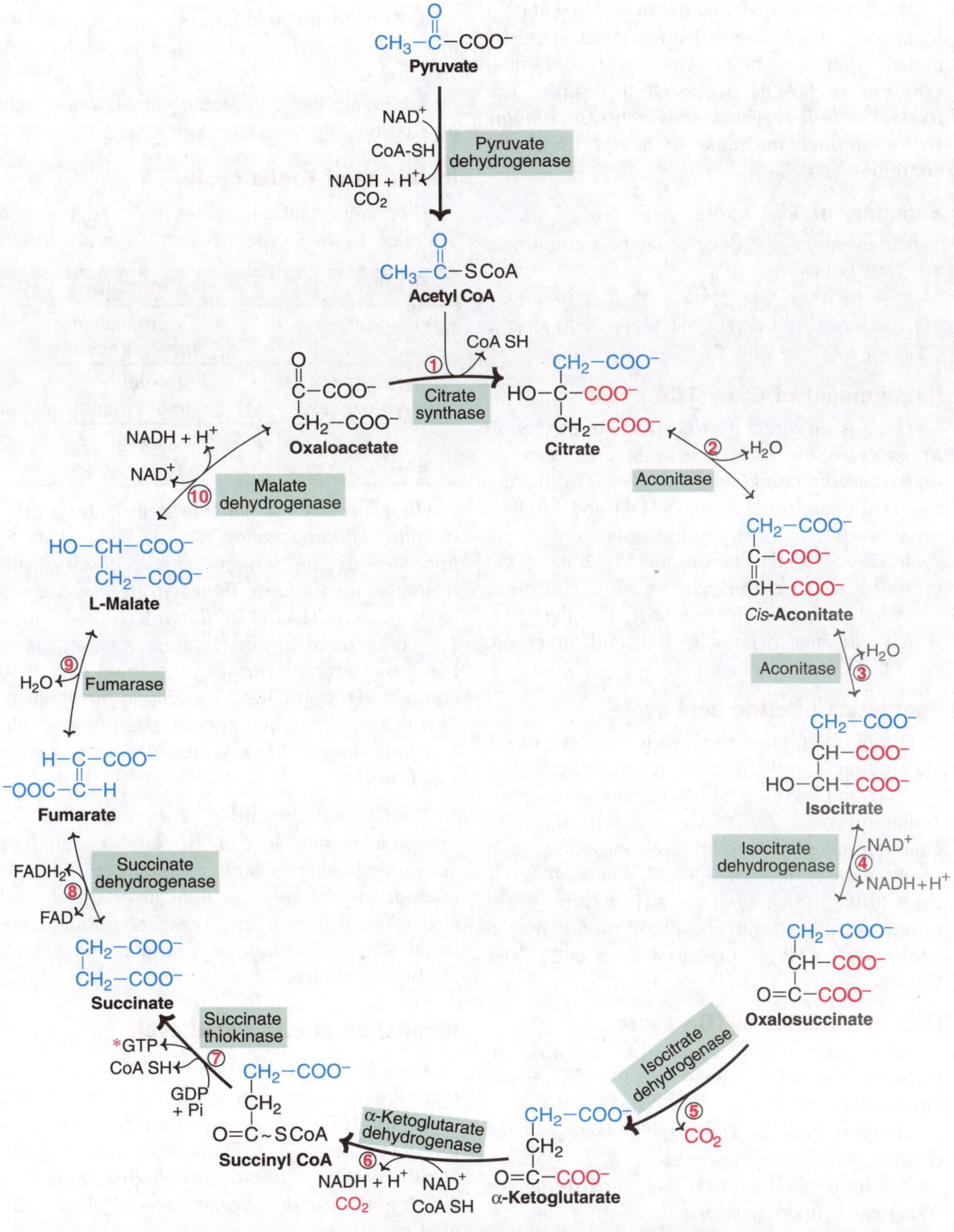

Fig. 13.11 : *The citric acid (Krebs) cycle. (**Irreversible** reactions shown by **thick** arrows).*
* *ATP, instead of GTP is produced in non-glucogenic tissues like brain and heart.*

10. **Conversion of malate to oxaloacetate :** Malate is then oxidized to oxaloacetate by malate dehydrogenase. The third and final synthesis of NADH occurs at this stage. The ***oxaloacetate is regenerated which can combine with another molecule of acetyl CoA***, and continue the cycle.

Summary of TCA cycle

The events of Krebs cycle may be summarized as given below

Acetyl CoA + 3 NAD^+ + FAD + GDP + Pi + $2H_2O \longrightarrow 2CO_2$ + 3NADH + $3H^+$ + $FADH_2$ + GTP + CoA

Requirement of O_2 by TCA cycle

There is ***no direct participation of oxygen in Krebs cycle***. However, the cycle operates only under ***aerobic conditions.*** This is due to the fact that NAD^+ and FAD (from NADH and $FADH_2$, respectively) required for the operation of the cycle can be regenerated in the ETC only in the presence of O_2. Therefore, citric acid cycle is strictly aerobic in contrast to glycolysis which operates in both aerobic and anaerobic conditions.

Energetics of citric acid cycle

During the process of oxidation of acetyl CoA via citric acid cycle, 4 reducing equivalents (3 as NADH and one as $FADH_2$) are produced. Oxidation of 3 NADH by electron transport chain coupled with oxidative phosphorylation results in the synthesis of 7.5 ATP, whereas $FADH_2$ leads to the formation of 1.5 ATP. Besides, there is one substrate level phosphorylation. Thus, a total of ***ten ATP*** are produced from one acetyl CoA.

Role of vitamins in TCA cycle

Four (rather five) B-complex vitamins are essential for Krebs cycle, and thus energy generation

1. **Thiamine** (as TPP) as a coenzyme for α-ketoglutarate dehydrogenase.

2. **Riboflavin** (as FAD) as a coenzyme for succinate dehydrogenase.

3. **Niacin** (as NAD^+) as electron acceptor for isocitrate dehydrogenase, α-ketoglutarate dehydrogenase and malate dehydrogenase.

4. **Pantothenic acid** (as coenzyme A) attached to active carboxylic acid residues i.e. acetyl CoA, succinyl CoA.

5. Vitamin like compound namely ***lipoic acid*** also participates in TCA cycle.

Inhibitors of Krebs cycle

The important enzymes of TCA cycle inhibited by the respective inhibitors are listed

Enzyme	Inhibitor
Aconitase	Fluoroacetate (non-competitive)
α-Ketoglutarate dehydrogenase	Arsenite (non-competitive)
Succinate dehydrogenase	Malonate (competitive)

Fluoroacetate – a suicide substrate : The inhibitor fluoroacetate is first activated to fluoroacetyl CoA which then ***condenses*** with ***oxaloacetate*** to form ***fluorocitrate***. TCA cycle (enzyme-***aconitase***) is ***inhibited*** by fluorocitrate. The compound fluoroacetate, as such, is a harmless substrate. But it is converted to a toxic compound (fluorocitrate) by cellular metabolism. This is a suicide reaction committed by the cell, and thus ***fluoroacetate*** is regarded as a ***suicide substrate***.

Fluoroacetate, regarded as a poison, is found in some plants. It can be fatal to animals consuming such plants. Certain fluorinated compounds are used as anticancer drugs, and also in some pesticides. These compounds get metabolized to produce fluoroacetate which inhibits aconitase.

Regulation of citric acid cycle

The cellular demands of ATP are crucial in controlling the rate of citric acid cycle. The regulation is brought about either by enzymes or the levels of ADP. Three enzymes—namely ***citrate synthase, isocitrate dehydrogenase*** and ***α-ketoglutarate dehydrogenase***—regulate citric acid cycle.

1. **Citrate synthase** is inhibited by ATP, NADH, and long chain fatty acyl CoA.

Medical Concepts/ Clinical Correlates

- Lactic acidosis is observed in patients with deficiency of the enzyme pyruvate dehydrogenase. It could also be due to collapse of circulatory system encountered in myocardial infarction and pulmonary embolism.
- Citric acid cycle is the final common oxidative pathway for carbohydrates, fats and amino acids. It utilizes (indirectly) about 2/3 of the total oxygen consumed by the body and generates about 2/3 of the total energy (ATP).
- Unlike the other metabolic pathways/cycles, very few genetic abnormalities of Krebs cycle are known. This may be due to the vital importance of this metabolic cycle for the survival of life.

2. **Isocitrate dehydrogenase** is activated by ADP, and inhibited by ATP and NADH.

3. **α-Ketoglutarate dehydrogenase** is inhibited by succinyl CoA and NADH.

4. **Availability of ADP** is very important for the citric acid cycle to proceed. This is due to the fact that unless sufficient levels of ADP are available, oxidation (coupled with phosphorylation of ADP to ATP) of NADH and $FADH_2$ through electron transport chain stops. The accumulation of NADH and $FADH_2$ will lead to inhibition of the enzymes (as stated above) and also limits the supply of NAD^+ and FAD which are essential for TCA cycle to proceed.

5. **Hyperammonemia** is a condition characterized by ***elevated ammonia*** in the circulation (***Refer Chapter 15***). It is usually found in severe liver diseases (acute hepatitis, severe cirrhosis), and genetic defects of urea synthesis. Accumulation of ammonia withdraws α-ketoglutarate (from Krebs cycle) to form glutamate and glutamine. This leads to a reduction in the levels of Krebs cycle intermediates, and therefore, decreased generation of ATP. Further, ***ammonia inhibits α-ketoglutarate and possibly pyruvate dehydrogenase***. The net result of hyperammonemia (due to very low ATP) is loss of consciousness, coma and convulsions, and may be even fatal.

The regulation of the enzyme ***pyruvate dehydrogenase*** (see ***page 252***) may also be considered as a part of the regulation of citric acid cycle.

Amphibolic nature of the citric acid cycle

The citric acid cycle provides various intermediates for the synthesis of many compounds needed by the body. Krebs cycle is ***both catabolic and anabolic in nature***, hence regarded as ***amphibolic***.

TCA cycle is actively ***involved in gluconeogenesis, transamination and deamination***. The most important synthetic (anabolic) reactions connected with TCA cycle are given (***Fig.13.12***)

1. Oxaloacetate and α-ketoglutarate, respectively, serve as precursors for the synthesis of aspartate and glutamate which, in turn, are required for the synthesis of other non-essential amino acids, purines and pyrimidines.

2. Succinyl CoA is used for the synthesis of porphyrins and heme.

3. Mitochondrial citrate is transported to the cytosol, where it is cleaved to provide acetyl CoA for the biosynthesis of fatty acids, sterols etc.

Anaplerosis or anaplerotic reactions

The synthetic reactions described above deplete the intermediates of citric acid cycle. The cycle will cease to operate unless the intermediates drawn out are replenished. ***The reactions concerned to replenish or to fill up the intermediates of citric acid cycle are called anaplerotic reactions or anaplerosis*** (*Greek* : fill up). In ***Fig.13.12***, the important synthetic pathways that draw the intermediates of TCA cycle and the anaplerotic reactions to fill them up are given.

The salient features of important anaplerotic reactions are described

1. Pyruvate carboxylase catalyses the conversion of pyruvate to oxaloacetate. This is an ATP dependent carboxylation reaction.

$$\text{Pyruvate} + CO_2 + \text{ATP} \longrightarrow \text{Oxaloacetate} + \text{ADP} + \text{Pi}$$

The details of the above reaction are described under gluconeogenesis.

2. Pyruvate is converted to malate by $NADP^+$ dependent malate dehydrogenase (malic enzyme).

$$\text{Pyruvate} + CO_2 + \text{NADPH} + H^+ \rightleftharpoons \text{Malate} + NADP^+ + H_2O$$

3. Transamination is a process wherein an amino acid transfers its amino group to a keto acid and itself gets converted to a keto acid. The formation of α-ketoglutarate and oxaloacetate occurs by this mechanism.

4. α-Ketoglutarate can also be synthesized from glutamate by glutamate dehydrogenase action.

$$\text{Glutamate} + NAD(P)^+ + H_2O \leftrightarrow \alpha\text{-Ketoglutarate} + NAD(P)H + H^+ + NH_4^+$$

Energetics of glucose oxidation

When a molecule of glucose (6 carbon) undergoes glycolysis, 2 molecules of pyruvate or lactate (3 carbon) are produced. Pyruvate is oxidatively decarboxylated to acetyl CoA (2 carbon) which enters the citric acid cycle and gets completely oxidized to CO_2 and H_2O. The overall process of glucose being completely oxidized to CO_2 and H_2O via glycolysis and citric acid cycle is as follows

$$C_6H_{12}O_6 + 6O_2 + 32ADP + 32Pi \longrightarrow 6CO_2 + 6H_2O + 32ATP$$

The enzymes of glucose metabolism responsible for generating ATP are given in ***Table 13.2*** (***page 248***).

GLUCONEOGENESIS

The ***synthesis of glucose from non-carbohydrate compounds*** is known as gluconeogenesis. The major ***substrates***/precursors for gluconeogenesis are ***lactate***, ***pyruvate***, ***glucogenic amino acids***, ***propionate*** and ***glycerol***.

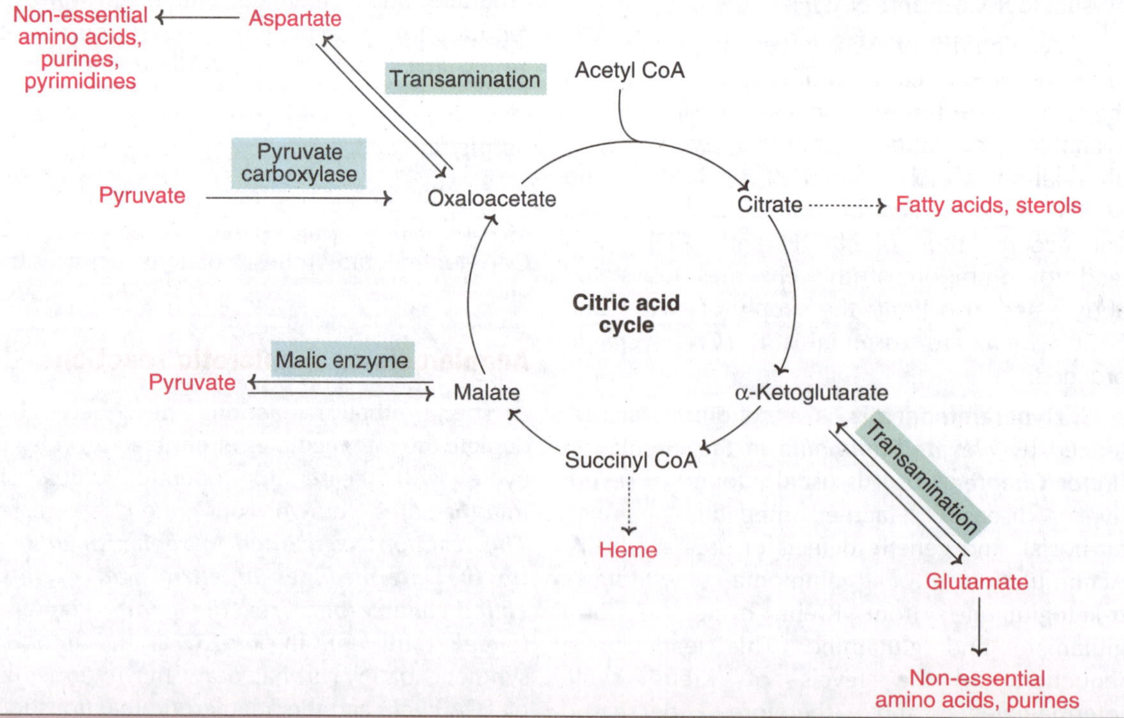

Fig. 13.12 : *Major synthetic and anaplerotic pathways of the intermediates of citric acid cycle.*

Location of gluconeogenesis

Gluconeogenesis occurs mainly in the ***cytosol***, although some precursors are produced in the mitochondria. Gluconeogenesis mostly takes place in ***liver*** (about 1 kg glucose synthesized everyday) and, to some extent, in ***kidney matrix*** (about one-tenth of liver capacity).

Importance of gluconeogenesis

Glucose occupies a key position in the metabolism and its continuous supply is absolutely essential to the body for a variety of functions

1. ***Brain*** and ***central nervous system, erythrocytes, testes and kidney medulla*** are dependent on glucose for continuous supply of energy. Human brain alone requires about 120 g of glucose per day, out of about 160 g needed by the entire body.

2. Glucose is the only source that supplies energy to the ***skeletal muscle, under anaerobic conditions***.

3. ***In fasting*** even more than a day, ***gluconeogenesis must occur*** to meet the basal requirements of the body for glucose and to maintain the intermediates of citric acid cycle. This is essential for the survival of humans and other animals.

4. Certain metabolites produced in the tissues accumulate in the blood, e.g. lactate, glycerol, propionate etc. Gluconeogenesis effectively clears them from the blood.

5. Increased gluconeogenesis is observed in critically ill patients (injury, infection) resulting in poor outcome. Elevated blood glucose (hyperglycemia) is associated with impaired immune function and blood coagulation, besides increased production of free radicals.

Reactions of gluconeogenesis

Broadly, gluconeogenesis involves glycolysis and Krebs cycle, besides some special reactions. It may be noted that complete reversal of glycolysis is not possible due to irreversible reactions at 3 stages, catalysed by hexokinase (or glucokinase) phosphofructokinase and pyruvate kinase. Thus, ***partial reversal of glycolysis*** besides other reactions is responsible for gluconeogenesis.

Gluconeogenesis closely resembles the reversed pathway of glycolysis, although it is ***not the complete reversal of glycolysis***. Essentially, 3 (out of 10) reactions of glycolysis are irreversible. The seven reactions are common for both glycolysis and gluconeogenesis (***Fig.13.13***). The ***three irreversible steps*** of glycolysis are catalysed by the enzymes, namely hexokinase, phosphofructokinase and pyruvate kinase. These three stages—bypassed by alternate enzymes specific to gluconeogenesis—are discussed

1. **Conversion of pyruvate to phosphoenolpyruvate :** This takes place in two steps (***Fig.13.14***). ***Pyruvate carboxylase*** is a ***biotin—dependent*** mitochondrial enzyme that converts pyruvate to oxaloacetate in presence of ATP and CO_2. This enzyme regulates gluconeogenesis and requires acetyl CoA for its activity.

Oxaloacetate is synthesized in the mitochondrial matrix. It has to be transported to the cytosol to be used in gluconeogenesis, where the rest of the pathway occurs. Due to membrane impermeability, oxaloacetate cannot diffuse out of the mitochondria. It is converted to malate and then transported to the cytosol. Within the cytosol, oxaloacetate is regenerated. The reversible conversion of oxaloacetate and malate is catalysed by malate dehydrogenase, an enzyme present in both mitochondria and cytosol.

In the cytosol, ***phosphoenolpyruvate carboxykinase*** converts oxaloacetate to phosphoenolpyruvate. GTP or ITP (not ATP) is used in this reaction and the CO_2 (fixed by carboxylase) is liberated. For the conversion of pyruvate to phosphoenolpyruvate, 2 ATP equivalents are utilized. This is in contrast to only one ATP that is liberated in glycolysis for this reaction.

2. **Conversion of fructose 1,6-bisphosphate to fructose 6-phosphate :** Phosphoenolpyruvate undergoes the reversal of glycolysis until fructose 1,6-bisphosphate is produced. The enzyme ***fructose 1,6-bisphosphatase*** converts fructose 1,6-bisphosphate to fructose 6-phosphate. This enzyme requires Mg^{2+} ions. Fructose 1,6-bisphosphatase is absent in smooth muscle and heart muscle. This enzyme is also regulatory in gluconeogenesis.

3. **Conversion of glucose 6-phosphate to glucose :** ***Glucose 6-phosphatase*** catalyses the conversion of glucose 6-phosphate to glucose. The presence or absence of this enzyme in a tissue determines whether the tissue is capable of contributing glucose to the blood or not. It is mostly present in liver and kidney but ***absent in muscle,*** brain and adipose tissue.

The overall summary of gluconeogenesis for the conversion of pyruvate to glucose is shown in the next column.

$$2\ \text{Pyruvate} + 4\text{ATP} + 2\text{GTP} + 2\text{NADH} + 2\text{H}^+ + 6\text{H}_2\text{O} \longrightarrow \text{Glucose} + 2\text{NAD}^+ + 4\text{ADP} + 2\text{GDP} + 6\text{Pi} + 6\text{H}^+$$

Gluconeogenesis consumes energy

As is evident from the summary of reactions, gluconeogenesis is associated with ***high ATP consumption.*** With restricted dietary carbohydrates and high consumption of fats and proteins, gluconeogenesis occurs from amino acids and ***energy is supplied by oxidation of fatty acids.***

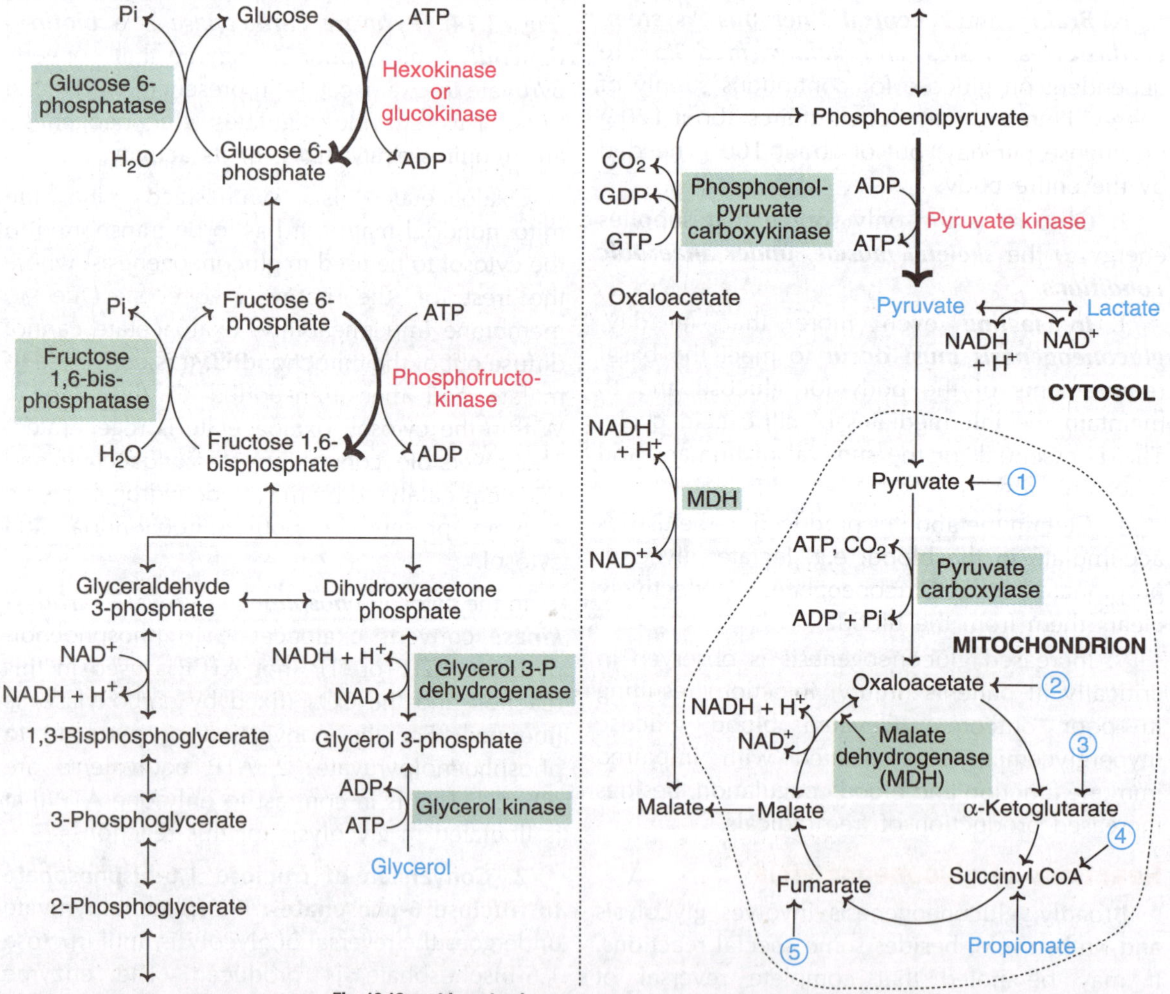

Fig. 13.13 contd. next column

Fig. 13.13 : The pathway of gluconeogenesis. [The enzymes catalysing irreversible steps in glycolysis are shown in red. The important enzymes participating in gluconeogenesis are shown in shaded green. The substrates for gluconeogenesis are in blue. The numbers represent the entry of glucogenic amino acids : (1) Alanine, glycine, serine, cysteine, threonine and tryptophan; (2) Aspartate and asparagine; (3) Arginine, glutamate, glutamine, histidine, proline; (4) Isoleucine, methionine, valine; (5) Phenylalanine, tyrosine].

In certain weight loss programmes, this type of dietary regime—very low carbohydrate and high fat and protein is effectively used.

Gluconeogenesis from amino acids

The carbon skeleton of ***glucogenic amino acids*** (***all except leucine and lysine***) results in the formation of pyruvate or the intermediates of citric acid cycle (***Fig.13.13***) which, ultimately, result in the synthesis of glucose.

Gluconeogenesis from glycerol

Glycerol is liberated mostly in the adipose tissue by the ***hydrolysis of fats*** (triacylglycerols). The enzyme glycerokinase (found in liver and kidney, absent in adipose tissue) activates glycerol to glycerol 3-phosphate. The latter is converted to dihydroxyacetone phosphate by glycerol 3-phosphate dehydrogenase. Dihydroxyacetone phosphate is an intermediate in glycolysis which can be conveniently ***used for glucose production***.

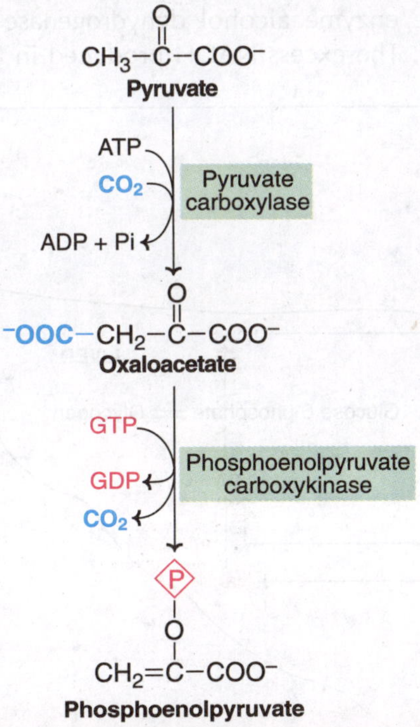

Fig. 13.14 : *Conversion of pyruvate to phosphoenolpyruvate.*

Gluconeogenesis from propionate

Oxidation of odd chain fatty acids and the breakdown of some amino acids (methionine, isoleucine) yields a three carbon propionyl CoA. Propionyl CoA carboxylase acts on this in presence of ATP and biotin and converts to methyl malonyl CoA which is then converted to succinyl CoA in presence of ***B_{12} coenzyme*** (***Refer Fig.7.38***). Succinyl CoA formed from propionyl CoA enters gluconeogenesis via citric acid cycle.

Gluconeogenesis from lactate (Cori cycle)

Lactate produced by active skeletal muscle is a ***major precursor for gluconeogenesis***. Under anaerobic conditions, pyruvate is reduced to lactate by lactate dehydrogenase (LDH)

$$\text{Pyruvate} + \text{NADH} + H^+ \xrightleftharpoons{\text{LDH}} \text{Lactate} + \text{NAD}^+$$

Lactate is a dead end in glycolysis, since it must be reconverted to pyruvate for its further metabolism. The very ***purpose of lactate production is to regenerate NADH*** so that glycolysis proceeds uninterrupted in skeletal muscle. Lactate or pyruvate produced in the muscle cannot be utilized for the synthesis of glucose due to the absence of the key enzymes of gluconeogenesis (glucose 6-phosphatase and fructose 1,6-bisphosphatase).

The plasma membrane is freely permeable to lactate. Lactate is carried from the skeletal muscle through blood and handed over to liver, where it is oxidized to pyruvate. Pyruvate, so produced, is converted to glucose by gluconeogenesis, which is then transported to the skeletal muscle.

The cycle involving the ***synthesis of glucose in liver from the skeletal muscle lactate*** and the reuse of glucose thus synthesized by the muscle for energy purpose is known as Cori cycle (***Fig.13.15***).

Glucose-alanine (Cahill) cycle

There is a continuous transport of amino acids from muscle to liver, which predominantly occurs during starvation. Alanine dominates among the transported amino acids. It is postulated that pyruvate in skeletal muscle undergoes transamination to produce alanine. Alanine is then transported to liver and used for gluconeogenesis. This cycle is referred to as glucose-alanine cycle or Cahill cycle (***Fig.13.15***).

Regulation of gluconeogenesis

The hormones ***glucagon***, glucocorticoids and insulin, and the availability of substrates mainly regulate gluconeogenesis, as discussed below

Influence of glucagon : This is a hormone, secreted by α-cells of the pancreatic islets. Glucagon ***stimulates gluconeogenesis*** by two mechanisms

1. Active form of pyruvate kinase is converted to inactive form through the mediation of cyclic AMP, brought about by glucagon. ***Decreased pyruvate kinase*** results in the reduced conversion of phosphoenol pyruvate to pyruvate and the former is diverted for the synthesis of glucose.

2. Glucagon reduces the concentration of ***fructose 2,6-bisphosphate***. This compound ***allosterically activates phosphofructokinase*** and ***inhibits fructose 1,6-bisphosphatase***. This favours gluconeogenesis.

Glucocorticoids ***promote*** the synthesis of hepatic aminotransferase, which provide substrates for gluconeogenesis. ***Insulin inhibits*** gluconeogenesis by reducing the activities of key enzymes of gluconeogenesis. An overview of hormonal regulation of gluconeogenesis is given in ***Fig.13.16***.

Availability of substrates : Among the various substrates, glucogenic amino acids have stimulating influence on gluconeogenesis. This is particularly important in a condition like diabetes mellitus (decreased insulin level) where amino acids are mobilized from muscle protein for the purpose of gluconeogenesis.

Acetyl CoA promotes gluconeogenesis : During starvation—due to excessive lipolysis in adipose tissue—acetyl CoA accumulates in the liver. Acetyl CoA allosterically activates pyruvate carboxylase resulting in enhanced glucose production.

Alcohol inhibits gluconeogenesis

Ethanol oxidation in the liver to acetaldehyde by the enzyme alcohol dehydrogenase utilizes NAD^+. The excess NADH produced in the liver

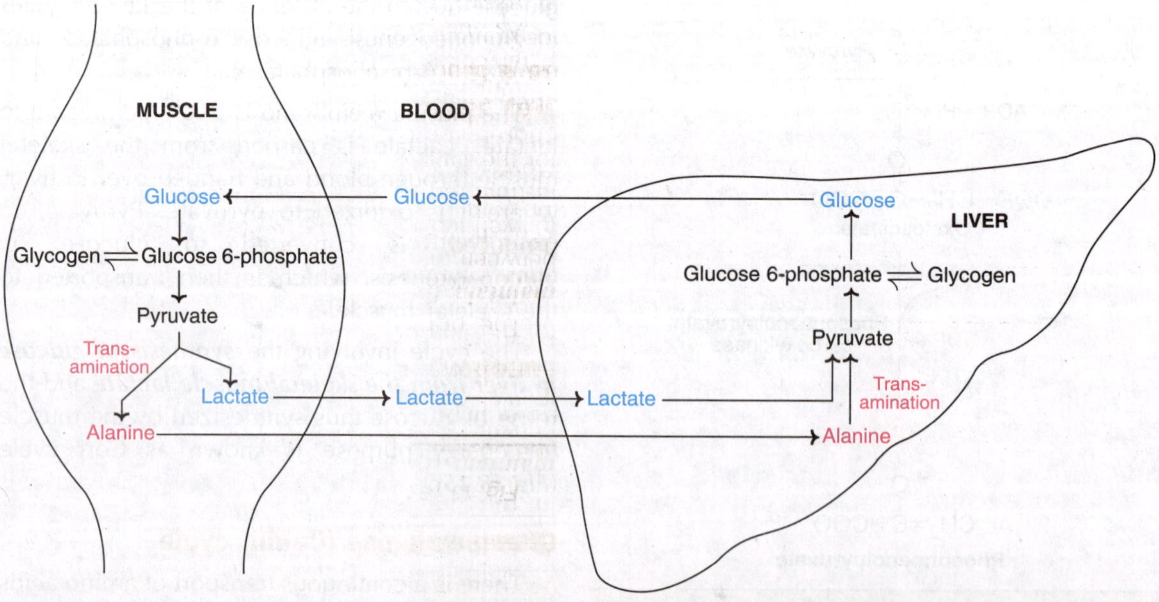

Fig. 13.15 : The Cori cycle (blue) and glucose-alanine (red) cycle (other reactions common for both cycles).

interferes with gluconeogenesis as illustrated below.

Ethanol + NAD^+ ⟶ Acetaldehyde + NADH + H^+

Pyruvate + NADH + H^+ ↔ Lactate + NAD^+

Oxaloacetate + NADH + H^+ ↔ Malate + NAD^+

It is evident from the above reactions that pyruvate and oxaloacetate, the predominant ***substrates for gluconeogenesis, are made unavailable by alcohol intoxication***. This happens due to overconsumption of NAD^+ and excessive production of NADH by alcohol.

Alcohol consumption ***increases the risk of hypoglycemia*** (reduced plasma glucose) due to lowered gluconeogenesis. Hypoglycemia is frequently observed in diabetic patients (particularly on insulin treatment), and under-nourished persons consuming alcohol.

Gluconeogenesis from fat?

It is often stated that glucose cannot be synthesized from fat. In a sense, it is certainly true, since the fatty acids (most of them being even chain), on oxidation, produce acetyl CoA which cannot be converted to pyruvate. Further, the two carbons of acetyl CoA disappear as 2 moles of CO_2 in TCA cycle. Therefore, even chain fatty acids cannot serve as precursors for glucose formation. The prime reason why animals cannot convert fat to glucose is the ***absence of glyoxylate cycle*** (described later).

However, the ***glycerol*** released from lipolysis and the ***propionate*** obtained from the oxidation of odd chain fatty acids ***are good substrates*** for gluconeogenesis, as discussed above.

GLYCOGEN METABOLISM

Glycogen is the ***storage form of glucose in animals***, as is starch in plants. It is stored mostly in liver (6-8%) and muscle (1-2%). Due to more muscle mass, the quantity of glycogen in muscle (250 g) is about three times higher than that in the liver (75 g). Glycogen is stored as granules in the cytosol, where most of the enzymes of glycogen synthesis and breakdown are present.

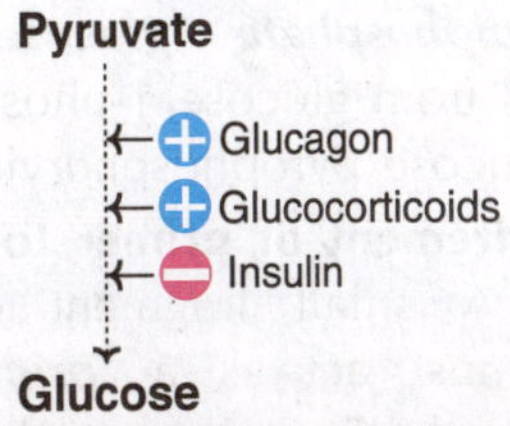

Fig. 13.16 : An overview of hormonal regulation of gluconeogenesis.

Functions of glycogen

The prime function of liver glycogen is to ***maintain the blood glucose levels***, particularly between meals. Liver glycogen stores increase in a well-fed state which are depleted during fasting. Muscle glycogen serves as a fuel reserve for the supply of ATP during muscle contraction.

Why store glycogen as a fuel reserve?

As such, fat is the fuel reserve of the body. However, fat is not preferred, instead glycogen is chosen for a routine, and day to day use of energy reserve for the following reasons

- Glycogen can be ***rapidly mobilized***
- Glycogen can generate energy in the absence of oxygen
- ***Brain*** depends on continuous ***glucose*** supply (which mostly comes ***from glycogen***.)

On the other hand, fat mobilization is slow, needs O_2 for energy production and cannot produce glucose (to a significant extent). Thus, ***fat*** may be considered as a ***fixed deposit*** while ***glycogen*** is in the ***current/saving account*** in a bank!

GLYCOGENESIS

The ***synthesis of glycogen*** from glucose is glycogenesis (***Fig.13.17***). Glycogenesis takes place in the cytosol and requires ATP and UTP, besides glucose.

1. **Synthesis of UDP-glucose :** The enzymes hexokinase (in muscle) and glucokinase (in liver) convert glucose to ***glucose 6-phosphate***. Phosphoglucomutase catalyses the conversion of glucose 6-phosphate to glucose 1-phosphate.

Uridine diphosphate glucose (UDPG) is synthesized from glucose 1-phosphate and UTP by UDP-glucose pyrophosphorylase.

2. **Requirement of primer to initiate glycogenesis :** A small fragment of pre-existing glycogen must act as a ***'primer'*** to initiate glycogen synthesis. It is recently found that in the absence of glycogen primer, a specific protein—namely ***'glycogenin'***—can accept glucose from UDPG. The hydroxyl group of the amino acid tyrosine of glycogenin is the site at which the initial glucose unit is attached. The enzyme glycogen initiator synthase transfers the first molecule of glucose to glycogenin. Then glycogenin itself takes up a few glucose residues to form a fragment of primer which serves as an acceptor for the rest of the glucose molecules.

3. **Glycogen synthesis by glycogen synthase :** Glycogen synthase is responsible for the ***formation of 1,4-glycosidic linkages.*** This enzyme transfers the glucose from UDP-glucose to the non-reducing end of glycogen to form α-1,4 linkages.

4. **Formation of branches in glycogen :** Glycogen synthase can catalyse the synthesis of a linear unbranched molecule with 1,4 α-glycosidic linkages. Glycogen, however, is a branched tree-like structure. The formation of branches is brought about by the action of a branching enzyme, namely ***glucosyl α-4-6 transferase.*** (amylo α 1,4 → 1,6 trans-glucosidase). This enzyme transfers a small fragment of five to eight glucose residues from the non-reducing end of glycogen chain (by breaking α-1,4 linkages) to another glucose residue where it is linked by α-1,6 bond. This leads to the formation of a new non-reducing end, besides the existing one. Glycogen is further elongated and branched, respectively, by the enzymes glycogen synthase and glucosyl 4-6 transferase.

The ***overall reaction of the glycogen synthesis*** for the addition of each glucose residue is

$$(\text{Glucose})_n + \text{Glucose} + 2\text{ATP} \longrightarrow (\text{Glucose})_{n+1} + 2\ \text{ADP} + \text{Pi}$$

Of the two ATP utilized, one is required for the phosphorylation of glucose while the other is needed for conversion of UDP to UTP.

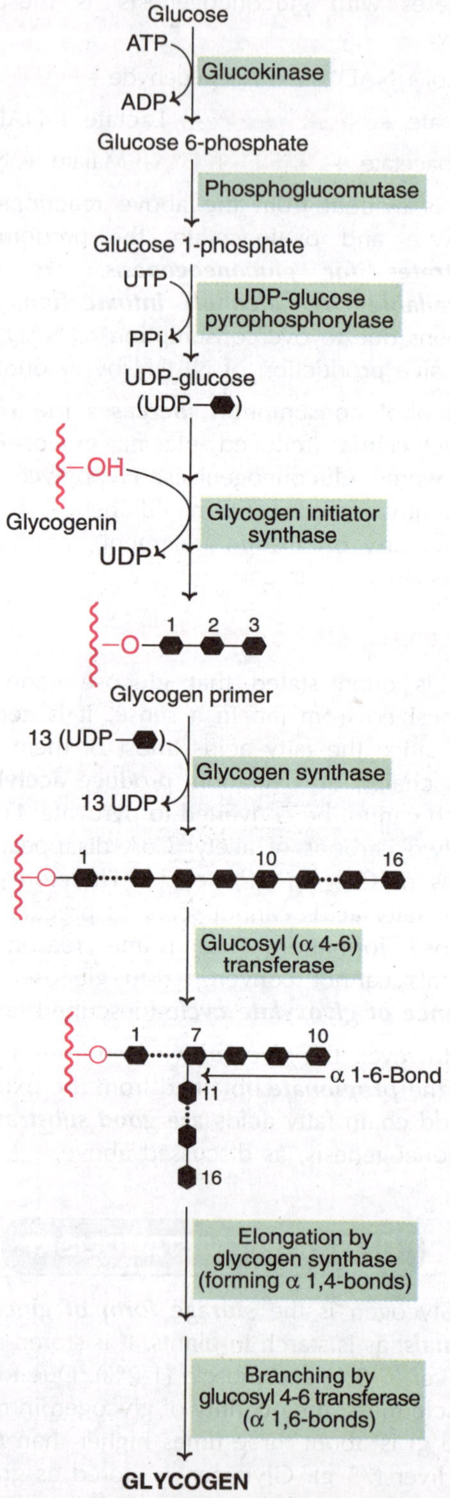

Fig. 13.17 : Glycogen synthesis from glucose (glycogenesis).

GLYCOGENOLYSIS

The ***degradation of stored glycogen*** in liver and muscle constitutes glycogenolysis. The pathways for the synthesis and degradation of glycogen are not reversible. An independent set of enzymes present in the cytosol carry out glycogenolysis. Glycogen is degraded by breaking α-1,4- and α-1,6-glycosidic bonds (***Fig.13.18***).

1. **Action of glycogen phosphorylase :** The α-1,4-glycosidic bonds (from the non-reducing ends) are cleaved sequentially by the enzyme glycogen phosphorylase to yield glucose 1-phosphate. This process—called ***phosphorolysis***—continues until four glucose residues remain on either side of branching point (α-1,6-glycosidic link). The glycogen so formed is known as ***limit dextrin*** which cannot be further degraded by phosphorylase. Glycogen phosphorylase possesses a molecule of pyridoxal phosphate, covalently bound to the enzyme.

2. **Action of debranching enzyme :** The branches of glycogen are cleaved by two enzyme activities present on a single polypeptide called ***debranching enzyme,*** hence it is a ***bifunctional enzyme.***

Glycosyl 4 : 4 transferase (oligo α-1,4 → 1,4 glucan transferase) activity removes a fragment of three or four glucose residues attached at a branch and transfers them to another chain. Here, one α-1,4-bond is cleaved and the same α-1,4 bond is made, but the places are different.

Amylo α-1,6-glucosidase breaks the α-1,6 bond at the branch with a single glucose residue and ***releases a free glucose.***

The remaining molecule of glycogen is again available for the action of phosphorylase and debranching enzyme to repeat the reactions stated in 1 and 2.

3. **Formation of glucose 6-phosphate and glucose :** Through the combined action of glycogen phosphorylase and debranching enzyme, glucose 1-phosphate and free glucose in a ratio of 8 : 1 are produced. Glucose 1-phosphate is converted to glucose 6-phosphate by the enzyme phosphoglucomutase.

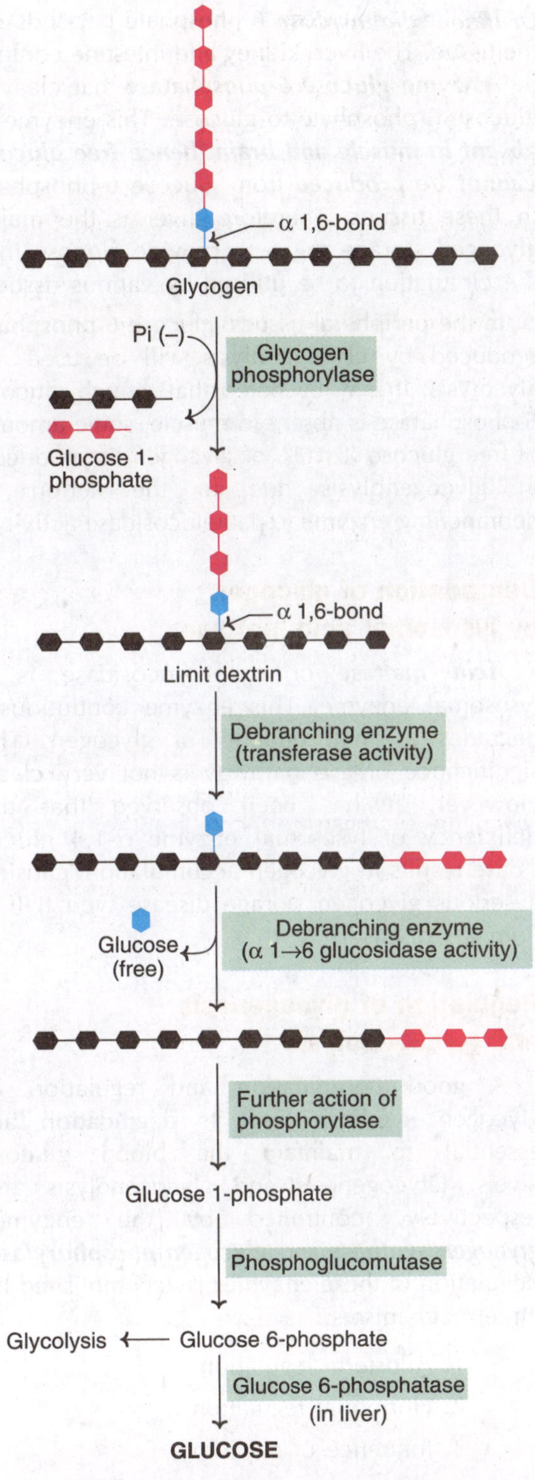

Fig. 13.18 : *Glycogen degradation to glucose-glycogenolysis. (The ratio of glucose 1-phosphate to glucose is 8 : 1).*

The fate of glucose 6-phosphate depends on the tissue. The liver, kidney and intestine contain the enzyme ***glucose 6-phosphatase*** that cleaves glucose 6-phosphate to glucose. This enzyme is ***absent in muscle and brain, hence free glucose cannot be produced*** from glucose 6-phosphate in these tissues. Therefore, liver is the major glycogen storage organ to provide glucose into the circulation to be utilised by various tissues.

In the peripheral tissues, glucose 6-phosphate produced by glycogenolysis will be used for glycolysis. It may be noted that though glucose 6-phosphatase is absent in muscle, some amount of free glucose (8-10% of glycogen) is produced in glycogenolysis due to the action of debranching enzyme (α-1,6-glucosidase activity).

Degradation of glycogen by lysosomal acid maltase

Acid maltase or α-1,4-glucosidase is a lysosomal enzyme. This enzyme continuously degrades a small quantity of glycogen. The significance of this pathway is not very clear. However, it has been observed that the deficiency of lysosomal enzyme α-1,4 glucosidase results in glycogen accumulation, causing a serious glycogen storage disease type II (i.e. Pompe's disease).

Regulation of glycogenesis and glycogenolysis

A good coordination and regulation of glycogen synthesis and its degradation are essential to maintain the blood glucose levels. Glycogenesis and glycogenolysis are, respectively, controlled by the enzymes ***glycogen synthase*** and ***glycogen phosphorylase***. Regulation of these enzymes is accomplished by three mechanisms

1. Allosteric regulation
2. Hormonal regulation
3. Influence of calcium.

1. **Allosteric regulation of glycogen metabolism :** There are certain metabolites that allosterically regulate the activities of glycogen synthase and glycogen phosphorylase. The control is carried out in such a way that glycogen synthesis is increased when substrate availability and energy levels are high. On the other hand, glycogen breakdown is enhanced when glucose concentration and energy levels are low. The allosteric regulation of glycogen metabolism is depicted in ***Fig.13.19***. In a well-fed state, the availability of ***glucose 6-phosphate*** is high which allosterically ***activates glycogen synthase*** for more glycogen synthesis. On the other hand, glucose 6-phosphate and ATP allosterically inhibit glycogen phosphorylase. Free ***glucose*** in liver also acts as an allosteric ***inhibitor of glycogen phosphorylase.***

2. **Hormonal regulation of glycogen metabolism :** The hormones, through a complex series of reactions, bring about covalent modification, namely phosphorylation and dephosphorylation of enzyme proteins which, ultimately control glycogen synthesis or its degradation.

cAMP as second messenger for hormones : Hormones like epinephrine and norepinephrine, and glucagon (in liver) activate adenylate cyclase to increase the production of cAMP. The enzyme phosphodiesterase breaks down cAMP. The hormone insulin increases the

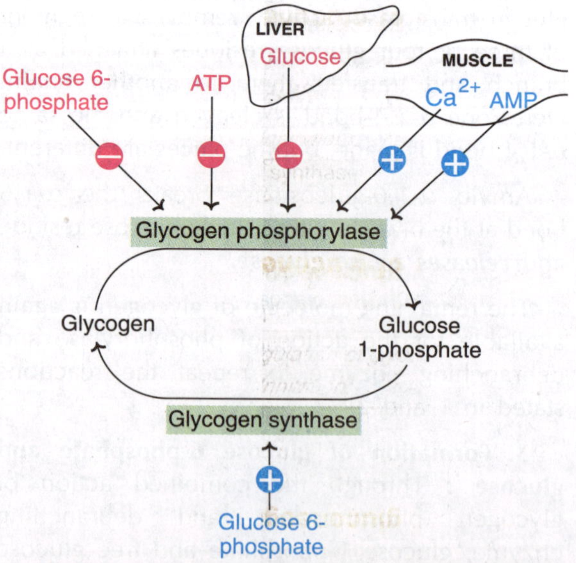

Fig. 13.19 : Allosteric regulation of glycogenolysis and glycogenesis (− : Inhibition; + : Activation).

Medical Concepts/ Clinical Correlates

- A continuous presence of glucose—supplied through diet or synthesized in the body (gluconeogenesis)—is essential for the survival of the organism. Alcohol intoxication reduces gluconeogenesis, particularly in diabetic patients and undernourished persons.
- Human brain consumes about 120 g of glucose per day out of the 160 g needed by the body. Insufficient supply of glucose to brain may lead to coma and death.
- Liver glycogen serves as an immediate source for maintaining blood glucose levels, particularly between the meals. The glycogen stores in the liver get depleted after 12-18 hours of fasting.

phosphodiesterase activity in liver and lowers the cAMP levels.

Regulation of glycogen synthesis by cAMP : The glycogenesis is regulated by glycogen synthase. This enzyme exists in two forms—***glycogen synthase 'a'***—which is not phosphorylated and most active, and secondly, ***glycogen synthase 'b'*** as phosphorylated inactive form. Glycogen synthase *'a'* can be converted to *'b'* form (inactive) by phosphorylation. The degree of phosphorylation is proportional to the inactive state of enzyme. The process of phosphorylation is catalysed by a cAMP-dependent protein kinase. The protein kinase phosphorylates and inactivates glycogen synthase by converting *'a'* form to *'b'* form. The glycogen synthase *'b'* can be converted back to synthase *'a'* by protein phosphatase I.

In the ***Fig.13.20***, the inhibition of glycogen synthesis brought by epinephrine (also norepinephrine) and glucagon through cAMP by converting active glycogen synthase *'a'* to inactive synthase *'b'*, is given.

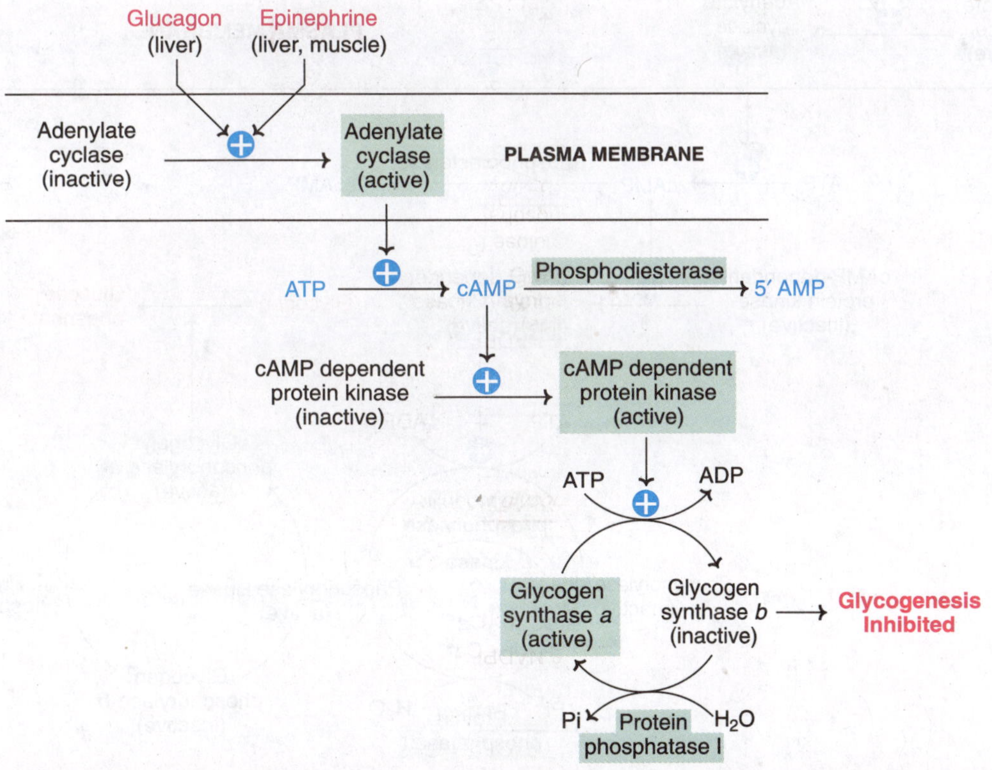

Fig. 13.20 : Hormonal regulation of glycogen synthesis (glycogenesis).

Regulation of glycogen degradation by cAMP : The hormones like epinephrine and glucagon bring about glycogenolysis by their action on ***glycogen phosphorylase*** through cAMP as illustrated in ***Fig.13.21***. Glycogen phosphorylase exists in two forms, an active *'a'* form and inactive form *'b'*.

The cAMP—formed due to hormonal stimulus—activates cAMP dependent protein kinase. This active protein kinase phosphorylates inactive form of glycogen phosphorylase kinase to active form. (The enzyme protein phosphatase removes phosphate and inactivates phosphorylase kinase). The active phosphorylase kinase phosphorylates inactive glycogen phosphorylase *'b'* to active glycogen phosphorylase *'a'* which degrades glycogen. The enzyme protein phosphatase I can dephosphorylate and convert active glycogen phosphorylase *'a'* to inactive *'b'* form.

3. **Effect of Ca^{2+} ions on glycogenolysis :** When the muscle contracts, Ca^{2+} ions are released from the sarcoplasmic reticulum. Ca^{2+} binds to ***calmodulin-calcium modulating protein*** and directly activates phosphorylase kinase without the involvement of cAMP-dependent protein kinase.

The overall effect of hormones on glycogen metabolism is that ***an elevated glucagon or epinephrine level increases glycogen degradation whereas an elevated insulin results in increased glycogen synthesis.***

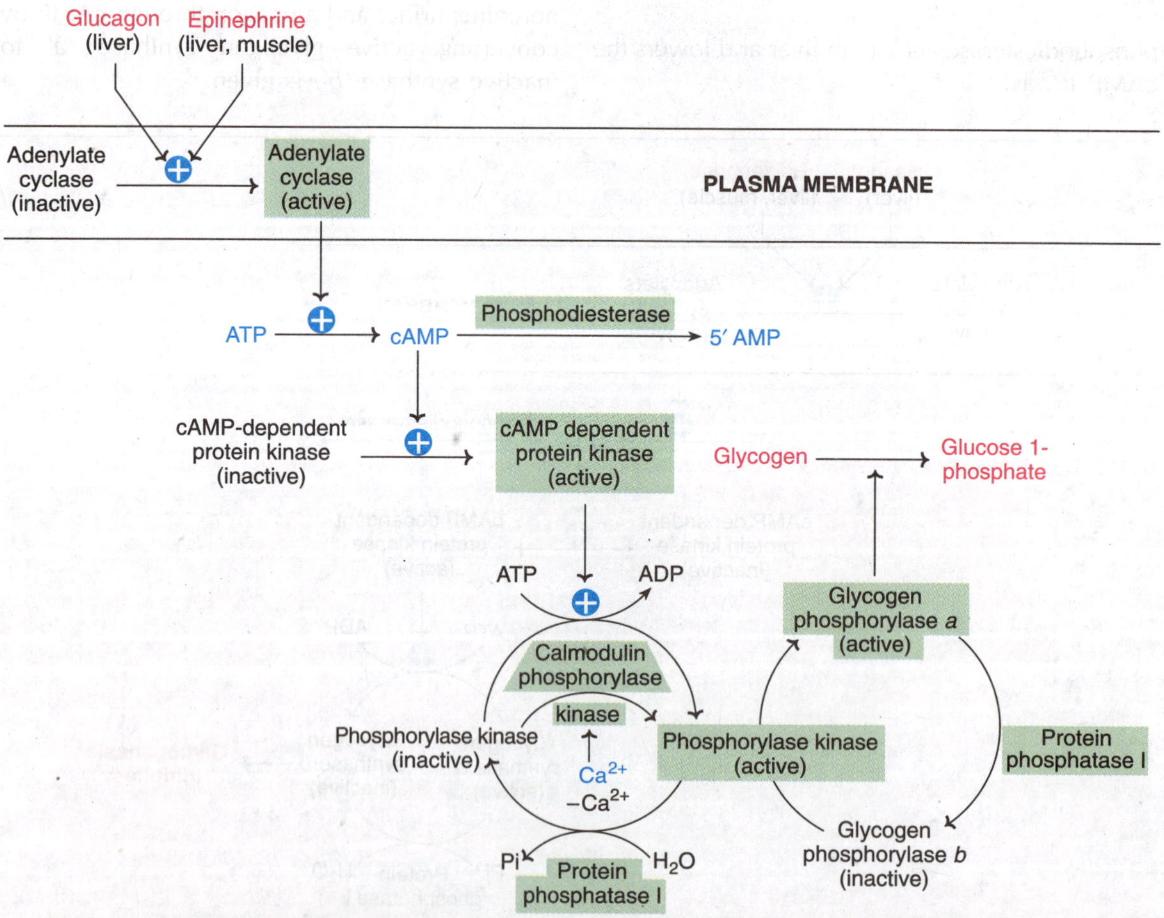

Fig. 13.21 : Hormonal regulation of glycogen degradation (glycogenolysis).

FUTILE (SUBSTRATE) CYCLES

The ***synthesis and degradative pathways*** of metabolism (particularly reactions involving phosphorylation and dephosphorylation utilizing ATP) are ***well regulated*** and subjected to fine tuning and rapid response to meet the body demands, ***with minimal wastage of energy and metabolites***. Thus, glycolysis and gluconeogenesis (breakdown of glucose to pyruvate, and conversion of pyruvate to glucose), glycogenolysis and glycogenesis operate in a selective fashion to suit the cellular demands. If on the other hand, the ***synthesis and degradative metabolic pathways*** of a particular substance (say gluconeogenesis and glycolysis related to glucose) ***operate to the same extent simultaneously***, this ***would result in futile cycles***. However, futile cycles, consuming energy (ATP) are ***wasteful metabolic exercises***. They are minimally operative due to a well coordinated metabolic machinery. The enzymes, namely ***protein kinases*** and ***phosphatases*** serve as ***molecular switches***. They can be ***turned on or off, as per the metabolic needs of the cells***.

Recent studies suggest that futile cycles may have some physiological (operative in arousal of hibernating animals) and pathological (malignant hyperthermia due to loss of control)significance.

GLYCOGEN STORAGE DISEASES

The ***metabolic defects*** concerned with the ***glycogen synthesis*** and ***degradation*** are collectively referred to as glycogen storage diseases (***GSD***) or ***glycogenosis***. The overall incidence of GSD is around 1 per 40,000 births. GSD are characterized by deposition of normal or abnormal type of glycogen in one or more tissues. A summary of glycogen metabolism along with the defective enzymes in the glycogen storage disorders is depicted in ***Fig.13.22***. The biochemical lesions and the characteristic features of the disorders are given in ***Table 13.4***.

Glycogen storage disease 0 is characterized by the deficiency of the enzyme ***glycogen synthase*** in the liver. Although ***GSD 0*** does not result in storage of extra glycogen in liver, it manifests in symptoms comparable to that of other GSD.

von Gierke's disease (type I)

The incidence of type I glycogen storage disease is 1 per 200,000 persons. It is transmitted by autosomal recessive trait. This disorder results in various biochemical manifestations.

1. **Fasting hypoglycemia :** Due to the defect in the enzyme ***glucose 6-phosphatase***, enough free glucose is not released from the liver.

2. **Lactic acidemia :** Glucose is not synthesized from lactate produced in muscle and liver. ***Lactate level in blood increases*** and the pH is lowered (acidosis).

3. **Hyperlipidemia :** There is a blockade in gluconeogenesis. Hence more fat is mobilized to meet energy requirements of the body. This results in ***increased plasma free fatty acids*** and ketone bodies.

4. **Hyperuricemia :** Glucose 6-phosphate that accumulates is diverted to pentose phosphate pathway, leading to ***increased synthesis of ribose phosphates*** which increase the cellular levels of phosphoribosyl pyrophosphate and enhance the metabolism of purine nucleotides to uric acid. ***Elevated*** plasma levels of ***uric acid*** (hyperuricemia; ***normal 4-7 mg/dl***) are often associated with ***gouty arthritis*** (painful joints).

The important features of the glycogen storage diseases are given in ***Table 13.4***.

HEXOSE MONOPHOSPHATE SHUNT

Hexose monophosphate pathway or ***HMP shunt*** is also called ***pentose phosphate pathway*** or phosphogluconate pathway or ***Dickens-Horecker pathway***. ***This is an alternative pathway to glycolysis and TCA cycle for the oxidation of glucose.*** However, HMP shunt is more anabolic in nature, since it is concerned with the biosynthesis of NADPH and pentoses.

HMP shunt—a unique multifunctional pathway

The pathway starts with glucose 6-phosphate. As such, ***no ATP is directly utilized or produced in HMP pathway***. It is a unique multifunctional pathway, since there are several interconvertible substances produced which may proceed in different directions in the metabolic reactions.

Location of the pathway

The enzymes of HMP shunt are located in the ***cytosol***. The tissues such as ***liver, adipose tissue, adrenal gland, erythrocytes, testes*** and ***lactating mammary gland*** are highly active in HMP shunt. Most of these tissues are involved in the biosynthesis of fatty acids and steroids which are ***dependent on the supply of NADPH***.

Reactions of the pathway

The sequence of reactions of HMP shunt (***Fig.13.23***) is divided into two phases—***oxidative*** and ***non-oxidative***.

1. **Oxidative phase :** ***Glucose 6-phosphate dehydrogenase (G6PD)*** is an NADP-dependent enzyme that converts glucose 6-phosphate to 6-phosphogluconolactone. The latter is then hydrolysed by the gluconolactone hydrolase to 6-phosphogluconate. The next reaction involving the synthesis of NADPH is catalysed by 6-phosphogluconate dehydrogenase to produce 3 keto 6-phosphogluconate which then undergoes decarboxylation to give ribulose 5-phosphate.

G6PD regulates HMP shunt : The first reaction ***catalysed by G6PD is most regulatory***

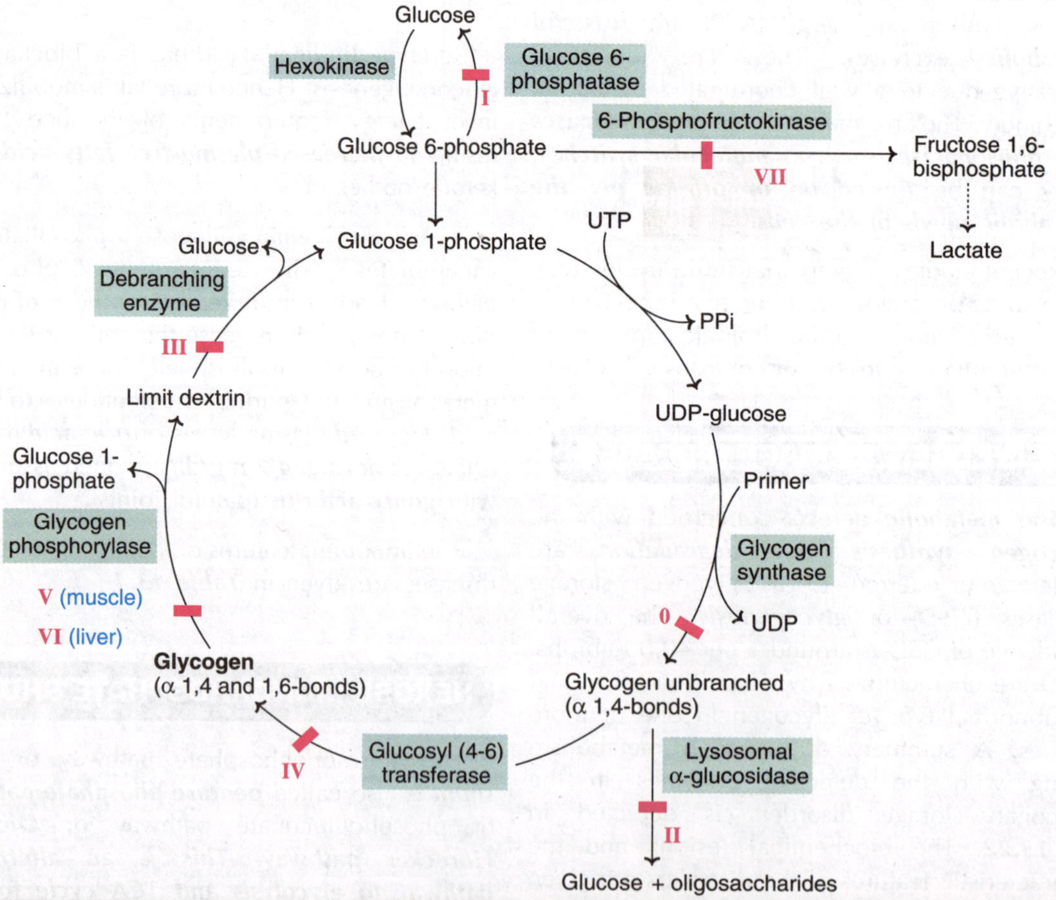

Fig. 13.22 : Summary of glycogen metabolism with glycogen storage diseases (Red blocks indicate storage disease, 0–glycogen storage disease 0; I–von Gierke's disease; II–Pompe's disease; III–Cori's disease; IV–Anderson's disease; V–Mc Ardle's disease; VI–Her's disease; VII–Tarui's disease).

in HMP shunt. This enzyme catalyses an irreversible reaction. NADPH competitively inhibits G6PD. It is the ratio of NADPH/NAD^+ that ultimately determines the flux of this cycle.

2. **Non-oxidative phase :** The non-oxidative reactions are concerned with the interconversion of three, four, five and seven carbon monosaccharides. Ribulose 5-phosphate is acted upon by an epimerase to produce xylulose 5-phosphate while ribose 5-phosphate ketoisomerase converts ribulose 5-phosphate to ribose 5-phosphate.

The enzyme ***transketolase*** catalyses the transfer of two carbon moiety from xylulose 5-phosphate to ribose 5-phosphate to give a 3-carbon glyceraldehyde 3-phosphate and a 7-carbon ***sedoheptulose 7-phosphate.*** Transketolase is dependent on the coenzyme ***thiamine pyrophosphate (TPP)*** and Mg^{2+} ions. Transaldolase brings about the transfer of a 3-carbon fragment (active dihydroxyacetone) from sedoheptulose 7-phosphate to glyceraldehyde 3-phosphate to give fructose 6-phosphate and four carbon erythrose 4-phosphate. Transketolase acts on xylulose 5-phosphate and transfers a 2-carbon fragment from it to erythrose 4-phosphate to generate fructose 6-phosphate and glyceraldehyde 3-phosphate.

Fructose 6-phosphate and glyceraldehyde 3-phosphate can be further catabolized through glycolysis and citric acid cycle. Glucose may also be synthesized from these two compounds.

An overview of HMP shunt is given in ***Fig.13.24***. For the complete oxidation of glucose 6-phosphate to $6CO_2$, we have to start with 6 molecules of glucose 6-phosphate. Of these 6,5

Table 13.4 Glycogen storage diseases – biochemical lesions and characteristic features

Type	Name	Enzyme defect	Organ(s) involved	Characteristic features
0	—	Glycogen synthase	Liver	Hypoglycemia, ketonemia, early death.
I	von Gierke's disease (type I glycogenosis)	Glucose 6-phosphatase	Liver, kidney and intestine	Glycogen accumulates in hepatocytes and renal cells, enlarged liver and kidney, fasting hypoglycemia, lactic acidemia; hyperlipidemia; ketosis; gouty arthritis.
II	Pompe's disease	Lysosomal α-1,4 glucosidase (acid maltase)	All organs	Glycogen accumulates in lysosomes in almost all the tissues; heart is mostly involved; enlarged liver and heart, nervous system is also affected; death occurs at an early age due to heart failure.
III	Cori's disease (limit dextrinosis, Forbe's disease)	Amylo α-1,6-glucosidase (debranching enzyme)	Liver, muscle, heart, leucocytes	Branched chain glycogen accumulates; liver enlarged; clinical manifestations are similar but milder compared to von Gierke's disease.
IV	Anderson's disease (amylopectinosis)	Glucosyl 4-6 transferase (branching enzyme)	Most tissues	A rare disease, glycogen with only few branches accumulate; cirrhosis of liver, impairment in liver function.
V	McArdle's disease (type V glycogenosis)	Muscle glycogen phosphorylase	Skeletal muscle	Muscle glycogen stores very high, not available during exercise; subjects cannot perform strenous exercise; suffer from muscle cramps; blood lactate and pyruvate do not increase after exercise; muscles may get damaged due to inadequate energy supply.
VI	Hers disease	Liver glycogen phosphorylase	Liver	Liver enlarged; liver glycogen cannot form glucose; mild hypoglycemia and ketosis seen.
VII	Tarui's disease	Phosphofructokinase	Skeletal muscle, erythrocytes	Muscle cramps due to exercise; blood lactate not elevated; hemolysis occurs.

Rare glycogen disorders VIII, IX, X and XI have been identified. They are due to defects in the enzymes concerned with activating and deactivating liver phosphorylase.

moles are regenerated with the production of 12 NADPH.

The overall reaction may be represented as

6 Glucose 6-phosphate + 12 $NADP^+$ + $6H_2O$ ⟶ $6CO_2$ + 12 NADPH + $12H^+$ + 5 Glucose 6-phosphate.

SIGNIFICANCE OF HMP SHUNT

HMP shunt is unique in generating two important products—***pentoses*** and ***NADPH***—needed for the biosynthetic reactions and other functions.

(**Note** : ATP is neither used nor generated in HMP shunt)

Importance of pentoses

In the HMP shunt, hexoses are converted into pentoses, the most important being ribose 5-phosphate. This pentose or its derivatives are useful for the ***synthesis of nucleic acids*** (RNA and DNA) and many ***nucleotides*** such as ATP, NAD^+, FAD and CoA. All tissues have to synthesize ribose they require for nucleotide and nucleic acid synthesis, since little or no ribose is found in the circulation.

Skeletal muscle is capable of synthesizing pentoses, although only the first few enzymes of HMP shunt are active. It, therefore, appears that the complete pathway of HMP shunt may not be required for the synthesis of pentoses.

Importance of NADPH

1. NADPH is required for the reductive ***biosynthesis of fatty acids and steroids***, hence HMP shunt is more active in the tissues concerned with lipogenesis, e.g. adipose tissue, liver etc.

2. NADPH is used in the synthesis of certain amino acids involving the enzyme ***glutamate dehydrogenase.***

3. There is a continuous production of H_2O_2 in the living cells which can chemically damage unsaturated lipids, proteins and DNA. This is, however, prevented to a large extent through ***antioxidant*** (free radical scavenging) ***reactions*** involving NADPH. Glutathione mediated reduction of H_2O_2 is described next.

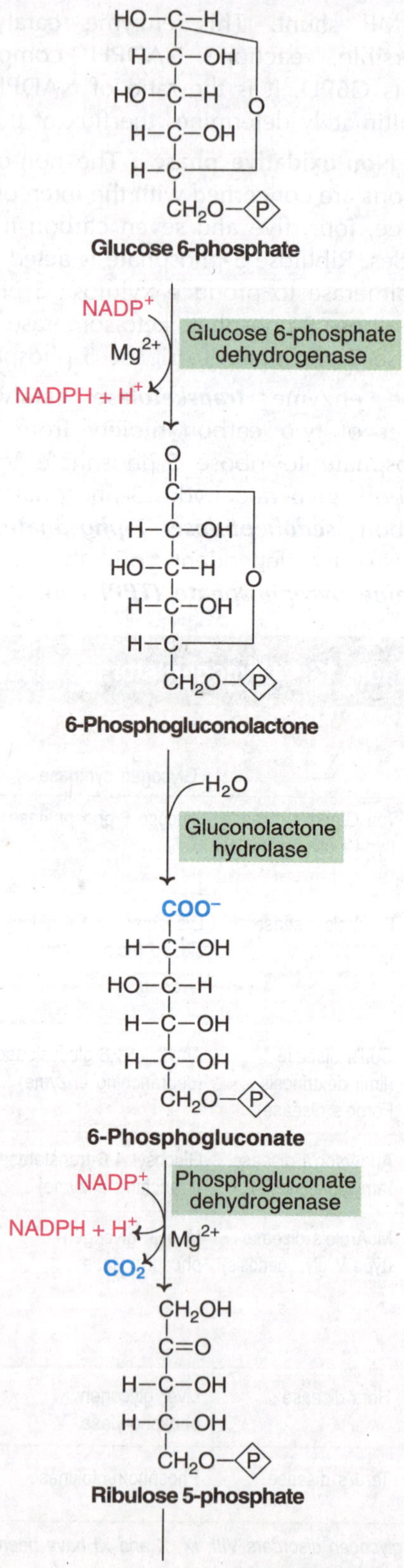

Fig. 13.23 contd. next page

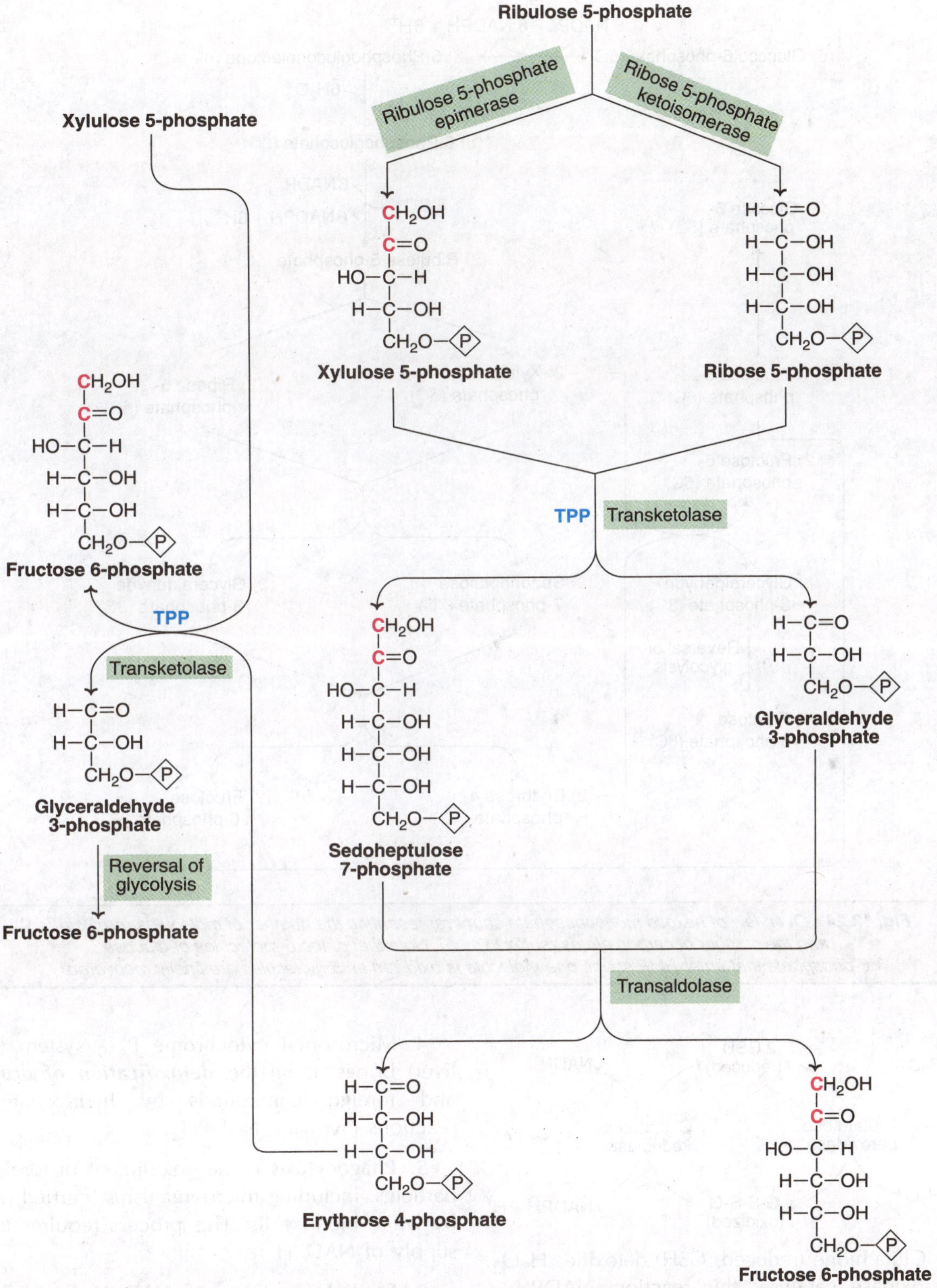

Fig. 13.23 : The hexose monophosphate shunt. (TPP – Thiamine pyrophosphate)

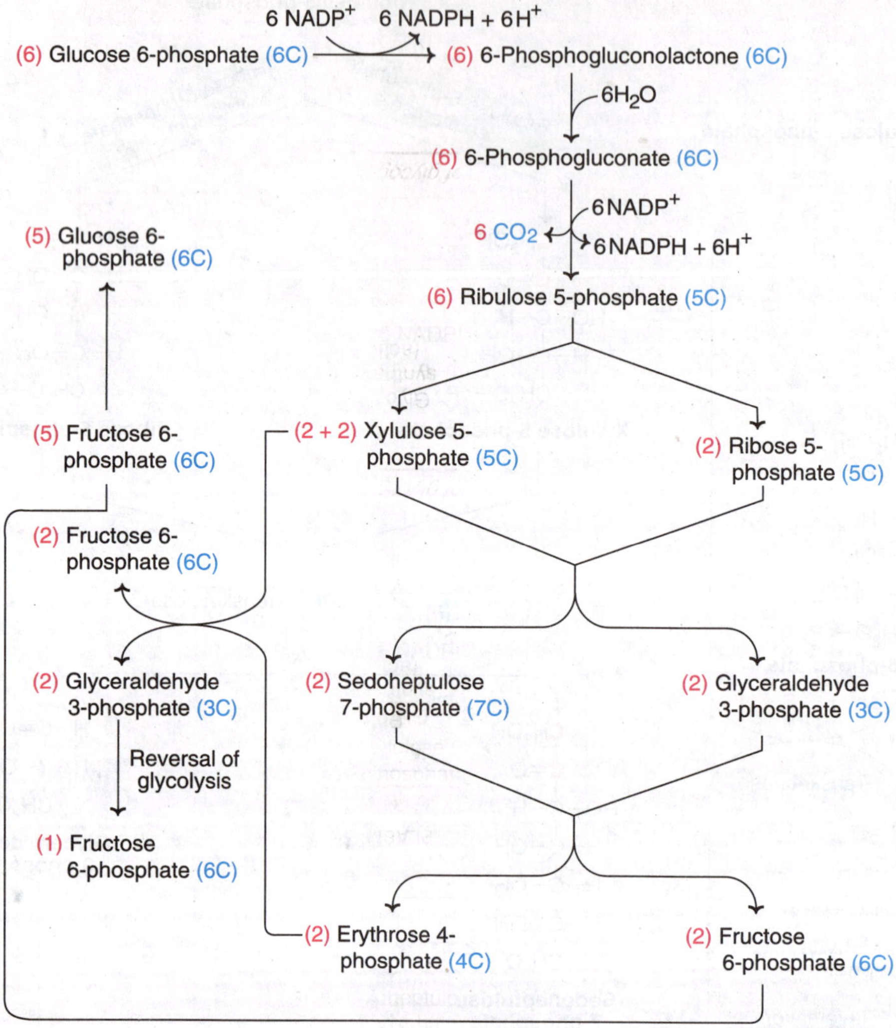

***Fig. 13.24 :** Overview of hexose monophosphate shunt representing the number of molecules (prefix in red) and the number of carbon atoms (suffix in blue). Note that of the 6-molecules of glucose 6-phosphate that enter HMP shunt, one molecule is oxidized as 5-molecules are finally recovered.*

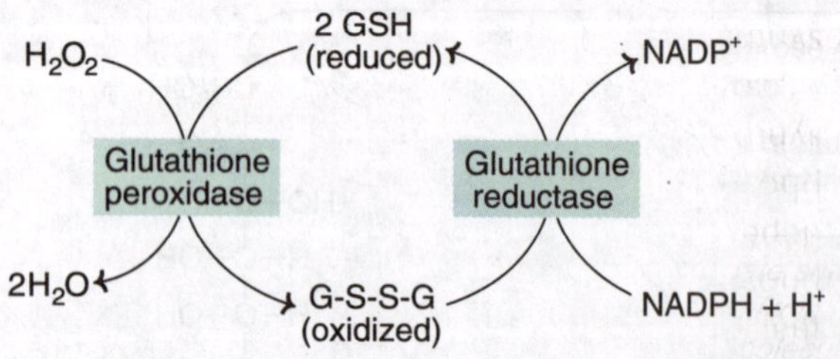

Glutathione (reduced, GSH) detoxifies H_2O_2, peroxidase catalyses this reaction. NADPH is responsible for the regeneration of reduced glutathione from the oxidized one.

4. Microsomal cytochrome P_{450} system (in liver) brings about the ***detoxification of drugs*** and foreign compounds by hydroxylation reactions involving NADPH.

5. **Phagocytosis** is the engulfment of foreign particles, including microorganisms, carried out by white blood cells. This process requires the supply of NADPH.

6. **Special functions of NADPH in RBC :** NADPH produced in erythrocytes has special functions to perform. It maintains the concen-

Medical Concepts/ Clinical Correlates

- Muscle glycogen is primarily concerned with the supply of hexoses that undergo glycolysis to provide energy during muscle contraction.
- Glycogen storage diseases—characterized by deposition of normal or abnormal type of glycogen in one or more tissues—result in muscular weakness, or even death.
- The occurrence of HMP shunt (NADPH production) in the RBC is necessary to maintain the integrity of erythrocyte membrane and to prevent the accumulation of methemoglobin.
- Deficiency of glucose 6-phosphate dehydrogenase results in hemolysis of RBC, causing hemolytic anemia. The subjects of G6PD deficiency are, however, resistant to malaria.

tration of reduced glutathione (reaction explained in 3) which is essentially required to preserve the ***integrity of RBC membrane.*** NADPH is also necessary to keep the ferrous iron (Fe^{2+}) of hemoglobin in the reduced state so that accumulation of methemoglobin (Fe^{3+}) is prevented.

7. High concentration of NADPH in lens of eyes is necessary to preserve the ***transparency of the lens.***

Glucose 6-phosphate dehydrogenase (G6PD) deficiency

G6PD deficiency is an ***inherited sex-linked*** trait. The gene of G6PD is on the X chromosome, hence males are affected. G6PD deficiency is regarded as the most common genetic defect, as about 400 million people worldwide carry mutated gene.

Although the deficiency occurs in all the cells of the affected individuals, it is more severe in RBC. HMP shunt is the only means of providing NADPH in the erythrocytes. Decreased activity of G6PD impairs the synthesis of NADPH in RBC. This results in the accumulation of methemoglobin and peroxides in erythrocytes leading to ***hemolysis.***

Clinical manifestations in G6PD deficiency : Most of the patients with G6PD deficiency do not usually exhibit clinical symptoms. Some of them, however, develop ***hemolytic anemia*** if they are administered oxidant drugs or exposed to a severe infection. The drugs such as primaquine (antimalarial), acetanilide (antipyretic), sulfamethoxazole (antibiotic) or ingestion of fava beans (***favism***) produce hemolytic jaundice in these patients. Severe infection results in the generation of free radicals (in macrophages) which can enter RBC and cause hemolysis (due to decreased NADPH and reduced GSH).

G6PD deficiency and resistance to malaria : It is interesting to note that G6PD deficiency is associated with resistance to malaria (caused by *Plasmodium falciparum*). This is explained from the fact that the ***parasites that cause malaria are dependent on HMP shunt and reduced glutathione for their optimum growth in RBC.*** Therefore, G6PD deficiency—which is seen frequently in Africans—protects them from malaria, a common disease in this region. It is regarded as an adaptability of the people living in malaria-infected regions of the world.

Biochemical diagnosis can be done by detecting reduced activity of G6PD in RBC. The ***management of G6PD deficiency*** includes avoiding oxidative stress and symptomatic treatment of hemolysis.

Wernicke-Korsakoff syndrome

This is a ***genetic disorder*** associated with ***HMP shunt***. An alteration in ***transketolase*** activity that reduces its affinity (by tenfold or so) with thiamine pyrophosphate is the biochemical lesion. The symptoms of Wernicke-Korsakoff syndrome include ***mental disorder***, loss of memory and partial paralysis. The symptoms are manifested in ***vitamin-deficient alcoholics.***

URONIC ACID PATHWAY

This is an alternative oxidative pathway for glucose and is also known as ***glucuronic acid*** *pathway* (*Fig.13.25*). It is concerned with the synthesis of glucuronic acid, pentoses, and vitamin, ascorbic acid (except in primates and guinea pigs). Dietary xylulose enters uronic acid pathway through which it can participate in other metabolisms. In most of the pathways of carbohydrate metabolism, phosphate esters participate, whereas, in uronic acid pathway, the free sugars or sugar acids are involved.

1. **Formation and importance of UDP-glucuronate :** Glucose 6-phosphate is first converted to glucose 1-phosphate. UDP-glucose is then synthesized by the enzyme UDP-glucose pyrophosphorylase. Till this step, the reactions are the same as described in glycogenesis (*Fig.13.17*). UDP-glucose dehydrogenase oxidizes UDP-glucose to UDP-glucuronate.

UDP-glucuronate is the metabolically active form of glucuronate which is utilized for conjugation with many substances like bilirubin, steroid hormones and certain drugs. Several insoluble compounds are converted to soluble ones through conjugation and, further, the ***drugs are detoxified***. UDP-glucuronate is also required for the synthesis of glycosaminoglycans and proteoglycans.

2. **Conversion of UDP-glucuronate to L-gulonate :** UDP-glucuronate loses its UDP moiety in a hydrolytic reaction and releases D-glucuronate which is reduced to L-gulonate by an NADPH-dependent reaction.

3. **Synthesis of ascorbic acid in some animals :** L-Gulonate is the precursor for the synthesis of ascorbic acid (vitamin C) in many animals. The enzyme ***L-gulonolactone oxidase***— which converts gulonate to ascorbic acid—is ***absent in man, other primates and guinea pigs***. Therefore, vitamin C has to be supplemented in the diet for these animals.

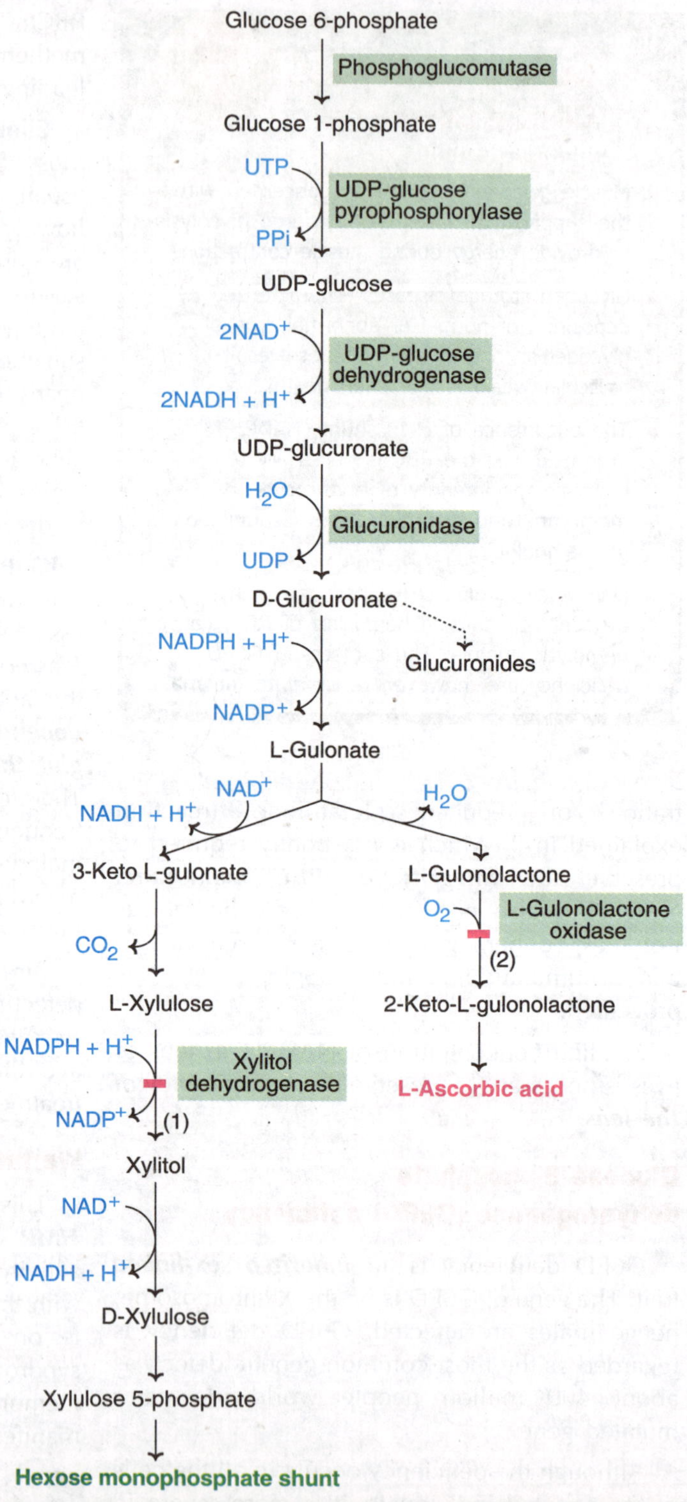

***Fig. 13.25 :** Uronic acid pathway [UDP—uridine diphosphate); (1) Block in essential pentosuria; (2) Enzyme absent in primates (including man) and guinea pigs].*

4. **Oxidation of L-gulonate :** L-Gulonate is oxidized to 3-ketogulonate and then decarboxylated to a pentose, L-xylulose. L-Xylulose is converted to D-xylulose via xylitol by a reduction (NADPH-dependent) followed by an oxidation (NAD^+-dependent) reaction. This is necessary since the D-xylulose (and not L-form)—after getting phosphorylated—can enter the hexose monophosphate shunt, for further metabolism.

Effect of drugs on uronic acid pathway

Administration of ***drugs*** (barbital, chlorobutanol etc.) ***significantly increases the uronic acid pathway*** to achieve more synthesis of glucuronate from glucose. Certain drugs (aminopyrine, antipyrine) were found to enhance the synthesis of ascorbic acid in rats.

Essential pentosuria

This is a rare genetic disorder related to the deficiency of an NADP-dependent enzyme ***xylitol dehydrogenase.*** Due to this enzyme defect, L-xylulose cannot be converted to xylitol. The affected individuals excrete large amounts of ***L-xylulose in urine.*** Essential pentosuria is asymptomatic and the individuals suffer from ***no ill-effects.*** It has been reported that the administration of drugs aminopyrine and antipyrine increases the excretion of L-xylulose in pentosuric patients.

METABOLISM OF GALACTOSE

The disaccharide ***lactose***, present in milk and milk products, is the principal dietary source of galactose. Lactase (β-galactosidase) of intestinal mucosal cells hydrolyses lactose to galactose and glucose. Galactose is also produced within the cells from the lysosomal degradation of glycoproteins and glycolipids. As is the case for fructose, galactose entry into the cells is not dependent on insulin.

The specific enzyme, namely ***galactokinase***, phosphorylates galactose to galactose 1-phosphate. This reacts with UDP-glucose in an exchange reaction to form UDP-galactose in presence of the enzyme galactose 1-phosphate uridyltransferase (***Fig.13.26***). UDP-galactose is an active donor of galactose for many synthetic reactions involving the formation of compounds like lactose, glycosaminoglycans, glycoproteins, cerebrosides and glycolipids. UDP-galactose can

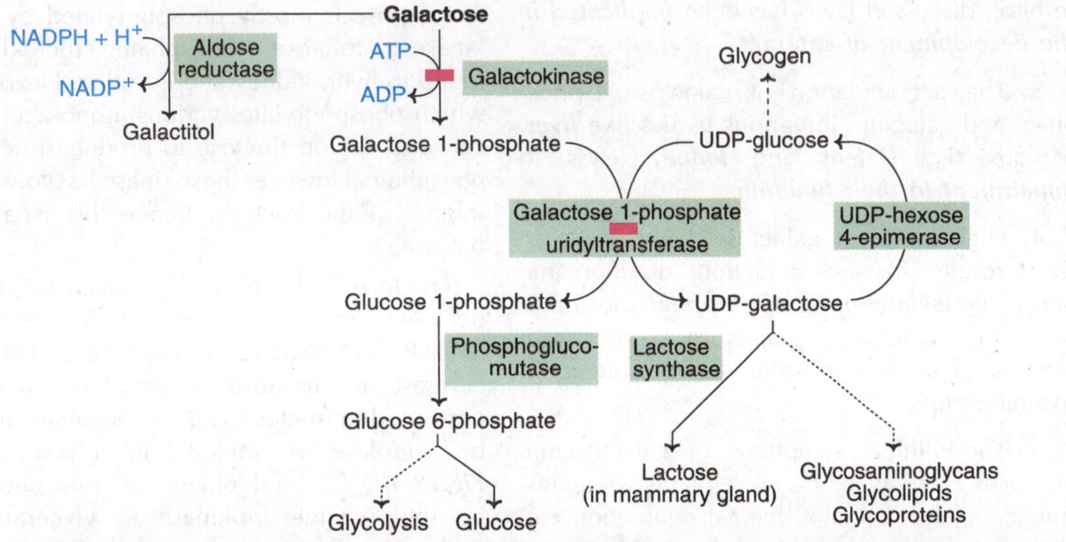

Fig. 13.26 : Metabolism of galactose (defects in galactosemia are indicated by red blocks).

be converted to UDP-glucose by UDP hexose 4-epimerase. In this way, ***galactose can enter the metabolic pathways of glucose***. It may be noted that ***galactose is not an essential nutrient*** since UDP-glucose can be converted to UDP-galactose by the enzyme UDP- hexose 4-epimerase.

DISORDERS OF GALACTOSE METABOLISM

Classical galactosemia

Galactosemia (incidence 1 : 30000) is due to the ***deficiency of*** the enzyme ***galactose 1-phosphate uridyltransferase.*** It is a rare congenital disease in infants, inherited as an autosomal recessive disorder. The salient features of galactosemia are listed.

1. Galactose metabolism is impaired leading to ***increased galactose levels*** in circulation (***galactosemia***) and ***urine*** (***galactosuria***).

2. The accumulated galactose is diverted for the production of ***galactitol*** (***dulcitol***) by the enzyme aldose reductase (the same enzyme that converts glucose to sorbitol). Aldose reductase is present in lens, nervous tissue, seminal vesicles etc. The conversion of galactose to galactitol is insignificant in routine galactose metabolism. However, with increased levels of galactose, this pathway assumes significance. Galactitol (like sorbitol, discussed later) has been ***implicated in the development of cataract***.

3. The accumulation of galactose 1-phosphate and galactitol in various tissues like ***liver***, nervous tissue, ***lens*** and ***kidney*** leads to ***impairment in their function***.

4. High levels of galactose 1-phosphate in liver results in the depletion of inorganic phosphate (***sequestering of phosphate***) for other metabolic functions. Galactose 1-phosphate inhibits glycogen phosphorylase resulting in ***hypoglycemia***.

5. The clinical symptoms of galactosemia are—loss of weight (in infants) hepatosplenomegaly, jaundice, mental retardation etc. In severe cases, ***cataract***, ***amino aciduria*** and albuminuria are also observed.

Diagnosis : Early detection of galactosemia is possible (biochemical diagnosis) by measuring the ***activity of galactose 1-phosphate uridyltransferase in erythrocytes***.

Treatment : The therapy includes the supply of ***diet deprived of galactose and lactose***.

Galactokinase deficiency : The defect in the enzyme ***galactokinase,*** responsible for phosphorylation of galactose, will also result in galactosemia and galactosuria. Here again galactose is shunted to the formation of galactitol. Generally, galactokinase-deficient individuals do not develop hepatic and renal complications. Development of cataract occurs at a very early age, sometimes within an year after birth. The treatment is the removal of galactose and lactose from the diet.

METABOLISM OF FRUCTOSE

The major dietary source of fructose is the disaccharide sucrose (cane sugar), and ***high-fructose corn syrups (HFCS)*** used in the manufactured foods and beverages. It is also found in free form in ***honey and many fruits***. In the body, entry of fructose into the cells is not controlled by the hormone insulin. This is in contrast to glucose which is regulated for its entry into majority of the tissues.

Fructose is mostly phosphorylated by fructokinase to fructose 1-phosphate. Fructokinase is found in liver, kidney and intestine. Hexokinase, which phosphorylates various monosaccharides, can also act on fructose to produce fructose 6-phosphate. However, hexokinase has low affinity (high K_m) for fructose, hence this is a minor pathway.

Fructose 1-phosphate is cleaved to glyceraldehyde and dihydroxyacetone phosphate (DHAP) by ***aldolase B (Fig.13.27)***. This is in contrast to fructose 6-phosphate which is converted to fructose 1, 6-bisphosphate and split by ***aldolase A*** (details in glycolysis—***See Fig.13.4***). Glyceraldehyde is phosphorylated by the enzyme triokinase to glyceraldehyde 3-phosphate which, along with DHAP, enters glycolysis or gluconeogenesis.

The fructose is more rapidly metabolized (via glycolysis) by the liver than glucose. This is due to the fact that the rate limiting reaction in glycolysis, catalysed by phosphofructokinase is bypassed. Increased dietary intake of fructose significantly elevates the production of acetyl CoA and lipogenesis (fatty acid, triacylglycerol and very low density lipoprotein synthesis). Ingestion of large quantities of fructose or sucrose is linked with many health complications.

Sorbitol / Polyol pathway

Polyol pathway (so termed since sorbitol is a polyhydroxy sugar) basically involves the ***conversion of glucose to fructose via sorbitol*** (***Fig.13.27***). This pathway is absent in liver. Sorbitol pathway is directly related to glucose concentration, and is ***higher in uncontrolled diabetes.***

The enzyme aldose reductase reduces glucose to sorbitol (glucitol) in the presence of NADPH. Sorbitol is then oxidized to fructose by sorbitol dehydrogenase and NAD^+. Aldose reductase is absent in liver but found in many tissues like lens and retina of the eye, kidney, placenta, Schwann cells of peripheral nerves, erythrocytes and seminal vesicles. The enzyme sorbitol dehydrogenase is present in seminal vesicle, spleen and ovaries. ***Fructose*** is a preferred carbohydrate for ***energy needs of sperm cells*** due to the ***presence of sorbitol pathway***.

Sorbitol pathway in diabetes mellitus

In uncontrolled diabetes (hyperglycemia), large amounts of glucose enter the cells which are not dependent on insulin. Significantly, the cells with increased intracellular glucose levels

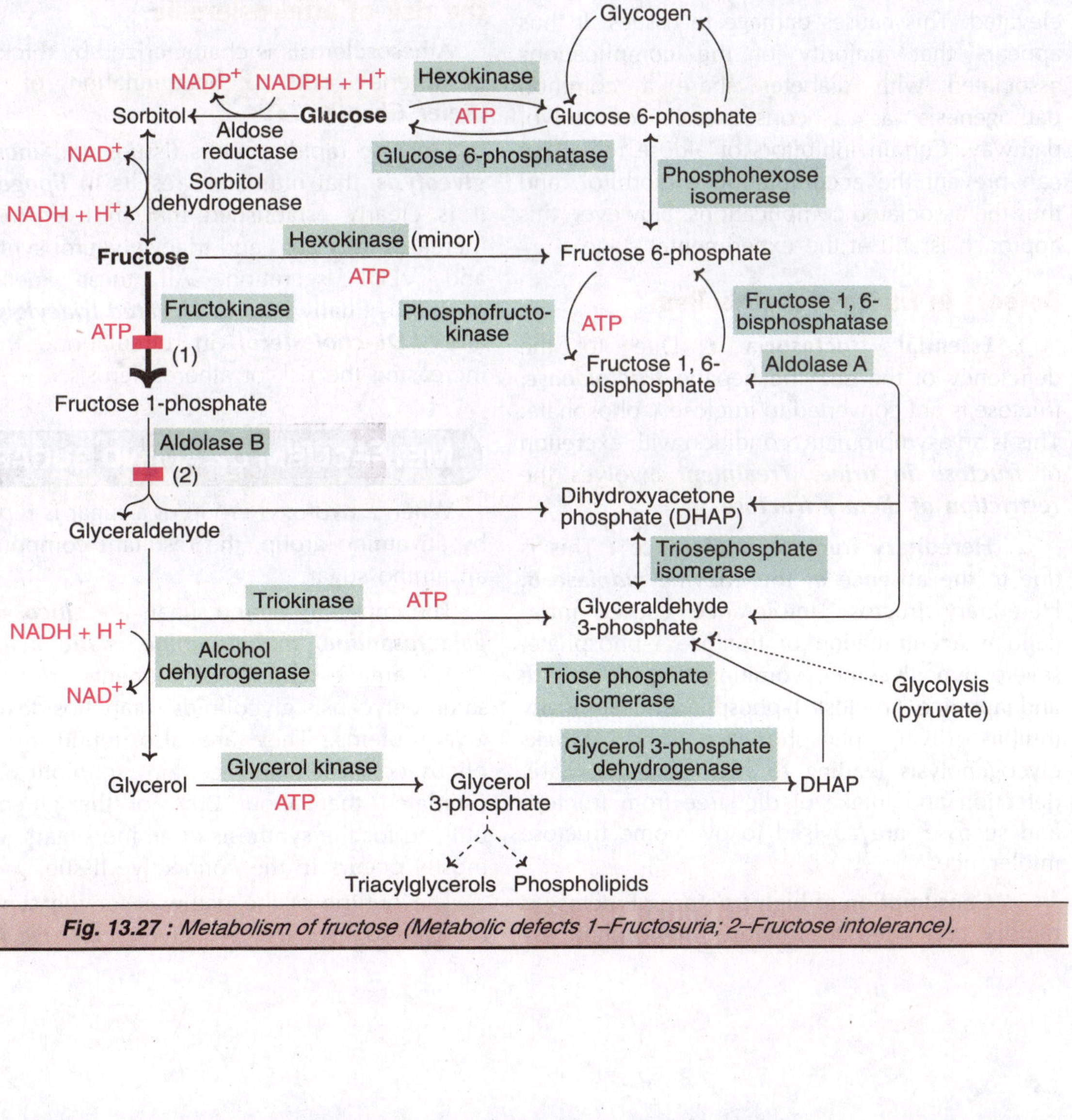

Fig. 13.27 : Metabolism of fructose (Metabolic defects 1–Fructosuria; 2–Fructose intolerance).

in diabetes (lens, retina, nerve cells, kidney etc.) possess high activity of aldose reductase and sufficient supply of NADPH. This results in a rapid and ***efficient conversion of glucose to sorbitol***. The enzyme sorbitol dehydrogenase, however, is either low in activity or absent in these cells, hence ***sorbitol is not converted to fructose***. Sorbitol cannot freely pass through the cell membrane, and ***accumulate in the cells*** where it is produced. Sorbitol—due to its hydrophilic nature—causes strong osmotic effects leading to swelling of the cells. Some of the ***pathological changes associated with diabetes*** (like ***cataract*** formation, peripheral ***neuropathy***, ***nephropathy*** etc.) are ***believed to be due to the accumulation of sorbitol***, as explained above.

It is clearly known that in diabetic animals sorbitol content of lens, nerve, and glomerulus is elevated. This causes damage to tissues. It thus appears that majority of the complications associated with diabetes share a common pathogenesis as a consequence of polyol pathway. Certain inhibitors of aldose reductase can prevent the accumulation of sorbitol, and thus the associated complications. However, this approach is still at the experimental stage.

Defects in fructose metabolism

1. **Essential fructosuria :** Due to the deficiency of the enzyme hepatic ***fructokinase***, fructose is not converted to fructose 1-phosphate. This is an asymptomatic condition with excretion of ***fructose in urine***. ***Treatment*** involves the ***restriction of dietary fructose***.

2. **Hereditary fructose intolerance :** This is due to the absence of the enzyme ***aldolase B***. Hereditary fructose intolerance causes intracellular accumulation of fructose 1-phosphate, severe hypoglycemia, vomiting, hepatic failure and jaundice. Fructose 1-phosphate allosterically inhibits liver phosphorylase and blocks glycogenolysis leading to hypoglycemia. Early detection and intake of diet free from fructose and sucrose, are advised to overcome fructose intolerance.

3. **Consumption of high fructose :** Fructose is rapidly converted to fructose 1-phosphate by fructokinase. The activity of the enzyme aldolase B is relatively less, and, due to this, fructose 1-phosphate accumulates in the cell. This leads to the ***depletion of intracellular inorganic phosphate*** (Pi) levels. The phenomenon of binding of Pi to the organic molecules (like fructose here)—that leads to the less availability of Pi for the essential metabolic functions—is known as ***sequestering of phosphate***. Due to the decreased availability of Pi, which happens in overconsumption of fructose, the liver metabolism is adversely affected. This includes the lowered synthesis of ATP from ADP and Pi. High consumption of fructose over a long period is associated with ***increased uric acid in blood leading to gout***. This is due to the excessive breakdown of ADP and AMP (accumulated due to lack of Pi) to uric acid.

High fructose consumption and the risk of atherosclerosis

Athesosclerosis is characterized by thickening of arteries due to accumulation of lipids (***Refer Chapter 14***).

Fructose rapidly enters tissues and ***increases glycolysis***, that ultimately results in ***lipogenesis***. It is clearly established that in liver fructose increases fatty acid and triacylglycerol synthesis, and VLDL secretion. All these metabolic processes finally lead to ***elevated triacylglycerol*** and ***LDL-cholesterol*** in circulation, thereby increasing the risk of atherosclerosis.

METABOLISM OF AMINO SUGARS

When a hydroxyl group of a sugar is replaced by an amino group, the resultant compound is an amino sugar.

The important amino sugars are ***glucosamine***, ***galactosamine***, ***mannosamine***, ***sialic acid*** etc. They are essential components of glycosaminoglycans, glycolipids (gangliosides) and glycoproteins. They are also found in some oligosaccharides and certain antibiotics. It is estimated that about 20% of the glucose is utilized for the synthesis of amino sugars, which mostly occurs in the connective tissue.

The outline of the pathway for the synthesis of amino sugars is given in ***Fig.13.28***.

Medical Concepts/ Clinical Correlates

- Uronic acid pathway is concerned with the production of glucuronic acid (involved in detoxification), pentoses and vitamin C. Man is incapable of synthesizing vitamin C due to the absence of a single enzyme—L-gulonolactone oxidase.
- The conversion of glucose to fructose is impaired in diabetes mellitus, causing accumulation of sorbitol. This compound has been implicated in the development of cataract, nephropathy, peripheral neuropathy etc.
- Severe cases of galactosemia are associated with the development of cataract, believed to be due to the accumulation of galactitol.

Fructose 6-phosphate is the major precursor for glucosamine, N-acetylgalactosamine and N-acetylneuraminic acid (NANA). The utilization of the amino sugars for the formation of glycosaminolgycans, glycoproteins and gangliosides is also indicated in this figure.

Mucopolysaccharidoses

The ***lysosomal storage diseases*** caused by enzyme defects in the degradation of glycosaminoglycans (GAGs) are known as mucopolysaccharidoses (***Table 13.5***). Mucopolysaccharidoses are characterized by the accumulation of GAGs in various tissues that may result in skeletal deformities, and mental retardation. Mucopolysaccharidoses are important for elucidating the role of lysosomes in health and disease.

GLYOXYLATE CYCLE

The animals, including man, cannot carry out the net synthesis of carbohydrate from fat. However, the ***plants*** and many ***microorganisms*** are equipped with the metabolic machinery—namely glyoxylate cycle—***to convert fat into carbohydrates***. This pathway is very significant

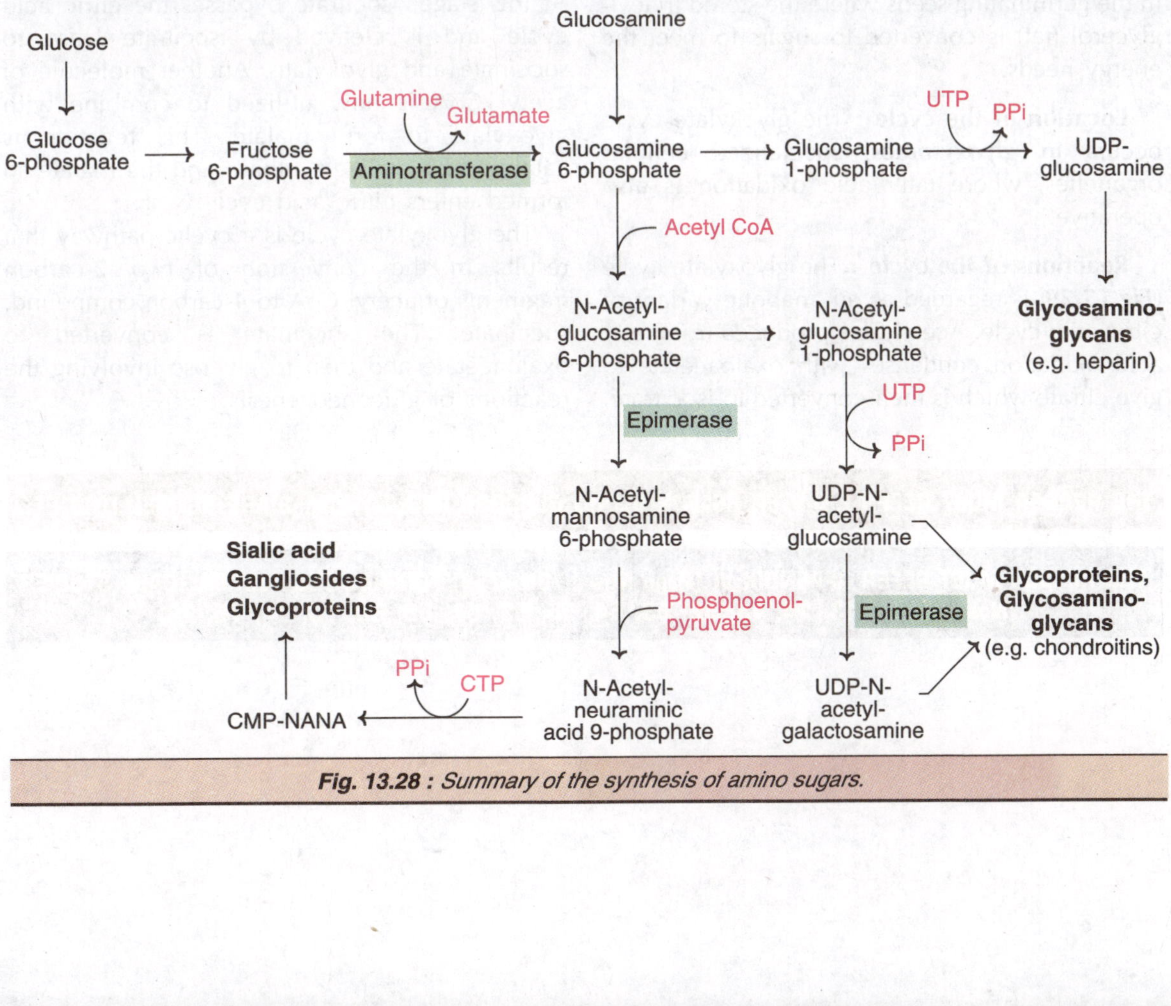

Fig. 13.28 : Summary of the synthesis of amino sugars.

Table 13.5 A selected list of mucopolysaccharidoses with defective enzymes and accumulated products

Syndrome	Enzyme defect	Accumulated product(s)
Hurler	α-Iduronidase	Heparan sulfate, Dermatan sulfate
Hunter	Iduronate sulfatase	Heparan sulfate Dermatan sulfate
Maroteauxlamy	N-Acetylgalactosamine sulfatase	Dermatan sulfate
Mucolipidosis VIII	β-Glucuironidase	Heparan sulfate Dermatan sulfate
Sanfilippo type A	Heparan sulfamidase	Heparan sulfate
Sanfilippo type B	N-Acetyl glucosaminase	Heparan sulfate
Sanfilippo type C	Acetyl CoA:α-glucosaminide acetyltransferase	Heparan sulfate
Sanfilippo type D	N-Acetyl glucosamine 6-sulfatase	Heparan sulfate
Morquio type A	Galactose 6-sulfatase	Keratan/chondroitin sulfate
Morquio type B	β-Galactosidase	Keratan sulfate

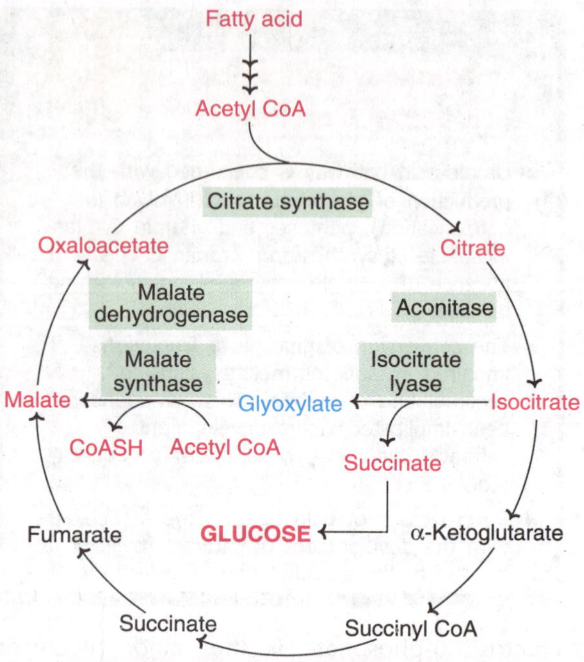

***Fig. 13.29 :** The glyoxylate cycle (enzymes represented in green shade).*

in the germinating seeds where the stored triacylglycerol (fat) is converted to sugars to meet the energy needs.

Location of the cycle : The glyoxylate cycle occurs in ***glyoxysomes***, specialized cellular organelles, where fatty acid oxidation is also operative.

Reactions of the cycle : The glyoxylate cycle (***Fig.13.29***) is regarded as an anabolic variant of citric acid cycle. Acetyl CoA produced from fatty acid oxidation condenses with oxaloacetate to give citrate which is then converted to isocitrate. At this stage, isocitrate bypasses the citric acid cycle and is cleaved by isocitrate lyase to succinate and glyoxylate. Another molecule of acetyl CoA is now utilized to combine with glyoxylate to form malate. This reaction is catalysed by malate synthase and the malate so formed enters citric acid cycle.

The glyoxylate cycle is a cyclic pathway that results in the conversion of two 2-carbon fragments of acetyl CoA to 4-carbon compound, succinate. The succinate is converted to oxaloacetate and then to glucose involving the reactions of gluconeogenesis.

THE READER MUST REFER CHAPTER 36 FOR BLOOD GLUCOSE REGULATION, DIABETES MELLITUS ETC.

THE READER MUST REFER APPENDIX VI (Page Nos : 738 - 756) FOR RELATED CLINICAL CASE STUDIES (Nos : 6, 12, 13, 14, 15, 52, 53, 54, 55, 56, 57, 58, 62, 73, 74, 75, 76)

SUMMARY

- *Carbohydrates are the major source of energy for the living cells. Glucose (normal fasting blood level 70-100 mg/dl) is the central molecule in carbohydrate metabolism, actively participating in a number of metabolic pathways—glycolysis, gluconeogenesis, glycogenesis, glycogenolysis, hexose monophosphate shunt, uronic acid pathway etc.*
- *Glucose is oxidized in glycolysis, either in anaerobic (2 ATP formed) or aerobic (7 ATP formed) conditions, resulting in the formation of 2 moles of lactate or pyruvate, respectively.*
- *Acetyl CoA is produced from pyruvate which is completely oxidized in citric acid cycle, the final common oxidative pathway for all foodstuffs. The complete oxidation of one mole of glucose generates 32 ATP.*
- *Gluconeogenesis is the synthesis of glucose from noncarbohydrate precursors like amino acids (except leucine and lysine), lactate, glycerol, propionate etc. The reversal of glycolysis with alternate arrangements made at three irreversible reactions of glycolysis constitutes gluconeogenesis.*
- *Glycogen is the storage form of glucose. The degradation of glycogen (glycogenolysis) in muscle meets the immediate fuel requirements, whereas the liver glycogen maintains the blood glucose level. Enzyme defects in synthesis or degradation of glycogen lead to storage disorders. von Gierke's disease (Type I) is due to the defect in the enzyme glucose 6-phosphatase.*
- *Hexose monophosphate shunt (HMP shunt) is the direct oxidative pathway of glucose. HMP shunt assumes significance since it generates NADPH and pentoses, respectively required for the synthesis of lipids and nucleic acids.*
- *Glucuronate—involved in the conjugation of bilirubin, steroid hormones and detoxification of drugs—is synthesized in uronic acid pathway. Due to a single enzyme defect (gulonolactone oxidase) in this pathway, man cannot synthesize ascorbic acid (vitamin C) whereas some animals can.*
- *Galactosemia is mostly due to the defect in the enzyme galactose 1-phosphate uridyltransferase. This results in the diversion of galactose to produce galactitol which has been implicated in the development of cataract.*
- *Glucose can be converted to fructose via sorbitol pathway. In prolonged hyperglycemia (uncontrolled diabetes), sorbitol accumulates in the tissues, resulting in cataract, nephropathy, peripheral neuropathy etc.*
- *Amino sugars (glucosamine, galactosamine, mannosamine etc.), synthesized from fructose 6-phosphate are essential components of glycosaminoglycans, glycolipids and glycoproteins.*

Self-Assessment Exercises

I. Essay questions

1. Describe briefly the metabolism of glucose 6-phosphate.
2. Give an account of glycogen metabolism.
3. Justify that citric acid cycle is the final common metabolic pathway for the oxidation of foodstuffs.
4. Discuss the synthesis of glucose from non-carbohydrate sources.
5. Describe the hexose monophosphate shunt and add a note on its significance.

II. Short notes

(a) Glycogenolysis, (b) UDPG, (c) Galactosemia, (d) Cori cycle, (e) 2, 3- BPG, (f) Glycogen storage diseases, (g) Essential fructosuria, (h) Conversion of pyruvate to acetyl CoA, (i) Energetics of TCA cycle, (j) TPP in carbohydrate metabolism.

III. Fill in the blanks

1. Name the five vitamins required by pyruvate dehydrogenase or α-ketoglutarate dehydrogenase complex ____________.
2. Muscle glycogen does not directly contribute to blood glucose due to absence of the enyme ____________.
3. Ascorbic acid is not synthesized in man due to lack of the enzyme ____________.
4. The compound implicated in the development of cataract in diabetic patients is ____________.
5. Galactosemia is mostly due to the deficiency of the enzyme ____________.
6. The two amino acids that are never glucogenic are ____________ and ____________.
7. Substrate level phosphorylation in citric acid cycle is catalysed by the enzyme ____________.
8. The metabolic pathway concerned with the conversion of L-xylulose to D-xylulose is ____________.
9. The name of the protein that has been identified to serve as a primer for glycogen synthesis is ____________.
10. The metabolite among the citric acid cycle intermediates performing a catalytic role ____________.

IV. Multiple choice questions

11. One of the following enzymes in glycolysis catalyses an irreversible reaction.
    (a) Hexokinase (b) Phosphofructokinase (c) Pyruvate kinase (d) All of them.
12. Synthesis of 2, 3-bisphosphoglycerate occurs in the tissue namely.
    (a) Liver (b) Kidney (c) Erythrocytes (d) Brain.
13. The hormone that lowers cAMP concentration in liver cells is
    (a) Glucagon (b) Insulin (c) Epinephrine (d) Thyroxine.
14. The number of ATP produced when a molecule of acetyl CoA is oxidized through citric acid cycle
    (a) 12 (b) 24 (c) 38 (d) 15.
15. The connecting link between HMP shunt and lipid synthesis is
    (a) Ribose (b) NADPH (c) Sedoheptulose 7-phosphate (d) NADH.

Metabolism of Lipids 14

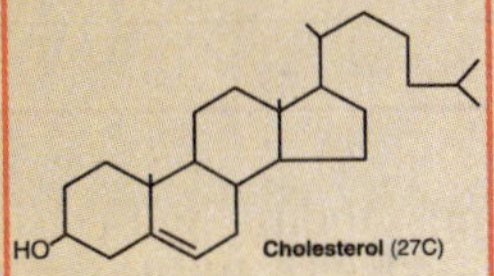
Cholesterol (27C)

Cholesterol, the most feared among lipids, speaks

"Consumed through diet and produced in the body;
Participate in innumerable cellular functions;
Implicated in several health complications;
And blamed I am, for no fault of mine!"

Lipids are indispensable for cell structure and function. Due to their hydrophobic and non-polar nature, lipids differ from rest of the body compounds and are unique in their action.

Triacylglycerols — the body fuel reserve

Lipids constitute about 15-20% of the body weight in humans. Triacylglycerols ***(formerly triglycerides)*** are the most abundant lipids comprising 85-90% of body lipids. Most of the triacylglycerols (TG; also called ***neutral fat*** or depot fat) are stored in the ***adipose tissue*** and serve as energy reserve of the body. This is in contrast to carbohydrates and proteins which cannot be stored to a significant extent for energy purposes. Fat also acts as an insulating material for maintaining the body temperature of animals.

Why should fat be the fuel reserve of the body?

Triacylglycerols are the most predominant storage form of energy. There are two main reasons for fat being the fuel reserve of the body

1. Triacylglycerols (TG) are highly ***concentrated form of energy***, yielding ***9 Cal/g***, in contrast to carbohydrates and proteins that produce only 4 Cal/g. This is because fatty acids found in TG are in the reduced form.

2. The triacylglycerols are non-polar and hydrophobic in nature, hence ***stored in*** pure form without any association with water ***(anhydrous form)***. On the other hand, glycogen and proteins are polar. One gram of glycogen combines with 2 g of water for storage.

For the two reasons stated above, ***one gram of fat stored in the body yields nearly six times as much energy as one gram of (hydrated) glycogen***. In a healthy adult individual (weighing 70 kg), about 10-11 kg of fat is stored (mostly in adipose tissue) which corresponds to a fuel reserve of 100,000 Cals. If this much of energy were to be ***stored as glycogen*** (instead of fat), then the ***weight*** of the person would ***increase by at least 55 kg!*** This explains why fat has been chosen as a fuel reserve during evolution.

Long chain fatty acids (of fat) are the ideal ***storage fuel reserves*** of the body. Fats can support the body's energy needs for long periods of food deprivation. In extreme cases, humans can fast and survive for 60–90 days, and the obese persons can survive even longer (6 months to one year!) without food.

Hibernating animals provide good example for utilizing fat reserve as fuel. For instance, bears go on hibernation for about 7 months and, during this entire period, the energy is derived from the degradation of fat stores. The ruby-throated humming birds fly, non-stop between New England and West Indies (2,400 km!), at a speed of 40 km/hr for 60 hours! This is possible only due to the stored fat.

Other important body lipids

Phospholipids, ***glycolipids*** and ***cholesterol*** are major components of cell membranes. Cholesterol is also a precursor for bile acids and steroid hormones. ***Arachidonic acid***—an unsaturated fatty acid—is the substrate for the synthesis of certain intercellular regulators—prostaglandins, thromboxanes, prostacyclins etc.

Transport of lipids

The insoluble lipids are solubilized in association with proteins to form ***lipoproteins*** in which form lipids are transported in the blood stream. Free lipids are undetectable in blood.

Chylomicrons, very low density lipoproteins (***VLDL***), low density lipoproteins (***LDL***), high density lipoproteins (***HDL***) and albumin-free fatty acids are the different lipoprotein complexes that transport lipids in the blood stream.

Transport of free fatty acids (FFAs) : FFAs are also called as ***non-esterified (NEFAs)*** or ***unesterified (UFAs) fatty acids***. In the plasma, long chain FFAs are bound to albumin and in the cells, they are attached to ***fatty acid binding protein***. Thus, long chain FFAs are never found in free form. However, short chain fatty acids, being relatively more water soluble, exist as fatty acid ion or sometimes unionized form.

Plasma lipids

The various fractions of lipids in the plasma can be estimated by different methods after extracting them with lipid solvents. The plasma levels of lipids (***Table 14.1***) are often useful for assessing the health of the individuals.

Table 14.1 The plasma concentration of lipids (lipid profile) in humans

Lipid fraction	Reference values (mg/dl)
Total lipid	400-600
Total cholesterol	150–200
LDL-cholesterol	<100
HDL-cholesterol	40–60
VLDL-cholesterol	20–40
Triglycerides	75–150
Phospholipids	150–200
Free fatty acids	5–15

Dynamic state of body lipids

It was earlier thought that the lipids are inert storage compounds and are less significant metabolically. However, later experiments with isotope studies have proved that the body lipids are ***continuously being degraded and resynthesized***. As already stated, fat stored in the adipose tissue is the fuel reserve of the body. This is in a dynamic state.

The triacylglycerols transported from intestine (as chylomicrons) and liver (as VLDL) are stored in the adipose tissue. Besides, they are also utilized by muscle, liver, heart etc., as per the needs of the body. An overview of fat metabolism is depicted in ***Fig.14.1***.

Mobilization of fat from adipose tissue

Triacylglycerol (***TG***) is the stored fat in the adipose tissue. The enzyme, namely ***hormone-sensitive triacylglycerol lipase,*** removes the fatty acid either from carbon 1 or 3 of the triacylglycerol to form diacylglycerol. The other two fatty acids of TG are cleaved by additional lipases specific for diacylglycerol and monoacylglycerol. The complete degradation of triacylglycerol to glycerol and free acids is known as ***lipolysis*** (***Fig.14.2***).

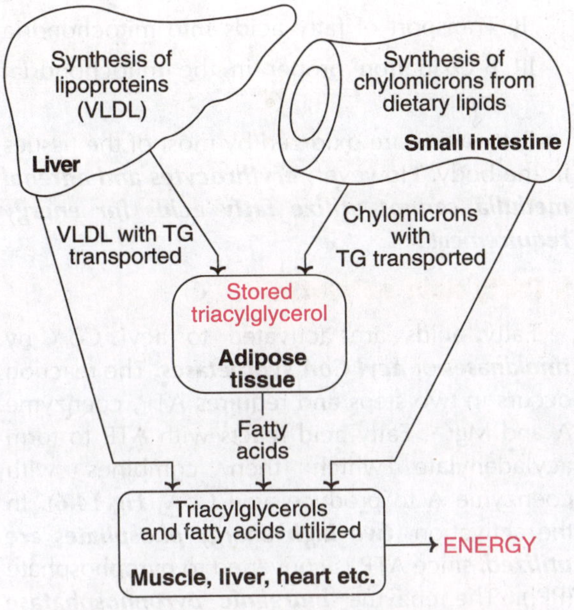

***Fig. 14.1 :** Overview of fat metabolism.*

Regulation of hormone-sensitive TG-lipase

Hormone-sensitive TG-lipase is so named because its activity is mostly controlled by hormones. Lipase is present in an inactive form '*b*' and is activated (phosphorylated) by a cAMP dependent protein kinase to lipase '*a*'. Several hormones—such as ***epinephrine*** (most effective), norepinephrine, glucagon, thyroxine, ACTH etc.— enhance the activity of adenylate cyclase and, thus, ***increase lipolysis***. On the other hand, insulin decreases cAMP levels and thereby inactivates lipase. Caffeine promotes lipolysis by increasing cAMP levels through its inhibition on phosphodiesterase activity. The control of cAMP mediated lipolysis is illustrated in ***Fig.14.3***.

Fate of glycerol : The adipose tissue lacks the enzyme glycerol kinase, hence glycerol produced in lipolysis cannot be phosporylated here. It is ***transported to liver*** where it is ***activated to glycerol 3-phosphate***. The latter may be used for the synthesis of triacylglycerols and phospholipids. Glycerol 3-phosphate may also enter glycolysis by getting converted to dihydroxyacetone phosphate (***Fig.14.4***).

Fate of free fatty acids : The fatty acids released in the adipocytes enter the circulation and are transported in a bound form to albumin. The free fatty acids enter various tissues and are utilized for the energy. About ***95% of the energy*** obtained from fat ***comes from the oxidation of fatty acids***. Brain (astrocytes) has very limited ability to oxidize fatty acids. Certain tissues, however, cannot oxidize fatty acids, e.g. erythrocytes, adrenal medulla.

Wolman disease

Wolman disease is a rare genetic disease characterized by ***accumulation of lipids*** (TG, cholesteryl esters etc.) in spleen, liver, small intestine, adrenal gland etc. This occurs due to the ***deficiency of*** the enzyme ***lysosomal acid lipase***. Children affected by Wolman disease do not survive long.

Triacylglycerol / fatty acid cycle

During starvation, TG stored in adipose tissue is hydrolysed to free fatty acids (for oxidation) to provide energy to skeletal and cardiac muscle. However, about 65% of these FFA are converted

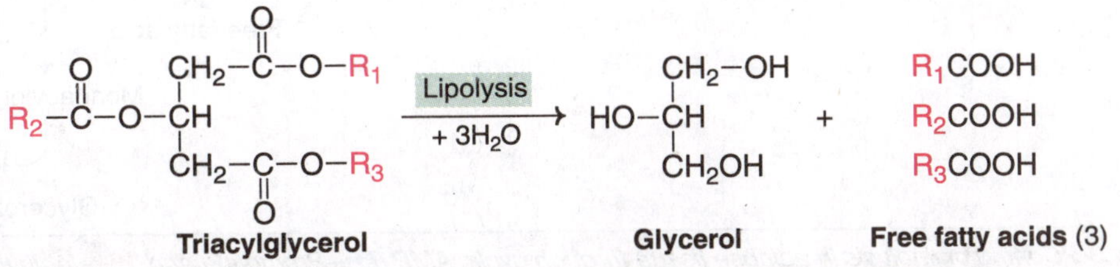

***Fig. 14.2 :** Complete hydrolysis (lipolysis) of triacylglycerol.*

to TG, and sent back to adipose tissue for deposition. This ***process of lipolysis of TG and reesterification of FFA to TG*** is termed as triacylglycerol/fatty acid cycle.

An overview of TG and FFA metabolism showing major pathways and products is shown in Fig.14.5.

FATTY ACID OXIDATION

The fatty acids in the body are mostly oxidized by ***β-oxidation***. β-Oxidation may be defined as the ***oxidation of fatty acids on the β-carbon atom***. This results in the sequential ***removal of*** a two carbon fragment, ***acetyl CoA***.

Fatty acid oxidation—stages and tissues

The β-oxidation of fatty acids involves three stages

I. Activation of fatty acids occurring in the cytosol

II. Transport of fatty acids into mitochondria

III. β-Oxidation proper in the mitochondrial matrix.

Fatty acids are oxidized by most of the tissues in the body. However, ***erythrocytes and adrenal medulla cannot utilize fatty acids for energy requirement***.

I. Fatty acid activation

Fatty acids are activated to acyl CoA by ***thiokinases*** or ***acyl CoA synthetases.*** The reaction occurs in two steps and requires ATP, coenzyme A and Mg^{2+}. Fatty acid reacts with ATP to form acyladenylate which then combines with coenzyme A to produce acyl CoA (***Fig.14.6***). In the activation, ***two high energy phosphates are utilized***, since ATP is converted to pyrophosphate (PPi). The enzyme ***inorganic pyrophosphatase*** hydrolyses PPi to phosphate (Pi). The immediate elimination of PPi makes this reaction totally irreversible. Thiokinase reaction in the oxidation of fatty acids is the only step that requires ATP.

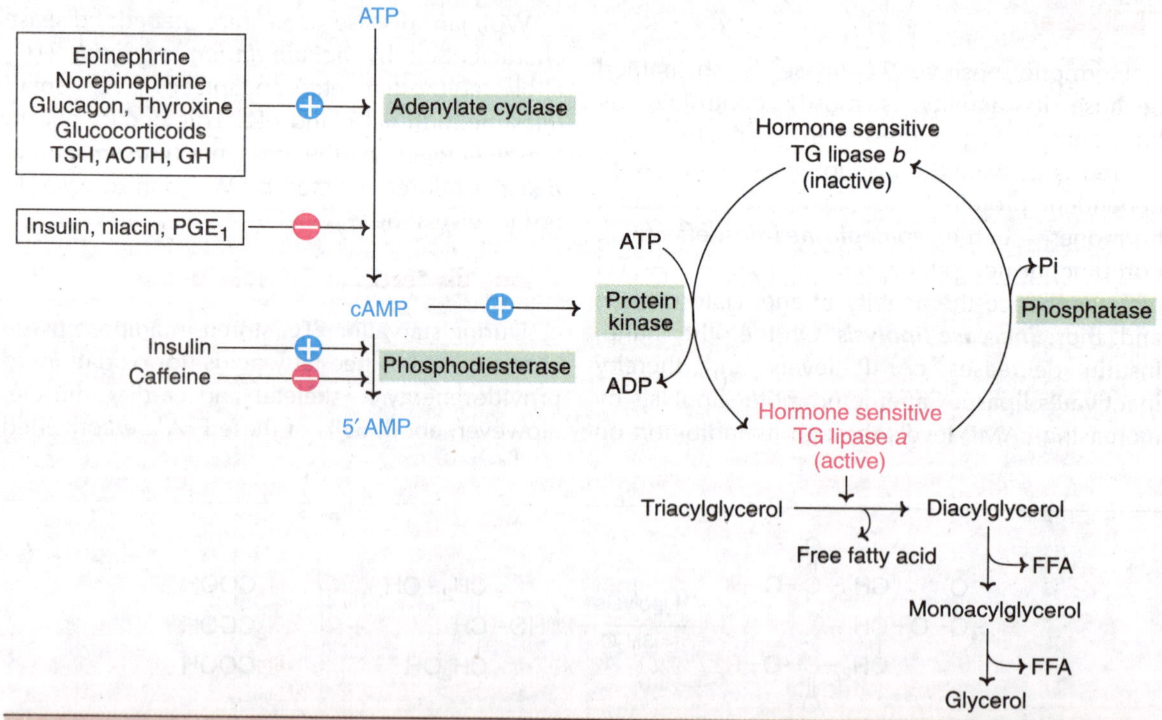

***Fig. 14.3 :** Control of lipolysis in adipose tissue through cyclic AMP (⊕–Promoting and ⊖–Inhibiting effect; TSH–Thyroid stimulating hormone; ACTH–Adrenocorticotrophic hormone; GH–Growth hormone; PGE_1–Prostaglandin E_1; TG–Triacylglycerol; FFA–Free fatty acid).*

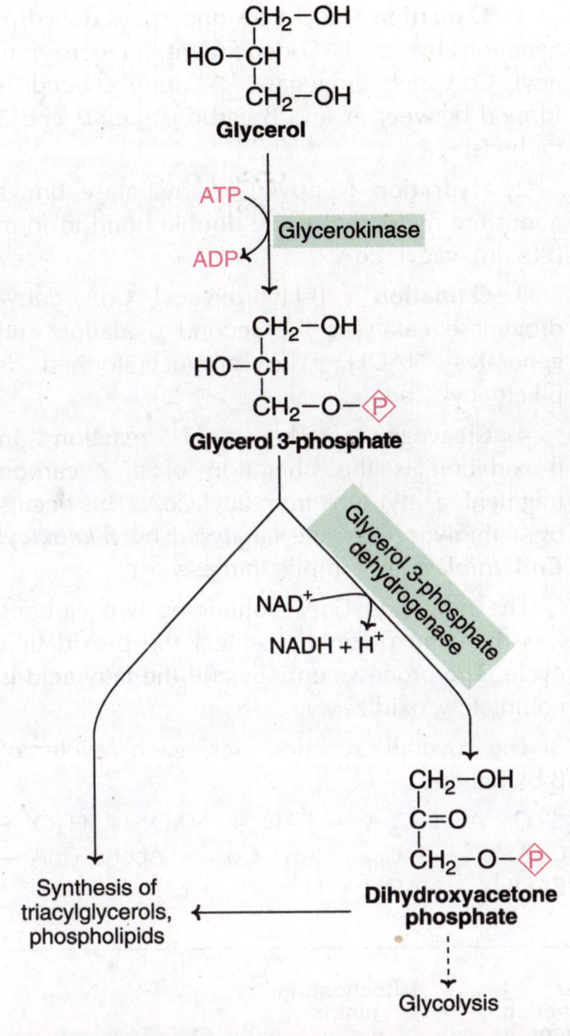

Fig. 14.4 : *Fate of glycerol.*

Three different thiokinases to activate long chain (10-20 carbon), medium chain (4-12 carbon) and short chain (< 4 carbon) fatty acids have been identified.

II. Transport of acyl CoA into mitochondria

The inner mitochondrial membrane is impermeable to fatty acids. A specialized ***carnitine carrier system*** (***carnitine shuttle***) operates to transport activated fatty acids from cytosol to the mitochondria. This occurs in four steps (***Fig.14.7***).

1. Acyl group of acyl CoA is transferred to ***carnitine*** (***β-hydroxy γ-trimethyl amino butyrate***), catalysed by carnitine palmitoyl-transferase I (present on the outer surface of inner mitochondrial membrane).

2. The acyl-carnitine is transported across the membrane to mitochondrial matrix by a specific carrier protein.

3. Carnitine palmitoyl transferase II (CPT II found on the inner surface of inner mitochondrial membrane) converts acyl-carnitine to acyl CoA.

4. The carnitine released returns to cytosol for reuse.

It should be noted that the coenzyme A used for activation is different from the one that finally combines with fatty acid in the mitochondria to form acyl CoA. Thus, the ***cell has two separate pools (cytosolic and mitochondrial) of coenzyme A.***

Inhibitor of carnitine shuttle : Carnitine acyl transferase I is inhibited by ***malonyl CoA***, a key metabolite involved in fatty acid synthesis that occurs in cytosol (details given later). In other words, while the fatty acid synthesis is in progress (reflected by high concentration of malonyl CoA), their oxidation does not occur, since carnitine shuttle is impaired.

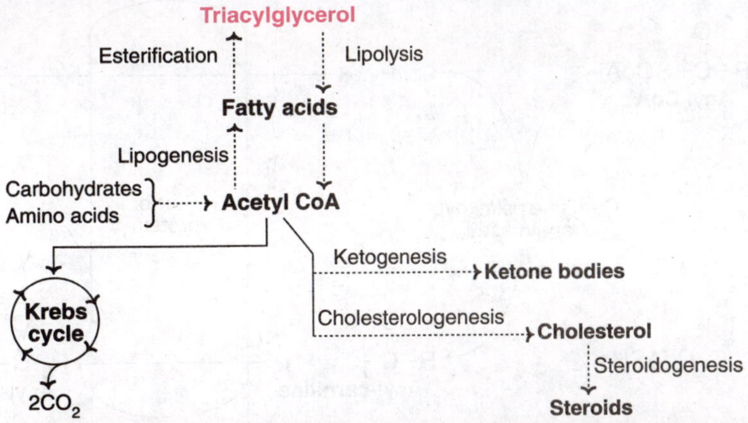

Fig. 14.5 : *An outline of triacylglycerol and fatty acid metabolism depicting major pathways and products.*

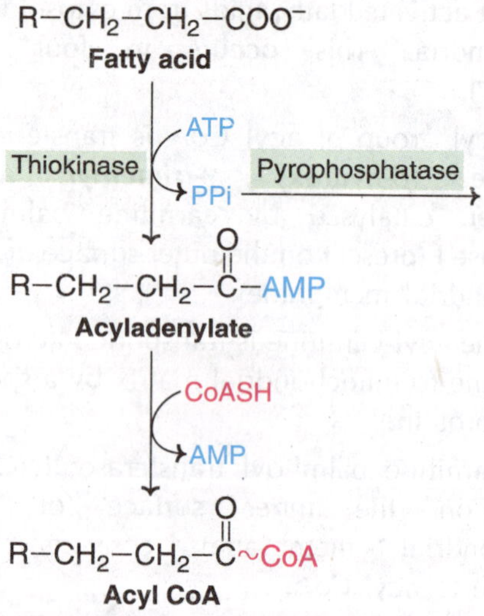

Fig. 14.6 : Activation of fatty acid to acyl CoA by the enzyme thiokinase.

III. β-Oxidation proper

The enzyme machinery responsible for β-oxidation in the mitochondrial matrix is collectively referred to as ***fatty acid oxidase system***. These enzymes, in a sequence of 4 reactions, catalyse the oxidation of acyl CoA to acetyl CoA (***Fig.14.8***).

1. **Oxidation :** Acyl CoA undergoes dehydrogenation by an FAD-dependent flavoenzyme, acyl CoA dehydrogenase. A double bond is formed between α and β carbons (i.e., 2 and 3 carbons).

2. **Hydration :** Enoyl CoA hydratase brings about the hydration of the double bond to form β-hydroxyacyl CoA.

3. **Oxidation :** β-Hydroxyacyl CoA dehydrogenase catalyses the second oxidation and generates NADH. The product formed is β-ketoacyl CoA.

4. **Cleavage :** The final reaction in β-oxidation is the liberation of a 2 carbon fragment, acetyl CoA from acyl CoA. This occurs by a thiolytic cleavage catalysed by ***β-ketoacyl CoA thiolase*** (or simply ***thiolase***).

The new acyl CoA, containing two carbons less than the original, reenters the β-oxidation cycle. The process continues till the fatty acid is completely oxidized.

The overall reaction for each cycle of β-oxidation

C_n Acyl CoA + FAD + NAD^+ + H_2O + CoASH ⟶ $C_{(n-2)}$ Acyl CoA + Acetyl CoA + $FADH_2$ + NADH + H^+.

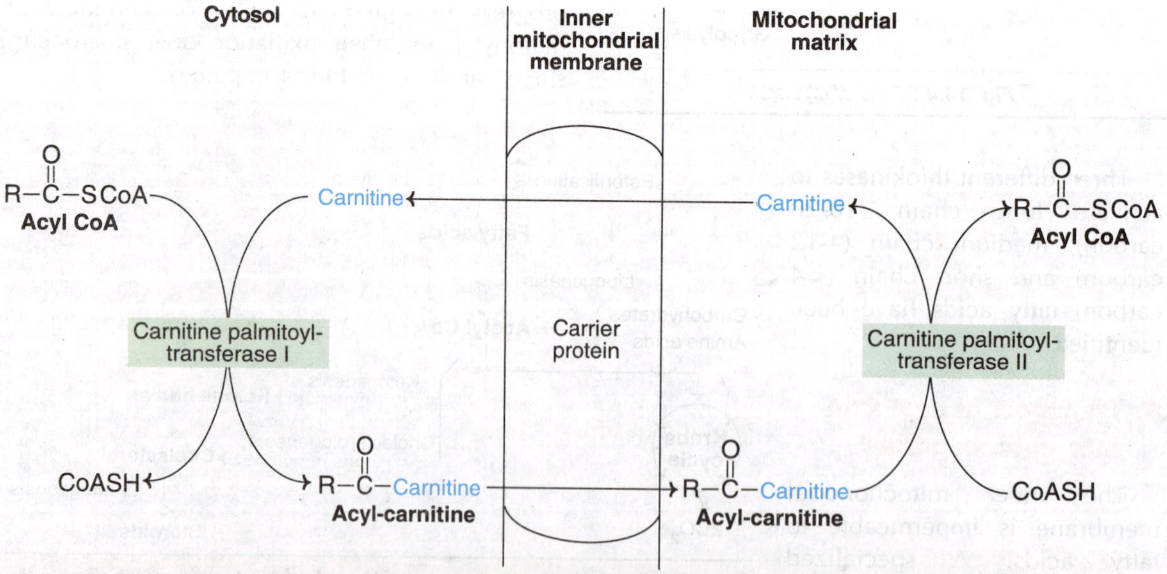

Fig. 14.7 : Carnitine shuttle for transport of activated fatty acid (acyl CoA) into mitochondria.

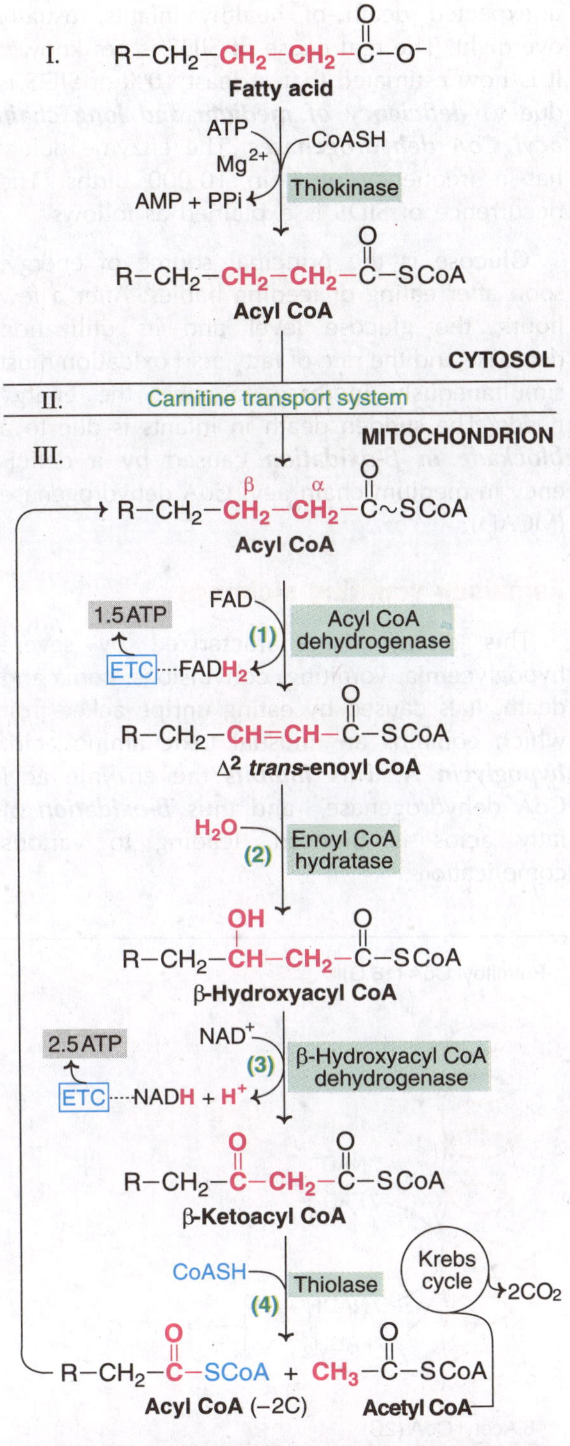

Fig. 14.8 : *β-Oxidation of fatty acids : Palmitoyl CoA (16 carbon) undergoes seven cycles to yield 8 acetyl CoA [I–Activation; II–Transport; III–β Oxidation proper— (1) Oxidation, (2) Hydration, (3) Oxidation and (4) Cleavage].*

Table 14.2 Energetics of palmitic acid oxidation

Mechanism	ATP yield
I. β-Oxidation 7 cycles	
7 $FADH_2$ [oxidized by electron transport chain (ETC), each $FADH_2$ gives 1.5 ATP]	*10.5*
7 NADH (oxidized by ETC, each NADH liberates 2.5 ATP)	*17.5*
II. From 8 acetyl CoA	
Oxidized by citric acid cycle, each acetyl CoA provides 10 ATP	80
Total energy from one mole of palmitoyl CoA	***108***
Energy utilized for activation (formation of palmitoyl CoA)	–2
Net yield for one molecule of palmitate	***106***

The scheme of fatty acid oxidation discussed above corresponds to saturated (no double bond) and even carbon fatty acids. This occurs most predominantly in biological system.

Oxidation of palmitoyl CoA

The summary of β-oxidation of palmitoyl CoA is shown below

Palmitoyl CoA + 7 CoASH + 7 FAD + 7 NAD^+ + 7H_2O $\longrightarrow$ 8 Acetyl CoA + 7 $FADH_2$ + 7 NADH + 7H^+

Palmitoyl CoA undergoes 7 cycles of β-oxidation to yield 8 acetyl CoA. Acetyl CoA can enter citric acid cycle and get completely oxidized to CO_2 and H_2O.

Energetics of β-oxidation

Fatty acid oxidation generates large quantity of energy. The ATP obtained from the complete oxidation of palmitic acid (16 carbon) is given in ***Table 14.2*** and ***Fig.14.9.***

Carnitine deficiency

The newborn, especially ***pre-term infants,*** are ***susceptible*** to ***carnitine*** deficiency. This may

Medical Concepts/ Clinical Correlates

- An adult human body contains about 10–11 kg of fat reserve corresponding to about 100,000 Cal. This can meet the energy requirements for several weeks of food deprivation in man.
- The sudden infant death syndrome (SIDS)—an unexpected overnight death of healthy infants—is attributed to a blockade in β-oxidation of fatty acids, caused by a deficiency of medium chain acyl CoA dehydrogenase (MCAD).
- Pre-term infants are susceptible to carnitine deficiency, and this causes hypoglycemia. Oral supplementation of carnitine is advocated.

be due to decreased synthesis or renal leakage of carnitine. In some ***patients undergoing hemodialysis***, carnitine deficiency has been reported. Hypoglycemia, due to impaired fatty acid oxidation, is the most significant manifestation of carnitine deficiency. In addition, accumulation of TG and muscular weakness are also observed. Oral supplementation with carnitine is the mode of treatment.

Carnitine palmitoyl transferase (CPT) deficiency

Inherited defects of CPT-I and CPT-II are known. CPT-I deficiency affects liver, reduces fatty acid oxidation and promotes ketone body synthesis, besides causing hypoglycemia.

CPT-II deficiency mainly affects skeletal muscle, and to some extent liver. In type 2 diabetic patients, on treatment with sulfonylurea drugs (e.g. glibenclamide) and tolbutamide. CPT-I is inhibited. This results in lowered fatty acid oxidation and also reduced gluconeogenesis (i.e., hyperglycemia is prevented).

SIDS—a disorder due to blockade in β-oxidation

The ***sudden infant death syndrome*** (SIDS), also known as ***cot death*** or ***crib death***, is an unexpected death of healthy infants, usually overnight. The real cause of SIDS is not known. It is now estimated that at least 10% of SIDS is due to ***deficiency of medium and long chain acyl CoA dehydrogenases***. The enzyme defect has a frequency of 1 in 10,000 births. The occurrence of SIDS is explained as follows

Glucose is the principal source of energy, soon after eating or feeding babies. After a few hours, the glucose level and its utilization decrease and the rate of fatty acid oxidation must simultaneously increase to meet the energy needs. The sudden death in infants is due to a ***blockade in β-oxidation*** caused by a deficiency in medium chain acyl CoA dehydrogenase (MCAD).

Jamaican vomiting sickness

This disease is characterized by severe hypoglycemia, vomiting, convulsions, coma and death. It is caused by eating unripe ackee fruit which contains an unusual toxic amino acid, ***hypoglycin A***. This ***inhibits*** the enzyme acyl CoA dehydrogenase and thus ***β-oxidation*** of fatty acids is blocked, leading to various complications.

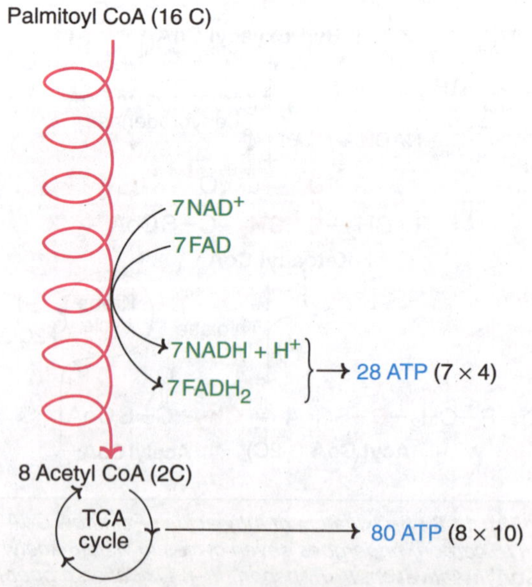

Fig. 14.9 : *An overview of oxidation of palmitic acid.*

Oxidation of odd carbon chain fatty acids

The β-oxidation of saturated fatty acids containing odd number of carbon atoms proceeds in the same manner, as described above for even carbon fatty acids. The only difference is that in the last and final β-oxidation cycle, a ***three-carbon fragment*** is left behind (in place of 2 carbon unit for saturated fatty acids). This compound is ***propionyl CoA*** which is converted to succinyl CoA as follows (***Fig.14.10***)

1. Propionyl CoA is carboxylated in the presence of ATP, CO_2 and vitamin ***biotin*** to D-methylmalonyl CoA.

2. Methylmalonyl CoA racemase converts the methylmalonyl CoA to L-form. This reaction (D → L) is essential for the entry of this compound into the metabolic reactions of the body.

3. The next enzyme, ***methylmalonyl CoA mutase***, is dependent on ***vitamin*** B_{12} (deoxyadenosyl cobalamin). It catalyses the conversion of methylmalonyl CoA (a branched compound) to succinyl CoA (a straight chain compound), which can enter citric acid cycle.

Methylmalonic acidemia

Two types of methylmalonic acidemias are known

1. Due to ***deficiency of vitamin*** B_{12};

2. Due to defect in the enzyme methylmalonyl CoA mutase.

In either case, there is an accumulation of methylmalonic acid in body, followed by its increased excretion in urine. This causes severe metabolic acidosis, damages the central nervous system and retards the growth. It is often fatal in the early years of life.

Oxidation of unsaturated fatty acids

Due to the presence of double bonds, the unsaturated fatty acids are not reduced to the same extent as saturated fatty acids. Therefore, ***oxidation of unsaturated fatty acids***, in general, provides ***less energy*** than that of saturated fatty acids.

Most of the reactions involved in the oxidation of unsaturated fatty acids are the same as found in the β-oxidation of saturated fatty acids. However, the presence of double bonds poses problem for β-oxidation to proceed. This is overcome by two additional enzymes—an ***isomerase*** and a ***reductase***.

β-Oxidation of long chain fatty acids in peroxisomes

Peroxisomes are organelles present in most eukaryotic cells. The β-oxidation occurs in a modified form in peroxisomes. Acyl CoA dehydrogenase (a flavoenzyme) leads to the formation of $FADH_2$, as in β-oxidation. The reducing equivalents from $FADH_2$ are not transferred to the electron transport chain, but handed over directly to O_2. This results in the formation of H_2O_2, which is cleaved by catalase.

$$\text{E-FADH}_2 + O_2 \longrightarrow \text{E-FAD} + H_2O_2$$

$$H_2O_2 \xrightarrow{\text{Catalase}} H_2O + \tfrac{1}{2}O_2$$

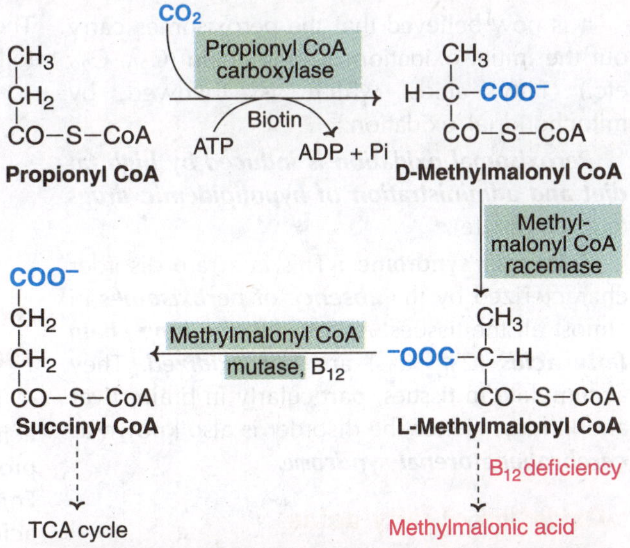

Fig. 14.10 : Conversion of propionyl CoA to succinyl CoA.

Medical Concepts/ Clinical Correlates

- Certain drugs used in the treatment of diabetes (e.g. tolbutamide, glibenclamide) inhibit carnitine palmitoyl transferase and reduce fatty acid oxidation.
- Methylmalonic acidemia occurs either due to a deficiency of the vitamin B_{12} or a defect in an enzyme methyl malonyl CoA mutase. This disorder retards growth and damages central nervous system.
- Zellweger syndrome is caused by the absence of peroxisomes in tissues; as a result, the long chain fatty acids cannot be oxidized.
- Refsum disease is due to a defect in α-oxidation of fatty acids. The patients are advised not to consume diets containing chlorophyll.
- Triclosan, an antibacterial agent, used in cosmetics, toothpaste etc., inhibits bacterial fatty acid synthesis. However, it is linked with reduced immunity and allergy in some individuals.

There is ***no ATP synthesized in peroxisomal β-oxidation of fatty acids***, since the reducing equivalents do not pass through ETC. However, heat is liberated.

It is now believed that the peroxisomes carry out the initial oxidation of long chain (C_{20}, C_{22} etc.) fatty acids which is followed by mitochondrial oxidation.

Peroxisomal oxidation is induced by high fat diet and administration of hypolipidemic drugs (e.g. clofibrate).

Zellweger syndrome : This is a rare disorder characterized by the ***absence of peroxisomes*** in almost all the tissues. As a result, the ***long chain fatty acids*** ($C_{26}-C_{38}$) are ***not oxidized***. They accumulate in tissues, particularly in brain, liver and kidney. Hence the disorder is also known as ***cerebrohepatorenal syndrome.***

α-Oxidation of fatty acids

β-Oxidation is the most predominant pathway for fatty acid degradation. However, the ***removal of one carbon unit at a time*** by the oxidation of α-carbon atom of fatty acid is known. α-Oxidation does not involve the binding of fatty acid to coenzyme A and no energy is produced.

Refsum disease is a rare but severe ***neurological disorder*** characterized by cerebral ataxia and peripheral neuropathy. The patients of this disease accumulate large quantities of an unusual fatty acid, ***phytanic acid.*** It is derived from phytol, a constituent ***of chlorophyll.*** Hence it is found mostly in plant foods. However, it is also present in milk lipids and animal fats. Phytanic acid ***cannot undergo β-oxidation*** due to the presence of a methyl group on carbon-3. This fatty acid undergoes ***initial α-oxidation*** (to remove α-carbon as carbon dioxide) and this is followed by β-oxidation.

Refsum disease is caused by a defect in the α-oxidation due to the ***deficiency of the enzyme phytanic acid α-oxidase***. The result is that phytanic acid cannot be converted to a compound that can be degraded by β-oxidation. The patients ***should not consume*** diets containing ***chlorophyll*** (i.e., green leafy vegetables).

ω-Oxidation of fatty acids

This is a minor pathway. It involves hydroxylation followed by oxidation of ω-carbon present as a methyl group at the other end (at one end carboxyl group is present) of fatty acid. The overall reaction may be represented as follows.

$$CH_3-(CH_2)_n-COO^-$$
$$\downarrow$$
$$HO-H_2C-(CH_2)_n-COO^-$$
$$\downarrow$$
$$^-OOC-(CH_2)_n-COO^-$$

Oxidation of fatty acids and metabolic water

Fatty acid oxidation (even other forms of aerobic respiration) is accompanied by the production of water, referred to metabolic water. For instance, when one molecule of palmitic acid is oxidized, it releases ***16 molecules*** of water. This metabolic water has great significance in some animals. ***Camel*** can store

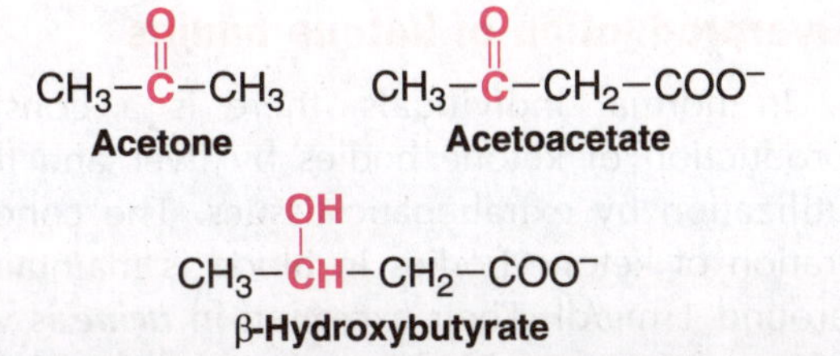

Fig. 14.11 : Structures of ketone bodies.

lipids in its ***hump*** which is ***good source of water***, besides ***energy*** supply. For this reason, camel can travel in deserts for long periods even without food and water supply. Kangaroo rat is a small animal that is believed to live indefinitely without water. It consumes only oil rich seeds, and the metabolic water produced is adequate to meet its water needs. The use of metabolic water is an adaptation, and is accompanied by reduced output of urine.

KETONE BODIES

The compounds namely ***acetone, acetoacetate*** and ***β-hydroxybutyrate*** (or 3-hydroxybutyrate) are known as ketone bodies or ***acetone bodies*** (***Fig.14.11***). Only the first two are true ketones while β-hydroxybutyrate does not possess a keto (C=O) group. ***Ketone bodies are water-soluble*** and ***energy yielding***. Acetone, however, is an exception, since it cannot be metabolized. Acetone is a volatile waste product, and can be excreted via lungs.

Ketogenesis

The synthesis of ketone bodies occurs in the ***liver***. The enzymes for ketone body synthesis are located in the ***mitochondrial matrix***. ***Acetyl CoA***, formed by oxidation of fatty acids, pyruvate or some amino acids, is the ***precursor*** for ketone bodies. Ketogenesis occurs through the following reactions (***Fig.14.12***).

1. Two moles of acetyl CoA condense to form acetoacetyl CoA. This reaction is catalysed by thiolase, an enzyme involved in the final step of β-oxidation. Hence, acetoacetate synthesis is appropriately regarded as the reversal of thiolase reaction of fatty acid oxidation.

2. Acetoacetyl CoA combines with another molecule of acetyl CoA to produce β-hydroxy β-methyl glutaryl CoA (HMG CoA). ***HMG CoA synthase***, catalysing this reaction, ***regulates the synthesis of ketone bodies.***

3. HMG CoA lyase cleaves HMG CoA to produce acetoacetate and acetyl CoA.

4. Acetoacetate can undergo spontaneous decarboxylation to form acetone.

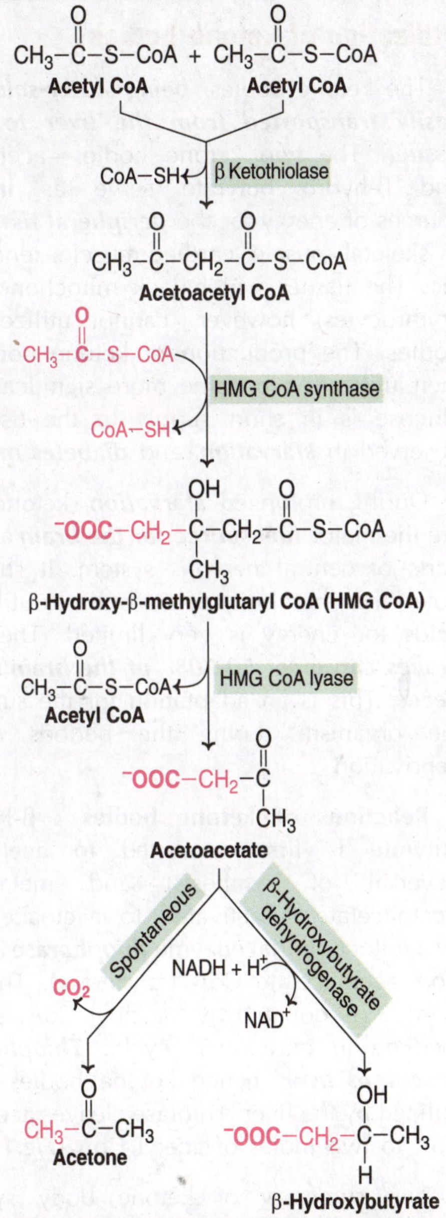

Fig. 14.12 : Synthesis of ketone bodies (ketogenesis).

5. Acetoacetate can be reduced by a dehydrogenase to β-hydroxybutyrate. Quantitatively, ***β-hydroxybutyrate*** is the most ***predominant ketone body*** present in circulation and urine, in ketosis.

The carbon skeleton of some amino acids (ketogenic) is degraded to acetoacetate or acetyl CoA and, therefore, to ketone bodies, e.g. leucine, lysine, phenylalanine etc.

Utilization of ketone bodies

The ketone bodies, ***being water-soluble***, are ***easily transported from the liver to various tissues***. The two ketone bodies—acetoacetate and β-hydroxybutyrate serve as important sources of energy for the ***peripheral tissues*** such as skeletal muscle, cardiac muscle, renal cortex etc. The tissues which lack mitochondria (e.g. erythrocytes) however, cannot utilize ketone bodies. The production of ketone bodies and their utilization become more significant when glucose is in short supply to the tissues, as observed in ***starvation***, and ***diabetes mellitus***.

During prolonged ***starvation***, ketone bodies are the major ***fuel source for the brain*** and other parts of central nervous system. It should be noted that the ability of the brain to utilize fatty acids for energy is very limited. The ***ketone bodies can meet 50-70% of the brain's energy needs***. This is an adaptation for the survival of the organism during the periods of food deprivation.

Reactions of ketone bodies : β-Hydroxybutyrate is first converted to acetoacetate (reversal of synthesis) and metabolized. Acetoacetate is activated to acetoacetyl CoA by a mitochondrial enzyme ***thiophorase*** (succinyl CoA acetoacetate CoA transferase). The coenzyme A is donated by succinyl CoA, an intermediate in citric acid cycle. ***Thiophorase is absent in liver***, hence ketone bodies are not utilized by the liver. Thiolase cleaves acetoacetyl CoA to two moles of acetyl CoA (***Fig.14.13***).

The summary of ketone body synthesis, utilization and excretion is depicted in ***Fig.14.14***.

Overproduction of ketone bodies

In normal individuals, there is a constant production of ketone bodies by liver and their utilization by extrahepatic tissues. The concentration of ketone bodies in blood is maintained around 1 mg/dl. Their excretion in ***urine*** is very low and undetectable by routine tests (***Rothera's test***).

When the rate of synthesis of ketone bodies exceeds the rate of utilization, their concentration in blood increases, this is known as ***ketonemia.*** Ketonemia is predominantly due to increased production of ketone bodies rather than the deficiency in their utilization. The term ***ketonuria*** represents the excretion of ketone bodies in urine. The overall picture of ketonemia and ketonuria is referred to as ***ketosis.*** Smell of

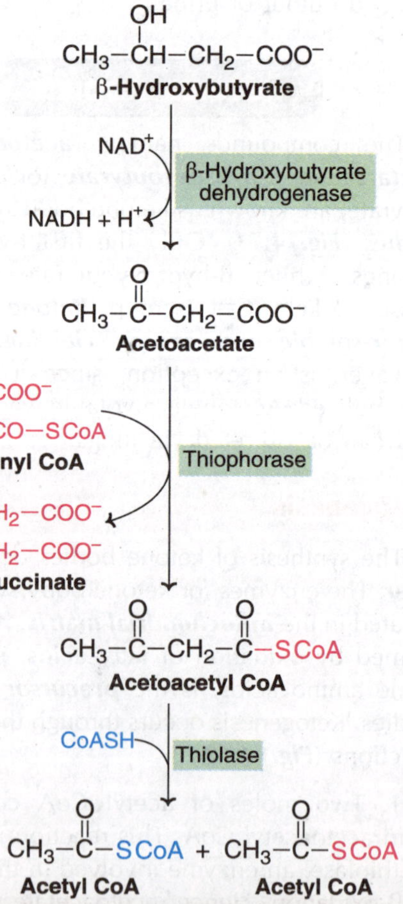

Fig. 14.13 : Metabolism (utilization) of ketone bodies to acetyl CoA.

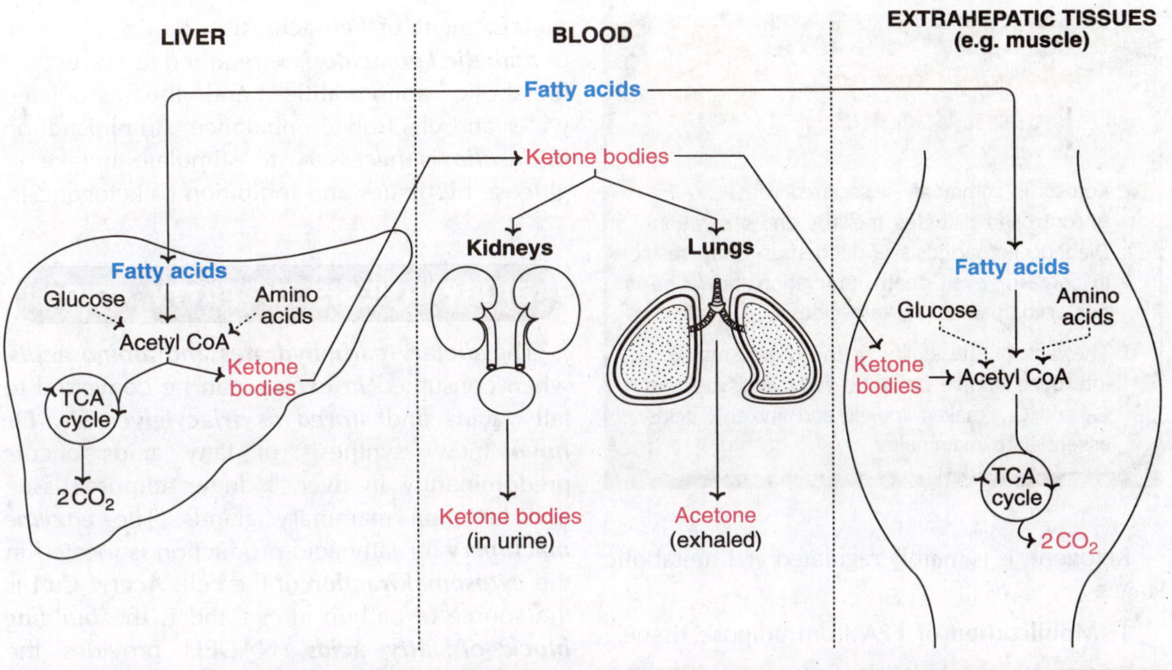

Fig. 14.14 : Summary of ketone body synthesis, utilization and excretion.

acetone in breath is a common feature in ketosis. Ketosis is most commonly ***associated with starvation and severe uncontrolled diabetes mellitus***.

Starvation : Starvation is accompanied by ***increased degradation of fatty acids*** (from the fuel reserve triacylglycerol) to meet the energy needs of the body. This causes an overproduction of acetyl CoA which cannot be fully handled by citric acid cycle. Furthermore, TCA cycle is impaired due to deficiency of oxaloacetate, since most of it is diverted for glucose synthesis to meet the essential requirements (often unsuccessful) for tissues like brain. The result is an accumulation of acetyl CoA and its diversion for ***overproduction of ketone bodies***. ***Hypermesis*** in early pregnancy (nausea, vomiting) may lead to starvation like condition and result in ketosis.

Ketonuria and weight loss programs : The appearance of ketone bodies in urine is an indication of active fat metabolism. Some programs of body weight loss encourage ***reduction in carbohydrate*** and total calorie intake until ketone bodies appear in urine.

Diabetes mellitus : Diabetes mellitus is associated with insulin deficiency. This results in impaired carbohydrate metabolism and ***increased lipolysis***, both of them ultimately leading to the accumulation of acetyl CoA and its conversion to ***ketone bodies***. In severe diabetes, the ketone body concentration in blood plasma may reach 100 mg/dl and the urinary excretion may be as high as 500 mg/day.

Non-pathologic ketosis : This is sometimes observed after consumption of high fat diet, and after severe excersise in a post absorptive state.

Regulation of ketogenesis

The ketone body formation (particularly overproduction) occurs primarily due to nonavailability of carbohydrates to the tissues. This is an outcome of excessive utilization of fatty acids to meet the energy requirements of the cells. The hormone ***glucagon stimulates ketogenesis*** whereas ***insulin inhibits***. The increased ratio of glucagon/insulin in diabetes mellitus promotes ketone body formation. This is due to disturbances caused in carbohydrate and lipid metabolisms in diabetes, as discussed elsewhere (***Chapter 36***).

Medical Concepts/ Clinical Correlates

- Ketosis is commonly associated with uncontrolled diabetes mellitus and starvation. Diabetic ketoacidosis is dangerous—may result in coma or even death. Starvation, however, is not accompanied by ketoacidosis.
- The lack of the ability of the organisms to introduce double bonds in fatty acids beyond C_9 and C_{10} makes linoleic and linolenic acids essential to mammals.

Ketogenesis is mainly regulated at 3 metabolic stages

1. **Mobilization of FFA** from adipose tissue.

2. **Carnitine palmitoyltransferase I (CPT-I) :** Activity is low in fed state, and high in starvation. Starvation results in elevated fatty acid oxidation.

3. **Concentration of oxaloacetate :** Low levels of oxaloacetate, found in starvation and uncontrolled diabetes, impair oxidation of acetyl CoA via Krebs cycle. This leads to overproduction of ketone bodies.

Ketogenic and antiketogenic substances

The ketogenic substances (promote ketogenesis) include fatty acids and certain amino acids (***leucine***, ***lysine***, tyrosine etc.). The antiketogenic substances (inhibit ketogenesis) are glucose, glycerol and glucogenic amino acids (e.g. ***glycine***, ***alanine***, serine, glutamate etc.)

Ketoacidosis

Both acetoacetate and β-hydroxybutyrate are strong acids. Increase in their concentration in blood would cause acidosis. The carboxyl group has a p^{Ka} around 4. Therefore, the ketone bodies in the blood dissociate and release H^+ ions which lower the pH. Diabetic ketoacidosis is dangerous—may result in coma, and even death, if not treated. Ketosis due to starvation is not usually accompanied by ketoacidosis.

Treatment of ketoacidosis : Rapid treatment of ***diabetic ketoacidosis*** is required to correct the metabolic abnormalities and the associated water and electrolyte imbalance. Administration of ***insulin*** is necessary to stimulate uptake of glucose by tissues and inhibition of ketogenesis.

BIOSYNTHESIS OF FATTY ACIDS

The ***dietary carbohydrates and amino acids***, when consumed ***in excess***, can be converted to fatty acids and ***stored as triacylglycerols***. ***De novo*** (new) synthesis of fatty acids occurs predominantly in liver, kidney, adipose tissue and lactating mammary glands. The ***enzyme machinery*** for fatty acid production is located in the ***cytosomal fraction*** of the cell. ***Acetyl CoA*** is the source of carbon atoms and is the ***building block of fatty acids***. NADPH provides the reducing equivalents and ATP supplies energy for fatty acid formation. The fatty acid synthesis may be learnt in 3 stages

I. Production of acetyl CoA and NADPH

II. Conversion of acetyl CoA to malonyl CoA

III. Reactions of fatty acid synthase complex.

I. Production of acetyl CoA and NADPH

Acetyl CoA and NADPH are the ***prerequisites for fatty acid synthesis***. Acetyl CoA is produced in the mitochondria by the oxidation of pyruvate and fatty acids, degradation of carbon skeleton of certain amino acids, and from ketone bodies. Mitochondria, however, are not permeable to acetyl CoA. An alternate or a bypass arrangement is made for the transfer of acetyl CoA to cytosol. Acetyl CoA condenses with oxaloacetate in mitochondria to form citrate. ***Citrate*** is freely transported to cytosol where it is cleaved by ***citrate lyase*** to liberate acetyl CoA and oxaloacetate. Oxaloacetate in the cytosol is converted to malate (***Fig.14.15***). The main ***source of NADPH*** for fatty acid biosynthesis is ***HMP shunt***.

Malic enzyme converts malate to pyruvate. NADPH and CO_2 are generated in this reaction. Both of them are utilized for fatty acid synthesis.

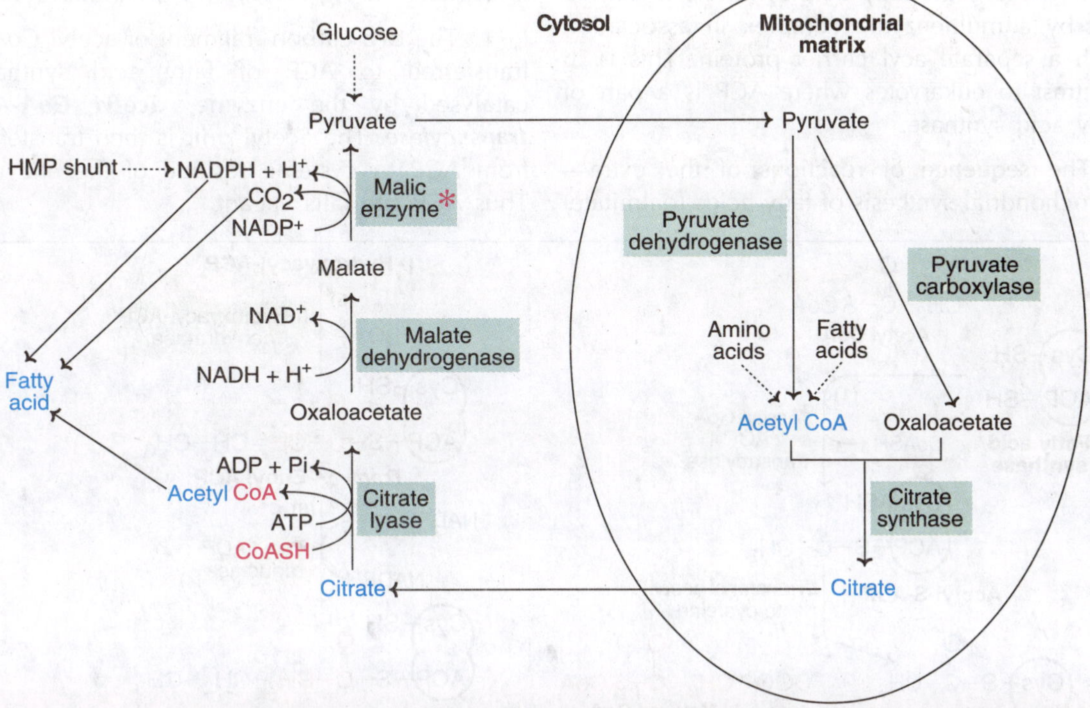

*Fig. 14.15 : Transfer of acetyl CoA from mitochondria to cytosol (HMP shunt—Hexose monophosphate shunt; *—also known as malate dehydrogenase).*

Advantages of coupled transport of acetyl CoA and NADPH : The transport of acetyl CoA from mitochondria to cytosol is coupled with the cytosomal production of NADPH and CO_2 which is highly advantageous to the cell for optimal synthesis of fatty acids.

II. Formation of malonyl CoA

Acetyl CoA is carboxylated to malonyl CoA by the enzyme ***acetyl CoA carboxylase*** (***Fig.14.16***). This is an ATP-dependent reaction and requires ***biotin*** for CO_2 fixation. The mechanism of action of acetyl CoA carboxylase is similar to that of pyruvate carboxylase (***Refer Chapter 7, Fig.7.29***). Acetyl CoA carboxylase is a regulatory enzyme in fatty acid synthesis (details given later).

III. Reactions of fatty acid synthase complex

The remaining reactions of fatty acid synthesis are catalysed by a ***multifunctional enzyme*** known as ***fatty acid synthase (FAS) complex.*** In eukaryotic cells, including man, the fatty acid synthase exists as a dimer with two identical units. Each monomer possesses the activities of seven different enzymes and an ***acyl carrier protein (ACP)*** bound to 4′-phosphopantetheine. Fatty acid synthase functions as a single unit catalysing all the seven reactions. Dissociation of the synthase complex results in loss of the enzyme activities. In the lower organisms

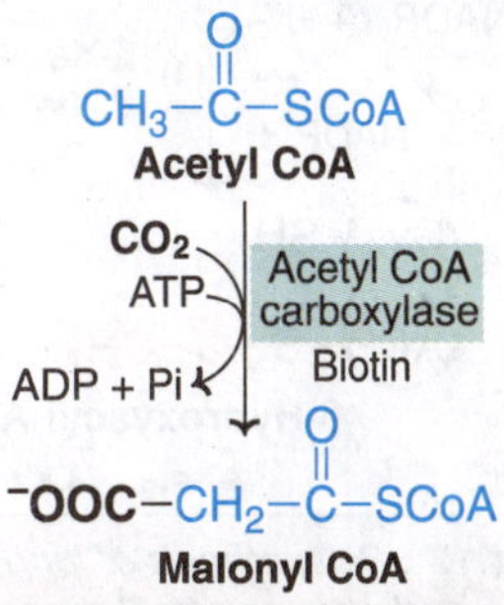

Fig. 14.16 : Conversion of acetyl CoA to malonyl CoA.

(prokaryotes), the fatty acid synthesis is carried out by a multienzyme complex in association with a separate acyl carrier protein. This is in contrast to eukaryotes where ACP is a part of fatty acid synthase.

The sequence of reactions of the extra—mitochondrial synthesis of fatty acids (palmitate) is depicted in ***Fig.14.17***, and described below.

1. The two carbon fragment of acetyl CoA is transferred to ACP of fatty acid synthase, catalysed by the enzyme, ***acetyl CoA-ACP transacylase***. The acetyl unit is then transferred from ACP to cysteine residue of the enzyme. Thus ACP site falls vacant.

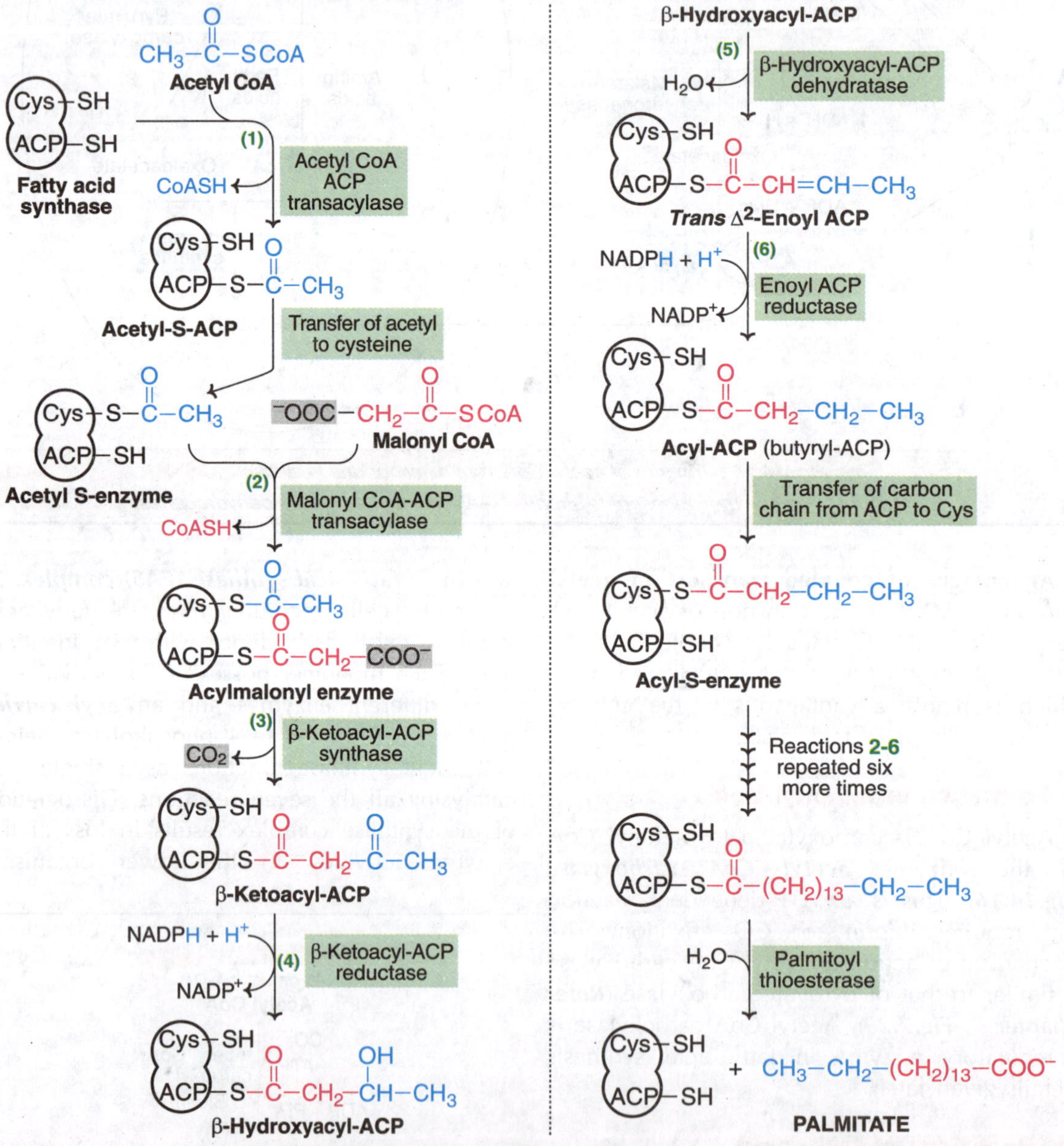

Fig. 14.17 : *Biosynthesis of long chain fatty acid-palmitate. (Cys–Cysteine; ACP–Acyl carrier protein; The pathway repeats 7 times to produce palmitate; the first two carbons at the methyl end are directly from acetyl CoA, the rest of the carbons come from malonyl CoA).*

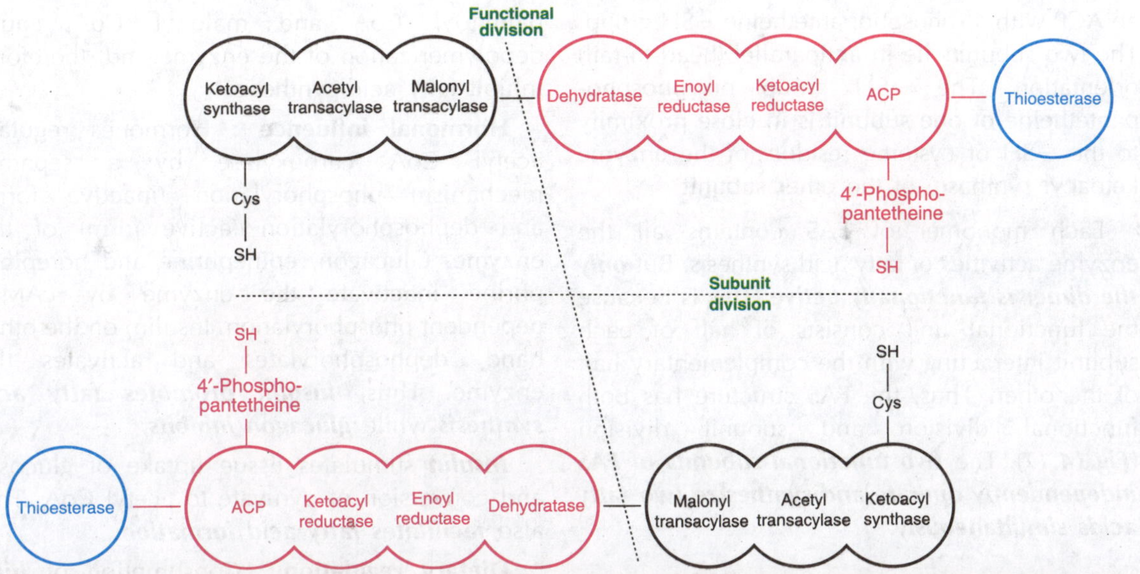

Fig. 14.18 : Fatty acid synthase multienzyme complex (ACP–Acyl carrier protein; FAS has two identical subunits which organize into two functional subunits to simultaneously synthesize two fatty acids).

2. The enzyme ***malonyl CoA-ACP transacylase*** transfers malonate from malonyl CoA to bind to ACP.

3. The acetyl unit attached to cysteine is transferred to malonyl group (bound to ACP). The malonyl moiety ***loses*** CO_2 which was added by acetyl CoA carboxylase. Thus, CO_2 is never incorporated into fatty acid carbon chain. The decarboxylation is accompanied by loss of free energy which allows the reaction to proceed forward. This reaction is catalyzed by ***β-ketoacyl ACP synthase.***

4. β-Ketoacyl ACP reductase reduces ketoacyl group to hydroxyacyl group. The reducing equivalents are supplied by NADPH.

5. β-Hydroxyacyl ACP undergoes dehydration. A molecule of water is eliminated and a double bond is introduced between α and β carbons.

6. A second NADPH-dependent reduction, catalysed by ***enoyl-ACP reductase*** occurs to produce acyl-ACP. The four-carbon unit attached to ACP is butyryl group.

The carbon chain attached to ACP is transferred to cysteine residue and the reactions 2-6 are repeated 6 more times. Each time, the fatty acid chain is lengthened by a two-carbon unit (obtained from malonyl CoA). At the end of 7 cycles, the fatty acid synthesis is complete and a 16-carbon fully saturated fatty acid—namely palmitate—bound to ACP is produced.

7. The enzyme palmitoyl thioesterase separates palmitate from fatty acid synthase. This completes the synthesis of palmitate.

Summary of palmitate synthesis

Of the 16 carbons present in palmitate, only two come from acetyl CoA directly. The remaining 14 are from malonyl CoA which, in turn, is produced by acetyl CoA. The overall reaction of palmitate synthesis is summarized

$$8 \text{ Acetyl CoA} + 7 \text{ ATP} + 14 \text{ NADPH} + 14 \text{ H}^+ \longrightarrow \text{Palmitate} + 8 \text{ CoA} + 7 \text{ ADP} + 7 \text{ Pi} + 6H_2O$$

Fatty acid synthase complex

The diagrammatic representation of the model for fatty acid synthase (*FAS*) ***multienzyme complex*** is depicted in ***Fig.14.18***. This model is tentative and is largely based on the work of Wakil.

Fatty acid synthase is a ***dimer*** composed of two identical subunits (monomers), each with a molecular weight of 240,000. Each subunit contains the activities of 7 enzymes of FAS and an ACP with 4′-phosphopantetheine –SH group. The two subunits lie in antiparallel (head-to-tail) orientation. The –SH group of phosphopantetheine of one subunit is in close proximity to the –SH of cysteine residue (of the enzyme ketoacyl synthase) of the other subunit.

Each monomer of FAS contains all the enzyme activities of fatty acid synthesis. But ***only the dimer is functionally active***. This is because the functional unit consists of half of each subunit interacting with the complementary half of the other. Thus, the FAS structure has both functional division and subunit division (***Fig.14.17***). The ***two functional subunits of FAS independently operate and synthesize two fatty acids simultaneously***.

Functional significance of FAS complex

The organization of different enzymes of a metabolic pathway into a single multienzyme functional unit has distinct advantages for cellular function

1. The FAS complex offers ***great efficiency*** that is free from interference of other cellular reactions for the synthesis of fatty acids.

2. Since the entire process of the metabolic pathway is confined to the complex, there are ***no permeability barriers*** for the various intermediates.

3. The multienzyme polypeptide complex is coded by a single gene. Thus, there is a ***good coordination*** in the synthesis of all enzymes of the FAS complex.

Regulation of fatty acid synthesis

Fatty acid production is controlled by enzymes, metabolites, end products, hormones and dietary manipulations. Some of the important regulatory mechanisms are discussed hereunder.

Acetyl CoA carboxylase : This enzyme controls a ***committed step in fatty acid synthesis.*** Acetyl CoA carboxylase exists as an inactive protomer (monomer) or an active polymer. Citrate promotes polymer formation, hence increases fatty acid synthesis. On the other hand, palmitoyl CoA and malonyl CoA cause depolymerization of the enzyme and, therefore, inhibit fatty acid synthesis.

Hormonal influence : Hormones regulate acetyl CoA carboxylase by a separate mechanism—phosphorylation (inactive form) and dephosphorylation (active form) of the enzyme. Glucagon, epinephrine and norepinephrine inactivate the enzyme by cAMP-dependent phosphorylation. Insulin, on the other hand, dephosphorylates and activates the enzyme. Thus, ***insulin promotes fatty acid synthesis*** while ***glucagon inhibits***.

Insulin stimulates tissue uptake of glucose, and conversion of pyruvate to acetyl CoA. This also ***facilitates fatty acid formation***.

Dietary regulation : Consumption of ***high carbohydrate or fat-free diet increases*** the synthesis of acetyl CoA carboxylase and fatty acid synthase, which promote ***fatty acid formation***. On the other hand, ***fasting*** or ***high fat diet decreases*** fatty acid production by reducing the synthesis of these two enzymes. Fatty acid synthesis is ***elevated when sucrose is fed*** (instead of glucose), as fructose bypasses phosphofructokinase step in glycolysis and increases fatty acid synthesis.

Availability of NADPH : The reducing equivalents for fatty acid synthesis are provided by NADPH which come either from citrate (acetyl CoA) transport or hexose monophosphate shunt. About 50-60% of required NADPH is obtained from ***HMP shunt***, which significantly influences fatty acid synthesis.

Serum free fatty acids : An inverse relationship is generally observed between hepatic fatty acid synthesis and the concentration of serum FFA.

Desaturation of fatty acid chains

A microsomal enzyme system called ***fatty acyl CoA desaturase*** is responsible for the formation of unsaturated fatty acids. This reaction also involves flavin-dependent cytochrome b_5 reductase, NADH and molecular O_2. The monounsaturated fatty acids—namely oleic acid and palmitoleic acid—are, respectively, synthesized from stearate and palmitate.

Table 14.3 Comparison of fatty acid synthesis and oxidation

		Fatty acid synthesis	β-Oxidation
1.	Major tissues	Liver, adipose tissue	Muscle, liver
2.	Subcellular site	Cytosol	Mitochondria
3.	Precursor/substrate	Acetyl CoA	Acyl CoA
4.	End product	Palmitate	Acetyl CoA
5.	Intermediates are bound to	Acyl carrier protein	Coenzyme A
6.	Coenzyme requirement	NADPH (supplying reducing equivalents)	FAD and NAD^+ (get reduced)
7.	Carbon units added/degraded	Malonyl CoA	Acetyl CoA
8.	Transport system	Citrate (mitochondria ⟶ cytosol)	Carnitine (cytosol ⟶ mitochondria)
9.	Inhibitor	Long chain acyl CoA (inhibits acetyl CoA carboxylase)	Malonyl CoA (inhibits carnitine acyltransferase I)
10.	The pathway increased	After rich carbohydrate diet	In starvation
11.	Hormonal status that promotes	High ratio of insulin/glucagon	Low ratio of insulin/glucagon
12.	Status of enzyme(s)	Multifunctional enzyme complex	Individual enzymes

Mammals ***lack the ability to introduce double bonds in fatty acids between carbon 10 and methyl terminal (ω) end.*** Hence, linoleic acid (18 : 2; 9, 12) and linolenic acid (18 : 3; 9, 12, 15) are essential for man in the diet. However, arachidonic acid (20 : 4; 5, 8, 11, 14) can be synthesized from linoleic acid by desaturation and chain elongation. Arachidonic acid is the precursor for eicosanoids (prostaglandins and thromboxanes), a group of compounds with diversified functions, discussed elsewhere (***Chapter 32***).

SYNTHESIS OF LONG CHAIN FATTY ACIDS FROM PALMITATE

Palmitate is the end product of the reactions of fatty acid synthase system that occurs in cytosol. Further, ***chain elongation*** can take place either in ***mitochondria*** or in ***endoplasmic reticulum*** (microsomes), by separate mechanisms. The microsomal chain elongation is more predominant and involves successive additions of malonyl CoA with the participation of NADPH. These reactions are similar to that catalysed by fatty acid synthase. A specific group of enzymes, namely ***elongases***, bring about fatty acid chain elongation.

The mitochondrial chain elongation is almost a reversal of β-oxidation of fatty acids. Acetyl CoA molecules are successively added to fatty acid to lengthen the chain. The reducing equivalents are derived from NADPH.

Comparison between fatty acid synthesis and oxidation

The synthesis of fatty acids and their oxidation are two distinct and independent pathways. A comparison between these two metabolic pathways in given in ***Table 14.3***.

SYNTHESIS OF TRIACYLGLYCEROLS

Triacylglycerol (TG) synthesis mostly occurs in ***liver*** and ***adipose tissue***, and to a lesser extent in other tissues. Fatty acids and glycerol must be activated prior to the synthesis of triacylglycerols. Conversion of fatty acids to acyl CoA by thiokinase is already described (***See Fig.14.6***).

Synthesis of glycerol 3-phosphate

Two mechanisms are involved for the synthesis of glycerol 3-phosphate

1. In the liver, glycerol is activated by glycerol kinase. This enzyme is absent in adipose tissue.

2. In both liver and adipose tissue, glucose serves as a precursor for glycerol 3-phosphate. Dihydroxyacetone phosphate (DHAP) produced in glycolysis is reduced by glycerol 3-phosphate dehydrogenase to glycerol 3-phosphate.

Addition of acyl groups to form TG

Glycerol 3-phosphate acyltransferase catalyses the transfer of an acyl group to produce lysophosphatidic acid. DHAP can also accept acyl group, ultimately resulting in the formation of lysophosphatidic acid. Another acyl group is added to lysophosphatidic acid to form phosphatidic acid (1,2-diacylglycerol phosphate). The enzyme phosphatase cleaves off phosphate of phosphatidic acid to produce diacylglycerol. Incorporation of another acyl group finally results in synthesis of triacylglycerol (***Fig.14.19***).

The ***three fatty acids found in triacylglycerol are not of the same type***. A saturated fatty acid is usually present on carbon 1, an unsaturated fatty acid is found on carbon 2, and carbon 3 may have either.

The intermediates of TG synthesis phosphatidic acid and diacylglycerol are also utilized for phospholipid synthesis (described later).

METABOLISM OF PHOSPHOLIPIDS

Phospholipids are a specialized group of lipids performing a variety of functions. These include the membrane structure and functions, involvement in blood clotting, and supply of arachidonic acid for the synthesis of prostaglandins (for details ***Refer Chapter 32***).

Synthesis of phospholipids

An overview of acylglycerol and phospholipid synthesis is given in ***Fig.14.20***. Almost all cells (except RBC) can synthesize phospholipids.

Phospholipids are synthesized from phosphatidic acid and 1,2-diacylglycerol, intermediates in the production of triacylglycerols (***Fig.14.19***). Phospholipid synthesis occurs in the smooth endoplasmic reticulum.

1. **Formation of lecithin and cephalin :** Choline and ethanolamine first get phosphorylated and then combine with CTP to form, respectively, CDP-choline and CDP-ethanolamine (***Fig.14.21***).

Phosphatidylcholine (lecithin) is synthesized when CDP-choline combines with 1,2-diacylglycerol. Phosphatidyl ethanolamine (cephalin) is produced when CDP-ethanolamine reacts with 1,2-diacylglycerol. Phosphatidyl ethanolamine can be converted to phosphatidyl choline on methylation.

Choline and ethanolamine, used for phospholipid synthesis, are mostly derived from the preexisting phospholipids. Thus, the ***phospholipid synthesis starting with choline or ethanolamine is regarded as salvage pathway***.

2. **Synthesis of phosphatidylserine :** Phosphatidyl ethanolamine can exchange its ethanolamine group with free serine to produce phosphatidylserine. The latter, on decarboxylation, gives the former.

3. **Formation of phosphatidylinositol :** CDP-diacylglycerol produced from phosphatidic acid combines with inositol to form phosphatidyl inositol (***PI***). This phospholipid contains ***arachidonic acid*** on carbon 2 of glycerol which serves as a substrate for ***prostaglandin synthesis***. Further, ***PI*** is important for ***signal transmission across membranes***.

4. **Synthesis of phosphatidyl glycerol and cardiolipin :** CDP-diacylglycerol combines with glycerol 3-phosphate to form phosphatidyl glycerol 3-phosphate, which then forms phosphatidylglycerol. The latter combines with another molecule of phosphatidylglycerol to finally produce cardiolipin (***Fig.14.21***). ***Cardiolipin*** is the only phospholipid possessing ***antigenic*** properties.

5. **Formation of plasmalogens :** These are phospholipids with fatty acid at carbon 1 bound by an ether linkage instead of ester linkage. An important plasmalogen, 1-alkenyl 2-acetyl glycerol 3-phosphocholine, causes blood platelet aggregation and is referred to as ***platelet-activating factor*** (PAF). The outline of the synthesis of plasmalogens is depicted in ***Fig.14.22***.

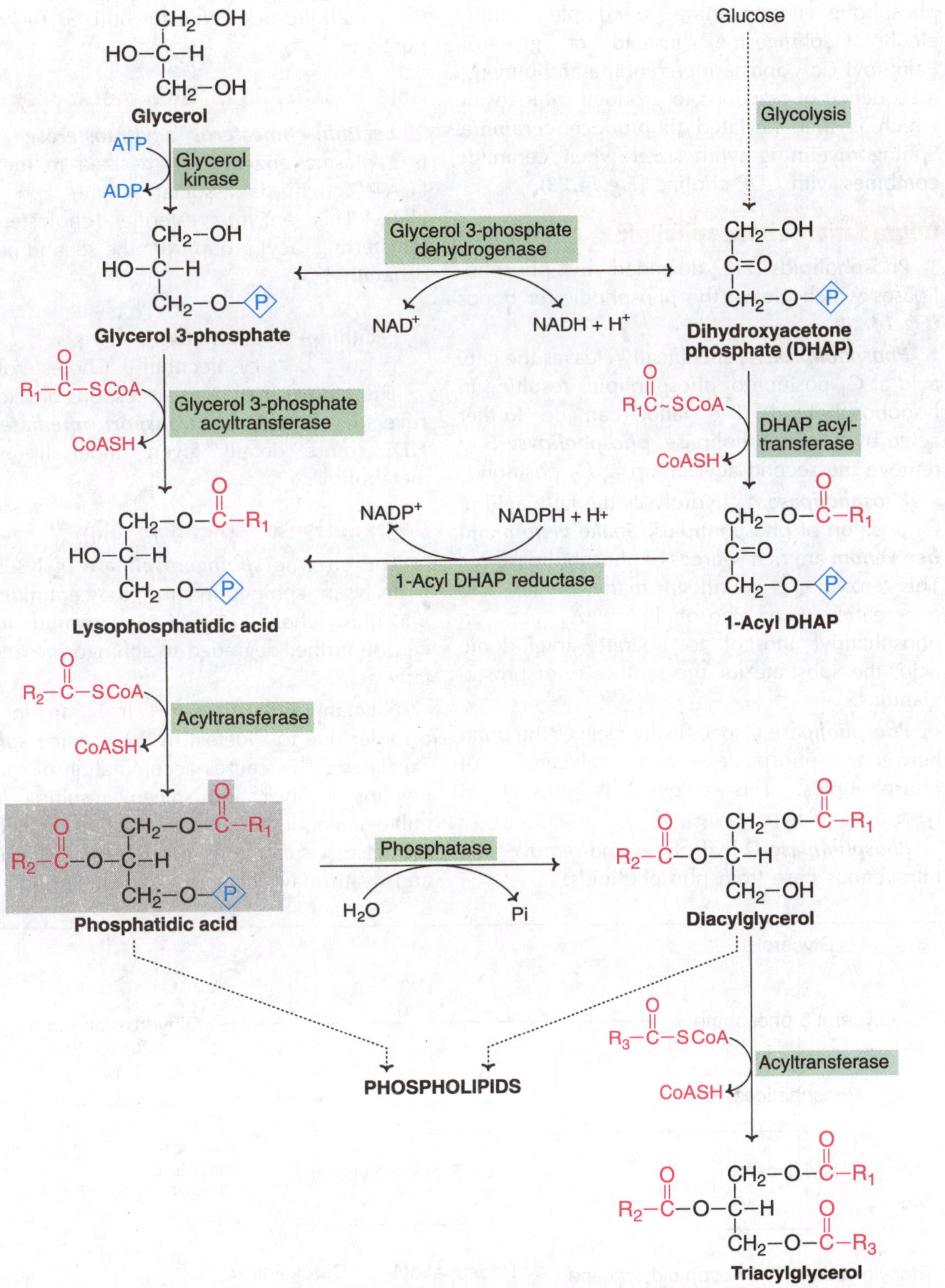

Fig. 14.19 : Synthesis of triacylglycerol.

6. **Synthesis of sphingomyelins :** These are phospholipids containing a complex amino alcohol, sphingosine, instead of glycerol. Palmitoyl CoA and serine combine and undergo a sequence of reactions to produce sphingosine which is then acylated to produce ***ceramide.*** Sphingomyelin is synthesized when ceramide combines with CDP-choline (***Fig.14.23***).

Degradation of phospholipids

Phospholipids are degraded by phospholipases which cleave the phosphodiester bonds (***Fig.14.24***).

Phospholipase A_1 specifically cleaves the fatty acid at C_1 position of phospholipids resulting in lysophospholipid. The latter can be further acted by lysophospholipase, ***phospholipase B*** to remove the second acyl group at C_2 position.

Phospholipase A_2 hydrolyses the fatty acid at C_2 position of phospholipids. ***Snake venom*** and ***bee venom*** are rich sources of phospholipase A_2. This enzyme is found in many tissues and pancreatic juice. Phospholipase A_2 acts on phosphatidyl inositol to liberate arachidonic acid, the substrate for the synthesis of prostaglandins.

Phospholipase C specifically cleaves the bond between phosphate and glycerol of phospholipids. This enzyme is present in lysosomes of hepatocytes.

Phospholipase D hydrolyses and removes the nitrogenous base from phospholipids.

The degraded products of phospholipids enter the metabolic pool and are utilized for various purposes.

Role of LCAT in lecithin metabolism

Lecithin-cholesterol acyltransferase (LCAT) is a plasma enzyme, synthesized in the liver. LCAT activity is associated with apo A_1 of HDL. This enzyme esterifies cholesterol by transferring acyl group from the second position of lecithin

$$\text{Lecithin} + \text{Cholesterol} \xrightarrow{\text{LCAT}} \text{Lysolecithin} + \text{Cholesterol ester}$$

The above reaction is responsible for the ***reverse cholesterol transport mediated by HDL*** (more details given under lipoprotein metabolism).

Degradation of sphingomyelins

The enzyme ***sphingomyelinase*** of lysosomes hydrolyses sphingomyelins to ceramide and phosphorylcholine (***Fig.14.25***). Ceramide formed can be further degraded to sphingosine and free fatty acid.

Niemann-Pick disease : It is an inherited disorder due to a defect in the enzyme ***sphingomyelinase***. This causes accumulation of sphingomyelins in liver and spleen, resulting in the enlargement of these organs. Victims of Niemann-Pick disease suffer from ***severe mental retardation***, and death may occur in early childhood.

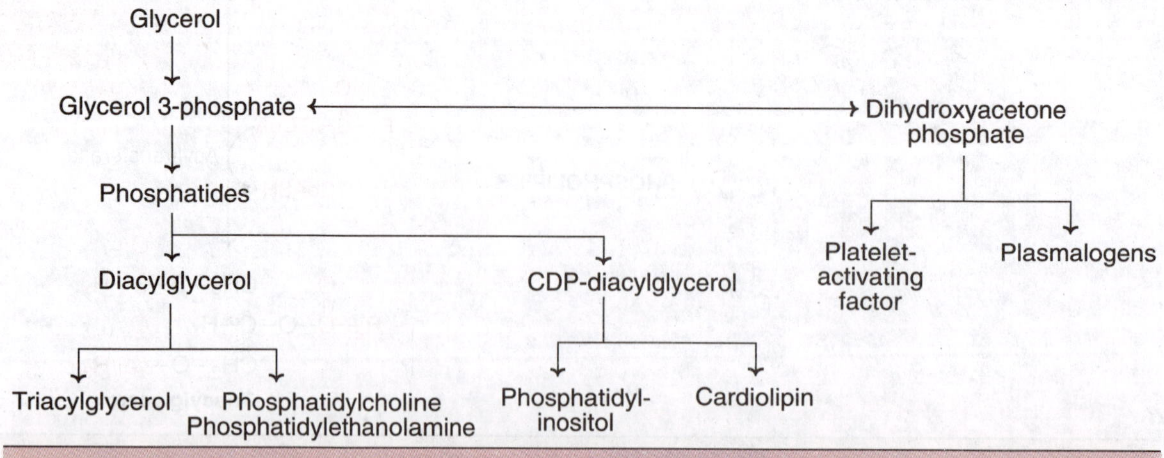

***Fig. 14.20 :** Overview of acylglycerol and phospholipid synthesis.*

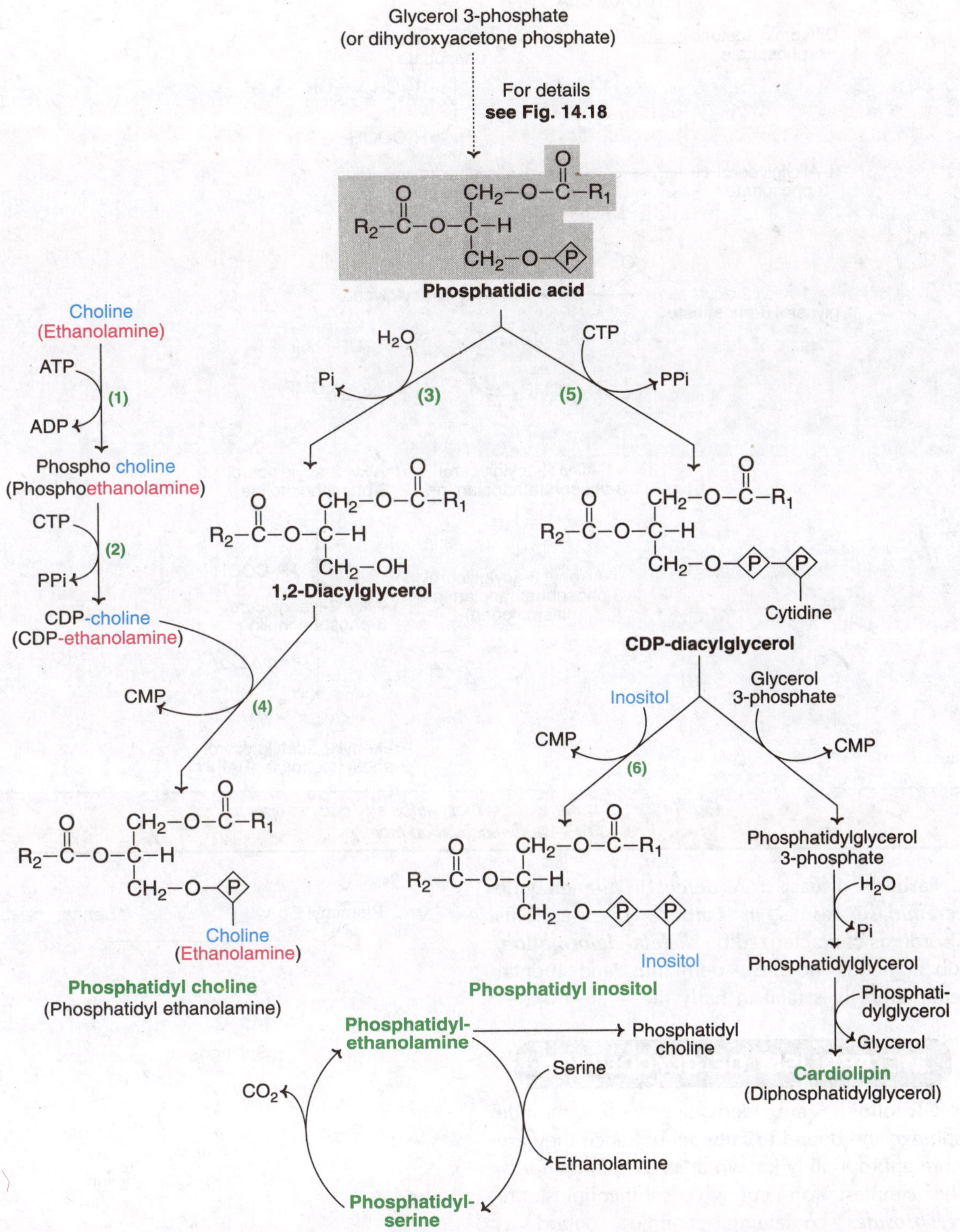

Fig. 14.21 : *Biosynthesis of phospholipids [The enzymes are numbered—(1) Choline kinase, (2) Phosphocholine cytidyltransferase, (3) Phosphatidate phosphohydrolase, (4) Phosphocholine diacylglycerol transferase, (5) CTP–Phosphatidate cytidyltransferase, (6) CDP–Diacylglycerol inositol transferase].*

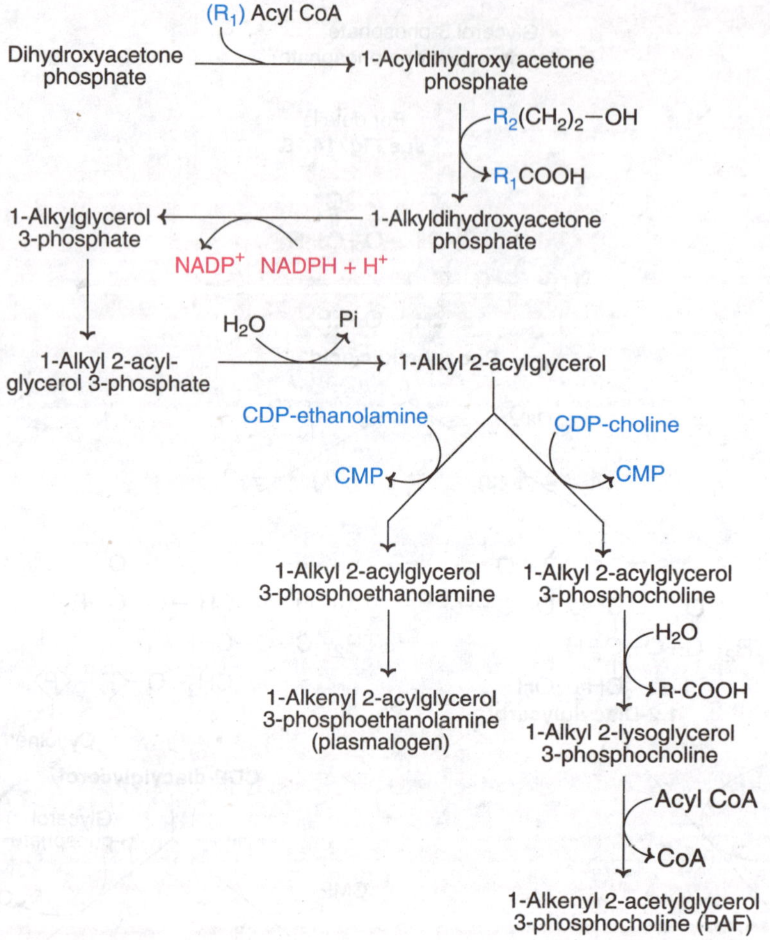

***Fig. 14.22 :** Summary of the biosynthesis of plasmalogens (PAF–Platelet activation factor).*

Farber's disease : A ***defect*** in the enzyme ***ceramidase*** results in Farber's disease. This disorder is characterized by ***skeletal deformation***, subcutaneous nodules, dermatitis and mental retardation. It is fatal in early life.

METABOLISM OF GLYCOLIPIDS

Glycolipids are derivatives of ceramide (sphingosine bound to fatty acid), hence they are more appropriately known as ***glycosphingolipids.*** The simplest form of glycosphingolipids are ***cerebrosides*** containing ceramide bound to monosaccharides. ***Galactocerebroside*** (Gal-Cer) and glucocerebroside (Glu-Cer) are the common glycosphingolipids. Galactocerebroside is a major component of membrane lipids in the nervous

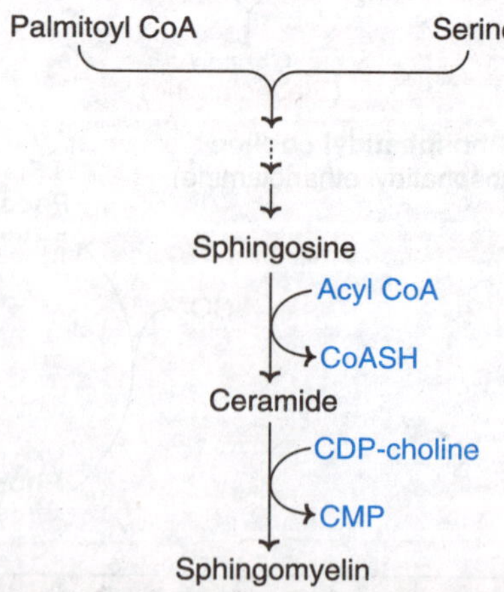

***Fig. 14.23 :** An outline of the synthesis of sphingomyelin.*

tissue (high in myelin sheath). Glucocerebroside is an intermediate in the synthesis and degradation of complex glycosphingolipids.

Synthesis of cerebrosides

The outline of the synthesis of cerebrosides and sulfatide is given in ***Fig.14.26***.

Metabolic disorders of cerebrosides

The degradation of cerebrosides along with the associated inborn errors is depicted in ***Fig.14.27***.

Gaucher's disease : This is due to a defect in the enzyme ***β-glucosidase.*** As a result, tissue glucocerebroside levels increase. This disorder is commonly associated with ***enlargement of liver and spleen***, osteoporosis, pigmentation of skin, anemia, mental retardation etc. Sometimes, Gaucher's disease is fatal.

Krabbe's disease : Defect in the enzyme ***β-galactosidase*** results in the accumulation of galactocerebrosides. A total ***absence of myelin in the nervous tissue*** is a common feature. Severe mental retardation, convulsions, blindness, deafness etc. are seen. Krabbe's disease is fatal in early life.

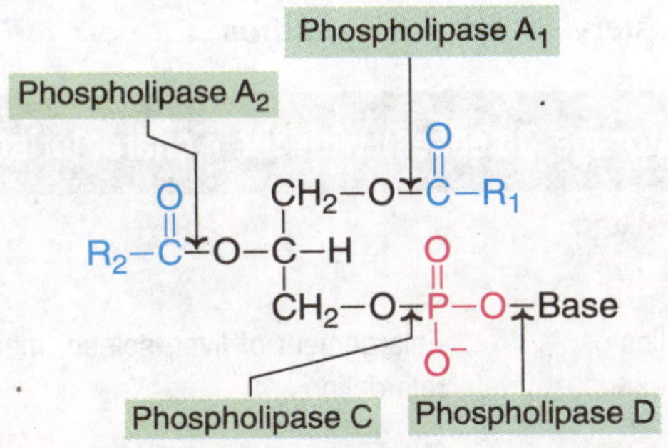

***Fig. 14.24 :** Degradation of phospholipids by phospholipases.*

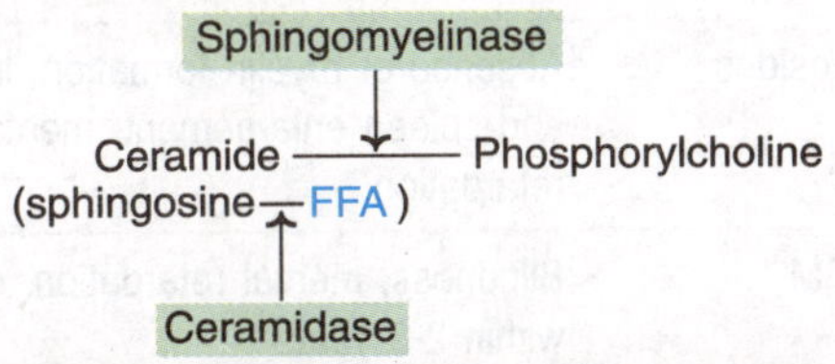

***Fig. 14.25 :** Site of action of sphingomyelinase and ceramidase on sphingomyelin.*

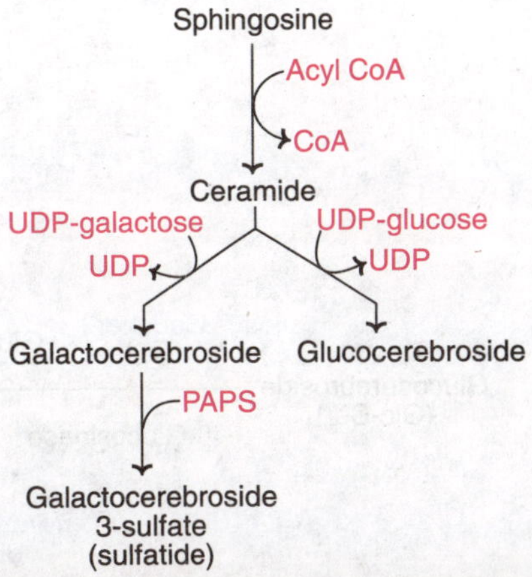

***Fig. 14.26 :** Biosynthesis of cerebrosides and sulfatides (PAPS–Phosphoadenosyl phosphosulfate).*

Niemann-Pick disease and Farber's disease connected with sphingomylein metabolism are already described. They are also depicted in ***Fig.14.27***.

Gangliosides are complex glycosphingolipids mostly found in ganglion cells. They contain one or more molecules of N-acetylneuraminic acid (NANA) bound ceramide oligosaccharides. Defect in the degradation of ***gangliosides*** causes gangliosidosis, Tay-Sach's disease etc.

Sphinogolipidoses

Lipid storage diseases, representing lysosomal storage defects, are inherited disorders. They are characterized by the accumulation of complex lipids.

The term ***sphingolipidoses*** is often used to collectively refer to the genetic disorders that lead to the ***accumulation*** of any one of the sphingolipids (***glycosphingolipids and sphingomyelins***). Some examples of sphiogolipidoses (***lipid storage diseases***) with important features are summarized in ***Table 14.4***.

METABOLISM OF CHOLESTEROL

Cholesterol is found exclusively in animals, hence it is often called as ***animal sterol***. The total body content of cholesterol in an adult man weighing 70 kg is around 140 g i.e., around 2 g/kg body weight. Cholesterol is

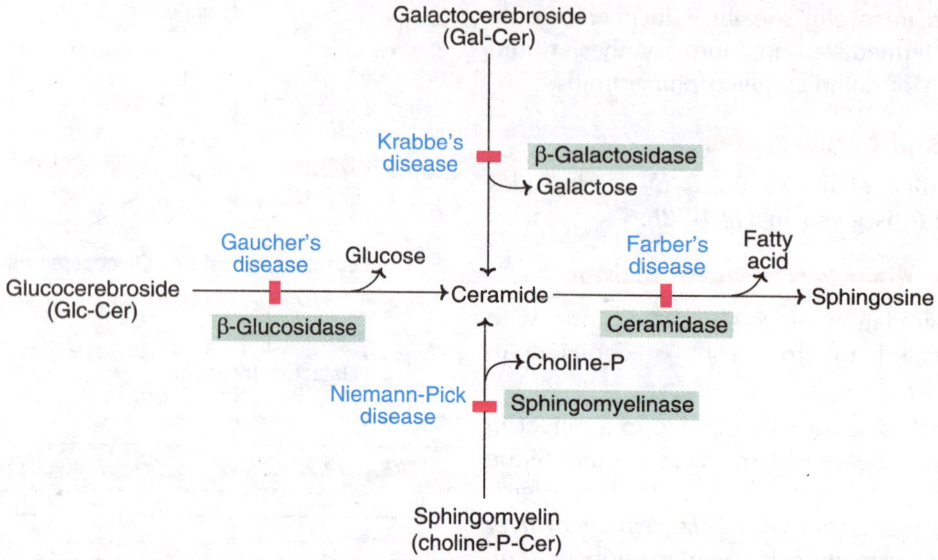

Fig. 14.27 : Degradation of cerebrosides and sphingomyelins with metabolic disorders.

amphipathic in nature, since it possesses both hydrophilic and hydrophobic regions in the structure.

Functions of cholesterol

Cholesterol is essential to life, as it performs a number of important functions

1. It is a ***structural component*** of cell membrane.

2. Cholesterol is the precursor for the ***synthesis of all other steroids*** in the body. These include steroid hormones, vitamin D and bile acids.

3. It is an essential ingredient in the structure of ***lipoproteins*** in which form the lipids in the body are transported.

4. ***Fatty acids are transported to liver*** as cholesteryl esters for ***oxidation***.

Table 14.4 Some examples of sphingolipidoses (lipid storage diseases) with their characteristics

Disease	Missing/defective enzyme	Major storage compound	Symptoms
Niemann-Pick disease	Sphingomyelinase	Sphingomyelins	Enlargement of liver, spleen, mental retardation.
Farber's disease	Ceramidase	Ceramide	Painful and deformed joints.
Gaucher's disease	β-Glucosidase	Glucocerebroside	Enlargement of liver and spleen, osteoporosis, mental retardation.
Krabbe's disease	β-Galactosidase	Galactocerebrosides	Absence of myelin formation, liver and spleen enlargement, mental retardation.
Tay-Sachs disease	Hexosaminidase A	Ganglioside GM_2	Blindness, mental retardation, death within 2-3 years.
Fabry's disease	α-Galactosidase	Ceramide trihexoside	Renal failure, skin rash, pain in lower extremities.

CHOLESTEROL BIOSYNTHESIS

About 1 g of cholesterol is synthesized per day in adults. Almost all the tissues of the body participate in cholesterol biosynthesis. The largest contribution is made by ***liver*** (***20%***), ***intestine*** (***15%***), skin, adrenal cortex, reproductive tissue etc.

The enzymes involved in cholesterol synthesis are found in the ***cytosol*** and ***microsomal fractions*** of the cell. Acetate of ***acetyl CoA*** provides all the carbon atoms in cholesterol.

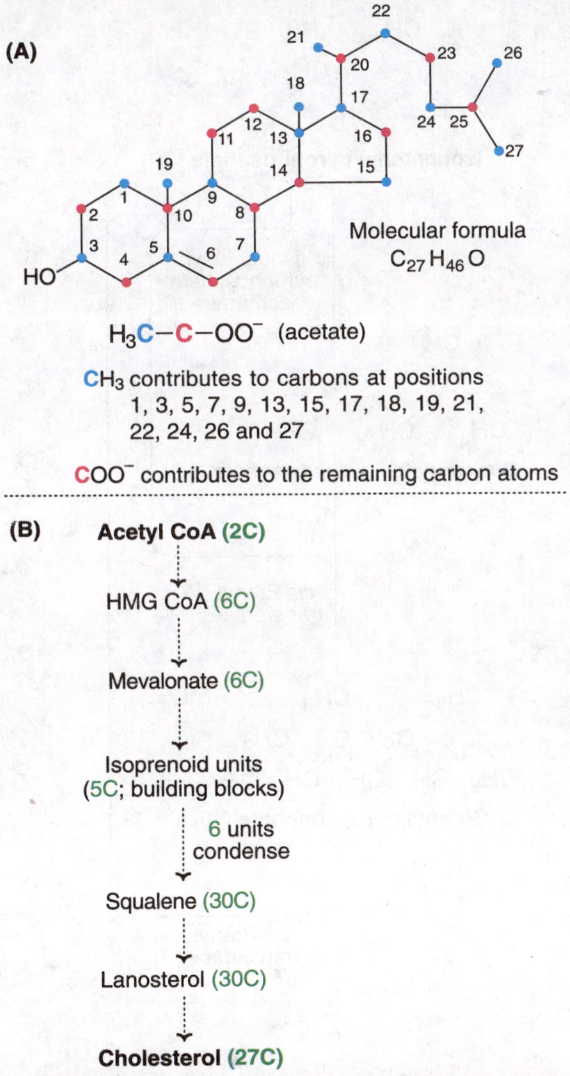

Fig. 14.28 : Outline of cholesterol biosynthesis— (A) Derivation of carbon atoms from acetate, (B) Key intermediates with the carbon atoms.

The reducing equivalents are supplied by ***NADPH*** while ***ATP*** provides energy. For the production of one mole of cholesterol, 18 moles of acetyl CoA, 36 moles of ATP and 16 moles of NADPH are required.

By administering acetate with ^{14}C isotope label either on the methyl ($-CH_3$) group or carboxyl ($-COO$) group, the origin of carbon atoms in the entire molecule of cholesterol has been established. The sources of carbon atoms and the ***key intermediates of cholesterol formation*** are depicted in ***Fig.14.28***, and the detailed reactions are given in ***Fig.14.29***.

The synthesis of cholesterol may be learnt in ***5 stages***

1. Synthesis of ***HMG CoA (6C)***
2. Formation of ***mevalonate (6C)***
3. Production of ***isoprenoid units (5C)***
4. Synthesis of ***squalene (30C)***
5. Conversion of ***squalene to cholesterol (27C)***.

1. **Synthesis of β-hydroxy β-methylglutaryl CoA (HMG CoA) :** Two moles of acetyl CoA condense to form acetoacetyl CoA. Another molecule of acetyl CoA is then added to produce HMG CoA. These reactions are similar to that of ketone body synthesis. However, the two pathways are distinct, since ketone bodies are produced in ***mitochondria*** while cholesterol synthesis occurs ***in cytosol***. Thus, there exist ***two pools of HMG CoA*** in the cell. Further, two isoenzymes of HMG CoA synthase are known. The cytosomal enzyme is involved in cholesterol synthesis whereas the mitochondrial HMG CoA synthase participates in ketone body formation.

2. **Formation of mevalonate :** ***HMG CoA reductase*** is the ***rate limiting enzyme*** in cholesterol biosynthesis. This enzyme is present in endoplasmic reticulum and catalyses the reduction of HMG CoA to mevalonate. The reducing equivalents are supplied by NADPH.

3. **Production of isoprenoid units :** In a three-step reaction catalysed by kinases, mevalonate is converted to 3-phospho 5-pyrophosphomevalonate which on decarboxylation forms isopentenyl pyrophosphate (IPP). The latter isomerizes to dimethylallyl pyrophosphate (DPP). Both IPP and DPP are 5-carbon isoprenoid units.

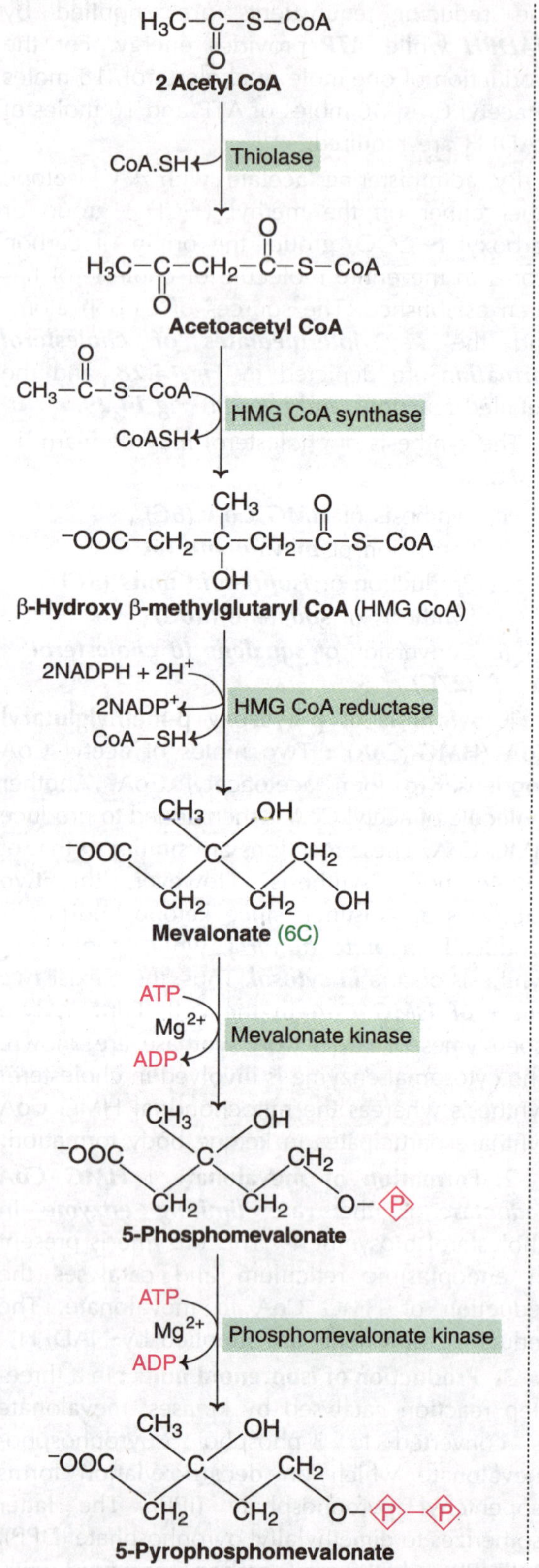

Fig. 14.29 contd. next coloumn

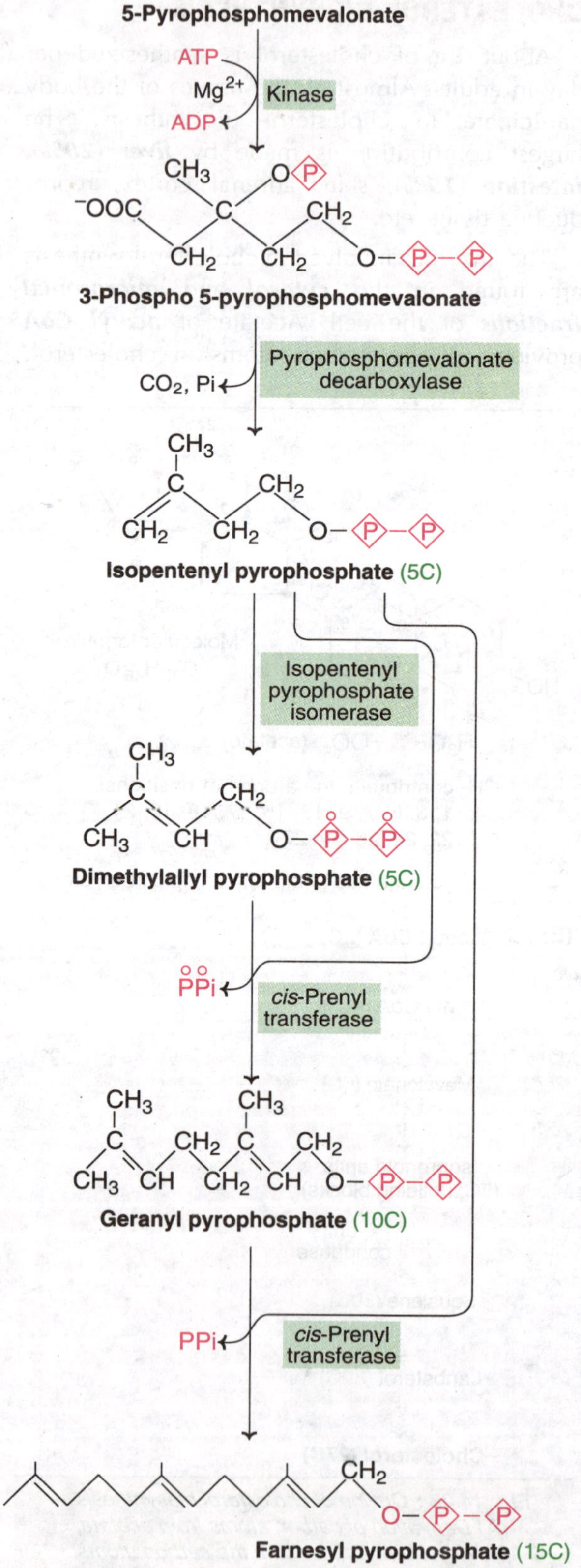

Fig. 14.29 contd. next page

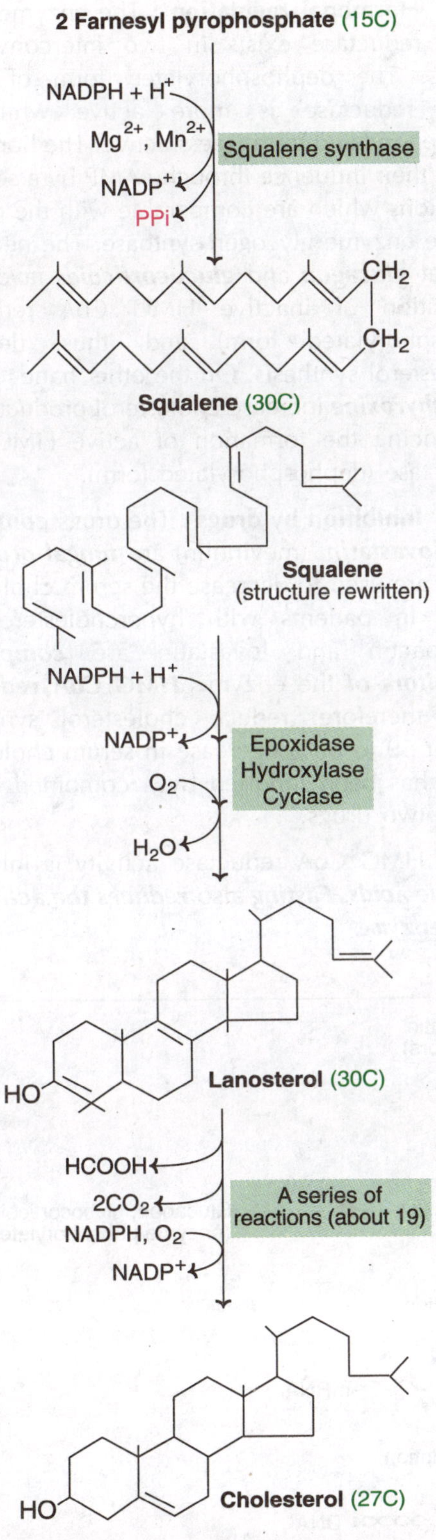

Fig. 14.29 : Biosynthesis of cholesterol.

4. **Synthesis of squalene** : IPP and DPP condense to produce a 10-carbon geranyl pyrophosphate (GPP). Another molecule of IPP condenses with GPP to form a 15-carbon farnesyl pyrophosphate (FPP). Two units of farnesyl pyrophosphate unite and get reduced to produce a 30-carbon squalene.

Farnesyl pyrophosphate, besides its participation in cholesterol synthesis is involved in the ***formation of ubiquinone, dolichol*** and ***prenylation of proteins***. Protein prenylation is known to facilitate protein-protein interaction, and membrane structure.

5. **Conversion of squalene to cholesterol** : Squalene undergoes hydroxylation and cyclization utilizing O_2 and NADPH and gets converted to lanosterol. The formation of cholesterol from lanosterol is a multistep process with a series of about 19 enzymatic reactions. The following are the most important reactions

- Reducing the carbon atoms from 30 to 27.
- Removal of two methyl groups from C_4 and one methyl group from C_{14}.
- Shift of double bond from C_8 to C_5.
- Reduction in the double bond present between C_{24} and C_{25}.

The enzymes (about 19?) involved in the conversion of lanosterol to cholesterol are associated with endoplasmic reticulum. 14-Desmethyl lanosterol, zymosterol, cholestadienol and desmosterol are among the intermediates in the cholesterol biosynthesis. ***The penultimate product is 7-dehydrocholesterol which, on reduction, finally yields cholesterol.***

Cholesterol biosynthesis is now believed to be a part of a major metabolic pathway concerned with the ***synthesis of*** several other isoprenoid compounds. These include ubiquinone (***coenzyme Q of electron transport chain***) and ***dolichol*** (found in glycoprotein). Both of them are derived from farnesyl pyrophosphate.

Regulation of cholesterol synthesis

Cholesterol biosynthesis is controlled by the rate limiting enzyme ***HMG CoA reductase,*** at the beginning of the pathway (***Fig.14.30***). HMG

Medical Concepts/ Clinical Correlates

- Niemann-Pick disease, caused by a defect in the enzyme sphingomyelinase, results in the accumulation of sphingomyelins in liver and spleen.
- About a dozen glycolipid storage diseases are known. These include Gaucher's disease and Krabbe's disease.

CoA reductase is found in association with endoplasmic reticulum, and is subjected to different metabolic controls.

1. **Feedback control** : The end product cholesterol controls its own synthesis by a feedback mechanism. Increase in the cellular concentration of ***cholesterol*** reduces the synthesis of the enzyme HMG CoA reductase. This is achieved by ***decreasing the transcription*** of the gene responsible for the production of ***HMG CoA reductase***. Feedback regulation has been investigated with regard to LDL cholesterol taken up by the cells, and the same mechanism is believed to operate whenever cellular cholesterol level is elevated.

2. **Hormonal regulation** : The enzyme HMG CoA reductase exists in two interconvertible forms. The dephosphorylated form of HMG CoA reductase is more active while the phosphorylated form is less active. The hormones exert their influence through cAMP by a series of reactions which are comparable with the control of the enzyme glycogen synthase. The net effect is that ***glucagon*** and ***glucocorticoids*** favour the formation of inactive HMG CoA reductase (phosphorylated form) and, thus, decrease cholesterol synthesis. On the other hand, ***insulin*** and ***thyroxine*** increase cholesterol production by enhancing the formation of active HMG CoA reductase (dephosphorylated form).

3. **Inhibition by drugs** : The drugs ***compactin*** and ***lovastatin*** (mevinolin) are ***fungal products***. They are used to decrease the serum cholesterol level in patients with hypercholesterolemia. Compactin and lovastatin are ***competitive inhibitors of*** the enzyme ***HMG CoA reductase*** and, therefore, reduce cholesterol synthesis. About 50 to 60% decrease in serum cholesterol level has been reported by a combined use of these two drugs.

4. HMG CoA reductase activity is inhibited by ***bile acids***. ***Fasting*** also ***reduces the activity of this enzyme***.

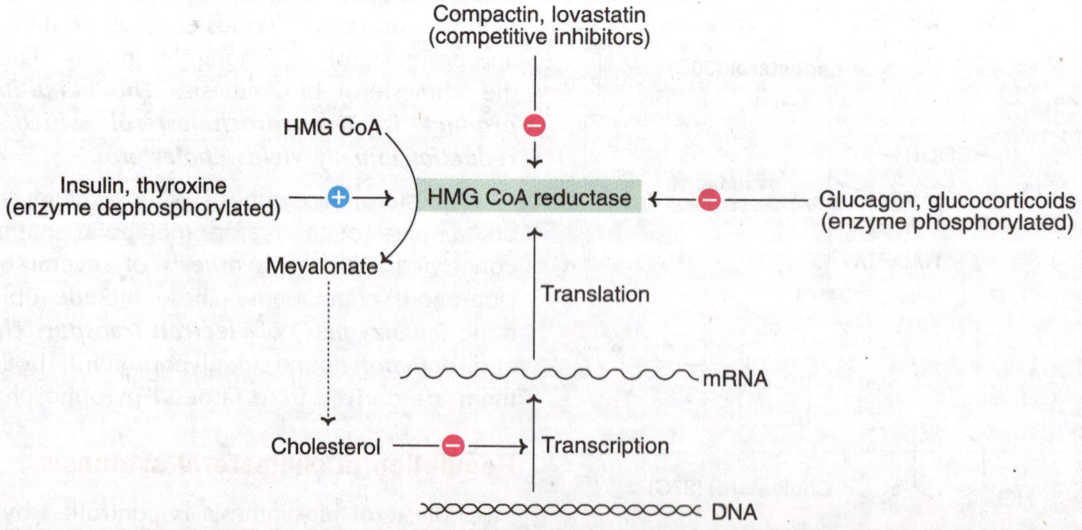

***Fig. 14.30 :** Regulation of cholesterol biosynthesis by HMG CoA reductase (+ –Promoting effect; – –Inhibitory effect).*

DEGRADATION OF CHOLESTEROL

The steroid nucleus (ring structure) of the ***cholesterol cannot be broken down*** in humans. Cholesterol (50%) is converted to ***bile acids***, excreted in feces, serves as a precursor for the synthesis of ***steroid hormones***, ***vitamin D***, ***coprostanol*** and ***cholestanol***. The latter two are the fecal sterols, besides cholesterol.

I. Synthesis of bile acids

The bile acids possess 24 carbon atoms, 2 or 3 hydroxyl groups in the steroid nucleus and a side chain ending in carboxyl group. The bile acids are amphipathic in nature since they possess both polar and non-polar groups. They serve as emulsifying agents in the intestine and actively participate in the digestion and absorption of lipids.

The synthesis of primary bile acids takes place in the liver and involves a series of reactions (***Fig.14.31***). The step catalysed by ***7 α-hydroxylase*** is inhibited by bile acids and this is the ***rate limiting*** reaction. ***Cholic acid and chenodeoxycholic acid*** are the ***primary bile acids*** and the former is found in the largest amount in bile. On conjugation with glycine or taurine, ***conjugated bile acids*** (glycocholic acid, taurocholic acid etc.) are formed which are more ***efficient in their function*** as surfactants. In the bile, the conjugated bile acids exist as ***sodium*** and ***potassium*** salts which are known as ***bile salts***.

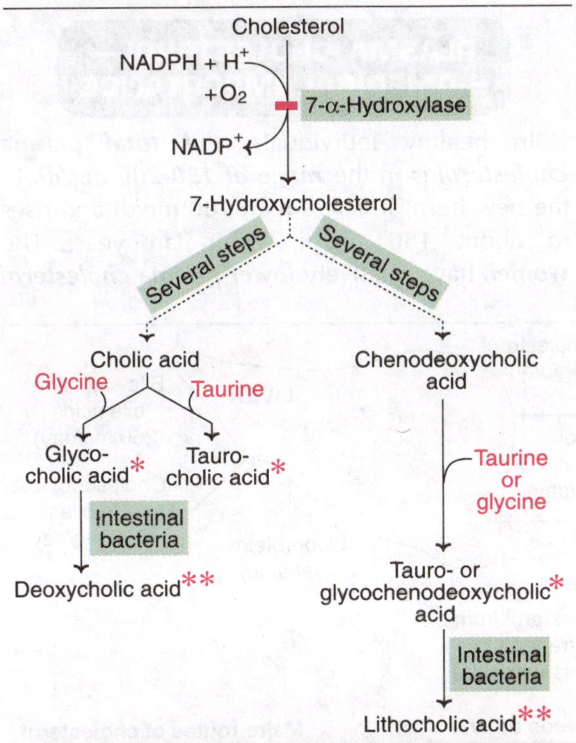

Fig. 14.31 : *Outline of bile acid synthesis (▬ inhibited by bile acids and, in vitamin C deficiency. *–Primary bile acids, **–Secondary bile acids).*

In the intestine, a portion of primary bile acids undergoes deconjugation and dehydroxylation to form ***secondary bile acids*** (***deoxycholic acid and lithocholic acid***). These reactions are catalysed by bacterial enzymes in the intestine.

Enterohepatic circulation : The conjugated bile salts synthesized in the liver accumulate in gall bladder. From there they are secreted into the small intestine where they serve as emulsifying agents for the digestion and absorption of fats and fat soluble vitamins. A large portion of the ***bile salts*** (primary and secondary) are ***reabsorbed and returned to the liver*** through portal vein. Thus the ***bile salts are recycled and reused several times in a day***. This is known as enterohepatic circulation. About 15-30 g of bile salts are secreted into the intestine each day and reabsorbed. However, a small portion of about 0.5 g/day is lost in the feces. An equal amount (0.5 g/day) is synthesized in liver to replace the lost bile salts. The ***fecal excretion of bile salts*** is the only ***route for the removal of cholesterol from the body***.

Cholelithiasis : Bile salts and phospholipids are responsible for keeping the cholesterol in bile in a soluble state. Due to their deficiency (particularly bile salts), cholesterol crystals precipitate in the gall bladder often resulting in cholelithiasis—***cholesterol gall stone disease***. Cholelithiasis may be due to defective absorption of bile salts from the intestine, impairment in liver function, obstruction of biliary tract etc.

The patients of cholelithiasis respond to the administration of bile acid chenodeoxy cholic acid, commonly known as ***chenodiol.*** It is believed that a slow but gradual dissolution of gall stones occurs due to chenodiol. For severe cases of cholelithiasis, surgical removal of gall bladder is the only remedy.

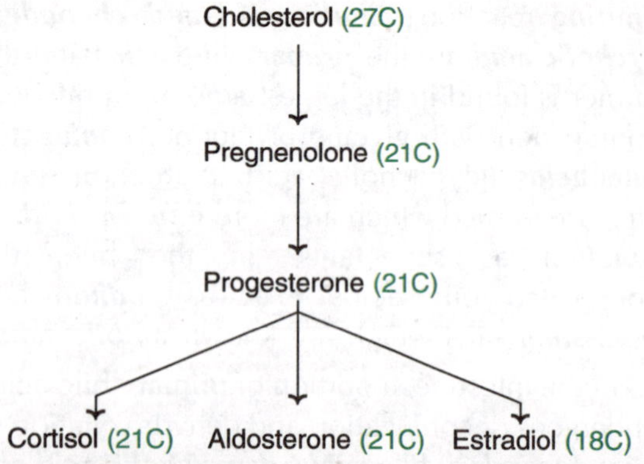

Fig. 14.32 : *Outline of steroid hormone synthesis from cholesterol (Numbers in the brackets represent the number of carbon atoms).*

II. Synthesis of steroid hormones from cholesterol

Cholesterol is the precursor for the synthesis of all the five classes of steroid hormones

(a) ***Glucocorticoids*** (e.g. cortisol)

(b) ***Mineralocorticoids*** (e.g. aldosterone)

(c) ***Progestins*** (e.g. progesterone)

(d) ***Androgens*** (e.g. testosterone)

(c) ***Estrogens*** (e.g. estradiol).

A brief outline of steroid hormonal synthesis is given in ***Fig.14.32*** and more details are discussed under ***'Hormones'*** (***Chapter 19***).

III. Synthesis of vitamin D

7-Dehydrocholesterol, an intermediate in the synthesis of cholesterol, is converted to cholecalciferol (vitamin D_3) by ultraviolet rays in the skin.

A brief summary of prominent sources and the major pathways for utilization of cholesterol with the liver as the central metabolic organ is depicted in ***Fig.14.33.***

Transport of cholesterol

Cholesterol is present in the plasma lipoproteins in two forms

1. About ***70-75%*** of it is in an ***esterified*** form with long chain fatty acids.

2. About 25-30% as free cholesterol. This form of cholesterol readily exchanges between different lipoproteins and also with the cell membranes.

Role of LCAT : High density lipoproteins (HDL) and the enzyme ***lecithin-cholesterol acyltransferase (LCAT)*** are responsible for the transport and elimination of cholesterol from the body. LCAT is a plasma enzyme, synthesized by the liver. It catalyses the transfer of fatty acid from the second position of phosphatidyl choline (lecithin) to the hydroxyl group of cholesterol (***Fig.14.34***). HDL-cholesterol is the real substrate for LCAT and this reaction is freely reversible. ***LCAT activity*** is associated with ***apo-A_1 of HDL.***

The cholesterol (cholesteryl) ester forms an integral part of HDL. In this manner, the ***cholesterol from the peripheral tissues is trapped in HDL***, by a reaction catalysed by LCAT and then ***transported to liver for degradation*** and excretion. This mechanism is commonly known as ***reverse cholesterol transport.***

PLASMA CHOLESTEROL— BIOMEDICAL IMPORTANCE

In healthy individuals, the ***total plasma cholesterol*** is in the ***range of 150-200 mg/dl***. In the new born, it is less than 100 mg/dl and rises to about 150 mg/dl within an year. The ***women*** have relatively ***lower plasma cholesterol***

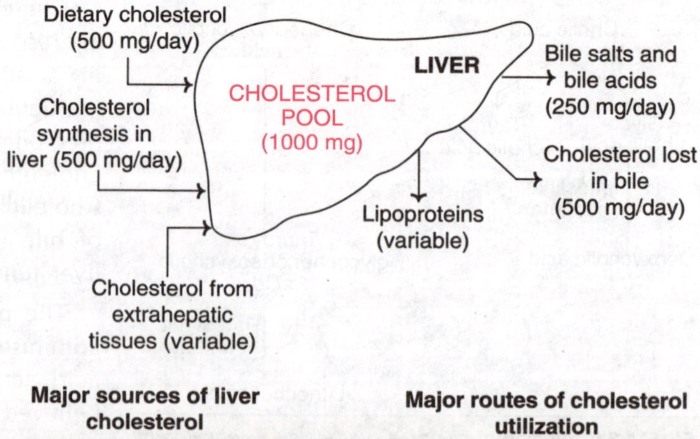

Fig. 14.33 *Summary of major sources of liver cholesterol and its utilization (values given in brackets are variable).*

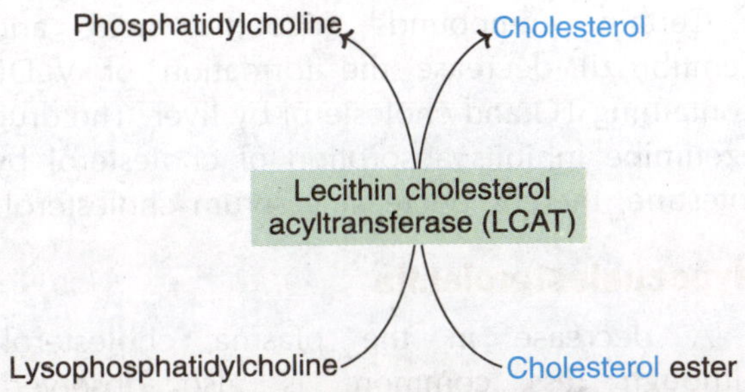

Fig. 14.34 : Reaction catalysed by LCAT.

which is attributed to the hormones-***estrogens***. Cholesterol level increases with increasing age (in women particularly after menopause), and also in pregnancy.

Plasma cholesterol is associated with different lipoprotein fractions (LDL, VLDL and HDL).

Total cholesterol can be estimated by many methods such as Libermann-Burchard reaction, Carr and Dructor method and, more recently, cholesterol oxidase method. HDL- cholesterol can be determined after precipitating LDL and VLDL by polyethylene glycol (PEG). VLDL cholesterol is equivalent to 1/5th of plasma triacylglycerol (TG) in a fasting state. LDL-cholesterol can be calculated by ***Friedewald formula*** given below.

LDL-cholesterol = Total cholesterol – (HDL-cholesterol + TG/5).

The above formula is not valid if TG concentration is above 400 mg/dl.

In adults, the normal LDL-cholesterol is about <100 mg/dl while HDL-cholesterol is around 40-60 mg/dl. Elevation in plasma HDL-cholesterol is beneficial to the body, since it protects the body from atherosclerosis and coronary heart diseases (CHD). On the other hand, increase in LDL-cholesterol is harmful to the body as it may lead to various complications, including CHD.

HYPERCHOLESTEROLEMIA

Increase in plasma cholesterol (> 200 mg/dl) concentration is known as hypercholesterolemia and is observed in many disorders

1. **Diabetes mellitus :** Due to increased ***cholesterol synthesis*** since the availability of acetyl CoA is increased.

2. **Hypothyroidism (myxoedema) :** This is believed to be due to ***decrease in the HDL receptors*** on hepatocytes.

3. **Obstructive jaundice :** Due to an ***obstruction in the excretion*** of cholesterol through bile.

4. **Nephrotic syndrome :** Increase in plasma globulin concentration is the characteristic feature of nephrotic syndrome. Cholesterol elevation is due to increase in plasma lipoprotein fractions in this disorder.

Hypercholesterolemia is associated with ***atherosclerosis*** and ***coronary heart disease*** (CHD). More specifically, ***LDL-cholesterol is positively correlated,*** whereas HDL-cholesterol is negatively correlated with CHD.

Bad cholesterol and good cholesterol : Cholesterol is a natural metabolite performing a wide range of functions (membrane structure, precursor for steroid hormones, bile acids). The usages good and bad to cholesterol, although inappropriate, are still in use. The cholesterol in high concentration, present in ***LDL,*** is considered ***bad*** due to its involvement in altherosclerosis and related complications. Thus, LDL may be regarded as ***l***ethally ***d***angerous ***l***ipoprotein. Small dense LDL (**sdLDL**) is considered to be the ***most dangerous fraction of LDL*** associated with CHD. On the other hand, ***HDL cholesterol*** is ***good*** since its high concentration counteracts athero-genesis. HDL may be considered as ***h***ighly ***d***esirable ***l***ipoprotein.

Effects of lifestyles on serum cholesterol level : Individual lifestyles and habits certainly influence serum cholesterol, and thus play a significant role in the development coronary heart disease. The parametres such as ***high blood pressure***, ***emotional stress***, ***smoking***, drinking of soft water (against hard water), coffee drinking, lack of exercise, obesity (particlarly of abdomen) ***elevate serum cholesterol level***.

Control of hypercholesterolemia

Several measures are advocated to lower the plasma cholesterol (particularly LDL) level

1. **Consumption of PUFA :** Dietary intake of polyunsaturated fatty acids (PUFA) ***reduces the plasma cholesterol level***. PUFA will help in

transport of cholesterol by LCAT mechanism (described earlier) and its excretion from the body. The oils with rich PUFA content include cottonseed oil, soya bean oil, sunflower oil, corn oil, fish oils etc. Ghee and coconut oil are poor sources of PUFA.

2. **Dietary cholesterol :** Dietary cholesterol ***influence*** on plasma cholesterol is ***minimal***. However, avoidance of cholesterol-rich foods is advocated, and a dietary intake of <300 mg/day is advised. Certain drugs (e.g. ezetimibe) inhibit intestinal cholesterol absorption.

3. **Plant sterols :** Certain plant sterols and their esters (e.g. sitostanol esters) ***reduce*** plasma cholesterol levels. They inhibit the intestinal absorption of dietary cholesterol.

4. **Dietary fiber :** Fiber present in vegetables ***decreases the cholesterol absorption*** from the intestine.

5. **Avoiding high carbohydrate diet :** Diets rich in carbohydrates (e.g. sucrose) should be avoided to control hypercholesterolemia.

6. **Impact of lifestyles :** Elevation in plasma cholesterol is obseved in people with smoking, abdominal obesity, lack of exercise, stress, high blood pressure, consumption of soft water etc. Therefore, adequate ***changes in the lifestyles*** will bring down plasma cholesterol.

7. **Moderate alcohol cosumption :** The beneficial effects of moderate alcohol intake are masked by the ill effects of chronic alcoholism. ***Red wine*** is particularly ***beneficial*** due to its antioxidants, besides low alcohol content.

8. **Use of drugs :** Drugs such as ***lovastatin*** which inhibit HMG CoA reductase and decrease cholesterol synthesis are used. Statins currently in use include atorvastatin, simvastatin, fluvastatin and pravastatin. ***Statins*** are usually ***taken at night to ensure maximum effect*** (HMG CoA reductase activity at peak about 6 hours after dark). Certain drugs—***cholestyramine*** and ***colestipol***—bind with bile acids and decrease their intestinal reabsorption. This helps in the conversion of more cholesterol to bile acids and its excretion through feces. ***Clofibrate*** increases the activity of lipoprotein lipase and reduces the plasma cholesterol and triacylglycerols.

Certain compounds nicotinic acid and gemfibrozil decrease the formation of VLDL containing TG and cholesterol by liver. The drug ezetimibe inhibits absorption of cholesterol by intestine, thereby decreasing serum cholesterol.

Hypocholesterolemia

A decrease in the plasma cholesterol, although less common, is also observed. Hyperthyroidism, pernicious anemia, malabsorption syndrome, hemolytic jaundice etc., are some of the disorders associated with hypocholesterolemia.

Dietary cholesterol—safe to consume!

In the dietary guidelines, cholesterol was treated as a nutrient of concern and its daily intake was restricted to not more than 300 mg. The US Government in 2015 has removed ***cholesterol from the list of nutrient of concern taking a U-turn on its earlier recommendations***.

The earlier warning (since 1970) world over was to stay away from high cholesterol foods to avoid heart diseases and clogging of arteries. Research findings repeatedly indicated that there is ***no significant relationship between consumption of dietary cholesterol and serum cholesterol***. Further, majority of people that have heart attacks were found to have normal cholesterol. Thus, ***cholesterol has become the biggest medical scam of all time.***

With the recent developments, ***consumption of cholesterol—rich foods such as eggs, sea foods***, animal fats and oils, fish and fish oils, cheese etc. are now ***considered safe*** (with regard to cholesterol).

Cholesterol is a friend, not an enemy

Cholesterol is a vital and universal ***component of every cell*** membrane on earth. Further, cholesterol is the precursor for the ***synthesis of steroid hormones, bile acids, vitamin D*** etc. Thus, there is no life without cholesterol. Therefore, cholesterol should be treated as one of our best friends, not an evil.

The focus now is on dietary transfat and sugars as culprits in the causation of heart diseases.

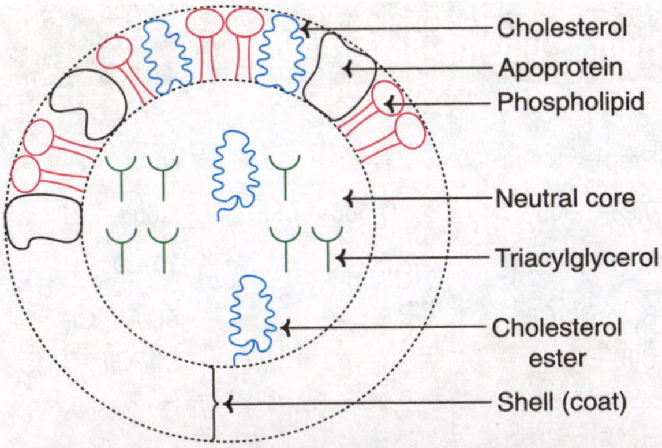

Fig. 14.35 : *A general structure of lipoprotein complex. (Note : For the sake of clarity, only a part of the shell and core are filled with the constituents).*

LIPOPROTEINS

Lipoproteins are ***molecular complexes*** that consist ***of lipids and proteins*** (conjugated proteins). They function as transport vehicles for lipids in blood plasma. Lipoproteins deliver the lipid components (cholesterol, triacylglycerol etc.) to various tissues for utilization.

Structure of lipoproteins

A lipoprotein basically consists of a neutral lipid core (with triacylglycerol and/or cholesteryl ester) surrounded by a coat shell of phospholipids, apoproteins and cholesterol (***Fig.14.35***). The polar portions (amphiphilic) of phospholipids and cholesterol are exposed on the surface of lipoproteins so that lipoprotein is soluble in aqueous solution.

Classification of lipoproteins

Five major classes of lipoproteins are identified in human plasma, based on their separation by electrophoresis (***Fig.14.36***).

1. **Chylomicrons :** They are ***synthesized in the intestine***, and transport exogenous (dietary) triacylglycerol to various tissues. They consist of highest (***99%***) quantity of ***lipid*** and lowest (1%) concentration of protein. The chylomicrons are the least in density and the largest in size, among the lipoproteins.

2. **Very low density lipoproteins (VLDL) :** They are ***produced in liver and intestine*** and are responsible for the transport of endogenously synthesized triacylglycerols.

3. **Low density lipoproteins (LDL) :** They are ***formed from VLDL in the blood circulation***. They transport cholesterol from liver to other tissues.

4. **High density lipoproteins (HDL) :** They are ***mostly synthesized in liver***. Three different fractions of HDL (1, 2 and 3) can be identified by ultracentrifugation. HDL particles transport cholesterol from peripheral tissues to liver (reverse cholesterol transport).

5. **Free fatty acids—albumin :** Free fatty acids in the circulation are in a bound form to albumin. Each molecule of albumin can hold about 20-30 molecules of free fatty acids. This lipoprotein cannot be separated by electrophoresis.

Apolipoproteins (apoproteins)

The protein components of lipoproteins are known as apolipoproteins or, simply, apoproteins. They perform the following functions

1. Act as structural components of lipoproteins.
2. Recognize the cell membrane surface receptors.
3. Activate enzymes involved in lipoprotein metabolism.

The comparative characteristic features of different lipoproteins with regard to electro-

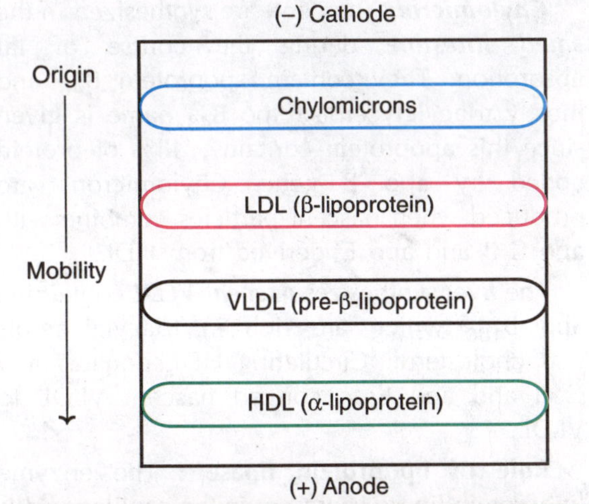

Fig. 14.36 : *Electrophoresis of plasma (serum) lipoproteins.*

Table 14.5 Characteristics of human plasma lipoproteins

Characteristic	Chylomicrons	VLDL	LDL	HDL
Electrophoretic mobility	Origin	Pre-β	β	α
Density	<0.96	0.96–1.006	1.006–1.063	1.063–1.21
Diameter (nm)	100–1,000	30–90	20–25	10–20
Apoproteins	AI, AII B_{48}	B_{100}, CI, CII CIII, E	B_{100}	AI, AII, CI, CII, CIII, D, E
Composition (%, approximate)				
Protein	2	10	20	40
Lipid (total)	98	90	80	60
Lipid components (%)				
Triacylglycerol	88	55	12	12
Cholesterol (free and ester)	4	24	59	40
Phospholipids	8	20	28	47
Free fatty acids	—	1	1	1

(VLDL : Very low density lipoproteins; LDL : Low density lipoproteins; HDL : High density lipoproteins).

phoretic patterns, size, composition etc. are given in *Table 14.5*.

Metabolism of lipoproteins —a general view

A general picture of lipoprotein metabolism is depicted in ***Fig.14.37***.

Chylomicrons (nascent) are synthesized in the ***small intestine*** during the course of fat absorption. They contain apoprotein B_{48} and mostly triacylglycerols. Apo B_{48} name is given since this apoprotein contains 48% of protein coded by apo B gene. Chylomicrons are produced when nascent particles combine with apo C II and apo E, derived from HDL.

The ***liver*** synthesizes ***nascent VLDL*** containing apo B_{100} which are rich in triacylglycerols and cholesterol. Circulating HDL donates apo C II and apo E to convert nascent VLDL to VLDL.

Role of lipoprotein lipase : The enzyme lipoprotein lipase is present in the capillary walls of adipose tissue, cardiac and skeletal muscle. It hydrolyses a portion of triacylglycerols present in chylomicrons and VLDL to liberate free fatty acids and glycerol. Lipoprotein lipase is activated by apo C II.

Uptake of chylomicron remnants by liver : As the triacylglycerols of chylomicrons and VLDL are degraded, they lose the apo C II which is returned to HDL. The chylomicron remnants are taken up by receptors present on the hepatocytes of liver.

Conversion of VLDL to LDL

During the course of VLDL metabolism, intermediate density lipoprotein (IDL) is formed which lose apo-E and get converted to LDL. The apo E is returned to HDL. LDL contains high cholesterol (free and esterified) and less triacylglycerol. The conversion of VLDL to IDL and then to LDL is referred to as ***lipoprotein cascade pathway***.

Cholesterol ester transfer protein (CETP) : CETP is synthesized in the liver, and it facilitates the exchange of components between different lipoproteins. CETP can transfer cholesterol esters from HDL to VLDL or LDL, in exchange for TG.

LDL receptors and supply of cholesterol to tissues

The most important ***function of LDL*** is to ***supply cholesterol to the extrahepatic tissues.*** The LDL particles bind to the specific receptor pits (identified as glycoprotein) on the cell membrane. The shape of the pit is stabilized by a protein called ***clathrin***. Apo B_{100} is responsible for the recognition of LDL receptor sites.

Deficiency of LDL receptors : A defect in LDL receptors ***results in the elevation of plasma LDL***, hence plasma cholesterol. However, plasma triacylglycerol concentration remains normal. Deficiency of LDL receptors is observed in

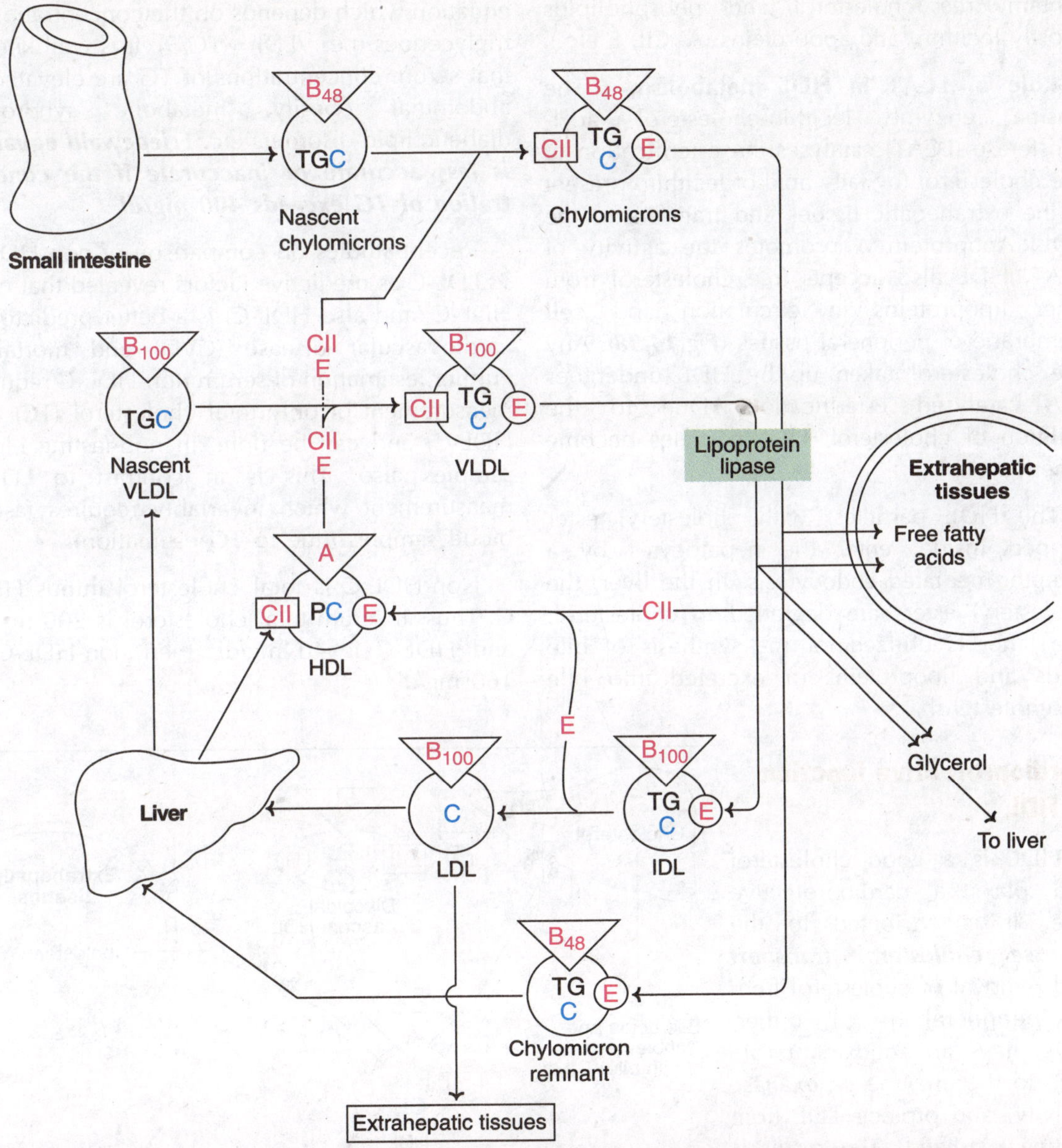

Fig. 14.37 : Summary of metabolism of lipoproteins (Apoproteins–A, B_{48}, B_{100}, CII and E; TG–Triacylglycerol; C–Cholesterol; P–Phospholipid; VLDL–Very low density lipoprotein; IDL–Intermediate density lipoprotein; LDL–Low density lipoprotein; HDL–High density lipoprotein).

type IIa hyperbetalipoproteinemia. This disorder is associated with a very high ***risk of atherosclerosis*** (particularly of coronary artery).

METABOLISM OF HDL

High density lipoproteins are ***synthesized in liver*** and ***intestine*** as – ***nascent HDL***. They contain free cholesterol and phospholipids (mostly lecithin) and apoproteins (A, CII, E etc.).

Role of LCAT in HDL metabolism : The plasma enzyme lecithin-cholesterol acyltransferase (LCAT) catalyses the esterification of free cholesterol (by fatty acid of lecithin) present in the extrahepatic tissues and transfers to the HDL. Apoprotein A promotes the activity of LCAT. HDL also accepts free cholesterol from other lipoproteins in circulation and cell membrane of peripheral tissues (***Fig.14.38***). Any free cholesterol taken up by HDL undergoes LCAT-catalysed esterification. Due to the addition of cholesterol, HDL particles become spherical.

The HDL particles, with cholesteryl ester trapped inside, enter the hepatocytes by a receptor-mediated endocytosis. In the liver, the cholesteryl esters are degraded to cholesterol. The latter is utilized for the synthesis of bile acids and lipoproteins or excreted into bile (as cholesterol).

Cardioprotective function of HDL

HDL is a good cholesterol and plays a cardioprotective role. It is attributed to the ***reverse cholesterol transport*** and removal of cholesterol from the peripheral tissue. Further, HDL plays an antioxidant role (due to the enzyme ***paroxanase*** activity) and protects LDL from getting oxidized. The result is that atherogenesis and related complications like heart attack are reduced.

NON-HDL CHOLESTEROL

Non-HDL cholesterol (***non-HDL-C***) represents all ***cholesterol present*** in potentially atherogenic lipoproteins namely ***VLDL, IDL, LDL*** and ***lipoprotein a***. In recent years, non-HDL-C is considered as a ***better predictor of cardiovascular risk*** than LDL-C for the reasons explained below.

At present, LDL-C is calculated by Friedewald equation which depends on the concentration of triglycerides (i.e. VLDl = TG/5). It is well known that serum concentrations of TG are elevated in abdominal obesity, metabolic syndrome, diabetic lipid disorders etc. ***Friedewald equation is less accurate or inaccurate if the concentration of TG exceeds 400 mg/dl.***

Recent studies on comparison of non-HDL-C *vs* LDL-C as predictive factors revealed that non-HDL-C (and also HDL-C) is a better predictor of cardiovascular disease (CVD) and mortality. Further, estimation of serum non-HDL-C requires measurement of only total cholesterol (TC) and HDL-C, and can be done in non-fasting blood samples also. This is in contrast to LDL-C measurement which invariably requires fasting blood samples (due to TG estimation).

Non-HDL-C is total cholesterol minus HDL-C. Thus, if serum total cholesterol is 200 mg/dl, and HDL-C is 40 mg/dl, then non-HDL-C is 160 mg/dl.

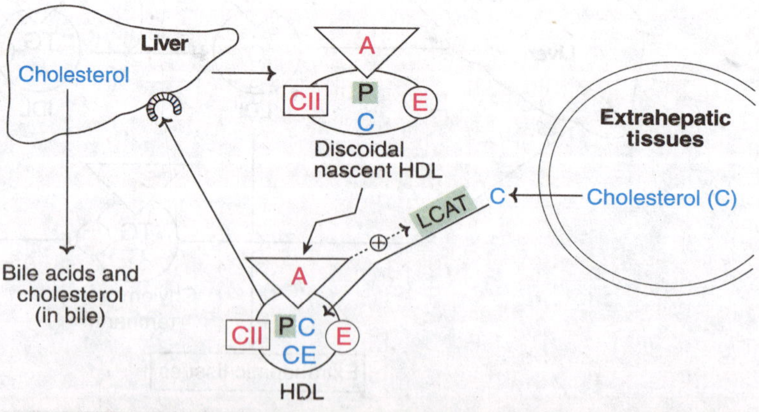

***Fig. 14.38 :** Metabolism of high density lipoproteins (P–Phospholipid; C–Cholesterol; CE–Cholesteryl ester; A, CII, E–Apoproteins; LCAT–Lecithin cholesterol acyltransferase).*

Desirable and risk levels of serum non-HDL-C

- ***Ideal levels—130–159 mg/dl (3.4-4.0 mmol/l)***
- Border line risk—160–189 mg/dl (4.1-4.8 mmol/l)
- High risk—190–219 mg/dl (4.9-5.6 mmol/l)
- Very high risk—above 220 mg/dl (6.7 mmol/l).

For people with known risk for heart diseases, non-HDL-C must be still lower i.e.

- Below 130 mg/dl (3.4 mmol/l) for high risk.
- Below 100 mg/dl (2.6 mmol/l) for very high risk.

Measures to lower non-HDL-C

Reduction in consumption of total calories, especially ***saturated and trans fatty acids*** and sugar in association with ***lifestyle therapy*** (mainly exercise) lowers non-HDL-C. Foods with high ω-3 fatty acid are beneficial. The drugs, namely statins lowering LDL-C will reduce non-HDL-C.

DISORDERS OF PLASMA LIPOPROTEINS

Inherited disorders of lipoproteins are encountered in some individuals resulting in ***primary*** hyper- or hypolipoproteinemias. These are due to ***genetic defects*** in lipoprotein metabolism and transport. The ***secondary*** acquired lipoprotein disorders are ***due to some other diseases*** (e.g. diabetes mellitus, nephrotic syndrome, atherosclerosis, hypothyrodism etc.), resulting in abnormal lipoprotein pattern which often resembles the primary inherited condition.

Hyperlipoproteinemias

Elevation in one or more of the lipoprotein fractions constitutes hyperlipoproteinemias. These disorders may be either primary or secondary. Some authors use ***hyperlipidemias*** or ***dyslipidemias*** instead of hyperlipoproteinemias. ***Frederickson's classification of hyperlipoproteinemias***—based on the electrophoretic patterns (***Fig.14.36***) of plasma lipoproteins—is widely accepted to understand these disorders. It is given in ***Table 14.6*** and briefly discussed hereunder.

1. **Type I :** This is due to familial ***lipoprotein lipase deficiency.*** The enzyme defect causes increase in plasma chylomicron and triacylglycerol levels.

2. **Type IIa :** This is also known as hyperbetalipoproteinemia and is caused by ***a defect in LDL receptors.*** Secondary type IIa hyperlipoproteinemia is observed in association with diabetes mellitus, hypothyroidism, nephrotic syndrome etc. This disorder is characterized by hypercholesterolemia.

3. **Type IIb :** Both LDL and VLDL increase along with elevation in plasma cholesterol and triacylglycerol. This is believed to be due to ***overproduction of apo B.***

Table 14.6 Classification and characteristics of hyperlipoproteinemias (hyperlipidemias)

Hyperlipoproteinemia Type	Increased plasma lipoprotein(s)	Increased plasma lipid (most)	Probable metabolic defect	Risk of atherosclerosis	Suggested treatment
I	Chylomicrons	Triacylglycerols	Deficiency of lipoprotein lipase	May increase	Low fat diet
IIa	LDL	Cholesterol	Deficiency of LDL receptors	Very high (mostly in coronary artery)	Low cholesterol fat diet; cholestyramine
IIb	LDL and VLDL	Triacylglycerols and cholesterol	Overproduction of apo-B	— do —	— do —
III	IDL	Triacylglycerols and cholesterol	Abnormality in apo-E	Very high (mostly in peripheral vessels)	Low fat and low caloric diet; clofibrate
IV	VLDL	Triacylglycerols	Overproduction of TG	May or may not increase	Low fat and low caloric diet; niacin
V	Chylomicrons and VLDL	Triacylglycerols	—	— do —	— do —

Medical Concepts/ Clinical Correlates

- Hypercholesterolemia is associated with atherosclerosis and coronary heart diseases. Consumption of polyunsaturated fatty acids and fiber decreases cholesterol in circulation. Drugs—such as lovastatin, cholestyramine, compactin and clofibrate—reduce plasma cholesterol.
- It has been proved that there is no significant correlation between dietary cholesterol and serum cholesterol. Thus, consumption of cholesterol-rich foods (eggs, sea foods etc.) are now considered safe.
- Non-HDL cholesterol (rather than LDL-cholesterol) is a better predictor of CVD and mortality.
- High LDL/HDL or high apoB/apoA ratio indicates high risk for atherosclerosis.

4. **Type III :** This is commonly known as ***broad beta disease*** and characterized by a broad β-band corresponding to ***intermediate density lipoprotein*** (IDL) on electrophoresis.

5. **Type IV :** This is due to overproduction of endogenous triacylglycerols with a concomitant ***rise in VLDL***. Type IV disorder is usually associated with obesity, alcoholism, diabetes etc.

6. **Type V :** Both ***chylomicrons and VLDL are elevated***. This is mostly a secondary condition, due to disorders such as obesity, diabetes and excessive alcohol consumption etc.

Hypolipoproteinemias

Although low levels of plasma lipids (not HDL!) within the normal range may be beneficial to the body, very low lipid levels are undesirable. These are commonly associated with certain abnormalities

1. **Familial hypobetalipoproteinemia :** It is an inherited disorder probably due to an ***impairment in the synthesis of apoprotein B***. The plasma LDL concentration in the affected individuals is between 10 to 50% of normal values. This disorder is harmless, and the individuals have healthy and long life.

2. **Abetalipoproteinemia :** This is a rare disorder due to a defect in the synthesis of apoprotein B. It is characterized by a total ***absence of β-lipoprotein (LDL)*** in plasma. Triacylglycerols are not found in plasma, but they accumulate in liver and intestine. Serum cholesterol level is low. Abetalipoproteinemia is associated with decreased absorption of fat and fat-soluble vitamins. Impairment in physical growth and mental retardation are commonly observed.

Familial alpha-lipoprotein deficiency (Tangier disease) : The plasma ***HDL particles are almost absent***. Due to this, the reverse transport of cholesterol is severely affected leading to the accumulation of cholesteryl esters in tissues. An absence of apoprotein C II—which activates lipoprotein lipase—is also found. The plasma triacylglycerol levels are elevated. The affected individuals are at an increased risk for atherosclerosis.

FATTY LIVER

The normal concentration of ***lipid*** (mostly phospholipid) ***in liver is around 5%***. Liver is not a storage organ for fat, unlike adipose tissue. However, in certain conditions, lipids—especially the ***triacylglycerols—accumulate*** excessively in liver, resulting in fatty liver (***Fig.14.39***). In the normal liver, ***Kupffer cells*** contain lipids in the form of droplets. In fatty liver, droplets of triacylglycerols are found in the entire cytoplasm of hepatic cells. This causes impairment in metabolic functions of liver. Fatty liver is associated with fibrotic changes and cirrhosis, Fatty liver may occur due to two main causes.

1. Increased synthesis of triacylglycerols
2. Impairment in lipoprotein synthesis.

1. **Increased triacylglycerol synthesis :** Mobilization of free fatty acids from adipose tissue and their influx into liver is much higher than their utilization. This leads to the over-production of triacylglycerols and their accumulation in liver. ***Diabetes mellitus***, ***starvation***, ***alcoholism*** and ***high fat diet*** are associated with

increased mobilization of fatty acids that often cause fatty liver. ***Alcohol*** also inhibits fatty acid oxidation and, thus, ***promotes fat synthesis and its deposition***.

2. **Impaired synthesis of lipoproteins** : The synthesis of very low density lipoproteins (VLDL) actively takes place in liver. VLDL formation requires phospholipids and apoprotein B. Fatty liver caused by impaired lipoprotein synthesis may be due to :

- a defect in phospholipid synthesis;
- a block in apoprotein formation;
- a failure in the formation/secretion of lipoprotein.

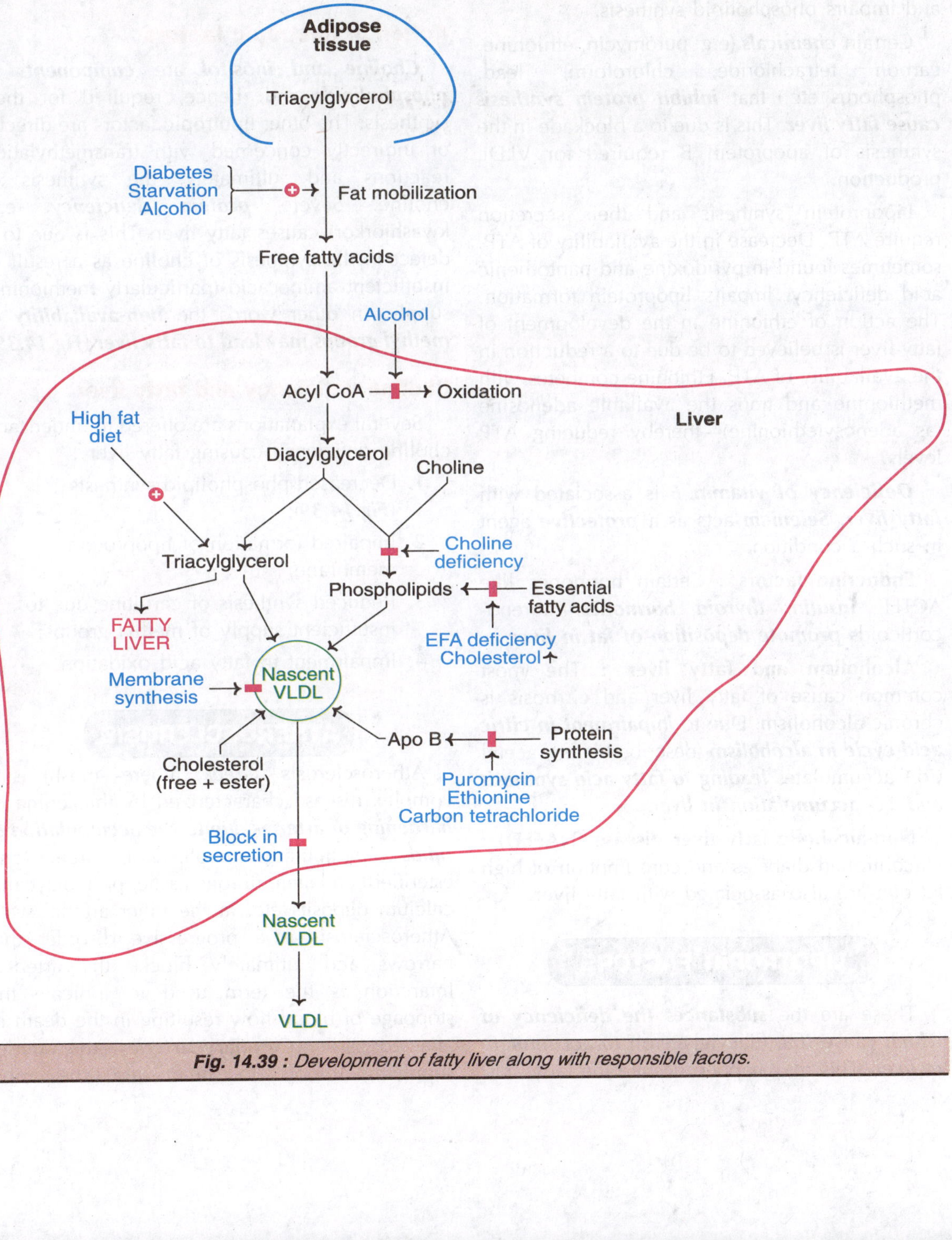

Fig. 14.39 : Development of fatty liver along with responsible factors.

Among the three causes, fatty liver due to impairment in phospholipid synthesis has been studied in some detail. This is usually associated with the dietary ***deficiency of lipotropic factors*** such as choline, betaine, inositol etc. (more details given later). Deficiency of essential fatty acids leads to a decreased formation of phospholipids. Further, excessive consumption of cholesterol competes with essential fatty acids and impairs phospholipid synthesis.

Certain ***chemicals*** (e.g. puromycin, ethionine, carbon tetrachloride, chloroform, lead, phosphorus etc.) that ***inhibit protein synthesis cause fatty liver***. This is due to a blockade in the synthesis of apoprotein B required for VLDL production.

Lipoprotein synthesis and their secretion require ATP. Decrease in the availability of ATP, sometimes found in pyridoxine and pantothenic acid deficiency, impairs lipoprotein formation. The action of ethionine in the development of fatty liver is believed to be due to a reduction in the availability of ATP. Ethionine competes with methionine and traps the available adenosine (as adenosylethionine)—thereby reducing ATP levels.

Deficiency of vitamin E is associated with ***fatty liver***. ***Selenium*** acts as a ***protective*** agent in such a condition.

Endocrine factors : Certain hormones like ACTH, ***insulin***, ***thyroid hormones***, adreno-corticoids ***promote deposition of fat in liver***.

Alcoholism and fatty liver : The most common cause of fatty liver and cirrhosis is chronic alcoholism. Due to ***impairment in citric acid cycle in alcoholism*** (described later), acetyl CoA accumulates ***leading to fatty acid synthesis and TG accumulation in liver***.

Non-alcoholic fatty liver disease (NAFLD) : Uncontrolled diabetes and consumption of high fat diet are also associated with fatty liver.

LIPOTROPIC FACTORS

These are the ***substances the deficiency of which causes fat*** (triacylglycerol) ***to accumulate in liver***. This may happen despite the fatty acid synthesis and uptake by liver being normal. Lipotropic factors are necessary for normal ***mobilization of fat*** from liver.

Important lipotropic factors

These include ***choline***, ***betaine***, ***methionine*** and ***inositol***. ***Folic acid***, ***vitamin*** B_{12}, glycine and serine also serve as lipotropic factors to some extent.

Action of lipotropic factors

Choline and inositol are ***components of phospholipids*** and, hence, required for their synthesis. The other lipotropic factors are directly or indirectly concerned with transmethylation reactions and, ultimately, the synthesis of choline. Severe ***protein deficiency*** (e.g. kwashiorkor) causes fatty liver. This is due to a defect in the synthesis of choline as a result of insufficient amino acid (particularly methionine) supply. In other words the ***non-availability of methyl groups may lead to fatty liver*** (***Fig.14.39***).

Choline deficiency and fatty liver

Several explanations are offered to understand choline deficiency causing fatty liver :

1. Decreased phospholipid synthesis (***Fig.14.39***);
2. Impaired formation of lipoprotein membrane;
3. Reduced synthesis of carnitine due to insufficient supply of methyl groups;
4. Impairment in fatty acid oxidation.

ATHEROSCLEROSIS

Atherosclerosis (*Greek*: athere—mush) is a complex disease characterized by thickening or ***hardening of arteries due to the accumulation of lipids*** (particularly cholesterol, free, and esterified) collagen, fibrous tissue, proteoglycans, calcium deposits etc. in the inner arterial wall. Atherosclerosis is a progressive disorder that narrows and ultimately blocks the arteries. Infarction is the term used to indicate the stoppage of blood flow resulting in the death of affected tissue. ***Coronary arteries***—the arteries supplying blood to heart—are the most

> **Medical Concepts/ Clinical Correlates**
>
> - Cholelithiasis, a cholesterol gall stone disease, is caused by a defect in the absorption of bile salts from the intestine or due to biliary tract obstruction.
> - High density lipoproteins are responsible for the transport and elimination of cholesterol from the body.
> - Niacin is known to boost HDL-cholesterol and lower LDL-cholesterol.
> - Hyperlipoproteinemias are a group of disorders caused by the elevation of one or more of plasma lipoprotein fractions.
> - Excessive accumulation of triacylglycerols causes fatty liver which can often be prevented by the consumption of lipotropic factors (choline, betaine, methionine).

commonly affected leading to myocardial infarction or heart attacks.

Causes of atherosclerosis and CHD : The development of atherosclerosis and the risk for the coronary heart disease (CHD) is directly correlated with plasma cholesterol and LDL. On the other hand, plasma HDL is inversely correlated with CHD.

Disorders that may cause atherosclerosis

Certain diseases are associated with atherosclerosis. These include ***diabetes mellitus***, ***hyperlipoproteinemias***, ***nephrotic syndrome***, ***hypothyroidism*** etc. Many other factors like obesity, high consumption of saturated fat, excessive smoking, ***lack of physical exercise***, ***hypertension***, stress etc., are the probable causes of atherosclerosis.

Relation between HDL and CHD

The increased levels of plasma HDL (good cholesterol) are correlated with a low incidence of cardiovascular disorders. ***Women*** have higher HDL and are ***less prone to heart diseases*** compared to men. This is attributed to ***estrogens*** in women. Strenuous physical exercise, moderate alcohol intake, consumption of unsaturated fatty acids (vegetable and fish oils), reduction in body weight—all tend to increase HDL levels and reduce the risk CHD (***see hyper-cholesterolemia, p-317***).

Lipoprotein a and CHD

Lipoprotein a (Lp-a) is almost identical in structure to LDL. Lp-a contains an additional apoprotein, apo-a. Lp-a inhibits fibrinolysis. Recent studies have shown that elevation of lipoprotein-a in the plasma (>30 mg/dl) suggests ***increased risk of CHD***. It is hypothesized that elevated Lp-a reduces the breakdown of blood clots by interfering with plasminogen activation. This results in intravascular thrombosis, and increased risk of heart attacks. ***Indians*** have ***higher levels of Lp-a*** compared to Western population.

Antioxidants and atherosclerosis

Antioxidants, in general, decrease the oxidation of LDL. There is some evidence, based on the epidemiological studies that taking of antioxidants (***vitamins E and C, or β-carotene***) reduces the risk of atherosclerosis, and CHD.

ALCOHOL METABOLISM

Walker has rightly said 'alcohol can be a food, a drug or a poison depending on the dose.' In small quantities, alcohol relieves tension and anxiety. Unfortunately, consumption of alcohol seldom ends with small doses, hence the beneficial effects are over-shadowed by the harmful effects.

Alcohol (***ethanol or ethyl alcohol***) is readily absorbed by the stomach and intestine. Consequently, less than 2% of the alcohol consumed is excreted through lungs, urine and sweat.

Medical Concepts/ Clinical Correlates

- Atherosclerosis is characterized by hardening of arteries due to the accumulation of lipids and other compounds. The probable causes of atherosclerosis include hyperlipoproteinemias, diabetes mellitus, obesity, high consumption of saturated fat, lack of exercise and stress.
- Atherosclerosis and coronary heart disease are directly correlated with plasma cholesterol and LDL, inversely with HDL. Elevation of plasma lipoprotein a suggests increased risk of CHD.
- Alcoholism is associated with fatty liver, hyperlipidemia and atherosclerosis.

Alcohol gets oxidized in the liver by alcohol dehydrogenase to acetaldehyde.

$$\underset{\textbf{Alcohol}}{CH_3-CH_2-OH} \xrightarrow[NAD^+ \;\rightarrow\; NADH + H^+]{\text{Alcohol dehydrogenase}} \underset{\textbf{Acetaldehyde}}{CH_3CHO}$$

Besides ADH, microsomal ethanol oxidizing system (MEOS) is also involved in the metabolism of alcohol. ***Aldehyde,*** produced by the action of either ADH or MEOS, is ***responsible for the manifestations of alcohol***. The enzyme aldehyde dehydrogenase converts aldehyde to acetic acid which then enters Krebs cycle in the form of acetyl CoA.

$$\underset{\text{Acetaldehyde}}{CH_3-CHO} \xrightarrow[NAD^+ \;\rightarrow\; NADH + H^+]{\text{Aldehyde dehydrogenase}} \underset{\text{Acetic acid}}{CH_3COOH}$$

Since the activity of aldehyde dehydrogenase is less than that of alcohol dehydrogenase, acetaldehyde accumulates leading to various complications. ***Disulfiram***, a drug used for the treatment of alcoholism, ***inhibits aldehyde dehydrogenase***.

Alcohol and MEOS

Microsomal ethanol oxidizing system (MEOS; belogs belongs to the family of cytochrome P_{450}) is an inducible enzyme by alcohol. Thus, MEOS is more active in metabolizing alcohol in alcoholics than alcohol dehydrogenase. MEOS is also involved in metabolizing certain drugs e.g. barbiturates. Therefore, care must be taken in administration of certain drugs to alcoholics with regard to dosage and effectiveness.

Biochemical changes in alcoholism

The metabolism of alcohol (by both dehydrogenases) involves the consumption of NAD^+, and consequently a high $NADH/NAD^+$ ratio. This is mostly responsible for the metabolic alterations observed in alcoholism. Some of them are listed.

1. High concentration of NADH favours the conversion of pyruvate to lactate which may lead to ***lactic acidosis***.

2. ***Hypoglycemia*** due to reduced gluconeogenesis is observed. This happens as a result of decreased availability of pyruvate and oxaloacetate (the latter gets converted to malate by high NADH).

3. ***Citric acid cycle is impaired*** since the availability of oxaloacetate and NAD^+ is reduced. As a result, acetyl CoA accumulates which gets diverted towards ketogenesis, cholesterologenesis, and fatty acid synthesis. Accumulation of fats leads to fatty liver and hyperlipidemia.

4. ***Increased*** concentration of ***serum uric acid*** due to its reduced excretion is observed in alcoholism. This is due to lactic acidosis.

5. Acetaldehyde interferes with the functioning of neurotransmitters, with an overall effect of neurological depression.

6. ***Acetaldehyde causes headache, nausea***, tachycardia, reduced blood pressure etc.

7. Alcoholics are more prone to ***vitamin deficiencies*** e.g. vitamins B_1, B_6 and C. ***Wernicke-Korsakoff syndrome*** due to deficiency of thiamine, is frequently observed in chronic alcoholics. The symptoms are comparable to beri-beri.

Effects of chronic alcoholism

Chronic alcoholism is associated with ***cirrhosis of liver, neurodegenerative changes***, cardiomyopathy, diuresis, impotence etc.

SUMMARY

- *Triacylglycerols (TG) are the highly concentrated form of energy, stored in adipose tissue. Hormone-sensitive lipase hydrolyses TG to free fatty acids which are transported as albumin-FFA complexes.*
- *Fatty acids are activated (acyl CoA) and transported by carnitine to mitchondria where they get oxidized (mostly by β-oxidation) to liberate energy. Complete oxidation of one mole palmitate liberates 106 ATP.*
- *Excessive utilization of fatty acids occurs in uncontrolled diabetes mellitus and starvation. This results in the overproduction of ketone bodies (in liver), namely acetone, acetoacetic acid and β-hydroxy butyric acid. The last two ketone bodies serve as energy source for peripheral tissues.*
- *Fatty acid biosynthesis occurs from acetyl CoA in the cytosol through the involvement of a multienzyme complex associated with acyl carrier protein (ACP). The reducing equivalents (NADPH + H^+) are supplied mostly by HMP shunt.*
- *Synthesis of triacylglycerols and phospholipids (PL) occurs from glycerol 3-phosphate and dihydroxyacetone phosphate with the addition of acyl CoA, and activated nitrogenous bases (for PL).*
- *Cholesterol is synthesized from acetyl CoA in a series of reactions involving HMG CoA, mevalonate, isoprenoid units and squalene as the intermediates. Cholesterol serves as a precursor for bile acids, steroid hormones and vitamin D.*
- *Lipoproteins are the transport vehicles for lipids in the plasma. Lipoprotein disorders are associated with abnormalities in their plasma levels. Elevation in LDL and VLDL—in association with cholesterol and TG—poses a serious health problem with increased risk of atherosclerosis and CHD.*
- *Excessive accumulation of triacylglycerols in liver causes fatty liver, which may be due to increased production of TG or impairment in lipoprotein (VLDL) synthesis. The latter is mostly associated with the deficiency of certain substances called lipotropic factors (e.g. choline, betaine, methionine etc.)*
- *Atherosclerosis is a complex disease characterized by thickening of arteries due to the accumulation of lipids. Atherosclerosis and CHD are directly correlated with LDL and inversely with HDL of plasma.*
- *Alcoholism is associated with hypoglycemia, impaired Krebs cycle, hyperuricemia etc. Chronic alcoholism may lead to cirrhosis of liver, cardiomyopathy, impotence etc.*

THE READER MUST REFER APPENDIX VI (Page Nos : 738 - 756) FOR RELATED CLINICAL CASE STUDIES (Nos : 17, 18, 19, 62, 71, 76)

Self-Assessment Exercises

I. Essay questions

1. Describe the functions and metabolism of phospholipids.
2. Give an account of cholesterol biosynthesis. Add a note on the significance of plasma cholesterol estimation.
3. Describe in detail the extramitochondrial synthesis of fatty acids.
4. Write about the types, characteristics and metabolism of lipoproteins. Add a note on lipoprotein disorders.
5. Give an account of fatty acid oxidation.

II. Short notes

(a) Carnitine, (b) LCAT, (c) Fatty liver, (d) Ketone bodies, (e) Lipotropic factors, (f) Acyl carrier protein, (g) Degradation of cholesterol, (h) HDL, (i) Lipoprotein lipase, (j) Brown adipose tissue.

III. Fill in the blanks

1. The most predominant lipid component of chylomicrons ____________.
2. Cholesterol synthesis is controlled by feedback inhibition of the enzyme ____________.
3. A compound possessing hydrophobic and hydrophilic groups in its structure is known as ____________.
4. Niemann-Pick disease is due to a defect in the enzyme ____________.
5. The lipoprotein involved in the reverse cholesterol transport is ____________.
6. The total number of ATP produced by the oxidation of a molecule of palmitic acid is ____________.
7. The long chain fatty acids (C_{26}—C_{35}) are not oxidized due to the absence of peroxisomes. This disorder is known as ____________.
8. Acetyl CoA from the mitochondria is transported into the cytosol after its conversion to ____________.
9. Plasma lipoprotein that is inversely correlated with coronary heart disease is ____________.
10. The fatty acid that is commonly found in the C_2 of triacylglycerols is ____________.

IV. Multiple choice questions

11. The following substance(s) is (are) ketogenic
    (a) Fatty acids (b) Leucine (c) Lysine (d) All of them.
12. The lipoprotein possessing the highest quantity of phospholipid
    (a) HDL (b) LDL (c) VLDL (d) Chylomicrons.
13. Hypercholesterolemia is observed in the disorder(s)
    (a) Hypothyroidism (b) Diabetes mellitus (c) Nephrotic syndrome (d) All of them.
14. The two final products in the β-oxidation of odd chain fatty acids are
    (a) Acetyl CoA and malonyl CoA (b) Acetyl CoA and acetyl CoA (c) Acetyl CoA and propionyl CoA (d) Acetyl CoA and succinyl CoA.
15. Hormone sensitive lipase activity is inhibited by the hormone
    (a) Epinephrine (b) Insulin (c) Thyroxine (d) Glucocorticoids.

Metabolism of Amino Acids 15

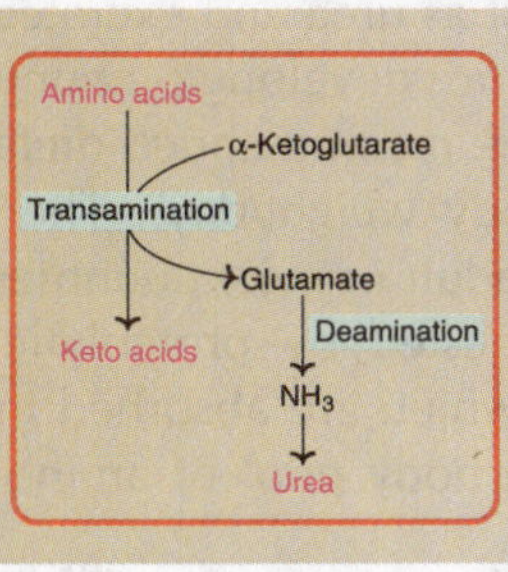

The amino acids speak

"We transaminate and deaminate to liberate ammonia;
That is detoxified in the liver to end product urea;
Greatly important to body are our nitrogen products;
Carbon skeleton is for glucose, fat or fuel."

Proteins (polymers of amino acids) are the ***most abundant organic compounds*** and constitute a major part of the body dry weight (10-12 kg in adults). They perform a wide variety of static (structural) and dynamic (enzymes, hormones, clotting factors, receptors etc.) functions. About half of the body protein (predominantly collagen) is present in the supportive tissue (skeleton and connective) while the other half is intracellular.

Life cycle of proteins : The blueprint of life of proteins lies in genes. Proteins are ***born at translation*** (biosynthesis of proteins), ***mature*** to function after ***posttranslational modifications*** and ***die*** (degraded) to release component amino acids.

Functions and fate of amino acids : Amino acids are not just the structural components of proteins. Each one of the 20 naturally occurring amino acids undergoes its own metabolism and performs specific functions. Some of the amino acids also serve as precursors for the synthesis of many biologically important compounds (e.g. melanin, serotonin, creatine etc.). Certain amino acids may directly act as neurotransmitters (e.g. glycine aspartate, glutamate). ***Protein metabolism is more appropriately learnt as metabolism of amino acids.***

AMINO ACID POOL

An adult has about ***100 g of free amino acids*** which represent the amino acid pool of the body. The amino acid pool may be an oversimplification of the facts, since there is no single compartment—rather, several compartments exist.

Glutamate and ***glutamine*** together constitute ***about 50%***, and essential amino acids about 10% of the body pool (100 g). The concentration of intracellular amino acids is always higher than the extracellular amino acids. Amino acids enter the cells against a concentration gradient by active transport.

The amino acid pool of the body is maintained by the sources that contribute (input) and the metabolic pathways that utilize (output) the amino acids (***Fig.15.1***).

It may be noted that everyday the kidneys filter around 50g of free amino acids, which are reabsorbed. Consequently, ***only traces of free amino acids appear in normal urine***.

I. Sources of amino acid pool

Turnover of body protein, intake of dietary protein and the synthesis of non-essential amino acids contribute to the body amino acid pool.

(a) **Protein turnover :** The protein present in the body is in a dynamic state. It is estimated that about ***300-400 g of protein per day*** is constantly degraded and synthesized which represents body protein turnover. There is a wide variation in the ***lifespan*** of individual ***proteins***. For instance, the plasma proteins and digestive enzymes are rapidly degraded, their half-lives being in hours or days. The structural proteins (e.g. collagen) have long half-lives, often in months and years.

Control of protein turnover : The turnover of the protein is influenced by many factors. A small polypeptide called ***ubiquitin*** (mol. wt. 8,500) tags with the proteins and facilitates degradation. Certain proteins with amino acid sequence proline, glutamic acid (one letter code E), serine and threonine (PEST sequence) are rapidly degraded.

(b) **Dietary protein :** There is a regular loss of nitrogen from the body due to degradation of amino acids. In healthy adults, it is estimated that about 30-50 g of protein is lost everyday from the body. This amount of protein (30-50 g/day) must, therefore, be supplied daily in the diet ***to maintain nitrogen balance***. The purpose of dietary protein is to supply amino acids (particularly the essential ones) for the synthesis of proteins and other nitrogen compounds.

There is ***no storage form of amino acids*** as is the case for carbohydrates (glycogen) and lipids (triacylglycerols). The excess intake of amino acids are metabolized—oxidized to provide energy, converted to glucose or fat. The amino groups are lost as urea and excreted. The protein consumption in developed countries is much higher than the recommended dietary allowance (i.e. 1g/kg body weight/day). The daily protein intake by an adult in most countries is 40-100 g. Protein is digested by proteolytic enzymes to amino acids which are absorbed in the intestine and enter the body pool of amino acids.

(c) **Synthesis of non-essential amino acids :** Ten out of the 20 naturally occurring amino acids can be synthesized by the body which contribute to the amino acid pool.

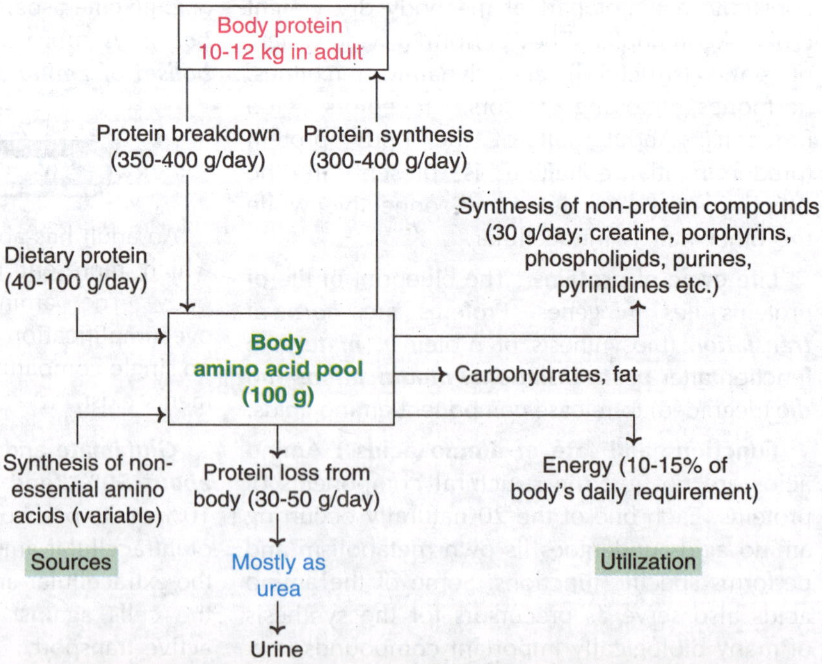

Fig. 15.1 : *Overview of body's amino acid pool—sources and utilization.*

II. Utilization of amino acids from body pool

(a) Most of the ***body proteins*** (300-400 g/day) degraded are synthesized from the amino acid pool. These include enzymes, hormones, immunoproteins, contractile proteins etc.

(b) Many important ***nitrogenous compounds*** (porphyrins, purines, pyrimidines, etc.) are produced from the amino acids. About 30 g of protein is daily utilized for this purpose.

(c) Generally, ***about 10-15% of body energy*** requirements are met from the amino acids.

(d) The amino acids are ***converted to carbohydrates and fats***. This becomes predominant when the protein consumption is in excess of the body requirements.

METABOLISM OF AMINO ACIDS —GENERAL ASPECTS

The amino acids undergo certain common reactions like ***transamination*** followed by ***deamination*** for the liberation of ***ammonia***. The amino group of the amino acids is utilized for the formation of ***urea*** which is an excretory ***end product*** of protein metabolism. The carbon ***skeleton*** of the amino acids is first converted to ***keto acids*** (by transamination) which meet one or more of the following fates.

1. Utilized to ***generate energy***.
2. Used for the ***synthesis of glucose***.
3. Diverted for the ***formation of fat or ketone bodies***.
4. Involved in the ***production of non-essential amino acids***.

A general picture of amino acid metabolism is depicted in ***Fig.15.2***.

The details of general and specific metabolic reactions of amino acids are described in the following pages.

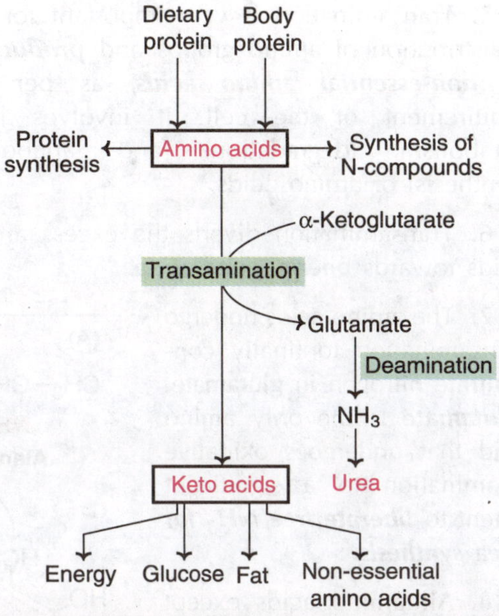

Fig. 15.2 : *An overview of amino acid metabolism.*

TRANSAMINATION

The ***transfer of an amino ($-NH_2$) group from an amino acid to a keto acid*** is known as transamination. This process involves the interconversion of a pair of amino acids and a pair of keto acids, catalysed by a group of enzymes called ***transaminases*** (recently, ***aminotransferases***). Transamination, as such, is the ***initiating point for amino acid metabolism.***

Salient features of transamination

1. All transaminases require ***pyridoxal phosphate*** (***PLP***), a coenzyme derived from vitamin B_6.

2. Specific transaminases exist for each pair of amino and keto acids. However, only two—namely, aspartate transaminase and alanine transaminase—make a significant contribution for transamination.

3. There is no free NH_3 liberated, only the transfer of amino group occurs.

4. Transamination is ***reversible*** (***Fig.15.3***).

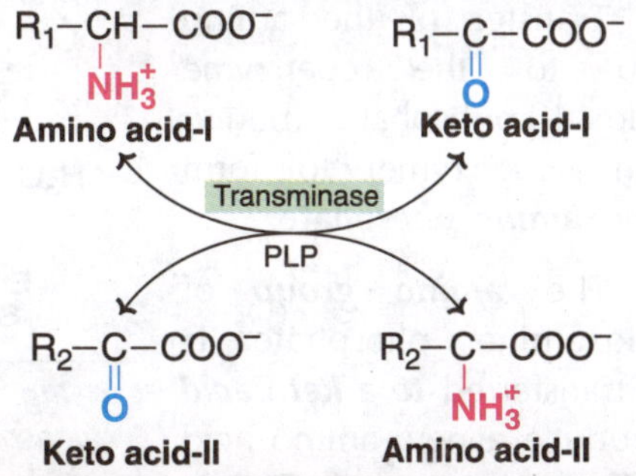

Fig. 15.3 : *Transamination reaction.*

5. Transamination is very important for the redistribution of amino groups and ***production of non-essential amino acids***, as per the requirement of the cell. It involves both catabolism (degradation) and anabolism (synthesis) of amino acids.

6. Transamination diverts the excess amino acids towards ***energy generation***.

7. The amino acids undergo transamination to finally concentrate nitrogen in glutamate. ***Glutamate*** is the only amino acid that undergoes oxidative deamination to a significant extent to ***liberate free NH_3 for urea synthesis***.

8. All amino acids except lysine, threonine, proline and hydroxyproline participate in transamination.

9. Transamination is not restricted to α-amino groups only. For instance, δ-amino group of ornithine is transaminated.

10. Serum transaminases are important for diagnostic and prognostic purposes. (***Refer Chapter 6***).

Mechanism of transamination

Transamination occurs in two stages (***Fig.15.4***)

1. Transfer of the amino group to the coenzyme pyridoxal phosphate (bound to the coenzyme) to form ***pyridoxamine phosphate***.

2. The ***amino group*** of pyridoxamine phosphate is then transferred ***to a keto acid*** to produce a new amino acid and the enzyme with PLP is regenerated.

All the transaminases require ***pyridoxal phosphate*** (PLP), a derivative of vitamin B_6. The aldehyde group of PLP is linked with ε-amino group of lysine residue, at the active site of the enzyme forming a ***Schiff base*** (imine linkage). When an amino acid (substrate) comes in contact with the enzyme, it displaces lysine and a new Schiff base linkage is formed. The amino

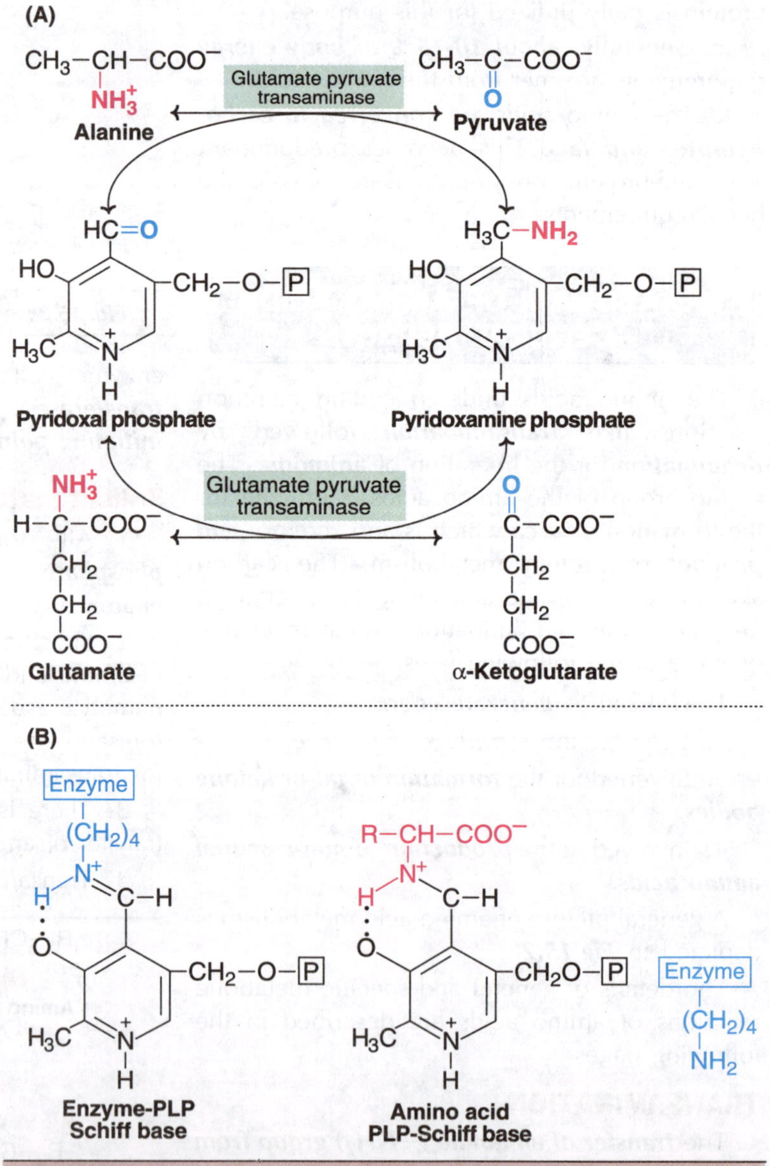

***Fig. 15.4 :** Mechanism of transamination—(**A**) Involvement of pyridoxal phosphate (PLP) in the transfer of amino group, (**B**) Formation of enzyme-PLP-Schiff base and amino acid-PLP-Schiff base. Note that when the amino acid binds, enzyme separates.*

acid-PLP-Schiff base tightly binds with the enzyme by non-covalent forces. Snell and Braustein proposed a ***Ping Pong Bi Bi*** mechanism involving a series of intermediates (aldimines and ketimines) in transamination reaction.

Clinical significance of transaminases

Aspartate transaminase (AST) and ***alanine transaminase*** (ALT) are very important as diagnostic enzymes. Serum AST is mostly elevated in cardiac disorders (myocardial infarction) while ALT is increased in liver diseases (viral hepatitis). For more details ***refer Chapter 6***.

DEAMINATION

The ***removal of amino group*** from the amino acids as NH_3 is deamination. Transamination (discussed above) involves only the shuffling of amino groups among the amino acids. On the other hand, deamination results in the liberation of ammonia ***for urea synthesis***. Simultaneously, the carbon skeleton of amino acids is converted to keto acids. Deamination may be either oxidative or non-oxidative.

Although transamination and deamination are separately discussed, they occur simultaneously, often involving glutamate as the central molecule. For this reason, some authors use the term ***transdeamination*** while describing the ***reactions of transamination and deamination***, particularly involving glutamate.

I. Oxidative deamination

Oxidative deamination is the ***liberation of free ammonia*** from the amino group of amino acids ***coupled with oxidation***. This takes place mostly in liver and kidney. The purpose of oxidative deamination is to provide NH_3 for urea synthesis and α-keto acids for a variety of reactions, including energy generation.

Role of glutamate dehydrogenase : In the process of transamination, the amino groups of most amino acids are transferred to α-ketoglutarate to produce glutamate. Thus, ***glutamate*** serves as a ***'collection centre' for amino groups*** in the biological system. Glutamate rapidly undergoes oxidative deamination, catalysed by glutamate dehydrogenase (GDH) to liberate ammonia. This enzyme is unique in that it can utilize ***either*** NAD^+ ***or*** $NADP^+$ ***as a coenzyme***. Conversion of glutamate to α-ketoglutarate occurs through the formation of an intermediate, α-iminoglutarate (***Fig.15.5***).

Glutamate dehydrogenase catalysed reaction is important as it reversibly links up glutamate metabolism with TCA cycle through α-ketoglutarate. GDH is involved in both catabolic and anabolic reactions.

Regulation of GDH activity : Glutamate dehydrogenase is a zinc containing mitochondrial enzyme. It is a complex enzyme consisting of six identical units with a molecular weight of 56,000 each. GDH is controlled by allosteric regulation. ***GTP*** and ***ATP inhibit***—whereas ***GDP*** and ***ADP activate***—glutamate dehydrogenase. Steroid and thyroid hormones inhibit GDH.

After ingestion of a protein-rich meal, liver glutamate level is elevated. It is converted to α-ketoglutarate with liberation of NH_3. Further, when the cellular energy levels are low, the degradation of glutamate is increased to provide α-ketoglutarate which enters TCA cycle to liberate energy.

Oxidative deamination by amino acid oxidases : L-Amino acid oxidase and D-amino acid oxidase are flavoproteins, possessing FMN and

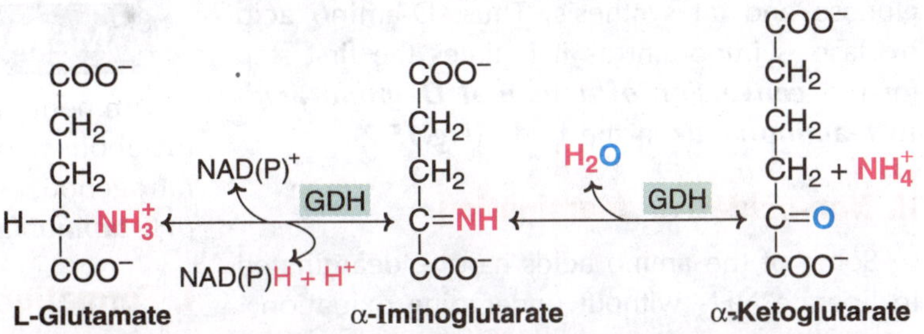

Fig. 15.5 : Oxidation of glutamate by glutamate dehydrogenase (GDH).

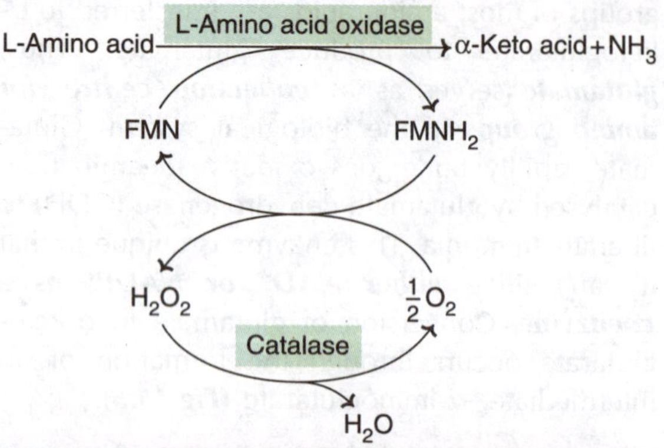

Fig. 15.6 : *Oxidative deamination of amino acids.*

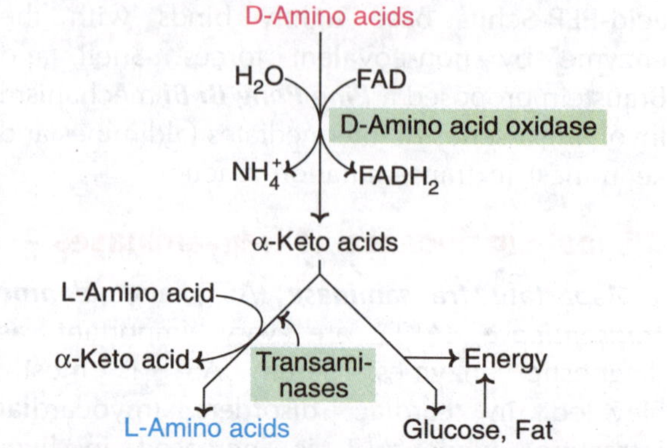

Fig. 15.7 : *Metabolic fate of D-amino acids.*

FAD, respectively. They act on the corresponding amino acids (L or D) to produce α-keto acids and NH_3. In this reaction, oxygen is reduced to H_2O_2, which is later decomposed by catalase (***Fig.15.6***).

The activity of ***L-amino acid oxidase*** is much low while that of ***D-amino acid oxidase*** is high in tissues (mostly liver and kidney). L-Amino acid oxidase does not act on glycine and dicarboxylic acids. This enzyme, due to its very low activity, does not appear to play any significant role in the amino acid metabolism.

Fate of D-amino acids : D-Amino acids are found in plants and microorganisms. They are, however, not present in the mammalian proteins. But D-amino acids are regularly taken in the diet and metabolized by the body. D-Amino acid oxidase converts them to the respective α-keto acids by oxidative deamination. The α-keto acids so produced undergo transamination to be converted to L-amino acids which participate in various metabolisms. Keto acids may be oxidized to generate energy or serve as precursors for glucose and fat synthesis. Thus, D-amino acid oxidase is important as it initiates the first step for the ***conversion of unnatural D-amino acids to L-amino acids*** in the body (***Fig.15.7***).

II. Non-oxidative deamination

Some of the amino acids can be deaminated to liberate NH_3 without undergoing oxidation

(a) **Amino acid dehydrases :** Serine, threonine and homoserine are the hydroxy amino acids. They undergo non-oxidative deamination, catalysed by PLP-dependent dehydrases (dehydratases).

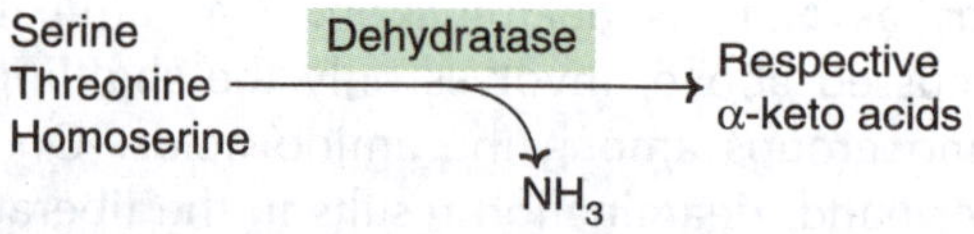

(b) **Amino acid desulfhydrases :** The sulfur amino acids, namely cysteine and homocysteine, undergo deamination coupled with desulfhydration to give keto acids.

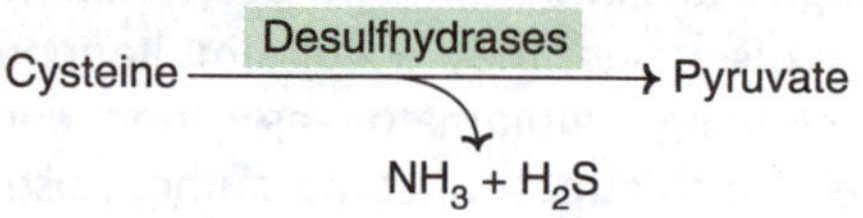

(c) **Deamination of histidine :** The enzyme histidase acts on histidine to liberate NH_3 by a non-oxidative deamination process.

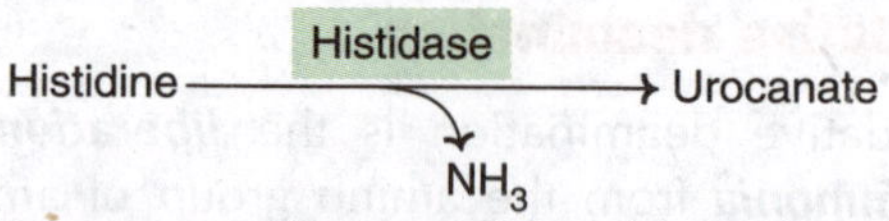

METABOLISM OF AMMONIA

Ammonia is constantly being liberated in the metabolism of amino acids (mostly) and other nitrogenous compounds. At the physiological pH, ***ammonia exists as ammonium (NH_4^+) ion.***

I. Formation of ammonia

The production of NH_3 occurs from the amino acids (transamination and deamination), biogenic amines, amino group of purines and pyrimidines and by the action of intestinal bacteria (urease) on urea.

II. Transport and storage of NH_3

Despite a regular and constant production of NH_3 from various tissues, its concentration in the circulation is surprisingly low (normal plasma 15-45 μg/dl). This is mostly because the body has an efficient mechanism for NH_3 transport and its immediate utilization for urea synthesis. The transport of ammonia between various tissues and the liver mostly occurs ***in the form of glutamine*** or ***alanine and not as free ammonia***. Alanine is important for NH_3 transport from muscle to liver by glucose-alanine cycle (***Refer Fig.13.15***).

Role of glutamine : Glutamine is a ***storehouse of NH_3***. It is present at the highest concentration (8 mg/dl in adults) in blood among the amino acids. Glutamine serves as a storage and transport form of NH_3. Its synthesis mostly occurs in liver, brain and muscle. Ammonia is removed from the brain predominantly as glutamine. Glutamine is ***freely diffusible*** in tissues, hence ***easily transported***.

Glutamine synthetase (a mitochondrial enzyme) is responsible for the synthesis of glutamine from glutamate and ammonia. This reaction is unidirectional and requires ATP and Mg^{2+} ions.

Glutamine can be ***deaminated*** by hydrolysis ***to release ammonia*** by ***glutaminase*** (***Fig.15.8***) an enzyme mostly found in kidney and intestinal cells.

III. Functions of ammonia

Ammonia is not just a waste product of nitrogen metabolism. It is involved (directly or via glutamine) for the ***synthesis*** of many compounds in the body. These include ***non-essential amino acids, purines, pyrimidines,*** amino sugars, asparagine etc. Ammonium ions (NH_4^+) are very important to maintain ***acid-base balance*** of the body.

IV. Disposal of ammonia

The organisms, during the course of evolution, have developed different mechanisms for the disposal of ammonia from the body. The animals in this regard are of three different types

(a) **Ammoniotelic** : The aquatic animals dispose off ***NH_3 into the surrounding water***.

(b) **Uricotelic** : Ammonia is converted mostly to ***uric acid*** e.g. reptiles and birds.

(c) **Ureotelic** : The mammals including man convert NH_3 to ***urea***. Urea is a non-toxic and soluble compound, hence easily excreted.

V. Toxicity of ammonia

Even a marginal elevation in the blood ammonia concentration is ***harmful to the brain.*** Ammonia, when it accumulates in the body, results in ***slurring of speech and blurring of the vision*** and causes tremors. It may lead to coma and, finally death, if not corrected.

Hyperammonemia : Elevation in plasma NH_3 (***reference range 15-45 μg/dl***) level may be genetic or acquired. Impairment in urea synthesis due to a defect in any one of the five enzymes is described in urea synthesis. All these disorders lead to hyperammonemia and cause ***hepatic coma*** and ***mental retardation***. The acquired hyperammonemia may be due to hepatitis, alcoholism etc. where the urea synthesis becomes defective, hence NH_3 accumulates.

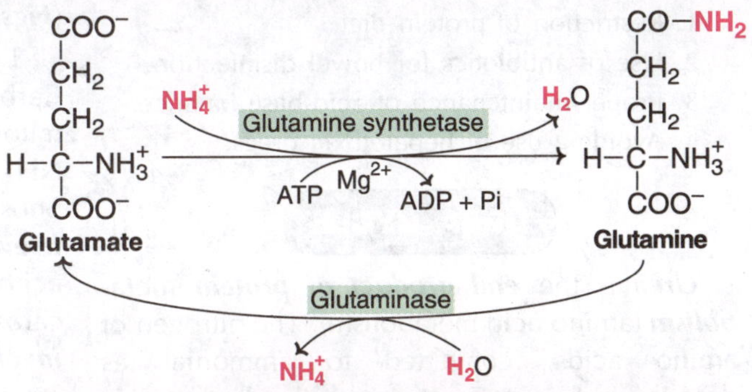

Fig. 15.8 : *Synthesis of glutamine and its conversion to glutamate.* (**Note :** *The reactions are independent and irreversible*).

Explanation for NH_3 toxicity : The reaction catalysed by glutamate dehydrogenase probably explains the toxic affects of NH_3 in brain

$$\alpha\text{-Ketoglutarate} + NH_3 \xrightleftharpoons[]{NADPH + H^+ \;\; NADP^+} \text{Glutamate}$$

Accumulation of NH_3 shifts the equilibrium to the right with more glutamate formation, hence more utilization of α-ketoglutarate. ***α-Ketoglutarate*** is a key intermediate in TCA cycle and its ***depleted levels impair the TCA cycle***. The net result is that production of energy (ATP) by the brain is reduced. The ***toxic effects of NH_3*** on brain are, therefore, ***due to impairment in ATP formation***.

Trapping and elimination of ammonia : When the plasma level of ammonia is highly elevated, intravenous administration of sodium benzoate and phenyllactate is done. These compounds can respectively condense with glycine and glutamate to form water soluble products that can be easily excreted. By this way, ammonia can be trapped and removed from the body. In some instances of toxic hyperammonemia, hemodialysis may become necessary.

Hepatic coma and management

Liver failure causes hyperammonemia, that finally results in hepatic coma. The manifestations of hepatic coma include dysfunction of central nervous system (***convulsions***) and impairment in liver function. The latter may result in hepatomegaly, jaundice, edema, hemorrhage etc.

The management of hepatic coma includes the following

1. Restriction of protein diet.
2. Use of antibiotics for bowel disinfection.
3. Proper maintenance of acid-base balance.
4. Avoiding use of hepatotoxic drugs.

UREA CYCLE

Urea is the ***end product of protein metabolism*** (amino acid metabolism). The nitrogen of amino acids, converted to ammonia (as described above), is toxic to the body. It is converted to urea and detoxified. As such, ***urea*** accounts for ***80-90% of the nitrogen containing substances excreted in urine***.

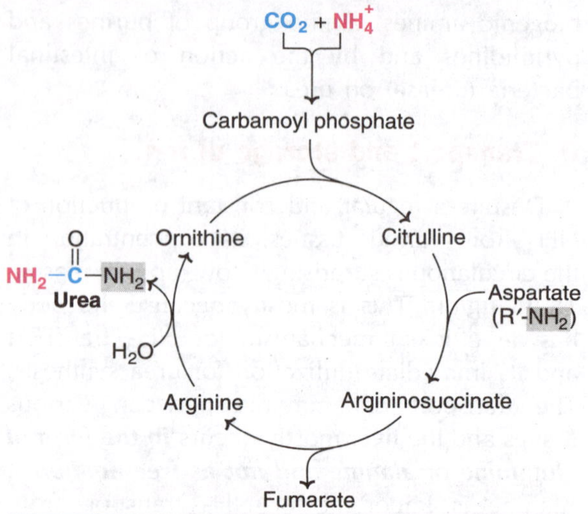

***Fig. 15.9 :** Outline of urea cycle. (**Note :** In the synthesis of urea one amino group comes from ammonium ion while the other is from aspartate; carbon is derived from CO_2. This is represented in colours.)*

Urea is ***synthesized in liver*** and transported to kidneys for excretion in urine. Urea cycle is the ***first metabolic cycle*** that was elucidated by Hans Krebs and Kurt Henseleit (1932), hence it is known as ***Krebs-Henseleit cycle or ornithine cycle***. The individual reactions, however, were described in more detail later on by Ratner and Cohen.

Urea has ***two amino ($-NH_2$) groups***, one derived ***from NH_3*** and the other ***from aspartate.*** Carbon atom is supplied by CO_2. Urea synthesis is a five-step cyclic process, with five distinct enzymes. The first ***two enzymes*** are present in ***mitochondria*** while the ***rest*** are localized in ***cytosol***. The details of urea cycle are described (***Figs.15.9 and 15.10***).

1. **Synthesis of carbamoyl phosphate :** Carbamoyl phosphate synthase I (***CPS I***) of mitochondria catalyses the condensation of NH_4^+ ions with CO_2 to form carbamoyl phosphate. This step consumes two ATP and is ***irreversible***, and ***rate-limiting***. CPS I requires ***N-acetylglutamate*** for its activity. Another enzyme, carbamoyl phosphate synthase II (***CPS II***)—***involved in pyrimidine synthesis***—is present in cytosol. It accepts amino group from glutamine and does not require N-acetylglutamate for its activity. The differences between CPS I and CPS II are given in ***Table 15.1***.

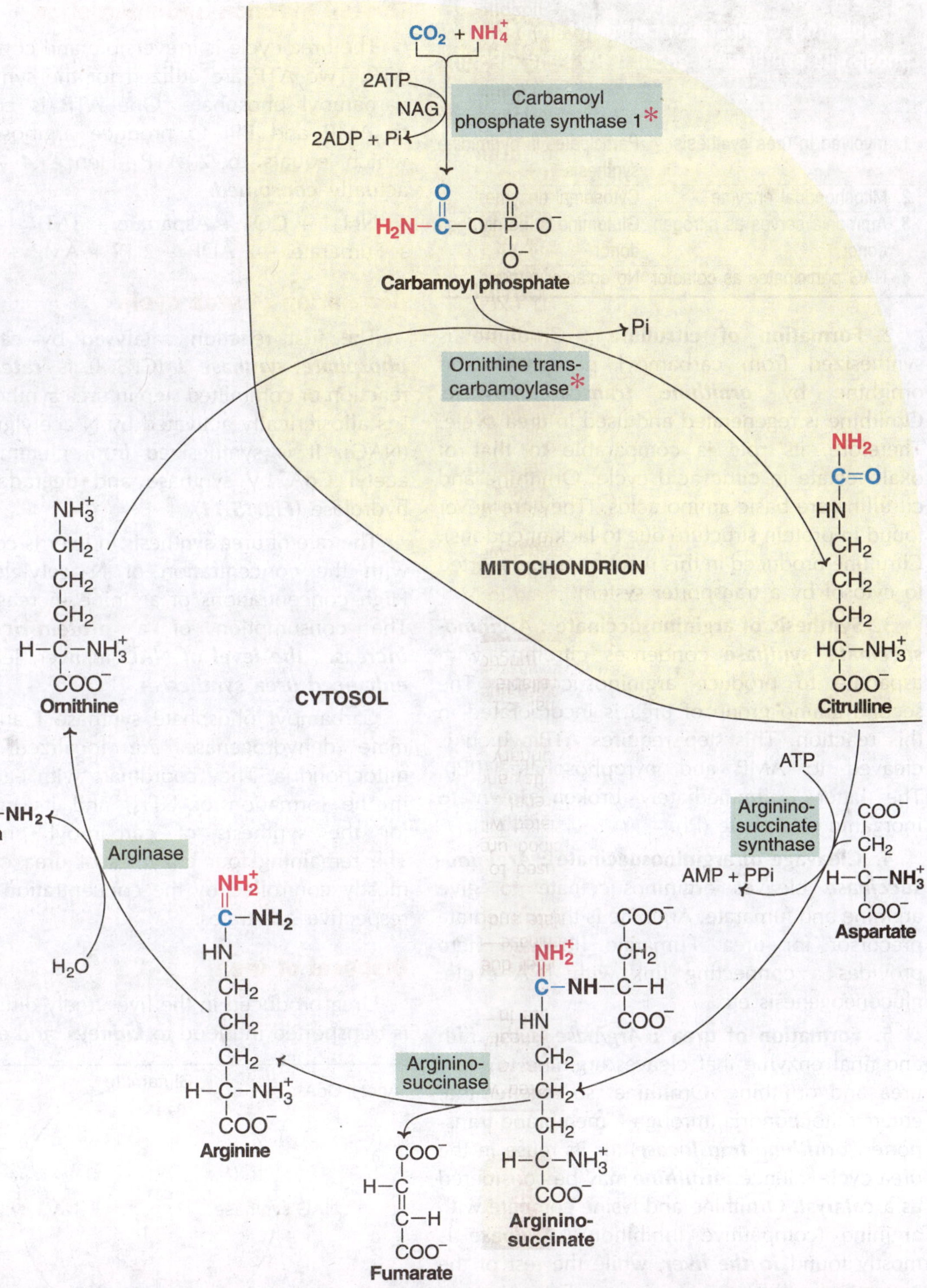

*Fig. 15.10 : Reactions of urea cycle (NAG—N-acetylglutamate; in the formation of urea, one amino group is derived from free ammonium ion while the other is from aspartate; carbon is obtained from CO_2; * mitochondrial enzymes, the rest of the enzymes are cytosomal).*

Table 15.1 Comparison between carbamoyl phosphate synthases I and II (CPS I and CPS II)

CPS I	CPS II
1. Involved in urea synthesis	Participates in pyrimidine synthesis
2. Mitochondrial enzyme	Cytosomal enzyme
3. Ammonia serves as nitrogen donor	Glutamine is the nitrogen donor
4. NAG participates as cofactor	No cofactor needed.

2. **Formation of citrulline :** Citrulline is synthesized from carbamoyl phosphate and ornithine by ***ornithine transcarbamoylase***. Ornithine is regenerated and used in urea cycle. Therefore, its role is comparable to that of oxaloacetate in citric acid cycle. Ornithine and citrulline are basic amino acids. (They are never found in protein structure due to lack of codons). Citrulline produced in this reaction is transported to cytosol by a transporter system.

3. **Synthesis of argininosuccinate :** ***Argininosuccinate synthase*** condenses citrulline with aspartate to produce argininosuccinate. The second amino group of urea is incorporated in this reaction. This step requires ATP which is cleaved to AMP and pyrophosphate (PPi). The latter is immediately broken down to inorganic phosphate (Pi).

4. **Cleavage of argininosuccinate :** ***Argininosuccinase*** cleaves argininosuccinate to give arginine and fumarate. Arginine is the immediate precursor for urea. Fumarate liberated here provides a connecting link with TCA cycle, gluconeogenesis etc.

5. **Formation of urea :** ***Arginase*** is the fifth and final enzyme that cleaves arginine to yield urea and ornithine. Ornithine, so regenerated, enters mitochondria (through a membrane transporter, ***ornithine translocase***) for its reuse in the urea cycle. Hence, ***ornithine*** may be considered as a ***catalyst***. Ornithine and lysine compete with arginine (competitive inhibition). ***Arginase*** is mostly found ***in the liver***, while the rest of the enzymes (four) of urea cycle are also present in other tissues. For this reason, arginine synthesis may occur to varying degrees in many tissues. But only the liver can ultimately produce urea.

Overall reaction and energetics

The urea cycle is irreversible and consumes 4 ATP. Two ATP are utilized for the synthesis of carbamoyl phosphate. One ATP is converted to AMP and PPi to produce arginosuccinate which equals to 2 ATP. Hence ***4 ATP*** are actually ***consumed***.

$NH_4^+ + CO_2$ + Aspartate + 3ATP $\longrightarrow$ Urea + Fumarate + 2 ADP + 2 Pi + AMP + PPi

Regulation of urea cycle

The first reaction catalysed by ***carbamoyl phosphate synthase I*** (CPS I) is ***rate-limiting*** reaction or committed step in urea synthesis. CPS I is allosterically activated by N-acetylglutamate (NAG). It is synthesized from glutamate and acetyl CoA by synthase and degraded by a hydrolase (***Fig.15.11***).

The rate of urea synthesis in liver is correlated with the concentration of N-acetylglutamate. High concentrations of arginine increase NAG. The consumption of a ***protein-rich meal increases the level of NAG*** in liver, leading to ***enhanced urea synthesis.***

Carbamoyl phosphate synthase I and glutamate dehydrogenase are localized in the mitochondria. They coordinate with each other in the formation of NH_3, and its utilization for the synthesis of carbamoyl phosphate. The remaining four enzymes of urea cycle are mostly controlled by the concentration of their respective substrates.

Disposal of urea

Urea produced in the liver freely diffuses and is transported in blood to ***kidneys***, and excreted.

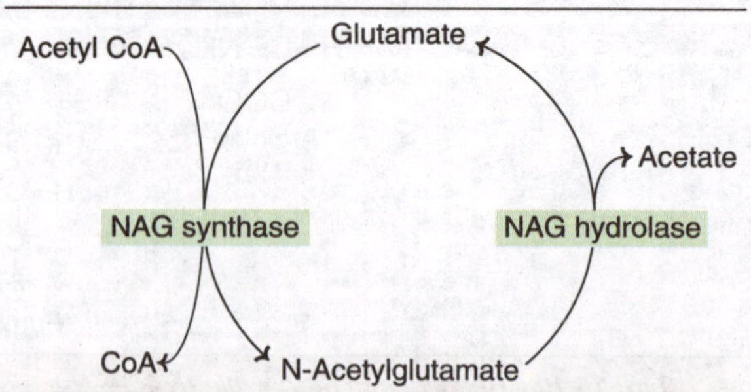

Fig. 15.11 : Formation and degradation of N-acetylglutamate.

A small amount of urea enters the intestine where it is broken down to CO_2 and NH_3 by the bacterial enzyme urease. This ammonia is either lost in the feces or absorbed into the blood. In renal failure, the blood urea level is elevated (uremia), resulting in diffusion of more urea into intestine and its breakdown to NH_3. ***Hyperammonemia*** (increased blood NH_3), linked to uremia, is commonly seen ***in patients of kidney failure***.

Integration between urea cycle and TCA cycle

Urea cycle is linked with TCA cycle in three different ways (***Fig.15.12***). This is regarded as ***bicyclic integration*** between the two cycles.

1. The production of ***fumarate*** in urea cycle is the most important integrating point with TCA cycle. Fumarate is converted to malate and then to oxaloacetate in TCA cycle. Oxaloacetate undergoes transamination to produce aspartate which enters urea cycle. Here, it combines with citrulline to produce arginosuccinate. Oxaloacetate is an important metabolite which can combine with acetyl CoA to form citrate and get finally oxidized. Oxaloacetate can also serve as a precursor for the synthesis of glucose.

2. ***ATP*** (10) are generated in the TCA cycle while ATP (4) are utilized for urea synthesis.

3. Citric acid cycle is an important metabolic pathway for the complete oxidation of various metabolites to CO_2 and H_2O. The ***CO_2*** liberated in TCA cycle (in the mitochondria) can be utilized in urea cycle.

Metabolic disorders of urea cycle

The clinical symptoms associated with defect in urea cycle enzymes include vomiting, lethargy, irritability, ataxia and mental retardation.

Metabolic defects of urea cycle are given in ***Table 15.2***. All disorders invariably result in ***hyperammonemia*** leading to toxicity.

Hyperornithinemia-hyperammonemia—homocitrullinuria syndrome (HHH syndrome) : It is an autosomal recessive disorder due to a defect in the enzyme ***ornithine translocase*** (mitochondrial membrane transporter).

Blood urea—clinical importance

In healthy people, the ***normal blood urea concentration is 10-40 mg/dl.*** Higher protein intake marginally increases blood urea level, however this is well within normal range. About 15-30 g of urea (7-15 g nitrogen) is excreted in urine per day.

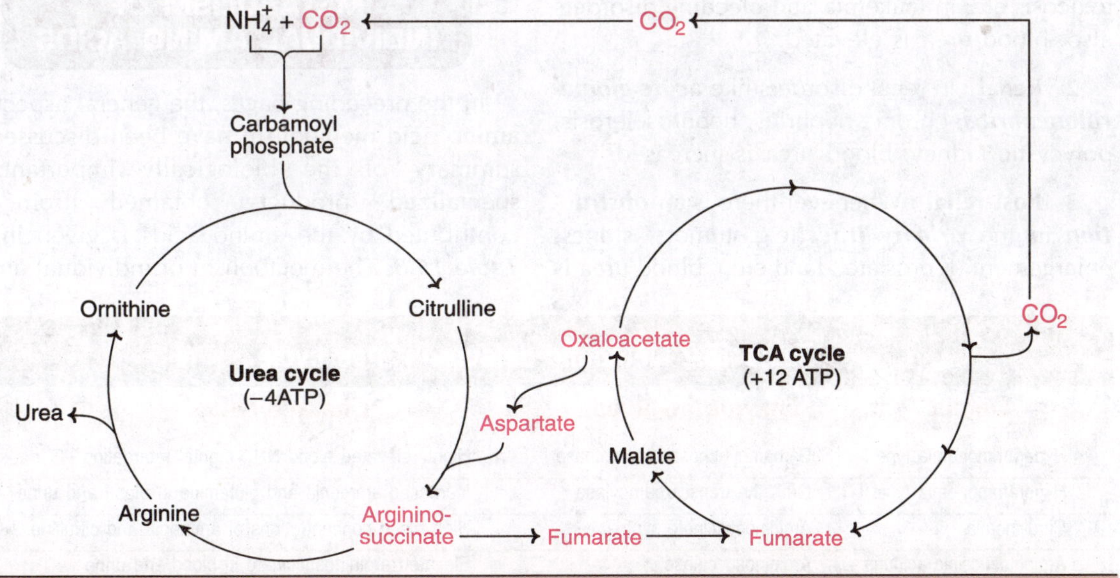

***Fig. 15.12 :** Interrelation between urea and tricarboxylic acid (TCA) cycle (Depicted in blue colour).*

> **Medical Concepts/ Clinical Correlates**
>
> - The amino acids are mainly utilized for protein biosynthesis, production of specialized products (creatine, porphyrin, amines, purines, pyrimidines) and generation of energy.
> - Ammonia accumulation in blood is toxic to brain causing slurring of speech, blurring of vision, tremors and even death. Mammals convert NH_3 to urea, a non-toxic excretory product. Metabolic defects in urea cycle enzymes result in hyperammonemia.
> - Serum alanine transaminase (ALT) and aspartate transaminase (AST) estimations are respectively useful for the diagnosis of liver diseases and heart diseases.

Blood urea estimation is widely used as a screening test for the ***evaluation*** of ***kidney*** (***renal***) ***function***. It is estimated in the laboratory either by urease method or diacetyl monoxime (DAM) procedure. Elevation in blood urea may be broadly classified into three categories.

1. **Pre-renal :** This is associated with ***increased protein breakdown***, leading to a negative nitrogen balance, as observed after major surgery, prolonged fever, burns, thyrotoxicosis etc. In leukemia and bleeding disorders also, blood urea is elevated.

2. **Renal :** In renal disorders like ***acute glomerulonephritis***, chronic nephritis, nephrosclerosis, polycystic kidney, blood urea is increased.

3. **Post-renal :** Whenever there is an ***obstruction*** in the ***urinary tract*** (e.g. tumors, stones, enlargement of prostate gland etc.), blood urea is elevated. This is due to increased reabsorption of urea from the renal tubules.

The term ***'uremia'*** is used to indicate ***increased blood urea levels due to renal failure***. ***Azotemia*** represents an ***elevation in blood urea/ or other nitrogen metabolites*** which may or may not be associated with renal diseases.

Non-protein nitrogen (NPN)

As is obvious from the name, the term NPN refers to all the nitrogen-containing substances other than proteins. These include ***urea*** (***most abundant***), ***creatinine***, ***creatine***, ***uric acid***, ***peptides, amino acids*** etc. In healthy persons, NPN ***concentration in blood is 20-40 mg/dl***.

The molecular weight of urea is 60 and about half of it (28) is contributed by the two nitrogen atoms. Thus, if blood urea concentration is 60 mg, then about half of it—28 mg—is ***blood urea nitrogen (BUN)***. Therefore,

$$\text{BUN} = \frac{1}{2}\,\text{NPN}$$

$$\text{NPN} = 2\ \text{BUN}$$

In some countries, estimations of BUN or NPN are used rather than blood urea for assessing kidney function. The normal range for ***ratio*** of ***BUN*** to serum ***creatinine*** is ***10:1 to 15:1***.

METABOLISM OF INDIVIDUAL AMINO ACIDS

In the preceding pages, the general aspects of amino acid metabolism have been discussed. A summary of the biologically important or specialized products obtained from or contributed by the amino acids is given in the ***Table 15.3***. The metabolism of individual amino

Table 15.2 Metabolic defects in urea cycle

Defect	Enzyme involved	Characteristics
Hyperammonemia type I	Carbamoyl phosphate synthase I	Highly elevated blood NH_3; mental retardation
Hyperammonemia type II	Ornithine transcarbamoylase	Increased ammonia and glutamine in blood and urine
Citrullinemia	Argininosuccinate synthase	High blood concentrations of ammonia and citrulline
Argininosuccinic aciduria	Argininosuccinase	Elevated argininosuccinate in blood and urine
Hyperargininemia	Arginase	Increased arginine in blood, cerebrospinal fluid and urine

Table 15.3 A summary of the specialized products formed/contributed by amino acids

Amino acid	Specialized product(s)
Glycine	Creatine, glutathione, heme, purines, conjugated bile acids.
Tyrosine	Thyroxine, triiodothyronine, epinephrine, norepinephrine, dopamine, melanin.
Tryptophan	NAD^+, $NADP^+$ (coenzymes of niacin), serotonin, melatonin.
Methionine	Active methionine, creatine, epinephrine, polyamines.
Cysteine	Glutathione, taurine, coenzyme A, active sulfate.
Histidine	Histamine
Arginine	Creatine, nitric oxide
Lysine	Carnitine
Glutamate	γ-Amino butyric acid, glutathione, γ-carboxyglutamate.
Glutamine	Purines, pyrimidines, amino sugars.
Aspartate	Purines, pyrimidines
Serine	Phosphatidylserine, sphingomyelins, choline.
β-Alanine	Coenzyme A

acids with special emphasis on the specialized products is described next.

GLYCINE

Glycine (Gly, G) is a ***non-essential***, optically inactive and ***glycogenic*** (precursor for glucose) amino acid. It is indispensable for chicks. The outline of glycine metabolism is depicted in ***Fig.15.13***. Glycine is actively involved in the synthesis of many specialized products (heme, purines, creatine etc.) in the body, besides its incorporation into proteins, synthesis of serine and glucose and participation in one-carbon metabolism. Glycine is an ***inhibitory neurotransmitter*** in central nervous system. Glycine is the most abundant amino acid normally excreted into urine (50–300 mg/day).

Glycine in proteins

Glycine is one among the commonest amino acids found in protein structure. Being small and non-polar, glycine is mostly present in the interior structure of protein. Collagen contains very high (about 30%) content of glycine.

Synthesis of glycine

Glycine is synthesized ***from serine*** by the enzyme serine hydroxymethyl transferase which is dependent on tetrahydrofolate (THF). Glycine can also be obtained from threonine, catalysed by threonine aldolase. Glycine synthase can convert a one-carbon unit (N^5, N^{10}-methylene THF), CO_2 and NH_3 to glycine (***Fig.15.4***).

Degradation of glycine

Glycine undergoes oxidative deamination by glycine synthase to liberate NH_4^+, CO_2 and one-carbon fragment as N^5, N^{10}-methylene THF. This provides a major route for glycine breakdown in mammals. Glycine synthase is a multienzyme complex and requires PLP, NAD^+ and THF for its activity. This reaction is reversible to generate glycine.

Glycine is reversibly converted to serine by THF dependent serine hydroxymethyl transferase. Pyruvate produced from serine by serine dehydratase, serves as a precursor for glucose. Serine is degraded to glyoxylate which undergoes transamination to give back glycine. Glyoxylate is also converted to oxalate, an excretory product and formate which enters one-carbon pool (***Fig.15.14***).

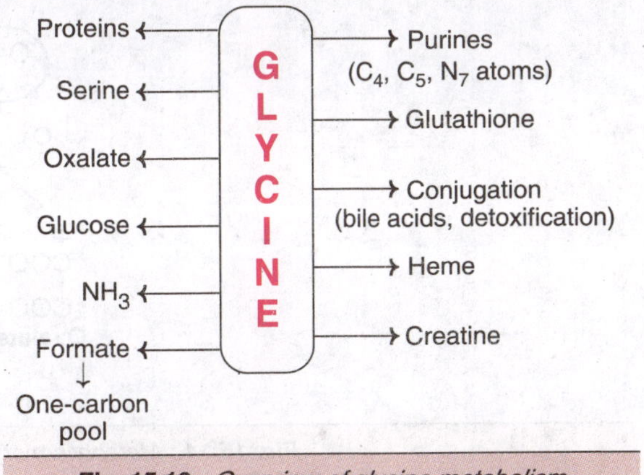

***Fig. 15.13 :** Overview of glycine metabolism.*

Synthesis of specialized products

1. **Formation of purine ring :** The entire molecule of glycine is utilized for the formation of positions 4 and 5 of carbon and position 7 of nitrogen of purines.

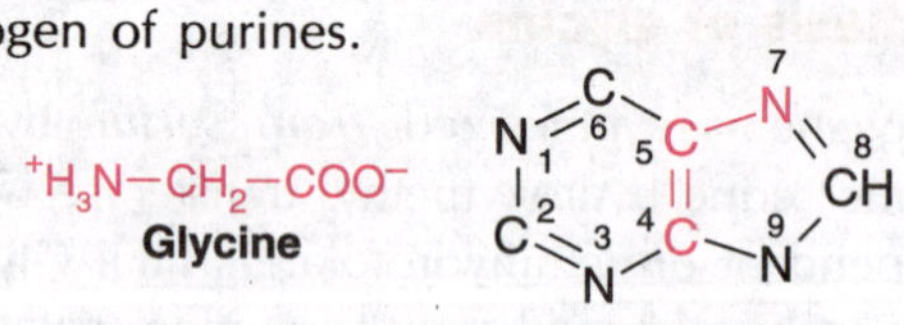

Purine ring

2. **Synthesis of glutathione :** Glutathione is a ***tripeptide*** (γ-glutamyl-cysteinyl-glycine) and requires three amino acids for its formation (***Fig.15.15***).

3. **Conjugation reactions :** As a conjugating agent, glycine performs two important functions

(a) The bile acids—cholic acid and chenodeoxy cholic acid—are conjugated with glycine.

Cholic acid + Glycine ⟶ Glycocholic acid

Chenodeoxycholic acid + Glycine ⟶ Glycochenodeoxy cholic acid

(b) Glycine is important for ***detoxification*** of benzoic acid (commonly used as a food preservative) to hippuric acid.

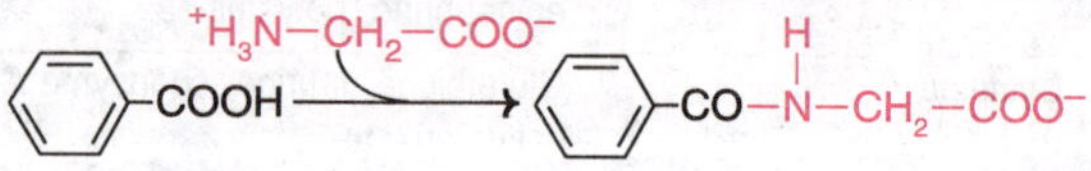

Benzoic acid **Hippuric acid**

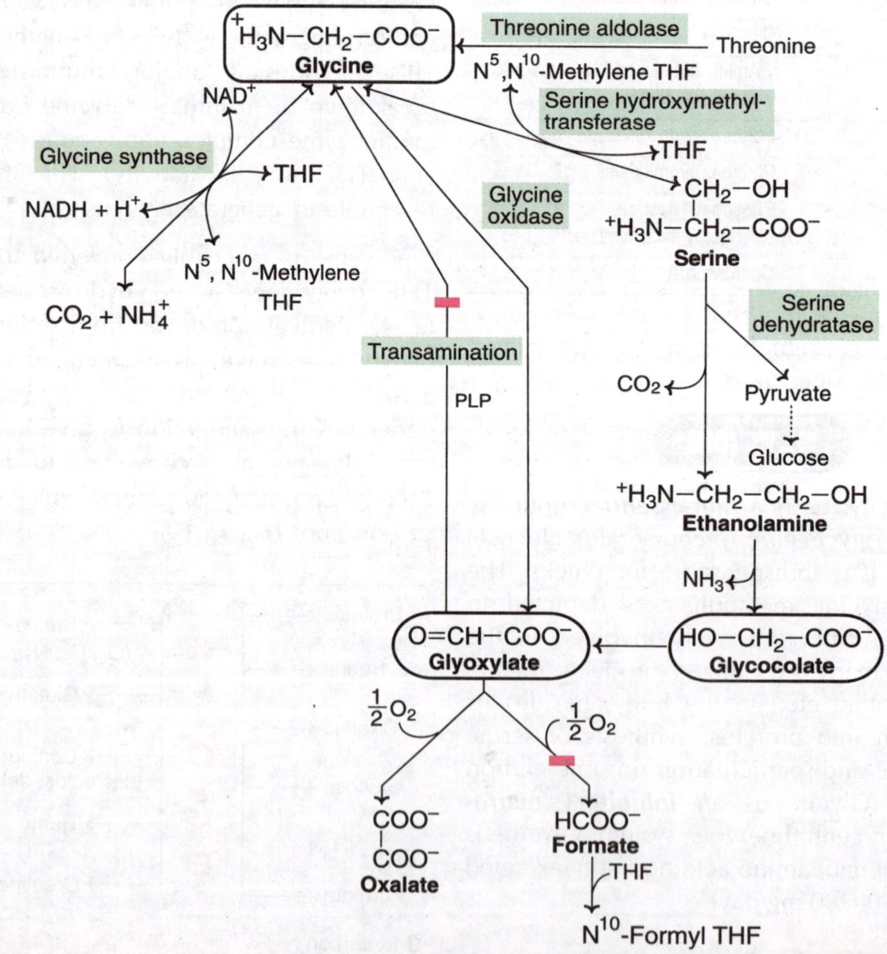

Fig. 15.14 : Metabolism of glycine (THF–Tetrahydrofolate; PLP–Pyridoxal phosphate; ▬–Block in primary hyperoxaluria).

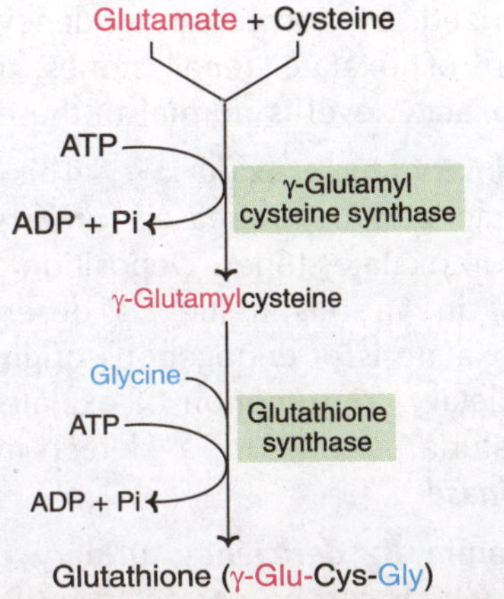

Fig. 15.15 : Outline of glutathione synthesis.

4. **Synthesis of heme :** Glycine condenses with succinyl CoA to form δ-amino levulinate which serves as a precursor for heme synthesis (details given in porphyrin metabolism—***Chapter 10***).

Succinyl CoA + Glycine —(ALA synthase)→ δ-Amino levulinate (ALA)

5. **Biosynthesis of creatine :** Creatine is present in the tissues (muscle, brain, blood etc.) as the high energy compound, phosphocreatine and as free creatine. Three amino acids—***glycine***, ***arginine*** and ***methionine***—are required for creatine formation (***Fig.15.16***). The first reaction occurs in the kidney. It involves the transfer of guanidino group of arginine to glycine, catalysed by arginine-glycine transamidinase to produce guanidoacetate (glycocyamine). ***S-Adenosylmethionine*** (active methionine) donates methyl group to glycocyamine to produce creatine. This reaction occurs in liver. Creatine is reversibly phosphorylated to phosphocreatine (***Lohmann reaction***) by creatine kinase. It is stored in muscle as high energy phosphate.

Creatinine is an anhydride of creatine. It is formed by spontaneous cyclization of creatine or creatine phosphate. Creatinine is excreted in urine.

Creatine and creatinine—clinical importance : The normal concentrations of creatine and creatinine in human serum and urine are as follows

Serum

Creatine — 0.2–0.6 mg/dl
Creatinine — 0.6–1 mg/dl

Urine

Creatine — 0–50 mg/day
Creatinine — 1–2 g/day

Estimation of ***serum creatinine*** (along with blood urea) is used as a ***diagnostic test to assess kidney function.*** Serum creatinine concentration

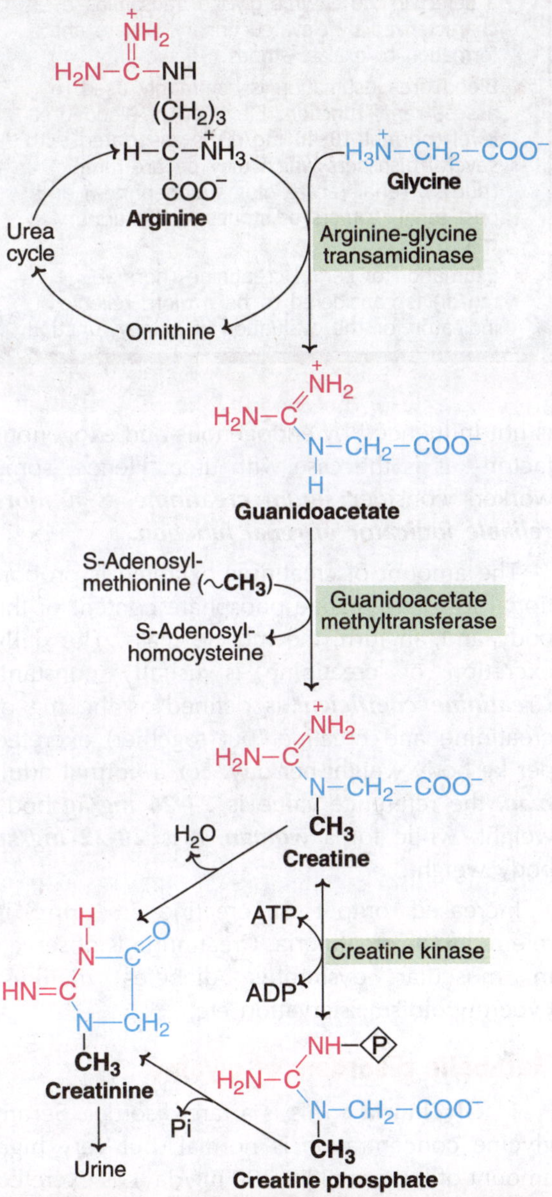

Fig. 15.16 : Metabolism of creatine.

Medical Concepts/ Clinical Correlates

- Hyperammonemia is sometimes associated with hepatic coma and liver failure.
- Dietary consumption of a protein-rich meal increases the level of N-acetylglutamate in liver which enhances urea production.
- Primary hyperoxaluria—a metabolic disorder due to a defect in the enzyme glycine transaminase—is characterized by elevated urinary oxalate and formation of oxalate stones.
- Blood urea estimation is commonly used to assess renal function. Elevation of blood urea level (normal 10–40 mg/dl) is associated with several disorders which may be pre-renal (burns), renal (acute glomerulonephritis) and post-renal (tumors or stones in the urinary tract).
- Estimation of serum creatinine (normal < 1 mg/dl) is considered to be a more reliable indicator for the evaluation of kidney function.

is not influenced by endogenous and exogenous factors, as is the case with urea. Hence, some workers consider ***serum creatinine*** as a ***more reliable indicator of renal function.***

The amount of creatinine excreted is proportional to total creatine phosphate content of the body and, in turn, the muscle mass. The daily excretion of creatinine is usually constant. ***Creatinine coefficient*** is defined as the mg of creatinine and creatine (put together) excreted per kg body weight per day. For a normal adult ***man***, the reference value is ***24-26 mg/kg*** body weight, while for a ***woman***, it is ***20-22 mg/kg*** body weight.

Increased output of creatine in urine is referred to as creatinuria. Creatinuria is observed in muscular dystrophy, diabetes mellitus, hyperthyroidism, starvation etc.

Metabolic disorders of glycine

1. **Glycinuria :** This is a rare disorder. Serum glycine concentration is normal, but very high amount of it (normal 50-300 mg/day) is excreted in urine. It is believed that glycinuria is due to a ***defective renal reabsorption.*** Glycinuria is characterized by increased tendency for the formation of oxalate renal stones. However, urinary oxalate level is normal in these patients.

2. **Primary hyperoxaluria :** This disorder is characterized by increased ***urinary oxalate*** resulting in oxalate stones. Deposition of oxalate (***oxalosis***) in various tissues is observed. The urinary oxalate is of endogenous origin and not due to dietary consumption of oxalate. Primary hyperoxaluria is due to a defect in ***glycine transaminase***.

In vitamin ***B_6 deficiency,*** urinary ***oxalate*** is ***elevated*** which can be corrected by B_6 supplementation. However, B_6 administration has no effect on endogenous hyperoxaluria.

Therapeutic uses of glycine

In recent years, glycine is being used to ***protect against heart diseases, liver damage***, improvement of insulin action (in type 2 diabetes), ***induction of sleep***, treatment of schizophrenia, stroke etc. However, some of these therapeutic uses of glycine are at experimental stages.

PHENYLALANINE AND TYROSINE

Phenylalanine (Phe, F) and tyrosine (Tyr, Y) are structurally related ***aromatic amino acids. Phenylalanine*** is an ***essential amino acid*** while tyrosine is non-essential. Besides its incorporation into proteins, the only function of phenylalanine is its conversion to tyrosine. For this reason, ingestion of tyrosine can reduce the dietary requirement of phenylalanine. This phenomenon is referred to as ***'sparing action' of tyrosine on phenylalanine.***

The predominant metabolism of phenylalanine occurs through tyrosine. Tyrosine is incorporated into proteins and is involved in the synthesis of a variety of biologically important compounds—***epinephrine, norepinephrine, dopamine*** (catecholamines), ***thyroid hormones***—and the pigment ***melanin*** (***Fig.15.17***). During the course of degradation, phenylalanine and tyrosine are converted to metabolites which can serve as precursors for the synthesis of glucose

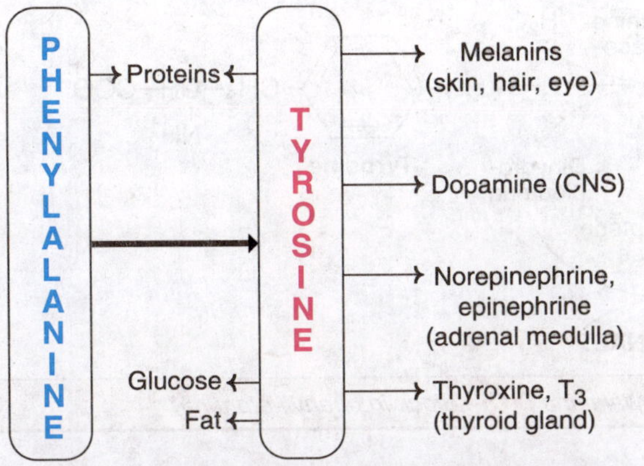

Fig. 15.17 : Overview of phenylalanine and tyrosine metabolism (CNS–Central nervous system; T_3–Triiodothyronine).

and fat. Hence, these amino acids are both ***glucogenic*** and ***ketogenic***. Biochemists attach special significance to phenylalanine and tyrosine metabolism for two reasons—synthesis of biologically important compounds and the metabolic disorders due to enzyme defects.

Conversion of phenylalanine to tyrosine

Under normal circumstances, the degradation of phenylalanine mostly occurs through tyrosine. Phenylalanine is hydroxylated at para-position by ***phenylalanine hydroxylase*** to produce tyrosine (p-hydroxy phenylalanine). This is an ***irreversible reaction*** and requires the participation of a specific coenzyme ***biopterin*** (containing pteridine ring) which is structurally related to folate. The active form of biopterin is tetrahydrobiopterin (H_4-biopterin). In the phenylalanine hydroxylase reaction, tetrahydrobiopterin is oxidized to dihydrobiopterin (H_2-biopterin). Tetrahydrobiopterin is then regenerated by an NADPH-dependent dihydrobiopterin reductase (***Fig.15.18***).

The enzyme phenylalanine hydroxylase is present in the liver. In the conversion of phenylalanine to tyrosine, the reaction involves the incorporation of one atom of molecular oxygen (O_2) into the para position of phenylalanine while the other atom of O_2 is reduced to form water. It is the ***tetrahydrobiopterin*** that supplies the reducing equivalents which, in turn, are provided by NADPH. Due to a ***defect in phenylalanine hydroxylase***, the conversion of phenylalanine to tyrosine is blocked resulting in the disorder ***phenylketonuria*** (PKU).

DEGRADATION OF TYROSINE (PHENYLALANINE)

The metabolism of phenylalanine and tyrosine is considered together. The sequence of the reactions in the degradation of these amino acids, depicted in ***Fig.15.19***, is described hereunder

1. As phenylalanine is ***converted to tyrosine*** (details in ***Fig.15.18***), a single pathway is responsible for the degradation of both these amino acids, which occurs mostly in liver.

2. Tyrosine first undergoes transamination to give p-hydroxyphenylpyruvate. This reaction is catalysed by ***tyrosine transaminase*** (PLP dependent).

3. ***p-Hydroxyphenylpyruvate hydroxylase*** (or dioxygenase) is a copper-containing enzyme. It catalyses oxidative decarboxylation as well as hydroxylation of the phenyl ring of p-hydroxyphenylpyruvate to produce ***homogentisate***. This reaction involves a shift in hydroxyl group from para position to meta position, and incorporates a new hydroxyl group at para position. This step in tyrosine metabolism requires ascorbic acid.

4. ***Homogentisate oxidase*** (iron metalloprotein) cleaves the benzene ring of homogentisate to form 4-maleylacetoacetate. Molecular oxygen is required for this reaction to break the aromatic ring.

5. Maleylacetoacetate undergoes isomerization to form 4-fumaryl acetoacetate and this reaction is catalysed by ***maleylacetoacetate isomerase***.

6. ***Fumaryl acetoacetase*** (fumaryl acetoacetate hydrolase) brings about the hydrolysis of fumaryl acetoacetate to liberate fumarate and acetoacetate.

Fumarate is an intermediate of citric acid cycle and can also serve as precursor for gluconeogenesis. Acetoacetate is a ketone body from which fat can be synthesized. Phenylalanine and tyrosine are, therefore, both glucogenic and ketogenic.

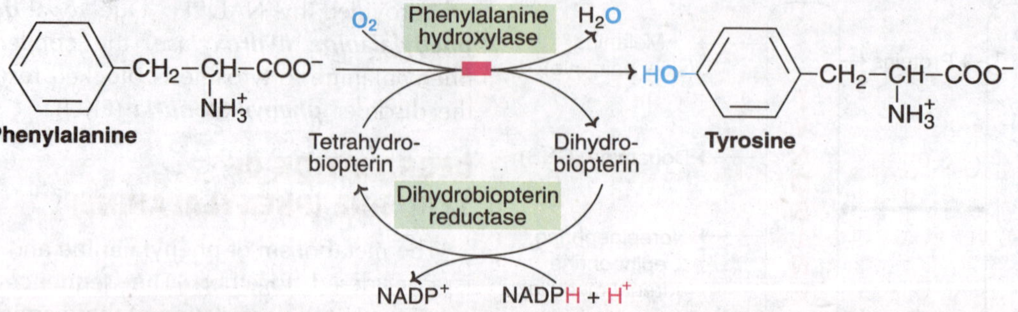

Fig. 15.18 : Synthesis of tyrosine from phenylalanine (▬—Block in phenylketonuria).

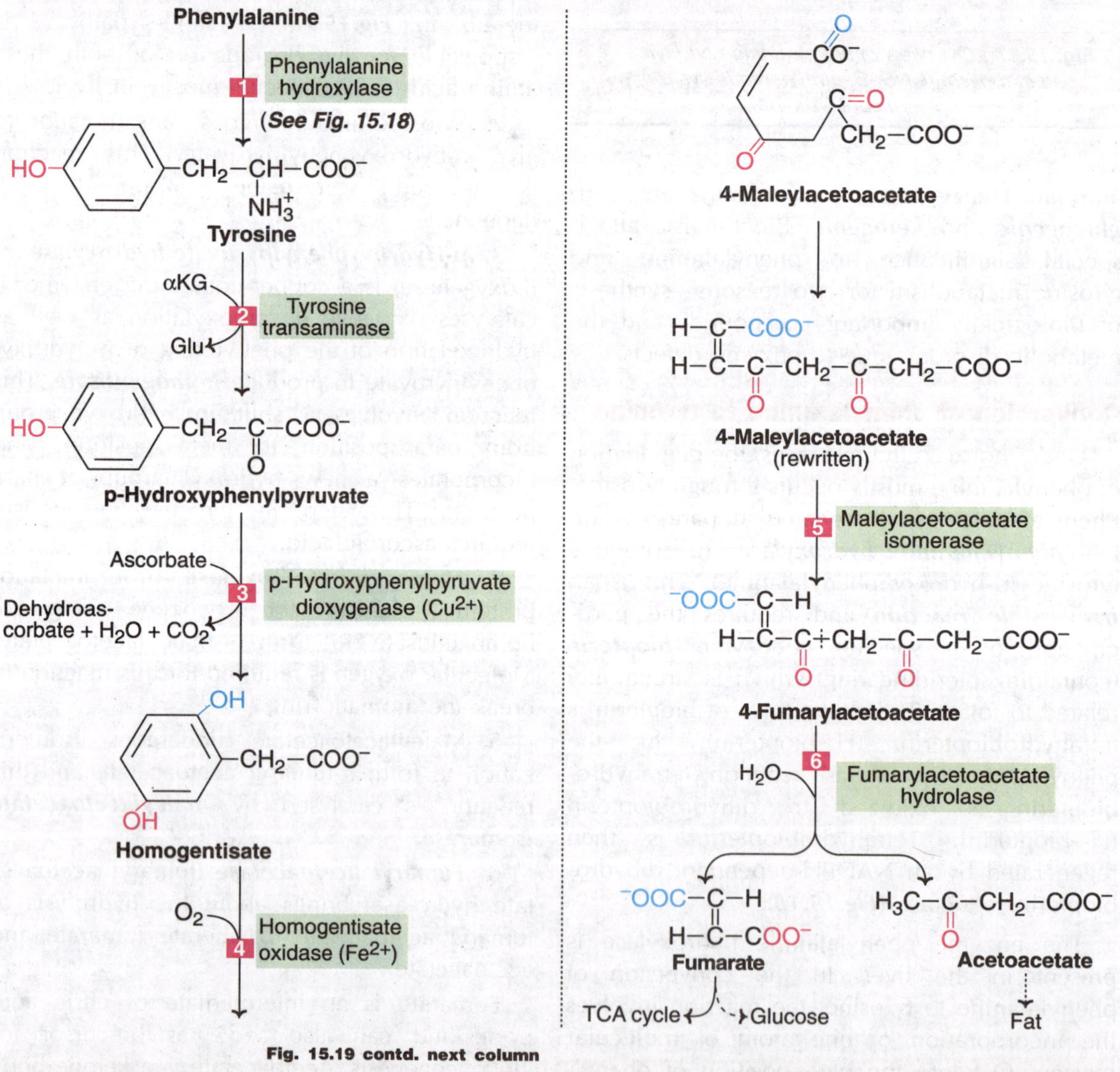

Fig. 15.19 : Tyrosine metabolism—degradative pathway [αKG-α-Ketoglutarate; Glu-Glutamate; The circled numbers indicate metabolic defects (1) Phenylketonuria; (2) Tyrosinemia type II; (3) Neonatal tyrosinemia; (4) Alkaptonuria; (5) and (6) Tyrosinosis (tyrosinemia, type I)].

Medical Concepts/ Clinical Correlates

- Melanin—the pigment of skin, hair and eyes—is produced from tyrosine. Lack of melanin synthesis (mostly due to a deficiency of tyrosinase) causes albinism.
- Defect in melanin synthesis, due to disappearance of melanocytes from hair roots, results in greying of hair.
- Parkinson's disease—a common disorder of the elderly—is linked with decreased synthesis of dopamine. It is characterized by muscular rigidity, tremors, lethargy etc.

The inborn errors of phenylalanine and tyrosine metabolism are indicated in ***Fig.15.19***. Detailed information on these disorders is given later.

Synthesis of melanin

Melanin (*Greek* : melan—black) is the ***pigment of skin, hair and eye***. The synthesis of melanin occurs in melanosomes present in melanocytes, the pigment-producing cells. Tyrosine is the precursor for melanin and only one enzyme, namely ***tyrosinase*** (a copper-containing oxygenase), is involved in its formation. Tyrosinase hydroxylates tyrosine to form 3,4-dihydroxy-phenylalanine (DOPA) (***Fig.15.20***). DOPA can act as a cofactor for tyrosinase. The next reaction is also catalysed by tyrosinase in which DOPA is converted to dopaquinone. It is believed that the subsequent couple of reactions occur spontaneously, forming leucodopachrome followed by 5,6-dihydroxyindole. The oxidation of 5,6-dihydroxyindole to indole 5,6-quinone is catalysed by tyrosinase, and DOPA serves as a cofactor. This reaction, inhibited by tyrosine regulates the synthesis of melanin. ***Melanochromes*** are formed from indole quinone, which on ***polymerization*** are converted ***to black*** melanin.

Another pathway from dopaquinone is also identified. Cysteine condenses with dopaquinone and in the next series of reactions results the synthesis of red melanins. The ***structure of melanin pigments is not clearly known***.

Melanin—the colour pigment : The ***skin colour*** of the individual is ***determined by the relative concentrations of black and red melanins***. This, in turn, is dependent on many factors, both genetic and environmental. These include the activity of tyrosinase, the density of melanocytes, availability of tyrosine etc.

The presence of ***moles*** on the body represents a localized severe ***hyperpigmentation*** due to hyperactivity of melanocytes. On the other hand, localized absence or degeneration of melanocytes results in ***white patches*** on the skin commonly known as ***leucoderma***. ***Greying of hair*** is due to lack of melanocytes at hair roots. ***Albinism*** is an ***inborn error*** with generalized lack of melanin synthesis (details described later).

Biosynthesis of thyroid hormones

Thyroid hormones—***thyroxine*** (tetraiodothyronine) and ***triiodothyronine***—are synthesized from the tyrosine residues of the protein ***thyroglobulin*** and activated iodine (***Fig.15.21***). Iodination of tyrosine ring occurs to produce mono- and diiodotyrosine from which triiodothyronine (T_3) and thyroxine (T_4) are synthesized. The protein thyroglobulin undergoes proteolytic breakdown to release the free hormones, T_3 and T_4.

Biosynthesis of catecholamines

The name catechol refers to the ***dihydroxylated phenyl ring***. The amine derivatives of catechol are called catecholamines.

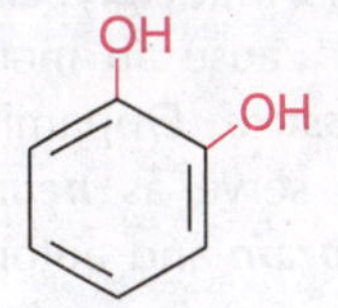

Catechol

Tyrosine is the precursor for the synthesis of catecholamines, namely ***dopamine***, ***norepinephrine*** (noradrenaline) and ***epinephrine*** (adrenaline).

The conversion of tyrosine to catecholamines occurs in adrenal medulla and central nervous system involving the following reactions (***Fig.15.22***).

Tyrosine is hydroxylated to 3,4-dihydroxyphenylalanine (DOPA) by ***tyrosine hydroxylase***.

This enzyme catalyses the ***rate limiting*** reaction and requires tetrahydrobiopterin as coenzyme (like phenylalanine hydroxylase). In contrast to this enzyme, tyrosinase present in melanocytes converts tyrosine to DOPA. Hence, two different enzyme systems exist to convert tyrosine to DOPA.

DOPA undergoes PLP-dependent decarboxylation to give dopamine which, in turn, is hydroxylated to produce norepinephrine. Methylation of norepinephrine by S-adenosylmethionine gives epinephrine. The difference between epinephrine and norepinephrine is only a ***methyl group*** (remember that ***nor***epinephrine has ***no*** methyl group).

There exists tissue specificity in the formation of catecholamines. In adrenal medulla, synthesis of the hormones, norepinephrine and epinephrine is prominent. Norepinephrine is produced in certain areas of the brain while dopamine is predominantly synthesized in substantia nigra and locus coeruleus of brain.

Functions of catecholamines : Norepinephrine and epinephrine ***regulate carbohydrate and lipid metabolisms***. They stimulate the degradation of triacylglycerol and glycogen. They cause an increase in the ***blood pressure***. Dopamine and norepinephrine serve as ***neurotransmitters in the brain*** and autonomous nervous system.

Dopamine and Parkinson's disease

Parkinson's disease is a common ***disorder*** in many ***elderly people***, with about 1% of the population above 60 years being affected. It is described as ***shaking palsy*** and characterized by muscular rigidity, tremors, expressionless face, lethargy, involuntary movements etc.

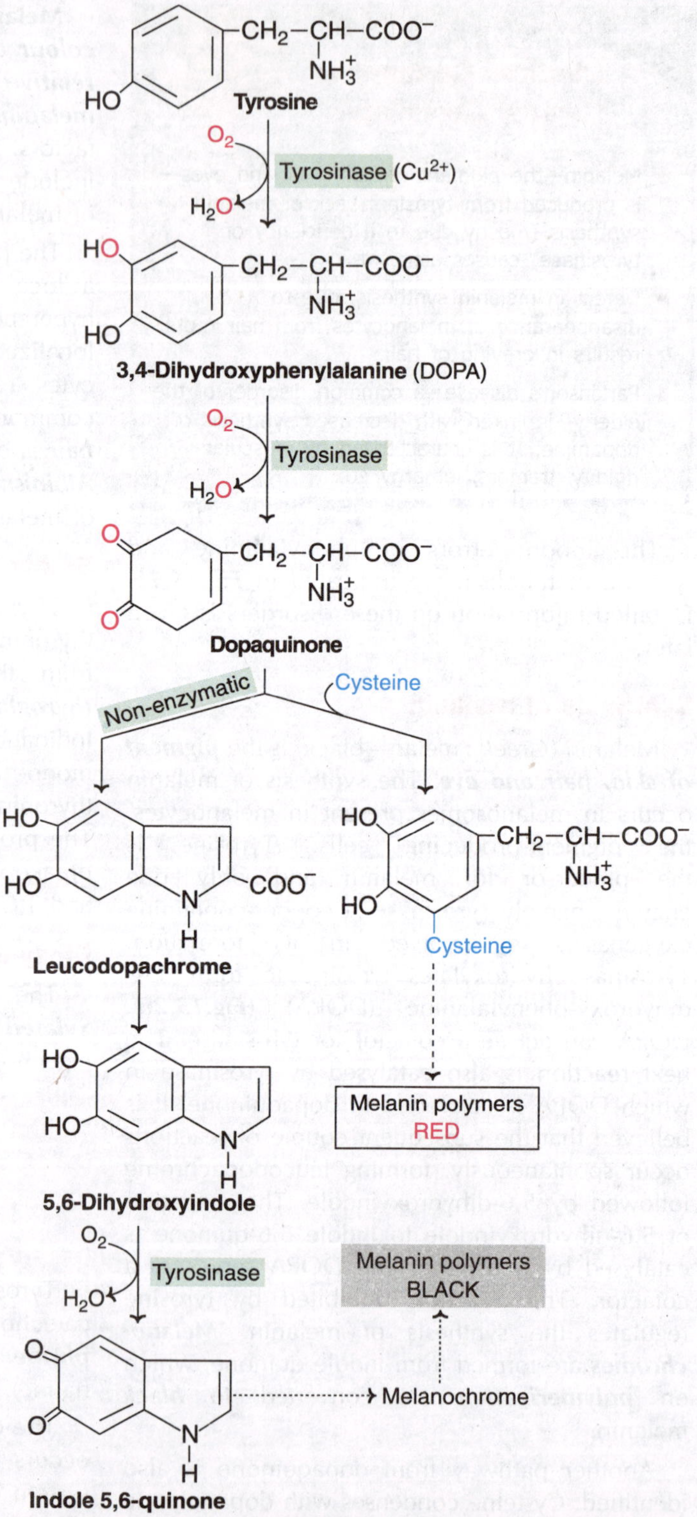

Fig. 15.20 : Metabolism of tyrosine—biosynthesis of melanin (Defect in tyrosinase causes albinism).

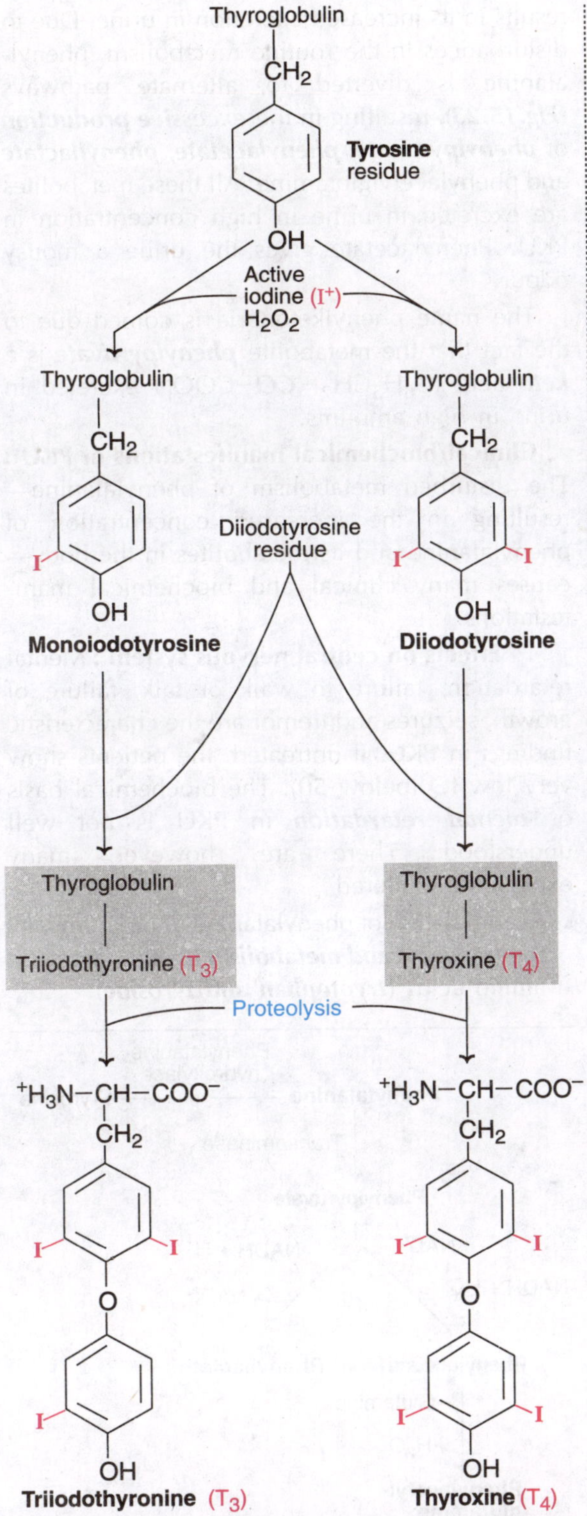

***Fig. 15.21 :** Metabolism of tyrosine-synthesis of thyroid hormones.*

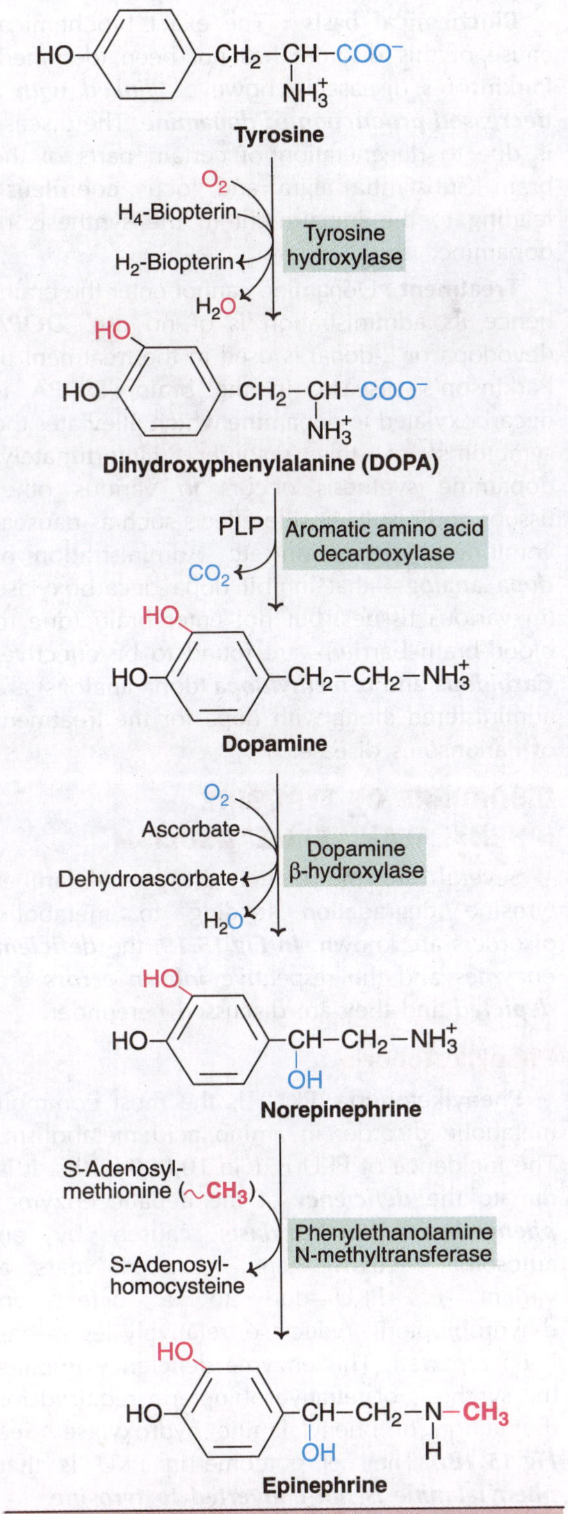

***Fig. 15.22 :** Metabolism of tyrosine-synthesis of catecholamines (dopamine, norepinephrine, epinephrine; PLP–pyridoxal phosphate).*

Biochemical basis : The exact biochemical cause of this disorder has not been identified. Parkinson's disease is, however, ***linked with a decreased production of dopamine***. The disease is due to degeneration of certain parts of the brain (substantia nigra and locus coeruleus), leading to the impairment in the synthesis of dopamine.

Treatment : Dopamine cannot enter the brain, hence its administration is of no use. ***DOPA*** (levodopa or L-dopa) is used in the treatment of Parkinson's disease. In the brain, DOPA is decarboxylated to dopamine which alleviates the symptoms of this disorder. Unfortunately, dopamine synthesis occurs in various other tissues and results in side-effects such as nausea, vomiting, hypertension etc. Administration of ***dopa analogs***—that inhibit dopa decarboxylase (in various tissues) but not enter brain (due to blood-brain barrier)—are found to be effective. ***Carbidopa*** and ***α-methyldopa*** (dopa analogs) are administered along with dopa for the treatment of Parkinson's disease.

DISORDERS OF TYROSINE (PHENYLALANINE) METABOLISM

Several enzyme defects in phenylalanine/tyrosine degradation leading to metabolic disorders are known. ***In Fig.15.19***, the ***deficient enzymes*** and the respective ***inborn errors*** are ***depicted*** and they are discussed hereunder.

Phenylketonuria

Phenylketonuria (PKU) is the most common metabolic disorder in amino acid metabolism. The incidence of PKU is 1 in 10,000 births. It is due to the ***deficiency*** of the hepatic enzyme, ***phenylalanine hydroxylase***, caused by an autosomal recessive gene. In recent years, a variant of PKU—due to a defect in dihydrobiopterin reductase (relatively less)—has been reported. This enzyme deficiency impairs the synthesis of tetrahydrobiopterin required for the action of phenylalanine hydroxylase (***See Fig.15.18***). The net outcome in PKU is that ***phenylalanine is not converted to tyrosine.***

Phenylalanine metabolism in PKU : Phenylketonuria primarily causes the accumulation of phenylalanine in tissues and blood, and results in its increased excretion in urine. Due to disturbances in the routine metabolism, phenylalanine is diverted to alternate pathways (***Fig.15.23***), resulting in the ***excessive production*** of ***phenylpyruvate, phenylacetate, phenyllactate*** and phenylacetylglutamine. All these metabolites are excreted in urine in high concentration in PKU. Phenylacetate gives the urine a mousy odour.

The name phenylketonuria is coined due to the fact that the metabolite ***phenylpyruvate*** is a ***keto*** acid ($C_6H_5CH_2-\boldsymbol{CO}-COO^-$) excreted in urine in high amounts.

Clinical/biochemical manifestations of PKU : The disturbed metabolism of phenylalanine—resulting in the increased concentration of phenylalanine and its metabolites in the body—causes many clinical and biochemical manifestations.

1. **Effects on central nervous system :** Mental retardation, failure to walk or talk, failure of growth, seizures and tremor are the characteristic findings in PKU. If untreated, the patients show very low IQ (below 50). The biochemical basis of ***mental retardation*** in PKU is not well understood. There are, however, many explanations offered

- Accumulation of phenylalanine in brain ***impairs*** the ***transport and metabolism*** of other aromatic amino acids (***tryptophan and tyrosine***).

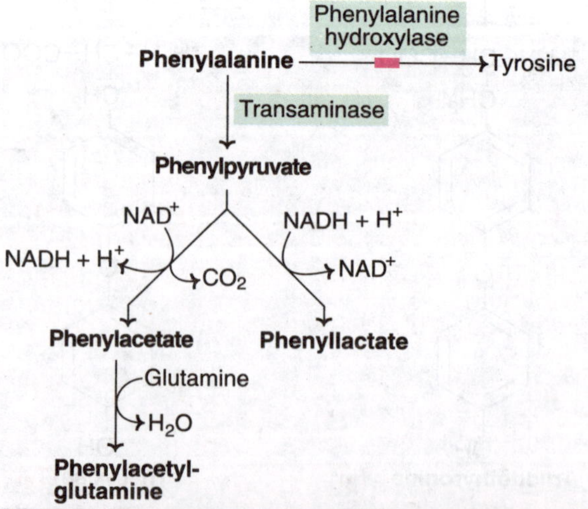

***Fig. 15.23 :** Metabolites that accumulate in phenylketonuria (▬–Block in phenylketonuria).*

- The synthesis of ***serotonin*** (an excitatory neurotransmitter) from tryptophan is ***insufficient***. This is due to the competition of phenylalanine and its metabolites with tryptophan that impairs the synthesis of serotonin.
- Defect in ***myelin*** formation is observed in PKU patients.

2. **Effect on pigmentation :** Melanin is the pigment synthesized from tyrosine by tyrosinase. Accumulation of phenylalanine competitively inhibits tyrosinase and impairs melanin formation. The result is ***hypopigmentation*** that causes light skin colour, fair hair, blue eyes etc.

Diagnosis of PKU : PKU is mostly detected by screening the newborn babies for the ***increased plasma levels of phenylalanine*** (PKU, 20–65 mg/dl; normal 1–2mg/dl). This is usually carried out by ***Guthrie test***, which is a bacterial *(Bacillus subtilis)* bioassay for phenylalanine. The test is usually performed after the baby is fed with breast milk for a couple of days by testing elevated levels of phenylalanine. Phenylpyruvate in urine can be detected by ***ferric chloride test*** (a green colour is obtained). This test is not specific, since many other compounds give a false positive test. Prenatal diagnosis of PKU can also be done by using cultured amniotic cells.

Treatment of PKU : The maintenance of ***plasma phenylalanine*** concentration ***within*** the ***normal range*** is a challenging task in the treatment of PKU. This is done by selecting ***foods with low phenylalanine content*** and/or feeding synthetic amino acid preparations, low in phenylalanine. Dietary intake of phenylalanine should be adjusted by measuring plasma levels. Early diagnosis (in the first couple of months of baby's life) and treatment for 4–5 years can prevent the damage to brain. However, the restriction to protein diet should be continued for many more years in life. Since the amino acid ***tyrosine*** cannot be synthesized in PKU patients, it ***becomes essential*** and should be provided in the diet in sufficient quantity.

In some seriously affected PKU patients, treatment includes administration of 5-hydroxytryptophan and dopa to restore the synthesis of serotonin and catecholamines. PKU patients with tetrahydrobiopterin deficiency require tetrahydrobiopterin supplementation.

Tyrosinemia type II

This disorder—also known as ***Richner-Hanhart syndrome***, is due to a defect in the enzyme ***tyrosine transaminase.*** The result is a blockade in the routine degradative pathway of tyrosine. Accumulation and excretion of tyrosine and its metabolites—namely p-hydroxyphenylpyruvate, p-hydroxyphenyllactate, phydroxyphenylacetate, N-acetyltyrosine—and tyramine are observed.

Tyrosinemia type II is characterized by ***skin*** (dermatitis) and ***eye lesions*** and, rarely, mental retardation. A disturbed self-coordination is seen in these patients.

Neonatal tyrosinemia

The absence of the enzyme ***p-hydroxyphenylpyruvate dioxygenase*** causes neonatal tyrosinemia. This is mostly a temporary condition and usually responds to ascorbic acid. It is explained that the substrate inhibition of the enzyme is overcome by the presence of ascorbic acid.

Alkaptonuria (Black urine disease)

Alkaptonuria has great historical importance. It was first described by Lusitanus in 1649 and characterized in 1859. Garrod conceived the idea of inborn errors of metabolism from his observation on alkaptonuria. The prevalance of this autosomal recessive disorder is 1 in 25,000.

Enzyme defect : The defective enzyme in alkaptonuria is ***homogentisate oxidase*** in tyrosine metabolism (***See Fig.15.19***). ***Homogentisate accumulates*** in tissues and blood, and is ***excreted*** into urine. Homogentisate, on standing, gets oxidized to the corresponding quinones, which polymerize to give black or brown colour. For this reason, the ***urine*** of alkaptonuric patients resembles ***coke in colour.***

Biochemical manifestations : Homogentisate gets oxidized by polyphenol oxidase to benzo-

Medical Concepts/ Clinical Correlates

- Phenylketonuria, due to a defect in the enzyme phenylalanine hydroxylase, is characterized by failure of growth, seizures and mental retardation (low IQ).
- Alkaptonuria causes the accumulation of homogentisate which undergoes oxidation followed by polymerization to produce the pigment alkapton. Deposition of alkapton in connective tissue, causes ochronosis which is associated with arthritis.
- Neurotransmitters namely serotonin, dopamine, epinephrine, nonepinephrine and γ-amino-butyrate are derived from amino acids. Many drugs used to treat psychiatric and neurologic disorders act by altering the metabolism of neurotransmitters.

quinone acetate which undergoes polymerization to produce a pigment called ***alkapton*** (***Fig.15.24***). Alkapton deposition occurs in connective tissue, bones and various organs (nose, ear etc.) resulting in a condition known as ***ochronosis.*** Many alkaptonuric patients suffer from arthritis and this is believed to be due to the deposition of pigment alkapton (in the joints), produced from homogentisate.

Diagnosis : ***Change in colour of the urine*** on standing to ***brown or dark*** has been the simple traditional method to identify alkaptonuria. The urine gives a positive test with ferric chloride and silver nitrate. This is due to the strong reducing activity of homogentisate. Benedict's test—employed for the detection of glucose and other reducing sugars—is also positive with homogentisate.

Treatment : Alkaptonuria is ***not*** a ***dangerous disorder*** and, therefore, does not require any specific treatment. However, consumption of protein diet with relatively low phenylalanine content is recommended.

Tyrosinosis or tyrosinemia type I

This is due to the deficiency of the enzymes ***fumarylacetoacetate hydroxylase*** and/or ***maleylacetoacetate isomerase.*** Tyrosinosis is a rare but serious disorder. It causes liver failure, rickets, renal tubular dysfunction and polyneuropathy. Tyrosine, its metabolites and many other amino acids are excreted in urine.

In acute tyrosinosis, the infant exhibits diarrhea, vomiting, and 'cabbage-like' odor. Death may even occur due to liver failure within one year. For the treatment, diets low in tyrosine, phenylalanine and methionine are recommended.

Albinism

Albinism (*Greek*: albino—white) is an inborn error, due to the ***lack of*** synthesis of the pigment ***melanin***. It is an autosomal recessive disorder with a frequency of 1 in 20,000.

Biochemical basis : The colour of skin and hair is controlled by a large number of genes. About 150 genes have been identified in mice. The melanin synthesis can be influenced by a variety of factors. Many possible causes (rather explanations) for albinism have been identified

1. Deficiency or ***lack of the enzyme tyrosinase***.
2. Decrease in melanosomes of melanocytes.
3. Impairment in melanin polymerization.
4. Lack of protein matrix in melanosomes.
5. Limitation of substrate (tyrosine) availability.
6. Presence of inhibitors of tyrosinase.

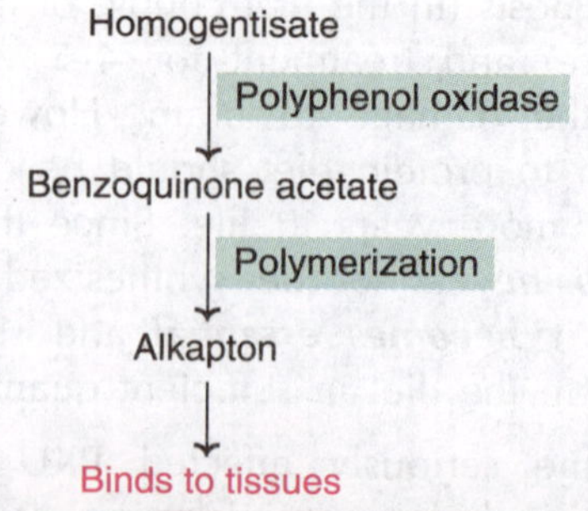

Fig. 15.24 : Conversion of homogentisate to alkapton.

The most common cause of albinism is a ***defect in tyrosinase***, the enzyme most responsible for the synthesis of melanin (***See Fig.15.20***).

Clinical manifestations : The most important function of melanin is the protection of the body from sun radiation. Lack of melanin in albinos makes them ***sensitive to sunlight***. Increased susceptibility to ***skin cancer*** (carcinoma) is observed. ***Photophobia*** (intolerance to light) is associated with lack of pigment in the eyes. However, there is no impairment in the eyesight of albinos.

Hypopigmentation

In some individuals, a reduced synthesis of melanin (instead of total lack) is often observed. Hypopigmentation disorders may be either diffuse or localized.

A good example of diffuse hypopigmentation is ***oculocutaneous albinism*** which is mostly due to mutations in the tyrosinase gene. The degree of hypopigmentation depends on the type and severity of mutated genes.

Vitiligo and ***leukoderma*** are the important among the localized hypopigmentation disorders. ***Vitiligo*** is an acquired progressive disease with ***loss of pigmentation around mouth***, ***nose***, eyes and nipples. ***Leukoderma*** is comparable with vitiligo, but ***lack of pigmentation*** usually ***begins with hands*** and then spreads.

Piebaldism is a disorder related to ***melanocytes.*** It is characterized by ***grey forelock of hair***, ***scattered normal pigmented*** and ***hypopigmented macules*** and ***patches on forehead***. Piebaldism is less common in man, and more frequently found in some animals e.g. mouse rabbit, dog, deer, horse. Human piebaldism is frequently associated with endocrine disorders.

Greying of hair is due to lack of melanin synthesis which usually occurs as a result of disappearance of melanocytes from the hair roots.

Localized ***hyperpigmentation*** due to hyperactivity of melanocytes results in the formation of ***moles***.

TRYPTOPHAN

Tryptophan (Trp, W) was the first to be identified as an ***essential amino acid***. It contains an ***indole ring*** and chemically, it is α-amino β-indole propionic acid. Tryptophan is both ***glucogenic*** and ***ketogenic*** in nature. It is a precursor for the synthesis of important compounds, namely ***NAD⁺*** and ***NADP⁺*** (coenzymes of ***niacin***), ***serotonin*** and ***melatonin*** (***Fig.15.25***).

The metabolism of tryptophan is divided into

I. Kynurenine (kynurenine-anthranilate) pathway;

II. Serotonin pathway.

I. Kynurenine pathway

This pathway mostly occurs in liver leading to oxidation of tryptophan and the synthesis of NAD^+ and $NADP^+$ (***Fig.15.26***).

Tryptophan pyrrolase or oxygenase cleaves the five-membered ring of the indole nucleus to produce formylkynurenine. Tryptophan pyrrolase is a metalloprotein containing an iron porphyrin ring. It is a ***substrate inducible enzyme*** and is controlled by feedback regulation (by NADPH and other niacin derivatives). Tryptophan pyrrolase activity is also elevated by corticosteroids. Formamidase hydrolyses formylkynurenine and liberates formate which enters the one carbon pool. Kynurenine formed in this reaction is a branch point with different fates. In the prominent pathway, kynurenine undergoes NADPH-dependent hydroxylation to give 3-hydroxykynurenine.

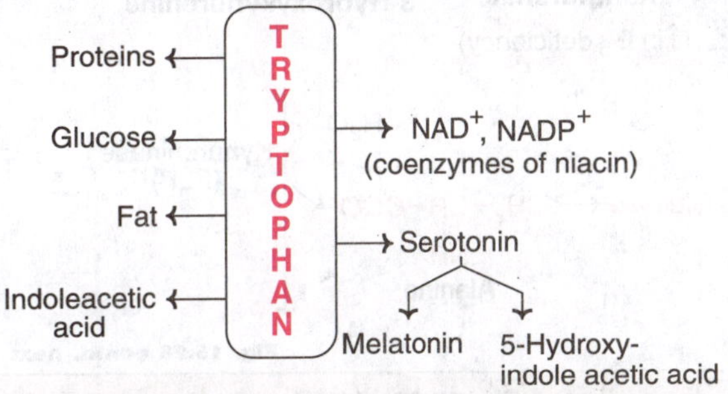

Fig. 15.25 : Overview of tryptophan metabolism.

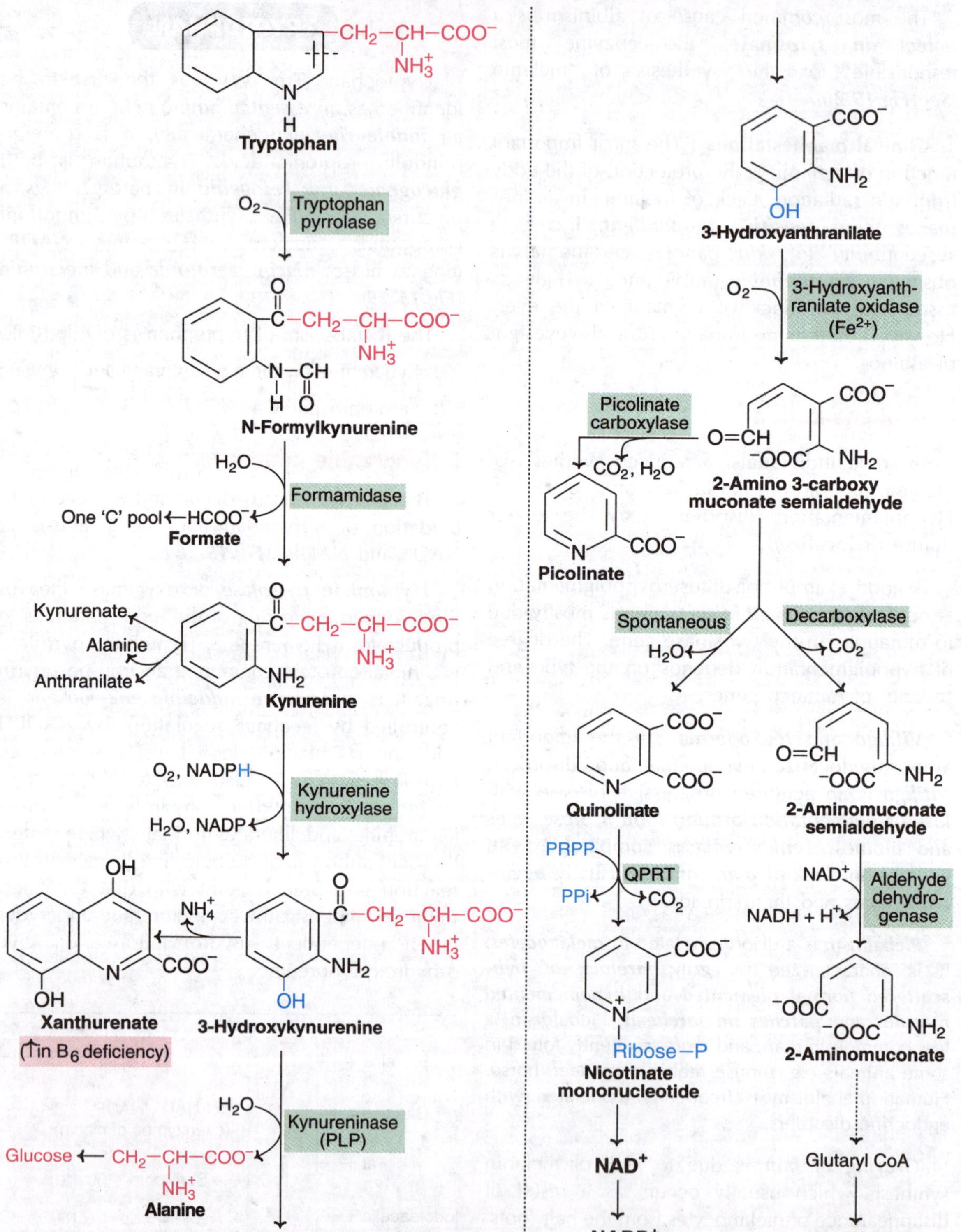

Fig. 15.26 contd. next column

Fig. 15.26 : Metabolism of tryptophan-kynurenine pathway (PLP–Pyridoxal phosphate; QPRT–Quinolinate phosphoribosyl transferase; PRPP–Phosphoribosyl pyrophosphate).

Kynureninase, a pyridoxal phosphate (PLP)-dependent enzyme acts on the 3-hydroxy-kynurenine and splits off alanine. Tryptophan is glucogenic, since alanine is a good precursor for glucose. The enzyme kynureninase is very sensitive to vitamin B_6 deficiency. Due to the lack of PLP, kynureninase reaction is blocked and 3-hydroxykynurenine is diverted to form xanthurenate. Elevated excretion of ***xanthurenate*** serves as an ***indication of B_6 deficiency.*** Administration of ***isoniazid***, an antituberculous drug—induces B_6 deficiency and results in xanthurenate excretion in urine. Defects in the activity of kynureninase (in B_6 deficiency) cause reduced synthesis of NAD^+ and $NADP^+$ from tryptophan. The symptoms of pellagra —observed in B_6 deficiency—are explained on this basis.

The enzyme kynurenine hydroxylase is inhibited by estrogen, hence ***women***, in general, are ***more susceptible to pellagra***.

3-Hydroxyanthranilate is cleaved by an oxidase (Fe^{2+} dependent) to form an unstable intermediate, 2-amino 3-carboxy muconate semialdehyde. This compound has three fates.

1. It undergoes spontaneous cyclization to form quinolinate for NAD^+ synthesis.

2. ***Picolinate carboxylase*** decarboxylates the intermediate which cyclizes to produce picolinate. This enzyme competes with the formation of quinolinate. High activity of picolinate carboxylase in some animals (e.g. cat) deprives them of NAD^+ synthesis from tryptophan. In other words, cat is exclusively dependent on niacin for its coenzymes (NAD^+, $NADP^+$), since tryptophan cannot serve as a precursor.

3. The intermediate undergoes decarboxylation, catalysed by amino carboxysemialdehyde decarboxylase to produce 2-aminomuconate semialdehyde that enters glutarate pathway. The semialdehyde is converted to 2-aminomuconate by a dehydrogenase. The aminomuconate, in a series of reactions involving reduction, deamination, decarboxylation etc., is converted to glutaryl CoA and finally to acety CoA. The latter is either completely oxidized via TCA cycle or converted to fat (hence tryptophan is ketogenic).

NAD^+ Pathway : Tryptophan is not a precursor for the synthesis of free niacin. Quinolinate undergoes decarboxylation and is converted to nicotinate mononucleotide by the enzyme quinolinate phosphoribosyl transferase (QPRT). The synthesis of NAD^+ and $NADP^+$ from nicotinate mononucleotide is similar to that from niacin as described in vitamins (***Refer Fig.7.21***).

Conversion of tryptophan to indole acetate : Tryptophan undergoes deamination and decarboxylation to produce indolepyruvate and tryptamine, respectively. Both these compounds are converted to indoleacetate (***Fig.15.27***) and excreted in urine.

II. Serotonin pathway

Serotonin or ***5-hydroxytryptamine*** (5HT) is a neurotransmitter, synthesized from tryptophan. Normally, about ***1% of the tryptophan is converted to serotonin***. The production of 5HT occurs in the target tissues.

Synthesis of serotonin : In mammals, the ***largest*** amount of serotonin is synthesized in the ***intestinal cells***. The formation of serotonin is comparable with the production of catecholamines. Tryptophan is first hydroxylated at 5th carbon by tryptophan hydroxylase. This enzyme requires tetrahydrobiopterin as a cofactor. 5-Hydroxytryptophan is decarboxylated by aromatic amino acid decarboxylase (PLP-dependent) to give serotonin (***Fig.15.27***).

Platelets contain ***high*** concentration of ***5HT***, the significance of which is not clear. As such, platelets cannot synthesize serotonin.

Degradation of serotonin : ***Monoamine oxidase*** (***MAO***) degrades serotonin to 5-hydroxy-indoleacetate (5HIA) which is excreted in urine.

Functions of serotonin : Serotonin is ***a neuro-transmitter*** and performs a variety of functions.

1. Serotonin is a powerful ***vasoconstrictor*** and results in smooth muscle contraction in bronchioles and arterioles.

2. It is closely involved in the ***regulation of cerebral activity*** (excitation).

3. Serotonin, regarded as a ***happiness hormone***, ***controls the behavioural patterns***, sleep, blood pressure and body temperature.

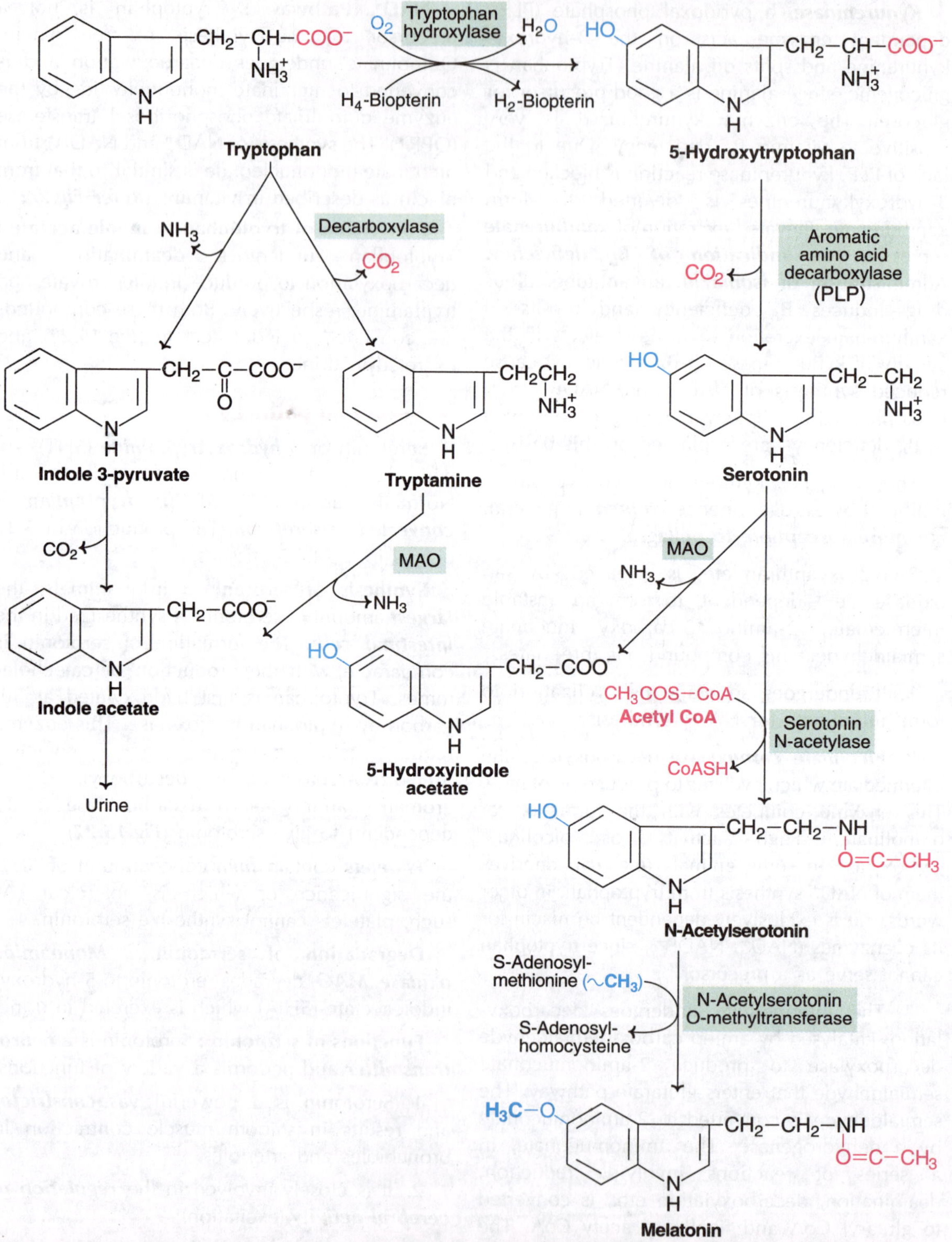

Fig. 15.27 : Metabolism of tryptophan – serotonin and melatonin synthesis (PLP–Pyridoxal phosphate; MAO–Monoamine oxidase).

Medical Concepts/ Clinical Correlates

- Serotonin, an excitatory neurotransmitter, is synthesized from tryptophan. Iproniazid elevates serotonin levels while reserpine decreases.
- Carbohydrate-rich diets promote sleep by enhancing tryptophan entry into brain and increasing the synthesis of serotonin and melatonin.
- Malignant carcinoid syndrome, a tumor of argentaffin cells of gastrointestinal tract, is characterized by tremendously increased production of serotonin. This disorder can be diagnosed by the elevated levels of 5-hydroxy-indoleacetate in urine.

4. Serotonin evokes the release of peptide hormones from gastrointestinal tract.

5. It is also necessary for the motility of GIT (peristalsis).

Serotonin and brain : The brain itself synthesizes 5HT which is in a bound form. The outside serotonin cannot enter the brain due to blood-brain barrier. Primarily, serotonin is a ***stimulator*** (excitation) ***of brain activity***, hence ***its deficiency causes depression. Serotonin level is decreased in psychosis patients.***

Defects in monoamine oxidase gene (lowered MAO activity) are linked to ***violent behaviour*** and slight mental retardation.

Effect of drugs on serotonin : The drug, ***iproniazid*** (isopropyl isonicotinyl hydrazine) inhibits MAO and elevates serotonin levels, therefore, this drug is a an antidepressent. On the other hand, ***reserpine*** increases the degradation of serotonin, hence it is a depressant drug. Lysergic acid diethylamide (**LSD**) acts as a psychic stimulant by activating serotonin receptors..

Serotonin syndrome : This disorder is caused by excess levels of serotonin which ***occurs due to serotonin boosting medication*** in the form of antidepressants and migraine medicines. Serotonin syndrome is characterized by restlessness, confusion, headache, sweating, increased blood pressure, loss of muscle coordination, seizures etc. The treatment mainly involves stopping the medication to restore normal serotonin levels.

Malignant carcinoid syndrome : Serotonin is produced by argentaffin cells of gastrointestinal tract. When these cells undergo uncontrolled growth, they develop into a tumor called malignant carcinoid or ***argentaffinomas.*** The patients exhibit symptoms like respiratory distress, sweating, hypertension etc.

Normally about 1% of the tryptophan is utilized for serotonin synthesis. In case of carcinoid syndrome, very high amount (up to 60%) of tryptophan is diverted for serotonin production. This disturbs the normal tryptophan metabolism and ***impairs the synthesis of NAD^+ and $NADP^+$***. Hence, the patients of carcinoid syndrome develop symptoms of ***pellagra*** (niacin deficiency). Further, ***negative nitrogen balance*** is also observed.

Diagnosis : The excretion of ***5-hydroxy indole acetate*** in urine is tremendously ***elevated*** (upto 500mg/day against normal <5 mg/day) in carcinoid syndrome. The estimation of 5 HIA in urine is used for the diagnosis of this disorder. In general, urine concentration of 5 HIA above 25 mg/day should be viewed with caution as it may be suggestive of carcinoid syndrome. Sufficient precaution should, however, be taken for sample collection. During the course of urine collection, the patients should not ingest certain foods (banana, tomato etc.) that increase urine 5 HIA.

Melatonin

Melatonin is a ***hormone***, mostly synthesized by the ***pineal gland.*** Serotonin—***produced from tryptophan***—is acted upon by serotonin N-acetylase (the rate limiting enzyme), to give N-acetylserotonin. The latter undergoes methylation, S-adenosylmethionine being the methyl group donor to produce melatonin or N-acetyl 5-methoxyserotonin (***Fig.15.27***). The synthesis and secretion of melatonin from pineal gland is controlled by light.

Functions of melatonin

1. Melatonin is involved in ***circadian rhythms*** or ***diurnal variations*** (24 hr cyclic process) of the body. It plays a significant role in ***sleep and wake process***.

2. Melatonin inhibits the production of melanocyte stimulating hormone (MSH) and adrenocorticotropic hormone (ACTH).

3. It has some inhibitory effect on ovarian functions.

4. Melatonin also performs a ***neuro-transmitter*** function.

Tryptophan metabolism and sleep

As described above, tryptophan is a precursor for sleep inducing compounds—serotonin and melatonin. Thus, ***tryptophan promotes sleep***, and is found in variable amounts in all protein foods. To have sedative effect, tryptophan should enter the brain by crossing the blood-brain barrier. Tryptophan shares the same transport system as other amino acids, and thus it has to compete against them to enter the brain.

Eating ***rich carbohydrate diets, along with proteins is benefical***. Carbohydrates stimulate the release of insulin, and this clears other amino acids from circulation and facilitates the entry of tryptophan into brain. The net result is increased synthesis of sleep inducer serotonin from tryptophan. Thus, ***tryptophan is considered as a sleep-inducing amino acid***.

Hartnup's disease

This disorder was first described in the family of Hartnup, hence the name—Hartnup's disease. It is a hereditary disorder of tryptophan metabolism. The clinical symptoms include dermatitis, ataxia, mental retardation etc. Hartnup's disease is characterized by ***low plasma levels of tryptophan and other neutral amino acids*** and their elevated urinary excretion. Increased urinary output of indoleacetic acid and indolepyruvic acid is also observed.

Pellagra-like symptoms are common in these patients. There is an impairment in the synthesis of NAD^+ and serotonin from tryptophan. Some authors (earlier) attributed Hartnup's disease to a defect in the enzyme tryptophan pyrrolase. This, however, does not appear to be true. Hartnup's disease is now believed to be due to an ***impairment*** in the ***absorption*** and/or ***transport*** of tryptophan and other ***neutral amino acids*** from the intestine, renal tubules and, probably brain. Some more details on Hartnup's disease are given under digestion and absorption (***Chapter 8***).

SULFUR AMINO ACIDS

The sulfur-containing amino acids are methionine, cysteine and cystine. Among these, only ***methionine*** is ***essential***. It serves as a precursor for the synthesis of cysteine and cystine which are, therefore, non-essential. ***Cysteine can spare the requirement of methionine*** in the diet. Cysteine and cystine are interconvertible. Cystine is found exclusively in proteins. Methionine and cysteine, besides being present in proteins, are involved in many important metabolic reactions (***Fig.15.28***). Methionine is also required for the initiation of protein biosynthesis. The sulfur-containing amino acids are almost an exclusive ***dietary source of sulfur*** to the body.

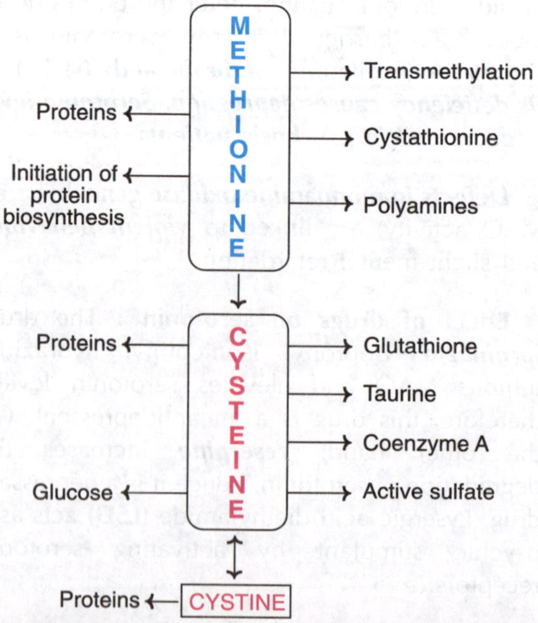

Fig. 15.28 : Overview of the metabolism of sulfur amino acids.

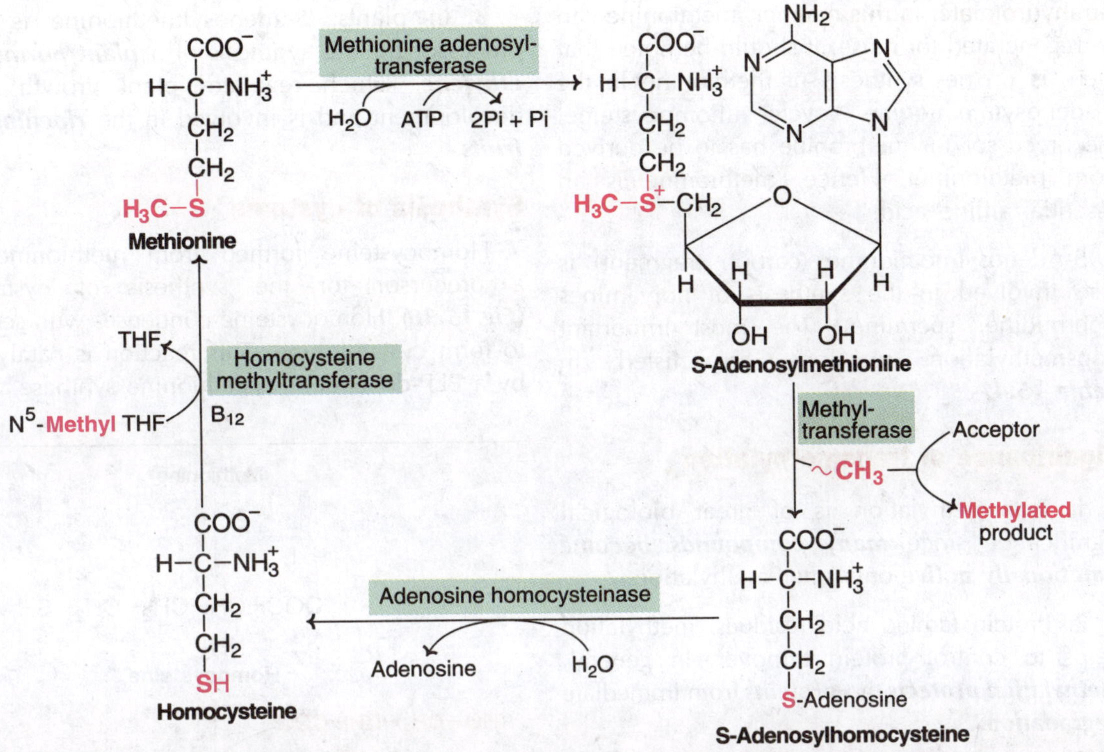

Fig. 15.29 : *S-Adenosylmethionine cycle – synthesis, utilization and regeneration.*

METABOLISM OF METHIONINE

Methionine (or sulfur amino acids) metabolism may be divided into three parts.

1. Utilization of methionine for transmethylation reactions.

2. Conversion of methionine to cysteine and cystine.

3. Degradation of cysteine and its conversion to specialized products.

Transmethylation

The ***transfer of methyl group*** ($-CH_3$) ***from active methionine*** to an acceptor is known as transmethylation. Methionine has to be activated to ***S-adenosylmethionine*** (SAM) or active methionine to donate the methyl group.

Synthesis of S-adenosylmethionine

The synthesis of S-adenosylmethionine occurs by the transfer of an adenosyl group from ATP to sulfur atom of methionine (***Fig.15.29***). This reaction is catalysed by methionine S-adenosyltransferase. The activation of methionine is unique as the sulfur becomes a sulfonium atom (SAM is a sulfonium compound) by the addition of a third carbon. This reaction is also unusual since all the three phosphates of ATP are eliminated as pyrophosphates (PPi) and inorganic phosphates (Pi). Three high energy phosphates (***3 ATP***) are ***consumed in the formation of SAM.***

Functions of S-adenosylmethionine

S-Adenosylmethionine is highly reactive due to the presence of a positive charge. The enzymes involved in the transfer of methyl group are collectively known as ***methyltransferases.*** S-Adenosylmethionine ***transfers the methyl group to an acceptor*** and gets itself converted to S-adenosylhomocysteine. The loss of free energy in this reaction makes the methyl transfer essentially irreversible. S-Adenosylhomocysteine is hydrolysed to homocysteine and adenosine. Homocysteine can be remethylated to methionine by N^5-methyl tetrahydrofolate. In this manner, methionine can be regenerated for reuse. It should be noted that there is no net synthesis of methionine in the S-adenosylmethionine cycle (homocysteine, the precussor for methionine has to be derived from methionine). Hence, methionine is an essential amino acid.

S-Adenosylmethionine (carbon fragment) is also involved in the synthesis of polyamines (spermidine, spermine). The most important transmethylation reactions are listed in *Table 15.4*.

Significance of transmethylation

1. Transmethylation is of great biological significance since ***many compounds become functionally active*** only after methylation.

2. Protein (amino acid residues) methylation helps to control protein turnover. In general, ***methylation protects*** the ***proteins*** from immediate ***degradation***.

3. In plants, S-adenosylmethionine is the precursor for the synthesis of a ***plant hormone, ethylene***, which regulates plant growth and development and is involved in the ***ripening of fruits***.

Table 15.4 Transmethylation reactions

S-Adenosylmethionine	*S-Adenosylhomocysteine*

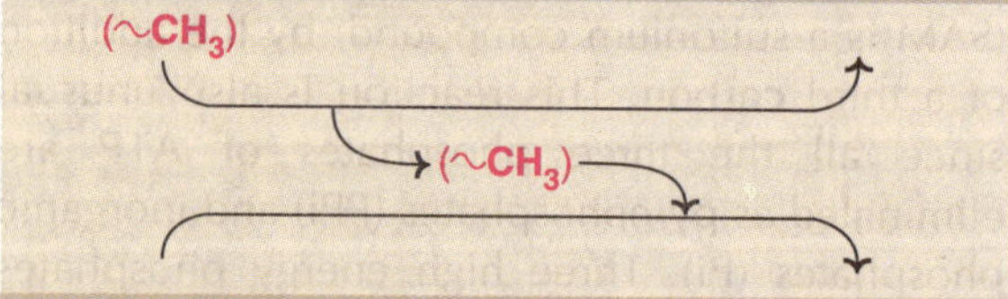

Methyl group acceptor	*Methylated product*
Guanidoacetate	Creatine
Norepinephrine	Epinephrine
Epinephrine	Metanephrine
Ethanolamine	Choline
Nicotinamide	N-Methylnicotinamide
Acetyl serotonin	Melatonin
Phosphatidylethanolamine	Phosphatidylcholine
Serine	Choline
Carnosine	Anserine
tRNA bases	Methylated tRNA bases
Protein-amino acid residues (histidine, lysine, arginine etc.)	Protein-methylated amino acids (methylhistidine, methyl-lysine, methylarginine etc.)

Synthesis of cysteine

Homocysteine formed from methionine is a precursor for the synthesis of cysteine (***Fig.15.30***). Homocysteine condenses with serine to form cystathionine. This reaction is catalysed by a PLP-dependent cystathionine synthase. The

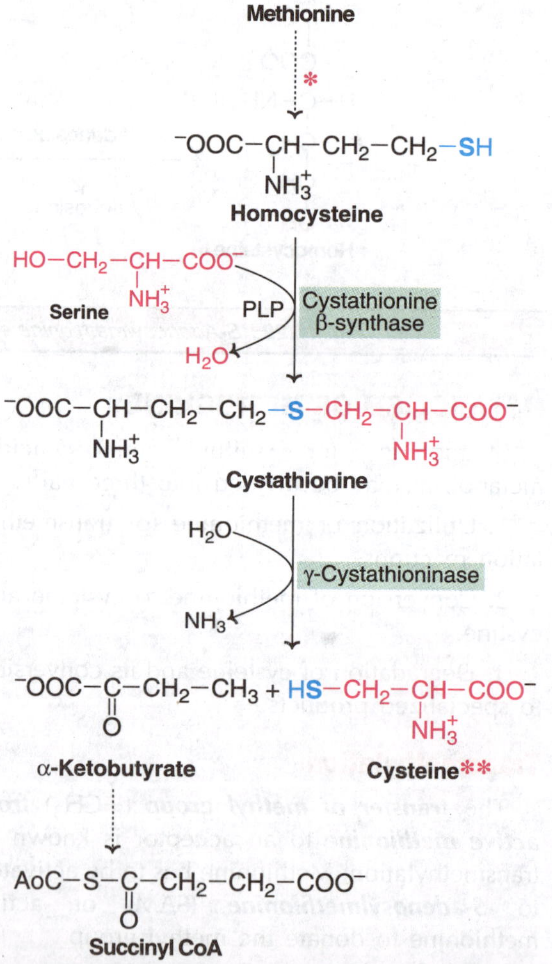

Fig. 15.30 : *Synthesis of cysteine from methionine (*–For reactions of methionine conversion to homocysteine see Fig. 15.29) (**–In cysteine synthesis, only sulfur is obtained from homocysteine, the rest of the molecule is from serine).*

Medical Concepts/ Clinical Correlates

- Melatonin, produced from serotonin, is involved in circadian rhythms or diurnal variations, i.e., maintenance of body's biological clock.
- Tryptophan, being a precursor of serotonin and melatonin, is considered as a sleep inducing amino acid.
- Homocysteine has been implicated as a risk factor in heart diseases, and neural tube defects (in fetus).
- Histidine loading test, characterized by elevated excretion of N-formiminoglutamate (FIGLU), is commonly employed to assess the deficiency of the vitamin, folic acid.

enzyme cystathioninase (PLP-dependent) cleaves and deaminates cystathionine to cysteine and α-ketobutyrate. The sum of the reactions catalysed by cystathionine synthase and cystathioninase is a good example of ***transsulfuration*** (transfer of sulfur from one compound to another). It should be noted that only the ***sulfur atom of cysteine*** comes from homocysteine (originally methionine) while the ***rest of the molecule is from serine.***

Homocysteine and heart attacks

Homocysteine is an intermediate in the synthesis of cysteine from methionine (***Fig.15.30***), ***Elevation*** in ***plasma homocysteine*** (normal < 15 μ mol/l) has been ***implicated in coronary artery diseases***, although the mechanism is not known. It is believed that homocysteine reacts with collagen to produce reactive free radicals, besides interfering with collagen cross links. Homocysteine is also involved in the aggregation of LDL particles. All this leads to an increased tendency for ***atherogenesis***, and consequently heart complications. ***Hyperhomocysteinemia*** in pregnancy increases the risk of ***neural tube defects in fetus.***

Supplementation of diet with ***folic acid, vitamin*** B_{12} and ***vitamin*** B_6 have some ***beneficial*** affects in lowering plasma homocysteine levels (***Refer Chapter 7***).

Degradation of cysteine

Cystine and cysteine are interconvertible by an NAD^+-dependent cystine reductase. Cysteine on decarboxylation produces ***mercapto-ethanolamine*** which is involved in the biosynthesis of ***coenzyme A*** from the vitamin pantothenic acid.

The enzyme cysteine dioxygenase oxidizes cysteine to cysteine sulfinate which, on further oxidation, is converted to cysteic acid. The latter undergoes decarboxylation to produce ***taurine*** which conjugates with ***bile acids***. Cysteic acid can also be degraded to pyruvate, which is glycogenic (***Fig.15.31***).

Cysteine sulfinate cleaves off alanine to produce sulfite which is converted to sulfate and excreted in urine. Some amount of sulfate condenses with ATP to form active sulfate or ***3′-phosphoadenosine 5′-phosphosulfate*** (***PAPS***). Active sulfate (PAPS) is utilized for the ***synthesis of mucopolysaccharides*** (sulfation), besides being used in ***detoxification***. Sulfate is also a structural component of some proteins, lipids etc.

Cysteine can be degraded by desulfhydrase to liberate sulfur (as H_2S), ammonia and pyruvate.

Cysteine is a component of tripeptide glutathione (synthesis described in glycine metabolism).

Inborn errors of sulfur amino acid metabolism

Cystinuria (cystine-lysinuria) : Cystinuria is one of the most ***common inherited diseases*** with a frequency of 1 in 7,000. It is primarily characterized by increased excretion of cystine (25-40 times normal). Elevation in the urinary output of lysine, arginine and ornithine is also observed. A specific carrier system exists in kidney tubules for the reabsorption of amino acids, namely cysteine, ornithine, arginine and lysine (remember ***COAL to recall***). In cystinuria, this ***carrier system*** becomes ***defective*** leading to the excretion of all these four amino acids in urine.

Cystine is relatively insoluble and increase in its concentration leads to precipitation and formation of cystine stones in kidney and urinary tract. Cystinuria is usually identified in the laboratory by cyanide nitroprusside test. The treatment includes restricted ingestion of dietary cystine and high intake of fluids.

Cystinosis (cystine storage disease) : Cystine crystals are deposited in many tissues and organs of reticuloendothelial system throughout the body. These include spleen, lymph nodes, liver,

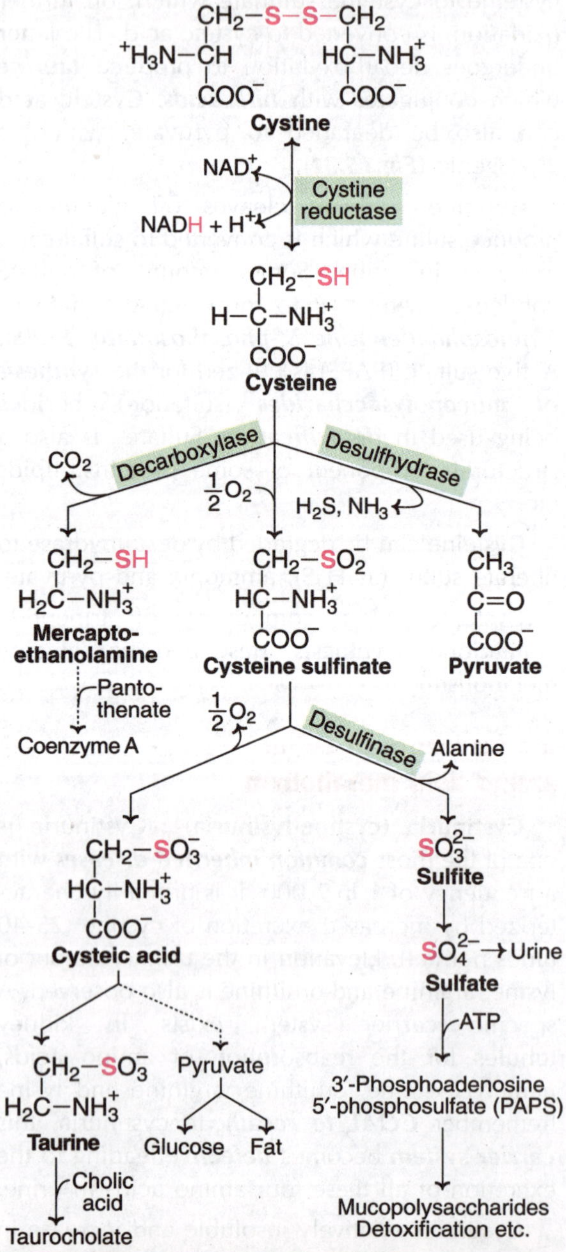

Fig. 15.31 : Metabolism of cysteine and cystine.

kidney, bone marrow etc. A ***defect in the lysosomal function*** is said to be the primary cause of this disorder. In fact, cystine accumulates in the lysosomes of various tissues. Impairment in renal function is commonly seen in cystinosis. It is characterized by generalized amino aciduria. The affected patients die usually within 10 years, mostly due to renal failure. Although the underlying metabolic defect in cystinosis is not clearly known, some authors attribute this to the defect in the enzyme ***cystine reductase.*** This is, however, not accepted by others.

Homocystinurias : Homocystinurias are a group of ***metabolic disorders*** characterized by the accumulation and ***increased urinary excretion of homocysteine and S-adenosylmethionine.*** Plasma concentration of methionine is increased.

Homocystinuria type I has been more thoroughly investigated. It is due to a defect in the enzyme ***cystathionine synthase. Accumulation of homocysteine*** results in various complications—***thrombosis***, osteoporosis and, very often, mental retardation. Further, the deficiency of cystathionine is associated with damage to endothelial cells which might lead to atherosclerosis. Two forms of type I homocystinurias are known, one of them can be corrected with vitamin B_6 supplementation (B_6 responsive) while the other does not respond to B_6. The treatment includes consumption of diet low in methionine and high in cystine.

The patients of homocystinuria have ***high*** levels of ***homocysteine,*** and usually die of ***myocardial infarction, stroke, or pulmonary embolism.***

The other homocystinurias are associated with enzyme defects (as stated below) in the conversion of homocysteine to methionine by remethylation.

Homocystinuria II : N^5-N^{10}—Methylene THF reductase.

Homocystinuria III : N^5—Methyl THF-homocysteine methyltransferase. This is mostly due to impairment in the synthesis of methylcobalamin.

Homocystinuria IV : This is primarily due to a defect in the intestinal absorption of vitamin B_{12}.

ONE-CARBON METABOLISM

Amino acid metabolism is particularly important for the transfer or exchange of one-carbon units. The following ***one-carbon fragments*** are encountered in the biological reactions, which constitute one-carbon pool

Methyl	$(-CH_3)$
Hydroxymethyl	$(-CH_2OH)$
Methylene	$(=CH_2)$
Methenyl	$(-CH=)$
Formyl	$(-CH=O)$
Formimino	$(-CH=NH)$

[**Note :** It may be stated here that CO_2 is also a one-carbon unit. Carbon dioxide is involved (carboxylation) in many biochemical reactions, which are dependent on biotin. For instance, conversion of pyruvate to oxaloacetate in gluconeogenesis. Most of the authors, however, ignore CO_2 as one-carbon unit and do not even consider it worth mentioning. This would be unfair to CO_2!]

Tetrahydrofolate (THF) is a versatile coenzyme that ***actively participates in one-carbon metabolism.*** With regard to the transfer of methyl groups from S-adenosylmethionine, vitamin B_{12} is also involved besides THF.

The ***one-carbon unit covalently binds*** with THF at position ***N^5 or N^{10}*** or ***on both N^5*** and ***N^{10}*** of pteroyl structure of folate. The details of different one-carbon units binding with THF and the structures of THF derivatives are given under vitamin-folic acid (***Chapter 7***).

The one-carbon metabolism is rather complex, involving many reactions. For the sake of better understanding, it is divided into generation and utilization of one-carbon units, and the role of methionine and vitamin B_{12}.

I. Generation of one-carbon units

Many compounds (particularly amino acids) act as donors of one-carbon fragments

1. The formate released from glycine and tryptophan metabolism combines with THF to form N^{10}-formyl THF.

2. Histidine contributes formimino fragment to produce N^5-formimino THF.

3. When serine is converted to glycine, N^5, N^{10}-methylene THF is formed. This is the most predominant entry of one carbon units into one carbon pool.

4. Choline and betaine contribute to the formation of N^5-methyl THF.

The ***different derivatives of THF carrying one-carbon units are interconvertible***, and this is metabolically significant for the continuity of one- carbon pool (***Fig.15.32***).

II. Utilization of one-carbon moieties

One-carbon fragments from THF are used for the synthesis of a wide variety of compounds. These include ***purines***, ***formylmethionine tRNA*** (for initiation of protein synthesis in prokaryotes), ***glycine, pyrimidine nucleotide*** (thymidylate) etc.

III. Role of methionine and B_{12} in one-carbon metabolism

Methyl $(-CH_3)$ group is an important one-carbon unit. The role of active methionine as methyl donor in transmethylation reactions is already described. After the release of methyl group, methionine is converted to homocysteine. For the regeneration of methionine, free homocysteine and N^5-methyl THF are required and this reaction is dependent on methyl-cobalamin (vitamin B_{12}). The one-carbon pool, under the control of THF, is linked with methionine metabolism (transmethylation) through vitamin B_{12}. Hence vitamin B_{12} is also involved in one-carbon metabolism.

BRANCHED CHAIN AMINO ACIDS

Valine, leucine and ***isoleucine*** are the branched chain and ***essential amino acids***. These three amino acids initially undergo a common pathway and then diverge to result in different end products. Based on the products obtained from the carbon skeleton, the branched chain amino acids are either glycogenic or ketogenic

Valine – ***glycogenic***

Leucine – ***ketogenic***

Isoleucine – ***glycogenic and ketogenic.***

The first three metabolic reactions are common to the branched chain amino acids (***Fig.15.33***).

1. **Transamination :** The three amino acids undergo a reversible transamination to form their respective keto acids.

2. **Oxidative decarboxylation : *α-Keto acid dehydrogenase*** is a complex mitochondrial enzyme. It is comparable in function to pyruvate dehydrogenase complex and employs 5 coenzymes—TPP, lipoamide, FAD, coenzyme A and NAD^+—for its action. α-Keto acid dehydrogenase catalyses oxidative decarboxylation of the keto acids to the corresponding acyl CoA thioesters. This is a regulatory enzyme in the catabolism of branched chain amino acids.

3. **Dehydrogenation :** The dehydrogenation is similar to that in fatty acid oxidation. FAD is the coenzyme and there is an incorporation of a double bond. It is now believed that there are two enzymes responsible for dehydrogenation.

After the initial three common reactions, the metabolism of branched chain amino acids diverges and takes independent routes. In a

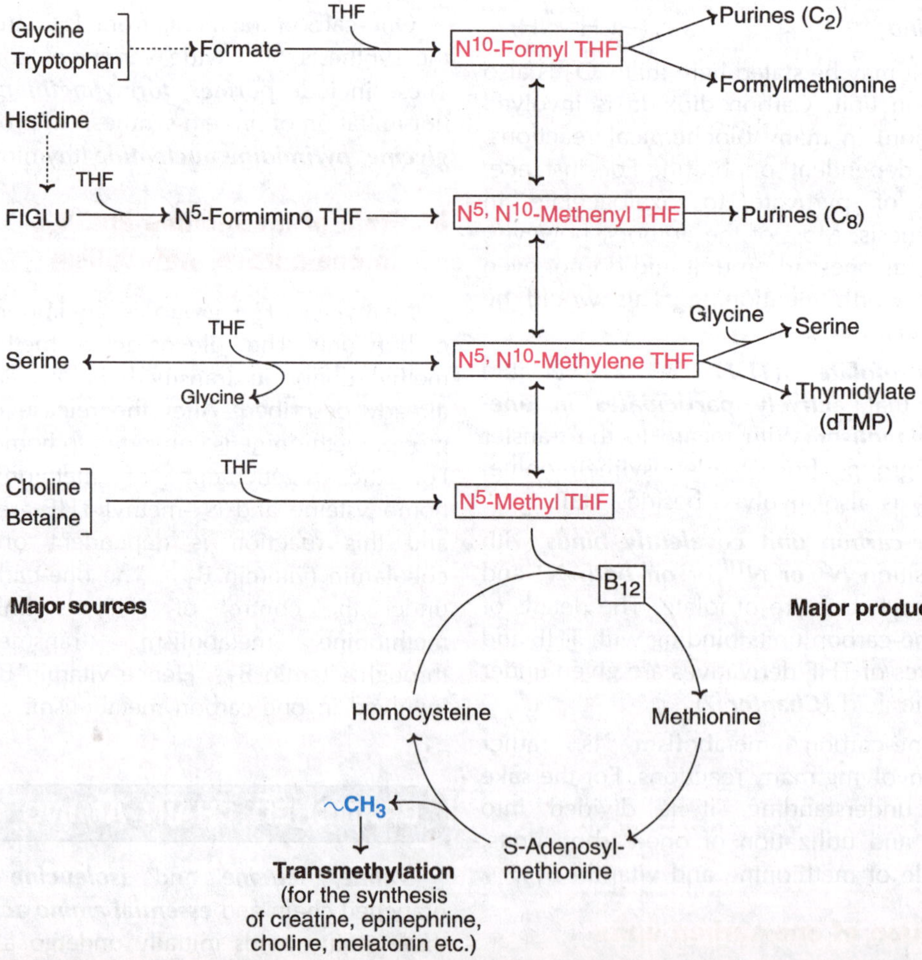

Fig. 15.32 : *Summary of one-carbon metabolism (THF–Tetrahydrofolate; $\sim CH_3$–Active methyl group).*

series of reactions that follow, valine is converted to propionyl CoA, a precursor for glucose. Leucine produces acetyl CoA and acetoacetate, the substrates for fatty acid synthesis. Isoleucine is degraded to propionyl CoA and acetyl CoA. Thus, valine is glycogenic and leucine is ketogenic while isoleucine is both glycogenic and ketogenic.

Metabolic defects of branched chain amino acids

1. **Maple syrup urine disease (MSUD) :** This is a metabolic disorder of branched chain amino acids. The ***urine*** of the affected individuals ***smells*** like maple syrup or ***burnt sugar***—hence the name.

Enzyme defect : Maple syrup urine disease is due to a defect in the enzyme ***branched chain α-keto acid dehydrogenase.*** This causes a blockade in the conversion of α-keto acids to the respective acyl CoA thioesters. The plasma and urine concentrations of branched amino acids and their keto acids are highly elevated. This disease is also known as ***branched chain ketonuria.***

Biochemical complications and symptoms :

- Accumulation of branched chain amino acids causes an ***impairment in transport and function of other amino acids.***
- Protein biosynthesis is reduced.
- Branched chain amino acids competitively inhibit glutamate dehydrogenase.
- The disease results in ***acidosis, lethargy, convulsions,*** mental retardation, coma and, finally, death within one year after birth.

Diagnosis and treatment : An early diagnosis by enzyme analysis—preferably within the first week of life—is ideal. Estimation of urinary branched amino acids and keto acids will also help in diagnosis.

The treatment is to feed a diet with low (or no) content of branched amino acids. Mild variants of MSUD respond to high doses of thiamine. In severe cases of MSUD, liver transplantation is required.

2. **Intermittent branched chain ketonuria :** This is a less severe variant form of maple syrup urine disease. The enzyme defect is the same—***α-keto acid dehydrogenase.*** As such, there is an impairment and no total blockade in the conversion of α-keto acids to their respective

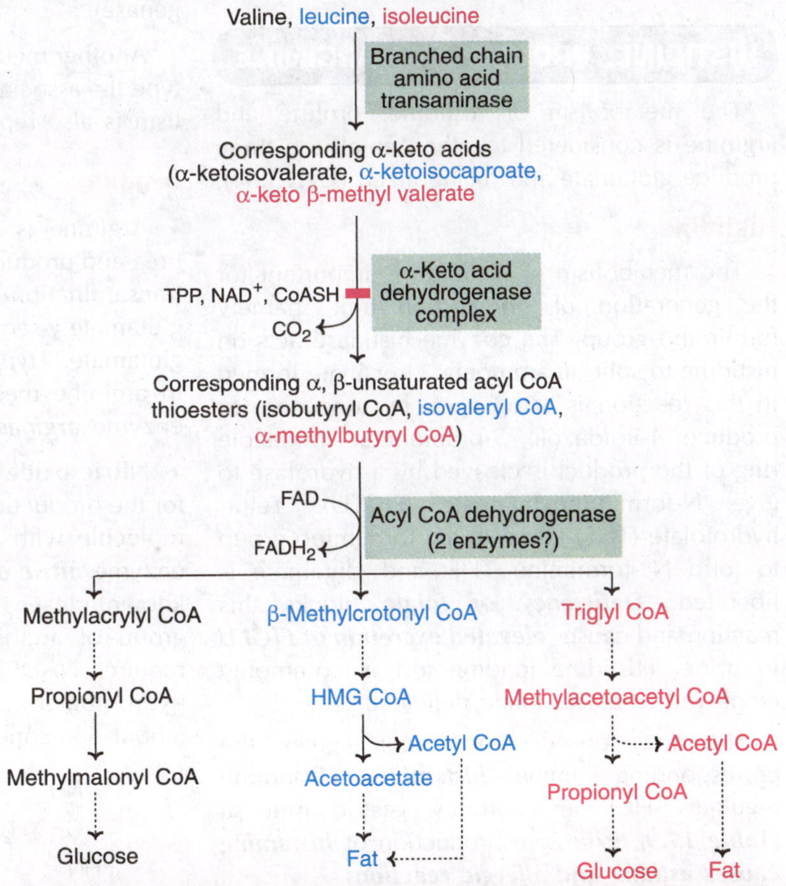

***Fig. 15.33 :** Summary of branched chain amino acid metabolism (▬ –Defect in α-keto acid dehydrogenase results in maple syrup urine disease; HMG CoA–β-Hydroxy β-methyl-glutaryl CoA; Valine is glycogenic, leucine is ketogenic, while isoleucine is both).*

acyl CoA thioesters. The symptoms are not as severe as in MSUD. Careful diet planning is adequate to overcome this disorder.

3. **Isovaleric acidemia :** This is a specific inborn error of leucine metabolism. It is due to a defect in the enzyme ***isovaleryl CoA dehydrogenase.*** The conversion of isovaleryl CoA to methylcrotonyl CoA is impaired. The excretion of isovalerate is high in urine. The affected individuals exhibit a 'cheesy' odor in the breath and body fluids. The symptoms include acidosis and mild mental retardation.

4. **Hypervalinemia :** This inborn error is characterized by increased plasma concentration of valine while leucine and isoleucine levels remain normal. The transamination of valine alone is selectively impaired.

HISTIDINE, PROLINE AND ARGININE

The metabolism of histidine, proline and arginine is considered together, as all the three produce glutamate and metabolized (***Fig.15.34***).

Histidine

The metabolism of histidine is important for the generation of ***one-carbon*** unit, namely formimino group. The enzyme histidase acts on histidine to split off ammonia. Urocanate formed in this reaction is acted upon by urocanase to produce 4-imidazole 5-propionate. Imidazole ring of the product is cleaved by a hydrolase to give N-formiminoglutamate (FIGLU). Tetrahydrofolate (THF) takes up the formimino group to form N^5-formimino THF, and glutamate is liberated. ***Deficiency of folate*** blocks this reaction and causes ***elevated excretion of FIGLU*** in urine. Histidine loading test is commonly employed to assess folate deficiency.

Histidine, on decarboxylation, gives the corresponding amine—***histamine.*** Histamine regulates HCl secretion by gastric mucosa (***Table 15.9***). Excessive production of ***histamine causes asthma and allergic reactions.***

Antihistamines are drugs used to treat allergies, gastric ulcers (e.g. omeprazole, ranitidine, cimetidine). Antihistamines bind to histamine receptors and block histamine action.

Histidinemia : The frequency of histidinemia is 1 in 20,000. It is due to a defect in the enzyme ***histidase.*** Histidinemia is characterized by elevated plasma histidine levels and increased excretion of imidazole pyruvate and histidine in urine. Most of the patients of histidinemia are mentally retarded and have defect in speech.

Proline

Proline is oxidized to pyrroline 5-carboxylate which undergoes a non-enzymatic conversion to glutamate 5-semialdehyde. The latter is converted to glutamate and then transaminated to α-ketoglutarate. The five carbons of proline are converted to α-ketoglutarate.

Hyperprolinemia type I : It is due to a defect in the enzyme ***proline oxidase*** (proline dehydrogenase).

Another metabolic disorder—hyperprolinemia type II—associated with hydroxyproline metabolism is also reported.

Arginine

Arginine is cleaved by arginase to liberate urea and produce ornithine. Ornithine undergoes transamination of δ-amino group to form glutamate γ-semialdehyde which is converted to glutamate. ***Hyperargininemia*** is an inborn error in arginine metabolism due to a defect in the enzyme ***arginase.***

Nitric oxide (NO) : Arginine is the substrate for the production of nitric oxide (NO), a wonder molecule with a wide range of functions. The enzyme ***nitric oxide synthase*** (three isoenzymes known) cleaves the nitrogen from the guanidino group of arginine to form NO. This reaction requires NADPH, FMN, FAD, heme and tetrahydrobiopterin. NO has a very short half-life (about 5 seconds).

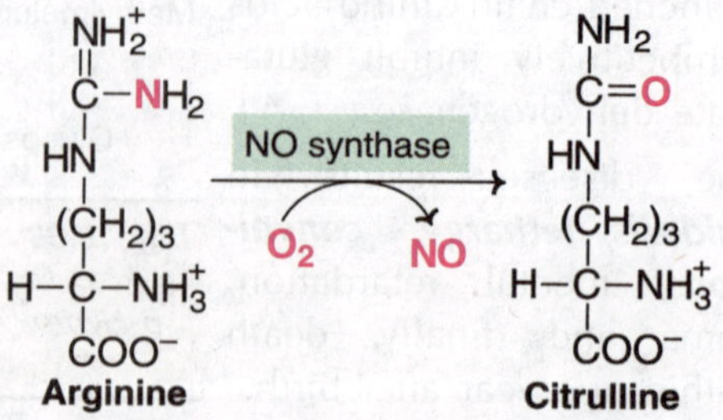

The occurrence of high concentrations of citrulline in human brain has been known for several years. Only recently it is realized that the citrulline is formed during the course of NO synthesis [**Note :** Nitric oxide (NO) should not be confused with nitrous oxide (NO_2)—laughing gas—used as an anesthetic].

Functions of NO : The role of nitric oxide as a therapeutic drug (in the form of nitroglycerine and amyl nitrate) for the treatment of ***angina pectoris*** has been known since 1867. However, it is only recently that *in vivo* production and the biological importance of NO are recognized. In fact, the ***endothelium derived releasing factor***

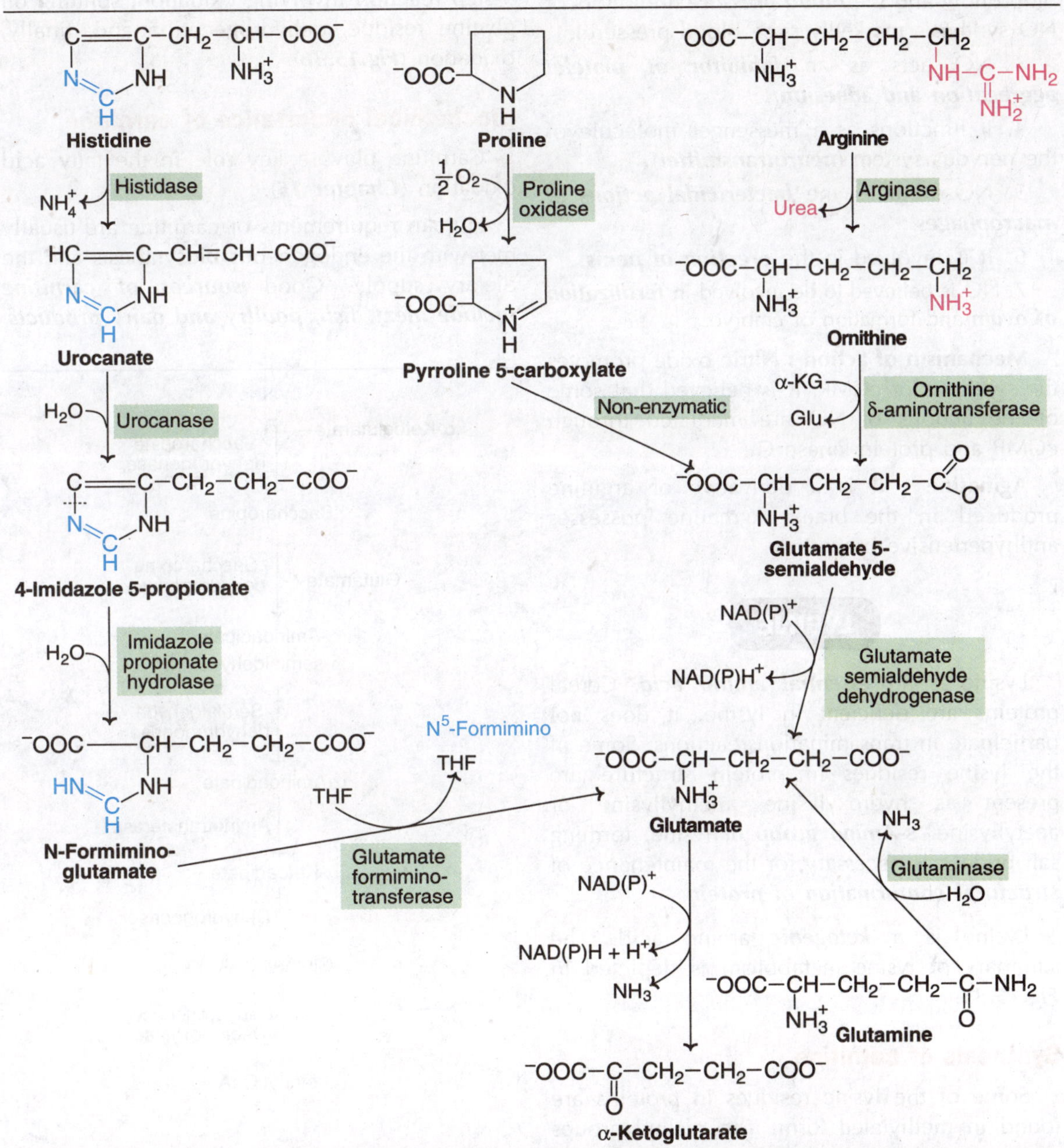

Fig. 15.34 : *Metabolism of histidine, proline, arginine, glutamate and glutamine (THF–Tetrahydrofolate; α-KG–α-Ketoglutarate; Glu–Glutamate).*

(EDRF) which causes smooth muscle relaxation is none other than a gas, NO.

Nitric oxide acts as a mediator for several biological functions.

1. NO functions as a vasodilator and causes ***relaxation of smooth muscles***.

2. It is a key molecule in the ***regulation of*** blood flow and the ***blood pressure*** (inhibitors of NO synthesis markedly raise blood pressure).

3. NO acts as an ***inhibitor of platelet aggregation and adhesion.***

4. It functions as a messenger molecule of the nervous system (***neurotransmitter***).

5. NO mediates the ***bactericidal actions of macrophages***.

6. It is involved in the ***erection of penis***.

7. NO is believed to be involved in ***fertilization of ovum*** and formation of embryo.

Mechanism of action : Nitric oxide promotes the synthesis of cGMP. It is believed that some of the actions of NO are mediated through cGMP and protein kinase G.

Agmatine : It is a derivative of arginine produced in the brain. Agmatine possesses antihypertensive properties.

LYSINE

Lysine is an ***essential amino acid***. Cereal proteins are deficient in lysine. It does not participate in transamination reactions. Some of the lysine residues in protein structure are present as hydroxylysine, methyllysine or acetyllysine. ***ε-Amino group*** of lysine, forming salt bridges is necessary for the maintenance of ***structural conformation of protein***.

Lysine is a ***ketogenic*** amino acid. The summary of lysine metabolism is depicted in ***Fig.15.35***.

Synthesis of carnitine

Some of the lysine residues in proteins are found in methylated form. The methyl groups are obtained from active methionine (SAM). Such ***proteins on degradation*** (by proteolysis) will release the methyllysines. The ***trimethyllysine*** serves as a ***precursor*** for the synthesis of ***carnitine***, a compound ***involved in the transport of fatty acids to mitochondria for oxidation***. It should be noted that free lysine is not methylated, hence it will not be a substrate for carnitine formation.

Synthesis of carnitine from trimethyllysine is a 4-step reaction involving oxidation, splitting off glycine residue, dehydrogenation and, finally, oxidation (***Fig.15.36***).

Biochemical importance of carnitine

Carnitine plays a key role in the fatty acid oxidation (***Chapter 15***).

Human requirements of carnitine are usually met with the endogeneous biosynthesis and the dietary supply. Good ***sources of carnitine include meat, fish, poultry and dairy products.***

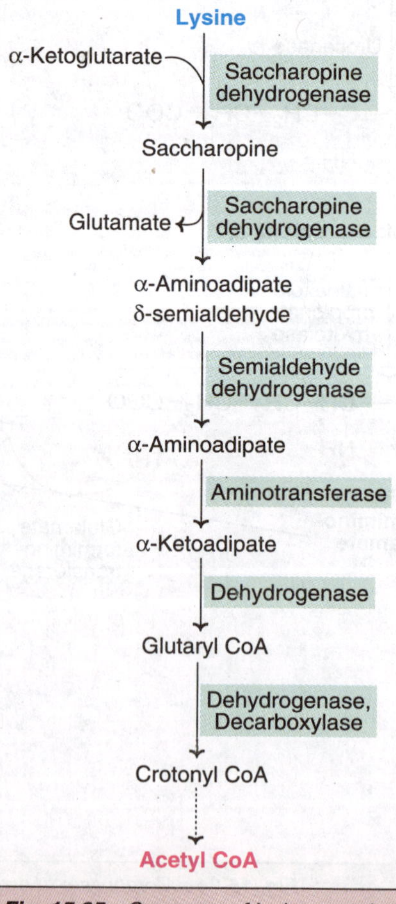

Fig. 15.35 : *Summary of lysine metabolism.*

Some research findings suggest that carnitine supplementation has some beneficial effects in the treatment of myocardial dysfunctions, AIDS, etc.

GLUTAMATE AND GLUTAMINE

Glutamate and glutamine are ***non-essential glycogenic*** amino acids. Both of them play a predominant role in the amino acid metabolism, and are directly involved in the final transfer of amino group for urea synthesis.

The amino acids—histidine, proline and arginine—are converted in their metabolism to glutamate (***See Fig.15.34***). α-Ketoglutarate—an intermediate in TCA cycle—serves as an immediate precursor for glutamate formation.

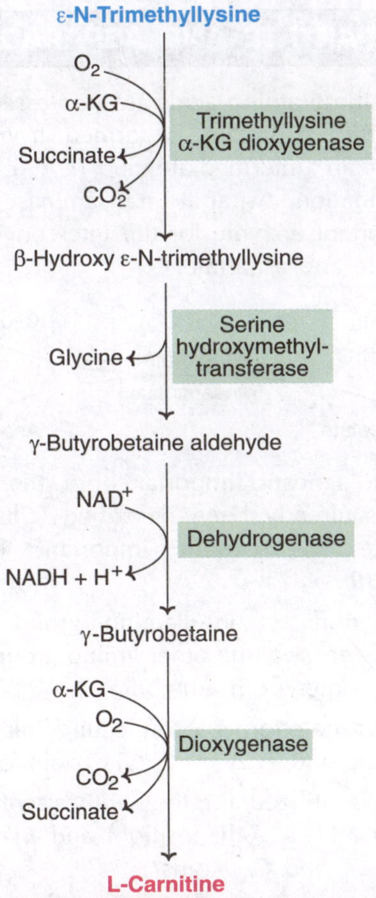

***Fig. 15.36 :** Synthesis of carnitine (α-KG—α-Ketoglutarate).*

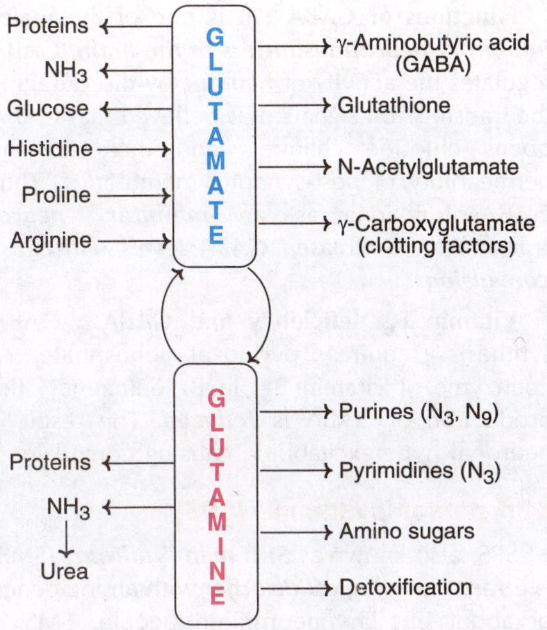

***Fig. 15.37 :** Overview of glutamate and glutamine metabolism.*

Glutamate—besides being converted to glutamine—is involved in the synthesis of certain specialized products (***Fig.15.37***).

1. **Glutathione** is a tripeptide that contains glutamate. Its formation is described under glycine metabolism.

2. **N-Acetylglutamate** is an allosteric regulator of carbamoyl phosphate synthase I, the first enzyme in urea synthesis.

3. Glutamate is present in the ***clotting factors*** (II, VII, IX, X) as γ-carboxyglutamate and is involved in coagulation.

4. **γ-Aminobutyric acid (GABA) :** Glutamate is decarboxylated to GABA—mostly in the brain. This is a sensitive and irreversible reaction, ***catalysed by a pyridoxal phosphate-dependent glutamate decarboxylase***.

GABA undergoes transamination followed by oxidation to form succinate which enters TCA cycle (***Fig.15.38***). The reactions involving the fate of GABA constitute a bypass route for glutamate to enter TCA cycle, which is known as ***GABA shunt.***

Functions of GABA : It is one of the major ***inhibitory neurotransmitters in the brain***. GABA regulates the activity of neurons by discouraging the transmission signals. It is believed that GABA opens chloride channels and increases the permeability of post-synaptic membranes. Thus GABA functions as an ***inhibitory neuro-transmitter. Decreased GABA levels will cause convulsions***.

Vitamin B_6 deficiency and GABA : GABA synthesis requires pyridoxal phosphate, a coenzyme of vitamin B_6. In B_6 deficiency, the production of GABA is ***reduced***. The result is neuronal hyperexcitability, causing convulsions.

Stiff person syndrome (SPS)

SPS, also known as ***Stiff man syndrome*** (SMS) is a ***rare neurological disorder*** with an incidence of about one in one million people. SMS is characterized by ***muscular rigidity***, ***stiffness*** and ***painful muscle spasms***. Biochemically, SMS appears to be due to ***impairment in the synthesis and function of GABA***. This occurs due to high amounts of glutamic acid decarboxylase (GAD) antibodies. GAD antibodies are found to inhibit GAD, and thereby reduce the synthesis of GABA.

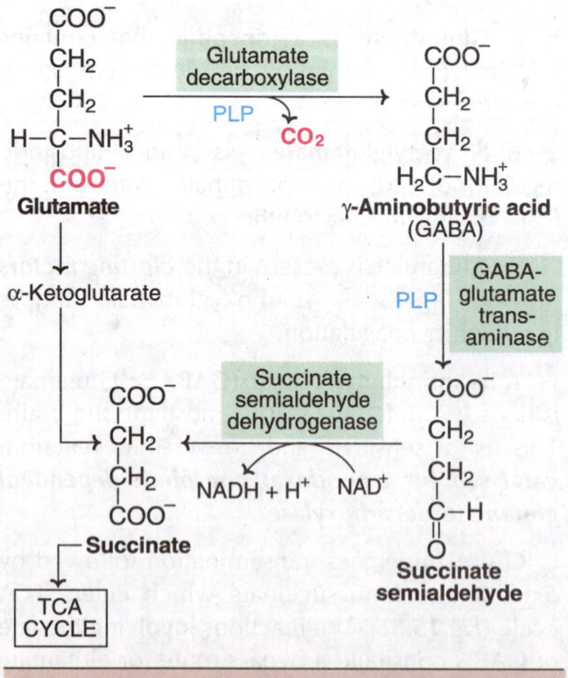

Fig. 15.38 : *Metabolism of γ-aminobutyric acid (PLP–Pyridoxal phosphate).*

Glutamine

Glutamine is a versatile amino acid. ***Ammonia*** is temporarily ***stored in the form of glutamine***. Glutamine is freely diffusible and, hence, easily transported. The synthesis and degradation of glutamine are described (***See Fig.15.8***).

Glutamine is the donor of nitrogen atoms for purine and pyrimidine synthesis. It is the chief source of ammonia in kidneys. The NH_3 production is elevated in acidosis to maintain ***acid-base balance.*** Glutamine also takes part in ***conjugation reactions.***

ASPARTATE AND ASPARAGINE

Both these amino acids are ***non-essential*** and ***glycogenic***. Aspartate is formed from oxalo-acetate (an intermediate in TCA cycle) by transamination. Aspartate transaminase (AST) is an important enzyme for the interconversion of glutamate and aspartate.

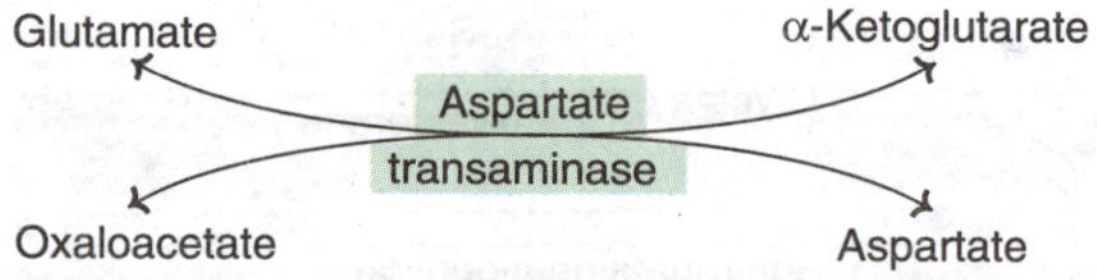

The diagnostic importance of the enzyme AST has already been described (***Chapter 6***). Aspartate has certain important functions (***Fig.15.39***).

1. It donates one amino group for the ***synthesis of urea*** (the other amino group in urea directly comes from ammonia).

2. Aspartate forms a connecting link between urea cycle and TCA cycle (via oxaloacetate).

3. It is utilized for the synthesis of ***purines*** (N_1 and NH_2 at 6th position) and ***pyrimidines*** (N_3, C_4 C_5 and C_6 atoms).

4. ***Malate-aspartate shuttle*** is important for the transfer of reducing equivalents (NADH) from the cytosol to mitochondria (***Refer Fig.11.13***).

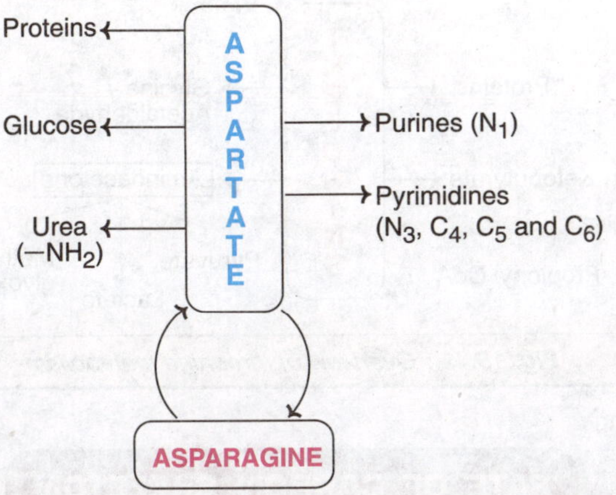

Fig. 15.39 : Overview of aspartate and asparagine metabolism.

Asparagine is synthesized from aspartate by a synthetase in an irreversible ATP-dependent reaction. Asparaginase hydrolyses asparagine and liberates ammonia (***Fig.15.40***). These reactions are comparable to glutamine synthesis and its breakdown.

ALANINE

The ***non-essential*** amino acid alanine performs two important functions in the body—incorporation into proteins and participation in transamination and NH_3 transport.

As already discussed, ammonia is toxic to the body, hence it cannot be transported in free form. Glutamate and glutamine shoulder the major burden of ammonia transport. Alanine is also important in this regard. In the peripheral tissues (most predominantly—muscle), pyruvate produced in glycolysis gets converted to alanine (by transamination) and is transported to liver. Pyruvate can be regenerated from alanine in liver and the pyruvate so produced serves as a precursor for glucose. Amino group is diverted for transamination or urea formation. This is an ***alanine-pyruvate shuttle for carrying nitrogen*** to be reutilized or converted to urea.

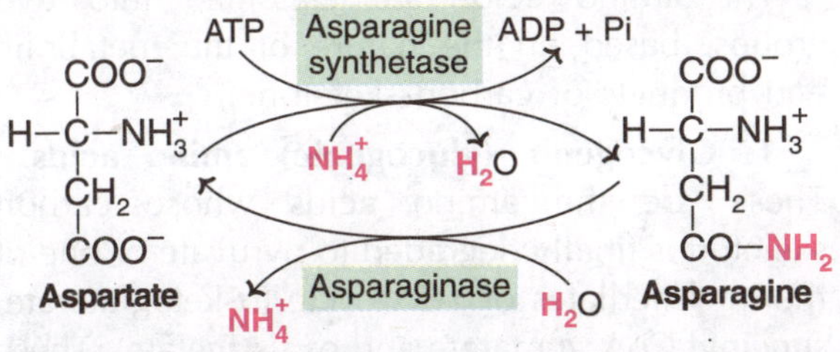

*Fig. 15.40 : Synthesis of asparagine and its conversion to aspartate (**Note :** The reactions are independent and irreversible).*

It is recently reported that people with higher levels of ***alanine in urine*** have increased risk for ***higher blood pressure***.

The amino acid ***β-alanine*** is a constituent of the vitamin pantothenic acid, and thus the ***coenzyme A***.

SERINE

Serine is a ***non-essential glycogenic amino acid***. As described in glycine metabolism, ***serine and glycine are interconvertible***. Serine can be synthesized from the intermediates of glycolysis (3-phosphoglycerate). The metabolic reactions of serine are described hereunder (***Fig.15.41***)

1. Serine undergoes deamination to form pyruvate.

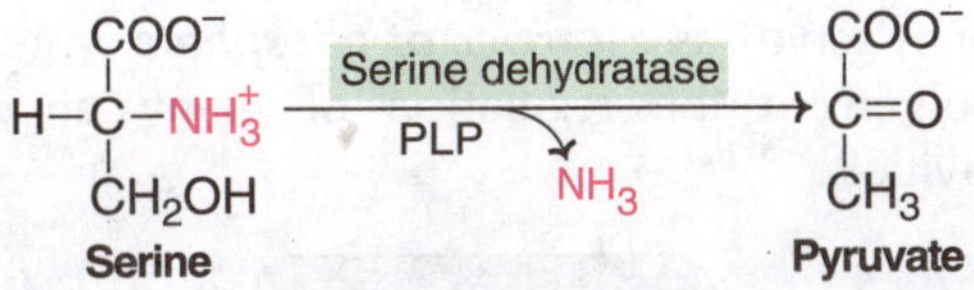

2. Serine is involved in ***one-carbon metabolism***. It donates methylene ($-CH_2$) moiety to tetrahydrofolate (THF).

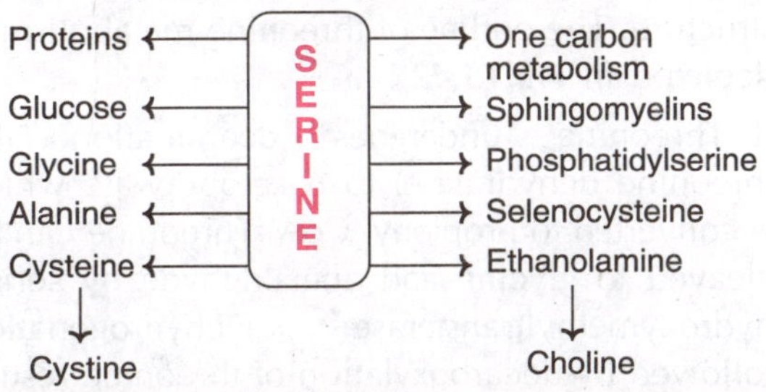

Fig. 15.41 : Overview of serine metabolism.

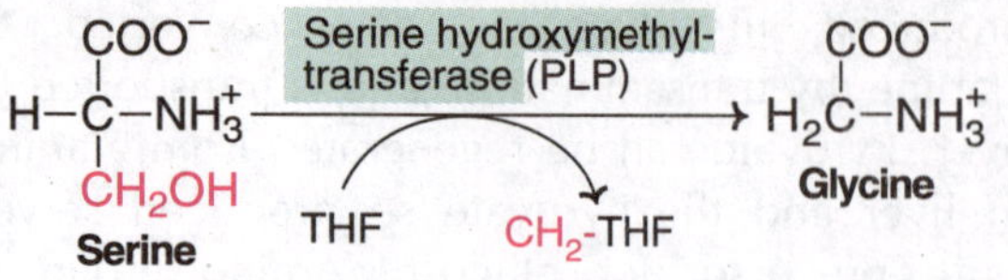

3. On decarboxylation (PLP-dependent) serine forms ***ethanolamine*** which is the precursor ***for choline synthesis***.

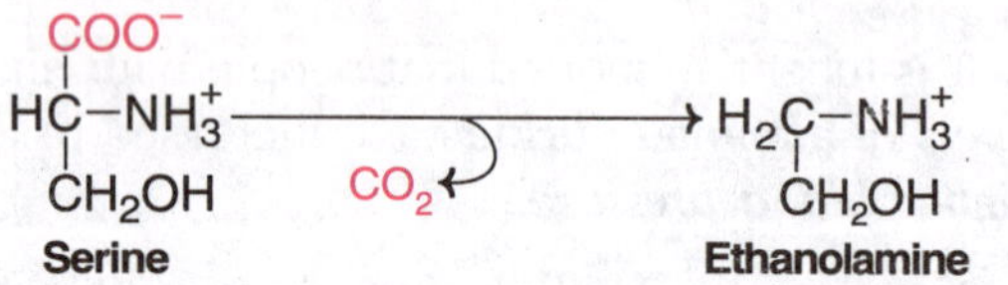

4. Serine is utilized for the ***synthesis of cysteine*** (***See Fig.15.30***). It may be noted that the entire cysteine molecule is derived from serine except the sulfur that comes from homocysteine.

5. Serine is involved in the formation of ***selenocysteine***, the 21st amino acid found in certain proteins.

6. Serine directly participates in the synthesis of phospholipid-phosphatidyl serine (details described in lipid metabolism, ***Chapter 14***).

7. Serine is also involved in the synthesis of ***sphingomyelins*** and ***cephalins***.

8. In the ***structure of proteins***, serine (–OH group) serves as a ***carrier of phosphate*** which is involved in the regulation of many enzyme activities.

THREONINE

Threonine is an ***essential hydroxy amino acid***. It is ***glycogenic*** and does not participate in transamination reactions. Threonine is often a carrier of phosphate group in the protein structure. The outline of threonine metabolism is depicted in ***Fig.15.42***.

Threonine undergoes deamination (by threonine dehydratase) to α-ketobutyrate which is converted to propionyl CoA. Threonine can be cleaved to glycine and acetaldehyde by serine hydroxymethyltransferase. Dehydrogenation followed by decarboxylation of threonine results in aminoacetone which may be converted to pyruvate or lactate.

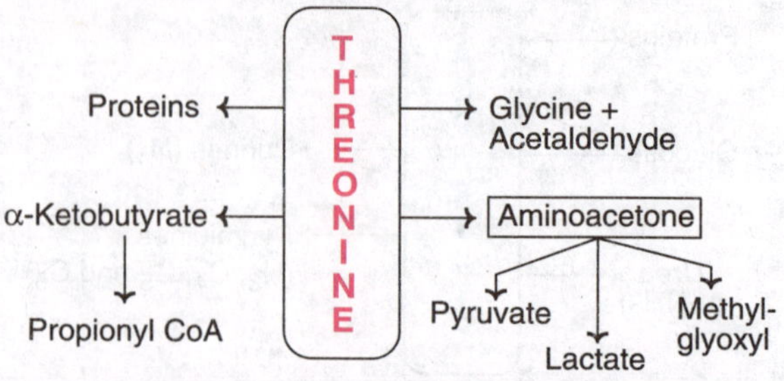

Fig. 15.42 : Overview of threonine metabolism.

FATE OF CARBON SKELETON OF AMINO ACIDS

The metabolic reactions of individual amino acids are described above. After the removal of amino groups, the carbon skeleton of amino acids is converted to intermediates of TCA cycle or their precursors. The ***carbon skeleton*** finally has one or more of the following ***fates***

1. Oxidation via TCA cycle to produce ***energy*** (about 10-15% of body needs).
2. Synthesis of ***glucose***.
3. Formation of lipids—***fatty acids and ketone bodies***.
4. Synthesis of ***non-essential amino acids***.

The carbon skeletons of the 20 standard (or more) amino acids (or the amino acids of proteins) are degraded to one of the following seven products—***pyruvate***, ***α-ketoglutarate***, ***succinyl CoA***, ***fumarate***, ***oxaloacetate***, ***acetyl CoA*** and ***acetoacetate***. Some authors use the term amphibolic (*Greek*: amphiboles—uncertain) intermediates to these compounds due to their multiple metabolic functions.

The amino acids are classified into two groups, based on the nature of the metabolic end products of carbon skeleton.

1. **Glycogenic (glucogenic) amino acids :** These are the amino acids whose carbon skeleton is finally degraded to pyruvate or one of the intermediates of TCA cycle (α-ketoglutarate, succinyl CoA, fumarate and oxaloacetate). These intermediates serve as good substrates for gluconeogenesis leading to the ***formation of glucose*** or glycogen.

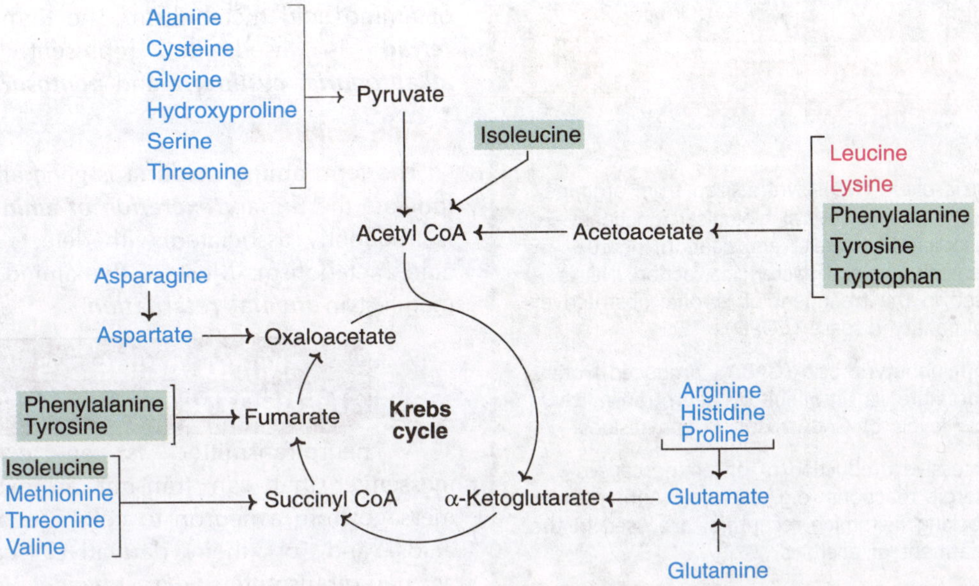

Fig. 15.43 : *Summary of the products formed from carbon skeleton of amino acids (colour indication, Blue – glucogenic; Green shade – glucogenic and ketogenic; Red – ketogenic).*

2. **Ketogenic amino acids :** The amino acids whose carbon skeleton is ***metabolized to acetyl CoA or acetoacetate*** can be ***converted to fat*** (i.e., fatty acids or ketone bodies). Acetoacetate is a ketone body (besides acetone and β-hydroxybutyrate).

Some of the amino acids are both glycogenic and ketogenic since they serve as precursors for glucose as well as fat.

The classification of amino acids (glycogenic, ketogenic, or both) is given in ***Table 15.5***. The various products obtained from the carbon skeleton of amino acids and their connection with the citric acid cycle is depicted in ***Fig.15.43***.

The details on the formation of amphibolic intermediates by the degradation of amino acids are given in the metabolism of respective amino acids. They are summarized hereunder

- **Pyruvate :** Alanine, cysteine, glycine, hydroxyproline, serine and threonine.
- **α-Ketoglutarate :** Glutamine, glutamate, arginine, histidine and proline.
- **Succinyl CoA :** Isoleucine, methionine, threonine and valine.

Table 15.5 Classification of amino acids based on the fate of carbon skeleton

Glycogenic (glucogenic)	*Glycogenic and ketogenic*	*Ketogenic*
Alanine	Phenylalanine*	Leucine*
Arginine*	Isoleucine*	Lysine*
Aspartate	Tyrosine	
Cysteine	Tryptophan*	
Glutamine		
Glutamate		
Glycine		
Histidine*		
Hydroxyproline		
Methionine*		
Proline		
Serine		
Threonine*		
Valine*		

* *Essential amino acids; (Helpful tips to recall—ketogenic amino acids start with letter 'L'; PITT for glyco- and ketogenic amino acids; rest of the 20 amino acids are only glycogenic).*

Medical Concepts/ Clinical Correlates

- Nitric oxide (NO), synthesized from arginine, is involved in several biological functions—vasodilation, platelet aggregation, neurotransmission and bactericidal action. NO is used in the treatment of chronic obstructive pulmonary disease (COPD).
- γ-Aminobutyric acid (GABA), produced from glutamate, is an inhibitory neurotransmitter. Low levels of GABA result in convulsions.
- Excessive production of histamine causes allergic reactions (e.g. asthma). Antihistamines, blocking histamine receptors, are used in the treatment of allergies.
- Polyamines (spermine, putrescine) are involved in the synthesis of DNA, RNA and proteins and, thus, they are essential for cell growth and differentiation.

- **Fumarate :** Phenylalanine and tyrosine.
- **Oxaloacetate :** Asparagine and aspartate.
- **Acetyl CoA and acetoacetate :** Phenylalanine, tyrosine, tryptophan, isoleucine, leucine and lysine.

Leucine and lysine are only ketogenic, since they produce acetoacetate or acetyl CoA.

Biosynthesis of non-essential amino acids

Of the ***20*** amino acids, ***about half of them*** are non-essential in the diet, as they ***can be synthesized*** in human body. This is carried out by the biosynthesis of carbon skeleton, followed by the addition of amino group via transamination. In the ***Table 15.6***, the sources of carbon skeleton for the synthesis of non-essential amino acids are given.

Inborn errors of amino acid metabolism—a summary

The details of amino acid metabolic disorders are described in the respective amino acids. ***Table 15.7*** gives a summary of the inborn errors of amino acid metabolism. The term ***Garrod's tetrad*** is used to represent ***albinism, alkaptonuria, cystinuria*** and ***pentosuria***.

Amino aciduria

The term amino aciduria is generally used to indicate the urinary ***excretion of amino acids***. It is frequently associated with defects in amino acid metabolism. Most of the amino acidurias manifest in ***mental retardation***.

AMINO ACIDS AS NEUROTRANSMITTERS

A neurotransmitter is an extracellular messenger that can transmit an extracellular message from a neuron to cells. Certain amino acids and or their derivatives can serve as neurotransmitters e.g. ***glycine***, ***glutamate***, ***serotonin***, ***GABA*** (***Table 15.8***).

BIOGENIC AMINES

In general, the ***decarboxylation*** of amino acids or their derivatives results in the formation of amines.

$$\underset{\text{Amino acid}}{R{-}\overset{NH_2}{\overset{|}{C}}H{-}COOH} \xrightarrow[\searrow CO_2]{\text{Decarboxylase (PLP)}} \underset{\text{Amine}}{R{-}CH_2{-}NH_2}$$

A summary of biogenic amines and their major functions are given in ***Table 15.9***.

Table 15.6 The sources of carbon skeleton for the biosynthesis of non-essential amino acids

Amino acid	Source(s) of carbon skeleton
Glycine	Serine
Alanine	Pyruvate
Serine	3-Phosphoglycerate
Aspartic acid Asparagine Glutamic acid Glutamine Proline	Intermediates of Krebs cycle
Cysteine	Serine (sulfur donated by methionine)
Tyrosine	Phenylalanine

Table 15.7 Inborn errors of amino acid metabolism

Disorder	Metabolic defect (enzyme/other)
I. Defects in urea synthesis —Refer *Table 15.2*	
II. Phenylalanine and tyrosine – *Figs.15.19* and *15.20*	
1. Phenylketonuria	Phenylalanine hydroxylase
2. Tyrosinemia type II	Tyrosine transaminase
3. Neonatal tyrosinemia	p-Hydroxy phenylpyruvate dioxygenase
4. Alkaptonuria	Homogentisate oxidase
5. Tyrosinosis (tyrosinemia type I)	Maleyl acetoacetate isomerase or fumaryl acetoacetate hydrolase
6. Albinism	Tyrosinase
III. Sulfur amino acids (methionine, cysteine and cystine)	
7. Cystinuria	Defect in renal reabsorption
8. Cystinosis	Impairment in cystine utilization (defect in lysosomal function)
9. Homocystinuria type I	Cystathionine synthetase
10. Homocystinuria type II	N^5, N^{10}-Methylene THF reductase
11. Homocystinuria type III	N^5-Methyl THF-homocysteine methyltransferase
12. Cystathioninuria	Cystathioninase
IV. Glycine	
13. Glycinuria	Defect in renal reabsorption
14. Primary hyperoxaluria	Glycine transaminase
V. Tryptophan	
15. Hartnup's disease	Defective intestinal absorption
VI. Branched chain amino acids (valine, leucine and isoleucine)	
16. Maple syrup urine disease	Branched chain α-keto acid dehydrogenase
17. Intermittent branched chain ketonuria	Variant of the above enzyme (less severe)
18. Hypervalinemia	Valine transaminase
19. Isovaleric acidemia	Isovaleryl CoA dehydrogenase
VII. Histidine	
20. Histidinemia	Histidase
VIII. Proline	
21. Hyperprolinemia type I	Proline oxidase

Table 15.8 Amino acids and their derivatives as neurotransmitters

Amino acid/derivative	Major function(s)
Glycine	Inhibitory neurotransmitter in spinal cord
Glutamate	Excitatory neurotransmitter
Dopamine	Increases blood pressure
Norepinephrine and epinephrine	Hormonal neurotransmitters, increase cardiac output and blood pressure
Serotonin	Regulates cerebral activity and behaviour
γ-Aminobutyric acid (GABA)	Inhibitory neurotransmitter in brain

Table 15.9 A summary of the biogenic amines and their functions

Amino acid	Amine	Function(s)
Serine	Ethanolamine	Forms choline
Glutamate	γ-Aminobutyric acid	Inhibitory neurotransmitter
Histidine	Histamine	Vasodilator, promotes gastric HCl and pepsin synthesis
Phenylalanine	Dopamine	For the synthesis of norepinephrine and epinephrine
Tyrosine	Tyramine	Vasoconstrictor (increases blood pressure)
Tryptophan	Tryptamine	Elevates blood pressure
	Serotonin	Stimulates cerebral activity
	Melatonin	Circadian rhythms
Cysteine	Taurine	Constituent of bile acid (taurocholic acid)

POLYAMINES

Polyamines (*Greek*: poly—many) possess multiple amino groups. ***Putrescine, spermine*** and ***spermidine*** are the biologically important polyamines. Spermine and spermidine were originally detected in human semen (sperms), hence they are so named.

Biosynthesis

Ornithine and S-adenosylmethionine are the precursors for polyamine synthesis. It should, however, be noted that only the ***four-carbon moiety of SAM*** (not the methyl group) is involved in polyamine formation. Ornithine decarboxylase acts on ornithine to split off CO_2 and produce putrescine (***Fig.15.44***). The enzyme ***ornithine decarboxylase has the shortest half-life*** (about 10 minutes) among the known mammalian enzymes. It ***regulates polyamine synthesis***. The activity of this enzyme is increased by hormones like corticosteroids, and growth hormone.

Putrescine is converted to spermidine and then spermine with the involvement of SAM. S-Adenosylmethionine is first decarboxylated to give decarboxylated SAM. SAM decarboxylase is a rare example of an enzyme that does not require pyridoxal phosphate as coenzyme. An amino acid residue bound to pyruvate is believed to function as a cofactor. The propylamino group of decarboxylated SAM is transferred to putrescine to give spermidine. Synthesis of spermine requires one more molecule of decarboxylated SAM, and is catalysed by spermine synthase.

Therapeutic inhibition of polyamine biosynthesis : The drug difluoro methyl ornithine (DFMO) is an irreversible suicide inhibitor of ornithine decarboxylase (ODC). DFMO is used in the treatment of African sleeping sickness (trypanosomiasis), as it can selectively inhibit ODC of the parasite, trypanosoma. In case of humans, by inhibiting ODC and reducing the synthesis of polyamines, cell division and growth can be blocked. This approach, although at an experimental stage, may be useful in the control of growth of cancerous cells.

Functions of polyamines

1. Polyamines are basic in nature and possess multiple positive charges. Hence they are readily ***associated with nucleic acids*** (DNA and RNA).

2. They are involved in the ***synthesis of DNA, RNA and proteins***.

Medical Concepts/ Clinical Correlates

- Dietary supplements containing branched chain amino acids are used by body builders and athletes for boosting muscle growth and performance.
- Screening of all new born babies is advised for detection of inborn errors of metabolism, at an early stage. Early diagnosis is useful to prevent irrepairable damage, including mental retardation.
- Biogenic amines have a wide range of functions. e.g. histamine-inflammatory response and itching, serotonin-modulating mood, learning, memory.
- Urinary excretion of polyamines is increased in all types of cancers.

3. They are essential for ***cell growth and proliferation***.

4. Some enzymes are inhibited by polyamines, e.g. protein kinase.

5. They are believed to be involved in the ***stabilization of the membrane structure*** (cell and cellular organelles).

Clinical importance and polyamines

The ***excretion of polyamines*** is found to be ***elevated in*** almost all types of ***cancers***, e.g. leukemias; carcinoma of lungs, bladder, kidney etc. Diagnostically, putrescine is an ideal marker for cell proliferation whereas spermidine is suitable for the assessment of cell destruction.

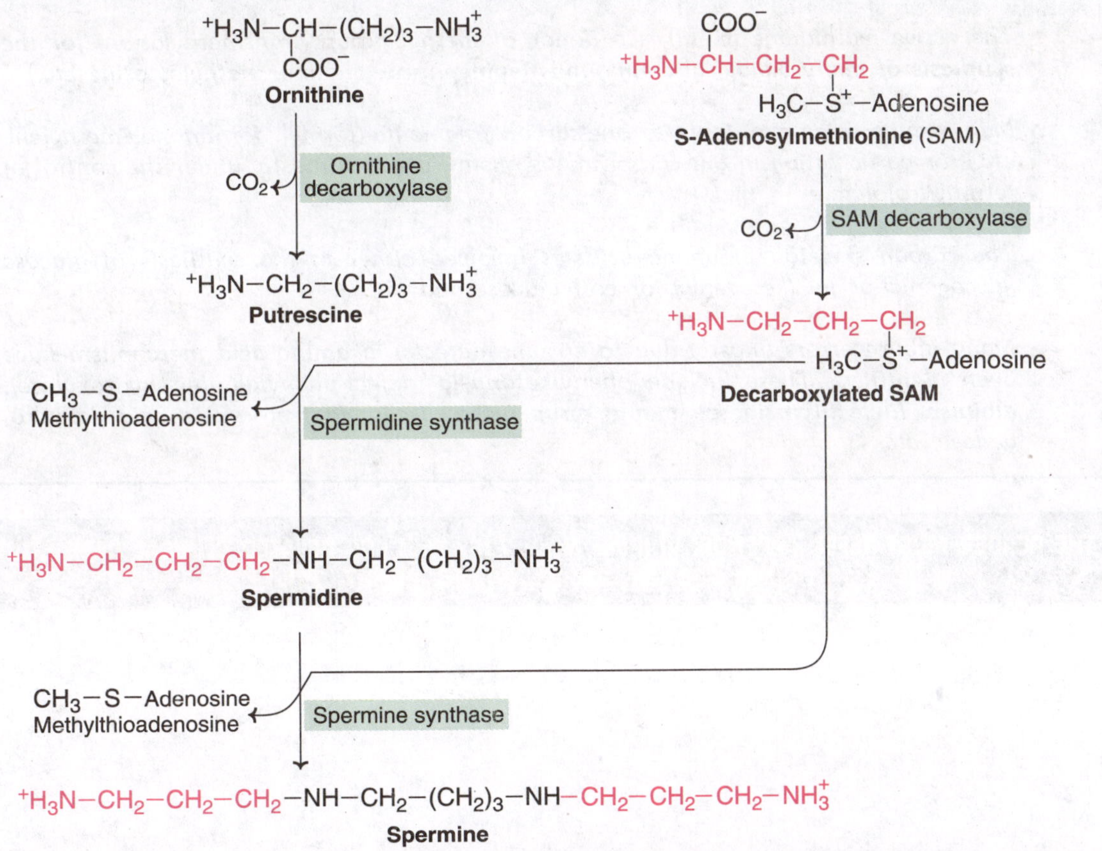

***Fig. 15.44 :** Biosynthesis of polyamines — putrescine, spermidine and spermine.*

SUMMARY

- *The body proteins are in a dynamic state (degradation and synthesis) and there is an active amino acid pool (100 g) maintained for this purpose.*
- *The amino acids undergo transamination and deamination to liberate ammonia for the synthesis of urea, the end product of protein metabolism.*
- *Besides being present as structural components of proteins, amino acids participate in the formation of several biologically important compounds.*
- *Glycine is involved in the synthesis of creatine, heme, purines, glutathione etc.*
- *Phenylalanine is hydroxylated to tyrosine, which is a precursor for the production of skin pigment (melanin), catecholamines (dopamine, epinephrine and norepinephrine) and thyroid hormones (T_3 and T_4).*
- *Tryptophan is converted to NAD^+ and $NADP^+$, the coenzymes of niacin, serotonin (a neurotransmitter) and melatonin.*
- *The active methionine (SAM) is a donor of methyl group (transmethylation) for the synthesis of many biological compounds (epinephrine, choline, methylcytosine etc.).*
- *Many amino acids contribute to one-carbon fragments (formyl, formimino, methylene etc.) for participation in one-carbon metabolism—which is mostly under the control of tetrahydrofolate.*
- *The carbon skeleton of amino acids is involved either in the synthesis of glucose (glycogenic) or fat (ketogenic), or both-glucose and fat.*
- *Many inborn errors (mostly due to enzyme defects) in amino acid metabolism have been identified. These include phenylketonuria (defect-phenylalanine hydroxylase), albinism (defect-tyrosinase), maple syrup urine disease (defect–α-keto acid dehydrogenase) etc.*

THE READER MUST REFER APPENDIX VI (Page Nos : 738 - 756) FOR RELATED CLINICAL CASE STUDIES (Nos : 20, 21, 22, 23, 24, 25, 26, 27, 71)

Self-Assessment Exercises

I. Essay questions

1. Describe the reactions in the synthesis of urea.
2. Give an account of the formation of specialized products from glycine.
3. Discuss the metabolism of phenylalanine and tyrosine.
4. Describe the fate of carbon skeleton of amino acids.
5. Write briefly on various inborn errors of amino acid metabolism.

II. Short notes

(a) Amino acid pool, (b) Transmethylation, (c) Transamination, (d) Deamination, (e) Ammonia toxicity, (f) One-carbon metabolism, (g) Albinism, (h) Serotonin, (i) Glutamate and glutamine, (j) Polyamines.

III. Fill in the blanks

1. The coenzyme that participates in transamination reactions is ____________.
2. The most important enzyme involved in oxidative deamination is ____________.
3. N-Acetylglutamate is required for the activation of the enzyme ____________.
4. Primary hyperoxaluria is due to a defect in the enzyme ____________.
5. The cofactor required for the conversion of phenylalanine to tyrosine is ____________.
6. Parkinson's disease is linked with decreased synthesis of ____________.
7. The metabolite excreted (elevated) in alkaptonuria is ____________.
8. The disease in which very high amount of tryptophan (nearly 60%) is converted to serotonin is ____________.
9. The mammalian enzyme with the shortest half-life (about 10 minutes) is ____________.
10. The branched chain amino acid that is only ketogenic is ____________.

IV. Multiple choice questions

11. The synthesis of urea occurs in
    (a) Kidney (b) Liver (c) Muscle (d) Brain.
12. The amino acid required for the formation of glutathione
    (a) Glycine (b) Cysteine (c) Glutamate (d) All of them.
13. In the synthesis of cysteine, the carbon skeleton is provided by
    (a) Serine (b) Methionine (c) Glutamate (d) Alanine.
14. The amino acids are said to be ketogenic when the carbon skeleton is finally degraded to
    (a) Succinyl CoA (b) Fumarate (c) Acetyl CoA (d) Pyruvate.
15. The amino acid that does not participate in transamination
    (a) Lysine (b) Glutamate (c) Alanine (d) Tryptophan.

Integration of Metabolism

16

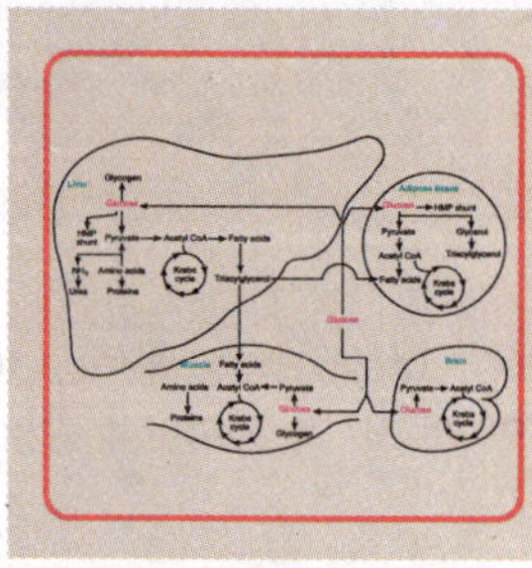

The energy metabolism speaks

"Carbohydrate, fat, protein metabolisms integrate;
Cells, tissues and organs coordinate;
To meet the body's fuel demands;
The essential requisite for existence."

Metabolism is a continuous process, with thousands of reactions, simultaneously occurring in the living cell. However, biochemists prefer to present metabolism in the form of reactions and metabolic pathways. This is done for the sake of convenience in presentation and understanding.

In the preceeding ***three chapters (13-15)***, we have learnt the ***metabolism of carbohydrates, lipids and amino acids***. We shall now consider the organism as a whole and integrate the metabolism with particular reference to energy demands of the body.

Energy demand and supply

The organisms possess variable energy demands, hence the supply (input) is also equally variable. The consumed metabolic fuel may be burnt (oxidized to CO_2 and H_2O) or stored to meet the energy requirements as per the body needs. ***ATP*** serves as the ***energy currency of the cell*** in this process (***Fig.16.1***).

The humans possess enormous capacity for food consumption. It is estimated that one can consume as much as 100 times his/her basal requirements! Obesity, a disorder of over-nutrition mostly prevalent in affluent societies, is primarily a consequence of overconsumption.

Integration of major metabolic pathways of energy metabolism

An overview of the interrelationship between the important metabolic pathways, concerned

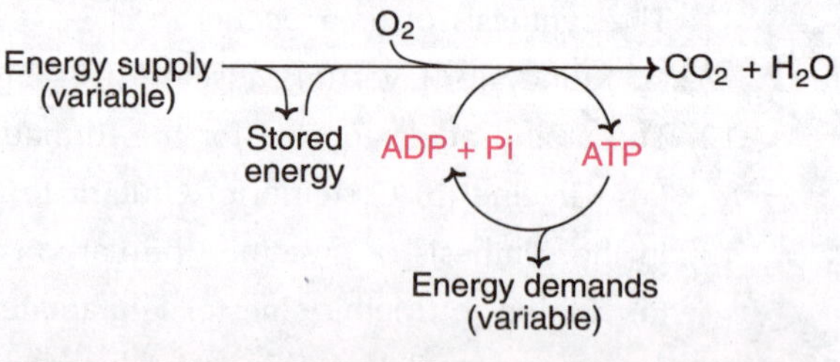

***Fig. 16.1 :** A summary of body's energy supply and demands. (**Note :** ATP serves as the energy currency).*

with fuel metabolism depicted in ***Fig.16.2***, is briefly described here. For detailed information on these metabolic pathways, the reader must refer the respective chapters.

1. **Glycolysis :** The degradation of glucose to pyruvate (lactate under anaerobic condition) generates 8 ATP. Pyruvate is converted to acetyl CoA.

2. **Fatty acid oxidation :** Fatty acids undergo sequential degradation with a release of 2-carbon fragment, namely acetyl CoA. The energy is trapped in the form of NADH and $FADH_2$.

3. **Degradation of amino acids :** Amino acids, particularly when consumed in excess than required for protein synthesis, are degraded and utilized to meet the fuel demands of the body. The glucogenic amino acids can serve as precursors for the synthesis of glucose via the formation of pyruvate or intermediates of citric acid cycle. The ketogenic amino acids are the precursors for acetyl CoA.

4. **Citric acid cycle :** Acetyl CoA is the key and common metabolite, produced from different fuel sources (carbohydrates, lipids, amino acids). ***Acetyl CoA*** enters citric acid (Krebs) cycle and gets oxidized to CO_2. Thus, citric acid cycle is the ***final common metabolic pathway for the oxidation of all foodstuffs.*** Most of the energy is trapped in the form of NADH and $FADH_2$.

5. **Oxidative phosphorylation :** The NADH and $FADH_2$, produced in different metabolic pathways, are finally oxidized in the electron transport chain (ETC). The ETC is coupled with oxidative phosphorylation ***to generate ATP.***

6. **Hexose monophosphate shunt :** This pathway is primarily concerned with the liberation of NADPH and ribose sugar. NADPH is utilized for the biosynthesis of several compounds, including fatty acids. Ribose is an essential component of nucleotides and nucleic acids (***Note :*** DNA contains deoxyribose).

7. **Gluconeogenesis :** The synthesis of glucose from non-carbohydrate sources constitutes gluconeogenesis. Several compounds (e.g. pyruvate, glycerol, amino acids) can serve as precursors for gluconeogenesis.

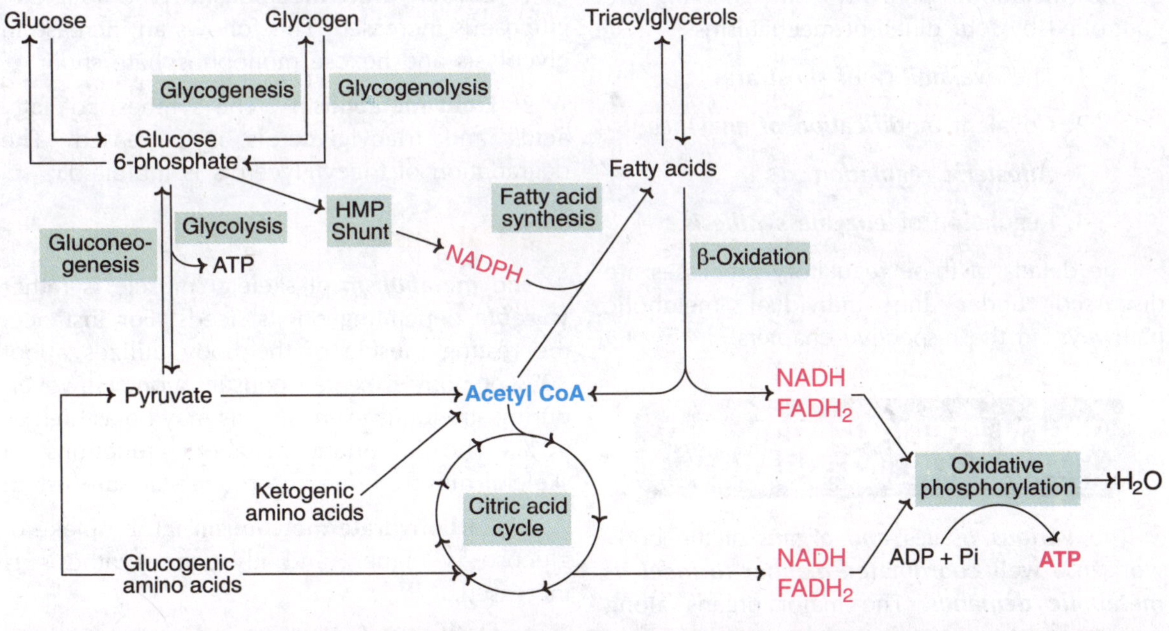

Fig. 16.2 : *An overview of integration of metabolic pathways of energy metabolism (HMP shunt–Hexose monophosphate shunt).*

Medical Concepts/ Clinical Correlates

- Biochemists, for their convenience, learn body chemical processes in terms of individual metabolic reactions and pathways, although thousands of reactions simultaneously occur in a living cell.
- The metabolic pathways in various tissues and organs are well coordinated to meet the demands of the body.
- Liver is appropriately regarded as the body's 'central metabolic clearing house' while adipose tissues constitute the energy (fat) storehouse.

8. **Glycogen metabolism :** Glycogen is the storage form of glucose, mostly found in liver and muscle. It is degraded (glycogenolysis) and synthesized (glycogenesis) by independent pathways. Glycogen effectively serves as a ***fuel reserve*** to meet body needs, ***for a brief period*** (between meals).

Regulation of metabolic pathways

The metabolic pathways, in general, are controlled by four different mechanisms

1. The ***availability of substrates***
2. Covalent ***modification of enzymes***
3. ***Allosteric regulation***
4. Regulation of ***enzyme synthesis.***

The details of these regulatory processes are discussed under the individual metabolic pathways, in the respective chapters.

ORGAN SPECIALIZATION AND METABOLIC INTEGRATION

The ***various tissues and organs*** of the body work in a well ***coordinated manner to meet its metabolic demands.*** The major organs along with their most important metabolic functions, in a well-fed absorptive state (usually 2-4 hours after food consumption), are described.

Liver

The liver is specialized to serve as the body's ***central metabolic clearing house.*** It processes and distributes the nutrients to different tissues for utilization. After a meal, the liver takes up the carbohydrates, lipids and most of the amino acids, processes them and routes to other tissues. The major metabolic functions of liver, in an absorptive state, are listed

1. **Carbohydrate metabolism :** Increased glycolysis, glycogenesis and hexose monophosphate shunt and decreased gluconeogenesis.

2. **Lipid metabolism :** Increased synthesis of fatty acids and triacylglycerols.

3. **Protein metabolism :** Increased degradation of amino acids and protein synthesis.

Adipose tissue

Adipose tissue is regarded as the ***energy storage tissue.*** As much as 15 kg. of triacylglycerol (equivalent to 135,000 Cal) is stored in a normal adult man. The major metabolic functions of adipose tissue in an absorptive state are listed here.

1. **Carbohydrate metabolism :** The uptake of glucose is increased. This follows an increase in glycolysis and hexose monophosphate shunt.

2. **Lipid metabolism :** The synthesis of fatty acids and triacylglycerols is increased. The degradation of triacylglycerols is inhibited.

Skeletal muscle

The ***metabolism*** of skeletal muscle is rather ***variable*** depending on its needs. For instance, the resting muscle of the body utilizes about 30% of body's oxygen consumption. However, during strenuous exercise, this may be as high as 90%. The important metabolic functions of skeletal muscle in an absorptive state are listed.

1. **Carbohydrate metabolism :** The uptake of glucose is higher, and glycogen synthesis is increased.

2. **Lipid metabolism :** Fatty acids taken up from the circulation are also important fuel sources for the skeletal muscle.

Table 16.1 Energy relationship in major mammalian organs (tissues)

Organ/Tissue	Energy compound(s) preferably utilized	Energy compound(s) exported
Liver	Amino acids, glucose, fatty acids	Glucose, fatty acids, ketone bodies
Adipose tissue	Fatty acids	Fatty acids, glycerol.
Skeletal muscle	Fatty acids Glucose	None Lactate
Brain	Glucose, ketone bodies (in starvation)	None

3. **Protein metabolism :** Incorporation of amino acids into proteins is higher.

Brain

The human brain constitutes about 2% of the body's weight. But it ***utilizes*** as much as ***20% of the oxygen*** consumed by the body. Being a vital organ, special priority is given to the metabolic needs of the brain.

1. **Carbohydrate metabolism :** In an absorptive state, glucose is the only fuel source to the brain. About 120 g of glucose is utilized per day by an adult brain. This constitutes about 60% of the glucose consumed by the body at rest. It is estimated that about 50% of the energy consumed by brain is utilized by plasma membrane Na^+-K^+-ATPase to maintain membrane potential required for nerve impulse transmission.

2. **Lipid metabolism :** The free fatty acids cannot cross the blood-brain barrier, hence their contribution for the supply of energy to the brain is insignificant. Further, in a fed state, ketone bodies are almost negligible as fuel source to the brain. However, ***brain*** predominantly ***depends on ketone bodies during prolonged starvation*** (details given later).

The metabolic interrelationship among the major tissues in an absorptive state are given in ***Fig.16.3***. The fuel sources that are preferably utilized by the major organs and the compounds exported from them are listed in ***Table 16.1***.

METABOLISM IN STARVATION

Starvation may be due to food scarcity or the desire to rapidly lose weight or certain clinical conditions (e.g. surgery, burns etc.). Starvation is a metabolic stress which imposes certain metabolic compulsions on the organism. The metabolism is reorganized to meet the new demands of starvation.

Glucose is the fuel of choice for brain and muscle. Unfortunately, the carbohydrate reserve of the body is so low that it cannot meet the energy requirements even for a day. The fuel stores (or energy reserves) of a 70 kg normal man are given in ***Table 16.2***. Triacylglycerol (fat) of adipose tissue is the predominant energy reserve of the body. The survival time of an individual on starvation is mostly dependent on his/her fat stores. And for this reason, obese individuals can survive longer than lean individuals without consuming food.

Protein is basically a structural constituent, mostly present in the muscle. However, during starvation, protein can also meet the fuel demands of the body. It is estimated that about $^1/_3$rd of the body's protein can be utilized towards energy needs without compromising the vital functions.

Starvation is associated with a decrease in insulin level and an increase in glucagon. The

Table 16.2 Energy reserves of a normal 70 kg man

Energy source (main storage tissue)	Weight (kg)	Energy equivalent (in Cal)
Triacylglycerol (adipose tissue)	15	135,000
Protein (muscle)	6	24,000
Glycogen (muscle, liver)	0.2	800

Medical Concepts/ Clinical Correlates

- Brain is a vital metabolic organ that consumes about 20% of body's oxygen, although it constitutes only 2% of body weight.
- The metabolism in starvation is reorganized to meet the body's changed demands and metabolic compulsions.
- Under normal circumstances, glucose is the only fuel source to brain. However, during starvation, the brain slowly gets adapted to use ketone bodies for energy needs.

metabolic changes during starvation are discussed with reference to the major organs/ tissues.

Liver in starvation

1. **Carbohydrate metabolism :** An important function of liver is to act as a ***blood glucose buffering organ.*** The action of liver is to suit the metabolic needs of the body. During starvation, increased gluconeogenesis and elevated glycogen degradation furnish glucose to the needy tissues (mostly brain).

2. **Lipid metabolism :** Fatty acid oxidation is increased with an elevated synthesis of ketone bodies. This is due to the fact that TCA (Krebs) cycle cannot cope up with the excess production of acetyl CoA, hence the latter is diverted for ketone body synthesis.

Ketone bodies (primarily β-hydroxybutyrate) effectively serve as fuel source for the peripheral tissues. The brain slowly adapts itself to use ketone bodies. Thus, after a 3-day fast, about

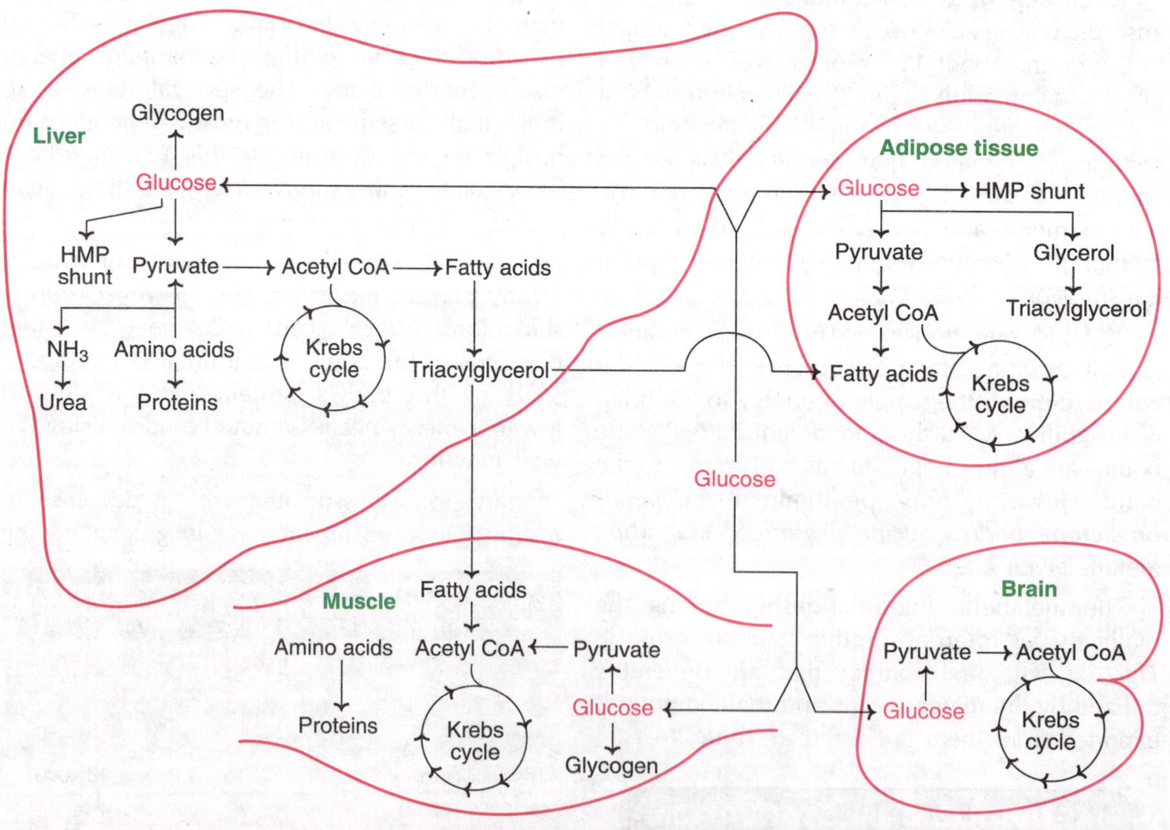

Fig. 16.3 : Metabolic interrelationship among the major tissues in a well fed state (HMP shunt–Hexose monophosphate shunt).

$^1/_3$rd of the brain's fuel demands are met by ketone bodies, while, after 40 days' starvation, they countribute to about 70% of energy needs.

Adipose tissue in starvation

1. **Carbohydrate metabolism** : Glucose uptake and its metabolism are lowered.

2. **Lipid metabolism** : The degradation of triacylglycerol is elevated, leading to an increased release of fatty acids from the adipose tissue which serve as fuel source for various tissues (brain is an exception). The glycerol liberated in lipolysis serves as a precursor for glucose synthesis by liver. The synthesis of fatty acids and triacylglycerols is totally stopped in adipose tissue.

Skeletal muscle in starvation

1. **Carbohydrate metabolism** : Glucose uptake and its metabolism are depressed.

2. **Lipid metabolism** : Both fatty acids and ketone bodies are utilized by the muscle as fuel source. However, on prolonged starvation beyond 3 weeks, the muscle adapts to exclusively utilize fatty acids.

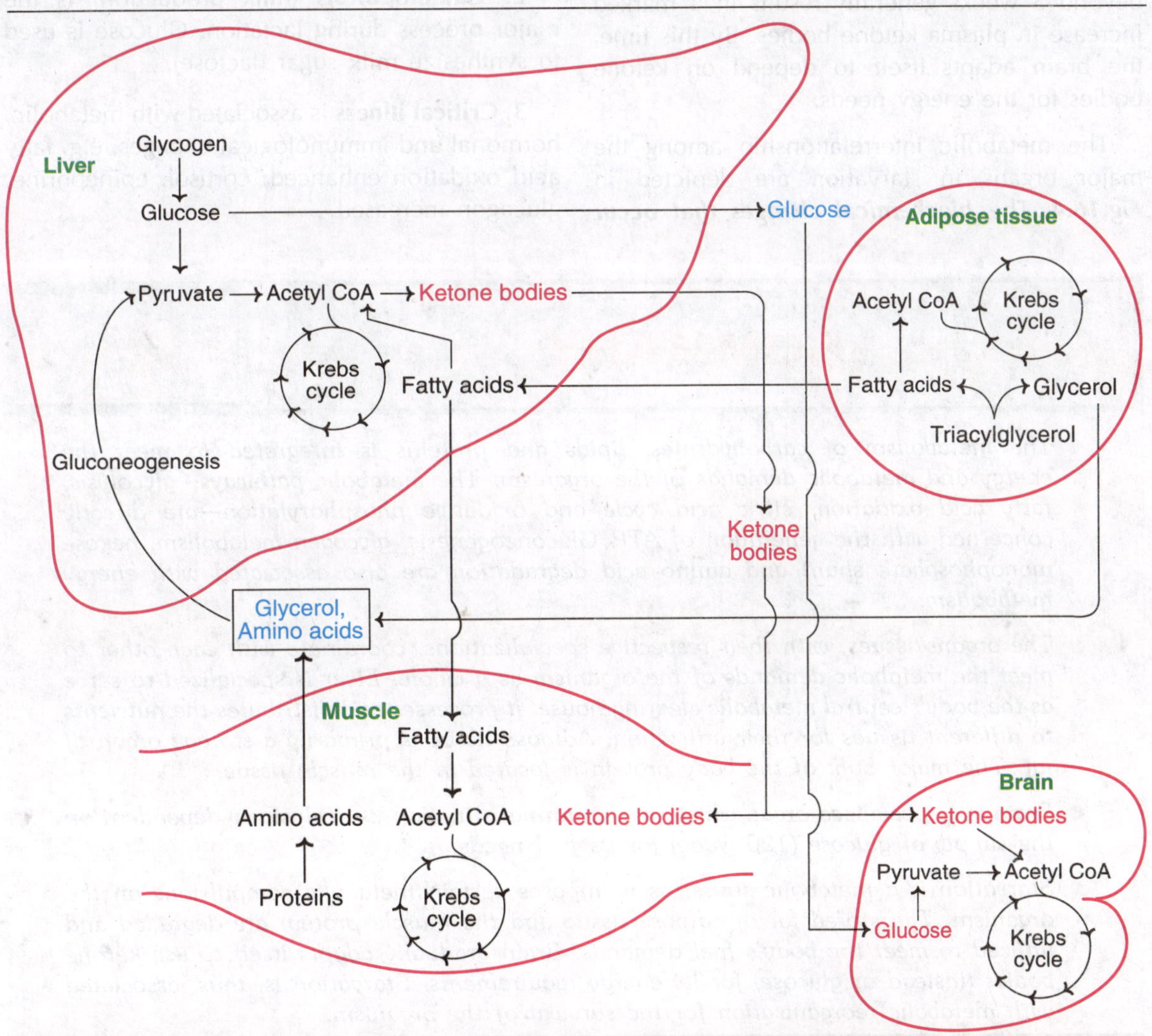

Fig. 16.4 : Metabolic interrelationship among major tissues during starvation.

3. **Protein metabolism :** During the early period of starvation, muscle proteins are degraded to liberate the amino acids which are effectively utilized by the liver for glucose synthesis. On prolonged starvation, however, protein breakdown is reduced.

Brain in starvation

As already stated, glucose is the preferred fuel source by brain. During the first 2 weeks of starvation, the brain is mostly dependent on glucose, supplied by liver gluconeogenesis. This, in turn, is dependent on the amino acids released from the muscle protein degradation. Starvation beyond 3 weeks generally results in a marked increase in plasma ketone bodies. By this time, the brain adapts itself to depend on ketone bodies for the energy needs.

The metabolic interrelationship among the major organs in starvation are depicted in ***Fig.16.4***. The ***biochemical changes that occur during starvation are such that an adequate supply of fuel molecules is maintained to various tissues to meet the energy demands.***

METABOLISM IN PREGNANCY, LACTATION, ILLNESS ETC

Metabolism gets geared up and gets adapted to meet the physiological and pathological demands of the body. Some examples are listed

1. **Pregnancy** is associated with increased availability of blood glucose, deposition of protein (positive nitrogen balance), etc.

2. **Galactopoiesis** (milk production) is the major process during lactation. Glucose is used to synthesize milk sugar (lactose).

3. **Critical illness** is associated with metabolic, hormonal and immunological changes. e.g. fatty acid oxidation enhanced; cortisol, epinephrine, glucagon increased.

SUMMARY

- *The metabolism of carbohydrates, lipids and proteins is integrated to meet the energy and metabolic demands of the organism. The metabolic pathways—glycolysis, fatty acid oxidation, citric acid cycle and oxidative phosphorylation—are directly concerned with the generation of ATP. Gluconeogenesis, glycogen metabolism, hexose monophosphate shunt and amino acid degradation are also associated with energy metabolism.*

- *The organs/tissues, with their respective specializations, coordinate with each other to meet the metabolic demands of the organism as a whole. Liver is specialized to serve as the body's central metabolic clearing house. It processes and distributes the nutrients to different tissues for their utilization. Adipose tissue is primarily a storage organ of fat. The major bulk of the body protein is located in the muscle tissue.*

- *Brain is a specialized organ which, in the normal situation, is exclusively dependent on the supply of glucose (120 g/day) for its fuel needs.*

- *Starvation is a metabolic stress, as it imposes certain metabolic compulsions on the organism. The stored fat of adipose tissue and the muscle protein are degraded and utilized to meet the body's fuel demands. Brain gradually adapts itself to use ketone bodies (instead of glucose) for its energy requirements. Starvation is, thus, associated with metabolic reorganization for the survival of the organism.*

Metabolism of Nucleotides 17

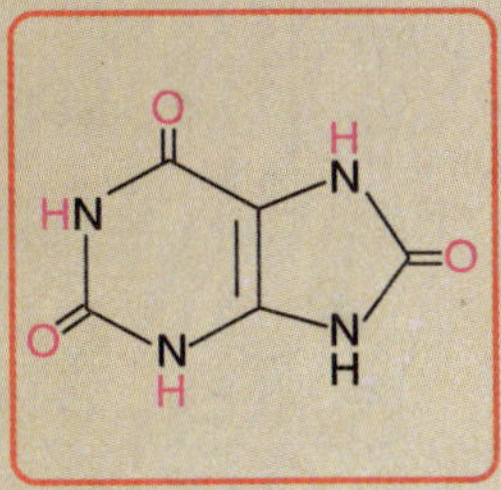

The uric acid speaks

"I am the end product of purines;
An increase in my production causes gout;
Inflammation of joints is the symptom,
And administration of allopurinol a relief.

Nucleotides consist of a nitrogenous base, a pentose and a phosphate. The pentose sugar is D-ribose in ribonucleotides of RNA while in deoxyribonucleotides (deoxynucleotides) of DNA, the sugar is 2-deoxy D-ribose. Nucleotides participate in almost all the biochemical processes, either directly or indirectly. They are the structural components of nucleic acids (DNA, RNA), coenzymes, and are involved in the regulation of several metabolic reactions.

Purines and pyrimidines are non-essential in the diet

Humans can synthesize purines and pyrimidines to meet the variable demands of the body. Thus, dietary nucleic acids (digested to nucleotides) and nucleotides are non-essential. The nucleotides and the corresponding bases (purines and pyrimidines) are absorbed and metabolized to uric acid. ***Very little dietary purines and pyrimidines are incorporated into nucleic acids.***

BIOSYNTHESIS OF PURINE RIBONUCLEOTIDES

Many compounds contribute to the purine ring of the nucleotides (***Fig.17.1***).

1. N_1 of purine is derived from amino group of aspartate.

2. C_2 arise from formate of N^{10}-formyl THF and C_8 from N^5,N^{10}-methenyl THF.

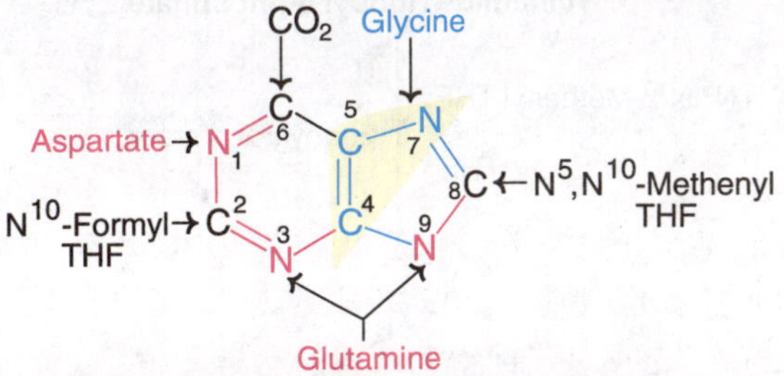

Fig. 17.1 : *The sources of individual atoms in purine ring. (**Note :** Same colours are used in the synthetic pathway **Fig. 17.2**).*

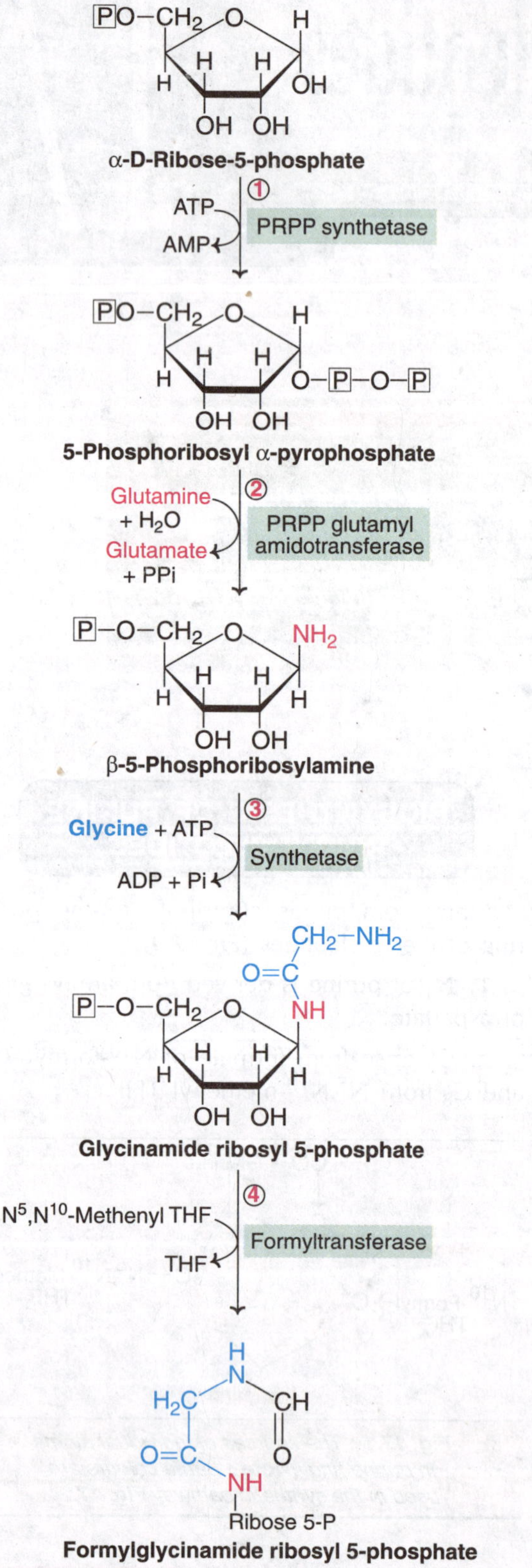

Fig. 17.2 contd. next column

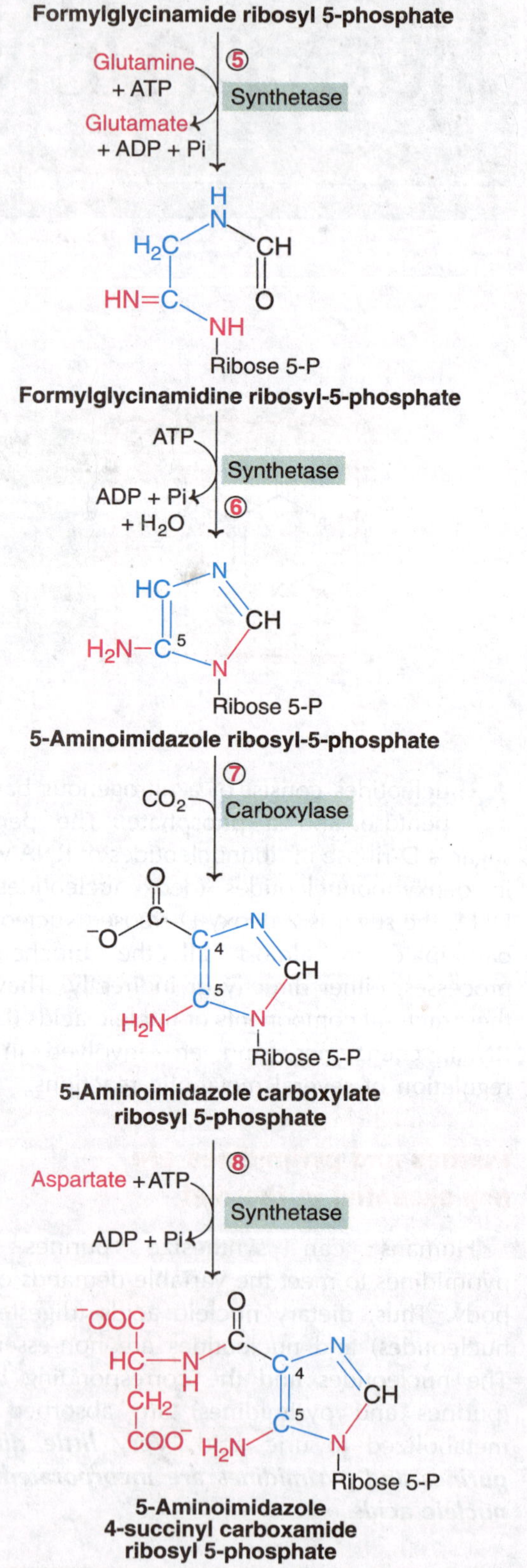

Fig. 17.2 contd. next page

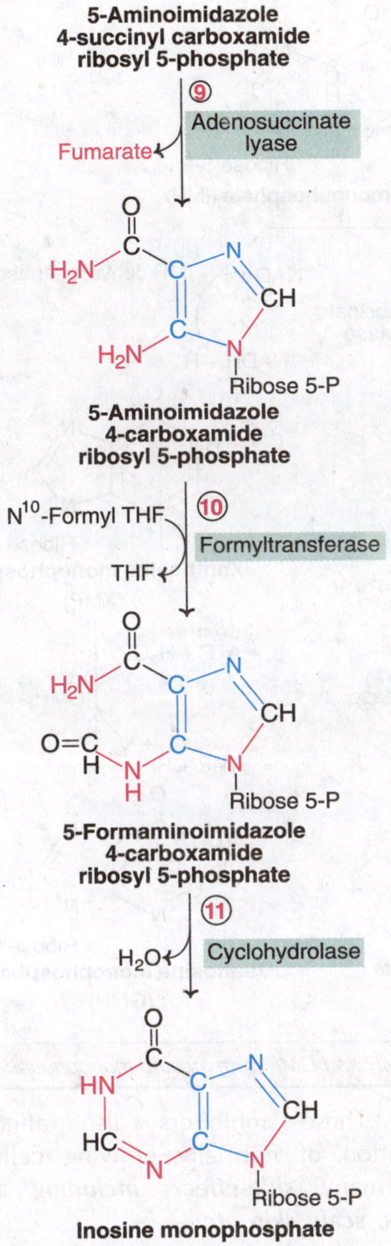

***Fig. 17.2 :** The metabolic pathway for the synthesis of inosine monophosphate, the parent purine nucleotide (PRPP–Phosphoribosyl pyrophosphate; PPi–Pyrophosphate).*

3. N_3 and N_9 are obtained from amide group of glutamine.

4. C_4, C_5 and N_7 are contributed by glycine.

5. C_6 directly comes from CO_2.

It should be remembered that purine bases are not synthesized as such, but they are formed as ribonucleotides. ***The purines are built upon a pre-existing ribose 5-phosphate.*** Liver is the major site for purine nucleotide synthesis. Erythrocytes, polymorphonuclear leukocytes and brain cannot produce purines.

The pathway for the synthesis of ***inosine monophosphate*** (IMP or inosinic acid), the 'parent' purine nucleotide is given in ***Fig.17.2.*** The reactions are briefly described.

1. ***Ribose 5-phosphate***, produced in the hexose monophosphate shunt of carbohydrate metabolism is the ***starting material*** for purine nucleotide synthesis. It reacts with ATP to form phosphoribosyl pyrophosphate (PRPP).

2. Glutamine transfers its amide nitrogen to PRPP to replace pyrophosphate and produce 5-phosphoribosylamine. The enzyme ***PRPP glutamyl amidotransferase*** is controlled by ***feedback inhibition*** of nucleotides (IMP, AMP and GMP). This reaction is the ***'committed step'*** in purine nucleotide biosynthesis.

3. Phosphoribosylamine reacts with glycine in the presence of ATP to form glycinamide ribosyl 5-phosphate or glycinamide ribotide (GAR).

4. N^5,N^{10}-Methenyl tetrahydrofolate donates the formyl group and the product formed is formyl- glycinamide ribosyl 5-phosphate.

5. Glutamine transfers the second amido amino group to produce formylglycinamidine ribosyl 5-phosphate.

6. The imidazole ring of the purine is closed in an ATP dependent reaction to yield 5-amino-imidazole ribosyl 5-phosphate.

7. Incorporation of CO_2 (carboxylation) occurs to yield aminoimidazole carboxylate ribosyl 5-phosphate. This reaction ***does not require*** the vitamin ***biotin*** and/or ATP which is the case with most of the ***carboxylation*** reactions.

8. Aspartate condenses with the product in reaction 7 to form aminoimidazole 4-succinyl carboxamide ribosyl 5-phosphate.

9. Adenosuccinate lyase cleaves off fumarate and only the amino group of aspartate is retained to yield aminoimidazole 4-carboxamide ribosyl 5-phosphate.

10. N^{10}-Formyl tetrahydrofolate donates a one-carbon moiety to produce formaminoimidazole 4-carboxamide ribosyl 5-phosphate. With this reaction, all the carbon and nitrogen atoms of purine ring are contributed by the respective sources.

11. The final reaction catalysed by cyclohydrolase leads to ring closure with an elimination of water molecule. The product obtained is ***inosine monophosphate (IMP)***, the ***parent purine nucleotide*** from which other purine nucleotides can be synthesized.

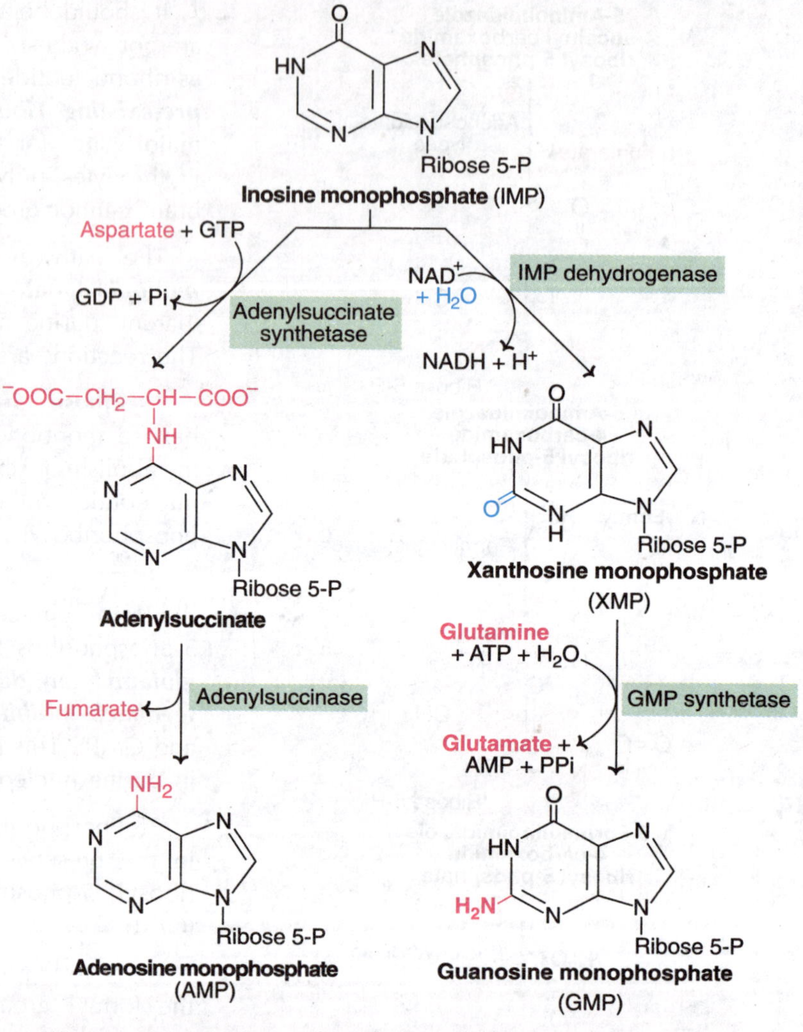

Fig. 17.3 : *Synthesis of AMP and GMP from inosine monophosphate.*

Inhibitors of purine synthesis

Inhibition of microbial growth : Folic acid (THF) is essential for the synthesis of purine nucleotides (reactions 4 and 10). ***Sulfonamides*** are the structural analogs of para-aminobenzoic acid (PABA) (see ***page no. 162*** for structures). These sulfa drugs can be used to ***inhibit the synthesis of folic acid by microorganisms.*** This indirectly reduces the synthesis of purines and, therefore, the nucleic acids (DNA and RNA). Sulfonamides have no influence on humans, since folic acid is not synthesized and is supplied through diet.

Treatment of cancer : The structural ***analogs of folic acid*** (e.g. ***methotrexate***) are widely ***used to control cancer.*** They inhibit the synthesis of purine nucleotides (reactions 4 and 10) and, thus, nucleic acids. Both these reactions are concerned with the transfer of one-carbon moiety (formyl group). These inhibitors also affect the proliferation of normally growing cells. This causes many ***side-effects including anemia, baldness, scaly skin etc.***

Azaserine, an antagonist of glutamine inhibits reactions (2 and 5) involving glutamine and blocks purine synthesis.

Synthesis of AMP and GMP from IMP

Inosine monophosphate is the immediate precursor for the formation of AMP and GMP (***Fig.17.3***). Aspartate condenses with IMP in the presence of GTP to produce adenylsuccinate which, on cleavage, forms AMP.

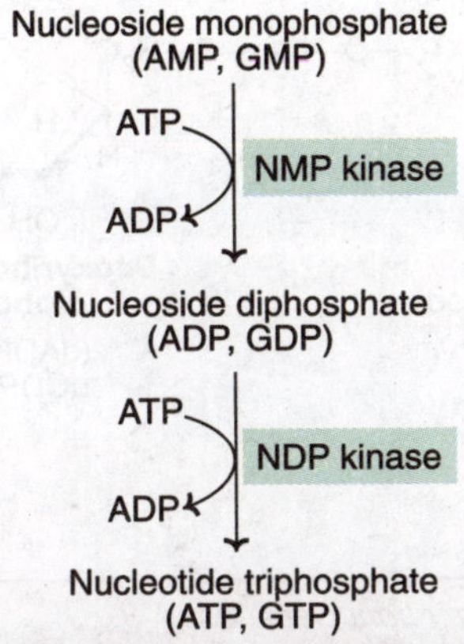

***Fig. 17.4 :** Conversion of nucleoside monophosphates to di- and triphosphates (NMP–Nucleoside monophosphate; NDP–Nucleoside diphosphate).*

For the synthesis of GMP, IMP undergoes NAD^+ dependent dehydrogenation to form xanthosine monophosphate (XMP). Glutamine then transfers amide nitrogen to XMP to produce GMP.

6-Mercaptopurine is an ***inhibitor*** of the synthesis of AMP and GMP. It acts on the enzyme ***adenylsuccinase*** (of AMP pathway) and IMP dehydrogenase (of GMP pathway).

Formation of purine nucleoside diphosphates and triphosphates

The nucleoside monophosphates (AMP and GMP) have to be converted to the corresponding di- and triphosphates to participate in most of the metabolic reactions. This is achieved by the transfer of phosphate group from ATP, catalysed by nucleoside monophosphate (NMP) kinases and nucleoside diphosphate (NDP) kinases (***Fig 17.4***).

Salvage pathway for purines

The free purines (adenine, guanine and hypoxanthine) are formed in the normal turnover of nucleic acids (particularly RNA), and also obtained from the dietary sources. The ***purines*** can be ***directly converted to the corresponding nucleotides***, and this process is known as 'salvage pathway' (***Fig.17.5***).

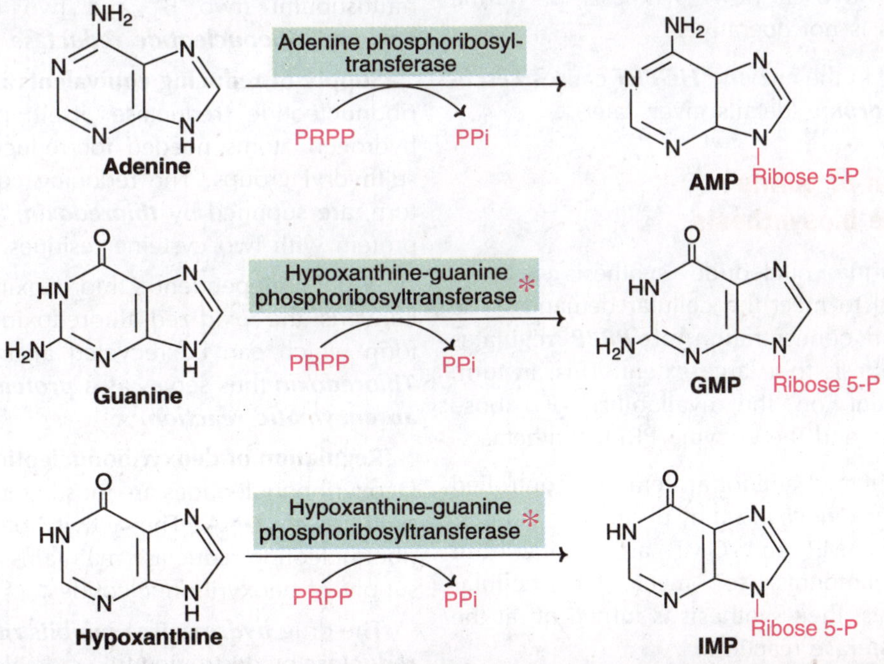

***Fig. 17.5 :** Salvage pathways of purine nucleotide synthesis (PRPP–Phosphoribosyl pyrophosphate; PPi–Inorganic pyrophosphate; AMP–Adenosine monophosphate; GMP–Guanosine monophosphate; IMP–Inosine monophosphate;* * *Deficiency of HGPRT causes Lesch-Nyhan syndrome).*

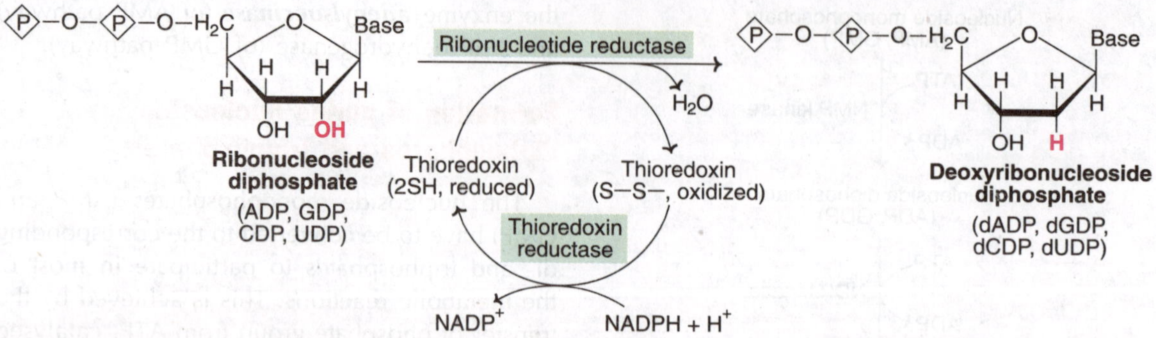

Fig. 17.6 : Formation of deoxyribonucleotides from ribonucleotides.

Adenine phosphoribosyl transferase catalyses the formation of AMP from adenine. Hypoxanthine-guanine phosphoribosyl transferase (HGPRT) converts guanine and hypoxanthine, respectively, to GMP and IMP. Phosphoribosyl pyrophosphate (PRPP) is the donor of ribose 5-phosphate in the salvage pathway.

The ***salvage pathway*** is particularly ***important*** in certain tissues such as ***erythrocytes and brain*** where *de novo* (a new) synthesis of purine nucleotides is not operative.

A ***defect*** in the enzyme ***HGPRT*** causes ***Lesch-Nyhan syndrome*** (details given later).

Regulation of purine nucleotide biosynthesis

The purine nucleotide synthesis is well coordinated to meet the cellular demands. The intracellular concentration of ***PRPP*** regulates purine synthesis to a large extent. This, in turn, is dependent on the availability of ribose 5-phosphate and the enzyme PRPP synthetase.

PRPP glutamyl amidotransferase is controlled by a ***feedback mechanism*** by purine nucleotides. That is, if AMP and GMP are available in adequate amounts to meet the cellular requirements, their synthesis is turned off at the ***amidotransferase*** reaction.

Another important stage of regulation is in the conversion of IMP to AMP and GMP. AMP inhibits adenylsuccinate synthetase while GMP inhibits IMP dehydrogenase. Thus, AMP and GMP control their respective synthesis from IMP by a feedback mechanism.

Conversion of ribonucleotides to deoxyribonucleotides

The synthesis of purine and pyrimidine deoxyribonucleotides occurs from ribonucleotides ***by a reduction at the C_2 of ribose moiety*** (***Fig.17.6***). This reaction is catalysed by a multisubunit (two B_1 and two B_2 subunits) enzyme, ***ribonucleotide reductase.***

Supply of reducing equivalents : The enzyme ribonucleotide reductase itself provides the hydrogen atoms needed for reduction from its sulfhydryl groups. The reducing equivalents, in turn, are supplied by ***thioredoxin***, a monomeric protein with two cysteine residues.

NADPH-dependent thioredoxin reductase converts the oxidized thioredoxin to reduced form which can be recycled again and again. ***Thioredoxin*** thus serves as a ***protein cofactor in an enzymatic reaction.***

Regulation of deoxyribonucleotide synthesis : Deoxyribonucleotides are mostly required for the synthesis of DNA. The activity of the enzyme ribonucleotide reductase maintains the adequate supply of deoxyribonucleotides.

The drug ***hydroxyurea inhibits ribonucleotide reductase*** by destroying free radicals required by this enzyme. Hydroxyurea is used in the ***treatment of cancers*** such as chronic myelogenous leukemia.

DEGRADATION OF PURINE NUCLEOTIDES

The ***end product*** of purine metabolism in humans is ***uric acid***. The sequence of reactions in purine nucleotide degradation is given in ***Fig.17.7***.

1. The nucleotide monophosphates (AMP, IMP and GMP) are converted to their respective nucleoside forms (adenosine, inosine and guanosine) by the action of ***nucleotidase***.

2. The amino group, either from AMP or adenosine, can be removed to produce IMP or inosine, respectively.

3. Inosine and guanosine are, respectively, converted to hypoxanthine and guanine (purine bases) by purine nucleoside phosphorylase. Adenosine is not degraded by this enzyme, hence it has to be converted to inosine.

4. Guanine undergoes deamination by guanase to form ***xanthine***.

5. ***Xanthine oxidase*** is an important enzyme that converts hypoxanthine to xanthine, and xanthine to uric acid. This enzyme contains FAD, molybdenum and iron, and is exclusively found in liver and small intestine. Xanthine oxidase liberates H_2O_2 which is harmful to the tissues. ***Catalase*** cleaves H_2O_2 to H_2O and O_2.

Uric acid (2,6,8-trioxypurine) is the final excretory product of purine metabolism in humans. Uric acid can serve as an important ***antioxidant*** by getting itself converted (non-enzymatically) to allantoin. It is believed that the antioxidant role of ascorbic acid in primates is replaced by uric acid, since these animals have lost the ability to synthesize ascorbic acid.

Most animals (other than primates) however, oxidize uric acid by the enzyme uricase to allantoin, where the purine ring is cleaved. Allantoin is then converted to allantoic acid and excreted in some fishes (***Fig.17.8***). Further degradation of allantoic acid may occur to produce urea (in amphibians, most fishes and some molluscs) and, later, to ammonia (in marine invertebrates).

DISORDERS OF PURINE METABOLISM

Hyperuricemia and gout

Uric acid is the end product of purine metabolism in humans. The normal concentration of ***uric acid in the serum*** of adults is in the ***range of 3-7 mg/dl***. In women, it is slightly lower (by about 1 mg) than in men. The daily excretion of uric acid is about 500-700 mg.

Hyperuricemia refers to an ***elevation in the serum uric acid*** concentration. This is sometimes associated with increased uric acid excretion (uricosuria).

Gout is a metabolic disease associated with ***overproduction of uric acid***. At the physiological pH, uric acid is found in a more soluble form as sodium urate. In severe hyperuricemia, crystals of sodium urate get deposited in the soft tissues, particularly in the joints. Such deposits are commonly known as ***tophi***. This causes ***inflammation in the joints*** resulting in a painful ***gouty arthritis***. Sodium urate and/or uric acid may also precipitate in kidneys and ureters that results in renal damage and stone formation.

Historically, gout was found to be often associated with high living, over-eating and ***alcohol consumption***. In the previous centuries, alcohol was contaminated with lead during its manufacture and storage. ***Lead poisoning*** leads to ***kidney damage*** and decreased uric acid excretion causing ***gout***. In general, a diet rich is ***meat*** and ***seafoods*** is associated with increased ***risk of gout***.

The ***prevalence*** of gout is about ***3 per 1,000*** persons, ***mostly affecting males***. Post-menopausal women, however, are as susceptible as men for this disease. Gout is of two types—primary and secondary.

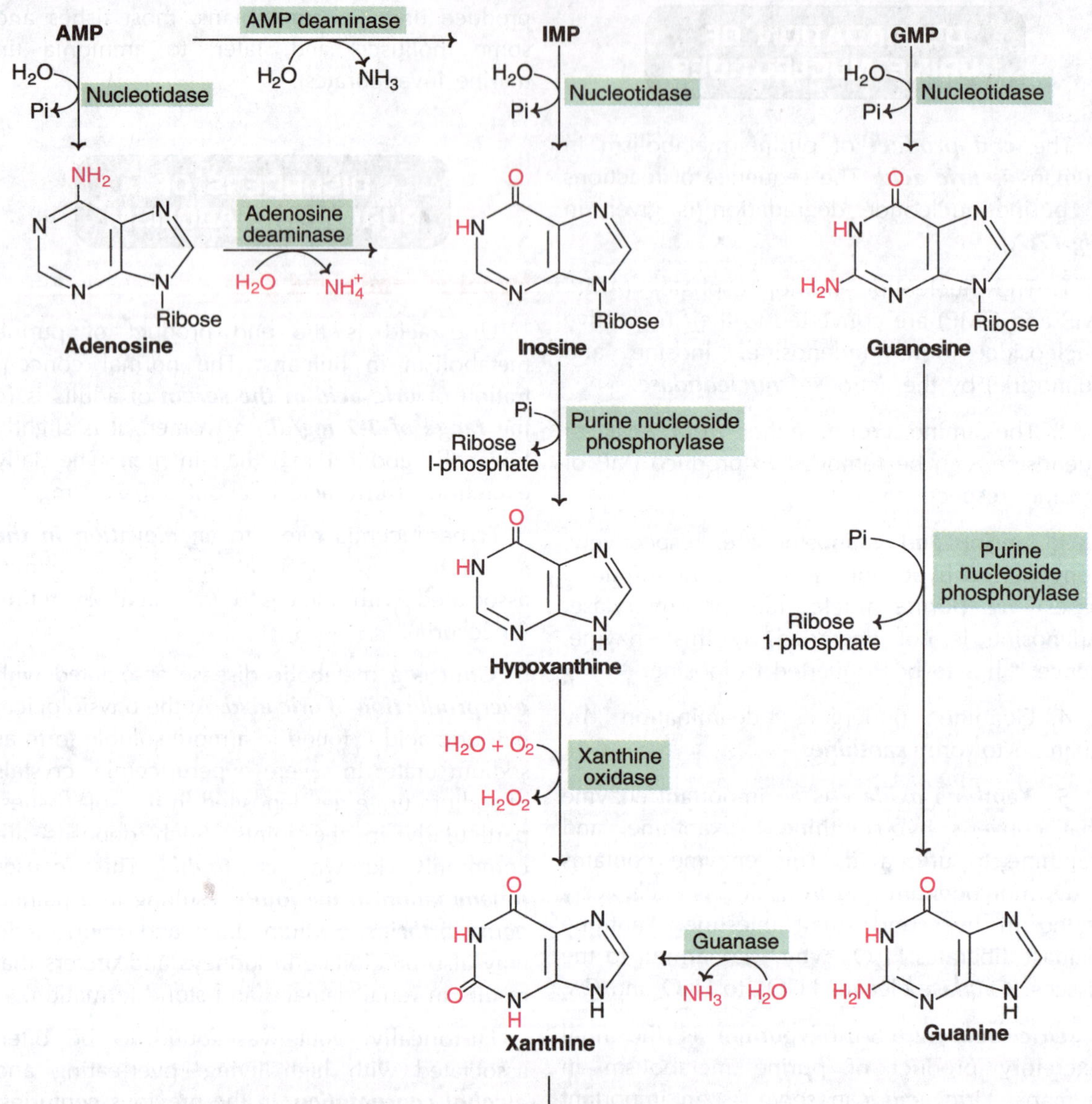

Fig. 17.7 : Degradation of purine nucleotides to uric acid (AMP–Adenosine monophosphate; IMP–Inosine monophosphate; GMP–Guanosine monophosphate).

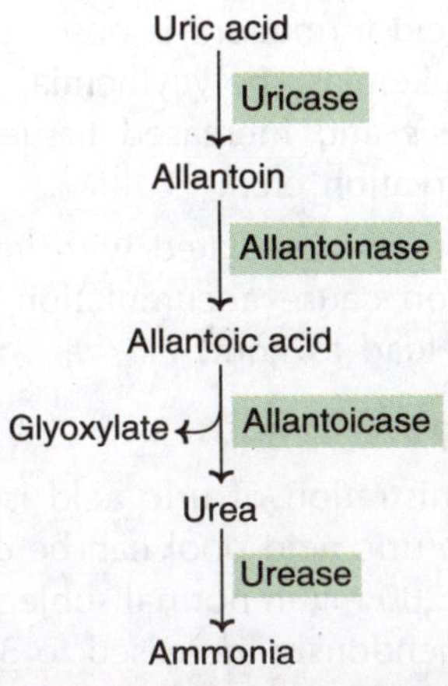

***Fig. 17.8 :** Degradation of uric acid in animals other than man.*

1. **Primary gout :** It is an inborn error of metabolism due to ***overproduction of uric acid.*** This is mostly related to increased synthesis of purine nucleotides. The following are the important metabolic defects (enzymes) associated with primary gout (***Fig.17.9***)

- **PRPP synthetase :** In normal circumstances, PRPP synthetase is under feedback control by purine nucleotides (ADP and GDP). However, variant forms of PRPP synthetase—which are not subjected to feedback regulation—have been detected. This leads to the increased production of purines.
- **PRPP glutamylamidotransferase :** The lack of feedback control of this enzyme by purine nucleotides also leads to their elevated synthesis.
- **HGPRT deficiency :** This is an enzyme of purine salvage pathway, and its defect causes ***Lesch-Nyhan syndrome.*** This disorder is associated with increased synthesis of purine nucleotides by a two-fold mechanism. Firstly, decreased utilization of purines (hypoxanthine and guanine) by salvage pathway, resulting in the accumulation and diversion of PRPP for purine nucleotides. Secondly, the defect in salvage pathway leads to decreased levels of IMP and GMP causing impairment in the tightly controlled feedback regulation of their production.

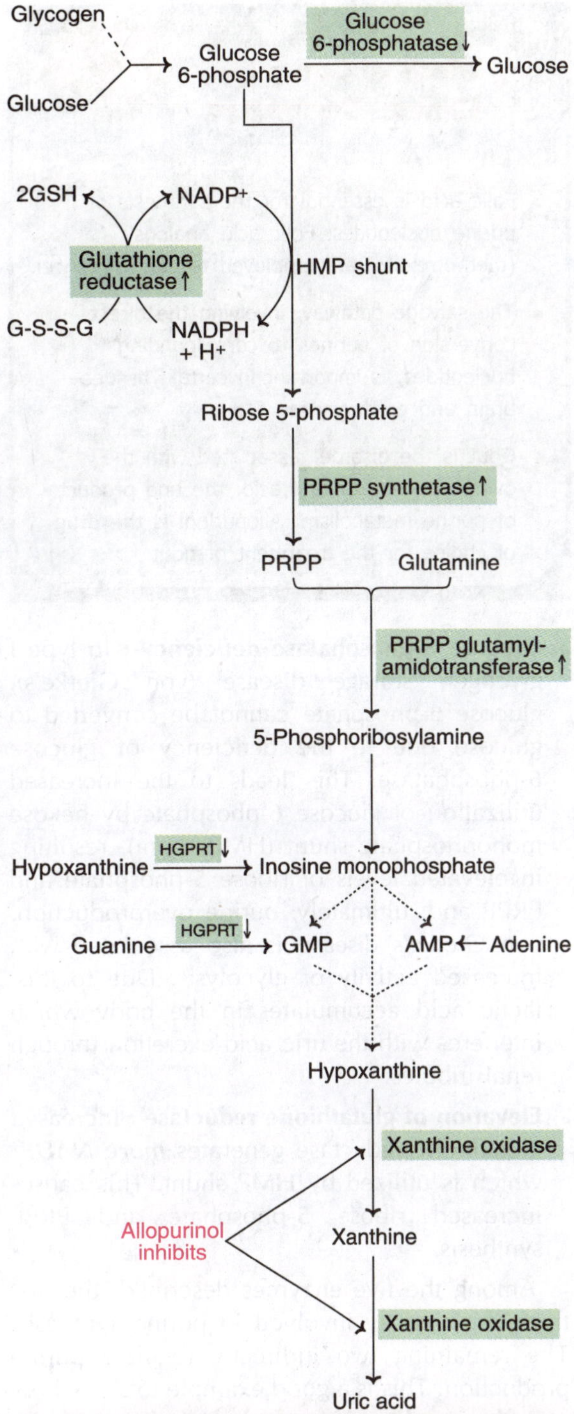

***Fig. 17.9 :** Summary of possible enzyme alterations causing gout (↑–Increased enzyme activity; ↓–Decreased enzyme activity; GSH–Reduced glutathione; G-S-S-G–Oxidized glutathione; PRPP–Phosphoribosyl pyrophosphate; HGPRT–Hypoxanthine-guanine phosphoribosyltransferase).*

Medical Concepts/ Clinical Correlates

- Folic acid is essential for the synthesis of purine nucleotides. Folic acid analogs (methotrexate) are employed to control cancer.
- The salvage pathway, involving the direct conversion of purines to corresponding nucleotides, is important in certain tissues—brain and erythrocytes.
- Gout is the disorder associated with the overproduction of uric acid, the end product of purine metabolism. Allopurinol is the drug of choice for the treatment of gout.

- **Glucose 6-phosphatase deficiency :** In type I glycogen storage disease (von Gierke's), glucose 6-phosphate cannot be converted to glucose due to the deficiency of glucose 6-phosphatase. This leads to the increased utilization of glucose 6-phosphate by hexose monophosphate shunt (HMP shunt), resulting in elevated levels of ribose 5-phosphate and PRPP and, ultimately, purine overproduction. von Gierke's disease is also associated with increased activity of glycolysis. Due to this, lactic acid accumulates in the body which interferes with the uric acid excretion through renal tubules.
- **Elevation of glutathione reductase :** Increased glutathione reductase generates ***more*** $NADP^+$ which is utilized by HMP shunt. This causes increased ribose 5-phosphate and PRPP synthesis.

Among the five enzymes described, the first three are directly involved in purine synthesis. The remaining two indirectly regulate purine production. This is a good example to show how an ***abnormality in one metabolic pathway influences the other***.

2. **Secondary gout :** Secondary hyperuricemia is due to various diseases causing increased synthesis or decreased excretion of uric acid. Increased degradation of nucleic acids (hence more uric acid formation) is observed in various ***cancers*** (leukemias, polycythemia, lymphomas, etc.) ***psoriasis*** and increased tissue breakdown (trauma, starvation etc.).

The disorders associated with impairment in renal function cause accumulation of uric acid which may lead to gout.

Uric acid pool in gout

By administration of uric acid isotope (N^{15}), the miscible uric acid pool can be calculated. It is around 1,200 mg in normal subjects. Uric acid pool is tremendously increased to 3,000 mg. or even more, in patients suffering from gout.

Treatment of gout

The drug of choice for the treatment of primary gout is ***allopurinol***. This is a structural analog of hypoxanthine that competitively inhibits the enzyme ***xanthine oxidase***. Further, allopurinol is oxidized to alloxanthine by xanthine oxidase (***Fig.17.10***). ***Alloxanthine***, in turn, is a more effective inhibitor of xanthine oxidase. This type of inhibition is referred to as ***suicide inhibition*** (***Refer Chapter 6***).

Inhibition of xanthine oxidase by allopurinol leads to the accumulation of hypoxanthine and xanthine. These two compounds are more soluble than uric acid, hence easily excreted.

Uricosuric drugs (e.g. probenecid) decrease renal reabsorption of uric acid and increase its excretion into urine. The drug ***halofenate*** is uricosuric, besides its hypolipidemic action.

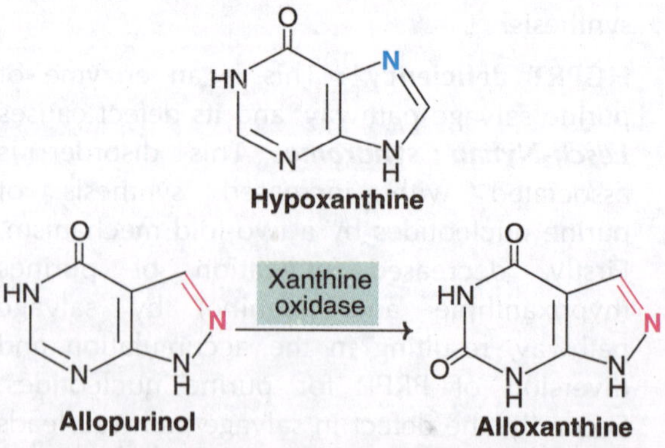

Fig. 17.10 : Structures of hypoxanthine and its structural analogs.

Besides the drug therapy, ***restriction in dietary intake of purines and alcohol is advised.*** Consumption of plenty of water will also be useful.

The ***anti-inflammatory drug colchicine*** is used for the treatment of gouty arthritis. Other anti-inflammatory drugs—such as ***phenylbutazone***, indomethacin, oxyphenbutazone, corticosteroids—are also useful.

Pseudogout

The clinical manifestations of pseudogout are similar to gout. But this disorder is caused by the ***deposition of calcium pyrophosphate crystals*** in joints. Further, serum uric acid concentration is normal in pseudogout.

Lesch-Nyhan syndrome

This disorder is due to the deficiency of ***hypoxanthine-guanine phosphoribosyltransferase*** (***HGPRT***), an enzyme of purine salvage pathway (***See Fig.17.5***). It was first described in 1964 by Michael Lesch (a medical student) and William L. Nyhan (his teacher).

Lesch-Nyhan syndrome is a ***sex-linked metabolic disorder*** since the structural gene for HGPRT is located on the X-chromosome. It ***affects*** only the ***males*** and is characterized by excessive uric acid production (often gouty arthritis), and ***neurological abnormalities*** such as ***mental retardation***, aggressive behavior, learning disability etc. The patients of this disorder have an ***irresistible urge to bite their fingers and lips***, often causing self-mutilation.

The overproduction of uric acid in Lesch-Nyhan syndrome is explained. HGPRT deficiency results in the accumulation of PRPP and decrease in GMP and IMP, ultimately leading to increased synthesis and degradation of purines (more details given under primary gout).

The biochemical basis for the neurological symptoms observed in Lesch-Nyhan syndrome is not clearly understood. This may be related to the dependence of brain on the salvage pathway for *de novo* synthesis of purine nucleotides. Uric acid is not toxic to the brain, since patients with severe hyperuricemia (not related to HGPRT deficiency) do not exhibit any neurological symptoms. Further, allopurinol treatment that helps to decrease uric acid production, has no affect on the neurological manifestations in these patients.

Immunodeficiency diseases associated with purine metabolism

Two different immunodeficiency disorders associated with the degradation of purine nucleosides are identified. The enzyme defects are ***adenosine deaminase*** and ***purine nucleoside phosphorylase,*** involved in uric acid synthesis (***See Fig.17.7***).

The ***deficiency of adenosine deaminase (ADA) causes severe combined immunodeficiency (SCID)*** involving T-cell and usually B-cell dysfunction. It is explained that ADA deficiency results in the accumulation of dATP which is an inhibitor of ribonucleotide reductase and, therefore, DNA synthesis and cell replication.

The deficiency of purine nucleotide phosphorylase is associated with impairment of T-cell function but has no effect on B-cell function. Uric acid synthesis is decreased (***hypouricemia***) and the tissue levels of purine nucleosides and nucleotides are higher. It is believed that dGTP inhibits the development of normal T-cells.

Hypouricemia

Decreased uric acid levels in the serum (< 2 mg/dl) represent hypouricemia. This is mostly associated with a rare genetic defect in the enzyme xanthine oxidase. It leads to the increased excretion of xanthine and hypoxanthine. Xanthinuria frequently causes the formation of xanthine stones in the urinary tract.

BIOSYNTHESIS OF PYRIMIDINE RIBONUCLEOTIDES

The synthesis of pyrimidines is a much simpler process compared to that of purines. ***Aspartate***, ***glutamine*** (amide group) and ***CO_2 contribute*** to atoms in the formation of ***pyrimidine ring*** (***Fig.17.11***). Pyrimidine ring is first synthesized and then attached to ribose 5-phosphate. This is in contrast to purine nucleotide synthesis wherein purine ring is built upon a pre-existing ribose 5-phosphate. The pathway of pyrimidine synthesis is depicted in ***Fig.17.12***, and the salient features are described below.

Glutamine transfers its amido nitrogen to CO_2 to produce carbamoyl phosphate. This reaction is ATP-dependent and is catalysed by cytosomal enzyme ***carbamoyl phosphate synthetase II*** (CPS II).

CPS II is activated by ATP and PRPP and inhibited by UTP. Carbamoyl phosphate synthetase I (CPS I) is a mitochondrial enzyme which synthesizes carbamoyl phosphate from ammonia and CO_2 and, in turn urea (***Refer*** protein metabolism, ***Chapter 15***, for more details). Prokaryotes have only one carbamoyl phosphate synthetase which is responsible for the biosynthesis of arginine and pyrimidines.

Carbamoyl phosphate condenses with aspartate to form carbamoyl aspartate. This reaction is catalysed by aspartate transcarbamoylase. Dihydroorotase catalyses the pyrimidine ring closure with a loss of H_2O.

The three enzymes—***CPS II***, ***aspartate transcarbamoylase*** and ***dihydroorotase*** are the domains (functional units) of the ***same protein***. This is a good example of a ***multifunctional enzyme.***

The next step in pyrimidine synthesis is an NAD^+ dependent dehydrogenation, leading to the formation of orotate.

Ribose 5-phosphate is now added to orotate to produce orotidine monophosphate (OMP). This reaction is catalysed by orotate phosphoribosyltransferase, an enzyme comparable with HGPRT in its function. OMP undergoes decarboxylation to uridine monophosphate (UMP).

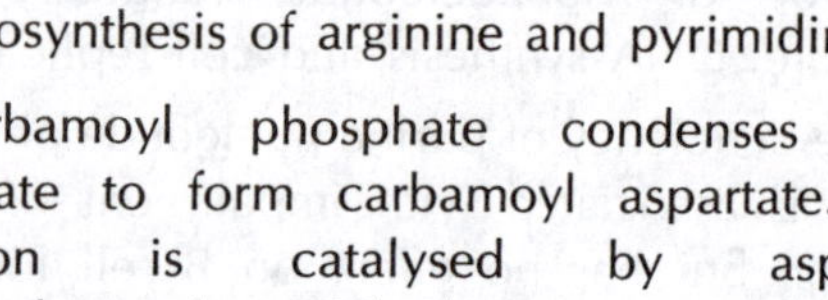

Fig. 17.11 : Sources of individual atoms in pyrimidine ring (Note : Same colours are used in the synthetic pathway, Fig. 17.12).

Orotate phosphoribosyltransferase and OMP decarboxylase are ***domains*** of a single protein. A defect in this ***bifunctional enzyme*** causes orotic aciduria (details given later).

By an ATP-dependent kinase reaction, UMP is converted to UDP which serves as a precursor for the synthesis of dUMP, dTMP, UTP and CTP.

Ribonucleotide reductase converts UDP to dUDP by a thioredoxin-dependent reaction. Thymidylate synthetase catalyses the transfer of a methyl group from N^5,N^{10}-methylene tetrahydrofolate to produce deoxythymidine monophosphate (dTMP).

UDP undergoes an ATP-dependent kinase reaction to produce UTP. Cytidine triphosphate (CTP) is synthesized from UTP by amination. CTP synthetase is the enzyme and glutamine provides the nitrogen.

Regulation of pyrimidine synthesis

In bacteria, ***aspartate transcarbamoylase*** (ATCase) catalyses a ***committed step*** in pyrimidine biosynthesis. ATCase is a good example of an enzyme controlled by ***feedback mechanism*** by the end product CTP. In certain bacteria, UTP also inhibits ATCase. ATP, however, stimulates ATCase activity.

Carbamoyl phosphate synthetase II (CPS II) is the regulatory enzyme of pyrimidine synthesis in animals. It is activated by PRPP and ATP and inhibited by UDP and UTP. OMP decarboxylase, inhibited by UMP and CMP, also controls pyrimidine formation.

Inhibition of pyrimidine synthesis

The enzyme ***thymidylate synthetase*** requires tetrahydrofolate derivative (N^5,N^{10}-methylene THF). This reaction is inhibited by methotrexate and therefore methotrexate serves as a good anticancer drug.

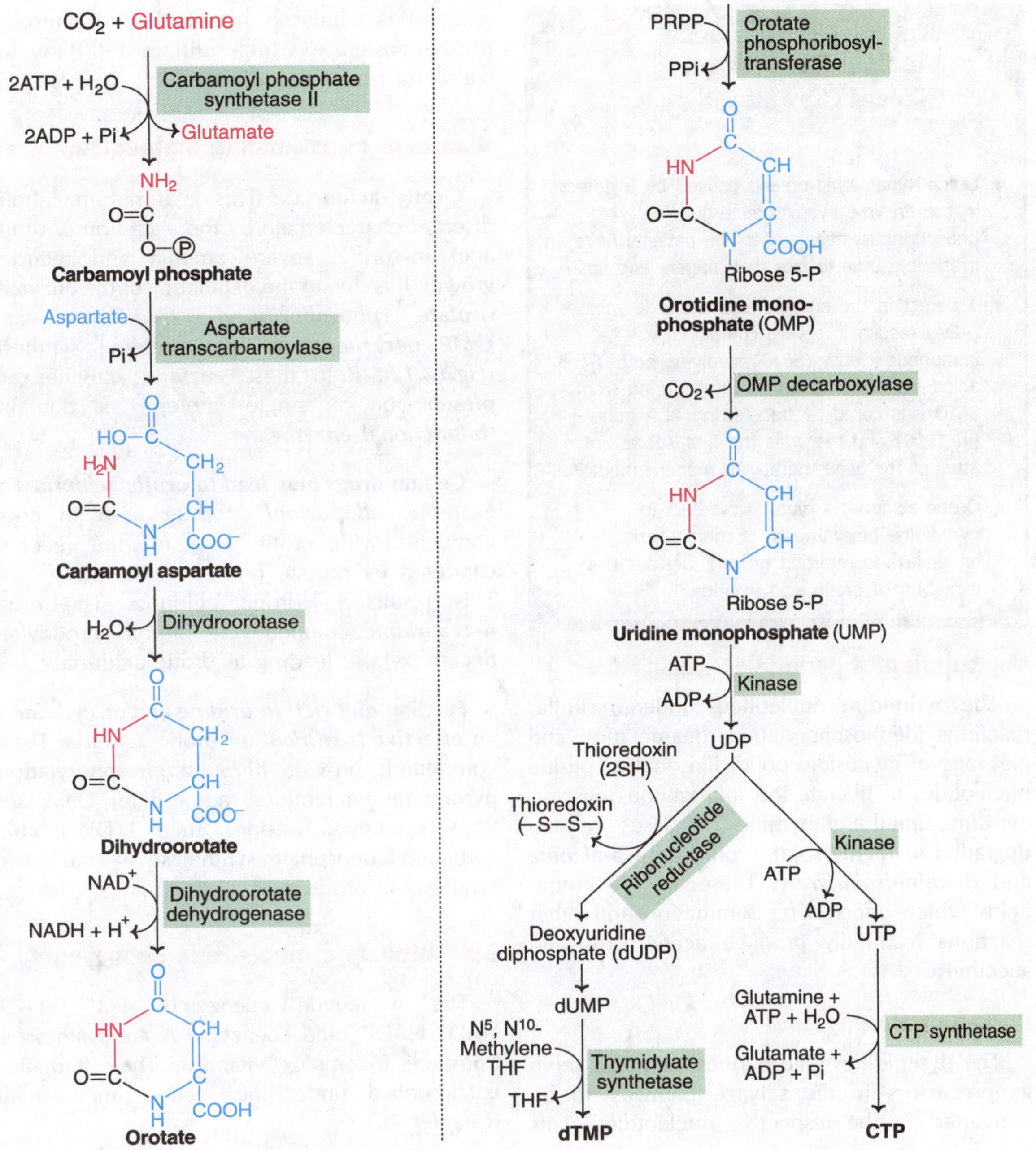

Fig. 17.12 contd. next column

Fig. 17.12 : Metabolic pathway for the synthesis of pyrimidine nucleotides.

Coordination of purine and pyrimidine nucleotide biosynthesis

There is a ***well coordinated*** and regulated control for the quantitative biosynthesis of purine and pyrimidine nucleotides. This is obvious from the fact that there is a ***mole for mole purines and pyrimidines ratio at the cellular level.*** The enzyme ***PRPP synthetase*** (I st reaction in purine synthesis), responsible for PRPP synthesis (needed for purine and pyrimidine formation) is feedback inhibited by both purine and pyrimidine nucleotides.

Medical Concepts/ Clinical Correlates

- Lesch-Nyhan syndrome is caused by a defect in the enzyme hypoxanthine-guanine phosphoribosyltransferase. The patients have an irresistible urge to bite their fingers and lips.
- A defect in the enzyme adenosine deaminase (ADA) results in severe combined immunodeficiency (SCID) involving both T-cell and B-cell dysfunction. A girl suffering from SCID was cured by transferring ADA gene (in 1990) and that was the first successful attempt for gene therapy in modern medicine.
- Orotic aciduria, a metabolic defect in pyrimidine biosynthesis, is characterized by anaemia and retarded growth, besides the excretion of orotic acid in urine.

Degradation of pyrimidine nucleotides

The pyrimidine nucleotides undergo similar reactions (dephosphorylation, deamination and cleavage of glycosidic bond) like that of purine nucleotides to liberate the nitrogenous bases—cytosine, uracil and thymine. The bases are then degraded to highly soluble products—***β-alanine*** and ***β-aminoisobutyrate.*** These are the amino acids which undergo transamination and other reactions to finally produce acetyl CoA and succinyl CoA.

Salvage pathway

The pyrimidines (like purines) can also serve as precursors in the salvage pathway to be converted to the respective nucleotides. This reaction is catalysed by pyrimidine phosphoribosyltransferase which utilizes PRPP as the source of ribose 5-phosphate.

Disorders of pyrimidine metabolism

Orotic aciduria : This is a rare metabolic disorder characterized by the excretion of orotic acid in urine, severe anemia and retarded growth. It is due to the deficiency of the enzymes ***orotate phosphoribosyl transferase*** and ***OMP decarboxylase*** of pyrimidine synthesis (***Fig.17.12***). Both these enzyme activities are present on a single protein as domains (***bifunctional enzyme***).

Certain ***drugs may lead to orotic aciduria***. For example, ***allopurinol*** (a drug used in gout) competes with orotic acid for the reaction catalysed by orotate phosphoribosyltransferase. This results in orotic aciduria. The drug, ***6-azauridine*** competitively inhibits orotidylate decarboxylase, leading to orotic aciduria.

Feeding ***diet rich in uridine*** and/or ***cytidine*** is an ***effective treatment*** for orotic aciduria. These compounds provide (through phosphorylation) pyrimidine nucleotides required for DNA and RNA synthesis. Besides this, UTP inhibits carbamoyl phosphate synthetase II and blocks synthesis of orotic acid.

Biosynthesis of nucleotide coenzymes

The nucleotide coenzymes FMN, FAD, NAD^+ $NADP^+$ and coenzyme A are synthesized from the B-complex vitamins. Their formation is described under the section on vitamins (***Chapter 7***).

SUMMARY

- *Nucleotides participate in a wide variety of reactions in the living cells—synthesis of DNA and RNA; as constituents of many coenzymes; in the regulation of metabolic reactions etc.*

- *Purine nucleotides are synthesized in a series of reactions starting from ribose 5-phosphate. Glycine, glutamine, aspartate, formate and CO_2 contribute to the synthesis of purine ring.*

- *Purine nucleotides can also be synthesized from free purines by a salvage pathway. The defect in the enzyme HGPRT causes Lesch-Nyhan syndrome.*

- *Deoxyribonucleotides are formed from ribonucleotides by a reduction process catalysed by ribonucleotide reductase. Thioredoxin is the protein cofactor required for this reaction.*

- *Purine nucleotides are degraded to uric acid, the excretory product in humans. Uric acid serves as a natural antioxidant in the living system.*

- *Uric acid in many animal species (other than primates) is converted to more soluble forms such as allantoin, allantoic acid etc., and excreted.*

- *Gout is a metabolic disease associated with overproduction of uric acid. This often leads to the accumulation of sodium urate crystals in the joints, causing painful gouty arthritis. Allopurinol, an inhibitor of xanthine oxidase, is the drug used for the treatment of gout.*

- *Pyrimidine nucleotides are synthesized from the precursors aspartate, glutamine and CO_2, besides ribose 5-phosphate.*

- *Orotic aciduria is a defect in pyrimidine synthesis caused by the deficiency of orotate phosphoribosyltransferase and OMP decarboxylase. Diet rich in uridine and/or cytidine is an effective treatment for orotic aciduria.*

- *Pyrimidines are degraded to amino acids, namely β-alanine and β-aminoisobutyrate which are then metabolized.*

THE READER MUST REFER APPENDIX VI (Page Nos : 738 - 756) FOR RELATED CLINICAL CASE STUDIES (Nos : 28, 29, 30)

Self-Assessment Exercises

I. Essay questions

1. Describe the catabolism of purine nucleotides and the associated metabolic disorders.
2. Write an account of the biosynthesis of inosine monophosphate.
3. Discuss the synthesis and degradation of pyrimidines.
4. Describe the role of PRPP in purine and pyrimidine synthesis.
5. Write an account of salvage pathway in purine nucleotide synthesis. Add a note on Lesch-Nyhan syndrome.

II. Short notes

(a) Gout, (b) PRPP, (c) Synthesis of deoxyribonucleotides, (d) Functions of nucleotides, (e) Immuno-deficiency diseases in purine metabolism, (f) Orotic aciduria, (g) Carbamoyl phosphate synthetase II, (h) HGPRT, (I) Degradation of uric acid in different animals, (j) Regulation of purine synthesis (k) Inhibitors of purine synthesis.

III. Fill in the blanks

1. The amino acids required for the synthesis of purines and pyrimidines are ________________.
2. The enzyme xanthine oxidase is inhibited by ________________.
3. Tophi are mostly made up of ________________.
4. Hypouricemia is due to the deficiency of the enzyme ________________.
5. The disorder in which the patients have an irresistible urge to bite their fingers and lips is ________________.
6. The cofactor required by the enzyme ribonucleotide reductase is ________________.
7. The 'parent' nucleotide synthesized in the biosynthesis of purines is ________________.
8. Xanthine oxidase converts allopurinol to ________________.
9. The amino acid that contributes to the synthesis of more than half of the pyrimidine ring ________________.
10. The regulatory enzyme in the pyrimidine biosynthesis in animals is ________________.

IV. Multiple choice questions

11. Name the enzyme associated with hyperuricemia
    (a) PRPP synthetase (b) HGPRT (c) Glucose 6-phosphatase (d) All of them.
12. An enzyme of purine metabolism associated with immunodeficiency disease
    (a) Adenosine deaminase (b) Xanthine oxidase (c) PRPP synthetase (d) HGPRT.
13. Orotic aciduria can be treated by a diet rich in
    (a) Adenine (b) Guanine (c) Uridine (d) Any one of them.
14. The end product of purine metabolism in humans is
    (a) Xanthine (b) Uric acid (c) Urea (d) Allantoin.
15. The nitrogen atoms in the purine ring are obtained from
    (a) Glycine (b) Glutamine (c) Aspartate (d) All of them.

Mineral Metabolism

18

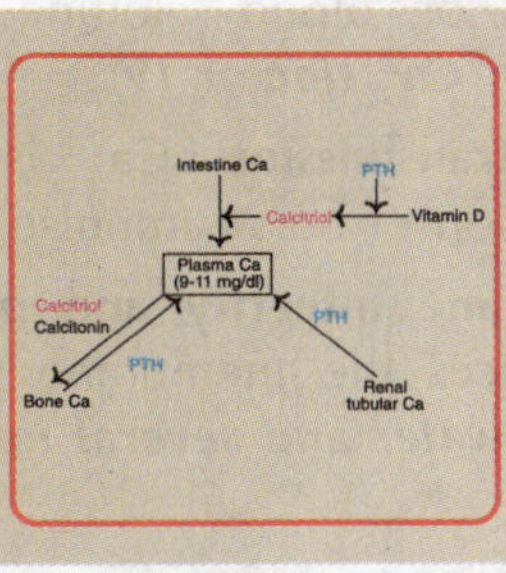

The mineral, calcium speaks

"I am the most abundant mineral;
Calcify and strengthen bones, teeth......
Coagulate blood and contract muscle;
Regulated by calcitriol, PTH and calcitonin."

The ***mineral (inorganic) elements*** constitute ***only a small proportion of the body weight***. There is a wide variation in their body content. For instance, calcium constitutes about 2% of body weight while cobalt about 0.00004%.

General functions

Minerals perform several vital functions which are absolutely ***essential for the very existence of the organism***. These include ***calcification*** of bone, ***blood coagulation***, neuromuscular irritability, ***acid-base equilibrium***, fluid balance and osmotic regulation.

Certain minerals are ***integral components*** of biologically important compounds such as ***hemoglobin*** (***Fe***), ***thyroxine*** (***I***), ***insulin*** (***Zn***) and vitamin B_{12} (Co). Sulfur is present in thiamine, biotin, lipoic acid and coenzyme A. Several minerals participate as cofactors for enzymes in metabolism (e.g. Mg, Mn, Cu, Zn, K). Some elements are essential constituents of certain enzymes (e.g. Co, Mo, Se).

Classification

The minerals are classified as principal elements and trace elements.

The seven **principal elements** (***macro-minerals***) constitute 60-80% of the body's inorganic material. These are ***calcium***, ***phosphorus***, ***magnesium***, ***sodium***, ***potassium***, ***chloride*** and ***sulfur***.

The principal elements are required in amounts greater than 100 mg/day.

The ***microminerals*** are required in amounts less than 100 mg/day. They are subdivided into three categories

1. **Essential trace elements :** ***Iron***, ***copper***, ***iodine***, ***manganese***, ***zinc***, ***molybdenum***, ***cobalt***, ***fluorine***, ***selenium*** and ***chromium***.

2. **Possibly essential trace elements :** Nickel, vanadium, cadmium and barium.

3. **Non-essential trace elements :** Aluminium, lead, mercury, boron, silver, bismuth etc.

A summary of the major characteristics of principal elements and trace elements is respectively given in ***Tables 18.1*** and ***18.2***. The individual elements are described next.

CALCIUM

Calcium is the ***most abundant among the minerals in the body***. The total content of calcium in an adult man is about 1 to 1.5 kg. As much as ***99%*** of it is ***present in the bones and teeth***. A small fraction (1%) of the calcium, found outside the skeletal tissue, performs a wide variety of functions.

Biochemical functions

1. **Development of bones and teeth :** Calcium, along with phosphate, is required for the formation (***of hydroxyapatite***) and physical strength of skeletal tissue. Bone is regarded as a mineralized connective tissue. Bones which are in a dynamic state serve as ***reservoir of Ca***. Osteoblasts are responsible for bone formation while osteoclasts result in demineralization.

2. **Muscle contraction :** Ca^{2+} interacts with troponin C to ***trigger muscle contraction***. Calcium also activates ATPase, increases the interaction between actin and myosin.

3. **Blood coagulation :** Several reactions in the cascade of blood clotting process are dependent on ***Ca^{2+}(factor IV)***.

4. **Nerve transmission :** Ca^{2+} is ***necessary*** for the transmission of nerve impulse.

5. **Membrane integrity and permeability :** Ca^{2+} ***influences*** the membrane structure and ***transport of water and several ions*** across it.

Table 18.1 A summary of the major characteristics of principal elements (macroelements)

Element	Major functions	Deficiency disease/symptoms	Recommended dietary allowance	Major sources
Calcium	Constituent of bones and teeth; muscle contraction, nerve transmission	Rickets; osteomalacia, osteoporosis	0.8–1.0 g/d	Milk and milk products, leafy vegetables, beans
Phosphorus	Constituent of bones and teeth; in the formation of high energy phosphates, nucleic acids, nucleotide coenzymes.	Rickets, osteomalacia	0.8–1.0 g/d	Milk, cereals, leafy vegetables
Magnesium	Constituent of bones and teeth; cofactor for enzymes e.g. kinases.	Neuromuscular weakness, irritation	300–350 mg/d	Cereals, vegetables, fruits, milk
Sodium	Chief cation of extracellular fluids; acid-base balance, osmotic pressure; nerve and muscle function	Almost unknown on normal diet	5–10 g/d	Table salt, salt added foods
Potassium	Chief cation of intracellular fluids; acid-base balance; osmotic pressure; muscle function	Muscular weakness, mental confusion	3–4 g/d	Fruits, nuts, vegetables
Chloride	Regulation of acid-base balance; formation of HCl	Almost unknown on normal diet	5–10 g/d	Table salt
Sulfur	Constituent of sulfur containing amino acids, certain vitamins (thiamine, biotin) and other compounds (heparin, chondroitin sulfate).	Almost unknown	—	Sulfur containing amino acids

6. **Activation of enzymes :** Ca^{2+} is needed for the direct activation of enzymes such as ***lipase*** (pancreatic), ***ATPase*** and succinate dehydrogenase.

7. **Calmodulin mediated action of Ca^{2+} :** Calmodulin (mol. wt. 17,000) is a calcium binding regulatory protein. ***Ca-calmodulin complex activates certain enzymes*** e.g. adenylate cyclase, Ca^{2+} dependent protein kinases, myosin kinase, phospholipase C, glycogen synthase. The mechanism of action of calmodulin is given in ***Fig.18.1***.

8. **Calcium as intracellular messenger :** Certain hormones exert their action through the mediation of Ca^{2+} (instead of cAMP). Calcium is regarded as a ***second messenger*** for such hormonal action e.g. epinephrine in liver glycogenolysis. Calcium serves as a third messenger for some hormones e.g. antidiuretic hormone (ADH) acts through cAMP, and then Ca^{2+}.

9. **Release of hormones :** The release of certain hormones (insulin, PTH, calcitonin) from the endocrine glands is facilitated by Ca^{2+}.

Table 18.2 A summary of the major characteristics of trace elements (microelements)

Element	Major functions	Deficiency disease/symptoms	Recommended dietary allowance	Major sources
Iron	Constituent of heme e.g. hemoglobin, myoglobin, cytochromes; involved in O_2 transport and biological oxidation.	Hypochromic, microcytic anemia	10–15 mg/d	Organ meats (liver, heart), leafy vegetables, iron cookware
Copper	Constituent of enzymes e.g. cytochrome C oxidase, catalase, tyrosinase; in iron transport.	Anemia, Menke's disease	2–3 mg/d	Organ meats, cereals, leafy vegetables
Iodine	Constituent of thyroxine and triiodothyronine	Cretinism, goiter, myxedema	150–200 μg/d	Iodized salt, sea foods
Manganese	Cofactor for enzymes e.g. arginase, pyruvate carboxylase; glycoprotein synthesis.	Almost unknown	2–9 mg/d	Cereals, leafy vegetables
Zinc	Cofactor for enzymes e.g. alcohol dehydrogenase, carbonic anhydrase, lactate dehydrogenase.	Growth retardation, poor wound healing, hypogonadism	10–15 mg/d	Meat, fish, milk
Molybdenum	Constituent of enzymes e.g. xanthine oxidase	Almost unknown	75–250 μg/d	Vegetables
Cobalt	Constituent of vitamin B_{12}, required for the formation of erythrocytes	Pernicious anemia (as in vitamin B_{12} deficiency)	5–8 μg/d	Foods of animal origin
Fluorine	Helps in the proper formation of bones and teeth	Dental caries, osteoporosis	2–4 mg/d	Drinking water
Selenium	Involved in antioxidant function along with vitamin E; constituent of glutathione peroxidase and selenocysteine	Muscular degeneration, cardiomyopathy	50–200 μg/d	Organ meats, sea foods
Chromium	Promotes insulin function (as glucose tolerance factor)	Impaired glucose tolerance	10–100 μg/d	Brewer's yeast, meat, whole grains

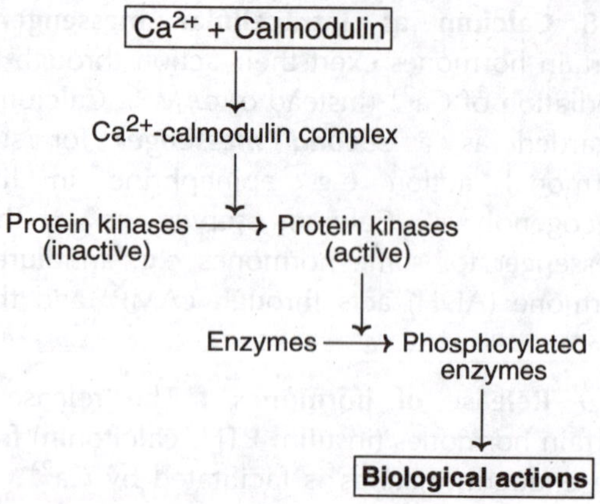

Fig. 18.1 : Mechanism of action of calcium mediated calmodulin.

10. **Secretory processes :** Ca^{2+} regulates microfilament and microtubule mediated processes such as endocytosis, exocytosis and cell motility.

11. **Contact inhibition :** Calcium is believed to be involved in cell to cell contact and adhesion of cells in a tissue (**Refer page 663** also). The cell to cell communication may also require Ca^{2+}.

12. **Action on heart :** Ca^{2+} acts on myocardium and ***prolongs systole***.

Dietary requirements

Adult men and ***women***	—	***800 mg/day***
Women during pregnancy, lactation and post-menopause	—	1.5 g/day
Children (1-18 yrs.)	—	0.8–1.2 g/day
Infants (<1 year)	—	300–500 mg/day

Sources

Best sources	—	***Milk*** and ***milk products***
Good sources	—	Beans, leafy vegetables, fish, cabbage, egg yolk.

Absorption

The absorption of calcium mostly occurs in the duodenum by an energy dependent active process. It is influenced by several factors.

Factors promoting Ca absorption

1. Vitamin D (through its active form ***calcitriol***) induces the synthesis of calcium binding protein in the intestinal epithelial cells and ***promotes Ca absorption***.

2. ***Parathyroid hormone enhances Ca*** absorption through the increased synthesis of calcitriol.

3. ***Acidity*** (low pH) is more ***favourable*** for Ca absorption.

4. ***Lactose promotes*** calcium uptake by intestinal cells.

5. The amino acids ***lysine*** and ***arginine facilitate*** Ca absorption.

Factors inhibiting Ca absorption

1. ***Phytates and oxalates*** form insoluble salts and interfere with Ca absorption.

2. High content of dietary phosphate results in the formation of insoluble calcium phosphate and prevents Ca uptake. The dietary ratio of ***Ca and P—between 1 : 2 and 2 : 1—is ideal for*** optimum Ca absorption by intestinal cells.

3. The free fatty acids react with Ca to form insoluble calcium soaps. This is particularly observed when the fat absorption is impaired.

4. ***Alkaline condition*** (high pH) is ***unfavourable*** for Ca absorption.

5. ***High content of dietary fiber interferes*** with Ca absorption.

Plasma calcium

Most of the blood Ca is present in the plasma since the blood cells contain very little of it. The normal concentration of plasma or serum Ca is ***9-11 mg/dl*** (***4.5-5.5 mEq/l***). About half of this (5 mg/dl) is in the ionized form which is functionally the most active (***Fig.18.2***). At least 1 mg/dl serum Ca is found in association with citrate and/or phosphate. About 40% of serum Ca (4-5mg/dl) is bound to proteins, mostly albumin and, to a lesser extent, globulin. Ionized and citrate (or phosphate) bound Ca is diffusible from blood to the tissues while protein bound Ca is non-diffusible. In the usual laboratory

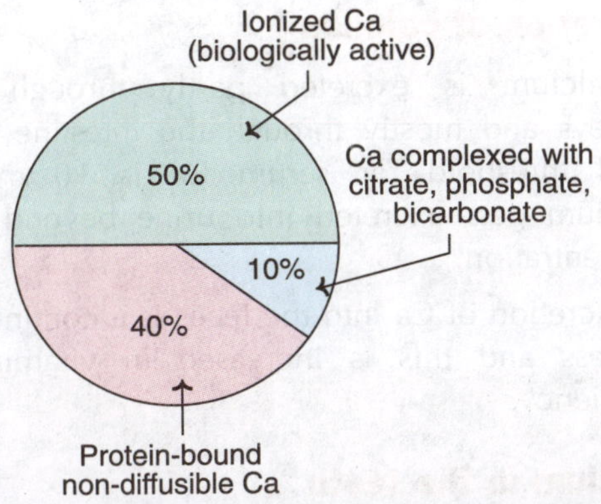

***Fig. 18.2 :** Different forms of circulating calcium.*

determination of serum Ca, all the three fractions are measured together.

FACTORS REGULATING PLASMA Ca LEVEL

As already stated, calcium is almost exclusively present in blood plasma (or serum). The hormones—***calcitriol, parathyroid hormone*** (PTH) and ***calcitonin*** are the major factors that regulate the plasma calcium *(**homeostasis of Ca; Fig.18.3**)* within a narrow range (9-11 mg/dl).

Calcitriol

The physiologically active form of vitamin D is a hormone, namely calcitriol or ***1,25-dihydroxy-cholecalciferol*** (***1,25 DHCC***). The synthesis of calcitriol and its wide range of biochemical actions are described under ***Vitamins*** (***Chapter 7***).

Calcitriol induces the synthesis of a specific calcium binding protein in the intestinal cells. This protein increases the intestinal absorption of calcium as well as phosphate. Thus blood Ca level is increased by calcitriol (the active vitamin D). Furthermore, calcitriol ***stimulates calcium uptake by osteoblasts of bone and promotes calcification*** or mineralization (deposition of calcium phosphate) and remodelling.

Parathyroid hormone

Parathyroid hormone (PTH) is secreted by two pairs of parathyroid glands that are closely associated with thyroid glands. Parathyroid hormone (mol. wt. 95,000) is a single chain polypeptide, containing 84 amino acids. It is originally synthesized as preproPTH which is degraded to proPTH and, finally, to active PTH. The rate of formation (by degradation of proPTH) and the secretion of PTH are promoted by low Ca^{2+} concentration. Thus, the ***release of PTH*** from parathyroid glands is ***under the negative feedback regulation of serum*** Ca^{2+}.

Mechanism of action of PTH : PTH binds to a membrane receptor protein on the target cell and activates adenylate cyclase to liberate cAMP. This, in turn, increases intracellular calcium that promotes the phosphorylation of proteins (by kinases) which, finally brings about the biological actions. PTH has 3 independent tissues—bone, kidneys and intestine—to exert its action. The ***prime function of PTH is to elevate serum calcium level.***

Action on the bone : ***PTH*** causes decalcification or ***demineralization of bone***, a process carried out ***by osteoclasts***. This is brought out by PTH stimulated increased activity of the enzymes pyrophosphatase and collagenase. These enzymes result in bone resorption. Demineralization ultimately leads to an increase in the blood Ca level. The action of PTH on bone is quantitatively very significant to maintain Ca homeostasis. It must, however, be noted that this is being done at the expense of loss of Ca from bone, particularly in dietary Ca deficiency.

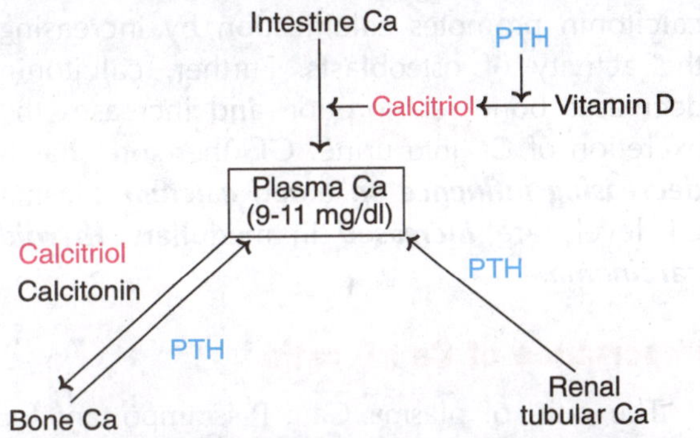

***Fig. 18.3 :** Overview of calcium homeostasis (PTH–Parathyroid hormone).*

Medical Concepts/ Clinical Correlates

- Serum calcium level is increased (normal 9–11 mg/dl) in hyperparathyroidism. This condition is also associated with elevated urinary excretion of Ca and P, often leading to stone formation.
- Tetany, caused by a drastic reduction in serum Ca, is characterized by neuromuscular irritability and convulsions.
- Rickets is due to defective calcification of bones. This may be caused by deficiency of Ca and P or vitamin D or both.

Action on the kidney : ***PTH increases the Ca reabsorption*** by kidney tubules. This is the most rapid action of PTH to elevate blood Ca levels. However, quantitatively, this is less important compared to the action of PTH on bone. PTH promotes the production of calcitriol (1,25 DHCC) in the kidney by stimulating 1-hydroxylation of 25-hydroxycholecalciferol.

Action on the intestine : The action of PTH on the intestine is indirect. It ***increases the intestinal absorption of Ca*** by promoting the synthesis of calcitriol.

Calcitonin (CT)

Calcitonin is a peptide containing 32 amino acids. It is secreted by parafollicular cells of thyroid gland. The ***action of CT on calcium metabolism is antagonistic to that of PTH***. Thus, calcitonin promotes calcification by increasing the activity of osteoblasts. Further, calcitonin decreases bone reabsorption and increases the excretion of Ca into urine. CT, therefore, has a ***decreasing influence on blood calcium***. Plasma CT levels are ***increased*** in medullary ***thyroid carcinoma***.

Importance of Ca : P ratio

The ratio of plasma Ca : P is important for calcification of bones. The product of Ca × P (in mg/dl) in children is around 50 and in adults 40. This product is less than 30 in rickets.

Excretion of calcium

Calcium is excreted partly through the kidneys and mostly through the intestine. The renal threshold for serum Ca is 10 mg/dl. Calcium gets excreted into urine beyond this concentration.

Excretion of Ca into the feces is a continuous process and this is increased in vitamin D deficiency.

Calcium in the teeth

The teeth calcium is not subjected to regulation as observed for bone calcium. Thus the adult teeth, once formed, ***do not undergo decalcification*** to meet the body needs of calcium. However, proper calcification of teeth is important in the growing children.

DISEASE STATES

The blood Ca level is maintained within a narrow range by the ***homeostatic control***, most predominantly by PTH. Hence ***abnormalities*** in Ca metabolism are mainly ***associated with alterations in PTH***.

Hypercalcemia

Elevation in serum Ca level (***normal 9–11 mg/dl***) is hypercalcemia. Hypercalcemia is associated with ***hyperparathyroidism*** caused by increased activity of parathyroid glands. Decrease in serum phosphate (due to increased renal losses) and increase in alkaline phosphatase activity are also found in hyperparathyroidism. Elevation in the urinary excretion of Ca and P, often resulting in the formation of urinary calculi, is also observed in these patients.

The determination of ionized serum calcium (elevated to 6-9 mg/dl) is more useful for the diagnosis of hyperparathyroidism. It has been observed that some of the patients may have normal levels of total calcium in the serum but differ with regard to ionized calcium.

The symptoms of hypercalcemia include lethargy, muscle weakness, loss of appetite, constipation, nausea, increased myocardial contractility and susceptibility to fractures.

Hypocalcemia

Hypocalcemia is a more serious and life threatening condition. It is characterized by a fall in the serum Ca to ***below 7 mg/dl***, causing ***tetany***. The symptoms of tetany include neuromuscular irritability, and convulsions.

Hypocalcemia is mostly due to ***hypoparathyroidism***. This may happen after an accidental surgical removal of parathyroid glands or due to an autoimmune disease.

Treatment : Supplementation of ***oral calcium*** with ***vitamin D*** is commonly employed. In severe cases of hypocalcemia, ***calcium gluconate*** is intravenously administered.

Rickets

Rickets is a disorder of defective calcification of bones. This may be due to a ***low levels of vitamin D*** in the body or due to a dietary deficiency of Ca and P — or both. The concentrations of serum Ca and P may be low or normal. An ***increase*** in the activity of ***alkaline phosphatase*** is a characteristic feature of rickets.

Renal rickets

Renal rickets is associated with damage to renal tissue, causing impairment in the synthesis of calcitriol. It does not respond to vitamin D in ordinary doses, therefore, some workers regard this as ***vitamin D resistant rickets***. Renal rickets can be treated by administration of calcitriol.

Osteoporosis

Osteoporosis is characterized by ***demineralization of bone*** resulting in the progressive loss of bone mass.

Occurrence : The ***elderly people*** (over 60 yr.) of both sexes are ***at risk*** for osteoporosis. However, it more predominantly occurs in the post-menopausal women. Osteoporosis results in ***frequent bone fractures*** which are a major cause of disability among the elderly. It is estimated that more than 50% of the fractures in USA are due to this disorder. Osteoporosis may be regarded as a silent thief.

Etiology : The etiology of osteoporosis is largely unknown, but it is believed that several causative factors may contribute to it. The ability to produce ***calcitriol*** from vitamin D is ***decreased with age***, particularly in the postmenopausal women. Immobilized or sedentary individuals tend to decrease bone mass while those on regular exercise tend to increase bone mass. ***Deficiency of sex hormones*** (in women) has been implicated in the development of osteoporosis.

Treatment : ***Estrogen*** administration along with ***calcium*** supplementation (in combination with vitamin D) to postmenopausal women reduces the risk of fractures. Higher dietary intake of Ca (about 1.5 g/day) is recommended for elderly people.

Osteopetrosis (marble bone disease)

Osteopetrosis is characterized by ***increased bone density***. This is primarily due to inability to resorb bone. This disorder is mostly ***observed in*** association with ***renal tubular acidosis*** (due to a defect in the enzyme carbonic anhydrase) and cerebral calcification.

PHOSPHORUS

An adult body contains about ***1 kg phosphate*** and it is found in every cell of the body. Most of it (about 80%) occurs in combination with Ca in the bones and teeth. About 10% of body P is found in muscles and blood in association with proteins, carbohydrates and lipids. The remaining 10% is widely distributed in various chemical compounds.

Biochemical functions

1. Phosphorus is ***essential for*** the development of ***bones and teeth***.

2. It plays a central role in the formation and utilization of ***high-energy phosphate compounds*** e.g. ***ATP***, GTP, creatine phosphate etc.

3. Phosphorus is required for the formation of ***phospholipids***, phosphoproteins and nucleic acids (DNA and RNA).

4. It is an essential component of several ***nucleotide coenzymes*** e.g. NAD^+, $NADP^+$, pyridoxal phosphate, ADP, AMP.

5. Several ***proteins and enzymes are activated*** by phosphorylation.

6. ***Phosphate buffer*** system is important for the maintenance of pH in the blood (around 7.4) as well as in the cells.

Dietary requirements

The recommended dietary allowance (RDA) of phosphate is based on the intake of calcium. The ratio of ***Ca : P of 1 : 1 is recommended*** (i.e. ***800 mg/day***) for an adult. For infants, however, the ratio is around 2 : 1, which is based on the ratio found in human milk. Calcium and phosphate are distributed in the majority of natural foods in 1 : 1 ratio. Therefore, adequate intake of Ca generally takes care of the P requirement also.

Sources

Milk, ***cereals***, leafy vegetables, meat, eggs.

Absorption

Phosphate absorption occurs from jejunum

1. Calcitriol promotes phosphate uptake along with calcium.

2. Absorption of phosphorus and calcium is optimum when the dietary Ca : P is between 1 : 2 and 2 : 1.

3. Acidity favours while phytate decreases phosphate uptake by intestinal cells.

Serum phosphate

The phosphate level of the whole blood is around 40 mg/dl while serum contains about ***3-4 mg/dl***. This is because the RBC and WBC have very high content of phosphate.

The serum phosphate may exist as free ions (40%) or in a complex form (50%) with cations such as Ca^{2+}, Mg^{2+}, Na^+, K^+. About 10% of serum phosphate is bound to proteins. It is interesting to note that the ***fasting serum phosphate*** levels are ***higher*** than the post-prandial. This is attributed to the fact that following the ingestion of carbohydrate (glucose), the phosphate from the serum is drawn by the cells for metabolism (phosphorylation reactions).

Excretion

About 500 mg phosphate is excreted in urine per day. The renal threshold is 2 mg/dl. The reabsorption of phosphate by renal tubules is inhibited by PTH.

Disease states

1. Serum ***phosphate*** level is ***increased in hypoparathyroidism*** and decreased in hyperparathyroidism.

2. In severe renal diseases, serum phosphate content is elevated causing acidosis.

3. Vitamin D deficient rickets is characterized by decreased serum phosphate (1–2 mg/dl).

4. Renal rickets is associated with low serum phosphate levels and increased alkaline phosphatase activity.

5. In diabetes mellitus, serum content of organic phosphate is lower while that of inorganic phosphate is higher.

MAGNESIUM

The adult body contains about 20 g magnesium, 70% of which is found in bones in combination with calcium and phosphorus. The remaining 30% occurs in the soft tissues and body fluids.

Biochemical functions

1. Magnesium is required for the ***formation of bones and teeth***.

2. Mg^{2+} serves as a ***cofactor for several enzymes*** requiring ATP e.g. ***hexokinase, glucokinase***, phosphofructokinase, ***adenylate cyclase***.

3. Mg^{2+} is necessary for proper ***neuro-muscular*** function. Low Mg^{2+} levels lead to neuromuscular irritability.

Dietary requirements

Adult man — ***350 mg/day***
Adult woman — 300 mg/day

Sources

Cereals, nuts, beans, vegetables (cabbage, cauliflower), meat, milk, fruits.

Absorption

Magnesium is absorbed by the intestinal cells through a specific carrier system. About 50% of the dietary Mg is normally absorbed. Consumption of large amounts of calcium, phosphate and alcohol diminishes Mg absorption. PTH increases Mg absorption.

Serum Mg

Normal serum concentration of Mg is ***2–3 mg/dl***. It is present in the ionized form (60%), in combination with other ions (10%) and bound to proteins (30%).

Disease states

1. Magnesium deficiency causes neuromuscular irritation, weakness and convulsions. These symptoms are similar to that observed in tetany (Ca deficiency) which are relieved only by Mg. Malnutrition, alcoholism, hyperthyroidsm and cirrhosis of liver may lead to Mg deficiency.

2. Low levels of Mg may be observed in uremia, rickets and abnormal pregnancy.

3. A ***raise*** in Mg levels is observed in ***uncontrolled diabetes and hypothyroidism***.

SODIUM

Sodium is the ***chief cation*** of the ***extracellular fluid***. About 50% of body sodium is present in the bones, 40% in the extracellular fluid and the remaining (10%) in the soft tissues.

Biochemical functions

1. In association with chloride and bicarbonate, sodium ***regulates*** the body's ***acid-base balance***.

2. Sodium is required for the maintenance of ***osmotic pressure and fluid balance***.

3. It is necessary for the normal muscle irritability and cell permeability.

4. Sodium is involved in the intestinal ***absorption of glucose***, galactose and amino acids.

5. It is necessary for initiating and maintaining ***heart beat***.

Dietary requirements

For normal individuals, the requirement of sodium is about 5-10 g/day which is mainly consumed as NaCl. For persons with a family history of hypertension, the daily NaCl intake should be less than 5 g. For patients of hypertension, around 1 g/day is recommended. It may be noted that 10 g of NaCl contains 4 g of sodium. The daily consumption of Na is generally higher than required due to its flavour.

Sources

The ***common salt*** (***NaCl***) used in the cooking medium is the major source of sodium. The ingested foods also contribute to sodium. The good sources of sodium include bread, whole grains, leafy vegetables, nuts, eggs and milk.

Absorption

Sodium is readily absorbed in the gastrointestinal tract and, therefore, very little of it (< 2%) is normally found in feces. However, in diarrhea, large quantities of sodium is lost in feces.

Plasma sodium

In the plasma (serum), the normal concentration of sodium is ***135-145 mEq/l***. Sodium is an ***extracellular cation,*** therefore, the blood cells contain much less (35 mEq/l). The mineralocorticoids, secreted by adrenal cortex, influence sodium metabolism. A decrease in plasma sodium and an increase in its urinary excretion are observed in adrenocortical insufficiency.

Excretion

Kidney is the major route of sodium excretion from the body. As much as 800 g Na/day is filtered by the glomeruli, 99% of this is reabsorbed by the renal tubules by an active process. This is ***controlled by aldosterone***. Extreme sweating also causes considerable amount of sodium loss from the body. There is, however, individual variation in sodium loss through sweat.

Disease states

1. **Hyponatremia** : This is a condition in which the ***serum sodium level falls*** below the normal. Hyponatremia may occur due to diarrhea, vomiting, chronic renal diseases, adrenocortical insufficiency (***Addison's disease***). Administration of salt free fluids to patients may also cause hyponatremia. This is due to overhydration. Decreased serum sodium concentration is also observed in edema which occurs in cirrhosis or congestive heart failure.

The manifestations of hyponatremia include reduced blood pressure and circulatory failure.

2. **Hypernatremia** : This condition is characterized by an ***elevation in the serum sodium level***. The symptoms include increase in blood volume and blood pressure. It may occur due to hyperactivity of adrenal cortex (***Cushing's syndrome***), prolonged administration of cortisone, ACTH and/or sex hormones. Loss of water from the body causing dehydration, as it occurs in diabetes insipidus, results in hypernatremia. Rapid administration of sodium salts also increases serum sodium concentration. It may be noted that in pregnancy, steroid and placental hormones cause sodium and water retention in the body, leading to edema.

In ***edema***, along with ***water***, ***sodium*** concentration in the body is also ***elevated***. Administration of diuretic drugs increases the urinary output of water along with sodium. In the patients of ***hypertension*** and congestive cardiac failure salt (***Na^+***) ***restriction*** is advocated.

POTASSIUM

Potassium is the principal ***intracellular cation***. It is equally important in the extracellular fluid for specific functions.

Biochemical functions

1. Potassium maintains ***intracellular osmotic pressure***.

2. It is required for the regulation of acid-base balance and water balance in the cells.

3. The enzyme pyruvate kinase (of glycolysis) is dependent on K^+ for optimal activity.

4. Potassium is required for the transmission of nerve impulse.

5. Adequate intracellular concentration K^+ is necessary for proper biosynthesis of proteins by ribosomes.

6. Extracellular K^+ influences cardiac muscle activity.

Dietary requirements

About ***3-4 g/day***.

Sources

Banana, orange, pineapple, potato, beans, chicken, and liver. Tender coconut water is a rich source of potassium.

Absorption

The absorption of K^+ from the gastrointestinal tract is very efficient (90%) and very little is lost through feces. However, in subjects with diarrhea, a good proportion of K^+ is lost in the feces.

Plasma potassium

The plasma (serum) concentration of potassium is ***3.4-5.0 mEq/l***. The whole blood contains much higher level of K^+ (50 mEq/l), since it is predominantly an intracellular cation. Care should, therefore, be taken to avoid hemolysis of RBC for the estimation of serum K^+.

Excretion

Potassium is mainly excreted through urine. The maintenance of body acid-base balance influences K^+ excretion. Aldosterone increases excretion of potassium.

Disease states

Serum potassium concentration is maintained within a narrow range. Either high or low concentrations are dangerous since potassium effects the contractility of heart muscle.

Hypokalemia : ***Decrease*** in the concentration of ***serum potassium*** is observed due to overactivity of adrenal cortex ***(Cushing's syndrome)***, prolonged cortisone therapy, intravenous administration of K^+-free fluids, treatment of diabetic coma with insulin, prolonged diarrhea and vomiting.

The symptoms of hypokalemia include irritability, muscular weakness, tachycardia, cardiomegaly and cardiac arrest. Changes in the ECG are observed (flattened waves with inverted T wave).

Hyperkalemia : ***Increase*** in the concentration of ***serum potassium*** is observed in renal failure, adrenocortical insufficiency ***(Addison's disease)***, diabetic coma, severe dehydration, intravenous administration of fluids with excessive potassium salts.

The manifestations of hyperkalemia include depression of central nervous system, mental confusion, numbness, bradycardia with reduced heart sounds and, finally, cardiac arrest. Changes in ECG are also observed (elevated T wave).

CHLORIDE

Chlorine is a constituent of sodium chloride. Hence, the metabolism of chlorine and sodium are intimately related.

Biochemical functions

1. Chloride is involved in the regulation of ***acid-base equilibrium***, fluid balance and osmotic pressure. These functions are carried out by the interaction of chloride with Na^+ and K^+.

2. Chloride is necessary for the formation of ***HCl in the gastric juice***.

3. Chloride shift involves the active participation of Cl^-.

4. The enzyme ***salivary amylase is activated*** by chloride.

Dietary requirements

The daily requirement of chloride as ***NaCl is 5-10 g***. Adequate intake of sodium will satisfy the chloride requirement of the body.

Sources

Common salt as cooking medium, whole grains, leafy vegetables, eggs and milk.

Absorption

In normal circumstances, chloride is almost totally absorbed in the gastrointestinal tract.

Plasma chloride

The normal plasma concentration of chloride is 95-105 mEq/l. Cerebrospinal fluid (CSF) contains higher level of Cl^- (125 mEq/l). This is due to the fact that protein content is low in CSF and, therefore, Cl^- is higher in order to maintain Donnan membrane equilibrium.

Excretion

There exists a parallel relationship between excretion of chloride and sodium. The renal threshold for Cl^- is about 110 mEq/l.

Disease states

1. **Hypochloremia :** A reduction in the serum Cl^- level may occur due to vomiting, diarrhea, respiratory alkalosis, Addison's disease and excessive sweating.

2. **Hyperchloremia :** An increase in serum Cl^- concentration may be due to dehydration, respiratory acidosis and Cushing's syndrome.

SULFUR

Sulfur of the body is mostly present in the organic form. Methionine, cysteine and cystine are the three sulfur-containing amino acids present in the proteins. Generally, proteins contain about 1% sulfur by weight.

Biochemical functions

1. ***Sulfur-containing amino acids*** are very essential for the structural conformation and

> **Medical Concepts/ Clinical Correlates**
>
> - Osteoporosis is a bone disorder of the elderly, characterized by demineralization resulting in a progressive loss of bone mass. It is the major cause of bone fractures and disability in the old people.
> - Decreased levels of serum Na (hyponatremia) is observed in diarrhea and vomiting, besides Addison's disease, while increased serum Na (hypernatremia) is found in Cushing's syndrome.
> - Iron deficiency anemia is the most prevalent nutritional disorder worldover. It is most commonly observed in pregnant and lactating women.

biological functions of proteins (enzymes, hormones, structural proteins etc.). The disulfide linkages (—S—S—) and sulfhydryl groups (—SH) are largely responsible for this.

2. The vitamins ***thiamine***, ***biotin***, ***lipoic acid***, and coenzyme A of pantothenic acid contain sulfur.

3. ***Heparin***, ***chondroitin sulfate***, glutathione, taurocholic acid are some other important sulfur-containing compounds.

4. ***Phosphoadenosine phosphosulfate*** (PAPS) is the active sulfate utilized for several reactions e.g. synthesis of glycosaminoglycans, detoxification mechanism.

5. The sulfur-containing amino acid methionine (as S–adenosylmethionine) is actively involved in ***transmethylation reactions***.

Dietary requirements and sources

There is no specific dietary requirement for sulfur. Adequate intake of sulfur-containing essential amino acid methionine will meet the body needs. Food proteins rich in methionine and cysteine are the sources of sulfur.

Excretion

The sulfur from different compounds is oxidized in the liver to sulfate and excreted in urine. The urine contains inorganic sulfate (80%), organic or conjugated or ethereal sulfate (10%) and unoxidized sulfur (10%). The unoxidized sulfur is in the form of sulfur-containing amino acids, thiocyanates etc.

IRON

The ***total content*** of iron in an adult body is ***3-5 g***. About 70% of this occurs in the erythrocytes of blood as a constituent of hemoglobin. At least 5% of body iron is present in myoglobin of muscle. Heme is the most predominant iron-containing substance. It is a constituent of several proteins/enzymes ***(hemoproteins)—hemoglobin***, ***myoglobin***, ***cytochromes***, xanthine oxidase, catalase, tryptophan pyrrolase, peroxidase. Certain other proteins contain ***non-heme iron*** e.g. ***transferrin***, ***ferritin***, hemosiderin.

Biochemical functions

1. Iron mainly exerts its functions through the compounds in which it is present. ***Hemoglobin*** and ***myoglobin*** are required for the ***transport of*** O_2 ***and*** CO_2.

2. Cytochromes and certain non-heme proteins are necessary for ***electron transport chain*** and oxidative phosphorylation.

3. Peroxidase, the lysosomal enzyme, is required for ***phagocytosis*** and killing of bacteria by neutrophils.

4. Iron is associated with effective ***immunocompetence*** of the body.

Dietary requirements

Adult ***man***	— ***10 mg/day***
Menstruating ***woman***	— ***18 mg/day***
Pregnant and lactating woman	— 40 mg/day

Sources

Rich sources — ***Organ meats*** (liver, heart, kidney).

Good sources — Leafy vegetables, pulses, cereals, fish, apples, dried fruits, molasses, jaggery.

Poor sources — ***Milk***, wheat, polished rice.

ABSORPTION, TRANSPORT AND STORAGE

Iron is mainly absorbed in the stomach and duodenum and upper jejunum. In normal people, ***about 10%*** of dietary iron is usually ***absorbed***. However, in iron deficient (anemic) individuals and growing children, a much higher proportion of dietary iron is absorbed to meet the increased body demands.

Iron is mostly found in the foods in ferric form (Fe^{3+}), bound to proteins or organic acids. In the acid medium provided by gastric HCl, the Fe^{3+} is released from foods. Reducing substances such as ascorbic acid (vitamin C) and cysteine convert ferric iron (Fe^{3+}) to ferrous form (Fe^{2+}). Iron in the ***ferrous form is soluble and readily absorbed.***

Factors affecting Fe absorption

1. ***Acidity***, ***ascorbic acid***, alcohol, fructose and ***cysteine promote*** iron absorption.

2. In ***iron deficiency anemia***, Fe absorption is ***increased*** to 2-10 times that of normal.

3. Small peptides and amino acids favour iron uptake.

4. ***Phytate*** (found in cereals) and ***oxalate*** (found in leafy vegetables) ***interfere*** with Fe absorption.

5. A diet with high phosphate content decreases Fe absorption while low phosphate promotes.

6. Impaired absorption of iron is observed in malabsorption syndromes such as steatorrhea.

7. In patients with partial or total surgical removal of stomach and/or intestine, iron absorption is severely impaired.

Mucosal block theory

Homeostasis of iron in the body is maintained by ***regulating iron absorption***, and not by its excretion. When iron stores in the body are less its absorption is increased. On the other hand, if iron stores are adequate, then absorption is decreased. This unique ***phenomenon of regulating iron absorption*** is referred to as mucosal block theory. It is interesting to note that no other nutrient in the body is regulated in this fashion.

Iron in the mucosal cells : The iron (Fe^{2+}) entering the mucosal cells by absorption is oxidized to ferric form (Fe^{3+}) by the enzyme ferroxidase. Fe^{3+} then combines with ***apoferritin to form ferritin*** which is the temporary storage form of iron. From the mucosal cells, iron may enter the blood stream (which mainly depends on the body needs) or lost when the cells are desquamated.

Transport of Fe in the plasma

The iron liberated from the ferritin of mucosal cells enters the plasma in ferrous state. Here, it is oxidized to ferric form by a copper-containing protein, ***ceruloplasmin*** which possesses ***ferroxidase activity***. Another cuproprotein ferroxidase II also helps for the conversion of Fe^{2+} to Fe^{3+}.

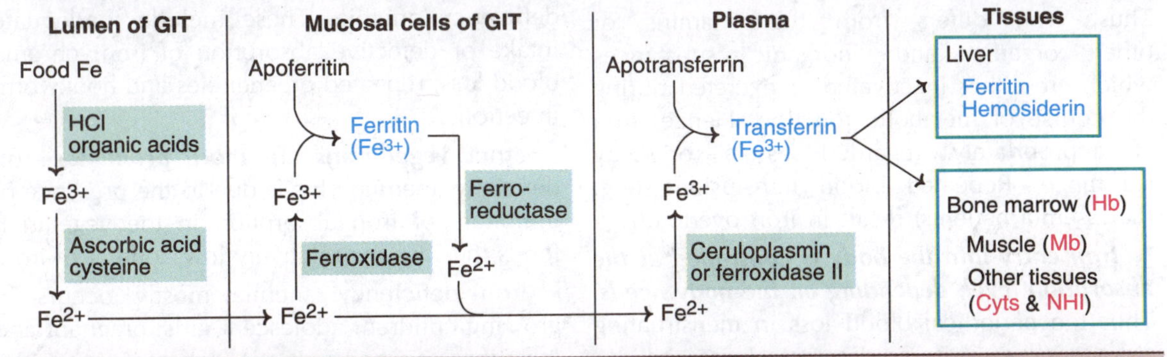

Fig. 18.4 : *Iron absorption and transport (GIT–Gastrointestinal tract; Hb–Hemoglobin; Mb–Myoglobin; Cyts–Cytochromes; NHI–Non-heme iron).*

Ferric iron then binds with a specific iron-binding protein, namely ***transferrin*** or siderophilin (a glycoprotein with mol. wt. 90,000). Each transferrin molecule can bind with two atoms of ferric iron (Fe^{3+}). The plasma transferrin (concentration 250 mg/dl) can bind with 400 mg of iron/dl plasma. This is known as ***total iron binding capacity (TIBC)*** of plasma.

Storage of iron

Iron is ***stored in liver, spleen*** and bone marrow in the form of ***ferritin***. In the mucosal cells, ferritin is the temporary storage form of iron. A molecule of apoferritin (mol. wt. 500,000) can combine with 4,000 atoms of iron. The maximum iron content of ferritin on weight basis is around 25%.

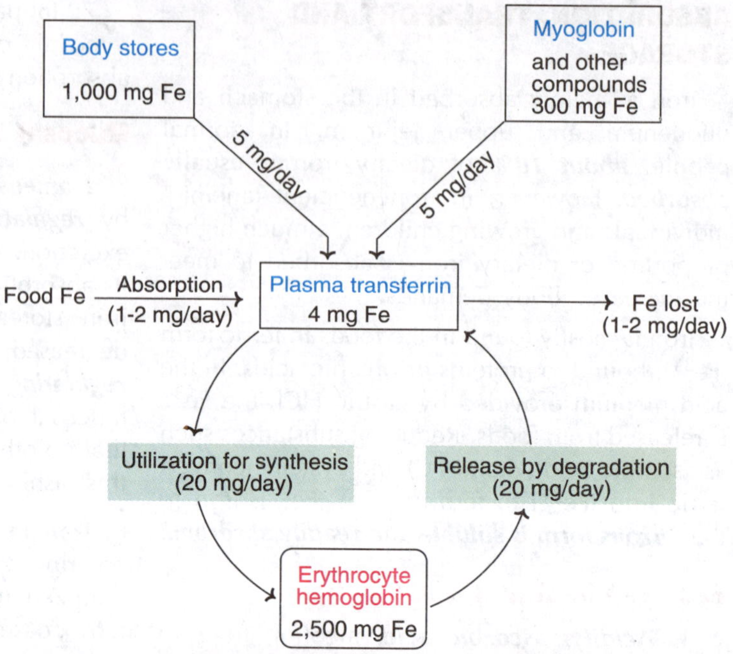

Fig. 18.5 : A general overview of iron metabolism.

Hemosiderin is another ***iron storage protein*** which can hold about 35% of iron by weight. Hemosiderin accumulates in the body (spleen, liver) when the supply of iron is in excess of body demands.

Iron is a one-way substance

Iron metabolism is unique as it operates in a closed system. It is very efficiently ***utilized and reutilized by the body***. Further, iron losses from the body are minimal (< 1 mg/day) which may occur through bile, sweat, hair loss etc. Iron is ***not excreted into urine***. Thus, iron differs from the vitamins or other organic and inorganic substances which are either inactivated or excreted during the course of metabolic function. Hence, iron is appropriately regarded as a one-way substance. Repeated blood transfusions (e.g. thalassemia patients) result in iron overload.

Iron entry into the body is controlled at the absorption level, depending on the body needs. Thus the periodical blood loss in menstruating women increases its requirements. Increased iron demands are also observed in pregnancy, lactation, and in growing children.

Iron metabolism

A general overview of iron metabolism is depicted in ***Fig.18.5***. It shows the distribution of iron in the body and its efficient reutilization. It may be noted that 1-2 mg of iron is absorbed per day to replace the loss.

Disease states

1. **Iron deficiency anemia :** This is the ***most prevalent nutritional disorder worldover***, including the well developed countries (e.g. USA). Several factors may contribute to iron deficiency anemia. These include inadequate intake or defective absorption of iron, chronic blood loss, repeated pregnancies and hookworm infections.

Strict ***vegetarians*** are ***more prone*** for iron deficieny anemia. This is due to the presence of inhibitors of iron absorption in the vegetarian foods, besides the relatively low content of iron.

Iron deficiency anemia mostly occurs in growing children, adolescent girls, pregnant and lactating women. It is characterized by ***microcytic hypochromic anemia*** with reduced blood hemoglobin levels (***<12 g/dl***). The other manifestations include apathy (dull and inactive), sluggish metabolic activities, retarded growth and loss of appetite.

Treatment : Iron deficiency is treated by supplementation of iron along wih folic acid and vitamin C.

2. **Hemosiderosis :** This is a less common disorder and is due to ***excessive iron in the body***, ***frequently referred*** to as ***iron toxicity***. It is commonly observed in subjects receiving repeated blood transfusions over the years, e.g. patients of hemolytic anemia, hemophilia. As already stated, iron is a one-way compound, once it enters the body, it cannot escape. Excessive iron is deposited as ferritin and hemosiderin.

Hemosiderosis is commonly observed among the Bantu tribe in South Africa. This is attributed to a high intake of iron from their staple diet corn and their habit of cooking foods in iron pots.

3. **Hemochromatosis :** This is a rare disease in which ***iron is directly deposited in the tissues*** (liver, spleen, pancreas and skin). Hemosiderosis is sometimes accompanied by hemochromatosis. Bronzed-pigmentation of the skin, cirrhosis of liver, pancreatic fibrosis are the manifestations of this disorder. Hemochromatosis causes a condition known as ***bronze diabetes***.

COPPER

The body contains about 100 mg copper distributed in different organs. It is involved in several important functions.

Biochemical functions

1. Copper is an essential constituent of several ***enzymes***. These include ***cytochrome oxidase***, ***catalase***, ***tyrosinase***, superoxide dismutase, monoamine oxidase, ascorbic acid oxidase, ALA synthase, phenol oxidase and uricase. Due to its presence in a wide variety of enzymes, copper is involved in many metabolic reactions.

2. Copper is necessary for the ***synthesis of hemoglobin*** (Cu is a constituent of ALA synthase, needed for heme synthesis).

3. Lysyl oxidase (a copper-containing enzyme) is required for the conversion of certain lysine residues of ***collagen and elastin*** to allysine which are necessary for cross-linking these structural proteins.

4. Ceruloplasmin serves as ferroxidase and is involved in the conversion of iron from Fe^{2+} to Fe^{3+} in which form ***iron*** (transferrin) is ***transported in plasma***.

5. Copper is necessary for the ***synthesis of melanin*** and phospholipids.

6. Development of bone and nervous system (myelin) requires Cu.

7. Certain copper-containing non-enzymatic proteins have been identified, although their functions are not clearly known. These include ***hepatocuprein*** (storage form in liver), ***cerebrocuprein (in brain)*** and ***hemocuprein*** (in RBC).

8. ***Hemocyanin***, a copper protein complex in invertebrates, functions like hemoglobin for O_2 transport.

Dietary requirements

Adults	—	***2-3 mg/day***
Infants and children	—	0.5-2 mg/day

Sources

Liver, ***kidney***, ***meat***, egg yolk, cereals, nuts and green leafy vegetables. Milk is a poor source.

Absorption

About 10% of dietary copper is absorbed, mainly in the duodenum. ***Metallothionein*** is a transport protein that facilitates copper absorption. Phytate, zinc and molybdenum decrease copper uptake.

Plasma copper

The copper concentration of plasma is about 100-200 mg/dl. Most of this (95%) is tightly bound to ceruloplasmin while a small fraction (5%) is loosely held to albumin. Normal concentration of serum ***ceruloplasmin*** is 25-50 mg/dl. It contains about 0.34% copper (6-8 atoms of Cu per molecule, half in Cu^{2+} state and the other half in Cu^{+} state).

Medical Concepts/ Clinical Correlates

- Wilson's disease is due to disturbed copper metabolism. It is characterized by abnormal deposition of copper in liver and brain, besides the low levels of plasma copper and ceruloplasmin.
- Endemic goitre, due to dietary iodine deficiency, is very common. Consumption of iodized salt is advocated to overcome this problem.
- Fluorosis is caused by excessive intake of fluoride. The manifestations include mottling of enamel and discoloration of teeth. In the advanced stages, hypercalcification of limb bones and ligaments of spine get calcified, ultimately crippling the individual.

Disease states

1. **Copper deficiency :** Severe deficiency of copper causes demineralization of bones, demyelination of neural tissue, anemia, fragility of arteries, myocardial fibrosis, hypopigmentation of skin, greying of hair.

2. **Menke's disease :** This disorder is due to a defect in the intestinal absorption of copper. It is possible that copper may be trapped by metallothionein in the intestinal cells. The symptoms include decreased copper in plasma and urine, anemia and depigmentation of hair.

Wilson's disease

Wilson's disease (hepatolenticular degeneration) is a rare disorder (1 : 50,000) of abnormal copper metabolism and is characterized by the following manifestations.

- Copper is ***deposited*** in abnormal amounts in ***liver and lenticular nucleus of brain***. This may lead to hepatic cirrhosis and brain necrosis.
- Deposition of Cu occurs around cornea to form ***Kayser-Fleischer (KF) rings***.
- ***Low levels of copper and ceruloplasmin*** in plasma (reference range 20–50 mg/dl) with increased excretion of copper in urine.
- Copper deposition in kidney causes renal damage. This leads to increased excretion of amino acids, glucose, peptides and hemoglobin in urine.
- Intestinal ***absorption of copper*** is very ***high***, about 4-6 times higher than normal.

Probable causes of Wilson's disease : The following explanations are offered to understand the etiology of this disease.

1. A failure to synthesize ceruloplasmin or an impairment in the binding capacity of copper to this protein or both. As a result of this, copper is free in the plasma which easily enters the tissues (liver, brain, kidney), binds with the proteins and gets deposited. The albumin bound copper is either normal or increased.

2. A mutation in the gene encoding ***copper binding ATPase*** is believed to be responsible for Wilson's disease. Defect in this ATPase reduces intestinal excretion of Cu through bile.

Treatment : Administration of penicillamine, a naturally occurring copper chelating agent, is used for the treatment of Wilson's disease.

IODINE

The total body contains about 20 mg iodine, most of it (80%) being present in the ***thyroid gland***. Muscle, salivary glands and ovaries also contain some amount of iodine.

Biochemical functions

The only known function of iodine is its requirement for the synthesis of thyroid hormones namely, ***thyroxine*** (T_4) and ***triiodothyronine*** (T_3). These hormones are involved in several biochemical functions (***Chapter 19***). Functionally, ***T_3 is more active than T_4.***

Dietary requirements

Adults	—	100-150 µg/day
Pregnant women	—	200 µg/day

Sources

Seafoods, drinking water, vegetables, fruits (grown on seaboard). High altitudes are deficient in iodine content in water as well as soil. Plant and animal foods of these areas, therefore, contain lesser amount of iodine. In these regions, iodine is added to drinking water or to table salt.

Absorption, storage and excretion

Iodine as iodide is mainly absorbed from the small intestine. Normally, about 30% of dietary iodine is taken up by the intestinal cells. Iodine absorption also occurs through skin and lungs.

About 80% of body's iodine is stored in the organic form as ***iodothyroglobulin*** (a glycoprotein) in the thyroid gland. This protein contains thyroxine, diiodotyrosine and triiodothyronine in different proportions.

Excretion of iodine mostly occurs through kidney. It is also excreted through saliva, bile, skin, and milk (in lactating women).

Plasma iodine

The normal concentration of plasma iodine is ***4-10 μg/dl***. Most of this is present as ***protein bound iodine (PBI)*** and represents the iodine contained in the circulating thyroid hormones. PBI level decreases in hypothyroidism and increases in hyperthyroidism. RBC do not contain iodine.

Disease states

The disorders of iodine metabolism—***simple goiter*** and ***toxic goiter***—are discussed in detail under thyroid hormones (***Chapter 19***).

MANGANESE

The total body content of manganese is about 15 mg. The liver and kidney are rich in Mn. Within the cells, Mn is mainly found in the nuclei in association with nucleic acids.

Biochemical functions

1. Mn serves as a ***cofactor*** for several ***enzymes***. These include arginase, pyruvate carboxylase, isocitrate dehydrogenase, superoxide dismutase (mitochondrial) and peptidase.

2. Mn is required for the ***formation of bone***, proper reproduction and normal functioning of nervous system.

3. Mn is necessary for the synthesis of mucopolysaccharides and glycoproteins.

4. Hemoglobin synthesis involves Mn.

5. Mn inhibits lipid peroxidation.

6. Mn is necessary for cholesterol biosynthesis.

Dietary requirements

The exact requirement of Mn is not known. About 2-9 mg/day is recommended for an adult.

Sources

Cereals, nuts, leafy vegetables and fruits. Tea is a rich source of Mn.

Absorption

About 3-4% of dietary Mn is normally absorbed in the small intestine. Iron inhibits Mn absorption.

Serum Mn

Manganese in the serum is bound to a specific carrier protein—***transmagnanin*** (a β-globulin). The normal plasma contains about 5-20 ng/dl.

Disease states

Mn deficiency in animals causes

1. Retarded growth, bone deformities and, in severe deficiency, sterility.

2. Accumulation of fat in liver.

3. Increased activity of serum alkaline phosphatase, and

4. Diminished activity of β-cells of pancreas (low insulin production).

ZINC

The total content of zinc in an adult body is about 2 g. ***Prostate gland*** is very ***rich*** in Zn (100 mg/g). Zinc is mainly an intracellular element.

Biochemical functions

1. Zn is an essential component of several enzymes e.g. carbonic anhydrase, alcohol dehydrogenase, alkaline phosphatase, carboxypeptidase, superoxide dismutase (cytosolic).

2. Zinc may be regarded as an ***antioxidant*** since the enzyme superoxide dismutase (Zn containing) protects the body against free radical damage.

3. The storage and ***secretion of insulin*** from the β-cells of pancreas ***require Zn.***

4. Zn is necessary to ***maintain the normal levels of vitamin A*** in serum. Zn promotes the synthesis of retinol binding protein.

5. It is required for ***wound healing.*** Zn enhances cell growth and division, besides stabilizing biomembranes.

6. ***Gusten,*** a zinc containing protein of the saliva, is important for taste sensation.

7. Zn is essential for proper reproduction.

Dietary requirements

Zinc requirement for an adult is 10-15 mg/ day. It is increased (by about 50%) in pregnancy and lactation.

Sources

Meat, fish, eggs, milk, beans, nuts.

Absorption

Zinc is absorbed mainly in the duodenum. Zn from the animal sources is better absorbed than the vegetable sources. Zn absorption appears to be dependent on a transport protein—***metallothionein***. Phytate, calcium, copper and iron interfere while small peptides and amino acids promote Zn absorption.

Serum Zn

The concentration of Zn in serum is about 100 mg/dl. Erythrocytes contain higher content of Zn (1.5 mg/dl) which is found in association with the enzyme carbonic anhydrase.

Disease states

1. Zinc deficiency is associated with growth retardation, poor wound healing, anemia, loss of appetite, loss of taste sensation, impaired spermatogenesis etc. It is reported that Zn deficiency in pregnant animals causes congenital malformations of the fetus. Deficiency of Zn may result in depression, dementia and other psychiatric disorders. The neuropsychiatric manifestations of chronic alcoholism may be partly due to zinc deficiency.

Acrodermatitis enteropathica is a rare inherited metabolic disease of zinc deficiency caused by a defect in the absorption of Zn from the intestine.

2. Zinc toxicity is often observed in welders due to inhalation of zinc oxide fumes. The manifestations of Zn toxicity include nausea, gastric ulcer, pancreatitis, anemia and excessive salivation.

MOLYBDENUM

Molybdenum is a constituent of the enzymes ***xanthine oxidase***, aldehyde oxidase and sulfite oxidase. Nitrite reductase (containing Mo) is a plant enzyme, required for nitrogen fixation.

The requirements of Mo are not clearly known. However, it is widely distributed in the natural foods. Dietary Mo is effectively (60%-70%) absorbed by the small intestine.

Some workers have reported that Mo decreases the mobilization and utilization of copper in the body.

Molybdenosis is a rare disorder caused by ***excessive consumption of Mo***. Its manifestations include impairment in growth, diarrhea and anemia. Intestinal absorption of copper is diminished.

COBALT

Cobalt is only important as a ***constituent of vitamin*** $\boldsymbol{B_{12}}$ ***(cobalamin)*** Cobalt content of vitamin B_{12} is about 4% by weight. The functions of cobalt are the same as that of vitamin B_{12} (***Chapter*** *7*). Administration of cobalt stimulates the production of the hormone erythropoietin, which promotes erythropoiesis.

Prolonged administration of cobalt is toxic as it results in polycythemia (increased RBC in blood).

FLUORINE

Fluoride is mostly found in bones and teeth. The beneficial effects of fluoride in trace amounts are overshadowed by its harmful effects caused by excess consumption.

Biochemical functions

1. Fluoride ***prevents*** the development of ***dental caries***. It forms a protective layer of acid resistant fluoroapatite with hydroxyapatite of the enamel and prevents the tooth decay by bacterial acids. Further, fluoride inhibits the bacterial enzymes and reduces the production of acids.

2. Fluoride is necessary for the proper ***development of bones***.

3. It inhibits the activities of certain enzymes. Sodium fluoride ***inhibits enolase (of glycolysis)*** while fluoroacetate inhibits aconitase (of citric acid cycle).

Dietary requirements and sources

An intake of less than 2 ppm of fluoride will meet the daily requirements. Drinking water is the main source.

Disease states

1. **Dental caries :** It is clearly established that drinking water containing ***less than 0.5 ppm of fluoride*** is associated with the development of dental caries in children (Refer ***Chapter 13***).

2. **Fluorosis :** ***Excessive*** intake of ***fluoride*** is harmful to the body. An intake above 2 ppm (particularly > 5 ppm) in children causes ***mottling of enamel and discoloration of teeth***. The teeth are weak and become rough with characteristic brown or yellow patches on their surface. These manifestations are collectively referred to as ***dental fluorosis***.

An intake of fluoride above 20 ppm is toxic and causes pathological changes in the bones. Hypercalcification, increasing the density of the bones of limbs, pelvis and spine, is a characteristic feature. Even the ligaments of spine and collagen of bones get calcified. Neurological disturbances are also commonly observed. The manifestations described here constitute ***skeletal fluorosis***. In the advanced stages, the individuals are crippled and cannot perform their daily routine work due to stiff joints. This condition of ***advanced fluorosis*** is referred to as ***genu valgum***.

The fluoride content of water in some parts of Andhra Pradesh, Punjab and Karnataka is quite high. Fluorosis is prevalent in these regions, causing concern to government and health officials.

3. **Fluoridation of water and use of fluoride tooth-pastes :** In order to prevent the dental caries in children, some advanced countries like USA have started fluoridation of water. Further, the consumer markets till recently were flooded with fluoride toothpastes. There is some rethinking on these aspects due to the toxic effects of excess fluoride.

SELENIUM

Selenium was originally identified as an element that causes toxicity to animals (alkali disease) in some parts of USA, containing large amounts of Se in the soil. Later work, however, has shown that Se in smaller amounts is biologically important.

Biochemical functions

1. Selenium, along ***with vitamin E***, prevents the development of hepatic necrosis and muscular dystrophy.

2. Se is involved in maintaining ***structural integrity of biological membranes***.

3. Se as ***selenocysteine*** is an essential component of the enzyme ***glutathione peroxidase***. This enzyme protects the cells against the damage caused by H_2O_2. It appears from recent studies that selenocysteine is directly incorporated during protein biosynthesis. Therefore, selenocysteine is considered as a separate (21st) amino acid.

4. Se ***prevents lipid peroxidation*** and protects the cells against the ***free radicals***, including superoxide (O_2^-).

5. Se protects animals from carcinogenic chemicals. However, the precise role of Se in humans with regard to cancer prevention is not clearly identified.

6. Se binds with certain heavy metals (Hg, Cd) and protects the body from their toxic effects.

7. A selenium containing enzyme ***5′-deiodinase*** converts thyroxine (T_4) to triiodothyronine in the thyroid gland.

8. ***Thioredoxin reductase***, involved in purine nucleotide metabolism, is also a selenoprotein.

Requirements and sources

A daily intake of 50-200 mg of Se has been recommended for adults. The good sources of Se are organ meats (liver, kidney) and sea foods.

Disease states

Deficiency : Se deficiency in animals leads to muscular dystrophy, pancreatic fibrosis and reproductive disorders. In humans, ***Keshan disease***, an endemic cardiomyopathy (in China) is attributed to the deficiency of Se. Epidemiological studies reveal that low serum Se levels are associated with increased risk of cardiovascular disease, and various cancers.

Kaschin-Beck disease is associated with severe Se deficiency. It mostly occurs in the age group 5–15 years, and is characterized by osteoarthropathy (a disease of bones and joints) and necrosis of joints. Kaschin-Beck disease is also linked to iodine deficiency and/or presence of mycotoxins.

Toxicity

Selenosis is toxicity due to very excessive intake of Se. The manifestations of selenosis include weight loss, emotional disturbances, diarrhea, hair loss and garlic odor in breath. The compound ***dimethyl selenide*** is responsible for the ***garlic odor***.

CHROMIUM

The total human body contains about 6 mg chromium. The Cr content of blood is about 20 μg/dl. Cr performs several biochemical functions.

1. In association with insulin, Cr promotes the utilization of glucose. Cr is a component of a protein namely ***chromodulin*** which facilitates the binding of insulin to cell receptor sites.

2. Cr lowers the total serum cholesterol level.

3. It is involved in lipoprotein metabolism. Cr decreases serum low density lipoproteins (LDL) and increases high density lipoproteins (HDL) and, thus, promotes health.

4. It is believed that Cr participates in the transport of amino acids into the cells (heart and liver).

The dietary requirement of Cr is not known. It is estimated that an adult man consumes about 10 to 100 mg/day. The good sources of Cr include brewer's yeast, grains, cereals, cheese and meat.

Chromium deficiency causes disturbances in carbohydrate, lipid and protein metabolisms. Excessive intake of Cr results in toxicity, leading to liver and kidney damage.

SUMMARY

- *The minerals or inorganic elements are required for normal growth and maintenance of the body. They are classified as principal elements and trace elements. There are seven principal elements—Ca, P, Mg, Na, K, Cl and S. The trace elements include Fe, Cu, I, Zn, Mn, Mo, Co, F, Se and Cr.*
- *Calcium is required for the development of bones and teeth, muscle contraction, blood coagulation, nerve transmission etc. Absorption of Ca from the duodenum is promoted by vitamin D, PTH and acidity while it is inhibited by phytate, oxalate, free fatty acids and fiber. The normal level of serum Ca (9-11 mg/dl) is controlled by an interplay of PTH, calcitriol and calcitonin.*
- *Serum Ca level is elevated in hyperparathyroidism and diminished in hypoparathyroidism. Hypocalcemia causes tetany, the symptoms of which include neuromuscular irritability, spasm and convulsions.*
- *Phosphorus, besides being essential for the development of bones and teeth, is a constituent of high energy phosphate compounds (ATP, GTP) and nucleotide coenzymes (NAD^+, $NADP^+$).*
- *Sodium, potassium and chlorine are involved in the regulation of acid-base equilibrium, fluid balance and osmotic pressure in the body. Sodium is the principal extracellular cation (serum level 135-145 mEq/l), while potassium is the chief intracellular cation (serum level 3.5-5.0 mEq/l).*
- *Iron is mainly required for O_2 transport and cellular respiration. Absorption of iron is promoted by ascorbic acid, cysteine, acidity and small peptides while it is inhibited by phytate, oxalate and high phosphate.*
- *Iron (Fe^{3+}) is transported in the plasma in a bound form to transferrin. It is stored as ferritin in liver, spleen and bone marrow. Iron deficiency anemia causes microcytic hypochromic anemia. Excessive consumption of iron results in hemosiderosis which is due to the tissue deposition of hemosiderin.*
- *Copper is an essential constituent of several enzymes (e.g catalase, cytochrome oxidase, tyrosinase). Ceruloplasmin is a copper containing protein required for the transport of iron (Fe^{3+}) in the plasma. Wilson's disease is an abnormality in copper metabolism, characterized by the deposition of copper in liver, brain and kidney.*
- *Iodine is important as a component of thyroid hormones (T_4 and T_3) while cobalt is a constituent of vitamin B_{12}. Zinc is necessary for the storage and secretion of insulin and maintenance of normal vitamin A levels in serum, besides being a component of several enzymes (e.g. carbonic anhydrase, alcohol dehydrogenase).*
- *Fluorine in trace amounts (<2 ppm) prevents dental caries while its higher intake leads to fluorosis. Selenium is assigned an antioxidant role as it protects the cells from free radicals. Chromium promotes the utilization of glucose and reduces serum cholesterol.*

THE READER MUST REFER APPENDIX VI (Page Nos : 738 – 756) FOR RELATED CLINICAL CASE STUDIES (Nos. 10, 33, 35, 49, 50, 59, 69)

Self-Assessment Exercises

I. Essay questions

1. Write briefly on the trace elements and their metabolism in the body.
2. Discuss the biochemical functions, dietary requirements, sources and absorption of calcium.
3. Write an essay on the iron metabolism in the body.
4. Describe the metabolism of copper, zinc and manganese.
5. Write on the biochemical importance and disease states of fluorine and selenium.

II. Short notes

(a) Homeostasis of calcium, (b) Osteoporosis, (c) Phosphorus, (d) Sodium and chlorine, (e) Potassium, (f) Factors affecting Fe absorption, (g) Hemosiderosis, (h) Wilson's disease, (i) Iodine, (j) Magnesium.

III. Fill in the blanks

1. The normal concentration of serum calcium ____________.
2. The vitamin derived hormone that regulates calcium homeostasis ____________.
3. The inorganic element found in the structure of majority of high-energy compounds ____________.
4. Several kinase enzymes require the mineral cofactor ____________.
5. The principal cation of extracellular fluid ____________.
6. The normal concentration of serum potassium ____________.
7. Iron is transported in the plasma in a bound form to a protein ____________.
8. The copper containing protein involved for the conversion of ferrous iron (Fe^{2+}) to ferric iron (Fe^{3+}) in the plasma ____________.
9. The zinc containing protein in the saliva involved in taste sensation ____________.
10. The element involved in the protection of cells against the damage of H_2O_2 and other free radicals ____________.

IV. Multiple choice questions

11. The following substance(s) is(are) involved in the regulation of plasma calcium level
    (a) Calcitriol (b) Parathyroid hormone (c) Calcitonin (d) All of them.
12. The following is a sulfur containing essential amino acid
    (a) Methionine (b) Cysteine (c) Cystine (d) All of them.
13. Iron in the mucosal cells binds with the protein
    (a) Transferrin (b) Ferritin (c) Ceruloplasmin (d) Hemosiderin.
14. The following element is involved in wound healing
    (a) Calcium (b) Sodium (c) Zinc (d) Magnesium.
15. Pick up element that prevents the development of dental caries
    (a) Fluorine (b) Calcium (c) Phosphorus (d) Sodium.

CLINICAL BIOCHEMISTRY AND NUTRITION

SECTION FOUR

Hormones

19

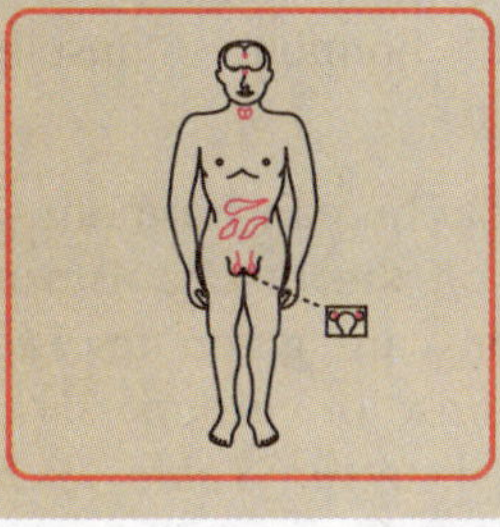

The hormones speak

"We are the chemical messengers of the body!
Diversified in our structure and function;
Act either directly or through messengers;
'Growth, health and welfare' is our motto."

The living body possesses a remarkable communication system to coordinate its biological functions. This is achieved by two distinctly organized functional systems.

1. The nervous system coordinates the body functions through the transmission of electrochemical impulses.

2. The ***endocrine system*** acts through a wide range of ***chemical messengers*** known as hormones.

Hormones are conventionally defined as ***organic substances, produced in small amounts by specific tissues (endocrine glands), secreted into the blood stream to control the metabolic and biological activities in the target cells***. Hormones may be regarded as the ***chemical messengers*** involved in the transmission of information from one tissue to another and from cell to cell. The major endocrine organs in human body are depicted in ***Fig.19.1***).

CLASSIFICATION OF HORMONES

Hormones may be classified in many ways based on their characteristics and functions. Two types of classification are discussed here

I. Based on the chemical nature

The hormones can be categorized into three groups considering their chemical nature.

1. **Protein** or **peptide hormones** e.g. ***insulin***, glucagon, antidiuretic hormone, oxytocin.

2. **Steroid hormones** e.g. glucocorticoids, mineralocorticoids, ***sex hormones***.

3. **Amino acid derivatives** e.g. ***epinephrine***, norepinephrine, thyroxine (T_4), triiodothyronine (T_3).

II. Based on the mechanism of action

Hormones are classified into two broad groups (I and II) based on the location of the receptors to which they bind and the signals used to mediate their action.

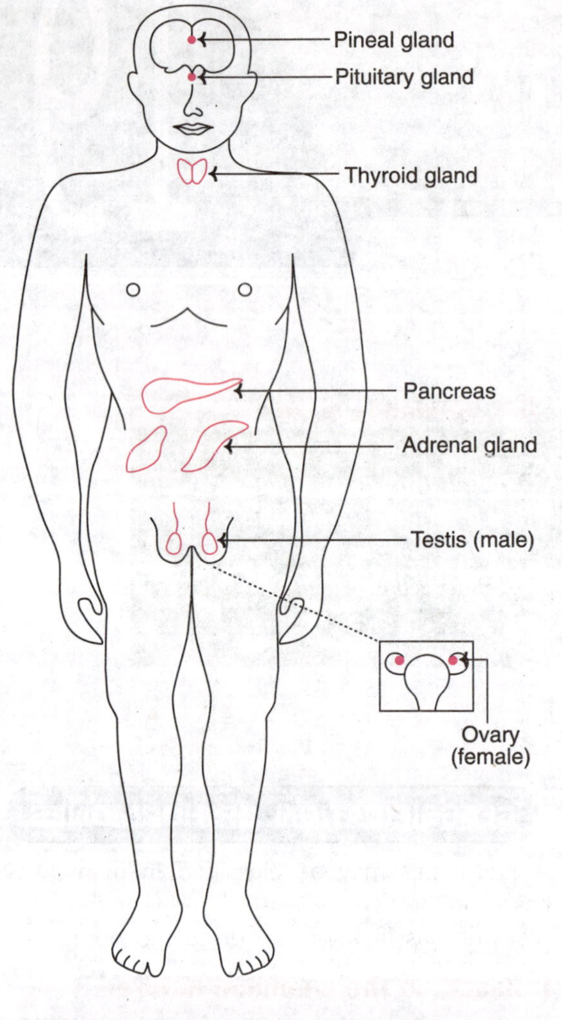

Fig. 19.1 : Diagrammatic representation of major endocrine glands.

1. **Group I hormones :** These hormones ***bind to intracellular receptors*** to form receptor-hormone complexes (the intracellular messengers) through which their biochemical functions are mediated. Group I hormones are lipophilic in nature and are mostly derivatives of cholesterol (exception–T_3 and T_4). e.g. estrogens, androgens, glucocorticoids, calcitriol.

2. **Group II hormones :** These hormones bind to cell surface (plasma membrane) receptors and stimulate the release of certain molecules, namely the ***second messengers*** which, in turn, perform the biochemical functions. Thus, hormones themselves are the first messengers.

Group II hormones are subdivided into three categories based on the chemical nature of the second messengers.

(a) The second messenger is ***cAMP*** e.g. ACTH, FSH, LH, PTH, glucagon, calcitonin.

(b) The second messenger is ***phosphatidylinositol/calcium*** e.g. TRH, GnRH, gastrin, CCK.

(c) The second messenger is ***unknown*** e.g. growth hormone, insulin, oxytocin, prolactin.

The principal human hormones, their classification based on the mechanism of action, and major functions are given in ***Table 19.1***. The general features of group I and group II hormones are listed in ***Table 19.2***.

Mechanism of action of group I hormones

These hormones are lipophilic in nature and can easily pass across the plasma membrane. They act through the ***intracellular receptors*** located either in the cytosol or the nucleus. The hormone-receptor complex binds to specific regions on the DNA called hormone responsive element (HRE) and causes ***increased expression of specific genes*** (***Fig.19.2***). It is believed that the interaction of hormone receptor complex with HRE promotes initiation and, to a lesser extent, elongation and termination of RNA synthesis (transcription). The ultimate outcome is the production of specific proteins (translation) in response to hormonal action.

Mechanism of action of group II hormones

These ***hormones*** are considered as ***the first messengers***. They exert their action through ***mediatory molecules***, collectively called ***second messengers***.

Table 19.1 Principal human hormones—classification (by mechanism of action), origin and major functions

Hormone(s)	Origin	Major Function(s)
Group I. HORMONES THAT BIND TO INTRACELLULAR RECEPTORS		
Estrogens	Ovaries and adrenal cortex	Female sexual characteristics, menstrual cycle.
Progestins	Ovaries and placenta	Involved in menstrual cycle and maintenance of pregnancy.
Androgens	Testes and adrenal cortex	Male sexual characteristics, spermatogenesis.
Glucocorticoids	Adrenal cortex	Affect metabolisms, suppress immune system.
Mineralocorticoids	Adrenal cortex	Maintenance of salt and water balance.
Calcitriol (1, 25–DHCC)	Kidney (final form)	Promotes absorption of Ca^{2+} from intestine, kidney and bone.
Thyroid hormones (T_3, T_4)	Thyroid	Promote general metabolic rate.
Group II. HORMONES THAT BIND TO CELL SURFACE RECEPTORS		
A. The second messenger is cAMP		
Adrenocorticotropic hormone (ACTH)	Anterior pituitary	Stimulates the release of adrenocorticosteroids.
Follicle stimulating hormone (FSH)	Anterior pituitary	In females, stimulates ovulation and estrogen synthesis. In males, promotes spermatogenesis.
Luteinizing hormone (LH)	Anterior pituitary	Stimulates synthesis of estrogens and progesterone and causes ovulation. Promotes androgen synthesis by testes.
Chorionic gonadotropin (hCG)	Anterior pituitary	Stimulates progesterone release from placenta.
Thyroid stimulating hormone (TSH)	Anterior pituitary	Promotes the release of thyroid hormones (T_3, T_4).
β-Endorphins and enkephalins	Anterior pituitary	Natural endogenous analgesics (pain relievers).
Antidiuretic hormone (ADH)	Posterior pituitary (stored)	Promotes water reabsorption by kidneys.
Glucagon	Pancreas	Increases blood glucose level, stimulates glycogenolysis and lipolysis.
Parathyroid hormone (PTH)	Parathyroid	Increases serum calcium, promotes Ca^{2+} release from bone.
Calcitonin	Thyroid	Lowers serum calcium. Decreases Ca^{2+} uptake by bone and kidney.
Epinephrine	Adrenal medulla	Increases heart rate and blood pressure. Promotes glycogenolysis in liver and muscle and lipolysis in adipose tissue.
Norepinephrine	Adrenal medulla	Stimulates lipolysis in adipose tissue.
B. The second messenger is phosphatidyl inositol/calcium		
Thyrotropin-releasing hormone (TRH)	Hypothalamus	Promotes TSH release.
Gonadotropin-releasing hormone (GnRH)	Hypothalamus	Stimulates release of FSH and LH.
Gastrin	Stomach	Stimulates gastric HCl and pepsinogen secretion.
Cholecystokinin (CCK)	Intestine	Stimulates contraction of gall bladder and secretion of pancreatic enzymes.
C. The second messenger is unknown/unsettled		
Growth hormone (GH)	Anterior pituitary	Promotes growth of the body (bones and organs).
Prolactin (PRL)	Anterior pituitary	Growth of mammary glands and lactation.
Oxytocin	Posterior pituitary (stored)	Stimulates uterine contraction and milk ejection.
Insulin	Pancreas	Lowers blood glucose (hypoglycemic effect), promotes protein synthesis and lipogenesis.
Somatomedins (insulin-like growth factors, IGF-I, IGF-II)	Liver	Growth related functions of GH are mediated. Stimulates growth of cartilage.

cAMP—THE SECOND MESSENGER

Cyclic AMP (cAMP, cyclic adenosine 3′,5′-monophosphate) is a ubiquitous nucleotide. It consists of adenine, ribose and a phosphate (linked by 3′,5′ linkage). cAMP acts as a second messenger for a majority of ***polypeptide hormones***.

The membrane-bound enzyme adenylate cyclase converts ATP to cyclic AMP. cAMP is hydrolysed by phosphodiesterase to 5′-AMP (***Fig.19.3***).

Adenylate cyclase system

A series of events occur at the membrane level that influence the activity of adenylate cyclase leading to the ***synthesis of cAMP***. This process is mediated by ***G-proteins***, so designated due to their ability to bind to ***g***uanine nucleotides.

Action of cAMP—a general view

Once produced, cAMP performs its role as a second messenger in eliciting biochemical responses (***Fig.19.4***). cAMP activates protein kinase **A** (A stands for cAMP). This enzyme is a heterotetramer consisting of 2 regulatory subunits (R) and 2 catalytic subunits (C).

cAMP binds to inactive protein kinase and causes the dissociation of R and C subunits.

$$\underset{\text{(inactive)}}{4\text{cAMP} + R_2C_2} \longrightarrow \underset{\text{(inactive)}}{R_2(4\ \text{cAMP})} + \underset{\text{(active)}}{2C}$$

The active subunit (C) catalyses phosphorylation of proteins (transfer of phosphate group to serine and threonine residues). It is the phosphoprotein that ultimately causes the biochemical response.

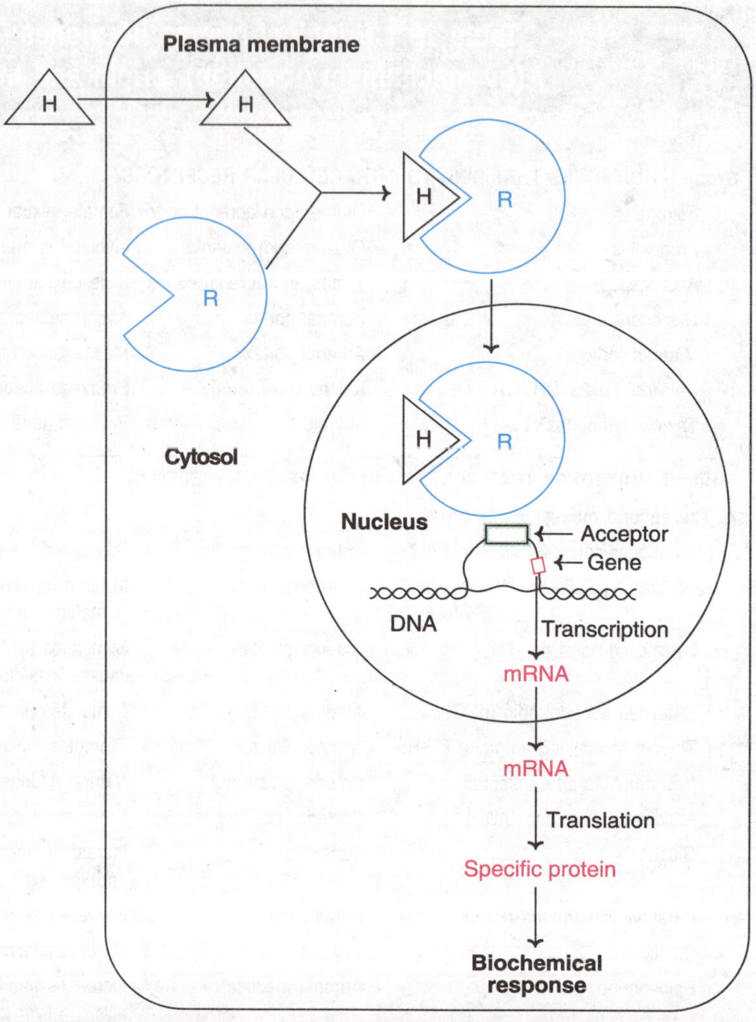

Fig. 19.2 : *Mechanism of action of steroid hormones (H–Hormone; R–Receptor; HR–Hormone-receptor complex).*

Table 19.2 General characteristics of group I and group II hormones

Character	Group I	Group II
Solubility	Lipophilic	Hydrophilic
Transport	Protein bound	Free
Half-life	Hours to days (usually long)	Minutes (usually short)
Receptor location	Intracellular	Plasma membrane
Mediator(s)	Receptor-hormone complex	cAMP, cGMP, Ca^{2+}, phosphatidylinositol kinase cascade.
Selected examples	Steroids, T_3, T_4, calcitriol	Proteins, polypeptides, catecholamines.

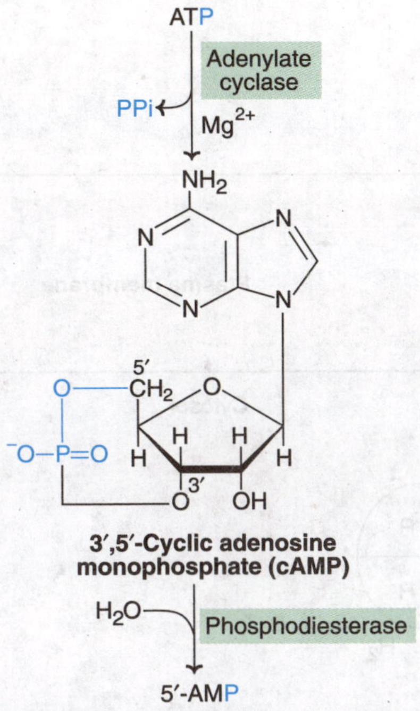

Fig. 19.3 : Synthesis and degradation of cAMP.

It should, however, be remembered that cAMP does not act on all protein kinases. For instance, on ***protein kinase C*** (the second messenger is diacylglycerol).

Dephosphorylation of proteins : A group of enzymes called protein phosphatases hydrolyse and remove the phosphate group added to proteins.

Degradation of cAMP : cAMP undergoes rapid hydrolysis, catalysed by the enzyme ***phosphodiesterase to 5′ AMP*** which is ***inactive***. Hence, the effect of cAMP will be shortlived if the hormone stimulating adenylate cyclase is removed. Caffeine and theophylline (methylxanthine derivatives) can inhibit phosphodiesterases and increase the intracellular levels of cAMP.

HYPOTHALAMIC AND PITUITARY HORMONES

The pituitary gland or hypophysis (weighing about 1 g) is located below the hypothalamus of the brain. It consists of two distinct parts—the anterior pituitary (***adenohypophysis***) and the posterior pitutitary (***neurohypophysis***) connected by pars intermedia (***Fig.19.5***). The latter is almost absent in humans, although found in lower organisms.

Hypothalamus is a specialized center in the brain that functions as a ***master coordinator of hormonal action***. In response to the stimuli of central nervous system, hypothalamus liberates certain releasing factors or hormones. These factors stimulate or inhibit the release of corresponding tropic hormones from the anterior pituitary. Tropic hormones stimulate the target endocrine tissues to secrete the hormones they synthesize. The relationship between hypothalamus and pituitary with endocrine glands is illustrated in ***Fig.19.6***. ***In general, the hormonal system is under feedback control***. For instance, adrenocorticotropic hormone (ACTH) inhibits the release of corticotropin releasing hormone (CRH).

HYPOTHALAMIC HORMONES

Hypothalamus produces at least six releasing factors or hormones.

1. **Thyrotropin-releasing hormone (TRH) :** It is a tripeptide consisting of glutamate derivative (pyroglutamate), histidine and proline. TRH stimulates anterior pituitary to release thyroid-stimulating hormone (TSH or thyrotropin) which, in turn, stimulates the release of thyroid hormones (T_3 and T_4).

2. **Corticotropin-releasing hormone (CRH) :** It stimulates anterior pituitary to release adrenocorticotropic hormone (ACTH) which in turn, acts on adrenal cortex to liberate adrenocorticosteroids. CRH contains 41 amino acids.

3. **Gonadotropin-releasing hormone (GnRH) :** It is a decapeptide. GnRH stimulates anterior pituitary to release ***gonadotropins,*** namely luteinizing hormone (LH) and follicle stimulating hormone (FSH).

4. **Growth hormone-releasing hormone (GRH)** with 44 amino acids stimulates the release of growth hormone (GH or somatotropin) which promotes growth.

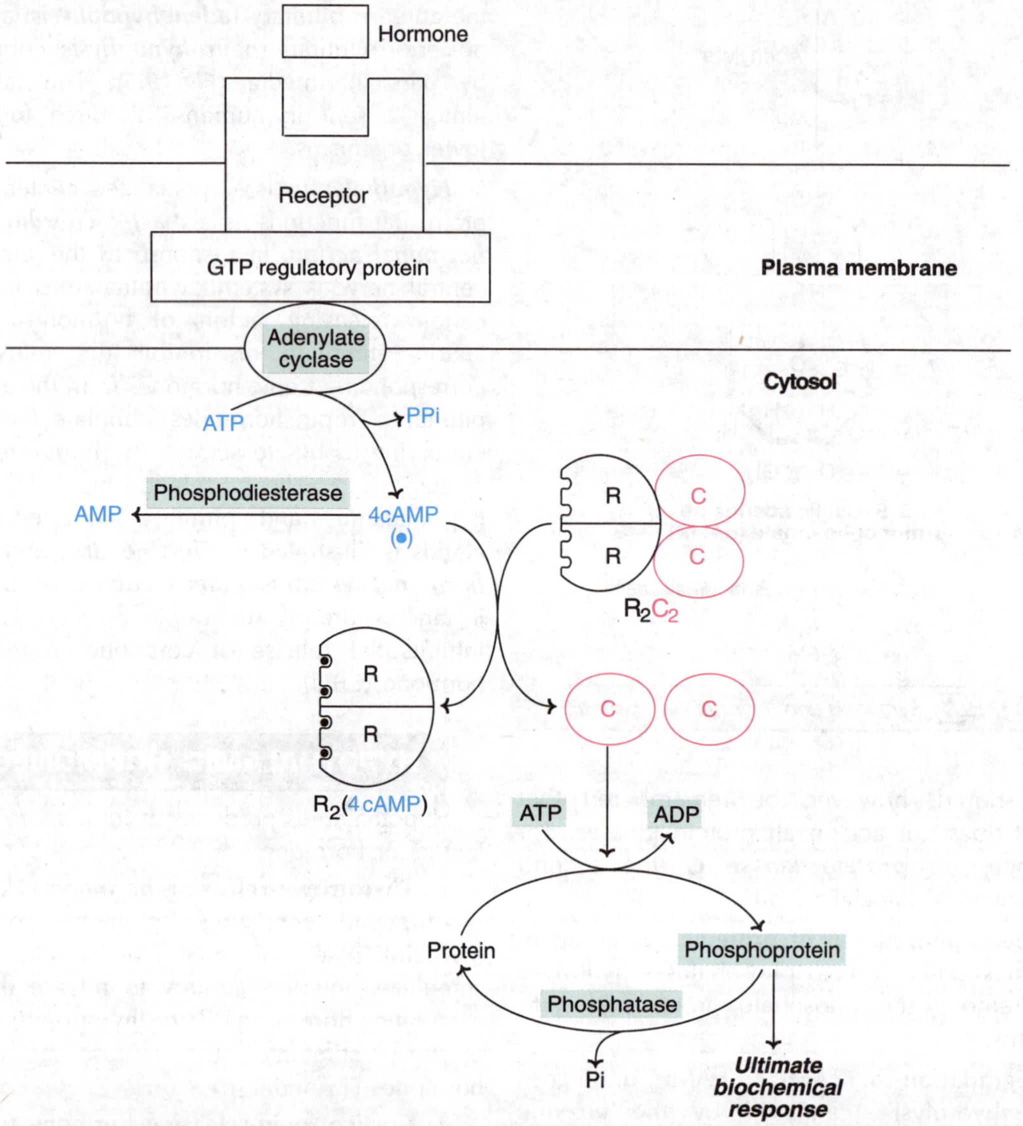

***Fig. 19.4 :** Overview of synthesis and action of cAMP (R_2C_2–cAMP dependent protein kinase A; R_2–Regulatory subunits; C_2–Catalytic subunits; C–Active catalytic unit of R_2C_2).*

5. **Growth hormone release-inhibiting hormone (GRIH) :** It contains 14 amino acids and is also known as ***somatostatin***. GRIH inhibits the release of growth hormone from the anterior pituitary.

6. **Prolactin release-inhibiting hormone (PRIH) :** It is believed to be a dopamine and/or a small peptide that inhibits the release of prolactin (PRL) from anterior pituitary.

ANTERIOR PITUITARY HORMONES

Anterior pituitary or adenohypophysis is truly the ***master endocrine organ***. It produces several hormones that influence—either directly or indirectly—a variety of biochemical processes in the body. The hormones of adenohypophysis are broadly classified into three categories.

I. The growth hormone-prolactin group.

II. The glycoprotein hormones.

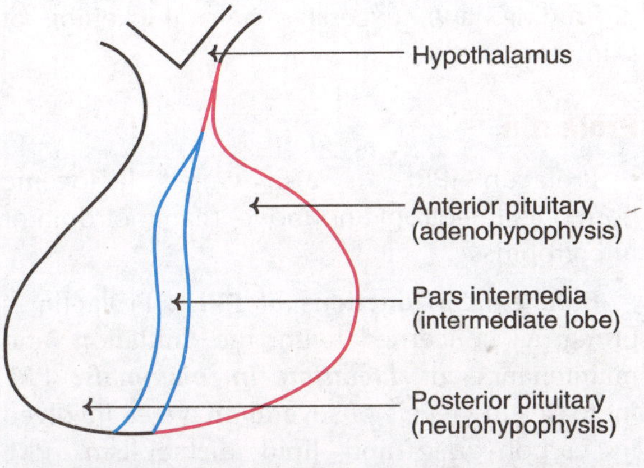

Fig. 19.5 : A diagrammatic view of pituitary gland.

III. The pro-opiomelanocortin peptide family.

I. The growth hormone-prolactin group

Growth hormone (GH), prolactin (PRL) and chorionic somatomammotropin (CS; placental lactogen) are protein hormones with many striking similarities in their structure.

Growth hormone (GH)

The growth hormone (or ***somatotropin***) is produced by somatotropes, a special group of acidophilic cells of anterior pituitary.

Regulation of GH release : Two hypothalamic factors play a prominent role in the release of growth hormones. These are the growth hormone-releasing hormone (GRH) that stimulates and the growth hormone release-inhibiting hormone (GRIH, ***somatostatin***) that inhibits. This, in turn, is regulated by a feedback mechanism.

Growth hormone production is influenced by many factors such as sleep, stress (pain, cold, surgery), exercise, food intake etc. It is observed that the largest increase in the production of GH occurs after the onset of sleep. This supports the ***adage "If you don't sleep, you won't grow."***

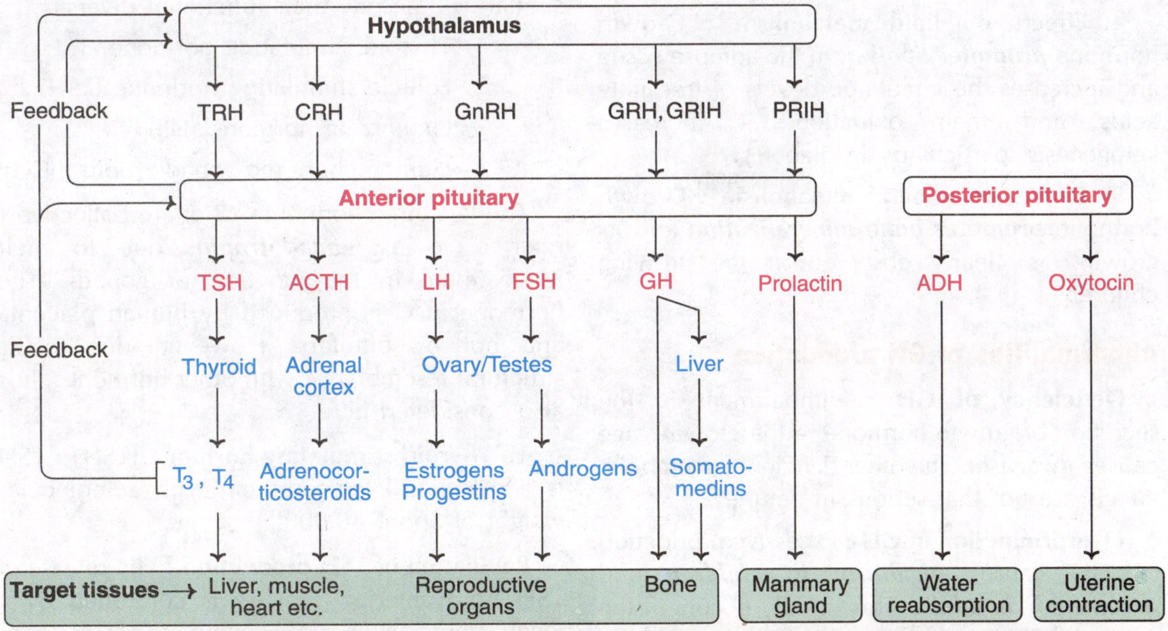

***Fig. 19.6 :** Hormonal hierarchy relationships between hypothalamus and pituitary with other endocrine glands [TRH–Thyrotropin releasing hormone; CRH–Corticotropin releasing hormone; GnRH–Gonadotropin releasing hormone; GRH–Growth hormone releasing hormone; GRIH–Growth hormone release inhibiting hormone; TSH–Thyroid stimulating hormone; ACTH–Adrenocorticotropic hormone; LH–Luteinizing hormone; FSH–Follicle stimulating hormone; GH–Growth hormone; ADH–Antidiuretic hormone; T_3–Triiodothyronine; T_4–Tetraiodothyronine (thyroxine)].*

Biochemical functions of GH : Growth hormone promotes growth, and also influences the normal metabolisms (protein, carbohydrate, lipid and mineral) in the body.

1. **Effects on growth :** As is obvious from the name, GH is ***essential for the growth***. The growth-related effects of GH are mediated through ***insulin like growth factor I (IGF-I)*** which is also known as ***somatomedin C*** (formerly sulfation factor), produced by liver.

2. **Effects on protein metabolism :** Growth hormone has an ***anabolic effect*** on protein metabolism. It promotes the uptake of amino acids into the tissues and increases the protein synthesis. The overall effect of GH is a ***positive nitrogen balance*** that leads to increase in body weight.

3. **Effects on carbohydrate metabolism :** Growth hormone is antagonistic to insulin and ***causes hyperglycemia***. GH increases gluconeogenesis, decreases glucose utilization, impairs glycolysis and reduces the tissue uptake of glucose.

4. **Effects on lipid metabolism :** Growth hormone ***promotes lipolysis*** in the adipose tissue and increases the circulatory levels of free fatty acids and their oxidation. It increases ketogenesis, particularly in diabetes.

5. **Effects on mineral metabolism :** Growth hormone ***promotes bone mineralization*** and its growth, as clearly observed in the growing children.

Abnormalities of GH production

Deficiency of GH : Impairment in the secretion of growth hormone in the growing age causes ***dwarfism***. The other deficiency metabolic effects are not that serious in nature.

Overproduction of GH : Excessive production of GH causes ***gigantism in children*** and ***acromegaly in adults***. This usually occurs in the acidophil tumor of pituitary gland. Gigantism is characterized by increased growth of long bones and this is observed before the epiphyseal plates close. Acromegaly occurs after epiphyseal closure and is characterized by increase in the size of hands, facial changes (enlarged nose, protruding jaw), excessive hair, thickening of skin etc.

Prolactin

Prolactin (PRL) is also called lactogenic hormone, luteotropic hormone, mammotropin or luteotropin.

Biochemical functions of PRL : Prolactin is primarily concerned with the initiation and maintenance of ***lactation in mammals***. PRL increases the levels of several enzymes involved in carbohydrate and lipid metabolism. PRL promotes HMP shunt, increases lipid biosynthesis and stimulates lactose production in mammary glands.

Prolactin promotes the growth of corpus luteum (hence also known as luteotropic hormone) and stimulates the production of progesterone.

II. The glycoprotein hormones

The following four hormones are glycoprotein in nature and possess certain structural similarities, despite their functional diversity.

1. Thyroid stimulating hormone (TSH)
2. Follicle stimulating hormone (FSH)
3. Luteinizing hormone (LH)
4. Human chorionic gonadotropin (hCG).

The last three hormones (***2-4***) are collectively referred to as ***gonadotropins*** due to their involvement in the function of gonads. The hormone hCG is produced by human placenta and not by pituitary. However, due to its structural resemblance with other hormones, it is also considered here.

1. **Thyroid stimulating hormone (TSH) :** TSH is a dimer (αβ) glycoprotein with a molecular weight of about 30,000.

Regulation of TSH production : The release of TSH from anterior pituitary is controlled by a feedback mechanism. This involves the hormones of thyroid gland (T_3 and T_4) and thyrotropin-releasing hormone (TRH) of hypothalamus.

Functions of TSH : The biochemical effects of TSH on thyroid gland are briefly discussed here. TSH binds with plasma membrane receptors and stimulates adenylate cyclase with a consequent increase in cAMP level. TSH, through the mediation of cAMP, exerts the following effects.

- ***Promotes the uptake of iodide*** (iodide pump) from the circulation ***by thyroid gland.***
- Enhances the ***conversion of iodide*** (I^-) to ***active iodide*** (I^+), a process known as organification.
- Increases the proteolysis of thyroglobulin to ***release*** T_3 ***and*** T_4 into the circulation.

TSH increases the synthesis of proteins, nucleic acids and phospholipids in thyroid gland.

Gonadotropins : The follicle-stimulating hormone (***FSH***), luteinizing hormone (***LH***) and human chorionic gonadotropin (***hCG***) are commonly known as gonadotropins. All three are glycoproteins.

The release of FSH and LH from the anterior pituitary is controlled by gonadotropin-releasing hormone (GnRH) of hypothalamus.

2. **Biochemical functions of FSH :** In females, FSH stimulates follicular growth, increases the weight of the ovaries and enhances the production of estrogens.

In males, FSH stimulates testosterone production, required for spermatogenesis. FSH also promotes growth of seminiferous tubules.

3. **Biochemical functions of LH :** Luteinizing hormone stimulates the production of progesterone from corpus luteum cells in females and testosterone from Leydig cells in males. LH and FSH are collectively responsible for the development and maintenance of secondary sexual characters in males.

4. **Human chorionic gonadotropin (hCG) :** hCG is a glycoprotein (mol. wt. 100,000), produced by syncytiotrophoblast cells of placenta. The structure of hCG closely resembles that of LH.

The levels of hCG in plasma and urine increase almost immediately after the implantation of fertilized ovum. The detection of ***hCG in urine*** is conveniently used for the ***early detection*** (within a week after missing the menstrual cycle) ***of pregnancy***.

III. The pro-opiomelanocortin (POMC) peptide family

This family consists of the hormones—adrenocorticotropic hormone (***ACTH***), lipotropin (***LPH***) and melanocyte stimulating hormone (***MSH***) and several (about 24) neuromodulators such as endorphins and enkephalins.

The synthesis of POMC family. is very interesting. All the members of POMC are produced from a single gene of the anterior and intermediate lobes of pituitary. It is fascinating that a single polypeptide—pro-opiomelanocortin—is the precursor (approximately 285 amino acids) that contains multiple hormones. The name ***pro-opiomelano-cortin*** is derived since it is a ***pro***hormone to ***opio***ids, ***melano***cyte-stimulating hormone and ***corti***cotropi***n***.

Products of POMC : The pituitary multihormone precursor is synthesized as prepro-opiomelanocortin from which POMC is formed. The POMC consists of 3 peptide groups.

1. ACTH that can give rise to α-MSH and corticotropin like intermediate lobe peptide (CLIP).

2. ***β-Lipotropin*** (β-LPH) that can produce γ-LPH, β-MSH and β-endorphin. The latter yields γ- and α-endorphins.

3. An N-terminal peptide that forms γ-MSH.

The products obtained from POMC are depicted in ***Fig.19.7***. These products undergo many modifications such as glycosylation, acetylation etc.

1. **Adrenocorticotropic hormone (ACTH) :** ACTH is a polypeptide with 39 amino acids and a molecular weight of 4,500. This hormone is primarily concerned with the growth and functions of adrenal cortex.

Regulation of ACTH production : The release of ACTH from the anterior pituitary is under the regulation of hypothalamic hormone, namely corticotropin releasing hormone (CRH).

Medical Concepts/ Clinical Correlates

- Growth hormone deficiency causes dwarfism while its excessive production results in gigantism (in children) or acromegaly (in adults).
- Identification of hCG in urine is employed for the early detection of pregnancy.
- Cushing's syndrome is due to overproduction of ACTH that results in the increased synthesis of adrenocorticosteroids. The symptoms of this syndrome include hypertension, edema and negative nitrogen balance.

Biochemical functions of ACTH

- ACTH promotes the conversion of cholesterol to pregnenolone in the adrenal cortex.
- It enhances RNA and protein synthesis and thus promotes adrenocortical growth.
- ACTH increases lipolysis by activating lipase of adipose tissue.

Overproduction of ACTH : *Cushing's syndrome* is caused by an excessive production of ACTH which may be due to a tumor. This syndrome is characterized by hyperpigmentation and increased production of adreno-corticosteroids. The associated symptoms include negative nitrogen balance, impaired glucose tolerance, hypertension, edema, muscle atrophy etc.

2. **β-Lipotropin (β-LPH) :** β-LPH is derived from POMC and contains 93 carboxy terminal amino acids. This polypeptide consists of γ-LPH and β-endorphin from which β-MSH and γ-endorphin are, respectively, formed. γ-Endorphin can be converted to α-endorphin and then to enkephalins (***Fig.19.7***). β-LPH is found only in the pituitary and not in other tissues since it is rapidly degraded.

The biochemical functions of β-LPH, as such, are limited. It promotes lipolysis and increases the mobilization of fatty acids. The most important function of β-LPH is its precursor role for the formation of β-endorphin and enkephalins.

Endorphins and enkephalins : These are the ***natural analgesics*** that control pain and emotions. They were discovered after an unexpected finding of opiate receptors in the human brain.

Synthesis : Endorphins and enkephalins are produced from β-endorphin which, in turn, is derived from POMC (***Fig.19.7***). β-Lipotropin has 31 amino acids while its modified products α and γ-endorphins have 15 and 14 amino acids, respectively. Methionine enkephalin (Tyr–Gly–Gly–Phe–***Met***) and leucine enkephalin (Tyr-Gly-Gly-Phe-***Leu***) are the two important pentapeptide derivatives of β-endorphin.

Biochemical actions : Endorphins and enkephalins are ***peptide neurotransmitters*** that produce opiate-like effects on the central nervous system, hence they are also known as ***opioid-peptides***. They bind to the same receptors as the ***morphine*** opiates and are believed to

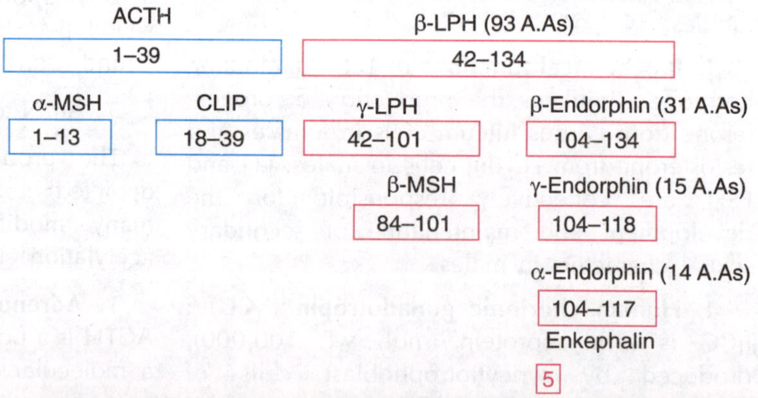

***Fig. 19.7 :** The members of the pro-opiomelanocortin (POMC) family derived from POMC cleavage. (Numbers in blocks represent amino acids in sequence; In the brackets are the number of amino acids–AAs; (ACTH–Adrenocorticotropic hormone; LPH–Lipotropin; MSH–Melanocyte-stimulating hormone; CLIP–Corticotropin like intermediate lobe peptide).*

control the ***endogenous pain perception***. Endorphins and enkephalins are more potent (20-30 times) than morphine in their function as analgesics.

It is ***believed that the pain relief*** through ***acupuncture and placebos is mediated through opioid peptides***.

3. **Melanocyte-stimulating hormone (MSH) :** Three types of MSH (α, β and γ) are present in the precursor POMC molecule. In humans, γ MSH is important while in some animals α and β are functional. The activity of γ-MSH is contained in the molecule γ-LPH or its precursor β-LPH (***Fig.19.7***).

The functions of MSH has been clearly established in some animals. MSH ***promotes the synthesis of skin pigment melanin*** (melanogenesis) and disperses melanin granules that ultimately leads to darkening of the skin. In humans, MSH does not appear to play any role in melanin synthesis.

POSTERIOR PITUITARY HORMONES

Two hormones namely ***oxytocin*** and ***antidiuretic hormone*** (ADH, vasopressin) are produced by the posterior pituitary gland (neurohypophysis). Both of them are non-apeptides (9 amino acids). Their structures are depicted in ***Fig.19.8***.

Oxytocin

The release of oxytocin from posterior pituitary gland is caused by the neural impulses of nipple stimulation. The other stimuli responsible for oxytocin release include vaginal and uterine distention.

(A) Cys[1]—Tyr—Ile—Gln—Asn—Cys—Pro—Leu—Gly[9] (S—S bridge between Cys[1] and Cys[6])

(B) Cys[1]—Tyr—Phe—Gln—Asn—Cys—Pro—Arg—Gly[9] (S—S bridge between Cys[1] and Cys[6])

Fig. 19.8 : *Structures of* ***(A)*** *Human oxytocin and* ***(B)*** *Human antidiuretic hormone (ADH).*

Biochemical functions

1. **Effect on uterus :** Oxytocin causes the contraction of pregnant uterus (smooth muscles) and ***induces labor***.

2. **Effect on milk ejection :** In mammals, oxytocin causes contraction of myoepithelial cells (look like smooth muscle cells) of breast. This stimulates the squeezing effect, ***causing milk ejection from the breast***.

3. Oxytocin synthesized in the ovary appears to ***inhibit the synthesis of steroids***.

Antidiuretic hormone (ADH)

The release of ADH (also called ***vasopressin***) is mostly controlled by osmoreceptors (of hypothalamus) and baroreceptors (of heart). Any increase in the osmolarity of plasma stimulates ADH secretion.

Biochemical functions : ADH is primarily concerned with the ***regulation of water balance*** in the body. It stimulates kidneys ***to retain water*** and, thus, increases the blood pressure.

In the absence of ADH, the urine output would be around 20 l/day. ADH acts on the distal convoluted tubules of kidneys and causes water reabsorption with a result that the urine output is around 0.5-1.5 l/day.

Mechanism of action : ADH stimulates adenylate cyclase causing production of cAMP. Water reabsorption is promoted by cAMP. Inhibitors of adenylate cyclase (e.g. calcium) inhibit the activity of ADH. This supports the view that ADH action is mostly mediated through cAMP.

Diabetes insipidus : This disorder is characterized by the excretion of large volumes of dilute urine ***(polyuria)***. It may be ***due to insufficient levels of ADH*** or a defect in the receptors of target cells.

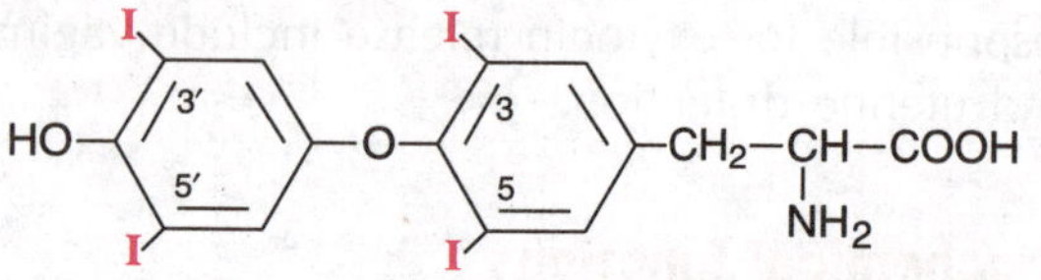

3, 5, 3′, 5′-Tetraiodothyronine (thyroxine, T_4)

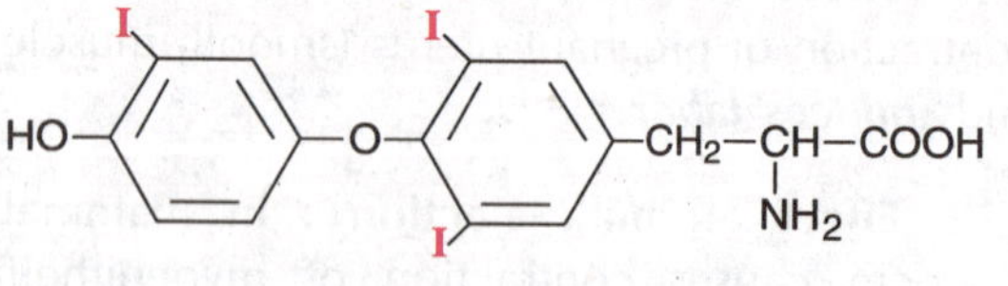

3,5,3′-Triiodothyronine (T_3)

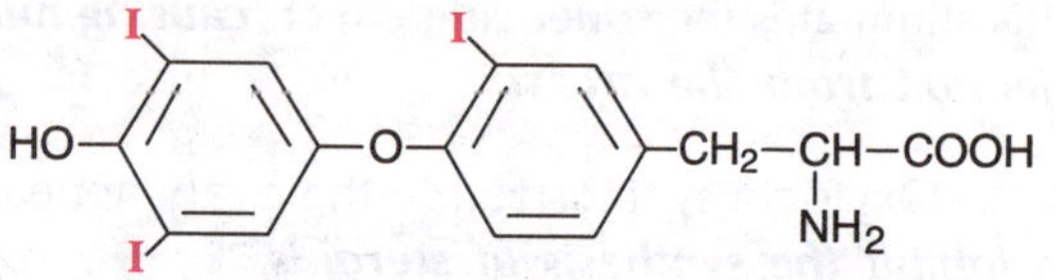

3, 3′, 5′-Triiodothyronine (reverse T_3, rT_3)

Fig. 19.9 : Structures of thyroid hormones (Refer Fig. 15.21 for their biosynthesis).

THYROID HORMONES

Thyroid gland (weighs about 30 g in adults) is located on either side of the trachea below the larynx. It produces two principal hormones (*Fig.19.9*)—***thyroxine*** (T_4; 3,5,3′,5′-tetraiodothyronine) and ***3,5,3′-triiodothyronine*** (T_3)—which regulate the metabolic rate of the body. Thyroid gland also secretes calcitonin, a hormone concerned with calcium homeostasis (discussed under calcium metabolism, ***Chapter 18***).

Biosynthesis of thyroid hormones

Iodine is essential for the synthesis of thyroid hormones. More than half of the body's total iodine content is found in the thyroid gland.

Uptake of iodide : The uptake of iodide by the thyroid gland occurs against a concentration gradient (about 20 : 1). It is an energy requiring process and is linked to the ATPase dependent Na^+-K^+ pump. Iodide uptake is primarily controlled by TSH, and inhibited by thiocyanate and perchlorate.

Formation of active iodine : The conversion of iodide (I^-) to active iodine (I^+) is an essential step for its incorporation into thyroid hormones. Thyroid is the only tissue that can oxidize I^- to a higher valence state I^+. This reaction requires H_2O_2 and is catalysed by the enzyme thyroperoxidase.

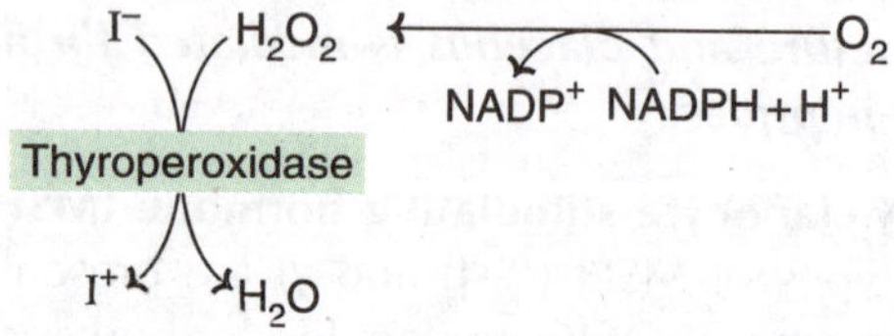

TSH promotes the oxidation of iodide to active iodine while the antithyroid drugs (thiourea, thiouracil, methimazole) inhibit.

Thyroglobulin and synthesis of T_3 and T_4 : Thyroglobulin is a glycoprotein and precursor for the synthesis of T_3 and T_4.

Tyrosine (of thyroglobulin) is first iodinated at position 3 to form monoiodotyrosine (MIT) and then at position 5 to form diiodotyrosine (DIT). Two molecules of DIT couple to form thyroxine (T_4). One molecule of MIT, when coupled with one molecule of DIT, triiodothyronine (T_3) is produced. The mechanism of coupling is not well understood. The details of synthesis of T_3 and T_4 are given under tyrosine metabolism (***Chapter 15***). A diagrammatic representation is depicted in ***Fig.19.10***.

Biochemical functions of thyroid hormones

Triiodothyronine (***T_3***) is ***about four times more active*** in its biological functions ***than*** thyroxine (***T_4***). The following are the biochemical functions attributed to thyroid hormones (T_3 and T_4).

1. **Influence on the metabolic rate :** Thyroid hormones ***stimulate the metabolic activities*** and increases the oxygen consumption in most of the tissues of the body (exception—brain, lungs, testes and retina).

Na^+-K^+ ATP pump : This is an energy dependent process which consumes a major share of cellular ATP. Na^+-K^+ ATPase activity is directly correlated to thyroid hormones and this, in turn, with ATP utilization. ***Obesity*** in some individuals

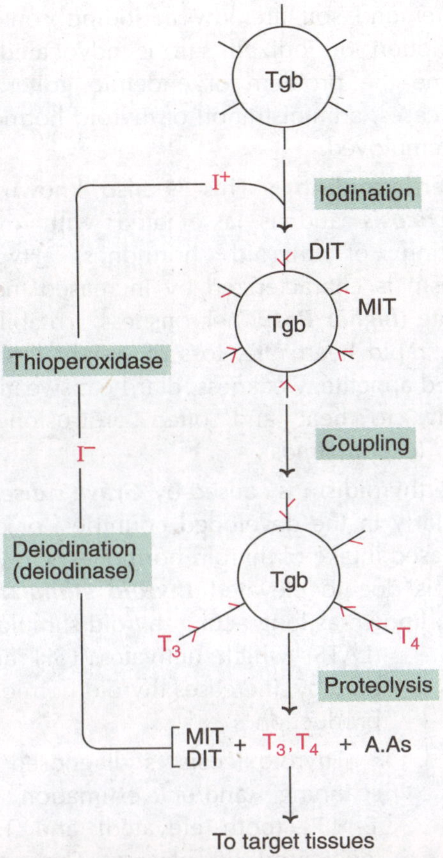

Fig. 19.10 : *Biosynthesis of thyroid hormones—diagrammatic representation [Note : Refer* **Fig. 15.21** *for synthesis with structures; Tgb–Thyroglobulin; I^+–Active iodine; T_3–Triiodothyronine; T_4–Thyroxine; MIT–Monoiodotyrosine; DIT–Diiodotyrosine; A. As–Amino acids].*

is attributed to a ***decreased energy utilization*** and heat production due to ***diminished Na^+-K^+ ATPase activity***.

2. **Effect on protein synthesis :** Thyroid hormones act like steroid hormones in promoting protein synthesis by acting at the transcriptional level (activate DNA to produce RNA). Thyroid hormones, thus, function as anabolic hormones and cause ***positive nitrogen balance and promote growth and development***.

3. **Influence on carbohydrate metabolism :** Thyroid hormones promote intestinal absorption of glucose and its utilization. These hormones increase ***gluconeogenesis and glycogenolysis***, with an overall effect of enhancing blood glucose level (hyperglycemia).

4. **Effect on lipid metabolism :** Lipid turnover and utilization are stimulated by thyroid hormones. ***Hypothyroidism*** is associated with ***elevated plasma cholesterol levels*** which can be reversed by thyroid hormone administration.

Regulation of T_3 and T_4 synthesis

The synthesis of thyroid hormones is controlled by ***feedback regulation*** (***Fig.19.11***). T_3 appears to be more actively involved than T_4 in the regulation process. The production of thyroid stimulating hormone (TSH) by pituitary, and thyrotropin releasing hormone (TRH) by hypothalamus are inhibited by T_3 and, to a lesser degree, by T_4. The increased synthesis of TSH and TRH occurs in response to decreased circulatory levels of T_3 and T_4. As already discussed, the body has sufficient stores of hormones to last for several weeks. Hence it takes some months to observe thyroid functional deficiency.

Abnormalities of thyroid function

Among the endocrine glands, ***thyroid is the most susceptible for hypo- or hyperfunction***.

Three abnormalities associated with thyroid functions are known.

Goiter : Any abnormal increase in the size of the thyroid gland is known as goiter. Enlargement of thyroid gland is mostly to compensate the decreased synthesis of thyroid hormones and is associated with ***elevated TSH***. Goiter is primarily due to a failure in the autoregulation of T_3 and T_4 synthesis. This may be ***caused by deficiency or excess of iodide***.

Goitrogenic substances (goitrogens) : These are the substances that interfere with the production of thyroid hormones. These include thiocyanates, nitrates and perchlorates and the drugs such as thiourea, thiouracil, thiocarbamide etc. Certain plant foods—cabbage, cauliflower and turnip—contain goitrogenic factors (mostly thiocyanates).

Simple endemic goiter : This is due to iodine deficiency in the diet. It is mostly found in the geographical regions away from sea coast where

Medical Concepts/ Clinical Correlates

- Endorphins and enkephalins are the natural painkillers in the brain. It is believed that the pain relief through acupuncture and placebos is mediated through these compounds.
- Deficiency of ADH causes diabetes insipidus, a disorder characterized by excretion of large volumes of dilute urine (polyuria).
- Thyroid hormones directly influence $Na^+ - K^+$ ATP pump which consumes a major share of cellular ATP. Obesity in some individuals is attributed to decreased energy utilization (heat production) due to diminished $Na^+ - K^+$ ATPase activity.

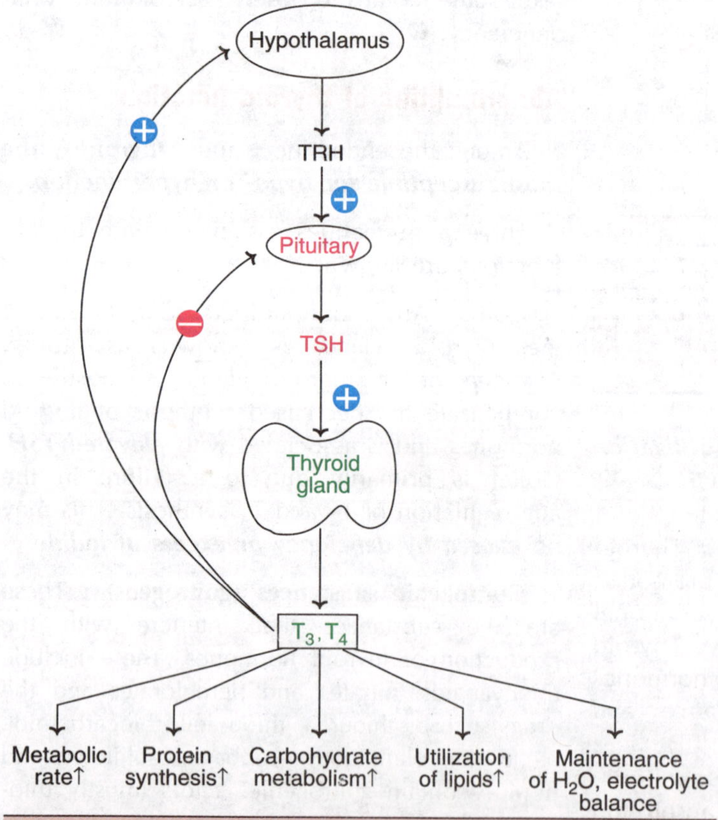

Fig. 19.11 : Regulation of synthesis and functions of thyroid hormones—an overview (TRH–Thyrotropin-stimulating hormone; TSH–Thyroid stimulating hormone; T_3–Triiodothyronine; T_4–Thyroxine; (+)–Promoting effect; (−)–Inhibitory effect).

the water and soil are low in iodine content. Consumption of iodized salt is advocated to overcome the problem of endemic goiter. In certain cases, administration of thyroid hormone is also employed.

Hyperthyroidism : This is also known as ***thyrotoxicosis*** and is associated with over-production of thyroid hormones. Hyperthyroidism is characterized by increased metabolic rate (***higher BMR***) nervousness, irritability, anxiety, rapid heart rate, loss of weight despite increased appetite, weakness, diarrhea, sweating, sensitivity to heat and often protrusion of eyeballs (exopthalmos).

Hyperthyroidism is caused by ***Grave's disease*** (particularly in the developed countries) or due to increased intake of thyroid hormones. Grave's disease is due to elevated ***thyroid stimulating IgG*** also known as long acting thyroid stimulator (LATS) which activates TSH and, thereby, increases thyroid hormonal production.

Thyrotoxicosis is diagnosed by scanning and/or estimation of T_3, T_4 (both elevated) and TSH (decreased) in plasma. The treatment includes administration of ***antithyroid drugs***. In severe cases, thyroid gland is surgically removed.

Hypothyroidism : This is due to an impairment in the function of thyroid gland that often causes decreased circulatory levels of T_3 and T_4. Disorders of pituitary or hypothalamus also contribute to hypothyroidism. Women are more susceptible than men. Hypothyroidism is characterized by ***reduced BMR***, slow heart rate, weight gain, sluggish behaviour, constipation, sensitivity to cold, dry skin etc.

Hypothyroidism in children is associated with physical and mental retardation, collectively known as ***cretinism***. Early diagnosis and proper treatment are essential. Hypothyroidism in adult causes

myxoedema, characterized by bagginess under the eyes, puffiness of face, slowness in physical and mental activities.

Laboratory diagnosis of thyroid function

Measurement of ***basal metabolic rate*** (BMR) was once used to reflect thyroid activity. The estimation of serum ***protein bound iodine*** (PBI), representing the circulating thyroid hormones, was employed for a long time to assess thyroid function. The normal serum PBI concentration is 3-8 μg/100 ml.

Hypothyroidism is associated with decreased PBI and hyperthyroidism with increased PBI.

In recent years, more sensitive and reliable tests have been developed to assess thyroid activity. The concentration of free $\boldsymbol{T_3}$ and $\boldsymbol{T_4}$, and ***TSH*** are measured (by RIA or ELISA) and their serum normal concentrations are

Free triiodothyronine (T_3)	—	80–220 ng/dl
Free thyroxine (T_4)	—	0.8–2.4 ng/dl
Total thyroxine (T_4)	—	5–12 μg/dl
Thyroid stimulating hormone (TSH)	—	<4.5 mIU/ml

Radioactive iodine uptake (RAIU) and scanning of thyroid gland are also used for diagnosis.

Primary hypothyroidism : Serum T_3 and T_4 levels are low while TSH is elevated.

Primary hyperthyroidism : Serum T_3 and T_4 are increased while TSH is reduced.

Thyroid activity and serum cholesterol

Serum cholesterol level is ***increased in hypothyroidism*** and decreased in hyperthyroidism. Unfortunately, cholesterol estimation will be of no value in the assessment of thyroid function. This is due to the fact that serum cholesterol level is elevated in many other disorders (diabetes, obstructive jaundice, nephrotic syndrome etc.). However, ***cholesterol estimation may be utilized for monitoring thyroid therapy***.

HORMONES OF ADRENAL CORTEX

The adrenal glands are two small organs (each weighing about 10 g), located above the kidneys. Each adrenal consists of two distinct tissues—an outer cortex (with 3 zones) and inner medulla (***Fig.19.12***).

As many as 50 steroid hormones (namely adrenocorticosteroids), produced by adrenal cortex, have been identified. However, only a few of them possess biological activity.

Adrenocorticosteroids are classified into three groups according to their dominant biological action. However, there is some overlap in their functions.

1. **Glucocorticoids :** These are 21-carbon steroids, produced mostly by zona fasciculata. They affect glucose (hence the name), amino acid and fat metabolism in a manner that is opposite to the action of insulin. ***Cortisol*** (also known as hydrocortisone) is the most important glucocorticoid in humans. Corticosterone is predominantly found in rats.

2. **Mineralocorticoids :** These are also 21-carbon containing steroids produced by zona glomerulosa. They regulate water and electrolyte balance. ***Aldosterone*** is the most prominent mineralocorticoid.

3. **Androgens and estrogens :** The innermost adrenal cortex zona reticularis produces small quantities of androgens (19-carbon) and ***estrogens*** (18-carbon). These hormones affecting ***sexual development*** and functions are mostly produced by gonads. Dehydroepiandrosterone—a precursor for androgens—is synthesized in adrenal cortex.

Synthesis of adrenocorticosteroids

Cholesterol undergoes cleavage with an elimination of a 6-carbon fragment to form pregnenolone. ***Pregnenolone is the common***

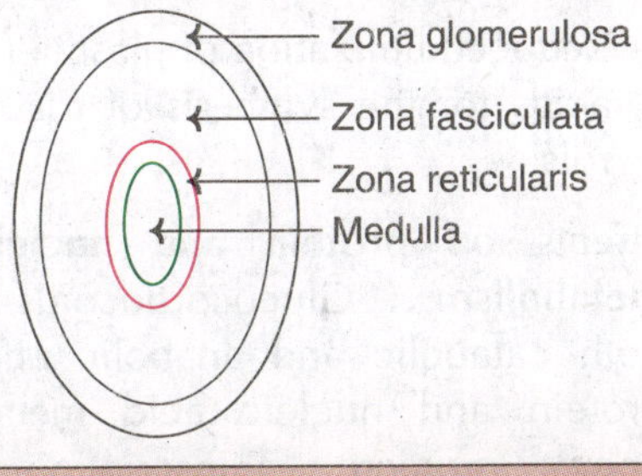

Fig. 19.12 : Adrenal gland with zones (3) and medulla.

precursor for the synthesis of all steroid hormones.

Conversion of cholesterol to pregnenolone is catalysed by cytochrome P_{450} side chain cleavage enzyme. This reaction is promoted by ACTH. The enzymes—hydroxylases, dehydrogenases/ isomerases and lyases associated with mitochondria or endoplasmic reticulum—are responsible for the synthesis of steroid hormones. The metabolic pathway for the formation of major adrenocorticosteroids is given in ***Fig.19.13***.

Biochemical functions of adrenocorticosteroids

1. **Glucocorticoid hormones :** The important glucocorticoids are—***cortisol***, ***cortisone*** and ***corticosterone***. They bring about several biochemical functions in the body.

(a) **Effects on carbohydrate metabolism :** Glucocorticoids ***promote*** the synthesis of glucose (***gluconeogenesis***). This is brought about by increasing the substrates (particularly amino acids) and enhancing the synthesis of phosphoenolpyruvate carboxykinase, the rate limiting enzyme in gluconeogenesis.

The overall influence of glucocorticoids on carbohydrate metabolism is to increase blood glucose concentration. The biological actions of glucocorticoids generally oppose that of insulin.

(b) **Effects on lipid metabolism :** Glucocorticoids ***increase the circulating free fatty acids***. This is caused by two mechanisms.

(i) Increased breakdown of storage triacylglycerol (lipolysis) in adipose tissue.

(ii) Reduced utilization of plasma free fatty acids for the synthesis of triacylglycerols.

(c) **Effects on protein and nucleic acid metabolism :** Glucocortiocoids exhibit both catabolic and anabolic effects on protein and nucleic acid metabolism. They promote transcription (RNA synthesis) and protein biosynthesis in liver. These anabolic effects of glucocorticoids are caused by the stimulation of specific genes.

Glucocorticoids (particularly at high concentration) cause catabolic effects in extrahepatic tissues (e.g. muscle, adipose tissue, bone etc.). This results in enhanced degradation of proteins.

(d) **Effects on water and electrolyte metabolism :** The influence of glucocorticoids on water metabolism is ***mediated through antidiuretic hormone (ADH)***. Deficiency of glucocorticoids causes increased production of ADH. ADH decreases glomerular filtration rate causing water retention in the body.

(e) **Effects on the immune system :** Glucocorticoids (particularly cortisol), in high doses, suppress the host immune response. The steroid hormones act at different levels—damaging lymphocytes, impairment of antibody synthesis, suppression of inflammatory response etc.

(f) **Other physiological effects of glucocorticoids :** Glucocorticoids are involved in several physiological functions.

(i) ***Stimulate the fight and flight response*** (to face sudden emergencies) of catecholamines.

(ii) Increase the production of gastric HCl and pepsinogen.

(iii) Inhibit the bone formation, hence the subjects are at a risk for osteoporosis.

Mechanism of action of glucocorticoids : Glucocorticoids bind to specific receptors on the target cells and bring about the action. These hormones mostly act at the transcription level and control the protein synthesis.

2. **Mineralocorticoid hormones :** The most active and potent mineralocorticoid is ***aldosterone***. It promotes Na^+ reabsorption at the distal convoluted tubules of kidney. Na^+ retention is accompanied by corresponding excretion of K^+, H^+ and NH_4^+ ions.

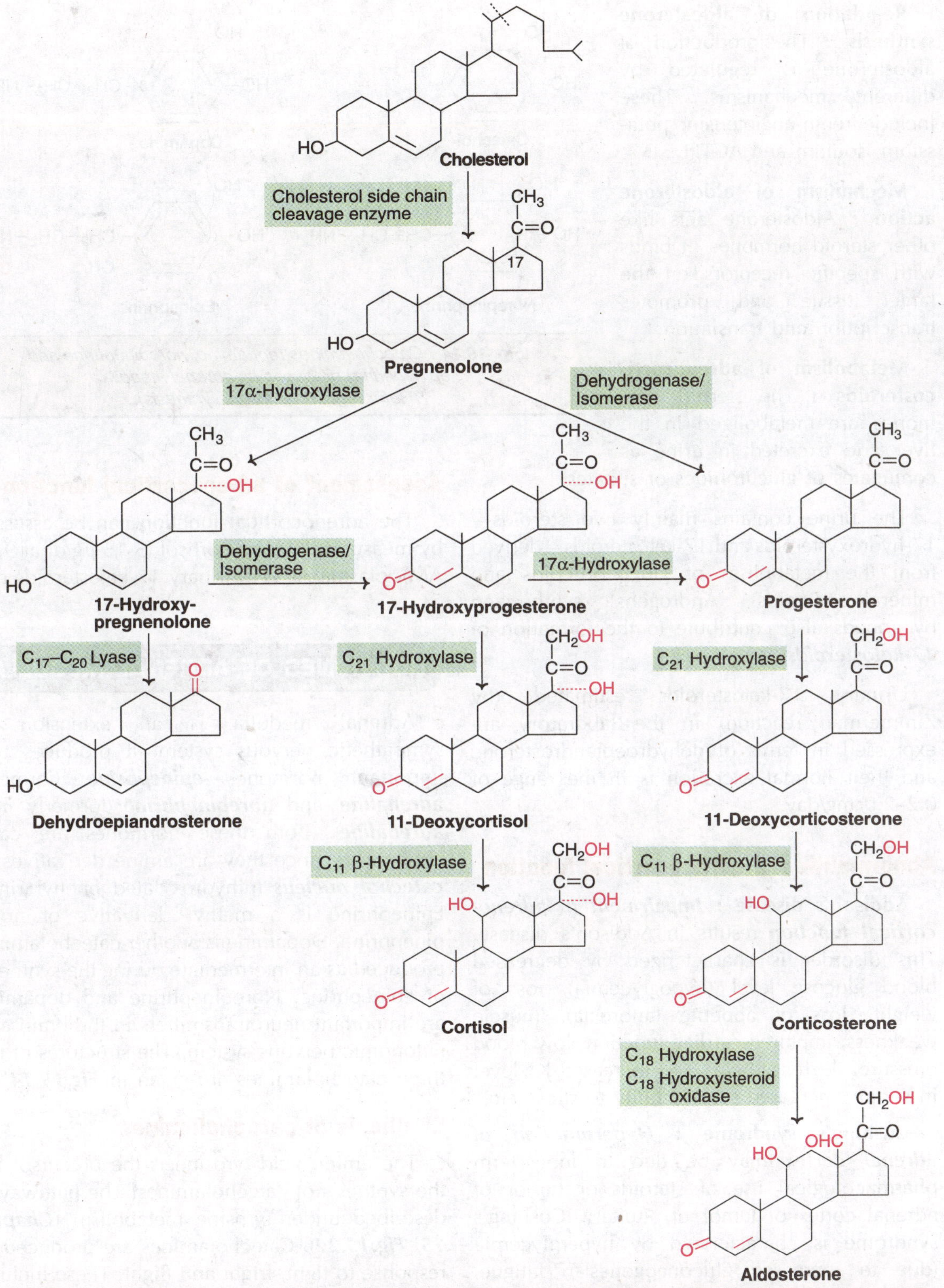

Fig. 19.13 : Biosynthesis of major adrenocorticosteroids.

Regulation of aldosterone synthesis : The production of aldosterone is regulated by different mechanisms. These include renin-angiotensin, potassium, sodium and ACTH.

Mechanism of aldosterone action : Aldosterone acts like other steroid hormones. It binds with specific receptors on the target tissue and promotes transcription and translation.

Metabolism of adrenocorticosteroids : The steroid hormones are metabolized in the liver and excreted in urine as conjugates of glucuronides or sulfates.

The urine contains mainly two steroids—17-hydroxysteroids and 17-ketosteroids—derived from the metabolism of glucocorticoids and mineralocorticoids. Androgens synthesized by gonads also contribute to the formation of ***17-ketosteroids***.

Urinary 17-ketosteroids estimated (by Zimmerman reaction) in the laboratory are expressed in terms of dehydroepiandrosterone and their normal excretion is in the range of 0.2–2.0 mg/day.

Abnormalities of adrenocortical function

Addison's disease : ***Impairment in adrenocortical function*** results in Addison's disease. This disorder is characterized by decreased blood glucose level (hypoglycemia), loss of weight, loss of appetite (anorexia), muscle weakness, impaired cardiac function, low blood pressure, decreased Na^+ and increased K^+ level in serum, increased susceptibility to stress etc.

Cushing's syndrome : ***Hyperfunction of adrenal cortex*** may be due to long term pharmacological use of steroids or tumor of adrenal cortex or tumor of pituitary. Cushing's syndrome is characterized by hyperglycemia (due to increased gluconeogenesis), fatigue, muscle wasting, edema, osteoporosis, negative nitrogen balance, hypertension, moon-face etc.

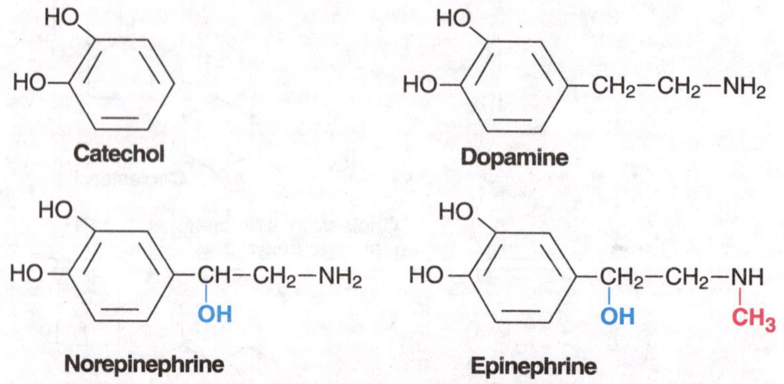

***Fig. 19.14 :** Catecholamines (dopamine, norepinephrine and epinephrine) produced by adrenal medulla (Refer **Fig. 15.22** for biosynthesis).*

Assessment of adrenocortical function

The adrenocortical function can be assessed by measuring plasma cortisol (5-15 μg/dl at 9.00 AM), plasma ACTH, urinary 17-ketosteroids etc.

HORMONES OF ADRENAL MEDULLA

Adrenal medulla is an extension of sympathetic nervous system. It produces two important hormones—***epinephrine*** (formerly ***adrenaline***) and ***norepinephrine*** (formerly ***noradrenaline***). Both these hormones are catecholamines since they are amine derivatives of ***catechol nucleus*** (dihydroxylated phenyl ring). Epinephrine is a methyl derivative of norepinephrine. Dopamine is another catecholamine, produced as an intermediate during the synthesis of epinephrine. Norepinephrine and dopamine are important neurotransmitters in the brain and autonomic nervous system. The structures of the three catecholamines are given in ***Fig.19.14***.

Synthesis of catecholamines

The amino acid tyrosine is the precursor for the synthesis of catecholamines. The pathway is described under tyrosine metabolism (***Chapter 15, Fig.15.22***). Catecholamines are produced in response to fight, fright and flight. These include the emergencies like shock, cold, fatigue, emotional conditions like anger etc.

Biochemical functions of catecholamines

Catecholamines cause diversified biochemical effects on the body. The ultimate goal of their action is to mobilize energy resources and prepare the individuals ***to meet emergencies*** (e.g. shock, cold, low blood glucose etc.).

1. **Effects on carbohydrate metabolism :** Epinephrine and norepinephrine in general ***increase the degradation of glycogen*** (glycogenolysis), synthesis of glucose (gluconeogenesis) and decrease glycogen formation (glycogenesis).

The overall effect of catecholamines is to ***elevate blood glucose levels*** and make it available for the brain and other tissues to meet the emergencies.

2. **Effects on lipid metabolism :** Both epinephrine and norepinephrine ***enhance the breakdown of triacylglycerols*** (lipolysis) in adipose tissue. This causes increase in the free fatty acids in the circulation which are effectively utilized by the heart and muscle as fuel source.

The metabolic effects of catecholamines are mostly related to the increase in adenylate cyclase activity causing elevation in cyclic AMP levels (refer carbohydrate and lipid metabolisms for more details).

3. **Effects on physiological functions :** In general, catecholamines (most predominantly epinephrine) ***increase cardiac output, blood pressure and oxygen consumption***. They cause smooth muscle relaxation in bronchi, gastrointestinal tract and the blood vessels supplying skeletal muscle. On the other hand, catecholamines stimulate smooth muscle contraction of the blood vessels supplying skin and kidney. Platelet aggregation is inhibited by catecholamines.

Metabolism of catecholamines

Catecholamines are rapidly inactivated and metabolized. The enzymes—catechol-O methyltransferase (COMT) and ***monoamine oxidase (MAO)***, found in many tissues act on catecholamines. The metabolic products metanephrine and ***vanillylmandelic acid (VMA)*** are excreted in urine.

Abnormalities of catecholamine production

Pheochromocytomas : These are the tumors of adrenal medulla. The diagnosis of pheochromocytoma is possible only when there is an ***excessive production of epinephrine and norepinephrine*** that causes severe hypertension. In the individuals affected by this disorder, the ratio of norepinephrine to epinephrine is increased. The measurement of ***urinary VMA*** (***normal <8 mg/day***) is helpful in the ***diagnosis*** of pheochromocytomas.

HORMONES OF GONADS

The gonads (testes in males, ovaries in females) perform closely related dual functions.

1. Synthesize sex hormones;
2. Produce germ cells.

The steroid sex hormones are responsible for growth, development, maintenance and regulation of reproductive system. Sex hormones are essentially required for the development of germ cells.

The ***sex hormones*** are categorized into three groups

1. **Androgens** or ***male sex hormones*** which are C-19 steroids.

2. **Estrogens** or ***female sex hormones*** which are C-18 steroids. Ring A of steroid nucleus is phenolic in nature and is devoid of C-19 methyl group.

3. **Progesterone** is a C-21 steroid produced during the luteal phase of menstrual cycle and also during pregnancy.

ANDROGENS

The male sex hormones or androgens are produced by the Leydig cells of the testes and to a minor extent by the adrenal glands in both the sexes. Ovaries also produce small amounts of androgens.

Biosynthesis of androgens

Cholesterol is the ***precursor*** for the synthesis of androgens. It is first converted to pregnenolone which then forms androstenedione by two

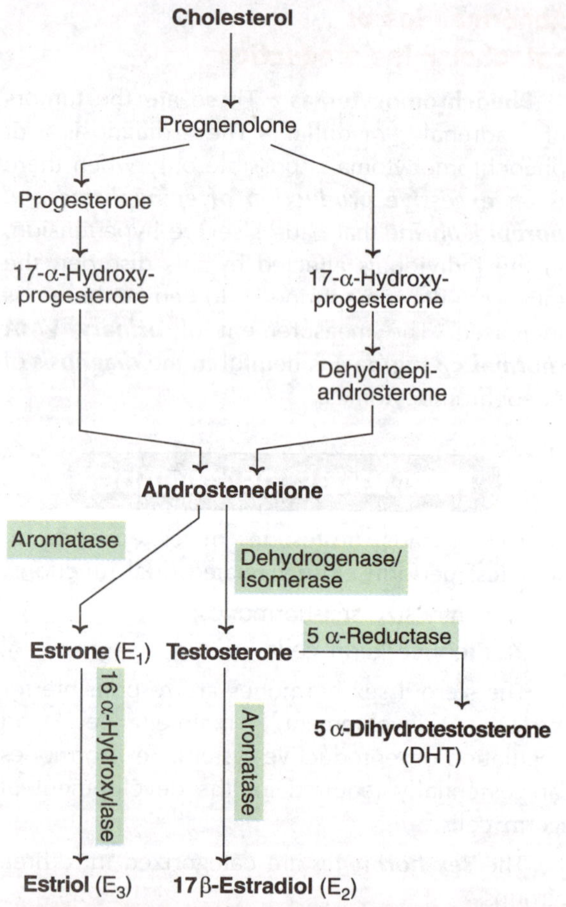

Fig. 19.15 : *Biosynthesis of steroid sex hormones from cholesterol*
(Note : *Male and female sex hormones are given together).*

pathways—either through progesterone or through 17-hydroxypregnenolone (***Fig.19.15***). Testosterone is produced from androstenedione, and this is under the control of LH and FSH.

Active form of androgen : The primary product of testes is testosterone. However, the active hormone in many tissues is not testosterone but its metabolite ***dihydrotestosterone*** (DHT). Testosterone, on reduction by the enzyme 5 α-reductase, forms DHT. Some workers consider testosterone as a prohormone and dihydrotestosterone, the more potent form as the hormone. DHT is implicated in ***benign prostate hypertrophy*** (BPH). Finasteride inhibits 5α-reductase, and hence used in the treatment of BPH.

Physiological and biochemical functions of androgens

1. **Sex-related physiological functions :** The androgens, primarily DHT and testosterone, influence :

- Growth, development and maintenance of male reproductive organs.
- Sexual differentiation and secondary sexual characteristics.
- Spermatogenesis.
- Male pattern of aggressive behavior.

2. **Biochemical functions :** Many specific biochemical effects of androgens that ultimately influence the physiological functions stated above are identified. Androgens are anabolic in nature.

- **Effects on protein metabolism :** Androgens promote RNA synthesis (transcription) and protein synthesis (translation). Androgens cause positive nitrogen balance and increase the muscle mass.
- **Effects on carbohydrate and fat metabolisms :** Androgens increase glycolysis fatty acid synthesis and citric acid cycle.
- **Effects on mineral metabolism :** Androgens promote mineral deposition and bone growth before the closure of epiphyseal cartilage.

ESTROGENS

Estrogens are predominantly ovarian hormones, synthesized by the follicles and corpus luteum of ovary. These hormones are responsible for maintenance of menstrual cycle and reproductive process in women.

Synthesis of estrogens

Estrogen synthesis occurs from the precursor cholesterol (***Fig.19.15***). Estrogens are produced by aromatization (formation of aromatic ring) of androgens. The ovary produces estradiol (E_2) and estrone (E_1) while the placenta synthesizes these two steroid hormones and estriol (E_3). The synthesis of estrogens is under the control of LH and FSH.

Physiological and biochemical functions of estrogens

1. **Sex-related physiological functions :** The estrogens are primarily concerned with

- Growth, development and maintenance of female reproductive organs.
- Maintenance of menstrual cycles.
- Development of female sexual characteristics.

2. **Biochemical functions :** Estrogens are involved in many metabolic functions.

- **Lipogenic effect :** Estrogens increase lipogenesis in adipose tissue and, for this reason, women have relatively more fat (about 5%) than men.
- **Hypocholesterolemic effect :** Estrogens lower the plasma total cholesterol. The LDL fraction of lipoproteins is decreased while the HDL fraction is increased. This explains the low incidence of atherosclerosis and coronary heart diseases in the women during reproductive age.
- **Anabolic effect :** Estrogens in general promote transcription and translation. The synthesis of many proteins in liver is elevated e.g. transferrin, ceruloplasmin.
- **Effect on bone growth :** Estrogens like androgens promote calcification and bone growth. It is believed that decalcification of bone in the postmenopausal women leading to osteoporosis is due to lack of estrogens.
- **Effect on transhydrogenase :** Transhydrogenase is an enzyme activated by estrogen. It is capable of ***transferring reducing equivalents from NADPH to NAD⁺***. The NADH so formed can be oxidized. It is explained that in the women ***after menopause, due to deficiency of estrogens***, the transhydrogenase activity is low. This results in the diversion of NADPH towards lipogenesis—causing ***obesity***.

PROGESTERONE

Progesterone is synthesized and secreted by corpus luteum and placenta. Progesterone, as such, is an intermediate in the formation of steroid hormones from cholesterol (***See Fig.20.13***). LH controls the production of progesterone.

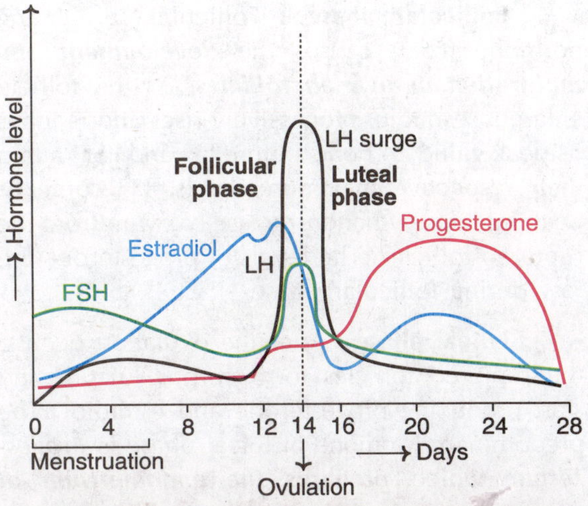

Fig. 19.16 : Hormonal pattern in women during mestrual cycle (FSH–Follicle stimulating hormone; LH–Luteinizing hormone).

Biochemical functions of progesterone

1. Progesterone is essentially required for the implantation of fertilized ovum and maintenance of pregnancy.

2. It promotes the growth of glandular tissue in uterus and mammary gland.

3. Progesterone increases the body temperature by 0.5–1.5 F°. The exact mechanism of this thermogenic effect is not clearly known. The measurement of temperature was used as an indicator for ovulation.

THE MENSTRUAL CYCLE

The occurrence of menstrual cycle is a good example of coordination among the hormonal functions. In humans, the menstrual cycle is under the control of FSH, LH, estrogens and progesterone. The cycle normally varies between 25 and 35 days in length, with a mean of 28 days. The menstrual cycle can be divided into two phases—follicular phase and luteal phase (***Fig.19.16***).

1. **Follicular phase :** Follicular stimulating hormone (FSH) causes the ***development and maturation of ovarian follicles***. As the follicle enlarges, estradiol progessively rises and reaches its peak value 24 hours before LH and FSH attain their respective maximum levels. LH surge or peak initiates ovulation-release of ovum from the ruptured follicles. The levels of progesterone are low during follicular phase

2. **Luteal phase :** After the ovulation occurs, the ruptured follicles form corpus luteum and start producing progesterone and estradiol. The predominant hormone of luteal phase is ***progesterone*** which ***prepares the endometrium of uterus for implantation of the fertilized ovum***. LH maintains the corpus for a few days. In the absence of implantation, the ***corpus luteum*** regresses and sheds endometrium causing menstruation. And another new cycle begins.

The luteal phase is always fixed, with 14 ± 2 days in length. The observed variations in the length of menstrual cycle are due to changes in the follicular phase. ***In case of implantation of the fertilized ovum***, human chorionic g***onadotropin (hCG) is produced by the cells of implanted early embryo***. hCG stimulates corpus luteum to synthesize progesterone. This continues till the plancenta starts making high quantities of progesterone.

Menopause

The ***menstrual cycles*** which begin in the women after puberty, continue till the age of 45-50 years. The cycles ***cease*** around this age which coincides with the loss of ovarian function. The progesterone and estrogen levels are very low in these women. However, the concentration of LH and FSH are elevated due to lack of feedback inhibition by estrogens.

Post-menopausal women are susceptible to two complications associated with insufficient levels of sex hormones.

1. **Atrophy of secondary sex tissues :** Mainly the epithelial tissue of vagina and lower urinary tract.

2. **Osteoporosis :** Decreased density of bones and increased susceptibility to fractures.

GASTROINTESTINAL (OR GUT) HORMONES

The digestion and absorption of nutrients (***Chapter 8***) is a complicated process which is regulated by the autonomic nervous system. This occurs in association with peptide hormones of gastrointestinal tract (GIT).

The specialized cells lining the GIT are responsible for the production of GIT hormones, Hence GIT may be considered as the largest mass of cells that secrete hormones. A large number of GIT hormones have been identified. However, only four GIT hormones have been well characterised.

1. **Gastrin :** This hormone contains 17 amino acids and is produced by gastric mucosa. It ***stimulates the secretion of gastric HCl and pepsinogen*** (proenzyme of pepsin). The release of gastrin is stimulated by vagus nerve of stomach and partially digested proteins. HCl and certain other hormones inhibit gastrin release.

2. **Secretin :** It is a 27-amino acid containing polypeptide and resembles glucagon in many ways. Secretin is synthesized by the mucosa of the upper small intestine. It is released in response to the presence of HCl in chyme in the duodenum which is passed on from the stomach. Secretin ***stimulates pancreatic cells to produce bicarbonate*** (HCO_3^-) in order to neutralize HCl.

3. **Cholecystokinin (CCK) :** It contains 33 amino acids and is produced by the upper part of small intestine. The secretion of CCK is stimulated by the products of protein and lipid digestion, namely peptides, amino acids, mono- or diacylglycerols, fatty acids and glycerol.

Cholecystokinin ***stimulates the contraction of gall bladder and increases the flow of bile into duodenum***. It also promotes the secretion of digestive enzymes and HCO_3^- from pancreas.

4. **Gastric inhibitory peptide (GIP) :** It contains 43 amino acids and is produced by duodenal mucosa. The release of GIP is stimulated by the presence of glucose in the gut. The most important function of GIP is to stimulate the release of insulin from pancreas. This is evident from the fact that the plasma

Medical Concepts/ Clinical Correlates

- Catecholamines are produced in response to fight, fright and flight. The ultimate goal of catecholamine function is to mobilize energy resources and prepare the individual to meet emergencies such as shock, cold, fatigue, anger etc.
- Pheochromocytomas are the tumors of adrenal medulla, characterized by excessive production of epinephrine and norepinephrine, associated with severe hypertension.
- Sex hormones are primarily responsible for growth, development, maintenance and regulation of reproductive system.
- The low incidence of atherosclerosis and coronary heart disease in the women during reproductive age is due to estrogens.

insulin level is elevated much before the increase in blood glucose. GIP also inhibits gastric HCl secretion, gastric motility and its emptying.

GIT hormones show certain structural relations and may be considered under two families.

(i) **Gastrin family :** Some of the C-terminal amino acids are identical. This family includes gastrin and CCK.

(ii) **Secretin family :** Secretin, GIP and glucagon are structurally related, hence may be considered under this family.

Besides the hormones described above, several other hormones (in hundreds!) from the GIT have been identified. These hormones are often known as ***candidate hormones,*** since their biological functions are yet to be precisely identified. The candidate hormones include ***vasoactive intestinal peptide*** (VIP), ***motilin, enteroglucagon, substance P, neurotensin, somatostatin*** and ***enkephalins.***

Mechanism of action of GIT hormones

Many of the GIT hormones have receptor sites specific for their action. At least two distinct mechanisms have been identified through which these hormones act.

1. ***Production of cAMP*** through the activation of adenylate cyclase e.g. secretin, VIP etc.

2. ***Stimulation of intracellular Ca^{2+}*** usually mediated through the metabolism of phosphatidylinositol e.g. gastrin , CCK.

Both these mechanisms ultimately influence the enzyme secretions/other biological effects.

Other hormones

Besides the hormones discussed above, there are a few other important hormones which are not referred to in this chapter. ***Insulin and glucagon*** are described under diabetes mellitus (***Chapter 36***) while ***parathyroid hormone*** and ***calcitonin*** are discussed under calcium metabolism (***Chapter 18***) These hormones are not given here to avoid repetition.

THE READER MUST REFER CHAPTER 36 FOR INSULIN AND GLUCAGON; CHAPTER 18 FOR PARATHYROID HORMONE AND CALCITONIN

THE READER MUST REFER APPENDIX VI (Page Nos : 738 - 756) FOR RELATED CLINICAL CASE STUDIES (Nos : 47, 48, 51, 60, 61)

SUMMARY

- *Hormones are the organic substances, produced in minute quantities by specific tissues (endocrine glands) and secreted into the blood stream to control the biological activities in the target cells. They may be regarded as the chemical massengers involved in the regulation and coordination of body functions.*

- *Hormones are classified based on their chemical nature or mechanism of action. Chemically, they may be proteins or peptides (insulin, oxytocin), steroids (glucocorticoids, sex hormones) and amino acid derivatives (epinephrine, thyroxine). By virtue of the function, group I hormones bind to the intracellular receptors (estrogens, calcitriol), while group II hormones (ACTH, LH) bind to the cell surface receptors and act through the second messengers.*

- *Cyclic AMP (cAMP) is an intracellular second messenger for a majority of polypeptide hormones. Membrane bound adenylate cyclase enzyme, through the mediation of G proteins, is responsible for the synthesis of cAMP. cAMP acts through protein kinases that phosphorylate specific proteins which, in turn, cause the ultimate biochemical response. Phosphatidylinositol/calcium system also functions as a second messenger for certain hormones (TRH, gastrin).*

- *Hypothalamus is the master coordinator of hormonal action as it liberates certain releasing factors or hormones (TRH, CRH, GRH, GRIH) that stimulate or inhibit the corresponding trophic hormones from the anterior pituitary.*

- *Anterior pituitary gland is the master endocrine organ that produces several hormones which influence either directly or indirectly (through the mediation of other endocrine organs) a variety of biochemical processes in the body. For instance, growth hormone is directly involved in growth promoting process while TSH, FSH and ACTH, respectively influence thyroid gland, gonads and adrenal cortex to synthesize hormones.*

- *Thyroid gland produces two principal hormones—thyroxine (T_4) and triiodothyronine (T_3)—which are primarily concerned with the regulation of the metabolic activity of the body. Goiter is a disorder caused by enlargement of thyroid gland and is mainly due to iodine deficiency in the diet.*

- *Adrenal cortex synthesizes glucocorticoids (e.g. cortisol) that influence glucose, amino acid and fat metabolism, and mineralocorticoids (e.g. aldosterone) that regulate water and electrolyte balance. Androgens and estrogens (sex hormones) in small quantities are also synthesized by the adrenal cortex.*

- *Adrenal medulla produces two important hormones—epinephrine and norepinephrine (catecholamines). They influence diversified biochemical functions with an ultimate goal to mobilize energy resources and prepare the individual to meet emergencies (shock, anger, fatigue etc.)*

- *The steroid sex hormones, primarily androgens in males and estrogens in females, are respectively synthesized by the testes and ovaries. These hormones are responsible for growth, development, maintenance and regulation of reproductive system in either sex.*

- *Several gastrointestinal hormones (e.g. gastrin, secretin) have been identified that are closely involved in the regulation of digestion and absorption of foodstuffs.*

Self-Assessment Exercises

I. Essay questions

1. Describe the role of second messengers in hormonal action.
2. Write an account of the anterior pituitary hormones.
3. Discuss in detail the synthesis and biochemical functions of thyroid hormones.
4. Describe the hormones of adrenal cortex with special reference to glucocorticoids.
5. Write briefly on the synthesis and biochemical functions of sex hormones.

II. Short notes

(a) 'G'-Proteins, (b) Inositol triphosphate, (c) Hypothalamic hormones, (d) ACTH, (e) Goiter, (f) Epinephrine, (g) Cortisol, (h) Gastrin, (i) ADH, (j) Aldosterone.

III. Fill in the blanks

1. The enzyme that catalyses the formation of cAMP from ATP is ______________.
2. The inorganic ion that can act as a second messenger for certain hormones is ______________.
3. The endocrine organ responsible for the synthesis of trophic hormones is ______________.
4. The compounds that produce opiate-like effects on the central nervous system are ______________.
5. The enzyme that converts iodide (I^-) to active iodine (I^+) ______________.
6. The most predominant mineralocorticoid synthesized by adrenal cortex ______________.
7. The major urinary excretory product of catecholamines ______________.
8. The male sex hormone, testosterone, is converted to a more active form, namely ______________.
9. The precursor for the synthesis of steroid hormones ______________.
10. The gastrointestinal hormone that increases the flow of bile from the gall bladder ______________.

IV. Multiple choice questions

11. Impairment in the synthesis of dopamine by the brain is a major causative factor for the disorder
    (a) Parkinson's disease (b) Addison's disease (c) Cushing's syndrome (d) Goiter.
12. One of the following hormones is an amino acid derivative
    (a) Epinephrine (b) Norepinephrine (c) Thyroxine (d) All of them.
13. The most active mineralocorticoid hormone is
    (a) Cortisol (b) Aldosterone (c) 11-Deoxycorticosterone (d) Corticosterone.
14. Name the hormone, predominantly produced in response to fight, fright and flight
    (a) Thyroxine (b) Aldosterone (c) Epinephrine (d) ADH.
15. The hormone essentially required for the implantation of fertilized ovum and maintenance of pregnancy
    (a) Progesterone (b) Estrogen (c) Cortisol (d) Prolactin.

20 Organ Function Tests

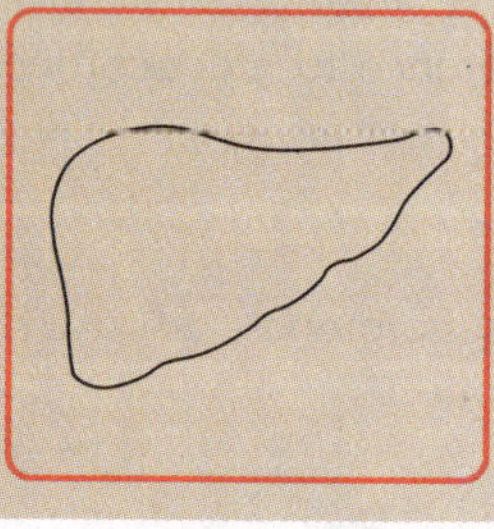

The liver speaks

"Master organ I am, for the body's metabolism!
Damage to my cells causes malfunction;
Raising serum bilirubin and certain enzymes markedly,
Tested in lab for my functional measurement."

Each organ of the body has to perform its biochemical functions to keep the body, as a whole, in a healthy state. This is possible only when the cells of the organ are intact in structure and function. Any abnormality in the tissue, caused by exogenous or endogenous factors, will seriously impair the organ function which, in turn, influences the health of the organism.

Based on the functional capabilities of the organs, ***specific biochemical investigations*** have been developed ***in the laboratory***, to assess their function. In this chapter, the biochemical investigations to assess the functioning of liver, kidney, stomach and pancreas are discussed. The tests to evaluate the ***functions of endocrine organs*** are discussed elsewhere (***Chapter 19***).

LIVER FUNCTION TESTS

Liver performs several diversified functions. It is the central organ of body's metabolism.

Major functions of liver

1. **Metabolic functions :** Liver actively participates in carbohydrate, lipid, protein, mineral and vitamin metabolisms.

2. **Excretory functions :** Bile pigments, bile salts and cholesterol are excreted in the bile into intestine.

3. **Protective functions and detoxification :** Kupffer cells of liver perform phagocytosis to eliminate foreign compounds. Ammonia is detoxified to urea. Liver is responsible for the metabolism of xenobiotics (detoxification).

4. **Hematological functions :** Liver participates in the formation of blood (particularly in the embryo), synthesis of plasma proteins (including blood clotting factors) and destruction of erythrocytes.

5. **Storage functions :** Glycogen, vitamins A, D and B_{12} and trace element, iron are stored in liver.

Causes of liver damage

Hepatocellular damage may occur due to ***viruses*** (hepatitis A virus, hepatitis B virus), ***toxins*** (carbon tetrachloride, aflatoxin), ***alcohol***, hepatocellular ***carcinoma***, ***autoimmune hepatitis*** etc.

Tests to assess liver function

The liver function tests (***LFT***) are the biochemical investigations to assess the capacity of the liver to carry out any of the functions it performs. LFT will ***help to detect the abnormalities and the extent of liver damage***.

Two important facts should be borne in mind while carrying out LFT.

1. Liver is a large-size factory of safety. Therefore, it ***can perform many of its functions almost normally, despite the damage***.

2. Selection of the ***right test*** is important in LFT. This is due to the fact that since liver participates in several functions, the function that is measured in LFT may not be the one that is adversely affected.

The major liver function tests may be ***classified*** as follows

1. Tests based on ***excretory function***—Measurement of ***bile pigments***, bile salts, bromosulphthalein.

2. Tests based on ***serum enzymes*** derived from liver—Determination of ***transaminases***, alkaline phosphatase, 5′-nucleotidase, γ-glutamyl-transpeptidase.

3. Tests based on ***metabolic capacity***—Galactose tolerance, antipyrine clearance.

4. Tests based on ***synthetic functions***—Prothrombin time, serum albumin.

5. Tests based on ***detoxification***—Hippuric acid synthesis.

This above list, contains the most important biochemical investigations to assess LFT. Among these, the commonly used tests are described in the following pages.

Markers of liver function

The important liver functions and the common plasma/serum markers for the impaired functions are listed in ***Table 20.1***. The most important markers namely, bilirubin, enzymes, albumin, prothrombin time and drug metabolism with special reference to jaundice and other liver diseases are described.

BILIRUBIN

Bilirubin is ***a bile pigment***, and is the excretory ***end product of heme degradation***. It is conjugated in the liver to form bilirubin diglucuronide, and excreted in bile. The details of bilirubin metabolism are discussed elsewhere (***Chapter 10***).

Serum bilirubin

The normal concentration of serum bilirubin is in the range of ***0.2-1.0 mg/dl***. Of this, the conjugated bilirubin (diglucuronide 75%; mono-glucuronide 25%) is 0.2-0.4 mg/dl, while the unconjugated bilirubin is 0.2-0.6 mg/dl.

Icterus index

This is a simple test to ***measure*** the ***yellow colour of serum due to bilirubin***. It is rather crude and almost outdated. However, it is often useful for a rapid assessment of neonatal jaundice.

Table 20.1 A list of liver (hepatic) functions and the common markers in plasma for the impaired function

Hepatic function	Common plasma/serum marker(s) for impaired function
Heme catabolism	↑Bilirubin
Enzymes	↑Alanine transaminase ↑Aspartate transaminase ↑γ-Glutamyltranspeptidase
Protein synthesis	↓Albumin ↑Prothrombin time
Protein catabolism	↑Urea ↑Ammonia
Lipid metabolism	↑Cholesterol ↑Triglycerides
Drug metabolism	↑Half-lives of drugs
Bile acid metabolism	↑Bile acids

van den Bergh reaction

This is a specific reaction ***to identify*** the increase in ***serum bilirubin*** (above the reference level). ***Normal serum gives a negative van den Bergh reaction.***

Mechanism of the reaction : van den Bergh reagent is a mixture of equal volumes of sulfanilic acid (in dilute HCl) and sodium nitrite. The principle of the reaction is that diazotised sulfanilic acid (in the above mixture) reacts with bilirubin to form a purple coloured azobilirubin.

Direct and indirect reactions : Bilirubin as such is insoluble in water while the conjugated bilirubin is soluble. van den Bergh reagent reacts with ***conjugated bilirubin*** and gives a purple colour immediately (normally within 30 seconds). This is referred to as a ***direct positive*** van den Bergh reaction. ***Addition of methanol*** (or alcohol) ***dissolves*** the ***unconjugated bilirubin*** which then gives the van den Bergh reaction (normally within 30 minutes) positive and this is referred to as ***indirect positive***. If the serum contains both unconjugated and conjugated bilirubin in high concentration, the purple colour is produced immediately (direct positive) which is further intensified by the addition of alcohol (indirect positive). This type of reaction is known as ***biphasic***.

van den Bergh reaction and jaundice : This reaction is highly useful in understanding the nature of jaundice. This is due to the fact that the type of jaundice is characterized by increased serum concentration of unconjugated bilirubin (hemolytic), conjugated bilirubin (obstructive) or both of them (hepatic). Therefore, the response of van den Bergh reaction can differentiate the jaundice as follows

Indirect positive — ***Hemolytic jaundice***
Direct positive — ***Obstructive jaundice***
Biphasic — ***Hepatic jaundice***.

Bilirubin in urine

The ***conjugated bilirubin***, being water soluble, is ***excreted in urine***. This is in contrast to unconjugated bilirubin which is not excreted. Bilirubin in urine can be detected by ***Fouchet's test*** or ***Gmelin's test***. Urobilinogen can be detected by ***Ehrlich test***.

Bromosulphthalein (BSP) test

Bromosulphthalein is a ***dye*** used ***to assess*** the ***excretory function of liver***. It is a non-toxic compound and almost exclusively excreted by the liver (through bile). BSP is administered intravenously (5 mg/kg body weight) and its serum concentration is measured at 45 min and at 2 hrs. In normal individuals, less than 5% of the dye is retained at the end of 45 min. Any impairment in liver function causes an ***increased retention of the dye***. This test is quite sensitive to assess liver abnormality with particular reference to excretory function.

SERUM ENZYMES DERIVED FROM LIVER

Liver cells contain several enzymes which may be released into the circulation in liver damage. Measurement of ***selected enzymes in serum*** is often ***used to assess the liver function***. It must, however, be noted that there is no single enzyme that is absolutely specific to liver alone. Despite this fact, serum enzymes provide valuable information for LFT. Some of these enzymes are discussed hereunder.

Transaminases or aminotransferases

The activities of two enzymes—namely serum glutamate pyruvate transaminase (SGPT; recently called as ***alanine transaminase***—ALT) and serum glutamate oxaloacetate transaminase (SGOT; recently known as ***aspartate transaminase***—AST) —are widely used to assess the liver function. ALT is a cytoplasmic enzyme while AST is found in both cytoplasm and mitochondria. The activity of these enzymes is low in normal serum (***ALT 10-45 U/l; AST 10-50 U/l***). Serum ***ALT*** and AST are increased in liver damage. However, alanine transaminase is ***more sensitive*** and reliable for the assessment of LFT.

The normal ***AST/ALT ratio*** is around ***0.8***. This ratio is ***increased*** (>2) in ***myocardial infarction***, alcoholic hepatitis, and cirrhosis. AST/ALT ratio is decreased (i.e. ALT higher) ***in acute hepatocellular damage and cholestasis***.

Alkaline phosphatase

Alkaline phosphatase (ALP) is mainly derived from bone and liver (the cells lining the bile canaliculi). A rise in serum ALP (normal 3-13 KA units/dl), usually associated with elevated serum bilirubin is an indicator of biliary obstruction (***obstructive/posthepatic jaundice***). ALP is also elevated in cirrhosis of liver and hepatic tumors.

Liver is not the sole source of alkaline phosphatase. Therefore, its measurement has to be carefully viewed (along with others) before arriving at any conclusion. The liver and bone isoenzymes of ALP can be separated by electrophoresis.

γ-Glutamyl transpeptidase

This is a microsomal enzyme widely distributed in body tissues, including liver. Measurement of γ-glutamyl transpeptidase (***GGT***) activity provides a ***sensitive index to assess liver abnormality***. The activity of this enzyme almost parallels that of transaminases in hepatic damage. Serum GGT is highly elevated (normal ***5-50 U/l***) in ***biliary obstruction*** and ***alcoholism***. Further, several drugs (e.g. phenytoin) induce (liver synthesis) and increase this enzyme in circulation.

5′-Nucleotidase

The serum activity of 5′-nucleotidase (normal ***2-10 U/l***) is ***elevated in hepatobiliary disease*** and this parallels ALP. The advantage with 5′-nucleotidase is that it is not altered in bone disease (as is the case with ALP).

Other enzymes

Serum isocitrate dehydrogenase and isoenzymes of lactate dehydrogenase (LDH_4 and LDH_5) are also useful in LFT.

Enzyme combinations

Very often, a ***panel of liver enzymes*** involving a combination of serum enzyme estimations (instead of a single one) is ***used for a better understanding of liver functions***. For instance, a large increase in transaminases (particularly ALT) relative to a small increase in alkaline phosphatase indicates hepatocellular damage. On the other hand, a small increase in transaminases and a large increase of alkaline phosphatase show biliary obstruction.

JAUNDICE

Jaundice (*French* : jaune—yellow) is characterized by ***yellow coloration of sclera (of eyes)*** and skin. This is due to the elevated ***serum bilirubin*** level, usually ***beyond 2 mg/dl*** (normal < 1 mg/dl).

The metabolism of heme to produce bilirubin and its conjugated derivatives and the types of jaundice have already been described. The reader must refer this (***Chapter 10***) now. The biochemical changes and the related parameters for the differential diagnosis of the three types of jaundice (hemolytic, obstructive and hepatic) are given in ***Table 20.2***.

In the ***Fig.20.1***, the normal and abnormal bilirubin metabolism (along with the associated

Table 20.2 Biochemical changes for the differential diagnosis of three types of jaundice

Parameter	Hemolytic jaundice (preheptic jaundice)	Obstructive jaundice (posthepatic jaundice)	Hepatic jaundice (Intrahepatic jaundice)
Serum bilirubin	Unconjugated bilirubin ↑	Conjugated bilirubin ↑	Both ↑
van den Bergh reaction	Indirect positive	Direct positive	Biphasic
Serum enzymes	ALT, AST and ALP →	ALP ↑↑, ALT and AST marginal ↑	ALT and AST ↑↑, ALP marginal ↑
Bilirubin in urine	Not excreted	Excreted	Excreted
Urobilinogen in urine	Excretion ↑	→ or ↓	→ or ↓

ALT : Alanine transaminase; AST : Aspartate transaminase; ALP : Alkaline phosphatase; ↑ : Increase; ↓ : Decrease; → : Normal.

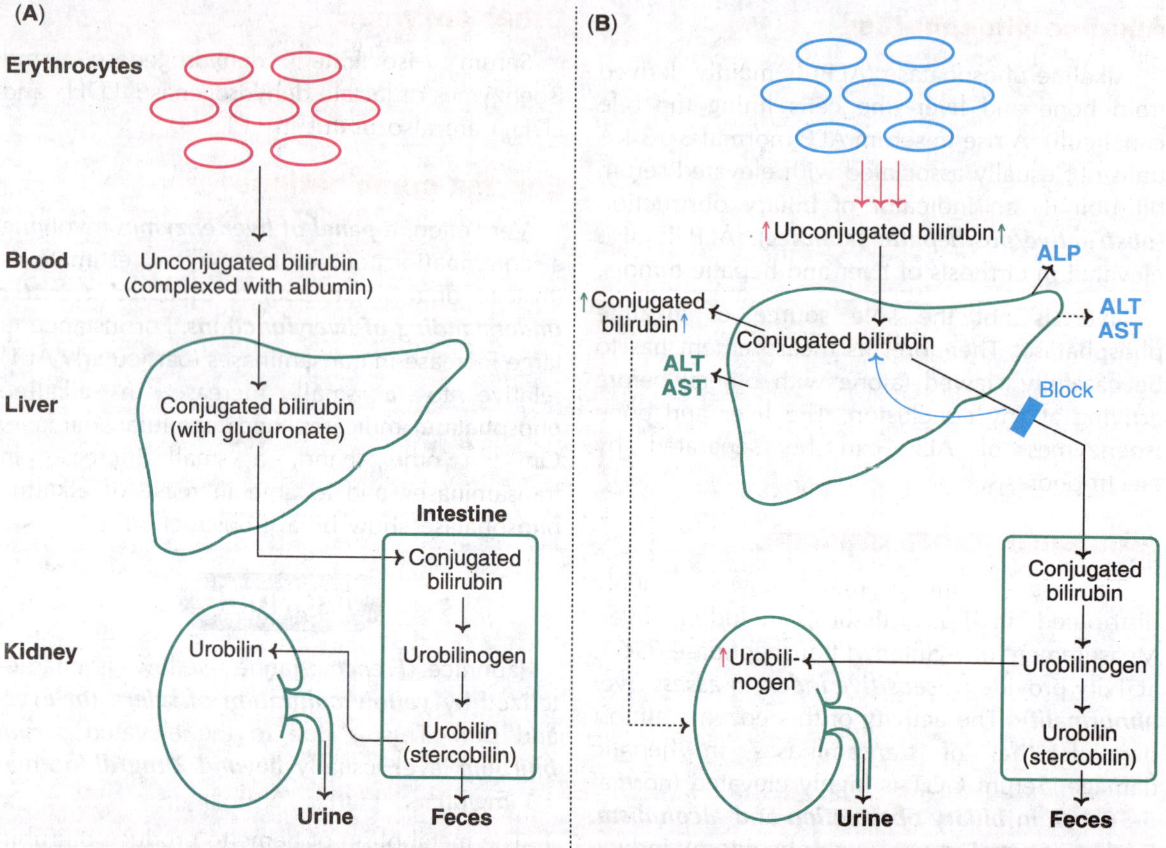

Fig. 20.1 : *Normal and abnormal bilirubin metabolism* **(A)** *Normal bilirubin metabolism* **(B)** *Alterations in bilirubin metabolism along with enzymes in three types of jaundice* (**Note :** *Colours indicate major changes; Red–changes in hemolytic jaundice; Green–changes in hepatic jaundice; Blue–changes in obstructive jaundice; Dotted lines indicate minor pathways; ALT–Alanine transaminase; AST–Aspartate transaminase; ALP–Alkaline phosphatase).*

enzyme changes) are depicted. The major changes in the 3 types of jaundice are listed below

Hemolytic jaundice : Elevated serum ***unconjugated bilirubin***, and increased urinary excretion of urobilinogen.

Obstructive jaundice : Elevated serum ***conjugated bilirubin*** and increased activities of ***alkaline phosphatase*** (ALP), alanine transaminase (ALT) and aspartate transaminase (AST).

Hepatic jaundice : Elevated serum ***unconjugated*** and ***conjugated bilirubin***, and increased activities of ***ALT*** and AST.

The pattern of rise in the serum alanine transaminase, aspartate transaminase and bilirubin in acute viral hepatitis is depicted in ***Fig.20.2***. It may be noted that the transaminase activities (more predominantly ALT) are elevated much before the bilirubin starts increasing.

OTHER LIVER FUNCTION TESTS

The most commonly used laboratory LFT are described above. In addition, some other tests to assess liver function are discussed hereunder.

Galactose tolerance

Galactose is a monosaccharide, almost exclusively metabolized by the liver. The liver function can be ***assessed by measuring the utilization of galactose***. This is referred to galactose tolerance test. The subject is given intravenous administration of galactose (about 300 mg/kg body weight). Blood is drawn at 10 minute intervals for the next 2 hours and galactose estimated. In the normal individuals,

Medical Concepts/ Clinical Correlates

- The impairment in the functions of any organ in the body will adversely influence the health of the organism. Organ function tests are the laboratory tools to biochemically evaluate the working of a given organ.
- Acute viral hepatitis is associated with elevated alanine transaminase (predominantly), aspartate transaminase and bilirubin.
- Increase in serum γ-glutamyl transpeptidase is observed in biliary obstruction and alcoholism.

the ***half-life of galactose*** is about 10-15 minutes. This is markedly ***elevated*** in ***hepatocellular damage*** (***infective hepatitis, cirrhosis***).

Serum albumin

Albumin is solely synthesized by the liver. It has a half-life of about 20-25 days, therefore, it is a good marker to assess chronic (and not acute) liver damage. ***Low serum albumin*** is commonly observed in ***patients with severe liver damage***. It must, however, be noted that the serum albumin concentration is also ***decreased*** due to other factors, such as ***malnutrition***.

Functional impairment of liver is frequently associated with increased synthesis of globulins. Cirrhosis of the liver causes a ***reversal*** of albumin/globulin ratio (***A/G ratio***). ***Serum electrophoresis*** of proteins reveals increased albumin and decreased γ-globulin concentration. This, however, may not have much diagnostic importance since several diseases are associated with altered electrophoretic pattern of serum proteins.

α-Fetoprotein

Alpha-fetoprotein (***α-AFP***) is an important protein produced by yolk sac and liver during fetal development. Circulatory AFP levels decrease after birth, and stabilize to normal range (< 5 ng/ml serum) within one year. Elevated serum AFP is found in hepatocellular carcinoma, and metastatic cancers of liver. AFP is used as a ***biomarker to detect tumors of liver, germ cells etc.***

Prothrombin time

The liver synthesizes all the factors concerned with blood clotting. A decrease in the concentration of plasma clotting factors is found in the impairment of liver function. This can be assessed in the laboratory by measuring prothrombin time which is ***prolonged in*** patients with ***liver damage***, compared to normal. The half-lives of clotting factors are relatively short (5-72 hrs.), therefore, changes in prothrombin time occur quickly. Hence, this test is useful to assess acute as well as chronic liver damages; besides its help in the prognosis.

Vitamin K is required for the synthesis of blood clotting factors II, VII, IX and X. Therefore, vitamin K deficiency can also cause prolonged prothrombin time which must be ruled out, before drawing conclusions on the liver functions. This is done by measuring prothrombin time before and after administration of vitamin K.

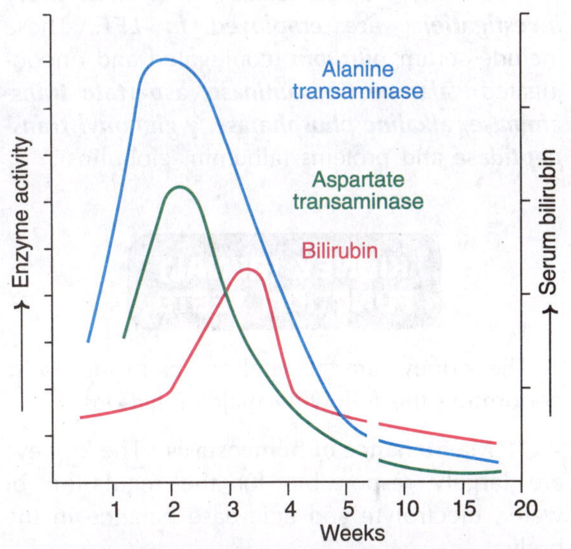

Fig. 20.2 : *Pattern of rise in serum enzymes and bilirubin in viral hepatitis.*

Hippuric acid synthesis

The liver is the major site for the metabolism of xenobiotics (detoxification). Measurement of hippuric acid synthesis is an ideal ***test for*** assessing the ***detoxification function of liver***. Hippuric acid is produced in the liver when benzoic acid combines with glycine.

About 6 g of sodium benzoate dissolved in (about 250 ml) water, is orally given to the subject, after a light breakfast (usually 2 hrs later) and after emptying the bladder. Urine collections are made for the next 4 hours and the amount of hippuric acid excreted is estimated. Theoretically, 6 g of sodium benzoate should yield 7.5 g of hippuric acid. In the healthy persons, about 60% of sodium benzoate (equivalent to 4.5 g hippuric acid) is excreted in urine. A ***reduction*** in hippuric acid excretion (particularly < 3 g) ***indicates hepatic damage***.

Choice of liver function tests

The choice of biochemical tests to measure liver functions, mostly depends on the purpose of the investigation. The clinical history of the subject is often a guiding factor in this regard. A single test in isolation may have a little diagnostic value.

Frequently, a ***combination of laboratory investigations are employed in LFT***. These include serum ***bilirubin*** (conjugated and unconjugated), ***alanine transaminase***, ***aspartate transaminase***, ***alkaline phosphatase***, ***γ-glutamyl transpeptidase*** and proteins (albumin, globulins).

KIDNEY (RENAL) FUNCTION TESTS

The kidneys are the vital organs of the body, performing the following major functions.

1. **Maintenance of homeostasis :** The kidneys are largely responsible for the regulation of water, electrolyte and acid-base balance in the body.

2. **Excretion of metabolic waste products :** The end products of protein and nucleic acid metabolism are eliminated from the body. These include urea, creatinine, creatine, uric acid, sulfate and phosphate.

3. **Retention of substances vital to body :** The kidneys reabsorb and retain several substances of biochemical importance in the body e.g. glucose, amino acids etc.

4. **Hormonal functions :** The kidneys also function as endocrine organs by producing hormones.

- **Erythropoietin,** a peptide hormone, stimulates hemoglobin synthesis and formation of erythrocytes.
- 1,25-Dihydroxycholecalciferol ***(calcitriol)*** – the biochemically active form of vitamin D – is finally produced in the kidney. It regulates calcium absorption from the gut.
- **Renin,** a proteolytic enzyme liberated by kidney, stimulates the formation of angiotensin II which, in turn, leads to aldosterone production. Angiotensin II and aldosterone are the hormones involved in the regulation of electrolyte balance.

The formation of urine

Nephron is the functional unit of kidney. Each kidney is composed of approximately one million nephrons. The structure of a nephron, as depicted in ***Fig.20.3***, consists of a ***Bowman's capsule*** (with blood capillaries), proximal convoluted tubule (PCT), loop of Henle, distal convoluted tubule (DCT) and collecting tubule.

The blood supply to kidneys is relatively large. About 1200 ml of blood (650 ml plasma) passes through the kidneys, every minute. From this, about ***120-125 ml is filtered per minute by the kidneys*** and this is referred to as ***glomerular filtration rate*** (GFR). With a normal GFR (120-125 ml/min), the glomerular filtrate formed in an adult is about 175-180 litres per day, out of which only 1.5 litres is excreted as urine. Thus, more than 99% of the glomerular filtrate is reabsorbed by the kidneys.

The process of urine formation basically involves two steps—glomerular filtration and tubular reabsorption.

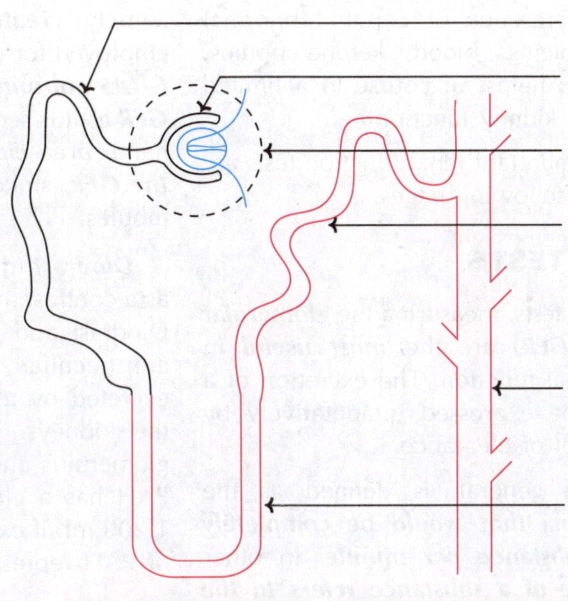

Fig. 20.3 : Diagrammatic representation of a nephron.

1. **Glomerular filtration :** This is a passive process that results in the formation of ultrafiltrate of blood. All the (unbound) constituents of plasma, with a molecular weight less than about 70,000, are passed into the filtrate. Therefore, the glomerular filtrate is almost similar in composition to plasma.

2. **Tubular reabsorption :** The renal tubules (PCT, DCT and collecting tubules) retain water and most of the soluble constituents of the glomerular filtrate by reabsorption. This may occur either by passive or active process. The excreted urine has an entirely different composition compared to glomerular filtrate from which it is derived. The normal composition of urine is given elsewhere (***Refer inside backcover***).

Renal threshold substances

There are certain ***substances in the blood whose excretion in urine is dependent on their concentration***. Such substances are referred to as renal threshold substances. At the normal concentration in the blood, they are completely reabsorbed by the kidneys, with a result that their excretion in urine is almost negligible.

The ***renal threshold*** of ***a substance*** is defined as its ***concentration in blood (or plasma) beyond which*** it is ***excreted into urine***. The renal threshold ***for glucose is 180 mg/dl; for ketone bodies 3 mg/dl***; for calcium 10 mg/dl and for bicarbonate 30 mEq/l. While calculating the renal threshold of a particular compound, it is assumed that both the kidneys are optimally functioning, without any abnormality. But this is not always true—in which case the renal threshold is altered. For instance, ***renal glycosuria*** is associated with reduced threshold for glucose due to its diminished tubular reabsorption.

The term ***tubular maximum*** (***Tm***) is used to indicate the maximum capacity of the kidneys to absorb a particular substance. For instance, tubular maximum for ***glucose (TmG) is 350 mg/min.***

Tests to assess renal function

In view of the important and sensitive functions the kidney performs (described already), it is essential that the abnormalities (renal damages), if any, must be detected at the earliest. Several tests are employed in the laboratory to assess kidney (renal) function. It must, however, be remembered that ***about two-thirds of the renal tissue must be functionally damaged to show any abnormality by these tests***. The kidney function tests may be divided into four groups.

1. **Glomerular function tests :** All the ***clearance tests*** (inulin, creatinine, urea) are included in this group.

2. **Tubular function tests :** ***Urine concentration or dilution test***, urine acidification test.

3. **Analysis of blood/serum :** Estimation of ***blood urea, serum creatinine, protein and electrolyte*** are often useful to assess renal function.

4. **Urine examination :** Simple routine examination of urine for volume, pH, specific gravity,

osmolality and presence of certain abnormal constituents (proteins, blood, ketone bodies, glucose etc.) also helps, of course to a limited degree, to assess kidney functioning.

Some of the important renal function tests are discussed in the following pages.

CLEARANCE TESTS

The clearance tests, measuring the ***glomerular filtration rate (GFR)*** are the ***most useful*** in assessing the renal function. The excretion of a substance can be expressed quantitatively by using the concept of clearance.

Clearance, in general, is defined as the ***volume of plasma that would be completely cleared of a substance per minute***. In other words, ***clearance of a substance refers to the milliliters of plasma which contains the amount of that substance excreted by kidney per minute.*** Clearance (C), expressed as ml/minute, can be calculated by using the ***formula***

$$C = \frac{U \times V}{P}$$

where U = Concentration of the substance in urine.

V = Volume of urine in ml excreted per minute.

P = Concentration of the substance in plasma.

Care should be taken to express the concentrations of plasma and urine in the same units (mmol/l or mg/dl).

The clearance of a given substance is determined by its mode of excretion. The maximum rate at which the plasma can be cleared of any substance is equal to the GFR. This can be easily calculated by measuring the clearance of a plasma compound which is freely filtered by the glomerulus and is neither absorbed nor secreted in the tubule. ***Inulin*** (a plant carbohydrate, composed of fructose units) and ^{51}Cr-EDTA satisfy this criteria. Inulin is intravenously administered ***to measure GFR.***

In practice, however, measurement of clearance for the substances already present in the blood is preferred. The two compounds, namely ***creatinine*** and ***urea***, are commonly employed for this purpose. ***Creatinine clearance (~145 ml/min) is marginally higher than the GFR*** as it is secreted by the tubules. On the other hand, ***urea clearance (~75 ml/min) is less than the GFR***, since it is partially reabsorbed by the tubules.

Diodrast (diiodopyridone acetic acid) is used as a contrast medium to take urinary tract X-rays. Diodrast and ***para amino hippuric acid (PAH)*** are peculiar substances as they are entirely excreted by a single passage of blood through the kidneys. They are partly filtered by the glomerulus and mostly excreted by the tubules. PAH has a clearance of about 700 ml/min (or 1,200 ml, if expressed as blood). Thus clearance of PAH represents the ***renal plasma flow.***

Creatinine clearance test

Creatinine is an excretory product derived from creatine phosphate (largely present in muscle). The ***excretion of creatinine*** is ***rather constant*** and is not influenced by body metabolism or dietary factors. As already stated, creatinine is filtered by the glomeruli and only marginally secreted by the tubules. The value of creatinine clearance is close to GFR, hence its measurement is a sensitive and good approach to assess the renal glomerular function. Creatinine clearance may be defined as ***the volume (ml) of plasma that would be completely cleared of creatinine per minute.***

Procedure : In the traditional method, creatinine content of a 24 hr urine collection and the plasma concentration in this period are estimated. The creatinine clearance (C) can be calculated as follows :

$$C = \frac{U \times V}{P}$$

where U = Urine concentration of creatinine

V = Urine output in ml/min (24 hr urine volume divided by 24 × 60)

P = Plasma concentration of creatinine.

As already stated, creatinine concentration in urine and plasma should be expressed in the same units (mg/dl or mmol/l).

Modified procedure : *Instead of a 24 hr urine* collection, the procedure is modified to collect ***urine for 1 hr***, after giving water. The volume of urine is recorded. Creatinine contents in plasma and urine are estimated. The creatinine clearance can be calculated by using the formula referred above.

Reference values : The ***normal*** range of ***creatinine clearance is around 120-145 ml/min.*** These values are slightly lower in women. In recent years, creatinine clearance is expressed in terms of body surface area.

Diagnostic importance : A ***decrease in creatinine clearance*** value (< 75% normal) serves as ***sensitive indicator of a decreased GFR***, due to renal damage. This test is useful for an early detection of impairment in kidney function, often before the clinical manifestations are seen.

Urea clearance test

Urea is the end product of protein metabolism. After being filtered by the glomeruli, it is partially reabsorbed by the renal tubules. Hence, urea clearance is less than the GFR and, further, it is influenced by the protein content of the diet. For these reasons, ***urea clearance is not as sensitive as creatinine clearance*** for assessing renal function. Despite this fact, several laboratories traditionally use this test.

Urea clearance is defined as the volume (ml) of plasma that would be completely cleared of urea per minute. It is calculated by the formula

$$C_m = \frac{U \times V}{P}$$

where C_m = Maximum urea clearance
U = Urea concentration in urine (mg/ml)
V = Urine excreted per minute in ml
P = Urea concentration in plasma (mg/ml).

The above calculation is applicable if the ***output of urine is more than 2 ml per minute.*** This is referred to as ***maximum urea clearance*** and the normal value is around 75 ml/min.

Standard urea clearance : It is observed that the urea clearance drastically changes when the volume of urine is less than 2 ml/min. This is known as ***standard urea clearance (C_s)*** and the normal value is around 54 ml/min. It is calculated by a modified formula

$$C_s = \frac{U \times \sqrt{V}}{P}$$

Diagnostic importance : A urea clearance value below 75% of the normal is viewed seriously, since it is an indicator of renal damage. Blood urea level as such is found to increase only when the clearance falls below 50% normal. As already stated, creatinine clearance is a better indicator of renal function.

Urine concentration test

This is a test to assess the renal tubular function. It is a simple test and involves the accurate measurement of specific gravity which depends on the concentration of solutes in urine. ***A specific gravity of 1.020 in the early morning urine sample is considered to be normal.***

Several measures are employed to concentrate urine and measure the specific gravity. These include overnight water deprivation and administration of antidiuretic hormone. If the specific gravity of urine is above 1.020 for at least one of the samples collected, the tubular function is considered to be normal.

Osmolality and specific gravity : The osmolality of urine is variable. In normal individuals, it may range from 500-1,200 milliosmoles/kg. The plasma osmolality is around 300 milliosmoles/kg. The normal ratio of the osmolality between urine and plasma is around 2-4. It is found that the ***urine*** (without any protein or high molecular weight substance) with an ***osmolality of 800 mosm/kg has a specific gravity of 1.020***. Therefore, measurement of urine osmolality will also help to assess tubular function.

Analysis of blood (or serum)

Estimation of serum ***creatinine*** and blood ***urea*** are often used to ***assess the overall kidney function***, although these tests are less sensitive than the clearance tests. ***Serum creatinine*** is a ***better*** indicator than urea in this regard. The diagnostic importance of urea and creatinine estimations are discussed elsewhere (***Refer Chapter 15***).

The relationship between GFR and serum creatinine levels is depicted in ***Fig.20.4***. It is observed that the GFR must fall to about 50% of its normal value before a significant increase in serum creatinine occurs. Therefore, a normal serum creatinine level does not necessarily mean that all is well with the kidney. It is estimated that a loss of 50% of the functions of nephrons leads to (approximate) doubling of serum creatinine concentration.

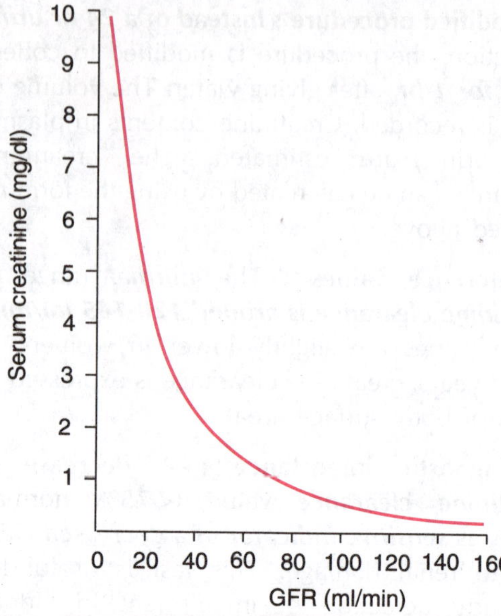

Fig. 20.4 : *The relationship between glomerular filtration rate (GFR) and serum creatinine concentration.*

Cystatin C

Cystatin C is mainly used as a ***biomarker of kidney function***. Normally, serum concentration of cystatin C is in the range of ***0.8-1.2 mg/dl***.

Because of its low molecular weight (13,300), cystatin C is easily filtered by glomerular filtration of the kidneys. If kidney function declines, cystatin C levels rise in serum. Even minor changes in GFR in the early stages of kidney disease are associated with increase in cystatin C. Thus, serum ***cystatin C*** estimation is a more precise test, as it is a ***better indicator of a kidney function than creatinine***.

In recent years, cystatin C has been found to be associated with the risk of death, and linked with cardiovascular diseases, cancer and AIDS. Therefore, cystatin C levels should be carefully interpreted with clinical findings.

Urine examination

The routine urine examination is undoubtedly a guiding factor for renal function. The volume of urine excreted, its pH, specific gravity, osmolality, the concentration of abnormal constituents (such as proteins, ketone bodies, glucose and blood) may help to have some preliminary knowledge of kidney function. More information on ***urine laboratory tests is given in the appendix***.

Choice of renal function tests

In general, the assessment of kidney function starts with the routine urine examination, followed by serum creatinine and/or blood urea estimations and, finally, the specific tests to measure the tubular and glomerular functions (clearance tests).

GASTRIC FUNCTION TESTS

The stomach is a major organ of digestion and performs the following functions

1. Stomach is a reservoir of ingested foodstuffs.

2. It has a great churning ability which promotes digestion.

3. Stomach elaborates HCl and proteases (pepsin) which are responsible for the initiation of digestive process.

4. The products obtained in the stomach (peptides, amino acids) stimulate the release of pancreatic juice and bile.

Secretion of gastric HCl

The parietal (oxyntic) cells of gastric glands produce HCl. The pH in the gastric lumen is as low as 0.8 (against the blood pH 7.4). Therefore, the protons are transported against the concentration gradient by an active process.

Medical Concepts/ Clinical Correlates

- A combination of laboratory investigations—instead of a single one—are commonly employed in assessing organ function. Kidney function can be accurately assessed by clearance tests, measuring glomerular filtration rate. A reduction in clearance reflects renal damage.
- Cystatin C is a more reliable and better biomarker of kidney function than creatinine.
- Zollinger-Ellison syndrome, a tumor of gastrin secreting cells of the pancreas, is associated with increased gastric HCl production.

A unique enzyme—namely K^+ activated ATPase—present in the parietal cells is connected with the mechanism of HCl secretion (***Fig.20.5***). The process involves an exchange of H^+ ions (of the parietal cells) for K^+ ions (of the lumen). This is coupled with the consumption of energy, supplied by ATP. The H^+ are continuously generated in the parietal cells by the dissociation of carbonic acid which, in turn, is produced from CO_2. The bicarbonate ions (HCO_3^-), liberated from the carbonic acid (H_2CO_3) dissociation, enter the blood in exchange for Cl^- ions. The latter diffuse into the gastric lumen to form HCl. ***Gastrin***—a peptide hormone of gastrointestinal tract—stimulates HCl secretion.

Following a meal, there is a slight ***elevation in the plasma bicarbonate concentration*** which is linked to the gastric HCl secretion. This is referred to as ***alkaline tide***.

TESTS TO ASSESS GASTRIC FUNCTION

There are several tests for gastric function evaluation, some of the important ones are briefly discussed.

Fractional test meal (FTM)

This is rather old and not used these days. Fractional test meal involves the collection of stomach contents by ***Ryle's tube*** in fasting. This is followed by a gastric stimulation, giving a test meal (rice gruel, black coffee etc.) The stomach contents are aspirated by Ryle's tube at different time periods (usually every 15 min for 2 hrs.) The samples are analysed for free and total acidity in the laboratory. The results are normally represented by a graph.

Alcohol test meal

In this case, the test meal in the form of 100 ml of 7% alcohol is administered. The response to alcohol test meal is more rapid, and the test time can be reduced to $1^1/_2$ hour. Clear specimens can be collected by this test, and the free acidity levels are relatively higher compared to FTM.

Pentagastrin stimulation test

Pentagastrin is a synthetic peptide which stimulates the gastric secretion in a manner similar to the natural gastrin. The test procedure adapted is as follows

The stomach contents are aspirated by Ryle's tube in a fasting condition. This is referred to as residual juice. The gastric juice elaborated for the next one hour is collected and pooled which represents the basal secretion. Pentagastrin (5 mg/kg body weight) is now given to stimulate gastric secretion. The gastric juice is collected at

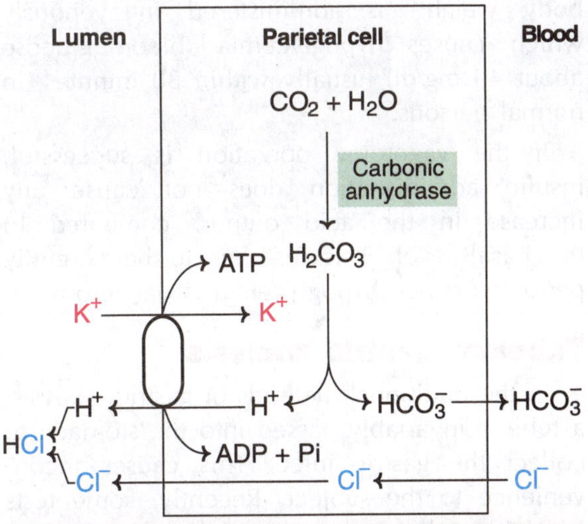

***Fig. 20.5 :** Mechanism of HCl secretion (0—represents K^+ activated ATPase).*

15 minute intervals for one hour. This represents the maximum secretion.

Each sample of the gastric secretion collected is measured for acidity by titrating the samples with N/10 NaOH to pH 7.4. The end point may be detected by an indicator (phenol red) or a pH meter.

Basal acid output (BAO) refers to the acid output (millimol per hour) under the basal conditions i.e. basal secretion.

Maximal acid output (MAO) represents the acid output (millimol per hour) after the gastric stimulation by pentagastrin i.e. maximum secretion.

In normal individuals, the BAO is 4-10 mmol/hr while the MAO is 20-50 mmol/hr.

Augmented histamine test meal

Histamine is a powerful stimulant of gastric secretion. The basal gastric secretion is collected for one hour. Histamine (0.04 mg/kg body weight) is administered subcutaneously and the gastric contents are aspirated for the next one hour (at 15 minute intervals). The acid content is measured in all these samples.

Insulin test meal

This is also known as ***Hollander's test.*** It is mainly done to assess the completeness of vagotomy (vagal resection). Insulin (0.1 unit/kg body weight) is administered intravenously which causes hypoglycemia (blood glucose about 40 mg/dl), usually within 30 minutes, in normal persons.

If the vagotomy operation is successful, insulin administration does not cause any increase in the acid output, compared to the basal level. This test has to be carefully perfomed, since hypoglycemia is dangerous.

Tubeless gastric analysis

In the traditional methods of gastric analysis, a tube is invariably passed into the stomach to collect the gastric juice. This causes inconvenience to the subject. Recently, some tests involving tubeless gastric analysis have been developed. Such tests, however, are mostly useful for preliminary screening.

The principle of tubeless gastric analysis involves administration of a cation exchange resin that gets quantitatively exchanged with the H^+ ions of the gastric juice. The resin is then excreted into urine which can be estimated for an indirect measure of gastric acidity (concentration of H^+ ions).

Diagnex blue containing ***azure-A-resin*** is employed in the tubeless gastric analysis.

Abnormalities of gastric function

Increased gastric HCl secretion is found in Zollinger-Ellison syndrome (a tumor of gastrin secreting cells of the pancreas), chronic duodenal ulcer, gastric cell hyperplasia, excessive histamine production etc.

A ***decrease in gastric HCl*** is observed in ***gastritis, gastric carcinoma, pernicious anemia*** etc.

OTHER ORGAN FUNCTION TESTS

PANCREATIC FUNCTION TESTS

The pancreas is a specialized organ with exocrine and endocrine functions. The endocrine functions are discussed under the topic diabetes mellitus (***Chapter 36***).

The exocrine functions involve the synthesis of pancreatic juice containing several enzymes (for the digestion of foodstuffs) and bicarbonate. The major enzymes of pancreatic juice are trypsin, chymotrypsin, elastase, carboxypeptidase, amylase and lipase.

Pancreatic enzymes in serum : ***Serum amylase*** and ***lipase measurements*** are commonly employed to assess the pancreatic function. Both these enzyme activities are ***elevated in acute pancreatitis, obstruction in the intestine*** and/or pancreatic duct.

THYROID FUNCTION TESTS

Thyroid gland produces two principal hormones—thyroxine (T_4) and triiodothyronine (T_3) which regulate the metabolic rate of the body. The laboratory tests employed for the diagnosis of thyroid function are described in the ***Chapter 19*** on ***hormones***.

SUMMARY

- *Specific laboratory biochemical investigations are employed to assess the functioning of the organs such as liver, kidney, stomach and pancreas.*
- *The liver function can be evaluated by the tests based on its excretory function (serum bilirubin), serum enzymes (transaminases), metabolic capability (galactose tolerance test) and synthetic functions (prothrombin time).*
- *Serum bilirubin (normal < 1mg/dl) is derived from heme degradation. It is mostly (75%) found in the conjugated form. van den Bergh reaction is a specific test to identify the increased serum bilirubin. Conjugated bilirubin gives a direct positive test while the unconjugated bilirubin gives an indirect positive test.*
- *The serum enzymes—namely ALT, AST, ALP and GGT—are frequently used for LFT. Increase in the activities of these enzymes indicates an impairment in liver function.*
- *Jaundice is due to elevated serum bilirubin level (>2 mg/dl). The three types of jaundice can be differentially diagnosed by biochemical tests. Thus, unconjugated bilirubin (indirect positive) is increased in hemolytic jaundice, conjugated bilirubin (direct positive) in obstructive jaundice and both of them (biphasic) are increased in hepatic jaundice.*
- *Impaired galactose tolerance test, diminished serum albumin concentration and prolonged prothrombin time are also associated with liver malfunction.*
- *The renal (kidney) function is usually assessed by evaluating either the glomerular (clearance tests) or tubular function (urine concentration test). This is often guided by blood analysis (for urea, creatinine) and/or urine examination.*
- *The clearance is defined as the volume of the plasma that would be completely cleared of a substance per minute. Inulin clearance represents glomerular filtration rate (GFR). Creatinine clearance and urea clearance tests are often used to assess renal function. A decrease in their clearance is an indication of renal damage.*
- *Impairment in renal function is often associated with elevated concentration of blood urea, serum creatinine, decrease in osmolality and specific gravity of urine.*
- *The tests to evaluate gastric function include fractional test meal, pentagastrin stimulation test, augmented histamine test and tubeless gastric analysis. Gastric HCI secretion is elevated in chronic duodenal ulcer and gastric hyperplasia. Gastritis and pernicious anemia are associated with decreased gastric HCI. Pancreatic function is assessed by serum amylase and lipase. Both of them are elevated in acute pancreatitis.*

THE READER MUST REFER CHAPTER 19 FOR THYROID FUNCTION TESTS AND ADRENOCORTICAL FUNCTION TESTS.

THE READER MUST REFER APPENDIX VI (Page Nos : 738 - 756) FOR RELATED CLINICAL CASE STUDIES (Nos : 36, 37, 38, 40, 43, 45, 46)

Self-Assessment Exercises

I. Essay questions

1. Write briefly on the different laboratory investigations employed to assess liver function.
2. Discuss the biochemical parameters for the differential diagnosis of jaundice.
3. Give an account of the serum enzymes derived from liver and their importance in LFT.
4. Describe the renal function tests.
5. Discuss the different laboratory investigations to evaluate gastric function.

II. Short notes

(a) Serum bilirubin, (b) van den Bergh reaction, (c) Galactose tolerance test, (d) Prothrombin time as LFT, (e) Renal threshold substances, (f) Glomerular filtration rate, (g) Creatinine clearance, (h) Standard urea clearance, (i) Urine concentration test, (j) Gastric function tests.

III. Fill in the blanks

1. Bilirubin is the excretory end product of __________.
2. The laboratory reaction most commonly employed to detect the elevation of serum bilirubin is __________.
3. The serum enzyme most predominantly elevated in viral hepatitis is __________.
4. Obstructive jaundice is characterized by an increase in the serum enzyme __________.
5. The excretory function of liver can be evaluated by using a dye __________.
6. The renal threshold for glucose is __________.
7. The exogenous substance used to measure glomerular filtration rate (GFR) is __________.
8. Standard urea clearance is calculated when the volume of urine output is less than __________.
9. Name the stomach tube used to aspirate gastric juice __________.
10. Name the synthetic peptide used to stimulate gastric secretion for evaluation of gastric function __________.

IV. Multiple choice questions

11. In hemolytic jaundice, van den Bergh reaction is
    (a) Indirect positive (b) Direct positive (c) Biphasic (d) None of these.
12. The serum enzyme elevated in alcoholic cirrhosis of liver is
    (a) Alanine transaminase (b) Aspartate transaminase (c) Alcohol dehydrogenase (d) γ-Glutamyl transpeptidase.
13. Bilirubin is not excreted in urine in
    (a) Obstructive jaundice (b) Hepatic jaundice (c) Hemolytic jaundice (d) All three.
14. Urea clearance is less than GFR because it is
    (a) Partially secreted by the renal tubules (b) Partially reabsorbed by the tubules (c) Only filtered by glomeruli (d) None of these.
15. The serum enzyme used to evaluate pancreatic function is
    (a) Alkaline phosphatase (b) Amylase (c) Aspartate transaminase (d) Lactate dehydrogenase.

Water, Electrolyte and Acid-base Balance

21

The acid-base homeostasis speaks

$$pH = pK_a + \log \frac{[HCO_3^-]}{[H_2CO_3]}$$

"We maintain the blood pH at 7.4!
Regulated by buffers, lungs and kidneys;
Increased hydrogen ion causes acidosis;
Decreased hydrogen ion leads to alkalosis."

The ***organism*** possesses tremendous capacity to survive against odds and to ***maintain homeostasis***. This is particularly true with regard to water, electrolyte and acid-base status of the body. These three are interrelated, hence they are considered together for the discussion in this chapter. Kidney actively participates in the regulation of water, electrolyte and acid-base balance. The general functions of kidney have already been described (***Chapter 20***).

WATER AND LIFE

Water is the ***solvent of life***. Undoubtedly, water is more important than any other single compound to life. It is involved in several body functions.

Functions of water

1. Water provides the ***aqueous medium*** to the organism which is ***essential for*** the various ***biochemical reactions to occur***.

2. Water directly ***participates as a reactant*** in several metabolic reactions.

3. It serves as a vehicle for ***transport of solutes***.

4. Water is closely associated with the regulation of body temperature.

Distribution of water

Water is the major body constituent. An adult human contains about 60% water (***men 55-70%, women 45-60%***). The women and obese individuals have relatively less water which is due to the higher content of stored fat in an anhydrous form.

A 70 kg normal man contains about 42 litres of water. This is distributed in intracellular (inside the cells 28l) and extracellular (outside the cells 14l) compartments, respectively known as ***intracellular fluid*** (***ICF***) and ***extracellular fluid*** (***ECF***). The ECF is further divided into interstitial fluid (10.5l) and plasma (3.5l). The distribution of water in man is given in ***Table 21.1***.

Table 21.1 Distribution of water in an adult man, weighing 70 kg.

Compartment	% Body weight	Volume (l)
Total	60	42
Intracellular fluid (ICF)	40	28
Extracellular fluid (ECF)	20	14
Interstitial fluid	15	10.5
Plasma	5	3.5

WATER TURNOVER AND BALANCE

The body possesses tremendous capacity to regulate its water content. In a healthy individual, this is achieved by balancing the daily water intake and water output.

Water intake

Water is supplied to the body by ***exogenous*** and ***endogenous*** sources.

Exogenous water : Ingested water and beverages, water content of solid foods-constitute the exogenous source of water. Water intake is highly variable which may range from 0.5-5 litres. It largely depends on the social habits and climate. In general, people living in hot climate drink more water. Ingestion of water is mainly controlled by a ***thirst centre*** located in the ***hypothalamus***. Increase in the osmolality of plasma causes increased water intake by stimulating thirst centre.

Endogenous water : The ***metabolic water*** produced within the body is the endogenous water. This water (300-350 ml/day) is ***derived from the oxidation of foodstuffs***. It is estimated that 1 g each of carbohydrate, protein and fat, respectively, yield 0.6 ml, 0.4 ml and 1.1 ml of water. On an average, about 125 ml of water is generated for 1,000 Cal consumed by the body.

Water output

Water losses from the body are variable. There are four routes for the elimination of water from the body—urine, skin, lungs and feces.

Urine : This is the ***major route*** for water loss from the body. In a healthy individual, the urine output is about ***1-2 l/day***. Water loss through kidneys, although highly variable, is well regulated to meet the body demands—to get rid of water or to retain. It should, however, be remembered that man cannot completely shut down urine production, despite there being no water intake. This is due to the fact that some amount of water (about 500 ml/day) is essential as the medium to eliminate the waste products from the body.

Hormonal regulation of urine production : It is indeed surprising to know that about 180 litres of water is filtered by the glomeruli into the renal tubules everyday. However, most of this is reabsorbed and only 1-2 litres is excreted as urine. Water excretion by the kidney is tightly controlled by ***vasopressin*** also known as ***antidiuretic hormone (ADH)*** of the posterior pituitary gland. The secretion of ADH is regulated by the osmotic pressure of plasma. An increase in osmolality promotes ADH secretion that leads to an increased water reabsorption from the renal tubules (less urine output). Plasma osmolality is largely dependent on the sodium concentration, hence sodium indirectly controls the amount of water in the body.

Diabetes insipidus is a disorder characterized by the ***deficiency of ADH*** which results in an ***increased loss of water*** from the body.

Aquaporins are membrane proteins that play a key role in water balance of the body. ADH translocates aquaporins from cytoplasm to membrane (of collecting duct cells) to promote reabsorption of water from the renal tubules.

Skin : Loss of water (450 ml/day) occurs through the body surface by perspiration. This is an unregulated process by the body which mostly depends on the atmospheric temperature and humidity. The loss is more in hot climate. Fever causes increased water loss through the skin. For every 1°C rise in body temperature, about 15% increase is observed in the loss of water (through skin).

Medical Concepts/ Clinical Correlates

- Existence of life is unimaginable in the absence of water.
- Kidneys play a predominant role in the regulation of water, electrolyte and acid-base balance.
- Electrolyte and water balance regulation occurs through the involvement of hormones—aldosterone, ADH and renin-angiotensin.
- Severe dehydration is characterized by low blood pressure, sunken eyeballs, lethargy, confusion and coma.

Lungs : During respiration, some amount of water (about 400 ml/day) is lost through the expired air. The latter is saturated with water and expelled from the body. In hot climates and/or when the person is suffering from fever, the water loss through lungs is increased.

The loss of water by perspiration (via skin) and respiration (via lungs) is collectively referred to as ***insensible water loss***.

Feces : Most of the water entering the gastrointestinal tract is reabsorbed by the intestine. About 150 ml/day is lost through feces in a healthy individual. Fecal loss of water is tremendously increased in diarrhea.

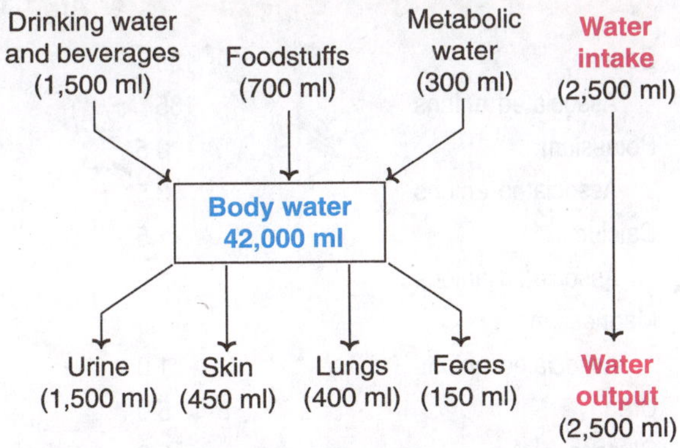

Fig. 21.1 : Water balance in the body, represented by daily intake and output (values are variable).

A summary of the water intake and output in the body is depicted in ***Fig.21.1***. It may be noted that water balance of the body is regulated predominantly by controlling the urine output. This happens after an obligatory water loss via skin, lungs and feces.

The abnormalities associated with water balance—dehydration and overhydration—will be described, following a discussion on electrolyte balance.

ELECTROLYTE BALANCE

Electrolytes are the compounds which readily dissociate in solution and ***exist as ions*** i.e. positively and negatively charged particles. For instance, NaCl does not exist as such, but it exists as cation (Na^+) and anion (Cl^-). The concentration of electrolytes are ***expressed as milliequivalents (mEq/l)*** rather than milligrams. A gram equivalent weight of a compound is defined as its weight in grams that can combine or displace 1 g of hydrogen. One gram equivalent weight is equivalent to 1,000 milliequivalents.

The following formula is employed to convert the concentration mg/l to mEq/l.

$$\text{mEq/l} = \frac{\text{mg per litre} \times \text{Valency}}{\text{Atomic weight}}$$

Electrolyte composition of body fluids

Electrolytes are well distributed in the body fluids in order to maintain the osmotic equilibrium and water balance. A comparison of electrolytes present in extracellular (plasma) and intracellular (muscle) fluids is given in ***Table 21.2***. The total concentration of cations and anions in each body compartment (ECF or ICF) is equal to maintain electrical neutrality.

There is a marked difference in the concentration of electrolytes (cations and anions) between the extracellular and intracellular fluids. ***Na^+*** is the principal ***extracellular cation*** while ***K^+*** is the ***intracellular cation***. This difference in the concentration is essential for the cell survival which is maintained by ***Na^+ – K^+ pump*** (for details, ***Refer Chapter 33***). As regards anions,

Table 21.2 Composition of electrolytes in the body fluids (expressed as mEq/l)

Extracellular fluid (plasma)				Intracellular fluid (muscle)			
Cations		Anions		Cations		Anions	
Na^+	142	Cl^-	103	K^+	150	HPO_4^{2-}	140
K^+	5	HCO_3^-	27	Na^+	10	HCO_3^-	10
Ca^{2+}	5	HPO_4^{2-}	2	Mg^{2+}	40	Cl^-	2
Mg^{2+}	3	SO_4^{2-}	1	Ca^{2+}	2	SO_4^{2-}	5
		Proteins	16			Proteins	40
		Organic acids	6			Organic acids	5
	155		155		202		202

Cl^- and HCO_3^- predominantly occur in extracellular fluids, while HPO_4^-, proteins and organic acids are found in the intracellular fluids.

Osmolarity and osmolality of body fluids

There are two ways of expressing the concentration of molecules with regard to the osmotic pressure.

1. **Osmolarity :** The number of moles (or millimoles) per liter of solution.

2. **Osmolality :** The number of moles (or millimoles) per kg of solvent.

If the solvent is pure water, there is almost no difference between osmolarity and osmolality. However, for biological fluids (containing molecules such as proteins), the osmolality is more commonly used. This is about 6% greater than osmolarity.

Osmolality of plasma

Osmolality is a measure of the solute particles present in the fluid medium. The osmolality of plasma is in the range of ***285-295 milliosmoles/kg*** (***Table 21.3***). Sodium and its associated anions make the largest contribution (~90%) to plasma osmolality. Osmolality is generally measured by osmometer.

For practical purposes, ***plasma osmolality*** can be computed from the concentrations (mmol/l) of Na^+, K^+, urea and glucose as follows

$$2(Na^+) + 2(K^+) + \text{Urea} + \text{Glucose}$$

The factor 2 is used for Na^+ and K^+ ions to account for the associated anion concentration (assuming complete ionization of the molecules). Since plasma Na^+ is the most predominant contributor to osmolality, the above calculation is further simplified as follows

$$\underset{\text{(mmol/kg)}}{\textbf{Plasma osmolality}} = 2 \times \underset{\text{(mmol/l)}}{\textbf{Plasma } Na^+}$$

The above calculation holds good preferably when plasma concentrations of glucose and urea are in the normal range. This calculation, however, will not be valid in severe hyperproteinemia and lipemia.

Table 21.3 Distribution of constituents in plasma osmolality

Constituent (solute)	Osmolality (mosm/kg)
Sodium	135
Associated anions	135
Potassium	3.5
Associated anions	3.5
Calcium	1.5
Associated anions	1.5
Magnesium	1.0
Associated anions	1.0
Urea	5.0
Glucose	5.0
Protein	1.0
Total	293

Osmolality of ECF and ICF

Movement of water across the biological membranes is dependent on the osmotic pressure differences between the intracellular fluid (ICF) and extracellular fluid (ECF). In a healthy state, the osmotic pressure of ECF, mainly due to Na^+ ions, is equal to the osmotic pressure of ICF which is predominantly due to K^+ ions. As such, there is no net passage of water molecules in or out of the cells, due to this osmotic equilibrium.

Regulation of electrolyte balance

Electrolyte and water balance are regulated together and the kidneys play a predominant role in this regard. The regulation is mostly achieved through the hormones aldosterone, ADH and renin-angiotensin.

Aldosterone : It is a ***mineralocorticoid*** produced by adrenal cortex. Aldosterone increases Na^+ reabsorption by the renal tubules at the expense of K^+ and H^+ ions. The net effect is the retention of Na^+ in the body.

Antidiuretic hormone (ADH) : An increase in the plasma osmolality (mostly due to Na^+) stimulates hypothalamus to release ADH. ADH effectively ***increases water reabsorption*** by renal tubules.

Renin-angiotensin : The ***secretion of aldosterone is controlled*** by renin-angiotensin system. Decrease in the blood pressure (due to a fall in ECF volume) is sensed by juxtaglomerular apparatus of the nephron which secrete renin. Renin acts on angiotensinogen to produce angiotensin I. The latter is then converted to angiotensin II which stimulates the release of aldosterone.

The relation between renin, angiotensin and aldosterone in the regulation of Na^+ balance is depicted in ***Fig.21.2***. Aldosterone and ADH coordinate with each other to maintain the normal fluid and electrolyte balance.

Atrial natriuretic factor (ANF) : ***ANF*** or ***atriopeptin*** is a 28-amino acids containing peptide. It is produced in the atrium of heart in response to increased blood volume, elevated blood pressure and high salt intake. ANF acts on kidneys to increase GFR, sodium excretion and urine output. Thus ANF opposes the actions of renin and aldosterone (which increase salt retention and blood pressure).

Na^+ concentration and ECF

It is important to realise that Na^+ and its anions (mainly Cl^-) are confined to the extracellular fluid. And the retention of water in the ECF is directly related to the osmotic effect of these ions (Na^+ and Cl^-). Therefore, the amount of Na^+ in the ECF ultimately determines its volume.

Dietary intake and electrolyte balance

Generally, the consumption of a well-balanced diet supplies the body requirement of electrolytes. ***Humans do not possess the ability to distinguish between the salt hunger and water hunger***. Thirst, however, may regulate electrolyte intake also. In hot climates, the loss of electrolyte is usually higher. Sometimes it may be necessary to supplement drinking water with electrolytes.

Dehydration

Dehydration is a condition characterized by ***water depletion in the body***. It may be due to insufficient intake or excessive water loss or

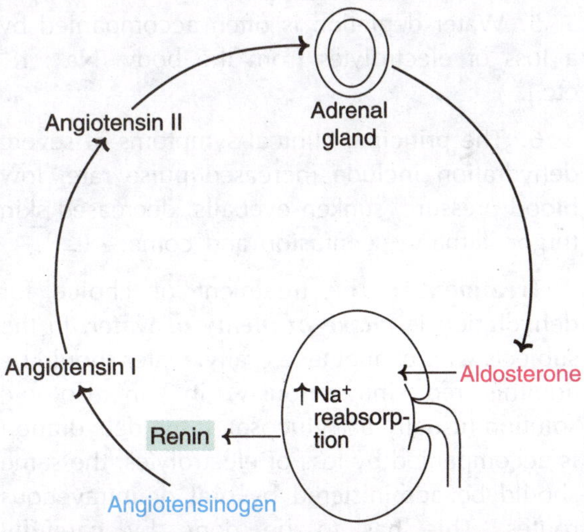

Fig. 21.2 : Hormonal regulation of Na^+ balance by the kidney.

both. Dehydration is generally classified into two types.

1. Due to loss of water alone.

2. Due to deprivation of water and electrolytes.

Causes of dehydration : Dehydration may occur as a result of ***diarrhea, vomiting, excessive sweating, fluid loss*** in burns, adrenocortical dysfunction, kidney diseases (e.g. renal insufficiency), deficiency of ADH (diabetes insipidus) etc.

Characteristic features of dehydration : There are three degrees of dehydration—mild, moderate and severe.

The salient features of dehydration are given hereunder

1. The volume of the ***extracellular fluid*** (e.g. plasma) is ***decreased with*** a concomitant ***rise in*** electrolyte concentration and ***osmotic pressure***.

2. Water is drawn from the intracellular fluid that results in shrunken cells and disturbed metabolism e.g. increased protein breakdown.

3. ADH secretion is increased. This causes increased water retention in the body and consequently urine volume is very low.

4. ***Plasma protein and blood urea*** concentrations are ***increased.***

5. Water depletion is often accompanied by a loss of electrolytes from the body (Na^+, K^+ etc.).

6. The principal clinical symptoms of severe dehydration include increased pulse rate, low blood pressure, sunken eyeballs, decreased skin turgor, lethargy, confusion and coma.

Treatment : The treatment of choice for dehydration is ***intake of plenty of water***. In the subjects who cannot take orally, water should be administered intravenously in an isotonic solution (usually 5% glucose). If the dehydration is accompanied by loss of electrolytes, the same should be administered by oral or intravenous routes. This has to be done by carefully monitoring the water and electrolyte status of the body.

Osmotic imbalance and dehydration in cholera

Cholera is transmitted through water and foods, contaminated by the bacterium ***Vibrio cholerae***. This bacterium produces a toxin which stimulates the intestinal cells to secrete various ions (Cl^-, Na^+, K^+, HCO_3^- etc.) into the intestinal lumen. These ions collectively raise the osmotic pressure and suck the water into lumen. This results in diarrhea with a heavy loss of water (5–10 liters/day). If not treated in time, the victims of cholera will die due to dehydration and loss of dissolved salts. Thus, cholera and other forms of severe diarrhea are the major killers of young children, in many developing countries.

Oral rehydration therapy (ORT) is commonly used ***to treat cholera*** and other ***diarrheal diseases***.

Overhydration

Overhydration or ***water intoxication*** is caused by excessive ***retention of water in the body***. This may occur due to excessive intake of large volumes of salt free fluids, renal failure, overproduction of ADH etc. Overhydration is observed after major trauma or operation, lung infections etc.

Water intoxication is associated with ***dilution of ECF and ICF with a decrease in osmolality***. The clinical symptoms include headache, lethargy and convulsions. The treatment advocated is stoppage of water intake and administration of hypertonic saline.

Water tank model

The distribution of body water (in the ECF and ICF), dehydration and overhydration can be better understood by a water tank model (***Fig.21.3***). The tank has an inlet and outlet, respectively, representing the water intake (mostly oral) and water output (mainly urine) by the body.

Dehydration is caused when the water output exceeds the intake. On the other hand, overhydration is due to more water intake and less output.

Medical Concepts/ Clinical Correlates

- Sodium is the principal extracellular cation while K^+ is intracellular. The maintenance of the differential concentration of these electrolytes is essential for the survival of life which is brought about by Na^+-K^+ pump.
- The body metabolism is accompanied by the production of acids such as carbonic acid, sulfuric acid, phosphoric acid etc.
- Vegetarian diet has an alkalizing effect on the body. This is attributed to the formation of organic acids such as sodium lactate which can deplete H^+ ions by combining with them.

Metabolism of electrolytes

The body distribution, dietary intake, intestinal absorption and biochemical functions of individual electrolytes are discussed under the section mineral metabolism (***Chapter 18***). The electrolyte disorders, particularly hypernatremia and hyponatremia (of sodium); hyperkalemia and hypokalemia (of potassium) must also be referred.

Diuretics in the treatment of edema and hypertension

Diuretics are the drugs that stimulate water and sodium excretion, so that ***urine volume is increased***. The commonly used diuretics are bendrofluazide, ***frusemide***, spironolactone and ***mannitol***. Diuretics are important in the treatment of edema, heart failure and hypertension.

ACID-BASE BALANCE

The ***normal pH*** of the ***blood*** is maintained in the narrow range of ***7.35-7.45***, i.e. slightly alkaline. The pH of intracellular fluid is rather variable. Thus, for erythrocytes the pH is 7.2, while for skeletal muscle, it may be as low as 6.0.

Maintenance of blood pH is an important homeostatic mechanism of the body. In normal circumstances, the ***regulation is so effective that the blood pH varies very little***. Changes in blood pH will alter the intracellular pH which, in turn, influences the metabolism e.g. distortion in protein structure, enzyme activity etc. It is estimated that the ***blood pH compatible to life is 6.8-7.8***. (For a good understanding of acid-base balance, adequate knowledge on acids, bases, pH and buffers is essential (refer ***Chapter 40***).

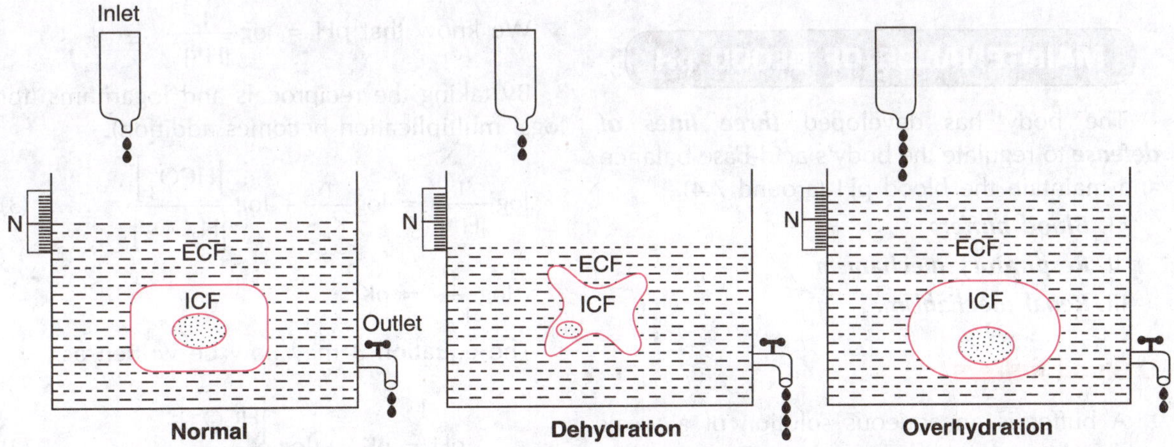

Fig. 21.3 : *Water tank model representing body fluid compartments (N–Normal level; ECF–Extracellular fluid; ICF–Intracellular fluid).*

Production of acids by the body

The metabolism of the body is accompanied by an overall production of acids. These include the ***volatile acids like carbonic acid*** (most predominant, about 20,000 mEq/day) or ***non-volatile acids*** (about 80 mEq/day) such as ***lactic acid, sulfuric acid, phosphoric acid*** etc. Carbonic acid is formed from the metabolic product CO_2; lactic acid is produced in anaerobic metabolism; sulfuric acid is generated from proteins (sulfur containing amino acids); phosphoric acid is derived from organic phosphates (e.g. phospholipids). All these acids add up H^+ ions to the blood. A ***diet rich in animal proteins results in more acid production*** by the body that ultimately leads to the excretion of ***urine*** which is profoundly ***acidic***.

Production of bases by the body

The formation of basic compounds in the body, in the normal circumstances, is negligible. Some amount of bicarbonate is generated from the organic acids such as lactate and citrate. A vegetarian diet has a tendency for a net production of bases. This is due to the fact that vegetarian diet produces salts of organic acids such as sodium lactate which can utilize H^+ ions produced in the body. For this reason, a ***vegetarian diet has an alkalizing effect on the body***. This is reflected by the excretion of neutral or slightly alkaline urine by these subjects.

MAINTENANCE OF BLOOD pH

The body has developed ***three lines of defense*** to regulate the body's acid-base balance and maintain the blood pH (around 7.4).

I. ***Blood buffers***

II. ***Respiratory mechanism***

III. ***Renal mechanism.***

I. Blood buffers

A buffer is an aqueous solution of a weak acid and its salt with a strong base. Or a solution of weak base and its salt of strong acid. The buffer resists the change in pH by the addition of acid or alkali and the buffering capacity is dependent on the absolute concentration of salt and acid. It should be borne in mind that the buffer cannot remove H^+ ions from the body. It temporarily acts as a shock absorbant to reduce the free H^+ ions. The H^+ ions have to be ultimately eliminated by the renal mechanism (described later).

The blood contains 3 buffer systems.

1. Bicarbonate buffer
2. Phosphate buffer
3. Protein buffer.

1. **Bicarbonate buffer system :** Sodium bicarbonate and carbonic acid ($NaHCO_3 - H_2CO_3$) is the most ***predominant buffer*** system of the extracellular fluid, particularly the plasma. Carbonic acid dissociates into hydrogen and bicarbonate ions.

$$H_2CO_3 \rightleftharpoons H^+ + HCO_3^-$$

By the law of mass action, at equilibrium

$$K_a = \frac{[H^+][HCO_3^-]}{[H_2CO_3]} \qquad \text{...... (1)}$$

(K_a = Dissociation constant of H_2CO_3).

The equation may be rewritten as follows

$$[H^+] = K_a \frac{[H_2CO_3]}{[HCO_3^-]}. \qquad \text{...... (2)}$$

We know that $pH = \log\frac{1}{[H^+]}$.

By taking the reciprocals and logarithms (for logs, multiplication becomes addition).

$$\log\frac{1}{[H^+]} = \log\frac{1}{K_a} + \log\frac{[HCO_3^-]}{[H_2CO_3]} \qquad \text{...... (3)}$$

$$\log\frac{1}{K_a} = pK_a$$

The equation 3 may now be written as

$$pH = pK_a + \log\frac{[HCO_3^-]}{[H_2CO_3]}. \qquad \text{...... (4)}$$

The above equation is valid for any buffer pair. The general equation referred to as

Henderson-Hasselbalch equation for any buffer is written as

$$pH = pK_a + \log \frac{[Base]}{[Acid]}. \quad \ldots\ldots (5)$$

It is evident from this equation that the pH is dependent on ratio of the concentration of the base to acid (HCO_3^- and H_2CO_3 in equation 4).

Blood pH and the ratio of HCO_3^- to H_2CO_3 : The ***plasma bicarbonate*** (*HCO_3^-*) concentration is around ***24 mmol/l*** (range 22-26 mmol/l). Carbonic acid is a solution of CO_2 in water. Its concentration is given by the product of pCO_2 (arterial partial pressure of CO_2 = 40 mm Hg) and the solubility constant of CO_2 (0.03).

Thus H_2CO_3 = 40 × 0.03 = 1.2 mmol/l.

The Henderson-Hasselbalch equation for bicarbonate buffer is

$$pH = pK_a + \log \frac{[HCO_3^-]}{[H_2CO_3]}.$$

Substituting the values (blood pH = 7.4; pK_a for H_2CO_3 = 6.1; HCO_3^- = 24 mmol/l; H_2CO_3 = 1.2 mmol/l), in the above equation

$$7.4 = 6.1 + \log \frac{24}{1.2}$$

$$= 6.1 + \log 20$$

$$= 6.1 + 1.3$$

$$= 7.4$$

It is evident that at a blood pH 7.4, the ***ratio of bicarbonate to carbonic acid is 20 : 1***. Thus, the bicarbonate concentration is much higher (20 times) than carbonic acid in the blood. This is referred to as ***alkali reserve*** and is responsible for the effective buffering of H^+ ions, generated in the body. In normal circumstances, the concentration of bicarbonate and carbonic acid determines the pH of blood. Further, the bicarbonate buffer system serves as an index to understand the disturbances in the acid-base balance of the body.

2. **Phosphate buffer system :** Sodium dihydrogen phosphate and disodium hydrogen phosphate ($NaH_2PO_4 - Na_2HPO_4$) constitute the phosphate buffer. It is ***mostly an intracellular buffer*** and is of less importance in plasma due to its low concentration. With a pK of 6.8 (close to blood pH 7.4), the phosphate buffer would have been more effective, had it been present in high concentration. It is estimated that the ratio of base to acid for phosphate buffer is 4 compared to 20 for bicarbonate buffer.

3. **Protein buffer system :** The ***plasma proteins and hemoglobin*** together constitute the protein buffer system of the blood. The buffering capacity of proteins is dependent on the pK of ionizable groups of amino acids. The imidazole group of histidine (pK = 6.7) is the most effective contributor of protein buffers. The plasma proteins account for about 2% of the total buffering capacity of the plasma.

Hemoglobin of RBC is also an important buffer. It mainly buffers the fixed acids, besides being involved in the transport of gases (O_2 and CO_2). More details on hemoglobin are given under respiratory mechanism for regulation of pH.

II. Respiratory mechanism for pH regulation

Respiratory system provides a ***rapid mechanism for the maintenance of acid-base balance***. This is achieved by regulating the concentration of carbonic acid (H_2CO_3) in the blood i.e. the denominator in the bicarbonate buffer system. The details of CO_2 transport and the role of hemoglobin in this process are described elsewhere (***Chapter 10, Refer Fig.10.5***).

The large volumes of CO_2 produced by the cellular metabolic activity endanger the acid-base equilibrium of the body. But in normal circumstances, all of this CO_2 is eliminated from the body in the expired air via the lungs, as summarized below.

$$H_2CO_3 \xleftrightarrow{\text{Carbonic anhydrase}} CO_2 + H_2O.$$

The rate of respiration (or the rate of removal of CO_2) is controlled by a respiratory centre, located in the medulla of the brain. This centre is highly sensitive to changes in the pH of blood. Any decrease in blood pH causes hyperventilation

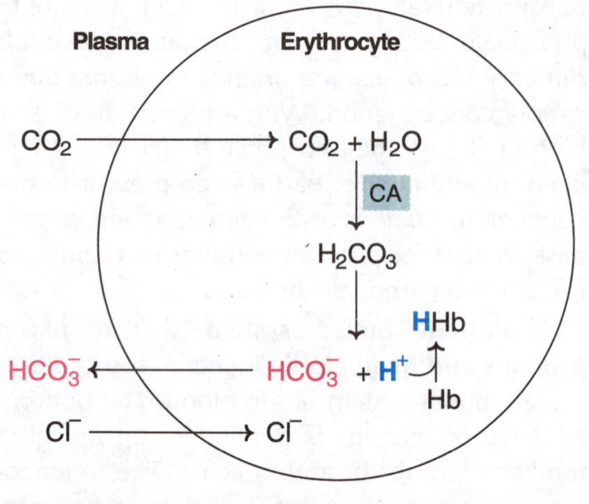

Fig. 21.4 : *Generation of bicarbonate by the erythrocyte (CA–Carbonic anhydrase; Hb–Hemoglobin).*

to blow off CO_2, thereby reducing the H_2CO_3 concentration. Simultaneously, the H^+ ions are eliminated as H_2O.

Respiratory control of blood pH is rapid but only a short term regulatory process, since hyperventilation cannot proceed for long.

Hemoglobin as a buffer : Hemoglobin of erythrocytes is also important in the respiratory regulation of pH. At the tissue level, hemoglobin binds to H^+ ions and helps to transport CO_2 as HCO_3^- with a minimum change in pH (referred to as ***isohydric transport***). In the lungs, as hemoglobin combines with O_2, H^+ ions are removed which combine with HCO_3^- to form H_2CO_3. The latter dissociates to release CO_2 to be exhaled (***Refer Fig.10.5***).

Generation of HCO_3^- by RBC : Due to lack of aerobic metabolic pathways, RBC produce very little CO_2. The plasma CO_2 diffuses into the RBC along the concentration gradient where it combines with water to form H_2CO_3. This reaction is catalysed by ***carbonic anhydrase*** (also called carbonate dehydratase). In the RBC, H_2CO_3 dissociates to produce H^+ and HCO_3^-. The H^+ ions are trapped and buffered by hemoglobin. As the concentration of HCO_3^- increases in the RBC, it diffuses into plasma along with the concentration gradient, in exchange for Cl^- ions, to maintain electrical neutrality. This phenomenon, referred to as ***chloride shift***, helps to generate HCO_3^- (***Fig.21.4***).

III. Renal mechanism for pH regulation

The role of kidneys in the maintenance of acid-base balance of the body (blood pH) is highly significant. The ***renal mechanism*** tries to provide a ***permanent solution to the acid-base disturbances***. This is in contrast to the temporary buffering system and a short term respiratory mechanism, described above.

The kidneys regulate the blood pH by maintaining the alkali reserve, besides excreting or reabsorbing the acidic or basic substances, as the situation demands.

Urine pH normally lower than blood pH : The pH of urine is normally acidic (***~6.0***). This clearly indicates that the ***kidneys have contributed to the acidification of urine***, when it is formed from the blood plasma (pH 7.4). In other words, the H^+ ions generated in the body in the normal circumstances, are eliminated by acidified urine. Hence the pH of urine is normally acidic (~6.0), while that of blood is alkaline (7.4). Urine pH, however, is variable and may range between 4.5-9.5, depending on the concentration of H^+ ions.

Carbonic anhydrase and renal regulation of pH : The enzyme carbonic anhydrase (inhibited by ***acetazolamide***) is of central importance in the renal regulation of pH which occurs by the following mechanisms.

1. Excretion of H^+ ions
2. Reabsorption of bicarbonate
3. Excretion of titratable acid
4. Excretion of ammonium ions.

1. **Excretion of H^+ ions :** Kidney is the ***only route through which the H^+ can be eliminated from the body***. H^+ excretion occurs in the proximal convoluted tubules (renal tubular cells) and is coupled with the regeneration of HCO_3^-. The process depicted in ***Fig.21.5***, occurs as described next.

Carbonic anhydrase catalyses the production of carbonic acid (H_2CO_3) from CO_2 and H_2O in the renal tubular cell. H_2CO_3 then dissociates to H^+ and HCO_3^-. The H^+ ions are secreted into the tubular lumen in exchange for Na^+. The Na^+ in association with HCO_3^- is reabsorbed into the blood. This is an effective mechanism to eliminate acids (H^+) from the body with a simultaneous generation of HCO_3^-. The latter adds up to the alkali reserve of the body. The H^+ combines with a non-carbonate base and is excreted in urine.

2. **Reabsorption of bicarbonate :** This mechanism is primarily ***responsible to conserve the blood HCO_3^-***, with a simultaneous excretion of H^+ ions. The normal urine is almost free from HCO_3^-. This is explained as follows (***Fig.21.6***).

Bicarbonate freely diffuses from the plasma into the tubular lumen. Here HCO_3^- combines with H^+, secreted by tubular cells, to form H_2CO_3. H_2CO_3 is then cleaved by carbonic anhydrase (of tubular cell membrane) to form CO_2 and H_2O. As the CO_2 concentration builds up in the lumen, it diffuses into the tubular cells along the concentration gradient. In the tubular cell, CO_2 again combines with H_2O to form H_2CO_3 which then dissociates into H^+ and HCO_3^-. The H^+ is secreted into the lumen in exchange for Na^+. The ***HCO_3^- is reabsorbed into plasma in association with Na^+***. Reabsorption of HCO_3^- is a cyclic process with the net excretion of H^+ or generation of new HCO_3^-. This is because the H^+ is derived from water. This mechanism helps to maintain the steady state and will not be effective for the elimination of H^+ or generation of new HCO_3^-.

3. **Excretion of titratable acid :** Titratable acidity is a ***measure of acid excreted into urine by the kidney***. This can be estimated by titrating urine back to the normal pH of blood (7.4). In quantitative terms, titratable acidity refers to the number of milliliters of N/10 NaOH required to titrate 1 liter of urine to pH 7.4. Titratable acidity ***reflects the H^+ ions excreted into urine which resulted in a fall of pH from 7.4*** (that of blood). The excreted H^+ ions are actually buffered in the urine by phosphate buffer as depicted in ***Fig.21.7***, and briefly described hereunder.

As already discussed, H^+ ion is secreted into the tubular lumen in exchange for Na^+ ion. This Na^+ is obtained from the base, disodium hydrogen phosphate (Na_2HPO_4). The latter in turn combines with H^+ to produce the acid, sodium dihydrogen phosphate (NaH_2PO_4), in which form the major quantity of titratable acid in urine is present. As the tubular fluid moves down the renal tubules, more and more H^+ ions are added, resulting in the acidification of urine. This causes a fall in the pH of urine to as low as 4.5. Any further fall in the pH will cause depletion of Na^+ ions.

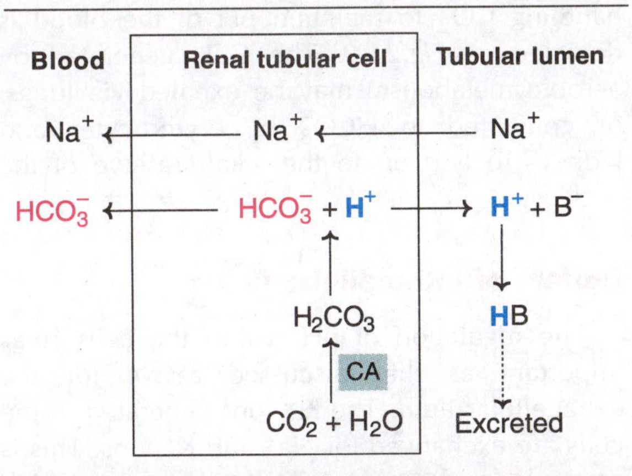

***Fig. 21.5 :** Renal regulation of blood pH–Excretion of H+ ions (CA–Carbonic anhydrase).*

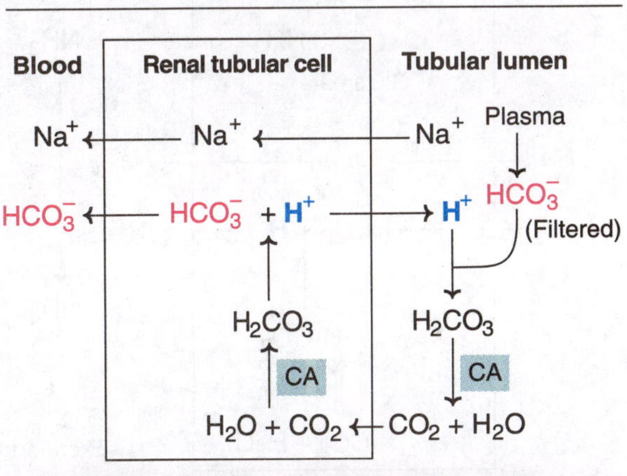

***Fig. 21.6 :** Renal regulation of blood pH–Reabsorption of bicarbonate (CA–Carbonic anhydrase).*

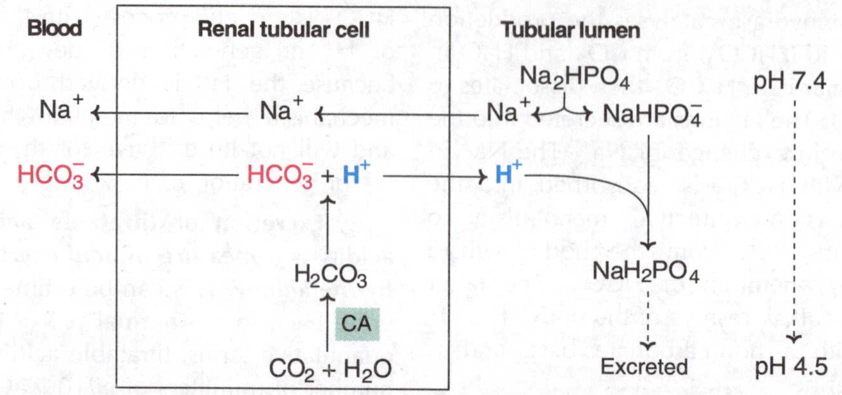

Fig. 21.7 : *Renal regulation of blood pH–Excretion of titratable acid by phosphate buffer mechanism (CA–Carbonic anhydrase).*

4. **Excretion of ammonium ions :** This is another mechanism to buffer H^+ ions secreted into the tubular fluid. The ***H^+ ion combines with NH_3 to form ammonium ion (NH_4^+).*** The renal tubular cells deamidate glutamine to glutamate and NH_3. This reaction is ***catalysed by*** the enzyme ***glutaminase***. The NH_3, liberated in this reaction, diffuses into the tubular lumen where it combines with H^+ to form NH_4^+ (***Fig.21.8***). ***Ammonium*** ions cannot diffuse back into tubular cells and, therefore, are ***excreted into urine***.

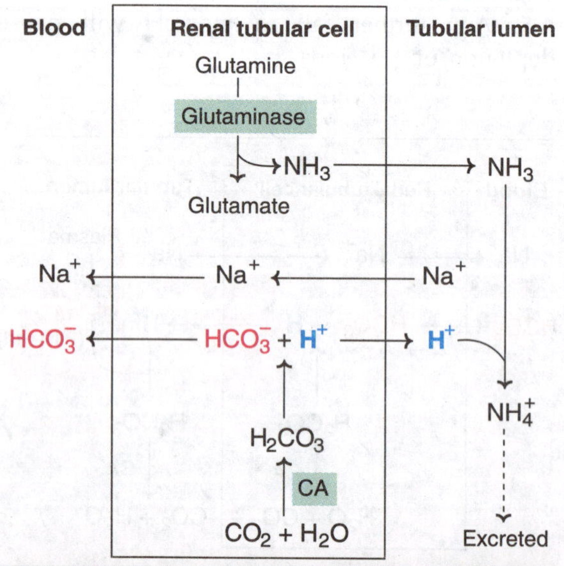

Fig. 21.8 : *Renal regulation of blood pH–Excretion of ammonium ions (CA–Carbonic anhydrase).*

NH_4^+ is a major urine acid. It is estimated that about half to two-thirds of body acid load is eliminated in the form of NH_4^+ ions. For this reason, renal regulation via NH_4^+ excretion is very effective to eliminate large quantities of acids produced in the body. This mechanism becomes predominant, particularly in acidosis.

Carbon dioxide—the central molecule of pH regulation

As is observed from the foregoing discussion, CO_2 is of central importance in the acid-base balance of the body. It has the ability to combine with H_2O to from H_2CO_3 which can dissociate to HCO_3^- and H^+. A summary of the interaction between the lungs, erythrocytes and kidneys in handling CO_2 to maintain pH of the blood is depicted in ***Fig.21.9***. The CO_2 generated by aerobic metabolism may be exhaled via lungs, or converted to HCO_3^- by erythrocytes and kidneys to add up to the alkali reserve of the body.

Buffers of intracellular fluids

The regulation of pH within the cells is as important as that discussed above for the extracellular fluid. The H^+ ions generated in the cells are exchanged for Na^+ and K^+ ions. This is particularly observed in skeletal muscle which reduces the potential danger of H^+ accumulation in the cells.

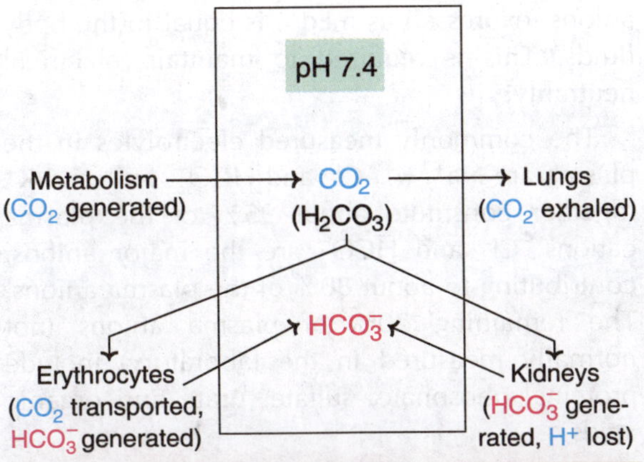

Fig. 21.9 : *Carbon dioxide—the central molecule of blood pH regulation.*

DISORDERS OF ACID-BASE BALANCE

The body has developed an efficient system for the maintenance of acid-base equilibrium with a result that the pH of blood is almost constant (7.4). The ***blood pH compatible to life is 6.8-7.8,*** beyond which life cannot exist.

For a better understanding of the disorders of acid-base balance, the ***Henderson-Hasselbalch equation*** must be frequently consulted.

$$pH = pK_a + \log \frac{[HCO_3^-]}{[H_2CO_3]}$$

It is evident from the above equation that the blood pH (H^+ ion concentration) is dependent on the relative concentration (ratio) of bicarbonate (HCO_3^-) and carbonic acid (H_2CO_3).

The acid-base disorders are mainly classified as

1. **Acidosis—a decline in blood pH**
   (a) ***Metabolic acidosis***—due to a ***decrease in bicarbonate.***
   (b) ***Respiratory acidosis***—due to an ***increase in carbonic acid.***
2. **Alkalosis—a rise in blood pH**
   (a) ***Metabolic alkalosis***—due to an ***increase in bicarbonate.***
   (b) ***Respiratory alkalosis***—due to a ***decrease in carbonic acid.***

The four acid-base disorders referred above are primarily due to alterations in either bicarbonate or carbonic acid. It may be observed that the metabolic acid-base balance disorders are caused by a direct alteration in bicarbonate concentration while the respiratory disturbances are due to a change in carbonic acid level (i.e. CO_2). This type of classification is more theoretical. In the ***actual clinical situations, mixed type of disorders are common.***

The terms ***acidemia*** and ***alkalemia,*** respectively, refer to an ***increase*** or ***a decrease*** in ***[H⁺]*** ion concentration in blood. They are, however, not commonly used.

Clinical causes of acid-base disorders

The most important clinical causes/disease states that result in acid-base disorders are listed in ***Table 21.4.*** Metabolic acidosis could occur due to ***diabetes mellitus*** (***ketoacidosis***), ***lactic acidosis, renal failure*** etc. Respiratory acidosis is common in ***severe asthma*** and ***cardiac arrest.*** Vomiting and hypokalemia may result in metabolic alkalosis while hyperventilation and severe anemia may lead to respiratory alkalosis.

Table 21.4 Major clinical causes of acid-base disorders

Metabolic acidosis	Respiratory acidosis
Diabetes mellitus (ketoacidosis)	Severe asthma Pneumonia Cardiac arrest
Renal failure	Obstruction in airways
Lactic acidosis	Chest deformities
Severe diarrhea Renal tubular acidosis	Depression of respiratory center (by drugs e.g. opiates)
Metabolic alkalosis	**Respiratory alkalosis**
Severe vomiting	Hyperventilation
Hypokalemia	Anemia High altitude
Intravenous administration of bicarbonate	Salicylate poisoning

Compensation of acid-base disorders

To counter the acid-base disturbances, ***the body gears up*** its homeostatic mechanism and ***makes every attempt to restore the pH to normal level (7.4).*** This is referred to as compensation which may be partial or full. Sometimes the acid-base disorders may remain uncompensated.

The principal acid-base disturbances, along with the blood concentration of HCO_3^- and H_2CO_3, in acute and compensated states are given in the ***Table. 21.5***.

For the acute metabolic disorders (due to changes in HCO_3^-), respiratory compensation sets in and regulates the H_2CO_3 (i.e. CO_2) by hyper- or hypoventilation. As regards acute respiratory disorders (due to changes in H_2CO_3), the renal compensation occurs to maintain the HCO_3^- level, by increasing or decreasing its excretion.

In the ***Table 21.6***, a summary of the acid-base disorders with primary changes and compensatory mechanisms is given.

Anion gap

For a better understanding of acid-base disorders, adequate knowledge of anion gap is essential. The total concentration of cations and anions (expressed as mEq/l) is equal in the body fluids. This is required to maintain electrical neutrality.

The commonly measured electrolytes in the plasma are Na^+, K^+, Cl^- and HCO_3^-. Na^+ and K^+ together constitute about 95% of the plasma cations. Cl^- and HCO_3^- are the major anions, contributing to about 80% of the plasma anions. The remaining 20% of plasma anions (not normally measured in the laboratory) include proteins, phosphate, sulfate, urate and organic acids.

Anion gap is defined as ***the difference between the total concentration of measured cations (Na^+ and K^+) and that of measured anion (Cl^- and HCO_3^-). The anion gap (A^-) in fact represents the unmeasured anions in the plasma*** which may be calculated as follows, by substituting the normal concentration of electrolytes (mEq/l).

$$Na^+ + K^+ = Cl^- + HCO_3^- + A^-$$
$$136 + 4 = 100 + 25 + A^-$$
$$A^- = 15 \text{ mEq/l}$$

The anion gap in a healthy individual is around 15 mEq/l (range 8-18 mEq/l). ***Acid-base disorders are often associated with alterations in the anion gap.***

Table 21.5 Acid-base disorders along with the concentrations of bicarbonate (HCO_3^-) and carbonic acid (H_2CO_3) in plasma

Disorder	Blood pH	$[HCO_3^-]$	$[H_2CO_3]$
Metabolic acidosis			
Acute	↓	↓	→
Compensated (by ↑ ventilation)	↘ or →	↓	↓
Respiratory acidosis			
Acute	↓	→	↑
Compensated (HCO_3^- retained by kidney)	↘ or →	↑	↑
Metabolic alkalosis			
Acute	↑	↑	→
Compensated (by ↓ ventilation)	↗ or →	↑	↑
Respiratory alkalosis			
Acute	↑	→	↓
Compensated (↑HCO_3^- excretion by kidney)	↗ or →	↓	↓

↑: *Increased*; ↓: *Decreased*; →: *Normal*; ↘: *Marginally decreased*; ↗: *Marginally increased.*

Metabolic acidosis

The primary defect in metabolic acidosis is a ***reduction in bicarbonate concentration*** which leads to a fall in blood pH. The bicarbonate concentration may be decreased due to its utilization in buffering H^+ ions, loss in urine or gastrointestinal tract or failure to be regenerated.

The most important cause of metabolic acidosis is due to an ***excessive production of organic acids which combine with $NaHCO_3$ and deplete the alkali reserve***.

$$NaHCO_3^- + \text{Organic acids} \longrightarrow \text{Na salts of organic acids} + CO_2$$

Metabolic acidosis is commonly seen in ***severe uncontrolled diabetes mellitus*** which is associated with excessive production of acetoacetic acid and β-hydroxybutyric acid (both are organic acids).

Anion gap and metabolic acidosis : Increased production and accumulation of organic acids causes an ***elevation*** in the anion gap. This type of picture is seen in metabolic acidosis associated with ***diabetes*** (***ketoacidosis***).

Compensation of metabolic acidosis : The acute metabolic acidosis is usually compensated ***by hyperventilation of lungs***. This leads to an increased elimination of CO_2 from the body (hence $H_2CO_3\downarrow$). but respiratory compensation is only short-lived. Renal compensation sets in within 3-4 days and the H^+ ions are excreted as NH_4^+ ions.

Respiratory acidosis

The primary defect in respiratory acidosis is due to a ***retention of CO_2*** ($H_2CO_3\uparrow$). There may be several causes for respiratory acidosis which include depression of the respiratory centre (overdose of drugs), pulmonary disorders (bronchopneumonia) and breathing air with high content of CO_2.

The renal mechanism comes for the rescue to compensate respiratory acidosis. ***More HCO_3^- is generated and retained by the kidneys*** which adds up to the alkali reserve of the body. The excretion of ***titratable acidity*** and NH_4^+ is ***elevated in urine***.

Metabolic alkalosis

The primary abnormality in metabolic alkalosis is an ***increase in HCO_3^- concentration***. This may occur due to excessive vomiting (resulting in loss of H^+) or an excessive intake of sodium bicarbonate for therapeutic purposes (e.g. control of gastric acidity). Cushing's syndrome (hypersecretion of aldosterone) causes increased retention of Na^+ and loss of K^+ from the body. Metabolic alkalosis is commonly associated with low K^+ concentration (hypokalemia). In severe K^+ deficiency, H^+ ions are retained inside the cells to

Table 21.6 Acid-base disorders with primary changes and compensatory mechanisms

Disorder	Primary change	Compensatory mechanism	Timescale for compensation
Metabolic acidosis	Decreased plasma bicarbonate	Hyperventilation (decrease in pCO_2)	Minutes to hours
Metabolic alkalosis	Increased plasma bicarbonate	Hypoventilation (increase in pCO_2)	Minutes to hours
Respiratory acidosis	Increased pCO_2	Elevation in plasma bicarbonate; increase in renal reabsorption of bicarbonate	Days
Respiratory alkalosis	Decreased pCO_2	Reduction in plasma bicarbonate; decrease in renal reabsorption of bicarbonate	Days

Medical Concepts/ Clinical Correlates

- The blood pH is maintained by blood buffers, respiratory and renal mechanisms.
- Carbon dioxide is the central molecule of acid-base regulation.
- Disturbances in acid-base regulation result in acidosis (decreased blood pH) or alkalosis (raised blood pH).
- Uncontrolled diabetes mellitus is associated with metabolic acidosis, commonly referred to as ketoacidosis (due to the overproduction of ketone bodies).

replace missing K^+ ions. In the renal tubular cells, H^+ ions are exchanged (instead of K^+) with the reabsorbed Na^+. Paradoxically, the patient excretes acid urine despite alkalosis.

The respiratory mechanism initiates the ***compensation by hypoventilation*** to retain CO_2 (hence $H_2CO_3\uparrow$). This is slowly taken over by renal mechanism which excretes more HCO_3^- and retains H^+.

Respiratory alkalosis

The primary abnormality in respiratory alkalosis is a ***decrease in H_2CO_3 concentration***. This may occur due to prolonged hyperventilation resulting in increased exhalation of CO_2 by the lungs. Hyperventilation is observed in conditions such as hysteria, hypoxia, raised intracranial pressure, excessive artificial ventilation and the action of certain drugs (salicylate) that stimulate respiratory centre.

The ***renal mechanism tries to compensate by increasing the urinary excretion of HCO_3^-***.

Mixed acid-base disorders

Sometimes, the ***patient may have two or more acid-base disturbances occurring simultaneously***. In such instances, both HCO_3^- and H_2CO_3 are altered. In general, if the biochemical data (of blood gas analysis) cannot be explained by a specific acid-base disorder, it is assumed that a mixed disturbance is occurring. Many a times, ***compensatory mechanisms may lead to mixed acid-base disorders***.

Acid-base disorders and plasma potassium

Plasma potassium concentration (***normal 3.5-5.0 mEq/l***) is very important as it affects the contractility of the heart. Hyperkalemia (high plasma K^+) or hypokalemia (low plasma K^+) can be life-threatening. The relevance of potassium balance in certain acid-base disorders is discussed briefly.

Potassium and diabetic ketoacidosis : The hormone insulin increases K^+ uptake by cells (particularly from skeletal muscle). The patient of severe uncontrolled diabetes (i.e. with metabolic acidosis) is usually with hypokalemia. When such a patient is given insulin, it stimulates K^+ entry into cells. The result is that plasma K^+ level is further depleted. Hypokalemia affects heart functioning, and is life threatening. Therefore, ***in the treatment of diabetic ketoacidosis, potassium has to be given*** (unless the patients have high plasma K^+ concentration).

Potassium and alkalosis : ***Low plasma*** concentration of K^+ (hypokalemia) leads to an ***increased excretion of hydrogen ions, and thus may cause metabolic alkalosis***. Conversely, metabolic alkalosis is associated with increased renal excretion of K^+.

In view of the importance discussed above, the measurement of plasma K^+ concentration assumes significance in the acid-base disorders. In cases of these disorders associated with hypokalemia, potassium supplementation (with carefull monitoring of plasma K^+) needs to be considered.

BLOOD GAS MEASUREMENT

Arterial blood gas (ABG) analysis is one of the most common investigations in intensive care units (ICUs). ***ABG analysis*** is necessary ***to assess acid-base balance in critical illness***, document severity of respiratory failure and monitor response to therapeutic interventions. Based on the ABG results obtained, oxygen treatment or artificial ventilation is carried out.

For blood gas analysis, a sample of ***arterial blood*** collected from (most commonly) radial artery in the forearm, or (less commonly) from the femoral artery in the leg is used. The biochemical profile measured include ***pO_2***, ***pCO_2***, and ***pH*** (H^+ ion concentration). The concentration of ***bicarbonate*** is calculated by using Henderson-Hasselbalch equation. In fact, the blood gas analysers employed in the hospitals are designed to perform the various calculations automatically and give the final results. The reference ranges of blood gas analysis are given in ***Table 21.7***.

Table 21.7 Reference ranges of arterial blood gas (ABG) analysis

Parameter	Reference range	
	Non-SI Units	SI Units
pH	7.35–7.45	7.35–7.45
$[H^+]$	35–43 mEq/l	35–43 mmol/l
pCO_2	35–45 mm Htg	4.5-6.0 kPa
pO_2	85–100 mm Hg	10.5–13.5 kPa
HCO_3^-*	22–26 mEq/l	22–26 mmol/l

*Bicarbonate concentration is calculated from pH and pCO_2 values.

SUMMARY

- *Water is the solvent of life and constitutes about 60% of the total body weight, distributed in intracellular and extracellular fluids. The daily water intake and output maintain the body balance of water.*
- *Electrolytes are distributed in the intracellular and extracellular fluids to maintain the osmotic equilibrium and water balance, Na^+ is the principal extracellular cation while K^+ is the intracellular cation.*
- *The osmolality of plasma is about 285 milliosmoles/kg, which is predominantly contributed by Na^+ and its associated anions. Thus, for practical purposes, plasma osmolality can be calculated from Na^+ concentration ($2 \times Na^+$ in mmol/l).*
- *Water and electrolyte balance are usually regulated together and this is under the control of hormones—aldosterone, antidiuretic hormone and renin.*
- *Dehydration of the body may be due to insufficient water intake or its excessive loss or both. Depletion of water in the ICF causes disturbance in metabolism. The manifestations of severe dehydration include increased pulse rate, low blood pressure, sunken eyeballs, decreased skin turgor, lethargy and coma.*
- *The normal pH of blood is maintained in the narrow range of 7.35–7.45. The metabolism of the body is accompanied by an overall production of acids. The body has developed three lines of defense (blood buffers, respiratory and renal mechanisms) to regulate the acid-base balance and maintain the blood pH.*
- *Among the blood buffers, bicarbonate buffer (with a ratio of HCO_3^- to H_2CO_3 as 20 : 1) is the most important in regulating blood pH. Phosphate and protein buffer systems also contribute in this regard. The respiratory system regulates the concentration of carbonic acid by controlling the elimination of CO_2 via lungs.*
- *The renal (kidney) mechanism regulates blood pH by excreting H^+ and NH_4^+ ions besides the reabsorption of HCO_3^-. The pH of urine is normally acidic which indicates that the kidneys have contributed to the acidification of urine.*
- *The acid-base disorders are classified as acidosis (metabolic or respiratory) and alkalosis (metabolic or respiratory), respectively, due to a rise or fall in blood pH.*
- *Blood gas measurement includes the parameters pO_2, pCO_2, pH and bicarbonate, and it is very important to evaluate and treat acid-base disorders.*

THE READER MUST REFER APPENDIX VI (Page Nos : 738 - 756) FOR RELATED CLINICAL CASE STUDIES (Nos : 62, 63, 64, 65)

Self-Assessment Exercises

I. Essay questions

1. Describe the role of kidney in the regulation of blood pH.
2. Give an account of the water distribution and its balance in the body.
3. Campare the composition of electrolytes in the extracellular and intracellular fluids. Discuss the regulation of electrolyte balance.
4. Describe the role of blood buffers in the acid-base balance.
5. Classify acid-base disorders and discuss them with compensatory mechanisms.

II. Short notes

(a) Dehydration, (b) Vasopressin and water balance, (c) Osmolality of plasma, (d) Acids produced in the body, (e) Henderson-Hasselbalch equation, (f) Bicarbonate buffer, (g) Excretion of H^+ by kidney, (h) Titratable acidity, (i) Metabolic acidosis, (j) Anion gap.

III. Fill in the blanks

1. The hormone controlling water excretion via kidneys is ___________.
2. The principal cation of extracellular fluid is ___________.
3. The normal osmolality of plasma is ___________.
4. Na^+ reabsorption by renal tubules is increased by the hormone ___________.
5. The most predominant volatile acid generated in the body is ___________.
6. The most important buffer system regulating blood pH is ___________.
7. At a normal blood pH 7.4, the ratio of bicarbonate to carbonic acid is ___________.
8. The body acid load is predominantly eliminated in the form of ___________.
9. The primary defect in metabolic acidosis is a reduction in the plasma concentration of ___________.
10. The respiratory alkalosis is primarily associated with a decrease in the plasma concentration of ___________.

IV. Multiple choice questions

11. The metabolic (endogenous) water is derived by the oxidation of
    (a) Carbohydrate (b) Protein (c) Fats d) All of them.
12. The most predominant anion in the extracellular fluids
    (a) Cl^- (b) HCO_3^- (c) HPO_4^{2-} (d) Protein.
13. The only route through which H^+ ions are eliminated from the body
    (a) Lungs (b) Stomach (c) Kidneys (d) None of them.
14. Name the amino acid from which ammonia is derived in the renal tubular cells which is finally excreted as NH_4^+
    (a) Asparagine (b) Glutamine (c) Glutamate (d) Aspartate.
15. The anion gap refers to the unmeasured plasma anion concentration (in the laboratory) and is represented by
    (a) Proteins and organic acids (b) Phosphate and sulfate (c) Urate (d) All of them.

Tissue Proteins and Body Fluids

22

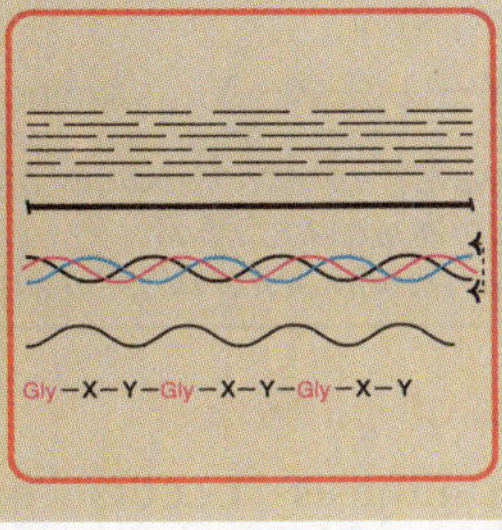

The protein, collagen, speaks

"I am the most abundant protein in mammals;
Triple helical in structure, with distinct types;
Predominantly composed of glycine and proline;
I give strength, support and shape to tissues."

The body possesses a vast number of proteins designed with specific structures to perform specialized functions. A selected few of the most important proteins that are intimately connected with the tissue structure and functions are briefly described in this chapter. In addition, the body fluids are also discussed.

CONNECTIVE TISSUE PROTEINS

The connective tissue or ***extracellular matrix (ECM)*** refers to the complex material surrounding the mammalian cells in tissues. The ***major protein*** components of ECM include ***collagen***, ***elastin***, ***fibrillin***, fibronectin, laminin and proteoglycans. Besides these proteins, the ***structural proteins***, ***namely keratins*** are also described.

COLLAGEN

Collagen is the most abundant protein in mammals, comprising approximately one-third of the total body protein. Collagen is the predominant component of the connective tissue, although its distribution varies in different tissues. For instance, collagen forms 90% of the organic matrix of bones, 85% of tendons, 70% of skin, and 4% of liver.

Functions of collagen

1. Being a major component of the connective tissue, collagen ***gives strength***, ***support*** and ***shape to the tissues***.

2. Collagen contributes to proper ***alignment of cells***, which in turn helps in ***cell proliferation***, and their differentiation to different tissues and organs.

3. Collagen (that is exposed in blood vessels) contributes to thrombus formation.

Types of collagen

Collagen is not a single homogeneous protein, but a group of structurally related and genetically distinct proteins. In humans, ***at least 19 different***

types of collagens, composed of 30 distinct polypeptide chains (encoded by separate genes), have been identified. The types of collagen are numbered (by Roman numerals) as ***I, II...XIX***. The different types of collagen are suited to perform specialized functions in tissues. For instance, collagens type I and type II are respectively found in skin and bone.

Structure of collagen

In principle, all types of collagen are ***triple helical structures***. The triple helix may occur throughout the molecule, or only a part of the molecule.

Type I mature collagen, containing about 1000 amino acids (for each polypeptide chain) possesses triple helical structure throughout the molecule. It is composed of ***three similar polypeptide chains twisted around each other to form a rod like molecule*** of 1.4 nm diameter, and about 300 nm length (***Fig.22.1***). The amino acid composition of collagen is unique. Approximately $^1/_3$rd of the amino acids are contributed by glycine i.e. ***every third amino acid in collagen is glycine***. Hence, the repetitive amino acid sequence of collagen is represented by $(Gly\text{-}X\text{-}Y)_n$, where X and Y represent other amino acids. Thus, collagen may be regarded as a polymer of glycine-led tripeptide. Among the other amino acids, ***proline*** and ***hydroxyproline*** are present in large quantities (about 100 residues each). These two amino acids ***confer rigidity*** to the collagen molecule.

The triple helical structure of collagen is stabilized by an extensive network of hydrogen bonds, covalent cross-links, electrostatic and hydrophobic interactions, and van der Waals forces.

The strength of the collagen fibers is contributed by the covalent cross links formed between lysine and hydroxylysine residues. The degree of collagen cross-linking increases with age. Thus, ***in older people***, the ***collagen containing tissues*** (e.g. skin, blood vessels) ***become less elastic and more stiff, contributing to health complications***.

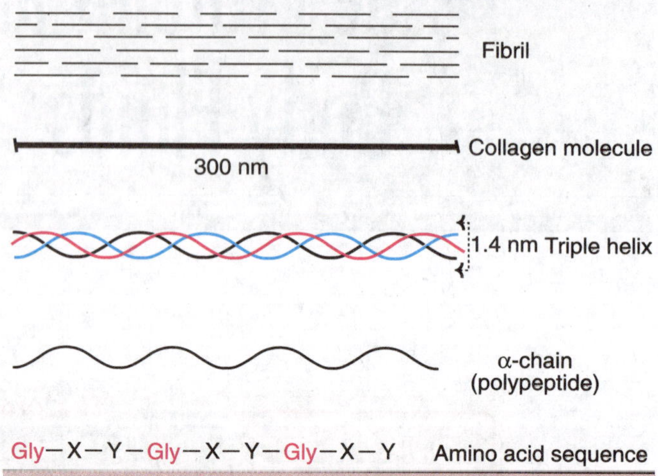

Fig. 22.1 : *A diagrammatic representation of the structure of collagen and fibril (X and Y represent amino acids other than glycine)*

Biosynthesis of collagen

Collagen synthesis occurs in fibroblasts, and related cells e.g. osteoblasts in bones, chondroblasts in cartilage, odontoblasts in teeth.

Collagen is synthesized on the ribosomes in a precursor form, namely ***preprocollagen***. This contains a signal peptide which directs the protein to reach the endoplasmic reticulum (ER). In the ER, the signal peptide is cleaved to form ***procollagen***. The latter undergoes extensive post-translational modifications (hydroxylation and glycosylation) and disulfide bonds formation. The procollagen so formed is secreted into the extracellular medium, and subjected to the action of aminoproteinase and carboxy-proteinase to remove the terminal amino acids. This is followed by a spontaneous assembly of the polypeptide chains to form triple helical structures of collagen.

Collagen and aging : Collagen production declines with increasing age. This results in wrinkles and sagging skin, besides muscle weakness.

Abnormalities associated with collagen

The biosynthesis of collagen is a complex process, involving at least ***30 genes*** (in humans), and about ***8 post-translational modifications***. ***Collagenopathies*** are inherited diseases due to gene mutations, linked with collagen formation. A few of them are listed next.

- **Ehlers-Danlos syndrome (EDS)**–a group of inherited disorders characterized by ***hyper-extensibility of skin***, and abnormal tissue fragility.
- **Alport syndrome**–due to a defect in the formation of type IV collagen fibres found in the basement membrane of renal glomeruli. The patients exhibit hematuria.
- **Osteogenesis imperfecta (brittle bone disease)**–characterized by ***abnormal fragility of bones*** due to decreased formation of collagen.
- **Epidermolysis bullosa**–due to alteration in the structure of type VII collagen. The victims exhibit ***skin breaks and blisters*** formation even for a minor trauma.

Scurvy : This is a disease due to the ***deficiency of vitamin C*** (ascorbic acid). Although not a genetic disease, scurvy is related to the ***improper formation of collagen***, hence referred here (***vitamin C is needed for the post-translational modifications of collagen***).

Lathyrism : It is a ***disease of bone deformities*** caused by the consumption of Kesari dal (*Lathyrus sativa*) in some parts of India. The toxic compound namely β-oxalyl aminoalanine (BOAA), found in kesari dal, interferes with the cross-linking of lysine amino acids in collagen. BOAA is found to inhibit enzyme lysyl oxidase.

Pharmaceutical applications of collagen

Collagen is used in cosmetic surgeries as a healing aid for patients of burns, and in bone graft. Collagen is also employed in tissue regeneration, wound healing etc.

ELASTIN

Elastin is another important (besides collagens) ***connective tissue protein***. It is mainly responsible for the extensibility and ***elasticity of tissues***. Elastin is found in large quantities in lungs, arterial blood vessels, elastic ligments etc.

Abnormalities associated with elastin

- **Williams syndrome** is a genetic disease due to impairment in elastin synthesis. The connective tissue and central nervous system are affected.
- ***Decreased synthesis of elastin*** is found in ***aging of skin*** and pulmonary emphysema.

FIBRILLIN

Fibrillin is a ***structural component of myofibrils*** found in various tissues.

Marfan syndrome is a genetic disorder due to a mutation in the gene for fibrillin. It is characterized by ***hyperextensibility of joints*** and skeletal system. Consequently, the patients of Marfan syndrome are ***tall***, and have ***long digits***. They may also have cardiovascular complications. Some researchers believe that Abraham Lincoln was a victim of Marfan syndrome.

FIBRONECTIN

Fibronectin, a ***glycoprotein***, is closely involved in the interaction of cells with extracellular matrix. It actively participates in cell adhesion and cell migration. In general, ***tumor cells are deficient*** in fibronectin which results in the lack of adhesion among the tumor cells that may often lead to metastasis.

KERATINS

Keratins are structural ***proteins*** found ***in hair, skin, nails, horns*** etc. The 3 polypeptides of keratin form α-helical structure and are held together by disulfide bonds. The toughness and strength of keratin are directly related to the number of disulfide bonds. Thus, the ***harder keratin possesses more disulfide bonds***.

Hair waving (curling)

When the hair is exposed to moist heat, the α-helices of α-keratin can be stretched. This results in the formation of β-conformation from α-helices. On cooling, the hair structure is reverted back to α-conformation. This property of ***α- and β-conformations of keratin*** is exploited in hair waving or curling.

The hair to be curled is first bent to appropriate shape. By applying a reducing agent, the disulfide bonds (of cystine) are converted to sulfhydyl groups (cysteine). This results in the uncoiling of α-helical structure. After some time, th reducing agent is removed, and an oxidizing agent is added. This allows the formation of

Medical Concepts/ Clinical Correlates

- Improper formation of collagen is associated with certain genetic diseases e.g. Ehlers-Danlos syndrome (abnormal tissue fragility), osteogenesis imperfecta (abnormal fragility of bones).
- Defective formation of collagen is observed in scurvy, caused by vitamin C deficiency. This results in bleeding of gums and poor wound healing.
- Hair waving (curling) through artificial means is possible with suitable alterations in the structure of keratins.

some new disulfide bonds between cysteine residues (***Fig.22.2***). The hair is now washed and cooled. The desired curls are formed on the hair due to new disulfide bonds and altered α-helical structure of keratin. It may however, be noted that a permanent curling of hair is not possible. The new hair that grows will be the native original hair only (without curls).

CONTRACTILE PROTEINS

The proteins that are involved in the ***movement of body organs*** (e.g. muscle, heart, lung) are regarded as contractile proteins. It is worthwhile to understand the basic structure of muscle before learning the contractile proteins.

STRUCTURE OF MUSCLE

Muscle is the ***single largest tissue*** of the human body. Muscle constitutes about 20% of body mass at birth, 40% in young adults and 30% in aged adults. Three types of muscles are found in vertebrates-skeletal, cardiac and smooth. The skeletal and cardiac muscles are ***striated*** while the smooth muscles are ***non-striated.***

The structure of striated muscle is represented in ***Fig.22.3***. It is composed of bundles of multinucleated muscle fibre cells. Each cell is surrounded by an electrically excitable plasma membrane, the sarcolemma. The muscle fibre cells are long which may extend the entire length of the muscle. The intracellular fluid of fibre cells is the sarcoplasm (i.e. cytoplasm) into which the ***myofibrils*** are embedded. The sarcoplasm is rich in glycogen, ATP, creatine phosphate, and the enzymes of glycolysis.

When the myofibril is examined under electron microscope, alternating dark bands (anisotropic or A bands) and light bands (isotropic or I bands) are observed. The less dense central region of A band is referred to as H band (or H line). A narrow and dense Z line bisects the I band. The region of the muscle fibre between two Z lines is termed as sarcomere. ***Sarcomere is the functional unit of muscle.***

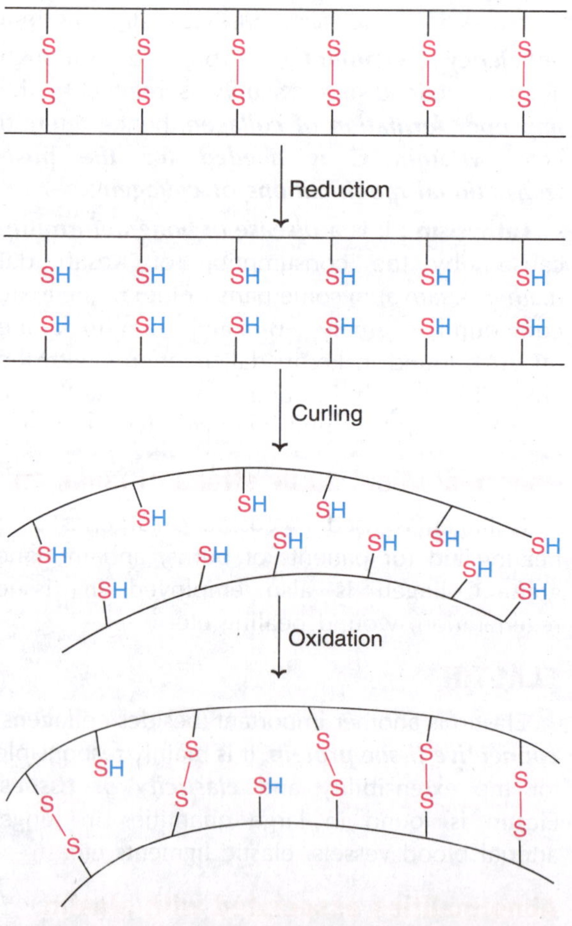

Fig. 22.2 : *A diagrammatic representation of hair waving with suitable alterations in keratin structure (— S— S corresponds to disulfide bonds of cystine; — SH indicates sulfhydryl groups of cysteine)*

During the course of ***muscle contraction***, the thick and thin filaments slide over each other (***sliding filament model*** of muscle contraction). Consequently, the H bands and I bands shorten. However, there is no change in the length of thick and thin filaments. The length of sarcomere which is around 2300 nm in an extended form of myofibril is reduced to 1500 nm in a contracted form (***Fig.22.3B***).

MUSCLE PROTEINS

More than 20% of the muscle mass is composed of proteins. This is largely contributed by structural proteins namely actin, myosin, and the actin cross-linking proteins, tropomyosin and troponin. Muscle also contains other proteins – myoglobin, collagen, enzymes etc. The term ***sarcopenia*** is used to indicate the ***loss of skeletal muscle mass with age***.

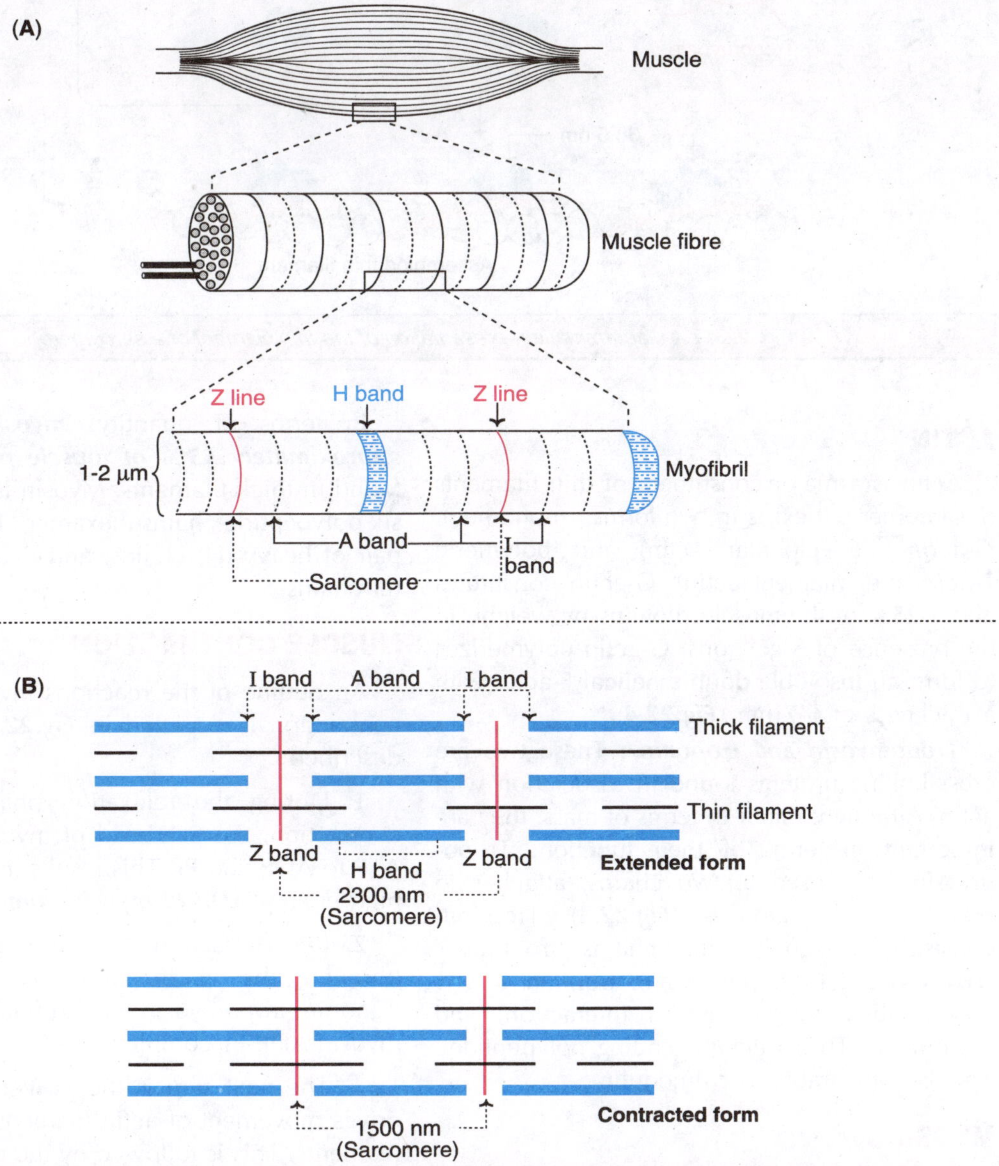

Fig. 22.3 : (A) Structure of myofibril of a straited muscle (B) Arrangement of filaments of myofibril in extended and contracted form. (Note : The length of sarcomere is reduced from 2300 nm to 1500 nm during contraction)

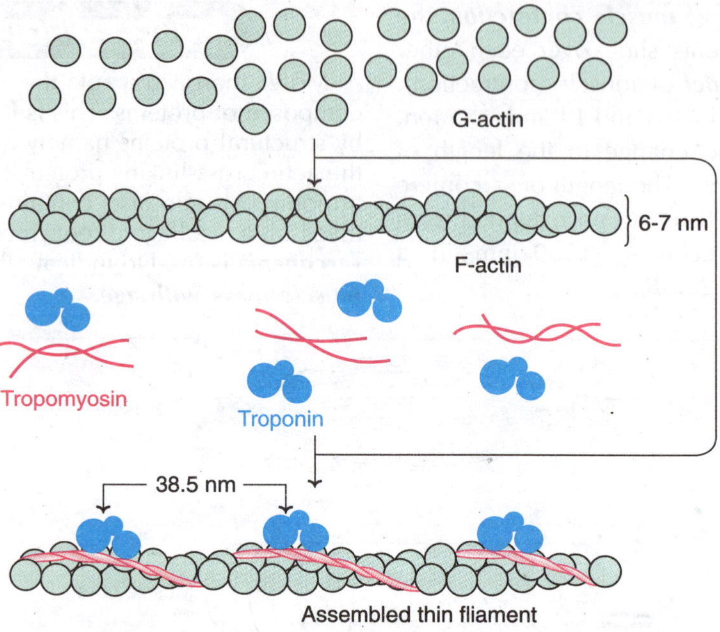

***Fig. 22.4 :** A diagrammatic representation of the thin filament of a sarcomere.*

ACTIN

Actin is a major constituent of thin filaments of sarcomere. It exists in two forms – monomeric ***G-actin*** (i.e. ***g***lobular actin) and polymeric ***F-actin*** (i.e. ***f***ilament actin). G-actin constitutes about 25% of the muscle proteins by weight. In the presence of Mg^{2+} ions, G-actin polymerizes to form an insoluble double helical F-actin with a thickness of 6-7 nm (***Fig.22.4***).

Tropomyosin and troponin **:** These two are cross-linking proteins found in association with actin. Although, minor in terms of mass, they are important in terms of their function. Tropomyosin, composed of two chains, attaches to F-actin in the grooves (***Fig.22.4***). Troponin consists of three polypeptide chains – troponin T (TpT binding to ***t***ropomyosin), tropinin I (TpI that ***i***nhibits F-actin myosin interaction) and troponin C (TpC, ***c***alcium binding polypeptide). TpC is comparable to calmodulin.

MYOSINS

Myosins are actually a family of proteins with about 15 members. The myosin that is predominantly present in muscle is myosin II.

In terms of quantity, ***myosin constitutes approximately 55% of muscle protein***, and is found in thick filaments. Myosin is composed of six polypeptide chains (hexamer). It contains one pair of heavy (H) chains, and two pairs of light (L) chains.

MUSCLE CONTRACTION

An outline of the reactions involving muscle contraction is depicted in ***Fig.22.5***, and briefly described

1. During the relaxation phase of muscle contraction, the S-1 head of myosin hydrolyses ATP to ADP and Pi. This results in the formation high energy ***ADP-Pi myosin complex***.

2. On contraction, the muscle gets stimulated (through the participation of actin, Ca^{2+}, troponin, tropomyosin etc.) to finally form actin-myosin-ADP-Pi complex.

3. The next step is the ***power stroke*** which drives movement of actin filaments over myosin filaments. This is followed by the release of ADP and Pi, and a conformation change in myosin. The actin-myosin complex is in a low energy state.

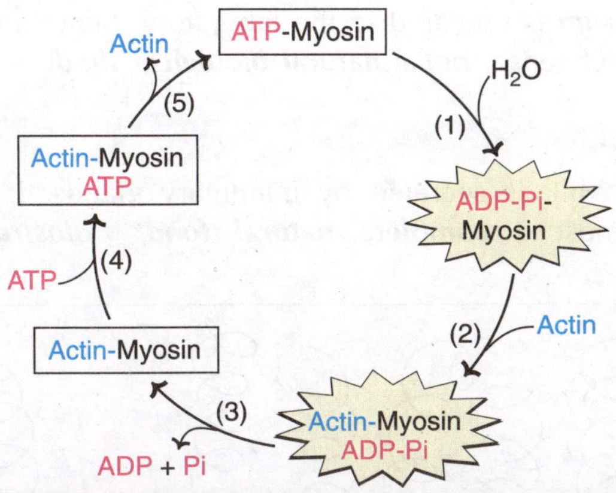

Fig. 22.5 : *Major biochemical events occurring during a cycle of muscle contraction and relaxation (The numbers 1-5 represent the steps in muscle contraction; The zig zag rounds indicate high energy states).*

4. A fresh molecule of ATP now binds to form actin-myosin ATP complex.

5. Actin is released, as myosin-ATP has low affinity for actin. This step is crucial for relaxation which is dependent on the binding of ATP to actin-myosin complex.

A fresh cycle of muscle contraction and relaxation now commences with the hydrolysis of ATP and the formation of ADP-Pi-myosin complex. It has to be noted that it is ultimately the ATP that is the immediate source of energy for muscle contraction.

Sources of ATP for muscle contraction

ATP is a constant source of energy for muscle contraction and relaxation cycle. ATP can be generated from the following ways.

- By substrate level phosphorylation of glycolysis using glucose or glycogen.
- By oxidative phosphorylation.
- From creatine phosphate.

OTHER PROTEINS OF MUSCLE

There are a large number of other proteins that are involved in the structure and functions of muscle. These include titin, nebulin, dystrophin, calcineurin and desmin. ***Titin is the largest protein known***. The gene coding for dystrophin is the largest gene (2,300 bp).

Muscular dystrophy

Muscular dystrophy is a ***hereditary disease*** in which ***muscles progressively deteriorate***. This is caused by mutations in the gene (located on X-chromosome) coding for the protein ***dystrophin***.

PROTEIN MISFOLDING AND DISEASES

The process of protein folding is complex and has been briefly described in ***Chapter 25***. Sometimes, improperly folded proteins may be formed (either spontaneous or by gene mutations). Such misfolded proteins usually get degraded within the cell. However, as the individuals age, the misfolded proteins accumulate and cause a number of diseases. Prion diseases and amyloidosis, two groups of diseases due to protein misfolding are briefly discussed.

Prion diseases

The term prion represents ***pro***teinous ***in***fectious agents. ***Prion proteins*** (***PrP***) are the altered forms of normal proteins. However, no differences in the primary structure (i.e. amino acid sequence) and post-translational modifications are observed.

Certain changes in three-dimensional structure are seen in prion proteins. The major alteration is the replacement of α-helices by β-sheets in PrP. This confers resistance to proteolytic digestion of prion proteins. PrP are highly infectious, and can act as template to convert non-infectious proteins (with α-helices) to infectious forms (***Fig.22.6***). This process continues in an exponential manner to accumulate a large number of prion proteins in tissues.

Prion proteins are implicated as causative agents in the following diseases.

- Transmissible spongiform encephalopathies (TSEs) and ***Creutzfeldt Jacob disease*** in humans.
- ***Scrapie*** in sheep
- ***Bovine spongiform encephalopathy*** (popularly known as ***mad cow disease***) in cattle.

Kuru is an interesting prion disease. It was first reported in Papau New Guinea in the tribal people who practice ***cannibalism*** (they eat the brains of the dead people).

As of now, there is no treatment for prion diseases. Transmissible spongiform encephalopathies are invariably fatal in humans.

Amyloidosis

The term amyloids is used to refer to the ***altered proteins*** (with ***β-sheets***) that ***accumulate in the body***, particularly in the nervous system. Amyloids are formed by protein misfolding or due to gene mutations. They are not infectious agents as prion proteins. However, as the age advances, amyloids accumulate, and they have been implicated in many degenerating diseases. A total of at least 15 different proteins are involved in amyloidosis.

Alzheimer's disease is a ***neurodegenerative disorder***, affecting about 5-10% of the people above 60 years of age. It is characterized by memory loss, confusion, hallucinations, personality changes with abnormal behaviour. As the disease progresses, the patient may enter a vegetative state, and may die after 10 years after the onset of the disease manifestations. The accumulation of amyloids (in the form of amyloid plaque) has been clearly demonstrated in the patients of Alzheimer's disease.

A specific protein, namely β-amyloid which is prone for self aggregation is believed to be the causative agent of Alzheimer's disease, β-Amyloid is formed from a conformational transformation of α-helix. Apolipoprotein E promotes the conformational change of α-amyloid to β-amyloid.

BODY FLUIDS

The specialized fluids of the body are ***milk, cerebrospinal fluid, amniotic fluid, pleural fluid, aqueous humor, sweat and tears. In a broader perspective, blood and plasma also biological fluids***. Their biochemical importance is discussed elsewhere (Hemoglobin, ***Chapter 10***; Plasma proteins, ***Chapter 9***; Diagnostic enzymes, ***Chapter 6***; Acid-base balance, ***Chapter 21***). Urine is an excretory biological fluid. (**Note** : ***Serum*** is prepared in the laboratory, hence in a strict sense, ***not a natural biological fluid***).

MILK

Milk is secreted by mammary glands. It is almost a ***complete natural food***. ***Colostrum***

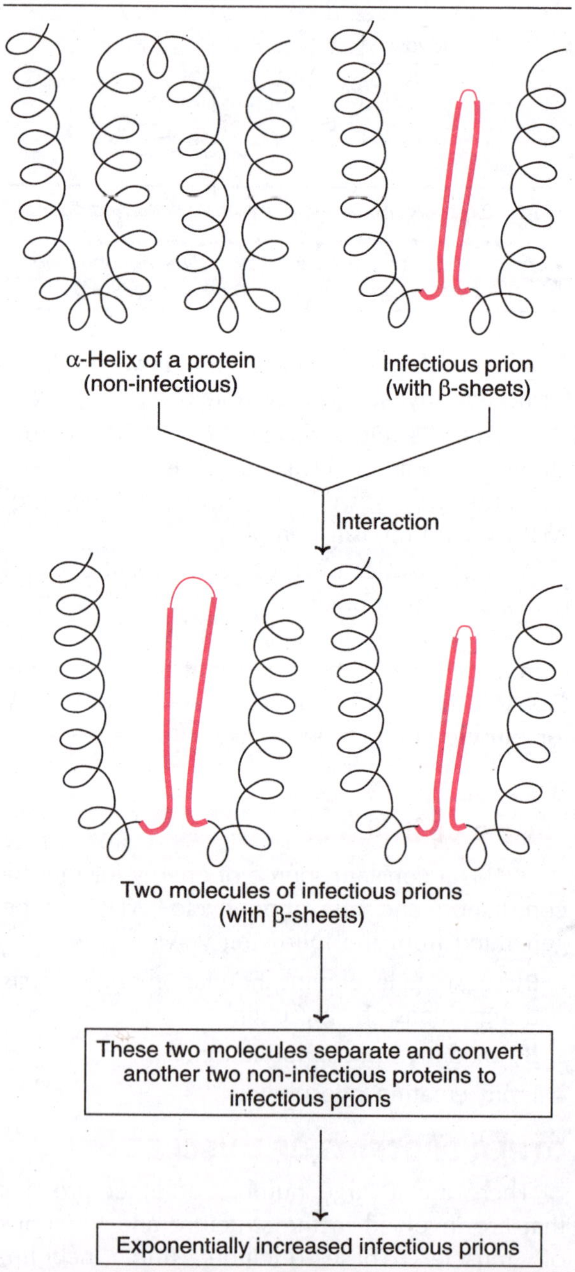

Fig. 22.6 : A model for the formation of infectious prions (Red thick lines represent β-sheets in protein).

refers to the mother's milk secreted during the first few days after delivery. Milk is the only food for the offsprings of mammals on their birth.

COMPOSITION OF MILK

The major constituents of milk in different species—human, cow, buffalo and goat are given in ***Table 22.1***. Water is the major constituent, with a concentration in the range of 83–87%, depending on the species. The remaining 13–17% is made up of solids–carbohydrates, lipids, proteins, minerals and vitamins.

Carbohydrates in milk

Milk contains the disaccharide ***lactose*** which imparts sweetness. Human milk has a higher concentration of lactose (7.5%) compared to milk of other species. Thus, human milk is sweet enough for the babies to relish. Milk sugar (lactose) serves two major functions.

1. It provides galactose, a structural unit for the growing infant.

2. In the intestine, it gets metabolized to lactic acid which eliminates harmful bacteria.

Lipids in milk

The lipids in the milk are dispersed as small globules. Milk fat is mainly composed of ***triacylglycerols***. Mono- and diacylglycerols are also present in trace quantities. The fatty acids found in milk (i.e. in TG) are mostly medium or short chain, and saturated e.g. palmitic acid, myristic acid, stearic acid, lauric acid and butyric acid. Oleic acid, an unsaturated fatty acid, is also present.

Proteins in milk

The major milk proteins are ***casein*** (about 80%) and ***lactalbumin***. Small concentrations of enzymes (proteases, lipase, lysozyme) and immunoglobulins are also found.

Milk casein (a phosphoprotein) is almost a complete protein (next to egg albumin), containing all the essential amino acids. It is present in milk in the form of aggregates called micelles. The ***white colour of milk*** is due to the dispersion of ***calcium caseinate micelles***.

Whey proteins : If milk is acidified, casein gets precipitated at isoelectric point (pH 4.7). The supernatant fluid contains whey proteins (20% of milk proteins). These include lactalbumin, lactoglobulin and various enzymes.

Minerals in milk

Milk is ***rich in calcium, magnesium***, phosphorus, sodium, potassium and chlorine. However, milk is a ***poor source of iron*** and ***copper***.

Table 22.1 Composition of milk in different species

Constituent	Human	Cow	Buffalo	Goat
Water	87.6	87.2	83.5	87.0
Total solids (g/dl)	12.4	12.8	16.5	13.0
Carbohydrates (g/dl)	7.5	4.4	5.4	4.6
Lipids (g/dl)	3.8	3.8	6.5	3.5
Proteins (g/dl)	1.1	3.3	4.3	3.7
Calcium (mg/dl)	35	150	160	175
Magnesium (mg/dl)	2.2	13	10	8
Phosphorus (mg/dl)	16	100	100	70
Sodium (mg/dl)	15	60	60	50
Potassium (mg/dl)	55	140	130	85

Vitamins in milk

Both fat soluble and water soluble vitamins are found in good concentration in milk. However milk is ***deficient in vitamin C***.

Calorific value of milk

Due to variability in the nutrient composition (carbohydrates, fats and proteins), the calorific value of milk from different species varies. Thus, human milk can provide about 70 Cal/100 ml, while for ***buffalo milk, it is around 95 Cal/100 ml***.

CEREBROSPINAL FLUID (CSF)

Cerebrospinal fluid is a clear, colourless liquid formed within the cavities (ventricles) of brain and around the spinal cord. CSF originates in the choroid plexus (as an ultrafiltrate of plasma) and returns to blood through arachnoid villi. About 500 ml of CSF is formed everyday. However, at any given time, there is about 120–150 ml CSF in the system. Further, CSF is completely replaced about three times a day.

Functions of CSF

As the brain has no lymphatic system, CSF drains into the ventricular system and moves into spaces surrounding the brain and spinal cord. The major functions of CSF are listed.

Table 22.2 Normal composition of cerebrospinal fluid

Parameter	Description/concentration
Volume	90–150 ml
Appearance	Clear and colourless
Specific gravity	1.006–1.008
Osmolality	280–290 mOsm/kg
Total cell count (lymphocytes)	0–5 × 10^6/l
pH	7.3–7.4
Protein	15–45 mg/dl
A/G ratio (albumin/globulin)	8 : 1
Glucose	45–85 mg/dl
Chloride	118–130 mEq/l
Calcium	2.1–2.7 mEq/l
Sodium	145–155 mEq/l
Potassium	2.0–3.5 mEq/l

- CSF ***serves as a hydraulic shock absorber***. It can diffuse the force from a hard blow to the skull that might otherwise cause severe injury.
- It helps in the regulation of intracranial pressure.
- It is believed that CSF influences the hunger sensation and eating behaviours.

Collection of CSF

Cerebrospinal fluid is usually collected by a spinal puncture for the purpose of biochemical analysis. The puncture is performed in the lumbar region, between the third and fourth, or between the fourth and fifth lumbar vertebrae.

Composition of CSF in health and disease

The normal composition of cerebrospinal fluid is given in the ***Table 22.2***. From the diagnostic point of view, the total cell count of lymphocytes (Reference : 0-5 × 10^6/1), protein concentration (15–45 mg/dl) and glucose concentration (45–85 mg/dl) are important.

In the ***Table 22.3***, the major alterations in the CSF in the disease states are given. The total ***cell count*** and ***protein content*** are ***increased*** while ***glucose*** concentration is ***reduced in tuberculosis meningitis***. In case of brain tumors, there is no change in total cell count while the protein concentration may be marginally increased.

The colour and appearance of CSF is sometimes a guiding factor in the disease diagnosis. For instance, CSF is opalescent and slightly yellow coloured in tuberculosis meningitis.

AMNIOTIC FLUID

Amniotic fluid is a ***liquid produced by the membranes and the fetus. It surrounds the fetus throughout pregnancy***. The volume of amniotic fluid increases with the gestational age. Thus, the volume increases from 30 ml (at 2 weeks of gestation) to 350 ml (at 20 weeks), and thereafter to 500–1000 ml. Amniotic fluid is almost clear with some desquamated fetal cells and a little lipid.

Medical Concepts/ Clinical Correlates

- Muscular dystrophy occurs due a mutation in the gene coding for the protein dystrophin.
- Protein misfolding results in prion diseases (e.g. mad cow disease) and amyloidosis (Alzheimer's disease).
- Biochemical analysis of CSF is useful for the diagnosis of certain diseases – tuberculous meningitis (increased protein and decreased glucose concentrations).
- Amniotic fluid is analysed to assess fetal maturity (creatinine concentration > 1.6 mg/dl), lung maturity (lecithin–sphingomyelin ratio > 2 : 1) and for the prenatal diagnosis of congenital disorders (e.g. hemophilia, Down's syndrome).

Functions of amniotic fluid

- It provides physical protection to the fetus.
- Amniotic fluid is a medium for the exchange of various chemicals.

Diagnostic importance of amniotic fluid

The term ***amniocentesis*** is used for the process by which amniotic fluid is collected for analysis. The diagnostic importance of amniotic fluid is given in the next column.

Assessment of fetal maturity : Fetal maturity can be assessed by cytological staining of fat cells, and estimation of ***creatinine*** concentration (> ***1.6 mg/dl indicates fetal maturity***).

Lung maturity : The fetal lung maturity is evaluated by measuring ***lecithin–sphingomyelin (L/S) ratio***. A L/S ratio of 2 : 1 or more indicates lung maturity. If L/S ratio is less than 1.2 : 1, it is better to delay the induced delivery until the lung has become more mature.

Diagnosis of congenital disorders : Amniotic fluid analysis is useful for the prenatal diagnosis of congenital disorders. Some of the important ones are listed.

- ***Chromosomal disorders*** such as Down's syndorme.
- ***Metabolic disorders*** e.g. cystic fibrosis.
- ***Sex-linked disorders*** e.g. hemophilia.
- ***Enzyme defects*** e.g. Tay-Sachs disease.

Assessment of hemolytic diseases : Estimation of bilirubin in amniotic fluid is useful to evaluate the severity of hemolytic diseases.

Measurement of α-fetoprotein : ***Increased*** levels of α-fetoprotein (normal 15–40 μg/ml during gestation; at 40 weeks < 1.0 μg/ml) are associated with ***neural tube defects, fetal distress***, Turner syndrome. Elevated α-fetoprotein may also indicate a possible death of the fetus.

PLEURAL FLUID

Pleural fluid is the filtrate of plasma, and is present in a minimal quantity in pleural cavity.

Table 22.3 Changes in cerebrospinal fluid in the disease states

Disease	Colour and appearance	Total cell count	Protein	Glucose
Normal	Clear and colourless	$0–5 \times 10^6$/l	15–45 mg/dl	45–85 mg/dl
Tuberculosis meningitis	Opalescent and slightly yellow	Increased	Increased	Relatively low
Bacterial meningitis	Opalescent and turbid	Markedly increased	Markedly increased	Markedly decreased
Brain tumour	Clear and colourless	No change	Increased	Low
Subarachnoid hemorrhage	Slightly blood colour	RBC and WBC present	Increased	Almost normal

The amount of pleural fluid increases in disease states due to ***pleural effusion***.

Pleural fluid may be ***transudate*** or ***exudate***, based on the composition. If the ratio of protein content between pleural fluid and plasma is less than 0.5, it is transudate. If this ratio is more than 0.5, the pleural fluid is exudate.

Transudate is fluid-like, with ***specific gravity <1.015***. It will not clot, and is not associated with inflammation. Transudate is observed in nephrotic syndrome (due to low osmotic pressure of plasma), and congestive cardiac failure (increased hydrostatic pressure).

Exudate is viscous in nature ***with specific gravity >1.015***. It may clot, and is associated with inflammation or malignancy. Exudate accumulates in various infections (tuberculosis, pneumonia, rheumatoid arthritis) and some cancers (e.g. lung cancer).

Biochemical analysis of exudate is useful in certain disease states

- Amylase activity increased in pancreatitis.
- Rheumatoid factor elevated in rheumatoid arthritis.
- Carcinoembryonic antigen (CEA) increased in malignancy.

AQUEOUS HUMOR

Aqueous humor is the ***fluid that fills the anterior chamber of the eye***. This fluid is responsible for maintaining the intraocular tension. Aqueous humor, secreted by the ciliary body, enters the anterior chamber. Blockade in the flow of aqueous humor causes ***glaucoma due to increased intraocular pressure***.

SUMMARY

- *The major proteins of connective tissue are collagen, elastin, fibrillin, laminin and proteoglycans. Among these, collagen is the most abundant, constituting one-third of the total body proteins.*
- *Type I mature collagen is a triple helical structure i.e. contains three polypeptide chains each with about 1000 amino acids. The repetitive amino acid sequence of collagen is* $(Gly\text{-}X\text{-}Y)_n$. *Glycine constitutes about 1/3 rd of the amino acids while X and Y represent other amino acids.*
- *Actin, myosin, tropomyosin and troponin are the major contractile proteins found in muscles. The muscle contraction and relaxation occur due to the active involvement of these proteins. ATP is the immediate source of energy for muscle contraction.*
- *Proper folding of proteins is essential for their structure. Misfolding of proteins results in certain diseases e.g. mad cow disease, Alzheimer's disease.*
- *Milk is almost a complete food with various nutrients—carbohydrates, lipids, proteins, vitamins and minerals. However, milk is deficient in vitamin C, iron and copper.*
- *CSF is an ultrafiltrate of plasma. In the disease, tuberculosis meningitis, the total cell count and protein concentration are increased, while glucose concentration is decreased.*
- *Amniotic fluid is a liquid produced by the fetus. Its biochemical analysis is important for the diagnostic purpose – assessment of fetal maturity.*

THE READER MUST REFER APPENDIX VI (Page Nos : 738 - 756) FOR RELATED CLINICAL CASE STUDIES (Nos : 68, 72)

Self-Assessment Exercises

I. Essay questions

1. Give an account of the structure and functions of collagen. Add a note on the abnormalities associated with collagen.
2. Describe the muscle proteins, and muscle contraction.
3. Discuss the protein misfolding and various diseases related to it.
4. Give an account of the composition of milk.
5. Describe the functions and composition of cerebrospinal fluid. Add a note on the alterations in CSF in diseased states.

II. Short notes

(a) Biosynthesis of collagen, (b) Collagen and scurvy, (c) Elastin, (d) Light and heavy meromyosins, (e) Prion diseases, (f) Amyloidosis, (g) Hair waving, (h) Vitamins and minerals in milk, (i) Collection of CSF, (j) Amniotic fluid.

III. Fill in the blanks

1. The most abundant protein in mammals ____________.
2. The amino acid that contributes to one-third of the total number of amino acids in collagen ____________.
3. The toxic compound that interferes with the cross-linking of lysine in collagen, causing lathyrism ____________.
4. Marfan syndrome is a genetic disorder due to a mutation of the gene coding for ____________.
5. Name the carbohydrates associated with the structure of proteoglycans ____________.
6. The region of the muscle fibre between two Z lines is termed as ____________.
7. Name the major protein found in the structure of thin filaments of sarcomere ____________.
8. The white colour of milk is due to the dispersion of ____________.
9. Name the vitamin deficient in milk ____________.
10. The fetal lung maturity is evaluated by measuring ____________ ratio.

IV. Multiple choice questions

11. The number of polypeptide chains present in collagen
    (a) 1 (b) 2 (c) 3 d) 4.
12. The functional unit of muscle
    (a) Fibre cell (b) Myofibril (c) H band d) Sarcomere.
13. The immediate source of energy for muscle contraction
    (a) ATP (b) Creatine phosphate (c) GTP d) Phosphoenol pyruvate.
14. One of the following minerals is lacking in milk
    (a) Calcium (b) Sodium (c) Iron d) Potassium.
15. One of the following biochemical parameters is increased in tuberculosis meningitis
    (a) Glucose (b) Protein (c) Sodium d) Chloride.

Nutrition

23

The nutrition speaks

"Some eat to live,
And some live to eat!
My function is
To cater for all."

Whether a man eats for living or lives for eating, food is his prime concern. ***Nutrition*** may be defined as the ***utilization of food by living organisms.*** Biochemists have largely contributed to the science of nutrition.

Nutrition significantly promotes man's development, his health and welfare. The subject nutrition, perhaps, is the most controversial. This is due to the fact that nutrition is concerned with food, and everyone feels competent enough to talk like an expert on nutrition. Further, ***high public awareness*** and the controversial reports by scientists also contribute to the controversy.

Methodology in nutrition : Most of the existing knowledge on nutrition is originally derived from ***animal experimentation***. This is despite the fact that there may exist several differences in the biochemical composition between man and animals! For instance, some animals can synthesize ascorbic acid while man cannot do so.

Study of human nutrition : The study of nutrition may be logically divided into three areas—***ideal nutrition***, ***undernutrition*** and ***overnutrition***. Ideal nutrition is the concern of everyone. Undernutrition is the prime concern of developing countries while overnutrition is a serious concern of developed countries. However, the fact remains that ***worldwide***, there are ***more overweight and obese people than undernourished people***.

Macronutrients and micronutrients

The nutrients necessary to support growth and maintenance of living organisms are broadly divided into two groups—macronutrients and micronutrients. The macronutrients forming the ***main bulk of food are carbohydrates, proteins and fats***. The ***micronutrients*** are ***required in small quantities*** and they include the various ***vitamins*** and ***minerals***.

Essential nutrients

A nutrient is said to be essential when it is ***required for normal physiological functions*** of the body and ***cannot be synthesized***, and therefore must be supplied through the diet. In case of humans, there are approximately ***8 amino acids, 2 fatty acids, 13 vitamins and 15 minerals*** that are considered ***as essential nutrients***. Besides these, there are certain ***conditionally essential nutrients***, as they are indispensable in developmental stages and pathological states.

A sound knowledge of chemistry and metabolism of foodstuffs (carbohydrates, lipids, proteins, vitamins and minerals) is an essential prerequisite for a better understanding of nutrition. The reader must, therefore, first refer these chapters. The ***principles of nutrition with special reference to energy demands, carbohydrates, fats, proteins, recommended dietary/daily allowances (RDA), balanced diet and nutritional disorders are discussed*** in the following pages.

NUTRITION AND ENERGY SUPPLY

Food is the fuel source of the body. The ingested food undergoes metabolism to liberate energy required for the vital activities of the body.

Energy content of foods

The calorific value (energy content) of a food is calculated from the heat released by the total combustion of food in a calorimeter.

Unit of heat : ***Calorie*** is the unit of heat. One calorie represents the amount of ***heat required to rise the temperature of one gram of water by 1°C*** (i.e. from 15° to 16°C). A calorie is too small a unit. Therefore, it is more conveniently expressed as kilocalories (1,000 times calorie) which is represented by kcal or simply ***Cal*** (with capital ***'C'***).

Table 23.1 Calorific values of foodstuffs

Foodstuff	Energy value (Cal/g) In bomb calorimeter	In the body
Carbohydrate	4.1	4
Fat	9.4	9
Protein	5.4	4
Alcohol	7.1	7

The ***joule*** is also a unit of energy used in some countries. The relationship between calories and joules (J) is

$$1 \text{ Cal } (1 \text{ kcal}) = 4.128 \text{ KJ}$$

The joule is less commonly used by nutritionists.

Caloric value of foods : The energy values of the three principal foodstuffs—carbohydrate, fat and protein—measured in a bomb calorimeter and in the body are given in the ***Table 23.1***. The carbohydrates and fats are completely oxidized (to CO_2 and H_2O) in the body; hence their fuel values, measured in the bomb calorimeter or in the body, are almost the same. Proteins, however, are not completely burnt in the body as they are converted to products such as urea, creatinine and ammonia, and excreted. Due to this reason, calorific value of protein in the body is less than that obtained in a bomb calorimeter. The energy values of ***carbohydrates, fats*** and ***proteins*** (when utilized in the body) respectively, are ***4, 9*** and ***4 Cal/g***.

Alcohol is a recent addition to the calorie (***7 Cal/g***) contribution, as it is a significant dietary component for some people.

It must be noted that the nutrients, namely ***vitamins and minerals***, have ***no calorific value***, although they are involved in several important body functions, including the generation of energy from carbohydrates, fats and proteins.

Respiratory quotient of foodstuffs

The respiratory quotient (***R. Q.***) is the ***ratio of the volume of CO_2 produced to the volume of O_2 utilized in the oxidation of foodstuffs.***

Carbohydrates : The carbohydrates are completely oxidized and their R. Q. is close to 1, as represented below for glucose.

$$C_6H_{12}O_6 + 6O_2 \longrightarrow 6CO_2 + 6H_2O$$

$$\text{R. Q. for carbohydrate} = \frac{CO_2}{O_2} = \frac{6}{6} = 1.$$

Fats : Fats have relatively lower R.Q. since they have a low oxygen content. For this reason, fats require more O_2 for oxidation. The R.Q.

for the oxidation of the fat, tristearin is given below.

$$2\,C_{57}H_{110}O_6 + 163\,O_2 \longrightarrow 114\,CO_2 + 110\,H_2O$$

$$\text{R. Q. for fat} = \frac{CO_2}{O_2} = \frac{114}{163} = 0.7.$$

Proteins : The chemical nature of proteins is highly variable, and this cannot be represented by any specific formula. By indirect measurements, the R.Q. of protein is found to be around 0.8.

Mixed diet : The R. Q. of the diet consumed is dependent of the relative composition of carbohydrates, fats and proteins. For a normally ingested diet, it is around 0.8. R.Q of alcohol is around 0.7 (as that of fat).

UTILIZATION OF ENERGY IN MAN

Man consumes energy to meet the fuel demands of the three ongoing processes in the body.

1. Basal metabolic rate
2. Specific dynamic action
3. Physical activity.

Besides the above three, additional energy supply is needed during growth, pregnancy and lactation.

BASAL METABOLIC RATE

Basal metabolism or basal metabolic rate (BMR) is defined as the ***minimum amount of energy required by the body*** to maintain life ***at complete physical and mental rest*** in the post-absorptive state (i.e. 12 hours after the last meal). It may be noted that ***resting metabolic rate*** (***RMR***) is in recent use for BMR.

Under the basal conditions, although the body appears to be at total rest, several functions within the body continuously occur. These include working of heart and other organs, conduction of nerve impulse, reabsorption by renal tubules, gastrointestinal motility and ion transport across membranes (***Na^+-K^+ pump*** consumes about 50% of basal energy).

Measurement of BMR

Preparation of the subject : For the measurement of BMR the subject should be awake, at complete physical and mental rest, in a post-absorptive state and in a comfortable surrounding (at 25°C).

Measurement : The BMR is determined either by the ***apparatus of Benedict and Roth*** (closed circuit device) or by the ***Douglas bag*** method (open circuit device). The former is more frequently used.

By Benedict-Roth method, the volume of O_2 consumed (recorded on a graph paper) by the subject for a period of 2-6 minutes under basal conditions is determined. Let this be A liters for 6 minutes. The standard calorific value for one liter O_2 consumed is 4.825 Cal.

Heat produced in 6 min = 4.825 × A

Heat produced in one hour = 4.825A × 10

Units of BMR : BMR is expressed as Calories per square meter of body surface area per hour i.e. ***Cal/sq.m/hr***.

For the calculation of body ***surface area***, the simple ***formula*** devised by ***Du Bois and Du Bois*** is used.

$$A = H^{0.725} \times W^{0.425} \times 71.84$$

where A = Surface area in cm^2

H = Height in cm

W = Weight in kg.

To convert the surface area into square meters (m^2), divide the above value (cm^2) by 10,000. Nomograms of body surface area (directly in m^2) from heights and weights are readily available in literature.

Normal values of BMR : ***For an adult man 35–38 Cal/sq. m/hr; for an adult woman 32-35 Cal/sq.m/hr.*** A BMR value between –15% and +20% is considered as normal.

Some authors continue to express BMR as Cal/day. ***For an adult man BMR is around 1,600 Cal/day, while for an adult woman around 1,400 Cal/day.*** This is particularly important for easily calculating energy requirements per day.

Factors affecting BMR

1. **Surface area :** The BMR is ***directly proportional to the surface area***. Surface area is related to weight and height.

2. **Sex :** ***Men*** have marginally ***higher*** (about 5%) BMR than women. This is due to the higher proportion of lean muscle mass in men.

3. **Age :** In infants and growing children, with lean muscle mass, the BMR is higher. ***In adults, BMR decreases*** at the rate of about 2% per decade of life.

4. **Physical activity :** BMR is ***increased*** in persons (notably athletes) with ***regular exercise***, due to increase in body surface area.

5. **Hormones :** Thyroid hormones (T_3 and T_4) have a stimulatory effect on the metabolism of the body and, therefore, BMR. Thus, BMR is ***raised in hyperthyroidism*** and ***reduced in hypothyroidism***. In fact, the measurement of BMR was earlier used to assess thyroid function.

The other hormones such as epinephrine, cortisol, growth hormone and sex hormones increase BMR.

6. **Environment :** In cold climates, the BMR is higher compared to warm climates.

7. **Starvation :** During the periods of starvation, the energy intake has an inverse relation with BMR, a ***decrease*** up to 50% has been reported. This may be an adaptation by the body.

8. **Fever :** Fever causes an ***increase*** in BMR. An elevation by more than 10% in BMR is observed for every 1°C rise in body temperature.

9. **Disease states :** BMR is ***elevated*** in various infections, leukemias, polycythemia, cardiac failure, hypertension etc. In Addison's disease (adrenal insufficiency), BMR is marginally lowered.

10. **Racial variations :** The BMR of Eskimos is much higher. The BMR of Oriental women living in USA is about 10% less than the average BMR of American women.

11. **Drugs :** Certain drugs like caffeine, nicotine, alcohol increase BMR while most anesthetics are known to decrease.

Significance of BMR

BMR is important to calculate the ***calorie requirement*** of an individual and ***planning of diets***. Determination of BMR is useful for the assessment of thyroid function. In hypothyroidism, BMR is lowered (by about – 40%) while in hyperthyroidism, it is elevated (by about + 70%). Certain disease conditions also influence BMR (described above).

SPECIFIC DYNAMIC ACTION

The phenomenon of the ***extra heat production by the body***, over and above the calculated caloric value, when a given ***food is metabolized*** by the body, is known as specific dynamic action (SDA). It is also known as ***calorigenic action*** or ***thermogenic action*** or thermic action (effect) of food.

SDA for different foods : For a food containing 25 g of protein, the heat production from the caloric value is 100 Cal (25 × 4 Cal). However, when 25 g protein is utilized by the body, 130 Cal of heat is liberated. The extra 30 Cal is the SDA of protein. Likewise, consumption of 100 Cal of fat results in 113 Cal and 100 Cal of carbohydrate in 105 Cal, when metabolized in the body. SDA for protein, fat and carbohydrate are 30%, 13% and 5%, respectively. Thus, ***proteins*** possess the ***highest SDA*** while ***carbohydrates*** have the ***lowest***.

SDA for mixed diet : For a mixed diet, the SDA is not an additive value of different foods but it is much less. The presence of fats and carbohydrates reduces the SDA of proteins. Fats are most efficient in reducing SDA of foodstuffs. For a regularly consumed mixed diet, the SDA is ***around 10%***.

Significance of SDA : For the utilization of foods by the body, certain amount of energy is consumed from the body stores. This is actually an expenditure by the body for the utilization of foodstuffs. It is the highest for proteins (30%) and lowest for carbohydrates (5%), and for a mixed diet around 10%. It is, therefore, essential that an additional 10% calories should be added to the total energy needs (of the body) towards SDA. And the diet should be planned, accordingly. (SDA is quite comparable to the handling charges levied by a bank for an outstation cheque).

The higher SDA for protein indicates that it is not a good source of energy. Fat is the best source of energy due to its lowering effect on SDA. However, excessive utilization of fat leads to ketosis.

Mechanism of SDA : The exact cause of SDA is not known. It is generally believed that SDA of foods is due to the energy required for digestion, absorption, transport, metabolism and storage of foods in the body.

Intravenous administration of amino acids or the oral ingestion of proteins gives the same SDA. This shows that the SDA of proteins is not due to their digestion and absorption. Hepatectomy abolishes SDA, thereby indicating that SDA is closely ***connected with the metabolic functions of liver***. The SDA of proteins is primarily to meet the energy requirements for deamination, synthesis of urea, biosynthesis of proteins, synthesis of triacylglycerol (from carbon skeleton of amino acids). It has been demonstrated that certain amino acids (phenylalanine, glycine and alanine) increase the SDA. It is a common experience that consumption of a ***protein-rich diet makes us feel warm and comfortable in cold weather***. This is due to the high SDA of proteins.

Table 23.2 Type of physical activity and energy expenditure (over and above BMR, about 65 Cal/hr).

Physical activity	Energy requirement (Cal/hr)
Sitting (quietly)	25
Standing (quietly)	30
Writing/eating/reading	30
Car driving	60
Typing	75
Household work (dish washing)	80
Walking (slow)	130
Sexual intercourse	140
Cycling (slow)	150
Running (moderate)	500
Swimming	600
Walking upstairs	800

The SDA of carbohydrates is attributed to the energy expenditure for the conversion of glucose to glycogen.

As regards fat, the SDA may be due to its storage, mobilization and oxidation.

PHYSICAL ACTIVITY OF THE BODY

The physical activity of the individual is highly variable. The amount of energy needed for this depends mainly on the duration and intensity of muscular activity. The expenditure of energy for the various physical activities has been calculated (***Table 23.2***).

For the sake of convenience, the individuals are grouped into four categories with regard to their physical activity and the requirement of energy.

Light work — 30–40% of BMR
(teachers, office workers, doctors)

Moderate work — 40–50% of BMR
(housewives, students)

Heavy work — 50–60% of BMR
(agricultural labourers, miners)

Very heavy work — 60–100% of BMR
(construction workers, rickshaw pullers)

Energy requirements of man

As already stated, the three factors—basal metabolic rate, specific dynamic action and physical activity—determine the energy needed by the body. In an individual with light work, ***about 60% of the calories are spent towards BMR, about 30% for physical activity and about 10% to take care of the SDA.***

The daily requirement of energy is rather variable which depends on the BMR (in turn depends on age, sex, body size etc.) and physical activity. As per some rough calculation, caloric requirements of adults per day (Cal/day) are in the following ranges.

Light work — 2,200–2,500
Moderate work — 2,500–2,900
Heavy work — 2,900–3,500
Very heavy work — 3,500–4,000

Non-exercise activity thermogenesis (NEAT)

NEAT, also called as ***non-exercise physical activity (NEPA)*** refers to the ***energy expended for all our routine physical activities other than sleeping*** or ***sports-like exercises***. These include the energy used for walking to work, climbing steps, performing yard work, kitchen work and fidgeting. All these are trivial physical activities, but when put together they substantially add to increase in metabolic rate and burning of calories.

In recent years the daily energy requirements (particularly for people who are not doing any specific and regular exercise), are calculated considering NEAT as follows.

Daily energy requirement = BMR + SDA + NEAT.

Normally, about 60–70% of calories are spent towards BMR, 10–15% towards SDA, and at least 15–30% towards NEAT. In individuals, with very sedentary life, NEAT may be as low as 15% while in active individuals, it may go up to 50%.

NEAT plays an important role in keeping the body in order. ***Defect in NEAT significantly contributes to obesity.***

NUTRITIONAL IMPORTANCE OF CARBOHYDRATES

Dietary carbohydrates are the ***chief source of energy***. They contribute to 60-70% of total caloric requirement of the body. Incidentally, carbohydrate-rich foods cost less.

Carbohydrates are the most abundant dietary constituents, despite the fact that they are ***not essential nutrients to the body***. From the nutritional point of view, carbohydrates are grouped into 2 categories.

1. Carbohydrates ***utilized*** by the body—starch, glycogen, sucrose, lactose, glucose, fructose etc.

2. Carbohydrates ***not utilized*** (***not digested***) by the body—cellulose, hemicellulose, pectin, gums etc.

Among the carbohydrates utilized by the body, starch is the most abundant. The consumption of starch has distinct advantages due to its bland taste, satiety value and slow digestion and absorption. Sucrose (the table sugar), due to its sweetness, can be consumed to a limited extent. ***Excessive intake of sucrose*** causes dental caries, and an increase in plasma lipid levels that is associated with many health complications.

Functions of carbohydrates

1. **Major sources of energy :** Carbohydrates are the principal source of energy, supplying 60–80% of the caloric requirements of the body.

2. **Protein sparing action :** Proteins perform a specialized function of body building and growth. The wasteful expenditure of proteins to meet the energy needs of the body should be curtailed. Carbohydrates come to the rescue and spare the proteins from being misused for caloric purpose.

3. **Absolute requirement by brain :** The brain and other parts of central nervous system are dependent on glucose for energy. Prolonged hypoglycemia may lead to irreversible brain damage.

4. **Required for the oxidation of fat :** Acetyl CoA is the product formed in fatty acid oxidation. For its further oxidation via citric acid cycle, acetyl CoA combines with oxaloacetate, the latter is predominantly derived from carbohydrates. It may therefore be stated ***'Fat burns in a fuel of carbohydrate'***.

5. **Synthesis of pentoses :** Pentoses (e.g. ribose) are the constituents of several compounds in the body e.g. nucleic acids (DNA, RNA), coenzymes (NAD^+, FAD). These pentoses are produced in carbohydrate metabolism.

6. **Synthesis of fat :** Excess consumption of carbohydrates leads to the formation of fat which is stored in the adipose tissue.

7. **Importance of non-digestible carbohydrates :** These are the carbohydrates not utilized by the body. Yet, they are important since they improve bowel motility, prevent constipation, lower cholesterol absorption and improve glucose tolerance (details discussed later).

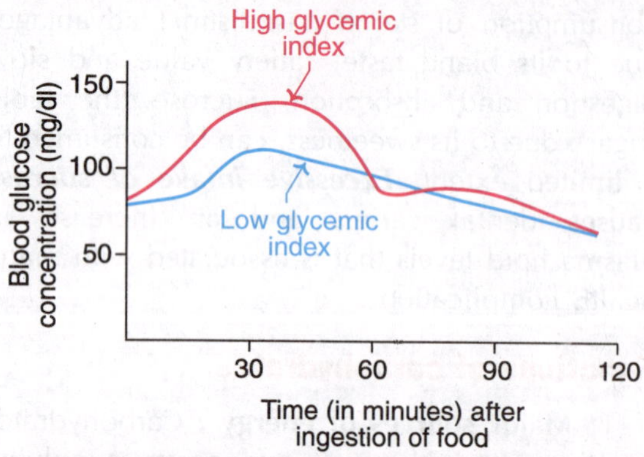

Fig. 23.1 : *The glycemic index curve after the ingestion of two different foods.*

High fructose corn syrups (HFCS)

HFCS are ***produced from glucose*** by employing enzymatic processes that convert glucose into fructose. HFCS contain approximately equal amount of glucose and fructose. They are commonly used as ***substitutes for sucrose in beverages***, including ***soft drinks***, and processed foods.

The composition and metabolism of HFCS and sucrose are similar except that HFCS is ingested as a mixture of monosaccharides. Further, most studies have shown that there is no significant difference between sucrose and HFCS with regard to post-prandial glucose and insulin response.

Glycemic index

There are variations in the increase and fall of blood glucose levels after the ingestion of different carbohydrate containing foods. These quantitative differences are assayed by glycemic index which measures the ***time course of post-prandial glucose concentrations*** from a graph. Glycemic index may be defined as the area under the blood glucose curve after the ingestion of a food compared with the area under the blood glucose curve after taking the same amount of carbohydrate as glucose. It is expressed as percentage.

$$\text{Glycemic index} = \frac{\text{Area under the blood glucose curve after ingestion of a test meal}}{\text{Area under the curve after ingestion of glucose}} \times 100$$

A graphic representation of high and low glycemic indices is depicted in ***Fig. 23.1***.

The glycemic index of a ***complex carbohydrate (i.e. starch) is lower than a refined carbohydrate (i.e. glucose)***. This is explained on the basis of slow digestion and absorption of complex carbohydrates. Further, the glycemic index of carbohydrate is usually lower when it is combined with protein, fat or fiber. The glycemic index of some selected foods is given in ***Table 23.3***.

The food item like ice cream has relatively lower glycemic index. This may be explained on the basis of high fat content which lowers the glucose absorption.

The nutritional importance of glycemic index is controversial. This is due to the fact that the foods with low glycemic index need not be good for health. However, low glycemic index foods usually have higher satiety value (creating a sense of stomachfulness), and thus may be helpful in limiting the caloric intake. Nutritionists are of the opinion that ***foods with high fiber content and low glycemic index*** (e.g. whole grains, fruits, vegetables) should be ***preferred for consumption.***

Sources of carbohydrates

Carbohydrates are abundant in several naturally occurring foods. These include table

Table 23.3 Glycemic index of some selected foods

Food item	Glycemic index
Glucose	100
Carrots	90–95
Honey	80–90
Bread, rice	70–80
Banana, potato	60–70
Sweet potato	50–60
Oranges, apples	40–45
Ice cream, milk	35–40
Fructose	20–25
Soy beans	15–20

Medical Concepts/ Clinical Correlates

- The dietary essential nutrients for humans are approximately 8 amino acids, 2 fatty acids, 13 vitamins and 15 minerals.
- The body at total rest (physical and mental) requires energy to meet the basal requirements such as working of heart, conduction of nerve impulse, membrane transport etc.
- Non-exercise physical activity (NEPA) is important in keeping the body in order. Defect in NEPA significantly contributes to obesity.
- Adequate intake of dietary fiber prevents constipation, eliminates bacterial toxins, reduces GIT cancers, improves glucose tolerance and reduces plasma cholesterol.

sugar (99%), cereals (60–80%), pulses (50–60%), roots and tubers (20–40%) and bread (50–60%).

Requirement of carbohydrates

In a well balanced diet, at least 40% of the caloric needs of the body should be met from carbohydrates.

FIBER IN NUTRITION

The ***complex carbohydrates*** that are ***not digested*** by the human enzymes are collectively referred to as dietary fiber. ***Soluble fibers***, mostly found in fruits and legumes, dissolve in water and form gels (e.g., ***pectins***, gums, mucilages). ***Insoluble fibers***, present in vegetables and grains, adsorb water and swell up (e.g. ***cellulose hemicellulose***, lignin). Certain fibers (e.g. pectins, gums) are digestible by intestinal bacterial ezymes. It may be stated that once regarded as a nutritional waste, a lot of importance is now given to dietary fiber in human health.

Beneficial effects of fiber

1. **Prevents constipation :** Fiber can absorb 10–15 times its own weight of water, by drawing fluid into the lumen of the intestine. This increases ***bowel motility***, and prevents constipation, besides decreasing the risk of hemorrhoids and diverticulosis.

2. **Eliminates bacterial toxins :** Fiber also ***adsorbs toxic compounds*** produced by intestinal bacteria and helps in their easier expulsion.

3. **Decreases GIT cancers :** The ***lower*** incidence of cancers of gastrointestinal tract (e.g. colon and rectum) ***in vegetarians*** compared to non-vegetarians is attributed to dietary fiber.

4. **Improves glucose tolerance :** Fiber improves glucose tolerance by the body. This is mainly done by a diminished rate of glucose absorption from the intestine.

5. **Reduces plasma cholesterol level :** Fiber decreases the absorption of dietary cholesterol from the intestine. Further, fiber binds with the bile salts and reduces their enterohepatic circulation. Thus, degradation of cholesterol to bile salts and its disposal from the body is increased.

6. **Satiety value :** Dietary fiber adds to the weight of the foodstuff ingested and gives a sensation of stomachfulness, giving satiety without consumption of excess calories.

Adverse affects of fiber

Some of the food fads went to the extent of ingesting huge quantities of rice bran to achieve all the benefits of fiber. This led to several complications. In general, the harmful effects are mostly observed in people consuming ***large quantities*** of dietary fiber.

1. Digestion and ***absorption of protein*** are adversely ***affected***.

2. The intestinal absorption of certain minerals (e.g. Ca, P, Mg) is decreased.

3. Intestinal bacteria ferment some fibers, causing ***flatulence*** and often discomfort.

Drinking plenty of water along with fiber is advocated to reduce adverse effects of fiber.

Sources of dietary fiber

Fruits, ***leafy vegetables***, vegetables, whole wheat legumes, rice bran etc., are rich sources of fiber. The ideal way to increase fiber intake is to reduce intake of refined carbohydrates, besides eating vegetables, fresh fruits and whole grains. In general, ***vegetarians consume more fiber*** than non-vegetarians. An average ***daily intake of*** about ***30 g fiber*** is recommended.

NUTRITIONAL IMPORTANCE OF LIPIDS

Triacylglycerols (fats and oils) are the concentrated dietary source of fuel, contributing 15-50% of the body energy requirements. Phospholipids and cholesterol (from animal sources) are also important in nutrition. The nutritional and biochemical functions of fat, phospholipids and cholesterol have already been discussed in detail and the reader must invariably refer them now (***Chapters 3*** and ***14***).

Major nutritional functions of lipids

Dietary lipids have two major nutritive functions.

1. Supply triacylglycerols that normally constitute about 90% of dietary lipids which is a concentrated source of fuel to the body.

2. Provide essential fatty acids and fat soluble vitamins (A, D, E and K).

ESSENTIAL FATTY ACIDS

The unsaturated fatty acids which the body ***cannot synthesize*** and, therefore, must be ***consumed in the diet*** are referred to as essential fatty acids (EFA).

The fatty acids—***linoleic*** and ***linolenic acid***—cannot be synthesized by humans. In a strict sense, only these two are essential fatty acids. Arachidonic acid can be synthesized from linoleic acid in some animal species, including man. However, the conversion efficiency of linoleic acid to arachidonic acid is not clearly known in man. And for this reason, some nutritionists recommend that it is better to include ***some amount of arachidonic acid*** also in the diet.

Functions of EFA

1. Essential fatty acids are the ***structural components of biological membranes***.

2. Participate in the ***transport and utilization of cholesterol***.

3. ***Prevent fat*** accumulation in the ***liver***.

4. Required for the synthesis of prostaglandins.

5. Maintain proper growth and reproduction of the organisms.

Deficiency of EFA

The EFA deficiency in humans is characterized by a scaly dermatitis on the posterior and lateral parts of limbs and buttocks. This condition is referred to as ***phrynoderma*** or ***toad skin***.

EFA content of foods

The essential fatty acids, more frequently called ***polyunsaturated fatty acids (PUFA)***, are predominantly present in ***vegetable oils*** and ***fish oils***. The rich vegetable sources include sunflower oil, cotton seed oil, corn oil, soyabean oil etc. The fat of animal origin (exception—fish), contain less PUFA e.g. butter, fat of meat, pork.

Dietary intake of EFA

Nutritionists recommend that at least 30% of the dietary fat should contain PUFA. Very high intake of PUFA (i.e. totally replacing saturated fatty acids) may not be advisable. This is due to the fact that excess PUFA, unless accompanied by ***antioxidants*** (vitamin E, carotenes), is believed to be injurious to the cells due to the overproduction of ***free radicals***.

Normally, from the nutrition point of view, a ***high*** ratio of polyunsaturated fatty acids to saturated fatty acids (***P.S. ratio***) is believed ***to reduce the risk of coronary heart diseases.***

ω-3 and ω-6 fatty acids

These are long chain PUFA with double bond beginning at 3rd (ω-3) and 6th (ω-6) position from the methyl end. ω-3 fatty acids (e.g. linolenic acid, ***docosahexaenoic acid (DHA), eicosapentaenoic acid (EPA)*** and ω-6 fatty acids (e.g. arachidonic acid) are known to ***reduce serum cholesterol*** and ***triacylglycerols***, thereby decrease the tendency for thrombosis, lower blood presure and reduce the risk of CHD. In recent years, ω-3 fatty acids are included in infant formulas to ***promote brain development***.

Conditionally essential fatty acids

The fatty acids that become essential in developmental or disease conditions are regarded as conditionally essential fatty acids. Supplementation of DHA to meet the needs of developing brain in children, and ***DHA and EPA*** to reduce the risk of heart diseases are good examples of conditionally EFA (for some more details ***Refer Chapter 3***).

TRANS FATTY ACIDS (TFA)

TFA possess double bonds and are formed during partial hydrogenation of vegetable oils. TFA are ***widely used in food industry due to long shelf-life***. They increase LDL and decrease HDL, and thus promote altherogenesis and ***heart diseases***. Therefore, TFA ***should be avoided*** in the diet, as far as possible. WHO has recommended to eliminate TFA from the global food supply by 2023.

CHOLESTEROL IN NUTRITION

Animal foods are the only dietary source of cholesterol. However, the role of dietary cholesterol on plasma cholesterol is less important than the amount and types of fatty acids consumed.

REQUIREMENT OF DIETARY FAT

Consumption of dietary fats and oils is considered in terms of their contribution towards the energy needs of the body. There is a wide variation in fat intake. It is much higher (up to 50% of daily calories) in affluent societies compared to the poorer sections of the people (about 15% of calories). The recommended fat intake is around 20–30% of the daily calorie requirement, containing about 50% of PUFA.

NUTRITIONAL IMPORTANCE OF PROTEINS

Proteins have been traditionally regarded as '***body-building foods***'. However, 10-15% of the total body energy is derived from proteins. As far as possible, carbohydrates spare proteins and make the latter available for body-building process. A few functions carried out by proteins in a living cell are listed hereunder.

Functions of proteins

1. Proteins are the ***fundamental basis of cell structure and its function***.

2. All the ***enzymes***, several hormones, immunoglobulins, etc., are proteins.

3. Proteins are involved in the maintenance of ***osmotic pressure***, clotting of blood, muscle contraction etc.

4. During starvation, proteins (amino acids) serve as the major suppliers of energy. It may be noted that the structural proteins themselves serve as 'storage proteins' to meet the emergency energy needs of the body. This is in contrast to lipids and carbohydrates which have storage forms.

Essential amino acids

The nutritional importance of proteins is based on the content of essential amino acids (***Chapter 4***). There are ten essential amino acids—***arginine, valine, histidine, isoleucine, leucine, lysine, methionine, phenylalanine, tryptophan and threonine (code to recall—AV HILL MP TT)***. Of these two—namely arginine and histidine—are semi-essential. The requirement of 8 essential amino acids per kg body weight per day is given in ***Table 23.4***. Cysteine and tyrosine can respectively spare the requirement of methionine and phenylalanine.

Conditionally essential amino acids

The synthesis of certain non-essential amino acids is limited to the extent of meeting body's requirements in a healthy state. In ***patho-physiological conditions*** like surgery, wound healing, sepsis, injury etc., the non-essential amino acids become conditionally essential (indispensable), and need to be supplemented in the diet. e.g. ***arginine, cysteine, glycine, glutamine, proline, tyrosine*** (for some more details ***Refer Chapter 4***).

NITROGEN BALANCE

Dietary protein is almost an exclusive source of nitrogen to the body. Thus, nitrogen balance

Medical Concepts/ Clinical Correlates

- In general, vegetable oils are good sources for essential fatty acids while animal proteins are superior for the supply of essential amino acids.
- Trans fatty acids, widely used in food industry promote atherogenesis and heart diseases.
- The recommended dietary allowance (RDA) of nutrients depends on the sex and age, besides pregnancy and lactation in the women.
- Conditionally essential amino acids (e.g, glycine, cysteine) and conditionally essential fatty acids (e.g. decosahexaenoic acid, eicosapentaenoic acid) are required in the diet in certain pathophysiological conditions.

truly represents the protein (16% of which is nitrogen) utilization and its loss from the body.

Nitrogen balance is determined by comparing the intake of nitrogen (chiefly by proteins) and the excretion of nitrogen (mostly undigested protein in feces; urea and ammonia in urine). A normal healthy adult is in a nitrogen equilibrium since the daily dietary ***intake*** (***I***) is equal to the loss through ***urine*** (***U***), ***feces*** (***F***) and ***sweat*** (***S***).

$$I = U + F + S$$

The term ***fudge factor*** (approximately 3g) is used to represent nitrogen lost in feces, sweat and nails etc.

Thus, an individual is said to be in a nitrogen balance if the intake and output of nitrogen are the same (***Fig. 23.2***). There are two other situations—a positive and a negative nitrogen balance.

Positive nitrogen balance : This is a state in which the ***nitrogen intake is higher than the output.*** Some amount of nitrogen is retained in the body causing a net increase in the body protein. Positive nitrogen balance is observed in growing children, pregnant women or during recovery after serious illness.

Negative nitrogen balance : This is a situation in which the ***nitrogen output is higher than the input.*** The result is that some amount of nitrogen is lost from the body depleting the body protein. Prolonged negative nitrogen balance may even lead to death. This is sometimes observed in children suffering from kwashiorkor or marasmus.

Table 23.4 Requirements of essential amino acids

Amino acid	Requirement (mg/kg body weight/day)
Valine	14–26
Isoleucine	12–20
Leucine	16–39
Lysine	12–30
Methionine*	10–15
Phenylalanine*	16–25
Tryptophan	3–41
Threonine	8–15

* *Cysteine and tyrosine can, respectively, spare (partly) the requirement of methionine and phenylalanine.*

Negative nitrogen balance may occur due to inadequate dietary intake of protein (deficiency of even a single essential amino acid) or destruction of tissues or serious illness. In all these cases, the body adapts itself and increases the breakdown of tissue proteins causing loss of nitrogen from the body.

Other factors influencing nitrogen balance

Besides the major factors discussed above (growth, pregnancy, protein deficiency, injury,

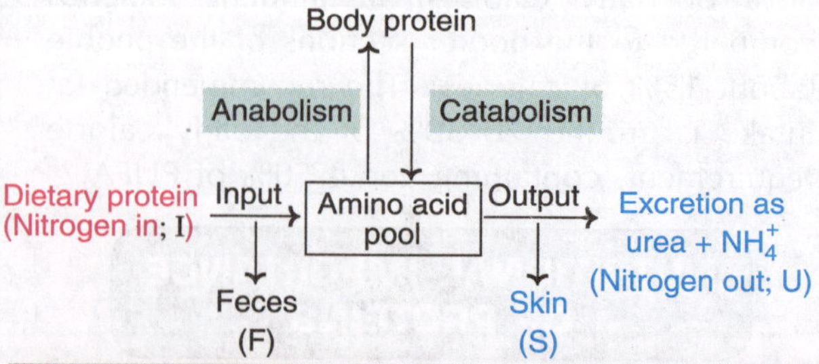

Fig. 23.2 : *Overview of nitrogen balance (At equilibrium N input = N output; For positive N balance, N input > N output; for negative N balance N input < N output).*

illness) several other factors influence nitrogen balance.

Hormones : Growth hormone and insulin promote positive nitrogen balance while corticosteroids result in negative nitrogen balance.

Disease states : Cancer and uncontrolled diabetes cause negative nitrogen balance.

ASSESSMENT OF NUTRITIVE VALUE OF PROTEINS

Knowledge on the quantity of dietary protein alone is not sufficient to evaluate the nutritional importance of proteins. This is in contrast to dietary carbohydrates and lipids. The ***quality*** of the proteins which ***depends on the composition of essential amino acids*** is more important. Several laboratory methods are in use to assess the nutritive value of proteins. Of these, four methods—protein efficiency ratio, biological value, net protein utilization and chemical score—are discussed briefly.

Protein efficiency ratio (PER)

This test consists of feeding weaning (21 day old) albino rats with a 10% test protein diet and recording the gain in body weight for a period of 4 weeks. PER is represented by ***gain in the weight of rats per gram protein ingested.***

$$\text{PER} = \frac{\text{Gain in body weight (g)}}{\text{Protein ingested (g)}}.$$

The PER for egg protein is 4.5; for milk protein 3.0; for rice protein 2.2.

Biological value (BV)

The biological value of protein is defined as the ***percentage of absorbed nitrogen retained by the body.***

$$\text{BV} = \frac{\text{Nitrogen retained}}{\text{Nitrogen absorbed}} \times 100$$

For the measurement of BV, the experimental animals, namely weaning albino rats are chosen. They are first fed with a protein-free diet for 10 days. Then they are kept on a 10% protein diet to be tested for BV. Urine and feces are collected for both the periods i.e. protein-free diet and protein diet. Nitrogen is estimated in the diet, feces and urine samples. Biological value can be calculated by the following formula

$$\text{BV} = \frac{(\text{N absorbed} - \text{N lost in metabolism})}{\text{N absorbed}} \times 100$$

$$\text{BV} = \frac{[I_n - (F_n - F_c)] - (U_n - U_c)}{I_n - (F_n - F_c)} \times 100$$

where I_n = Nitrogen ingested

F_n = Nitrogen in feces (on protein diet)

F_c = Nitrogen in feces (on protein-free diet)

U_n = Nitrogen in urine (on protein diet)

U_c = Nitrogen in urine (on protein-free diet)

For the calculation of BV of proteins, experiments can be done even in human subjects. The BV for different protein sources is given in ***Table 23.5.***

The biological value provides a reasonably good index for the nutritive value of proteins. But unfortunately this method has an inherent drawback. It cannot take into account the nitrogen that might be lost during the digestion process. For instance, if the ingested nitrogen is 100 mg, absorbed is 10 mg and retained is 8 mg, the BV 8/10 × 100 = 80. This figure is erroneous, since the major part of the protein (90 mg) did not enter the body at all for utilization.

Net protein utilization (NPU)

NPU is a better nutritional index than biological value, since it takes into account the digestibility factor. The experimental procedure for NPU is similar to that of BV. Net protein utilization can be calculated as

$$\text{NPU} = \frac{\text{Nitrogen retained}}{\text{Nitrogen ingested}} \times 100$$

Chemical score

This is ***based on the chemical analysis of the protein for*** the composition of ***essential amino acids*** which is then compared with a reference protein (usually egg protein). The chemical score is defined as the ratio between the quantity of the most limiting essential amino acid in the test protein to the quantity of the same amino acid in the egg protein, expressed as percentage.

Chemical score =

$$\frac{\text{mg of the limiting amino acid / g test protein}}{\text{mg of the same amino acid / g egg protein}} \times 100$$

The chemical score of egg protein, for any one of the essential amino acids, is taken as 100 and the rest of the proteins are compared.

In the ***Table 23.5***, the four methods employed (***PER, BV, NPU and chemical score***) for the assessment of nutritive value of proteins are compared with regard to the different sources of dietary proteins. Although there are certain variations, any one of these methods provides sufficient information on the nutritive value of proteins.

Mutual supplementation of proteins.

As is observed from the ***Table 23.5***, the ***animal proteins are superior*** in their nutritive value compared to the proteins of vegetable origin. Further, some of the essential amino acids are limiting in certain foods. For instance, ***rice*** and ***wheat*** proteins are ***limiting in lysine*** and ***threonine*** while the protein of Bengal gram is limited in sulfur-containing amino acids (methionine and cystine).

It is fortunate that humans (world over) have the habit of consuming a mixed diet, with different foods, simultaneously. This helps to overcome the deficiency of certain essential amino acids in one food by being supplemented from the others. This phenomenon is referred to as ***mutual supplementation***. For instance, an Indian diet with cereals (wheat, rice) is taken along with pulses (dal). The limitation of lysine and threonine in cereal proteins is overcome by their supplementation from dal proteins. Simultaneously, the limitation of sulfur-containing amino acids in dal is also compensated by the cereals which are rich in them.

The nutritive value of protein of a particular food can be enhanced by appropriate combination with other foods. Due to the consumption of mixed diets, ***dietary deficiency of essential amino acids is most uncommon***. Further, the principle of mixed diet takes care to supply adequate quantities of essential amino acids to the people subsisting on pure vegetarian diets. It has to be remembered that the effect of mutual supplementation in proteins is best observed with the same meal (or at least on the same day).

Requirement of proteins

The requirement of protein is dependent on its nutritive value, caloric intake and physiological states (growth, pregnancy, lactation) of the individual. For an ***adult, 0.8-1.0 g protein/kg body weight/day*** is adequate. The requirement

Table 23.5 Nutritive value of food proteins, assessed by PER, BV, NPU and chemical score

Source of protein	PER	BV	NPU	Chemical score	Limiting amino acid(s)
Egg	4.5	94	90	100	Nil
Milk	3.0	84	75	65	S-Containing amino acids
Fish	3.0	85	70	60	Tryptophan
Meat	2.7	75	76	70	S-Containing amino acids
Rice	2.2	68	60	60	Lysine, threonine
Wheat	1.5	58	47	42	Lysine, threonine
Bengal gram	1.7	58	47	45	S-Containing amino acids
Red gram	1.5	57	46	45	S-Containing amino acids
Groundnut	1.7	55	45	44	Lysine, threonine, S-amino acids
Soyabean	2.1	65	55	55	S-Containing amino acids

PER–Protein efficiency ratio; BV–Biological value; NPU–Net protein utilization; S–Sulfur.

should be nearly double for growing children, pregnant and lactating women.

Dietary sources of proteins

The protein content of foods is variable, cereals have 6-12%; pulses 18-22%; meat 18-25%, egg 10-14%; milk 3-4% and leafy vegetables 1-2%. In general, the ***animal proteins*** are ***superior*** than vegetable proteins as the dietary source.

NUTRITIONAL IMPORTANCE OF VITAMINS AND MINERALS

The nutritional aspects including metabolism, biochemical functions, dietary sources, requirements and associated disorders for vitamins (***Chapter 7***) and for minerals (***Chapter 18***) have already been discussed in much detail.

RECOMMENDED DIETARY ALLOWANCES (RDA)

The recommended dietary/daily allowances (RDA) represents the ***quantities of the nutrients to be provided in the diet daily*** for maintaining good health and physical efficiency of the body. It must be remembered that RDA is not the minimum amount to just meet the body needs, but allowance is given for a safe margin.

Factors affecting RDA

1. **Sex :** The RDA for ***men*** is about ***20% higher*** than that for ***women***. Iron is an exception as the requirement is greater in menstruating women. Additional requirements (20-30% above normal) are needed by pregnant and lactating women.

2. **Age :** In general, the nutrient requirement is much ***higher in the growing age***. For instance, the protein requirement for a growing child is about 2 g/kg body wt/day compared to 1 g/kg body wt/day for adults.

RDA an for adult man

The details of RDA for each of the nutrients in relation to age, sex and physiological status is described in the respective chapters. For a quick recapitulation, the RDA of macronutrients (carbohydrate, fat and protein) and selected micronutrients (vitamins and minerals) for an adult man weighing 70 kg are given in ***Table 23.6***.

BALANCED DIET

After discussing the nutritional aspects of dietary ingredients and their RDA, it is worthwhile to formulate a diet for man. A balanced diet or ***prudent diet*** is defined as the diet which ***contains different types of foods, possessing the nutrients—carbohydrates, fats, proteins, vitamins and minerals—in a proportion to meet the requirements of the body***. A balanced diet invariably supplies a little more of

Table 23.6 Recommended dietary allowance (RDA) of important nutrients for an adult man, weighing 70 kg.

Nutrient(s)	RDA
Carbohydrates	400 g
Fats	70 g
Proteins	56 g*
Essential fatty acids	4 g
Vitamin A	600 µg **
Vitamin D	5 µg***
Vitamin E	10 mg
Vitamin K	70 µg
Ascorbic acid	50 mg
Thiamine	1.5 mg
Riboflavin	1.5 mg
Niacin	15 mg
Pyridoxine	2 mg
Folic acid	200 µg
Cobalamin	1.5 µg
Calcium	800 mg
Phosphorus	800 mg
Iron	10 mg

* *0.8 g/kg body weight/day,* ** *Retinol equivalents;* *** *As cholecalciferol*
(***Note :*** *RDA variable with ± 10–20% values given*)

each nutrient than the minimum requirement to withstand the short duration of leanness and keep the body in a state of good health.

The basic composition of balanced diet is highly variable, as it differs from country to country, depending on the availability of foods. Social and cultural habits, besides the economic status, age, sex and physical activity of the individual largely influence the intake of diet.

The Nutrition Expert Group, constituted by the Indian Council of Medical Research (ICMR) has recommended the composition of balanced diets for Indians (***Table 23.7***).

The Indian balanced diet is composed of cereals (rice, wheat, jowar), pulses, vegetables, roots and tubers, fruits, milk and milk products, fats and oils, sugar and groundnuts. Meat, fish and eggs are present in the non-vegetarian diets. In case of vegetarians, an additional intake of milk and pulses is recommended. The nutritional aspects of milk are given in the ***Chapter 22***.

Flavor of foods–Maillard reaction

The acceptability of foods is mainly based on the taste and flavor (satisfying the taste buds) rather than the balanced composition of nutrients. The ***distinct flavor of browned foods*** (cookies, biscuits, bread, coffee roasting, fried foods) is due to a chemical reaction, namely ***Maillard reaction***. It is a ***non-enzymatic*** reaction between amino acids and reducing sugars that takes place at a temperature of 140–160 °C. Maillard reaction that occurs during the course of cooking process can produce hundreds of different flavored compounds. This depends on the constituents of the food, temperature, cooking time, presence of air etc.

NUTRITIONAL DISORDERS

Some of the important nutritional diseases are discussed hereunder.

Protein-energy malnutrition

Protein-energy malnutrition (***PEM***)—sometimes called ***protein-calorie malnutrition (PCM)***—is the most common nutritional disorder of the developing countries. PEM is widely prevalent in the infants and pre-school children. ***Kwashiorkor*** and ***marasmus*** are the two extreme forms of protein-energy malnutrition.

Table 23.7 Balanced diet for an adult man*

	Sedentary work		Moderate work		Heavy work	
	Vegetarian (g)	Non-vegetarian (g)	Vegetarian (g)	Non-vegetarian (g)	Vegetarian (g)	Non-vegetarian (g)
Cereals	400	400	475	475	650	650
Pulses	70	55	80	65	80	65
Green leafy vegetables	100	100	125	125	125	125
Other vegetables	75	75	75	75	100	100
Roots and tubers	75	75	100	100	100	100
Fruits	30	30	30	30	30	30
Milk	200	100	200	100	200	100
Fats and oils	35	40	40	40	50	50
Meat and fish	...	30	...	30	...	30
Eggs	...	30	...	30	...	30
Sugar and jaggery	30	30	40	40	55	55
Groundnuts	...	...	...	...	50	50

* *Formulations based on the recommended dietary (daily) allowances (RDA) of the Indian Council of Medical Research.*

Kwashiorkor

The term kwashiorkor was introduced by Cicely Williams (1933) to a nutritional disease affecting the people of Gold Coast (modern Ghane) in Africa. Kwashiorkor literally means ***sickness of the deposed child*** i.e. a disease the child gets when the next baby is born.

Occurrence and causes : Kwashiorkor is predominantly found in ***children between 1-5 years of age***. This is primarily due to insufficient intake of proteins, as the diet of a weaning child mainly consists of carbohydrates.

Clinical symptoms : The major clinical manifestations of kwashiorkor include ***stunted growth, edema***, diarrhea, discoloration of hair and skin, anemia, apathy and moonface.

Biochemical manifestations : Kwashiorkor is associated with a ***decreased plasma albumin*** concentration (< 2 g/dl against normal 3–4.5 g/dl), fatty liver, deficiency of K^+ due to diarrhea. Edema occurs due to lack of adequate plasma proteins to maintain water distribution between blood and tissues. Disturbances in the metabolisms of carbohydrate, protein and fat, besides vitamin deficiencies occur.

Treatment : Ingestion of ***protein-rich foods*** or the dietary combinations to provide about 3–4 g of protein/kg body weight/day will control kwashiorkor. The treatment can be monitored by measuring plasma albumin concentration, disappearance of edema and gain in body weight.

Marasmus

Marasmus literally means ***'to waste'***. It mainly occurs in children under one year of age. Marasmus is predominantly due to the ***deficiency of calories.*** This is usually observed in children given watery gruels (of cereals) to supplement the mother's breast milk.

The symptoms of marasmus include growth retardation, muscle wasting (emaciation), anemia and weakness. A marasmic child does not show edema or decreased concentration of plasma albumin, a major difference to distinguish marasmus from kwashiorkor. In ***Table 23.8***, a comparison between kwashiorkor and marasmus is given.

Signs comparable to marasmus in advanced cancer and AIDS

The patients of certain chronic diseases like cancer and AIDS are fequently undernourished, resulting in a condition called ***cachexia***. This is mainly due to the loss of body proteins as a result of ***hypermetabolism***, particularly increased basal metabolic rate. Further, increased metabolisms leading to thermogenesis is also observed in cancer and AIDS.

Nutritional anemias

Anemia is characterized by lower concentration of hemoglobin (reference 14–16 g/dl). Nutritional anemias are classified based on the size of erythrocytes.

- **Microcytic anemia**—most common, with reduced RBC size. Occurs due to the ***deficiency of iron***, copper and pyridoxine.
- **Macrocytic anemia**—RBC are large and immature. Mostly due to the ***deficiency of folic acid and vitamin*** B_{12}.
- **Normocytic anemia**—Size of the RBC is normal, but their quantity in blood is low. Mostly found in ***protein-energy malnutrition***.

OBESITY

Obesity is an abnormal increase in the body weight due to excessive fat deposition. It mainly occurs as a result of imbalance between energy intake and energy expenditure.

Nutritional basis

Men and women are considered as obese if their weight due to fat (in adipose tissue), respectively, exceeds more than 20% and 25% of body weight. Obesity is basically a disorder of excess calorie intake, in simple language—***overeating***. It has to be remembered that every 7 calories of excess consumption leads to 1 g fat deposit and increase in body weight. Overeating—coupled with ***lack of physical exercise***—contribute to obesity.

Table 23.8 Comparison between kwashiorkor and marasmus

Clinical/biochemical parameter	Kwashiorkor	Marasmus
Age of onset	Pre-school children (1-5 yr)	Weaned infants (< 1 yr)
Main nutritional cause	Low protein intake	Low calorie intake
Body weight	60–80% of normal	Less than 60% of normal
Growth	Mild retardation	Severe retardation
Oedema	Present	Absent
Facial appearance	Moon face	Like old man's face
Abdomen	Protruding	Shrunken
Skin	Dermatitis	Dry and atrophic
Muscles	Undergo wasting	Weak and atrophic
Subcutaneous fat	Present	Absent
Vitamin deficiencies	Present	Present
Serum albumin	0.5–2 g/dl	2–3 g/dl
Serum cortisol	Normal or decreased	Increased
Fasting blood glucose	Decreased	Decreased
Serum K^+	Decreased	Normal

Other causes of obesity

Obesity can occur due to certain diseases-***polycystic ovary syndrome***, ***hypothyroidism***, Cushing's syndrome. Some medications may also result in obesity e.g. anticonvulsants, oral contraceptives, corticosteroids. Recently, ***obesity due to a virus*** (adenovirus-36) infection has been identified.

Body mass index (BMI)

Clinical obesity is represented by body mass index. BMI is calculated as the weight (in kilograms) divided by the height (in meters²).

$$BMI\ (kg/m^2) = \frac{Weight\ (kg)}{[height\ (m)^2]}$$

Healthy ***reference range*** for BMI is between ***18.5–24.9 kg/m²***.

- Grade I obesity or overweight – BMI 25–30 kg/m^2
- Grade II or clinical obesity – BMI > 30 kg/m^2
- Grade III or morbid obesity – BMI > 40 kg/m^2

Obesity is associated with many health complications e.g. type II diabetes, CHD, hypertension, stroke, arthritis, gall bladder disease.

In recent years, the ***ratio between waist and hip sizes*** (for men < 0.9 and for women < 0.85) is considered as ***more effective than BMI***, particularly with regard to the risk of heart diseases. The lower is the waist to hip ratio, the lower the risk for health complications, and therefore better is the health.

Genetics, obesity and leptin

There is strong evidence to suggest that obesity has genetic basis. Thus, a child born to two obese people has about 75% chances of being obese. One gene namely ***ob gene***, expressed in adipocytes (of white adipose tissue) producing a protein called ***leptin*** (mol. wt. 16,000 daltons), is associated with obesity.

Leptin is regarded as a ***body weight regulatory hormone***. It binds to a specific receptor in the brain and functions as a ***lipostat***.

Medical Concepts/ Clinical Correlates

- Obesity is linked to ob gene that synthesises a protein called leptin, regarded as the body weight regulatory hormone. Genetic defect in leptin is associated with obesity.
- In general, illness and metabolic stress increase nutritional demands. e.g. RDA of vitamin D increased in liver and kidney diseases.
- Kwashiorkor and marasmus, the two extreme forms of protein-energy malnutrition in infants and preschool children, are highly prevalent in developing countries.
- Use of oral contraceptives may lead to deficiencies of vitamins B_6, B_{12}, and folic acid.

When the fat stores in the adipose tissue are adequate, leptin levels are high. This signals to restrict the feeding behaviour and limit fat deposition. Further, leptin stimulates lipolysis and inhibits lipogenesis. Any genetic defect in leptin or its receptor will lead to extreme overeating and obesity. Treatment of such obese individuals with leptin has been shown to reverse obesity. During starvation, leptin levels fall which promote feeding, and fat production and its deposition.

Obesity and adipose tissue

Obesity is due to an increase in both the number and size of adipocytes (of adipose tissue). There are two types of adipose tissues

1. **White adipose tissue :** The fat is mostly stored and this tissue is metabolically less active.

2. **Brown adipose tissue :** The stored fat is less but the tissue is metabolically very active.

Brown adipose tissue possesses high proportion of mitochondria and cytochromes but low activity of ATP synthase. This is an active centre for the oxidation of fat and glucose and is responsible for the ***diet-induced thermogenesis***.

The peculiarity of mitochondria of brown adipose tissue is that the oxidation and phosphorylation are not coupled. Mitochondrial oxidation produces ***more heat*** and ***less ATP***. A specific protein—namely ***thermogenin***—has been isolated in the inner membrane of these mitochondria. Thermogenin functions like an uncoupler and dissipates the energy in the form of heat, and thus blocks the formation of ATP.

Brown adipose tissue is mostly found in hibernating animals, and the animals exposed to cold, besides the newborn. In adult humans, though not a prominent tissue, it is located in the thoracic region. It is significant to note that ***brown adipose tissue is almost absent in obese persons***. Some individuals are fortunate to have active brown adipose tissue. They eat and liberate it as heat, and therefore do not become obese.

Pharmacological treatment of obesity : In recent years, synthetic lipids such as ***Olestra*** and Orlistat are used to treat obesity. They taste like natural lipids but cannot be digested, and excreted unchanged.

METABOLIC SYNDROME

Metabolic syndrome (MS) is a cluster of different conditions that adversely affect the health. The ***components*** contributing to MS include ***abdominal obesity, insulin resistance, dyslipidemia, elevated blood pressure, over-nutrition, sedentary lifestyles*** etc. As per WHO criteria, metabolic syndrome has the following characteristics

1. **Insulin resistance** – identified either as type 2 diabetes or elevated fasting blood glucose (>100 mg/dl) or impaired glucose tolerance.

2. And ***any two*** of the following
   (i) ***Hypertension*** (> = 140/90 mm Hg)
   (ii) ***Dyslipidemia*** (serum TG > = 150 mg/dl or HDL cholesterol <40 mg/dl in men or <50 mg/dl in women.
   (iii) ***BMI*** >30 kg/m^2 or waist : hip ratio of >0.9 in men or 0.85 in women.

Metabolic syndrome can be ***managed by healthy habits*** and change in lifestyles—restricted balanced diet, adequate intake of fiber, and antioxidants, exercise, avoiding smoking, stress-free life etc.

CACHEXIA

This is opposite of what is seen in obesity. Cachexia is characterized by a failure to maintain normal lipid stores in the body. It involves higher rate of fat mobilization than deposition.

Anorexia nervosa is a total loss of appetite. This is mostly seen in females in the age group 10-30 years. Surprisingly, majority of the affected individuals are from wealthy families where food is aplenty. And some members in these families may be even obese! Anorexia nervosa is more a psychiatric disease.

OTHER NUTRITIONAL DISORDERS

There are several other nutritional disorders which have been discussed elsewhere. These include atherosclerosis (*Chapter 14*); vitamin deficiency disorders—xerophthalmia, rickets, beri-beri, pellagra, scurvy, pernicious anemia etc. (*Chapter 7*); goiter and other disorders of minerals (*Chapter 18*). The biochemical ramifications of starvation are discussed along with the integration of metabolism (*Chapter 16*).

THERAPEUTIC DIETS

Diet therapy in disease states is a part of nutrition. Therapeutic diets are ***usually not palatable***. However, they possess high or low amounts of specific nutrients to meet the body demands as per the situation. Selected examples of therapeutic diets are listed

- Liquid diets — for post-operative patients
- Low sodium diets — for hypertensive people
- Low fat diet — for patients of malabsorption syndrome
- Low protein diet — for patients of hepatic encephalopathy, renal failure
- High fiber diet — for patients of constipation, CHD, diabetes mellitus.

Atkins diet

In Atkins diet, ***fat and protein are high***, and ***carbohydrate very low*** (<50g/day; <10% of a 2000 Cal/day). Atkins diet is ***advocated for weight loss programmes of obese people***. High fat diet reduces appetite, and thus food intake. However, long term human consumption of Atkins diet is controversial.

NUTRITIONAL STATUS AND CLINICAL PRACTICE

The nutritional status of an individual is important in clinical practice. The ***dietary requirements of nutrients are variable***, and are mostly related to age and sex, and physiological status. Some examples are listed

- Infants and young children have increased needs of protein, iron and calcium.
- During teenage, high calcium and magnesium are recommended.
- In pregnancy, and lactation, the requirements of iron, calcium, magnesium, folic acid and vitamin B_6 and B_{12} are increased.
- Elderly people have to take more of vitamins-B_6 and B_{12}, folic acid and vitamin D, and minerals chromium, zinc etc. However, vitamin A intake should be restricted in the elderly to avoid toxicity.

In general, illness and metabolic stress increase the nutritional demands. For instances, liver and kidney diseases reduce the formation of active vitamin D (calcitriol), and storage and utilization of vitamins—folic acid, vitamin B_{12} and vitamin D.

Drug and nutrient interactions

Many drugs are known to lead to potential nutrient deficiencies (***Table 23.9***). For instance, oral contraceptives may result in deficiency manifestations of vitamin B_6, B_{12} and folic acid.

NUTRIGENOMICS

Nutrigenomics, a new field of nutrition, relates to 3 distinct areas of nutrient-gene interactions

1. **Nutritional genetics** involves the individual genetic differences and their influence on the nutrient intake.

2. **Nutritional epigenetics** describes the nutrient induced changes in DNA, such as ***DNA methylation***, histone post-translational changes etc.

3. **Nutritional transcriptomics** involves the effects of ***nutrients on gene expression***. The role vitamins A and D on nuclear receptors, and in turn on gene expression is well known.

Nutrigenomics may soon revolutionize the clinical and nutritional practice, and result in individualized RDA for disease prevention and treatment.

TABLE 23.9 Drug and nutrient interactions

Drug	Risk of nutrient deficiencies
Oral contraceptives	Vitamin B_6, vitamin B_{12}, folic acid
Diuretics	Potassium, zinc
Anticonvulsants	Folic acid, vitamin D, vitamin K
Isoniazid	Vitamin B_6
Corticosteroids	Vitamin D, calcium, potassium, zinc
Alcohol	Thiamine, vitamin B_6, folic acid

SUMMARY

- *The calorific values of carbohydrates, fats and proteins respectively are 4, 9 and 4 Cal/g. These three nutrients (macronutrients) supply energy to the body to meet the requirements of basal metabolic rate, SDA and physical activity.*
- *Basal metabolic rate (BMR) represents the minimum amount of energy required by the body to maintain life at complete physical and mental rest, in the post-absorptive state. The normal BMR for an adult man is 35-38 Cal/m^2 body surface/hr.*
- *Specific dynamic action (SDA) is the extra heat produced by the body over and above the calculated calorific value of foodstuff. It is higher for proteins (30%), lower for carbohydrates (5%), and for a mixed diet, it is around 10%.*
- *Carbohydrates are the major source of body fuel supplying about 40-70% of body calories. The non-digested carbohydrates (cellulose, pectin), referred to as fiber, prevent constipation, improve glucose tolerance and reduce plasma cholesterol.*
- *Lipids are the concentrated source of energy. They also provide essential fatty acids (linoleic and linolenic acids) and fat-soluble vitamins (A, D, E and K).*
- *Proteins are the body building foods that supply essential amino acids, besides meeting the body energy requirement partly (10-15%).*
- *Several methods are employed to assess the nutritive value of proteins. These include protein efficiency ratio, biological value, net protein utilization and chemical score.*
- *The RDA represents the quantities of nutrients to be provided daily in diet for maintaining good health and physical efficiency. RDA for protein is 1g/kg body weight/day.*
- *A balanced diet is the diet which contains different types of foods with the nutrients, namely carbohydrates, fats, proteins, vitamins and minerals, in a proportion to meet the body requirements.*
- *Protein energy malnutrition (kwashiorkor and marasmus) obesity, atherosclerosis, vitamin deficiencies (e.g. beri-beri, rickets, scurvy) are important nutrition-related disorders.*

THE READER MUST REFER APPENDIX VI (Page Nos : 738 - 756) FOR RELATED CLINICAL CASE STUDIES (Nos : 1, 2, 3, 4, 5, 6, 7, 8, 9, 10, 50, 59)

Self-Assessment Exercises

I. Essay questions

1. Define BMR. Discuss the factors affecting BMR.
2. Describe the different methods employed for the nutritional evaluation of proteins.
3. Define a balanced diet. Formulate a diet for a medical student.
4. Discuss the protein-energy malnutrition with special reference to kwashiorkor.
5. Give an account of the recommended dietary allowance (RDA) for macro- and micronutrients.

II. Short notes

(a) Essential amino acids, (b) Mutual supplementation of proteins, (c) Caloric value of foods, (d) Specific dynamic action, (e) Energy requirements of man, (f) Fiber in nutrition, (g) Kwashiorkor, (h) Limiting amino acids, (i) Nitrogen balance, (j) Biological value of proteins.

III. Fill in the blanks

1. One calorie of energy is equivalent to ____________ Joules (KJ).
2. The endocrine organ most predominantly associated with BMR is ____________.
3. The non-digestible carbohydrates are collectively known as ____________.
4. The major source of energy to the body is supplied by ____________.
5. The nutritional assessment method used to know the most limiting essential amino acid in relation to a standard protein is ____________.
6. The daily normal requirement of protein in an adult is ____________.
7. The percentage of absorbed nitrogen retained in the body represents ____________.
8. The proteins of Bengal gram are limiting in the amino acids ____________.
9. The nutrient required in greater amounts in menstruating women compared to men is ______.
10. The biochemical parameter often used as an index for monitoring the recovery from kwashiorkor is ______.

IV. Multiple choice questions

11. The specific dynamic action (SDA) is the greatest for the following foodstuff
    (a) Protein (b) Carbohydrate (c) Fat (d) Vitamins.
12. The reference protein for the calculation of chemical score
    (a) Meat protein (b) Fish protein (c) Milk protein (d) Egg protein.
13. The essential amino acid limiting in rice
    (a) Methionine (b) Tryptophan (c) Lysine (d) Histidine.
14. A continuous supply of energy to the body is necessary to meet the requirements of
    (a) Basal metabolic rate (b) Specific dynamic action (c) Physical activity (d) All of them.
15. One of the following is the most important essential fatty acid in the diet
    (a) Linoleic acid (b) Arachidonic acid (c) Oleic acid (d) Palmitic acid.

MOLECULAR BIOLOGY AND BIOTECHNOLOGY

SECTION FIVE

DNA-Replication, Recombination, and Repair

24

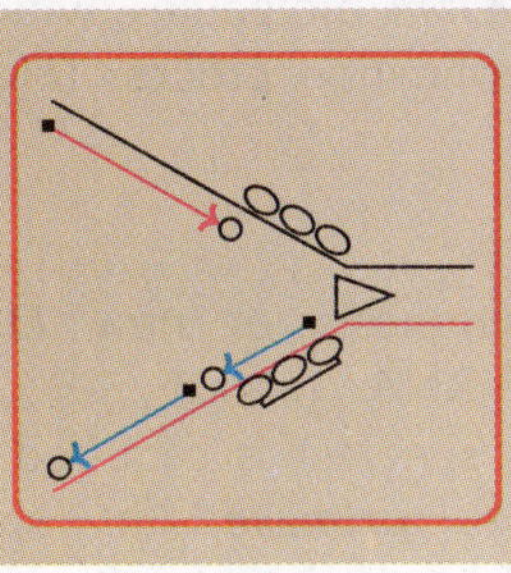

The hereditary molecule, DNA speaks

"I replicate and recombinate,
To permit the cells to proliferate,
Environmental insults try to damage me,
But I protect myself with adequate repairs"

Deoxyribonucleic acid (DNA) is a macromolecule that carries genetic information from generation to generation. It is responsible to preserve the identity of the species over millions of years. ***DNA*** may be regarded as a ***reserve bank of genetic information*** or a memory bank.

A single mammalian fetal cell contains only a few picograms (10^{-12} g) of DNA. It is surprising that this little quantity of DNA stores information that will determine the differentiation and every function of an adult animal.

Why did DNA evolve as genetic material?

RNA molecules, in principle, can perform the cellular functions that are carried out by DNA. In fact, many viruses contain RNA as the genetic material. Chemically, ***DNA is more stable*** than RNA. Hence, during the course of evolution, DNA is preferred as a more suitable molecule for long-term repository of genetic information.

The central dogma of life

The ***biological information flows from DNA to RNA, and from there to proteins***. This is the central dogma of life (***Fig.24.1***). It is ultimately the DNA that controls every function of the cell through protein synthesis.

As the carrier of genetic information, DNA in a cell must be duplicated (replicated), maintained and passed down accurately to the daughter cells. Three distinct processes are designed for this purpose. The ***'three Rs' of DNA-r***eplication, ***r***ecombination, and ***r***epair, are dealt with in this chapter. There are certain common features between the three Rs.

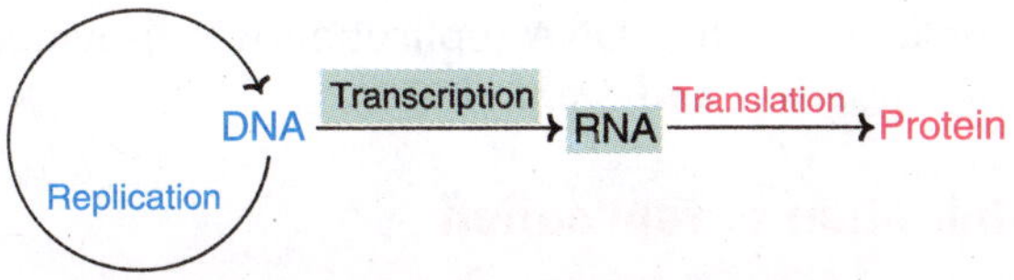

Fig. 24.1 : *The central dogma of life.*

- They act on the same substrate (DNA).
- They are primarily concerned with the making and breaking of phosphodiester bonds (the backbone of DNA structure).
- Enzymes used in the three processes are mostly similar/comparable.

REPLICATION OF DNA

DNA is the genetic material. When the cell divides, the daughter cells receive an identical copy of genetic information from the parent cell.

Replication is a ***process in which DNA copies itself to produce identical daughter molecules of DNA***. Replication is carried out with high fidelity which is essential for the survival of the species. Synthesis of a new DNA molecule is a complex process involving a series of steps.

The salient features of replication in prokaryotes are described first. This is followed by some recent information on the eukaryotic replication.

REPLICATION IN PROKARYOTES

Replication is semiconservative

The parent DNA has two strands complementary to each other. Both the strands undergo simultaneous replication to produce two daughter molecules. Each one of the newly synthesized DNA has one-half of the parental DNA (one strand from original) and one-half of new DNA (***Fig.24.2***). This type of replication is known as semiconservative since ***half of the original DNA is conserved in the daughter DNA***. The first experimental evidence for the semiconservative DNA replication was provided by ***Meselson and Stahl*** (1958).

Initiation of replication

The initiation of DNA synthesis occurs at a site called ***origin of replication***. In case of prokaryotes, there is a single site whereas in ***eukaryotes***, there are ***multiple sites of origin***. These sites mostly consist of a short sequence of A-T base pairs. A specific protein called ***dna A*** (20-50 monomers) binds with the site of origin for replication. This causes the double-stranded DNA to separate.

Replication bubbles

The two complementary strands of DNA separate at the site of replication to form a bubble. Multiple replication bubbles are formed in eukaryotic DNA molecules, which is essential for a rapid replication process (***Fig.24.3***).

RNA primer

For the synthesis of new DNA, a short fragment of RNA (about 5-50 nucleotides, variable with species) is required as a primer. The enzyme ***primase*** (a specific RNA polymerase) in association with single-stranded binding proteins forms a complex called ***primosome***, and produces RNA primers. A constant synthesis and supply of RNA primers should occur on the lagging strand of DNA. This is in contrast to the leading strand which has almost a single RNA primer.

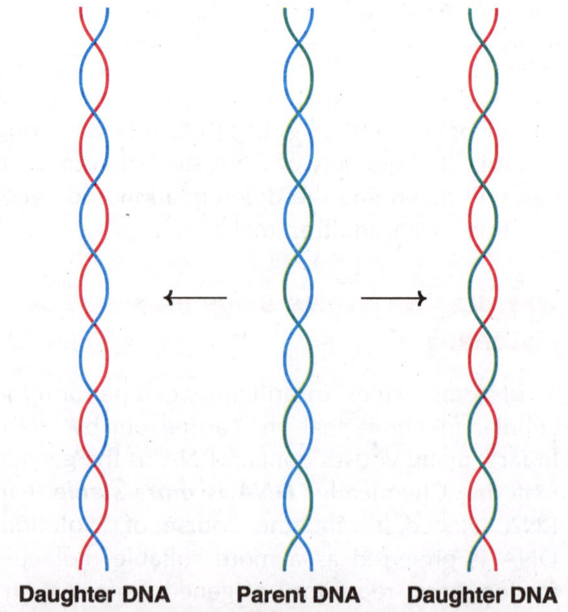

Fig. 24.2 : DNA replication—semiconservative.

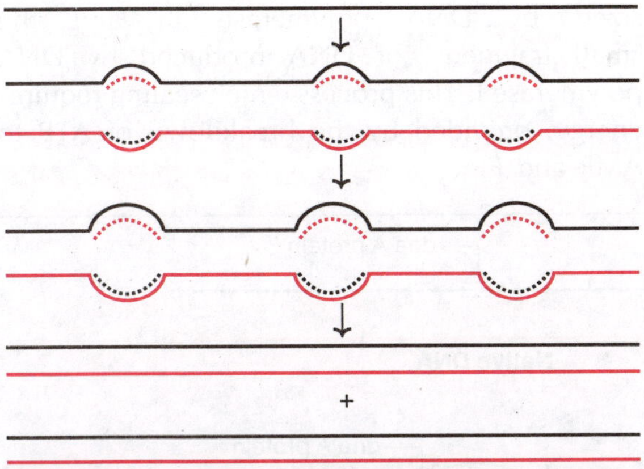

Fig. 24.3 : Schematic representation of multiple replication bubbles in DNA replication.

DNA synthesis is semidiscontinuous and bidirectional

The replication of DNA occurs in 5′ to 3′ direction, simultaneously, on both the strands of DNA. On one strand, the ***leading (continuous or forward) strand—the DNA synthesis is continuous***. On the other strand, the ***lagging (discontinuous or retrograde) strand—the synthesis of DNA is discontinuous***. Short pieces of DNA (15-250 nucleotides) are produced on the lagging strand.

In the replication bubble, the DNA synthesis occurs in both the directions (bidirectional) from the point of origin.

Replication fork and DNA synthesis

The separation of the two strands of parent DNA results in the formation of a replication fork. The active synthesis of DNA occurs in this region. The replication fork moves along the parent DNA as the daughter DNA molecules are synthesized.

DNA helicases : These enzymes bind to both the DNA strands at the replication fork. Helicases move along the DNA helix and separate the strands. Their function is comparable with a ***zip opener***. Helicases are dependent on ATP for energy supply.

Single-stranded DNA binding (SSB) proteins : These are also known as DNA helix-destabilizing proteins. They possess no enzyme activity. SSB proteins bind only to single-stranded DNA (separated by helicases), keep the two strands separate and provide the template for new DNA synthesis. It is believed that SSB proteins also protect the single-stranded DNA degradation by nucleases.

DNA synthesis catalysed by DNA polymerase III

The synthesis of a new DNA strand, catalysed by DNA polymerase III, occurs ***in 5′→3′ direction***. This is antiparallel to the parent template DNA strand. The presence of all the four deoxyribonucleoside triphosphates (dATP, dGTP, dCTP and dTTP) is an essential prerequisite for replication to take place.

The ***synthesis of two new DNA strands, simultaneously, takes place in the opposite direction***—one is in a direction (5′→3′) towards the replication fork which is continuous, the other in a direction (5′→3′) away from the replication fork which is ***discontinuous*** (***Fig.24.4***).

The incoming ***deoxyribonucleotides*** are added one after another, to 3′ end of the growing DNA chain (***Fig.24.5***). A molecule of pyrophosphate (PPi) is removed with the addition of each nucleotide. The template DNA strand (the parent) determines the base sequence of the newly synthesized complementary DNA.

Polarity problem

The DNA strand (leading strand) with its 3′-end (3′-OH) oriented towards the fork can be elongated by sequential addition of new nucleotides. The other DNA strand (lagging strand) with 5′-end presents some problem, as there is no DNA polymerase enzyme (in any organism) that can catalyse the addition of nucleotides to the 5′ end (i.e. 3′→5′ direction) of the growing chain. This problem however is solved by synthesizing this strand as a series of small fragments. These pieces are made in the normal 5′→3′ direction, and later joined together.

Okazaki pieces : The small ***fragments of the discontinuously synthesized DNA*** are called

Okazaki pieces. These are ***produced on the lagging strand of the parent DNA***. Okazaki pieces are later joined to form a continuous strand of DNA. DNA polymerase I and ***DNA ligase*** are responsible for this process (details given later).

Proof-reading function of DNA polymerase III

Fidelity of replication is the most important for the very existence of an organism. Besides its 5′→3′ directed catalytic function, DNA polymerase III also has a ***proof-reading activity.*** It checks the incoming nucleotides and allows only the correctly matched bases (i.e. complementary bases) to be added to the growing DNA strand. Further, DNA polymerase edits its mistakes (if any) and removes the wrongly placed nucleotide bases.

Replacement of RNA primer by DNA

The synthesis of new DNA strand continues till it is in close proximity to RNA primer. Now the DNA polymerase I comes into picture. It removes the RNA primer and takes its position. DNA polymerase I catalyses the synthesis (5′→3′ direction) of a fragment of DNA that replaces RNA primer (***Fig.24.6***).

The enzyme ***DNA ligase*** catalyses the formation of a phosphodiester linkage between the DNA synthesized by DNA polymerase III and the small fragments of DNA produced by DNA polymerase I. This process—nick sealing requires energy, provided by the breakdown of ATP to AMP and PPi.

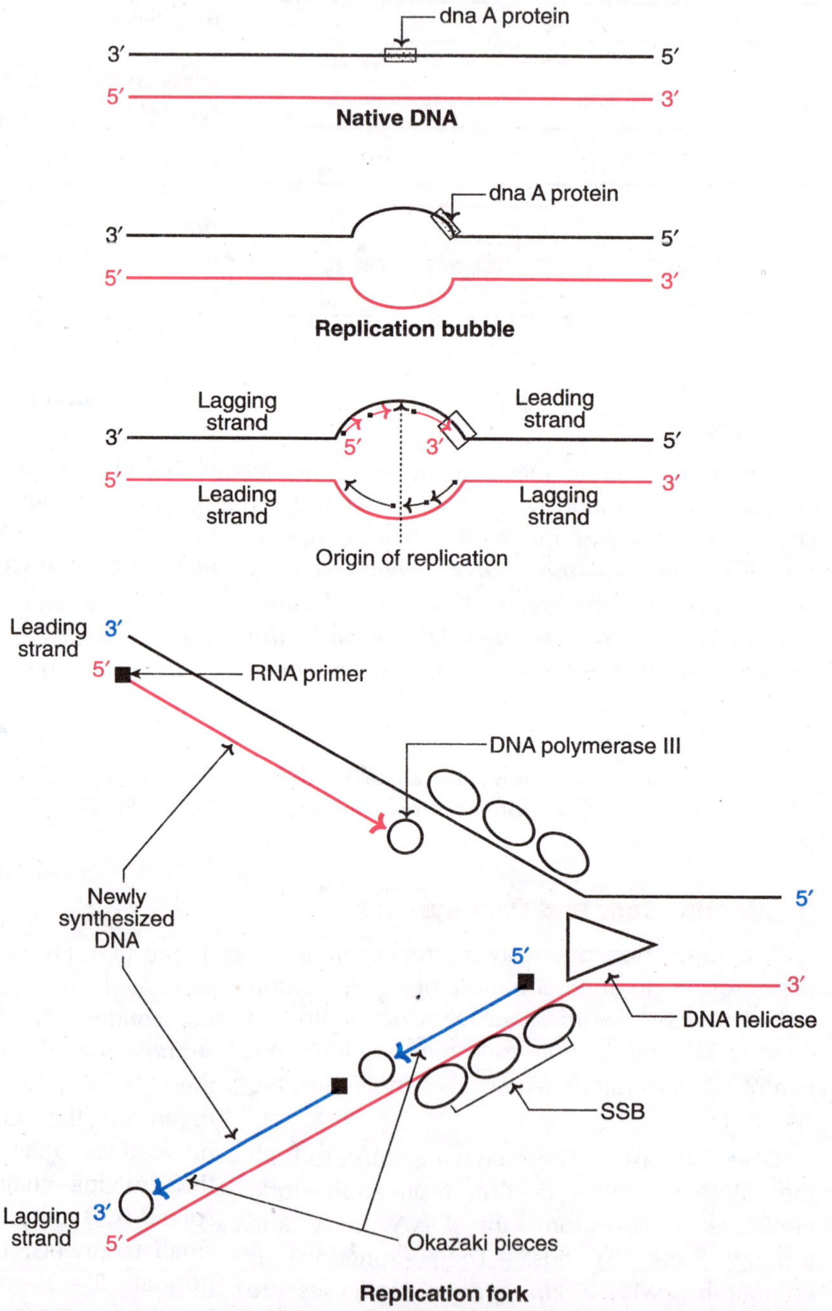

Fig. 24.4 : *Overview of DNA replication process (SSB–Single-stranded binding proteins).*

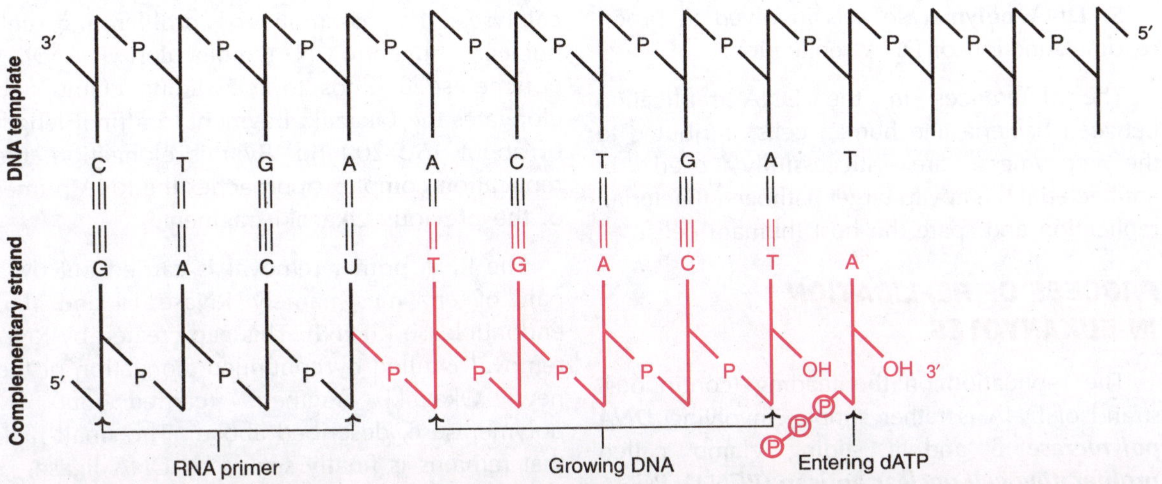

Fig. 24.5 : DNA replication with a growing complementary strand.

Another enzyme—***DNA polymerase II***—has been isolated. It participates in the ***DNA repair process.***

Supercoils and DNA topoisomerases

As the double helix of DNA separates from one side and replication proceeds, ***supercoils are formed at the other side.*** The formation of supercoils can be better understood by comparing DNA helix with two twisted ropes tied at one end. Hold the ropes at the tied end in a fixed position. And let your friend pull the ropes apart from the other side. The formation of supercoils is clearly observed.

Type I DNA topoisomerase cuts the single DNA strand (nuclease activity) ***to overcome*** the problem of ***supercoils*** and then reseals the strand (ligase activity). Type II DNA topoisomerase (also known as DNA gyrase) cuts both strands and reseals them to overcome the problem of supercoils. DNA topoisomerases are targeted by drugs (campthoterin for topoisomerase I, and amsacrime and etoposide for topoisomerase I, and amsacrime and etoposide for topoisomerase II) in the treatment of cancers.

REPLICATION IN EUKARYOTES

Replication of DNA in eukaryotes closely resembles that of prokaryotes. Certain differences, however, exist. ***Multiple origins of replication*** is a characteristic feature of eukaryotic cell. Further, at least ***five distinct DNA polymerases*** are known in eukaryotes. Greek letters are used to number these enzymes.

1. **DNA polymerase α** is responsible for the ***synthesis of RNA primer*** for both the leading and lagging strands of DNA.

2. **DNA polymerase β** is involved in the ***repair of DNA.*** Its function is comparable with DNA polymerase I found in prokaryotes.

3. **DNA polymerase γ** participates in the replication of ***mitochondrial DNA.***

4. **DNA polymerase δ** is responsible for the replication on the ***leading and lagging strands of DNA.*** It also possesses proof-reading activity.

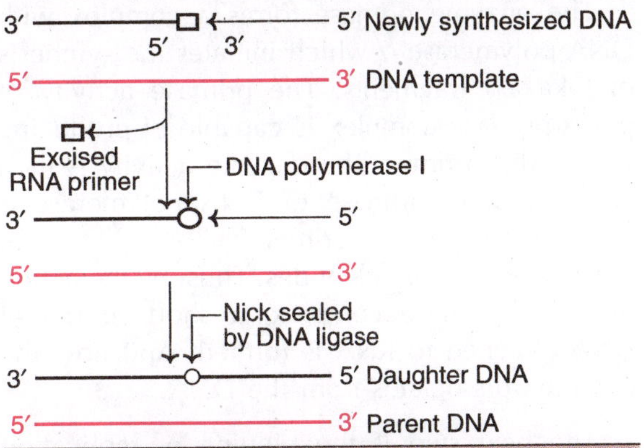

Fig. 24.6 : Overview of the action of DNA polymerase I and DNA ligase.

5. **DNA polymerase *ε*** is involved in proof-reading function of DNA replication.

The differences in the DNA replication between bacteria and human cells, attributed to the enzymes, are successfully used in antibacterial therapy to target pathogen (bacterial) replication and spare the host (human) cells.

PROCESS OF REPLICATION IN EUKARYOTES

The replication on the leading (continuous) strand of DNA is rather simple, involving ***DNA polymerase δ*** and a sliding clamp called ***proliferating cell nuclear antigen (PCNA)***. PCNA is so named as it was first detected as an antigen in the nuclei of replicating cells. PCNA forms a ring around DNA to which DNA polymerase δ binds. Formation of this ring also requires another factor namely ***replication factor C (RFC)***.

The ***replication on the lagging*** (discontinuous) ***strand in eukaryotes is more complex*** when compared to prokaryotes or even the leading strand of eukaryotes. This is depicted in ***Fig.24.7***, and briefly described hereunder.

The parental strands of DNA are separated by the enzyme helicase. A single-stranded DNA binding protein called ***replication protein A (RPA)*** binds to the exposed single-stranded template. This strand has been opened up by the replication fork (a previously formed Okazaki fragment with an RNA primer is also shown in ***Fig.24.4***).

The enzyme primase forms a complex with DNA polymerase α which initiates the synthesis of Okazaki fragments. The primase activity of pol α-primase complex is capable of producing 10-bp RNA primer. The enzyme activity is then switched from primase to DNA polymerase α which elongates the primer by the addition of 20–30 deoxyribonucleotides. Thus, by the action of pol α-primase complex, a short stretch of DNA attached to RNA is formed. And now the complex dissociates from the DNA.

The next step is the binding of replication factor C (RFC) to the elongated primer (short RNA-DNA). RFC serves as a clamp loader, and catalyses the assembly of proliferating cell nuclear antigen (PCNA) molecules. The DNA polymerase δ binds to the sliding clamp and elongates the Okazaki fragment to a final length of about 150–200 bp. By this elongation, the replication complex approaches the RNA primer of the previous Okazaki fragment.

The RNA primer removal is carried out by a pair of enzymes namely RNase H and flap endonuclease I (FENI). This gap created by RNA removal is filled by continued elongation of the new Okazaki fragment (carried out by polymerase δ, described above). The small nick that remains is finally sealed by DNA ligase.

Eukaryotic DNA is tightly bound to histones (basic proteins) to form nucleosomes which, in turn, organize into chromosomes. During the course of replication, the chromosomes are relaxed and the nucleosomes get loosened. The DNA strands separate for replication, and the parental histones associate with one of the parental strands. As the synthesis of new DNA strand proceeds, histones are also produced simultaneously, on the parent strand. At the end of replication, of the two daughter chromosomal DNAs formed, one contains the parental histones while the other has the newly synthesized histones.

INHIBITORS OF DNA REPLICATION

Bacteria contain a specific type II topoisomerase namely ***gyrase***. This enzyme cuts and reseals the circular DNA (of bacteria), and thus overcomes the problem of supercoils. Bacterial gyrase is ***inhibited*** by the antibiotics ***ciprofloxacin, novobiocin and nalidixic acid***. These are widely used as antibacterial agents since they can effectively block the replication of DNA and multiplication of cells. These antibacterial agents have almost ***no effect on human enzymes***.

Certain compounds that ***inhibit human topoisomerases*** are used as anticancer agents e.g. adriamycin, etoposide, doxorubicin. The nucleotide analogs that inhibit DNA replication are also used as anticancer drugs e.g. 6-mercaptopurine, 5-fluorouracil.

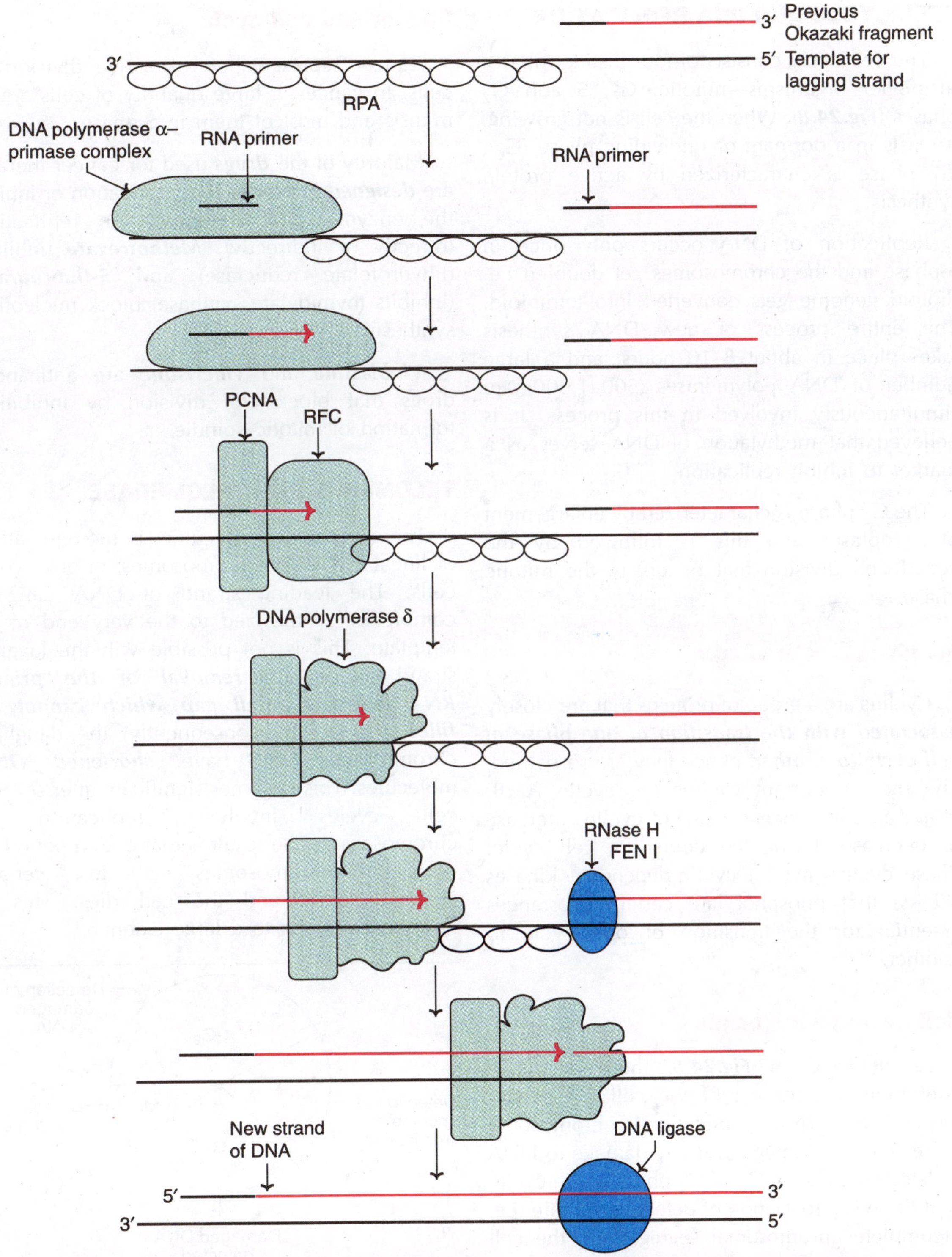

Fig. 24.7 :** An outline of DNA replication on the lagging strand in eukaryotes (RPA–Replication protein A; PCNA–Proliferating cell nuclear antigen; RFC–Replication factor C; RNase H–Ribonuclease H; FENI–Flap endonuclease I;* ***Note : *Leading strand not shown).*

CELL CYCLE AND DNA REPLICATION

The cell cycle consists of four distinct phases in higher organisms—mitotic, G_1, S and G_2 phases (***Fig.24.8***). When the cell is not growing, it exists in a dormant or undividing phase (G_0). G_1 phase is characterized by active protein synthesis.

Replication of DNA occurs only once in S-phase and the chromosomes get doubled i.e. diploid genome gets converted into tetraploid. The entire process of new DNA synthesis takes place in about 8–10 hours, and a large number of DNA polymerases (500–1,000) are simultaneously involved in this process. It is believed that methylation of DNA serves as a marker to inhibit replication.

The G_2 phase is characterized by enlargement of cytoplasm and this is followed by the actual cell division that occurs in the mitotic phase.

Cyclins and cell cycle

Cyclins are a group of proteins that are closely ***associated with the transition of one phase of cell cycle to another***, hence they are so named. The most important cyclins are cyclin A, B, D and E. The concentrations of cyclins increase or decrease during the course of cell cycle. These cyclins act on cyclin-dependent kinases (CDKs) that phosphorylate certain substances essential for the transition of one cycle to another.

Cell cycle check points

As depicted in ***Fig.24.8***, there occurs a continuous monitoring of the cell cycle with respect to DNA replication, chromosome segregation and integrity. If any damage to DNA is detected either in G_1 or G_2 phase of the cycle, or if there is a formation of defective spindle (i.e. incomplete chromosomal segregation), the cell cycle will not progress until appropriately corrected. If it is not possible to repair the damage done, the cells undergo ***apoptosis*** (programmed cell death).

Cancer and cell cycle

Cancer represents an excessive division of cells. In cancer, a large quantity of cells are in mitosis and most of them in S-phase.

Majority of the ***drugs*** used for cancer therapy are ***designed to block DNA*** replication or inhibit the enzymes that participate in replication (directly or indirectly). ***Methotrexate*** (inhibits dihydrofolate reductase) and ***5-fluorouracil*** (inhibits thymidylate synthase) block nucleotide synthesis.

Vinblastine and ***vincristine*** are anticancer drugs that block cell division by inhibiting formation of mitotic spindle.

TELOMERES AND TELOMERASE

There are certain difficulties in the replication of linear DNAs (or chromosomes) of eukaryotic cells. The leading strand of DNA can be completely synthesized to the very end of its template. This is not possible wih the lagging strand, since the ***removal of the primer RNA leaves a small gap which cannot be filled*** (***Fig.24.9A***). Consequently, the daughter chromosomes will have ***shortened DNA*** molecules. This becomes significant after several cell cycles involving replication of chromosomes. The result is that over a period of time, the chromosomes may lose certain essential genes and the cell dies. This is however, avoided to a large extent.

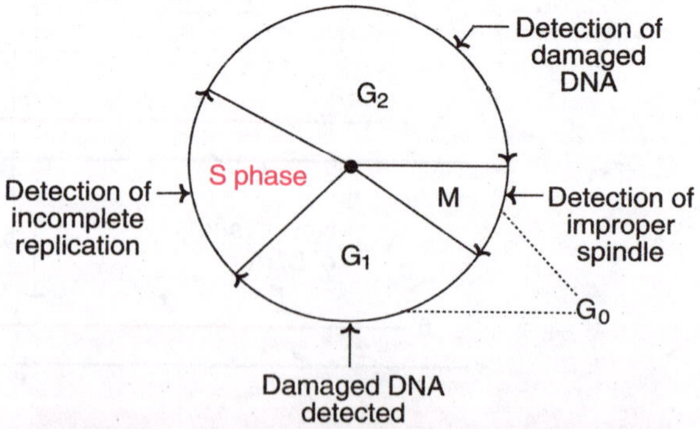

Fig. 24.8 : *The cell cycle of a mammalian cell (M–Mitotic phase; G_1–Gap1 phase; G_0–Dormant phase; S phase – Period of replication; G_2–Gap 2 phase).*

Telomeres are the special structures that prevent the continuous loss of DNA at the end of the chromosomes during the course of replication. Thus, they protect the ends of the chromosomes, and are also responsible to prevent the chromosomes from fusing with each other. Telomeres are many repeat sequences of six nucleotides present at the ends of eukaryotic chromosomes. Human telomeres contain thousands of ***repeat TTAGGG sequences***, which can be up to a length of 1500 bp.

Role of telomerase

Telomeres are maintained by the enzyme telomerase, also called as ***telomere terminal transferase***. Telomerase is an unusual enzyme as it is composed of both protein and RNA. In case of humans, the RNA component is 450 nucleotides in length, and at the 5′-terminal and it contains the sequence 5′-CUAACCCUAAC-3′. It may be noted that the central region of this sequence is complementary to the telomere repeat sequence 5′-TTAGGG-3′. The telomerase RNA sequence can be used as a template for extension of telomeres (***Fig.24.9B***).

The telomerase RNA base pairs to the end of the DNA molecule with telomeres and extends to a small distance. Then translocation of telomerase occurs and a fresh extension of DNA takes place. This process of DNA synthesis and translocation is repeated several times until the chromosome gets sufficiently extended. The extension process gets completed through the participation of DNA polymerase and primase complex and sealing of the new DNA formed.

It may be noted here that as such the telomeres do not encode proteins. Hence, when extended by telomerase, they need not have to remain the same length, and some shortening will not pose any problem. During the course of repeated cell cycles, there occurs ***progressive shortening of telomeres***, and this has to be ***prevented***, which is appropriately carried out ***by telomerase***.

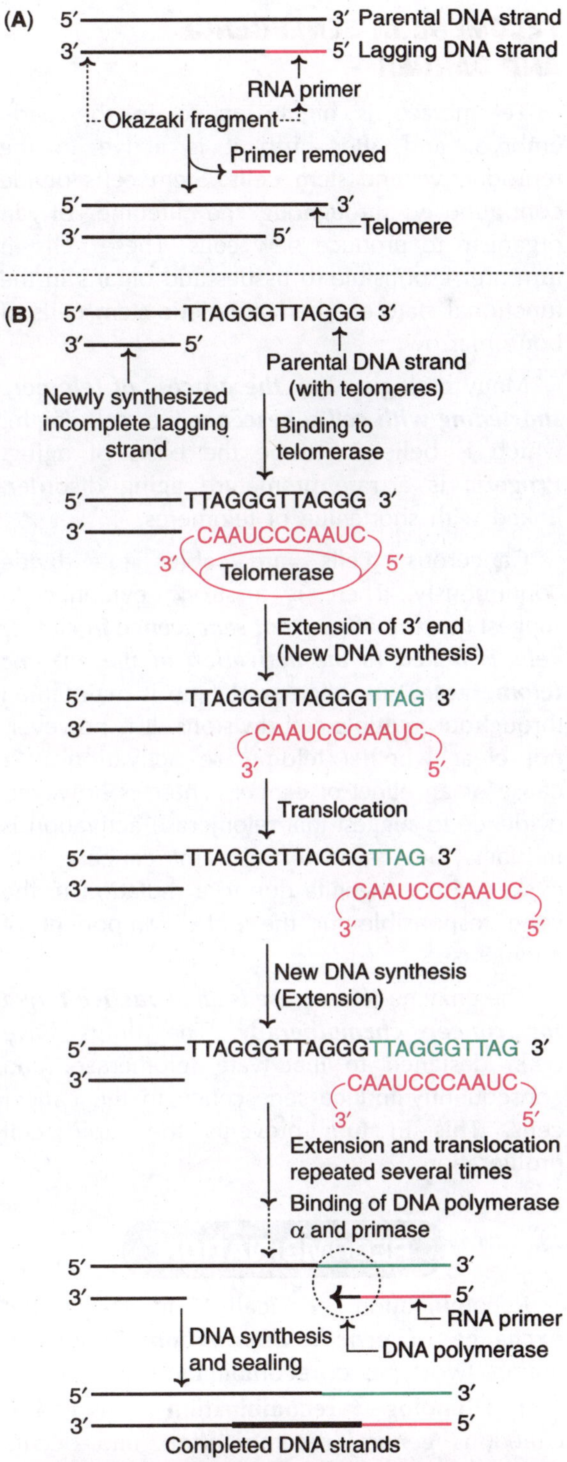

Fig. 24.9 : Replication of DNA with telomeres (A) Formation of telomere (B) Role of telomerase in the replication of telomere (Note : Only 2/3 repeat sequences of telomere shown for clarity).

TELOMERE IN SENESCENCE AND CANCER

Telomerase is highly active in the early embryo, and after birth it is active in the reproductive and stem cells. Stem cells divide continuously throughout the lifetime of an organism to produce new cells. These cells in turn are responsible to tissues and organs in the functional state e.g. hematopoietic stem cells of bone marrow.

Many biologists ***link the process of telomere shortening with cell senescence*** (i.e. cell death). which is believed to be the basis of aging. ***Progeria*** is a rare premature aging disorder, linked with shortening of telomeres.

Cancerous cells are able to divide continuously. There is a strong evidence to suggest that the ***absence of senescence in cancer cells is linked to the activation of the enzyme telomerase***. Thus, telomere length is maintained throughout multiple cell divisions. It is however, not clear whether telomerase activation is a cause or an effect of cancer. There is however, evidence to suggest that telomerase activation is in fact the cause of certain cancers e.g. dyskeratosis congenita due to a mutation in the gene responsible for the RNA component of telomerase.

The enzyme ***telomerase is an attractive target for cancer chemotherapy***. The drugs have been designed to inactivate telomerase, and consequently induce senescence in the cancer cells. This in turn prevents the rapid cell proliferation.

RECOMBINATION

Recombination basically involves the ***exchange of genetic information***. There are mainly two types of recombinations.

1. **Homologous recombination :** This is also called as ***general recombination***, and occurs between identical or nearly identical chromosomes (DNA sequences). The best example is the recombination between the paternal and maternal chromosomal pairs (***Fig.24.10***).

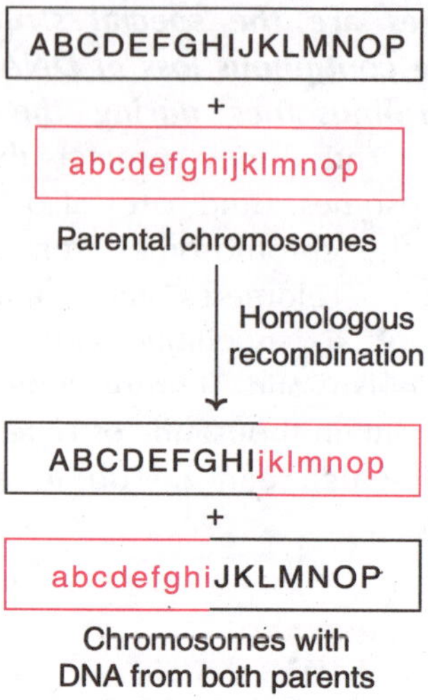

Fig. 24.10 : A diagrammatic representation of homologous recombination.

2. **Non-homologous recombination :** This is regarded as ***illegitimate recombination*** and does not require any special homologous sequences. ***Transposition*** is a good example of non-homologous recombination. Random integration of outside genes into mammalian chromosomes is another example.

HOMOLOGOUS RECOMBINATION

It is a known fact that the chromosomes are not passed on intact from generation to generation. Instead, they are inherited from both the parents. This is possible due to homologous recombination. Three models have been put forth to explain homologous recombinations.

- ***Holliday model***
- ***Meselson-Radding model***
- ***Double-strand break model.***

NON-HOMOLOGOUS RECOMBINATION

The recombination process without any special homologous sequences of DNA is regarded as non-homologous recombination.

Transposition

Transposition primarily involves the ***movement of specific pieces of DNA in the genome***. The mobile segments of DNA are called

Medical Concepts/ Clinical Correlates

- A few picograms (10^{-12} g) of DNA in a fetal cell stores the genetic information that will determine the differentiation and every function of an adult animal. This is the marvel of molecular biology.
- Topoisomerase inhibitors (e.g. adriamycin, etoposide) are useful to prevent DNA replication, and thus uncontrolled cell proliferation in cancer. These compounds block the unwinding of DNA strands.
- The progressive shortening of telomeres (DNA sequences at chromosomal ends) is prevented by telomerase. This enzyme is an attractive target for cancer therapy.

transposons or ***transposable elements***. They were first discovered by Barbara McClintock (in 1950) in maize, and their significance was ignored for about two decades by other workers.

Transposons are mobile and can move almost to any place in the target chromosome. There are two modes of transposition. One that involves an RNA intermediate, and the other which does not involve RNA intermediate.

Significance of transposition

It is now widely accepted that a large fraction of the human genome has resulted due to the accumulation of transposons. ***Short interspersed elements (SINEs)*** are repeats of DNA sequences which are present in about 500,000 copies per haploid human genome e.g. ***Alu sequences***.

Long interspersed elements (LINEs) are also repeated DNA sequences and are present in about 50,000 copies in the human genome e.g. L1 elements.

Some of the diseases caused by mutations are due to insertion of transposons into genes.

DAMAGE AND REPAIR OF DNA

Being the carrier of genetic information, the cellular DNA must be replicated (duplicated), maintained, and passed down to the daughter cells accurately. In general, the accuracy of replication is extremely high. However, there do occur replication errors. It is estimated that approximately one error is introduced per billion base pairs during each cycle of replication. The cells do possess the capability to repair damages done to DNA to a large extent.

Consequences of DNA damage

Despite an efficient repair system for the damaged DNA, replication errors do accumulate that ultimately result in mutations. The human body possesses 10^{14} nucleated cells, each with 3×10^9 base pairs of DNA. It is estimated that about 10^{16} cell divisions occur in a lifetime. If 10^{-10} mutations per base pair per cell generation escape repair, this results in about one mutation per 10^6 base pairs in genome.

Besides the possible errors in replication, the DNA is constantly subjected to attack by both physical and chemical agents. These include ***radiation, free radicals, chemicals*** etc., which also result in ***mutations***.

It is fortunate that a great majority of the mutations probably occur in the DNA that does not encode proteins, and consequently will not have any serious impact on the organism. This is not, however, all the time true, since mutations do occur in the coding regions of DNA also. There are situations in which the change in a single base pair in the human genome can cause a serious disease e.g. sickle-cell anemia.

TYPES OF DNA DAMAGES

The damages done to DNA by physical, chemical and environmental agents may be broadly classified into ***four categories*** with different types (***Table 24.1***).

The DNA damage may occur due to ***single-base alterations*** (e.g. depurination, deamination), ***two-base alterations*** (e.g. pyrimidine dimer) ***chain breaks*** (e.g. ionizing radiation) and ***cross-linkages*** (e.g. between bases). Some selected DNA damages are briefly described.

The production of reactive oxygen species is often associated with alteration of bases e.g.

Table 24.1 Major types of DNA damages

Category	Types
Single-base alteration	Deamination (C→U; A→hypoxanthine)
	Depurination
	Base alkylation
	Insertion or deletion of nucleotides
	Incorporation of base analogue
Two-base alteration	UV light induced pyrimidine dimer alteration (T–T)
Chain breaks	Oxidative free radical formation
	Ionizing radiation
Cross-linkage	Between bases in the same or opposite strands
	Between the DNA and protein molecules

formation of 8-hydroxy guanine. Free radical formation and oxidative damage to DNA increases with advancement of age.

Ultraviolet radiations result in the formation of covalent links between adjacent pyrimidines along the DNA strand to form ***pyrimidine dimers***. DNA chain breaks can be caused by ionizing radiations (e.g. X-rays).

MUTATIONS

The genetic macromolecule DNA is highly stable with regard to its base composition and sequence. However, DNA is not totally exempt from gradual change.

Mutation refers to ***a change in the DNA structure of a gene.*** The substances (chemicals) which can induce mutations are collectively known as ***mutagens***.

The changes that occur in DNA on mutation are reflected in replication, transcription and translation.

Types of mutations

Mutations are mainly of two major types—point mutations, frameshift mutations (***Fig.24.11***).

1. **Point mutations :** The replacement of one base pair by another results in point mutation. They are of two sub-types.

(a) **Transitions :** In this case, a purine (or a pyrimidine) is replaced by another.

(b) **Transversions :** These are characterized by replacement of a purine by a pyrimidine or vice versa.

2. **Frameshift mutations :** These occur when one or more base pairs are inserted in or deleted from the DNA, respectively, causing ***insertion or deletion mutations.***

Consequences of point mutations

The change in a single base sequence in point mutation may cause one of the following (***Fig.24.12***).

1. **Silent mutation :** The codon (of mRNA) containing the changed base may code for the same amino acid. For instance, *UCA* codes for serine and change in the third base *(UCU)* still codes for serine. This is due to degeneracy of the genetic code. Therefore, there are ***no detectable effects*** in silent mutation.

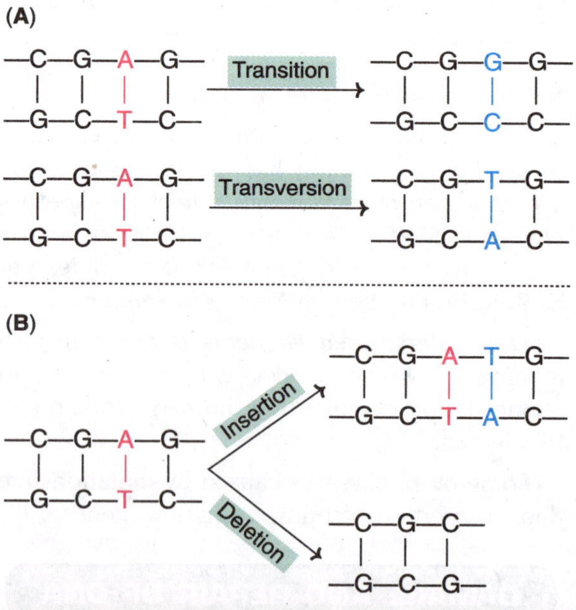

Fig. 24.11 :** An illustration of mutations* ***(A) *-Point mutations;* ***(B)*** *-Frameshift mutations.*

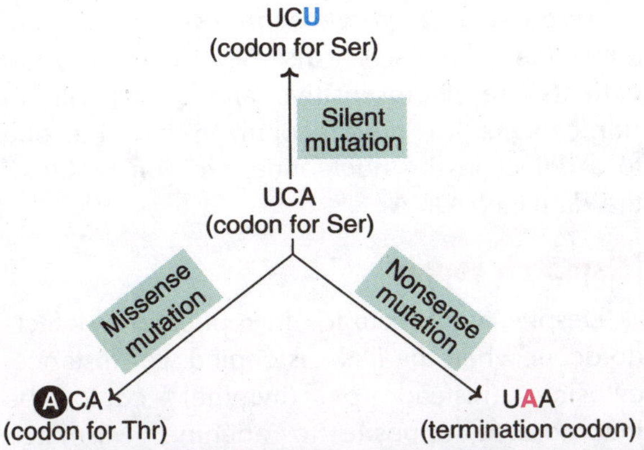

Fig. 24.12 : An illustration of point mutations (represented by a codon of mRNA).

2. **Missense mutation** : In this case, the changed base may code for a different amino acid. For example, *UCA* codes for serine while *ACA* codes for threonine. The mistaken (or missense) amino acid may be ***acceptable, partially acceptable*** or ***unacceptable*** with regard to the function of protein molecule. ***Sickle-cell anemia*** is a classical ***example of missense mutation***.

3. **Nonsense mutation** : Sometimes, the codon with the altered base may become a ***termination*** (or nonsense) ***codon***. For instance, change in the second base of serine codon *(UCA)* may result in *UAA*. The altered codon acts as a stop signal and causes termination of protein synthesis, at that point.

Consequences of frameshift mutations

The insertion or deletion of a base in a gene results in an ***altered reading frame of the mRNA*** (hence the name frameshift). The machinery of mRNA (containing codons) does not recognize that a base was missing or a new base was added. Since there are no punctuations in the reading of codons, translation continues. The result is that the protein synthesized will have several altered amino acids and/or prematurely terminated protein.

Mutations and cancer

Mutations are permanent alterations in DNA structure, which have been implicated in the etiopathogenesis of cancer.

REPAIR OF DNA

As already stated, damage to DNA caused by replication errors or mutations may have serious consequences. The cell possesses an inbuilt system to repair the damaged DNA. This may be achieved by four distinct mechanisms (***Table 24.2***).

1. ***Base excision-repair***
2. ***Nucleotide excision-repair***
3. ***Mismatch repair***
4. ***Double-strand break repair***.

Table 24.2 Major mechanisms of DNA repair

Mechanism	Damage to DNA	DNA repair
Base excision-repair	Damage to a single base due to spontaneous alteration or by chemical or radiation means.	Removal of the base by N–glycosylase; abasic sugar removal, replacement.
Nucleotide excision-repair	Damage to a segment of DNA by spontaneous, chemical or radiation means.	Removal of the DNA fragment (≃ 30 nt length) and replacement.
Mismatch repair	Damage due to copying errors (1-5 base unpaired loops).	Removal of the strand (by exonuclease digestion) and replacement.
Double-strand break repair	Damage caused by ionizing radiations, free radicals, chemotherapy etc.	Unwinding, alignment and ligation.

Medical Concepts/ Clinical Correlates

- Mutations may sometimes result in serious diseases. e.g. sickle-cell anemia, cancer.
- Xeroderma pigmentosum is a rare disease characterized by photosensitivity and risk for skin cancer. This is due to a defect in the nucleotide excision repair of the damaged DNA (caused by UV rays).
- Hereditary nonpolyposis colon cancer is a common inherited cancer, and is due to a faulty mismatch repair of defective DNA.

Base excision-repair

The bases cytosine, adenine and guanine can undergo spontaneous depurination to respectively form uracil, hypoxanthine and xanthine. These altered bases do not exist in the normal DNA, and therefore need to be removed. This is carried out by base excision repair (***Fig.24.13***).

A defective DNA in which cytosine is deaminated to uracil is acted upon by the enzyme uracil DNA glycosylase. This results in the removal of the defective base uracil. An endonuclease cuts the backbone of DNA strand near the defect and removes a few bases. The gap so created is filled up by the action of repair DNA polymerase and DNA ligase.

Nucleotide excision-repair

The DNA damage due to ultraviolet light, ionizing radiation and other environmental factors often results in the modification of certain bases, strand breaks, cross-linkages etc. Nucleotide excision-repair is ideally suited for such large-scale defects in DNA. After the identification of the defective piece of the DNA, the DNA double helix is unwound to expose the damaged part. An ***excision nuclease*** (exinuclease) cuts the DNA on either side (upstream and downstream) of the damaged DNA. This defective piece is degraded. The gap created by the nucleotide excision is filled up by DNA polymerase which gets ligated by DNA ligase (***Fig.24.14***).

Xeroderma pigmentosum (XP) is a rare autosomal recessive disease. The affected patients are photosensitive and susceptible to skin cancers. It is now recognized that XP is due to a defect in the nucleotide excision repair of the damaged DNA.

Mismatch repair

Despite high accuracy in replication, defects do occur when the DNA is copied. For instance, cytosine (instead of thymine) could be incorporated opposite to adenine. Mismatch repair corrects a single mismatch base pair e.g. C to A, instead of T to A.

The template strand of the DNA exists in a methylated form, while the newly synthesized strand is not methylated. This difference allows the recognition of the new strands. The enzyme

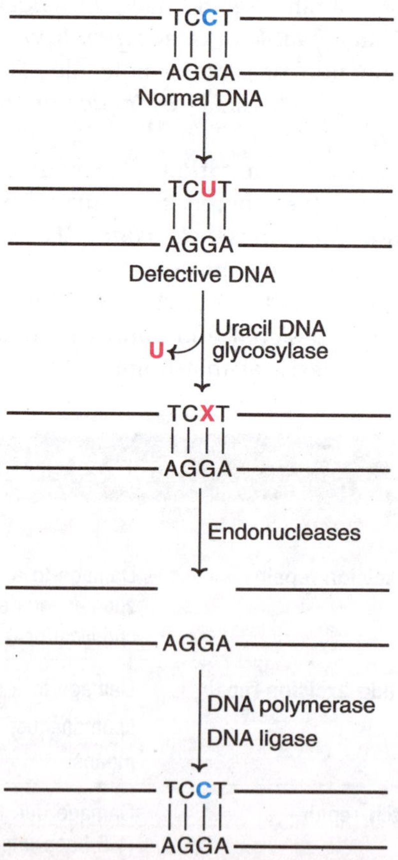

***Fig. 24.13 :** A diagrammatic representation of base excision-repair of DNA.*

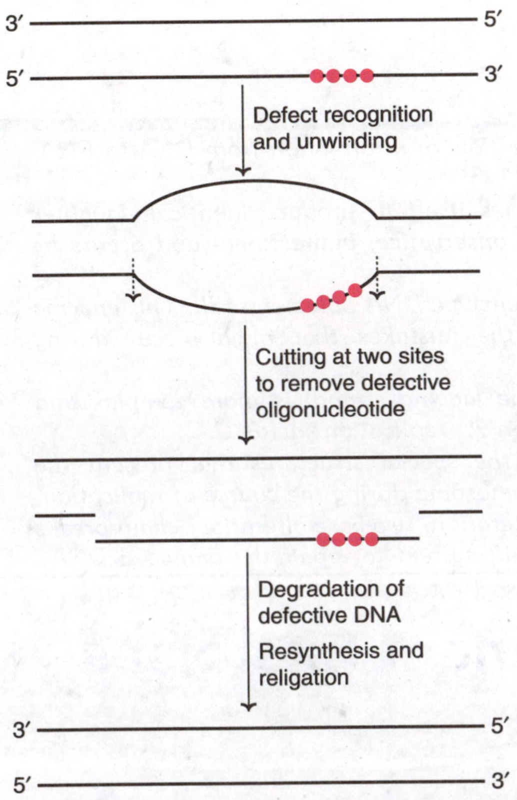

Fig. 24.14 : *A diagrammatic representation of nucleotide excision-repair of DNA.*

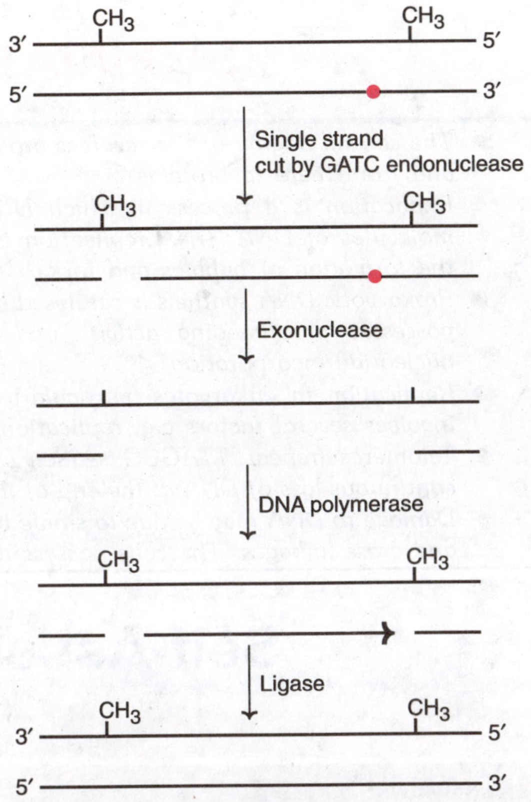

Fig. 24.15 : *A diagrammatic representation of mismatch repair of DNA.*

GATC endonuclease cuts the strand at an adjacent methylated GATC sequence. (***Fig.24.15***). This is followed by an exonuclease digestion of the defective strand, and thus its removal. A new DNA strand is now synthesized to replace the damaged one.

Hereditary nonpolyposis colon cancer (HNPCC) is one of the most common inherited cancers. This cancer is now linked with ***faulty mismatch repair*** of defective DNA.

Double-strand break repair

Double-strand breaks (DSBs) in DNA are dangerous. They result in genetic recombination which may lead to chromosomal translocation, broken chromosomes, and finally cell death. DSBs can be repaired by homologous recombination or non-homologous end joining. Homologous recombination occurs in yeasts while in mammals, non-homologous and joining dominates.

DEFECTS IN DNA REPAIR AND CANCER

Cancer develops when certain genes that regulate normal cell division fail or are altered. Defects in the genes encoding proteins involved in nucleotide-excision repair, mismatch repair and recombinational repair are linked to human cancers. For instance, as already referred above, ***HNPCC is due to a defect in mismatch repair.***

SUMMARY

- *The central dogma of life revolves around the flow of information from DNA to RNA, and from there to proteins.*
- *Replication is a process in which DNA copies itself to produce identical daughter molecules of DNA. DNA replication is semiconservative, bidirectional and occurs by the formation of bubbles and forks.*
- *Prokaryotic DNA syntheis is catalysed by the enzyme DNA polymerase III. This enzyme possesses proof-reading activity and edits the mistakes that might occur during nucleotide incorporation.*
- *Replication in eukaryotes (particularly on the lagging strand) is more complex and involves several factors e.g. replication protein A, replication factor C.*
- *Telomeres (repeat TTAGGG sequences) are the special structures that prevent the continuous loss of DNA at the end of the chromosome during the course of replication.*
- *Damage to DNA may be due to single base alteration, two-base alteration, chain breaks and cross linkages. The cells possess an inbuilt system to repair the damaged DNA*

Self-Assessment Exercises

I. Essay questions

1. Describe the replication of DNA.
2. Discuss different types of DNA damages, and the repair mechanisms.
3. What are mutations? Describe different types, and consequences of mutations.
4. Give an account of telomeres and their role in senescence and cancer.

II. Short notes

(a) Replication fork, (b) Okazaki pieces, (c) RNA primer, (d) DNA topoisomerases, (e) Inhibitors of DNA replication, (f) Telomerase, (g) Holliday model of DNA recombination, (h) Transposition.

III. Fill in the blanks

1. DNA strands for replication process are separated by the enzyme ______________.
2. The small fragments of DNA produced during replication are called ______________.
3. During the course of DNA replication, the proof-reading function is carried out by the enzyme ______________.
4. The problem of supercoils in DNA replication is overcome by a group of enzymes, namely ______________.
5. The proteins that are associated with the transition of one phase of cell cycle to another ______________.
6. The mobile segments of DNA are called ______________.
7. Sickle-cell anemia is a good example of ______________ mutation.

IV. Multiple choice questions

8. The chemical nature of the primer required for the synthesis of DNA
   (a) DNA (b) Histone (c) RNA (d) hnRNA.
9. The enzyme responsible for the synthesis of RNA primer in eukaryotes
   (a) DNA polymerase α (b) DNA polymerase β (c) DNA polymerase γ (d) Topisomerases.
10. The repeat sequence of nucleotides in telomeres
    (a) TTGGGA (b) TTAGGG (c) GGGATT (d) TTGAGG.

Transcription and Translation 25

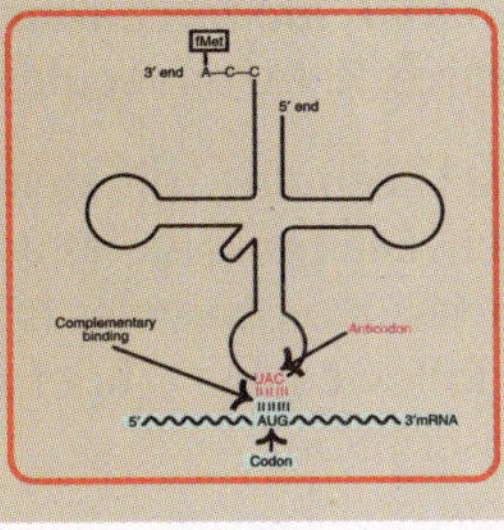

The genetic code speaks

"Triplet base sequence of messenger RNA, I am;
Universal, specific, non-overlapping, degenerate, in character;
Faithfully work under the dictates of DNA :
To execute my master's orders for protein synthesis."

The conventional concept of central dogma of life which in essence is "***DNA makes RNA makes protein***" is an oversimplification of molecular biology. With the advances in cell biology and rapid developments in bio-informatics, the terms ***genome***, ***transcriptome*** and ***proteome*** are ***in current use*** to represent the central dogma of life (***Fig.25.1***).

GENOME

The ***total DNA*** (genetic information) contained ***in an organism*** or ***a cell*** is regarded as the genome. Thus, the genome is the storehouse of biological information. It includes the chromosomes in the nucleus and the DNA in mitochondria, and chloroplasts.

Genomics : The study of the structure and function of genome is genomics. The term ***functional genomics*** is used to represent the gene expression and relationship of genes with gene products. ***Structural genomics*** refers to the structural motifs and complete protein structures. ***Comparative genomics*** involves the study of comparative gene function and phylogeny. ***Metagenomics*** refers to the study of genomes of whole communities of microscopic life (microorganisms, viruses).

TRANSCRIPTOME

The ***RNA copies of*** the active protein coding ***genes*** represent transcriptome. Thus,

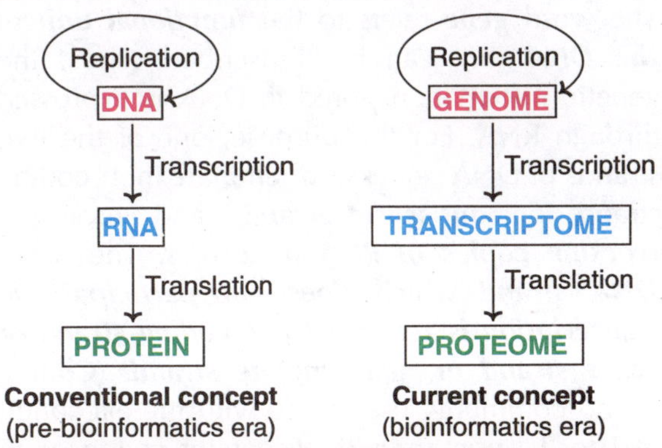

Fig. 25.1 : The central dogma of life (or molecular biology) represented in the form of conventional and current concepts.

transcriptome is the initial product of gene expression which directs the synthesis of proteins.

Transcriptomics : The study of transcriptome that involves ***all the RNA molecules*** made by a cell, tissue or an organism is transcriptomics.

PROTEOME

The ***cell's repertoire*** (repository/storehouse) of ***proteins*** with their nature ***and biological functions*** is regarded as proteome. Thus, proteome represents the entire range of proteins and their biological functions in a cell.

Proteomics : The study of the proteome.

Metabolomics : The study of chemical processes involving metabolites.

Whether the central dogma of life is represented in the conventional or more recent form, replication, transcription and translation are the key or core processes that ultimately control life. Replication of DNA has been described in ***Chapter 24***, while transcription and translation are discussed in this chapter.

TRANSCRIPTION

Transcription is a process in which ribonucleic acid ***(RNA) is synthesized from DNA***. The word ***gene*** refers to the ***functional unit of the DNA*** that can be transcribed. Thus, the genetic information stored in DNA is expressed through RNA. For this purpose, one of the two strands of DNA serves as a ***template*** (non-coding strand or ***antisense strand***) and produces ***working copies of RNA molecules***. The ***other DNA strand*** which ***does not participate in transcription is referred to as coding strand or sense strand or non-template strand***. (Coding strand commonly used since with the exception of T for U, primary mRNA contains codons with the same base sequence).

Transcription is selective

The entire molecule of DNA is not expressed in transcription. RNAs are synthesized only for some selected regions of DNA. For certain other regions of DNA, there may not be any transcription at all. The exact reason for the selective transcription is not known. This ***may be due to some inbuilt signals in the DNA molecule***.

The product formed in transcription is referred to as ***primary transcript.*** Most often, the primary RNA transcripts are inactive. They undergo certain alterations (splicing, terminal additions, base modifications etc.) commonly known as ***post-transcriptional modifications***, to produce functionally active RNA molecules.

There exist certain differences in the transcription between prokaryotes and eukaryotes. The RNA synthesis in prokaryotes is given in some detail. This is followed by a brief discussion on eukaryotic transcription.

TRANSCRIPTION IN PROKARYOTES

A single enzyme—DNA dependent RNA polymerase or simply ***RNA polymerase***—synthesizes all the RNAs in prokaryotes. RNA polymerase of *E. coli* is a complex holoenzyme (mol wt. 465 kDa) with five polypeptide subunits—2α, 1β and 1β′ and ***one sigma(s) factor*** (***Fig.25.2***). The enzyme without sigma factor is referred to as core enzyme ($\alpha_2\beta\beta'$).

An overview of RNA synthesis is depicted in ***Fig.25.3***. Transcription involves three different stages—initiation, elongation and termination (***Fig.25.4***).

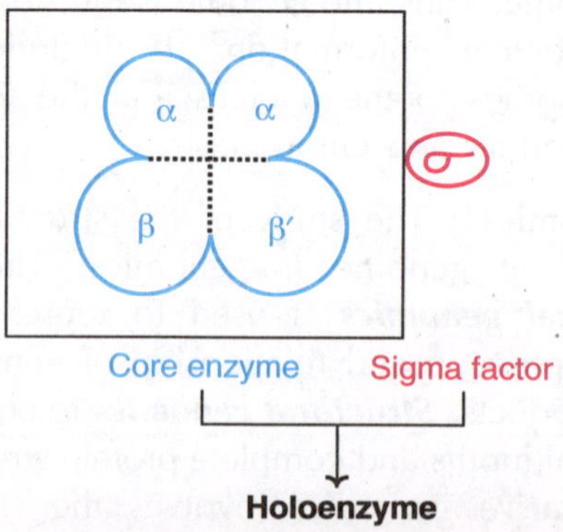

Fig. 25.2 : *RNA polymerase of* E. coli.

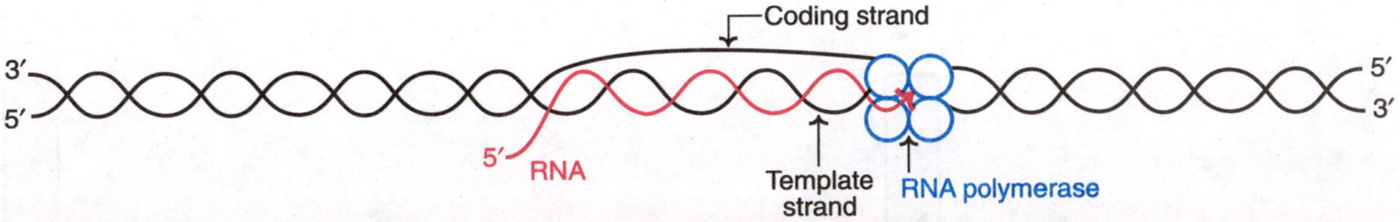

Fig. 25.3 : An overview of transcription.

Initiation

The binding of the enzyme RNA polymerase to DNA is the prerequisite for the transcription to start. The specific region on the DNA where the enzyme binds is known as ***promoter region.*** There are two base sequences on the ***coding DNA strand*** which the sigma factor of RNA polymerase can recognize for initiation of transcription (***Fig.25.5***).

1. **Pribnow box (TATA box) :** This consists of 6 nucleotide bases (TATAAT), located on the left side about 10 bases away (upstream) from the starting point of transcription.

2. **The '–35' sequence :** This is the second recognition site in the promoter region of DNA. It contains a base sequence TTGACA, which is located about 35 bases (upstream, hence –35) away on the left side from the site of transcription start.

Elongation

As the holoenzyme, RNA polymerase recognizes the promoter region, the sigma factor is released and transcription proceeds. ***RNA is synthesized from 5′ end to 3′ end (5′→3′)*** antiparallel to the DNA template. RNA polymerase ***utilizes*** ribonucleotide triphosphates (***ATP, GTP, CTP and UTP***) for the formation of RNA. For the addition of each nucleotide to the growing chain, a pyrophosphate moiety is released.

The sequence of nucleotide bases in the ***mRNA is complementary to the template DNA strand***. It is however, identical to that of coding strand except that RNA contains U in place of T in DNA (***Fig.25.6***).

RNA polymerase differs from DNA polymerase in two aspects. ***No primer is required*** for RNA polymerase and, further, this enzyme does not possess endo- or exonuclease activity. Due to lack of the latter function (proof-reading activity), RNA polymerase has no ability to repair the mistakes in the RNA synthesized. This is in contrast to DNA replication which is carried out with high fidelity. It is, however, fortunate that mistakes in RNA synthesis are less dangerous, since they are not transmitted to the daughter cells.

The double helical structure of DNA unwinds as the transcription goes on, resulting in supercoils. The problem of supercoils is overcome by topoisomerases (more details in ***Chapter 24***).

Termination

The process of transcription ***stops by termination signals***. Two types of termination are identified.

1. **Rho (ρ) dependent termination :** A specific protein, named ***ρ factor***, binds to the growing RNA (and not to RNA polymerase) or weakly to DNA, and in the bound state it acts as ATPase and terminates transcription and releases RNA. The ρ factor is also responsible for the dissociation of RNA polymerase from DNA.

2. **Rho (ρ) independent termination :** The termination in this case is brought about by the formation of ***hairpins*** of newly synthesized RNA. This occurs due to the presence of ***palindromes***. A palindrome is a word that reads alike forward and backward e.g. ***madam***, ***rotor***. The presence of palindromes in the base sequence of DNA template (same when read in opposite direction) in the termination region is known. As a result of this, the newly synthesized RNA folds to form hairpins (due to complementary base pairing) that cause termination of transcription.

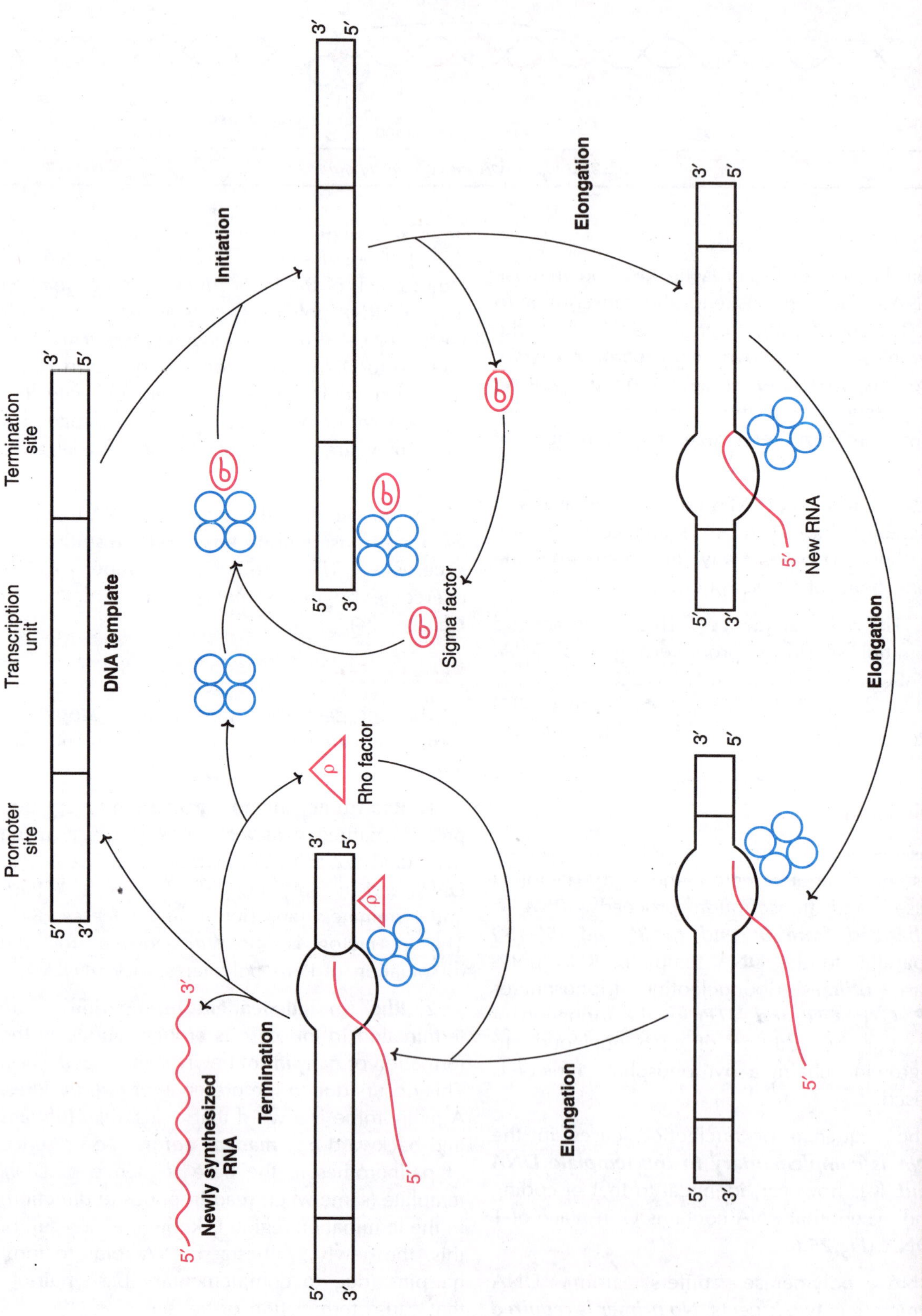

Fig. 25.4 : *Synthesis of RNA from DNA template (transcription).*

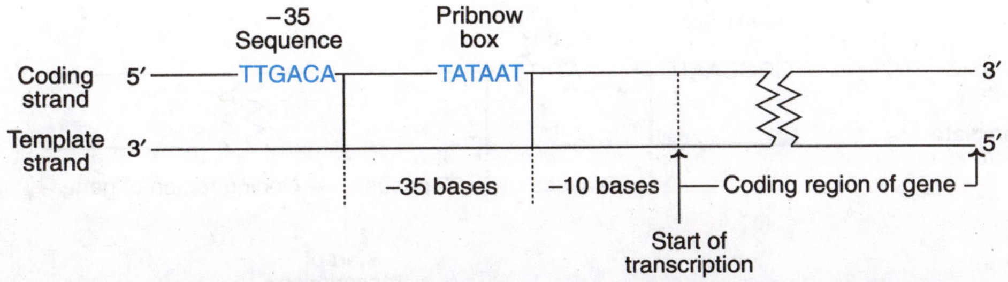

***Fig. 25.5 :** Promoter regions of DNA in prokaryotes.*

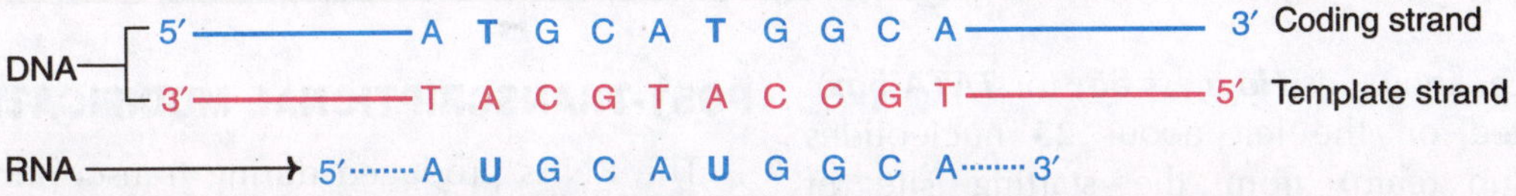

***Fig. 25.6 :** Transcription—Complementary base pair relationship.*

TRANSCRIPTION IN EUKARYOTES

RNA synthesis in eukaryotes is a much more complicated process than the transcription described above for prokaryotes. As such, all the details of eukaryotic transcription (particularly about termination) are not clearly known. The salient features of available information are given hereunder.

RNA polymerases

The nuclei of eukaryotic cells possess three distinct RNA polymerases (***Fig.25.7***).

1. **RNA polymerase I** is responsible for the synthesis of precursors for the large ***ribosomal RNAs***.

2. **RNA polymerase II** synthesizes the precursors for ***mRNAs and small nuclear RNAs***.

3. **RNA polymerase III** participates in the formation of ***tRNAs and small ribosomal RNAs***.

Besides the three RNA polymerases found in the nucleus, there also exists a mitochondrial RNA polymerase in eukaryotes. The latter resembles prokaryotic RNA polymerase in structure and function.

Promoter sites

In eukaryotes, a sequence of DNA bases—which is almost identical to ***pribnow box*** of prokaryotes—is identified (***Fig.25.8***). This

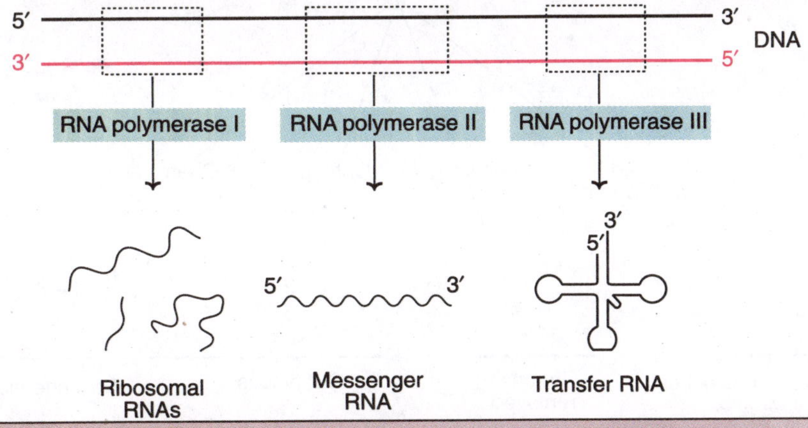

***Fig. 25.7 :** An overview of transcription in eukaryotes.*

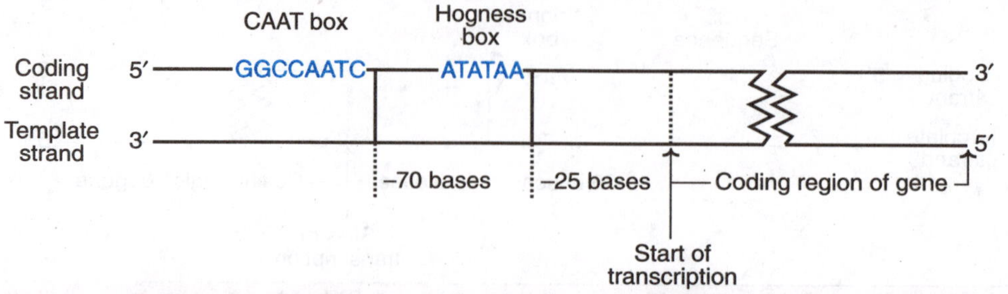

Fig. 25.8 : Promoter regions of DNA in eukaryotes.

sequence, known as ***Hogness box*** (or ***TATA box***), is located on the left about 25 nucleotides away (upstream) from the starting site of mRNA synthesis. There also exists another site of recognition between 70 and 80 nucleotides upstream from the start of transcription. This second site is referred to as ***CAAT box***. One of these two sites (or sometimes both) helps RNA polymerase II to recognize the requisite sequence on DNA for transcription.

Heterogeneous nuclear RNA (hnRNA)

The ***primary mRNA transcript*** produced by RNA polymerase II ***in eukaryotes*** is often referred to as heterogeneous nuclear RNA (hnRNA). This is then processed to produce mRNA needed for protein synthesis.

POST-TRANSCRIPTIONAL MODIFICATIONS

The RNAs produced during transcription are called primary transcripts. They undergo many alterations—***terminal base additions, base modifications, splicing*** etc., which are collectively referred to as post-transcriptional modifications. This process is required to convert the RNAs into the active forms. A group of enzymes, namely ribonucleases, are responsible for the processing of tRNAs and rRNAs of both prokaryotes and eukaryotes.

The prokaryotic mRNA synthesized in transcription is almost similar to the functional mRNA. In contrast, eukaryotic mRNA (i.e. hnRNA) undergoes extensive post-transcriptional changes.

An outline of the post-transcriptional modifications is given in ***Fig.25.9***, and some highlights are described.

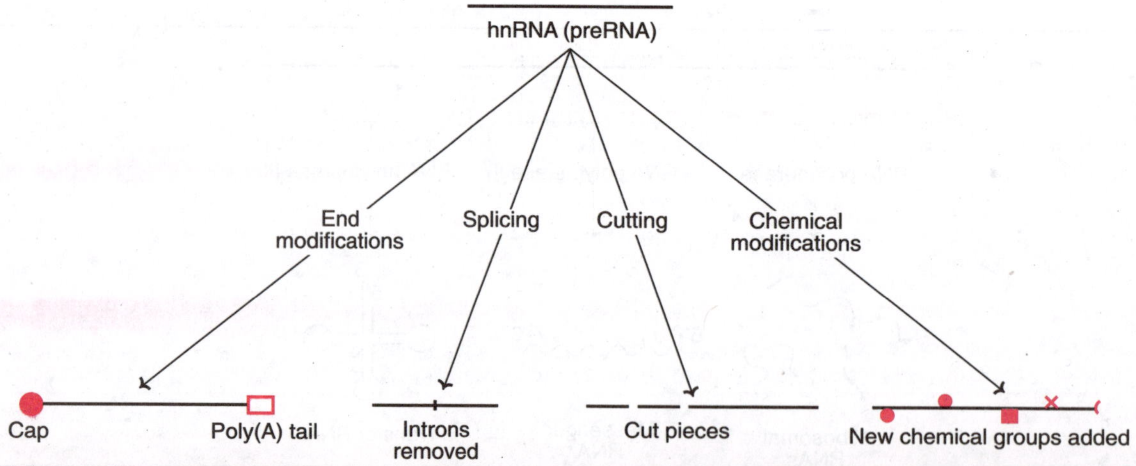

Fig. 25.9 : An outline of post-transcriptional modifications of RNA (hnRNA-Heterogeneous nuclear RNA).

Messenger RNA

The primary transcript of mRNA is the hnRNA in eukaryotes, which is subjected to many changes before functional mRNA is produced.

1. **The 5′ capping :** The 5′ end of mRNA is capped with 7-methylguanosine by an unusual 5′→ 5′ triphosphate linkage. S-Adenosylmethionine is the donor of methyl group. This cap is required for translation, besides stabilizing the structure of mRNA.

2. **Poly-A tail :** A large number of eukaryotic mRNAs possess an adenine nucleotide chain at the 3′-end. This poly-A tail, as such, is not produced during transcription. It is later added to stabilize mRNA. However, poly-A chain gets reduced as the mRNA enters cytosol.

3. **Introns and their removal :** Introns are the intervening nucleotide sequences in mRNA which do not code for proteins. On the other hand, ***exons of mRNA possess genetic code and are responsible for protein synthesis***. The splicing and excision of introns is illustrated in ***Fig.25.10***. The removal of introns is promoted by small nuclear ribonucleoprotein particles (snRNPs). snRNPs (pronounced as snurps) in turn, are formed by the association of small nuclear RNA (snRNA) with proteins.

The term ***spliceosome*** is used to represent the snRNP association with hnRNA at the exon-intron junction.

Post-transcriptional modifications of mRNA occur in the nucleus. The mature RNA then enters the cytosol to perform its function (translation).

Different mRNAs produced by alternate splicing

Alternate patterns of hnRNA splicing result in different mRNA molecules which can produce different proteins. Alternate splicing results in mRNA heterogeneity. In fact, the processing of hnRNA molecules becomes a site for the regulation of gene expression.

Transfer RNA

All the tRNAs of prokaryotes and eukaryotes undergo post-transcriptional modification. These include trimming, converting the existing bases into unusual ones, and addition of CCA nucleotides to 3′ terminal end of tRNAs.

Ribosomal RNA

The preribosomal RNAs originally synthesized are converted to ribosomal RNAs by a series of post-transcriptional changes.

Inhibitors of transcription

The synthesis of RNA is inhibited by certain antibiotics and toxins.

Actinomycin D : This is also known as dactinomycin. It is synthesized by *Streptomyces*. Actinomycin D binds with DNA template strand and ***blocks the movement of RNA polymerase***. This was the very first antibiotic used for the treatment of tumors.

Rifampin : It is an antibiotic widely used for the treatment of tuberculosis and leprosy. Rifampin binds with the β-subunit of ***prokaryotic RNA polymerase and inhibits its activity***.

α-Amanitin : It is a toxin produced by mushroom, *Amanita phalloides*. This mushroom is delicious in taste but poisonous due to the

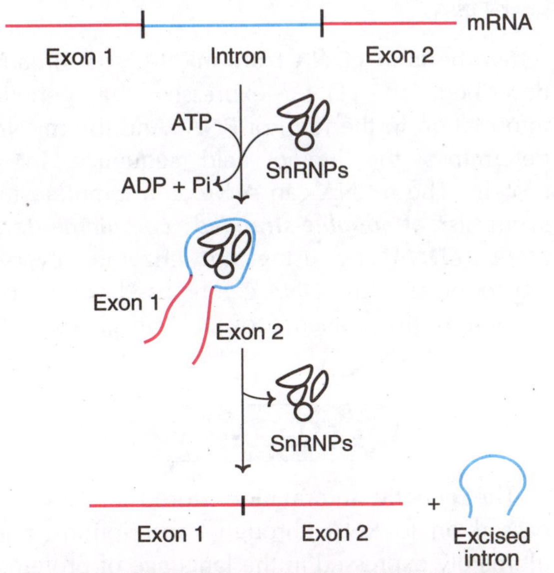

***Fig. 25.10 :** Formation of mature RNA from eukaryotic mRNA (SnRNPs–Small nuclear ribonucleoprotein particles).*

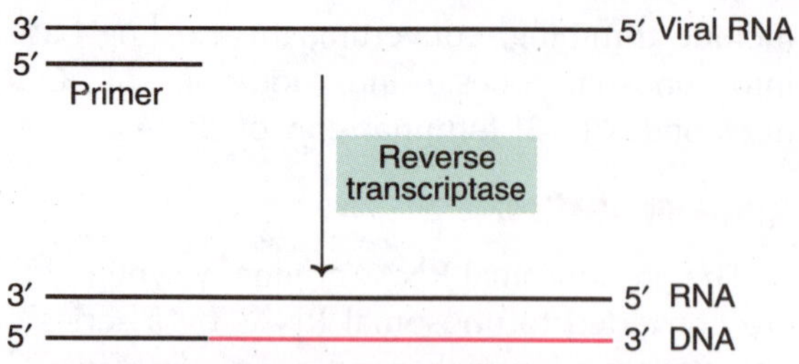

Fig. 25.11 : Reverse transcription of RNA virus.

toxin α-amanitin which tightly binds with RNA polymerase II of eukaryotes and inhibits transcription.

REVERSE TRANSCRIPTION

Some of the viruses—known as ***retroviruses***—possess RNA as the genetic material. These viruses cause cancers in animals, hence known as ***oncogenic***. They are actually found in the transformed cells of the tumors.

The enzyme RNA dependent DNA polymerase —or simply ***reverse transcriptase***— is responsible for the formation of ***DNA from RNA*** (***Fig.25.11***). This DNA is complementary (cDNA) to viral RNA and can be transmitted into host DNA.

Synthesis of cDNA from mRNA : As already described, the DNA expresses the genetic information in the form of RNA. And the mRNA determines the amino acid sequence in a protein. The mRNA can serve as a template for synthesis of ***double-stranded complementary DNA*** (***cDNA***) by using the enzyme reverse transcriptase. This cDNA can be used as a probe to identify the sequence of DNA in genes.

TRANSLATION

The genetic information stored in DNA is passed on to RNA (through transcription), and ultimately expressed in the language of proteins. The ***biosynthesis of a protein or a polypeptide in a living cell*** is referred to as translation. The term translation is used to represent the biochemical translation of four-letter language information from nucleic acids (DNA and then RNA) to 20 letter language of proteins. ***The sequence of amino acids in the protein synthesized is determined by the nucleotide base sequence of mRNA.***

Variability of cells in translation

There are wide variations in the cells with respect to the quality and quantity of proteins synthesized. This largely depends on the need and ability of the cells. ***Erythrocytes*** (red blood cells) lack the machinery for translation, and therefore ***cannot synthesize proteins***.

In general, the growing and dividing cells produce larger quantities of proteins. Some of the cells continuously synthesize proteins for export. For instance, liver cells produce albumin and blood clotting factors for export into the blood for circulation. The normal liver cells are very rich in the protein biosynthetic machinery, and thus the ***liver*** may be regarded as the ***protein factory in the human body***.

GENETIC CODE

The ***three nucleotide*** (triplet) ***base sequences in mRNA that act as code words for amino acids*** in protein constitute the genetic code or simply ***codons.*** The genetic code is regarded as a ***dictionary of nucleotide bases (A, G, C and U)*** that determines the sequence of amino acids in proteins.

The codons consist of the four nucleotide bases, the purines—adenine (A) and guanine (G), and the pyrimidines—cytosine (C) and uracil (U). These four bases produce 64 different combinations (4^3) of three base codons, as depicted in ***Table 25.1***. The nucleotide sequence of the codon on mRNA is written from the 5′-end to 3′ end. ***Sixty one codons code for the 20 amino acids found in protein.***

The three codons ***UAA, UAG*** and ***UGA*** do not code for amino acids. They act as ***stop signals*** in protein synthesis. These three codons are collectively known as ***termination codons*** or non-sense codons. The codons UAG, UAA and UGA are often referred to, respectively, as ***amber***, ***ochre*** and ***opal*** codons.

The codons ***AUG***—and, sometimes, ***GUG***—are the chain ***initiating codons***.

Other characteristics of genetic code

The genetic code is ***universal, specific, non-overlapping and degenerate***.

1. **Universality :** The same codons are used to code for the same amino acids in all the living organisms. Thus, the genetic code has been conserved during the course of evolution. Hence genetic code is appropriately regarded as universal. There are, ***however, a few exceptions***. For instance, AUA is the codon for methionine in mitochondria. The same codon (AUA) codes for isoleucine in cytoplasm. With some exceptions noted, the genetic code is universal.

2. **Specificity :** A particular ***codon always codes for the same amino acid***, hence the genetic code is highly specific or unambiguous e.g. UGG is the codon for tryptophan.

3. **Non-overlapping :** The genetic code is read from a fixed point as a continuous base sequence. It is non-overlapping, commaless and without any punctuations. For instance, UUUCUUAGAGGG is read as UUU/CUU/AGA/GGG. Addition or deletion of one or two bases will radically change the message sequence in mRNA. And the protein synthesized from such mRNA will be totally different. This is encountered in ***frameshift mutations*** which cause an alteration in the reading frame of mRNA.

4. **Degenerate :** Most of the ***amino acids have more than one codon***. The codon is degenerate

Table 25.1 The genetic code along with respective amino acids

First base 5′end	Second base (middle one) U		C		A		G		Third base 3′end
U	UUU	Phe	UCU	Ser	UAU	Try	UGU	Cys	U
	UUC	Phe	UCC	Ser	UAC	Try	UGC	Cys	C
	UUA	Leu	UCA	Ser	UAA	Stop	UGA	Stop	A
	UUG	Leu	UCG	Ser	UAG	Stop	UGG	Trp	G
C	CUU	Leu	CCU	Pro	CAU	His	CGU	Arg	U
	CUC	Leu	CCC	Pro	CAC	His	CGC	Arg	C
	CUA	Leu	CCA	Pro	CAA	Gln	CGA	Arg	A
	CUG	Leu	CCG	Pro	CAG	Gln	CGG	Arg	G
A	AUU	Ile	ACU	Thr	AAU	Asn	AGU	Ser	U
	AUC	Ile	ACC	Thr	AAC	Asn	AGC	Ser	C
	AUA	Ile	ACA	Thr	AAA	Lys	AGA	Arg	A
	AUG*	Met	ACG	Thr	AAG	Lys	AGG	Arg	G
G	GUU	Val	GCU	Ala	GAU	Asp	GGU	Gly	U
	GUC	Val	GCC	Ala	GAC	Asp	GGC	Gly	C
	GUA	Val	GCA	Ala	GAA	Glu	GGA	Gly	A
	GUG	Val	GCG	Ala	GAG	Glu	GGG	Gly	G

***** *AUG serves as initiating codon, besides coding for methionine residue in protein synthesis; UAA, UAG and UGA called as nonsense codons, are responsible for termination of protein synthesis.*

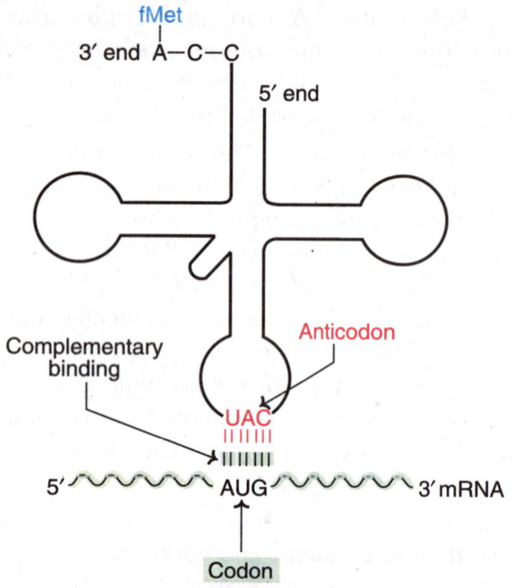

Fig. 25.12 : Complementary binding of codon (of mRNA) and anticodon (of tRNA).

or redundant, since there are ***61 codons*** available to code ***for*** only ***20 amino acids***. For instance, ***glycine has four codons***. The codons that designate the same amino acid are called ***synonyms***. Most of the synonyms differ only in the third (3′ end) base of the codon.

The Wobble hypothesis explains codon degeneracy (described later).

Codon-anticodon recognition

The codon of the mRNA is recognized by the anticodon of tRNA (***Fig.25.12***). They pair with each other in antiparallel direction (5′ → 3′ of mRNA with 3′ → 5′ of tRNA). The usual conventional complementary base pairing (A=U, C≡G) occurs between the first two bases of codon and the last two bases of anticodon. The third base of the codon is rather lenient or flexible with regard to the complementary base. The ***anticodon region*** of tRNA consists of seven nucleotides and it recognizes the three letter codon in mRNA.

Wobble hypothesis

Wobble hypothesis, put forth by Crick, is the phenomenon in which a ***single tRNA can recognize more than one codon***. This is due to the fact that the third base (3′-base) in the codon often fails to recognize the specific complementary base in the anticodon (5′-base). Wobbling is attributed to the difference in the spatial arrangement of the 5′-end of the anticodon. The possible pairing of 5′-end base of anticodon (of tRNA) with the 3′-end base of codon (mRNA) is given

Anticodon		Codon	
C	—	G	Conventional base pairing
A	—	U	
U	—	**G** or A	Non-conventional base (coloured) pairing
G	—	**U** or C	

Wobble hypothesis explains the degeneracy of the genetic code, i.e. existence of ***multiple codons for a single amino acid***. Although there are 61 codons for amino acids, the number of tRNAs is far less (around 40) which is due to wobbling.

Codon bias : Many amino acids have multiple codons. However, the organisms prefer to use one or two codons (not all of them), and this phenomenon is referred to as ***codon bias***. It is variable, depending the organism. As a result of codon bias, low amounts of tRNAs for the rarely used codons are made.

Mutations and genetic code

Mutations result in the change of nucleotide sequences in the DNA, and consequently in the RNA. The different types of mutations are described in ***Chapter 24***. The ultimate effect of mutations is on the translation through the alterations in codons. Some of the mutations are harmful.

The occurrence of the disease ***sickle-cell anemia*** due to a single base alteration (CTC → CAC in DNA, and GAG → GUG in RNA) is a classical example of the seriousness of mutations. The result is that glutamate at the 6th position of β-chain of hemoglobin is replaced by valine. This happens since the altered codon GUG of mRNA codes for valine instead of glutamate (coded by GAG in normal people).

Frameshift mutations are caused by deletion or insertion of nucleotides in the DNA that generate altered mRNAs. As the reading frame of mRNA is continuous, the codons are read in continuation, and amino acids are added. This results in proteins that may contain several altered amino acids, or sometimes the protein synthesis may be terminated prematurely.

PROTEIN BIOSYNTHESIS

The ***protein synthesis*** which involves the translation of nucleotide base sequence of mRNA into the language of amino acid sequence may be divided into the ***following stages*** for the convenience of understanding.

I. Requirement of the components

II. Activation of amino acids

III. Protein synthesis proper

IV. Chaperones and protein folding

V. Post-translational modifications.

I. REQUIREMENT OF THE COMPONENTS

The ***protein synthesis may be considered as a biochemical factory operating on the ribosomes.*** As a factory is dependent on the supply of raw materials to give a final product, the protein synthesis also requires many components.

1. **Amino acids :** Proteins are polymers of amino acids. Of the ***20 amino acids*** found ***in protein structure***, half of them (10) can be synthesized by man. About ***10 essential amino acids*** have to be provided through the diet. Protein synthesis can occur only when all the amino acids needed for a particular protein are available. If there is a deficiency in the dietary supply of any one of the essential amino acids, the translation stops. It is, therefore, necessary that a regular dietary supply of essential amino acids, in sufficient quantities, is maintained, as it is a prerequisite for protein synthesis.

As regards prokaryotes, there is no requirement of amino acids, since all the 20 are synthesized from the inorganic components.

2. **Ribosomes :** Ribosomes are the best examples of ***molecular machines***. They perform highly complex process of ***synthesizing proteins*** as per the instructions encoded in mRNA. The functionally active ribosomes are the ***centres*** or ***factories for protein synthesis***. Ribosomes may also be considered as workbenches of translation. Ribosomes are huge complex structures (70S for prokaryotes and 80S for eukaryotes) of proteins and ribosomal RNAs. Each ribosome consists of two subunits—one big and one small. The functional ribosome has two sites—A site and P site. Each site covers both the subunits. ***A site*** is for binding of aminoacyl tRNA and ***P site*** is for binding peptidyl tRNA, during the course of translation. Some authors consider ***A site as acceptor site***, and ***P site as donor site***. In case of eukaryotes, there is another site called **exit site** or ***E site***. Thus, eukaryotes contain three sites (A, P and E) on the ribosomes.

The ribosomes are located in the cytosomal fraction of the cell. They are found in association with rough endoplasmic reticulum (RER) to form clusters RER—ribosomes, where the protein synthesis occurs. The term ***polyribosome*** (***polysome***) is used when several ribosomes simultaneously translate on a single mRNA (***Fig.25.13***).

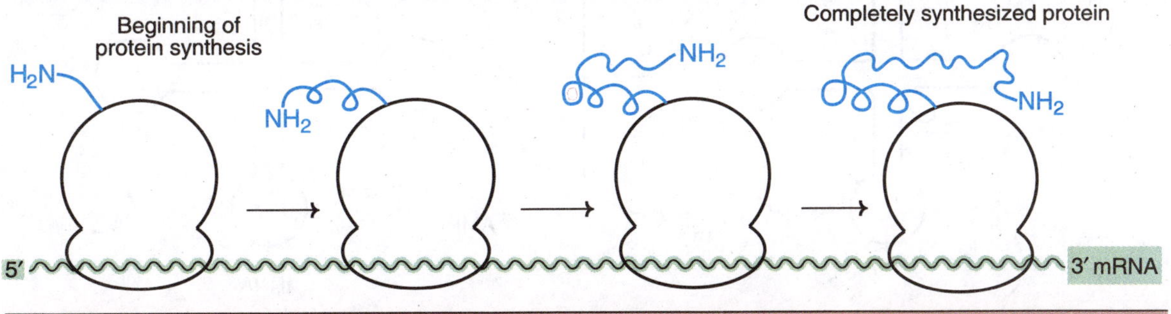

***Fig. 25.13 :** A polyribosome in protein synthesis.*

3. **Messenger RNA (mRNA) :** The specific information required for the synthesis of a given protein is present on the mRNA. The DNA has passed on the genetic information in the form of ***codons*** to mRNA to ***translate into a protein sequence.***

4. **Transfer RNAs (tRNAs) :** They ***carry the amino acids***, and hand them over to the growing peptide chain. The amino acid is covalently bound to tRNA at the 3′-end. Each tRNA has a three nucleotide base sequence—the ***anticodon,*** which is responsible to recognize the codon (complementary bases) of mRNA for protein synthesis.

In man, there are ***about 50 different tRNAs*** whereas in bacteria around 40 tRNAs are found. Some amino acids (particularly those with multiple codons) have more than one tRNA.

5. **Energy sources :** Both ***ATP*** and ***GTP*** are required for the supply of energy in protein synthesis. Some of the reactions involve the breakdown of ATP or GTP, respectively, to AMP and GMP with the liberation of pyrophosphate. Each one of these reactions consumes two high energy phosphates (equivalent to 2 ATP).

6. **Protein factors :** The process of translation involves a number of protein factors. These are needed for initiation, elongation and termination of protein synthesis. The protein factors are more complex in eukaryotes compared to prokaryotes.

II. ACTIVATION OF AMINO ACIDS

Amino acids are activated and attached to tRNAs in a two step reaction. A group of enzymes—namely aminoacyl tRNA synthetases—are required for this process. These enzymes are highly specific for the amino acid and the corresponding tRNA.

The amino acid is first attached to the enzyme utilizing ATP to form enzyme-AMP-amino acid complex. The amino acid is then transferred to the 3′ end of the tRNA to form aminoacyl tRNA (***Fig.25.14***).

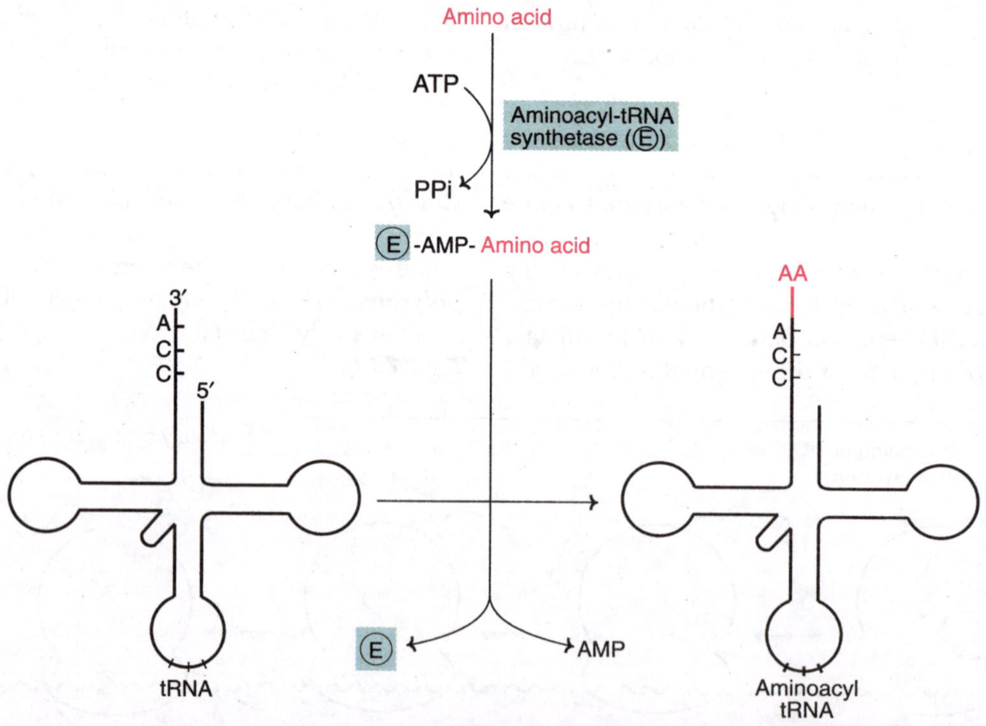

Fig. 25.14 : Formation of aminoacyl tRNA (AA–Amino acid; E–Enzyme).

III. PROTEIN SYNTHESIS PROPER

The protein or polypeptide synthesis occurs on the ribosomes (rather ***polyribosomes***). The ***mRNA is read in the 5′→3′ direction and the polypeptide synthesis proceeds from N-terminal end to C-terminal end***. Translation is directional and collinear with mRNA.

The prokaryotic mRNAs are ***polycistronic,*** since a single mRNA has many coding regions that code for different polypeptides. In contrast, eukaryotic mRNA is ***monocistronic,*** since it codes for a single polypeptide.

In case of prokaryotes, translation commences before the transcription of the gene is completed. Thus, simultaneous transcription and translation are possible. This is not so in case of eukaryotic organisms since transcription occurs in the nucleus whereas translation takes place in the cytosol. Further, the primary transcript (hnRNA) formed from DNA has to undergo several modifications to generate functional mRNA.

Protein synthesis is comparatively simple in case of prokaryotes compared to eukaryotes. Further, many steps in eukaryotic translation were not understood for quite sometime. For these reasons, majority of the textbooks earlier used to describe translation in prokaryotes in detail, and give most important and relevant information for eukaryotic translation. With the advances in molecular biology, the process of protein biosynthesis in eukaryotes is better understood now.

Translation in eukaryotes is briefly ***described here***, along with some relevant features of prokaryotic protein biosynthesis. Translation proper is divided into three stages—***initiation, elongation and termination*** (as it is done for transcription).

INITIATION OF TRANSLATION

The initiation of translation in eukaryotes is complex, involving at least ***ten eukaryotic initiation factors (eIFs)***. Some of the eIFs contain multiple (3-8) subunits. The process of translation initiation can be divided into four steps (***Fig.25.15***).

1. Ribosomal dissociation.
2. Formation of 43S preinitiation complex.
3. Formation of 48S initiation complex.
4. Formation of 80S initiation complex.

Ribosomal dissociation

The 80S ribosome dissociates to form 40S and 60S subunits. Two initiating factors namely eIF-3 and eIF-1A bind to the newly formed 40S subunit, and thereby block its reassociation with 60S subunit. For this reason, some workers name eIF-3 as ***anti-association factor***.

Formation of 43S preinitiation complex

A ternary complex containing met-tRNA^i and eIF-2 bound to GTP attaches to 40S ribosomal subunit to form 43S preinitiation complex. The presence of eIF-3 and eIF-1A stabilizes this complex (***Note*** : Met-tRNA is specifically involved in binding to the ***i***nitiation condon AUGs; hence the superscript^i is used in met-tRNA^i).

Formation of 48S initiation complex

The binding of mRNA to 43S preinitiation complex results in the formation of 48S initiation complex through the intermediate 43S initiation complex. This, however, involves certain interactions between some of the eIFs and activation of mRNA.

eIF-4F complex is formed by the association of eIF-4G, eIF-4A with eIF-4E. The so formed eIF-4F (referred to as cap binding protein) binds to the cap of mRNA. Then eIF-4A and eIF-4B bind to mRNA and reduce its complex structure. This mRNA is then transferred to 43S complex. For the appropriate association of 43S preinitiation complex with mRNA, energy has to be supplied by ATP.

Recognition of initiation codon : The ribosomal initiation complex scans the mRNA for the identification of appropriate initiation codon. 5′-AUG is the initiation codon and its recognition is facilitated by a specific sequence of nucleotides surrounding it. This marker sequence for the identification of AUG is called as ***Kozak consensus sequences***. In case of ***prokaryotes*** the recognition sequence of initiation codon is referred to as ***Shine-Dalgarno sequence***.

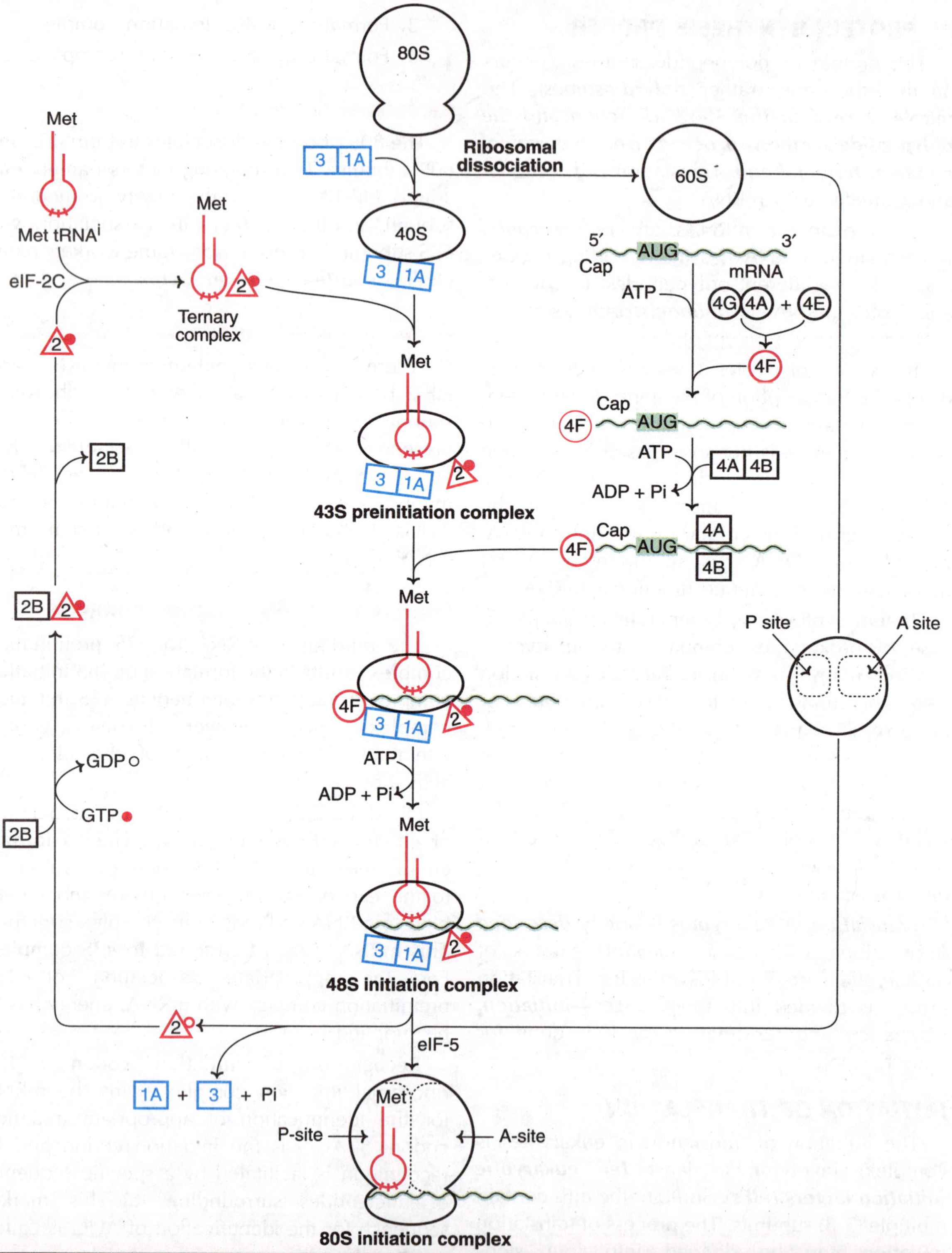

Fig. 25.15 : *A diagrammatic representation of initiation of protein biosynthesis (translation) in eukaryotic cells (The eukaryotic initiation factors are represented by symbols ▭, △, and ○. By prefixing with eIF, the full names of the factors are obtained e.g. 3 represents eIF–3).*

Formation of 80S initiation complex

48S initiation complex binds to 60S ribosomal subunit to form 80S initiation complex. The binding involves the hydrolysis of GTP (bound to eIF-2). This step is facilitated by the involvement of eIF-5.

As the 80S complex is formed, the initiation factors bound to 48S initiation complex are released, and recycled. The activation of eIF-2 requires eIF-2B (also called as guanine nucleotide exchange factor) and GTP. The activated eIF-2 (i.e. bound to GTP) requires eIF-2C to form the ternary complex.

Regulation of initiation

The eIF-4F, a complex formed by the assembly of three initiation factors controls initiation, and thus the translation process. eIF-4E, a component of eIF-4F is primarily responsible for the recognition of mRNA cap. And this step is the rate-limiting in translation.

eIF-2 which is involved in the formation of 43S preinitiation complex also controls protein biosynthesis to some extent.

Initiation of translation in prokaryotes

The formation of translation initiation complex in prokaryotes is less complicated compared to eukaryotes. The 30S ribosomal subunit is bound to initiation factor 3 (IF-3) and attached to ternary complex of IF-2, formyl met-tRNA and GTP. Another initiation factor namely IF-I also participates in the formation of preinitiation complex. The recognition of initiation codon AUG is done through Shine-Dalgarno sequence. A 50S ribosome unit is now bound with the 30S unit to produce 70S initiation complex in prokaryotes.

ELONGATION OF TRANSLATION

Ribosomes elongate the polypeptide chain by a sequential addition of amino acids. The amino acid sequence is determined by the order of the codons in the specific mRNA. Elongation, a cyclic process involving certain ***elongation factors (EFs)***, may be divided into three steps (***Fig.25.16***).

1. Binding of aminoacyl t-RNA to A-site.
2. Peptide bond formation.
3. Translocation.

Binding of aminoacyl—tRNA to A-site

The 80S initiation complex contains met-$tRNA^i$ in the P-site, and the A-site is free. Another aminoacyl-tRNA is placed in the A-site. This requires proper codon recognition on the mRNA and the involvement of elongation factor 1a (EF-Ia) and supply of energy by GTP. As the aminoacyl-tRNA is placed in the A-site, EF-1α and GDP are recycled to bring another aminoacyl-tRNA.

Peptide bond formation

The enzyme ***peptidyltransferase*** catalyses the formation of peptide bond (***Fig.25.17***). The activity of this enzyme lies on 28S RNA of 60S ribosomal subunit. It is therefore the ***rRNA*** (and not protein) referred to as ***ribozyme*** that ***catalyses the peptide bond formation***. As the amino acid in the aminoacyl-tRNA is already activated, no additional energy is required for peptide bond formation.

The net result of peptide bond formation is the attachment of the growing peptide chain to the tRNA in the A-site.

Translocation

As the peptide bond formation occurs, the ribosome moves to the next codon of the mRNA (towards 3′-end). This process called translocation, basically involves the ***movement of growing peptide chain from A-site to P-site***. Translocation requires EF-2 and GTP. GTP gets hydrolysed and supplies energy to move mRNA. EF-2 and GTP complex recycles for translocation.

In recent years, another site namely ***exit site*** (***E-site***) has been identified in eukaryotes. The deacylated tRNA moves into the E-site, from where it leaves the ribosome.

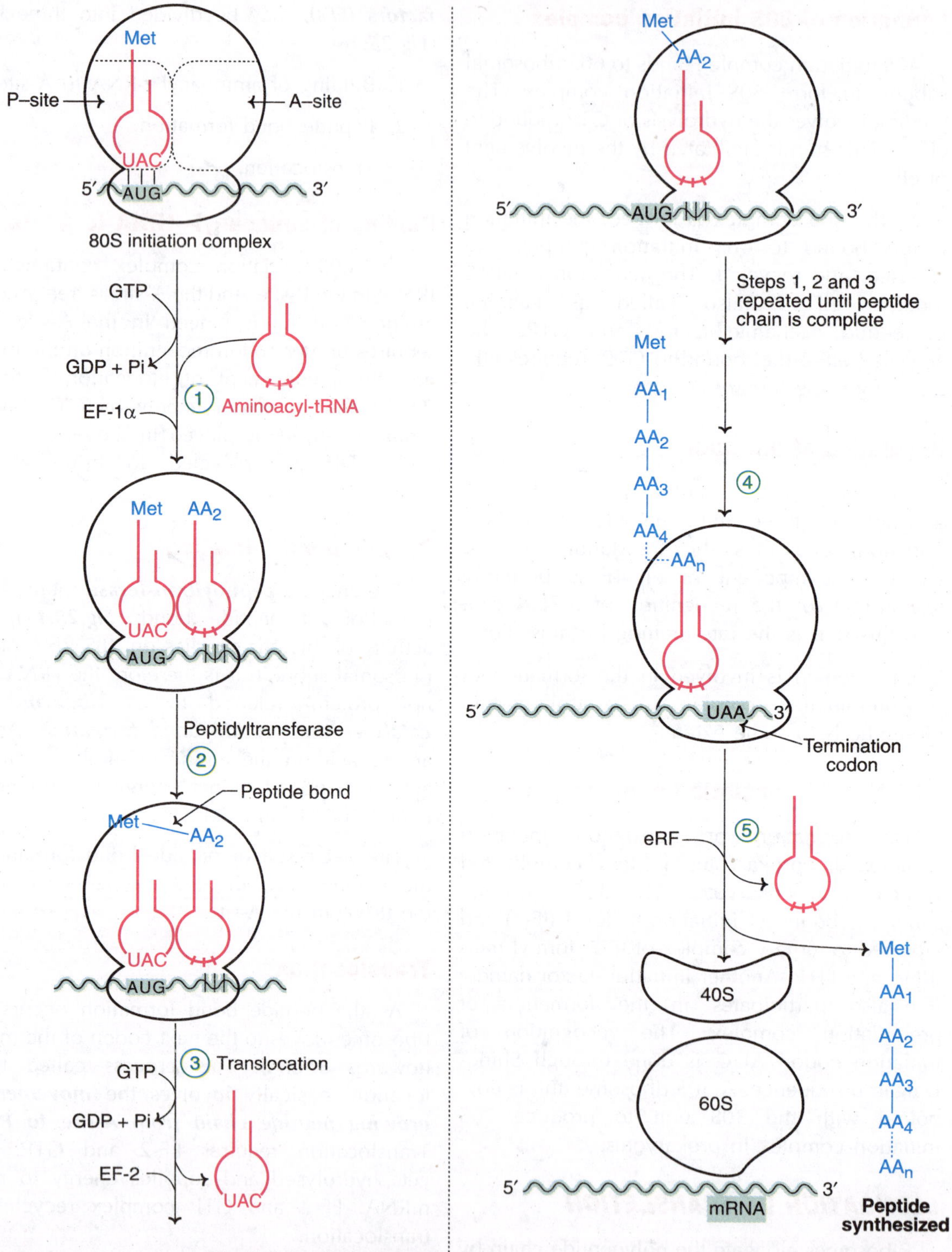

Fig. 25.16 contd. next column

Fig. 25.16 : *Protein biosynthesis — Elongation and termination (for initiation* **See Fig. 25.15**). *Met-Methionine; P-site — Peptidyl tRNA binding site; A-site — Aminoacyl tRNA binding site. AA-Amino acid; EF-Elongation factor; RF-Releasing factor.*

Medical Concepts/ Clinical Correlates

- Faulty splicing of hnRNA may result in certain diseases e.g. β-thalassemia.
- Inhibitors of transcription are used as therapeutic agents. Thus, actinomycin D was the first antibiotic used in the treatment of tumors. Rifampin is employed to treat tuberculosis and leprosy.
- Retroviruses (RNA is the genetic material) are oncogenic i.e. cause cancers.

In case of prokaryotes, the elongation factors are different, and they are EF-Tu, EF-Ts (in place of of EF-1a) and EF-G (instead of EF-2).

Incorporation of amino acids

It is estimated that about ***six amino acids per second*** are incorporated during the course of elongation of translation ***in eukaryotes***. In case of prokaryotes, as many as 20 amino acids can be incorporated per second. Thus the process of protein/polypeptide synthesis in translation occurs with ***great speed*** and ***accuracy***.

TERMINATION OF TRANSLATION

Termination is a simple process when compared to initiation and elongation. After several cycles of elongation, incorporating amino acids and the formation of the specific protein/ polypeptide molecule, one of the ***stop or termination signals (UAA, UAG and UCA)*** terminates the growing polypeptide. The termination codons which act as stop signals do not have specific tRNAs to bind. As the termination codon occupies the ribosomal A-site, the ***release factor*** namely eRF recognizes the stop signal. eRF-GTP complex, in association with the enzyme peptidyltransferase, cleaves the peptide bond between the polypeptide and the tRNA occupying P-site. In this reaction, a water molecule, instead of an amino acid is added. This hydrolysis releases the protein and tRNA from the P-site. The 80S ribosome dissociates to form 40S and 60S subunits which are recycled. The mRNA is also released.

INHIBITORS OF PROTEIN SYNTHESIS

Translation is a complex process and it has become a favourite target for inhibition by antibiotics. Antibiotics are the substances

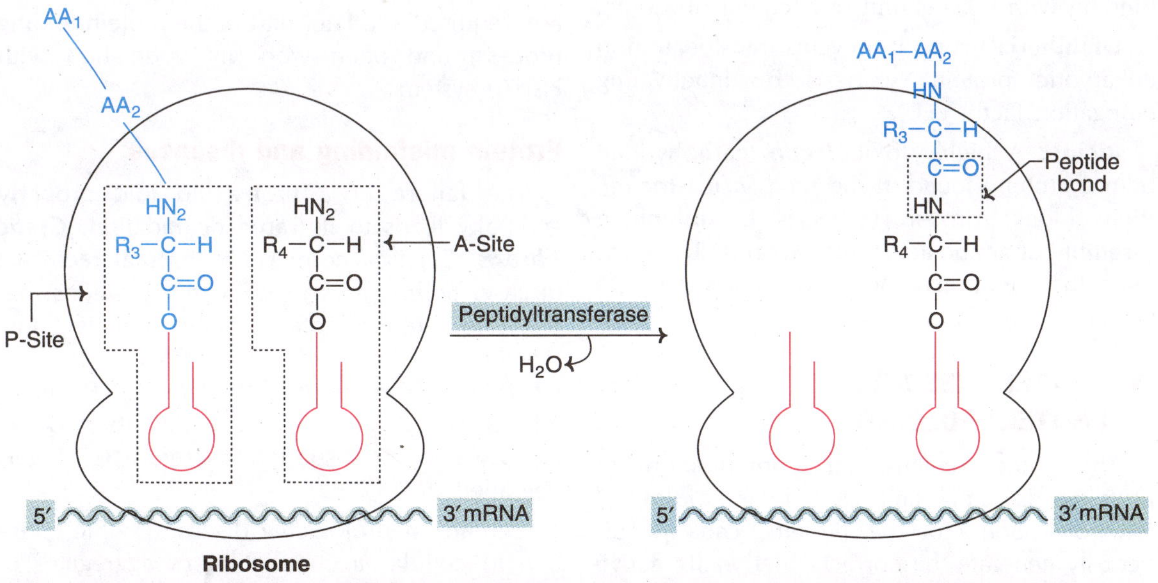

Fig. 25.17 : *Formation of peptide bond in translation (P-site — Peptidyl tRNA site; A-site — Aminoacyl tRNA site).*

produced by bacteria or fungi which inhibit the growth of other organisms. ***Majority*** of the ***antibiotics interfere with the bacterial protein synthesis and are harmless to higher organisms***. This is due to the fact that the process of translation sufficiently differs between prokaryotes and eukaryotes. The action of a few important antibiotics on translation is described next.

Streptomycin : ***Initiation of protein synthesis is inhibited by streptomycin***. It causes misreading of mRNA and interferes with the normal pairing between codons and anticodons.

Tetracycline : It ***inhibits the binding of aminoacyl tRNA to the ribosomal complex***. In fact, tetracycline can also block eukaryotic protein synthesis. This, however, does not happen since eukaryotic cell membrane is not permeable to this drug.

Puromycin : This has a structural resemblance to aminoacyl tRNA. ***Puromycin enters the A site and gets incorporated into the growing peptide chain*** and causes its release. This antibiotic prevents protein synthesis in both prokaryotes and eukaryotes.

Chloramphenicol : It acts as a ***competitive inhibitor of the enzyme peptidyltransferase*** and thus interferes with elongation of peptide chain.

Erythromycin : It inhibits translocation by binding with 50S subunit of bacterial ribosome.

Diphtheria toxin : It prevents translocation in eukaryotic protein synthesis by inactivating elongation factor eEF_2.

Ricin : A highly toxic ***lectin*** (carbohydrate bound protein) found in the ***seeds of castor oil***. Ricin blocks protein synthesis by inhibiting assembly of amino acids into proteins. There are fears that ricin may be used as a biological weapon in warfares.

IV. CHAPERONES AND PROTEIN FOLDING

The three dimensional conformation of proteins is important for their biological functions. Some of the proteins can spontaneously generate the correct functionally active conformation e.g. denatured pancreatic ribonuclease. However, a vast majority of proteins can attain correct conformation, only through the assistance of certain proteins referred to as chaperones. Chaperones are ***heat shock proteins*** (originally discovered in response to heat shock). They ***facilitate*** and favour the interactions on the polypeptide surfaces to finally give the ***specific conformation of a protein***. Chaperones can reversibly bind to hydrophobic regions of unfolded proteins and folding intermediates. They can stabilize intermediates, prevent formation of incorrect intermediates, and also prevent undesirable interactions with other proteins. All these activities of chaperones help the protein to attain compact and biologically active conformation.

Types of chaperones

Chaperones are categorized into two major groups

1. **Hsp70 system :** This mainly consists of Hsp70 (70-kDa ***h***eat ***s***hock ***p***rotein) and Hsp40 (40-kDa Hsp). These proteins can bind individually to the substrate (protein) and help in the correct formation of protein folding.

2. **Chaperonin system :** This is a large oligomeric assembly which forms a structure into which the folded proteins are inserted. The chaperonin system mainly has Hsp60 and Hsp10 i.e. 60 kDa Hsp and 10 kDa Hsp. Chaperonins are required at a later part of the protein folding process, and often work in association with Hsp70 system.

Protein misfolding and diseases

The failure of a protein to fold properly generally leads to its rapid degradation. ***Cystic fibrosis*** (CF) is a common autosomal recessive disease. Some cases of CF with mutations that result in altered protein (cystic fibrosis transmembrane conductance regulator or in short ***CFTR***) have been reported. Mutated CFTR cannot fold properly, besides not being able to get glycosylated or transported. Therefore, CFTR gets degraded.

Certain neurological diseases which are due to cellular accumulation of aggregates of misfolded proteins or their partially degraded products have been identified. The term ***prions*** (***pr***oteinous ***in***fectious agents) is used to collectively represent them.

Prions exhibit the characteristics of viral or microbial pathogens and ***have been implicated in many diseases***. e.g. ***mad cow disease***, Creutzfeldt-Jacob disease, ***Alzheimer's disease***, Huntington's disease (***Refer Chapter 22***).

It is now recognized that ***prion diseases are protein conformation diseases*** due to alteration in protein conformation resulting in altered protein properties.

V. POST-TRANSLATIONAL MODIFICATIONS OF PROTEINS

The proteins synthesized in translation are, as such, not functional. Many changes take place in the polypeptides after the initiation of their synthesis or, most frequently, ***after the protein synthesis is completed***. These ***modifications*** include protein folding (described already), trimming by proteolytic degradation, intein splicing and covalent changes which are collectively known as post-translational modifications (***Fig.25.18***).

Proteolytic degradation

Many proteins are synthesized as the precursors which are much bigger in size than the functional proteins. Some portions of precursor molecules are removed by proteolysis to liberate active proteins. This process is commonly referred to as trimming. The formation of insulin from preproinsulin, conversion of ***zymogens*** (inactive digestive enzymes e.g. trypsinogen) to the active enzymes are some examples of trimming.

Intein splicing

Inteins are ***intervening sequences in certain proteins***. These are comparable to introns in mRNAs. ***Inteins*** have to be ***removed***, *and* ***exteins ligated*** in the appropriate order for the protein to become active.

Covalent modifications

The proteins synthesized in translation are subjected to many covalent changes. By these modifications in the amino acids, the proteins may be converted to active form or inactive form. Selected examples of covalent modifications are described below.

1. **Phosphorylation :** The ***hydroxyl group*** containing amino acids of proteins, namely ***serine, threonine and tyrosine*** are subjected to phosphorylation. The phosphorylation may either increase or decrease the activity of the proteins. A group of enzymes called ***protein kinases*** catalyse phosphorylation while protein phosphatases are responsible for dephosphorylation (removal of phosphate group). Many enzymes that undergo phosphorylation or dephosphorylation are known in metabolisms (e.g. glycogen synthase).

2. **Hydroxylation :** During the formation of collagen, the amino acids ***proline and lysine are respectively converted to hydroxyproline and hydroxylysine***. This hydroxylation occurs in the endoplasmic reticulum and requires ***vitamin C***.

3. **Glycosylation :** The attachment of carbohydrate moiety is essential for some proteins to perform their functions. The complex carbohydrate moiety is attached to the amino acids, serine and threonine (O-linked) or to asparagine

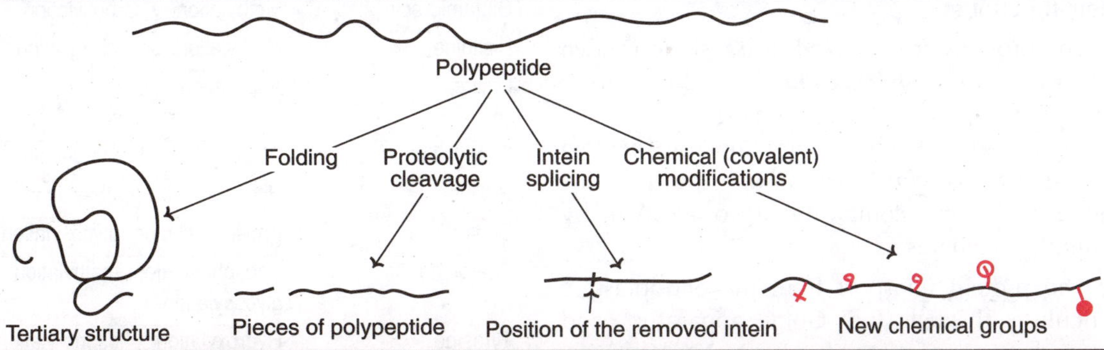

***Fig. 25.18 :** An outline of post-translational modifications of proteins.*

Medical Concepts/ Clinical Correlates

- Several antibiotics selectively block bacterial translation, and thus inhibit their growth e.g. streptomycin, tetracycline, puromycin.
- Protein misfolding often results in the formation of prions (proteinous infectious agents) which have been implicated in many diseases e.g. mad cow disese, Alzheimer's disease.
- Lebers' hereditary optic neuropathy is caused by mutation in mtDNA in males. The victims become blind due to loss of central vision as a result of neuroretinal degeneration.

(N-linked), leading to the synthesis of glycoproteins.

Vitamin K dependent ***carboxylation*** of glutamic acid residues in certain clotting factors is also a post-translational modification.

In the ***Table 25.2***, selected examples of post-translational modification of proteins through their amino acids are given.

PROTEIN TARGETING

The eukaryotic proteins (tens of thousands) are distributed between the cytosol, plasma membrane and a number of cellular organelles (nucleus, mitochondria, endoplasmic reticulum etc.). At the appropriate places, they perform their functions.

The proteins, synthesized in translation, have to ***reach*** their ***destination*** to exhibit their biological activities. This is carried out by a process called protein targeting or ***protein sorting*** or ***protein localization***. The proteins move from one compartment to another by multiple mechanisms.

The protein transport from the endoplasmic reticulum through the Golgi apparatus, and beyond uses carrier vesicles. It may be, however, noted that only the ***correctly folded proteins*** are recognized as the ***cargo for transport***. Protein targeting and post-translational modifications occur in a well coordinated manner.

Protein targeting to mitochondria

Most of the proteins of mitochondria are synthesized in the cytosol, and their transport to mitochondria is a complex process. Majority of the proteins are synthesized as larger preproteins with N-terminal presequences for the entry of these proteins into mitochondria. The ***transport of unfolded proteins*** is often ***facilitated by chaperones***.

One protein namely ***mitochondrial matrix targeting signal***, involved in protein targeting has been identified. This protein can recognize mitochondrial receptor and transport certain proteins from cytosol to mitochondria. This is an energy-dependent process.

Table 25.2 Selected examples of post-translational modifications of proteins through their amino acids

Amino acid	Post-translational modification(s)
Amino-terminal amino acid	Glycosylation, acetylation, myristoylation, formylation.
Carboxy terminal amino acid	Methylation, ADP-ribosylation
Aspartic acid	Phosphorylation, hydroxylation
Cysteine (—SH)	Cystine (—S—S—) formation, selenocysteine formation, glycosylation.
Glutamic acid	Methylation, γ-carboxylation.
Histidine	Methylation, phosphorylation.
Lysine	Acetylation, methylation, hydroxylation, biotinylation.
Phenylalanine	Glycosylation, hydroxylation.
Proline	Hydroxylation, glycosylation.
Serine	Phosphorylation, glycosylation.
Threonine	Phosphorylation, methylation glycosylation.
Tyrosine	Hydroxylation, phosphorylation, sulfonylation, iodination.

MITOCHONDRIAL DNA, TRANSCRIPTION AND TRANSLATION

The mitochondrial DNA (***mtDNA***) has ***structural and functional resemblances with prokaryotic DNA***. This fact supports the view that mitochondria are derivatives of prokaryotes. mtDNA is circular in nature and contains about 16,000 nucleotide bases.

A vast majority of structural and functional proteins of the mitochondria are synthesized in the cytosol, under the influence of nuclear DNA. However, ***certain proteins (around 13)***, most of them being the components of electron transport chain, are ***synthesized in the mitochondria*** (e.g. cytochrome b of complex III, two subunits of ATP synthase). Transcription takes place in the mitochondria leading to the synthesis of mRNAs, tRNAs and rRNAs. Two types of rRNA and about 22 species of tRNA have been so far identified. Transcription is followed by translation resulting in protein synthesis.

The mitochondria of the sperm cell do not enter the ovum during fertilization, therefore, ***mtDNA is inherited from the mother. Mitochondrial DNA*** is subjected to ***high rate of mutations*** (about 10 times more than nuclear DNA) that causes inherited defects in oxidative phosphorylation. The best known among them are certain mitochondrial myopathies and Leber's hereditary optic neuropathy. The latter is mostly found in males and is characterized by blindness due to loss of central vision as a result of neuroretinal degeneration. ***Leber's hereditary optic neuropathy*** is a consequence of single base mutation in mtDNA. Due to this, the amino acid histidine, in place of arginine, is incorporated into the enzyme NADH coenzyme Q reductase.

SUMMARY

- *Transcription is the process in which RNA is synthesized from DNA, which is carried out in 3 stages–initiation, elongation and termination.*
- *In case of prokaryotes, a single enzyme synthesizes all the RNAs. In eukaryotes, RNA polymerase I, II and III respectively produce rRNAs, mRNAs and tRNAs.*
- *The primary mRNA transcript (i.e. hnRNA) undergoes post-transcriptional modifications e.g. base modifications, splicing etc.*
- *Reverse transcription is the process of synthesizing DNA from RNA by the enzyme reverse transcriptase.*
- *Biosynthesis of a protein or a polypeptide is known as translation. The amino acid sequence of a protein is determined by the triplet codons of mRNA.*
- *The genetic code (codons)–composed of A, G, C and U–is universal, specific, non-overlapping and degenerate. Of the 64 codons, three (UAA, UAG, UGA) are termination codons while the rest code for amino acids.*
- *Ribosomes are the factories of protein biosynthesis. Translation involves activation of amino acids, protein synthesis proper (initiation, elongation and termination), protein folding and post-translational modifications.*
- *The post-translational modifications include proteolytic degradation, intein splicing and covalent modifications (phosphorylation, hydroxylation, glycosylation etc.). These modifications are required to make the proteins biologically active.*
- *The mitochondria possess independent DNA with the machinery for transcription and translation. However, only a few proteins (around 13) are actually synthesized in the mitochondria.*

Self-Assessment Exercises

I. Essay questions

1. Give an account of transcription. Compare the RNA synthesis between prokaryotes and eukaryotes.
2. Describe protein biosynthesis (translation).
3. Discuss the inhibitors of transcription and translation.
4. Give an account of post-transcriptional and post-translational modifications.
5. What is genetic code? Describe the characteristics of genetic code. Add a note on the effects of mutations on genetic code.

II. Short notes

(a) Genome, (b) Heterogeneous nuclear RNA (hnRNA), (c) Eukaryotic RNA polymerases, (d) Introns and exons, (e) Reverse transcription, (f) Wobble hypothesis, (g) Anticodon, (h) Shine-Dalgarno sequence, (i) Peptidyltransferase, (j) Chaperones, (k) Protein targeting.

III. Fill in the blanks

1. The total DNA (genetic information) contained in an organism (or a cell) is referred to as ______________.
2. The primary transcript produced by RNA polymerase II is eukaryotes ______________.
3. The intervening nucleotide sequences in mRNA that do not code for proteins ______________.
4. The synthesis of complementary DNA (cDNA) from mRNA is catalysed by the enzyme ______________.
5. A single tRNA is capable of recognizing more than one codon, and this phenomenon is referred to as ______________.
6. The factories for protein biosynthesis are ______________.
7. The enzyme peptidyltransferase calalyses the formation of peptide bond during translation. The chemical nature of this enzyme is ______________.
8. The proteins that facilitate the formation of specific conformation of proteins are ______________.
9. The common term used for the diseases due to misfolding of proteins _______.
10. The process of delivery of proteins in a cell to the site their biological activity is _______.

IV. Multiple choice questions

11. The codon(s) that terminate(s) protein biosynthesis
    (a) UAA (b) UAG (c) UGA (d) All of them.
12. The nitrogenous base that is never found in the genetic code
    (a) Adenine (b) Guanine (c) Thymine (d) Cytosine.
13. The total DNA (genetic information) contained in a living cell (or organism) is regarded as
    (a) Genome (b) Transciptome (c) Proteome (d) Gene.
14. The enzyme responsible for the synthesis of mRNAs in eukaryotic cells
    (a) RNA polymerase I (b) RNA polymerase II (c) RNA polymerase III (d) RNA polymerase α.
15. Mitochondrial DNA is inherited from
    (a) Mother only (b) Father only (c) Both of them (d) Either mother or father.

Regulation of Gene Expression 26

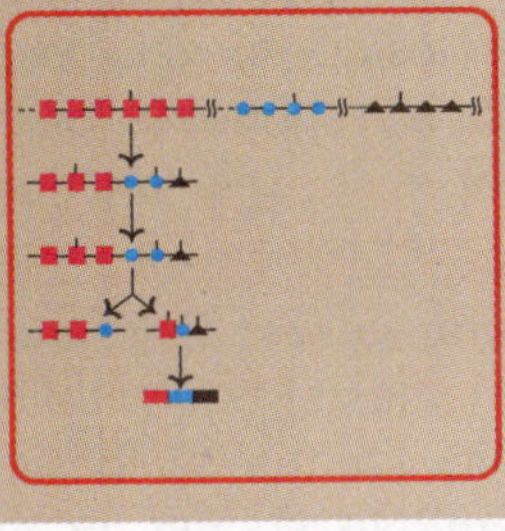

The genes speak

"Functional units of DNA, we are;
Ultimate for all cellular activities;
Tailored to express as per tissue demands;
Mystery of our molecular actions await unfolding."

DNA, the chemical vehicle of heredity, is composed of ***functional units***, namely ***genes***. The term ***genome*** refers to the total genetic information contained in a cell. The bacterium *Escherichia coli* contains about 4,400 genes present on a single chromosome. The ***genome of humans*** is more complex, with 23 pairs of (diploid) chromosomes containing 6 billion (6×10^9) base pairs of DNA, with an estimated ***19,000–20,000 genes***. At any given time, only a fraction of the genome is expressed.

The living cells possess a remarkable property to adapt to changes in the environment by regulating the gene expression. For instance, insulin is synthesized by specialized cells of pancreas and not by cells of other organs (say kidney, liver), although the nuclei of all the cells of the body contain the insulin genes. Molecular regulatory mechanisms facilitate the expression of insulin gene in pancreas (***tissue specific gene expression***), while preventing its expression in other cells.

GENE REGULATION—GENERAL

The ***regulation of the expression of genes*** is absolutely ***essential for the growth, development, differentiation*** and the very existence of an organism. There are two types of gene regulation-positive and negative.

1. **Positive regulation :** The gene regulation is said to be positive when its ***expression is increased by a regulatory element*** (positive regulator).

2. **Negative regulation :** A ***decrease in the gene expression due to the presence of a regulatory element*** (negative regulator) is referred to as negative regulation.

It may be noted here that double negative effect on gene regulation results in a positive phenomenon.

Constitutive and inducible genes

The genes are generally considered under two categories.

1. **Constitutive genes :** The products (proteins) of these genes are ***required all the time*** in a cell. Therefore, the constitutive genes (or ***house-keeping genes***) are expressed at more or less constant rate in almost all the cells and, further, they are not subjected to regulation e.g. the ***enzymes of citric acid cycle***.

2. **Inducible genes :** The concentration of the proteins synthesized by inducible genes is ***regulated by*** various ***molecular signals***. An inducer increases the expression of these genes while a repressor decreases, e.g. tryptophan pyrrolase of liver is induced by tryptophan.

The term ***pseudogenes*** is used to represent DNA sequences that have significant homology to a functional gene, but they cannot express due to mutations. Thus pseudogenes are non-functional. However, they significantly increase the size of the eukaryotic genome without any contribution to the expression of genes.

One cistron-one subunit concept

The chemical product of a gene expression is a protein which may be an enzyme. It was originally believed that each gene codes for a specific enzyme, leading to the popular concept, one gene-one enzyme. This however, is not necessarily valid due to the fact that several enzymes (or proteins) are composed of two or more nonidentical subunits (polypeptide chains).

The cistron is the smallest unit of genetic expression. It is the fragment of DNA coding for the subunit of a protein molecule. The original concept of ***one gene-one enzyme is replaced by one cistron-one subunit***.

Models to study gene expression

Elucidation of the regulation of gene expression in prokaryotes has largely helped to understand the principles of the flow of information from genes to mRNA to synthesize specific proteins. Some important features of ***prokaryotic gene*** expression are described first. This is followed by a brief account of eukaryotic gene expression.

THE OPERON CONCEPT

The operon is the ***coordinated unit of genetic expression in bacteria***. The concept of operon was introduced by ***Jacob and Monod in 1961*** (Nobel Prize 1965), based on their observations on the regulation of lactose metabolism in *E. coli*. This is popularly known as ***lac operon***.

LACTOSE (LAC) OPERON

Structure of lac operon

The lac operon (***Fig.26.1***) consists of a regulatory gene (I; I for inhibition), operator gene (O) and three structural genes (Z, Y, A). Besides these genes, there is a promoter site (P), next to the operator gene, where the enzyme RNA polymerase binds. The structural genes ***Z, Y and A respectively***, code for the enzymes ***β-galactosidase, galactoside permease and galactoside acetylase***. β-Galactosidase hydrolyses lactose (β-galactoside) to galactose and glucose while permease is responsible for the transport of lactose into the cell. The function of acetylase (coded by A gene) remains a mystery.

The structural genes Z, Y and A transcribe into a single large mRNA with 3 independent translation units for the synthesis of 3 distinct enzymes. An mRNA coding for more than one protein is known as ***polycistronic mRNA***. Prokaryotic organisms contain a large number of polycistronic mRNAs.

Repression of lac operon

The regulatory gene (I) is constitutive. It is expressed at a constant rate leading to the synthesis of lac repressor. Lac repressor is a tetrameric (4 subunits) regulatory protein (total mol. wt. 150,000) which specifically binds to the operator gene (O). This prevents the binding of the enzyme RNA polymerase to the promoter site (P), thereby blocking the transcription of structural genes (Z, Y and A). This is what happens in the absence of lactose in *E. coli*. The ***repressor molecule acts as a negative regulator of gene expression***.

Derepression of lac operon

In the presence of lactose (inducer) in the medium, a small amount of it can enter the *E. coli* cells. The repressor molecules have a high affinity for lactose. The lactose molecules bind and induce a conformational change in the repressor. The result is that the repressor gets inactivated and, therefore, cannot bind to the operator gene (O). The RNA polymerase attaches to the DNA at the promoter site and transcription proceeds, leading to the formation of polycistronic mRNA (for genes Z, Y and A) and, finally, the 3 enzymes. Thus, ***lactose induces the synthesis of the three enzymes*** β-galactosidase, galactoside permease and galactoside acetylase. ***Lactose acts by inactivating the repressor molecules,*** hence this process is known as derepression of lac operon.

Gratuitous inducers : There are certain structural analogs of lactose which can induce the lac operon but are not the substrates for the enzyme β-galactosidase. Such substances are known as gratuitous inducers. Isopropylthiogalactoside (IPTG) is a gratuitous inducer, extensively used for the study of lac operon.

The catabolite gene activator protein : The cells of *E. coli* utilize glucose in preference to lactose; when both of them are present in the medium. After the depletion of glucose in the medium, utilization of lactose starts. This

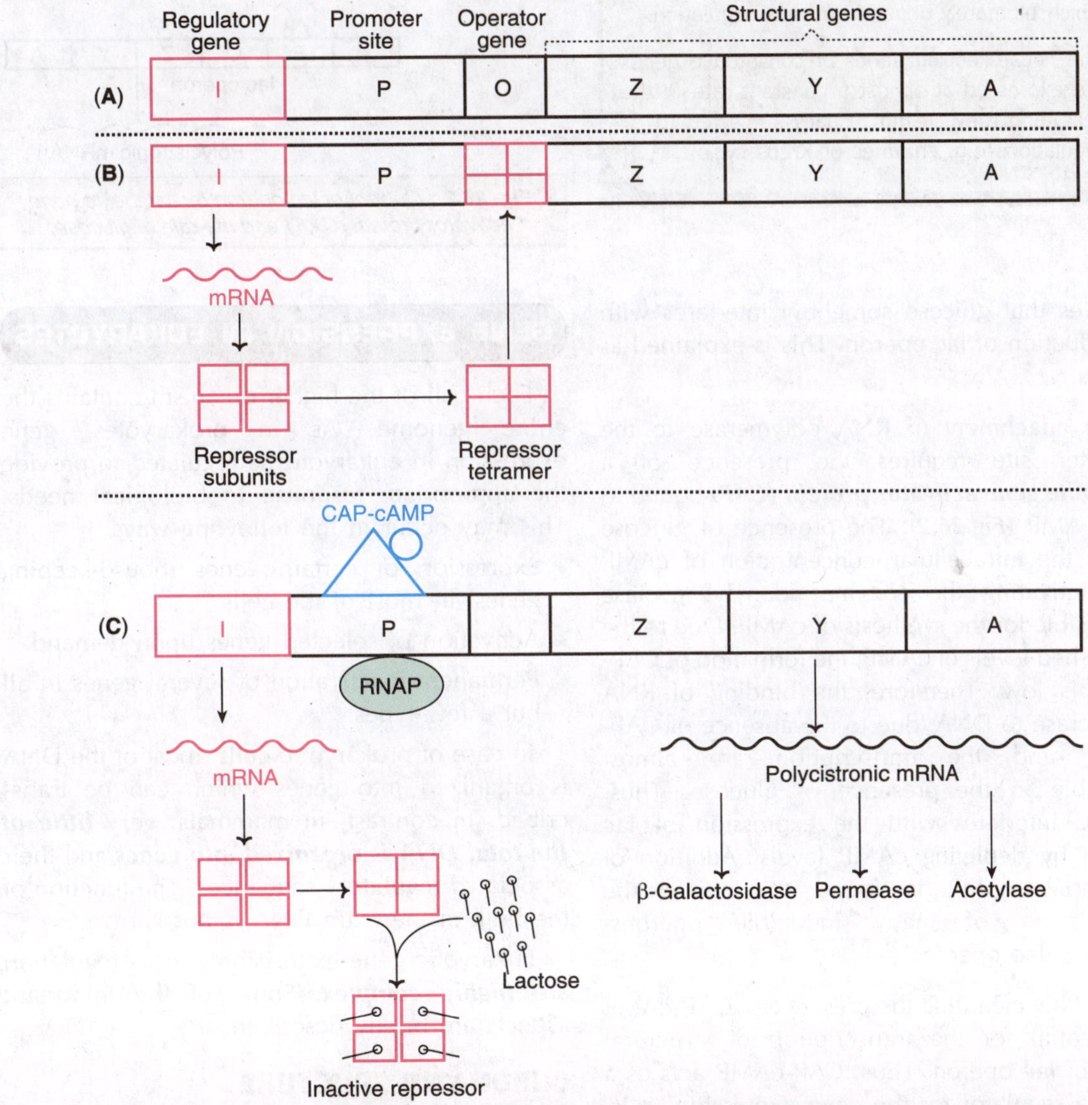

Fig. 26.1 : Model of lactose operon in* E.coli *(A) Structure of lac operon (B) Repression of lac operon (C) Derepression of lac operon. (CAP—cAMP–catabolite gene activator protein bound to cAMP; RNAP–RNA polymerase).

Medical Concepts/ Clinical Correlates

- Regulation of gene expression to adapt to the changes in the environment is a remarkable property of living cells e.g. synthesis of insulin by β-cells of pancreas and nowhere else.
- The growth, development and differentiation of an organism involves complex mechanisms which ultimately depend on gene regulation.
- The house-keeping genes or constitutive genes are expressed at almost a constant rate in the cells, and they are not usually subjected to regulations e.g. enzymes of Krebs cycle.

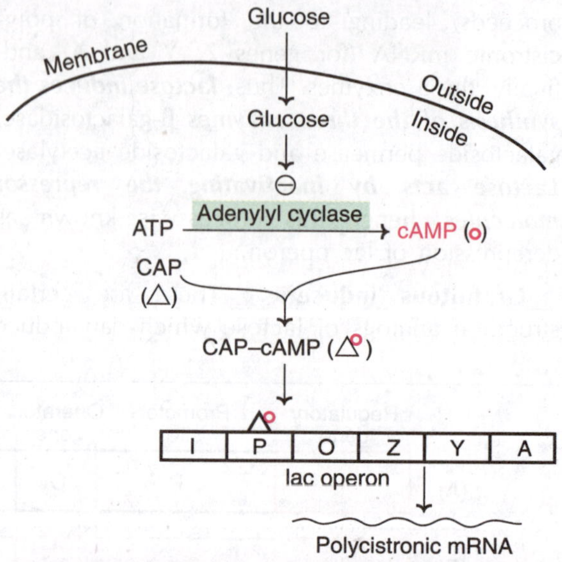

Fig. 26.2 : *Control of lac operon by catabolite gene activator protein (CAP) and the role of glucose.*

indicates that glucose somehow interferes with the induction of lac operon. This is explained as follows.

The attachment of RNA polymerase to the promoter site requires the presence of a ***catabolite gene activator protein (CAP)*** bound to cyclic AMP (***Fig.26.2***). The presence of glucose lowers the intracellular concentration of cAMP by inactivating the enzyme adenylyl cyclase responsible for the synthesis of cAMP. Due to the diminished levels of cAMP, the formation of CAP-cAMP is low. Therefore, the binding of RNA polymerase to DNA (due to the absence of CAP-cAMP) and the transcription are almost negligible in the presence of glucose. Thus, glucose interferes with the expression of lac operon by depleting cAMP levels. Addition of exogenous cAMP is found to initiate the transcription of many inducible operons, including lac operon.

It is now clear that the presence of CAP-cAMP is essential for the transcription of structural genes of lac operon. Thus, CAP-cAMP acts as a positive regulator for the gene expression. It is, therefore, evident that lac operon is subjected to both positive (by repressor, described above) and negative regulation.

GENE EXPRESSION IN EUKARYOTES

Each cell of the higher organism contains the entire genome. As in prokaryotes, gene expression in eukaryotes is regulated to provide the appropriate response to biological needs. This may occur in the following ways

- Expression of certain genes (house-keeping genes) in most of the cells.
- Activation of selected genes upon demand.
- Permanent inactivation of several genes in all but a few types.

In case of prokaryotic cells, most of the DNA is organized into genes which can be transcribed. In contrast, in mammals, ***very little of the total DNA is organized into genes*** and their associated regulatory sequences. The function of the bulk of the extra DNA is not known.

Eukaryotic gene expression and its regulation are ***highly complex***. Some of the important aspects are briefly described.

CHROMATIN SRUCTURE AND GENE EXPRESSION

The DNA in higher organisms is extensively folded and packed to form ***protein-DNA***

complex called chromatin. The structural organization of DNA in the form of chromatin plays an important role in eukaryotic gene expression. In fact, chromatin structure provides an additional level of control of gene expression.

A selected list of genes (represented by the products) along with the respective chromosomes on which they are located is given in ***Table 26.1***.

In general, the genes that are transcribed within a particular cell are less condensed and more open in structure. This is in contrast to genes that are not transcribed which form highly condensed chromatin.

Histone code

Histone code refers to ***histone covalent modifications*** that ***determine gene activity***. Eukaryotic DNA segments are wrapped around histone proteins to form nucleosome. Acetylation or deacetylation of histones is an ***important factor in determining the gene expression***. In general, ***acetylation of histones leads to activation of gene expression*** while deacetylation reverses the effect.

Histones are subjected to a wide range of ***post-translational modifications*** (***PTMs***). Acetylation predominantly occurs on the lysine residues in the amino terminal ends of histones. This modification in histones reduces the positive charges of terminal ends (tails), and decreases their binding affinity to negatively charged DNA. Consequently, nucleosome structure is disrupted to allow transcription.

Methylation of DNA and inactivation of genes

Cytosine in the sequence CG of DNA gets methylated to form 5′-methylcytosine. A major portion of CG sequences (about 20%) in human DNA exists in methylated form. In general, ***methylation leads to*** loss of transcriptional activity, and thus ***inactivation of genes***. This occurs due to binding of methylcytosine binding proteins to methylated DNA. As a result, methylated DNA is not exposed and bound to transcription factors. It is interesting to note that methylation of DNA correlates with deacetylation of histones. This provides a double means for repression of genes.

The activation and normal expression of genes, and gene inactivation by DNA methylation are depicted in ***Fig.26.3***.

Table 26.1 A selected list of genes (represented by the products) along with respective chromosomes

Genes	Chromosome number
Alkaline phosphatase	1
Apolipoprotein B	2
Transferrin	3
Alcohol dehydrogenase	4
HMG CoA reductase	5
Steroid 21-hydroxylase	6
Arginase	7
Carbonic anhydrase	8
Interferon	9
Parathyroid hormone	11
Glyceraldehyde 3-phosphate dehydrogenase	12
Adenosine deaminase	13
α_1-Antitrypsin	14
Cytochrome P_{450}	15
Hemoglobin α-chain	16
Growth hormone	17
Prealbumin	18
Creatine phosphokinase (M chain)	19
Adenosine deaminase	20
Superoxide dismutase	21
Immunoglobulin (λ chain)	22
Glucose 6-phosphate dehydrogenase	X
Steroid sulfatase	Y

ENHANCERS AND TISSUE-SPECIFIC GENE EXPRESSION

Enhancers (or activators) are DNA elements that facilitate or ***enhance gene expression***. The enhancers provide binding sites for specific proteins that regulate transcription. They facilitate binding of the transcription complex to promoter regions. It is believed that the

tissue-specific gene expression is largely mediated through the involvement of enhancers.

COMBINATION OF DNA ELEMENTS AND PROTEINS IN GENE EXPRESSION

Gene expression in mammals is a complicated process with several environmental stimuli on a single gene. The ultimate response of the gene which may be positive or negative is brought out by the association of DNA elements and proteins.

In the illustration given in the ***Fig.26.4***, gene I is activated by a combination of activators 1, 2 and 3. Gene II is more effectively activated by the combined action of 1, 3 and 4. Activator 4 is not in direct contact with DNA, but it forms a bridge between activators 1 and 3, and activates gene II. As regards gene III, it gets inactivated by a combination of 1, 5 and 3. In this case, protein 5 interferes with the binding of protein 2 with the DNA, and inactivates the gene.

MOTIFS IN PROTEINS AND GENE EXPRESSION

A motif literally means a dominant element. Certain motifs in proteins mediate the binding of regulatory proteins (transcription factors) to DNA. The specific control of transcription occurs by the binding of regulatory proteins with high affinity to the correct regions of DNA.

A great majority of specific ***protein-DNA interactions*** are brought out by four unique motifs—helix-turn-helix (HTH), zinc finger, leucine zipper, helix-loop-helix (HLH).

These ***amino acid motifs*** bind with high affinity to the specific site and low affinity to other parts of DNA. The motif-DNA interactions are maintained by hydrogen bonds and van der Waals forces.

Helix-turn-helix motif

The helix-turn-helix (***HTH***) motif is about 20 amino acids which represents a small part of a large protein. HTH is the domain part of the protein which specifically interacts with the DNA (***Fig.26.5A***). Examples of helix-turn-helix motif proteins include lactose repressor, and cyclic AMP catabolite activator protein (CAP) of *E. coli*, and several developmentally important transcription factors in mammals.

Zinc finger motif

Sometime ago, it was recognized that the transcription factor TFIIIA requires zinc for its activity. On analysis, it was revealed that each TFIIIA contains zinc ions as a repeating coordinated complex. This complex is formed by the closely spaced amino acids cysteine and cysteine, followed by a histidine—histidine pair. In some instances, His-His is replaced by a second Cys-Cys pair (***Fig.26.5B***).

The zinc fingers bind to the major groove of DNA, and lie on the face of the DNA. This binding makes a contact with 5 bp of DNA. The

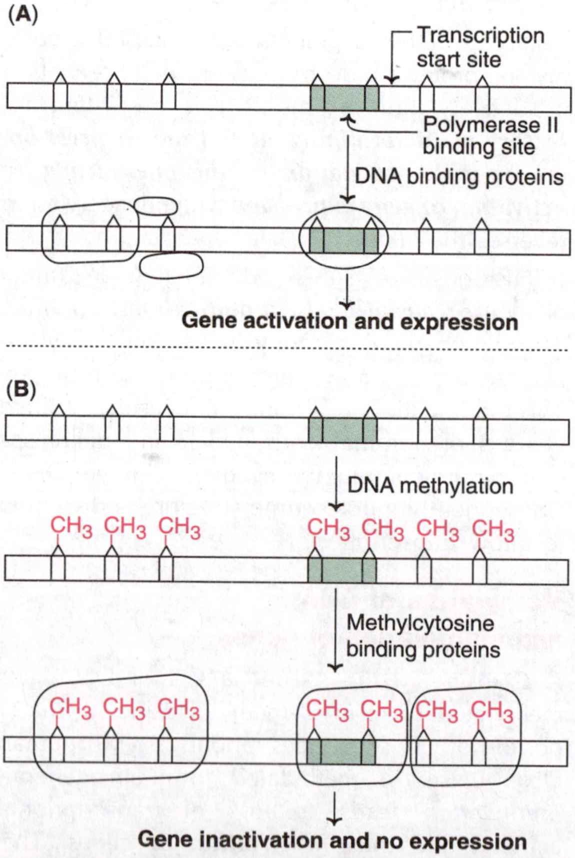

Fig. 26.3 : Methylation of DNA and inactivation of genes (A) Gene activation in the absence of DNA methylation (B) Gene inactivation due to methylation (⌂ represent CG sequences).

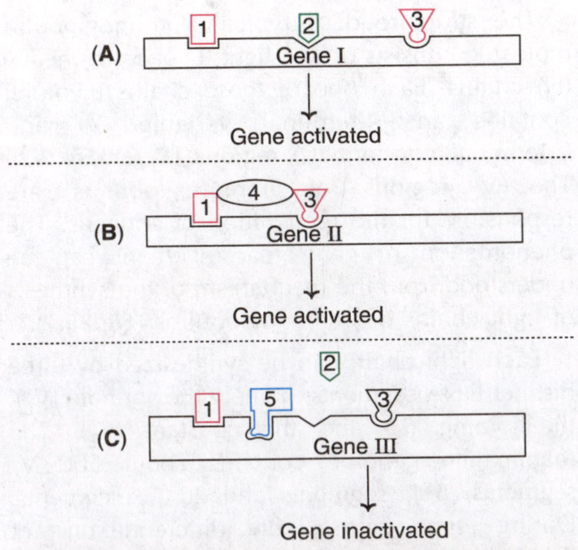

Fig. 26.4 : A diagrammatic representation of the association of DNA elements and proteins in gene regulation. **A, B** *and* **C** *represent genes I, II and III (1...5 represent proteins).*

steroid hormone receptor transcription factors use zinc finger motifs to bind to DNA.

Leucine zipper motif

The ***b***asic regions of leucine ***zip***per (***bZIP***) proteins are rich is the amino acid leucine. There occurs a periodic repeat of leucine residues at every seventh position. This type of repeat structure allows two identical monomers or heterodimers to ***zip together*** and form a dimeric complex. This protein-protein complex associates and interacts with DNA (***Fig.26.5C***). Good examples of leucine zipper proteins are the enhancer binding proteins (EBP)—fos and jun.

Helix-loop-helix motif

Two amphipathic (literally means a feeling of closeness) α-helical segments of proteins can form helix-loop-helix motif and bind to DNA. The dimeric form of the protein actually binds to DNA (***Fig.26.5D***).

GENE REGULATION IN EUKARYOTES

The important features of eukaryotic gene expression along with the regulatory aspects are described in the preceeding pages. Besides transcription, eukaryotic cells also employ variety of other mechanisms to regulate gene expression. The most important ones are listed below, and briefly described next.

1. Gene amplification
2. Gene rearrangement

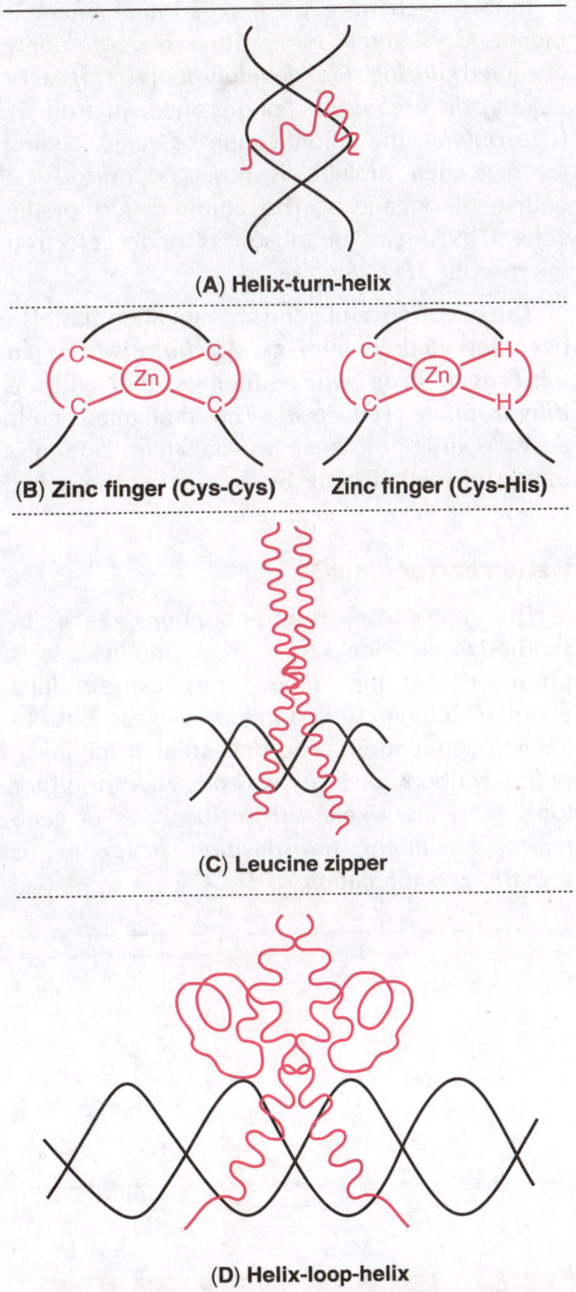

Fig. 26.5 : A diagrammatic representation of common motifs in proteins interacting with DNA.

3. Processing of RNA
4. Alternate mRNA splicing
5. Transport of mRNA from nucleus to cytoplasm
6. Degradation of mRNA.

Gene amplification

In this mechanism, the expression of a gene is increased several fold. This is commonly observed ***during the developmental stages of eukaryotic organisms***. For instance, in fruit fly (*Drosophila*), the amplification of genes coding for egg shell proteins is observed during the course of oogenesis. The amplification of the gene (DNA) can be observed under electron microscope (***Fig.26.6***).

The occurrence of gene amplification has also been reported in humans. ***Methotrexate is an anticancer drug*** which inhibits the enzyme ***dihydrofolate reductase***. The malignant cells develop drug resistance to long term administration of methotrexate by amplifying the genes coding for dihydrofolate reductase.

Gene rearrangment

The body possesses an enormous capacity to synthesize a wide range of antibodies. It is estimated that the human body can produce about 10 billion (10^{10}) antibodies in response to antigen stimulations. The molecular mechanism of this antibody diversity was not understood for long. It is now explained on the basis of gene rearrangement or ***transposition of genes*** or ***somatic recombination of DNA.***

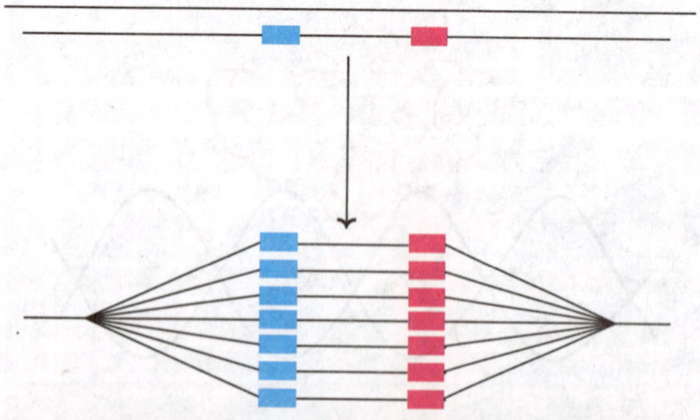

Fig. 26.6 : A diagrammatic representation of gene amplification (the genes are depicted in blue and red colours).

The structure of a typical immunoglobulin molecule consists of two light (L) and two heavy (H) chains. Each one of these chains (L or H) contains an N-terminal variable (V) and C-terminal constant (C) regions (***Refer Fig.9.3***). The V regions of immunoglobulins are responsible for the recognition of antigens. The phenomenon of gene rearrangement can be understood from the mechanism of the synthesis of light chains of immunoglobulins (***Fig.26.7***).

Each light chain can be synthesized by three distinct DNA segments, namely the variable (V_L), the joining (J_L) and the constant (C_L). The mammalian genome contains about 500 V_L segments, 6 J_L segments and 20 C_L segments. During the course of differentiation of B-lymphocytes, one V_L segment (out of the 500) is brought closer to J_L and C_L segments. This occurs on the same chromosome. For the sake of illustration, 100th V_L, 3rd J_L and 10th C_L segments are rearranged in ***Fig.26.7***. The rearranged DNA (with V_L, J_L and C_L fragments) is then transcribed to produce a single mRNA for the synthesis of a specific light chain of the antibody. By innumerable combinations of V_L, J_L and C_L segments, the body's immune system can generate millions of antigen specific immunoglobulin molecules.

Processing of RNA

The RNA synthesized in transcription undergoes modifications resulting in a functional RNA. The changes include intron-exon splicing, polyadenylation etc. (***Chapter 25***).

Alternate mRNA splicing

Eukaryotic cells are capable of carrying out alternate mRNA processing to control gene expression. Different mRNAs can be produced by alternate splicing which code for different proteins (for more details, ***Refer Chapter 25***).

Degradation of mRNA

The expression of genes is indirectly influenced by the stability of mRNA. Certain hormones regulate the synthesis and degradation of some mRNAs. For instance, estradiol prolongs the half-life of vitellogenin mRNA from a few hours to about 200 hours.

Medical Concepts/ Clinical Correlates

- The malignant cells develop drug resistance to long term administration of methotrexate. This occurs by amplification of the genes coding for dihydrofolate reductase.
- The human body has the capability to produce around 10 billion antigen-specific immunoglobulins. This is achieved by a process called gene rearrangement.
- Knowledge on the gene expression and its regulation helps in the understanding and control of several diseases, including cancer.
- Epigenetic therapy of cancer employing certain inhibitors holds a great promise.

EPIGENETIC REGULATION OF GENE EXPRESSION

The term epigenetics (literally means above genetics) is used to describe the ***changes in the characteristics*** of a cell or an organism that are ***not due to changes in the nucleotide sequence of the DNA***.

Epigenetics regulates gene expression by modulating chromatin structure via histone modification or modification of DNA via methylation. For instance, ***acetylation of histones leads to activation of gene expression***, while DNA methylation is associated with a reduction in gene expression.

The pattern of covalent modifications in histones that provide a tunable means of regulating gene expression is referred to as ***histone code.***

Although there is no change in DNA sequence, epigenetics is heritable due to coordination between histone modification and methylation of DNA.

Epigenetic therapy of cancers

Hypermethylation of DNA in some parts of tumor suppressor genes is found in certain cancers. The enzyme DNA methyltransferase (DNMT) responsible for DNA methylation is targeted for cancer therapy. Many inhibitors of DNMT (e.g. 5-azacytidine) have been approved by FDA, and are in use for the treatment of leukemia.

Likewise, inhibitors of histone deacetylase (HDAC) are also employed in the epigenetic therapy cancers. HDAC inhibitors stimulate tumor suppressor gene expression by allowing acetylation of histones in chromatin structure. FDA has aproved the use of vorinostat for the treatment of T-cell lymphoma.

Epigenetic therapy of cancer by employing various inhibitors (individually or in combination) holds a great promise.

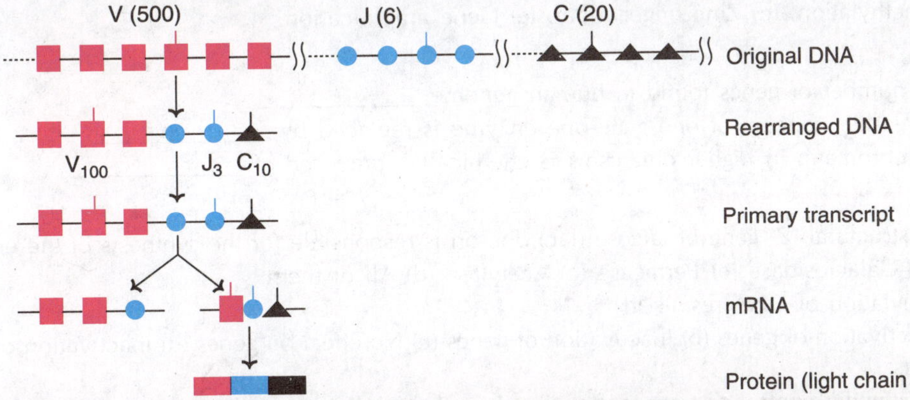

Fig. 26.7 : A diagrammatic representation of gene rearrangement for the synthesis of light chain of immunoglobulin.

SUMMARY

- *DNA, the chemical vehicle of heredity, is composed of genes. The regulation of gene expression is absolutely essential for the growth, development and differentiation of an organism. A positive regulation increases gene expression while a negative regulation decreases.*
- *The operon is the coordinated unit of gene expression. The lac operon of* E. coli *consists of regulatory genes and structural genes. The lac repressor binds to the DNA and halts the process of transcription of structural genes. However, the presence of lactose inactivates the repressor (derepression) leading to the expression of structural genes.*
- *Tryptophan operon is regulated by a repressor. Tryptophan repressor binds to tryptophan, and then to trp operator gene to turn off the transcription.*
- *Eukaryotic gene expression and its regulation are highly complex. Acetylation of histones leads to gene expression while deacetylation reverses the effect. In general, methylation of DNA results in the inactivation of genes.*
- *The protein-DNA interactions, brought out by motifs (helix-turn-helix, zinc finger, leucine zipper, helix-loop-helix), are involved in the control of gene expression.*
- *Eukaryotic cells have developed several mechanisms to regulate gene expression. These include gene amplification, gene rearrangement, and processing, transport and degradation of DNA.*

Self-Assessment Exercises

I. Essay questions

1. Describe lactose (lac) operon.
2. Write briefly on the gene expression and its regulation in eukaryotes.

II. Short notes

(a) One cistron-one subunit concept, (b) Catabolite gene activator protein, (c) Gene inactivation by DNA methylation, (d) Zinc finger motif, (e) Gene amplification.

III. Fill in the blanks

1. The number of genes found in human genome ______________.
2. The earlier concept of one gene-one enzyme is replaced by ______________.
3. The chromatin in higher organisms is chemically composed of ______________.

IV. Multiple choice questions

4. The structural 'Z' gene of lactose (lac) operon is responsible for the synthesis of the enzyme(s)
   (a) β-Galactosidase (b) Permease (c) Acetylase (d) All of them.
5. Methylation of DNA results in
   (a) Activation of genes (b) Inactivation of genes (c) No effect on genes (d) Inactivation of protein motifs.
6. The specific control of transcription involves the following motif(s)
   (a) Helix-turn-helix (b) Zinc finger (c) Leucine zipper (d) All of them.

Recombinant DNA and Biotechnology

27

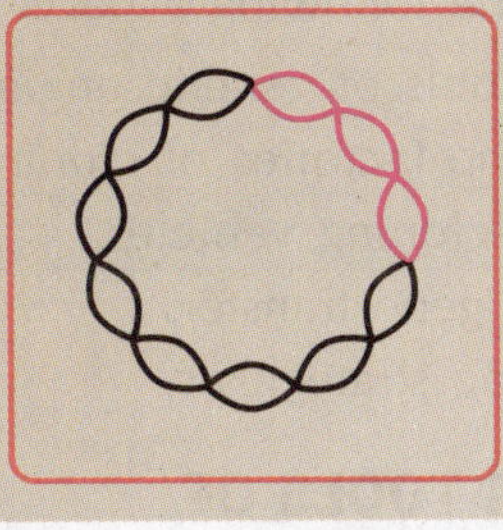

The recombinant DNA speaks

"I am the hybridized DNA molecule;
Created by cutting and sealing;
When introduced into host cells;
I multiply and code for desired proteins."

The term biotechnology represents a fusion or an alliance between ***bio**logy* and ***technology***. Frankly speaking, biotechnology is a newly discovered discipline for age-old practices e.g. preparation of wine, beer, curd, bread. These natural processes are regarded as ***old*** or ***traditional biotechnology***.

The ***new*** or ***modern biotechnology*** embraces all the ***genetic manipulations***, cell fusion techniques, and improvements made in the old biotechnological processes. The biotechnology with particular reference to recombinant DNA in human health and disease is briefly described in this chapter.

Genetic engineering primarily involves the ***manipulation of genetic material (DNA)*** to achieve the desired goal in a pre-determined way. Some ***other terms*** are also in common use to describe genetic engineering.

- ***Gene manipulation***
- ***Recombinant DNA (rDNA) technology***
- ***Gene cloning (molecular cloning)***
- ***Genetic modifications***
- ***New genetics.***

Brief history of recombinant DNA technology

The present day DNA technology has its roots in the experiments performed by ***Boyer and Cohen in 1973***. In their experiments, they successfully ***recombined two plasmids*** (pSC 101 and pSC 102) and ***cloned the new plasmid in E.coli.*** In the later experiments the genes of a frog could be successfully transplanted, and expressed in *E.coli*. This made the real beginning of modern rDNA technology and laid foundations for the present day ***molecular biotechnology***.

Some biotechnologists who admire Boyer-Cohen experiments divide the subject into two chronological categories.

1. ***BBC***-biotechnology ***B****efore* ***B****oyer* and ***C****ohen*.
2. ***ABC***-biotechnology ***A****fter* ***B****oyer* and ***C****ohen*.

Recombinant DNA technology is a vast field. The basic principles and techniques of rDNA technology along with the most important applications are briefly described in this chapter.

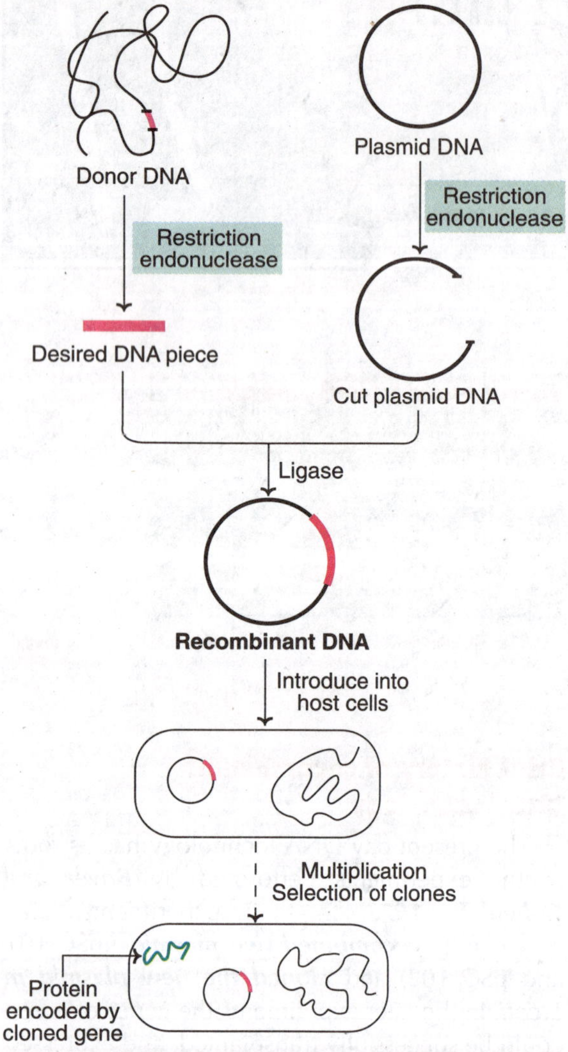

Fig. 27.1 : The basic principle of recombinant DNA technology.

BASIC PRINCIPLES OF rDNA TECHNOLOGY

There are many diverse and complex techniques involved in gene manipulation. However, the basic principles of recombinant DNA technology are reasonably simple, and broadly involve the following stages (***Fig.27.1***).

1. Generation of DNA fragments and selection of the ***desired piece of DNA*** (e.g. a ***human gene***).

2. Insertion of the selected DNA into a cloning vector (e.g. a plasmid) to ***create a recombinant DNA*** or ***chimeric DNA*** (Chimera is a monster in Greek mythology that has a lion's head, a goat's body and a serpent's tail. This may be comparable to Narasimha in Indian mythology).

3. Introduction of the recombinant vectors into ***host cells*** (e.g. bacteria).

4. Multiplication and ***selection of clones*** containing the recombinant molecules.

5. Expression of the gene to produce the ***desired product***.

Recombinant DNA technology with special reference to the following aspects is described

- Molecular tools of genetic engineering.
- Host cells-the factories of cloning.
- Vectors-the cloning vehicles.
- Methods of gene transfer.
- Gene cloning strategies.

MOLECULAR TOOLS OF GENETIC ENGINEERING

The term genetic engineer may be appropriate for an individual who is involved in genetic manipulations. The ***genetic engineer's toolkit*** or molecular tools namely the ***enzymes*** most commonly used in recombinant DNA experiments are briefly described.

Restriction endonucleases—DNA cutting enzymes

Restriction endonucleases are one of the most important groups of enzymes for the manipulation of DNA. These are the ***bacterial enzymes*** that can ***cut/split DNA*** (from any source) at specific sites. They were first discovered in *E.coli* restricting the replication of bacteriophages, by cutting the viral DNA (The host *E.coli* DNA is protected from cleavage by addition of methyl groups). Thus, the enzymes that restrict the viral replication are known as ***restriction enzymes*** or restriction endonucleases.

Nomenclature : Restriction endonucleases are named by a standard procedure, with particular reference to the bacteria from which they are isolated. The first letter (in italics) of the enzymes indicates the genus name, followed by the first two letters (also in italics) of the species, then comes the strain of the organism and finally a

Roman numeral indicating the order of discovery. A couple of examples are given below.

EcoRI is from *Escherichia* (***E***) *coli* (***co***), strain Ry13 (***R***), and first endonuclease (***I***) to be discovered. ***HindIII*** is from *Haemophilus* (***H***) *influenzae* (***in***), strain Rd (***d***) and, the third endonucleases (***III***) to be discovered.

Recognition sequences : Recognition sequence is the ***site where the DNA is cut*** by a restriction endonuclease. Restriction endonucleases can specifically recognize DNA with a particular sequence of 4-8 nucleotides and cleave.

Cleavage patterns : Majority of restriction endonucleases (particularly type II) cut DNA at defined sites within recognition sequence. A selected list of enzymes, recognition sequences, and their products formed is given in ***Table 27.1.***

The cut DNA fragments by restriction endonucleases may have mostly ***sticky ends*** (cohesive ends) or ***blunt ends***, as given in ***Table 27.1***. DNA fragments with sticky ends are particularly useful for recombinant DNA experiments. This is because the single-stranded sticky DNA ends can easily pair with any other DNA fragment having complementary sticky ends.

DNA ligases—DNA joining enzymes

The cut DNA fragments are covalently joined together by DNA ligases. These enzymes were originally isolated from viruses. They also occur in *E.coli* and eukaryotic cells. DNA ligases actively participate in cellular DNA repair process.

Many enzymes are used in the recombinant DNA technology/genetic engineering. A selected list of these enzymes and the reactions catalysed by them is given in ***Table 27.2***.

Table 27.1 Some restriction enzymes with sources, recognition sequences and the products formed

Enzyme (source)	Recognition sequence	Products
EcoRI (*Escherichia coli*)	5′......G✂A-A-T-T-C......3′ 3′......C-T-T-A-A✂G......5′	A-A-T-T-C...... G...... ......G C-T-T-A-A
BamHI (*Bacillus amyloliquefaciens*)	5′......G✂G-A-T-C-C......3′ 3′......C-C-T-A-G✂G......5′	G-A-T-C-C...... G...... ......G ......C-C-T-A-G
HaeIII (*Haemophilus aegyptius*)	5′......G-G✂C-C......3′ 3′......C-C✂G-G......5′	*C-C...... G-G...... ......*G-G ...... C-C
HindIII (*Haemophilus influenzae*)	5′......A✂A-G-C-T-T......3′ 3′......T-T-C-G-A✂A......5′	A-G-C-T-T...... A...... ......A ......T-T-C-G-A
NotI (*Nocardia otitidis*)	5′......G-C✂G-G-C-C-G-C......3′ 3′......C-G-C-C-G-G✂C-G......5′	G-G-C-C-G-C...... C-G...... ......G-C ......C-G-C-C-G-G......

(***Note :*** *Scissors indicate the sites of cleavage.* **The products are with blunt ends while for the rest, the products are with sticky ends*).

Table 27.2 The most commonly used enzymes in recombinant DNA technology/genetic engineering

Enzyme	Use/reaction
Alkaline phosphatase	Removes phosphate groups from 5′-ends of double/single-stranded DNA (or RNA).
Bal 31 nuclease	For the progressive shortening of DNA.
DNA ligase	Joins DNA molecules by forming phosphodiester linkages between DNA segments.
DNA polymerase I	Synthesizes DNA complementary to a DNA template.
DNase I	Produces single-stranded nicks in DNA.
Exonuclease III	Removes nucleotides from 3′-end of DNA.
λ exonuclease	Removes nucleotides from 5′-end of DNA.
Polynucleotide kinase	Transfers phosphate from ATP to 5′-OH ends of DNA or RNA.
Restriction enzymes	Cut double-stranded DNA with a specific recognition site.
Reverse transcriptase	Synthesizes DNA from RNA.
RNase A	Cleaves and digests RNA (and not DNA).
RNase H	Cleaves and digests the RNA strand of RNA-DNA heteroduplex.
Taq DNA polymerase	Used in polymerase chain reaction
SI nuclease	Degrades single-stranded DNA and RNA.
Terminal transferase	Adds nucleotides to the 3′-ends of DNA or RNA. Useful in homopolymer tailing.

HOST CELLS — THE FACTORIES OF CLONING

The hosts are the ***living systems or cells*** in which the carrier of recombinant DNA molecule or ***vector can be propagated***. There are different types of host cells-prokaryotic (bacteria) and eukaryotic (fungi, animals and plants). Some examples of host cells used in genetic engineering are given in ***Table 27.3***.

Host cells, besides effectively incorporating the vector's genetic material, must be conveniently cultivated in the laboratory to collect the products. In general, ***microorganisms*** are ***preferred*** as host cells, since they ***multiply faster*** compared to cells of higher organisms (plants or animals).

VECTORS–THE CLONING VEHICLES

Vectors are the DNA molecules, which can ***carry a foreign DNA fragment to be cloned***. They are self-replicating in an appropriate host cell. The most important vectors are plasmids, bacteriophages, cosmids and artificial chromosome vectors.

Choice of vector

Among the several factors, the size of the foreign DNA is very important in the choice of vectors. The efficiency of this process is often crucial for determining the success of cloning. The size of DNA inserts that can be accepted by different vectors is shown in ***Table 27.4***.

METHODS OF GENE TRANSFER

Introducing a foreign DNA (i.e. the gene) into the cells is an important task in biotechnology.

Table 27.3 Some examples of host cells used in genetic engineering

Group	Examples
Prokaryotic	
Bacteria	*Escherichia coli*
	Bacillus subtilis
	Streptomyces sp
Eukaryotic	
Fungi	*Saccharomyces cerevisiae*
	Aspergillus nidulans
Animals	Insect cells
	Oocytes
	Mammalian cells
	Whole organisms
Plants	Protoplasts
	Intact cells
	Whole plants

Table 27.4 The different cloning vectors with the corresponding hosts and the sizes of foreign insert DNAs

Vector	Host	Foreign insert DNA size
Phage λ	*E. coli*	5–25 kb
Cosmid λ	*E. coli*	35–45 kb
Plasmid artifical chromosome (PAC)	*E. coli*	100–300 kb
Bacterial artificial chromosome (BAC)	*E. coli*	100–300 kb
Yeast chromosome	*S. cerevisiae*	200–2000 kb

The efficiency of this process is often crucial for determining the success of cloning. The most commonly employed ***gene transfer methods*** are ***transformation, conjugation, electroporation and lipofection, and direct transfer of DNA.***

GENE CLONING STRATEGIES

A ***clone*** refers to ***a group of*** organisms, cells, ***molecules*** or other objects, ***arising from a single individual.*** Clone and colony are almost synonymous.

Gene cloning strategies in relation to recombinant DNA technology broadly involve the following aspects (***Fig.27.2***).

- Generation of desired DNA fragments.
- Insertion of these fragments into a cloning vector.
- Introduction of the vectors into host cells.
- Selection or screening of the recipient cells for the recombinant DNA molecules.

BASIC TECHNIQUES IN GENETIC ENGINEERING

There are several techniques used in recombinant DNA technology or gene manipulation. The most frequently used methods are listed.

- Isolation and purification of nucleic acids.
- Nucleic acid blotting techniques.
- DNA sequencing.
- Methods of gene transfer (described already).
- Polymerase chain reaction.
- Production of monoclonal antibodies (***Chapter 41***).
- Construction of gene library.
- Site-directed mutagenesis and protein engineering.

ISOLATION AND PURIFICATION OF NUCLEIC ACIDS

Almost all the experiments dealing with gene manipulations require pure forms of either DNA or RNA, or sometimes even both. Hence there is a need for the reliable isolation of nucleic acids from the cells. The purification of nucleic acids broadly involves three stages.

1. Breaking or opening of the cells to expose nucleic acids.
2. Separation of nucleic acids from other cellular components.
3. Recovery of nucleic acids in a pure form.

NUCLEIC ACID BLOTTING TECHNIQUES

Blotting techniques are very widely used analytical tools for the specific ***identification of desired DNA or RNA fragments*** from thousands of molecules. Blotting refers to the ***process of immobilization of sample nucleic acids on solid support*** (nitrocellulose or nylon membranes). The blotted nucleic acids are then used as targets in the hybridization experiments for their specific detection. An outline of the nucleic acid blotting technique is depicted in ***Fig.27.3.***

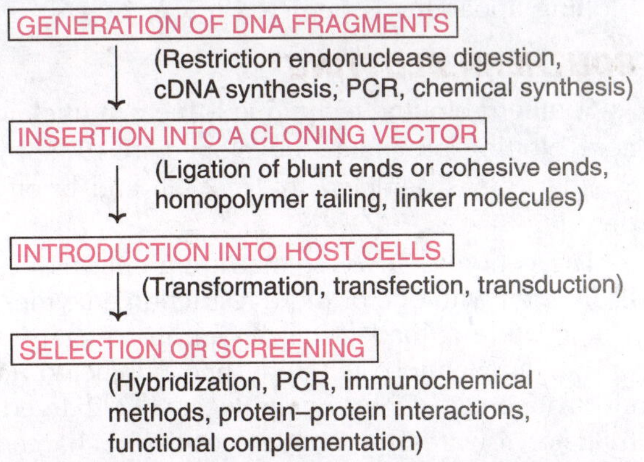

***Fig. 27.2 :** An overview of cloning strategies in recombinant DNA technology.*

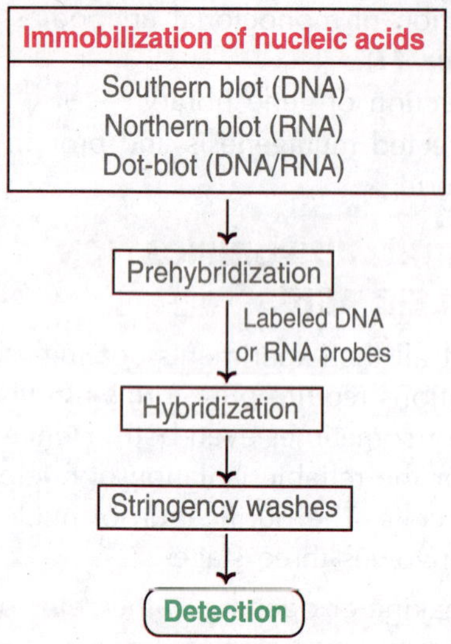

Fig. 27.3: An outline of the nucleic acid blotting techniques.

Types of blotting techniques

The most comonly used blotting techniques are listed below

- ***Southern blotting (for DNA)***
- ***Northern blotting (for RNA)***
- ***Dot blotting (DNA/RNA)***

The Southern blotting is named after the scientist Ed Southern (1975) who developed it. The other names Northern blotting and Western blotting are laboratory jargons which are now accepted. ***Western blotting involves the transfer of protein blots and their identification by using specific antibodies.***

A diagrammatic representation of a typical blotting apparatus is depicted in ***Fig.27.4.***

SOUTHERN BLOTTING

Southern blotting technique is the first nucleic acid blotting procedure developed in 1975 by Southern. It is depicted in ***Fig.27.5***, and briefly described.

The genomic DNA isolated from cells/tissues is digested with one or more restriction enzymes. This mixture is loaded into a well in an agarose or polyacrylamide gel and then subjected to electrophoresis. DNA, being negatively charged, migrates towards the anode (positively charged electrode); smaller DNA fragments move faster.

The separated DNA molecules are denatured by exposure to a mild alkali and transferred to nitrocellulose or nylon paper. This results in an exact replica of the pattern of DNA fragments on the gel. The DNA can be annealed to the paper on exposure to heat (80°C). The nitrocellulose or nylon paper is then exposed to labeled cDNA probes. These probes hybridize with complementary DNA molecules on the paper.

The paper after thorough washing is exposed to X-ray film to develop autoradiograph. This reveals specific bands corresponding to the ***DNA fragments recognized by cDNA probe.***

Zoo blot : This is a specialized Southern blot technique used ***to compare DNA sequences*** (genomes) ***between humans and other organisms.*** e.g. hemoglobin gene sequences in humans compared to that of chimpanzee, horse and pig. Zoo blot technique is also useful to distinguish between coding and non-coding regions and their evolution in different organisms.

Applications of Southern blotting

- It is an invaluable method in ***gene analysis.***
- Important for confirmation of DNA cloning.
- Forensically applied to detect minute quantities of DNA (to ***identify parenthood, thieves, rapists*** etc.).
- Highly useful for the determination of restriction fragment length polymorphism (RFLP) associated with ***pathological conditions.***

NORTHERN BLOTTING

Northern blotting is the technique for the ***specific identification of RNA molecules.*** The

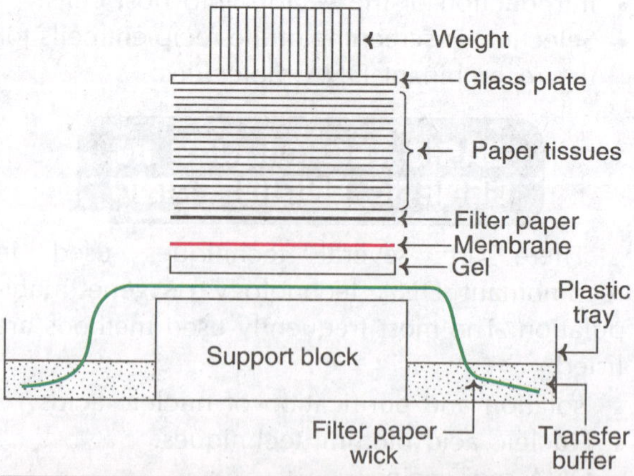

Fig. 27.4 : Diagrammatic representation of a typical blotting apparatus.

procedure adapted is almost similar to that described for Southern blotting and is depicted in ***Fig.27.6***. RNA molecules are subjected to electrophoresis, followed by blot transfer, hybridization and autoradiography.

RNA molecules do not easily bind to nitrocellulose paper or nylon membranes. Blot-transfer of RNA molecules is carried out by using a chemically reactive paper prepared by diazotization of aminobenzyloxymethyl to create diazobenzyloxymethyl (DBM) paper. The RNA can covalently bind to DBM paper.

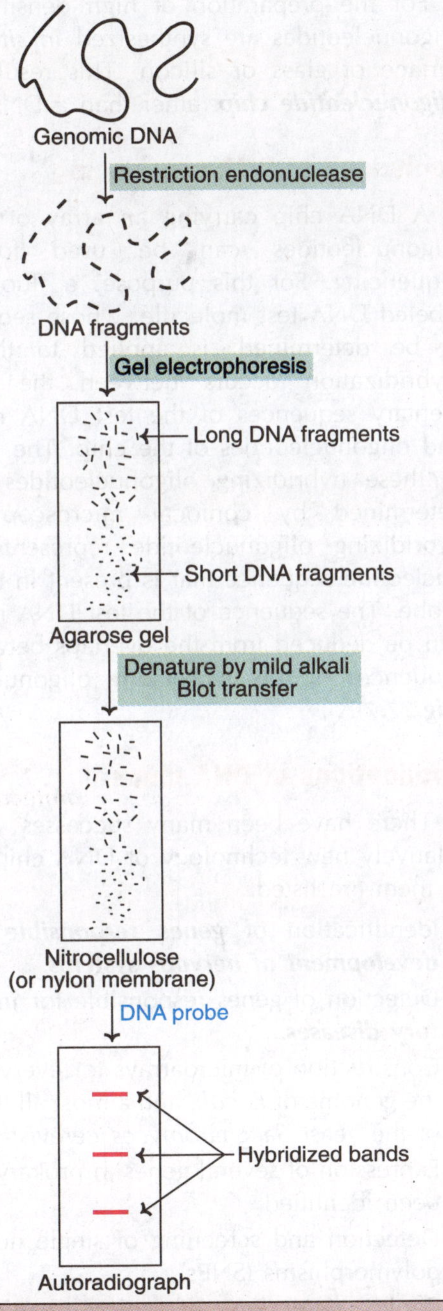

Fig. 27.5 : Southern blotting technique.

Northern blotting is theoretically, a good technique for determining the number of genes (through mRNA) present on a given DNA. But this is not really practicable since each gene may give rise to two or more RNA transcripts. Another drawback is the presence of exons and introns.

DOT-BLOTTING

Dot-blotting is a modification of Southern and Northern blotting techniques described above. In this approach, the ***nucleic acids (DNA or RNA) are directly spotted onto the filters***, and not subjected to electrophoresis. The hybridization procedure is the same as in original blotting techniques.

Dot-blotting technique is particularly useful in obtaining quantitative data for the evaluation of gene expression.

Western blotting

Western blotting involves the ***identification of proteins***. It is very useful to understand the

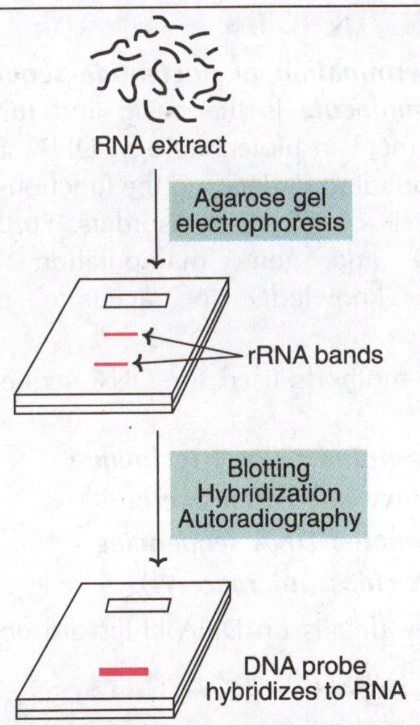

Fig. 27.6 : An outline of Northern blotting.

nucleic acid functions, particularly during the course of gene manipulations.

The technique of Western blotting involves the transfer of electrophoresed protein bands from polyacrylamide gel to nylon or nitrocellulose membrane. These proteins can be detected by specific protein-ligand interactions. Antibodies or lectins are commonly used for this purpose.

Autoradiography

Autoradiography is the process of ***localization and recording of a radiolabel*** within a solid specimen, with the production of an ***image in a photographic emulsion***. These emulsions are composed of silver halide crystals suspended in gelatin.

When a β-particle or a γ-ray from a radiolabel passes through the emulsions, silver ions are converted to metallic silver atoms. This results in the development of a visible image which can be easily detected.

Applications of autoradiography

As already described, autoradiography is closely associated with blotting techniques for the detection of DNA, RNA and proteins.

DNA SEQUENCING

Determination of nucleotide sequence in a DNA molecule is the basic and fundamental requirement in biotechnology. DNA sequencing is important to understand the functions of genes, and basis of inherited disorders. Further, DNA cloning and gene manipulation invariably require knowledge of accurate nucleotide sequence.

The mothods used for DNA sequencing are listed

- ***Maxam and Gilbert technique***
- ***Dideoxy nucleotide method***
- ***Automated DNA sequencing***
- ***DNA chips (microarrays)***

Some details on DNA chips are described.

DNA CHIPS (MICROARRAYS)

DNA chips or DNA microarrays are recent developments for DNA sequencing as result of advances made in automation and miniaturization. A large number of DNA probes, each one with different sequence, are immobilized at defined positions on the solid surface, made up of either nylon or glass. The ***probes*** can be short DNA molecules such as ***cDNAs or synthetic oligonucleotides.***

For the preparation of high density arrays, oligonucleotides are synthesized *in situ* on the surface of glass or silicon. This results in an ***oligonucleotide chip*** rather than a DNA chip.

Technique of DNA sequencing

A DNA chip carrying an array of different oligonucleotides can be used for DNA sequencing. For this purpose, a fluorescently labeled DNA test molecule, whose sequence is to be determined, is applied to the chip. Hybridization occurs between the complementary sequences of the test DNA molecule and oligonucleotides of the chip. The positions of these hybridizing oligonucleotides can be determined by confocal microscopy. Each hybridizing oligonucleotide represents an 8-nucleotide sequence that is present in the DNA probe. The sequence of the test DNA molecule can be deduced from the overlaps between the sequences of the hybridizing oligonucleotides (***Fig.27.7***).

Applications of DNA chips

There have been many successes with this relatively new technology of DNA chips. Some of them are listed.

- Identification of ***genes responsible*** for the ***development of nervous systems.***
- Detection of genes responsible for ***inflammatory diseases.***
- Construction of microarrays for every gene in the genome of *E. coli*, and almost all the genes of the yeast *Saccharomyces cerevisiae*.
- Expression of several genes in prokaryotes has been identified.
- Detection and screening of single nucleotide polymorphisms (SNPs).
- Rapid ***detection of microorganisms*** for environmental monitoring.

The future of DNA chips

The major limitation of DNA chips at present is the unavailability of complete genome arrays for higher eukaryotes, including humans. It is expected that within the next few years such DNA chips will be available. This will help the biotechnologists to capture the functional snapshots of the genome in action for higher organisms.

POLYMERASE CHAIN REACTION (DNA AMPLIFICATION)

The polymerase chain reaction (PCR) is a laboratory (*in vitro*) ***technique for generating large quantities of a specified DNA***. Obviously,

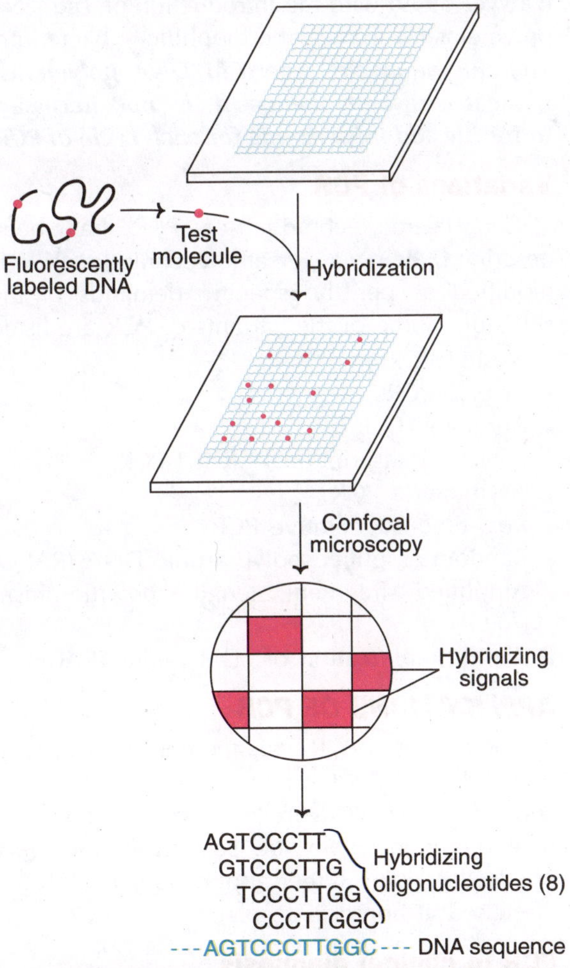

Fig. 27.7 : Microarray (or chip) technology in DNA sequencing.

PCR is a ***cell-free amplification technique*** for synthesizing multiple identical copies (billions) of ***any DNA*** of interest. Developed in 1984 by Karry Mullis (Nobel Prize, 1993), PCR is now considered as a basic tool for the molecular biologist. As is a photocopier a basic requirement in an office, so is the PCR machine in a molecular biology laboratory!

Principle of PCR

The double-stranded DNA of interest is denatured to separate into two individual strands. Each strand is then allowed to hybridize with a primer (renaturation). The primer-template duplex is used for DNA synthesis (the enzyme-DNA polymerase). These three steps—***denaturation, renaturation*** and ***synthesis*** are repeated again and again to generate multiple forms of target DNA.

Technique of PCR

The essential requirements for PCR are listed below

1. A target DNA (100–35,000 bp in length).
2. Two primers (synthetic oligonucleotides of 17–30 nucleotides length) that are complementary to regions flanking the target DNA.
3. Four deoxyribonucleotides (dATP, dCTP, dGTP, dTTP).
4. A DNA polymerase that can withstand at a temperature up to 95°C (i.e., thermostable).

The actual technique of PCR involves repeated cycles for amplification of target DNA. Each cycle has three stages.

1. **Denaturation :** On raising the temperature to about 95°C for about one minute, the DNA gets denatured and the two strands separate.
2. **Renaturation or annealing :** As the temperature of the mixutre is slowly cooled to about 55°C, the primers base pair with the complementary regions flanking target DNA strands. This process is called renaturation or annealing. High concentration of primer ensures annealing between each DNA strand and the primer rather than the two strands of DNA.
3. **Synthesis :** The initiation of DNA synthesis occurs at 3′-hydroxyl end of each primer. The primers are extended by joining the bases

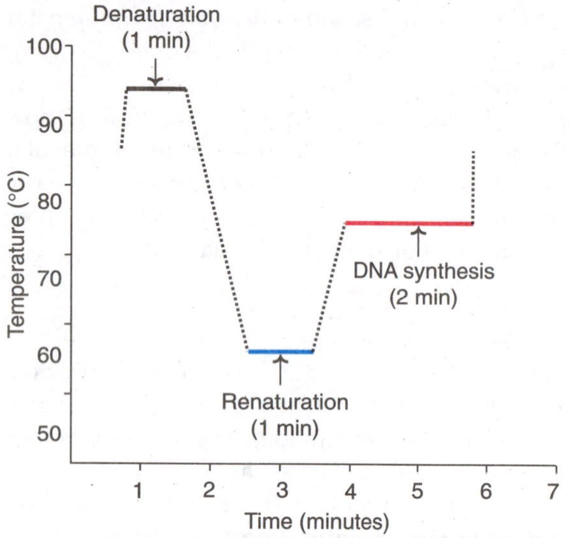

Fig. 27.8 : *The three stages in each cycle of PCR in relation to temperature and time (Each cycle takes approximately 3-5 minutes).*

complementary to DNA strands. The synthetic process in PCR is quite comparable to the DNA replication of the leading strand (***Refer Chapter 24***). However, the temperature has to be kept optimal as required by the enzyme DNA polymerase. For ***Taq DNA polymerase, the optimum temperature is around 75°C*** (for *E. coli* DNA polymerase, it is around 37°C). The reaction can be stopped by raising the temperature (to about 95°C).

The 3 stages of PCR in relation to temperature and time are depicted in ***Fig.27.8***. Each cycle of PCR takes about 3-5 minutes. In the normal practice, the PCR is carried out in an automated machine.

As is evident from the ***Fig.27.9*** (cycle I), the new DNA strand joined to each primer is beyond the sequence that is complementary to the second primer. These new strands are referred to as ***long templates***, and they will be used in the second cycle.

For the second cycle of PCR, the DNA strands (original + newly synthesized long template) are denatured, annealed with primers and subjected to DNA synthesis. At the end of second round, long templates, and short templates (DNA strands with primer sequence at one end, and sequence complementary to the other end primer) are formed.

In the third cycle of PCR, the original DNA strands along with long and short templates are the starting materials. The technique of ***denaturation, renaturation and synthesis are repeated***. This procedure is repeated ***again and again for each cycle***. It is estimated that at the end of 32nd cycle of PCR, about a million-fold target DNA is synthesized. The short templates possessing precisely the target DNA as double-stranded molecules accumulate.

Sources of DNA polymerase

In the original technique of PCR, Klenow fragment of *E. coli* DNA polymerase was used. This enzyme, gets denatured at higher temperature, therefore, fresh enzyme had to be added for each cycle. A breakthrough occurred (Lawyer 1989) with the introduction of *Taq* DNA polymerase from thermophilic bacterium, *Thermus aquaticus*. The ***Taq DNA polymerase is heat resistant, hence it is not necessary to freshly add this enzyme for each cycle of PCR.***

Variations of PCR

The basic technique of PCR has been described. Being a versatile technique, PCR is modified as per the specific demands of the situation. Some of the variants of PCR are listed

- Nested PCR
- Inverse PCR
- Anchored PCR
- Reverse transcription PCR (RT-PCR)
- Asymmetric PCR
- Real-time quantitative PCR
- Random amplified polymorphic DNA (RAPD)
- Amplified fragment length polymorphism (AFLP)
- Rapid amplification of cDNA ends (RACE).

APPLICATIONS OF PCR

The advent of PCR had, and continues to have tremendous impact on molecular biology. The applications of PCR are too many to be listed here. Some of them are selectively and very briefly described. Other applications of PCR are discussed at appropriate places.

PCR in clinical diagnosis

The specificity and sensitivity of PCR is highly useful for the diagnosis of various diseases in humans. These include diagnosis of inherited disorders (genetic diseases), viral diseases, bacterial diseases etc.

Prenatal diagnosis of inherited diseases : PCR is employed in the prenatal diagnosis of inherited diseases by using chorionic villus samples or cells from amniocentesis. Thus, diseases ***like sickle-cell anemia, β-thalassemia and phenylketonuria can be detected*** by PCR in these samples.

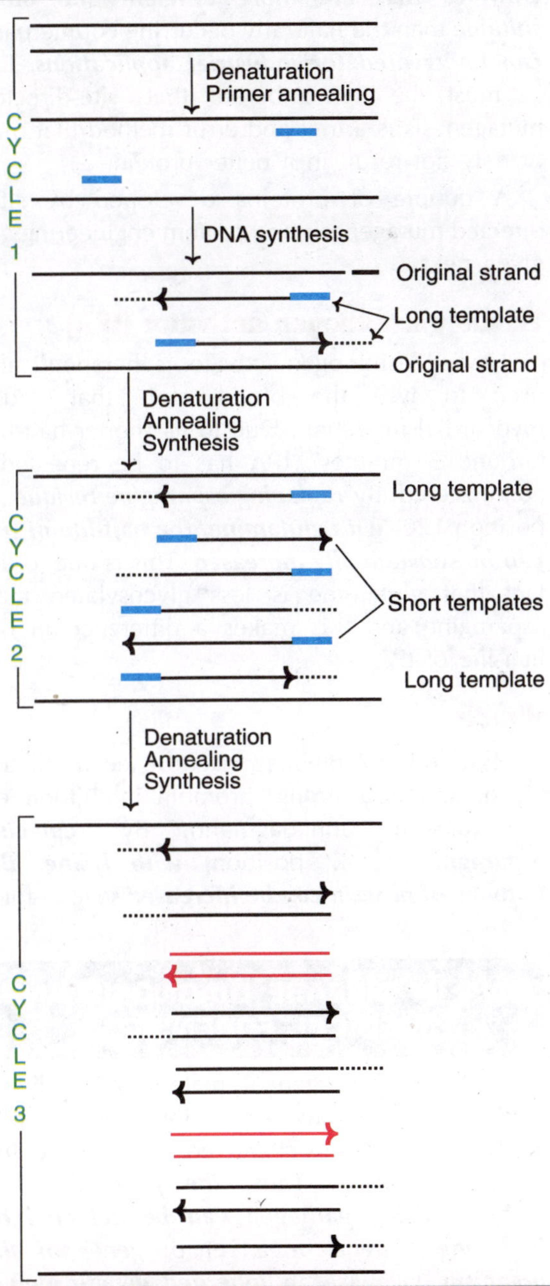

Fig. 27.9 : The polymerase chain reaction (PCR) representing the initial three cycles (— indicate primers).

Diagnosis of retroviral infections : PCR from cDNA is a valuable tool for diagnosis and monitoring of retroviral infections, e.g., HIV infection.

Diagnosis of bacterial infections : PCR is used for the detection of bacterial infections e.g., tuberculosis by *Mycobacterium tuberculosis.*

Diagnosis of cancers : Several virally-induced cancers (e.g., cervical cancer caused by human papilloma virus) can be detected by PCR. Further, some cancers which occur due to chromosomal translocation (chromosome 14 and 18 in follicular lymphoma) involving known genes are identified by PCR.

PCR in sex determination of embryos : Sex of human and live stock embryos fertilized *in vitro,* can be determined by PCR, by using primers and DNA probes specific for sex chromosomes. Further, this technique is also useful to detect sex — linked disorders in fertilized embryos.

PCR in the diagnosis of COVID-19 : The diagnosis of COVID-19 pandemic can be done ***by real-time PCR (RT-PCR)*** against the specific genes of severe acute respiratory syndrome coronavirus 2 (SARS-CoV_2).

PCR in comparative studies of genomes

The differences in the genomes of two organisms can be measured by PCR with random primers. The products are separated by electrophoresis for comparative identification. Two genomes from closely related organisms are expected to yield more similar bands.

PCR is very important in the study of ***evolutionary biology***, more specifically referred to as ***phylogenetics***. As a technique which can amplify even minute quantities of DNA from any source (hair, mummified tissues, bone, or any fossilized material), PCR has revolutionized the studies in palaentology and archaelogy. The movie 'Jurassic Park', has created public awareness of the potential applications of PCR!

Medical Concepts/ Clinical Correlates

- Biotechnology is a newly discovered discipline for age-old practices (e.g. preparation of curd, wine, beer), with special emphasis on genetic manipulations.
- Human artificial chromosome (HAC) is a synthetic vector, possessing the characteristics of human chromosome. HAC is capable of carrying large-sized human genes that may be useful in gene therapy.
- Southern blotting technique (that specifically detects DNA) is employed for the identification of thieves, rapists, and settlement of parenthood.

PCR in forensic medicine

A single molecule of DNA from any source (blood strains, hair, semen etc.) of an individual is adequate for amplification by PCR. Thus, PCR is very important for identification of criminals.

The reader may refer DNA finger printing technique described later in this chapter.

GENE LIBRARIES

The ***collection of DNA fragments*** (specifically genes) from a particular species represents gene libraries. The creation or construction of gene libraries (broadly ***genomic libraries***) is accomplished by isolating the complete genome (entire DNA from a cell) which is cut into fragments, and cloned in suitable vectors. Then the specific clone carrying the desired (target) DNA can be identified, isolated and characterized. In this manner, a library of genes or clones (appropriately considered as ***gene bank***) for the entire genome of a species can be constructed.

SITE-DIRECTED MUTAGENESIS AND PROTEIN ENGINEERING

Modifications in the DNA sequence of a gene are ideal to create a protein with desired properties. ***Site-directed mutagenesis is the technique for generating amino acid coding changes in the DNA (gene)***. By this approach ***specific (site-directed) change (mutagenesis)*** can be made in the base (or bases) of the gene ***to produce a desired enzyme***. The net result in site-directed mutagenesis is incorporation of a desired amino acid (of one's choice) in place of a specific amino acid in a protein or a polypeptide. By employing this technique, ***enzymes that are more efficient and more suitable*** than the naturally occurring counterparts ***can be created for industrial applications***. But it must be remembered that site-directed mutagenesis is a trial and error method that may or may not result in a better protein.

A couple of proteins developed by site-directed mutagenesis and protein engineering are given next.

Tissue plasminogen activator (tPA)

Tissue plasminogen activator is therapeutically used to lyse the blood clots that cause myocardial infarction. Due to its shorter half-life (around 5 minutes), tPA has to be repeatedly administered. By ***replacing asparagine residue*** (at position 120) ***with glutamine, the half-life of tPA can be substantially increased***. This is due to the fact that glutamine is less glycosylated than asparagine and this makes a difference in the half-life of tPA.

Hirudin

Hirudin is a protein secreted by leech salivary gland, and is a strong thrombin inhibitor (i.e., acts as an anticoagulant). By ***replacing asparagine*** (at 47 position) ***with lysine, the potency of hirudin can be increased*** several-fold.

DNA IN DISEASE DIAGNOSIS AND MEDICAL FORENSICS

DNA, being the genetic material of the living organisms, contains the information that contributes to various characteristic features of the specific organism. Thus, ***the presence of a disease-causing pathogen can be detected by identifing a gene or a set of genes of the organism***. Likewise an ***inherited genetic defect can be diagnosed by identifying the alterations in the gene***. In the modern laboratory diagnostics, DNA analysis is a very useful and a sensitive tool.

DNA FINGERPRINTING OR DNA PROFILING

DNA fingerprinting is the present day ***genetic detective*** in the practice ***of modern medical forensics***. The underlying principles of DNA fingerprinting are briefly described.

The structure of each person's genome is unique. The only exception being monozygotic identical twins (twins developed from a single fertilized ovum). The unique nature of genome structure provides a good opportunity for the specific identification of an individual.

It may be remembered here that in the traditional fingerprint technique, the individual is identified by preparing an ink impression of the skin folds at the tip of the person's finger. This is based on the fact that the nature of these skin folds is genetically determined, and thus the fingerprint is unique for an individual. In contrast, the ***DNA fingerprint is an analysis of the nitrogenous base sequence in the DNA of an individual***.

History and terminology

The original DNA fingerprinting technique was developed by Alec Jaffreys in 1985. Although the DNA fingerprinting is commonly used, a more general term ***DNA profiling*** is preferred. This is due to the fact that a wide range of tests can be carried out by DNA sequencing with improved technology.

Applications of DNA fingerprinting

The amount of DNA required for DNA fingerprint is remarkably small. The ***minute quantities of DNA*** from blood stains, body fluids, hair fiber or skin fragments ***are enough***. ***Polymerase chain reaction is used to amplify*** this DNA for use in fingerprinting. DNA profiling has wide range of applications—most of them related to medical forensics. Some important ones are listed below.

- Identification of criminals, rapists, thieves etc.
- Settlement of paternity disputes.
- Use in immigration test cases and disputes.

In general, ***the fingerprinting technique is carried out by collecting the DNA from a suspect*** (or a person in a paternity or immigration dispute) and ***matching it with that of a reference sample*** (from the victim of a crime, or a close relative in a civil case).

DNA MARKERS IN DISEASE DIAGNOSIS AND FINGERPRINTING

The ***DNA markers are highly useful for genetic mapping of genomes***. There are four types of DNA sequences which can be used as markers.

1. Restriction fragment length polymorphisms (***RFLPs***, pronounced as rif-lips).
2. Minisatellites or variable number tandem repeats (***VNTRs***, pronounced as vinters).
3. Microsatellites or simple tandem repeats (***STRs***).
4. Single nucleotide polymorphisms (***SNPs***, pronounced as snips).

The general aspects of the above DNA markers are described along with their utility in disease diagnosis and DNA fingerprinting.

RESTRICTION FRAGMENT LENGTH POLYMORPHISMS (RFLPs)

A RFLP represents a stretch of DNA that serves as a marker for mapping a specified gene. RFLPs are located randomly throughout a person's chromosomes and have no apparent function.

A DNA molecule can be cut into different fragments by a group of enzymes called restriction endonucleases (***See Table 27.1***). These fragments are called ***polymorphisms*** (literally means ***many forms***).

An outline of RFLP is depicted in ***Fig.27.10***. The DNA molecule 1 has three restriction sites (R_1, R_2, R_3), and when cleaved by restriction endonucleases forms 4 fragments. Let us now consider DNA 2 with an inherited mutation (or a genetic change) that has altered some base pairs. As a result, the site (R_2) for the recognition by restriction endonuclease is lost. This DNA molecule 2 when cut by restriction endonuclease forms only 3 fragments (instead of 4 in DNA 1).

As is evident from the above description, a stretch of DNA exists in ***fragments*** of various ***lengths*** (***polymorphisms***), derived by the action of ***restriction*** enzymes, hence the name restriction fragment length polymorphisms.

RFLPs in the diagnosis of diseases

If the RFLP lies within or even close to the locus of a gene that causes a particular disease, it is possible to ***trace the defective gene by the analysis of RFLP in DNA***. The person's cellular DNA is isolated and treated with restriction enzymes. The DNA fragments so obtained are separated by electrophoresis. The RFLP patterns of the disease suspected individuals can be compared with that of normal people (preferably with the relatives in the same family). By this approach, it is possible to determine whether the individual has the marker RFLP and the disease gene. With 95% certainity, RFLPs can ***detect single gene-based diseases***.

Methods of RFLP scoring : Two methods are in common use for the detection of RFLPs (***Fig.27.11***).

1. **Southern hybridization :** The DNA is digested with appropriate restriction enzyme, and separated by agarose gel electrophoresis. The so obtained DNA fragments are transferred to a nylon membrane. A DNA probe that spans the suspected restriction site is now added, and the hybridized bands are detected by autoradiograph. If the restriction site is absent, then only a single restriction fragment is detected. If the site is present, then two fragments are detected (***Fig.27.11A***).

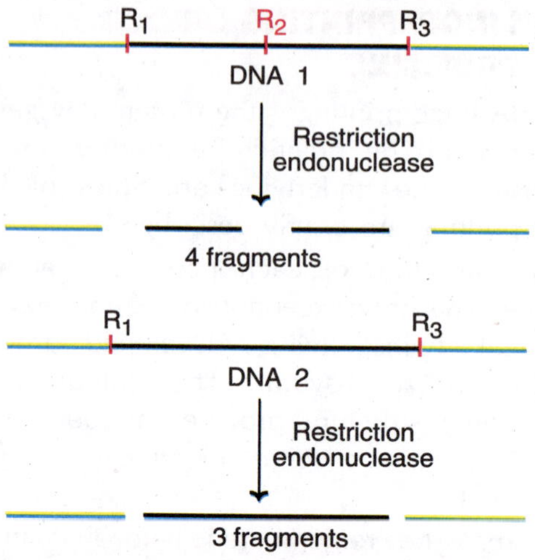

***Fig. 27.10 :** An outline of the restriction fragment length polymorphism (RFLP) (R_1, R_2, R_3 represent the sites for the action of restriction endonucleases).*

2. **Polymerase chain reaction :** RFLPs can also be scored by PCR. For this purpose, PCR primers that can anneal on either side of the suspected restriction site are used. After amplification by PCR, the DNA molecules are treated with restriction enzyme and then analysed by agarose gel electrophoresis. If the restriction site is absent only one band is seen, while two bands are found if the site is found (***Fig.27.11B***).

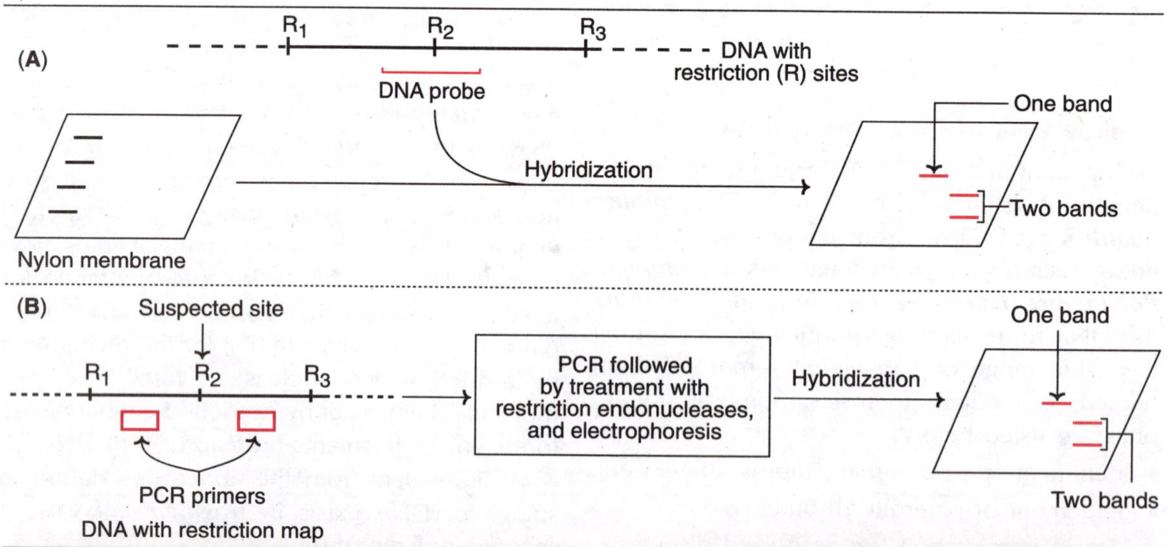

***Fig. 27.11 :** Two common methods used for scoring restriction fragment length polymorphism (RFLP) (**A**) RFLP by Southern hybridization (**B**) RFLP by polymerase chain reaction (PCR).*

Applications of RFLPs : The approach by RFLP is very powerful and has helped many genes to be mapped on the chromosomes. e.g. sickle-cell anemia (chromosome 11), cystic fibrosis (chromosome 7), Huntington's desease (chromosome 4), retinoblastoma (chromosome 13), Alzheimer's disease (chromosome 21).

VARIABLE NUMBER TANDEM REPEATS (VNTRs)

VNTRs, also known as ***minisatellites***, like RFLPs, are DNA fragments of different length. The main difference is that RFLPs develop from random mutations at the site of restriction enzyme activity while VNTRs are formed due to different number of base sequences between two points of a DNA molecule. In general, VNTRs are made up of tandem repeats of short base sequences (10–100 base pairs). The number of elements in a given region may vary, hence they are known as variable number tandem repeats.

An individual's genome has many different VNTRs and RFLPs which are unique to the individual. The ***pattern of VNTRs and RFLPs forms the basis of DNA fingerprinting*** or DNA profiling.

In the ***Fig.27.12***, two different DNA molecules with different number of copies (bands) of VNTRs are shown. When these molecules are subjected to restriction endonuclease action (at two sites R_1 and R_2), the VNTR sequences are released, and they can be detected due to variability in repeat sequence copies. These can be used in mapping of genomes, besides their utility in DNA fingerprinting.

VNTRs are useful for the detection of certain genetic diseases associated with alterations in the degree of repetition of microsatellites e.g. Huntington's chorea is a disorder which is found when the VNTRs exceed 40 repeat units.

Limitations of VNTRs : The major drawback of VNTRs is that they are not evenly distributed throughout the genome. VNTRs tend to be localized in the telomeric regions, at the ends of the chromosomes.

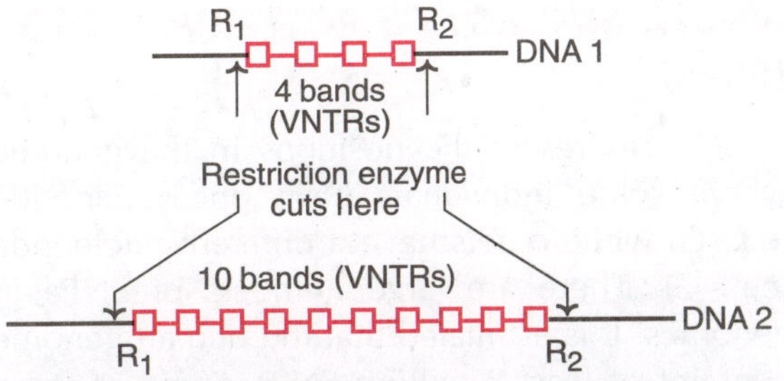

Fig. 27.12 : A diagrammatic representation of variable number tandem repeats (VNTRs). Each band (or copy) represents a repeating sequence in the DNA (e.g. 100 base pairs each). R_1 and R_2 indicate the sites cut by a restriction enzyme.

Use of RFLPs and VNTRs in genetic fingerprinting

RFLPs caused by variations in the number of VNTRs between two restriction sites can be detected (***Fig.27.13***). The DNAs from three individuals with different VNTRs are cut by the specific restriction endonuclease. The DNA fragments are separated by electrophoresis, and identified after hybridization with a probe complementary to a specific sequence on the fragments.

MICROSATELLITES (SIMPLE TANDEM REPEATS)

Microsatellites are short repeat units (10–30 copies) usually composed of dinucleotide or tetranucleotide units. These simple tandem repeats (STRs) are more popular than ***minisatellites (VNTRs)*** as DNA markers for two reasons.

1. Microsatellites are evenly ***distributed throughout the genome***.

2. PCR can be effectively and conveniently used to identify the length of polymorphism.

Two variants (alleles) of DNA molecules with 5 and 10 repeating units of a dimer nucleotides (GA) are depicted in ***Fig.27.14***.

By use of PCR, the region surrounding the microsatellites is amplified, separated by agarose gel electrophoresis and identified.

SINGLE NUCLEOTIDE POLYMORPHISMS (SNPs)

SNPs represent the positions in the genome where some individuals have one nucleotide (e.g. G) while others have a different nucleotide (e.g. C). There are large numbers of SNPs in genomes. It is estimated that the human genome contains at least 3 million SNPs. Some of these SNPs may give rise to RFLPs.

SNPs are highly ***useful as DNA markers*** since there is no need for gel electrophoresis and this saves a lot of time and labour. The detection of SNPs is based on the oligonucleotide hybridization analysis (***Fig.27.15***).

An oligonucleotide is a short single-stranded DNA molecule, synthesized in the laboratory with a length not usually exceeding 50 nucleotides. Under appropriate conditions, this nucleotide sequence will hybridize with a target DNA strand if both have completely base paired structure. Even a single mismatch in base pair will not allow the hybridization to occur.

DNA chip technology is most commonly used to screen SNPs hybridization with oligonucleotide (**See page 578**).

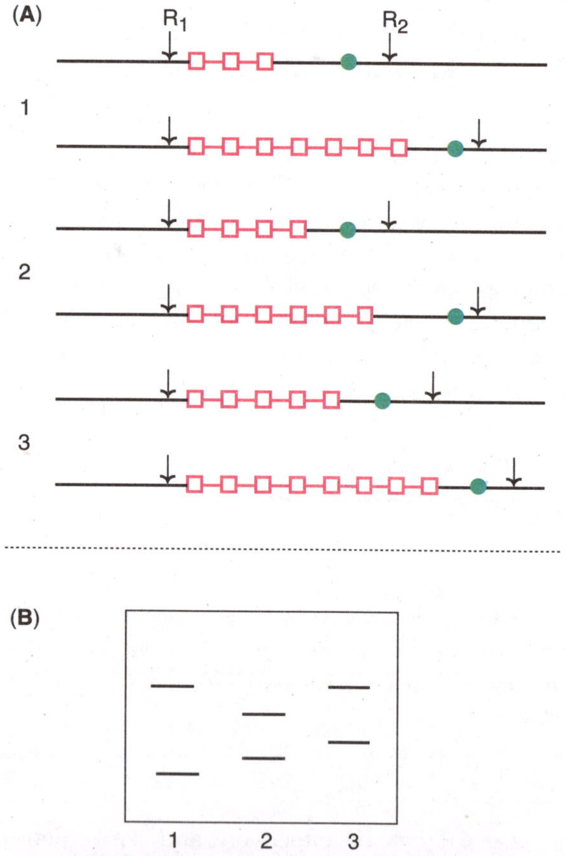

Fig. 27.13 : *Use of restriction fragment length polymorphisms (RFLPs) caused by variable number tandem repeats (VNTRs) in genetic fingerprinting (**A**) An illustration of DNA structure from three individuals (**B**) Hybridized pattern of DNA fragment with a probe complementary to the sequence shown in green circles (1, 2 and 3 represent the individuals; R_1 and R_2 indicate restriction sites; coloured squares are the number of VNTRs)*

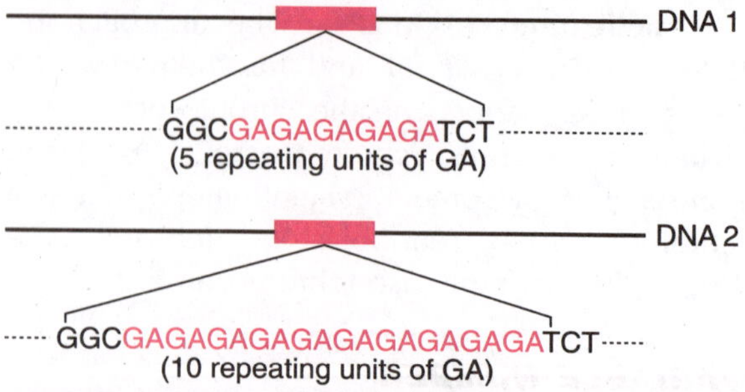

Fig. 27.14 : *Two alleles of DNA molecules representing 5 and 10 dimer repeating units.*

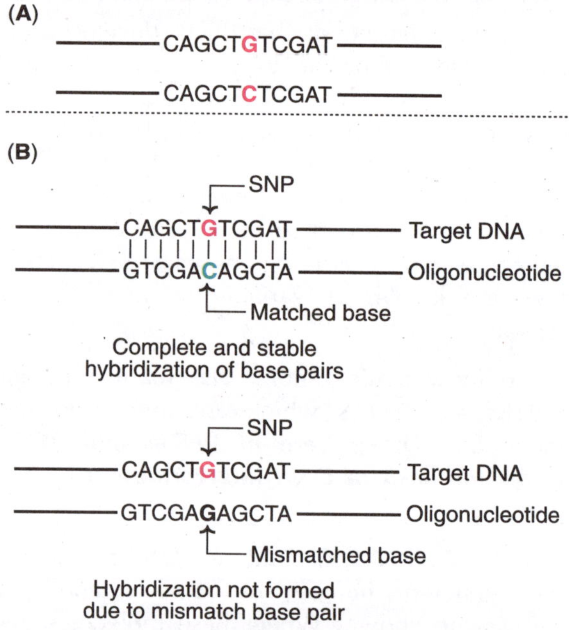

Fig. 27.15 : *(**A**) An illustration of single nucleotide polymorphism (SNP) (**B**) Oligonucleotide hybridization to detect SNP.*

PHARMACEUTICAL PRODUCTS OF DNA TECHNOLOGY

The advent of recombinant DNA technology heralded a new chapter for the production of a wide range of therapeutic agents in sufficient quantities for human use. The commercial exploitation of ***recombinant DNA (rDNA) technology*** began in late 1970s by a few biotechnological companies to produce proteins. There are at least 400 different proteins being produced (by DNA technology) which may serve as therapeutic agents for humans. A selected list of some important human proteins produced by recombinant DNA technology potential for the treatment of human disorders is given in ***Table 27.5***. As of now, only a selected few of them (around 30) have been approved for human use, and the most important among these are given in ***Table 27.6***.

Table 27.5 A selected list of human proteins produced by recombinant DNA technology for treatment of human disorders

Disorder	Recombinant protein(s)
Anemia	Hemoglobin, erythropoietin
Asthma	Interleukin-I receptor
Atherosclerosis	Platelet-derived growth factor
Delivery	Relaxin
Blood clots	Tissue plasminogen activator, urokinase
Burns	Epidermal growth factor
Cancer	Interferons, tumor necrosis factor, colony stimulating factors, interleukins, lymphotoxin, macrophage-activating factor
Diabetes	Insulin, insulin-like growth factor
Emphysema	α_1-Antitrypsin
Female infertility	Chorionic gonadotropin
Free radical damage (minimizing)	Superoxide dismutase
Growth defects	Growth hormone, growth hormone-releasing factor, somatomedin-C
Heart attacks	Prourokinase
Hemophilia A	Factor VIII
Hemophilia B	Factor IX
Hepatitis B	Hepatitis B vaccine
Hypoalbuminemia	Serum albumin
Immune disorders	Interleukins, β-cell growth factors
Kidney disorders	Erythropoietin
Lou Gehrig's disease (amytrophic lateral sclerosis)	Brain-derived neurotropic factor
Multiple sclerosis	Interferons (α, β, γ)
Nerve damage	Nerve growth factor
Osteomalacia	Calcitonin
Pain	Endorphins and enkephalins
Rheumatic disease	Adrenocorticotropic hormone
Ulcers	Urogastrone
Viral infections	Interferons (α, β, γ)

INSULIN AND DIABETES

Diabetes mellitus is characterized by increased blood glucose concentration (hyperglycemia) which occurs ***due to insufficient or inefficient insulin***. In the early years, insulin isolated and purified from the pancreases of pigs and cows was used for the treatment of severe diabetics. This often resulted in allergies. ***Recombinant DNA technology has become a boon to diabetic patients.***

Production of recombinant insulin

Attemps to produce insulin by recombinant DNA technology started in late 1970s. The ***basic technique consisted of inserting human insulin gene and the promoter gene of lac operon on to the plasmids of*** **E. coli**. By this method human insulin was produced. It was in July 1980, seventeen human volunteers were, for the first time, administered recombinant insulin for treatment of diabetes at Guy's Hospital, London. And in fact, insulin was the first ever pharmaceutical product of recombinant DNA technology administered to humans. Recombinant insulin worked well, and this gave hope to scientists that DNA technology could be successfully employed to produce substances of medical and commercial importance. An ***approval***, by the concerned authorities, for using

Table 27.6 A selected list of rDNA-derived therapeutic agents (approved by FDA) with trade names and their applications in humans

rDNA product	Trade name(s)	Applications/uses
Insulin	Humulin	Diabetes
Growth hormone	Protropin/Humatrope	Pituitary dwarfism
α-Interferon	Intron A	Hairy cell leukemia
Hepatitis B vaccine	Recombinax HB/Engerix B	Hepatitis B
Tissue plasminogen activator	Activase	Myocardial infarction
Factor VIII	Kogenate/Recombinate	Hemophilia
DNase	Pulmozyme	Cystic fibrosis
Erythropoietin	Epogen/Procrit	Severe anemia with kidney damage

recombinant insulin ***for the treatment of diabetes mellitus*** was ***given in 1982***. And in 1986, Eli Lilly company received approval ***to market hum***an ins***ulin*** under the trade name ***Humulin***.

Technique for production of recombinant insulin : The orginal technique (described briefly above) of insulin synthesis in *E. coli* has undergone several changes, for improving the yield. e.g. addition of signal peptide, synthesis of A and B chains separately etc.

The procedure employed for the synthesis of two insulin chains A and B is illustrated in ***Fig.27.16***. The genes for insulin A chain and B chain are separately inserted to the plasmids of two different *E. coli* cultures. The *lac* operon system (consisting of inducer gene, promoter gene, operator gene and structural gene Z for β-galactosidase) is used to express both the genes. The presence of lactose in the culture medium induces the synthesis of insulin A and B chains in separate cultures. The so formed insulin chains can be isolated, purified and joined together to give a full-fledged human insulin.

RECOMBINANT VACCINES

Recombinant DNA technology in recent years, has become a boon to produce ***new generation vaccines***. By this approach, some of the limitations (low yield, high cost, side effects) of traditional vaccine production could be overcome.

The list of diseases for which recombinant vaccines are developed or being developed is given in ***Table 27.7***. It may be stated here that due to very stringent regulatory requirements to use in humans, the new generation vaccines are first tried in animals, and it may ***take some more years before most of them are approved for use in humans***.

Hepatitis B vaccine — the first synthetic vaccine

In 1987, the recombinant vaccine for hepatitis B (i.e. ***HBsAg***) became the first synthetic vaccine for public use. It was marketed by trade names ***Recombivax*** and ***Engerix-B***. Hepatitis B vaccine is safe to use, very effective and produces no allergic reactions. For these reasons, this recombinant vaccine has been in use since 1987. The individuals must be administered three doses over a period of six months. Immunization against hepatitis B is strongly recommended to anyone coming in contact with blood or body secretions. All the health professionals—physicians, surgeons, medical laboratory technicians, nurses, dentists, besides police officers, firefighters etc., must get vaccinated against hepatitis B.

Hepatitis B vaccine in India

India is the ***fourth country*** (after USA, France and Belgium) in the world to develop an indigenous hepatitis B vaccine. It was launched in 1997, and is now being used.

RECOMBINANT VACCINES FOR COVID-19

A novel coronavirus (COVID-19) was identified in December 2019 in Wuhan, China.

Medical Concepts/ Clinical Correlates

- Polymerase chain reaction (PCR) is useful for the diagnosis of inherited diseases, in DNA sequencing, and in forensic medicine.
- By employing site-directed mutagenesis, it is possible to produce more efficient and more suitable enzymes for therapeutic and industrial purposes.
- The analysis of genetic material DNA (gene/genes) is employed for the diagnosis of certain diseases, and in medical forensics e.g. AIDS, sickle-cell anemia, certain cancers, DNA fingerprinting.

This virus rapidly spread throughout the world and resulted in millions of deaths. To contain COVID-19 pandemic, many pharmaceutical companies have come up with recombinant vaccines.

The most important vaccines are listed below.

- ***Covaxin*** (inactivated virus), developed by India.
- ***Covishield/Astrazenca*** (viral vector-modern chimpanzee adeno), developed by U.K, Sweden and India.
- ***Moderna*** (mRNA based) developed by US.
- ***Sputnik V*** (viral vector-modified adeno), developed by Russia.
- ***Prizer and Biotech*** (mRNA based) developed by US and Germany.

TRANSGENIC ANIMALS

With the advent of modern biotechnology, it is now possible to carry out manipulations at the genetic level to get the desired characteristics in animals. ***Transgenesis refers to the phenomenon of introduction of exogeneous DNA into the genome to create and maintain a stable heritable character***. The foreign DNA that is introduced is called ***transgene***. And the animal whose genome is altered by adding one or more transgenes is said to be transgenic. The transgenes behave like other genes present in the animals' genome, and are passed on to the offsprings. Thus, transgenic animals are ***genetically engineered*** or ***genetically modified organisms (GMOs)*** with a new heritable character.

Importance of transgenic animals—general

Transgenesis has now become a powerful tool for studying the gene expression and developmental processes in higher organisms, besides the improvement in their genetic characteristics. Transgenic animals serve as good ***models for understanding the human diseases***. Further, several proteins produced by transgenic animals are important for ***medical***

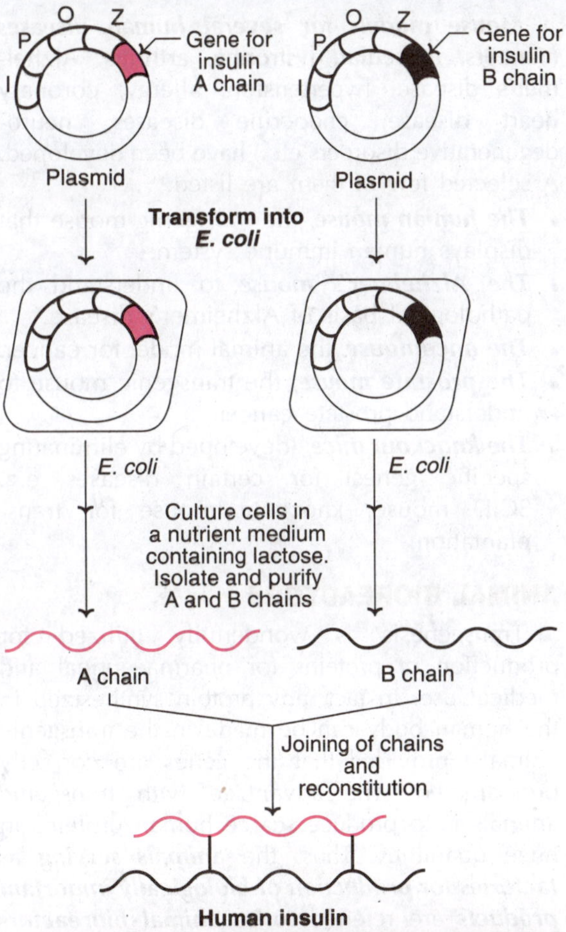

***Fig. 27.16 :** The production of recombinant insulin in* E. coli *(I-Inducer gene, P-Promoter gene, O-Operator gene, Z-β-Galactosidase gene; all these genes are of lac operon system).*

and pharmaceutical applications. Thus, the transgenic farm animals are a part of the lucrative world-wide biotechnology industry, with great benefits to mankind.

TRANSGENIC MICE AND THEIR APPLICATIONS

Mouse, although not close to humans in its biology, has been and continues to be the most exploited animal model in transgenesis experiments. The common feature between man and mouse is that both are mammals. Transgenic mice are extensively used as animal models for understanding human diseases, and for the production of therapeutic agents. Adequate care, however, must be exercised before extrapolating data of transgenic mice to humans.

Mouse models for several human diseases (cancers, muscular dystrophy, arthritis, Alzheimer's disease, hypertension, allergy, coronary heart disease, endocrine diseases, neurodegenerative disorders etc.) have been developed. A selected few of them are listed.

- ***The human mouse***, the transgenic mouse that displays human immune system.
- ***The Alzheimer's mouse*** to understand the pathological basis of Alzheimer's disease.
- ***The oncomouse***, the animal model for cancer.
- ***The prostate mouse***, the transgenic mouse to understand prostate cancer.
- ***The knockout mice***, (developed by eliminating specific genes) for certain diseases e.g. SCID mouse, knockout mouse for transplantation.

ANIMAL BIOREACTORS

Transgenesis is wonderfully utilized for production of proteins for pharmaceutical and medical use. In fact, any protein synthesized in the human body can be made in the transgenic animals, provided that the genes are correctly programmed. The advantage with transgenic animals is to produce scarce human proteins in huge quantities. Thus, the ***animals serving as factories for production of biologically important products*** are referred to as ***animal bioreactors or*** sometimes ***pharm animals***. Some transgenic animals that serve as bioreactors are listed

- ***Transgenic cow for the production of lactoferrin and interferons***.

Table 27.7 A selected list of diseases along with the pathogenic organisms for which recombinant vaccines are developed or being developed

Disease	Pathogenic organism
Viral diseases	
Acute infantile gastroenteritis	Rotavirus
Acute respiratory diseases	Influenza A and B viruses
AIDS	Human immunodeficiency virus
Chicken pox	Varicella–zoster virus
Encephalitis	Japanese encephalitis virus
Genital ulcers	Herpes simplex virus type–2
Hemorrhagic fever	Dengue virus
Liver damage	Hepatitis A virus
Liver damage	Hepatitis B virus
Upper and lower respiratory tract lesions	Yellow fever virus
Bacterial diseases	
Cholera	*Vibrio cholerae*
Diarrhea	*E. coli*
Dysentery	*Shigella* strain
Gonorrhea	*Niesseria gonorroheae*
Leprosy	*Mycobacterium leprae*
Meningitis	*Neisseria meningitidis*
Pneumonia	*Streptococcus pneumoniae*
Rheumatic fever	*Streptococcus* group A
Tetanus	*Clostridium tetani*
Tuberculosis	*Mycobacterium tuberculosis*
Typhoid	*Salmonella typhi*
Urogenital tract infection	*Streptococcus* group B
Parasitic diseases	
Filariasis	*Wuchereria bancrofti*
Malaria	*Plasmodium* sp
River blindness	*Onchocerca volvulus*
Schistosomiasis	*Schistosoma mansoni*
Sleeping sickness	*Trypanosoma* sp

Medical Concepts/ Clinical Correlates

- The pharmaceutical products of rDNA technology have revolutionized the treatment of certain diseases e.g. diabetes, asthma, atherosclerosis, heart attacks, hemophilia.
- Recombinant vaccine for hepatitis B is the first synthetic vaccine. It is effective, safe and produces no allergic reactions.
- Genetic immunization by using DNA vaccines is a novel concept. It has been shown that the immune response (humoral and cellular) of the body can be stimulated by a DNA molecule.

- Transgenic goat to synthesize tissue plasminogen activator, and antithrombin III.
- Transgenic mouse for the production of immunoglobulins, and urokinase.
- Transgenic pig to produce hemoglobin.

DOLLY – THE TRANSGENIC CLONE

Dolly, the first ever mammal clone was developed by Wilmut and Campbell in 1997. It is a ***sheep*** (female lamb) ***with a mother and no father.***

The ***technique primarily involves nuclear transfer and the phenomenon of totipotency***. The character of a cell to develop into different cells, tissues, organs, and finally an organism is referred to as ***totipotency*** or ***pluripotency***. Totipotency is the basic character of embryonic cells. As the embryo develops, the cells specialize to finally give the whole organism. As such, the cells of an adult lack totipotency. Totipotency was induced into the adult cells for developing Dolly.

The cloning of sheep for producing Dolly, illustrated in ***Fig.27.17***, is briefly described here. The mammary gland cells from a donor ewe were isolated. They were subjected to total nutrient deprivation (starvation) for five days. By this process, the mammary cells abandon their normal growth cycle, enter a dormant stage and regain totipotency character. An ovum (egg cell)

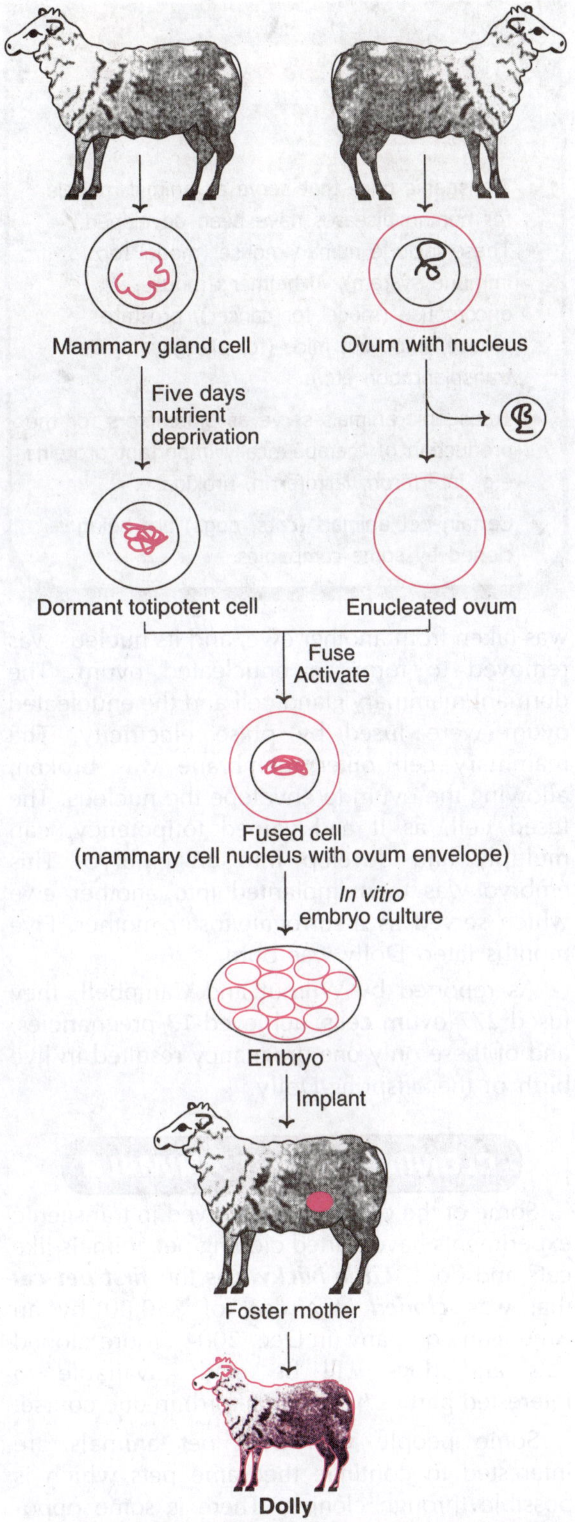

Fig. 27.17 : The cloning of sheep for developing Dolly.

Medical Concepts/ Clinical Correlates

- Transgenic mice that serve as animal models for human diseases have been developed. These include human mouse (model for immune system), Alzheimer's mouse, oncomouse (model for cancer), prostate mouse, knockout mice (for allergy, transplantation etc.).
- Transgenic animals serve as bioreactors for the production of therapeutically important proteins e.g. interferon, lactoferrin, urokinase.
- Certain pet animals (cats, dogs) are being cloned by some companies.

was taken from another ewe, and its nucleus was removed to form an enucleated ovum. The dormant mammary gland cell and the enucleated ovum were fused by pulse electricity. The mammary cell outer membrane was broken, allowing the ovum to envelope the nucleus. The fused cell, as it had gained totipotency, can multiply and develop into an embryo. This embryo was then implanted into another ewe which served as a surrogate/foster mother. Five months later, Dolly was born.

As reported by Wilmut and Campbell, they fused 277 ovum cells, achieved 13 pregnancies, and of these only one pregnancy resulted in live birth of the offspring-Dolly.

CLONING OF PET ANIMALS

Some of the companies involved in transgenic experiments have started cloning pet animals like cats and dogs. ***Little Nicky*** was the ***first pet cat*** that was ***cloned*** at a cost of $50,00 by an American company (in Dec. 2004). More cloned cats and dogs will be made available to interested parties (who can afford) in due course.

Some people who own pet animals are interested to continue the same pets which is possible through cloning. There is some opposition to this approach as the cloned animals are less healthy, and have shorter life span, besides the high cost factor.

BIOTECHNOLOGY AND SOCIETY

Advances in biotechnology, and their applications are most frequently associated with controversies. Based on their perception to biotechnology, the people may be grouped into three broad categories.

1. ***Strong opponents*** who oppose the new technology, as it will give rise to problems, issues and concerns humans have never faced before. They ***consider biotechnology as an unnatural manipulative technology***.

2. ***Strong proponents*** who consider that the biotechnology will ***provide untold benefits to society***. They argue that for centuries the society has safely used the products and processes of biotechnology.

3. A ***neutral group*** of people who have a ***balanced approach*** to biotechnology. This group believes that research on biotechnology (with regulatory systems), and extending its fruits to the society should be pursued with a cautious approach.

BENEFITS OF BIOTECHNOLOGY

The fruits of biotechnology are beneficial to the fields of ***healthcare***, ***agriculture***, ***food production***, ***manufacture of industrial enzymes*** and appropriate environmental management.

It is a fact that modern technology in various forms is woven tightly into the fabric of our lives. Our day-to-day life is inseparable from technology. Imagine life about 1-2 centuries ago where there was no electricity, no running water, sewage in the streets, unpredictable food supply and an expected life span of less than 40 years. Undoubtedly, technology has largely contributed to the present day world we live in. Many pepople consider biotechnology as a technology that will improve the quality of life in every country, besides maintaining living standards at a reasonably higher level.

ELSI OF BIOTECHNOLOGY

Why so much uproar and negativity to biotechnology? This is mainly because the major part of the modern biotechnology deals with genetic manipulations. These unnatural genetic manipulations, as many people fear, may lead to unknown consequences.

ELSI is the short form to represent the ***ethical, legal and social implications of biotechnology***. ELSI broadly covers the relationship between biotechnology and society with particular reference to ethical and legal aspects.

Risks and ethics of biotechnology

The modern biotechnology deals with genetic manipulations of viruses, bacteria, plants, animals, fish and birds. Introduction of foreign genes into various organisms raises concerns about the safety, ethics and unforeseen consequences. Some of the popular phrases used in the media while referring to experiments on recombinant DNA technology are listed.

- Manipulation of life
- Playing God
- Man-made evolution

The major apprehension of genetic engineering is that through recombinant DNA experiments, unique microorganisms or viruses (either inadvertently, or sometimes deliberately for the purpose of war) may be developed that would cause epidemics and environmental catastrophes. Due to these fears, the regulatory guidelines for research dealing with DNA manipulation were very stringent in the earlier years.

So far, risk assessment studies have failed to demonstrate any hazardous properties acquired by host cells/organisms due to transfer of DNA. Thus, ***the fears of genetic manipulations may be unfounded*** to a large extent. Consequently, there has been some relaxation in the regulatory guidelines for recombinant DNA research.

It is now widely accepted that biotechnology is certainly beneficial to humans. But it should not cause problems of safety to people and environment, and create unacceptable social, moral and ethical issues.

SUMMARY

- *Recombinant DNA (rDNA) technology is primarily concerned with the manipulation of genetic material (DNA) to achieve the desired goal in a pre-determined way.*
- *The procedure for rDNA technology involves molecular tools (enzymes e.g. restriction endonucleases), host cells (*E. coli, S. cerevisiae*), vectors (plasmids, bacteriophages), gene transfer (transformation, electroporation) and the strategies of gene cloning.*
- *Blotting techniques are employed for the identification of desired DNA (Southern blot), RNA (Northern blot), and protein (Western blot).*
- *Polymerase chain reaction is an* in vitro *technique for generating large quantities of a specified DNA i.e. cell-free amplification.*
- *Gene libraries or genomic libraries represents the collection of DNA fragments (i.e. genes) from a genome of a particular species.*
- *Site-directed mutagenesis is the technique for generating amino acid coding changes in the DNA (gene) to produce a desired protein/enzyme.*
- *Analysis of DNA (i.e. detection of gene/genes) can be used as a diagnostic system for the detection of many pathogenic and genetic diseases e.g. tuberculosis, malaria, AIDS, sickle-cell anemia, certain cancers.*
- *DNA fingerprinting or DNA profiling is the present day genetic detective in the practice of modern medical forensics. Four types of DNA markers are used in DNA fingerprinting–RFLFs, VNTRs, STRs, and SNPs.*
- *Many pharmaceutical compounds of health importance (for disease treatment) are being produced by rDNA technology e.g. insulin, growth hormone, interferons, erythropoietin, hepatitis B vaccine.*
- *Transgenic animals can be developed by introducing a foreign DNA (transgene). These animals are genetically modified or engineered with new heritable characters e.g. oncomouse, knockout mouse, prostate mouse.*

Self-Assessment Exercises

I. Essay questions

1. Describe the basic principles underlying the recombinant DNA technology.
2. Give an account of the nucleic acid blotting techniques. Add a note on their importance.
3. Describe the polymerase chain reaction along with its applications.
4. Write briefly on the utility of DNA in disease diagnosis and medical forensics.
5. Give an account of the pharmaceutical products of DNA technology.

III. Short notes

(a) Restriction endonucleases, (b) Plasmids, (c) Methods of gene transfer, (d) Purification of nucleic acids, (e) Western blotting, (f) DNA sequencing, (g) DNA chips (h) Gene libraries, (i) Restriction fragment length polymorphisms, (j) Recombinant vaccines.

III. Fill in the blanks

1. The most commonly used prokaryotic host in rDNA technology is ______________.
2. Northern blotting technique is used for the detection of ______________.
3. Name the blotting technique in which nucleic acids (DNA or RNA) are directly blotted onto the filters without electrophoresis______________.
4. The bacterial source of the enzyme *Taq* DNA polymerase, that is widely used in polymerase chain reaction ______________.
5. The collection of DNA fragments from the genome of a particular species represents ______________.
6. The technique for generating amino acid coding changes in the DNA (gene) is regarded as ______________.
7. The trade name for insulin produced by rDNA technology ______________.
8. The first synthetic vaccine developed by rDNA technology ______________.
9. The most commonly used animal model in transgenesis to represent humans ______________.
10. Name the first ever mammmal that has been cloned ______________.

IV. Multiple choice questions

11. One of the following enzyme produces single-stranded nicks in DNA
    (a) DNA ligase (b) DNA polymerase (c) DNase I (d) SI nuclease.
12. Western blotting is the technique for the identification of
    (a) DNA (b) RNA (c) Carbohydrates (d) Proteins.
13. The DNA markers used in the diagnosis of diseases and DNA fingerprinting
    (a) Restriction fragment length polymorphisms, (b) Minisatellites and microsatellites, (c) Single nucleotide polymorphisms, (d) Any one of the above.
14. The first pharmaceutical product of recombinant DNA technology approved for human use
    (a) Insulin (b) Growth hormone (c) Interferon (d) Hypatitis B vaccine.
15. Genetic immunization involves the administration of
    (a) Antigens (b) Antibodies (c) DNA (d) RNA.

CURRENT TOPICS

SECTION SIX

Human Genome Project

28

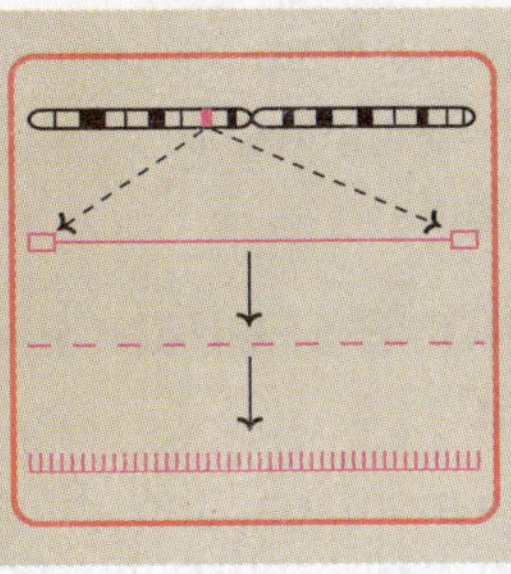

The Human Genome speaks

"I am the outcome of international collaborative research;
Composed of 3.2 billion nucleotide base pairs;
With 19,000 to 20,000 protein coding genes;
That represents only 1.1 to 1.5% of the genome"

The most important features of a DNA molecule are the nucleotide sequences, and the identification of genes and their activities. Since 1920, scientists have been working to determine the sequences of pieces of DNA.

THE BIRTH AND ACTIVITY OF HUMAN GENOME PROJECT

The human genome project (***HGP***) was ***conceived in 1984***, and officially begun in earnest in October 1990. The primary objective of HGP was to ***determine the nucleotide sequence of the entire human nuclear genome***. James Watson (who elucidated DNA structure) was the first Director of HGP.

In 1997, United States established the ***National Human Genome Research Institute*** (***NHGRI***). The HGP was an international venture involving research groups from six countries — USA, UK, France, Germany, Japan and China, and several individual laboratories and a large number of scientists and technicians from various disciplines. This collaborative venture was named as ***International Human Genome Sequencing Consortium*** (***IHGSC***) and was headed by Francis Collins. A total expenditure of $3 billion, and a time period of 10–15 years for the completion of HGP was expected. A second human genome project was set up by a private company — ***Celera Genomics***, of Maryland USA in 1998. This team was led by Craig Venter.

Announcement of the draft sequence of human genome

The date ***26th June 2000*** will be remembered as one of the most important dates in the history of science or even mankind. It was on this day, Francis Collins and Craig Venter, the leaders of the two human genome projects, in the presence of the President of U.S., jointly ***announced the working drafts of human genome sequence***. The detailed results of the teams were later published in February 2001 in scientific journals Nature (IHGSC) and Science (Celera Genomics).

The human genome project results attracted worldwide attention. This achievement was hailed with many descriptions in the media.

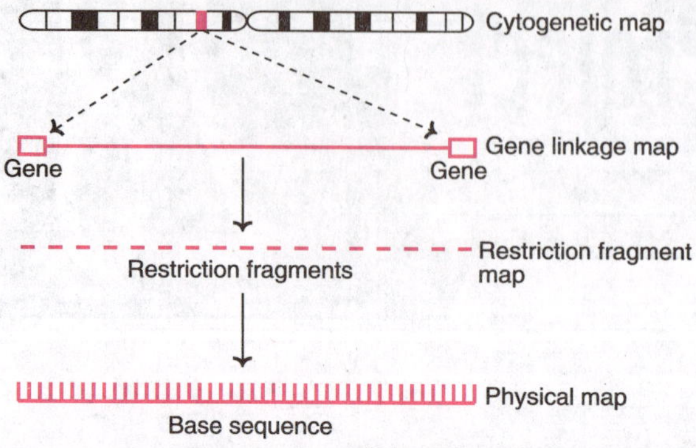

Fig. 28.1 : Different types of genome maps.

- The mystery of life unravelled.
- The library of life.
- The periodic table of life.

MAPPING OF THE HUMAN GENOME

The most important objective of human genome project was to construct a series of maps for each chromosome. In ***Fig.28.1***, an outline of the different types of maps is given.

1. **Cytogenetic map :** This is a map of the chromosome in which the active genes respond to a chemical dye and display themselves as ***bands on the chromosome***.

2. **Gene linkage map :** A chromosome map in which the ***active genes*** are identified by locating closely ***associated marker genes***. The most commonly used DNA markers are ***restriction fragment length polymorphism (RFLP), variable number tandem repeats (VNTRs) and short tandem repeats (STRs).*** VNTRs are also called as ***minisatellites*** while STRs are ***microsatellites.***

3. **Restriction fragment map :** This consists of the ***random DNA fragments*** that have been ***sequenced.***

4. **Physical map :** This is the ***ultimate map*** of the chromosome ***with highest resolution base sequence***. Physical map depicts the location of the active genes and the number of bases between the active genes.

APPROACHES FOR GENOME SEQUENCING

The principle methods used for mapping of genomes are listed below

- DNA sequencing
- Use of DNA probes
- Fluorescence *in situ* hybridization (FISH)
- Cloning in vectors
- Polymerase chain reaction
- Chromosome walking
- Chromosome jumping
- Data bases utility.

HUMAN GENOME SEQUENCE—RESULTS SUMMARISED

The information on the human genome projects is too vast, and only some highlights can be given (***Table 28.1***). Some of them are briefly described.

Most of the genome sequence is identified

About 90% of the human genome has been sequenced. It is composed of 3.2 billion base pairs (3200 Mb or 3.2 Gb). If written in the format of a telephone book, the base sequence of human genome would fill about 200 telephone books of 1000 pages each. Some other ***interesting analogs/sidelights of genome*** are given in ***Table 28.2.***

Individual differences in genomes : It has to be remembered that every individual, except identical twins, have their own versions of genome sequences. The differences between individuals are largely due to ***single nucleotide polymorphisms (SNPs).*** SNPs represent positions in the genome where some individuals have one nucleotide (i.e. an A), and others have a different nucleotide (i.e. a G). The frequency of occurrence of SNPs is estimated to be one per 1000 base pairs. About 3 million SNPs are believed to be present and at least half of them have been identified.

Table 28.1 Major highlights of human genome

• The draft represents about 90% of the entire human genome. It is believed that most of the important parts have been identified.
• The remaining 10% of the genome sequences are at the very ends of chromosomes (i.e. telomeres) and around the centromeres.
• Human genome is composed of 3200 Mb (or 3.2 Gb) i.e. 3.2 billion base pairs (3,200,000,000).
• Approximately 1.1 to 1.5% of the genome codes for proteins.
• Approximately 24% of the total genome is composed of introns that split the coding regions (exons), and appear as repeating sequences with no specific functions.
• The number of protein coding genes is in the range of 30,000–40,000. (now revised–19,000 to 20,000 genes)
• An average gene consists of 3000 bases, the sizes however vary greatly. Dystrophin gene is the largest known human gene with 2.4 million bases.
• Chromosome 1 (the largest human chromosome) contains the highest number of genes (2968), while the Y chromosome has the lowest. Chromosomes also differ in their GC content and number of transposable elements.
• Genes and DNA sequences associated with many diseases such as breast cancer, muscle diseases, deafness and blindness have been identified.
• Repeated sequences constitute about 50% of the human genome.
• A vast majority of the genome (~ 97%) has no known functions.
• Between the humans, the DNA differs only by 0.2% or one in 500 bases.
• More than 3 million single nucleotide polymorphisms (SNPs) have been identified.
• Human DNA is about 98% identical to that of chimpanzees.
• About 200 genes are close to that found in bacteria.

Organization of human genome

An outline of the organization of the human genome is given in ***Fig.28.2***. Of the 3200 Mb, only a small fraction (48 Mb) represents the actual genes, while the rest is due to gene-related sequences (introns, pseudogenes) and intergenic DNA (long interspersed nuclear elements, short interspread nuclear elements, microsatellites, DNA transposons etc.). ***Intergenic DNA*** represents the parts of the genome that lie between the genes which have no known function. This is appropriately regarded as ***junk DNA***.

Genes present in human genome

The two genome projects differ in their estimates of the total number of genes in humans. Their figures are in the range of 30,000–40,000 genes. The main reason for this variation is that it is rather difficult to specifically recognize the DNA sequences which are genes and which are not. (**Note :** As per one ***recent estimate***, there are ***19,00–20,000 protein–coding genes*** in humans)

Before the results of the HGP were announced, the best guess of human genes was in the range of 80,000–100,000. This estimate was based on the fact that the number of proteins in human cells is 80,000–100,000, and thus so

Table 28.2 Some interesting analogs/ sidelights about human genome

• The base sequence in human genome would fill about 200 telephone books of 1000 pages each.
• If the genome is recited at the rate of one base per second for 24 hours a day, it would take a century to recite the book of life.
• If a typist types at the rate of 60 words per minute (i.e. 360 letters) for 8 hours a day, he/she would take around 50 years to type human genome.
• If the DNA sequence is typed in lines 10 cm containing 60 nucleotide bases and printed, the human genome sequence (from a single cell) would stretch a distance of 5000 km.
• If the DNA in the entire human body is put end to end, it would reach to the sun and back over 600 times (**Note :** The human body contains 100 trillion cells; the length of DNA in a cell is 6 feet; the distance between the sun and earth is 93 million miles).
• The total expenditure for human genome project was $3 billion. The magnitude of this huge amount has to be appreciated. If one starts counting at a non–stop rate of a dollar per second, it would take about 90 years to complete.

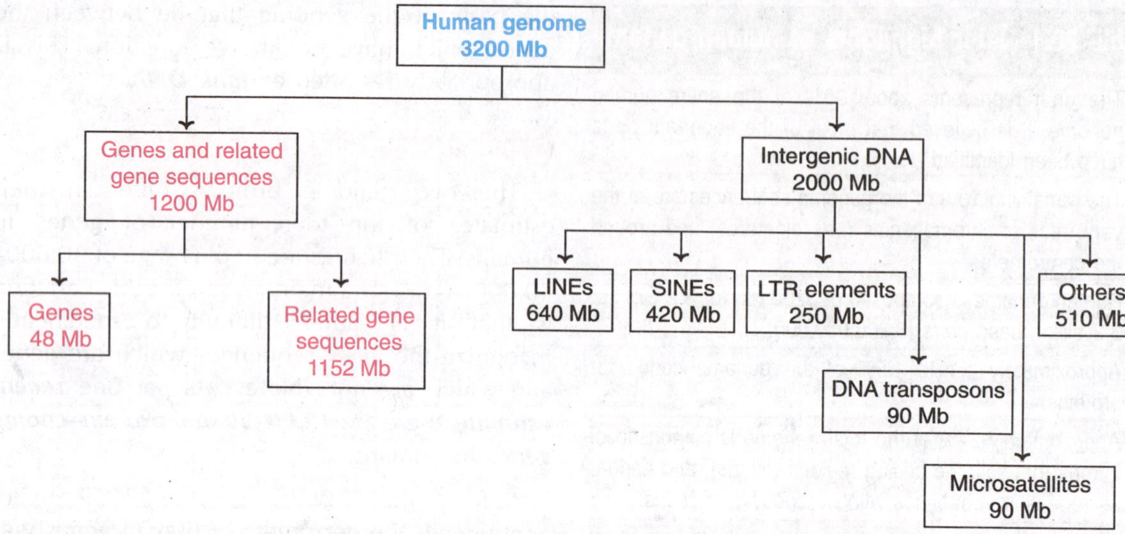

***Fig. 28.2 :** An overview of the organization of human genome (LINEs-Long interspersed nuclear elements; SINEs–Short interspersed nuclear elements; LTR-Long terminal repeats).*

many genes expected. The fact that the number of genes is much lower than the proteins suggests that the ***RNA editing*** (RNA processing) is widespread, so that a single mRNA may code for more than one protein.

A diagrammatic representation of a typical structure of an average human gene is given in ***Fig.28.3***. It has exons and introns.

A broad categorization of human gene catalog in the form of a pie chart is depicted in ***Fig.28.4***. About 17.5% of the genes participate in the general biochemical functions of the cells, 23% in the maintenance of genome, 21% in signal transduction while the remaining 38% are involved in the production of structural proteins, transport proteins, immunoglobins etc.

Human genes encoding proteins

It is now clear that only 1.1-1.5% of the human genome codes for proteins. Thus, this figure 1.1-1.5% represents exons of genome.

As already described, a huge portion of the genome is composed of introns, and intergenic sequences (***junk DNA***).

The major categories of the proteins encoded by human genes are listed in ***Table 28.3***. The functions of at least 40% of these proteins are not known.

BENEFITS/APPLICATIONS OF HUMAN GENOME SEQUENCING

It is expected that the sequencing of human genome, and the genomes of other organisms

Table 28.3 Different categories of proteins encoded by human genes (*based on the Human Genome Project report, 2001*)

Category of proteins	Percentage	Actual number of genes
Unknown functions	41.0%	12,809
Nucleic acid enzymes	7.5%	2,308
Transcription factors	6.0%	1,850
Receptors	5.0%	1,543
Hydrolases	4.0%	1,227
Regulatory proteins (G-proteins, cell cycle regulators etc.)	3.2%	988
Protooncogenes	2.9%	902
Structural proteins of cytoskeleton	2.8%	876
Kinases	2.8%	868

*(**Note :** This table is based on the rough draft of human genome reported by Celera Genomics. The percentages are derived from a total of 26,383 genes)*

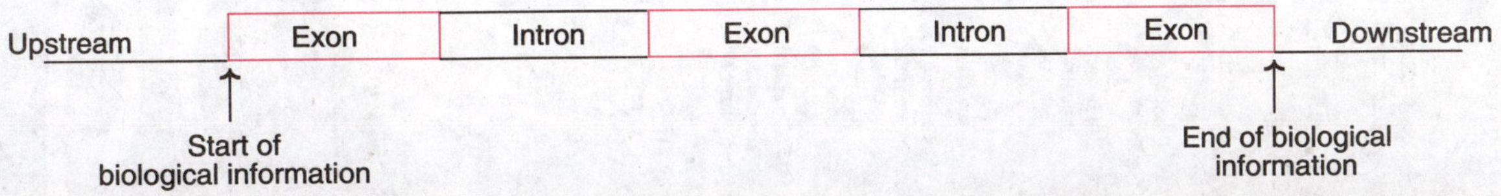

Fig. 28.3 : A diagrammatic representation of a typical structure of an average human gene.

will dramatically change our understanding and perceptions of biology and medicine. Some of the benefits of human genome project are given.

- ***Identification of*** human genes and their functions.
- ***Understanding*** of polygenic ***disorders*** e.g. ***cancer***, hypertension, ***diabetes***.
- Improvements in ***gene therapy***

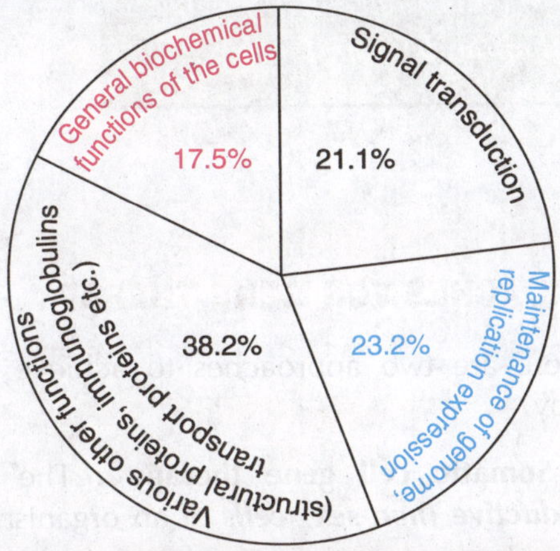

Fig. 28.4 : A pie chart showing a broad categorization of the human gene catalog (About 13000 genes whose functions are not known are not included).

- Improved diagnosis of diseases
- Development of pharmacogenomics
- Genetic basis of psychiatric disorders
- Understanding of complex social trait
- Improved knowledge on mutations
- Better understanding of developmental biology
- Comparative genomics
- Development of biotechnology

ETHICS AND HUMAN GENOME

The research on human genomes will make very sensitive data available that will affect the personal and private lives of individuals. For instance, once it is known that a person carries genes for an incurable disease, what would be the strategy of an insurance company? How will the society treat him/her? There is a possibility that ***individuals with substandard genome sequences may be discriminated.*** Human genome results may also promote racial discrimination categorizing the people with ***good*** and ***bad genome sequences.*** Considering the gravity of ethics related to a human genome, about 3% of the HGP budget was earmarked for ethical research.

SUMMARY

- *Human Genome Project is an international venture involving several laboratories, and a large number of scientists and technicians from various disciplines.*
- *About 90% of the human genome has been sequenced. It is composed of 3.2 billion base pairs.*
- *The total number of genes in the humans is believed to be in the range of 19,000–20,000.*
- *About 1.1–1.5% of the human genome codes for proteins while the remaining portion is regarded as junk DNA (composed of introns and intergenic sequences).*
- *Human genome sequencing has wide range of applications–better understanding of genetic diseases, improvements in gene therapy, development of pharmacogenomics, and advancement of biotechnology.*

Gene Therapy

29

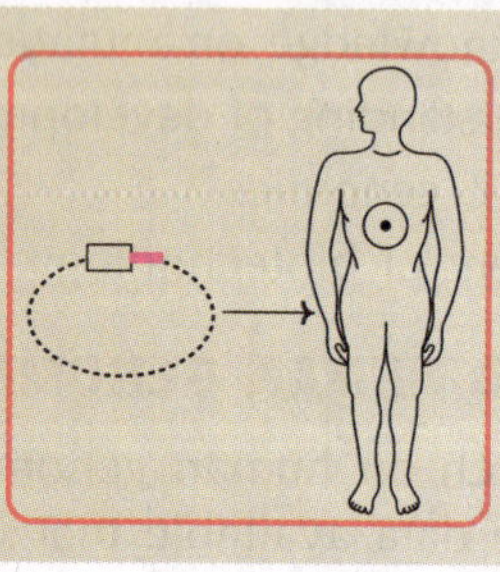

The gene therapy speaks

"I represent the insertion of genes into cells;
The preferred being somatic cells to treat diseases;
Although unsuccessful and unable to satisfy now;
I am highly optimistic about my future!"

Advances in biochemistry and molecular biology have helped to understand the genetic basis of inherited diseases. It was a dream of the researchers to replace the defective genes with good ones, and cure the genetic disorders.

Gene therapy is the process of inserting genes into cells to treat diseases. The newly introduced genes will encode proteins and correct the deficiencies that occur in genetic diseases. Thus, gene therapy primarily involves ***genetic manipulations*** in animals or humans ***to correct a disease***, and keep the organism in good health. The initial experiments on gene therapy are carried out in animals, and then in humans. Obviously, the goal of the researchers is to benefit the mankind and improve their health.

An overview of gene therapy strategies is depicted in ***Fig.29.1***. In ***gene augmentation therapy***, a DNA is inserted into the genome to replace the missing gene product. In case of ***gene inhibition therapy***, the antisense gene inhibits the expression of the dominant gene.

APPROACHES FOR GENE THERAPY

There are two approaches to achieve gene therapy.

1. **Somatic cell gene therapy** : The ***non-reproductive (non-sex) cells*** of an organism are

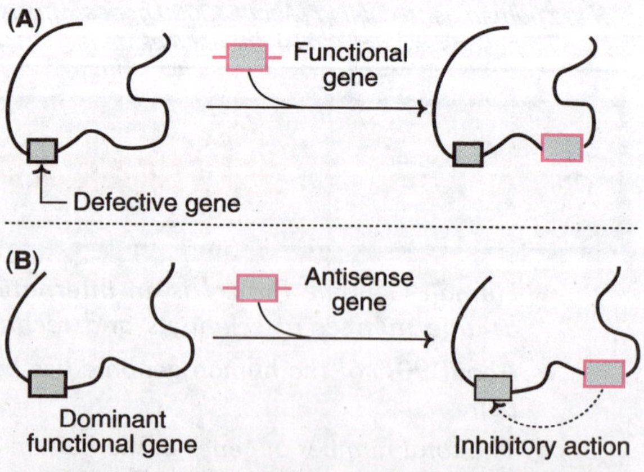

Fig. 29.1 : Overview of two major gene therapy strategies (A) Gene augmentation therapy (B) Gene inhibition therapy.

referred to as somatic cells. These are the cells of an organism other than sperm or egg cells, e.g., bone marrow cells, blood cells, skin cells, intestinal cells. At present, all the research on gene therapy is directed to correct the genetic defects in somatic cells. In essence, somatic cell gene therapy involves the ***insertion of a fully*** functional and ***expressible gene into a target somatic cell to correct a genetic disease permanently***.

2. **Germ cell gene therapy :** The ***reproductive (sex) cells*** of an organism constitute germ cell line. Gene therapy involving the introduction of DNA into germ cells is passed on to the successive generations. For safety, ethical and technical reasons, germ cell gene therapy is ***not being attempted at present***.

The genetic alterations in somatic cells are not carried to the next generations. Therefore, ***somatic cell gene therapy is preferred*** and extensively studied with an ultimate objective of correcting human diseases.

A large number of genetic disorders and other diseases are currently at various stages of gene therapy trials. A selected list of some important ones is given in ***Table 29.1***.

There are two types of gene therapies.

I. ***Ex vivo* gene therapy :** This involves the ***transfer of genes in cultured cells*** (e.g., bone marrow cells) which are then reintroduced into the patient.

II. ***In vivo* gene therapy :** The ***direct delivery of genes into the cells*** of a particular tissue is referred to as *in vivo* gene therapy.

Table 29.1 Human gene therapy trials

Disease	Gene therapy
Severe combined immunodeficiency (SCID)	Adenosine deaminase (ADA).
Cystic fibrosis	Cystic fibrosis transmembrane regulator (CFTR).
Familial hypercholesterolemia	Low density lipoprotein (LDL) receptor.
Emphysema	α_1-Antitrypsin
Hemophilia B	Factor IX
Thalassemia	α- or β-Globin
Sickle-cell anemia	β-Globin
Lesch-Nyhan syndrome	Hypoxanthine-guanine phosphoribosyltransferase (HGPRT).
Gaucher's disease	Glucocerebrosidase
Melanoma, renal cancer	Interleukin-2 (IL-2)
Glioblastoma (brain tumor), AIDS, ovarian cancer	Thymidine kinase (herpes simplex virus)
Head and neck cancer	p^{53}
Breast cancer	Multidrug resistance I
AIDS	*rev* and *env*
Colorectal cancer, melanoma, renal cancer	Histocompatability locus antigen-B_7 (HLA-B_7)
Duchenne muscular dystrophy	Dystrophin
Short stature*	Growth hormone
Diabetes*	Glucose transporter-2, (GLUT-2), glucokinase
Phenylketonuria*	Phenylalanine hydroxylase
Citrullinemia*	Arginosuccinate synthetase

* *Mostly confined to animal experiments*

EX VIVO GENE THERAPY

The *ex vivo* gene therapy can be applied to only ***selected tissues*** (e.g., bone marrow) ***whose cells can be cultured in the laboratory***.

The technique of *ex vivo* gene therapy involves the following steps (***Fig.29.2***).

1. Isolate cells with genetic defect from a patient.
2. Grow the cells in culture.
3. Introduce the therapeutic gene to correct gene defect.
4. Select the genetically corrected cells (stable transformants) and grow.
5. Transplant the modified cells to the patient.

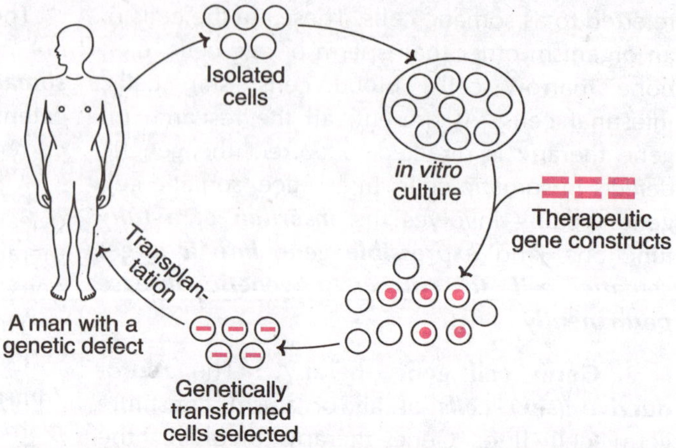

Fig. 29.2 : *The procedure for* ex vivo *gene therapy.*

The procedure basically involves the ***use of the patient's own cells for culture and genetic correction, and then their return back to the patient***. This technique is therefore, not associated with adverse immunological responses after transplanting the cells. *Ex vivo* gene therapy is efficient only if the therapeutic gene (remedial gene) is stably incorporated and continuously expressed. This can be achieved by use of vectors.

VECTORS IN GENE THERAPY

The ***carrier particles or molecules used to deliver genes to somatic cells*** are referred to as vectors. The important vectors employed in *ex vivo* gene therapy are listed below

- ***Viruses***
- Human artificial chromosome
- Bone marrow cells.

SELECTED EXAMPLES OF EX VIVO GENE THERAPY

THERAPY FOR ADENOSINE DEAMINASE DEFICIENCY

The ***first*** and the most publicised human ***gene therapy*** was carried out to correct the deficiency of the enzyme adenosine deaminase (ADA). This was done on September 14, 1990 by a team of workers led by Blaese and Anderson at the National Institute of Health, USA (The girl's name is Ashanti, 4 years old then).

Severe combined immunodeficiency (SCID)

SCID is rare inherited immune disorder associated with T-lymphocytes, and (to a lesser extent) B-lymphocytes dysfunction. About 50% of SCID patients have a defect in the gene (located on chromosome 20, and has 32,000 base pairs and 12 exons) that encodes for adenosine deaminase. In the deficiency of ADA, deoxyadenosine and its metabolites (primarily deoxyadenosine 5′-triphosphate) accumulate and destroy T-lymphocytes. T-Lymphocytes are essential for body's immunity. Besides participating directly in body's defense, they promote the function of B-lymphocytes to produce antibodies. Thus, the ***patients of SCID*** (lacking ADA) ***suffer from infectious diseases and die at an young age***. Previously, the children suffering from SCID were treated with conjugated bovine ADA, or by bone marrow transplantation.

Technique of therapy for ADA deficiency

The general scheme of gene therapy adopted for introducing a defective gene in the patient has been depicted in ***Fig.29.2***. The same procedure with suitable modifications can also be applied for other gene therapies.

A plasmid vector bearing a proviral DNA is selected. A part of the proviral DNA is replaced by the ADA gene and a gene (G 418) coding for antibiotic resistance, and then cloned. The antibiotic resistance gene will help to select the desired clones with ADA gene.

A diagrammatic representation of the treatment of ADA deficient patient is depicted in ***Fig.29.3***.

Circulating lymphocytes are removed from a patient suffering from ADA deficiency. These cells are transfected with ADA gene by exposing to billions of retroviruses carrying the said gene. The genetically-modified lymphocytes are grown in cultures to confirm the expression of ADA gene and returned to the patient. These lymphocytes persist in the circulation and synthesize ADA. Consequently, the ability of the patient to produce antibodies is increased. However, there is a limitation. The lymphocytes

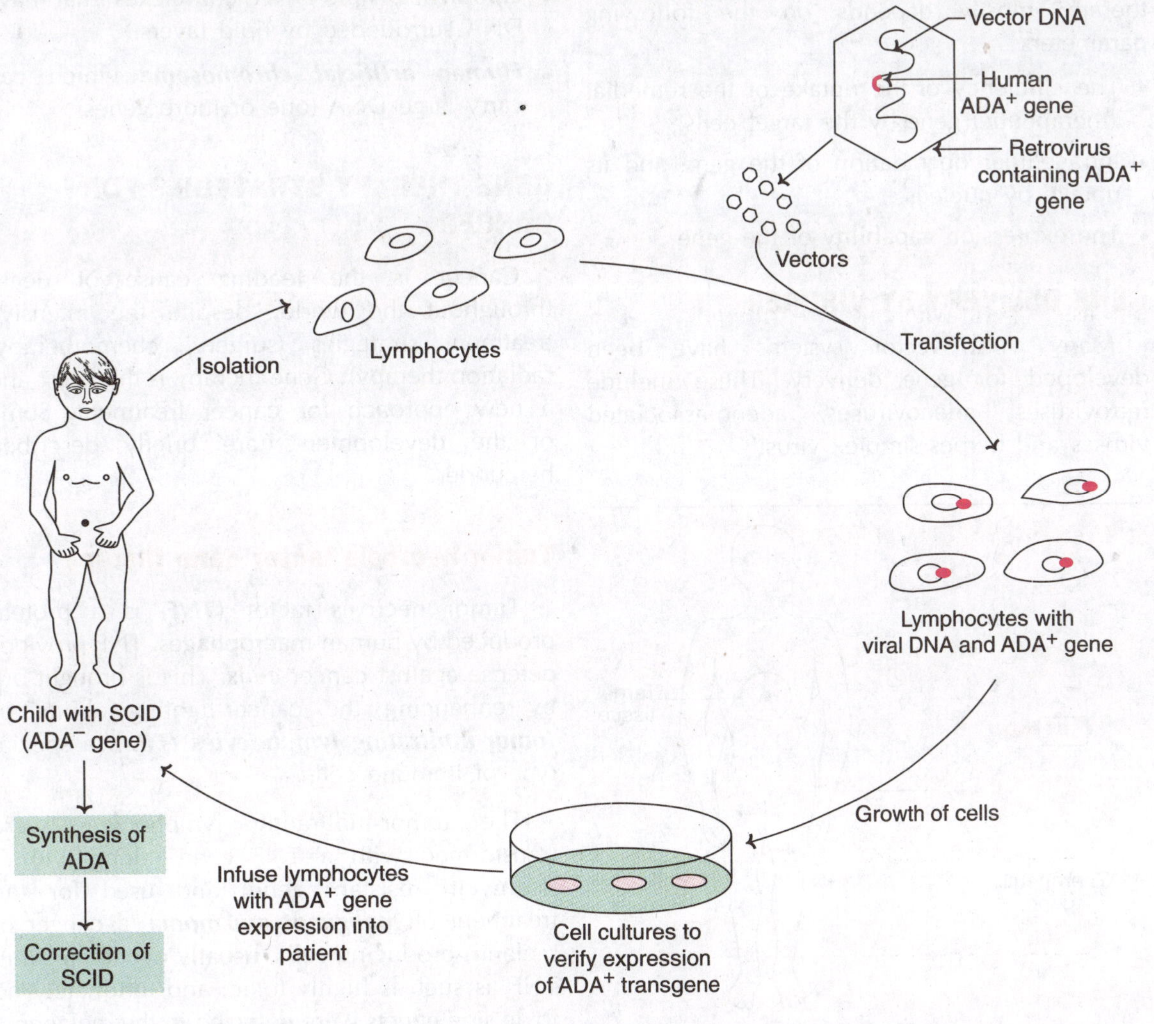

Fig. 29.3 : *Treatment of adenosine deaminase (ADA) deficient patient by somatic* ex vivo *gene therapy (SCID-Severe combined immunodeficiency).*

have a short life span (just live for a few months), hence the transfusions have to be carried out frequently.

IN VIVO GENE THERAPY

The ***direct delivery of the therapeutic gene*** (DNA) ***into the target cells*** of a particular tissue of a patient constitutes *in vivo* gene therapy (***Fig.29.4***). Many tissues are the potential candidates for this approach. These include liver, muscle, skin, spleen, lung, brain and blood cells. Gene delivery can be carried out by ***viral*** or ***non-viral vector systems***. The success of ***in vivo*** gene therapy mostly depends on the following parameters

- The efficiency of the uptake of the remedial (therapeutic) gene by the target cells.
- Intracellular degradation of the gene and its uptake by nucleus.
- The expression capability of the gene.

GENE DELIVERY BY VIRUSES

Many viral vector systems have been developed for gene delivery. These include retroviruses, adenoviruses, adeno-associated viruses and herpes simplex virus.

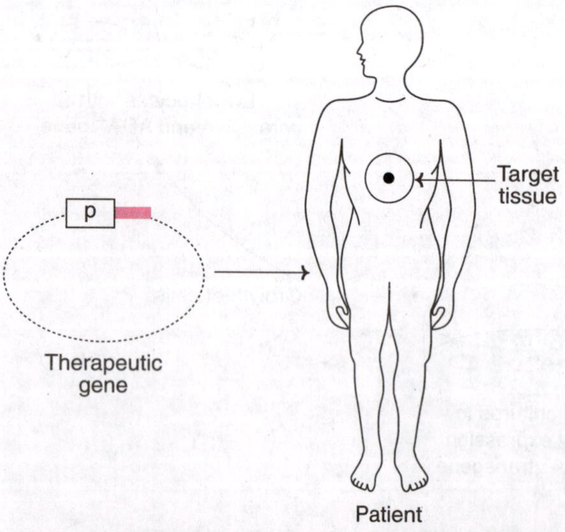

Fig. 29.4 : *Diagrammatic representation of* in vivo *gene therapy. (p-Promoter gene specific for therapeutic gene)*

GENE DELIVERY BY NON-VIRAL SYSTEMS

There are certain limitations in using viral vectors in gene therapy. In addition to the prohibitive cost of maintaining the viruses, the viral proteins often induce inflammatory responses in the host. Therefore, there is a continuous search by researchers to find alternatives to viral vector systems.

The non-viral gene delivery systems are listed

- ***Pure DNA constructs*** that can be directly introduced into target tissues.
- ***Lipoplexes***, lipid-DNA complexes that have DNA surrounded by lipid layers.
- ***Human artificial chromosome*** which can carry large DNA (one or more genes).

GENE THERAPY STRATEGIES FOR CANCER

Cancer is the leading cause of death throughout the world, despite the intensive treatment strategies (surgery, chemotherapy, radiation therapy). Gene therapy is the latest and a new approach for cancer treatment. Some of the developments are briefly described hereunder.

Tumor necrosis factor gene therapy

Tumor necrosis factor (***TNF***) is a protein produced by human macrophages. TNF provides defense against cancer cells. This is brought out by enhancing the cancer-fighting ability of ***tumor-infiltrating lymphocytes (TILs)***, a special type of immune cells.

The tumor-infiltrating lymphocytes were transformed with a TNF gene (along with a neomycin resistant gene) and used for the treatment of ***malignant melanoma*** (a cancer of melanin producing cells, usually occurs in skin). TNF as such is highly toxic, and fortunately no toxic side effects were detected in the melanoma patients injected with genetically altered TILs with TNF gene. Some improvement in the cancer patients was observed.

Medical Concepts/ Clinical Correlates

- Theoretically, gene therapy is the permanent solution for genetic diseases.
- A large number of genetic disorders and other diseases are at various stages of gene therapy trials e.g. sickle-cell anemia, cystic fibrosis, AIDS, cancer.
- Despite extensive research and trials, as of now, no disease has been permanently cured by gene therapy. However, a breakthrough may come at anytime.

Suicide gene therapy

The gene encoding the enzyme ***thymidine kinase*** is often referred to as suicide gene, and is used for the treatment of certain cancers.

Gene replacement therapy

A gene named $\boldsymbol{p^{53}}$ codes for a **p**rotein with a molecular weight of ***53*** kilodaltons (hence p^{53}). p^{53} is considered to be a ***tumor-suppressor gene***, since the protein it encodes binds with DNA and inhibits replication. The tumor cells of several tissues (breast, brain, lung, skin, bladder, colon, bone) were found to have altered genes of p^{53} (mutated p^{53}), synthesizing different proteins from the original. These altered proteins cannot inhibit DNA replication. It is believed that the damaged p^{53} gene may be a causative factor in tumor development.

Some workers have tried to replace the damaged p^{53} gene by a normal gene by employing adenovirus vector systems. There are some encouraging results in the patients with liver cancer.

ANTIGENE AND ANTISENSE THERAPY

In general, gene therapy is carried out by introducing a therapeutic gene to produce the defective or the lacking protein. But there are certain disorders (cancer, viral and parasitic infections, inflammatory diseases) which result in an overproduction of certain normal proteins. It is possible to treat these diseases by ***blocking transcription*** using a single-stranded nucleotide sequence (antigene oligonucleotide) that hybridizes with the specific gene, and this is called ***antigene therapy***. ***Antisense therapy*** refers to the ***inhibition of translation*** by using a single-stranded nucleotide (antisense oligonucleotide). Further, it is also possible to inhibit both transcription and translation by blocking (with oligonucleotides) the transcription factor responsbile for the specific gene expression.

Nucleic acid therapy refers to the ***use of DNA or RNA molecules for therapeutic purposes***, as stated above. The naturally occurring sequences of DNA and RNA (with suitable modifications) or the synthetic ones can be employed in nucleic acid therapy. Theoretically, there is a vast potential for use of nucleic acids as therapeutic agents. But most of the work that is being carried out relates to the use of RNA in antisense therapy. Some of these are described below.

ANTISENSE THERAPY FOR CANCER

Oncogenes are the genes responsible for the causation of cancer. The dominantly acting oncogenes can be targeted in antisense technology by using antisense transgenes or oligonucleotides. Antisense oligonucleotides are used for the treatment of myeloid leukemia in as early as 1991.

Antisense RNA molecules are more frequently used in cancer therapy. This approach is effective only if the antisense oligonucleotide (antisense mRNA) specifically binds to the target mRNA, and blocks protein biosynthesis (translation).

Peptide nucleic acid (PNA) therapy : *PNAs* are ***artificial analogs of nucleic acid with a polypeptide backbone***. They possess standard nucleic acid bases attached to a polypeptide in place of sugar-phosphate backbone. Antisense PNAs have been developed to inhibit translation of HIV viral transcript, ***gag-pol***, and to block translation of cancer genes e.g. ***Has-ras*** and ***bcl-2***.

THE FUTURE OF GENE THERAPY

Theoretically, gene therapy is ***the permanent solution for genetic diseases***. But it is not as simple as it appears since gene therapy has several inbuilt complexicities. Gene therapy broadly involves isolation of a specific gene, making its copies, inserting them into target tissue cells to make the desired protein. The story does not end here. It is absolutely essential to ensure that the gene is harmless to the patient and it is appropriately expressed (too much or too little will be no good). Another concern in gene therapy is the body's immune system which reacts to the foreign proteins produced by the new genes.

The public, in general, have ***exaggerated expectations*** on gene therapy. The researchers, at least for the present, are ***unable to satisfy them***. This is mainly because the gene therapists are unable to categorically claim that gene therapy has permanently cured any one patient! Some people in the media (leading news papers and magazines) have openly questioned whether it is worth to continue research on gene therapy!!

It may be true that as of now, gene therapy due to several limitations, has not progressed the way it should, despite intensive research. But ***a breakthrough may come anytime***, and of course, this is only possible with persistent research. And a day may come (it might take some years) when almost every disease will have a gene therapy, as one of the treatment modalities. And gene therapy will revolutionize the practice of medicine!

Good News!

It was recently reported (The New England Journal of Medicine, March 2017) that a 13 year old boy in France was ***cured of Sickle-cell disease by gene therapy***. The technique involved the remov of patient's bone marrow, altering the DNA, and inserting it back to his body to make normal erythrocytes.

SUMMARY

- *Gene therapy is the process of inserting genes into cells to treat diseases. Somatic cell gene therapy, involving the insertion of an expressible gene into somatic cells, is the preferred approach.*
- Ex vivo *gene therapy involves the transfer of genes in cultured cells which are then reintroduced into the patient. The direct delivery of genes into the cells of a particular tissue is regarded as* in vivo *gene therapy.*
- *Gene therapy was successfully carried out in a patient of severe combined immunodeficiency (caused by the deficiency of the enzyme adenosine deaminase).*
- *Antigene therapy involves blocking of transcription (by antigene oligonucleotide) while in antisense therapy, translation is inhibited (by antisense oligonucleotide). These approaches are in the experimental stages for the therapy of cancer and AIDS.*
- *Although as of now, gene therapy has not offered any permanent cure to any human patients, a breakthrough may come anytime. And gene therapy may revolutionize the practice of medicine.*

Bioinformatics

30

The bioinformatics speaks

"I am the product of biology and informational technology;
International computer network (internet) is my brain;
Biological databases represent my body;
I have revolutionized the advances in biology."

Bioinformatics is ***the combination*** (***or marriage!***) ***of biology and information technology***. Basically, bioinformatics is a recently developed science using information to understand biological phenomenon. It broadly involves the computational tools and methods used to manage, analyse and manipulate volumes and volumes of biological data.

Bioinformatics may also be regarded as a part of the ***computational biology***. The latter is concerned with the application of quantitative analytical techniques in modeling and solving problems in the biological systems. Bioinformatics is an interdisciplinary approach requiring advanced knowledge of computer science, mathematics and statistical methods for the understanding of biological phenomena at the molecular level.

History and relevance of bioinformatics

The term bioinformatics was first ***introduced in 1990s***. Originally, it dealt with the management and analysis of the data pertaining to ***DNA, RNA and protein sequences***. As the biological data is being produced at an unprecedented rate, their management and interpretation invariably requires bioinformatics. Thus, bioinformatics now includes many other types of biological data. Some of the most important ones are listed below

- Gene expression profiles
- Protein structure
- Protein interactions
- Microarrays (DNA chips)
- Functional analysis of biomolecules
- Drug designing.

Bioinformatics is largely (not exclusively) a computer-based discipline. Computers are in fact very essential to handle large volumes of biological data, their storage and retrieval.

We have to accept the fact that there is no computer on earth (however advanced) which

Medical Concepts/ Clinical Correlates

- Bioinformatics has largely benefited biological and medical sciences, particularly related to molecular biology and biotechnology. Some applications are listed :
  - Sequencing of macromolecules (proteins, DNA, RNA)
  - Human genome sequencing
  - Molecular modelling of biomolecules
  - Handling of vast biological data
  - Designing of drugs for the treatment of diseases
  - Development of models for the functioning of cells, tissues and organs
- As such, there is no field of biological science that is not benefited by bioinformatics.

can store information, and perform the functions like a living cell. Thus a highly complex information technology lies right within the cells of an organism. This primarily includes the organism's genes and their dictates for the organisms biological processes and behaviour.

BROAD COVERAGE OF BIOINFORMATICS

Bioinformatics covers many specialized and advanced areas of biology.

Functional genomics : Identification of genes and their respective functions.

Structural genomics : Predictions related to functions of proteins.

Comparative genomics : For understanding the genomes of different species of organisms.

DNA microarrays : These are designed to measure the levels of gene expression in different tissues, various stages of development and in different diseases.

Medical informatics : This involves the management of biomedical data with special reference to biomolecules, *in vitro* assays and clinical trials.

COMPONENTS OF BIOINFORMATICS

Bioinformatics comprises three components

1. **Creation of databases :** This involves the organizing, storage and management of the biological data sets. The databases are accessible to researchers to know the existing information and submit new entries. e.g. protein sequence data bank for molecular structure. Databases will be of no use until analysed.

2. **Development of algorithms and statistics :** This involves the development of tools and resources to determine the relationship among the members of large data sets e.g. comparison of protein sequence data with the already existing protein sequences.

3. **Analysis of data and interpretation :** The appropriate use of components 1 and 2 (given above) to analyse the data and interpret the results in a biologically meaningful manner. This includes DNA, RNA and protein sequences, protein structure, gene expression profiles, and biochemical pathways.

BIOINFORMATICS AND THE INTERNET

The internet is an ***inter**national* computer ***net**work*. A computer network involves a group of computers that can communicate (usually over a telephone system) and exchange data between users.

It is the ***internet protocol*** (IP) that determines how the packets of information are addressed and routed over the network. To access the internet, a computer must have the correct hardware (modem/network card), appropriate software and permission for access to network. For this purpose, one has to subscribe to an ***internet service provider*** (ISP).

World wide web (**www**) : www involves the exchange of information over the internet using a programme called ***browser***. The most widely used browsers are Internet explorer and Netscape navigator.

www works on the basis of ***Uniform resource locator*** (URL) which is a document with a unique address. URLs takes the format ***http.//*** (***h****yper****t****ext* ***t****ransfer* ***p****rotocol*) that can identify the protocol for communication over www.

BIOLOGICAL DATABASES

The collection of the ***biological data*** on a computer which can be manipulated to appear ***in varying arrangements and subsets*** is regarded as a database. The biological information can be stored in different databases. Each database has its own website with unique navigation tools.

The biological databases are, in general, publicly accessible. Selected examples of biological databases are briefly described (***Table 30.1***).

Nucleotide sequence databases

The nucleotide sequence data submitted by the scientists and genome sequencing groups is at the databases namely GenBank, EMBL (European Molecular Biology Laboratory) and DDBJ (DNA Data Bank of Japan). There is a good coordination between these three databases as they are synchronized on daily basis.

Besides the primary nucleotide databases (referred above), there are some other databases also to provide information on genes, genomes and ongoing research projects.

Table 30.1 Selected examples of biological databases in bioinformatics

Database(s)	Salient features
Primary nucleotide sequence databases	
GenBank (www.ncbi.nih.gov/GeneBank/)	Provides nucleotide sequence databases maintained by the National Center for Biotechnology Information (NCBI), USA.
Other nucleotide sequence databases	
UniGene (www.ncbi.nih.gov/UniGene/)	The nucleotide sequences of GenBank in the form of clusters, representing genes are available.
Genome Biology (www.ncbi.nlm.nih.gov/Genomes/)	The information about the completed genomes is available.
Protein sequence database	
SWISS-PROT (www.expasy.ch/sport)	Provides the description of the structure of a protein, its domains structure, post-translational modifications, variants etc. It has high level of integration with other databases and minimal level of redundancy.
Protein sequence motif databases	
PROSITE (www.expasy.ch/prosite/)	Provides information on protein families and domains. It also has patterns and profiles for sequences and biological functions.
Macromolecular databases	
PDB (www.rcsb.org/pdb)	This is the primary database for 3-dimensional (3-D) structures of biological macromolecules (determined by X-ray and NMR studies).
Other databases	
KEGG (www.genome.ad.jp/kegg/)	The Kyoto Encyclopedia of Genes and Genomes (KEGG) is a database with latest computerised information on biomolecules and cell biology. KEGG provides details on information pathways, interacting molecules and the connecting links with genes.

Protein sequence databases

Protein sequence databases are usually prepared from the existing literature and/or in consultation with the experts. In fact, these databases represent the translated DNA databases.

Molecular structure of databases

The three dimensional (3-D) structures of macromolecules are determined by X-ray crystallography and nuclear magnetic resonance (NMR). PDB and SCOP are the primary databases of 3-D structures of biological molecules.

Other databases

KEGG database is an important one that provides information on the current knowledge of molecular biology and cell biology with special reference to information on metabolic pathways, interacting molecules and genes.

APPLICATIONS OF BIOINFORMATICS

The advent of bioinformatics has revolutionized the advancements in biological science. And biotechnology is largely benefited by bioinformatics. The best example is the sequencing of human genome in a record time which would not have been possible without bioinformatics. A selected list of applications of bioinformatics is given below.

- Sequence mapping of biomolecules (DNA, RNA, proteins).
- Identification of nucleotide sequences of functional genes.
- Finding of sites that can be cut by restriction enzymes.
- Prediction of functional gene products.
- To trace the evolutionary trees of genes.
- For the prediction of 3-dimensional structure of proteins.
- Molecular modelling of biomolecules.
- Designing of drugs for medical treatment.
- Handling of vast biological data which otherwise is not possible.
- Development of models for the functioning various cells, tissues and organs.

The above list of applications however, may be treated as incomplete, since at present there is no field in biological sciences that does not involve bioinformatics.

SUMMARY

- *Bioinformatics (a computer-based discipline) represents an alliance between biology and information technology.*
- *The storage, management and interpretation of vast biological data invariably requires bioinformatics.*
- *Bioinformatics comprises three components-creation of data base, development of algorithms and statistics, and analysis of data and interpretation.*
- *Biological databases, containing the biological information, are publicly accessible e.g. GenBank (www.ncbi.nih.gov/GeneBank).*
- *Bioinformatics has revolutionized the advancements of biological and medical sciences e.g. sequencing of human genome.*

Metabolism of Xenobiotics (Detoxification)

31

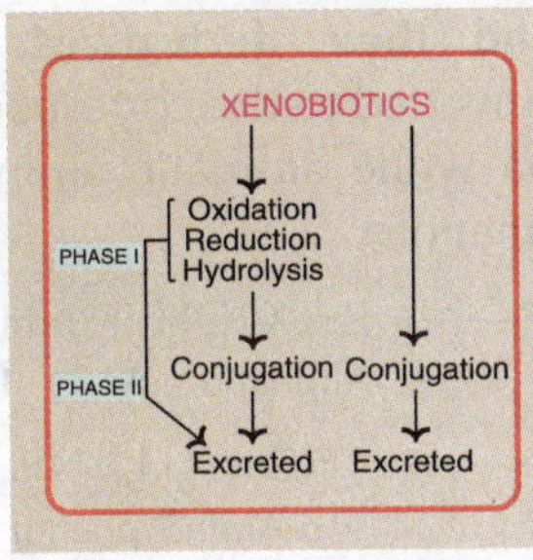

The detoxification speaks

"I deal with the metabolism of foreign compounds;
Through oxidation, reduction, hydrolysis and conjugation,
To convert xenobiotics into soluble forms;
For their effective elimination from the body."

Man is continuously exposed to several foreign compounds such as drugs, pollutants, food additives, cosmetics, pesticides etc. Certain unwanted compounds are produced in the large intestine by the bacteria which enter the circulation. These include indole from ***tryptophan, cadaverine*** from lysine, tyramine from tyrosine, phenol from phenylalanine etc. In the normal metabolism of the body, certain waste compounds (e.g. bilirubin) are formed. A vast majority of the foreign compounds or the unwanted substances, produced in the body, are toxic and, therefore, they should be quickly eliminated from the body.

The term ***detoxication or detoxification*** refers to ***the series of biochemical reactions occurring in the body to convert the foreign (often toxic) compounds to non-toxic or less toxic, and more easily excretable forms.***

Detoxification—a misnomer?

Detoxification is rather misleading, since sometimes a detoxified product is more toxic than the original substance (e.g. procarcinogens to carcinogens). It appears that the body tries to get rid of a foreign substance by converting it into a more soluble (often polar), and easily excretable compound, and this may be sometimes associated with increased toxicity (e.g. conversion of methanol to formaldehyde).

In recent years, the term detoxification is replaced by ***biotransformation*** or metabolism of xenobiotics (*Greek* : xenos—strange, foreign) or simply ***metabolism of foreign compounds***.

Site of detoxification

The detoxification reactions are carried out ***mainly in the liver*** which is equipped with the enzyme machinery. Kidney and other organs may sometimes be involved. The products formed by detoxification are mostly excreted by the kidneys, less frequently excreted via feces or expired air.

MECHANISM OF DETOXIFICATION

The metabolism of xenobiotics may be divided into two phases which may occur together or separately (***Fig.31.1***).

Phase I : The reactions of phase I are ***oxidation, reduction*** and ***hydrolysis.***

Phase II : These are the ***conjugation reactions,*** involving compounds such as glucuronic acid, amino acids (glycine), glutathione, sulfate, acetate and methyl group.

Generally, detoxification of a compound involves phase I as well as phase II reactions. For instance, oxidation followed by conjugation is the most frequent process in the metabolism of xenobiotics.

Oxidation

A large number of foreign substances are detoxified by oxidation. These include alcohols, aldehydes, amines, aromatic hydrocarbons and sulfur compounds. In general, aliphatic compounds are more easily oxidized than aromatic ones.

Alcohols : Aliphatic and aromatic alcohols undergo oxidation to form the corresponding acids.

$CH_3OH \longrightarrow HCOOH$
Methanol → Formic acid

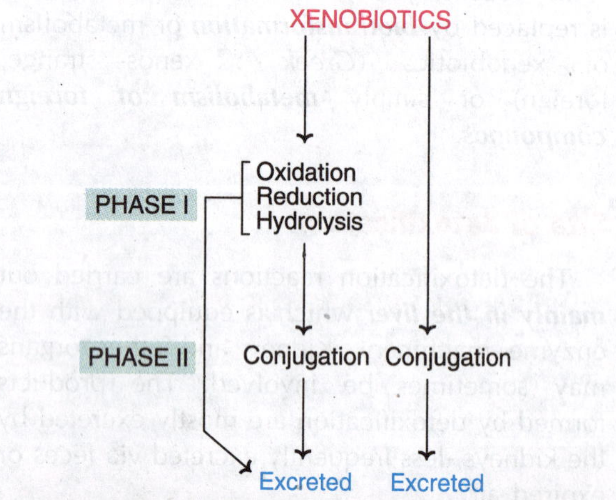

Fig. 31.1 : Phase I and phase II reactions in the metabolism of xenobiotics.

$C_2H_5OH \longrightarrow CH_3COOH$
Ethanol → Acetic acid

$C_6H_5CH_2OH \longrightarrow C_6H_5COOH$
Benzyl alcohol → Benzoic acid

Aldehydes : Aldehydes are oxidized to produce the corresponding acids.

$C_6H_5CHO \longrightarrow C_6H_5COOH$
Benzaldehyde → Benzoic acid

$C.Cl_3CHO \longrightarrow CCl_3COOH$
Chloral → Trichloroacetic acid

Amines and their derivatives : Alipahtic amines are converted to the corresponding acids, liberating urea while aromatic amino acids are oxidized to phenols.

$RCH_2NH_2 \longrightarrow R\text{–}COOH + NH_2\text{–}CO\text{–}NH_2$
Aliphatic amine → Aliphatic acid, Urea

$C_6H_5NH_2 \longrightarrow HO\text{–}C_6H_4\text{–}NH_2$
Aniline → p-Amino phenol

Aromatic hydrocarbons : Benzene may be oxidized to mono, di- and trihydroxy phenols as shown below

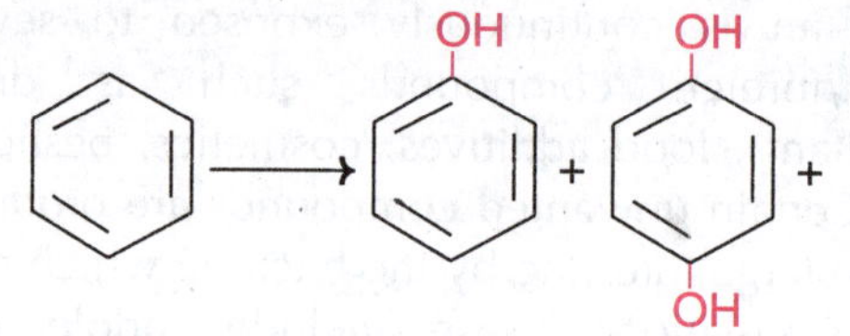

Benzene Phenol Quinol

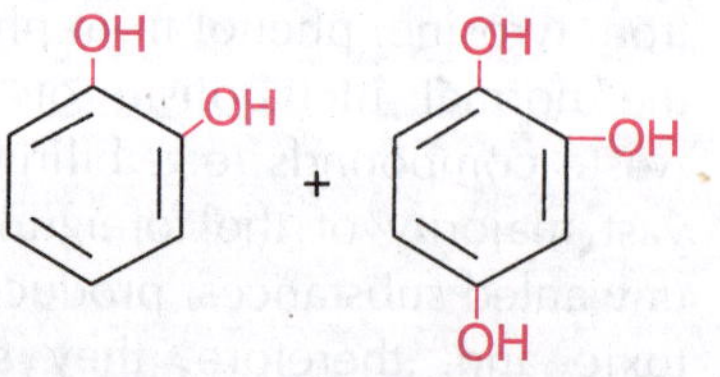

Catechol Hydroxyquinol

Sulfur compounds : Organic sulfur is oxidized to sulfuric acid.

Drugs : Meprobamate is a tranquilizer. It is oxidized to hydroxymeprobamate and excreted in urine.

Role of cytochrome P_{450}

Most of the oxidation reactions of detoxification are catalysed by monooxygenase or cytochrome P_{450}. This enzyme, also called ***mixed function oxidase,*** is associated with the microsomes. The usage P_{450} refers to the absorption peak (at 450 nm), exhibited by the enzyme when exposed to carbon monoxide.

Most of the reactions of cytochrome P_{450} involve the addition of a hydroxyl group to aliphatic or aromatic compounds which may be represented as

$$RH + O_2 + NADPH \longrightarrow ROH + H_2O + NADP^+$$

Salient features of cytochrome P_{450}

1. Multiple forms of cytochrome P_{450} are believed to exist, ranging from 20 to 200. At least 6 species have been isolated and worked in detail.

2. They are all ***hemoproteins***, containing heme as the prosthetic group.

3. Cytochrome P_{450} species are ***found in*** the highest concentration in the ***microsomes of liver***. In the adrenal gland, they occur in mitochondria.

4. The mechanism of action of cytochrome P_{450} is complex and is ***dependent on NADPH***.

5. The phospholipid-phosphatidylcholine is a constituent of cytochrome P_{450} system which is necessary for the action of this enzyme.

6. Cytochrome P_{450} is an ***inducible enzyme***. Its synthesis is increased by the administration of drugs such as ***phenobarbitol***.

7. A distinct species namely cytochrome P_{448} (with absorption peak at 448 nm) has been studied. It is specific for the metabolism of polycyclic aromatic hydrocarbons, hence it is also known as aromatic hydrocarbon hydroxylase.

Reduction

A few examples of detoxification by reduction are given.

$$C_6H_2OH(NO_2)_3 \longrightarrow C_6H_2OH(NO_2)_2NH_2$$
Picric acid → Picramic acid

$$CCl_3.CH(OH)_2 \longrightarrow CCl_3CH_2OH$$
Chloral → Trichloroethanol

$$C_6H_5NO_2 \longrightarrow C_6H_5NH_2$$
Nitrobenzene → Aminobenzene

Hydrolysis

The hydrolysis of the bonds such as ester, glycoside and amide is important in the metabolism of xenobiotics. Several compounds undergo hydrolysis during the course of their detoxification. These include aspirin, acetanilide, diisopropylfluorophosphate, atropine and procaine.

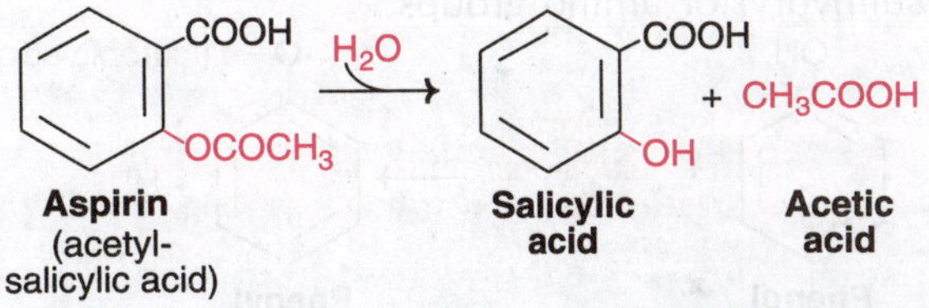

$$\text{Acetanilide} \xrightarrow{H_2O} \text{Aniline + Acetic acid}$$

$$(C_3H_7O)_2POF \xrightarrow{H_2O} (C_3H_7O)_2PO(OH) + HF$$
Diisopropyl fluoro-phosphate → Dialkyl phosphate

$$\text{Atropine} \xrightarrow{H_2O} \text{Tropic acid + Tropine}$$

$$\text{Procaine} \xrightarrow{H_2O} \text{p-Aminobenzoic acid + Diethylaminoethanol}$$

Conjugation

Several xenobiotics undergo detoxification by conjugation to produce less toxic and/or more easily excretable compounds. Conjugation is the process in which a ***foreign compound combines with a substance produced in the body***. The process of conjugation may occur either directly or after the phase I reactions. At least 8 different ***conjugating agents*** have been identified in the body. These are ***glucuronic acid, glycine, cysteine (of glutathione), glutamine, methyl group, sulfate, acetic acid and thiosulfate.***

Glucuronic acid : Conjugation with glucuronic acid is the most common. The active form of glucuronic acid is UDP-glucuronic acid produced in the uronic acid pathway (***Chapter 13***). The microsomal enzymes UDP-glucuronyltransferases participate in glucuronide formation. A general reaction of glucuronide conjugation is shown below (X-OH represents xenobiotic).

$$\text{X–OH + UDP–glucuronic acid} \xrightarrow{\text{UDP-glucuronyltransferase}} \text{XO–glucuronide + UDP}$$

Certain drugs (e.g. barbiturates) when administered induce glucuronyltransferase and this increases the glucuronide formation.

Glucuronic acid conjugation may occur with compounds containing hydroxyl, carbonyl, sulfhydryl or amino groups.

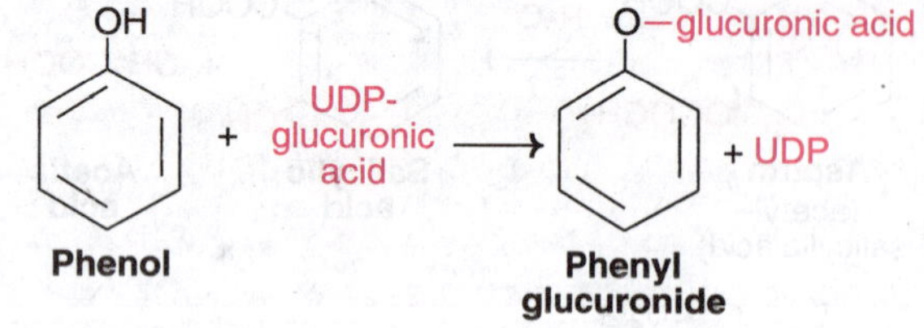

Benzoic acid + UDP-Glucuronic acid ⟶ Benzoyl glucuronide + UDP

Bilirubin + UDP-Glucuronic acid ⟶ Bilirubin-diglucuronide

In general, conjugation with glucuronic acid produces strongly acidic compounds which are more water soluble at physiological pH, and therefore, more easily excreted.

Glycine : Many aromatic carboxylic acids (e.g. benzoic acid, phenylacetic acid) are conjugated with glycine. Hippuric acid is formed when glycine is conjugated with benzyl CoA.

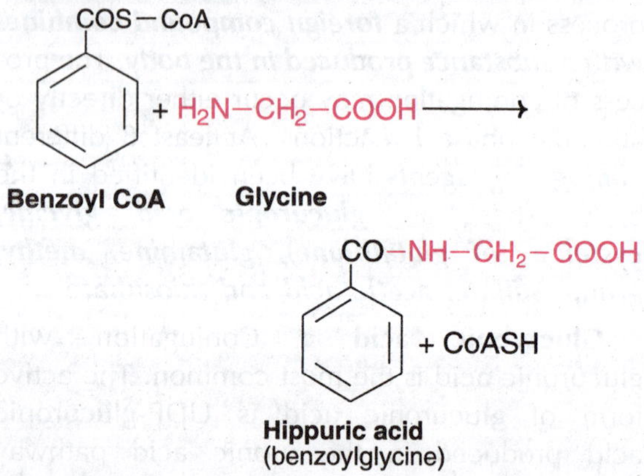

Phenylacetic acid + Glycine ⟶ Phenylaceturic acid

Cholic acid + Glycine ⟶ Glycocholic acid

Glutathione : The tripeptide glutathione (Glu-Cys-Gly), is the active conjugating agent. A wide range of organic compounds such as alkyl or aryl halides, alkenes, nitro compounds and epoxides get ***conjugated with cysteine of glutathione***. The formation of mercapturic acid is depicted in ***Fig.31.2***. The glutamate and glycine of glutathione are removed and an acetyl group is added to the cysteine residue.

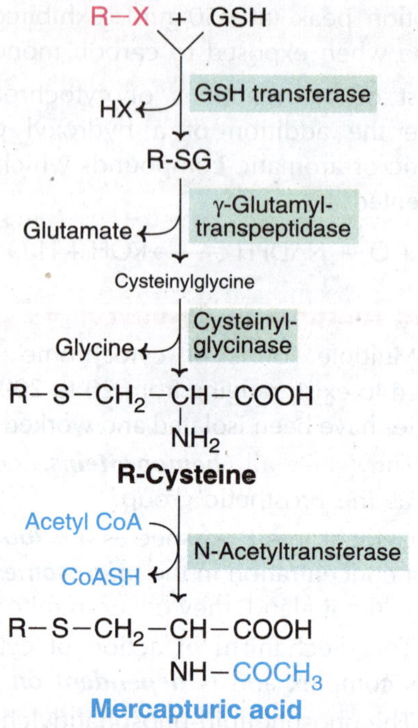

***Fig. 31.2 :** Role of glutathione in conjugation to form mercapturic acid (R-X–A xenobiotic; GSH–Glutathione).*

Glutamine : Phenylacetic acid is conjugated with glutamine to form phenylacetyl glutamine. Conjugation with glutamine is, however, relatively less important.

Methyl group : The methyl group ($-CH_3$) of ***S-adenosylmethionine*** is frequently used to methylate certain xenobiotics. This is catalysed by the enzyme methyltransferase.

S-Adenosylmethionine + X–OH $\xrightarrow{\text{Methyltransferase}}$ S-Adenosylhomocysteine + $XO-CH_3$

Sulfate : The active form of sulfate—***3′-phosphoadenosine 5-phosphosulfate*** (***PAPS***)—participates in conjugation reactions and the enzyme sulfotransferase is involved in this process. Several aliphatic and aromatic compounds undergo sulfation.

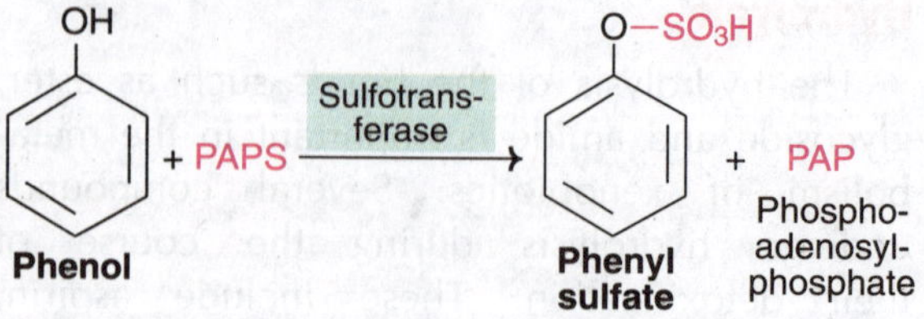

Medical Concepts/ Clinical Correlates

- The body possesses the capability to get rid of the foreign substances by converting them into more easily excretable forms.
- Detoxification is not necessarily associated with the conversion of toxic into non-toxic compounds. For instance, methanol is metabolized to a more toxic formaldehyde.
- Detoxification primarily occurs in the liver through one or more of the reactions, namely oxidation, reduction, hydrolysis and conjugation.
- British antilewisite (BAL), a compound developed during Second World War, was used to detoxify certain war poisons.

Acetic acid : Acetyl CoA is the active form of acetic acid that takes part in conjugating reactions. Drugs such as sulfanilamide are converted to acetyl derivatives.

Sulfanilamide + Acetyl CoA ⟶ Acetyl sulfanilamide + CoASH

Thiosulfate : The highly toxic cyanides are conjugated with thiosulfate to form less toxic thiocyanate.

Cyanide + Sodium thiosulfate ⟶ Thiocyanate + Sodium sulfate

Detoxification by drugs : It may be surprising to know that some drugs are administered to detoxify foreign substances. The toxic effects of certain metals such as arsenic, mercury and cadminum could be overcome by administering ***BAL*** (***British antilewisite***). This compound was developed during the World War II and was used as a detoxifying agent for certain war poisons. The mechanism of action of BAL is not clearly known. It is believed that BAL readily combines with metals and gets easily excreted into urine.

SUMMARY

- *Detoxification deals with the series of biochemical reactions occurring in the body to convert the foreign (often toxic) compounds to non-toxic or less toxic and more easily excretable forms. Liver is the major site of detoxification. In recent years, the term detoxification is replaced by biotransformation or metabolism of xenobiotics.*
- *Detoxification may be divided into phase I (oxidation, reduction, hydrolysis) and phase II reactions (conjugation). Oxidation is a major process of detoxification, involving the microsomal enzyme cytochrome P_{450} which is an inducible, NADPH dependent hemoprotein.*
- *Conjugation is a process in which a foreign compound combines with a substance produced in the body. The process of conjugation may occur either directly or after phase I reactions. At least 8 different conjugating agents have been identified in the body—glucuronic acid, glycine, cysteine, glutamine, methyl group, sulfate, acetic acid and thiosulfate.*

Prostaglandins and Related Compounds

32

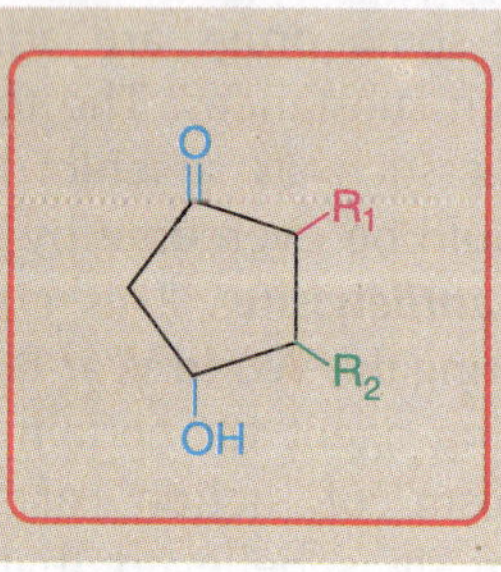

The prostaglandins speak

"Twenty carbon compounds are we!
Synthesized from arachidonic acid;
Act as local hormones in function;
Widely used as therapeutic agents."

Prostaglandins, and their related compounds—prostacyclins (PGI), thromboxanes (TXA), and leukotrienes (LT) are collectively known as ***eicosaniods,*** since they all ***contain 20 carbons*** (*Greek* : eikosi-twenty). Eicosanoids are considered as ***locally acting hormones*** with a wide range of biochemical functions.

History : Prostaglandins (PGs) were first discovered in human semen by Ulf von Euler (of Sweden) in 1930. These compounds were found to stimulate uterine contraction and reduce blood pressure. von Euler presumed that they were synthesized by prostate gland and hence named them as prostaglandins. It was later realized that PGs and other eicosanoids are synthesized in almost all the tissues (exception–erythrocytes). By then, however, the name prostaglandins was accepted worldwide, and hence continued.

The prostaglandins E and F were first isolated from the biological fluids. They were so named due to their solubility in ***ether*** (PG***E***) and phosphate buffer (PG***F***, F for fosfat, in Swedish). All other prostaglandins discovered later were denoted by a letter—PGA···PGH etc.

Structure of prostaglandins

Prostaglandins are derivatives of a hypothetical 20-carbon fatty acid namely ***prostanoic acid***, hence known as ***prostanoids.*** This has a cyclopentane ring (formed by carbon atoms 8 to 12) and two side chains, with carboxyl group on one side (***Fig. 32.1***). Prostaglandins differ in their structure due to substituent group and double bond on cyclopentane ring. e.g. PGA-double bond at $C_{10}-C_{11}$ and keto group at C_9; PGE-keto group at C_9 and OH group at C_{11}.

The structures of the most important prostaglandins (PGF_2 and $PGF_2\alpha$), ***prostacyclins (PGI_2), thromboxanes (TXA_2) and leukotrienes (LTA_4)*** along with arachidonic acid are depicted in ***Fig.32.2***. A subscript numeral indicates the number of double bonds in the two side chains. A subscript α-denotes that the hydroxyl group at C_9 of the ring and the carboxyl group are on the same side of the ring.

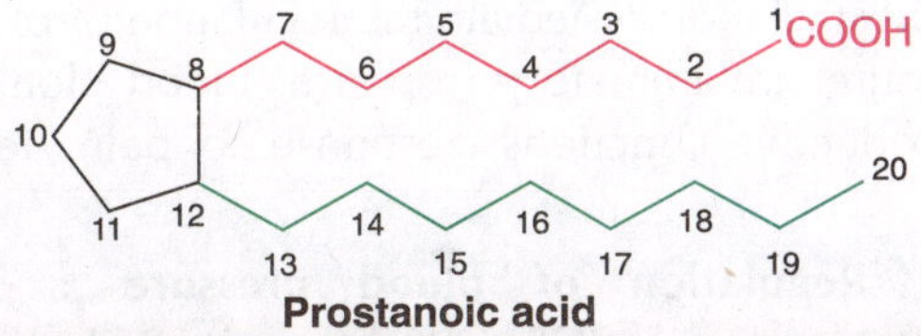

Fig. 32.1 : *Structure of prostanoic acid*

Synthesis of prostaglandins

Arachidonic acid (5,8,11,14-eicosatetraenoic acid) is the ***precursor*** for most of the prostaglandins in humans. The biosynthesis occurs by ***cyclooxygenase pathway*** in the endoplasmic reticulum in the following stages, as depicted in ***Fig.32.3***.

1. Release of ***arachidonic acid from membrane bound phospholipids*** by phospholipase A_2—this reaction occurs due to a specific stimuli by hormones such as epinephrine.

2. Oxidation and cyclization of arachidonic acid to PGG_2 which is then converted to PGH_2 by a reduced glutathione dependent peroxidase.

3. PGH_2 serves as the immediate precursor for the synthesis of a number of prostaglandins, including prostacyclins and thromboxanes.

Cyclooxygenase–a suicide enzyme : It is interesting to note that prostaglandin synthesis can be partly controlled by suicidal activity of the enzyme cyclooxygenase. This enzyme is ***capable of undergoing self-catalysed destruction to switch off PG synthesis***.

Inhibition of PG synthesis : A number of structurally unrelated compounds can inhibit prostaglandin synthesis. Corticosteroids (e.g. cortisol) prevent the formation of arachidonic acid by inhibiting the enzyme phospholipase A_2.

Many non-steroidal ***anti-inflammatory drugs*** inhibit the synthesis of prostaglandins, prostacyclins and thromboxanes. They do so by ***blocking*** the enzyme ***cyclooxygenase***.

Aspirin inhibits PG synthesis : Aspirin (acetyl salicylic acid) has been used since nineteenth century as an antipyretic (fever-reducing) and analgesic (pain relieving). The mechanism of action of aspirin however, was not known for a long period. It was only in 1971, John Vane discovered that aspirin inhibits the synthesis of PG from arachidonic acid. ***Aspirin acetylates*** serine at the active site of cyclooxygenase and irreversibly ***inhibits***. Other antiinflammatory drugs, such as indomethacin and phenylbutazone act as reversible inhibitors of the enzyme cyclooxygenase. ***Paracetamol*** is also a ***reversible inhibitor***.

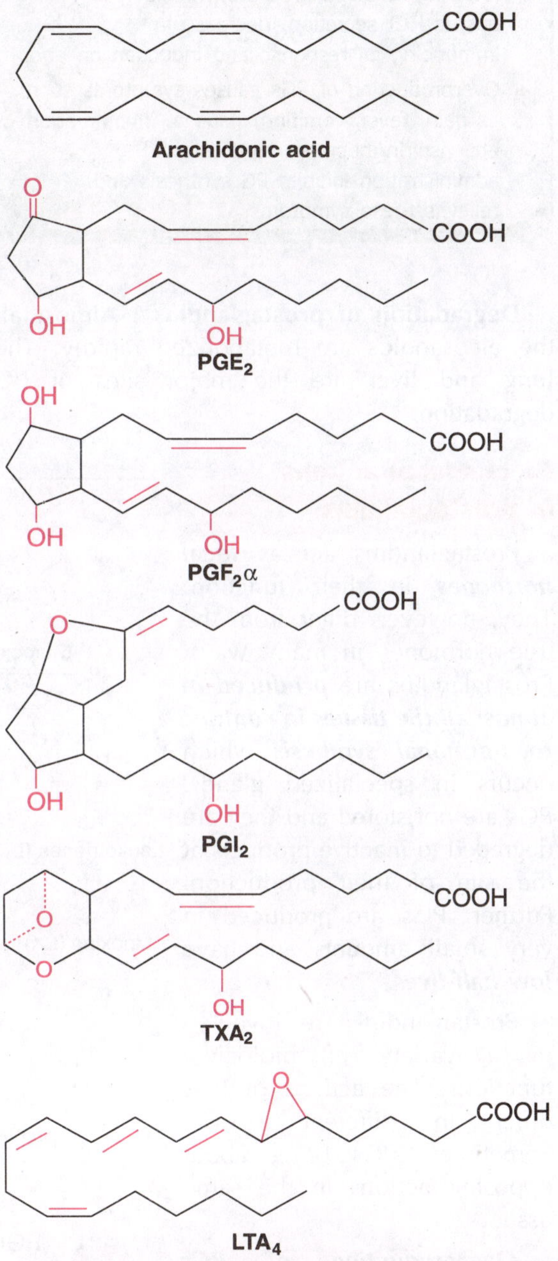

Fig. 32.2 : *The structures of arachidonic acid, common prostaglandins (PGE_2 and $PGF_2\alpha$), prostacyclins (PGI_2), thromboxanes (TXA_2) and leukotrienes (LTA_4).*

Medical Concepts/ Clinical Correlates

- PGs act as local hormones and perform diversified biochemical functions. These include lowering of blood pressure, inhibition of gastric HCI secretion, decrease in immunological response and induction of labor.
- Overproduction of PGs causes symptoms such as pain, fever, vomiting, nausea, inflammation etc. Aspirin/ibuprofen/corticosteroid administration inhibits PG synthesis and relieves these symptoms.

Degradation of prostaglandins : Almost all the eicosanoids are metabolized rapidly. The lung and liver are the major sites of PG degradation.

Biochemical actions of prostaglandins

Prostaglandins act as ***local hormones*** in their function. They, however, differ from the true hormones in many ways. Prostaglandins are ***produced in almost all the tissues in contrast to hormonal synthesis*** which occurs in specialized glands. PGs are not stored and they are degraded to inactive products at the site of their production. Further, PGs are produced in very small amounts and have ***low half-lives.***

Prostaglandins are involved in a variety of biological functions. The actions of PGs differ in different tissues. Sometimes, PGs bring about opposing actions in the same tissue.

Overproduction of PGs results in many symptoms which include ***pain, fever, nausea, vomiting and inflammation.***

Prostaglandins mediate the regulation of blood pressure, inflammatory response, blood clotting, reproductive functions, response to pain, fever etc.

1. **Regulation of blood pressure :** The prostaglandins (PGE, PGA and PGI_2) are vasodilator in function. This results in increased blood flow and decreased peripheral resistance to ***lower the blood pressure.*** PGs serve as agents in the treatment of hypertension.

2. **Inflammation :** The prostaglandins PGE_1 and PGE_2 ***induce*** the symptoms of ***inflammation*** (redness, swelling, edema etc.) due to arteriolar vasodilation. This led to the belief that PGs are natural mediators of inflammatory reactions of ***rheumatoid arthritis*** (involving joints), ***psoriasis*** (skin), ***conjunctivitis*** (eyes) etc. ***Corticosteroids*** are frequently used to treat these inflammatory reactions, since they inhibit prostaglandin synthesis.

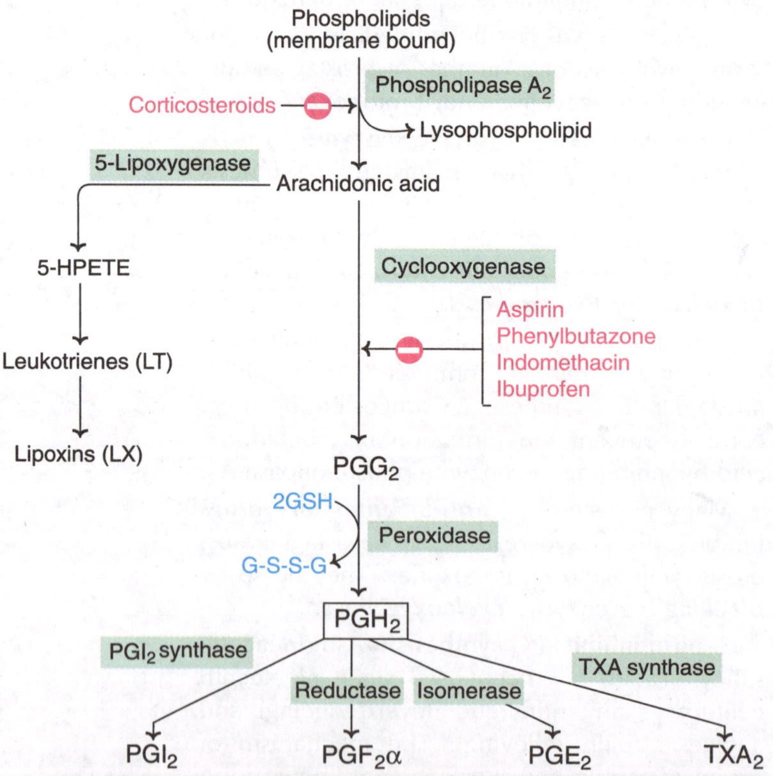

***Fig. 32.3 :** Overview of biosynthesis of prostaglandins and related compounds (5-HPETE–5-Hydroxyperoxyeicosatetraenoic acid; PG–Prostaglandins; PGI_2–Prostacyclin I_2; TXA_2–Thromboxane A_2).*

3. **Reproduction :** Prostaglandins have widespread applications in the field of reproduction. PGE_2 and PGF_2 are ***used for*** the medical ***termination of pregnancy*** and ***induction of labor***.

4. **Pain and fever :** It is believed that pyrogens (fever producing agents) promote prostaglandin biosynthesis leading to the formation of PGE_2 in the hypothalamus, the site of regulation of body temperature. ***PGE_2*** along with histamine and bradykinin ***cause pain***. ***Migraine is also due to PGE_2***. ***Aspirin*** and other non-steroidal drugs ***inhibit PG synthesis*** and thus control fever and relieve pain.

5. **Regulation of gastric secretion :** In general, prostaglandins (PGE) ***inhibit*** gastric secretion. ***PGs*** are used for the ***treatment of gastric ulcers***. However, PGs stimulate pancreatic secretion and increase the motility of intestine which often causes diarrhea.

6. **Influence on immune system :** Macrophages secrete PGE which ***decreases*** the immunological functions of B-and T-lymphocytes.

7. **Effects on respiratory function :** PGE is a bronchodilator whereas PGF acts as a constrictor of bronchial smooth muscles. Thus, PGE and PGF oppose the actions of each other in the lungs. PGE_1 and PGE_2 are used in the ***treatment of asthma***.

8. **Influence on renal functions :** PGE increases glomerular filtration rate (GFR) and ***promotes urine output***. Excretion of Na^+ and K^+ is also increased by PGE.

9. **Effects on metabolism :** Prostaglandins influence certain metabolic reactions, probably through the mediation of cAMP. ***PGE decreases lipolysis***, increases glycogen formation and promotes calcium mobilization from the bone.

10. **Platelet aggregation and thrombosis :** The prostaglandins, namely ***prostacyclins*** (***PGI_2***), ***inhibit platelet aggregation***. On the other hand, thromboxanes (TXA_2) and prostaglandin E_1 promote platelet aggregation and blood clotting that might lead to thrombosis. PGI_2, produced by endothelial cells lining the blood vessels, prevents the adherence of platelets to the blood vessels. ***TXA_2*** is released by the ***platelets*** and is responsible for their spontaneous ***aggregation***. Thus, prostacyclins and thromboxanes are antagonists in their action. In the overall effect, PGI_2 acts as a vasodilator, while TXA_2 is a vasoconstrictor. The ***balance between PGI_2 and TXA_2 is important in the regulation of hemostasis and thrombosis***.

Mechanism of action of PGs

The ***mechanism of action*** of prostaglandins is ***not known*** for certain. It is believed that PGs may act through the mediation of cyclic nucleotides. PGE increases cAMP levels whereas PGF elevates cGMP.

Low doses of aspirin reduce heart attacks

At low doses (80–325 mg/day), aspirin inhibits platelet cyclooxygenase, thereby reduces thromboxane (TXA_2) formation, and thus platelet aggregation and thrombus formation. This helps to prevent heart attacks to some extent.

Biomedical applications of PGs

Prostaglandins perform diversified functions. And for this reason, PGs (or other derivatives) are the most exploited in therapeutic applications. They are used in the ***treatment of gastric ulcers***, ***hypertension***, ***thrombosis***, ***asthma*** etc. Prostaglandins are also employed in the medical termination of pregnancy, prevention of conception, induction of labor etc.

Inhibitors of prostaglandin synthesis (e.g. aspirin, ibuprofen) are utilized in controlling fever, pain, migraine, inflammation etc.

LEUKOTRIENES

Leukotrienes are synthesized by leucocytes, mast cells, lung, heart, spleen etc., by ***lipoxygenase pathway of arachidonic*** acid. The synthesis of different leukotrienes (A_4, B_4, C_4, D_4 and E_4) through the intermediate, 5–hydroperoxyeicosatetraenoic acid (5–HPETE) is depicted in ***Fig.32.4***.

Anaphylaxis is a violent and fatal allergic reaction. It is now known that leukotrienes (C_4, D_4 and E_4) are the components of ***slow-reacting substances of anaphylaxis*** (SRS-A), released after immunological challenge. SRS–A is 100–1,000 times more potent than histamine

Medical Concepts/ Clinical Correlates

- Platelet aggregation that may lead to thrombosis is promoted by thromboxanes and prostaglandins E_1 and inhibited by prostacyclins.
- Leukotrienes are implicated in hypersensitivity (allergy) and asthma.
- Consumption of fish foods containing the unsaturated fatty acid namely eicosapentaenoic acid is advocated to prevent heart attacks.

or prostaglandins in its action as a stimulant of allergic reactions. Leukotrienes are implicated in asthma, inflammatory reactions, hypersensitivity (allergy) and heart attacks.

Leukotrienes cause contraction of smooth muscles, bronchoconstriction, vasoconstriction, adhesion of white blood cells and release of lysosomal enzymes.

Dietary marine lipids in relation to PGs, LTs and heart diseases

Eskimos of Greenland have a ***low incidence of coronary heart diseases***, despite the fact that they consume high quantities of fat and cholesterol. This is due to the high intake of marine lipids containing unsaturated fatty acids (UFA). The most predominant UFA in the fish foods consumed by Eskimos is 5, 8, 11, 14, 17-eicosapentaenoic acid (EPA). EPA is the precursor for leukotrienes-5 series which are much lower in their activity than the leukotriene-4 series, produced from arachidonic acid. Further, ***eicosapentaenoic acid inhibits the formation thromboxanes (TXA_2)***. As already described, TXA_2 promotes platelet aggregation and thrombosis.

The diet rich in marine lipids (with EPA) decreases plasma cholesterol and triacylglycerols. These factors, along with reduced synthesis of TXA_2 are believed to be responsible for the low incidence of heart attacks in Eskimos.

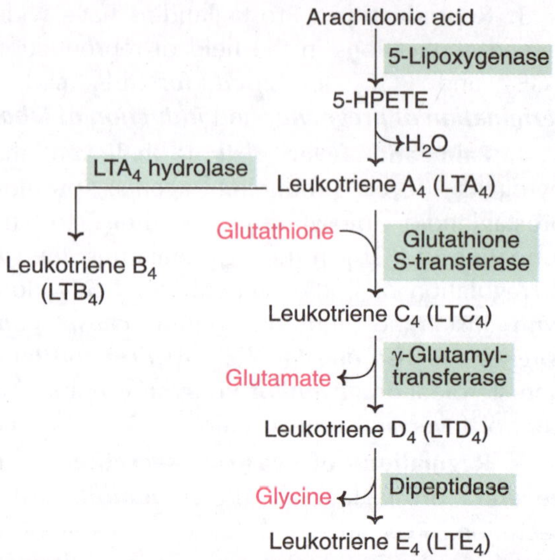

***Fig. 32.4 :** Synthesis of leukotrienes from arachidonic acid (5-HPETE–5-Hydroperoxyicosatetraenoic acid).*

SUMMARY

- *Prostaglandins (PGs) and related compounds prostacyclins (PGI), thromboxanes (TXA) and leukotrienes (LT) are collectively known as eicosanoids. They are the derivatives of a hypothetical 20 carbon fatty acid, namely prostanoic acid. Prostaglandins are synthesized from arachidonic acid, released from the membrane bound phospholipids. Corticosteroids and aspirin inhibit PG synthesis.*
- *Prostaglandins act as local hormones and are involved in a wide range of biochemical functions. In general, PGs are involved in the lowering of blood pressure, induction of inflammation, medical termination of pregnancy, induction of labor, inhibition of gastric HCl secretion, decrease in immunological response and increase in glomerular filtration rate. Thromboxanes (TXA_2) and prostaglandin E_1 promote while prostacyclins (PGI_2) inhibit platelet aggregation.*

Biological Membranes and Transport

33

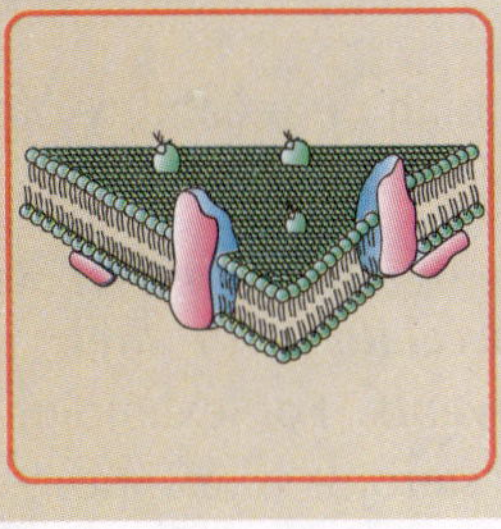

The plasma membrane speaks

"I earmark the cell territory;
For protection from hostile environment;
Regulate solute import and export
By passive or active transport."

The ***plasma membrane is an envelope surrounding the cell*** (***Refer Fig.1.1***). It separates and protects the cell from the external hostile environment. Besides being a protective barrier, plasma membrane ***provides a connecting system between the cell and its environment***. The subcellular organelles such as nucleus, mitochondria, lysosomes are also surrounded by membranes.

Chemical composition

The membranes are composed of lipids, proteins and carbohydrates. The actual composition differs from tissue to tissue. Among the lipids, ***amphipathic lipids*** (containing hydrophobic and hydrophilic groups) namely ***phospholipids***, ***glycolipids*** and ***cholesterol***, are found in the animal membranes.

Many animal cell membranes have thick coating of complex polysaccharides referred to as ***glycocalyx***. The oligosaccharides of glycocalyx interact with collagen of intercellular matrix in the tissues.

Structure of membranes

A ***lipid bilayer model*** originally proposed for membrane structure in 1935 by ***Davson and Danielle*** has been modified.

Fluid mosaic model, proposed by Singer and Nicolson, is a more recent and acceptable model for membrane structure. The biological membranes usually have a thickness of 5-8 nm. A membrane is essentially composed of a lipid bilayer. The hydrophobic (nonpolar) regions of the lipids face each other at the core of the bilayer while the hydrophilic (polar) regions face outward. Globular proteins are irregularly embedded in the lipid bilayer (***Fig.33.1***). Membrane proteins are categorized into two groups.

1. **Extrinsic (peripheral) membrane proteins** are loosely held to the surface of the membrane

and they can be easily separated e.g. cytochrome c of mitochondria.

2. **Intrinsic (integral) membrane proteins** are tightly bound to the lipid bilayer and they can be separated only by the use of detergents or organic solvents e.g. hormone receptors, cytochrome P_{450}.

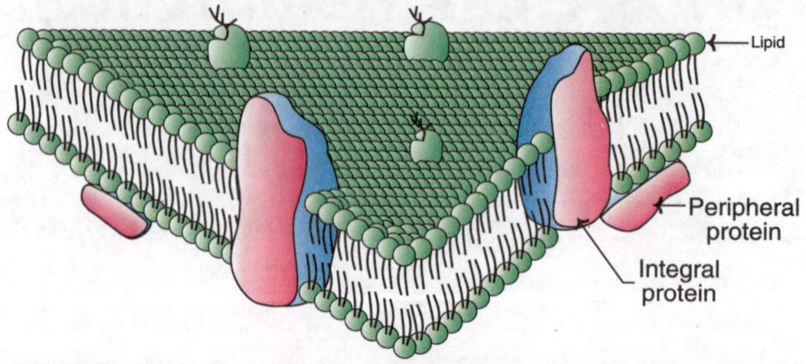

***Fig. 33.1** : The fluid mosaic model of membrane structure.*

The membrane is asymmetric due to the irregular distribution of proteins. The lipid and protein subunits of the membrane give an appearance of mosaic or a ceramic tile. Unlike a fixed ceramic tile, the ***membrane freely changes***, hence the structure of the membrane is considered as fluid mosaic.

Permeability of membranes

The biological membranes are composed of hydrophobic bilayers. Therefore, membranes are generally impermeable to water soluble molecules. However, gases such as O_2, CO_2 and nitrogen, with little interaction towards hydrophobic membranes, can readily diffuse.

Permeability coefficient : It represents the ***measure of the ability of a molecule to diffuse across a biological membrane***. Permeability coefficients are ***expressed as cm/second***. In the ***Fig.33.2***, the permeability coefficients of selected molecules and ions are depicted. The electrolytes namely Na^+, K^+ and Cl^- move through the membranes much slower than water. The relatively high permeability of water may be due to lack of charge, besides its small size.

In general, the ***permeability coefficients of molecules correlate with their solubilities in non-polar solvents***. For example, lipid soluble alcohol and steroids can more readily pass through membranes than electrolytes. Most of the drugs used in human healthcare can easily cross the membranes and enter cells.

The transport of lipid insoluble molecules is promoted through the participation of membrane proteins. The proteins spread over the biological membranes form either channels (for movement of ions and small molecules) or serve as transporters (for large molecules).

Transport across membranes

The biological membranes are relatively impermeable. The membrane, therefore, forms a barrier for the free passage of compounds across it. At least three distinct mechanisms have been

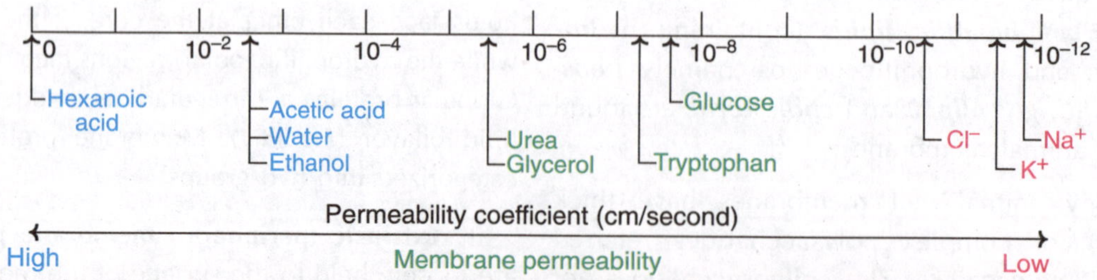

***Fig. 33.2** : Permeability coefficients of selected molecules and ions.*

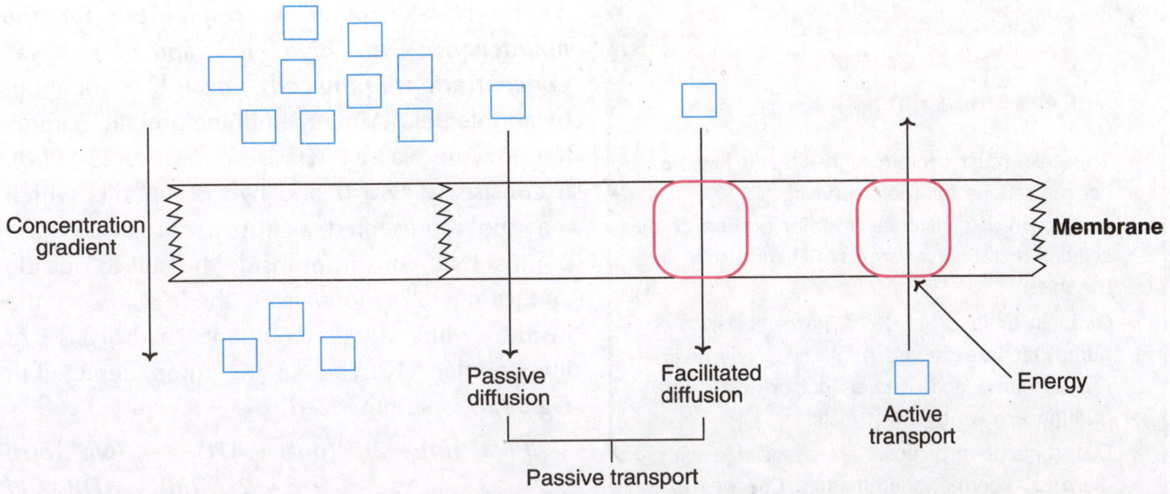

Fig. 33.3 : *Mechanism of transport across biological membrane*
*(**Note :** Transport molecule are represented in blue; the carrier proteins in red).*

identified for the transport of solutes (metabolites) through the membrane (***Fig.33.3***).

1. Passive diffusion
2. Facilitated diffusion
3. Active transport.

1. **Passive diffusion :** This is a simple process which ***depends on the concentration gradient of a particular substance*** across the membrane. ***Passage of water*** and gases through membrane occurs by passive diffusion. This process ***does not require energy***.

2. **Facilitated diffusion :** This is somewhat comparable with diffusion since the solute moves along the concentration gradient (from higher to lower concentration) and no energy is needed. But the most important distinguishing feature is that facilitated diffusion occurs ***through the mediation of carrier or transport proteins.*** Specific carrier proteins for the transport of ***glucose***, ***galactose***, leucine, phenylalanine etc. have been isolated and characterized.

Mechanism of facilitated diffusion : A ***ping-pong model*** is put forth to explain the occurrence of facilitated diffusion (***Fig.33.4***). According to this mechanism, a transport (carrier) protein exists in two conformations. In the pong conformation, it is exposed to the side with high solute concentration. This allows the binding of solute to specific sites on the carrier protein. The protein

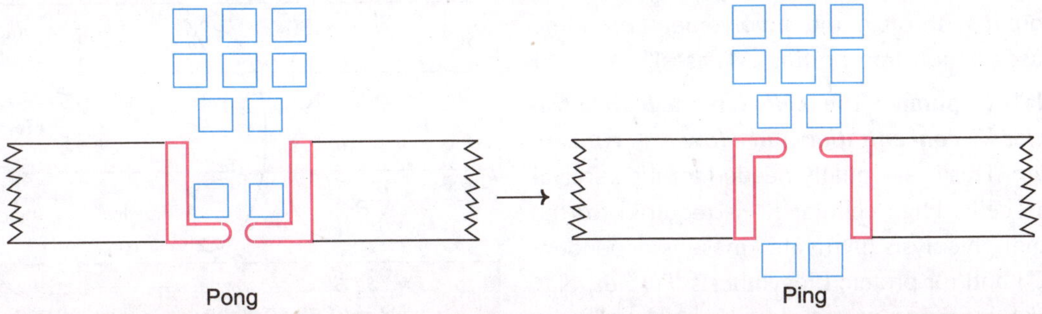

Fig. 33.4 : *A diagrammatic representation of 'ping-pong' model for facilitated diffusion.*

Medical Concepts/ Clinical Correlates

- The cells must contain high K^+ and low Na^+ concentrations for their survival. Na^+-K^+ pump, which consumes a major portion of the cellular metabolic energy (ATP), is responsible for this.
- Ouabain inhibits Na^+-K^+ ATPase (Na^+-K^+ pump). It is extracted from the seeds of an African shrub and used as poison to tip the hunting arrows by the tribals.
- Disturbances in osmosis are associated with diarrhea, edema, inflammation. Changes in membrane fluidity have been suggested to be involved in LCAT deficiency, hypertension and Alzheimer's disease.
- Corona virus (COVID-19) binds to angiotensin converting enzyme-2 (ACE2) membrane receptor and enters human cells.

then undergoes a conformational change (ping state) to expose to the side with low solute concentration where the solute molecule is released. ***Hormones regulate facilitated diffusion***. For instance, insulin increases glucose transport in muscle and adipose tissue; amino acid transport in liver and other tissues.

3. **Active transport :** Active transport occurs ***against a concentration gradient*** and this is dependent on the ***supply of metabolic energy (ATP)***. Active transport is also a carrier mediated process like facilitated diffusion. The most important ***primary active transport systems*** are ion-pumps (through the involvement of pump ATPases or ion transporting ATPases).

Na^+-K^+ pump : The ***cells have a high intracellular K^+ concentration and a low Na^+ concentration***. This is essentially needed for the survival of the cells. High cellular K^+ is required for the optimal glycolysis (pyruvate kinase is dependent on K^+) and for protein biosynthesis. Further, Na^+ and K^+ gradients across plasma membranes are needed for the transmission of nerve impulse.

The Na^+-K^+ pump is ***responsible for the maintenance of high K^+ and low Na^+ concentrations in the cells***. This is brought about by an integral plasma membrane protein, namely the enzyme Na^+-K^+ ATPase (mol. wt. 250,000). It consists of two α and two β subunits which may be represented as $(\alpha\beta)_2$. Na^+-K^+ ATPase pumps $3Na^+$ ions from inside the cell to outside and brings $2K^+$ ions from the outside to the inside with a concomitant hydrolysis of intracellular ATP. The Na^+-K^+ pump, depicted in ***Fig.33.5***, is summarized.

$$3\,Na^+\ (in) + 2K'\ (out) + ATP \longrightarrow 3Na^+\ (out) + 2K^+\ (in) + ADP + Pi$$

A major portion of the cellular ATP (up to 70% in nerve cells) is in fact utilized by Na^+-K^+ pump to maintain the requisite cytosolic Na^+ and K^+ levels. Ouabain (pronounced as Wah-báin) inhibits Na^+-K^+ ATPase. ***Ouabain*** is a steroid derivative extracted from the seeds of an African shrub. It is a poison used to tip the hunting arrows by the tribals in Africa. ***Digoxin*** a steroid glycoside and an inhibitor of Na^+-K^+ ATPase, is used in the treatment of congestive cardiac failure (digoxin improves cardiac contractility).

Na^+-cotransport system : The ***amino acids and sugars are transported into the cells*** by a Na^+-cotransport system. This process essentially consists of the passage of glucose (or amino acid) into the cell with a simultaneous movement of Na^+. ATP is required to pump out the intracellular Na^+ through the mediation of Na^+-K^+ ATPase. (***Refer Chapter 8***).

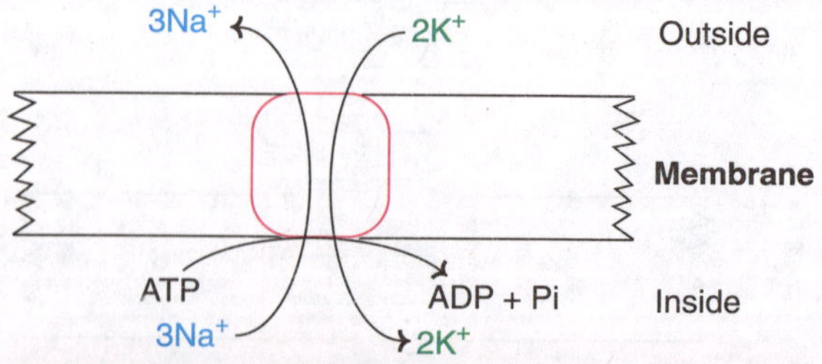

***Fig. 33.5 :** Diagrammatic representation of Na^+—K^+ pump (**Note :** Red colour block represents the enzyme Na^+—K^+ ATPase).*

Transport systems

The transport systems may be divided into 3 categories (***Fig.33.6***).

1. **Uniport system :** This involves the ***movement of a single molecule*** through the membrane e.g. transport of glucose to the erythrocytes.

2. **Symport system :** The ***simultaneous transport of two different molecules in the same direction*** e.g. transport of Na^+ and glucose to the intestinal mucosal cells from the gut.

3. **Antiport system :** The ***simultaneous transport of two different molecules in the opposite direction*** e.g exchange of Cl^- and HCO_3^- in the erythrocytes. Uniport, symport and antiport systems are considered as ***secondary active transport systems.***

Cotransport system : In cotransport, the ***transport of a substance*** through the membrane is ***coupled to the spontaneous movement of another substance***. The symport and antiport systems referred to above are good examples of cotransport system.

Proton pump in the stomach : This is an antiport transport system of gastric parietal cells. It is brought out by the enzyme ***H^+–K^+ ATPase*** to maintain highly acidic (pH≈1) conditions in the lumen of the stomach. Proton pump antiports two cytoplasmic protons ($2H^+$) and two extracellular potassium ($2K^+$) ions

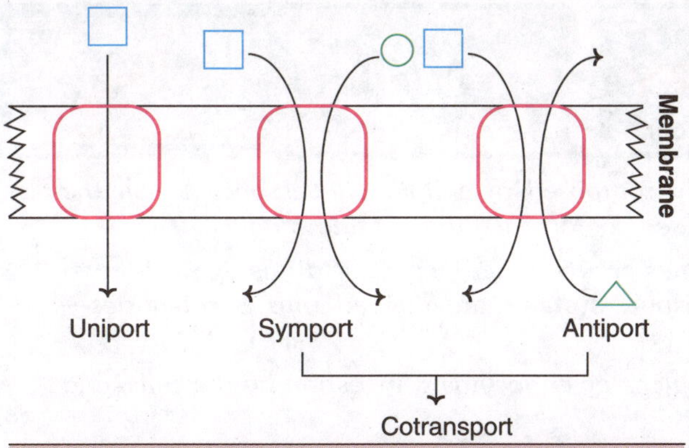

Fig. 33.6 : Diagrammatic representation of transport systems.

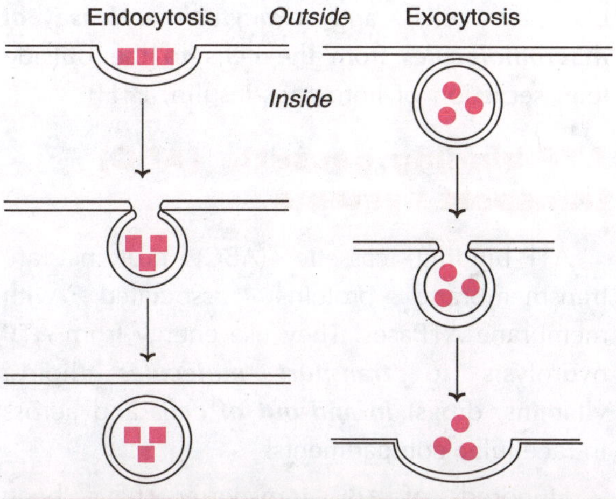

Fig. 33.7 : Diagrammatic representation of endocytosis and exocytosis (Note : The red coloured particles indicate the transport material).

for a molecule of ATP hydrolysed. The chloride ions secreted by Cl^- channels combine with protons to form gastric HCl. ***Omeprazole*** is a ***drug*** used in the treatment of ***peptic ulcer***. It ***inhibits H^+-K^+ ATPase*** and results in ***reduced secretion of HCl***.

Passive transport of water-osmosis

Osmosis is the phenomenon of movement of water from low osmotic pressure (dilute solution) to high osmotic pressure (concentrated solution) across biological membranes. The movement of water in the body occurs through osmosis, and this process ***does not require energy*** (ATP). Certain medical and health complications are due to disturbances in osmosis. e.g. edema, diarrhea, cholera, inflammation of tissues. The reader may refer ***Chapter 40*** for more information on osmosis, water and electrolyte imbalance in cholera/diarrhea.

Transport of macromolecules

The transport of macromolecules such as proteins, polysaccharides and polynucleotides across the membranes is equally important. This is brought about by two independent mechanisms namely ***endocytosis—intake of macromolecules*** by the cells (e.g. uptake of

LDL by cells) and ***exocytosis—release of macromolecules*** from the cells to the outside (e.g. secretion of hormones-insulin, PTH).

ATP-binding cassette (ABC) transport systems

ATP-binding cassette (ABC) proteins are transmembrane proteins, associated with membrane ATPases. They use energy from ATP hydrolysis to ***transport molecules*** (lipids, vitamins, drugs) ***in and out of cells*** and across intracellular compartments.

Hundreds of ABC transporters have been identified. They are involved in multidrug resistance, and over-expressed in cancer cells.

Defects in ABC proteins are linked with several disorders e.g. cystic fibrosis, Tangier disease, respiratory distress syndrome, macular degeneration.

Diseases due to loss of membrane transport systems

Alterations in transport systems result in a number of pathological conditions, selected examples are listed.

- ***Hartnup's disease*** due to a ***decrease*** in the ***transport of neutral amino acids*** in the intestinal cells and renal tubules.
- ***Cystinuria*** characterized by increased excretion of cystine, lysine, arginine and ornithine. This results in the formation of ***renal cystine stones***.

Renal reabsorption of phosphate is decreased in vitamin D resistant rickets.

Gene mutations associated with diseases of membranes

A selected list of membrane diseases along with the corresponding gene mutations/abnormalities is given in ***Table 33.1***.

Table 33.1 Selected list of diseases of membranes and the corresponding gene mutations/abnormalities

Disease	Abnormality
Familial hypercholesterolemia	LDL receptor gene mutations
Cystic fibrosis	Cystic fibrosis transmembrane regulator protein (CFTR) gene mutation
Wilson's disease	Copper-dependent ATPase gene mutation
Metastasis of cancer cells	Abnormalities in oligosaccharide chains of membrane glycoproteins
Leber hereditary optic neuropathy	Mutations in genes of mitochondrial membrane proteins involved in oxidative phosphorylation

SUMMARY

- *The biological membranes are primarily composed of a lipid bilayer onto which the globular proteins are irregularly embedded to form a fluid mosaic model.*
- *Transport of molecules through membranes occurs either by passive diffusion, facilitated diffusion or active transport. The transport systems are divided into 3 categories—uniport, symport and antiport.*
- *The transport of macromolecules takes place by endocytosis (ingestion by the cells) and exocytosis (release from the cells).*
- *Some membrane diseases as a result of gene mutations are known e.g. cystic fibrosis due to CFTR gene mutation.*

Free Radicals and Antioxidants

34

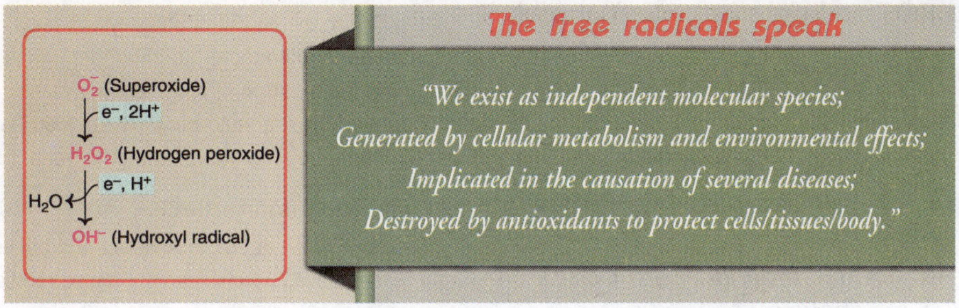

The supply of oxygen is absolutely essential for the existence of higher organisms. As the saying goes too much of even the best is bad. Very high concentrations of O_2 are found to be toxic, and can damage tissues. The present day concept of oxygen toxicity is due to the involvement of ***oxygen free radicals*** or ***reactive oxygen species (ROS)***. In fact, the generation of reactive metabolites of O_2 is an integral part of our daily life.

A ***free radical*** is defined as a molecule or a molecular species that ***contains one or more unpaired electrons***, and is ***capable of independent existence***.

Types of free radicals

Oxygen is required in many metabolic reactions, particularly for the release of energy. During these processes, molecular O_2 is completely reduced, and converted to water. However, if the reduction of O_2 is incomplete, a series of reactive radicals are formed, as shown in the next column.

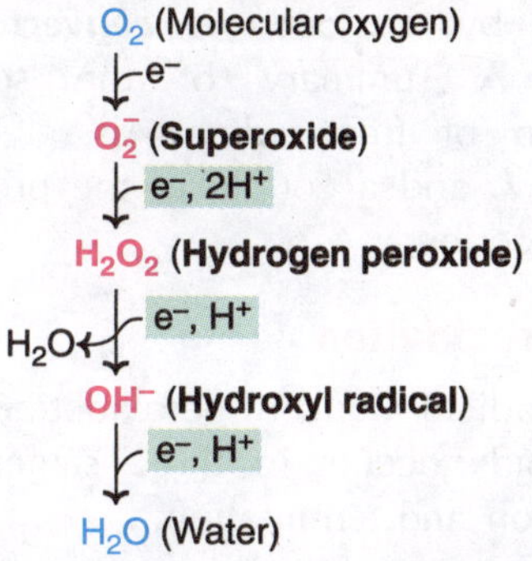

Besides the above (O_2^-, H_2O_2, OH^-), the other free radicals and reactive oxygen species of biological importance include singlet oxygen ($^1O_2$), hydroperoxy radical (HOO^-), lipid peroxide radical (ROO^-), nitric oxide (NO^-) and peroxynitrite ($ONOO^-$).

The common ***characteristic features*** of free radicals are listed

- ***Highly reactive***
- Very ***short half-life***
- Can generate new radicals by chain reaction
- Cause ***damage to biomolecules***, cells and tissues

Free radicals and reactive oxygen species (ROS)—not synonymous : By definition, a free radical contains one or more unpaired electrons. e.g. O_2^-, OH^-, ROO^-. There are certain non-radical derivatives of O_2 which do not contain unpaired electrons e.g. H_2O_2, $^1O_2$. The term reactive oxygen species is used in a broad sense to collectively represent free radicals, and non-free radicals (which are extremely reactive) of the biological systems. However, most authors do not make a clear distinction between free radicals and ROS, and use them interchangebly.

SOURCES AND GENERATION OF FREE RADICALS

The major sources responsible for the generation of free radicals may be considered under two categories

I. Due to ***normal biological processes*** (or cellular metabolism).

II. Due to ***environmental effects***.

It is estimated that about ***1-4% of the O_2*** taken up by the body is ***converted to free radicals***. A summary of the sources for generation of free radicals is given in the ***Table 34.1***, and a couple of the processes are briefly described.

Lipid peroxidation

Free radical-induced peroxidation of membrane lipids occurs in three stages-initiation, propagation and termination.

Malondialdehyde (MDA) as a marker for lipid peroxidation

Most of the products of lipid peroxidation are unstable e.g. carbonyls, esters, alkanes, alkenes, 2-alkenal, 2,4-alkadienal, MDA. Of these, malondialdehyde ($-CHO-CH_2-CHO-$) is the most extensively studied, and is used as a biochemical marker for the assessment of lipid peroxidation. MDA and other aldehydes react with thiobarbituric acid and produce red-coloured products namely ***thiobarbituric acid reactive substances (TBARS)*** which can be measured colorimetrically. The estimation of serum MDA is often used to assess oxidative stress, and free radical damage to the body.

Table 34.1 Sources along with some examples for generation of free radicals

I Cellular metabolism
• Leakage of electrons from the respiratory chain (ETC).
• Production of H_2O_2 or O_2^- by oxidase enzymes (e.g. xanthine oxidase, NADPH oxidase).
• Due to chain reactions of membrane lipid peroxidation.
• Peroxisomal generation of O_2 and H_2O_2.
• During the synthesis of prostaglandins.
• Production of nitric oxide from arginine.
• During the course of phagocytosis (as a part of bactericidal action).
• In the oxidation of heme to bile pigments.
• As a result of auto-oxidation e.g. metal ions [Fe^{2+}, Cu^{2+}]; ascorbic acid, glutathione, flavin coenzymes.
II Environmental effects
• As a result of drug metabolism e.g. paracetamol, halothane, cytochrome P_{450} related reactions.
• Due to damages caused by ionizing radiations (e.g. X-rays) on tissues.
• Photolysis of O_2 by light.
• Photoexcitation of organic molecules
• Cigarette smoke contains free radicals, and trace metals that generate OH^-.
• Alcohol, promoting lipid peroxidation.

Generation of ROS by macrophages

During the course of ***phagocytosis, macrophages produce superoxide (O_2^-)***, by a reaction catalysed by NADPH oxidase (***Fig.34.1***). This O_2^- radical gets converted to H_2O_2, and then to hypochlorous acid (HClO). The superoxide radical along with hypochlorous ions brings about ***bactericidal action***. This truly represents the ***beneficial affects of the free radicals*** generated by the body. A large amount of O_2 is consumed by macrophages during their bactericidal function, a phenomenon referred to as ***respiratory burst***. It is estimated that about 10% of the O_2 taken up by macrophages is utilized for the generation of free radicals.

Medical applications of ROS

1. **Radiation therapy** uring cobalt-60 or γ-rays to destroy tumor tissue involves ROS (hydroxyl radicals, organic radicals). Thus, the biochemical basis of radiation therapy is localized ***oxidative***

Medical Concepts/ Clinical Correlates

- Free radicals have been implicated in the causation and progress of several diseases e.g. atherosclerosis and CHD, cancer, respiratory diseases.
- The estimation of serum malondialdehyde is often used to assess oxidative stress and free radical damage to the body.

stress, causing damage to all biomolecules. Most important is the damage to DNA that ***prevents tumor cell replication and tumor growth***.

2. **Sterilization** of foods by ***irradiation*** destroying viral or bacterial contaminants also involves ROS.

3. **Nuclear accidents** explosions result in ionizing radiations. This causes oxidative ***damage to DNA and mutations which may lead to cancers***.

HARMFUL EFFECTS OF FREE RADICALS

Free radicals and biomolecules

Free radicals are highly reactive, and are ***capable of damaging*** almost all types of ***biomolecules*** (proteins, lipids, carbohydrates, nucleic acids). ***Free radicals beget free radicals*** i.e. generate free radicals from normal compounds which continues as a chain reaction.

Proteins : Free radicals cause ***oxidation of sulfhydryl groups***, and modification of certain amino acids (e.g. methionine, cysteine, histidine, tryptophan, tyrosine). ROS may damage ***proteins by fragmentation***, cross-linking and aggregation. The net result is that free radicals may often result in the ***loss of biological activity of proteins***.

Lipids : Polyunsaturated fatty acids (PUFA) are highly susceptible to damage by free radicals. (see ***lipid peroxidation***)

Carbohydrates : At physiological pH, oxidation of monosaccharides (e.g. glucose) can produce H_2O_2 and oxoaldehydes. It appears that the linkage of carbohydrates to proteins (glycation) increases the susceptibility of proteins to the attack by free radicals. This character assumes significance in diabetes mellitus where protein glycation is associated with many health complications e.g. diabetic microangiopathy, diabetic nephropathy.

Nucleic acids : Free radicals may cause DNA strand breaks, fragmentation of bases and deoxyribose. Such damages may be associated with cytotoxicity and mutations.

Free radicals and diseases

As discussed above, free radicals are harmful to biomolecules, and in turn cells and tissues. Free radicals have been implicated in the causation and progress of several diseases.

Cardiovascular diseases (CHD) : Oxidized low density lipoproteins (LDL), formed by the action of free radicals, ***promote atherosclerosis*** and CHD.

Cancer : Free radicals ***can damage DNA***, and cause mutagenicity and cytotoxicity, and thus play a ***key role in carcinogenesis***. It is believed that ROS can induce mutations, and inhibit DNA repair process, that results in the inactivation of certain tumor suppressor genes leading to cancer. Further, free radicals promote biochemical and molecular changes for rapid growth of tumor cells.

Inflammatory diseases : Rheumatoid arthritis is a chronic inflammatory disease. The free

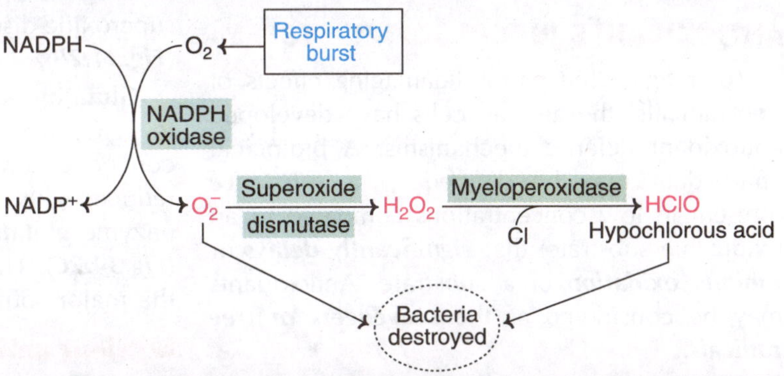

Fig. 34.1 : Generation of free radicals by macrophages and respiratory burst.

radicals produced by neutrophils are the predominant causative agents. The occurrence of other inflammatory disorders–***chronic glomerulonephritis*** and ulcerative colitis is also due to the ***damages caused by ROS*** on the extracellular components (e.g. collagen, hyaluronic acid).

Respiratory diseases : Direct exposure of lungs to 100% oxygen for a long period (more than 24 hrs) is known to destroy endothelium and cause lung edema. This is mediated by free radicals. ***ROS*** are also responsible for ***adult respiratory distress syndrome (ARDS)***, a disorder characterized by pulmonary edema.

Cigarette smoke, as such, contains free radicals, and further it promotes the generation of more free radicals. The damages caused to lungs in the smokers are due to ROS.

Diabetes : Destruction of islets of pancreas due to the accumulation of free radicals is one of the causes for the pathogenesis of insulin-dependent diabetes mellitus.

Cataract : Increased exposure to oxidative stress contributes to cataract formation, which is mostly related to aging.

Male infertility : Free radicals are known to reduce sperm motility and viability, and thus may contribute to male infertility.

Aging process : Free radicals are closely associated with the various biochemical and morphological changes that occur during normal aging.

Other diseases : Free radicals play a key role in ***Parkinson's disease***, ***Alzheimer's disease***, multiple sclerosis, ***liver cirrhosis***, muscular dystrophy, toxemia of pregnancy etc.

ANTIOXIDANTS IN BIOLOGICAL SYSTEM

To mitigate the harmful/damaging effects of free radicals, the aerobic cells have developed antioxidant defense mechanisms. A biological antioxidant may be defined as a ***substance*** (present in low concentrations compared to an oxidizable substrate) that ***significantly delays*** or ***inhibits oxidation*** of a ***substrate***. Antioxidants may be considered as the ***scavengers of free radicals***.

The production of free radicals and their neutralization by antioxidants is a normal bodily process. There are different ways of classifying antioxidants.

I. Antioxidants according to their location

1. ***Plasma antioxidants*** e.g. β-carotene, ascorbic acid, bilirubin, uric acid, ceruloplasmin, transferrin.

2. ***Cell membrane antioxidants*** e.g. α-tocopherol.

3. ***Intracellular antioxidants*** e.g. superoxide dismutase, catalase, glutathione peroxidase.

II. Antioxidants according to their nature and action

1. ***Enzymatic antioxidants*** e.g. superoxide dismutase, catalase, glutathione peroxidase, glutathione reductase.

2. ***Non-enzymatic antioxidants***

(a) ***Nutrient antioxidants*** e.g. carotenoids (β-carotene), α-tocopherol, ascorbic acid, selenium.

(b) ***Metabolic antioxidants*** e.g. glutathione, ceruloplasmin, albumin, bilirubin, transferrin, ferritin, uric acid

The antioxidant enzyme system

The antioxidant enzymes are truly the scavangers of free radicals. The major reactions of these enzymes are depicted in ***Fig.34.2***, some highlights are given below.

Superoxide dismutase : It converts superoxide (O_2^-) to hydrogen peroxide and O_2 (***Fig.34.2A***). This is the first line of defense to protect cells from the injurious effects of superoxide.

Catalase : Hydrogen peroxide, produced by superoxide dismutase, is metabolised by catalase (***Fig.34.2B***).

Glutathione peroxidase : It detoxifies H_2O_2 to H_2O, while reduced glutathione (G–SH) is converted to oxidized glutathione (GS–SG). The reduced glutathione can be regenerated by the enzyme ***glutathione reductase*** utilizing NADPH (***Fig.34.2C***). The hexose monophosphate shunt is the major source of NADPH.

Nutrient antioxidants

Tocopherols (vitamin E) : Vitamin E is fat soluble, and among the tocopherols, α-toco-

Medical Concepts/ Clinical Correlates

- The respiratory burst of macrophages, accompanied by the generation of ROS (H_2O_2 and HClO), brings about bactericidal action, and is beneficial to the body.
- Dietary consumption of a variety of nutrient antioxidants (vitamins C and E, β-carotene, lycopenes, Se, β-lipoic acid) is desirable since each antioxidant targets certain types of damaging free radicals.

pherol is biologically the most active. It is an ***antioxidant present in all cellular membranes***, and ***protects against lipid peroxidation***. α-Tocopherol can directly act on oxyradicals (e.g. O_2^-, OH^-, singlet oxygen), and thus serves as an important chain breaking antioxidant.

Ascorbic acid (vitamin C) : It is a vitamin that participates in many normal metabolic reactions of the body. Ascorbic acid is an important water-soluble antioxidant in biological fluids. Vitamin C ***efficiently scavenges free radicals, and inhibits lipid peroxidation***. It also promotes the regeneration of α-tocopherol (from α-toco-pheroxyl radical produced during scavenging of ROS).

Carotenoids : These are the natural compounds with lipophilic properties. About 500 different carotenoids have been identified, among them ***β-carotene is the most important***. It can act as an antioxidant under low partial pressure of O_2. β-Carotene usually functions in association with vitamins C and E. ***Lycopene***, a fat soluble pigment is a carotenoid. It is responsible for colour of certain fruits and vegetables (e.g. tomato). Lycopene possesses antioxidant property. ***Lutein*** and ***zeaxanthin*** are also carotenoid pigments that impart yellow or green colour to fruits and vegetables. These pigments can also serve as antioxidants.

Selenium : It is an ***essential trace element***, and is proved to be a significant antioxidant. Selenium ***works with vitamin E*** in fighting free radicals. It is also required for the function of an important antioxidant enzyme, namely ***glutathione peroxidase***.

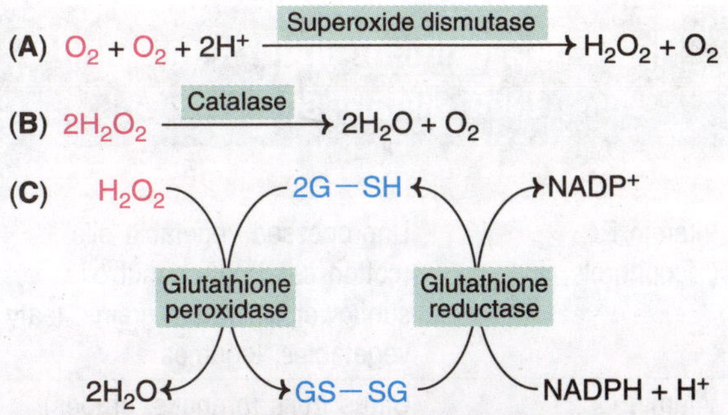

***Fig. 34.2 :** The antioxidant enzyme system (G−SH-reduced glutathione; GS−SG-oxidized glutathione).*

α-Lipoic acid : It is vitamin-like compound, produced in the body, besides the supply from plant and animal sources. α-Lipoic acid plays a key role in ***recycling*** other important ***antioxidants such as ascorbic acid, α-tocopherol and glutathione.***

Caffeine : Coffee contains flavonoids which are antioxidant in nature. Recent studies indicate that caffeine can directly act as an antioxidant.

Besides the above, there are many other important nutrient antioxidants, some of them are listed below

- Coenzyme Q_{10} of ubiquinone family
- ***Proanthocyanidins*** of grape seeds
- ***Catechins*** of green tea
- ***Curcuminoids*** of turmeric
- ***Quercetin*** of onions

In the ***Table 34.2***, some important nutrient antioxidants and their dietary sources are given. ***Consumption of a variety of nutrient antioxidants*** is important, since each antioxidant targets certain types of damaging free radicals.

Metabolic antioxidants

Glutathione : Reduced glutathione (GSH) plays a key role in the biological antioxidant enzyme system (***See Fig.34.2C***). GSH and H_2O_2 are the twin substrates for glutathione peroxidase. The reduced glutathione (GSH) gets regenerated from the oxidized glutathione (GS−SG) through the participation of glutathione reductase and NADPH. It is sugested that the ability to synthesize GSH decreases as age advances, and this has been implicated in certain diseases e.g. cataract.

Table 34.2 Nutrient antioxidants and their dietary sources

Antioxidant	Dietary Source
Vitamin E (tocopherols)	Unprocessed vegetable oils (cotton seed oil, peanut oil, sunflower oil) whole grains, leafy vegetables, legumes
Vitamin C (ascorbic acid)	Citrus fruits (oranges, grapes) gooseberry (amla), guava, green vegetables (cabbage, spinach), cauliflower, melons
β-Carotene (provitamin A)	Carrots, green fruits and vegetables, spinach, turnip, apricots.
Lycopene	Tomatoes, and their products (tomato sauce), papaya, pink guava, watermelon.
Leutein and zeaxanthin	Egg yolk, fruits, green leafy vegetables, corn, green peas.
Selenium	Sea foods, meats, organ meats, whole grains
α-Lipoic acid	Red meat, liver, yeast
Coenzyme Q_{10}	Organ meats (best-heart), beef, chicken.
N-Acetylcysteine	Available as supplement or drug.
Proanthocyanidins	Grape seeds
Catechins	Green tea
Curcuminoids	Turmeric
Quercetin	Onions, red wine, green tea
Ellagic acid	Berries, walnuts, pomegranates
Hesperidin	Citrus fruits (oranges), lemon.

Some more metabolic antioxidants of biological importance.

- **Uric acid**, a powerful scavenger of singlet oxygen ($^1O_2$) and OH^- radicals.
- **Ceruloplasmin** inhibits iron and copper dependent lipid peroxidation.
- **Transferrin** binds to iron and prevents iron-catalysed free radical formation.
- **Albumin** can scavange the free radicals formed on its surface.
- **Bilirubin** protects the albumin bound free fatty acids from peroxidation.
- **Haptoglobin** binds to free hemoglobin and prevents the acceleration of lipid peroxidation.

DIETARY SUPPLEMENTATION OF ANTIOXIDANTS

Free radicals damage biomolecules (proteins, nucleic acids, lipids), and are implicated in the causation and progress of several diseases (CHD, cancer, autoimmune diseases). To counter the action of free radicals, many protective antioxidant nutrients (vitamins E and C, β-carotene, selenium) are in use as dietary supplements.

Some recent studies show that antioxidant supplements may be beneficial to the people who are on deficient states, but not to all. On the other hand, some clinical trials indicate that supplementation of β-carotene and vitamin E is associated with increased mortality. However, this is controversial, and needs to be proved beyond doubt.

In any case, caution should be exercised in the supplementation of antioxidants, and their overuse should be avoided.

SUMMARY

- *Free radicals are the molecules or molecular species containing one or more unpaired electrons with independent existence. e.g.* O_2^-, H_2O_2, OH^-, $^1O_2$.
- *ROS are constantly formed during the normal cellular metabolism, (e.g. lipid peroxidation) and due to various environmental influences (e.g. ionizing radiations).*
- *Free radicals are highly reactive and are capable of damaging almost all types of biomolecules (proteins, lipids, carbohydrates, nucleic acids), and have been implicated in the causation of many diseases e.g. cardiovascular diseases, cancer.*
- *To mitigate the harmful effects of free radicals, the aerobic cells have developed antioxidant defense mechanisms-enzymatic antioxidants (superoxide dismutase, catalase) and non-enzymatic antioxidants (glutathione, Se, α-tocopherol, β-carotene).*

Environmental Biochemistry 35

The environment speaks

"Composed of living and non-living entities;
Co-existing and friendly with you all;
Continuously insulted by your activities of pollution;
That threatens only your healthy existence!"

Environment constitutes the ***non-living*** (***air, water, land, energy etc.***) as well as the ***living*** (***biological and social***) ***systems*** surrounding man. Environmental biochemistry primarily deals with the metabolic (biochemical) responses and adaptations in man (or other organisms) due to the environmental factors.

A healthy environment is required for a healthy life which is however, not really possible or practicable. This is mainly because of the atmospheric (climatic) changes and environmental pollution.

Environmental biochemistry is a very vast subject. The basic concepts regarding the atmospheric changes and environmental pollution on humans are dealt with here.

ATMOSPHERIC (CLIMATIC) CHANGES

The climatic changes include ***cold, heat*** etc. The body makes every effort to maintain its normal temperature (despite cold and heat surroundings) for optimal physiological and biochemical functions.

EXPOSURE TO COLD

Short-term exposure to cold ***causes shivering*** (mainly due to skeletal muscle) to produce extra ***heat***. Heat is generated by the hydrolysis of ATP.

Non-shivering phase

Chronic ***exposure to cold results in non-shivering phase*** which is characterized by several metabolic adaptations.

- **Energy metabolism :** Heat generation by a process called ***chemical thermogenesis*** occurs in non-shivering phase. The foodstuffs undergo oxidation to generate heat at the expense of growth and other anabolic processes. Elevation in BMR, and increased intake of foods are observed.
- **Lipid metabolism :** ***Stored fat*** (triacylglycerol) in the adipose tissue is ***mobilized*** to supply free fatty acids for oxidation and production of energy. Brown adipose tissue, particularly in neonatal life, significantly contributes to thermogenesis.

Medical Concepts/ Clinical Correlates

- The body makes every effort to maintain its normal temperature, despite cold and heat surroundings, for optimal physiological and biochemical functions.
- Failure of heat regulatory system (thermoregulation) leads to heat stroke characterized by high body temperature, convulsions etc.
- Sulfur dioxide (SO_2) is the most dangerous industrial pollutant gas to man. It primarily affects the respiratory system, and may result in bronchitis, and even lung cancer.

- **Hormonal changes :** Thyroxine, a hormone closely associated with energy metabolism, is elevated. Further, corticosteroids are increased on exposure to cold.

EXPOSURE TO HEAT

There is a continuous ***generation of heat*** by the body ***due to the ongoing biochemical processes***, referred to as ***metabolic heat***. This heat has to be exchanged with the environment to maintain a constant body temperature. On exposure to heat in surroundings, as happens in summer, the body is subjected to an uncomfortable situation (since temperature of the surroundings is much higher than that of the body). However, heat is still lost from the body through sweating and evaporation. Normally, the body (thermoregulation) gets acclimatized to higher temperature within 3-5 days.

Heat stroke : It is characterized by the ***failure of the heat regulatory system (thermoregulation)*** of the body. The manifestations of heat stroke include high body temperature, convulsions, partial (some times total) loss of consciousness. In extreme cases, heat stroke may cause irreversible damage to brain. The treatment for the heat stroke ***involves rapid cooling of the body***.

The milder form of heat stroke is referred to as ***heat syncope***. Although the body temperature is not raised much in this condition, the blood pressure falls and the person may collapse suddenly. Heat syncope is easily reversible.

ENVIRONMENTAL POLLUTION

Environmental pollution may be regarded as the addition of extraneous (foreign) materials to air, water or land which adversely affects the quality of life. ***Pollution may be caused by physical, chemical or biological processes***.

The term ***pollutant*** refers to a substance which ***increases in quantity due to human activity and adversely affects the environment*** (e.g. carbon monoxide, sulfur dioxide, lead). A substance which is not present in nature but released during human activity is the ***contaminant*** (e.g. methyl isocyanate, DDT, malathion). A contaminant however, is regarded as a pollutant when it exerts detrimental effects. Environmental pollution may be considered in different ways—industrial pollution; agricultural pollution; pollution due to gaseous wastes, liquid wastes and solid wastes. Environmental pollution with reference to air, water and foodstuffs is briefly discussed.

AIR POLLUTION

The major components of air include nitrogen (78.1%), oxygen (20.93%) and carbon dioxide (0.03%), along with water vapour and suspended particles. The ***rapid growth of industries coupled*** with changing lifestyles of man (urbanization, smoking, use of motor vehicles etc.) largely ***contribute to air pollution***. The major chemical constituents of air pollution are sulfur dioxide, oxides of carbon (CO_2 and CO), oxides of nitrogen, hydrocarbons and particulates. The biochemical affects of air pollution are described.

Sulfur dioxide

Sulfur dioxide (SO_2) is the ***most dangerous pollutant gas to man***. Industrial activities such as burning of coal and oil emit large quantities of SO_2.

Sulfur dioxide pollution primarily affects respiratory system in man. Irritation of the respiratory tract and increasing airway resistance (breathing difficulty) are observed. Lung tissue may get damaged due to acidic pH. Further, dipalmityl lecithin, the phospholipid acting as the lung surfactant, gets affected. Continuous exposure to SO_2 (> 1 ppm) for several days causes bronchitis and in some individuals lung cancer.

Carbon monoxide

Carbon monoxide (CO) is ***mostly produced by incomplete combustion of fuel or carbon-containing compounds***. Automobiles, aircrafts, rail engines and burning of coal in factories contribute to CO pollution.

Carbon monoxide combines with hemoglobin to form ***carboxyhemoglobin*** (***Refer Chapter 10***). This causes a drastic ***reduction in the supply of O_2 to tissues***. At a CO concentration around 1 ppm, impairment in mental performance and visual perception take place. With a further increase in CO level, headache, dizziness and loss of consciousness occur. Death may be inevitable in persons exposed to above 750 ppm of CO.

Carbon dioxide

Carbon dioxide (CO_2), constituting only a fraction (0.03%) of the atmospheric gases, plays a significant role in controlling the climate. This is ***done by trapping the heat radiation*** from the earth's surface. Without the presence of CO_2, the earth would be as cold as moon!

Carbon dioxide is often referred to as greenhouse gas. The term ***greenhouse effect*** refers to an ***elevation in CO_2*** near earth's surface that traps sunlight and ***increases atmospheric temperature***. Deforestation, burning of coal, oils etc., elevate atmospheric CO_2 resulting in greenhouse effect. Hence the global propaganda for increased plantation of trees!

Nitrogen dioxide

Nitrogen dioxide (NO_2) like carbon monoxide (CO), combines with hemoglobin and reduces the supply of O_2 to the tissues. NO_2 is more harmful to human health than CO. It is fortunate that the atmospheric concentration of NO_2 is relatively lower.

Nitrogen dioxide (in the form of HNO_3) along with SO_2 (as H_2SO_4) contributes to ***acid rain***.

Hydrocarbons

Many hydrocarbons polluting the environment affect human life. The aromatic hydrocarbons cause irritation to injuries.

Particulates

The ***solid dust particles*** suspended ***in the atmosphere*** constitute particulates. The sources of particulates are grinding, spraying, erosion, smoking etc.

The particulates have ill-affects on humans. These include interference in respiratory function (coughing, sneezing) and toxicity caused by the absorption particulate chemicals. Further, the ***dust particles carry microorganisms and other infective agents to spread diseases.***

Ozone layer

Ozone is formed from atmospheric oxygen during high energy radiations of electrical discharges. This ozone forms a layer above the earth's surface (15-35 km). It ***absorbs harmful ultraviolet radiations of sun*** which would otherwise cause skin diseases and mutations, besides increasing the temperature of earth.

In recent years, a decrease in the ozone layer is observed due to chemical pollution in the air. Nitrogen oxides (released from engines of aeroplanes) and chlorofluoro carbons (used in refrigerators and air conditioners) deplete the ozone layer.

WATER POLLUTION

Water is the most predominant constituent of living matter. The very existence of life is unimaginable without water.

As such, pure water does not exist in nature. The available water contains dissolved gases, minerals and some suspended particles. Pollution of water occurs due to waste disposal from industries, agriculture and municipalities. The pollutants may be organic, inorganic, sediments, radioactive, thermal etc., in nature.

ORGANIC POLLUTANTS

The organic pollutants include agents carrying water borne diseases, oxygen demanding wastes and organic chemicals.

Water-borne disease agents

Several pathogenic organisms find their entry into water and cause diseases. The water borne disease include ***typhoid, paratyphoid, cholera, amoebiasis, giardiasis and infectious hepatitis***. These diseases can be prevented by disinfection techniques employed for the treatment of water.

Oxygen demanding wastes

Sewage, and wastes from industries and agriculture provide good nutrients for algae. As the algae grow utilizing the wastes, oxygen depletion occurs. This phenomenon of ***water deoxygenation*** is technically referred to as ***eutrophication***. As a consequence of eutrophication, fish and other acquatic animals die (due to lack of O_2), causing foul smell.

Organic chemicals

The organic chemical pollutants of water include pesticides and several synthetic compounds (detergents, paints, plastics, pharmaceuticals, food additives etc.)

Pesticides

Pesticides is a broad term used for ***insecticides, herbicides, fungicides*** and ***rodenticides***. Based on their structure, pesticides are classified as follows.

(a) **Chlorinated hydrocarbons :** e.g. aldrin, dieldrin, endrin, dichlorodiphenyl trichloroethane (DDT).

(b) **Organophosphates :** e.g. malathion, diazinon.

(c) **Carbamates** e.g. baygon, carbaryl (sevin)

(d) **Chlorophenoxy** e.g. 2,4-dichlorophenoxy acetic acid.

The use of pesticides has helped in controlling certain diseases (malaria, typhus), besides boosting food production. However, pesticides pollute water and cause several health complications to humans, besides damaging acquatic life.

Dichloro-diphenyl trichloroethane (DDT) is a widely used pesticide to control cotton and peanut pests, besides malaria. However, continuous use of DDT leads to its accumulation in foods causing ill effects (hence banned in some countries like USA).

DDT, being fat soluble, ***accumulates in the adipose tissue*** and is not excreted. Thus, its concentration in the body goes on increasing. DDT causes nervous irritability, muscle twitching and convulsions.

Organophosphates and carbamates are powerful ***neurotoxic agents***. They prevent the transmission of nerve impulse by ***inhibiting the enzyme cholinesterase***.

INORGANIC POLLUTANTS

Heavy metals (lead, mercury, cadmium, aluminium, arsenic etc.) are the most dangerous among the inorganic pollutants.

Lead

Lead is the ***most common inorganic pollutant*** found in water, air, foods and soils. The sources of lead pollution include petrol, gasoline, paints, cigarettes, news papers, lead pipes and xerox copies. The plasma concentration of > 25 μg/dl in adults and > 10 μg/dl in children results in toxic manifestations.

The principal target of lead toxicity is ***central nervous system***. In the growing children, Pb causes learning disabilities, ***behavioural changes*** (hyperexcitability) and ***mental retardation***. In adults, confusion, irritability, abdominal colic and severe anemia are associated with lead toxicity.

Lead inhibits several enzymes, particularly, δ-aminolevulinate (ALA) synthase, ALA dehydratase and ferrochelatase of heme synthesis (***Refer Chapter 10*** also). This results in severe anemia. There has been an increasing awareness worldover on the toxic manifestations of lead. This has lead to the supply of ***unleaded petrol*** in most countries.

Medical Concepts/ Clinical Correlates

- Carbon monoxide (CO) combines with hemoglobin to form carboxyHb. This reduces O_2 supply to tissues.
- Pollution of water with pathogenic organisms causes many diseases e.g. typhoid, cholera, amoebiasis.
- Lead toxicity affects central nervous system—learning disabilities, mental retardation etc.
- Radioactive pollution may lead to gene mutations, cancer etc.

Mercury

Mercury is a ***common industrial*** (plastic, paints, electrical apparatus, fungicides) ***pollutant***. Acute mercuric poisoning causes gastritis, vomiting and pulmonary edema. Chronic manifestations of Hg include emotional changes, loss of memory and other neuropsychiatric disturbances. In addition, deposition of mercuric salts may cause renal failure.

Organic mercuric poisoning is commonly referred to as ***minamata disease*** (as it first occurred in Minamata, Japan in 1953-60 by consuming fish containing methyl mercury, as industrial pollutant).

Cadmium

The outbreak of cadmium toxicity was reported in Japan in the form of ***itai itai*** or ***ouch disease***. Cadmium poisoning causes fragile bones, anemia, bone marrow disorders and kidney damage. Biochemically, cadmium replaces zinc and adversely influences several metabolic reactions.

Aluminium

The sources of aluminium include cooking vessels, building materials, food additives and cosmetics. Aluminium toxicity is associated with Alzheimer's disease, anemia and osteomalacia.

Arsenic

Arsenic, commonly found in many insecticides and fungicides, is toxic to the body. Arsenic binds with–SH groups of several enzymes and inhibits biochemical reactions e.g. pyruvate dehydrogenase. Further, arsenic causes coagulation of proteins and blockage of ATP generation (functions as an uncoupler).

NOISE POLLUTION

The unwanted sound is noise, which is a major urban environmental pollutant. Man can ***tolerate noise upto 100 decibels*** (speaking – 60 decibels; telephone bell 70 decibels; motor cycle 110 decibels; rockets 170 decibels). A noise above 150 decibels is uncomfortable.

The affects of noise pollution include headache, increased blood pressure, irritability, neuromuscular tension, confusion, disturbed vision and digestion, depression and loss of hearing.

RADIOACTIVE POLLUTION

The pollution due to radioactive substances is the most dangerous to human life. The health hazards of radioactive pollution include ***gene mutations,*** cancer, destruction of living cells etc.

TOXIC COMPOUNDS IN FOODSTUFFS

The foodstuffs consumed by humans contain several toxic compounds which may be either normally present or enter foodstuffs during the course of cultivation, processing or storage.

Natural toxins in foodstuffs

Neurotoxins : Kesari dal (*Lathyrus sativus*) is a pulse grown in some parts of Madhya Pradesh, Bihar and Uttar Prodesh. Excessive consumption of kesari dal causes ***paralysis of lower limbs*** referred to as ***lathyrism***. This is due to a neuro-toxin namely ***β-oxalylaminoalanine*** (*BOAA*). BOAA damages upper motor neurons, and inhibits the enzyme lysyl oxidase (reduces collagen cross-linking). Cooking of kesari dal 2-3 times and removal of the supernatant water will eliminate the toxin.

Protease inhibitors : Certain legumes (soya bean, peanut) contain inhibitors of protease enzymes particularly trypsin. Normally, protease inhibitors are destroyed by cooking. However, partial cooking does not totally destroy them. In such a case, protease inhibitors ***can inhibit digestion and proteins***.

Goitrogens : These compounds ***prevent uptake and utilization of iodine by thyroid gland***. Goitrogens are found in cabbage and turnips (thioglycosides), mustard and rape seed oils (thiocyanates), ground nuts and almonds (polyphenolic glycosides).

Antivitamins : ***Avidin*** of raw egg is a good example of ***antivitamin of biotin***.

Toxic pollutants of foodstuffs

The foodstuffs may get polluted with several toxic chemicals which might occur during cultivation, processing or storage.

Cultivation : Pesticides and other unnatural chemicals used during cultivation do find an entry into the foodstuffs. It is fortunate that most of these chemicals can be removed by peeling the outer layers of vegetables and fruits, besides repeated washings.

Processing : Defects in freezing, and packing provide a suitable environment for the growth of several organisms which release toxic products e.g. milk contamination by *Salmonella*.

Several food additives are in use for preservation and enchancing flavour. Not all of them are safe e.g. aniline dyes used as colouring agents are carcinogenic; sweetening agent cyclamate may cause bladder cancer.

Storage : Contamination of stored foods occurs mostly due to fungal infections. ***Aflatoxins*** are produced by *Aspergillus favus* when ground nuts or coconuts are stored in moist conditions. Aflatoxins are ***hepatotoxic and carcinogenic***.

CARCINOGENS

The group of chemicals that cause cancer in man and animals are collectively referred to as carcinogens (***Refer Chapter 37***). Environmental pollution is undoubtedly associated with increased risk of cancer. The topic '***cancer***' may be considered as a part of environmental biochemistry for learning purpose.

SUMMARY

- *Environmental biochemistry deals with the biochemical responses and adaptations in man (and other organisms) due to environmental factors.*
- *The atmospheric (climatic) changes like cold and heat influence the body. Several metabolic adaptations occur to overcome the adverse affects.*
- *The major chemical constituents of air pollution include SO_2, CO, CO_2 and oxides of nitrogen. Among these, sulfur dioxide is the most dangerous.*
- *Water pollution occurs mainly due to waste disposal from industries, agriculture and municipalities. The pollutants may be organic (pathogenic organisms, pesticides), or inorganic (lead, mercury).*
- *The foodstuffs consumed by humans may contain several toxic compounds. These may be normally present (e.g. BOAA causing lathyrism) or enter the foodstuffs during the course of cultivation (e.g. pesticides), or storage (e.g. aflatoxins).*

Insulin, Glucose Homeostasis, and Diabetes Mellitus

36

Diabetes mellitus
- Insulin-dependent diabetes mellitus (IDDM)
- Non-insulin dependent diabetes mellitus (NIDDM)

The Diabetes speaks

"Dubbed as a disease of fuel scarcity in plenty;
Starving the cells bathed in ample glucose quantity;
Attributed to insufficient or inefficient insulin;
Affecting several tissues by metabolic complications."

Diabetes mellitus (World Diabetes Day-14th November) is the ***third leading cause of death*** (after heart disease and cancer) in many developed countries. It affects about ***6 to 8% of the general population***. The complications of diabetes affect the eye, kidney and nervous system. Diabetes is a ***major cause of blindness, renal failure, amputation, heart attacks and stroke***. (The term diabetes, whenever used, refers to diabetes mellitus. It should, however, be noted that ***diabetes insipidus*** is another disorder characterized by ***large volumes of urine excretion*** due to ***antidiuretic hormone deficiency***).

Diabetes mellitus is a clinical condition characterized by ***increased blood glucose level (hyperglycemia)*** due to insufficient or inefficient (incompetent) insulin. In other words, insulin is either not produced in sufficient quantity or inefficient in its action on the target tissues. As a consequence, the blood glucose level is elevated which spills over into urine in diabetes mellitus (*Greek* : diabetes—a siphon or running through; mellitus—sweet).

An important feature of diabetes is that the body cells are starved of glucose despite its very high concentration around i.e. scarcity in plenty. For a comprehensive understanding of diabetes, the relevant hormones, namely insulin and glucagon, homeostasis of blood glucose, besides the biochemical aspects of diabetes, are discussed in this chapter.

INSULIN

Insulin is a polypeptide hormone ***produced by the β-cells of islets of Langerhans of pancreas***. It has profound influence on the metabolism of carbohydrate, fat and protein. Insulin is considered as ***anabolic hormone***, as it promotes the synthesis of glycogen, triacylglycerols and proteins. This hormone has been implicated in the development of diabetes mellitus.

Insulin occupies a special place in the ***history*** of biochemistry as well as medicine. Insulin was the first hormone to be isolated, purified and synthesized; first hormone to be sequenced; first hormone to be produced by recombinant DNA technology.

Structure of insulin

Human insulin (mol. wt. 5,734) contains ***51 amino acids***, arranged in two polypeptide chains. The chain A has 21 amino acids while B has 30 amino acids. Both are held together by two interchain disulfide bridges, connecting A_7 to B_7 and A_{20} to B_{19}. In addition, there is an intrachain disulfide link in chain A between the amino acids 6 and 11.

Biosynthesis of insulin

Insulin is produced by the β-cells of the islets of Langerhans of pancreas. The gene for this protein synthesis is located on chromosome 11. The synthesis of insulin involves two precursors, namely ***preproinsulin*** with 110 amino acids (mol. wt. 11,500) and ***proinsulin*** with 86 amino acids (mol. wt. 9,000). They are sequentially degraded (***Fig.36.1***) to form the ***active hormone*** insulin and a connecting peptide (C-peptide). Insulin and C-peptide are produced in equimolar concentration. ***C-peptide*** has no biological activity, however its estimation in the plasma serves as a ***useful index*** for the endogenous production of insulin.

In the β-cells, ***insulin*** (and also proinsulin) ***combines with zinc to form complexes***. In this form, insulin is stored in the granules of the cytosol which is released in response to various stimuli (discussed below) by exocytosis.

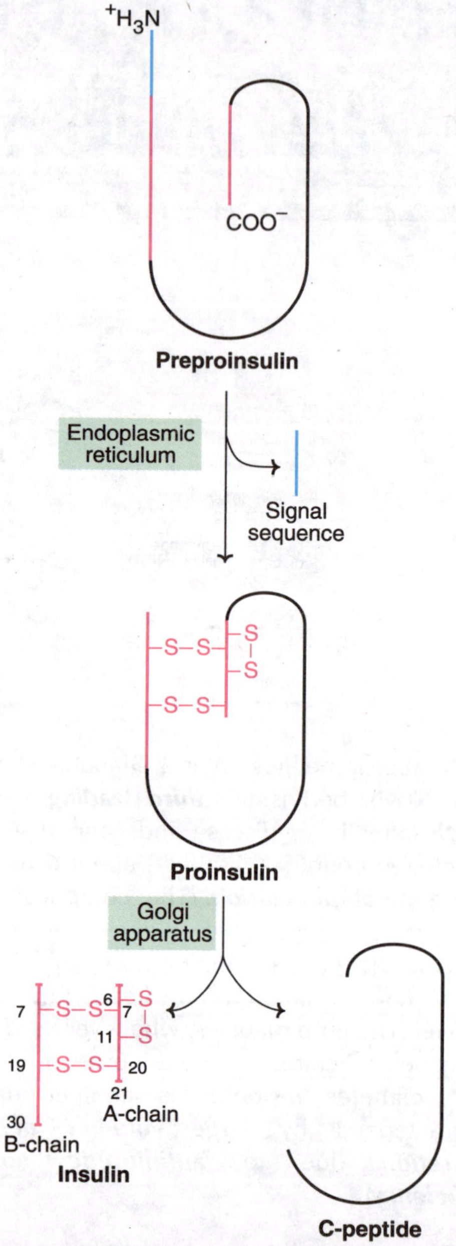

Fig. 36.1 : Formation of insulin from preproinsulin.

Regulation of insulin secretion

About 40-50 units of insulin is secreted daily by human pancreas. The ***normal insulin*** concentration in ***plasma*** is ***2-20 μU/ml***. The important factors that influence the release of insulin from the β-cells of pancreas are discussed hereunder.

1. **Factors stimulating insulin secretion :** These include glucose, amino acids and gastrointestinal hormones.

- **Glucose** is the ***most important stimulus for insulin release***. The effect is more predominant when glucose is administered orally (either direct or through a carbohydrate-rich meal). A rise in blood glucose level is a signal for insulin secretion.
- **Amino acids** induce the secretion of insulin. This is particularly observed after the ingestion of protein-rich meal that causes transient rise in plasma amino acid concentration. Among the amino acids, ***arginine and leucine are potent stimulators of insulin release***.

Table 36.1 Metabolic effects of insulin—a summary

Metabolism	Net effect	Effect on important enzyme(s)
Carbohydrate metabolism		
1. Glycolysis	Increased	Glucokinase ↑ Phosphofructokinase ↑ Pyruvate kinase ↑
2. Gluconeogenesis	Decreased	Pyruvate carboxylase ↓ Phosphoenol pyruvate carboxykinase ↓ Glucose 6-phosphatase ↓
3. Glycogenesis	Increased	Glycogen synthetase ↑
4. Glycogenolysis	Decreased	Glycogen phosphorylase ↓
5. HMP shunt	Increased	Glucose 6- phosphate dehydrogenase ↑
Lipid metabolism		
6. Lipogenesis	Increased	Acetyl CoA carboxylase ↑
7. Lipolysis	Decreased	Hormone sensitive lipase ↓
8. Ketogenesis	Decreased	HMG CoA synthetase ↓
Protein metabolism		
9. Protein synthesis	Increased	RNA polymerase ↑
10. Protein degradation	Decreased	Transaminases ↓ Deaminases ↓

- **Gastrointestinal hormones** (secretin, gastrin, pancreozymin) enhance the secretion of insulin. The GIT hormones are released after the ingestion of food.

2. **Factors inhibiting insulin secretion :** Epinephrine is the most potent inhibitor of insulin release. In emergency situations like stress, extreme exercise and trauma, the nervous system stimulates adrenal medulla to release epinephrine. ***Epinephrine*** suppresses insulin release and promotes energy metabolism by mobilizing energy-yielding compounds—glucose from liver and fatty acids from adipose tissue.

Degradation of insulin

In the plasma, insulin has a normal ***half-life*** of ***4-5 minutes***. This short half-life permits rapid metabolic changes in accordance to the alterations in the circulating levels of insulin. This is advantageous for the therapeutic purposes. A protease enzyme, namely ***insulinase*** (mainly found in liver and kidney), ***degrades insulin***.

Metabolic effects of insulin

Insulin plays a key role in the regulation of carbohydrate, lipid and protein metabolisms (***Table 36.1***). Insulin exerts ***anabolic*** and ***anticatabolic influences*** on the body metabolism.

1. **Effects on carbohydrate metabolism :** Insulin influences glucose metabolism in many ways. The net effect is that insulin ***lowers blood glucose level (hypoglycemic effect)*** by promoting its utilization and storage and by inhibiting its production.

- **Effect on glucose uptake by tissues :** ***Insulin*** is ***required*** for the uptake of glucose ***by muscle*** (skeletal, cardiac and smooth), ***adipose tissue***, leukocytes and mammary glands. Surprisingly, about 80% of glucose uptake in the body is not dependent on insulin. Tissues into which glucose can freely enter include brain, kidney, erythrocytes, retina, nerve, blood vessels and intestinal mucosa. As regards liver, glucose entry into hepatocytes does not require insulin. However, insulin stimulates glucose utilization in liver and, thus, indirectly promotes its uptake.

- **Effect on glucose utilization :** Insulin ***increases glycolysis*** in muscle and liver. The activation as well as the quantities of certain key enzymes of glycolysis, namely glucokinase (not hexokinase) phosphofructokinase and pyruvate kinase are increased by insulin.
- **Effect on glucose production :** Insulin ***decreases gluconeogenesis*** by suppressing the enzymes pyruvate carboxylase, phosphoenol pyruvate carboxykinase and glucose 6-phosphatase.

2. **Effects on lipid metabolism :** The net effect of insulin on lipid metabolism is to ***reduce the release of fatty acids from the stored fat and decrease the production of ketone bodies***. Among the tissues, adipose tissue is the most sensitive to the action of insulin.

- **Effect on lipogenesis :** Insulin ***favours*** the ***synthesis of triacylglycerols*** from glucose by providing more glycerol 3-phosphate (from glycolysis) and NADPH (from HMP shunt).
- **Effect on lipolysis :** Insulin decreases the activity of hormone-sensitive lipase and thus reduces the release of fatty acids from stored fat in adipose tissue. The mobilization of fatty acids from liver is also decreased by insulin. In this way, insulin keeps the circulating free fatty acids under a constant check.
- **Effect on ketogenesis :** Insulin ***reduces ketogenesis*** by decreasing the activity of HMG CoA synthetase.

3. **Effects on protein metabolism :** Insulin is an anabolic hormone. It stimulates the entry of amino acids into the cells, ***enhances protein synthesis*** and reduces protein degradation.

Besides the metabolic effects described above, ***insulin promotes cell growth and replication***. This is mediated through certain factors such as epidermal growth factor (EGF), platelet derived growth factor (PDGF) and prostaglandins.

Mechanism of action of insulin

It is now recognized that ***insulin binds to*** specific plasma ***membrane receptors*** present on the target tissues, such as muscle and adipose. This results in a series of reactions ultimately leading to the biological action. Three distinct mechanisms of insulin action are known. One concerned with the induction of transmembrane signals (signal transduction), second with the glucose transport across the membrane and the third with induction of enzyme synthesis.

1. **Insulin receptor mediated signal transduction**

Insulin receptor : It is a ***tetramer*** consisting of 4 subunits of two types and is ***designated as*** $\alpha_2\beta_2$. The subunits are in the glycosylated form. They are held together by disulfide linkages. The α-subunit (mol. wt. 135,000) is extracellular and it contains insulin binding site. The β-subunit (mol. wt. 95,000) is a transmembrane protein which is activated by insulin. The cytoplasmic domain of ***β-subunit has tyrosine kinase activity***.

Signal transduction : As the hormone insulin binds to the receptor, a conformational change is induced in the α-subunits of insulin receptor. This results in the generation of signals which are transduced to β-subunits. The net effect is that ***insulin*** binding ***activates tyrosine kinase activity*** of intracellular β-subunit of insulin receptor. This causes the autophosphorylation of tyrosine residues on β-subunit. It is believed that receptor ***tyrosine kinase also phosphorylates insulin receptor substrate (IRS)***. The phosphorylated IRS, in turn, promotes activation of other protein kinases and phosphatases, finally leading to biological action (***Fig.36.2***).

2. **Insulin-mediated glucose transport :** The binding of insulin to insulin receptors signals the translocation of vesicles containing glucose transporters from intracellular pool to the plasma membrane. The vesicles fuse with the membrane recruiting the glucose transporters. The ***glucose transporters are responsible for the insulin–mediated uptake of glucose by the cells***. As the insulin level falls, the glucose transporters move away from the membrane to the intracellular pool for storage and recycle (***Fig.36.3***).

3. **Insulin mediated enzyme synthesis :** Insulin promotes the synthesis of enzymes such as glucokinase, phosphofructokinase and pyruvate kinase. This is brought about by

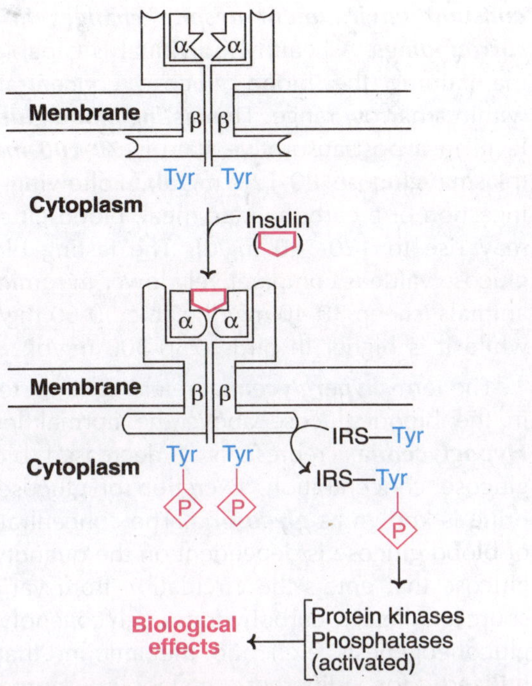

Fig. 36.2 : Insulin receptor mediated signal transduction (IRS—Insulin receptor substrate).

increased transcription (mRNA synthesis), followed by translation (protein synthesis).

GLUCAGON

Glucagon, ***secreted by α-cells*** of the pancreas, opposes the actions of insulin. It is a polypeptide hormone composed of 29 amino acids (mol. wt. 3,500) in a single chain. Glucagon is actually synthesized as proglucagon (mol. wt. 9,000) which on sequential degradation releases active glucagon. Unlike insulin, the ***amino acid sequence*** of glucagon is the ***same in all mammalian species*** (so far studied). Glucagon has a short half-life in plasma i.e. about 5 minutes.

Regulation of glucagon secretion

The secretion of glucagon is ***stimulated by low blood glucose*** concentration, amino acids derived from dietary protein and epinephrine. ***Increased blood glucose*** level markedly ***inhibits*** glucagon secretion.

Metabolic effects of glucagon

Glucagon influences carbohydrate, lipid and protein metabolisms. In general, the effects of this hormone ***oppose that of insulin.***

1. **Effects on carbohydrate metabolism :** Glucagon is the most potent hormone that

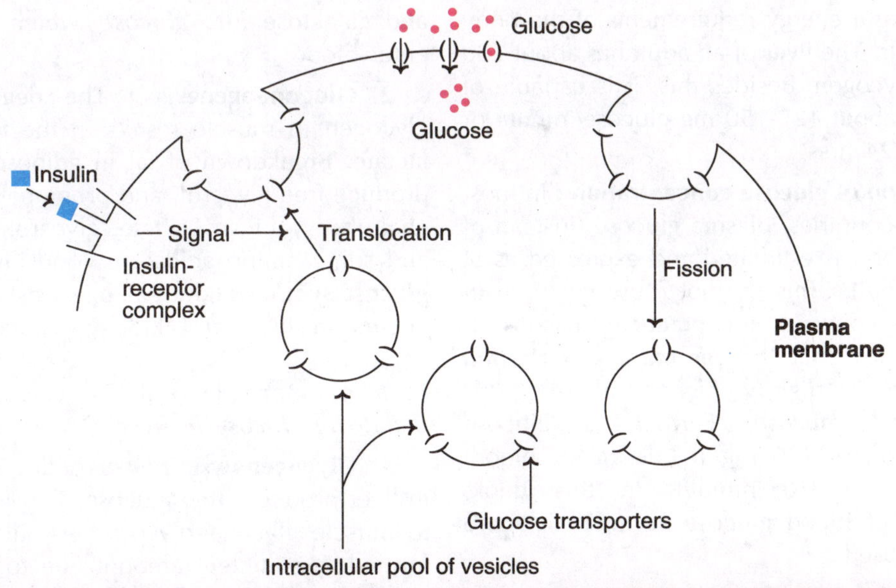

Fig. 36.3 : Insulin mediated glucose transport.

enhances the blood glucose level (***hyperglycemic***). Primarily, glucagon acts on liver to cause increased synthesis of glucose (gluconeogenesis) and enhanced degradation of glycogen (glycogenolysis). The actions of glucagon are mediated through cyclic AMP (***Chapter 13***).

2. **Effects on lipid metabolism :** Glucagon ***promotes fatty acid oxidation*** resulting in energy production and ketone body synthesis (ketogenesis).

3. **Effects on protein metabolism :** Glucagon increases the amino acid uptake by liver which, in turn, promotes gluconeogenesis. Thus, glucagon lowers plasma amino acids.

Mechanism of action of glucagon

Glucagon binds to the specific receptors on the plasma membrane and acts through the mediation of cyclic AMP, the second messenger. The details are given elsewhere (***Chapter 19***).

REGULATION OF BLOOD GLUCOSE LEVEL (HOMEOSTASIS OF BLOOD GLUCOSE)

Glucose is ***carbohydrate currency*** of the body. An adult human body contains about 18 g free glucose. This amount is just sufficient to meet the basal energy requirements of the body for one hour! The liver of an adult has about 100 g stored glycogen. Besides this, it is capable of producing about 125-150 mg glucose/minute or 180-220 g/24 hrs.

Expression of glucose concentration : In most developed countries, plasma glucose (instead of blood glucose) is estimated and expressed as SI units (mmol/l). This is not however so, in developing countries for practical reasons. It may be noted that the ***plasma concentration of glucose is slightly higher (about 15%) than blood glucose***. Further, a glucose concentration of 180 mg/dl (plasma or blood) corresponds to 10 mmol/l. In this book, expression of blood glucose as mg/dl is more frequently used.

The term ***homeostasis***, in general, refers to the ***livings organisms ability to maintain a constant environment despite changes in the surroundings***. A healthy individual is capable of maintaining the blood glucose concentration within a narrow range. The ***fasting blood glucose*** level in a post-absorptive state is ***70-100 mg/dl*** (plasma glucose 80-120 mg/dl). Following the ingestion of a carbohydrate meal, blood glucose may rise to 120-140 mg/dl. The fasting blood glucose value is comparatively lower in ruminant animals (sheep 30-40 mg/dl; cattle 50-60 mg/dl), while it is higher in birds (250-300 mg/dl).

The term ***hyperglycemia*** refers to an increase in the blood glucose above the normal level. ***Hypoglycemia*** represents a decreased blood glucose concentration. Excretion of glucose in urine is known as ***glycosuria.*** The concentration of blood glucose is dependent on the quantity of glucose that enters the circulation from various sources (dietary carbohydrates, glycogenolysis, gluconeogenesis etc.) and the amount that is utilized for different metabolic purposes (glycolysis, glycogenesis, fat synthesis etc.) as illustrated in ***Fig.36.4***.

Sources of blood glucose

1. **Dietary sources :** The dietary carbohydrates are digested and ***absorbed as monosaccharides*** (glucose, fructose, galactose etc.). The liver is capable of ***converting*** fructose and galactose ***into glucose***, which can readily enter blood.

2. **Gluconeogenesis :** The degradation of glycogen in muscle results in the formation of lactate. Breakdown of fat in adipose tissue will produce free glycerol, and propionate (from odd chain fatty acids). Lactate, glycerol, propionate and some amino acids are good precursors for ***glucose synthesis*** (gluco-neogenesis) that actively occurs in liver and kidney. Gluconeogenesis continuously adds glucose to the blood. ***Cori cycle*** is responsible for the conversion of ***muscle lactate to glucose in liver***.

3. **Glycogenolysis :** Degradation of glycogen in liver produces free glucose. This is in contrast to muscle glycogenolysis where glucose is not formed in sufficient amount due to lack of the enzyme glucose 6-phosphatase. However, the contribution of liver glycogenolysis to blood

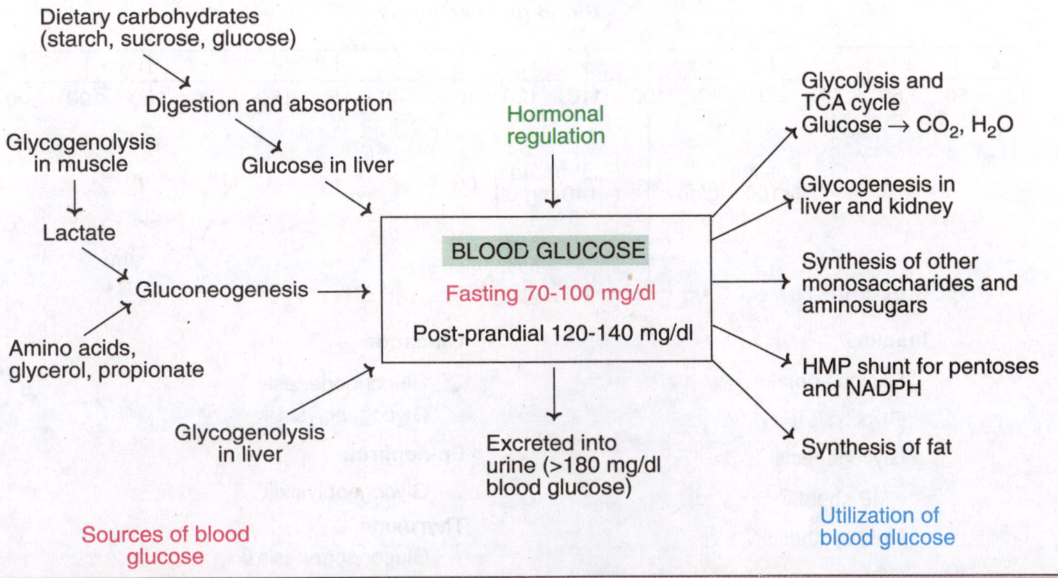

Fig. 36.4 : Overview of blood glucose homeostasis.

glucose is rather limited and can meet only the short intervals of emergency. This is due to the limited presence of glycogen in liver. An adult liver (weighing about 1.5 kg) can provide only 40-50 g of blood glucose from glycogen, that can last only for a few hours to meet the body requirements.

In the ***Fig.36.5***, the sources of blood glucose during a normal day (24 hours) are given. Glucose is primarily derived from ***glycogenolysis*** (of hepatic glycogen) ***between the meals. Gluconeogenesis*** becomes a predominant source of glucose in ***late night*** (after depletion of hepatic glycogen). During day time, gluconeogenesis may be more or less active, depending on the frequency of consumption of snacks, coffee, tea, fruit juices etc.

Utilization of blood glucose

Certain tissues like brain, erythrocytes, renal medulla and bone marrow are exclusively dependent on glucose for their energy needs. When the body is ***at total rest***, about ***two-thirds*** of the blood glucose is utilized ***by the brain***. The remaining one-third by RBC and skeletal muscle. A regular supply of glucose, by whatever means it may be, is absolutely required to keep the brain functionally intact.

The different metabolic pathways (glycolysis, glycogenesis, HMP shunt etc.) responsible for the utilization of blood glucose are already discussed (***Chapter 13***). The synthesis of fat from acetyl CoA and glycerol is described in lipid metabolism (***Chapter 14***).

Kidney plays a special role in the homeostasis of blood glucose. Glucose is continuously filtered by the glomeruli, reabsorbed and returned to the blood. If the level of glucose in

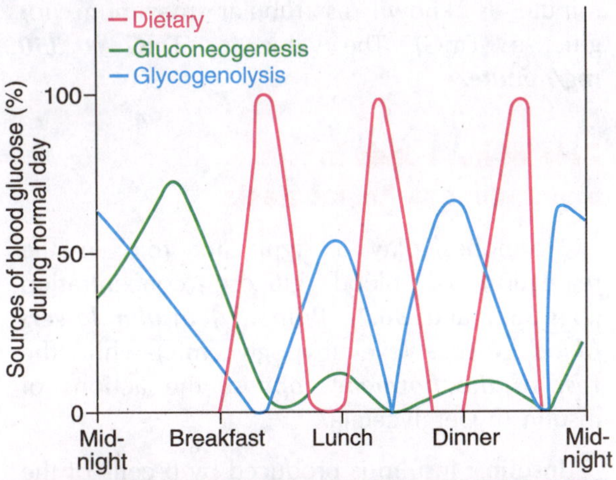

Fig. 36.5 : Sources of blood glucose during a normal day (24 hours).

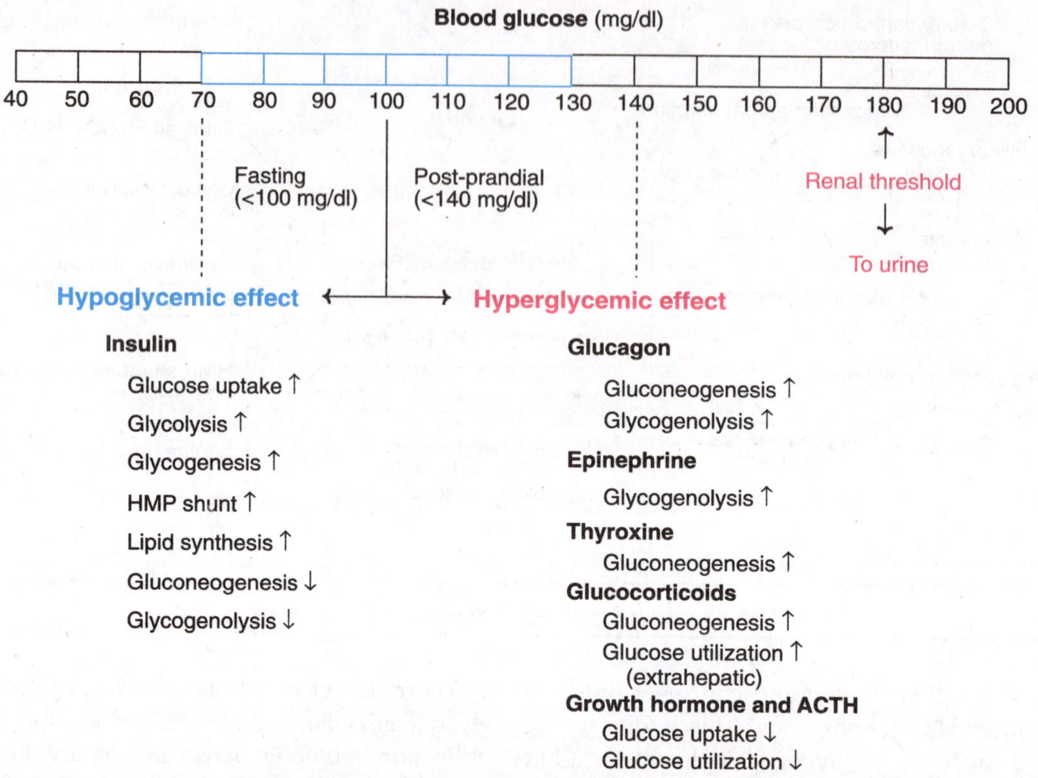

Fig. 36.6 : *Hormonal regulation of blood glucose.*

blood is above 160-180 mg/dl, glucose is excreted in urine (glycosuria). This value (160-180 mg/dl) is referred to as ***renal threshold for glucose.*** The maximum ability of the renal tubules to reabsorb glucose per minute is known as tubular maximum for glucose (TmG). The value for ***TmG is 350 mg/minute.***

Role of hormones in blood glucose homeostasis

Hormones play a significant role in the regulation of blood glucose concentration (***Figs.36.6*** and ***36.7***). Primarily, ***insulin lowers blood glucose*** level (hypoglycemic) while the ***rest of the hormones oppose*** the actions of insulin (hyperglycemia).

Insulin : Insulin is produced by β-cells of the islets of Langerhans in response to hyperglycemia (elevated blood glucose level). Some amino acids, free fatty acids, ketone bodies, drugs such as tolbutamide also cause the secretion of insulin.

Insulin is basically a hypoglycemic hormone that ***lowers blood glucose level*** through various means. It is an anti-diabetogenic hormone. For details of insulin action on glucose homeostasis refer metabolic effects of insulin (carbohydrate metabolism) in this chapter.

Incretins : These are a group of hormones that ***increase the release of insulin*** (from β-cells of pancreas) and thus decrease the blood glucose levels. Further, ***incretins also inhibit glucagon release*** from α-cells of pancreas. The two best examples of incretins are intestinal ***glucagon-like peptide I*** (GLP-I) and ***gastric inhibitory peptide*** (GIP).

Glucagon : Glucagon is synthesized by α-cells of the islets of Langerhans of the pancreas. Hypoglycemia (low blood glucose level) stimulates its production. Glucagon is basically involved in ***elevating blood glucose*** concentration. It enhances gluconeogenesis and glycogenolysis.

Epinephrine : This hormone is secreted by adrenal medulla. It acts both on muscle and liver to bring about glycogenolysis by increasing phosphorylase activity. The end product is glucose in liver and lactate in muscle. The net outcome is that epinephrine ***increases blood glucose level***.

Thyroxine : It is a hormone of thyroid gland. It ***elevates blood glucose*** level by stimulating hepatic glycogenolysis and gluconeogenesis.

Glucocorticoids : These hormones are produced by adrenal cortex. Glucocorticoids stimulate protein metabolism and increase gluconeogenesis (increase the activities of enzymes—glucose 6-phosphatase and fructose 1,6-bisphosphatase). The glucose utilization by extrahepatic tissues is inhibited by glucocorticoids. The overall effect of glucocorticoids is to ***elevate blood glucose concentration***.

Growth hormone and adrenocorticotropic hormone (ACTH) : The anterior pituitary gland secretes growth hormone and ACTH. The uptake of glucose by certain tissues (muscle, adipose tissue etc.) is decreased by growth hormone. ACTH decreases glucose utilization. The net effect of both these hormones is ***hyperglycemic***.

[In ***Fig.36.7***, regulation of blood glucose level by hormones is depicted as a game of tug of war with elephant (representing insulin) on one side and the other animals (as rest of the hormones) on the opposite side. This is just an illustration (a cartoon) for a quick understanding of glucose homeostasis].

HYPOGLYCEMIA

When the blood glucose concentration falls to ***less than 45 mg/dl***, the symptoms of hypoglycemia appear. The manifestations include ***headache, anxiety, confusion, sweating, slurred speech, seizures and coma***, and, if not corrected, death. All these symptoms are directly and indirectly related to the deprivation of glucose supply to the central nervous system (particularly the brain) due to a fall in blood glucose level.

The mammalian body has developed a well regulated system for an efficient maintenance of blood glucose concentration (details already described). Hypoglycemia, therefore, is not commonly observed. The following three types of hypoglycemia are encountered by physicians.

1. **Post-prandial hypoglycemia :** This is also called ***reactive hypoglycemia*** and is observed in subjects with an elevated insulin secretion following a meal. This causes transient hypoglycemia and is associated with mild symptoms. The patient is advised to eat frequently rather than the 3 usual meals.

2. **Fasting hypoglycemia :** Low blood glucose concentration in fasting is not very common. However, fasting hypoglycemia is observed in patients with pancreatic β-cell tumor and hepatocellular damage.

3. **Hypoglycemia due to alcohol intake :** In some individuals who are starved or engaged in prolonged exercise, alcohol consumption may cause hypoglycemia. This is due to the accumulation of NADH (during the course of alcohol metabolism by alcohol dehydrogenase) which diverts the pyruvate and oxaloacetate (substrates of gluconeogenesis) to form, respectively, lactate and malate. The net effect is that ***gluconeogenesis is reduced due to alcohol consumption***.

4. **Hypoglycemia due to insulin overdose :** The most common complication of ***insulin therapy in diabetic patients is hypoglycemia***. This is particularly observed in patients who are on intensive treatment regime.

5. **Hypoglycemia in pregnancy and in neonates :** During pregnancy, maternal and fetal glucose consumption are higher. Hence, pregnant women are at risk of hypoglycemia in case of long intervals between meals and at nights. ***Premature and low-birth weight babies are susceptible to hypoglycemia due to lower liver stores of glycogen***, and less developed enzyme machinery gluconeogenesis.

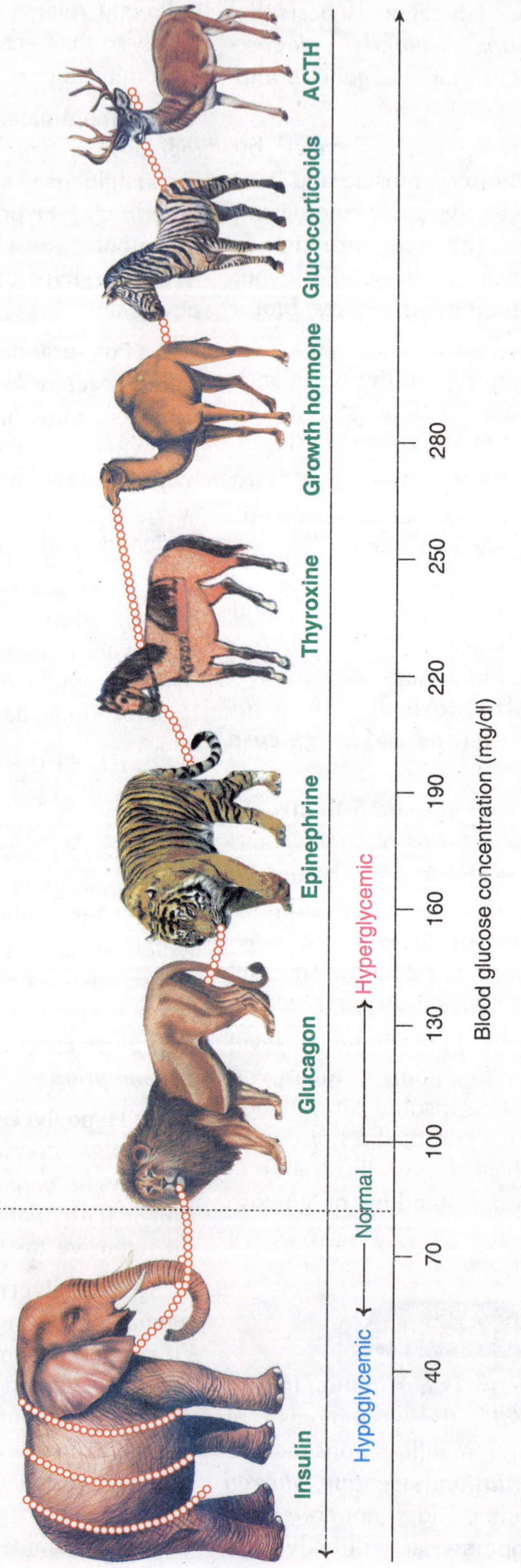

Fig. 36.7 : *A cartoon of tug of war illustrating hormonal action on blood glucose regulation.*

Medical Concepts/ Clinical Correlates

- Diabetes affects about 6-8% of the population and is a major cause of blindness, renal failure, heart attack and stroke. The hormone insulin has been implicated in the development of diabetes.
- Hypoglycemia (blood glucose <45 mg/dl) is a serious condition that may lead to coma and death, if not corrected.
- GTT is commonly used for confirmation of diabetes mellitus.
- Insulinoma, a rare tumor of β-cells of pancreas, leads to unregulated insulin secretion and hypoglycemia.

Whipple's triad-diagnosis of hypoglycemia

Whipple's triad refers to the collection of three criteria called ***Whipple's criteria*** to diagnose hypoglycemia

1. Symptoms of hypoglycemia
2. Low plasma glucose level at the time of symptoms.
3. Relief of symptoms on administration of glucose.

Whipple's criteria is used by endocrinologists to separate ***true hypoglycemia*** from other conditions e.g. idiopathic postprandial syndrome.

CLASSIFICATION OF DIABETES MELLITUS

Diabetes mellitus is a ***metabolic disease***, more appropriately a disorder of fuel metabolism. It is mainly characterized by ***hyperglycemia*** that leads to several long term complications.

Diabetes mellitus is broadly divided into 2 groups, namely ***insulin-dependent diabetes mellitus (IDDM) and non-insulin dependent diabetes mellitus (NIDDM)***. This classification is mainly based on the requirement of insulin for treatment.

Insulin-dependent diabetes mellitus (IDDM)

IDDM, also known as ***type 1 diabetes*** or (less frequently) ***juvenile onset diabetes,*** mainly occurs in childhood (particularly between 12-15 yrs age). IDDM accounts for about 5-10% of the known diabetics. This disease is characterized by almost total deficiency of insulin due to destruction of β-cells of pancreas. The β-cell destruction may be caused by drugs, viruses or autoimmunity. Due to certain genetic or environmental variations, the β-cells are recognized as non-self and they are destroyed by immune mediated injury. Usually, the symptoms of diabetes appear when 80-90% of the β-cells have been destroyed. The pancreas ultimately fails to secrete insulin in response to glucose ingestion. The patients of IDDM require insulin therapy.

Non-insulin dependent diabetes mellitus (NIDDM)

NIDDM, also called ***type 2 diabetes*** or (less frequently) ***adult-onset diabetes,*** is the most common, accounting for 90-95% of the diabetic population. NIDDM occurs in adults (usually above 35 years) and is less severe than IDDM. Type II diabetes is mainly due to ***insulin resistance***. This is characterized by decreased response of target cells (skeletal muscle) to insulin, caused by reduction in glucose transporter 4 (GLUT-4). ***Obesity acts as a diabetogenic factor and leads to a decrease in insulin receptors on*** the insulin responsive (target) cells. The patients of NIDDM may have either normal or even increased insulin levels. Many a times weight reduction by diet control alone is often sufficient to correct NIDDM.

Recent research findings on NIDDM suggest that increased levels of tumor necrosis factor-α (TNF-α) and resistin, and reduced secretion of adiponectin by adipocytes of obese people cause insulin resistance.

The comparison between IDDM and NIDDM is given in ***Table 36.2***. (For ***metabolic syndrome refer page no. 517***)

Table 36.2 Comparison of two types of diabetes mellitus

Character	Insulin-dependent diabetes mellitus (IDDM)	Non-insulin dependent diabetes mellitus (NIDDM)
General		
Prevalence	5-10% of diabetic population	90-95% of diabetic population
Age at onset	Usually childhood (<20 yrs)	Predominantly in adults (>30yrs)
Body weight	Normal or low	Obese
Genetic predisposition	Mild or moderate	Very strong
Biochemical		
Defect	Insulin deficiency due to destruction of β-cells	Impaired production of insulin and/or resistance of target cells
Plasma insulin	Decreased or absent	Normal or increased
Auto antibodies	Frequently found	Rare
Ketosis	Very common	Rare
Clinical		
Duration of symptoms	Weeks	Months to years
Diabetic complications at diagnosis	Rare	Found in 10-20% cases
Oral hypoglycemic drugs	Not useful for treatment	Suitable for treatment
Administration of insulin	Always required	Usually not necessary

GLUCOSE TOLERANCE TEST (GTT)

The oral glucose tolerance test (OGTT) is important for the diagnosis diabetes. ***OGTT*** is normally done for subjects, with doubtful ***diagnosis of diabetes for confirmation, and identification of gestational diabetes***.

Preparation of the subject for GTT

The person should have been taking ***carbohydrate-rich diet for at least 3 days prior to the test***. All drugs known to influence carbohydrate metabolism should be discontinued (for at least 2 days). The subject should avoid strenuous exercise on the previous day of the test. He/she should be in an ***overnight*** (at least 10 hr) ***fasting state***. During the course of GTT, the person should be comfortably seated and should refrain from smoking and exercise.

Procedure for GTT

Glucose tolerance test should be conducted ***preferably in the morning*** (ideal 9 to 11 AM). A fasting blood sample is drawn and urine collected. The subject is given 75 g glucose orally, dissolved in about 300 ml of water, to be drunk in about 5 minutes. Blood and urine samples are collected at 30 minute intervals for $2^1/_2$ hours. All blood samples are subjected to glucose estimation while urine samples are qualitatively tested for glucose.

Interpretation of GTT

The graphic representation of the GTT results is depicted in ***Fig.36.8***. The fasting plasma glucose level is 75–110 mg/dl in normal persons. On oral glucose load, the concentration increases and the peak value (140 mg/dl) is reached in less than an hour which returns to normal by 2 hours. Glucose is not detected in any of the urine samples.

In individuals with ***impaired glucose tolerance***, the fasting (110-125 mg/dl) as well as 2 hour (140-199 mg/dl) plasma glucose levels are elevated. These subjects slowly develop frank

Table 36.3 Diagnostic criteria for oral glucose tolerance test (WHO criteria)

Condition	Plasma glucose concentration as mmol/l (mg/dl)		
	Normal	Impaired glucose tolerance	Diabetes
Fasting	< 6.1 (< 110)	< 7.0 (< 126)	≥ 7.0 (≥ 126)
2 hours after glucose	< 7.8 (< 140)	< 11.1 (< 200)	≥ 11.1 (≥ 200)

diabetes at an estimated rate of 2% per year. Dietary restriction and exercise are advocated for the treatment of impaired glucose tolerance.

The WHO criteria for the diagnosis of diabetes by OGTT is presented in ***Table 36.3***. A person is said to be suffering from diabetes mellitus if his/her ***fasting plasma glucose exceeds*** 7.0 mmol/l (***126 mg/dl***) and, ***at 2 hrs.*** 11.1 mmol/l (***200 mg/dl***).

Other relevant aspects of GTT

1. For conducting GTT in children, oral glucose is given on the basis of weight (1.5 to 1.75 g/kg).

2. In case of pregnant women, 100 g oral glucose is recommended. Further, the diagnostic criteria for diabetes in pregnancy should be more stringent than WHO recommendations.

3. In the ***mini GTT***, carried out in some laboratories, fasting and 2 hrs. sample (instead of $^1/_2$ hr. intervals) of blood and urine are collected.

4. The GTT is rather unphysiological. To evaluate the glucose handling of the body under physiological conditions, fasting blood sample is drawn, the subject is allowed to take ***heavy breakfast***, blood samples are collected at ***1 hour*** and ***2 hrs*** (***post-prandial***—meaning after food). Urine samples are also collected. This type of test is commonly employed in established diabetic patients for monitoring the control.

5. For individuals with suspected malabsorption, ***intravenous GTT is carried out.***

6. **Corticosteroid stressed GTT** is employed to detect latent diabetes.

Glycosuria (glucosuria)

The commonest cause of ***glucose excretion in urine*** (glycosuria) is ***diabetes mellitus***. Therefore, glycosuria is the first line screening test for diabetes. Normally, glucose does not appear in urine until the plasma glucose concentration exceeds ***renal threshold (180 mg/dl)***. As age advances, renal threshold for glucose increases marginally.

Renal glycosuria : Renal glycosuria is a benign condition due to a ***reduced renal threshold for glucose***. It is unrelated to diabetes and, therefore, should not be mistaken as

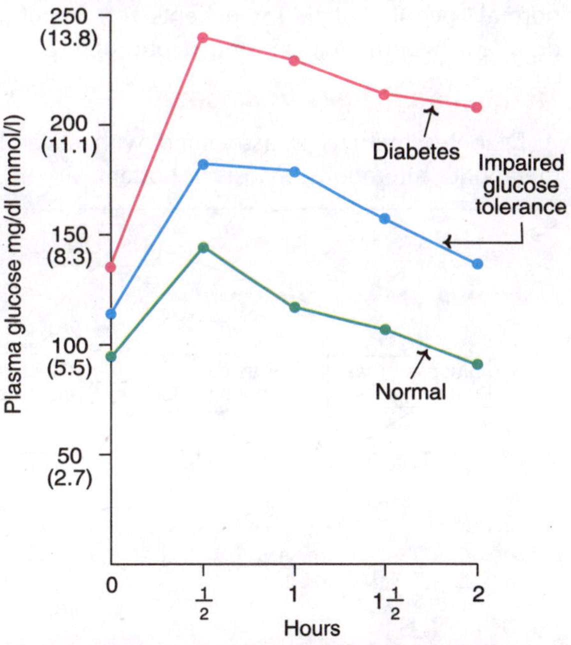

Fig. 36.8 : *Oral glucose tolerance test.*

Medical Concepts/ Clinical Correlates

- Diabetic ketoacidosis is frequently encountered in severe uncontrolled diabetics. The management includes administration of insulin, fluids and potassium.
- The oral antidiabetic drugs commonly used in diabetic patients include glibenclamide, glimepiride, and gliclazide.
- The long term complications of uncontrolled diabetes include atherosclerosis, retinopathy, nephropathy and neuropathy.
- Measurement of glycated hemoglobin (HbA_{1c}) serves as a marker for diabetic control.

diabetes. Further, it is not accompanied by the classical symptoms of diabetes.

Alimentary glycosuria : In certain individuals, blood glucose level rises rapidly after meals resulting in its spill over into urine. This condition is referred to as ***alimentary (lag storage) glycosuria.*** It is observed in some normal people, and in patients of hepatic diseases, hyperthyroidism and peptic ulcer.

Metabolic changes in diabetes

Diabetes mellitus is associated with several metabolic alterations. Most important among them are hyperglycemia, ketoacidosis and hypertriglyceridemia (***Fig.36.9***).

1. **Hyperglycemia :** Elevation of blood glucose concentration is the hallmark of uncontrolled diabetes. Hyperglycemia is primarily due to reduced glucose uptake by tissues and its increased production via gluconeogenesis and glycogenolysis. When the blood glucose level goes beyond the renal threshold, glucose is excreted into urine (glycosuria).

Glucose toxicity : High concentrations of glucose can be harmful causing ***osmotic effects/*** hypertonic effects (water drawn from cells into extracellular fluid and excreted into urine, resulting in dehydration), ***β-cell damage by free radicals*** (due to enhanced oxidative phosphorylation, oxidative stress, and increased free radicals) and ***glycation of proteins*** (associated with diabetic complications-neuropathy, nephropathy, retinopathy etc.).

2. **Ketoacidosis :** Increased mobilization of fatty acids results in overproduction of ketone bodies which often leads to ketoacidosis.

3. **Hypertriglyceridemia :** Conversion of fatty acids to triacylglycerols and the secretion of VLDL and chylomicrons is comparatively higher in diabetics. Further, the activity of the enzyme lipoprotein lipase is low in diabetic patients. Consequently, the plasma levels of ***VLDL, chylomicrons*** and triacylglycerols are ***increased***. ***Hypercholesterolemia*** is also frequently seen in diabetics.

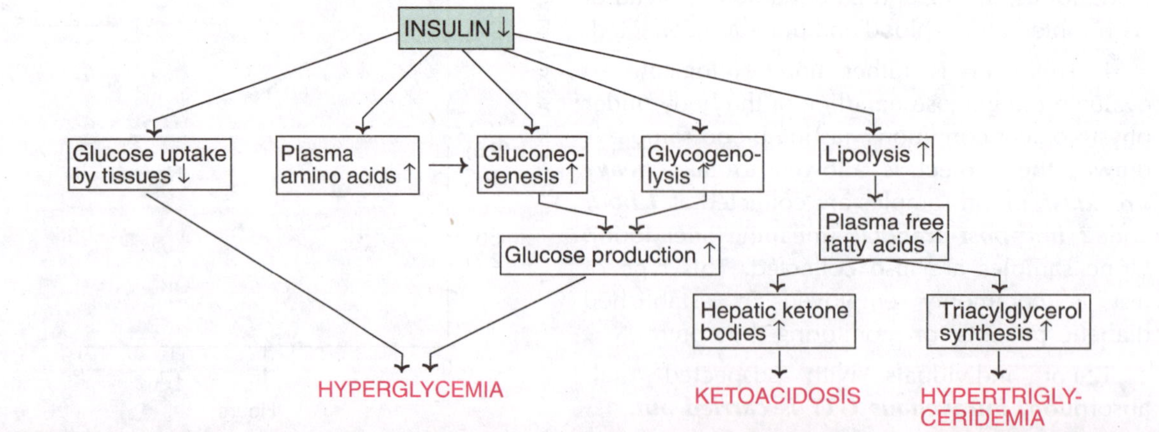

***Fig. 36.9 :** Major metabolic alterations in diabetes mellitus.*

Long term effects of diabetes

Hyperglycemia is directly or indirectly associated with several complications. These include ***atherosclerosis, retinopathy, nephropathy*** and ***neuropathy***. ***Diabetic retinopathy*** causes loss of transparency of the eye lens, resulting in impairment of vision and blindness. ***Diabetic neuropathy*** primarily affects and damages peripheral nervous system. ***Diabetic nephropathy*** progressively damages kidney and results in renal failure. The biochemical basis of these complications is not clearly understood. It is believed that at least some of them are related to microvascular changes caused by glycation of proteins, and elevated levels of sorbitol.

Management of diabetes

Diet, exercise, drug and, finally, insulin are the management options for diabetics. Approximately, 50% of the new cases of diabetes can be adequately controlled by diet alone, 20-30% need oral hypoglycemic drugs while the remaining 20-30% require insulin.

Dietary management : A diabetic patient is advised to consume ***low calories*** (i.e. low carbohydrate and fat), ***high protein and fiber rich diet***. Carbohydrates should be taken in the form of starches and complex sugars. As far as possible, ***refined sugars (sucrose, glucose) should be avoided***. Fat intake should be drastically reduced so as to meet the nutritional requirements of unsaturated fatty acids. ***Diet control and exercise*** will help to a large extent obese NIDDM patients.

Antidiabetic drugs : The oral antidiabetic drugs are broadly of two categories-***sulfonylureas*** and ***biguanides***.

Sulfonylureas such as acetohexamide, ***tolbutamide*** and ***glibenclamide*** promote secretion of endogenous insulin and help in reducing blood glucose level. Biguanides (***metformin***, ***phenoformin***) increase insulin sensitivity, resulting in enhanced glucose uptake, reduced gluconeogenesis, and thus lower blood glucose.

Management with insulin : Two types of insulin preparations are commercially available—***short acting*** and ***long acting.*** The short acting insulins are unmodified and their action lasts for about 6 hours. The long acting insulins are modified ones (such as adsorption to protamine) and act for several hours, which depends on the type of preparation.

The advent of ***genetic engineering*** is a boon to diabetic patients since ***bulk quantities*** of ***insulin*** can be produced in the laboratory.

Biochemical indices of diabetic control

For a diabetic patient who is on treatment (drug or insulin therapy), ***periodical assessment*** of the efficacy of the ***treatment is essential***. Urine glucose detection and blood glucose estimations are traditionally followed in several laboratories. In recent years, more reliable and long-term biochemical indices of diabetic control are in use.

Glycated hemoglobin : Glycated or ***glycosylated hemoglobin*** refers to the glucose derived products of normal adult hemoglobin (HbA). Glycation is a post-translational, non-enzymatic addition of sugar residue to amino acids of proteins. Among the glycated hemoglobins, the most abundant form is ***HbA_{1c}***. ***HbA_{1c}*** is produced by the condensation of glucose with N-terminal valine of each β-chain of HbA.

Diagnostic importance of HbA_{1c} : The rate of synthesis of HbA_{1c} is directly related to the exposure of RBC to glucose. Thus, the concentration of HbA_{1c} serves as an indication of the blood glucose concentration over a period, approximating to the half-life of RBC (hemoglobin) i.e. 6–8 weeks. A close correlation between blood glucose and HbA_{1c} concentrations has been observed when simultaneously monitored for several months.

Normally, ***HbA_{1c} concentration is about 4–5.6% of the total hemoglobin***. In diabetic patients, HbA_{1c} is elevated (to as high as 15%). Determination of HbA_{1c} is used for monitoring of diabetes control. HbA_{1c} reflects the mean ***blood glucose level over 3 months period prior to its measurement***.

In the routine clinical practice, if the HbA_{1c} concentration is less than 7%, the diabetic patient is considered to be in good control.

Estimated average glucose(eAG) : eAG is a term (introduced by American Diabetic Association) used in diabetic management. It is a laboratory tool to understand approximate ***relationship between HbA_{1c} and glucose concentrations***, and is given by the following formula

$$\text{eAG (mg/dl)} = (28.7 \times HbA_{1c}) - 46.7$$

(e.g. for HBA_{1c} values of 6%, 8% and 10%, the eAG values respectively are 126 mg/dl, 183 mg/dl and 240 mg/dl)

Fructosamine : Glycated serum proteins (fructosamine) can also be measured in diabetics. As albumin is the most abundant plasma protein, glycated albumin largely contributes to plasma fructosamine measurements. Albumin has shorter half-life than Hb. Thus, glycated albumin represents glucose status over 3 weeks prior to its determination.

Microalbuminuria : Microalbuminuria is defined as the excretion of ***30-300 mg of albumin in urine per day***. It may be noted that microalbuminuria represents an intermediary stage between normal albumin excretion (2.5–30 mg/d) and macroalbuminuria (> 300 mg/d). The small increase in albumin excretion ***predicts impairment in renal function*** in diabetic patients. Microalbuminuria serves ***as a signal of early reversible renal damage***.

Collection of 24 hours urine for albumin estimation is cumbersome. ***Albumin creatinine ratio (ACR)*** in spot urine samples is an alternative. Reference values of ACR: Normal-30 µg/mg creatinine; Albuminuria->300 µg/mg creatinine.

SUMMARY

- *Diabetes mellitus is a common metabolic disorder, characterized by insufficient or inefficient insulin.*
- *In a healthy person, the blood glucose level (fasting 70-110 mg/dl) is maintained by a well coordinated hormonal action regulating the sources that contribute to glucose, and the utilization pathways. Insulin is hypoglycemic while other hormones (glucagon, epinephrine, thyroxine, glucocorticoids) are hyperglycemic.*
- *Diabetes mellitus is broadly classified into 2 categories—insulin dependent diabetes mellitus (IDDM) and non-insulin dependent diabetes mellitus (NIDDM).*
- *The confirmatory diagnosis of diabetes is carried out by oral glucose tolerance test (GTT). As per WHO criteria, a person is said to be suffering from diabetes if his/her fasting blood glucose exceeds 126 mg/dl, and 2 hrs. after OGTT goes beyond 200 mg/dl.*
- *Diabetes is associated with several metabolic derangements such as ketoacidosis and hypertriglyceridemia, besides hyperglycemia. The chronic complications of diabetes include atherosclerosis, retinopathy, nephropathy and neuropathy.*
- *Estimation of glycated hemoglobin (HbA_{1C}), plasma fructosamine, microalbumin in urine, and serum lipids serve as biochemical indices to monitor diabetic control.*

THE READER MUST REFER APPENDIX VI (Page Nos : 738-756) FOR RELATED CLINICAL CASE STUDIES (Nos : 52, 53, 54, 55, 56, 57, 58, 62, 73, 74, 75, 76)

Cancer

37

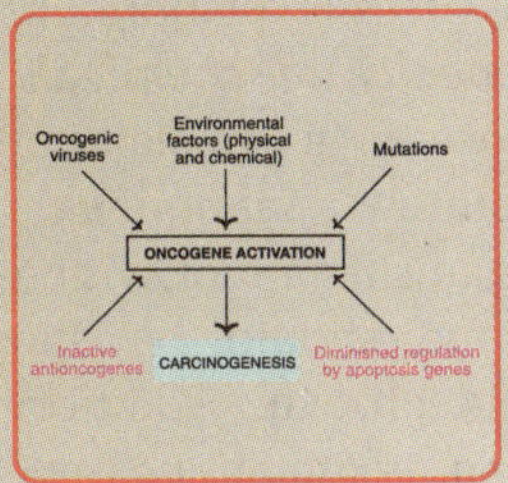

The dreaded disease cancer speaks

"I am the world's second largest killer;
Characterized by uncontrolled proliferation of cells;
Implicating environmental and genetic factors;
Perhaps due to oncogene and antioncogene imbalance."

In the normal circumstances, the proliferation of body cells is under strict control. The cells differentiate, divide and die in a sequential manner in a healthy organism. Cancer is characterized by ***loss of control of cellular growth*** and development leading to ***excessive proliferation and spread of cells***. Cancer is derived from a ***Latin word meaning crab***. It is presumed that the word cancer originated from the character of cancerous cells which can migrate and adhere and cause pain (like a crab) to any part of the body.

Neoplasia literally means new growth. Uncontrolled growth of cells results in tumors (a word originally used to represent swelling). ***Oncology*** (*Greek* : oncos—tumor) deals with the ***study of tumors***.

The tumors are of two types.

1. **Benign tumors :** They usually grow by expansion and remain encapsulated in a layer of connective tissue. Normally benign tumors are ***not life-threatening*** e.g. ***moles, warts***. These types of benign tumors are not considered as cancers.

2. **Malignant tumors or cancers :** They are characterized by ***uncontrolled proliferation and spread of cells to various parts of the body***, a process referred to as ***metastasis***. Malignant tumors are invariably life-threatening e.g. lung cancer, leukemia.

About 100 different types of human cancers have been recognized. Cancers arising from ***epithelial cells*** are referred to as ***carcinomas*** while that from ***connective tissues*** are known as ***sarcomas***. Methods for the early detection and treatment of cancers have been developed. However, little is known about the biochemical basis of cancer.

Incidence

Cancer is the ***second largest killer disease*** (the first being coronary heart disease) in the developed countries. It is estimated that cancer accounts for more than 20% of the deaths in United States. Based on the current rate of incidence, it is believed that one in every 3 persons will develop cancer at sometime during his life.

Although humans of all ages develop cancer, the incidence increases with advancement of age. More than 70% of the new cancer cases occur in persons over 60 years. Surprisingly, cancer is a leading cause of death in children in the age group 3-13 years, half of them die due to leukemia.

ETIOLOGY

In general, cancers are multifactorial in origin. The causative agents include ***physical***, ***chemical***, ***genetic*** and ***environmental factors***. A survey in USA has shown that about ***90% of all cancer deaths are due to avoidable factors such as*** tobacco, pollution, occupation, alcohol and diet.

Most of the cancers are caused by chemical carcinogens, radiation energy and viruses. These agents may damage DNA or interfere with its replication or repair.

Chemical carcinogens

It is estimated that ***almost 80% of the human cancers are caused*** by chemical carcinogens in nature. The chemicals may be ***organic*** (e.g. dimethylbenzanthracene, benzo (a) pyrene, dimethyl nitrosamine) ***or inorganic*** (arsenic, cadmium) in nature. Entry of the chemicals into the body may occur by one of the following mechanisms.

1. Occupation e.g. asbestos, benzene.

2. Diet e.g. aflatoxin B produced by fungus (*Aspergillus flavus*) contamination of foodstuffs, particularly peanuts.

3. Drugs—certain therapeutic drugs can be carcinogenic e.g. diethylstibesterol.

4. Life style e.g. cigarette smoking.

Mechanism of action : Although a few of the chemicals are directly carcinogenic, majority of them require prior metabolism to become carcinogenic. The enzymes such as cytochrome P_{450} responsible for the metabolism of xenobiotics (***Chapter 31***) are involved in dealing with the chemical carcinogens. In general, a chemically non-reactive procarcinogen is converted to an ultimate carcinogen by a series of reactions.

The carcinogens can covalently bind to purines, pyrimidines and phosphodiester bonds of DNA, often causing unrepairable damage. The chemical carcinogens frequently ***cause mutations*** (a change in the nucleotide sequence of DNA) which may finally lead to the development of cancer, hence they are regarded as mutagens.

Ames assay : This is a laboratory test ***to check*** the ***carcinogenecity of chemicals***. Ames assay employs the use of a special mutant strain of bacterium, namely ***Salmonella typhimurium*** (His^-). This organism cannot synthesize histidine; hence the same should be supplied in the medium for its growth. Addition of chemical carcinogens causes mutations (reverse mutation) restoring the ability of the bacteria to synthesize histidine (His^+). By detecting the strain of *Salmonella* (His^+) in the colonies of agar plates, the chemical mutagens can be identified. The Ames assay can detect about 90% of the chemical carcinogens. This test is regarded as a ***preliminary screening procedure***. Animal experiments are conducted for the final assessment of carcinogenecity.

Promoters of carcinogenesis : Some of the chemicals on their own are not carcinogenic. Certain substances known as promoting agents (***procarcinogens***) make them carcinogenic. The application of benzo- (a)pyrene to the skin, as such, does not cause tumor development. However, if this is followed by the application of croton oil, tumors will develop. In this case, benzo(a)pyrene is the initiating agent while croton oil acts as a promoting agent or promoter. Several compounds that act as promoting agents in various organs of the body have been identified. These include saccharin and phenobarbital.

Radiation energy

Ultraviolet rays, X-rays and γ-rays have been proved to be mutagenic in nature causing cancers. These rays ***damage DNA*** which is the ***basic mechanism to explain the carcinogenicity*** of radiation energy. For instance, exposure to

Medical Concepts/ Clinical Correlates

- About 80% of the human cancers are caused by chemical carcinogens.
- The products of oncogenes (e.g. growth factors) have been implicated in the development of cancer. Antioncogenes apply breaks and regulate the cell proliferation.
- The physical and chemical agents, viruses and mutations result in the activation of oncogenes causing carcinogenesis.

UV rays results in the formation of pyrimidine dimers in DNA while X-rays cause the production of free radicals. This type of molecular damages are responsible for the carcinogenic effects of radiations.

Carcinogenic viruses

The involvement of viruses in the etiology of cancer was first reported by ***Rous in 1911***. He demonstrated that the cell-free filtrates from certain chicken sarcomas (tumors of connective tissues) promote new sarcomas in chickens. Unfortunately, this epoch-making discovery of Rous was ignored for several years. This is evident from the fact that Rous was awarded the Nobel Prize in 1966 at the age of 85 for his discovery in 1911!

The presence of viral particles and the enzyme reverse transcriptase, besides the occurrence of base sequence in the DNA of malignant cells, complementary to tumor viruses indicate the involvement of viruses in cancer. The viruses involved in the development of cancer, commonly known as ***oncogenic viruses, may contain either DNA or RNA***. A selected list of tumor viruses is given in ***Table 37.1***.

DNA—the ultimate in carcinogenesis

DNA is the ultimate critical macromolecule in carcinogenesis. This fact is supported by several evidences.

1. Cancers are transmitted from mother to daughter cells. In other words, ***cancer cells beget cancer cells***.

2. Chromosomal abnormalities are observed in many tumor cells.

3. Damage to DNA caused by mutations often results in carcinogenesis.

4. Laboratory experiments have proved that purified oncogenes can transform normal cells into cancer cells.

MOLECULAR BASIS OF CANCER

Cancer is caused by a genetic change in a single cell resulting in its uncontrolled multiplication. Thus, ***tumors*** are ***monoclonal***. Two types of regulatory genes—***oncogenes*** and ***antioncogenes*** are involved in the development of cancer (carcinogenesis). In recent years, a third category of genes that control the cell death or ***apoptosis*** are also believed to be involved in carcinogenesis.

Oncogenes

The genes capable of causing cancer are known as oncogenes (*Greek* : oncos—tumor or mass). Oncogenes were originally discovered in tumor causing viruses. These viral oncogenes were found to be closely similar to certain genes present in the normal host cells which are referred to as protooncogenes. Now, about 40

Table 37.1 Selected tumor viruses

Class	Members
DNA viruses	
Adenovirus	Adenovirus 12 and 18
Herpesvirus	Epstein-Barr virus, Herpes simplex virus
Papovirus	Papilloma virus, Polyoma virus
RNA viruses	
Retrovirus type B	Mammary tumor virus of mouse
Retrovirus type C	Leukemia, sarcoma.

Table 37.2 Selected oncoproteins, protooncogenes and associated cancers

Oncoproteins	Protooncogene	Associated human cancer(s)
Growth factors		
Platelet derived growth factor (PDGF)	*sis*	Osteosarcoma
Epidermal growth factor (EGF)	*hst*-1	Cancers of stomach, breast and bladder
Growth factor receptors	*erb*-B_1	Lung cancer
	erb-B_2	Stomach cancer
	erb-B_3	Breast cancer
Signal—transducing proteins		
GTP— binding proteins	*ras*	Leukemias, cancers of lung, pancreas and colon
Non-receptor tyrosine kinase	*abl*	Leukemia

viral and cellular protooncogenes have been identified. Protooncogenes encode for growth-regulating proteins. The activation of protooncogenes to oncogenes is an important step in the causation of cancer.

In the ***Table 37.2***, a selected list of oncoproteins, protooncogenes and the associated human cancers is given.

Activation of protooncogenes to oncogenes

There are several mechanisms for converting the protooncogenes to oncogenes, some of the important ones are described next.

1. **Viral insertion into chromosome :** When certain retroviruses (genetic material RNA) infect cells, a complementary DNA (***cDNA***) is made from their RNA by the enzyme ***reverse transcriptase***. The cDNA so produced gets inserted into the host genome (***Fig.37.1***). The integrated double-stranded cDNA is referred to as provirus. This pro-viral DNA takes over the control of the transcription of cellular chromosomal DNA and transforms the cells. Activation of protooncogene *myc* to oncogene by viral insertion ultimately causing carcinogenesis is well known (e.g. avian leukemia).

Some DNA viruses also get inserted into the host chromosome and activate the proto-oncogenes.

2. **Chromosomal translocation :** Some of the tumors exhibit chromosomal abnormalities. This is due to the rearrangement of genetic material (DNA) by chromosomal translocation i.e. splitting off a small fragment of chromosome which is joined to another chromosome. Chromosomal translocation usually results in ***overexpression of protooncogenes***.

Burkitt's lymphoma, a cancer of human B-lymphocytes, is a good ***example of chromosomal translocation***. In this case, a fragment from chromosome 8 is split off and joined to chromosome 14 containing *myc* gene (***Fig.37.2***). This results in the activation of inactive *myc* gene leading to the increased synthesis of certain proteins which make the cell malignant.

3. **Gene amplification :** Severalfold amplifications of certain DNA sequences are observed

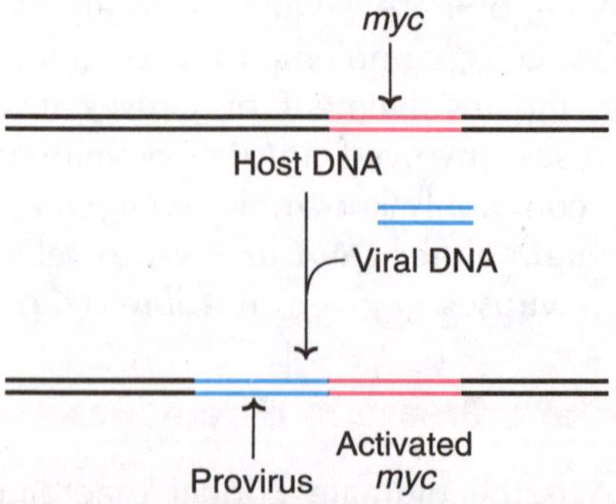

Fig. 37.1 : *Integration of viral DNA into host DNA.*

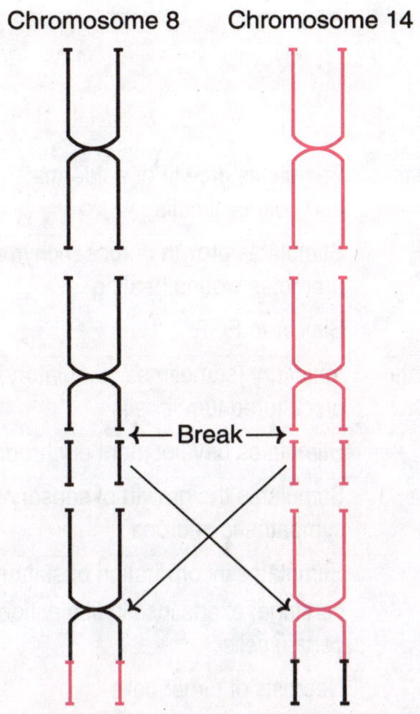

Fig. 37.2 : *Diagrammatic representation of reciprocal translocation occurring in Burkitt's lymphoma.*

in some cancers. Administration of anticancer drugs methotrexate (an inhibitor of the enzyme dihydrofolate reductase) is associated with gene amplification. The drug becomes inactive due to gene amplification resulting in a severalfold (about 400) increase in the activity of dihydrofolate reductase.

4. **Point mutation :** The ***ras protooncogene*** is the best ***example*** of activation by point mutation (change in a single base in the DNA). The mutated *ras* oncogene produces a protein (GTPase) which differs in structure by a single amino acid. This alteration diminishes the activity of GTPase, a key enzyme involved in the control of cell growth (details described later).

The presence of *ras* mutations is detected in several human tumors—90% of pancreatic, 50% of colon and 30% of lung. However, *ras* mutations have not been detected in the breast cancer.

Mechanism of action of oncogenes

Oncogenes encode for certain proteins, namely ***oncoproteins***. These proteins are the altered versions of their normal counterparts and are involved in the transformation and multiplication of cells. Some of the products of oncogenes are discussed below.

Growth factors : Several growth factors stimulating the proliferation of normal cells are known. They regulate cell division by transmitting the message across the plasma membrane to the interior of the cell (transmembrane signal transduction). It is believed that growth factors play a key role in carcinogenesis.

A selected list of polypeptide growth factors, their sources and major functions is given in ***Table 37.3***.

The ***cell proliferation*** is ***stimulated by growth factors***. In general, a growth factor binds to a protein receptor on the plasma membrane. This binding activates cytoplasmic protein kinases leading to the phosphorylation of intracellular target proteins. The phosphorylated proteins, in turn, act as intracellular messengers to stimulate cell division, the mechanism of which is not clearly known.

Transforming growth factor (TGF-α) is a protein synthesized and required for the growth of epithelial cells. ***TGF-α is produced in high concentration in individuals suffering from psoriasis***, a disease characterized by ***excessive proliferation of epidermal cells***.

Antioncogenes

A special category of genes, namely ***cancer suppressor genes*** (e.g. ***p^{53} gene***) or, more commonly, antioncogenes, have been identified. The products of these genes apply breaks and regulate cell proliferation. The loss of these suppressor genes removes the growth control of cells and is believed to be a key factor in the development of several tumors, e.g. retinoblastoma, one type of breast cancer, carcinoma of lung, Wilms' kidney tumor.

With the rapid advances in the field of genetic engineering, introducing antioncogenes to a normal chromosome to correct the altered growth rate of cells may soon become a reality.

Genes that regulate apoptosis

A new category of genes that regulate ***programmed cell death*** (apoptosis) have been discovered. These genes are also important in the development of tumors.

Table 37.3 Selected polypeptide growth factors

Growth factor	Source(s)	Major function(s)
Epidermal growth factor (EGF)	Salivary gland, fibroblasts	Stimulates growth of epidermal and epithelial cells
Platelet derived growth factor (PDGF)	Platelets	Stimulates growth of mesenchymal cells, promotes wound healing
Transforming growth factor-α (TGF-α)	Epithelial cell	Similar to EGF
Transforming growth factor-β (TGF-β)	Platelets, kidney, placenta	Inhibitory (sometimes stimulatory) effect on cultured tumor cells
Erythropoietin	Kidney	Stimulates development erythropoietic cells
Nerve growth factor (NGF)	Salivary gland	Stimulates the growth of sensory and sympathetic neurons
Insulin like growth factors (IGF-I and IGF-II, respectively known as somatomedins C and A)	Serum	Stimulates incorporation of sulfates into cartilage; exerts insulin-like action on certain cells
Tumor necrosis factor (TNF-α)	Monocytes	Necrosis of tumor cells
Interleukin-1 (IL-1)	Monocytes, leukocytes	Stimulates synthesis of IL-2.
Interleukin-2 (IL-2)	Lymphocytes (mainly T-helper cells).	Stimulates growth and maturation of T-cells

The gene, namely *bcl*-2, causes B-cell lymphoma by preventing programmed cell death. It is believed that overexpression of *bcl*-2 allows other mutations of protooncogenes that, ultimately, leads to cancer.

Unified hypothesis of carcinogenesis

The multifactorial origin of cancer can be suitably explained by oncogenes. The physical and chemical agents, viruses and mutations all lead to the activation of oncogenes causing carcinogenesis. The antioncogenes and the genes regulating apoptosis are intimately involved in development of cancer. A simplification of a unified hypothesis of carcinogenesis is depicted in ***Fig.37.3***.

TUMOR MARKERS

The biochemical indicators employed to ***detect the presence of cancers*** are collectively referred to as tumor markers. These are the ***abnormally produced molecules of tumor cells*** such as surface antigens, cytoplasmic proteins, enzymes and hormones. Tumor markers can be measured in serum (or plasma). In theory, the tumor markers must ideally be useful for screening the population to detect cancers. In practice, however, this has not been totally true. As such, the ***tumor markers support the diagnosis of cancers, besides being useful for monitoring the response to therapy and for the early detection of recurrence***.

A host of tumor markers have been described and the list is evergrowing. However, only a few of them have proved to be clinically useful. A

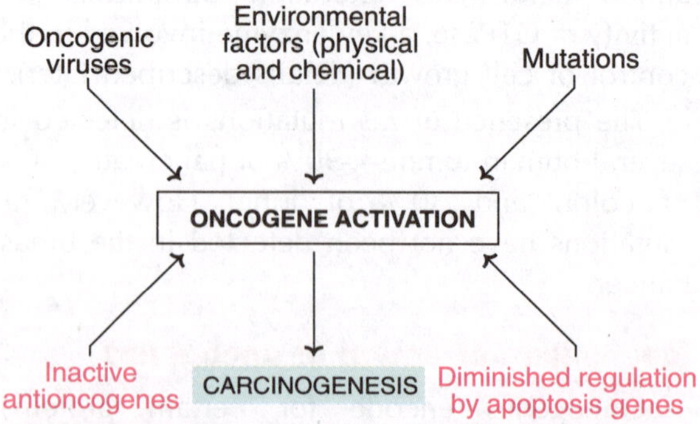

Fig. 37.3 : *A simplified hypothesis for the development of cancer.*

Table 37.4 Selected tumor markers and associated cancers

Tumor marker	Associated cancer(s)
Oncofetal antigens	
Carcinoembryonic antigen (CEA)	Cancers of colon, stomach, lung, pancreas and breast
Alpha fetoprotein (AFP)	Cancer of liver and germ cells of testis
Cancer antigen-125 (CA-125)	Ovarian cancer
Hormones	
Human chorionic gonadotropin (hCG)	Choriocarcinoma
Calcitonin	Carcinoma of medullary thyroid
Catecholamines and their metabolites (mainly vanillyl mandelic acid)	Pheochromocytoma and neuroblastoma
Enzymes	
Prostatic acid phosphatase	Prostate cancer
Neuron specific enolase	Neuroblastoma
Alkaline phosphatase	Bone secondaries
Specific proteins	
Prostate specific antigen (PSA)	Prostate cancer
Immunoglobulins	Multiple myeloma
Bence-Jones proteins	Multiple myeloma

selected list of tumor markers and the associated cancers are given in ***Table 37.4***.

A couple of the most commonly used tumor markers are discussed hereunder.

1. **Carcinoembryonic antigen (*CEA*) :** This is a complex glycoprotein, normally produced by the embryonic tissue of liver, gut and pancreas. The presence of CEA in serum is detected in several cancers (***colon, pancreas, stomach, lung***). In about 67% of the patients with ***colorectal cancer***, CEA can be identified. Unfortunately, serum CEA is also detected in several other disorders such as alcoholic cirrhosis (70%), emphysema (57%) and diabetes mellitus (38%). Due to this, CEA lacks specificity for cancer detection. However, in established cancer patients (particularly of colon and breast), the serum level of CEA is a useful indicator to detect the burden of tumor mass, besides ***monitoring the treatment***.

2. **Alpha-fetoprotein** (AFP) **:** It is chemically a glycoprotein, normally synthesized by yolk sac in early fetal life. Elevation in serum levels of AFP mainly ***indicates the cancers of liver and germ cells of testis*** and, to some extent, carcinomas of lung, pancreas and colon. As is the case with CEA, alpha-fetoprotein is not specific for the detection of cancers. Elevated levels of AFP are observed in cirrhosis, hepatitis and pregnancy. However, measurement of serum AFP provides a sensitive ***index for tumor therapy and detection of recurrence***.

CHARACTERISTICS OF GROWING TUMOR CELLS

The morphological and biochemical changes in the growing tumor cells are briefly described here. These observations are mostly based on the *in vitro* culture studies. Knowledge on the biochemical profile of tumor cells guides in the selection of chemotherapy of cancers.

1. **General and morphological changes**

- **Shape of cells :** The tumor cells are much ***rounder*** in shape compared to normal cells.
- **Alterations in cell structures :** The cytoskeletal structure of the tumor cells with regard to actin filaments is different.
- **Loss of contact inhibition :** The normal cells are characterized by contact inhibition i.e. they form monolayers and cannot move away from each other. The ***cancer cells form multilayers due to loss of contact inhibition*** (***Fig.37.4***). As a result, the cancer cells freely move and get deposited in any part of the body, a property referred to as ***metastasis***.

Medical Concepts/ Clinical Correlates

- The abnormal products of tumor cells, referred to as tumor markers (CEA, AFP, PSA) are useful for the diagnosis and prognosis of cancer.
- Anticancer drugs (e.g. methotrexate, cisplatin) are commonly used in the treatment of cancer. Antioxidants (vitamins E and C, β-carotene, Se) decrease the risk of carcinogenesis and hence their increased consumption is advocated.

- **Loss of anchorage dependence :** The cancer cells can grow without attachment to the surface. This is in contrast to the normal cells which firmly adhere to the surface.
- **Alteration in permeability properties :** The tumor cells have altered permeability and transport.

2. **Biochemical changes**

- **Increased replication and transcription :** The synthesis of DNA and RNA is increased in cancer cells.
- **Increased glycolysis :** The fast growing tumor cells are characterized by ***elevation in aerobic and anaerobic glycolysis*** due to increased energy demands of multiplying cells.
- **Reduced requirement of growth factors :** The tumor cells require much less quantities of growth factors. Despite this fact, there is an increased production of growth factors by these cells.
- **Synthesis of fetal proteins :** During fetal life, certain genes are active, leading to the synthesis of specific proteins. These genes are suppressed in adult cells. However, the tumor cells synthesize the fetal proteins e.g. carcino-embryonic antigen, alfa fetoprotein.
- **Alterations in the structure of molecules :** Changes in the structure of glycoproteins and glycolipids are observed.

Metastasis

Metastasis refers to the ***spread of cancer cells from the primary site of origin to other tissues of the body where they get deposited and grow as secondary tumors***. Metastasis is the major cause of cancer related morbidity and mortality. It is believed that the morphological changes in tumor cells, loss of contact inhibition, loss of anchorage dependence and alterations in the structure of certain macromolecules are among the important factors responsible for metastasis.

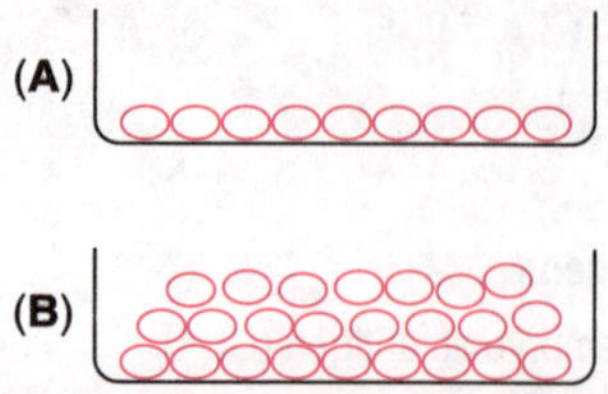

Fig. 37.4 : *Growth cells in culture **(A)** Normal cells forming monolayer (exhibiting contact inhibition); **(B)** Cancer cells forming multilayers (loss of contact inhibition).*

CANCER THERAPY

Chemotherapy, employing certain anticancer drugs, is widely used in the treatment of cancer. In the ***Table 37.5***, a selected list of the most commonly used drugs, and their mode of action is given. The effectiveness of anticancer drugs is inversely proportional to the size of the tumor i.e. the number of cancer cells. The major limitation of cancer chemotherapy is that the rapidly dividing normal cells (of hematopoietic system, gastrointestinal tract, hair follicles) are also affected. Thus, the ***use of anticancer drugs is associated with toxic manifestations***.

Cisplatin is used in the treatment of testicular, ovarian and several other cancers (bone, lung). The side effects of cisplatin include bone marrow depletion, loss of hearing and impairment in kidney function. About 80% of testicular cancer patients survive with a new combination therapy of cisplatin, etoposide, and bleomycin.

The term ***tumor lysis syndrome (TLS)*** represents all the metabolic consequences that occur during cancer treatment. These include increased uric acid levels in serum and urine, acute renal failure, hyperkalemia, hyperphosphatemia etc. ***Recombinant urate oxidase*** (converts uric acid to soluble allantion) is successfully used in the treatment of TLS.

Table 37.5 A selected list of the most commonly used anticancer drugs and their mode of action

Anticancer drug	Chemical nature	Mode of action
Methotrexate	Folic acid analogue	Blocks the formatin of tetrahydrofolate (inhibits the enzyme dihydrofolate reductase). THF is required for nucleotide synthesis.
6-Mercaptopurine	Purine analogue	Inhibits the formation of AMP from IMP.
6-Thioguanine	Purine analogue	Blocks thymidylate synthase reaction.
Mitomycin C	Antibiotic	Results in cross bridges between DNA base pairs.
Actinomycin D	Antibiotic	Blocks transcription
Vinblastine and vincristine	Alkaloids	Inhibit cell division and cytoskeleton formation
Cisplatin	Platinum compound	Results in the formation of intrastrand DNA adducts.
Cyclophosphamide	Alkylating agent	Cross-links bases and inhibits DNA strand separation
Imatinib	Monoclonal antibody	Tyrosine kinase inhibitor

PREVENTION OF CANCER

In recent years, certain precautionary measures are advocated to prevent or reduce the occurrence of cancer. The most important among them, from the biochemical perspective, are the ***antioxidants*** namely vitamin E, β-carotene, vitamin C and selenium. They ***prevent the formation or detoxify the existing free radicals*** (free radicals are known to promote carcinogenesis). In addition, ***antioxidants stimulate body's immune system***, and promote detoxification of various carcinogens.

In general, most of the vegetables and fruits are rich in antioxidants. Their increased consumption is advocated to prevent cancer. (For more details on free radicals and antioxidants, ***Refer Chapter 34***).

SUMMARY

- *Cancer is characterized by uncontrolled cellular growth and development, leading to excessive proliferation and spread of cells. Cancer is the second largest killer disease (next to heart disease) in the developed world.*
- *Regulatory genes—namely oncogenes, antioncogenes and genes controlling cell death—are involved in the development of cancer. Activation of oncogenes is a fundamental step in carcinogenesis. This may occur by insertion of viral DNA into host chromosome, translocation of chromosomes, gene amplification and point mutation.*
- *Tumor markers of cancers include carcinoembryonic antigen (CEA), alpha fetoprotein (AFP), cancer antigen-125 and prostate specific antigen (PSA). They are mainly useful to support diagnosis, monitor therapy and detect recurrence.*
- *There are several morphological and biochemical changes in the tumor cells which distinguish them from the normal cells. The cancer cells are characterized by loss of contact inhibition, altered membrane transport, increased DNA and RNA synthesis, increased glycolysis, alteration in the structure of certain molecules etc.*

THE READER MUST REFER APPENDIX VI (Page Nos : 738 - 756) FOR RELATED CLINICAL CASE STUDIES (Nos : 44, 66, 67)

Acquired Immunodeficiency Syndrome (AIDS)

38

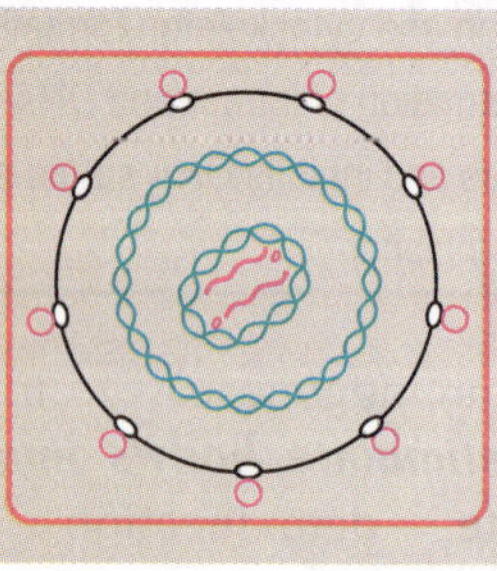

AIDS speaks

"I am the most feared disease of the world;
Due to a retrovirus, causing immunodeficiency;
With no cure in sight, except prevention;
I challenge the scientists worldwide to conquer me!"

Acquired immunodeficiency syndrome (AIDS) was ***first reported in 1981 in homosexual men***. AIDS is a retroviral disease caused by human immunodeficiency virus (HIV). The disease is ***characterized by immunosuppression, secondary neoplasma*** and ***neurological manifestations***. AIDS is invariably fatal, since there is no cure. In the USA, it is the fourth leading cause of death in men between the ages 15 to 55 years.

No other disease has attracted as much attention as AIDS by the governments, public and scientists. AIDS has stimulated an unprecedented amount of biomedical research which led to a major understanding of this deadly disease within a short period of time. So rapid is the research on AIDS (particularly relating to molecular biology), any review is destined to be out of date by the time it is published!

The isolation of human immunodeficiency virus (HIV) from lymphocytes of AIDS patients was independently achieved by Gallo (USA) and Montagnier (France) in 1984.

Epidemiology

AIDS was first described in USA and this country has the majority of reported cases. The prevalence of AIDS has been reported from almost every country. The number of people living with HIV worldwide is estimated to be around 40 million in the year 2020.At least one million deaths occurred in 2020, due to AIDS. AIDS is truely a global disease with an alarming increase in almost every country.

Transmission of HIV : Transmission of AIDS essentially requires the exchange of body fluids (semen, vaginal secretions, blood, milk) containing the virus or virus-infected cells. There are three major routes of HIV transmission—sexual contact, parenteral inoculation, and from infected mothers to their newborns.

The distribution of risk factors for AIDS transmission are as follows.

Sex between men (homosexuals)	— 60%
Sex between men and women	— 15%

Medical Concepts/ Clinical Correlates

- AIDS is a global disease with an alarming increase in the incidence of occurrence. By the year 2014, more than 40 million AIDS deaths occurred globally.
- Homosexuality (predominantly in men) and intravenous drug abuse are the major factors in the risk of AIDS transmission.

Intravenous drug abusers — 15%
Transfusion of blood and blood products — 6%
All others — 4%

The predominant methods of HIV transmission (about 75%) are through anal or vaginal intercourse. The risk for the transmission is much higher with anal than with vaginal intercourse. The practice of 'needle sharing' is mainly responsible for the transmission of HIV in drug abusers. Pediatric AIDS is mostly caused by vertical transmission (mother to infant).

It should, however, be noted that HIV cannot be transmitted by casual personal contact in the household or work place. Further, the transmission of AIDS from an infected individual to health personnel attending on him is extremely rare.

Virology of HIV

AIDS is caused by a retrovirus, namely ***human immunodeficiency virus*** (HIV), belonging to lentivirus family. ***Retroviruses*** contain RNA as the genetic material. On entry into the host cell, they transcribe DNA which is a complementary copy of RNA. The DNA, in turn is used, as a template to produce new viral RNA copies.

Two different forms of HIV, namely HIV-1 and HIV-2 have been isolated from AIDS patients. HIV-1 is more common, being found in AIDS patients of USA, Canada, Europe and Central Africa while HIV-2 is mainly found in West Africa. Both the viruses are almost similar except they differ in certain immunological properties.

HIV-1 is described in some detail.

Structure of HIV : The virus is spherical with a diameter of about 110 nm. It contains a core, surrounded by a lipid envelop derived from the host plasma membrane (***Fig.38.1***). The core of the HIV has two strands of genomic RNA and four core proteins, p_{24}, p_{18}, ***reverse transcriptase*** (p_{66}/p_{51}) and endonuclease (p_{32}). Note that the naming of the proteins is based on the molecular weight. For instance, a protein with a molecular weight of 24,000 is designated as p_{24}.

The lipid membrane of the virus is studded with two glycoproteins gp_{120} and gp_{41}. The surface antigen gp_{120} is very important for the viral infection and the detection of AIDS.

Immunological abnormalities in AIDS

As is evident from the name AIDS, immunodeficiency (or immunosuppression) is the hallmark of this disease. AIDS ***primarily affects*** the ***cell-mediated immune system*** which protects the body from intracellular parasites such as viruses, protozoa and mycobacteria. This is caused by a reduction in CD_4 (cluster determinant antigen 4) cells of T-lymphocytes, besides impairment in the functions of surviving CD_4 cells.

CD_4 cells may be regarded as ***master cells of cell mediated immunity***. They produce cytokines, macrophage chemotactic factors, hemopoietic

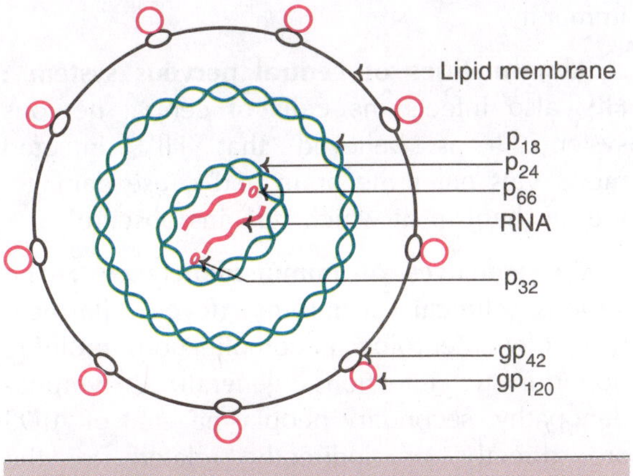

***Fig. 38.1 :** Diagrammatic representation of HIV (p represents protein with molecular weight e.g. p_{18} is with molecular weight 18,000; gp represents glycoprotein).*

growth factors, and others involved in the body immunity.

Entry of HIV and lysis of CD_4 cells : The virus enters the CD_4 T-lymphocytes. HIV binds to the specific receptors on CD_4 cells by using its surface membrane glycoprotein (gp_{120}). Following the entry into the host cells, RNA of HIV is transcribed into DNA by the viral enzyme reverse transcriptase. The viral DNA gets incorporated into the host genomic DNA. The virus may remain locked in the host genome for months or years and this is considered as the latent period. The viral DNA may undergo replication and translation, respectively, producing viral RNA and viral proteins. The latter two, on assembly, result in new viruses. The newly synthesized viruses leave the host cells by forming buds on plasma membrane. Extensive viral budding is associated with lysis and death of CD_4 cells (***Fig.38.2***). The new viral particles infect other host cells and repeat the whole process, ultimately resulting in a profound loss of CD_4 cells from the blood. ***Most of the immunodeficiency symptoms of AIDS are associated with the reduction in CD_4 cells.***

Other immunological abnormalities

The viral membrane protein gp_{120} binds with normal T-helper cells and kills them. AIDS patients also display abnormalities in antibody production by B-lymphocytes (humoral immunity).

Abnormalities of central nervous system : HIV also infects the cells of central nervous system. It is believed that HIV infected monocytes enter the brain and cause damage, the mechanism of which remains obscure.

Consequences of immunodeficiency : The various clinical symptoms (fever, diarrhea, weight loss, neurological complications, multiple opportunistic infections, generalized lymphadenopathy, secondary neoplasma etc.) of AIDS are directly or indirectly related to the ***immunosuppression*** caused by HIV. Due to the deficiency in the immune system, the body of AIDS patient is freely exposed to all sorts of infections (viral, bacterial, fungal).

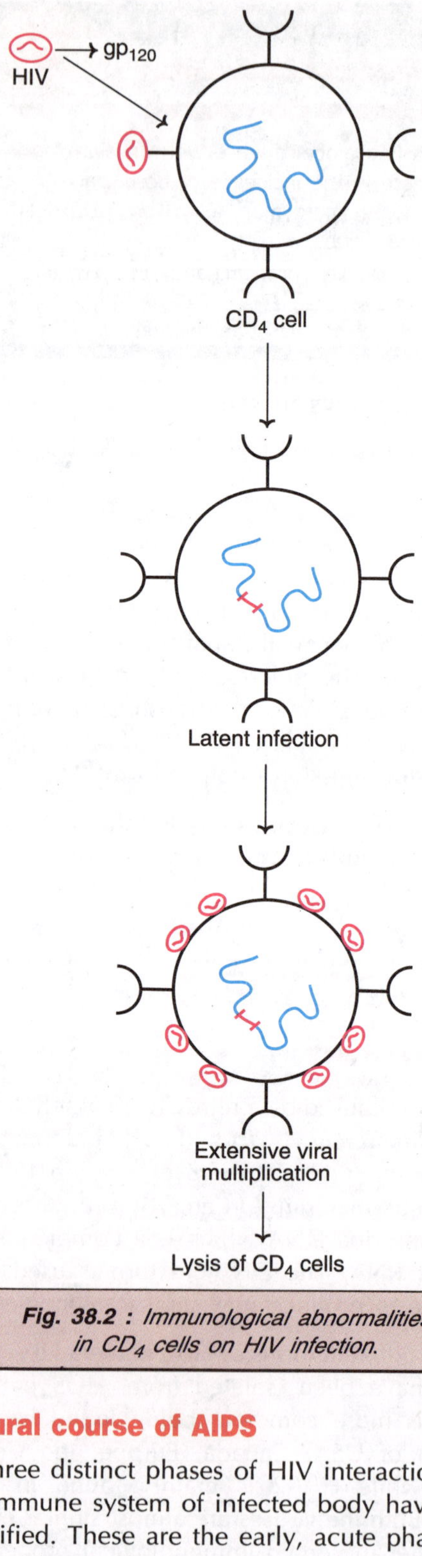

***Fig. 38.2 :** Immunological abnormalities in CD_4 cells on HIV infection.*

Natural course of AIDS

Three distinct phases of HIV interaction with the immune system of infected body have been identified. These are the early, acute phase; the intermediate, chronic phase; the final, crisis phase (***Fig.38.3***).

Medical Concepts/ Clinical Correlates

- The patients of AIDS are destined to die (within 5–10 years after infection), since there is no cure. However, administration of certain drugs (AZT, DDI) prolongs life.
- The clinical manifestations of AIDS are directly or indirectly related to immunosuppression (mostly due to reduced CD_4 cells). AIDS patients are highly susceptible to all sorts of infections (viral, bacterial, fungal).

1. **Acute phase :** This represents the initial body response to HIV infection. It is characterized by ***high rate of production of viruses*** which are lodged in the lymphoid tissues and the antiviral immune response of the body. This period may ***last for about 8-12 weeks.***

2. **Chronic phase :** During this period that may last for ***5 to 10 years or even more***, the ***body tries to contain the virus***. The immune system is largely intact. The person obviously appears normal, although he/she is the carrier of HIV which can be transmitted to others. Antibodies to HIV are found in the circulation, hence this phase is also referred to as ***seropositive period.***

3. **Crisis phase :** ***A failure in the defense system of the body, caused by immunosuppression by HIV***, represents the crisis phase. The plasma level of virus is tremendously increased. CD_4 T-lymphocyte concentration drastically falls. A patient with lower than 200 CD_4 T-lymphocytes/µl blood is considered to have developed AIDS. Crisis phase is characterized by opportunistic infections and the related clinical manifestations. In Western countries, a cancer—Kaposis sarcoma—is associated with AIDS.

In general, AIDS patients die between 5-10 years after HIV infection. Treatment may, however, prolong the life.

Laboratory diagnosis of AIDS

The following laboratory tests are employed to diagnose the HIV infection.

1. The detection of antibodies in the circulation by ***ELISA*** (enzyme-linked immunosorbant assay).

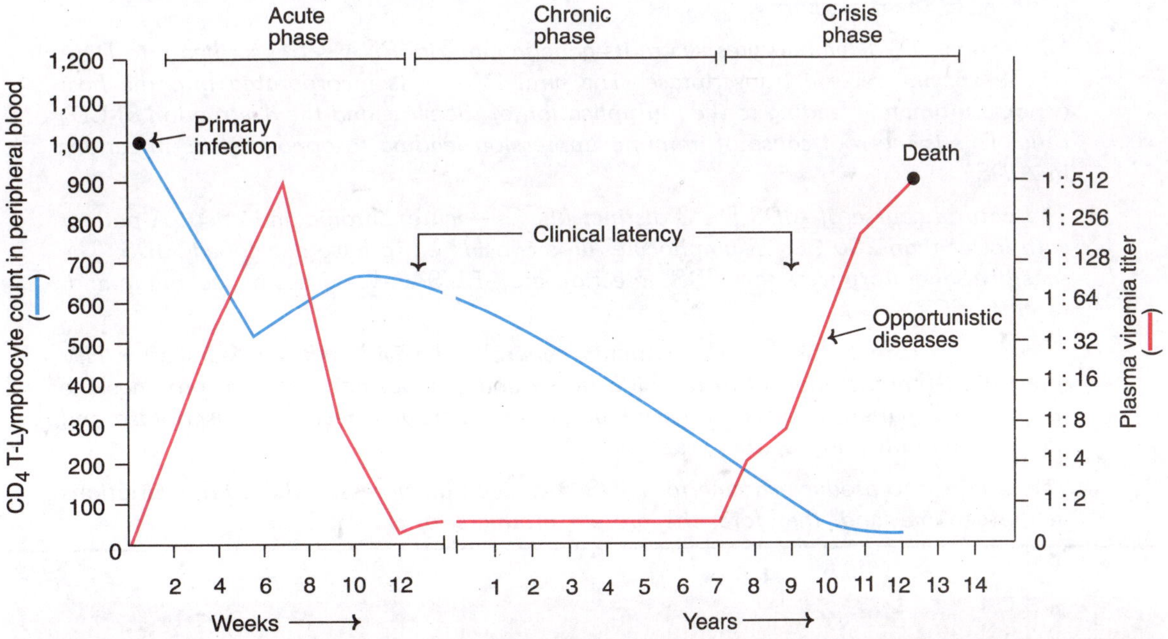

Fig. 38.3 : Graphic representation of a typical course of HIV infection.

2. ***Western blot technique***, a more specific test for the HIV antibodies, is employed for confirmation of ELISA positive cases.

3. A more recent and sophisticated ***PCR*** can be used to detect the presence of the HIV genome in the peripheral blood lymphocytes.

Drugs for the treatment of AIDS

Although there is ***no cure for AIDS***, use of certain ***drugs can prolong the life*** of AIDS patients. ***Zidovudine*** or ***AZT*** (3′-azido 2′, 3′-dideoxy thymidine), a structural analog of deoxythymidine was the first drug used and continues to be the drug of choice for the treatment of AIDS. ***Didanosine*** (dideoxyionosine, DDI) is another drug employed to treat AIDS.

Vaccine against AIDS—a failure so far

HIV exhibits genetic heterogenecity with a result that several species of virus may be found in the same AIDS patient. The principal cause for the ***genetic variation*** is the lack of proof-reading activity by the enzyme reverse transcriptase. This leads to ***very frequent alterations in the DNA base sequence*** synthesized from viral RNA which, in turn, influences the amino acid sequence of proteins. Thus, the protein products of HIV are highly variable in the amino acid composition and, therefore, the antigenic properties. For this reason, it has not been possible to develop a vaccine against AIDS. However, there have been some encouraging animal and *in vitro* experiments which raise fresh hopes for a vaccine in the near future.

SUMMARY

- *AIDS is a retorviral disease caused by human immunodeficiency virus (HIV). It is characterized by immunosuppression, secondary neoplasms and neurological manifestations. Transmission of HIV occurs by sexual contact (more in male homosexuals), parental inoculation (intravenous drug abusers) and from infected mothers to their newborns.*
- *HIV enters CD_4 T-lymphocytes where its genetic material RNA is transcribed into DNA by the enzyme reverse transcriptase. The viral DNA gets incorporated into the host genome ultimately leading to the multiplication of the virus and the destruction of CD_4 cells. This is the root cause of immunosuppression leading to opportunistic infections in AIDS.*
- *The natural course of AIDS has 3 distinct phases—acute, chronic and crisis. A patient with lower than 200 CD_4 T-lymphocytes/μl is considered to have developed AIDS. The sensitive laboratory tests for AIDS detection are—ELISA, Western blot technique and, recently PCR.*
- *There is no cure for AIDS. The patients generally die within 5–10 years after HIV infection. Administration of drugs (zidovudine and didanosine), however, prolongs the life of AIDS patients. These drugs inhibit the viral enzyme reverse transcriptase and halt the multiplication of the virus.*
- *The attempts to produce vaccine for AIDS have been unsuccessful due to the variations in the genome (and, therefore, the protein products) of the HIV.*

THE READER MUST REFER APPENDIX VI (Page Nos : 738 - 756) FOR RELATED CLINICAL CASE STUDY (No : 77)

BASICS TO LEARN BIOCHEMISTRY

SECTION SEVEN

Introduction to Bioorganic Chemistry

39

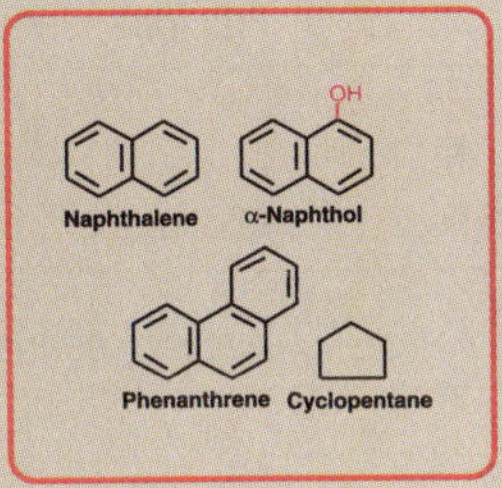

Carbon, the official spokesperson of organic chemistry, speaks

"With water, I say, 'Touch me not';
To the tongue, I am tasteful;
Within limits, I am dutiful;
In excess, I am dangerous!"

As life comes from the existing life, it was believed for a long that the carbon compounds of organisms (hence the name organic) arose from life only. This is referred to as ***vital force theory***. Friedrich Wohler (1825) first discovered that urea ($NH_2-CO-NH_2$), the organic compound, could be prepared by heating ammonium cyanate (NH_4NCO), in the laboratory. Thereafter, thousands and thousands of organic compounds have been synthesized outside the living system.

Organic chemistry broadly deals with the chemistry of carbon compounds, regardless of their origin. Biochemistry, however, is concerned with the ***carbon chemistry of life only***.

Most common organic compounds found in living system

The organic compounds, namely carbohydrates, lipids, proteins, nucleic acids and vitamins are the most common organic compounds of life. Their chemistry has been discussed in Section I (***Chapters 1-7***).

Common functional groups in biochemistry

Most of the physical and chemical properties of organic compounds are determined by their functional groups. Biomolecules possess certain functional groups which are their reactive centres. The common functional groups of importance in biomolecules are presented in ***Table 39.1***.

Common ring structures in biochemistry

There are many homocyclic and heterocyclic rings, commonly encountered in biomolecules. A selected list of them is given in ***Fig.39.1***.

Homocyclic rings : Phenyl ring derived from benzene is found in several biomolecules (phenylalanine, tyrosine, catecholamines). Phenanthrene and cyclopentane form the backbone of steroids (cholesterol, aldosterone).

Heterocyclic rings : Furan is the ring structure found in pentoses. Pyrrole is the basic unit of porphyrins found in many biomolecules (heme) while pyrrolidine is the ring present in the amino acid, proline. Thiophene ring is a part of the vitamin biotin. The amino acid histidine contains imidazole.

Pyran structure is found in hexoses. Pyridine nucleus is a part of the vitamins-niacin and pyridoxine. Pyrimidines (cytosine, thymine) and

Table 39.1 Common functional groups of importance in biomolecules

Functional group Name	Group	General structural formula	Type of compound	Examples of biomolecule(s)
Hydroxyl	$-OH$	$R-OH$	Alcohol	Glycerol, ethanol
Aldehyde	$-C(=O)-H$	$R-C(=O)-H$	Aldehyde	Glyceraldehyde, glucose
Keto	$-C(=O)-$	$R_1-C(=O)-R_2$	Ketone	Fructose, sedoheptulose
Carboxyl	$-C(=O)-OH$	$R-C(=O)-OH$	Carboxylic acid	Acetic acid, palmitic acid
Amino	$-NH_2$	$R-NH_2$	Amino acid	Alanine, serine
Imino	$-N(H)-$	$R-N(H)-$	Imino acid	Proline, hydroxyproline
Sulfhydryl	$-SH$	$R-SH$	Thiol	Cysteine, coenzyme A
Ether	$-O-$	R_1-O-R_2	Ether	Thromboxane A_2
Ester	$-C(=O)-O-R_1$	$R_2-C(=O)-O-R_1$	Ester	Cholesterol ester
Amido	$-C(=O)-N(R_1)(R_2)$	$R_3-C(=O)-N(R_1)(R_2)$	Amide	N-Acetylglucosamine

purines (adenine, guanine) are the constituents of nucleotides and nucleic acids. Indole ring is found in the amino acid tryptophan.

ISOMERISM

The compounds possessing identical molecular formulae but different structures are referred to as ***isomers.*** The phenomenon of existence of isomers is called isomerism (*Greek* : isos—equal; meros—parts). Isomers differ from each other in physical and chemical properties. ***Isomerism is partly responsible for the existence of a large number of organic molecules.***

Consider the molecular formula—C_2H_6O. There are two important isomers of this—ethyl alcohol (C_2H_5OH) and diethyl ether (CH_3OCH_3) depicted

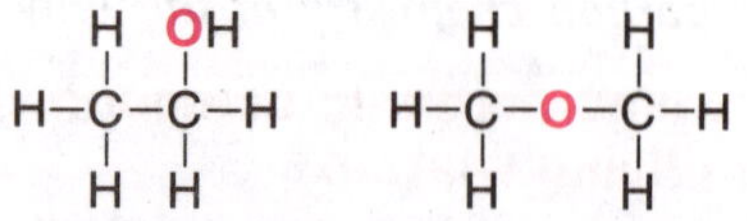

Ethyl alcohol **Diethyl ether**

If is broadly divided into two cate-gories-structural isomerism and stereoisomerism.

Structural isomerism

The difference in the arrangement of the atoms in the molecule (i.e. molecular framework) is responsible for structural isomerism. This may be due to variation in carbon chains (***chain isomerism***) or difference in the position of functional groups (***position isomerism***) or

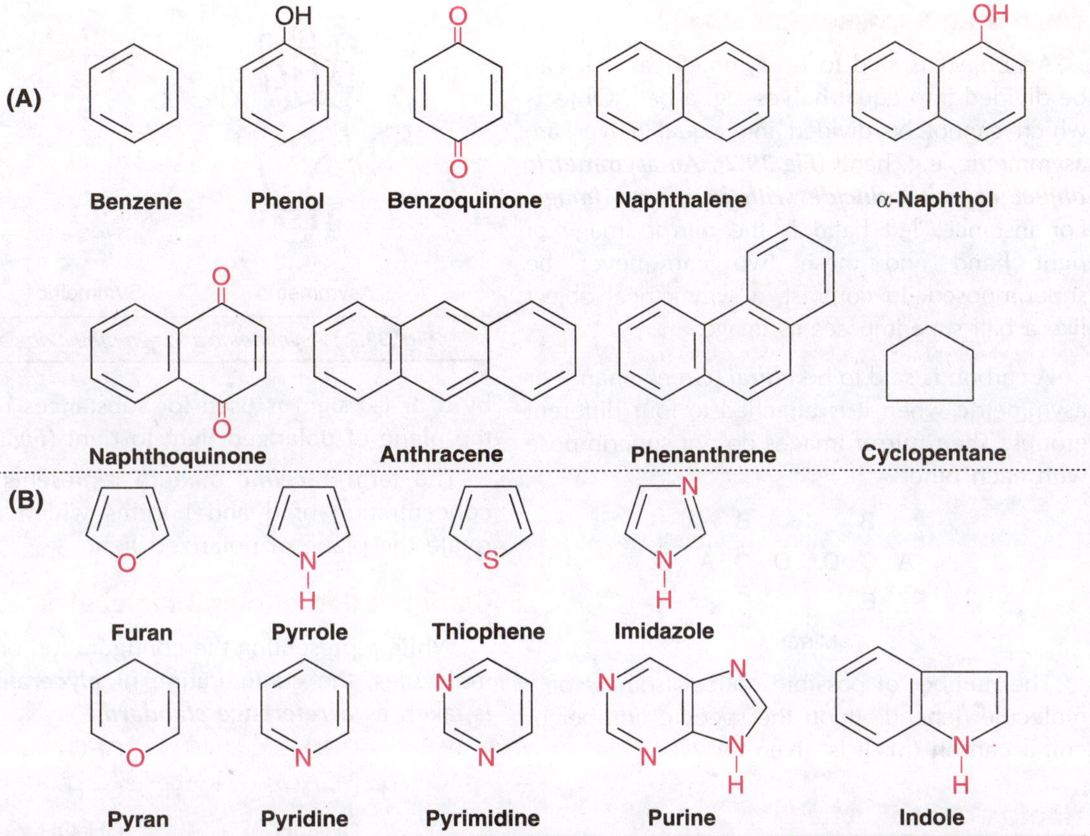

Fig. 39.1 : Common ring structures found in biomolecules (A) Homocyclic rings (B) Heterocyclic rings.

difference in both molecular chains and functional groups ***(functional isomerism).***

Structural isomerism, as such, is more common in general organic molecules. ***Tautomerism,*** a type of structural isomerism, occurs due to the migration of an atom or group from one position to the other e.g. ***purines and pyrimidines*** (***Chapter** 5*).

Stereoisomerism

Stereoisomerism (*Greek* : stereos—space occupying) is, perhaps, more relevant and important to biomolecules. The differential space arrangement of atoms or groups in molecules gives rise to stereoisomerism. Thus, stereoisomers have the same structural formula but differ in their spatial arrangement.

Stereoisomerism is of two types—***geometric isomerism*** and ***optical isomerism.***

Geometrical isomerism : This is also called ***cis-trans isomerism*** and is exhibited by certain molecules possessing double bonds. Geometrical isomerism is due to restriction of freedom of rotation of groups around a carbon-carbon double bond (C=C). Maleic acid and fumaric acid are classical examples of *cis-trans* isomerism.

Maleic acid *(cis)* **Fumaric acid** *(trans)*

When similar groups lie on the same side, it is called *cis* isomer (*Latin* : cis—on the same side). On the other hand, when similar groups lie on the opposite sides, it is referred to as *trans* isomer (*Latin* : trans—across). As is observed from the above structure, maleic acid is a *cis* form while fumaric acid is a *trans* form.

Geometric isomerism is also observed in sterols and porphyrins. *cis-trans* isomers differ in physical and chemical properties.

Optical isomerism : Optical isomers or enantiomers occur due to the presence of an ***asymmetric carbon*** (a chiral carbon).

What is an asymmetric carbon?

An object is said to be symmetrical if it can be divided into equal halves e.g. a ball. Objects which cannot be divided into equal halves are asymmetric, e.g. hand (***Fig.39.2***). ***An asymmetric object cannot coincide with its mirror image***. For instance, left hand is the mirror image of right hand and these two can never be superimposed. In contrast, a symmetrical object like a ball superimposes its image.

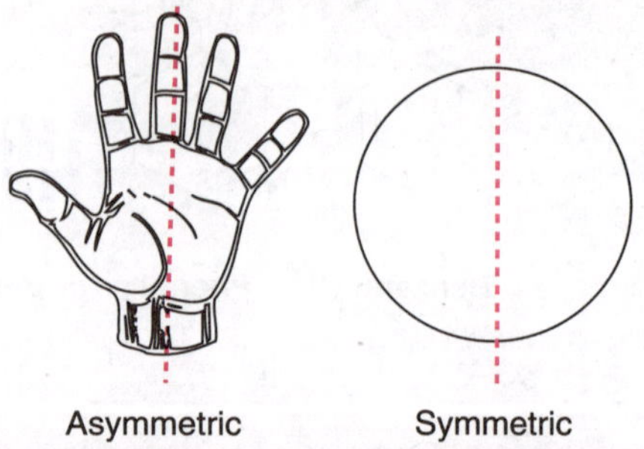

Fig. 39.2 : *Asymmetric and symmetric objects.*

A carbon is said to be ***chiral*** (*Greek* : hand) or asymmetric when it is attached to four different groups. Their mirror images do not superimpose with each other.

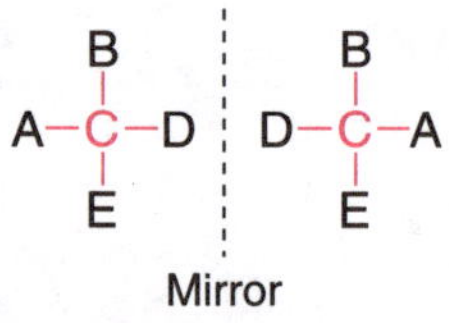

The number of possible optical isomers of a molecule depends upon the specific number of chiral carbon (n). It is given by 2^n.

What is optical activity?

The ordinary light propagates in all directions. However, on passing ordinary light through a Nicol prism, the plane of polarized light vibrates in one direction only (***Fig.39.3***).

Certain organic compounds (optical isomers) which are said to exhibit optical activity rotate the plane of polarized light either to the left or to the right.

The term ***levorotatory*** (indicated by 1 or (–) sign) is used for the substances which rotate the plane of polarized light to the left. On the other hand, the term ***dextrorotatory*** (indicated by d or (+) sign) is used for substances rotating the plane of polarized light to right (***Fig.39.3***).

The term ***racemic*** mixture represents equal concentration of d and l forms which cannot rotate the plane of polarized light.

Configuration of chiral molecules

While representing the configuration of chiral molecules, the configuration of ***glyceraldehyde*** is taken as a ***reference standard.***

D-Glyceraldehyde **L-Glyceraldehyde**

It must, however, be remembered that D- and L- do not represent the direction of the rotation of plane of polarized light.

Existence of chiral biomolecules

As you know, you can never come across anybody who is your mirror image. The same is true with biomolecules. Only one type of molecules (D or L) are found in the living system. Thus, the naturally occurring amino acids are of L-type while the carbohydrates are of D-type.

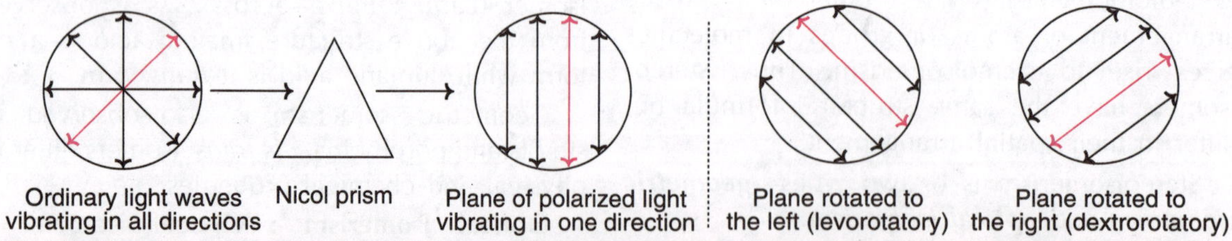

Fig. 39.3 : *Diagrammatic illustration of optical activity.*

Overview of Biophysical Chemistry

40

The radioactive isotopes speak

Isotope	Radiation	Half-life
^{3}H	β	12.2 years
^{14}C	β	5,700 years
^{22}Na	γ	2.5 years

"We are elements of same atomic numbers;
Undergo decay to emit α or β or γ rays;
Exhibit identical chemical properties;
Exploited as tracers in biomedical research."

The general laws and principles of chemistry and physics are applicable to biochemistry as well. It is, therefore, worthwhile to have a ***brief understanding of some of the basic chemical and physical principles that have direct relevance to life***.

WATER

Water is the most abundant fluid on earth. It is justifiably regarded as the ***solvent of life***. As much as 70% of a typical cell is composed of water. The unique physical and chemical properties of water have profound biological importance. The structures of biomolecules (proteins, nucleic acids, lipids and carbohydrates) are maintained due to their interaction with water, which forms an aqueous environment. This is essential for sustaining life.

Water expands on freezing : Water is one of the very few substances that expands on freezing. Thus, ice has a density of 0.92 g/ml, while water at 0°C has density of 1.0 g/ml. For this reason, ice floats on water. And this property is essential to maintain water equilibrium in the environment, and to sustain life.

(Imagine that water contracted on cooling and becomes denser. In such a case, ice would sink to the bottom of seas and lakes and would never get exposed to sun rays. Thus, frozen water would permanently remain as ice. If this were to happen, earth would have a permanent ice age!)

ACIDS AND BASES

According to Lowry and Bronsted, an acid is defined as a substance that gives off protons while base is a substance that accepts protons. Thus, ***an acid is a proton (H^+) donor and a base is a proton acceptor.*** A few examples of acids and their corresponding bases are given in the next page.

Acid	Base
HCl	$H^+ + Cl^-$
H_2CO_3	$H^+ + HCO_3^-$
H_2PO_4	$H^+ + HPO_4^-$
NH_4^+	$H^+ + NH_3$
H_2O	$H^+ + OH^-$
CH_3COOH	$H^+ + CH_3COO^-$
HA (general)	$H^+ + A^-$ (general)

It is evident that an acid dissociates to form proton and base. On the other hand, the base combines with proton to form acid. The ***difference between an acid and its corresponding base*** (more commonly referred to as ***conjugate base***) is the ***presence or absence of a proton***. In general, a strong acid has a weak base while a weak acid has a strong base. For instance, strong acid HCl has weak base Cl^-, weak acid HCN has a strong base CN^-.

Alkalies : The ***metallic hydroxides*** such as ***NaOH and KOH*** are commonly referred to as alkalies. These compounds do not directly satisfy the criteria of bases. However, they dissociate to form metallic ion and OH^- ion. The latter, being a base, accepts H^+ ions.

Ampholytes : The substances which can function both as acids and bases are referred to as ampholytes. ***Water is*** the ***best example for ampholytes.***

Dissociation of water

Water is a weak electrolyte and dissociates as follows.

$$H_2O \rightleftharpoons H^+ + OH$$

The proton reacts with another molecule of water to form hydronium ion (H_3O^+).

$$H^+ + H_2O \rightleftharpoons H_3O^+$$

For the sake of convenience, the presence of proton as H_3O^+ is ignored.

By applying the law of mass action for the dissociation of water.

$$K = \frac{[H^+][OH^-]}{[H_2O]}$$

Here K is a constant; the concentrations are expressed in molarity. Since the degree of dissociation is very small, the concentration of undissociated $[H_2O]$ may be taken as constant.

$$K_\omega = [H^+]\ [OH^-]$$

K_ω is the dissociation constant for water. Its value is 10^{-14} at 25°C.

$$[H^+]\ [OH^-] = 10^{-14}$$

In a neutral solution

$$[H^+] = [OH^-] = 10^{-7}$$

Hydrogen ion concentration (pH)

The acidic or basic nature of a solution is measured by H^+ ion concentration. The strength of H^+ ions in the biological fluids is exceedingly low. For this reason, the conventional units such as moles/l or g/l are not commonly used to express H^+ ion concentration.

Sorenson (1909) introduced the term pH to express H^+ ion concentration. ***pH is defined as the negative logarithm of H^+ ion concentration.***

$$pH = -\log [H^+]$$

The pH (may be considered as **p**otential of ***H^+*** ions) is a narrow scale, ranging from 0 to 14 which corresponds to 1 M solution to 10^{-14} M solution of $[H^+]$ concentration.

Pure water has a pH 7 which is neutral. Solutions with ***pH less than 7*** are said to be ***acidic*** while those with pH greater than 7 are alkaline. It must be remembered that the term acidic or alkaline are not absolute but only relative. Thus, a solution with pH 3.0 is more acidic when compared with a solution of pH 4.5.

A rise in H^+ concentration decreases pH while a fall in H^+ concentration increases pH. The reverse is true for OH^- concentration. The pH of a solution containing 1N $[H^+]$ is 0 while that containing 1N $[OH^-]$ is 14.

The pH of important biological fluids is presented in ***Table 40.1***.

BUFFERS

The pH of a given solution can be easily altered by the addition of acids or bases. ***Buffers***

Table 40.1 pH of important biological fluids

Fluid	pH
Pancreatic juice	7.5 – 8.0
Blood plasma (or whole blood)	7.35 – 7.45
Cerebrospinal fluid	7.2 – 7.4
Tears	7.2 – 7.4
Interstitial fluid	7.2 – 7.4
Human milk	7.2 – 7.4
Saliva	6.4 – 7.0
Intracellular fluid (cytosol)	6.5 – 6.9
Gastric juice	1.5 – 3.0
Urine	5.0 – 7.5

are defined as the solutions which resist change in pH by the addition of small amounts of acids or bases. A buffer usually consists of a weak acid and its salt (e.g. acetic acid and sodium acetate) or a weak base and its salt (e.g. ammonium hydroxide and ammonium chloride). Several buffers can be prepared in the laboratory. Nature has provided many buffers in the living system.

Mechanism of buffer action

Let us consider the buffer pair of acetic acid and sodium acetate. Acetic acid, being a weak acid, feebly ionizes. On the other hand, sodium acetate ionizes to a large extent.

$$CH_3COOH \rightleftharpoons CH_3COO^- + H^+$$

$$CH_3COONa \rightleftharpoons CH_3COO^- + Na^+$$

When an acid (say HCl) is added, the acetate ions of the buffer bind with H^+ ions (of HCl) to form acetic acid which is weakly ionizing. Therefore, the pH change due to acid is resisted by the buffer.

$$H^+ + CH_3COO^- \longrightarrow CH_3COOH$$

When a base (say NaOH) is added the H^+ ions of the buffer (acetic acid) combine with OH^- ions to form water, which is weakly dissociated. Thus, the pH change due to base addition is also prevented by the buffer.

$$OH^- + H^+ \longrightarrow H_2O$$

Buffering capacity : The ***efficiency of*** a buffer in ***maintaining a constant pH*** on the addition of acid or base is referred to as buffering capacity. It mostly depends on the concentration of the buffer components. The maximum buffering capacity is usually achieved by keeping the same concentration of the salt as well as the acid.

For a comprehensive discussion on blood buffers, refer ***Chapter 21.***

SOLUTIONS

Solutions may be regarded as mixtures of substances. In general, a solution is composed of two parts—***solute*** and ***solvent.*** The substance that is dissolved is solute and the medium that dissolves the solute is referred to as solvent. The particle size of a solute in solution is < 1 nm.

The relative concentrations of substances in a solution can be measured by several ways.

Per cent concentration : This represents parts per 100. The most frequently used is weight per volume (w/v) e.g. ***9% saline (9 g/100 ml solution)***. For expressing smaller concentration, mg (10^{-3}g), µg (10^{-6} g), ng (10^{-9} g) and pg (10^{-12} g) are used.

Parts per million (ppm) : This refers to the number of parts of a substance in one million parts of the solution. Thus ***10 ppm chlorine means 10 µg of chlorine in 1 g of water.***

Molarity (M) : It is defined as the ***number of moles of solute per liter solution***. NaCl has a molecular weight of 58.5. To get one molar (1 M) or one mole solution of NaCl, one gram molecular weight (58.5 g) of it should be dissolved in the solvent (H_2O) to make to a final total volume of 1 liter. For smaller concentrations, millimole and micromole are used.

Molality : It represents the number ***of moles of solute per 1,000 g of solvent***. One molal solution can be prepared by dissolving 1 mole of solute in 1,000 g of solvent.

Normality : Molarity is based on molecular weight while normality is ***based on equivalent***

weight. One gram equivalent weight of an element or compound represents its capacity to combine or replace 1 mole of hydrogen.

COLLOIDAL STATE

Thomas Graham (1861), regarded as the 'father of colloidal chemistry', divided substances into two classes—***crystalloids*** and ***colloids.***

Crystalloids are the substances which in solution can freely pass (diffuse) through parchment membrane e.g. sugar, urea, NaCl. Colloids (*Greek* : glue-like), on other hand, are the substances that are retained by parchment membrane e.g. gum, gelatin, albumin. The above classification of Graham is no longer tenable, since any substance can be converted into a colloid by suitable means. For instance, sodium chloride in benzene forms a colloid.

Colloidal state : As such, there are no group of substances as colloids, rather, substances can exist in the form of ***colloidal state*** or ***colloidal system.*** Colloidal state is characterized by the ***particle size of 1 to 100 nm***. When the particle size is <1 nm, it is in true solution. For the particle sizes >100 nm, the matter exists as a visible precipitate. Thus, the colloidal state is an intermediate between true solution and precipitate.

Phases of colloids : Colloidal state is heterogeneous with two phases.

1. **Dispersed phase** (internal phase) which constitutes the colloidal particles.

2. **Dispersion medium** (external phase) which refers to the medium in which the colloidal particles are suspended.

CLASSIFICATION OF COLLOIDS

Based on the affinity of dispersion medium with dispersed phase, colloids are classified as ***lyophobic*** and ***lyophilic*** colloids.

1. **Lyophobic** (*Greek* : solvent-hating) **:** These colloids ***do not have any attraction towards dispersion medium***. When water is used as dispersion medium, the colloids are referred to as ***hydrophobic.***

2. **Lyophilic** (*Greek* : solvent-loving) **:** These colloids have ***distinct affinity towards dispersion medium***. The term ***hydrophilic*** is used for the colloids when water is the dispersion medium.

The terms ***gel*** and ***sol*** are, respectively, used to jelly-like and solution-like colloids. ***Emulsions*** are the colloids formed by two immiscible liquids (e.g. oil + water). Frequently, emulsions can be stabilized by using agents known as ***emulsifiers.*** For instance, the protein casein acts as an emulsifier for milk.

Micelles are the aggregates of colloidal particles. Soap (sodium palmitate) in water is the classical example for the micelles formation.

Properties of colloids

1. **Brownian movement :** The continuous and ***haphazard motion of the colloidal particles*** is known as Brownian movement.

2. **Optical properties :** When light is passed through a colloidal solution, it gets scattered. This phenomenon is referred to as ***Tyndal effect***.

3. **Electrical properties :** The colloidal particles carry electrical charges, either positive or negative. The separation of charged colloids can be achieved by the analytical technique—electrophoresis (***Refer Chapter 41***).

4. **Osmotic pressure :** Since the colloidal particles are larger in size, their contribution to osmotic pressure is relatively less.

5. **Non-dialysable nature :** The colloidal particles, being larger in size, cannot pass through a membrane (cellophane or parchment). The membrane, however, allows dispersion medium and smaller particles to escape through the pores. This process is referred to as dialysis and is useful for the separation of colloids.

6. **Donnan membrane equilibrium :** The presence of non-diffusible colloidal particles (e.g. protein) in the biological systems influences the concentration of diffusible ions across the membrane. This is an important phenomenon, the details of which are given on page 682.

Biological importance of colloids

1. **Biological fluids as colloids :** These include blood, milk and cerebrospinal fluid.

2. **Biological compounds as colloidal particles :** The complex molecules of life, the high molecular weight proteins, complex lipids and polysaccharides exist in colloidal state.

3. **Blood coagulation :** When blood clotting occurs, the ***sol*** is converted finally into the ***gel.***

4. **Fat digestion and absorption :** The formation of ***emulsions***, facilitated by the emulsifying agents bile salts, ***promotes fat digestion and absorption*** in the intestinal tract.

5. **Formation of urine :** The filtration of urine is based on the principle of ***dialysis***.

DIFFUSION

The molecules in liquids or gases are in continuous motion. ***Diffusion may be regarded as the movement of solute molecules from a higher concentration to a lower concentration.*** Diffusion is more rapid in gases than in liquids. The smaller particles diffuse faster than the larger ones. The greater the temperature, the higher is the rate of diffusion.

Diffusion occurs in true solutions as well as in colloidal solutions.

Applications of diffusion

1. Exchange of O_2 and CO_2 in lungs and in tissues occurs through diffusion.

2. Certain nutrients are absorbed by diffusion in the gastrointestinal tract e.g. pentoses, minerals, water soluble vitamins.

3. Passage of the waste products namely ammonia, in the renal tubules occurs due to diffusion.

OSMOSIS

Osmosis (*Greek* : push) refers to the ***movement of solvent*** (most frequently water) ***through a semipermeable membrane***.

The flow of solvent occurs from a solution of low concentration to a solution of high concentration, when both are separated by a semipermeable membrane. In a strict sense, the semipermeable membrane is expected to be permeable to the solvent and not to the solute.

Osmotic pressure

Osmotic pressure may be defined as ***the excess pressure that must be applied to a solution to prevent the passage of solvent into the solution***, when both are separated by a semipermeable membrane.

Osmosis is a ***colligative property*** i.e. a character which ***depends on the number of solute particles and not their nature***. Osmotic pressure is directly proportional to the concentration (number) of the solute molecules or ions. Low molecular weight substances (e.g. NaCl, glucose) will have more number of molecules compared to high molecular weight substances (albumin, globulin) for unit mass. Therefore, the substances with low molecular weight, in general, exhibit greater osmotic pressure. Further, for ionizable compounds, the total osmotic pressure is equivalent to the sum of the individual pressures exerted by each ion. For instance, one molar solution of NaCl will exert double the osmotic pressure of one molar solution of glucose. This is because NaCl ionizes to Na^+ and Cl^- while glucose is non-ionizable.

The solutions that exert the same osmotic pressure are said to be ***isoosmotic.*** The term ***isotonic*** is used when a cell is in direct contact with an isoosmotic solution (0.9% NaCl) which does not change the cell volume and, thus, the cell tone is maintained. A solution with relatively greater osmotic pressure is referred to as ***hypertonic***. On the other hand, a solution with relatively lower pressure is ***hypotonic.***

The term ***oncotic pressure*** is commonly used to represent the ***osmotic pressure of colloidal substances*** (e.g. albumin, globulin).

Units of osmotic pressure : ***Osmole*** is the unit of osmotic pressure. Osmotic pressure of biological fluids is frequently expressed as ***milliosmoles.*** The osmotic pressure of plasma is 280–300 milliosmoles/l.

Applications of osmosis

1. **Fluid balance and blood volume :** The fluid balance of the different compartments of the body is maintained due to osmosis. Further, osmosis significantly contributes to ***the regulation of blood volume and urine excretion***.

2. **Red blood cells and fragility :** When RBC are suspended in an isotonic (0.9% NaCl) solution, the cell volume remains unchanged and they are intact. In hypertonic solution (say 1.5% NaCl), ***water flows out of RBC*** and the cytoplasm shrinks, a phenomenon referred to as ***crenation.***

On the other hand, when the RBC are kept in ***hypotonic solution*** (say 0.4% NaCl), the cells bulge due to entry of water which often causes rupture of plasma membrane of RBC (***hemolysis***).

Osmotic fragility test for RBC is employed in laboratories for diagnostic purposes. For a normal human blood, RBC begin to hemolyse in 0.45% NaCl and the hemolysis is almost complete in 0.33% NaCl. ***Increased fragility*** of RBC is observed ***in hemolytic jaundice*** while it is decreased in certain anemias.

3. **Transfusion :** ***Isotonic solutions*** of NaCl (0.9%) or glucose (5%) or a suitable combination of these two are commonly used in transfusion in hospitals for ***the treatment of dehydration, burns etc***.

4. **Action of purgatives :** The mechanism of action of purgatives is mainly due to ***osmotic phenomenon***. For instance, epson ($MgSO_4$ $7H_2O$) or Glauber's (Na_2SO_4 $10H_2O$) salts withdraw water from the body, besides preventing the intestinal water absorption.

5. **Osmotic diuresis :** The ***high blood glucose*** concentration causes osmotic diuresis resulting in the ***loss of water, electrolytes and glucose in the urine***. This is the basis of polyuria observed in diabetes mellitus. Diuresis can be produced by administering compounds (e.g. mannitol) which are filtered but not reabsorbed by renal tubules.

6. **Edema due to hypoalbuminemia :** Disorders such as kwashiorkor and glomerulonephritis are associated with lowered plasma albumin concentration and edema. Edema is caused by reduced oncotic pressure of plasma, leading to the accumulation of excess fluid in tissue spaces.

7. **Cerebral edema :** Hypertonic solutions of salts (NaCl, $MgSO_4$) are in use to reduce the volume of the brain or the pressure of cerebrospinal fluid.

8. **Irrigation of wounds :** Isotonic solutions are used for washing wounds. The pain experienced by the direct addition of salt or sugar to wounds is due to osmotic removal of water.

DONNAN MEMBRANE EQUILIBRIUM

When membrane is freely permeable to ions (say Na^+, Cl^-) and if the concentration of ions on both the sides is different, the ions freely diffuse to attain equal concentration. Gibbs-Donnan observed that the ***presence of a non-diffusible ion on one side of the membrane alters the diffusion of diffusible ions.***

In the molecule sodium proteinate (Na^+Pr^-), the protein (Pr^-) ion is non-diffusable through the membrane. Let us consider two sides of a compartment separated by a membrane. Initially, sodium proteinate is on side I while sodium chloride is on side II (***Fig.40.2***). Diffusible ions (Na^+, Cl^-) can freely pass through the membrane. On side I, Na^+ ions will balance the incoming Cl^- ions besides Pr^- ions, while on side II Na^+ ions have to balance only Cl^- ions. Therefore, the concentration of Na^+ on side I is greater than on side II. However, from the thermodynamical

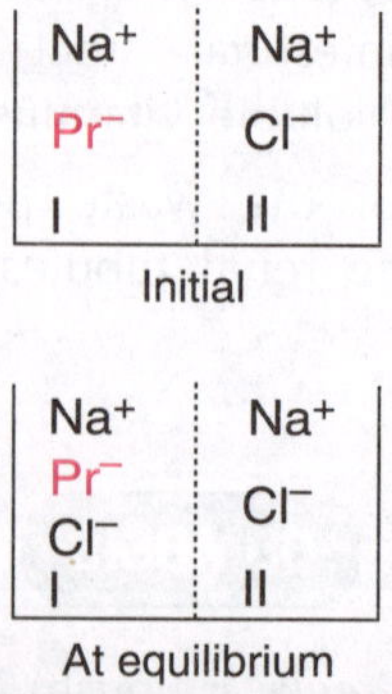

Fig. 40.1 : Diagrammatic representation of Donnan membrane equilibrium.

point of view, at equilibrium, the concentration of $Na^+ Cl^-$ on both the sides should be the same.

Thus $Na^+ Cl^-$ (I) = $Na^+ Cl^-$ (II)

Since Na^+ (I) > Na^+ (II)

Cl^- (I) < Cl^- (II)

Consequently, the concentration of Cl^- ions should be greater on side II. Further, the total concentration of ions on side I is higher than on side II.

The net concentration of total ions will be greater on the side of the membrane containing non-diffusible ions. This leads to a difference in the osmotic pressure on either side of the membrane.

Applications of Donnan membrane equilibrium

1. **Difference in the ionic concentrations of biological fluids :** The lymph and interstitial fluids have lower concentration of inorganic cations (Na^+, K^+) and higher concentration of anions (Cl^-) compared to plasma. This is due to the higher protein (Pr^-) content in plasma.

2. **Membrane hydrolysis :** The relative strength of H^+ and OH^- ions and, therefore, the acidic or alkaline nature on either side of a membrane, is influenced by the presence of non-diffusible ions. This phenomenon is referred to as membrane hydrolysis. Donnan membrane equilibrium explains the ***greater*** concentration of ***H^+ ions in the gastric juice.***

3. **Lower pH in RBC :** Hemoglobin of RBC is negatively charged and, this causes accumulation of positively charged ions including H^+. Therefore, the pH of RBC is slightly lower (7.25) than that of plasma (7.4).

4. **Osmotic imbalance :** Donnan membrane equilibrium—which results in the differential distribution of ions in different compartments of the body—partly explains the osmotic pressure differences.

5. **Dialysis in renal failure :** Donnan membrane equilibrium is the ***basic principle*** involved in the artificial means of ***purifying blood by dialysis*** in the patients of renal failure.

VISCOSITY

Liquid or fluid has a tendency to flow which is referred to as fluidity. The term viscosity may be defined as the ***internal resistance offered by a liquid or a gas to flow***. The property of viscosity is due to frictional forces between the layers while their movement occurs. Viscosity may be appropriately regarded as the internal friction of a liquid.

Liquids vary widely as regards their viscosity. For instance, ether has very low viscosity while honey and blood are highly viscous.

Viscosity of colloidal solutions, particularly lyophilic colloids, is generally higher than that of true solutions.

Units of viscosity : The unit of viscosity is ***poise***, after the scientist Poiseuille, who first systematically studied the flow of liquids. A poise represents ***dynes/cm²***.

Applications of viscosity

1. **Viscosity of blood :** Blood is about ***4 times more*** viscous than water. The viscosity of blood is mainly attributed to ***suspended blood cells*** and colloidal ***plasma proteins***. As the blood flows through capillaries the viscosity decreases to facilitate free flow of blood. Blood viscosity is increased in polycythemia (elevation of RBC), while it is reduced in anemia and nephritis. A more viscous blood increases cardiac work load. When dehydration occurs, the viscosity of the blood increases.

2. **Viscosity change in muscle :** Excitation of the muscle is associated with increase in the viscosity of the muscle fibres. This delays the change in the tension of the contracting muscle.

3. **Vitreous body :** This is an amorphous viscous body located in the posterior chamber of the eye. It is rich in albumin and hyaluronic acid.

4. **Synovial fluid :** It contains hyaluronic acid which imparts viscosity and helps in the lubricating function of joints.

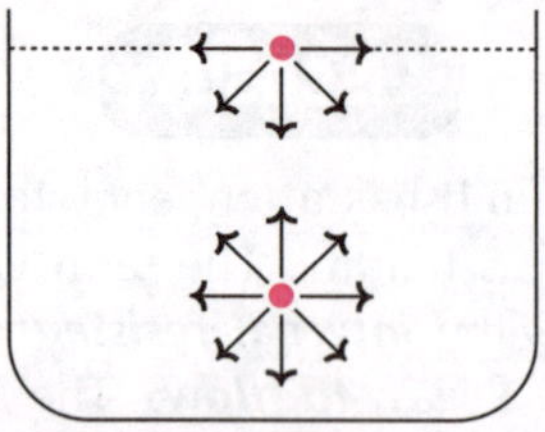

Fig. 40.2 : Surface tension of a liquid.

SURFACE TENSION

A molecule in the interior of a liquid is attracted by other molecules in all directions. In contrast, a molecule on the surface is attracted only downwards and sideways and not upwards (***Fig.40.2***). Due to this, the surface layer behaves like a stretched film. ***Surface tension is the force with which the molecules on the surface are held together.*** It is expressed as dynes/cm. Surface tension decreases with increase in temperature.

Due to the phenomenon of surface tension, any liquid occupies the minimum possible volume.

Applications of surface tension

1. **Digestion and absorption of fat :** Bile salts reduce the surface tension. They act as detergents and cause emulsification of fat, thereby allowing the formation of minute particles for effective digestion and absorption.

2. **Hay's sulfur test :** This is a common laboratory test employed for the detection of ***bile salts in urine of jaundice*** patients. Sulfur powder, when sprinkled on the surface of urine possessing bile salts, sinks. This is in contrast to a normal urine where sulfur powder floats. Hay's test is based on the principle that bile salts in urine lower surface tension which is responsible for sulfur to sink.

3. **Surfactants and lung function :** The low surface tension of the alveoli keeps them apart and allows an efficient exchange of gases in lungs. In fact, certain surfactants, predominantly dipalmitoyl phosphatidyl choline (***dipalmitoyl lecithin***) are responsible for maintaining low surface tension in the alveoli. ***Surfactant deficiency causes respiratory distress syndrome in the infants.***

4. **Surface tension and adsorption :** Adsorption, being a surface phenomenon, is closely related to surface tension. Due to the coupled action of these two processes, the formation of complexes of proteins and lipids occurs in the biological systems.

5. **Lipoprotein complex membranes :** The structure of plasma membrane is composed of surface tension reducing substances, namely lipids and proteins. This facilitates absorption of these compounds.

ADSORPTION

Adsorption is a ***surface phenomenon***. It is the process of accumulation of a substance (adsorbate) on the surface of another substance (adsorbent). Adsorption differs from absorption, as the latter involves the diffusion into the interior of the material.

The capacity of an adsorbent depends on the surface area. Therefore, porous substances serve as better adsorbents e.g. charcoal, alumina, silica gel. Adsorption is a dynamic and reversible process which decreases with rise in temperature.

Applications of adsorption

1. **Formation of enzyme-substrate complex :** For the catalysis to occur in biological system, formation of enzyme-substrate complex is a prerequisite. This happens by adsorption of substrate on the enzyme.

2. **Action of drugs and poisons :** On adsorption at the cell surface, drugs and poisons exert their action.

3. **Adsorption in analytical biochemistry :** The principle of adsorption is widely employed in the chromatography technique for the separation and purification of compounds (enzymes, immunoglobulins).

ISOTOPES

Isotopes have revolutionized biochemistry when they became available to investigators soon after Second World War. Isotopes are defined as the ***elements with same atomic number but different atomic weights.*** They possess the same number of protons but differ in the neutrons in their nuclei. Therefore, isotopes (*Greek* : iso—equal; tope—place) occupy the same place in the periodic table. The chemical properties of different isotopes of a particular element are identical.

Isotopes are of two types—***stable*** and ***unstable.*** The latter are more commonly referred to as radioactive isotopes and they are of particular interest to biochemists. Conventionally while representing isotopes, the atomic weight is written on upper left side of the element symbol.

Stable isotopes

They are naturally occurring and do not emit radiations (non-radioactive) e.g. ***deuterium (heavy hydrogen)*** ^2H; ^{13}C; ^{15}N; ^{18}O. Stable isotopes can be identified and quantitated by mass spectrometry or nuclear magnetic resonance (NMR). They are less frequently used in biochemical investigations.

Radioactive isotopes

The atomic nucleus of radioactive isotopes is unstable and, therefore, undergoes decay. The radioactive decay gives rise to one of the following 3 ionizing radiations.

1. **α-Rays**—an α particle possessing 2 protons i.e. helium nuclei.

2. **β-Rays**—due to the emission of electrons.

3. **γ-Rays**—due to emission of high energy photons.

The radiations emmitted by radioactive nuclei are characteristic of the isotope. For instance, ^3H, ^{14}C, and ^{32}P all emit β-particles in the respective energies of 0.018, 0.155 and 1.71 MeV.

The ***β and γ emitting radioisotopes are employed in biochemical research***. These isotopes are produced in nuclear reactors. The simple chemicals so produced are then converted to radiolabelled biochemicals by chemical or enzymatic synthesis.

Table 40.2 Commonly used radioisotopes in biochemistry

Isotope	Radiation	Half-life
^3H	β	12.2 years
^{14}C	β	5,700 years
^{22}Na	γ	2.5 years
^{32}P	β	14.5 days
^{35}S	β	87 days
^{45}Ca	β	164 days
^{59}Fe	β, γ	45 days
^{60}Co	γ	5.25 years
^{125}I	γ	60 days
^{131}I	β, γ	8.1 days

Units of radioactivity : *Curie* (Ci) is the basic unit of radioactive decay. It is defined as the amount of radioactivity equivalent to 1 g of radium i.e. 2.22 × 10^{12} disintegrations per minute (dpm). Millicurie (mCi) and microcurie (μCi), respectively, corresponding to 2.2 × 10^9 and 2.2 × 10^6 dpm, are more commonly used.

Half-lives of isotopes : The unstable radioisotopes undergo decay. The radioactivity gets reduced to half of the original within a fixed time. This represents the half-life which is characteristic for a given isotope.

Some of the commonly used radioactive isotopes in biochemical research with their characteristics are given in ***Table 40.2***.

Measurement of radioactivity of isotopes

Several techniques are in use for the detection of radioactivity of the isotopes. The most commonly employed in biochemical research are—***Geiger counters, liquid scintillation counter*** and ***autoradiography.*** Geiger counters are almost outdated. Liquid scintillation counters are now widely used.

In the liquid scintillation counter, the sample is dissolved or suspended in a solution containing one or two fluorescent organic compounds (fluors). The fluors emit a pulse of light when struck by radiation. The light, proportional to the radiation energy, can be detected. The advantage with liquid scintillation counter is that it can discriminate the particles of different energies. Thus, two or more isotopes can be simultaneously detected.

In autoradiography, the radiations are detected by its blackening of photographic film. This technique is commonly used for the detection of radioactive substances separated in polyacrylamide gel electrophoresis (PAGE).

Applications of radioisotopes in biochemistry

Radioactive isotopes have become indispensable tools of biochemistry. They can be conveniently used as ***tracers in biochemical research*** since the chemical properties of different isotopes of a particular element are identical. Therefore, the living cells cannot distinguish the radioactive isotope from a normal atom.

Radioisotopes are widely used in establishing the precursor-product relationships in metabolisms and ***understanding of the complex metabolic pathways***.

A few important application of radioisotopes are

1. By the use of isotope tracers, the metabolic origin of complex molecules such as heme, cholesterol, purines and phospholipids can be determined. As early as 1945, it was established that nitrogen atom of heme was derived from glycine. This was done by feeding rats with (^{15}N) glycine and detecting (^{15}N) heme.

2. The ***precursor-product relationship*** in several metabolic pathways has been ***investigated*** by radioisotopes. e.g. Krebs cycle, β-oxidation of fatty acids, urea cycle, fatty acid synthesis.

3. Radioisotopes are conveniently used in the ***study of metabolic pools*** (e.g. amino acid pool) and metabolic turnovers (e.g. protein turnover).

4. Certain endocrine and immunological studies also depend on the use of radioisotopes e.g. ***radioimmunoassay***.

5. Radioisotopes are employed in elucidating ***drug metabolism***.

Radioisotopes in diagnosis and treatment

Certain radioisotopes are used in the ***scanning of organs—thyroid gland (^{131}I), bone (^{90}Sr) and kidney (^{131}I hippuran)***.

Radioactivity has been employed in the ***treatment of cancers***. This is based on the principle that radiations produce ionizations which damage nucleic acids. Thus, the uncontrolled proliferation of cells is restricted.

Tools of Biochemistry

41

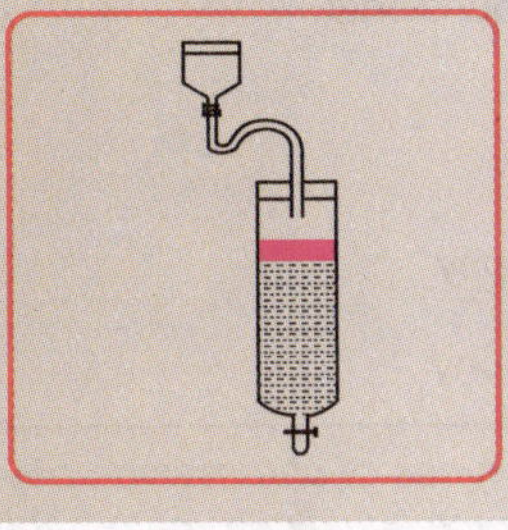

The chromotography speaks

"A technique I am, separating mixture of compounds;
To isolate, identify, characterize molecules, as desired;
Working on the principles of adsorption, partition, ion-exchange;
I am a key biochemical tool in laboratory experimentation."

Biochemistry is an experimental rather than a theoretical science. The understanding and development of concepts in biochemistry are a result of continuous experimentation and evidence obtained therein. It is no exaggeration to state that the foundations for the present (and the future, of course!) ***knowledge of biochemistry are based on the laboratory tools employed for biochemical experimentation***. Thus, the development of sensitive and sophisticated analytical techniques has tremendously contributed to our understanding of biochemistry.

A detailed discussion on the tools of biochemistry is beyond the scope of this book. The ***basic principles of some of the commonly employed tools are described in this chapter***. The reader must, however, ***refer Chapter 27***, for the following techniques related to molecular biology,and recombinant DNA technology

- Isolation and purification of nucleic acids
- Nucleic acid blotting techniques
- DNA sequencing
- Polymerase chain reaction
- Methods of DNA assay
- DNA fingerprinting or DNA profiling.

CHROMATOGRAPHY

Chromatography is one of the most useful and popular tools of biochemistry. It is an analytical technique dealing with the ***separation of closely related compounds from a mixture***. These include proteins, peptides, amino acids, lipids, carbohydrates, vitamins and drugs.

Historical perspective

The credit for the discovery of chromatography goes to the ***Russian botanist Mikhail Tswett***. It was in 1906, Tswett described the ***separation of plant leaf pigments*** in solution by passing through a column of solid adsorbents. He coined the term chromatography (*Greek* : chroma—colour; graphein—to write), since the technique dealt with the separation of colour compounds (pigments). Coincidently, the term Tswett means colour in Russian! Truly speaking, chromatography is a misnomer, since it is no longer limited to the separation of coloured compounds.

Principles and classification

Chromatography usually consists of a ***mobile phase*** and a ***stationary phase.*** The mobile phase refers to the mixture of substances (to be separated), dissoved in a liquid or a gas. The stationary phase is a porous solid matrix through which the sample contained in the mobile phase percolates. The interaction between the mobile and stationary phases results in the separation of the compounds from the mixture. These interactions include the physicochemical principles such as ***adsorption, partition, ion-exchange, molecular sieving*** and ***affinity.***

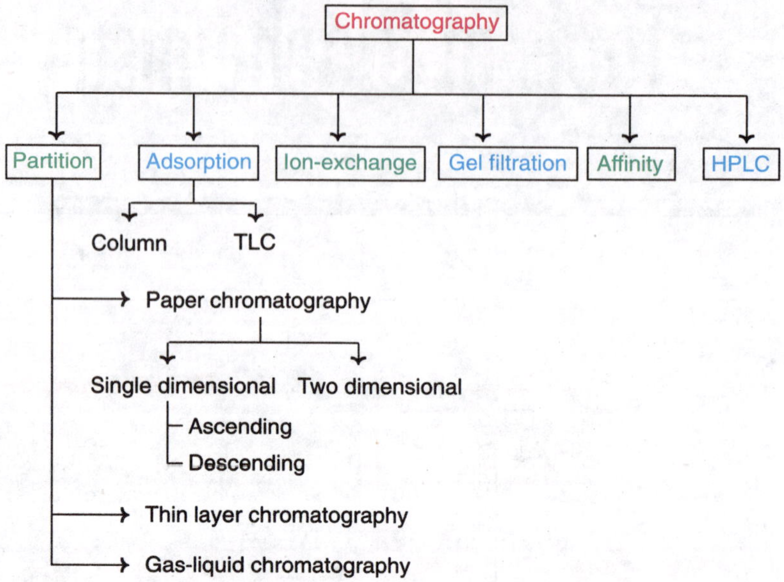

Fig. 41.1 : *Important types of chromatography (HPLC—High performance liquid chromatography; TLC—Thin layer chromatography).*

The interaction between stationary phase and mobile phase is often employed in the classification chromatography e.g. partition, adsorption, ion-exchange. Further, the classification of chromatography is also based either on the nature of the stationary phase (paper, thin layer, column), or on the nature of both mobile and stationary phases (gas-liquid chromatography). A summary of the different methods (classes) of chromatography is given in ***Fig.41.1***.

1. **Partition chromatography :** The molecules of a mixture get partitioned between the stationary phase and mobile phase depending on their relative affinity to each one of the phases.

(a) **Paper chromatography :** This technique is commonly used for the separation of amino acids, sugars, sugar derivatives and peptides. In paper chromatography, a few drops of solution containing a mixture of the compounds to be separated is applied (spotted) at one end, usually ~2 cm above, a strip of filter paper (Whatman No. 1 or 3). The paper is dried and dipped into a solvent mixture consisting of butanol, acetic acid and water in 4 : 1 : 5 ratio (for the separation of amino acids). The aqueous component of the solvent system binds to the paper and forms a stationary phase. The organic component that migrates on the paper is the mobile phase. When the migration of the solvent is upwards, it is referred to as ***ascending chromatography.*** *In* ***descending chromatography***, the solvent moves downwards (***Fig.41.2***). As the solvent flows, it takes along with it the unknown substances. The rate of migration of the molecules depends on the relative solubilities in the stationary phase (aqueous) and mobile phase (organic).

After a sufficient migration of the solvent front, the paper (chromatogram) is removed, dried and developed for the identification of the specific spots. ***Ninhydrin***, which forms purple complex with ***α-amino acids***, is frequently used as a colouring reagent.

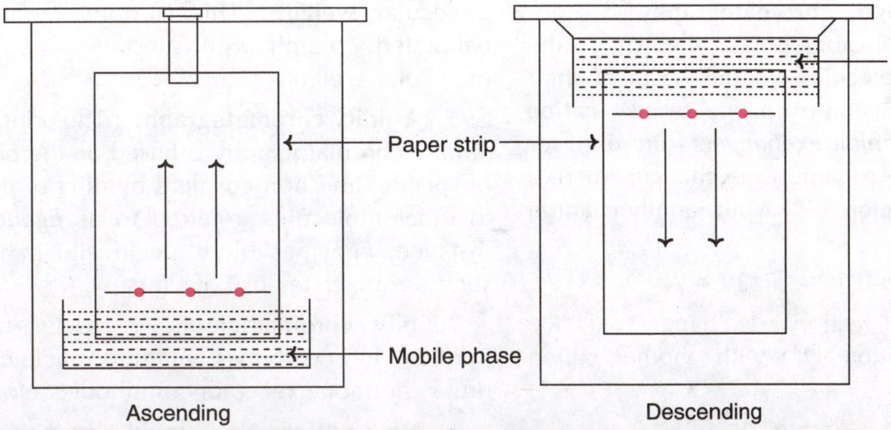

Fig. 41.2 : Paper chromatography—ascending and descending types.

The chemical nature of the individual spots can be identified by running known standards with the unknown mixture.

The migration of a substance is frequently expressed as ***R_f value*** (**r**atio of **f**ronts)

$$R_f = \frac{\textit{Distance travelled by the substance}}{\textit{Distance travelled by solvent front}}$$

(b) **Thin layer chromatography (TLC) :** The principle of TLC is the same as described for paper chromatography (partition). In place of a paper, an inert substance, such as cellulose, is employed as supporting material. Cellulose is spread as a thin layer on glass or plastic plates. The chromatographic separation is comparatively rapid in TLC.

(c) **Gas-liquid chromatography (GLC) :** This is the method of choice for ***the separation of volatile substances*** or the volatile derivatives of certain non-volatile substances.

Gas-liquid chromatography is sensitive, rapid and reliable. It is frequently used for the quantitative estimation of biological materials such as lipids, drugs and vitamins.

2. **Adsorption column chromatography :** The adsorbents such as silica gel, alumina, charcoal powder and calcium hydroxyapatite are packed into a column in a glass tube. This serves as the stationary phase. The sample mixture in a solvent is loaded on this column. The individual components get differentially adsorbed on to the adsorbent. The elution is carried out by a buffer system (mobile phase). The individual compounds come out of the column at different rates which may be separately collected and identified (***Fig.41.3***). For instance, amino acids can be identified by ninhydrin calorimetric method. An automated column chromatography apparatus—***fraction collector***—is frequently used nowadays.

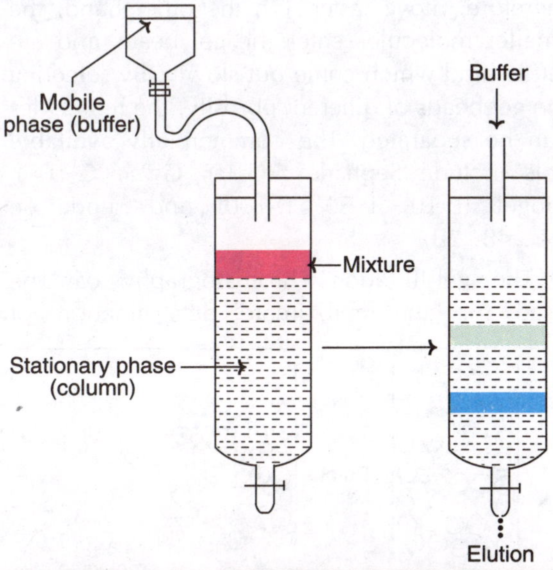

Fig. 41.3 : Diagrammatic representation of adsorption column chromatography.

3. **Ion-exchange chromatography :** Ion-exchange chromatography involves the separation of molecules on the basis of their electrical charges. Ion-exchange resins—***cation exchangers*** and ***anion exchangers***—are used for this purpose. An anion exchanger (R^+A^-) exchanges its anion (A^-) with another anion (B^-) in solution.

$$R^+A^- + B^- \rightleftharpoons R^+B^- + A^-$$

Similarly, a cation exchanger (H^+R^-) exchanges its cation (H^+) with another cation (C^+) in solution.

$$H^+R^- + C^+ \rightleftharpoons C^+R^- + H^+$$

4. **Gel filtration chromatography :** In gel filtration chromatography, the separation of molecules is based on their size, shape and molecular weight. This technique is also referred to as ***molecular sieve or molecular exclusion chromatography.*** The apparatus consists of a column packed with spongelike gel beads (usually cross-linked polysaccharides) containing pores. The gels serve as molecular sieves for the separation of smaller and bigger molecules (***Fig.41.4***).

The solution mixture containing molecules of different sizes (say proteins) is applied to column and eluted with a buffer. The larger molecules cannot pass through the pores of gel and, therefore, move faster. On the other hand, the smaller molecules enter the gel beads and are left behind which come out slowly. By selecting the gel beads of different porosity, the molecules can be separated. The commercially available gels include Sephadex (G-10, G-25, G-100), Biogel (P-10, P-30, P-100) and sepharose (6B, 4B, 2B).

The gel-filtration chromatography can be used for an approximate determination of molecular weights. This is done by using a calibrated column with substances of known molecular weight.

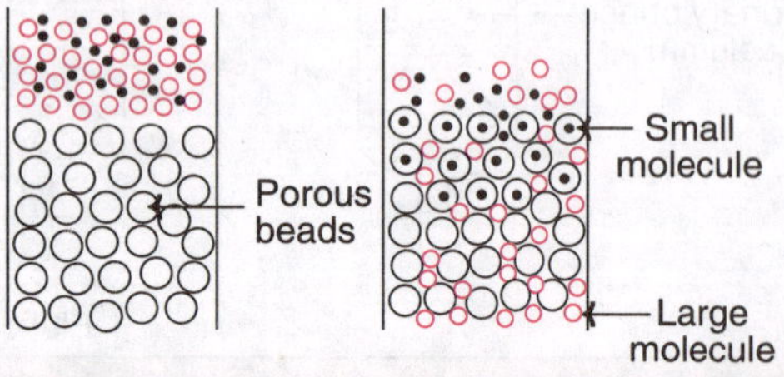

Fig. 41.4 : The principle of gel-filtration chromatography.

5. **Affinity chromatography :** The principle of affinity chromatography is based on the property of specific and non-covalent binding of proteins to other molecules, referred to as ***ligands.*** For instance, enzymes bind specifically to ligands such as substrates or cofactors.

Affinity chromatography is useful for the purification of enzymes, vitamins, nucleic acids, drugs, hormone receptors, antibodies etc.

6. **High performance liquid chromatography (HPLC) :** In general, the chromatographic techniques are slow and time consuming. The separation can be greatly improved by applying high pressure in the range of 5,000-10,000 psi (pounds per square inch), hence this technique is also referred (less frequently) to as high pressure liquid chromatography. HPLC requires the use of non-compressible resin materials and strong metal columns. The eluants of the column are detected by methods such as UV absorption and fluorescence.

ELECTROPHORESIS

The ***movement of charged particles*** (ions) ***in an electric field*** resulting in their migration towards the oppositely charged electrode is known as electrophoresis. Molecules with a net positive charge (cations) move towards the negative cathode while those with net negative charge (anions) migrate towards positive anode. Electrophoresis is a widely used analytical technique for the separation of biological molecules such as plasma proteins, lipoproteins and immunoglobulins.

The rate of migration of ions in an electric field depends on several factors that include shape, size, net charge and solvation of the ions, viscosity of the solution and magnitude of the current employed.

Different types of electrophoresis

Among the electrophoretic techniques, zone electrophoresis (paper, gel), isoelectric focussing

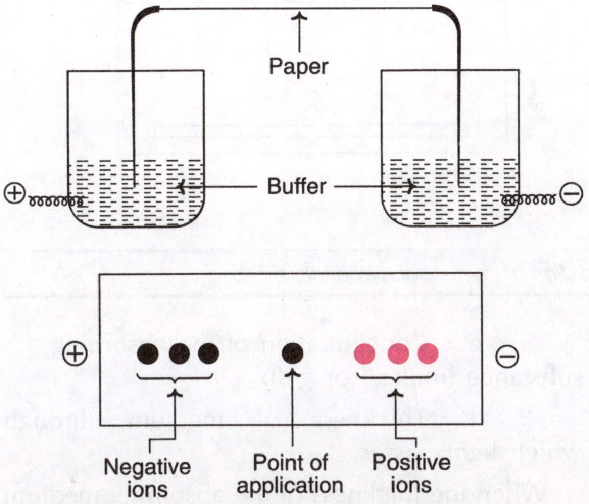

Fig. 41.5 : Diagrammatic representation of paper electrophoresis.

and immunoelectrophoresis are important and commonly employed in the laboratory. The original moving boundary electrophoresis, developed by Tiselius (1933), is less frequently used these days. In this technique, the U-tube is filled with protein solution overlaid by a buffer solution. As the proteins move in solution during electrophoresis, they form boundaries which can be identified by refractive index.

1. **Zone electrophoresis :** An inert supporting material such as paper or gel are used.

(a) **Paper electrophoresis :** In this technique, the sample is applied on a strip of filter paper wetted with desired buffer solution.

For the separation of serum proteins, Whatman No. 1 filter paper, veronal or tris buffer at pH 8.6 and the stains amido black or bromophenol blue are employed. The ***serum proteins are separated into five distinct bands***—albumin, α_1-, α_2-, β- and γ-globulins (Refer ***Fig.9.1***). For the electrophoretic pattern of serum lipoproteins, refer ***Fig.14.34***.

(b) **Gel electrophoresis :** This technique involves the separation of molecules based on their size, in addition to the electrical charge. The resolution is much higher in this technique. Thus, serum proteins can be separated to about 15 bands, instead of 5 bands on paper electrophoresis.

The gels commonly used in gel electrophoresis are agarose and polyacrylamide, sodium dodecyl sulfate (SDS). ***Polyacrylamide*** is employed for the determination of molecular weights of proteins in a popularly known electrophoresis technique known as SDS-PAGE.

2. **Isoelectric focussing :** This technique is primarily based on the ***immobilization of the molecules at isoelectric*** pH during electrophoresis. Stable pH gradients are set up (usually in a gel) covering the pH range to include the isoelectric points of the components in a mixture. As the electrophoresis occurs, the molecules (say proteins) migrate to positions corresponding to their isoelectric points, get immobilized and form sharp stationary bonds. The gel blocks can be stained and identified. By isoelectric focussing, serum proteins can be separated to as many as 40 bands. Isoelectric focussing can be conveniently used for the purification of proteins.

3. **Immunoelectrophoresis :** This technique involves combination of the principles of electrophoresis and immunological reactions. Immunoelectrophoresis is useful for the analysis of complex mixtures of antigens and antibodies.

The complex proteins of biological samples (say human serum) are subjected to electrophoresis. The antibody (antihuman immune serum from rabbit or horse) is then applied in a trough parallel to the electrophoretic separation. The antibodies diffuse and, when they come in contact with antigens, precipitation occurs, resulting in the formation of ***precipitin bands*** which can be identified (***Fig.41.6***).

PHOTOMETRY—COLORIMETER AND SPECTROPHOTOMETER

Photometry broadly deals with the study of the phenomenon of light absorption by

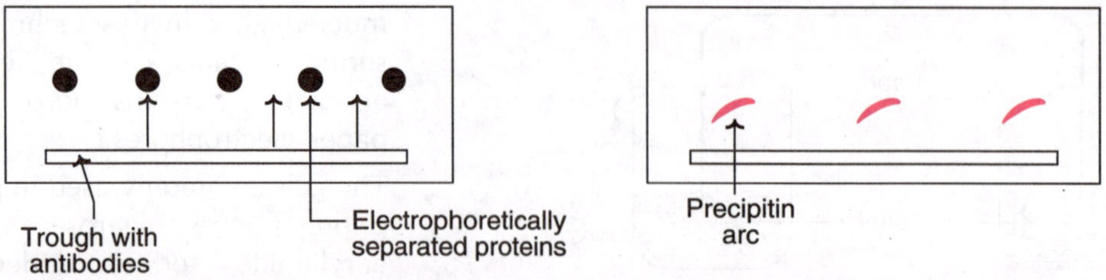

Fig. 41.6 : Diagrammatic representation of immunoelectrophoresis.

molecules in solution. The specificity of a compound to absorb light at a particular wavelength (***monochromatic light***) is exploited in the laboratory for quantitative measurements. From the biochemist's perspective, photometry forms an important laboratory tool for accurate estimation of a wide variety of compounds in biological samples. Colorimeter and spectrophotometer are the laboratory instruments used for this purpose. They work on the principles discussed below.

When a light at a particular wavelength is passed through a solution (incident light), some amount of it is absorbed and, therefore, the light that comes out (transmitted light) is diminished. The nature of light absorption in a solution is governed by ***Beer-Lambert law.***

Beer's law states that the amount of transmitted light decreases exponentially with an increase in the concentration of absorbing material (i.e. the amount of light absorbed depends on the concentration of the absorbing molecules). And according to Lambert's law, the transmitted light decreases exponentially with increase in the thickness of the absorbing molecules (i.e. the amount of light absorbed is dependent on the thickness of the medium).

By combining the two laws (Beer-Lambert law), the following mathematical derivation can be obtained

$$I = I_0^{\varepsilon ct}$$

where I = Intensity of the transmitted light

I_0 = Intensity of the incident light

ε = Molar extinction coefficient (characteristic of the substance being investigated)

c = Concentration of the absorbing substance (moles/l or g/dl)

t = Thickness of medium through which light passes.

When the thickness of the absorbing medium is kept constant (i.e. Lambert's law), the intensity of the transmitted light depends only on concentration of the absorbing material. In other words, the Beer's law is operative.

The ratio of transmitted light (I) to that of incident light (I_0) is referred to as transmittance (T).

$$T = \frac{I}{I_0}$$

Absorbance (A) or ***optical density (OD)*** is very commonly used in laboratories. The relation between absorbance and transmittance is expressed by the following equation.

$$A = 2 - \log_{10} \quad T = 2 - \log\% \, T$$

Colorimeter

Colorimeter (or photoelectric colorimeter) is the instrument used for the measurement of coloured substances. This instrument is operative in the visible range (400-800 nm) of the electromagnetic spectrum of light. The ***working of colorimeter*** is ***based on the principle of Beer-Lambert law*** (discussed above).

The colorimeter, in general consists of light source, filter sample holder and detector with display (meter or digital). A filament lamp usually serves as a light source.The filters allow the passage of a small range of wave length as incident light. The sample holder is a special glass cuvette with a fixed thickness. The photoelectric selenium cells are

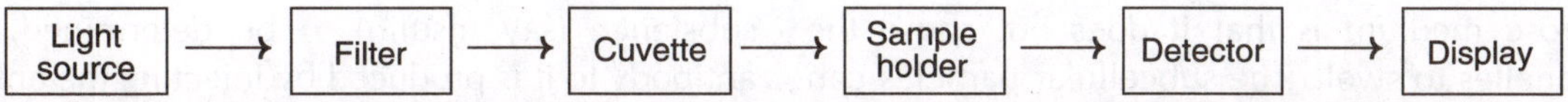

Fig. 41.7 : Diagrammatic representation of the components in a colorimeter.

the most common detectors used in colorimeter. The diagrammatic representation of a colorimeter is depicted in ***Fig.41.7***.

Spectrophotometer

The spectrophotometer primarily differs from colorimeter by covering the ultraviolet region (200-400 nm) of the electromagnetic spectrum. Further, the spectrophotometer is more sophisticated with several additional devices that ultimately increase the sensitivity of its operation severalfold when compared to a colorimeter. A precisely selected wavelength (say 234 nm or 610 nm) in both ultra-violet and visible range can be used for measurements. In place of glass cuvettes (in colorimeter), quartz cells are used in a spectrophotometer.

The spectrophotometer has similar basic components as described for a colorimeter (***Fig.41.7***), and its operation is also based on the Beer-Lambert law (already discussed).

FLUORIMETRY

When certain compounds are subjected to light of a particular wavelength, some of the ***molecules get excited***. These molecules, while they ***return to ground state***, ***emit light in the form of fluorescence*** which is proportional to the concentration of the compound. This is the principle in the operation of the instrument fluorometer.

FLAME PHOTOMETRY

Flame photometry primarily deals with the quantitative ***measurement of electrolytes*** such as sodium, potassium and lithium. The instrument, namely flame photometer, works on the following principle. As a solution in air is finally sprayed over a burner, it dissociates to give neutral atoms. Some of these atoms get excited and move to a higher energy state. When the excited atoms fall back to the ground state, they emit light of a characteristic wavelength which can be measured. The intensity of emission light is proportional to the concentration of the electrolyte being estimated.

ULTRACENTRIFUGATION

The ultracentrifuge was developed by a Swedish biochemist Svedberg (1923). The principle is based on the generation of centrifugal force to as high as 600,000 g (earth's gravity g = 9.81 m/s^2) that allows the sedimentation of particles or macromolecules. Ultracentrifugation is an indispensable tool for the ***isolation of subcellular organelles***, proteins and nucleic acids. In addition, this technique is also employed in the determination of molecular weights of macromolecules.

The rate at which the sedimentation occurs in ultracentrifugation primarily depends on the size and shape of the particles or macromolecules (i.e. on the molecular weight). It is expressed in terms of ***sedimentation coefficient***(s).

The sedimentation coefficient has the units of seconds. It was usually expressed in units of 10^{-13}s (since several biological macromolecules occur in this range), which is designated as one ***Svedberg unit.*** For instance, the sedimentation coefficient of hemoglobin is 4 × 10^{-13} s or 4S; ribonuclease is 2 × 10^{-13} s or 2S. Conventionally, the subcellular organelles are often referred to by their S value e.g. 70S ribosome.

Isolation of subcellular organelles by centrifugation

The cells are subjected to disruption by sonication or osmotic shock or by use of homogenizer. This is usually carried out in an isotonic (0.25 M) sucrose. The advantage with sucrose medium is that it does not cause the organelles to swell. The subcellular particles can be separated by differential centrifugation. The most commonly employed laboratory method separates subcellular organelles into 3 major fractions—nuclear, mitochondrial and microsomal (***Fig.41.8***).

When the homogenate is centrifuged at 700 g for about 10 min, the nuclear fraction (includes plasma membrane) gets sedimented. On centrifuging the supernatant (I) at 15,000 g for about 5 min mitochondrial fraction (that includes lysosomes, peroxisomes) is pelleted. Further centrifugation of the supernatant (II) at 100,000 g for about 60 min separates microsomal fraction (that includes ribosomes and endoplasmic reticulum). The supernatant (III) then obtained corresponds to the cytosol.

The ***purity*** (or contamination) of the subcellular fractionation can be ***checked by*** the use of ***marker enzymes. DNA polymerase is the marker enzyme for nucleus***, while ***glutamate dehydrogenase and glucose 6-phosphatase*** are ***the markers for mitochondria and ribosomes***, respectively. ***Hexokinase is the marker enzyme for cytosol***.

RADIOIMMUNOASSAY

Radioimmunoassay (***RIA***) was developed in 1959 by Solomon, Benson and Rosalyn Yalow for the estimation of insulin in human serum. This technique has revolutionized the estimation of several compounds in biological fluids that are found in exceedingly low concentrations (nanogram or picogram). RIA is a highly sensitive and specific analytical tool.

Principle

Radioimmunoassay combines the principles of ***radioactivity*** of isotopes and ***immunological*** reactions of antigen and antibody, hence the name.

The principle of RIA is primarily based on the competition between the labelled and unlabelled antigens to bind with antibody to form antigen-antibody complexes (either labelled or unlabelled). The unlabelled antigen is the substance (say insulin) to be determined. The antibody to it is produced by injecting the antigen to a goat or a rabbit. The specific antibody (Ab) is then subjected to react with unlabelled antigen in the presence of excess amounts of isotopically labelled (^{131}I) antigen (Ag^+) with known radioactivity. There occurs a competition between the antigens (Ag^+ and Ag) to bind the antibody. Certainly, the labelled Ag^+ will have an upper hand due to its excess presence.

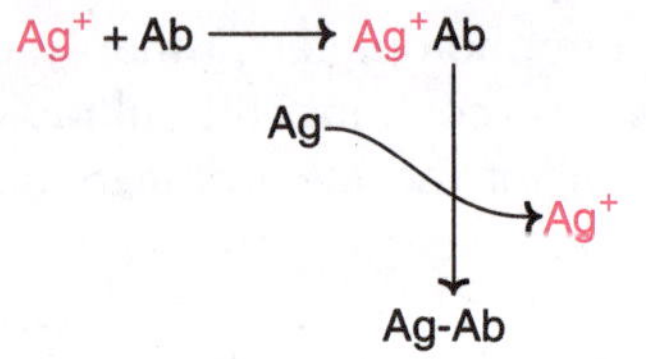

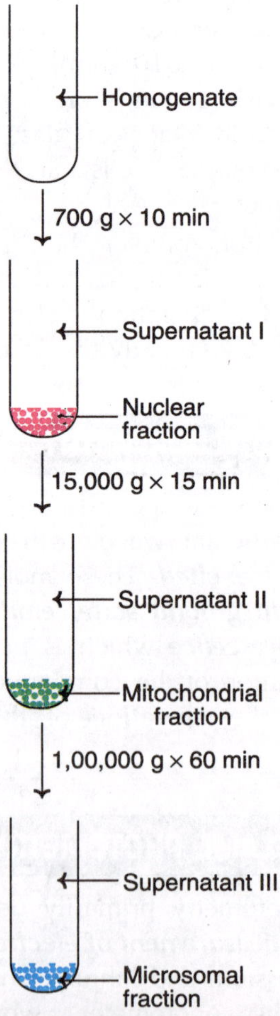

Fig. 41.8 : *Separation of subcellular fractions by differential centrifugation.*

As the concentration of unlabelled antigen (Ag) increases the amount of labelled antigen-antibody complex (Ag^+-Ab) decreases. Thus, the concentration of Ag^+-Ab is inversely related to the concentration of unlabelled Ag i.e. the substance to be determined. This relation is almost linear. A standard curve can be drawn by using different concentrations of unlabelled antigen and the same quantities of antibody and labelled antigen.

The labelled antigen-antibody (Ag^+-Ab) complex is separated by precipitation. The radioactivity of ^{131}I present is Ag^+-Ab is determined.

Applications

RIA is no more limited to estimating of hormones and proteins that exhibit antigenic properties. By the use of ***haptens*** (small molecules such as dinitrophenol, which, by themselves, are not antigenic), several substances can be made antigenic to elicit specific antibody responses. In this way, a wide variety of compounds have been brought under the net of RIA estimation. These include peptides, steroid hormones, vitamins, drugs, antibiotics, nucleic acids, structural proteins and hormone receptor proteins.

Radioimmunoassay has tremendous application in ***the diagnosis of hormonal disorders, cancers and therapeutic monitoring of drugs, besides being useful in biomedical research***.

ENZYME-LINKED IMMUNOSORBANT ASSAY

Enzyme-linked immunosorbant assay (***ELISA***) is a non-isotopic immunoassay. An enzyme is used as a label in ELISA in place of radioactive isotope employed in RIA. ELISA is as sensitive as or even more sensitive than RIA. In addition, there is no risk of radiation hazards (as is the case with RIA) in ELISA.

Principle

ELISA is based on the ***immunochemical principles of antigen-antibody reaction***. The stages of ELISA, depicted in ***Fig.41.9***, are summarized.

1. The antibody against the protein to be determined is fixed on an inert solid such as polystyrene.

2. The biological sample containing the protein to be estimated is applied on the antibody coated surface.

3. The protein antibody complex is then reacted with a second protein specific antibody to which an enzyme is covalently linked. These enzymes must be easily assayable and produce preferably coloured products. Peroxidase, amylase and alkaline phosphatase are commonly used.

4. After washing the unbound antibody linked enzyme, the enzyme bound to the second antibody complex is assayed.

5. The enzyme activity is determined by its action on a substrate to form a product (usually coloured). This is related to the concentration of the protein being estimated.

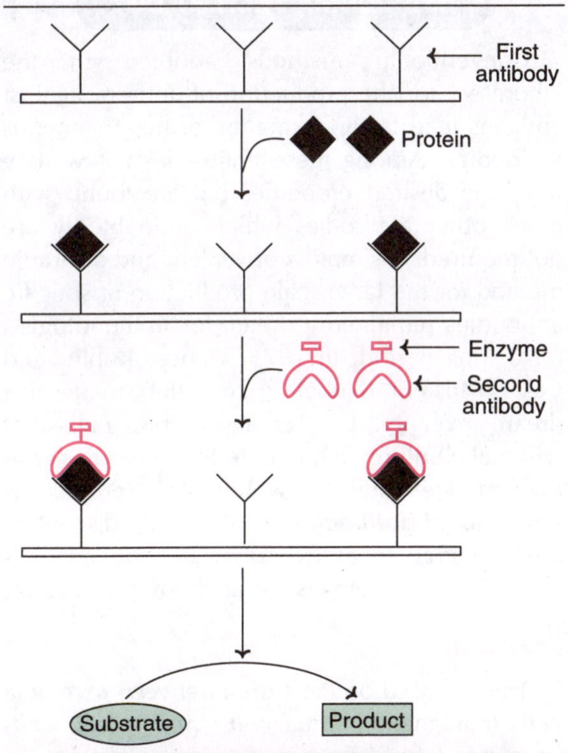

Fig. 41.9 : Diagrammatic representation of enzyme-linked immunosorbant assay (ELISA).

The principle for the use of the enzyme peroxidase in ELISA is illustrated next.

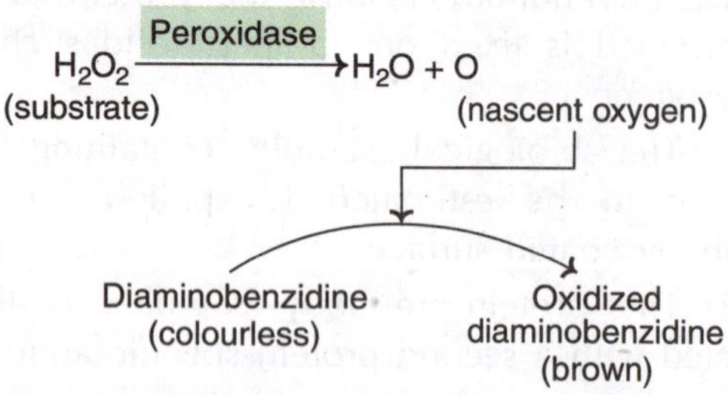

Applications

ELISA is widely used for the ***determination of*** small quantities of ***proteins*** (hormones, antigens, antibodies) and other biological substances. The most commonly used pregnancy test for the detection of human chorionic gonadotropin (hCG) in urine is based on ELISA. By this test, ***pregnancy*** can be detected within few days after conception. ELISA is also been used for the diagnosis of ***AIDS***.

HYBRIDOMA TECHNOLOGY

Conventional methods adopted in the laboratory for the production of antisera against antigens lead to the formation of heterogeneous antibodies. Among these antibodies a few may have the desired properties but are found with many other antibodies which undoubtedly are not required. A simple, convenient and desirable method for the large scale production of specific antibodies remained a dream for immunologists for a long period. In 1975, George Kohler and Cesar Milstein (Nobel Prize 1984) made this dream a reality. They created hybrid cells that will make unlimited quantities of antibodies with defined specificities, which are termed as ***monoclonal antibodies (McAb).*** This discovery, often referred to as hybridoma technology, has revolutionized methods for antibody production.

Principle

This is based on the ***fusion between myeloma cells*** (malignant plasma cells) ***and spleen cells*** from a suitably immunized animal. Spleen cells die in a short period under ordinary tissue culture conditions while myeloma cells are adopted to grow permanently in culture. Mutants of myeloma cells lack the enzyme hypoxanthine guanine phosphoribosyltransferase (azaquinine resistant) or thymidine kinase (bromodeoxy-uridine resistant). These mutants cannot grow in a medium containing aminopterin, supplemented with hypoxanthine and thymidine (HAT medium). Hybrids between the mutant myeloma cells and spleen cells can be selected and cultured in HAT medium.

From the growing hybrids, individual clones can be chosen that secrete the desired antibodies (monoclonal origin). The selected clones like ordinary myeloma cells can be maintained indefinitely. In short, the hybridoma technology for the production of monoclonal antibodies involves the following steps.

1. Immunization of appropriate animals with antigen (need not be pure) under study.

2. Fusion of suitable drug resistant myeloma cells with plasma cells, obtained from the spleen of the immunized animal.

3. Selection and cloning of the hybrid cells that grow in culture and produce antibody molecules of desired class and specificity against the antigen of interest.

Hybridoma technology can make available highly specific antibodies in abundant amounts. The clones once developed are far cheaper than the traditionally employed animals (horses, rabbits) for producing antibodies. The clones developed from the hybrids will also ensure constancy of the quality of the product and will also avoid the batch to batch variation inherent in the conventional methods.

Applications of monoclonal antibodies

The antibodies produced by hybridoma technology have been widely used for a variety of purposes. These include the ***early detection of pregnancy***, detection and treatment ***of cancer***, diagnosis of leprosy and treatment of autoimmune diseases.

42

Immunology

The immunology speaks

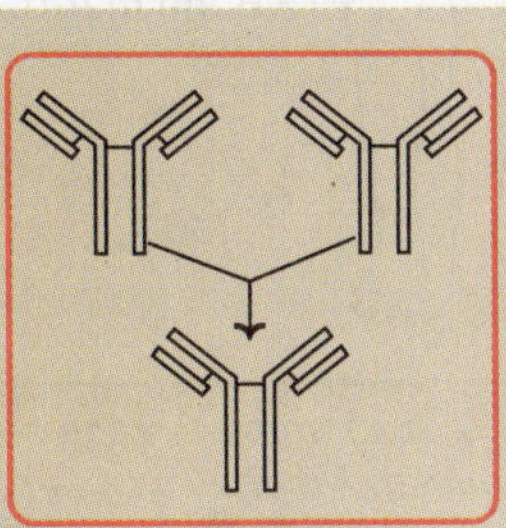

"I represent the defense system of the body;
Mainly composed of β-lymphocytes and T-lymphocytes;
Designed to eliminate invading microbes and moles;
My memory can distinguish self, and non-self."

Immunology deals with the study of immunity and immune systems of vertebrates. Immunity (immunis literally means exempt/free from burden) broadly involves the ***resistance shown***, and ***protection offered by the host organism against the infectious diseases***. The immune system consists of a complex network of cells and molecules, and their interactions. It is specifically designed to eliminate infectious organisms from the body. This is possible since the organism is capable of distinguishing the self from non-self, and eliminate non-self.

Immunity is broadly divided into two types — innate (non-specific) immunity and adaptive or acquired (specific) immunity.

INNATE IMMUNITY

Innate immunity is non-specific, and represents the inherent capability of the organism to offer resistance against diseases. It consists of defensive barriers.

First line of defense

The ***skin*** is the largest organ in the human body, constituting about 15% of the adult body weight. The skin provides mechanical barrier to prevent the entry of microorganisms and viruses. The acidic (pH 3-5) environment on the skin surface inhibits the growth of certain microorganisms. Further, the sweat contains an enzyme lysozyme that can destroy bacterial cell wall.

Second line of defense

Despite the physical barriers, the microorganisms do enter the body. The body defends itself and ***eliminates the invading organisms by non-specific mechanisms*** such as sneezing and secretions of the mucus. In addition, the body also tries to kill the pathogens by phagocytosis (involving macrophages and complement system). The inflammatory response and fever response of the body also form a part of innate immunity.

THE IMMUNE SYSTEM

The immune system represents the third and most potent defense mechanism of the body. Acquired (adaptive or specific) immunity is capable of specifically recognizing and eliminating the invading microorganisms and foreign molecules (antigens). In contrast to innate immunity, the acquired immunity displays four distinct characteristics

- Antigen specificity
- Recognition diversity
- Immunological memory
- Discrimination between self and non-self.

The ***body*** possess tremendous capability to specifically ***identify various antigens*** (antigen is a foreign substance, usually a protein or a carbohydrate that elicits immune response). Exposure to an antigen leads to the development of immunological memory. As a result, a second encounter of the body to the same antigen results in a heightened state of immune response. The immune system recognizes and responds to foreign antigens as it is capable of distinguishing self and non-self. Autoimmune diseases are caused due to a failure to discriminate self and non-self antigens.

ORGANIZATION OF IMMUNE SYSTEM

The immune system consists of several organs distributed throughout the body (***Fig. 42.1***). These lymphoid organs are categorized as primary and secondary.

Primary lymphoid organs

These organs provide appropriate micro-environment for the development and maturation of antigen-sensitive lymphocytes (a type of white blood cells). The ***thymus*** (situated above the heart) and bone marrow are the central or primary lymphoid organs. T-lymphocyte maturation occurs in the thymus while B-lymphocyte maturation takes place in the bone marrow.

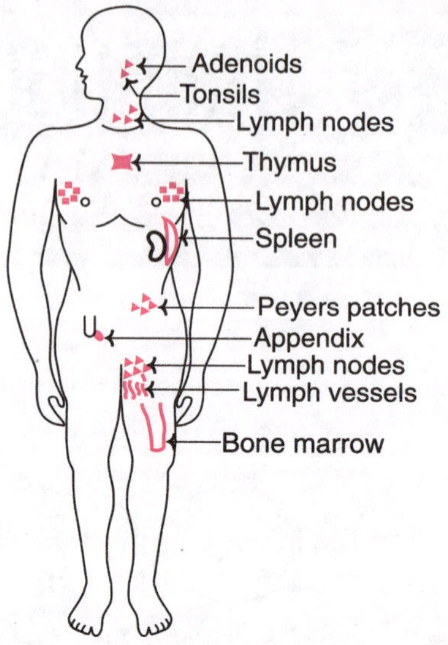

Fig. 42.1 : *A diagrammatic representation of human lymphatic system.*

Secondary lymphoid organs

These are the sites for the initiation of immune response. e.g. spleen, tonsils, lymph nodes, appendix, Peyers patches in the gut. Secondary lymphoid organs provide the microenvironment for interaction between antigens and mature lymphocytes.

CELLS OF THE IMMUNE SYSTEM

Two types of lymphocytes namely B-cells and T-cells are critical for the immune system. In addition, several accessory cells and effector cells also participate.

B-lymphocytes

The site of development and maturation of B-cells occurs in ***b***ursa fabricius in birds, and ***b***one marrow in mammals. During the course of immune response. B-cells mature into plasma cells and secrete antibodies (immunoglobulins).

The B-cells possess the capability to specifically recognize each antigen and produce antibodies (i.e. immunoglobulins) against it. B-lymphocytes are intimately associated with ***humoral immunity. Immunoglobulins are described in Chapter 9.***

T-lymphocytes

The maturation of T-cells occurs in the thymus, hence the name. The ***T-cells*** can identify viruses and microorganisms from the antigens displayed on their surfaces. There are at least four different types of T-cells.

- ***Inducer T-cells*** that mediate the development of T-cells in the thymus.
- ***Cytotoxic T-cells*** (T_C), capable of recognizing and killing the infected or abnormal cells.
- ***Helper T-cells*** (T_H) that initiate immune responses.
- ***Suppressor T-cells*** mediate the suppression of immune response.

T-lymphocytes are responsible for the ***cell-mediated immunity***.

MAJOR HISTOCOMPATIBILITY COMPLEX

The major histocompatibility complex (***MHC***) represents a special group of ***proteins, present on the cell surfaces of T-lymphocytes***. MHC is involved in the recognition of antigens on T-cells. It may be noted here that the B-cell receptors (antibodies) can recognize antigens on their own, while T-cells can do so through the mediation of MHC.

In humans, the MHC proteins are encoded by a cluster of genes located on chromosome 6 (it is on chromosome 17 for mice). The major histocompatibility complex in humans is referred to as ***human leukocyte antigen (HLA)***. Three classes of MHC molecules (chemically glycoproteins) are known in human. Class I molecules are found on almost all the nucleated cells of the body. Class II molecules are associated only with leukocytes involved in cell-mediated immune response. Class III molecules are the secreted proteins possessing immune functions e.g. complement components (C_2, C_4), tumor necrosis factor.

THE COMPLEMENT SYSTEM

The complement system is composed of about 20 plasma proteins that 'complement' the function of antibodies in defending the body from the invading antigens. The complementary factors are heat labile and get inactivated if heated at 56°C for about 30 minutes. The complement system helps the body immunity in 4 ways

1. **Complement fixation :** The complement system binds to the foreign invading cells and causes lysis of the cell membranes.
2. **Opsonization :** The process of promoting the phagocytosis of foreign cells is referred to as opsonization.
3. **Inflammatory reaction :** The complement system stimulates local inflammatory reaction and attracts phagocytic cells.
4. **Clearance of antigen-antibody complexes :** The complement system promotes the clearance of antigen-antibody complexes from the body.

Nomenclature of complement system : The complement proteins are designated by the letter 'C', followed by a component number—C_1, C_2, C_3 etc.

Types of reaction : The complement system brings about two sets of reactions :

1. Antibody dependent classical pathway.
2. Antibody independent alternative pathway.

Each one of the pathways consists of a series of reactions converting inactive precursors to active products by serine proteases which resembles blood coagulation.

THE IMMUNE RESPONSE

The immune response refers to the series of reactions carried out by the immune system in the body against the foreign invader. When an infection takes place or when an antigen enters the body, it is trapped by the macrophages in lymphoid organs. The phagocytic cells which are guarding the body by constant patrolling engulf and digest the foreign substance. However, the partially digested antigen (i.e. processed antigen) with antigenic epitopes attaches to lymphocytes.

T-helper cells (T_H) play a key role the immune response (***Fig. 42.2***). This is brought out through

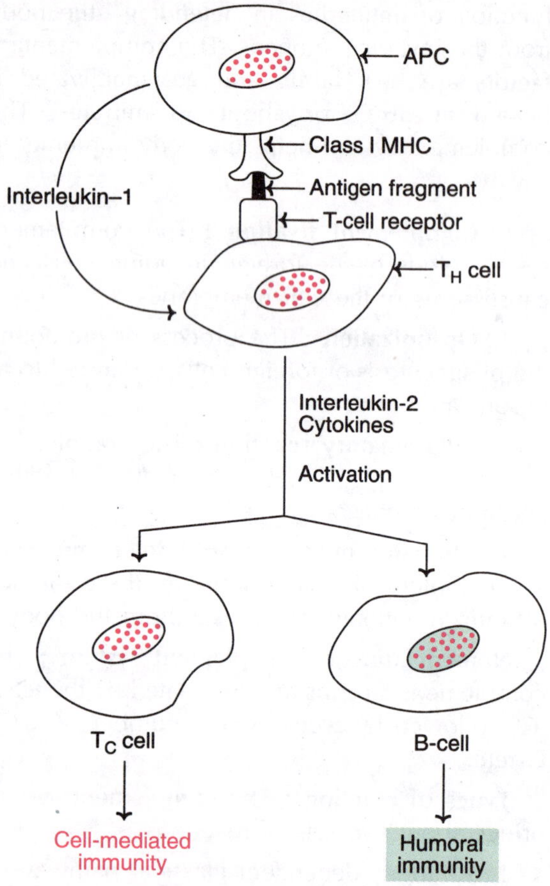

Fig. 42.2 : *A diagrammatic representation of the central role of helper T cells (T_H cells) in immune response (APC—Antigen presenting cell; T_C cell—Cytotoxic T-cell).*

the participation of antigen presenting cell (APC), usually a macrophage. Receptors of T_H cell bind to class II MHC-antigen complex displayed on the surface of APC. APC secretes interleukin-I, which activates the T_H cell. This activated T_H cell actively grows and divides to produce clones of T_H cells. All the T_H cells possess receptors that are specific for the MHC-antigen complex. This facilitates triggering of immune response in an exponential manner. The T_H cells secrete interleukin-2 which promotes the prolifiration of cytotoxic T cells (T_C cells) to attack the infected cells through cell-mediated immunity. Further, interleukin-2 also activates B-cells to produce immunoglobulins which perform humoral immunity.

CYTOKINES

Cytokines are a group of ***proteins*** that ***bring*** about ***communication between different cell types*** involved in immunity. They are low molecular weight glycoproteins and are produced by lymphoid and non-lymphoid cells during the course of immune response. Cytokines may be regarded as soluble messenger molecules of immune system. They can act as short messengers between the cells or long range messengers by circulating in the blood and affecting cells at far off sites. The latter function is comparable to that of hormones.

The term ***interleukin (IL)*** is frequently used to represent cytokines. There are more than a dozen interleukins (IL-I……IL_{12}), produced by different cells with wide range of functions. The main function (directly or indirectly) of cytokines is to ***amplify immune responses*** and inflammatory responses.

Therapeutic uses of cytokines

It is now possible to produce cytokines *in vitro*. Some of the cytokines have potential applications in the practice of medicine. For instance, IL-2 is used in cancer immunotherapy, and in the treatment of immunodeficiency diseases. IL-2 induces the proliferation and differentiation of T-and B-cells, besides increasing the cytotoxic capacity of natural killer cells.

A group of cytokines namely interferons can combat viral infection by inhibiting their replication.

IMMUNITY IN HEALTH AND DISEASE

The prime function of immune system is to protect the host against the invading pathogens. The body tries its best to overcome various strategies of infectious agents (bacteria, viruses), and provides immunity.

Some of the important immunological aspects in human health and disease are briefly described.

AUTOIMMUNE DISEASES

In general, the immune system is self-tolerant i.e. not responsive to cells or proteins of self. Sometimes, for various reasons, the ***immune system fails to discriminate between self and non-self***. As a consequence, the cells or tissues of the body are attacked. This phenomenon is referred to as ***autoimmunity*** and the diseases are regarded as autoimmune diseases. The antibodies produced to self molecules are regarded as ***autoantibodies***. Some examples of autoimmune diseases are listed.

- Insulin-dependent diabetes (pancreatic β-cell autoreactive T-cells and antibodies).
- Rheumatoid arthritis (antibodies against proteins present in joints).
- Myasthenia gravis (acetylcholine receptor autoantibodies).
- Autoimmune hemolytic anemia (erythrocyte autoantibodies).

Mechanism of autoimmunity : It is widely accepted that autoimmunity generally occurs as a consequence of ***body's response against bacterial, viral or any foreign antigen***. Some of the epitopes of foreign antigens are similar (homologous) to epitopes present on certain host proteins. This results in cross reaction of antigens and antibodies which may lead to autoimmune diseases.

ORGAN TRANSPLANTATION

The phenomenon of transfer of cells, tissues or organs from one site to another (in the same organism, ***autograft*** or from another organism ***allograft***) is regarded as organ transplantation. In case of humans, majority of organ transplantations are allografts (between two individuals). The term ***xenograft*** is used if tissues/organs are transferred from one species to another e.g. from pig to man.

Organ transplantation is associated with immunological complications, and tissue rejection. This is because the host body responds to the transplanted tissue in a similar way as if it were an invading foreign organism. Major histocompatibility complex (MHC) is primarily involved in allograft rejection. This is due to the fact that MHC proteins are unique to each individual, and the immune system responds promptly to foreign MHCs.

Organ transplantation between closely related family members is preferred, since their MHCs are also likely to be closely related. And major immunological complications can be averted.

CANCERS

Growth of tumors is often associated with the formation of novel antigens. These ***tumor antigens*** (also referred to as oncofetal antigens e.g. α-fetoprotein) are recognized as non-self by the immune systems. However, tumors have developed several mechanisms to evade immune responses.

AIDS

Acquired immunodeficiency syndrome (AIDS), caused by human immunodeficiency virus, is characterized by immunosuppression, secondary neoplasma and neurological manifestations. AIDS primarily affects the cell-mediated immune system which protects the body from intracellular parasites such as viruses, and bacteria. Most of the immunodeficiency symptoms of AIDS are associated with a reduction in CD_4 (cluster determinant antigen 4) cells.

Genetics

43

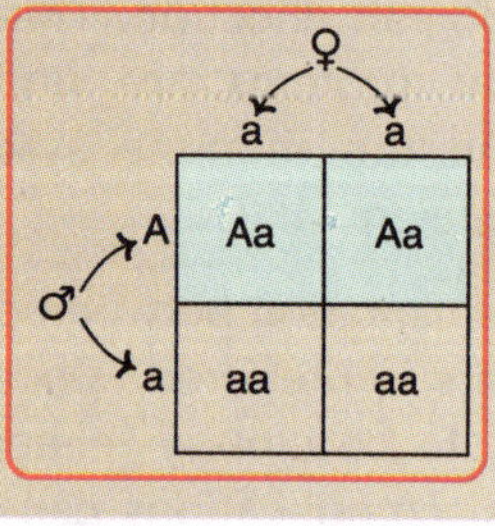

The genetics speaks

"I am the science for the study of heredity;
With DNA as the chemical of inheritance;
Transmitting characters from parents to offsprings;
Mutations and chromosomal abnormalities result in diseases."

Genetics is the ***study of heredity***. It is appropriately regarded as the science that explains the similarities and differences among the related organisms.

The blood theory of inheritance in humans

For many centuries, it was customary to explain inheritance in humans through blood theory. People used to believe that the children received blood from their parents, and it was the union of blood that led to the blending of characteristics. That is how the terms 'blood relations', 'blood will tell', and 'blood is thicker than water' came into existence. They are still used, despite the fact that blood is no more involved in inheritance. With the advances in genetics, the more appropriate terms should be as follows

- ***Gene relations*** in place of blood relations.
- ***Genes will tell*** instead of blood will tell.

BRIEF HISTORY AND DEVELOPMENT OF GENETICS

Genetics is relatively young, not even 150 years. The blood theory of inheritance was questioned in 1850s, based on the fact that the semen contained no blood. Thus, blood was not being transferred to the offspring. Then the big question was what was the hereditary substance.

Mendel's experiments : It was in 1866, an Austrian monk named Gregor Johann Mendel, for the first time reported the fundamental laws of inheritance. He conducted several experiments on the breeding patterns of pea plants. Mendel put forth the ***theory of transmissible factors*** which states that inheritance is controlled by ***certain factors*** passed from parents to offsprings. His results were published in 1866 in an obscure journal Proceedings of the Society of Natural Sciences.

For about 35 years, the observations made by Mendel went unnoticed, and were almost forgotten. Two European botanists (Correns and Hugo de Vries) in 1900, independently and simultaneously rediscovered the theories of Mendel. The year 1900 is important as it marks the beginning the modern era of genetics.

The origin of the word gene : In the early years of twentieth century, it was believed that the Mendel's inheritance factors are very closely related to chromosomes (literally coloured bodies) of the cells. It was in 1920s, the term ***gene*** (derived from a Greek word *gennan* meaning to produce) was introduced by Willard Johannsen. Thus, ***gene replaced the earlier terms inheritance factor or inheritance unit***.

Chemical basis of heredity : There was a controversy for quite sometime on the chemical basis of inheritance. There were two groups— the protein supporters and DNA supporters. It was in 1944, Avery and his associates presented convincing evidence that the chemical basis of heredity lies in DNA, and not in protein. Thus, ***DNA*** was finally identified as ***the genetic material***. Its structure was elucidated in 1952 by Watson and Crick.

Importance of genes in inheritance—studies on twins

Monozygotic or ***identical twins*** contain the ***same*** genetic material — ***DNA or genes***. Studies conducted on identical twins make startling revelations with regard to inheritance. One such study is described here.

Oskar Stohr and Jack Yufe were identical twins separated at birth. Oskar was taken to Germany where he was brought up by his grandmother as a Christian. Jack was raised by his father in Israel as a Jew. The two brothers were reunited at the age of 47. Despite the different environmental influences, their behavioural patterns and personalities were remarkably similar

- Both men had moustaches, wore two pocket shirts, and wire-rimmed glasses.
- Both loved spicy foods and tended to fall asleep in front of television.
- Both flushed the toilet before using.
- Both read maganizes from back to front.
- Both stored rubber bands on their wrists.
- Both liked to sneeze in a room of strangers.

Besides Oskar and Jack, many other studies conducted on identical twins point out the importance of genes on the inherited characters related to personality and mannerisms.

BASIC PRINCIPLES OF HEREDITY IN HUMANS

The understanding of how genetic characteristics are passed on from one generation to the next is based on the principles developed by Mendel.

As we know now, the human genome is organized into a diploid (2n) set of 46 chromosomes. They exist as ***22 pairs of autosomes*** and ***one pair of sex chromosomes (XX/XY)***. During the course of meiosis, the chromosome number becomes haploid (n). Thus, haploid male and female gametes — sperm and oocyte respectively, are formed. On fertilization of the oocyte by the sperm, the diploid status is restored. This becomes possible as the zygote receives one member of each chromosome pair from the father, and the other from the mother. As regards the sex chromosomes, the males have X and Y, while the females have XX. The sex of the child is determined by the father.

Monogenic and polygenic traits

The genetic traits or characters are controlled by single genes or multiple genes. The changes in genes are associated with genetic diseases.

Monogenic disorders : These are the ***single gene disease traits*** due to alterations in the corresponding gene e.g. sickle-cell anemia, phenylketonuria. Inheritance of monogenic disorders usually follows the Mendelian pattern of inheritance.

Polygenic disorders : The genetic traits conferred ***by more than on gene*** (i.e multiple genes), and the disorders associated with them are very important e.g. height, weight, skin colours, academic performance, blood pressure, aggressiveness, length of life.

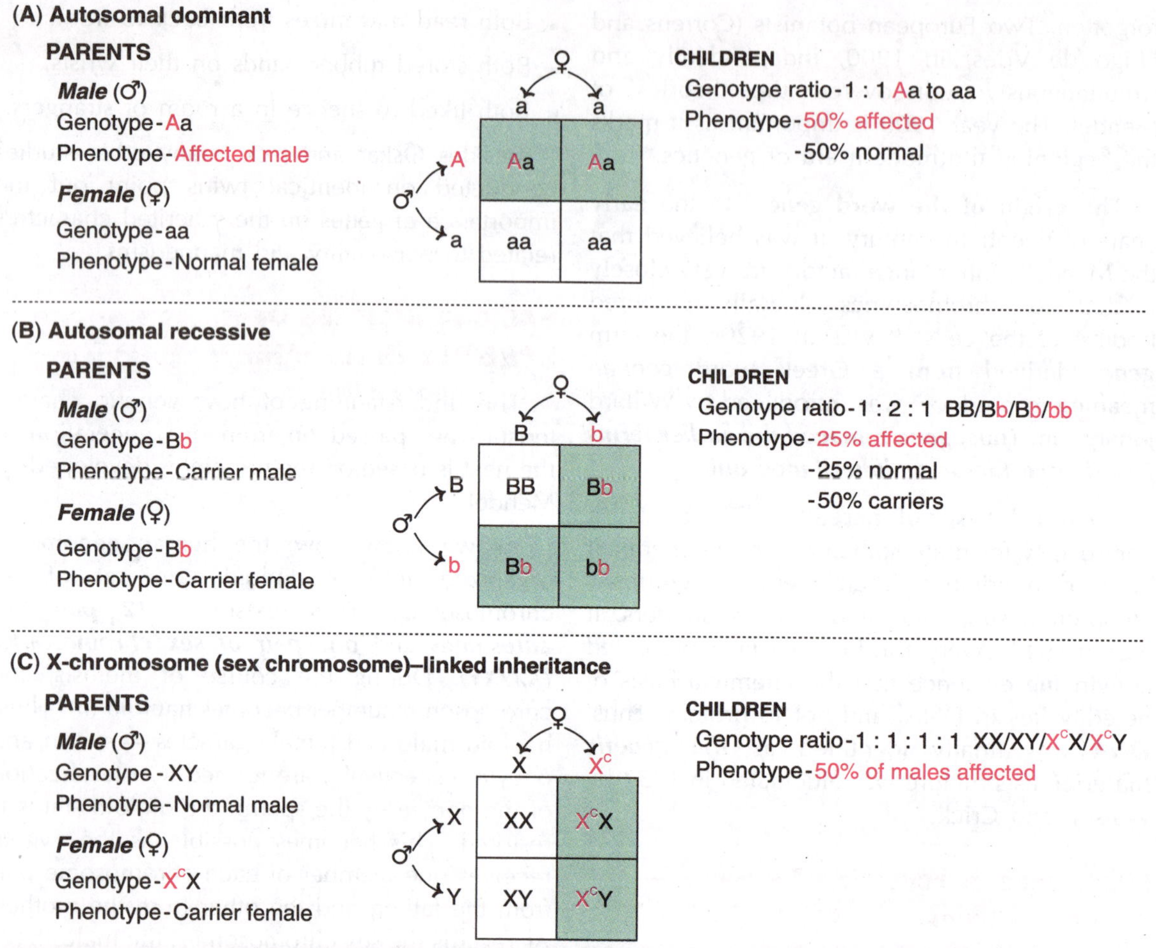

***Fig. 43.1 :** Patterns of inheritance-autosomal dominant, autosomal recessive and X-linked (**Note :** Genotype refers to the description of genetic composition, while phenotype represents the observable character displayed by an organism).*

PATTERNS OF INHERITANCE

The heredity is transmitted from parent to offspring as individual characters controlled by genes. The genes are linearly distributed on chromosomes at fixed positions called ***loci***.

A gene may have different forms referred to as ***alleles***. Usually one allele is transferred from the father, and the other from the mother. The allele is regarded as ***dominant*** if the trait is exhibited due to its presence. On the other hand, the allele is said to be ***recessive*** if its effect is masked by a dominant allele. The individuals are said to be ***homozygous*** if both the alleles are the same. When the alleles are different they are said to be ***heterozygous***.

The pattern of inheritance of monogenic traits may occur in the following ways (***Fig. 43.1***).

1. Autosomal dominant
2. Autosomal recessive
3. Sex-linked.

1. **Autosomal dominant inheritance** : A normal allele may be designated as ***a*** while an autosomal dominant disease allele as ***A*** (***Fig. 43.1A***). The male with Aa genotype is an affected one while the female with aa is normal. Half of the genes from the affected male will carry the disease allele. On mating, the male and female gametes are mixed in different combinations. The result is that ***half of the***

Table 43.1 Selected examples of genetic disorders (monogenic traits) in humans

Inherited pattern/disease	Estimated incidence	Salient features
Autosomal dominant		
Familial hypercholesterolemia	1 : 500	High risk for heart diseases
Huntington's disease	1 : 5000	Nervous disorders, dementia
Familial retinoblastoma	1 : 12000	Tumors of retina
Breast cancer genes *(BRAC 1 and 2)*	1 : 800	High risk for breast and ovarian cancers
β-Thalassemia	1 : 2500 (in people of Mediterranean descent)	A blood disorder; the blood appears to be blue instead of red
Autosomal recessive		
Sickle-cell anemia	1 : 100 (in Africans)	Severe life threatening anemia; confers resistance to malaria
Cystic fibrosis	1 : 2500 (in Caucasians)	Defective ion transport; severe lung infections and early death (before they reach 30 years)
Phenylketonuria	1 : 2000	Mental retardation due to brain damage
α_1-Antitrypsin deficiency	1 : 5000	Damage to lungs and liver
Tay-Sachs disease	1 : 3000 (in Ashkenazi Jews)	Nervous disorder; blindness and paralysis
Severe combined immunodeficiency disease (SCID)	Rare (only 100 cases reported worldwide)	Highly defective immune system; early death
Sex-linked		
Colour blindness	1 : 50 males	Unable to distinguish colours
Hemophilia (A/B)	1 : 10,000 males	Defective blood clotting
Duchenne muscular dystrophy	1 : 7000 males	Muscle wastage
Mitochondrial		
Leber hereditary optic neuropathy	Not known	Damage to optic nerves, may lead to blindness

children will be heterozygous (Aa) and ***have the disease***. Example of autosomal dominant inherited diseases are familial hypercholesterolemia, β-thalassemia, breast cancer genes.

2. **Autosomal recessive inheritance :** In this case, the normal allele is designated as ***B*** while the disease-causing one is ***a*** (***Fig. 43.1B***). The gametes of carrier male and carrier female (both with genotype Bb) get mixed. For these heterozygous carrier parents, there is one fourth chance of having an affected child. Cystic fibrosis, sickle-cell anemia and phenylketonuria are some good examples of autosomal recessive disorders.

3. **Sex (X)-linked inheritance :** In the ***Fig. 43.1C***, sex-linked pattern of inheritance is depicted. A normal male (XY) and a carrier female (X^cY) will produce children wherein, half of the male children are affected while no female child is affected. This is due to the fact that the male children possess only one X chromosome, and there is no dominant allele to mark its effects (as is the case with females). Colour blindness and hemophilia are good examples of X-linked diseases.

A selected list of genetic disorders (monogenic traits) due to autosomal and sex-linked inheritance in humans is given in ***Table 43.1***.

GENETIC DISEASES IN HUMANS

The pattern of inheritance and monogenic traits along with some of the associated disorders are described in previous page (***Table 43.1***). Besides ***gene mutations***, ***chromosomal abnormalities*** (aberrations) also result in genetic diseases.

Aneuploidy : The presence of ***abnormal number of chromosomes*** within the cells is referred to as aneuploidy. The most common aneuploid condition is ***trisomy*** in which three copies of a particular chromosome are present in a cell instead of the normal two e.g. trisomy-21 causing ***Down's syndrome***; trisomy-18 that results in ***Edward's syndrome***. These are the examples of autosomal aneuploidy.

In case of sex-linked aneuploidy, the sex chromosomes occur as three copies. e.g. phenotypically male causing Klinefelter's syndrome has XXY; trisomy-X is phenotypically a female with XXX.

ONLINE MENDELIAN INHERITANCE IN MAN (OMIM)

OMIM is a ***catalogue of human genes and genetic disorders*** with a focus on gene-phenotype relationship. OMIM data base has been designed for use by physicians and health care professionals dealing with genetic disorders, besides being freely available to public.

OMIM classification system deals with a ***unified index for genetic diseases***. Each OMIM entery is given a unique 6-digit identifying number. Some e.g. examples are listed

Phenylketonuria (enzyme defect –phenylalanine hydroxylase	261600
Galactosemia (enzyme defect-galactose 1-phosphate uridyltransferase)	230400

Some more details on OMIM are given in ***Appendix V***.

EUGENICS

Eugenics is a ***science of improving human race based on genetics***. Improving the traits of plants and animals through breeding programmes has been in practice for centuries.

Eugenics is a highly controversial subject due to social, ethical, and political reasons. The proponents of eugenics argue that people with desirable and good traits (good blood) should reproduce while those with undersirable characters (bad blood) should not. The advocates of eugenics, however, do not force any policy, but they try to convince the people to perform their duty voluntarily. The object of eugenics is to ***limit the production of people who are unfit to live in the society***.

Eugenics in Nazi Germany

Germany developed its own eugenic programme during 1930s. A law on eugenic sterilization was passed in 1933. In a span of three years, compulsory sterilization was done on about 250,000 people, who allegedly suffered from hereditary disabilities, feeble mindedness, epilepsy, schizophrenia, blindness, physical deformaties, and drug or alcohol addiction.

The German Government committed many atrocities in the name of racial purity. Other countries, however do not support this kind of eugenics.

APPENDICES

Answers to Self-assessment Exercises

CHAPTER 2

Answers to III and IV

1. Sucrose,
2. Glyceraldehyde,
3. Epimers,
4. Anomers,
5. Aglycone,
6. Streptomycin,
7. α-1,6-Glycosidic bond,
8. Inulin,
9. Hyaluronic acid,
10. N-Acetylneuraminic acid,
11. b,
12. d,
13. a,
14. d,
15. a.

CHAPTER 3

Answers to III and IV

1. Triacylglycerolds,
2. Geometric isomerism (*cis-trans* isomerism),
3. Chaulmoogric acid,
4. Triacylglycerols,
5. Stereospecific number,
6. Saponification number,
7. Dipalmitoyl lecithin,
8. Phosphatidylinositol,
9. Gangliosides,
10. Cyclopentanoperhydrophenanthrene,
11. a,
12. d,
13. d,
14. c,
15. b.

CHAPTER 4

Answers to III and IV

1. 16%,
2. L-α-Amino acids,
3. Methionine,
4. Zwitterion,
5. β-Alanine,
6. Peptide bonds,
7. Tryptophan,
8. 9,
9. 1-Fluro 2,4-dinitrobenzene (FDNB),
10. Denaturation,
11. b,
12. d,
13. b,
14. d,
15. a.

CHAPTER 5

Answers to III and IV

1. Gene,
2. RNA,
3. Nucleotides,
4. Thymine,
5. 2,
6. Base + sugar + phosphate,
7. Erwin Chargaff,
8. 3 Hydrogen bonds (in place of 2 in A-T),
9. β-Form,
10. CCA(5′ to 3′),
11. d,
12. b,
13. c,
14. d,
15. d.

CHAPTER 6

Answers to III and IV

1. In yeast,
2. Ligases,
3. Coenzyme,
4. Denaturation,
5. Alcohol dehydrogenase, carbonic anhydrase,
6. Active site,
7. $NADP^+$,
8. E.C. 1.1.1.1,
9. AMP/ADP,
10. Creatine phosphokinase (CPK),
11. c,
12. d,
13. b,
14. d,
15. b.

CHAPTER 7

Answers to III and IV

1. Acetylation,
2. Riboflavin,
3. Vitamin E (tocopherol),
4. Pyridoxine (B_6),
5. Avidin,
6. Pantothenic acid,
7. Cobalamin (B_{12}),
8. Dermatitis, diarrhea and dementia,
9. Vitamin K,
10. Folic acid,
11. b,
12. d,
13. a,
14. d,
15. a.

CHAPTER 8

Answers to III and IV

1. β-Glycosidic bonds,
2. Raffinose,
3. Lactase (β-galactosidase),
4. Fiber,
5. Parietal (oxyntic) cells,
6. Glutathione,
7. Hartnup's disease,
8. Arginine, lysine,
9. Colipase
10. Mixed micelles,
11. a,
12. d,
13. c,
14. b,
15. a.

CHAPTER 9

Answers to III and IV

1. Fibrinogen,
2. Electrophoresis,
3. Hemoglobin,
4. B-Lymphocytes,
5. IgG,
6. IgE,
7. 40–50°C,
8. C-reactive protein,
9. Staurt factor (Xa),
10. Plasmin,
11. c,
12. d,
13. a,
14. b,
15. b.

CHAPTER 10

Answers to III and IV

1. 574,
2. Methemoglobin,
3. Carbonic anhydrase,
4. 2,3-Bisphosphoglycerate,
5. Deoxyhemoglobin,
6. Thalassemias,
7. Succinyl CoA,
8. Uroporphyrinogen synthase I,
9. δ -Aminolevulinate synthase,
10. Biliverdin,
11. a,
12. a,
13. b,
14. d,
15. c.

CHAPTER 11

Answers to III and IV

1. $\Delta G = \Delta H - T\Delta S$ (T = Absolute temperature),
2. Exergonic or spontaneous,
3. Phosphoanhydride bonds,
4. Phosphoarginine,
5. Electrons,
6. Inner mitochondrial membrane,
7. Heme (porphyrin with iron),
8. Cytochrome oxidase (cyt a + a_3),
9. Cytochrome a + a_3,
10. Superoxide dismutase,
11. d,
12. a,
13. b,
14. d,
15. a.

CHAPTER 13

Answers to III and IV

1. Thiamine, riboflavin, lipoic acid, niacin, pantothenic acid,
2. Absence of glucose 6-phosphatase,
3. L-Gulonolactone oxidase,
4. Sorbitol,
5. Galactose 1-phosphate uridyltransferase,
6. Leucine and lysine,
7. Succinate thiokinase,
8. Uronic acid pathway,
9. Glycogenin,
10. Oxaloacetate,
11. d,
12. c,
13. b,
14. a,
15. b.

CHAPTER 14

Answers to III and IV

1. Triacylglycerols,
2. HMG CoA reductase,
3. Ampipathic,
4. Sphingomyelinase,
5. HDL,
6. 129 ATP,
7. Zellweger syndrome,
8. Citrate,
9. HDL,
10. Unsaturated fatty acid,
11. d,
12. a,
13. d,
14. c,
15. b.

CHAPTER 15

Answers to III and IV

1. Pyridoxal phosphate,
2. Glutamate dehydrogenase,
3. Carbamoyl phosphate synthase I,
4. Glycine transaminase,
5. Tetrahydrobiopterin,
6. Dopamine,
7. Homogentisate,
8. Malignant carcinoid syndrome,
9. Ornithine decarboxylase,
10. Leucine,
11. b,
12. d,
13. a,
14. c,
15. a.

CHAPTER 17

Answers to III and IV

1. Glutamine and aspartate,
2. Allopurinol,
3. Sodium urate,
4. Xanthine oxidase,
5. Lesch-Nyhan syndrome,
6. Thioredoxin,
7. Inosine monophosphate,
8. Alloxanthine,
9. Aspartate,
10. Carbamoyl phosphate synthetase II,
11. d,
12. a,
13. c,
14. b,
15. d.

CHAPTER 18

Answers to III and IV

1. 9–11 mg/dl. (4.5–5.5 mEq./l.),
2. Calcitriol,
3. Phosphorus,
4. Magnesium,
5. Sodium,
6. 3.5–5.0 mEq/l,
7. Transferrin,
8. Ceruloplasmin,
9. Gusten,
10. Selenium,
11. d,
12. a,
13. b,
14. c,
15. a.

CHAPTER 19

Answers to III and IV

1. Adenylate cyclase,
2. Ca^{2+},
3. Anterior pituitary,
4. Endorphins and enkephalins,
5. Thyroperoxidase,
6. Aldosterone,
7. Vanillyl mandelic acid (VMA),

8. Dihydrotestosterone (DHT),
9. Cholesterol,
10. Cholecystokinin (CCK),
11. a,
12. d,
13. b,
14. c,
15. a.

CHAPTER 20

Answers to III and IV

1. Heme,
2. van den Bergh reaction,
3. Alanine transaminase,
4. Alkaline phosphatase,
5. Bromosulphthalein (BSP),
6. 180 mg/dl,
7. Inulin,
8. 2ml/min,
9. Ryle's tube,
10. Pentagastrin,
11. a,
12. d,
13. c,
14. b,
15. b.

CHAPTER 21

Answers to III and IV

1. Antidiuretic hormone (ADH),
2. Na^+,
3. 285–295 milliosmoles /kg,
4. Aldosterone,
5. Carbonic acid (H_2CO_3),
6. Bicarbonate buffer,
7. 20 : 1,
8. Ammonium ion (NH_4^+),
9. Bicarbonate (HCO_3^-),
10. Carbonic acid (H_2CO_3) or CO_2,
11. d,
12. a,
13. c,
14. b,
15. d.

CHAPTER 22

Answers to III and IV

1. Collagen,
2. Glycine,
3. β-oxalyl aminoalamine,
4. Fibrillin,
5. Glycosaminoglycans,
6. Sarcomere,
7. Actin,
8. Calcium caseinate,
9. Vitamin C,
10. Lecithin/Sphingomyelin,
11. c,
12. d,
13. a,
14. c,
15. b.

CHAPTER 23

Answers to III and IV

1. 4.128,
2. Thyroid gland,
3. Fiber,
4. Carbohydrates,
5. Chemical score,
6. 1g/kg body weight/day,
7. Biological value (BV) of protein,
8. Sulfur containing amino acids,
9. Iron,
10. Plasma albumin,
11. a,
12. d,
13. c,
14. d,
15. a.

CHAPTER 24

Answers to III and IV

1. DNA helicase,
2. Okazaki pieces,
3. DNA polymerase III,
4. DNA topoisomerases,
5. Cyclins,
6. Transposons or transposable elements,
7. Missense,
8. c,
9. a,
10. b,

CHAPTER 25

Answers to III and IV

1. Genome,
2. hnRNA,
3. Introns,
4. Reverse transcriptase,
5. Wobble hypothesis,
6. Ribosomes,
7. rRNA,
8. Chaperones,
9. Prion diseases,
10. Protein targeting,
11. d,
12. c,
13. a,
14. b,
15. a.

CHAPTER 26

Answers to III and IV

1. 30,000–40,000,
2. One cistron-one subunit concept,
3. Protein-DNA complex,
4. a,
5. b,
6. d.

CHAPTER 27

Answers to III and IV

1. *Escherichia coli*,
2. RNA,
3. Dot-blotting,
4. *Thermus aquaticus*,
5. Genomic library/DNA library,
6. Site-directed mutagenesis,
7. Humulin,
8. Hepatitis B vaccine,
9. Mouse,
10. Sheep (Dolly),
11. c,
12. d,
13. d,
14. a,
15. c.

Appendix I

Abbreviations used in this Book

A	adenine, adenosine
ABC	ATP binding cassette
ABG	arterial blood gas
Ab	antibody
ACP	acyl carrier protein
ACTH	adrenocorticotropic hormone
Acyl CoA	fatty acid derivative of coenzyme A
ADA	adenosine deaminase
ADH	alcohol dehydrogenase
ADH	antidiuretic hormone
ADP	adenosine diphosphate
AFP	α-fetoprotein
AFLP	amplified fragment length polymorphism
Ag	antigen
A/G	albumin/globulin (ratio)
AIDS	acquired immunodeficiency syndrome
ALA	δ-aminolevulinic acid
ALP	alkaline phosphatase
ALT	alanine transaminase
AMS	acute mountain sickness
AMP	adenosine monophosphate
ANF	atrial natriuretic factor
APC	antigen presenting cell
APL	antiphospholipid
Apo-A	apoprotein-A
AP sites	apurinic sites
AST	aspartate transaminase
AT	α_1-antitrypsin
ATCase	aspartate transcarbamoylase
ATP	adenosine triphosphate
BAL	British antilewisite
BAO	basal acid output
BHA	butylated hydroxyanisole
BHT	butylated hydroxy toluene
BMR	basal metabolic rate
BNP	brain natriuretic peptide
BOAA	β-oxalylaminoalanine
bp	base pair
BP	blood pressure
BPG	bisphosphoglycerate (2,3-BPG, 1,3-BPG)
BPH	benign prostate hypertrophy
BSP	bromosulphthalein
BUN	blood urea nitrogen
BV	biological value
C	cytosine, cytidine
CA	carbonic anhydrase
Cal	calorie
Cam	calmodulin
cAMP	3′,5′-cyclic adenosine monophosphate (cyclic AMP)
CAP	catabolite activator protein
CBG	corticosteroid binding globulin
CCK	cholecystokinin
cDNA	complementary DNA
CD_4	cluster determinant antigen 4
CDP	cytidine diphosphate
CEA	carcinoembryonic antigen
CF	cystic fibrosis
CFTR	cystic fibrosis transmembrane regulator
cGMP	3′,5′-cyclic guanosine monophosphate
C_H	constant heavy chain
CHD	coronary heart disease
ChE	cholinesterase
Chl	chlorophyll
C_L	constant light chain
CLIP	corticotropin like intermediate lobe peptide
CMP	cytidine monophosphate
CNS	central nervous system
CoA or CoASH	coenzyme A
COHb	carboxyhemoglobin
COMT	catechol-o-methyltransferase
COPD	chronic obstructive pulmonary disease
CoQ	coenzyme Q (ubiquinone)

CPK (CK)	creatine phosphokinase (creatine kinase)
CPPP	cyclopentanoperhydrophenanthrene
CPS	carbamoyl phosphate synthase
CPT	carnitine palmitoyl transferase
CRH	corticotropin releasing hormone
CS	chorionic somatomammotropin
CSF	cerebrospinal fluid
CT	calcitonin
CTP	cytidine triphosphate
dA	deoxyadenosine
dADP	deoxyadenosine diphosphate
DG	diacylglycerol
DAM	diacetyl monoxime
dAMP	deoxyadenosine monophosphate
dATP	deoxyadenosine triphosphate
dCMP	deoxycytidine monophosphate
DCT	distal convoluted tubule
DEAE	diethyl aminoethylamine
DFP or DIFP	diidopropyl fluorophosphate
dGMP	deoxyguanosine monophosphate
DHA	docosahexaenoic acid
DHAP	dihydroxyacetone phosphate
DHCC	dihydroxycholecalciferol (1, 25-DHCC; 24, 25-DHCC)
DHEA	dehydroepiandrosterone
DHF	dihydrofolate
DHT	dihydrotestosterone
DIT	diiodotyrosine
dl	deciliter
DMB	dimethyl benzimidazole
DMS	dimethyl sulfate
DNA	deoxyribonucleic acid
DNase	deoxyribonuclease
DNMT	DNA methyltransferase
DNP	2, 4-dinitrophenol
DOPA	dihydroxy phenylalanine
DPG	diphosphoglycerate
DPP	dimethyl allyl pyrophosphate
DRV	daily reference values
DV	daily value
dTMP	deoxythymidine monophosphate
E_O	redox potential
eAG	estimated average glucose
EC	enzyme commission
ECF	extracellular fluid
EDRF	endothelium-derived releasing factor
EDS	Ehless-Dantos syndrome
EDTA	ethylene diamine tetraacetate
EF	elongation factor
EFA	essential fatty acids
eIFs	eukaryotic initiation factors
EGF	epidermal growth factor
ELISA	enzyme-linked immunosorbent assay
EM	Embden-Meyerhof
ER	endoplasmic reticulum
ES	enzyme-substrate complex
ES cells	embryonic stem cells
E-site	exist site
ETC	electron transport chain
FA	fatty acid
Fab	antigen binding fragment
FAD	flavin adenine dinucleotide
$FADH_2$	reduced FAD
FAS	fatty acid synthase
F 1,6-BP	fructose 1, 6-bisphosphate
F 2,6-BP	fructose 2, 6-bisphosphate
Fc	crystalline fragment
FDA	food and drug administration
FDNB	1-fluoro 2, 4-dinitrobenzene
FFA	free fatty acid
FGF	fibroblast growth factor
FH_4	tetrahydrofolate
FIGLU	formiminoglutamic acid
fMet	N-formylmethionine
FMN	flavin mononucleotide
$FMNH_2$	reduced FMN
F 1-P	fructose 1-phosphate
F 6-P	fructose 6-phosphate
Fp	flavoprotein
FSH	follicle stimulating hormone
FTM	fractional test meal
G	guanine, guanosine
g	gram
ΔG	free energy change
GABA	γ-aminobutyric acid
GAG	glycosaminoglycans
Gal-Cer	galactocerebroside
GAR	glycinamide ribotide
GDH	glutamate dehydrogenase
GDP	guanosine diphosphate
GFR	glomerular filtration rate
GGT (GT)	γ-glutamyl transpeptidase
GH	growth hormone
GHRH	growth hormone releasing hormone
GIP	gastric inhibitory peptide
GIT	gastrointestinal tract
Gla	γ-carboxy glutamate
GLC	gas liquid chromatography

Glu-Cer	glucocerebroside
Gly	glycine
GLUT	glucose transporters
GN	glucose-nitrogen (ratio)
GNA	glucol nucleic acid
GMP	guanosine monophosphate
GnRH	gonadotropin releasing hormone
GRH	growth hormone releasing hormone
GRIH	growth hormone release-inhibiting hormone
G 6-P	glucose 6-phosphate
G 6-PD	glucose 6-phosphate dehydrogenase
GPP	geranyl pyrophosphate
GSD	glycogen storage disease
GSH	glutathione (reduced form)
GSSG	glutathione (oxidized form)
GTP	guanosine triphosphate
GTT	glucose tolerance test
ΔH	change in enthalpy
HAC	human artificial chromosome
Hb	hemoglobin
HbA_1	adult hemoglobin
HbA_{1C}	glycosylated hemoglobin
HbF	fetal hemoglobin
HbO_2	oxyhemoglobin
HBsAg	hepatitis B surface antigen
HbS	sickle-cell hemoglobin
hCG	human chorionic gonadotropin
HDAC	histone deacetylase
HDL	high density lipoproteins
HFCS	high fructose–corn syrups
HGPRT	hypoxanthine guanine phosphoribosyltransferase
HGPS	Hutchinson-Gilford progeria syndrome
HIAA	hydroxy indole acetic acid
HIF	hypoxia inducible transcription factor
HIV	human immunodeficiency virus
HLA	human leukocyte antigen
HLH	helix-loop-helix
HMG CoA	β-hydroxy β-methylglutaryl CoA
HMP	hexose monophosphate
HNPCC	hereditary nonpolyposis colon cancer
hnRNA	heterogeneous nuclear RNA
Hp	haptoglobin
HPFH	hereditary persistence of fetal hemoglobin
HPLC	high performance liquid chromatography
HRE	hormone responsive element
hs-CRP	high sensitive C-reactive protein
Hsp	heat shock protein
5HT	5-hydroxytryptamine
HTH	helix-turn-helix
ICD	isocitrate dehydrogenase
IDDM	insulin dependent diabetes mellitus
IDL	intermediate density lipoproteins
IDP	inosine diphosphate
IF	initiation factor
Ig	immunoglobulin
IgG	immunoglobulin G
IGF	insulin-like growth factor
IL	interleukins
IMP	inosine monophosphate
INH	isonicotinic acid hydrazide (isoniazid)
$InsP_2$ (IP_2)	inositol 1, 4-bisphosphate
$InsP_3$ (IP_3)	inositol 1, 4, 5-triphosphate
IPP	isopentenyl pyrophosphate
IR	infrared
IRDS	infant respiratory distress syndrome
ITP	inosine triphosphate
IU	international unit
IV	intravenous
K	dissociation constant
KA	King Armstrong
K_a	dissociation constant of acid
Kbp	kilo base pair
KD	kilodalton
Keq	equilibrium constant
α-KG	α-ketoglutarate
Ki	inhibition constant
K_m	Michaelis constant
KJ	kilojoule
LATS	long acting thyroid stimulator
LCAT	lecithin cholesterol acyltransferase
LDH	lactate dehydrogenase
LDL	low density lipoproteins
LFT	liver function tests
LH	luteinizing hormone
LINES	long interspersed nuclear elements
LMWH	low molecular weight heparins
LPH	lipotrophic hormone (lipotropin)
LT	leukotrienes
Lp-a	lipoprotein-a
LSD	lysergic acid diethylamide
M	molar
MAO	maximal acid output
MAO	monoamine oxidase
Mb	myoglobin
MbO_2	oxymyoglobin
MCAD	medium chain acyl CoA dehydrogenase

MDH	malate dehydrogenase
MELAS	mitochondrial encephalopathy, lactic acidosis and stroke
mEq	milliequivalents
mg	milligram
MHC	major histocompatibility complex
MI	myocardial infarction
MIT	monoiodotyrosine
mol	mole(s)
mM	millimolar
mol. wt.	molecular weight
mRNA	messenger RNA
MSH	melanocyte stimulating hormone
mtCK	mitochondrial creatine kinase
mtDNA	mitochondrial DNA
MW	molecular weight
NAD^+	nicotinamide adenine dinucleotide
NADH	reduced NAD^+
$NADP^+$	nicotinamide adenine dinucleotide phosphate
NADPH	reduced $NADP^+$
NAG	N-acetylglutamate
NANA	N-acetylneuraminic acid
ncRNAs	non-protein coding RNAs
NDP	nucleoside diphosphate
NE	niacin equivalents
NEFA	non esterified fatty acid
ng	nanogram (10^{-9} g)
NGF	nerve growth factor
NIDDM	non-insulin dependent diabetes mellitus
NMP	nucleoside monophsphate
NMR	nuclear magnetic resonance
NPN	non-protein nitrogen
NPU	net protein utilization
OAA	oxaloacetate
Ob	obese
OD	optical density
OMP	orotidine monophosphate
Osm	osmoles
OXPHOS	oxidative phosphorylation
PABA	para amino benzoic acid
PAF	platelet-activating factor
PAGE	polyacrylamide gel electrophoresis
PAH	para amino hippurate
PAPS	phosphoadenosine phosphosulfate
PBG	porphobilinogen
PBI	protein bound iodine
PCM	protein-calorie malnutrition
PCNA	proliferating cell nuclear antigen
pCO_2	partial presence of CO_2
PCOS	polycystic overy syndrome
PCR	polymerase chain reaction
PCT	proximal convoluted tubule
PDGF	platelet derived growth factor
PDH	pyruvate dehydrogenase
PEG	polyethylene glycol
PEM	protein-energy malnutrition
PEP	phosphoenol pyruvate
PER	protein efficiency ratio
PEST	proline, glutamine, serine, threonine
PFK	phosphofructokinase
PG	prostaglandins
PGA	pteroyl glutamic acid
pH	negative log of (H^+)
PI	phosphatidyl inositol
Pi	inorganic phosphate
p^I	isoelectric pH
PIF	prolactin inhibitory factor
PIP_2	inositol 4, 5-bisphosphate
p^{Ka}	negative log of Ka
PKU	phenylketonuria
PL	phospholipid
PLP	pyridoxal phosphate
PNA	peptide nucleic acid
pO_2	partial pressure of O_2
POMC	pro-opiomelanocortin
PPi	inorganic pyrophosphate
ppm	parts per million
PRIH	prolactin release-inhibiting hormone
PRL	prolactin
PRPP	5-phosphoribosyl 1-pyrophosphate
PT	prothrombin time
PTH	parathyroid hormone
PTH	phenyl thiohydantoin
PUFA	polyunsaturated fatty acids
QPRT	quinolinate phosphoribosyltransferase
RACE	rapid amplification of cDNA ends
RAIU	radioactive iodine uptake
RAPD	random amplified polymorphic DNA
ras	rat sarcoma
RBC	red blood cells
RBP	retinol binding protein
RDA	recommended dietary (daily) allowance
RDI	recommended daily intake
rDNA	recombinant DNA
RE	retinol equivalents
RER	rough endoplasmic reticulum
RF	releasing factor

Rf	ratio of fronts
RFC	replication factor C
RFLP	restriction fragment length polymorphism
R-form	relaxed form
RIA	radioimmunoassay
RMR	resting metabolic rate
RNA	ribonucleic acid
RNAP	RNA polymerase
RNase	ribonuclease
R 5-P	ribose 5-phosphate
RPA	replication protein A
RQ	respiratory quotient
rRNA	ribosomal RNA
RSV	rous sarcoma virus
RT	reverse transcriptase
rT_3	reverse T_3
SAM	S-adenosylmethionine
SCID	severe combined immunodeficiency
SDA	specific dynamic action
SdLDL	small dense low density lipoproteins
Sf	Svedberg flotation
SGLT	Sodium glucose cotransporter
SGOT	serum glutamate oxaloacetate transaminase
SGPT	serum glutamate pyruvate transaminase
SHBG	sex hormone binding globulin
SIDS	sudden infant death syndrome
SINEs	short interspersed nuclear elements
SMS	stiff man syndrome
sn	stereospecific number
SNPs	single nucleotide polymorphisms
snRNA	small nuclear RNA
SPS	stiff person syndrome
snRNP	small nuclear ribonucleoprotein
sRNA	soluble RNA
SRS	slow reacting substance
STRs	simple tandem repeats
T	thymine, thymidine
T	thymus (T-lymphocyte)
T_3	3,5,3′-triiodothyronine
T_4	3,5,3′,5′-tetraiodothyronine (thyroxine)
TBG	thyroxine binding globulin
TBPA	thyroxine binding prealbumin
TCA	tricarboxylic acid
TF	tissue factor
TFA	trans fatty acids
T-form	taut or tense form
TG	triacylglycerol
Tgb	thyroglobulin
TGF	transforming growth factor
THF	tetrahydrofolate
TIBC	total iron binding capacity
TLC	thin layer chromatography
TLS	tumor lysis syndrome
Tm	tubular maximum
TMP	thymidine monophosphate
TNA	threose nucleic acid
TNF	tumor necrosis factor
tPA	tissue plasminogen activator
TPP	thiamine pyrophosphate
TRH	thyrotropin releasing hormone
tRNA	transfer RNA
TSH	thyroid stimulating hormone
TX	thromboxane
μm	micrometer (10^{-6} m)
UBG	urobilinogen
UCP	uncoupling protein
UDP	uridine diphosphate
UDPG	uridine diphosphate glucose
μl	microliter (10^{-6} l)
μM	micromoles (10^{-6} M)
UMP	uridine monophosphate
UTP	uridine triphosphate
UV	ultraviolet
V_H	variable heavy chain
VIP	vasoactive intestinal peptide
V_L	variable light chain
VLDL	very low density lipoproteins
VMA	vanillyl mandelic acid
V_{max}	velocity maximum
VNTRs	variable number tandem repeats
WBC	white blood cells
XMP	xanthosine monophosphate
XNA	xeno nucleic acid
XP	xeroderma pigmentosum
Xyl	xylose
YAC	yeast artificial chromosome

Origins of Important Biochemical Words

Appendix II

Acid *(Latin)* acidus–sour

Acidosis *(Latin)* acidus–sour; osis–condition

Albinism *(Greek)* albino–white

Alkali *(Arabic)* al-qite–ashes of saltwort

Allergy *(Greek)* allos–other; ergon–work

Alloseric *(Greek)* allo–the other

Amentia *(Latin)* amentis–mental deficiency

Amnesia *(Greek)* a–not; mnesis–memory

Amphipathic *(Greek)* amphi–both; pathos–feeling

Amphiphilic *(Greek)* amphi–both; philic–love

Anaerobe *(Greek)* a–not; aer–air; bios–life

Anaplerotic *(Greek)* ana–up; plerotikos–to fill

Androgen *(Greek)* aner–man; genesis–production

Anemia *(Greek)* a–not; haima–blood

Anorexia *(Greek)* a–not; orexis–appetite

Anticoagulant anti *(Greek)*–against; coagulare *(Latin)*–to curdle

Antimetabolite *(Greek)* anti-against; metabole-change

Arteriosclerosis arteria *(Latin)*–artery; sclerosis *(Greek)* hardening.

Arthritis *(Greek)* arthron–joint; itis–inflammation

Atherosclerosis *(Greek)* athere–porridge; sclerosis–hardening

Beri-beri *(Singhalese)*–I cannot (said twice)

Biochemistry *(Greek)* bios–life; chymos–juice

Biology *(Greek)* bios–life; logos–discourse

Bovine *(Latin)* bovinus–pertaining to cow or ox

Calorie *(Latin)* calor–heat

Cancer *(Latin)* crab

Carbohydrate carbo *(Latin)*–coal; hydor *(Greek)*–water

Caries *(Latin)*–decay

Casein *(Latin)* caseus–cheese

Catabolism *(Greek)* kata–down; ballein–to throw

Catalysis *(Greek)* kata–down; lysis–degradation

Cathepsin *(Greek)* to digest

Cephalins *(Greek)* kephale–head

Cheilitis *(Greek)* cheilos–lip; itis–inflammation

Cheilosis *(Greek)* cheilos–lip; osis–condition

Chirality *(Greek)* cheir–hand

Chlorophyll *(Greek)* chloros–pale green; phyllon–leaf

Cholelithiasis *(Greek)* chole–bile; lithos–stone; asis–condition

Cholesterol *(Greek)* chole–bile; sterol–solid alcohol

Chromatography *(Greek)* chroma–colour; graphein –to write

Chromosome *(Greek)* chroma–colour; soma–body

Chyle *(Greek)* chylos–juice

Chyluria *(Greek)* chylos–juice; auron–urine

Chyme *(Greek)* chymos–juice

Cirrhosis *(Greek)* kirrhos–orange-tawny; osis–condition

Cis *(Latin)* same side

Coagulation (*Greek*) coagulare–to curdle

Collagen *(Greek)* kolla–glue; genesthai–to be produced

Colloid *(Greek)* kolla–glue; eidos–form

Consanguinity *(Latin)* con-with; sanguis–blood

Creatine *(Greek)* kreas–flesh

Cristae *(Latin)* crests

Cutaneous *(Latin)* cutis–skin

Cytology *(Greek)* kytos–cell; logos–discourse

Cytoplasm *(Greek)* kytos–cell; plassein–to mould

Dermatitis *(Greek)* derma–skin; itis–inflammation

Diabetes mellitus *(Greek)* diabetes–running through (or a siphon); mellitus–sweet

Dialysis (*Greek*) dia—through; lysis–loosening

Doctor (*Latin*) docere—to teach

Eicosanoids *(Greek)* eikosi–twenty

Embolism *(Greek)* embolos–to plug

Emphysema *(Greek)* emphysan–to inflate

Enkephalin *(Greek)* in the brain

Enthalpy *(Greek)* to warm within

Entropy *(Greek)* in turning

Enzyme *(Greek)* in yeast

Erythrocyte *(Greek)* erythros–red; kytos–cell

Eukaryotes *(Greek)* eu-true; karyon–nucleus

Ferrous *(Latin)* ferrum–iron

Folate *(Latin)* folium–leaf

Galactose *(Greek)* gala–milk

Gastritis *(Greek)* gaster–belly; itis–inflammation

Gene *(Greek)* genesis–descent

Genome *(Greek)* genos–birth

Globin *(Latin)* globus–ball

Globulin *(Latin)* globulus—little ball

Glossitis *(Greek)* glossa–tongue; itis–inflammation

Glycolysis *(Greek)* glycos–sweet; lysis–dissolution

Goitre *(Latin)* gultur–throat

Gonadotrophin *(Greek)* gona–generation; trophe–nourishment

Hemoglobin haima *(Greek)*–blood; globus *(Latin)*–ball

Hepatitis *(Greek)* hepar–liver; itis–inflammation

Hippos *(Greek)* horse

Hormone *(Greek)* hormain–to excite

Hydrophilic *(Greek)* hydro–water; philic–living

Hydrophobic *(Greek)* hydro–water; phobic–hating

Hyperglycemia *(Greek)* hyper–above; glycos-sweet; haima–blood

Hypertonic *(Greek)* hyper-above; tonos–tension

Hypoglycemia *(Greek)* hypo-below; glycos-sweet; haima–blood

Hypotonic *(Greek)* hypo–below; tonos–tension

Icterus *(Greek)* ikteros–jaundice

Immunity *(Latin)* immunis–exempt from public burden

Inflammation *(Latin)* inflammare–to set on fire

In situ *(Latin)* in the correct position

Insulin (*Latin*) insula—island

In vitro *(Latin)* in a test tube

In vivo *(Latin)* in the living tissue

Isomerism *(Greek)* iso-equal; mesos–part

Isotonic *(Greek)* iso–equal; tonos–tension

Isotope *(Greek)* iso–equal; topos–place

Jaundice *(French)* jaune–yellow

Keratin *(Greek)* keras–horn

Kwashiorkor *(Ga-African)* sickness of the deposed child

Lactalbumin *(Greek)* lac-milk; albumin-white

Lecithin *(Greek)* lekithos–egg yolk

Lipids *(Greek)* lipos–fat

Lactosuria lac *(Latin)*–milk; ovron *(Greek)*–urine

Leukocytes (*Geek)* leukos–white; kytos–cell

Leukoderma *(Greek)* leukos–white; derma–skin

Ligase *(Greek)* ligate–to bind

Malaria *(Italian)* bad air

Malnutrition *(Latin)* malus–bad; nutrire–nourishment

Marasmus *(Greek)* to waste

Melanin *(Greek)* melan–black

Menopause *(Greek)* men–month; pausis–stopping

Metabolism *(Greek)* metabole–change

Mitochondria *(Greek)* mitos–thread; chondros–granule

Mitosis *(Greek)* mitos–thread; osis–condition

Monosaccharide *(Greek)*–mono–one; saccharin–sugar

Myeloma *(Greek)* myelos–marrow; oma–tumor

Myoglobin (*Greek*) myo–myscle

Nephritis *(Greek)* nephros–kidney; itis–inflammation

Neurosis *(Greek)* neuron–nerve; osis–condition

Oedema or edema *(Greek)* oidema–swelling

Oligosaccharides *(Greek)* oligo–few; saccharon–sugar

Osmosis *(Greek)*–push

Osteomalacia *(Greek)* osteon–bone; malakia–softness

Oxyntic *(Greek)* oxynein–to make acid

Oxytocin *(Greek)*–rapid birth

Palindrome *(Greek)*–to run back again

Pantothenic acid *(Greek)* pantos–everywhere

Pathogenesis *(Greek)* pathos–disease; genesis–producing

Pellagra *(Italian)*–rough skin

Pepsin *(Greek)* pepsis–digestion

Phagocytosis *(Greek)* phagein–to eat; kytos–cell; osis–condition

Phobia *(Greek)* phobos–fear

Polysaccharide *(Greek)* poly–many; saccharin–sugar

Porphyrin *(Greek)* porphyra–purple colour

Post-prandial (*Latin*)–after food

Prokaryotes *(Greek)* pro–before; karyon–nucleus

Proteins *(Greek)* proteios–holding first place

Rickets *(Old English)* wrickken–to twist

Serum *(Latin)*–whey

Sphingosine *(Greek)* sphingein–to bind tight

Steatorrhea *(Greek)* stear–fat; rheein–to flow

Stereoisomerism *(Greek)* stero-space

Sterol *(Greek)* steros–solid; ol- alcohol

Syndrome (*Greek*) syn–together; dramein–to run

Thalassemia *(Greek)* thalassa–sea

Thermodynamics *(Greek)* therme–heat; dynamics–power

Thermogenesis *(Greek)* therme–heat; genesis–production

Thrombosis *(Greek)* thrombos–clot; osis–condition

Thylakoid *(Greek)* thylakos–a sac or pouch

Tocopherol *(Greek)* tokos–child birth; pheros–to bear; ol–alcohol

Trans *(Latin)* across

Tumor *(Latin)* swelling

Vitamin (coined inappropriately in 1906*) (Latin)* vita–life; amine

Xanthoma *(Greek)* xanthos–yellow

Xenobiotics *(Greek)* xenos–strange

Zwitterion *(German)* zwitter–hybrid.

Common Confusables in Biochemistry

Appendix III

Acetone; acetate – Acetone is a ketone; acetate is a carboxylic acid.

Acetyl CoA; acyl CoA – Acetyl CoA is a specific compound containing acetate bound to coenzyme A; acyl CoA is a general term used to refer to any fatty acid (acyl group) bound to coenzyme A.

Albumin; albinism – Albumin is a serum protein; albinism is a genetic disease in tysosine metabolism.

Amino; imino – Amino group ($-NH_2$) is found in majority of amino acids; imino group ($=NH$) is present in a few amino acids like proline and hydroxyproline.

Anabolism; catabolism – Anabolism refers to the biosynthetic reactions involving the formation of complex molecules from simpler ones; catabolism is concerned with the degradation of complex molecules to simpler ones with a concomitant release of energy.

Anomers; epimers – Anomers refer to two stereoisomers of a sugar that differ in configuration around a single carbonyl atom; epimers are two stereoisomers that differ in configuration around one asymmetric carbon of a sugar possessing two or more asymmetric carbon atoms.

Apoenzyme; coenzyme – Apoenzyme is the protein part of the functional enzyme (holoenzyme); coenzyme is the non-protein organic part associated with enzyme activity.

Bile pigments; bile salts – Bile pigments (biliverdin, bilirubin) are the breakdown products of heme; bile salts are the sodium and potassium salts of bile acids (glycocholate, taurocholate) produced by cholesterol.

Biliverdin; bilirubin – Both are bile pigments. Biliverdin is produced from heme in the reticuloendothelial cells; bilirubin is formed by reduction of biliverdin.

Biotin; biocytin – Biotin is a B-complex vitamin; biocytin refers to the covalently bound biotin to enzymes (through ε-amino group of lysine).

B-Lymphocytes; T-lymphocytes – B-lymphocytes produce immunoglobulins (antibodies) and are involved in humoral immunity; T-lymphocytes are responsible for cellular immunity.

Bisphosphate; diphosphate – Bisphosphate has two phosphates held separately e.g. 2,3-BPG; diphosphate has two phosphates linked together e.g. ADP.

Calcitriol; calcitonin – Calcitriol (1,25-DHCC) is the physiologically active form of vitamin D; calcitonin is a peptide hormone, synthesized by thyroid gland.

Calorimetry; colorimetry – Calorimetry deals with the measurement of heat production by organism; colorimetry is concerned with the measurement of colour compounds.

Carboxyl; carbonyl – These two are functional groups found in organic substances; carboxyl group $-COOH$; carbonyl $-\overset{\|}{C}-$.

Carnitine; creatine; creatinine – Carnitine transports activated fatty acids (acyl CoA) from cytosol to mitochondria; creatine is mostly found in the muscle as creatine phosphate, a high energy compound; creatinine is the anhydride of creatine.

Choline; cholic acid – Choline is a trimethyl quaternary base and is a constituent of acetylcholine; cholic acid is an important bile acid.

Chyle; chyme – Chyle refers to lymph with milky appearance due to chylomicrons; chyme is the partially digested food in the stomach that passes to deodenum.

Configuration; conformation – Configuration is the geometric relationship between a given set of atoms (e.g. L- and D-amino acids). Conformation is the special relationship of every atom in a molecule (e.g. secondary structure of protein).

Cysteine; cystine – Both are sulfur containing non-essential amino acids. Cysteine contains sulfhydryl (—SH) group; cystine is formed by condensation of two cysteine residues and contains a disulfide (—S—S—) group.

Dextrins; dextrans; dextrose – The first two are polysaccharides composed of glucose. Dextrins are the breakdown products of starch; dextrans are gels produced by bacteria from glucose. Dextrose is glucose in solution (dextrorotatory) used in medical practice.

Diabetes mellitus; diabetes insipidus – Diabetes mellitus is primarily an impairment in glucose metabolism due to the deficiency of, or inefficient insulin; diabetes insipidus is characterized by excretion of large volumes of urine (polyuria), caused by the deficiency of antidiuretic hormone (ADH).

Endocytosis; exocytosis – Endocytosis is the intake of macromolecules by the cells; exocytosis refers to the release of macromolecules from the cells to the outside.

Epinephrine; norepinephrine – Both are catecholamines synthesized from tyrosine. Epinephrine is methylated while norepinephrine does not contain a methyl group.

Exons; introns – Exons are the DNA sequences coding for proteins; introns are the intervening DNA sequences that do not code for proteins.

GABA; PABA – γ-Aminobutyric acid (GABA) is a neurotransmitter; p-aminobenzoic acid (PABA) is a vitamin.

Gene; genome – A gene refers to the DNA fragment of a chromosome that codes for a single polypeptide; all the genes of a cell or an organism are collectively known as genome.

Glu; Gla – Glu is the code for glutamic acid; Gla is the code for γ-carboxy glutamic acid.

Glucuronic acid; gluconic acid – Both are derived from glucose; oxidation of C_6 results in glucuronic acid while oxidation of C_1 yields gluconic acid. Glucuronic acid is produced in uronic acid pathway; gluconic acid is formed in hexose monophosphate shunt.

Glutaric acid; glutamic acid – Glutaric acid is a dicarboxylic acid; glutamic acid (α-amino glutaric acid) is an amino acid.

Glycation; glycosylation – Glycation is a non-enzymatic process involving covalent binding of free sugars (glucose, fructose, galactose) to proteins forming glycated products; glycosylation is a post-translational modification, mediated by enzymes wherein a specific carbohydrate is added to protein in a predetermined manner.

Glycogen; glycogenin – Glycogen is a storage form of carbohydrate (polysaccharide) in the animal body; glycogenin is a protein which serves as a primer for the initiation of glycogen synthesis.

Glycoproteins; mucoproteins – Both are conjugated proteins containing carbohydrate as the prosthetic group. The term glycoprotein is used if the carbohydrate content is $<4\%$; mucoprotein contains $>4\%$ carbohydrate.

Hydrophilic; hydrophobic – Hydrophilic refers to affinity to water; hydrphobic means hatred towards water.

Hypercalcemia; hyperkalemia – Elevation in serum calcium is hypercalcemia; increase in serum potassium is hyperkalemia.

Hypocalcemia; hypokalemia – Hypocalcemia refers to a fall in serum calcium; hypokalemia is a decrease in serum potassium.

Insulin; inulin – Insulin is a peptide hormone; inulin is a polysaccharide composed of fructose.

In vivo; in vitro – In vivo refers to within the cell or organism; in vitro means in the test tube.

Isoniazid; iproniazid – Isoniazid is an anti-tuberculosis drug; iproniazid is an anti-depressant drug.

Lactam; lactim – These terms are used to represent tautomerism. Lactam indicates the existence of a molecule in keto form; lactim represents a molecule in enol form.

Lactose; lactase; lactulose – Lactose is a disaccharide; lactase is an enzyme that cleaves lactose to glucose and gulactose; lactulose is a disaccharide containing galactose and fructose.

Linoleic acid; linolenic acid – Both are 18 carbon unsaturated fatty acids. Linoleic acid has two double bonds; linolenic acid has three double bonds.

Lipoproteins; lipotropic factors – Lipoproteins are molecular complexes composed of lipids and proteins; lipotropic factors are the substances (e.g. choline, betaine), the deficiency of which causes accumulation of fat in liver.

β-Lipoprotein; β-lipotropin – β-Lipoprotein refers to the low density lipoproteins; β-lipotropin is a peptide hormone derived from pro-opiomelanocortin (POMC) peptide.

Lyases; ligases – Lyases are the enzymes that catalyse the addition or removal of water, ammonia, CO_2 etc.; ligases catalyse the synthetic reactions where two molecules are joined together.

Malate; malonate; mevalonate – Malate is an intermediate in the citric acid cycle; malonate is a competitive inhibitor of the enzyme succinate dehydrogenase; mevalonate is an intermediate in cholesterol biosynthesis.

Melanin; melatonin – Melanin is the pigment of skin and hair; melatonin is a hormone synthesized by pineal gland.

Maltose; maltase – Maltose is a disaccharide; maltase is an enzyme that cleaves maltose to two molecules of glucose.

Metabolomics; metabonomics – Metabolomics is the study of whole assay of metabolites; metabonomics is the study of metabolic profile of biological systems in response to stresses like disease, toxic exposure, dietary changes.

Methyl, methenyl; methylene – All the three are one-carbon fragments as shown in brackets, methyl ($-CH_3$); methenyl ($-CH=$); methylene ($-CH_2-$).

Molarity; molality – Molarity is defined as the number of moles of a solute per liter solution; molality represents the number of moles of a solute per 1,000 g of solvent.

Nicotinic acid; nicotine – Nicotinic acid is a B-complex vitamin; nicotine is an alkaloid present in tobacco leaves.

Nucleoside; nucleotide – A nucleoside is composed of a nitrogen base and a sugar; nucleotide contains one or more phosphate groups bound to nucleoside.

Oncogens; oncogenes – Oncogens are the chemicals that cause cancer; oncogenes are the genes causing cancer.

Osmolarity; osmolality – Osmolarity represents osmotic pressure exerted by the number of moles (milli moles) per liter solution; osmolality refers to the osmotic pressure exerted by the number of moles (milli moles) per kg solvent.

Oxidase; Oxygenase – Oxidase accepts O_2 but oxygen atoms are not incorporated into substrate; oxygenase incorporates one or both oxygen atoms into substrate.

Palmitate; palmitoleate – Both are even chain (16-carbon) fatty acids. Palmitate is a saturated fatty acid; palmitoleate is a monounsaturated fatty acid.

Phosphatase; phosphorylase – Phosphatase uses water to remove phosphoryl group; phosphorylase utilizes Pi to break a bond and produce a phosphorylated compound.

Phosphatidyl ethanolamine; phosphatidal ethanolamine – Both are phospholipids. In phosphatidyl ethanolamine, the fatty acid is bound by an ester linkage. The fatty acid is held by an ether linkage in phosphatidal ethanolamine.

Phytic acid; phytanic acid – Phytic acid is formed by the addition of six phosphate molecules to inositol, it is an inhibitor of the intestinal absorption of calcium and iron; phytanic acid is an unusal fatty acid derived from phytol, a constituent of chlorophyll.

Prokaryotes; eukaryotes – Prokaryotes are the cells that lack a well defined nucleus; eukaryotes possess a well-defined nucleus.

Prolamines; protamines – Both are simple proteins. Prolamines are soluble in alcohol; protamines are basic protein soluble in NH_4OH.

Pyridine; pyrimidine; pteridine – All the three are heterocyclic rings containing nitrogen, as depicted below.

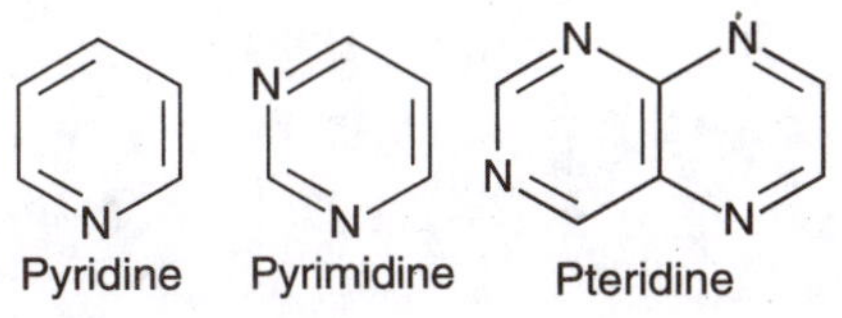

Pyridine ring is found in niacin and pyridoxine; pyrimidine is present in thiamine (vitamin B_1), thymine, cytosine and uracil; folic acid contains pteridine ring.

Pyridoxine; pyridoxal – Pyridoxine is the primary alcohol form of vitamin B_6; pyridoxal is the aldehyde form of B_6.

RDA; SDA – RDA (recommended dietary/daily allowance) represents the quantities of nutrients to be provided in the diet daily for maintenance of good health and physical efficiency; specific dynamic action (SDA) is the extra heat produced by the body over and above the caloric value of foodstuffs.

Renin; Rennin – Renin is synthesized by the kidneys and is involved in vasoconstriction causing hypertension; rennin is an enzyme found in gastric juice responsible for coagulation of milk.

Ribosomes; ribozymes – Ribosomes are the sites of protein biosynthesis; ribozymes refer to the RNA molecules which function as enzymes.

Retinol; retinal – Retinol is the alcohol form of vitamin A; retinal is the aldehyde form obtained by the oxidation of retinol.

Scleroproteins; selenoproteins – Scleroproteins are a group of fibrous proteins; selenoproteins contain the amino acid selenocysteine.

Serotonin; melatonin – Serotonin is a neuro-transmitter synthesized from tryptophan; melatonin is a hormone derived from serotonin in the pineal gland.

Somatotropin; somatostatin; somatomedin – Somatotropin is the other name for growth hormone (GH); growth hormone release inhibiting hormone (GRIH) is also called somatostatin; somatomedin refers to the insulin-like growth factor -I (IGF-I), produced by liver in response to GH action.

Sucrose; sucrase; sucralose – Sucrose is a disaccharide; sucrase is an enzyme that cleaves sucrose to glucose and fructose. Sucralose (trichlorosucrose) is an artifical sweetener.

Synthase; synthetase – Both the enzymes are concerned with biosynthetic reactions. Synthase does not require ATP; synthetase is dependent on ATP for energy supply. (***Note :*** This distinction between synthase and synthetase however, is not maintained strictly by most authors).

Thiamine; thymine – Thiamine is a vitamin (B_1); thymine is a pyrimidine base found in DNA structure.

Thiokinase; thiolase – Thiokinase activates fatty acids to acyl CoA; Thiolase catalyses the final reaction in β-oxidation to liberate acetyl CoA from acyl CoA.

Transcription; translation – Transcription refers to the synthesis of RNA from DNA; translation involves the protein synthesis from the RNA.

Uric acid; uronic acid – Uric acid is the end product of purine metabolism; uronic acids are formed by the oxidation of aldehyde group of monosaccharides (e.g. glucuronic acid).

Ureotelic; uricotelic – Ureotelic organisms (e.g. mammals) convert NH_3 to urea; uricotelic organisms (e.g. reptiles) convert NH_3 to uric acid.

Vitamin A; coenzyme A – Vitamin A is fat soluble vitamin; coenzyme A is derived from water soluble vitamin, pantothenic acid.

Practical Biochemistry—Principles

Appendix IV

QUALITATIVE EXPERIMENTS

Several laboratory qualitative experiments are performed to ***indentify the compounds of biochemical importance*** (carbohydrates, proteins/amino acids, non-protein nitrogenous substances) and to detect the ***abnormal constituents of urine***. The principles of the reactions pertaining to the most widely employed qualitative tests are described here.

I. REACTIONS OF CARBOHYDRATES

The carbohydrates used in the laboratory for the qualitative tests include glucose and fructose (monosaccharides), sucrose, lactose and maltose (disaccharides) and starch (polysaccharide). The principles of the reactions of carbohydrates are given :

1. **Molisch test :** It is a general test for the detection of ***carbohydrates***. The strong H_2SO_4 hydrolyses carbohydrates (poly- and disaccharides) to liberate monosaccharides. The monosaccharides get dehydrated to form furfural (from pentoses) or hydroxy methylfurfural (from hexoses) which condense with α-naphthol to form a violet coloured complex.

2. **Iodine test :** Polysaccharides combine with iodine to form a coloured complex. Thus, ***starch*** gives ***blue colour*** while dextrins give red colour with iodine.

3. **Benedict's test :** This is a test for the identification of ***reducing sugars***, which form enediols (predominantly under alkaline conditions). The enediol forms of sugars reduce ***cupric ions*** (***Cu^{2+}***) of copper sulfate ***to cuprous ions*** (***Cu^{+}***) which form a yellow precipitate of cuprous hydroxide or a red precipitate of cuprous oxide.

4. **Barfoed's test :** The principle of this test is the same as that of Benedict's test except that the reduction is carried out in ***mild acidic medium***. Since acidic medium is not favourable for reduction, only strong reducing sugars (monosaccharides) give this test positive. Thus, Barfoed's test serves as a key reaction to distinguish ***monosaccharides*** form disaccharides.

5. **Seliwanoff's test :** This is a specific test for ***ketohexoses***. Concentrated hydrochloric acid dehydrates ketohexoses to form furfural derivatives which condense with resorcinol to give a cherry red complex.

6. **Foulger's test :** This is also a test for ***ketohexoses***. The furfural derivatives formed from ketohexoses condense with urea in the presence of stannous chloride to give a blue colour.

7. **Rapid furfural test :** ***Ketohexoses*** are converted to furfural derivatives by HCl which form a purple colour complex with α-naphthol.

8. **Osazone test :** Phenylhydrazine in acetic acid, when boiled with reducing sugars forms osazones. The first two carbons (C_1 and C_2) are involved in this reaction. The sugars that differ in their configuration on these two carbons give the same type of osazones, since the difference is marked by binding with phenylhydrazine. Thus, glucose, ***fructose*** and ***mannose*** give the same type (***needle shaped***) of osazones. However, the osazones of reducing disaccharides differ — ***maltose gives sunflower***-shaped while ***lactose powder-puff shaped***.

9. **Sucrose hydrolysis test :** Sucrose is a non-reducing sugar, hence it does not give Benedict's and Barfoed's tests. ***Sucrose*** can be ***hydrolysed*** by concentrated HCl, to be converted ***to glucose and fructose*** (reducing monosaccharides) which answer the reducing reactions. However, after sucrose hydrolysis, the medium has to be made alkaline (by adding Na_2CO_3) for effective reduction process.

II. REACTIONS OF PROTEINS

The proteins employed in the laboratory for the qualitative tests include albumin, globulins, casein, gelatin and peptones. The principle of the most common reactions of proteins / amino acids performed in the laboratory are given hereunder.

A. PRECIPITATION REACTIONS

Proteins exist in colloidal solution due to hydration of polar groups (—COO^-, —NH_3^+, —OH). They can be precipitated by dehydration or neutralization of polar groups. Several methods are in use to achieve protein precipitation.

1. **Precipitation by neutral salts :** The process of protein precipitation by the addition of neutral salts such as ammonium sulfate or sodium sulfate is referred to as ***salting out***. This phenomenon is explained on the basis of dehydration of protein molecules by salts. This causes increased protein-protein interaction, resulting in molecular aggregation and precipitation.

The amount of salt required for protein precipitation depends on the size (molecular weight) of the protein molecule. In general, the ***higher is the protein molecular weight, the lower is the salt required for precipitation***. Thus, serum globulins are precipitated by half saturation with ammonium sulfate while albumin is precipitated by full saturation.

2. **Precipitation by salts of heavy metals :** Heavy metal ions like Pb^{2+}, Hg^{2+}, Fe^{2+}, Zn^{2+}, Cd^{2+} cause precipitation of proteins. These metals being positively charged, when added to protein solution (negatively charged) in alkaline medium result in precipitate formation.

3. **Precipitation by anionic or alkaloid reagents :** Proteins can be precipitated by ***trichloroacetic acid, sulphosalicylic acid***, phosphotungstic acid, picric acid, tannic acid, phosphomolybdic acid etc. By the addition of these acids, the proteins existing as cations are precipitated by the anionic form of acids to produce protein-sulphosalicylate, protein-tungstate, protein-picrate etc.

The anionic reagents such as phosphotungstic acid and trichloroacetic acid are ***used to prepare protein-free filtrate of blood*** needed for several estimations (e.g., urea, sugar) in the laboratory.

4. **Precipitation by organic solvents :** Organic solvents such as alcohol are good protein precipitating agents. They dehydrate the protein molecule by removing that water envelope and cause precipitation.

B. COLOUR REACTIONS

The proteins give several colour reactions which are often useful ***to identify the nature of the amino acids*** present in them as shown in the table.

Colour reactions of proteins/amino acids

Reaction	Specific group or amino acid
1. Biuret reaction	Two peptide linkages
2. Ninhydrin reaction	α-Amino acids
3. Xanthoproteic reaction	Benzene ring of aromatic amino acids (Phe, Tyr, Trp)
4. Millons reaction	Phenolic group (Tyr)
5. Hopkins-Cole reaction	Indole ring (Trp)
6. Sakaguchi reaction	Guanidino group (Arg)
7. Nitroprusside reaction	Sulfhydryl groups (Cys)
8. Sulfur test	Sulfhydryl groups (Cys)
9. Pauly's test	Imidazole ring (His)
10. Folin–Coicalteau's test	Phenolic groups (Tyr)

1. **Biuret reactions :** Biuret is a compound formed by heating urea to 180°C. When biuret is treated with dilute copper sulfate in alkaline medium, a purple colour is obtained. This is the basis of biuret test used for identification of proteins and peptides.

Biuret test is answered by compounds containing two or more CO—NH groups i.e., ***peptide bonds***. All proteins and peptides possessing atleast two peptide linkages i.e., tripeptides (with 3 amino acids) give positive biuret test. The principle of biuret test is conveniently used to detect the presence of proteins in biological fluids. The mechanism of biuret test is not clearly known. It is believed that the colour is due to the formation of a copper co-ordinated complex.

2. **Ninhydrin reaction :** The ***α-amino acids*** react with ninhydrin to form a purple, ***blue or pink colour complex (Ruhemann's purple)***.

$$\text{Amino acid + Ninhydrin} \longrightarrow \text{Keto acid} + NH_3 + CO_2 + \text{Hydrindantin}$$

$$\text{Hydrindantin} + NH_3 + \text{Ninhydrin} \longrightarrow \text{Ruhemann's purple}$$

3. **Xanthoproteic reaction :** Xanthoproteic reaction is due to nitration of ***aromatic amino acids*** (tryptophan, tyrosine and phenylalanine) on treatment with strong nitric acid at high temperature.

4. **Millon's test :** This test is given by the amino acid ***tyrosine***, or any other compound containing hydroxyphenyl ring. A red colour or precipitate is obtained in this reaction due to the formation of mercury complex of nitrophenol derivative.

5. **Hopkins-Cole reaction :** This reaction is specific for the indole ring of ***tryptophan***. It combines with formaldehyde in the presence of the oxidizing agent (sulfuric acid with mercuric sulfate) to form a violet or purple coloured compound.

6. **Sakaguchi reaction :** Arginine, containing guanidino group, reacts with α-naphthol and alkaline hypobromite to form a red colour complex.

7. **Sulfur test :** This is a test specific for sulfur containing amino acids namely ***cysteine*** and ***cystine***, but not methionine. When cysteine and cystine are boiled with sodium hydroxide, organic sulfur is converted to inorganic sodium sulfide. This reacts with lead acetate to form a black precipitate of lead sulfide. Methionine does not give this test, since sulfur of methionine is not split by alkali.

8. **Pauly's test :** This reaction is specific for ***histidine*** (imidazole ring). Diazotised sulfanilic acid reacts with imidazole ring in alkaline medium to form a red coloured complex.

9. **Molisch test :** This is a specific test for the detection of ***carbohydrates***. The proteins containing carbohydrates (e.g., glycoproteins) give this test positive. Albumin contains carbohydrate bound to it, hence answers Molisch test.

III. REACTIONS OF NON-PROTEIN NITROGENOUS SUBSTANCES

The non-protein nitrogenous (NPN) substances of biochemical importance include urea, uric acid and creatinine.

1. **Sodium hypobromite test :** This is a test for the detection of ***urea***. Sodium hypobromite decomposes urea to liberate nitrogen. The latter can be identified by brisk effervescence.

2. **Specific urease test :** The enzyme urease (source-horse gram) specifically acts on ***urea*** to liberate ammonium carbonate (alkali). The latter can be identified by a colour change in phenophthalein indicator (pink colour in alkaline medium).

3. **Benedict's uric acid test :** ***Uric acid***, being a strong reducing agent, reduces phosphotungstate to tungsten blue in alkaline medium.

4. **Murexide test :** ***Uric acid*** is oxidized by nitric acid to give purpuric acid (reddish yellow). This in turn combines with ammonia to form purple red colour ammonium purpurate (murexide).

5. **Jaffe's test :** ***Creatinine*** reacts with picric acid in alkaline medium to form orange red colour complex.

IV. ABNORMAL CONSTITUENTS OF URINE

Urine is the most important excretory fluid from the body. Some of the diseases are associated with an excretion of abnormal constituents in urine. The identification of such compounds in urine is of great diagnostic importance.

Urine abnormal constituent	Associated disorder(s)
Albumin	Kidney damage (glomerulonephritis)
Hemoglobin	Damage to kidneys or urinary tract.
Glucose	Diabetes mellitus, renal glycosuria.
Ketone bodies	Diabetes mellitus, starvation.
Bile salts	Obstructive jaundice
Bile pigments	Obstructive jaundice and hepatic jaundice.

1. **Sulfosalicylic acid test :** ***Proteins*** get precipitated by sulfosalicylic acid by forming protein-sulfosalicylate.

2. **Heat coagulation test :** This is a test for the detection of ***albumin*** and/or ***globulins*** in urine. Heat coagulation test is based on the principle of denaturation of proteins, followed by coagulation.

(**Note :** Small amounts of dilute acetic acid are added to dissolve the phosphates and sulfates that get precipitated on heating.)

3. **Benzidine test :** This test detects the presence of ***blood***. Hemoglobin (acts like peroxidase) decomposes hydrogen peroxide to liberate nascent oxygen (O–) which oxidises benzidine to a green or blue coloured complex.

(**Note :** Pus cells of urine possess peroxidase activity which interferes in benzidine test. This can be eliminated by boiling the urine prior to the test to inactivate the enzyme).

4. **Benedict's test :** This is a ***semiquantitative test*** for the detection of urine reducing sugars (primarily glucose). Benedict's test is based on the principle of reducing property of sugars (described in detail under reactions of carbohydrates). Colour of the precipitate formed indicates the approximate amount of ***glucose*** present in urine. Thus, green turbidity = traces; green precipitate = 0.5%; yellow precipitate = 1%; orange precipitate = 1.5% ***brick red precipitate = 2%***. (Note : Benedict's test is not specific to glucose, since it can be answered by any reducing substance like glucuronides, salicylates, homogentisic acid, vitamin C).

5. **Glucose oxidase test :** This is a strip test for the ***specific detection of glucose***. The enzyme glucose oxidase oxidizes glucose to liberate hydrogen peroxide which in turn is converted to nascent oxygen (O–) by peroxidase enzyme. The compound O-diansidine combines with nascent oxygen to form a coloured (yellow to red) complex.

6. **Rothera's test :** Nitroprusside in alkaline medium reacts with keto group of ***ketone bodies*** (acetone and acetoacetate) to form a purple ring. This test is not given by β-hydroxybutyrate.

7. **Hay's test :** This test is based on the surface tension lowering property of ***bile salts*** (sodium glycocholate and sodium taurocholate). Sulfur powder sprinkled on the surface of urine containing bile salts sinks to the bottom.

8. **Petternkofer's test :** This test is employed for the detection of ***bile salts***. The furfural derivatives (by reacting sugar with concentrated H_2SO_4) condense with bile salts to form a purple ring.

9. **Gmelin's test :** Nitric acid oxidizes the bile pigment bilirubin to biliverdin (green) or bilicyanin (blue). Gmelin's test gives a play of colours and is used for the identification of ***bile pigments***.

10. **Fouchet's test :** It is also employed for the detection of ***bile pigments***. Bile pigments are adsorbed on barium sulfate. Fouchet's reagents (containing ferric chloride in trichloroacetic acid) oxidizes bilirubin to biliverdin (green) and bilicyanin (blue).

QUANTITATIVE EXPERIMENTS

Quantitative experiments, dealing with the determination of concentrations of several biologically important compounds and the assay of many enzymes, are of great significance in the laboratory practice. Very often, the ultimate diagnosis and prognosis of a large number of diseases are guided by the quantitative biochemical investigations.

The principles involved in some of the quantitative experiments, commonly employed in the biochemistry laboratory by an undergraduate student, are briefly described here.

1. Blood glucose estimation

The quantitative determination of blood (plasma/serum) glucose is of great importance in the diagnosis and monitoring of diabetes mellitus.

(i) ***Folin Wu method*** **:** Alkaline copper (cupric ions) is reduced by glucose when boiled with protein free blood filtrate to cuprous oxide. The cuprous oxide in turn reacts with phosphomolybdic acid to form blue coloured oxides of molybdenum. The intensity of the colour can be measured in a colorimeter at a wavelength 680 nm. [Folin Wu method is rather ***old and is not specific for glucose determination***, since other substances (e.g., fructose, lactose, glutathione) also bring about reduction. Consequently the blood glucose level when estimated by Folin Wu method is higher i.e., normal fasting is 80-120 mg/dl against true glucose 60-100 mg/dl]

(ii) ***O-Toluidine method*** **:** Glucose combines with O-toluidine when boiled in acid medium to form a green coloured complex which can be measured in a colorimeter at a wavelength 630 nm. (This method ***determines glucose alone***).

(iii) ***Glucose oxidase-peroxidase (GOD—POD) method*** **:** This is an enzymatic determination of blood glucose. Glucose gets oxidized by glucose oxidase to gluconic acid and hydrogen peroxide. The enzyme peroxidase converts hydrogen peroxide to water and oxygen. The oxygen in turn reacts with 4-aminophenzone in the presence of phenol to form a pink coloured complex, the intensity of which can be measured at 530 nm.

2. Blood urea estimation

Determination of blood urea (reference range 10-40 mg/dl) is important for the evaluation of kidney (renal) function. Elevation of blood urea is associated with pre-renal (diabetic coma, thyrotoxicosis), renal (acute glomerulonephritis, polycystic kidney) and post-renal (obstruction in the urinary tract, due to tumors, stones) conditions.

Diacetyl monoxime (DAM) method **:** Urea when heated with diacetyl monoxime forms a yellow coloured complex of dioxime derivatives which can be measured at 520 nm.

3. Serum creatinine estimation

Estimation of serum creatinine (reference range 0.5-1.5 mg/dl) is used as a diagnostic test to assess kidney function. Serum creatinine is not influenced by endogenous and exogenous factors, as is the case with urea. Hence, some workers consider serum creatinine as a more reliable indicator of renal function.

Alkaline picrate method **:** This method is based on ***Jaffe's reaction***. Creatinine reacts with alkaline picrate to form creatinine picrate, an orange red coloured complex, which can be measured in a colorimeter at 530 nm.

(**Note :** Urinary creatinine can also be determined by employing the same principle given above).

4. Determination of serum proteins

The normal concentration of total serum proteins is in the range 6-8 g/dl (albumin 3.5-5.0 g/dl; globulins 2.5-3.5 g/dl; A/G ratio is 1.2 to 1.5 : 1). The A/G ratio is lowered either due to a decrease in albumin or an increase in globulins.

Serum albumin concentration is decreased in liver diseases, severe protein malnutrition, and excretion of albumin in urine (due to renal damage). Serum globulin concentration is elevated in chronic infections and multiple myeloma.

Biuret method **:** Peptide bonds (—CO—NH) of proteins react with cupric ions in alkaline medium to form a violet colour complex which is measured at a wavelength 530 nm. This method is suitable for total serum proteins with estimation.

Bromocresol green (BCG) dye method **:** This technique is employed for the estimation of serum albumin. BCG dye reacts with albumin to form an intense blue-green coloured complex which can be measured at 628 nm.

5. Estimation of serum bilirubin

The total bilirubin concentration in serum is 0.2-1 mg/dl (conjugated ~ 0.6 mg/dl; unconjugated ~ 0.4 mg/dl). Elevation in serum bilirubin concentration is observed in jaundice. Unconjugated bilirubin is increased in hemolytic jaundice, conjugated bilirubin in obstructive jaundice, while both of them are increased in hepatic jaundice.

van den Bergh reaction **:** Serum bilirubin estimation is based on van den Bergh reaction. The principle of the reaction is that diazotised sulfanilic acid (formed by mixing equal volumes of sulfanilic acid in HCl and sodium nitrite) reacts with bilirubin to form a purple coloured azobilirubin which can be measured at 540 nm.

6. Estimation of serum cholesterol

Serum cholesterol concentration (reference range 150-200 mg/dl) is elevated in atherosclerosis, diabetes mellitus, obstructive jaundice and hypothyroidism. Decreased levels are observed in hyperthyroidism.

Acetic anhydride method **:** Serum cholesterol reacts with acetic anhydride in the presence of glacial acetic acid and concentrated H_2SO_4 to form a green coloured complex. Intensity of this colour is measured at 560 nm.

7. Estimation of serum uric acid

Uric acid is the end product of purine metabolism. Its concentration in serum is increased (reference range - men 4-8 mg/dl; women 3-6 mg/dl) in gout.

Henry-Caraway's method **:** Uric acid in the protein-free filtrate when treated with phosphotungstic acid in the presence of sodium carbonate (alkaline solution) gives a blue coloured complex which can be measured at 660 nm.

8. Estimation of serum calcium

Serum calcium level is elevated (reference range 9-11 mg/dl) in hyperparathyroidism and decreased in hypothyroidism.

O-Cresolphthalein complexone method : Calcium reacts with the dye, O-cresolphthalein complexone (CPC) in alkaline solution to form a complex which can be measured at a wavelength 660 nm.

9. Estimation of serum phosphorus (inorganic)

Serum phosphate (reference range 3-4.5 mg/dl) is increased in hypoparathyroidism, and decreased in hyperparathyroidism and renal rickets.

For the determination of serum phosphate, serum proteins are precipitated by trichloroacetic acid. The protein-free filtrate containing inorganic phosphate is reacted with molybdic acid reagent to form phosphomolybdate. The latter in turn is reduced to molybdenum blue by treatment with 1-amino 2-naphthol-4 sulfonic acid (ANSA). The intensity of the blue colour is measured at 689 nm.

10. Determination of SGPT and SGOT

Serum glutamate pyruvate transaminase (SGPT; alanine transaminase) and serum glutamate oxaloacetate transaminase (SGOT; aspartate transaminase) are two important diagnostic enzymes. SGPT activity (reference range 5-40 IU/L) is more specifically increased in liver diseases (hepatic jaundice). SGOT activity is elevated (reference range 5-45 IU/L) in heart diseases (myocardial infarction).

Principle of assay : SGPT catalyses the following reaction

$$\text{L-Alanine} + \alpha\text{-ketoglutarate} \longrightarrow \text{L-glutamate} + \text{pyruvate}$$

SGOT brings about the following reaction

$$\text{L-Aspartic acid} + \alpha\text{-ketoglutarate} \longrightarrow \text{L-glutamate} + \text{oxaloacetate}$$

The keto acid (pyruvate or oxaloacetate), formed in the above reaction, when treated with 2, 4-dinitrophenyl hydrazine forms dinitrophenyl hydroazone (brown colour) in alkaline medium which can be measured at 505 nm.

11. Determination of serum alkaline phosphatase

The activity of the enzyme serum alkaline phosphatase (normal range 3-13 KA Units/dl) is elevated in rickets and obstructive jaundice.

Principle of assay : Alkaline phosphatase hydrolyses disodium phenylphosphate liberating phenol. On treatment with 4-amino antipyrine in alkaline medium, phenol gives ferricyanide (reddish colour) which can be measured at 520 nm.

12. Determination of serum amylase

Serum amylase activity is increased (reference range 80-180 Somogyi Units/dl) in acute pancreatitis.

Principle of assay : Amylase acts on starch and hydrolyses to dextrins and maltose. Starch forms blue coloured complex with iodine, a decrease in the colour (measured at 670 nm) is proportional to the activity of amylase.

13. Analysis of cerebrospinal fluid

Cerebrospinal fluid (CSF) is the aqueous medium surrounding the brain and spinal cord. From the biochemical perspective, estimation of ***proteins and glucose*** in CSF is ***important***. Increase in protein (reference range 15-40 mg/dl) and decrease in glucose (reference range 50-75 mg/dl) in the cerebrospinal fluid are observed in tuberculosis meningitis.

CSF protein estimation : Sulfosalicylic acid (in sodium sulfate solution) precipitates CSF proteins and the turbidity is measured at 680 nm.

CSF glucose estimation : Any one of the standard methods employed for the determination of blood glucose (already described) can be used for CSF glucose estimation.

Clinical Biochemistry Laboratory

Appendix V

The ultimate application of the biochemistry subject is for the health and welfare of mankind. Clinical biochemistry (also known as ***clinical chemistry*** or ***chemical pathology***) is the laboratory service absolutely essential for medical practice. The results of the biochemical investigations carried out in a clinical chemistry laboratory will help the clinicians to determine the diseases (diagnosis) and for follow-up of the treatment/recovery from the illness (prognosis). Biochemical investigations hold the key for the ***diagnosis and prognosis of diabetes mellitus, jaundice, myocardial infarction, gout, pancreatitis, rickets, cancers, acid-base imbalance*** etc. Successful medical practice is unimaginable without the service of clinical biochemistry laboratory.

The ***biological fluids*** employed in the clinical biochemistry laboratory include ***blood***, ***urine***, ***cerebrospinal fluid*** and pleural fluid. Among these, ***blood*** (directly or in the form of plasma or serum) is ***frequently used*** for the investigations in the clinical biochemistry laboratory.

On rare occassions, tissue biopsy samples may also be used for diagnosis.

COMMON ABNORMALITIES AND BIOCHEMICAL PROFILE

A large number of conditions, particularly diseases (diabetes mellitus, genetic diseases, MI, organ failure etc.), are associated with abnormal biochemical profiles. A selected list of them is given in ***Table V***, e.g. cancer can be diagnosed by estimating tumor markers.

Table V : Common abnormalities associated with blood/plasma biochemical profile (selected examples listed)

Abnormality	Biochemical profile
Organ function failure	***Renal failure***......serum creatinine and urea elevated ***Liver failure***......serum bilirubin and ALT increased.
Myocardial infarction	Cardiac troponin I and CPK in serum elevated.
Cancer	Tumor markers elevated e.g. PSA in prostate cancer.
Diabetes mellitus	Increased blood glucose and HbA_{1C}
Genetic diseases	Serum phenylalanine elevated in phenylketonuria; NH_3 increased in urea cycle disorders.
Nutritional disorders	Serum albumin low in kwashiorkor.
Endocrine dysfunction	TSH high in hypothyroidism and low in hyperthyrodism.
Changes in blood pH	Metabolic acidosis—serum bicarbonate low; metabolic alkalosis—serum bicarbonate high.
Inflammation/infection	C-reactive protein elevated.
Use of drugs	Increase in serum uric acid in cancer chemotherapy.
Poisons	Decrease in the activity of serum butylcholinesterase in organo-phosphorus poisoning.
Stress	Increase in serum cortisol and catecholamines.
Altered physiological conditions	Elevated levels of serum hCG in pregnancy; increased blood lactate after strenuous exercise

THE REFERENCE RANGE

There exists biological variations between normal and healthy individuals for any analyte/compound measured in the laboratory. This is referred to as reference range around a mean (or average) which is considered to be normal. The most important reference biochemical values are given inside back cover.

If may benoted that the reference values of some analytes differ between sexes, different age groups, and in different ethinic groups. Further, the biochemical resuls differ from laboratory to laboratory due to variations in the methods employed.

In general, for a healthy population (depending on the gender, age and ethenicity), the range that includes 95% of target population is considered to the reference range (for a given analyte). In statistical terms, a mean ±2 SD is considered to be a reference range. The values of the analytes outside the reference ranges are considered to be abnormal, and therefore need to be further investigated.

TECHNIQUES USED IN CLINICAL BIOCHEMISTRY LABORATORY

The reader may refer tools of biochemistry (**Chapter 41**) and principles of practical biochemistry (appendix IV) for a brief knowledge on the techniques employed in clinical biochemistry. Broadly, the following principles or techniques are used

- Absorption spectrophotometry
- Fluorescence spectrophotometry (or spectrophotofluorimetry)
- High pressure liquid chromatography
- Flame photometry
- Ion-specific electrodes
- Atomic absorption spectrometry
- Electrophoresis
- Radioimmunoassay
- Enzyme-linked immunosorbant assay

Dry chemistry dipstick

For certain assays, the reagents, enzymes or antibodies can be put together on ***specially designed plastic strip*** namely dipstick. Dipsticks are handy devices for use at the bedside in a hospital, doctors clinic or even at home by individuals. Some commonly used tests employing dry chemistry dipsticks are listed—

- Finger-prick blood glucose measurement by test strip and glucometer.
- Strips to semiquantitatively detect urinary levels of glucose, ketone bodies, proteins etc. (individual dipsticks or multidipsticks are available)
- Dipstick for home pregnancy test to detect human chorionic gonadotropin (hCG) in urine.

COLLECTION OF BLOOD

Venous blood is ***most commonly*** used for a majority of biochemical investigations. It can be drawn from any prominent vein (usually from a vein on the front of the elbow). ***Capillary blood*** (<0.2 ml) obtained from a finger or thumb, is less frequently employed. ***Arterial blood*** (usually drawn under local anesthesia) is used for ***blood gas determinations***.

Precautions for blood collection : Use of sterile (preferably disposable) needles and syringes, cleaning of patients skin, blood collection in clean and dry vials/tubes are some of the important precautions.

CHOICE OF BLOOD SPECIMENS

Biochemical investigations can be performed on 4 types of blood specimens-whole blood, plasma, serum and red blood cells. The selection of the specimen depends on the parameter to be estimated. ***Whole blood*** (usually mixed with an anticoagulant) is used for the estimation of hemoglobin, carboxyhemoglobin, pH, glucose, urea, non-protein nitrogen, pyruvate, lactate, ammonia etc. (Note : for glucose determination, plasma is prefered in recent years).

Plasma, obtained by centrifuging the whole blood collected with an anticoagulant, is employed for the parameters—fibrinogen, glucose, bicarbonate, chloride, ascorbic acid etc.

Serum is the supernatant fluid that can be ***collected after centrifuging the clotted blood.*** It is the most frequently used specimen in the clinical biochemistry laboratory. The parameters estimated in serum include proteins (albumin/globulins), creatinine, bilirubin, cholesterol, uric acid, electroylets (Na^+, K^+, Cl^-), enzymes (ALT, AST, LDH, CK, ALP, ACP, amylase, lipase) and vitamins.

It may be noted that plasma is physiologic fluid while serum is prepared in the laboratory.

Red blood cells are employed for the determination of abnormal hemoglobins, glucose 6-phosphate dehydrogenase, pyruvate kinase etc.

ANTICOAGULANTS

Certain biochemical tests require unclotted blood. Anticoagulants are employed for collecting such specimens.

Heparin : Heparin (inhibits the conversion of prothrombin to thrombin) is an ideal anticoagulant, since it does not cause any change in blood composition. However, other anticoagulants are prefered to heparin, due to the cost factor.

Potassium or sodium oxalate : These compounds precipitate calcium and inhibit blood coagulation. Being more soluble, potassium oxalate (5-10 mg per 5 ml blood) is prefered.

Potasium oxalate and sodium fluoride : These anticoagulants are employed for collecting blood to estimate glucose. Further sodium fluoride inhibits glycolysis and preserves blood glucose concentration.

Ammonium oxalate and potassium oxalate : A mixture of these two compounds in the ratio 3 : 2 is used for blood collection to carry out certain hematological tests.

Enthylene diaminetetracetic acid (EDTA) : It chelates with calcium and blocks coagulation. EDTA is employed to collect blood for hematological examinations.

HEMOLYSIS

The rupture or lysis of RBC, releasing the cellular constituents interferes with the laboratory investigations. Therefore, utmost care should be taken to avoid hemolysis when plasma or serum are used for biochemical tests. Use of dry syringes, needles and containers, allowing slow flow of blood into syringe are among the important precautions to avoid hemolysis.

PRESERVATION OF BLOOD SPECIMENS

Plasma or serum should be separated within 2 hours after blood collection. It is ideal and advisable to analyse blood, plasma or serum, immediately after the specimen collection. This however, may not be always possible. In such a case, the samples (usually plasma/serum) can be stored at 4°C until analysed. For enzyme analysis, the sample are preserved at –20°C.

TYPES OF LABORATORY TESTS

The biochemical investigations (on blood/plasma/serum) carried out in the clinical biochemistry laboratory may be grouped into different types.

1. **Discretionary or on-off tests :** Most common clinical biochemistry tests that are designed to answer specific questions. e.g., does the patient have increased blood urea/glucose concentration? Normally, these tests are useful to support the diagnosis.

2. **Biochemical profiles :** These tests are based on the fact that more useful information on the patients disease status can be obtained by analysing more constituents rather than one e.g., ***plasma electrolytes*** (Na^+, K^+, Cl^-, bicarbonate, urea); ***liven function tests*** (serum bilirubin, ALT, AST).

3. **Dynamic function tests :** These tests are designed to measure the body's response to external stimulus e.g., oral glucose tolerance test (to assess glucose homeostasis) : bromosulphthalein test (to assess liver function).

4. **Screening tests :** These tests are commonly employed to identify the inborn errors of metabolism, and to check the entry of toxic agents (pesticides, lead, mercury) into the body.

5. **Metabolic work-up tests :** The programmed intensive investigations carried out to identify the endocrinological disorders come under this category.

The term ***emergency tests*** is frequently used in the clinical laboratory. It refers to the tests to be performed immediately to help the clinician for proper treatment of the patient e.g., blood glucose, urea, serum electrolytes.

COLLECTION OF URINE

Urine, containing the metabolic waste products of the body in water is the most important excretory fluid. For biochemical investigations, urine can be collected as a single specimen or for 24 hours. Single specimens of urine, normally collected in the morning, are useful for qualitative tests e.g., sugar, proteins. Twenty four hour urine collections (done between 8 AM to 8 AM) are employed for quantitative estimation of certain urinary constituents e.g., proteins, hormones, metabolites.

Preservatives for urine : For the collection of 24 hr urine samples, preservatives have to be used or else urine undergoes changes due to bacterial action. Hydrochloric acid, toluene, light petroleum, thymol, formalin etc., are among the common preservatives used.

CEREBROSPINAL FLUID (CSF)

CSF is a fluid of the nervous system. It is formed by a process of selective dialysis of plasma by the choroid plexuses of the ventricles of the brain. The total volume CSF is 100-200 ml.

Collection of CSF : CSF is collected by puncturing the interspace between the 3rd and the 5th lumbar vertebrae, under aseptic conditions and local anesthesia.

Biochemical investigations on CSF : Protein, glucose and chloride estimations are usually performed in the clinical biochemistry laboratory.

ANALYSIS IN CLINICAL BIOCHEMISTRY LABORATORY

The details on the biochemistry of health and disease states in relation to normal and abnormal biochemical data have been described in this book. The reader must refer the following chapters for examples of laboratory tests

- Chapter 6 for enzyme alterations in disease states.
- Chapter 19 for hormonal tests.
- Chapter 20 for organ function tests.
- Chapter 21 for acid-base abnormalities.
- Chapter 37 for tumor markers.

For ready reference, the most common reference biochemical values are given on inside back cover .

Screening of neonates for genetic diseases

It is estimated that around 2% of the live births are associated with inborn errors of metabolism. Majority of genetic diseases related to metabolism lead to mental retardation. Screening of newborns (usually done within a week after birth) is useful for the early detection of possible occurrence of inborn errors of metabolism, and take steps for correction. It must be remembered that screening of neonates is not a confirmatory diagnosis. It only gives an indication for further investigations.

Although not that commonly done in India and other developing countries, the following desease screening tests are advised for the newborns

- Phenylketonuria
- Maple syrup urine disease
- Homocysteinuria
- Congenital hypothyroidism
- Organic acidurias
- Glucose 6-phosphate dehydrogenese deficiency
- Congenital adrenal hyperplasia
- Galactosemia.

ONLINE MENDELIAN INHERITANCE IN MAN (OMIM)

OMIM is a ***catalogue of human genes and genetic disorders*** with a focus on gene-phenotype relationship. The present OMIM Website (omim.org) has been developed and maintained by John Hopkins University School of Medicine (JHUSOM). OMIM database has been designed for use by physicians, health care professionals and researches dealing with genetic disorders. Further OMIM databse is freely available to public.

OMIM classification system

There is a unified index for genetic diseases with a specific numbering system that is widely used in medical literature.

Each OMIM entry is given a unique 6-digit identifying number as summarized below.

Autosomal loci or phenotypes (created before May 15, 1994)	100000–299999
X-linked loci or phenotypes	300000–399999
Y-linked loci or phenotypes	400000–499999
Mitochondrial loci or phenotypes	500000–599999
Autosomal loci or phenotypes	600000 and above

Some examples of OMIM are given

Phenylketonuria type I (enzyme defect-phenylalanine hydroxylase)	261600
Alkaptonuria (Enzyme defect-homogentisate oxidase)	607474
Galactosemia (Enzyme defect-galactose 1-phosphate uridyltransferase)	230400

QUALITY CONTROL

Quality control in clinical biochemistry laboratory refers to the reliability of investigative service. Any error in the laboratory will jeopardize the lives of patients. It is therefore utmost important that the laboratory errors are identified and rectified.

Quality control comprises of four interrelated factors namely precision, accuracy, specificity and sensitivity.

Precision refers to the reproducibility of the result when the same sample is analysed on different occasions (replicate measurements) by the same person. For instance, the precision is good, if the blood glucose level is 78, 80 and 82 mg/dl on replicates.

Accuracy means the closeness of the estimated result to the true value e.g., if true blood urea level is 50 mg/dl, the laboratory reporting 45 mg/dl is more accurate than the one reporting 35 mg/dl.

Specificity refers to the ability of the analytical method to specifically determine a particular parameter e.g., glucose can be specifically estimated by enzymatic glucose oxidase method.

Sensitivity deals with the ability of a particular method to detect small amounts of the measured constituent.

METHODS OF QUALITY CONTROL

Internal quality control refers to the analysis of the same pooled sample on different days in a laboratory, the results should vary within a narrow range.

External quality control deals with the analysis of a sample received from outside, usually from a national or regional quality control centre. The results obtained are then compared.

AUTOANALYSERS IN CLINICAL CHEMISTRY

The heavy work load in the clinical biochemistry laboratory has lead to the discovery of autoanalysers. These modern equipment are useful to analyse hundreds of samples in a short time. Single channel and multi-channel machines (autoanalysers) based on the principles of either continuous or discrete analysis are available on the market.

Case Studies with Biochemical Correlations

Appendix VI

I. Nutritional Disorders

CASE STUDY 1

A 3-year old female child had stunted growth, edema (particularly on legs and hands) discoloration of skin and hair, apathy and moon-face. She also had frequent respiratory infections and diarrhea. On enquiring, the mother informed the physician that the child was mostly breast-fed until 2 years of age, and for the past one year she was being given dilute buffalo milk and a small quantities of rice with ghee and dhal. The following are the laboratory data of the child

Parameter	Subject	Reference range
Hemoglobin	7 g/dl	12-14 g/dl
Serum proteins (total)	4 g/dl	6-8 g/dl
Serum albumin	2 g/dl	3-4.5 g/dl
Serum potassium	2.8 mEq/1	3.5-4.5 mEq/1

Diagnosis and discussion

The clinical manifestations and the nutritional history, supported by the laboratory data clearly indicate that the child was suffering from ***kwashiorkor***, a predominant nutritional disorder, in the developing countries. Edema occurred due to lack of adequate serum proteins to maintain water distribution between blood and tissues. The immunological response of the child to infections was very low. Deficiency of serum K^+ was observed due to diarrhea.

CASE STUDY 2

A one year old male baby had growth retardation, reduced physical activity, muscle wasting (emaciation), loose folds of skin wrapped over bones. The mother of the child, now in third month of gestation, informed the physician that she had been giving the child rice gruel, along with breast milk. The laboratory data of the child are given hereunder.

Parameter	Subject	Reference range
Hemoglobin	7 g/dl	12-14 g/dl
Serum proteins (total)	6 g/dl	6-8 g/dl
Serum albumin	3 g/dl	3-4.5 g/dl
Serum potassium	3.5 mEq/1	3.5-5 mEq/1

Diagnosis and discussion

The child was suffering from ***marasmus***, a nutritional disorder, predominantly due to the deficiency of calories. Marasmus mostly occurs in children less than one year of age. It can be distinguished from kwashiorkor (Case 1) by lack of edema and almost unaltered serum albumin level.

CASE STUDY 3

A 12-year old boy had difficulty to see in dim light (evenings and while he enters cine theatres). His plasma retinol level was very low.

Diagnosis and discussion

This is a case of ***night blindness (nyctalopia)*** due to vitamin A deficiency. Since vitamin A is closely associated with dim light vision, its deficiency causes prolonged dark adaptation time. Severe deficiency of vitamin A leads to dryness of conjunctiva and cornea (xerophthalmia) and the occurrence of triangular plaques known as Bitot's spots.

CASE STUDY 4

A 6-year old boy had bone deformities such as bow legs and pigeon chest. He had a history of delayed eruption of teeth. On enquiring, the mother informed the physician that the boy had been on a strict vegetarian diet with low intake of milk as well as fats and oils. The following are the laboratory findings of this boy.

Parameter	Subject	Reference range
Calcium	8.1 mg/dl	9-11 mg/dl
Phosphate	2.5 mg/dl	3-4.5 mg/dl
Alkaline phosphatase	300 U/l	50-100 U/l
25-Hydroxy vitamin D	10 ng/ml	25-50 ng/ml

Diagnosis and discussion

This case is classical example of ***rickets*** due to vitamin D deficiency. As a result of low serum calcitriol (the biochemically active form of vitamin D), calcium and phosphate levels are not maintained in the circulation, hence mineralization is impaired, ultimately causing bone deformities such as bow legs. Serum alkaline phosphatase is elevated in rickets in a vain attempt to result in bone formation.

CASE STUDY 5

A 25-year old working woman had been in the habit of eating freeze-stored vegetables and fruits for over 2 years. She developed the symptoms of spongy and sore gums, swollen joints. She used to have frequent infections (common cold and respiratory tract infections) and observed delay in the healing of her wounds. She was mildly anaemic (hemoglobin 10 g/dl against normal 12-14 g/dl). Her friend, who happened to be a nutritionist, advised her to eat fresh fruits and vegetables, besides the supplementation of vitamin C. The situation improved in about 3 months.

Diagnosis and discussion

The manifestations described along with the response of the woman to vitamin C supplementation clearly point that the woman was suffering from ***scurvy***, a vitamin C deficiency nutritional disorder. This is observed in people who do not eat vitamin C rich foods and/or fresh fruits and vegetables. Vitamin C (ascorbic acid) participates in the synthesis of collagen, besides its involvement as an antioxidant. The symptoms observed in scurvy are due to impairment in these two functions.

CASE STUDY 6

A 40-year old alcoholic man had the symptoms of mental confusion, loss of memory, rhythmical to and fro motion of eye balls. Enquiries revealed that he had not been taking diet with sufficient quantities of vitamins to meet his body demands. From the analysis of the cultured fibroblasts, the man was found to have the enzyme transketolase with much reduced affinity for thiamine pyrophosphate.

Diagnosis and discussion

This is a case of ***Wernicke-Korsakoff syndrome*** due to the deficiency of thiamine (vitamin B_1). It is predominantly observed in alcoholics consuming a thiamine-deficient diet. The transketolase enzyme of hexose monophosphate shunt is mainly affected.

CASE STUDY 7

A 55-year old woman, a patient of tuberculosis, was given the drug isoniazid (isonicotinic acid hydrazide, INH). At the end of the treatment regime, she developed neurological manifestations such as depression, irritability, nervousness and mental confusion. Her urine contained increased concentration of xanthurenic acid.

Diagnosis and discussion

The drug isoniazid combines with pyridoxal phosphate (PLP), the coenzyme of vitamin B_6, to form inactive hydrazone derivatives which inhibit PLP dependent enzymes. The elevated excretion of xanthurenic acid indicates a blockade in tryptophan metabolism (due to impairment in the activity of kynureninase). This case is an example of ***drug induced vitamin B_6 deficiency***. It is therefore, advised that for patients of tuberculosis, treatment with isoniazid should be accompanied by vitamin B_6 supplementation.

CASE STUDY 8

A 23-year old woman with 3 children had the complaints of weakness and lethargy. Her hemoglobin level was 7 g/dl (normal 12-14 g/dl). Her blood was found to contain large abnormal immature erythrocytes. This woman had a highly elevated excretion of formiminoglutamate (FIGLU), a metabolite of histidine, in urine.

Diagnosis and discussion

Elevated excretion of FIGLU in urine indicates that the woman was suffering from ***folate deficiency***, the most common vitamin deficiency observed in developing countries. Deficiency of folate results in ***macrocytic anemia*** with abnormally large erythrocytes. Due to a block in DNA synthesis, the maturation of RBC is slowed down leading to the increase in their size.

CASE STUDY 9

A 25-year old strict vegetarian woman, with a very low consumption of milk products, was found to be anemic (hemoglobin 7 g/dl against normal 12-14 g/dl). She was lethargic and had the complaints of numbness and tingling of fingers and toes. Her urine had an elevated excretion of methylmalonate.

Diagnosis and discussion

The strict vegetarians of low socioeconomic group in the developing countries are susceptible to anemia. In the case described above, the woman had low hemoglobin level along with an increased urinary excretion of methylmalonate. The latter character is prominent feature of ***vitamin*** B_{12} ***deficiency***.

CASE STUDY 10

A 25-year old strict vegetarian woman, with two children aged 5 years and 3 years, complained of tiredness and appeared pale. In the recent few months, the woman had heavy and prolonged menstrual flow. Enquiries revealed that her consumption of milk and milk products was reasonably good, but leafy vegetables was low. The laboratory investigations showed that the woman's hematocrit was 28% (reference range 40% to 50%), while her hemoglobin concentration was 8 g/dl (reference range 12-14 g/dl).

Diagnosis and discussion

The woman with a low hemoglobin concentration and reduced hematocrit depicts a very common nutritional disorder — ***iron deficiency anemia***. A strict vegetarian diet coupled with low consumption of leafy vegetables led her to iron deficiency.

CASE STUDY 11

A 20-year old medical student had a delicious dinner with mushroom dishes in a Chinese restaurant. The next day morning, he was admitted in a hospital with abdominal cramps, vomiting and diarrhea.

Diagnosis and discussion

This appears to be a case of ***food-poisoning***, caused by a dangerous species of ***mushroom*** (most probably *Amanita phalloides)*. The toxicity is due to toxin α-amanitin which is not inactivated by cooking. α-Amanitin is a cyclic octapeptide and it strongly inhibits RNA polymerase II and thus transcription (synthesis of RNA).

II. Metabolic Diseases/Inborn Errors

CASE STUDY 12

A boy, aged 12 years was given treatment for prolonged diarrhea. After improvement, he complained abdominal discomfort and diarrhea with a feeling of being bloated, after consumption of milk. He was taken to a physician who advised him to stop his intake of milk. He felt better in 3 days.

Diagnosis and discussion

The boy was suffering from ***acquired lactose intolerance***. This was precipitated by diarrhea where the intestinal mucosal cells were denuded faster. The brush border of the intestine houses the enzyme lactase which is lost due to diarrhea, and hence typical symptoms of flatulence.

CASE STUDY 13

A normal one month old baby had a history of vomiting and diarrhea that frequently occurred after breast feeding. The urine gave a positive test for reducing sugars (Benedict's test) while the test was negative by Glucostix (specific for glucose). The RBC were found to be totally deprived of activity of the enzyme galactose 1-phosphate uridyltransferase (Reference range 4-30 units/g of hemoglobin).

Diagnosis and discussion

The baby was suffering from ***galactosemia***, a metabolic disorder, due to the deficiency of the enzyme galactose 1-phosphate uridyltransferase. The disease is characterized by increase in blood galactose level, and its excretion into urine. For the babies suffering from galactosemia, milk has to be removed from the diet and replaced with infant formula containing sucrose.

CASE STUDY 14

An 8-year old girl with enlarged abdomen was irritable and lethargic, and was frequently hungry. She often had the symptoms of weakness and sweating which disappeared on eating. Enquiries revealed that her developments during childhood were rather slow compared to her elder brother. She was weak at school. On clinical examination, the girl was found to have enlarged liver. The laboratory data of the subject are given next.

Parameter	Subject	Reference range
Blood glucose (fasting)	40 mg/dl	70-100 mg/dl
Blood pH	7.25	7.35-7.45
Blood ketone bodies (total)	5 mg/dl	2-3 mg/dl
Serum lactate	25 mg/dl	3-8 mg/dl
Serum triglycerides	350 mg/dl	75-150 mg/dl
Serum uric acid	10 mg/dl	4-6 mg/dl

Diagnosis and discussion

From the girl's history, the clinical manifestations and the laboratory data, it is clear that this girl was a victim of ***von Gierke's disease*** (glycogen storage disease type I), due to a defect in the enzyme glucose 6-phosphatase in liver. The result is that enough free glucose is not liberated in the liver to maintain blood glucose level. Hence the girl had frequent episodes of hypoglycemia. The substrate glucose 6-phosphate is diverted to other pathways (HMP shunt, lipogenesis etc.) ultimately resulting in elevated triglycerides, ketone bodies, uric acid etc.

CASE STUDY 15

A 25-year old man was on treatment with antimalarial drug primaquine. As the treatment was in progress, he developed complications. This subject's laboratory investigative data are given hereunder.

Parameter	Subject	Reference range
Hematocrit	30 %	45-50 %
Hemoglobin		
before treatment	12 g/dl	12-14 g/dl
after treatment	8 g/dl	
Serum bilirubin	3.5 mg/dl	< 1.0 mg/dl

The laboratory tests also indicated increased fragility of erythrocytes.

Diagnosis and discussion

This subject obviously had the ***deficiency of the glucose 6-phosphate dehydrogenase*** (G6PD). It was severe in erythrocytes resulting in an impairment in the production of NADPH, a coenzyme required by glutathione peroxidase for the inactivation of free radicals (superoxide and H_2O_2). The drugs such as primaquine cause increased generation of free radicals which cannot be effectively inactivated due to G6PD deficiency. This leads to hemolysis of RBC and consequently jaundice, as is evident from the laboratory data-reduced hemoglobin and increased serum bilirubin.

CASE STUDY 16

A 20 year old man had a history of cramping abdominal pain, diarrhaea, vomiting, tachycardia, insomnia, anxiety and muscle weakness, following intake of low-calorie diet for weight reduction. His laboratory investigative data are given hereunder.

Parameter	Subject	Reference range
Urinary porphobilinogen (PBG)	100 mg/24 hr	0-4 mg/24 hr
Urinary δ-aminolevulinate (ALA)	124 mg/24 hr	1-7 mg/24 hr

Diagnosis and discussion

The subject was suffering from a rare metabolic disorder namely, ***acute intermittent porphyria (AIP)***. It is characterized by the deficiency of the enzyme uroporphyrinogen I synthase, and increased urinary excretion of porphobilinogen and δ-aminolevulinate. AIP is precipitated by exogenous and endogenous factors such as steroids, porphyrinogenic drugs (e.g. barbiturates), alcohol, low calorie diets etc. AIP can be treated by administration of hematin which inhibits the first enzyme of heme synthesis (δ-aminolevulinate synthase), and the accumulation of ALA and porphobilinogen.

CASE STUDY 17

A 30-year old man was admitted in a cardiology ward after he complained chest pain. His clinical and biochemical investigations (ECG changes, isoenzymes of creatine phosphokinase etc.) indicated that he suffered a mild myocardial infarction. Timely intervention and appropriate treatment saved him from death. The lipid profile data of this man are given here.

Parameter	Subject (mg/dl)	Reference range
Serum total cholesterol	450	150-200
LDL-cholesterol	380	<100
HDL-cholesterol	40	40-60
VLDL-cholesterol	30	20-40
Serum triglycerides	150	75-150

Diagnosis and discussion

This man had highly elevated total and LDL-cholesterol while other lipid profile parameters are within the normal limits. This is a case of ***familial hypercholesterolemia (hyperlipoproteinemia type IIa)***. The victims of this disorder have decreased number of LDL receptors. They have very high risk of coronary heart diseases in the 3rd and 4th decades of life. Lipid-lowering drugs, besides low cholesterol, low fat diet and regular exercise are useful for these people.

CASE STUDY 18

The laboratory data on the lipid profile of a 35-year old man are given here.

Parameter	Subject (mg/dl)	Reference range
Serum total cholesterol	350	150-200
LDL-cholesterol	225	<100
HDL-cholesterol	35	40-60
VLDL-cholesterol	90	20-40
Serum triglycerides	450	75-150

Further, the electrophoresis of plasma lipoproteins showed a thick band of intermediate density lipoproteins (IDL).

Diagnosis and discussion

The serum triglycerides and LDL-cholesterol were elevated in this subject, besides a distinct band of IDL on electrophoresis. These findings suggest that this is a case of ***type III hyperlipoproteinemia***. Although the etiopathology of this disorder is not known, it is attributed to an abnormality in apoprotein E. The sufferers of this disease are at a high risk for atherosclerosis and coronary heart diseases.

CASE STUDY 19

A 55-year old diabetic man had been on treatment with oral antidiabetic drugs for over a decade. Despite the physician's advice, he could not give up his drinking habit. The lipid profile data of this man are given below.

Parameter	Subject (mg/dl)	Reference range
Serum total cholesterol	250	150-200
LDL-cholesterol	110	<100
HDL-cholesterol	40	40-60
VLDL-cholesterol	100	20-40
Serum triglycerides	500	75-150

Further, the serum lipoprotein electrophoretic pattern indicated an increase in the concentration of chylomicrons and VLDL.

Diagnosis and discussion

The diabetic patient and an alcoholic had elevated levels of triglycerides and VLDL-cholesterol. This is a case of ***type V hyperlipoproteinemia***. This disorder is mostly found in adults and is due to overproduction of triglycerides. The risk for atherosclerosis may or may not increase. The sufferers of this disease are advised low calorie and low fat diet.

CASE STUDY 20

A six month old male infant had growth retardation, lethargy, frequent vomitings. His plasma ammonia concentration was 520 μg/dl (normal 15-45 μg/dl). He also had elevated plasma glutamate concentration and reduced citrulline. The infant's urine analysis revealed increased excretion of glutamine and uracil. Liver biopsy indicated that the activity of orinithine transcarbamoylase was only 15% that of normal.

Diagnosis and discussion

This is a case of ***hereditary hyperammonemia*** due to a defect in the enzyme ornithine transcarbamoylase. Elevated plasma levels of ammonia and glutamate along with reduced citrulline concentration strongly support the diagnosis. Due to a blockage in the urea cycle, carbamoyl phosphate which accumulates in the mitochondria leaks out into cytosol to be utilized for increased uracil formation (hence elevated uracil excretion in urine). Glutamate and glutamine are the carriers of ammonia, therefore they get elevated in circulation as well as urine.

CASE STUDY 21

A 35-year old man was operated thrice for renal stones in a span of 5 years. The chemical analysis of the stones revealed that they were primarily composed of oxalate. The man's urine contained elevated oxalate concentration — 140 mg/day against normal <60 mg/day. Liver biopsy revealed the deficiency of the enzyme glycine transaminase.

Diagnosis and discussion

This is a case of ***primary hyperoxaluria***, caused by a defect in the enzyme glycine transaminase. Due to a metabolic blockade in glycine metabolism glyoxylate accumulates and gets converted to oxalate. Elevated urine excretion of oxalate causes frequent stone formation.

CASE STUDY 22

A 5-year old boy had delayed developments and was unable to speak or walk properly. He was found to be mentally retarded with characteristic seizures and tremors. The boy had a plasma phenylalanine level of 30 mg/dl (reference range of 1-2 mg/dl). His urine gave a positive ferric chloride test, indicating an elevated excretion of phenylpyruvate.

Diagnosis and discussion

An elevation is plasma phenylalanine concentration with an increased excretion of phenylpyruvate in urine, supported by the clinical manifestations, clearly indicate that the boy was a victim of ***phenylketonuria (PKU)***. This is an inborn error of phenylalanine metabolism due to a defect in the enzyme phenylalanine hydroxylase. As a result of this, phenylalanine cannot be converted to tyrosine, hence gets diverted to alternate pathways, producing metabolites which are directly or indirectly inolved in growth failure and mental retardation. The treatment for PKU is to provide diet low or deficient in phenylalanine, so that their plasma phenylalanine levels are kept within the normal limits.

CASE STUDY 23

A 50-year old man had a complaint of arthritis. His urine gave a positive Benedict's test for reducing substances. On careful examination, his urine was found to turn to blackish (coke-like) colour on standing. The laboratory investigations revealed that the old man had normal fasting (90 mg/dl) and post-prandial (120 mg/dl) blood glucose and normal serum uric acid (4 mg/dl) concentrations.

Diagnosis and discussion

This is a case of ***alkaptonuria***, an inborn error of tyrosine metabolism, due to a defect in the enzyme homogentisate oxidase. The urine of these subjects contains large quantities of homogentisate — a strong reducing substance (hence Benedict's test positive). Homogentisate on standing gets oxidized to impart the characteristic coke colour to urine. Arthritis in alkaptonuric patients is believed to be due to the deposition of pigment alkapton (in the joints) produced from homogentisate.

CASE STUDY 24

A full term baby, born to normal and healthy parents, was found to have marked lack of pigmentation. The baby had white skin, hair and blue eyes. The paternal grandfather of the baby had similar problem.

Diagnosis and discussion

The generalized lack of pigmentation (melanin) of the baby is good example of ***albinism***. This is a metabolic disorder due to lack of the enzyme tyrosinase that is mainly responsible for the formation of the pigment melanin. The albinos are sensitive to light and are susceptible to skin cancers. Photophobia (intolerance to light) is associated with lack of pigmentation in eyes. However, there is no impairment in the eyesight of albinos.

CASE STUDY 25

A 12-year old boy was brought to the hospital with a complaint of severe pain in his lower right flank. His urine was found to have elevated levels of cystine, lysine, arginine and ornithine. Microscopic examination of urine revealed the presence of hexagonal crystals. Further, X-rays showed the occurrence of stones in the kidney.

Diagnosis and discussion

This is a case of ***cystinuria***, an inborn error of sulfur containing amino acids. Due to a defect in the carrier system in the renal tubules that is responsible for the absorption of cystine, ornithine, arginine and lysine, their excretion is increased in urine. Cystine is relatively insoluble and therefore gets precipitated leading to the formation of cystine stones in the kidney and urinary tract. The crystals observed in urine are due to cystine. The treatment for cystinuria includes restricted ingestion of dietary cystine and high intake of fluids.

CASE STUDY 26

A 5-year old girl was found to have stunted growth, delayed developments and mental retardation. X-rays indicated that she had signs of osteoporosis. Her urine gave a positive test for homocysteine. Vitamin B_6 supplementation did not make any improvement.

Diagnosis and discussion

Accumulation of homocysteine in the body along with its elevated urinary excretion results in ***homocysteinuria***. Homocysteinuria causes several complications such as osteoporosis, mental retardation and thrombosis. The biochemical lesion lies in the deficiency of the enzyme cystathionine synthase, as a result of which homocysteine cannot be converted to cystathionine.

CASE STUDY 27

A 2-week old infant was passing urine which smelled like burnt sugar. The plasma and urine concentrations of branched chain amino acids (valine, leucine and isoleucine) were found to be very high. Further, the enzyme activity of branched chain α-keto acid dehydrogenase in leucocytes was found to be very low.

Diagnosis and discussion

This is a case of ***maple syrup urine disease (MSUD)***, an inborn error associated with branched chain amino acid metabolism. The urine smells like burnt sugar (maple syrup). A defect in the enzyme branched chain α-keto acid dehydrogenase which catalyses the oxidative decarboxylation of the keto acids (of branched chain amino acids) to the corresponding acyl CoA thioesters, is absent in the disorder. MSUD presents in early infancy and should be properly treated by giving a diet low in branched chain amino acids. If not corrected, the disease results in acidosis, lethargy, convulsions, menal retardation, coma and finally death within one year after birth.

CASE STUDY 28

A 55-year old man complained severe pain in the joints. He was a non-vegetarian and consumed alcohol occasionally. His laboratory findings are as follows.

Parameter	Subject	Reference range
Serum uric acid	12 mg/dl	3-7 mg/dl
Blood urea	25 mg/dl	15-40 mg/dl
Urinary uric acid	2.5 g/day	0.5-0.7 g/day
Urinary pH	4.5	5.5-7.0

The subject responded to the treatment by the drug-allopurinol.

Diagnosis and discussion

Elevation of uric acid in serum and urine along with case history, further supported by the response of the subject to the drug allopurinol, clearly indicate that the patient was suffering from ***gout.*** Over-production of uric acid, the excretory end product of purine metabolism, causes gout. Crystals of sodium urate get deposited in the joints causing gouty arthritis. Allopurinol competitively inhibits the enzyme xanthine oxidase and lowers the formation of uric acid, hence the relief by administering this drug.

CASE STUDY 29

A 4-year old boy showed signs of learning disability and aggressive behaviour, besides pain in the joints. It was observed that he had an irresistible urge to bite his fingers and lips. The laboratory investigations revealed that the boy had serum uric acid concentration of 10 mg/dl (reference range 4-6 mg/dl).

Diagnosis and discussion

The clinical manfestations along with elevated serum uric acid level support that the boy was suffering from ***Lesch-Nyhan syndrome***. This is an inborn error of purine metabolism (salvage pathway). It is a sex-linked disorder affecting only males due to the deficiency of the enzyme hypoxanthine-guanine phosphoribosyl transferase (HGPRT). Lesch Nyhan syndrome is characterized by increased production of uric acid, often causing gouty arthritis, besides the neurological abnormalities.

CASE STUDY 30

A 2-year old male child with retarded growth was found to have anemia. This was non-responsive to treatment by iron, folic acid, pyridoxine and cobalamin. The following are his laboratory investigative data

Parameter	Subject	Reference range
Hemoglobin	7 g/dl	14-16 g/dl
Hematocrit	30 %	40-50 %
Orotic acid (urine)	800 mg/day	<2 mg/day

Diagnosis and discussion

The child was suffering from the metabolic disorder ***orotic aciduria***. This is due to deficiency of the enzymes orotate phosphoribosyl transferase and OMP decarboxylase of pyrimidine biosynthesis. Feeding a diet rich in uridine and/or cytidine is an effective treatment for orotic aciduria.

CASE STUDY 31

An African negro boy (aged 12 years), studying in an Indian High School, was admitted in a hospital with complaints of fever and severe pain in the arms and legs. His laboratory investigative data are given hereunder.

Parameter	Subject	Reference range
Hemoglobin	7 g/dl	14-16 g/dl
Hematocrit	20 %	40-50 %
Hemoglobin electrophoresis	A distinct HbS band with slower movement than adult hemoglobin (HbA) detected	—
Microscopic examination of blood smear	Crescent (sickle) shaped erythrocytes	—

Diagnosis and discussion

The laboratory data of the boy with sickle-shaped erythrocytes and altered hemoglobin electrophoretic pattern with a distinct HbS band supported by anemia and reduced hematocrit, along with the clinical manifestations support the diagnosis of ***sickle cell anemia***. This disorder primarily occurs in black population and the patients may die before they reach adulthood (< 20 years). Sickle cell anemia is caused by an abnormality in hemoglobin. It occurs due to the replacement of glutamate by valine at the 6th position of β-chain of hemoglobin.

CASE STUDY 32

A 30-year old man with a history of smoking for 15 years complained breathlessness and developed chronic bronchitis. His plasma concentration of α_1-antitrypsin was 20 mg/dl against the normal 200 mg/dl.

Diagnosis and discussion

α_1-Antitrypsin (α_1-AT) is a natural inhibitor of trypsin. The deficiency of α_1-antitrypsin causes ***emphysema*** (abnormal distension of lungs by air), which in turn is associated with lung infections. Further, α_1-AT deficiency causes destruction of lung tissues due to lack of inhibition of the enzymes elastase and proteases in lungs. Emphysema is more commonly associated with heavy smoking. This is because smoking causes oxidation of methionine (358th amino acid residue of α_1-AT) to methionine sulfoxide, which cannot bind and inactivate proteases.

CASE STUDY 33

A school boy aged 12 years complained of abdominal pain and was admitted in a hospital. He was weak and tired, and had a history of behavioural disturbance and epileptic form of seizures. Clinical examination revealed that the boy had an enlarged liver. His cornea showed the presence of Kayser Fleischer ring. The laboratory data of the boy are given below.

Parameter	Subject	Reference range
Serum copper	40 μg/dl	100-200 μg/dl
Serum ceruloplasmin	5 mg/dl	25-50 mg/dl
Urinary copper	200 μg/dl	< 25 μg/dl

Diagnosis and discussion

This boy had low serum level of copper and ceruloplasmin with an elevated copper excretion in urine. The clinical manifestations and the biochemical data indicate that the boy was suffering from ***Wilson's disease***. This disorder is associated with the deposition of copper in liver, kidney and brain.

CASE STUDY 34

A 16-year old boy had viral infection. He had mild symptoms of jaundice — abdominal pain, anorexia and malaise. His laboratory findings are given hereunder.

Parameter (serum)	Subject	Reference range
Bilirubin (total) (van den Bergh reaction	4.5 mg/dl Indirect positive	0.2-1 mg/dl Negative)
Conjugated bilirubin	0.4 mg/dl	0.4 mg/dl
Unconjugated bilirubin	4.1 mg/dl	0.6 mg/dl
Alanine transaminase	30 U/l	10-45 U/l
Aspartate transaminase	25 U/l	10-50 U/l
Alkaline phosphatase	80 U/l	50-100 U/l

The above investigations were repeated six months later and the data obtained were almost the same as above.

Diagnosis and discussion

The boy had an elevated unconjugated bilirubin while the other parameters remained normal. This is a case of ***Gilbert's syndrome*** mainly due to a genetic defect in the enzyme hepatic UDP-glucuronyl transferase. This is a common form of congenital non-hemolytic hyperbilirubinemia. Gilbert's syndrome is more common in men and occurs in about 5% of the population. This disorder is often mistaken for more serious liver diseases. It however, requires no treatment.

CASE STUDY 35

A 10-year old boy had mild manifestations of rickets which could not be corrected by supplementation of vitamin D as well calcium in the diet. His laboratory findings are given below.

Parameter (serum)	Subject	Reference range
Calcium	9.5 mg/dl	9-11 mg/dl
Phosphate	2.0 mg/dl	3-4.5 mg/dl
Alkaline phosphatase	300 U/l	50-100 U/l
25-Hydroxycholecalciferol	30 ng/ml	30-40 ng/ml

Diagnosis and discussion

As there was no response for vitamin D supplementation, the boy was suffering from vitamin D resistant rickets. The laboratory data with normal serum calcium and 25-Hydroxycholecalciferol concentrations, along with low serum phosphate and elevated alkaline phosphatase indicate that the boy was a victim of ***familial hypophosphatemic rickets***. This is a sex-linked disorder caused by a defect in the renal tubular reabsorption of phosphate.

III. CASES RELATED TO ORGAN FUNCTION

CASE STUDY 36

An 18-year old male medical student had the complaints of loss of appetite (anorexia), nausea, headache and malaise. On clinical examination, his liver was found to be slightly enlarged. He was passing pale coloured stools but dark coloured urine. The data of his laboratory investigations are as follows.

Parameter (serum)	Subject	Reference range
Bilirubin (total)	8 mg/dl	0.2-1 mg/dl
(van den Bergh reaction	Biphasic	Negative)
Conjugated bilirubin	4.5 mg/dl	0.4 mg/dl
Unconjugated bilirubin	3.5 mg/dl	0.6 mg/dl
Alanine transaminase (ALT)	180 U/l	10-45 U/l
Aspartate transaminase (AST)	80 U/l	10-50 U/l
Alkaline phosphatase (ALP)	10 U/l	50-100 U/l

Diagnosis and discussion

A biphasic van den Bergh positive reaction (elevated serum conjugated and unconjugated bilirubin), along with increased activity of ALT and AST clearly indicates that this is a case of ***hepatic jaundice***. This is due to impairment in liver cell function caused by viral infection (viral hepatitis), poisons and toxins or cirrhosis of liver. Damage to the liver cells adversely affects the bilirubin uptake and its conjugation (hence conjugated and unconjugated bilirubins are elevated). The pale coloured stools are due to the absence of stercobilinogen while the dark coloured urine is due to urobilinogen.

CASE STUDY 37

A 20-year old man was treated for malaria a month ago. For the past one week, he was excreting slightly dark coloured urine and dark brown coloured stools. His laboratory findings are given hereunder.

Parameter	Subject	Reference range
Bilirubin (total)	7 mg/dl	0.2-1 mg/dl
(van den Bergh reaction	Indirect positive	Negative)
Conjugated bilirubin	0.4 mg/dl	<0.4 mg/dl
Unconjugated bilirubin	6.6 mg/dl	<0.6 mg/dl
Alanine transaminase (ALT)	30 U/l	10-45 U/l
Aspartate transaminase (AST)	40 U/l	10-50 U/l
Alkaline phosphatase (ALP)	10 U/l	50-100 U/l

Diagnosis and discussion

Following a malarial attack, the man developed ***hemolytic jaundice***. This was due to an excessive breakdown of erythrocytes releasing hemoglobin. The heme degraded to bilirubin cannot be effectively conjugated, as a result unconjugated bilirubin in the serum was elevated. The laboratory data — indirect van den Bergh reaction positive with no increase in the activities of serum ALT, AST and ALP support the diagnosis of hemolytic jaundice. The urine was dark coloured due to increased excretion of urobilinogen while the feces were dark brown due to stercobilinogen.

CASE STUDY 38

A 30-year old man had fever and abdominal pain. His stools were clay in colour and contained fat (steatorrhea). The man's urine was found to be dark in colour. His laboratory findings are given in the next page.

Parameter	Subject	Reference range
Bilirubin (total)	12 mg/dl	0.2-1 mg/dl
(van den Bergh reaction	Direct positive	Negative)
Conjugated bilirubin	11.5 mg/dl	<0.4 mg/dl
Unconjugated bilirubin	0.5 mg/dl	<0.6 mg/dl
Alanine transaminase (ALT)	70 U/l	10-45 U/l
Aspartate transminase (AST)	75 U/l	10-50 U/l
Alkaline phosphatase (ALP)	450 U/l	50-100 U/l

In addition, the urine gave a positive test for bile salts (Hay's test) and bile pigments (Fouchet's test).

Diagnosis and discussion

A direct van den Bergh reaction reflecting an increase in conjugated bilirubin, along with an elevation in serum alkaline phosphatase indicates that the man was suffering from ***obstructive jaundice (cholestasis)***. A marginal increase in the activity of ALT and AST is also observed in this disorder. Cholestasis is due to an obstruction (caused by gall stones, tumors etc.) that prevents the passage of bile into the intestine. The urine was dark in colour due to the excretion of conjugated bilirubin while the stools were clay in colour due to the deficiency of stercobilinogen, a pigment derived from bilirubin. Feces contained excess fat indicating impairment in fat digestion and absorption due to the absence of bile sats.

CASE STUDY 39

A 3-day old male baby had yellow colouration of the sclerae of the eyes. His urine was also found to be yellow in colour. The laboratory investigations revealed that the serum total bilirubin concentration was 18 mg/dl (normal <1 mg/dl), most of it being unconjugated (van den Bergh indirect positive). The paediatrician advised phototherapy for the baby.

Diagnosis and discussion

About 50% of the normal new born babies develop ***physiological neonatal jaundice*** after 30 hours after birth, as was described in this case. This is due to increased hemolysis coupled with immature hepatic system for the uptake, conjugation and secretion of bilirubin. The activity of the enzyme UDP-glucuronyl transferase is low in the newborn. In severe forms of hyperbilirubinemia, phototherapy is advised. The ultraviolet rays of the light isomerize bilirubin into a non-toxic form. It may be noted that the neonatal physiological jaundice usually disappears within 10 days of birth.

CASE STUDY 40

A 50-year old man, an industrialist by profession, had a routine medical check up. He was found to have slightly enlarged liver. His laboratory findings are given below.

Parameter	Subject	Reference range
Blood glucose, fasting	100 mg/dl	70-100 mg/dl
Blood urea	30 mg/dl	15-40 mg/dl
Serum bilirubin	0.8 mg/dl	< 1 mg/dl
Serum alanine transaminase	250 U/l	10-45 U/l
Serum aspartate transaminase (AST)	550 U/l	10-50 U/l
Serum γ-glutamyl transpeptidase (GGT)	700 U/l	5-50 U/l

Diagnosis and discussion

The biochemical data indicate that this man had chronic liver disease. This is evident from the highly elevated activity of ALT, AST and GGT in the serum. These 3 enzymes leak into the circulation due to hepatocellular damage. The data indicate that the man was suffering from ***alcoholic liver disease***.

CASE STUDY 41

A 55-year old man was brought to the hospital with severe chest pain, breathlessness and vomiting. He could be rushed to the city hospital 5 hours after the onset of chest pain. His blood was immediately drawn, and the laboratory data are given below.

Parameter	Subject	Reference range
Creatine phosphokinase (CPK)	600 U/l	50-200 U/l
Aspartate transaminase (AST)	85 U/l	10-50 U/l
Lactate dehydrogenase (LDH)	315 U/l	50-200 U/l

Further laboratory evaluation revealed that the isoenzymes CPK_2(MB) and LDH_1 were highly elevated.

Diagnosis and discussion

The man obviously had an attack of ***myocardial infarction***. This is supported by the laboratory findings — elevation in CPK (MB), LDH (isoenzyme LDH_1) and aspartate transaminase. It may be noted that

CPK starts to rise at 4-6 hours and reaches a peak value within 24-30 hours. As regards LDH and AST, they rise from the second day onwards. Thus, CPK is the earliest marker enzyme for the diagnosis of myocardial infarction.

CASE STUDY 42

A 60-year old man had mild chest pain and breathlessness. He ignored and did not consider it worth to visit a hospital. His son came to know about this and could bring him to the city hospital only on the 3rd day after he had the symptoms. The man's laboratory findings are given hereunder.

Parameter (serum)	*Subject*	*Reference range*
Creatine phosphokinase (CPK)	120 U/l	50-200 U/l
Aspartate transaminase (AST)	350 U/l	10-50 U/l
Lactate dehydrogenase (LDH)	620 U/l	200-450 U/l
LDH electrophoresis	LDH_1 elevated	

Diagnosis and discussion

This is a case of ***myocardial infarction***. Elevation in the serum LDH (particularly LDH_1) and AST support the diagnosis. CPK is almost normal since the patient was brought to the hospital on the third day after he developed complications (Note : CPK rises at 4-6 hours and reaches peak value within 24-30 hours).

CASE STUDY 43

A 50-year old man, an occasional alcoholic, had the complaint of severe abdominal pain following a large meal or alcohol intake. He also had the symptoms of nausea and vomiting. His laboratory data are given hereunder.

Parameter	*Subject*	*Reference range*
Serum amylase	650 U/l	30-150 U/l
Serum lipase	450 U/l	30-150 U/l
Urine amylase	1350 U/l	200-800 U/l

Diagnosis and discussion

Elevation in the amylase activity is an indication of ***pancreatitis***. It may be noted that serum amylase activity is elevated on the first day of the disease and falls rapidly due to renal clearance. Very frequently, the urinary amylase activity is more important for the diagnosis of pancreatitis.

CASE STUDY 44

A 70-year old man had the complaints of frequent urination and pain in the lower abdomen. He had a sensation of obstruction in his urine flow. His rectal examination revealed the presence of hard mass. The laboratory findings of the man are as follows.

Parameter	*Subject*	*Reference range*
Serum acid phosphatase	30 U/l	0.5-5 U/l
Serum alkaline phosphatase	60 U/l	50-100 U/l
Serum prostatic specific antigen (PSA)	8 ng/ml	< 4 ng/ml

Diagnosis and discussion

The old man with an elevation in serum acid phosphatase and PSA, and rectal examination with prostate hard mass was suffering from ***cancer of prostate gland***. It may be noted that estimation of serum PSA (elevated) is highly specific for prostate carcinoma.

CASE STUDY 45

A 4-year old male child was passing less quantity of urine that was red in colour. He was found to have marked edema around the eyes. Enquiries revealed that the child had the history of repeated upper respiratory tract infections during the past 3 years. Microscopic examination of urine showed erythrocytes, leucocytes, epithelial cells and granular casts. The child had almost normal concentrations of blood urea, serum creatinine, serum total proteins and albumin. His 24 hour urine contained 1000 mg protein (normal < 150 mg).

Diagnosis and discussion

This child with oliguria (low urine output), hematuria and proteinuria, with characteristic erythrocyte casts in urine and marked edema around the eyes represents a case of ***acute glomerulonephritis***. This is supported by the history of respiratory tract infections. In most cases, acute glomerulonephritis is reversible.

CASE STUDY 46

A 20-year old man had generalized edema of the body with puffiness of the face in the mornings. His laboratory findings are given in the next page.

Parameter	Subject	Reference range
Serum total proteins	4.5 g/dl	6-8 g/dl
albumin	1.5 g/dl	3.5-5 g/dl
globulins	3.0 g/dl	2.5-3.5 g/dl
Serum cholesterol	350 mg/dl	150-200 mg/dl
Blood urea	30 mg/dl	15-40 mg/dl
Serum creatinine	1.2 mg/dl	0.5-1.5 mg/dl
Urinary proteins	4 g/day	< 100 mg/day

The serum electrophoresis of the subject showed a sharp and prominent α_2-globulin band.

Diagnosis and discussion

This is a case of ***nephrotic syndrome***, characterized by generalized edema heavy proteinuria and hypoproteinemia (particularly low albumin). Increased serum cholesterol level in this disorder is quite common. The serum electrophoretic pattern (a prominent α_2-globulin band) supports the diagnosis of nephrotic syndrome. Generalized edema is due to low plasma colloidal osmotic pressure, as a result of reduced albumin concentration.

IV. Cases Related to Endocrine Function

CASE STUDY 47

A 30-year old woman, married five years ago, had no children. She complained of tiredness, weight gain, cold intolerance, neuromuscular pains and constipation. She was found to be anemic. Her laboratory data are given hereunder.

Parameter (serum)	Subject	Reference range
Triiodothyronine (T_3)	100 ng/dl	110-180 ng/dl
Thyroxine, total (T_4)	4 μg/dl	5-12 μg/dl
Thyroid stimulating hormone (TSH)	50 μU/ml	0.5-4 μU/ml
Cholesterol	250 mg/dl	150-200 mg/dl

Diagnosis and discussion

The woman with the described clinical manifestations, elevated serum TSH and marginally lower T_4 and T_3 was a case of ***hypothyroidism***. Serum cholesterol level is also increased in hypothyroidism. (Note : This has no diagnositic importance, since serum cholesterol gets elevated in several other diseases e.g. diabetes, nephrotic syndrome).

CASE STUDY 48

A 35-year old woman complained of weight loss, despite normal or even increased appetite. She was found to have the symptoms of nervousness, irritability, anxiety, diarrhea, sweating, sensitive to heat and had increased pulse rate (tachycardia). Her laboratory findings are given here.

Parameter (serum)	Subject	Reference range
Triiodothyronine (T_3)	350 ng/dl	110-180 ng/dl
Thyroxine, total (T_4)	14 μg/dl	5-12 μg/dl
Thyroid stimulating hormone (TSH)	0-0.1 μU/ml	0.5-4 μU/ml

Diagnosis and discussion

The clinical manifestations of the woman, along with reduced serum TSH and increased T_3 and T_4 indicate that the woman was suffering from ***hyperthyroidism***.

CASE STUDY 49

A 50-year old post-menopausal woman had the symptoms of lethargy, depression, muscle weakness, loss of memory, loss of appetite, constipation, polyuria and polydipsia. Over the past 10 years, she had 2 kidney stones that passed spontaneously. Chemical analysis of the stones revealed that calcium phosphate was the major constituent. The following are the laboratory findings of this subject.

Parameter (serum)	Subject	Reference range
Calcium	13 mg/dl	9-11 mg/dl
Phosphate	2.5 mg/dl	3-4.5 mg/dl
Chloride	110 mEq/l	95-105 mEq/l
Alkaline phosphatase	110 U/l	50-100 U/l

Diagnosis and discussion

The clinical manifestations and the laboratory data — serum calcium elevation with decreased phosphate and a marginal increase in chloride and alkaline phosphatase strongly suggest that the woman was suffering from ***hyperparathyroidism***. This diagnosis was further supported from the fact that the woman had a recurrent stone formation (containing calcium and phosphate).

CASE STUDY 50

A 35-year old woman had the complaints of neuromuscular irritability, spasms and convulsions, besides tingling in the hands, feet and around mouth. Laboratory data showed that her serum calcium level was low (7 mg/dl against normal 9-11 mg/dl) while phosphate concentration was high (6 mg/dl against normal 3-4.5 mg/dl).

Diagnosis and discussion

The clinical symptoms and the laboratory findings — hypocalcemia and hyperphosphatemia indicate that woman was suffering from ***tetany***. This might be due to hypoparathyroidism/osteomalacia/ calcium malabsorption.

CASE STUDY 51

A 50-year old woman complained of increased thirst (polydipsia) and high frequency of urination (polyuria). She was passing about 10 litres of urine per day and had to wake up 4 to 5 times during night to urinate. The specific gravity of the morning urine sample was 1.001 (reference range 1.003-1.030) while the osmolality was 200 milliosmoles/kg (reference range 500-1,200 milliosmoles/kg). A computed tomogram (CT) of hypothalamus and pituitary gland revealed a mass lesion about 1.2 cm in diameter in the hypothalamus. The woman developed tuberculoma in the brain due to tuberculosis.

Diagnosis and discussion

The characteristic features — polyuria, polydipsia with a very dilute urine output (low specific gravity and osmolality) are indicative of the disorder ***diabetes insipidus***. The woman, suffering from tuberculosis, developed tuberculoma destroying the ability of the hypothalamus to synthesize vasopressin (antidiuretic hormone exclusively responsible for the reabsorption of water by renal tubules) ultimately resulting in diabetes insipidus.

CASE STUDY 52

Oral glucose tolerance test (mini GTT) was performed on a 45-year old obese women with a family history of diabetes mellitus (her mother was a diabetic patient). The following are the laboratory data given.

	Fasting	*2 hr*
Blood glucose (mg/dl)	120	160
(Normal	<110	<140)
Urine glucose	–ve	–ve
Urine ketone bodies	–ve	–ve

Diagnosis and discussion

The woman had marginally elevated fasting blood glucose and ***impaired oral glucose tolerance test***. Although she cannot be labelled as a diabetic patient at present, she requires careful periodical check up.

CASE STUDY 53

A 13-year old boy had the complaints of increased frequency of urination, and urinary tract infection. On routine examination, his urine was found to contain glucose and ketone bodies. He had a random blood glucose concentration 190 mg/dl. His laboratory data on the oral glucose tolerance test (mini GTT) are given below.

	Fasting	*2 hr*
Blood glucose (mg/dl)	150	240
(Normal	< 110	< 140)
Urine glucose	–ve	++
Ketone bodies	+	++

Diagnosis and discussion

The laboratory data of the boy-increase in blood glucose coupled with urinary excretion of glucose and ketone bodies, along with the results obtained in the OGTT point out that this is a good example of ***juvenile onset diabetes mellitus*** (type I diabetes or insulin dependent diabetes mellitus, IDDM). IDDM, which mainly occurs in childhood, is characterized by almost total deficiency of insulin due to the destruction of β-cells of pancreas. The boy has to be treated by insulin administration.

CASE STUDY 54

A 45-year old woman visited her physician with complaints of increased appetite and thirst with high frequency of urination. She complained of generalized weakness, fatigue and sweating. Her fasting blood glucose level was 160 mg/dl, with no presence of glucose or ketone bodies in urine. Her laboratory findings on the oral glucose tolerance test (mini GTT) are as follows.

	Fasting	*2 hr*
Blood glucose (mg/dl)	155	205
(Normal Reference	<110	<140)
Urine glucose	–ve	++
Ketone bodies	–ve	–ve

Diagnosis and discussion

The woman's fasting (elevated) blood glucose and the data obtained on OGTT indicate that she was suffering from ***non-insulin dependent diabetes mellitus*** (NIDDM or type II diabetes). NIDDM mainly occurs in adults (>30 yrs) with a strong genetic predisposition. The patients of NIDDM can be usually treated with oral antidiabetic drugs. Before prescribing the drugs, diet control and exercise are tried.

CASE STUDY 55

A healthy 25-year old pregnant woman (in 5th month of gestation), on routine examination, was found to contain glucose in her urine. Her random blood glucose was 150 mg/dl. She recalled that before pregnancy, her random blood glucose was <110 mg/dl and never she had glucose excreted into urine. Her mother was a diabetic patient. During her next visit a month later, her fasting and post-prandial blood glucose were respectively 130 mg/dl and 180 mg/dl, with traces of glucose in urine.

Diagnosis and discussion

The young woman with a family history of diabetes was found to be hyperglycemic during pregnancy. This is a case of ***gestational diabetes*** and needs treatment with insulin during pregnancy. This woman is likely to be a victim of diabetes in later years. Therefore, she has to be carefully monitored.

CASE STUDY 56

A 60-year old diabetic woman, who had been on insulin for the past 5 years, visited her physician for a routine check up. She informed that she was scrupulously following the dietary advise and had never missed her insulin injection. The following are the laboratory data of this woman.

Parameter	*Subject*	*Reference range*
Blood glucose, fasting	100 mg/dl	70-100 mg/dl
Blood glucose, postprandial	140 mg/dl	130-160 mg/dl
Blood urea	25 mg/dl	15-40 mg/dl
HbA_{Ic} (% of total Hb)	9 %	3-5 %

Urine did not contain glucose or ketone bodies.

On seeing her laboratory data, the physician stated that the patient was not taking insulin injection properly. The worried woman revealed that it was true that she did not take insulin injection for 10-15 days (as she went to her sister's place), but that was nearly one month back. She had been regular thereafter.

Diagnosis and discussion

The woman's normal blood glucose concentration (fasting and post prandial) indicate that she was under good diabetic control at the time of measurement. But ***elevated*** $\boldsymbol{HbA_{Ic}}$ (glycated hemoglobin) reflects ***poor diabetic control*** 6-8 weeks ago. This is frequently seen in diabetics who are irregular in the treatment regime. It may be noted that the concentration of HbA_{Ic} serves as an indication of blood glucose level over a period equivalent to the half-life of RBC (hemoglobin) i.e., 6-8 weeks.

CASE STUDY 57

A 60-year old man, a patient of diabetes for 20 years had been on insulin for the past 5 years. His laboratory data are given below.

Parameter	*Subject*	*Reference range*
Blood glucose, fasting	110 mg/dl	70-100 mg/dl
Blood glucose, post prandial	140 mg/dl	130-150 mg/dl
Blood urea	35 mg/dl	15-40 mg/dl
Serum creatinine	1.2 mg/dl	0.5-1.5 mg/dl
HbA_{Ic} (% Hb)	5 %	3-5 %
Urine albumin	250 mg/day	<30 mg/day
Urine glucose	–ve	–ve
Ketone bodies	–ve	–ve
Proteins (heat coagulation test)	–ve	–ve

Diagnosis and discussion

The laboratory data indicate that the patient was under control for diabetes. However, he was excreting slightly increased albumin in urine. This is a case of ***microalbuminuria***, which represents an intermediary stage between normal albumin excretion (< 30 mg/day) and macroalbuminuria (> 300 mg/day). The small increase in albumin excretion predicts the onset of impairment in renal function in diabetic patients. Thus, microalbuminuria serves as a signal of early reversible renal damage (diabetic nephropathy).

CASE STUDY 58

A 65-year old man had been on treatment for his diabetic control for two decades. Initially, he was on oral antidiabetic drugs and for the past 7 years he had been on insulin. This diabetic patient had not been strictly following the instructions of his physician as regards his treatment schedule and dietary regime. He first developed microalbuminuria and now full fledged proteinuria. The following are his laboratory findings.

Parameter	*Subject*	*Reference range*
Blood glucose, fasting	220 mg/dl	70-110 mg/dl
Blood urea	90 mg/dl	15-40 mg/dl
Serum creatinine	4 mg/dl	1-1.5 mg/dl
Creatinine clearance	20 ml/min	120-145 ml/min

Diagnosis and discussion

The diabetic patient was not on the proper treatment. Consequently, he had renal damage as is evident from the increased serum creatinine and drastically reduced creatinine clearance data, besides increase in blood urea level. This is a case of ***diabetic nephropathy***. The patient may require dialysis/renal replacement.

CASE STUDY 59

A 55-year old woman, who had menopause at the age of 48, fell down in bathroom and fractured her leg. She informed the orthopaedician that she had been experiencing severe back pain for the past 2 years. She was not on hormonal replacement therapy (HRT). Radiographic studies revealed that her bone density was decreased. Her laboratory data are given below.

Parameter (serum)	*Subject*	*Reference range*
Calcium	9 mg/dl	9-11 mg/dl
Phosphate	3.5 mg/dl	3-4.5 mg/dl
Alkaline phosphatase	70 KAU/dl	3-13 KAU/dl

Diagnosis and discussion

The post-menopausal woman was suffering from ***osteoporosis***, a common bone disease in the elderly (particularly woman). Osteoporosis is characterized by demineralization of bones resulting in progressive loss of bone mass (hence decreased bone density). Usually, the biochemical parameters in osteoporosis patients are in the normal range. But serum alkaline phosphatase may get elevated following a fracture. The etiology of osteoporosis is largely unknown. However, in post-menopausal woman, deficiency of sex hormones are implicated in the development of osteoporosis.

CASE STUDY 60

A 28-year old man complained constant headache and visual disturbance. He also had the symptoms of anxiety, tremors and was found to be hypertensive. His laboratory data are given below.

Parameter	*Subject*	*Reference range*
Serum epinephrine	10 ng/ml	0.1 ng/ml
Serum norepinephrine	26 μg/l	0.5 μg/l
Urine vanillylmandelic acid (VMA)	62 mg/day	<8 mg/day

Diagnosis and discussion

Elevation in the serum levels of catecholamines (epinephrine and norepinephrine) along with an increased urinary excretion of their principal metabolite - VMA, supported by the clinical manifestations (anxiety, hypertension) indicates that this man was suffering from ***pheochromocytoma***, a tumor of adrenal medulla.

CASE STUDY 61

A 25-year old woman visited her gynecologist as soon as she missed the menstrual cycle. The presence of human chorionic gonadotropin (hCG) hormone could be detected in her urine.

Diagnosis and discussion

Human chorionic gonadotropin is synthesized by the placenta. The levels of hCG in plasma and urine increase almost immediately after the implantation of fertilized ovum. Therefore, detection of hCG in urine is used for the ***detection of pregnancy***. Thus, this woman was pregnant, as confired by the hormonal test.

V. Miscellaneous

CASE STUDY 62

A 55-year old man was brought to the hospital in a confused and semiconscious state. He had low BP and feeble pulse. His breath had fruity odour. The data of his laboratory investigations are given below.

Parameter	Subject	Reference range
Blood pH	7.1	7.35-7.45
Plasma bicarbonate (HCO_3^-)	12 mmol/l	22-26 mmol/l
Plasma carbonic acid (H_2CO_3)	1.2 mmol/l	1.2 mmol/l
Blood glucose (random)	580 mg/dl	<140 mg/dl
Blood urea	40 mg/dl	15-40 mg/dl
Serum creatinine	1.5 mg/dl	0.5-1.5 mg/dl
Urine sugar	4+	negative
Urine ketone bodies	3+	negative

Diagnosis and discussion

The blood pH was reduced due to a decrease in plasma bicarbonate (HCO_3^-) concentration. This is case of ***metabolic acidosis***. The blood glucose level was highly increased along with urinary excretion of glucose and ketone bodies. The breath of the subject was fruity due to exhalation of acetone. Obviously, this diabetic patient was not under control. The acidosis was due to the accumulation of ketone bodies. This may be appropriately regarded as ***diabetic ketoacidosis***.

CASE STUDY 63

One patient with severe vomiting and hyokalemia was admitted in an intensive care unit. His laboratory data and acid-base status are given below.

Parameter	Subject	Reference range
Blood pH	7.7	7.35-7.45
Plasma bicarbonate (HCO_3^-)	42 mmol/l	22-26 mmol/l
Plasma carbonic acid (H_2CO_3) (pCO_2 40 mm Hg)	1.2 mmol/l	1.2 mmol/l

Diagnosis and discussion

This patient with increased blood pH due to an elevation in the plasma bicarbonate concentration is a typical case of ***metabolic alkalosis***.

CASE STUDY 64

A woman had severe pneumonia, and was on treatment. The following are the laboratory data on the acid-base status of this woman admitted in a nursing home.

Parameter	Subject	Reference range
Blood pH	7.1	7.4
Plasma bicarbonate (HCO_3^-)	24 mmol/l	24-30 mmol/l
Plasma carbonic acid (H_2CO_3) (pCO_2 80 mm Hg)	2.4 mmol/l	1.2 mmol/l

Diagnosis and discussion

The blood pH is lower due to an increased level of carbonic acid, hence this is a case of ***respiratory acidosis***.

CASE STUDY 65

An adult man with a history of anxiety and severe hyperventilation was admitted in an intensive care unit. The data on the acid-base status of this man are given below.

Parameter	Subject	Reference range
Blood pH	7.7	7.35-7.45
Plasma bicarbonate (HCO_3^-)	24 mmol/l	22-26 mmol/l
Plasma carbonic acid (H_2CO_3) (pCO_2 20 mm Hg)	0.6 mmol/l	1.2 mmol/l

Diagnosis and discussion

The blood pH of this man is increased due to a reduction in the concentration of carbonic acid. This is a good example of ***respiratory alkalosis***.

CASE STUDY 66

A 68 year old woman had complaints of bloating, abdominal discomfort, post-menopausal vaginal bleeding, loss of appetite and fatigue. Her pelvic examination and ultrasound indicated ovarian mass. Her laboratory data are given below.

Parameter	Subject	Reference range
CA-125	280 U/ml	0-35 U/ml

Diagnosis and discussion

The old woman with an elevation in serum CA-125 levels, and pelvic mass appears to be a victim of ***ovarian cancer***. This has to be however, confirmed by laparotomy and histopathology.

CASE STUDY 67

A 60-year old woman complained of back pain, non-responsive to drugs. The physical examination and enquiries revealed that a small marble-sized mass in the neck of the patient, with an existence of 15 years, had started increasing in size during the recent 4 months. However, there was no pain associated with this mass. But she had recurrent infections. Serum electrophoresis showed an abnormal protein pattern with a distinct band of monoclonal immunoglobin between β- and γ-globulins. The urine gave a positive test for Bence Jones proteins.

Diagnosis and discussion

The characteristic electrophoretic pattern of serum proteins with an 'M' band along with the excretion of Bence Jones proteins in urine and the clinical manifestations led to the diagnosis that the patient was suffering from ***multiple myeloma***. This is due to the malignancy of a single clone of plasma cells in the bone marrow resulting in an abnormal production of immunoglobulins, mostly IgG (75%) and in some cases IgA (25%). The patients of multiple myeloma have depressed immunity, hence recurrent infections are common in them.

CASE STUDY 68

A young boy aged 6 years was brought to a pediatric clinic. His mother was worried because of the muscular weakness in his legs. He was walking awkwardly, fell down frequently and could not climb stairs. A slight increase in the size of the calf muscle was noted by the pediatrician. The boy had an highly elevated serum creatine phosphokinase (550 IU/l against normal 50 IU/l). By employing Western blot technique, it was found that the muscle totally lacked the protein dystrophin. Muscle biopsy indicated some degree of necrosis.

Diagnosis and discussion

Elevated serum creatine phosphokinase and the total absence of dystrophin in muscle strongly indicate that the boy was a victim of ***Duchenne muscular dystrophy***. Dystropin is a high molecular weight (400 KDA) cytoskeletal protein that is essential for the proper muscle contraction and relaxation. Its absence causes progressive muscle weakness. Serum creatine phosphokinase is elevated, as it is released into circulation from the damaged muscle.

CASE STUDY 69

A 55-year old fair complexioned man had the complaints of weakness, fatigue and joint pains. He observed that his skin was becoming darker during the past one year. He had no history of exposure to toxic fumes or metals. His laboratory data are given below.

Parameter	Subject	Reference range
Hemoglobin	15 g/dl	14-16 g/dl
Hematocrit	55 %	40-50 %
Serum iron	330 μg/dl	50-150 μg/dl
Serum ferritin	400 μg/dl	3-30 μg/dl
Transferrin saturation with iron	85 %	25-50 %

Diagnosis and discussion

The subject had increased iron and ferritin in the body, the characteristic features of the disorder ***hemochromatosis***. This disorder is due to increased iron absorption from the intestine as a result of iron overload or repeated blood transfusions. The iron containing compounds, particularly hemosiderin gets deposited and damages the tissues — complications such as cirrhosis of liver, diabetes mellitus and cardiac problems are seen in advanced cases.

CASE STUDY 70

A 4-year old male child came along with his mother to his grand parents place. He accompanied his grand-father to the latter's work place, a coal factory. Despite the instructions of his grand-father, the child moved freely in the factory and inhaled smoke fumes. He developed lethargy and cyanosis. The spectrophotometric observations, of the blood showed absorption maxima at 500 nm and 631 nm.

Diagnosis and discussion

The child was exposed to industrial chemicals and the spectrophotometric analysis revealed the presence of methemoglobin (metHb). Here, the child suffered from ***toxic (acquired) methemoglobinemia***. Hemoglobin (Fe^{2+}) can be oxidized to methemoglobin (Fe^{3+}) by H_2O_2, free radicals, drugs, industrial chemicals and smoke (aniline, nitrobenzene etc.). The methemoglobin cannot bind to O_2 and therefore O_2 supply to the tissues is impaired, hence the symptoms of lethargy and cyanosis. In the normal circumstances, the metHb concentration in blood is <1% of total hemoglobin and this gets reduced to hemoglobin by methemoglobin reductase.

CASE STUDY 71

A normal healthy man aged 55 years had a routine check up by his family physician. His biochemical laboratory data are given below.

Parameter	Subject	Reference range
Blood glucose (fasting)	80 mg/dl	70-100 mg/dl
Blood urea	30 mg/dl	15-40 mg/dl
Serum creatinine	1.2 mg/dl	0.5-1.5 mg/dl
Serum cholesterol, total	320 mg/dl	150-200 mg/dl
HDL fraction	30 mg/dl	40-60 mg/dl
LDL fraction	230 mg/dl	<100 mg/dl
VLDL fraction	60 mg/dl	20-40 mg/dl
Serum triglycerides	180 mg/dl	75-150 mg/dl
Serum homocysteine	35 μmol/l	4-15 μmol/l

Diagnosis and discussion

This subject had an elevated serum total cholesterol (both LDL and VLDL fractions raised) with a marginally increased triglycerides. In recent years, elevation in serum homocysteine level has been implicated in an increased risk of coronary heart diseases. The data on the lipid profile along with increased levels of ***homocysteine*** strongly suggest that this man is at an ***increased risk for coronary heart diseases***.

CASE STUDY 72

A 6-year old boy had high temperature with sweats during sleep, marked tachycardia and loss of weight. He also had the complaints of headache, neck stiffness and vomiting. The liver was found to be slightly enlarged. The following are the laboratory findings of this boy.

Parameter	Subject	Reference range
Cerebrospinal fluid (CSF)		
Glucose	30 mg/dl	50-75 mg/dl
Proteins	100 mg/dl	15-40 mg/dl
Chloride	105 mEq/l	120-130 mEq/l
Blood glucose (simultaneously collected)	100 mg/dl	70-100 mg/dl

Diagnosis and discussion

A decrease in the CSF glucose and chloride concentrations along with an increase in protein level indicates that the boy was a victim of ***tuberculous meningitis***. This is supported by the clinical manifestations.

CASE STUDY 73

A 30-year old man was found to excrete glucose in urine in a random sample. He did not have any symptoms of diabetes mellitus. His fasting and post-prandial blood glucose respectively were 90 mg/dl and 140 mg/dl. Urine contained glucose but there were no ketone bodies. Oral glucose tolerance test (OGTT) also revealed that the man had a normal glucose tolerance except the presence of glucose in urine at 1 hr.

Diagnosis and discussion

This is a case of ***renal glycosuria***. It is a benign condition due to reduced renal threshold for blood glucose (normal around 180 mg/dl). It may be noted that renal glycosuria is unrelated to diabetes mellitus and therefore not accompanied by the classical symptoms of diabetes. This disorder does not require any specific treatment.

CASE STUDY 74

A 13-year old boy was constantly found to excrete glucose (by glucostrip method) in urine after meals, although no glucose was detected in the fasting state. A village doctor diagnosed the boy as a case of type I diabetes and prescribed 5 units of insulin administration. Following insulin injection, the boy had headache, sweating with anxiety, confusion and seizures. He became unconscious and was immediately shifted to the nearby town hospital. He was given intravenous glucose and the above symptoms disappeared. When he became normal, his fasting and post-prandial blood glucose estimations were done and they were respectively found to be 90 mg/dl and 140 mg/dl.

Diagnosis and discussion

This boy, a case of ***renal glycosuria, mistaken as diabetes mellitus***, without proper investigation (i.e., blood glucose, GTT), was given insulin. He developed hypoglycemia symptoms which were corrected on glucose administration. For more details on renal glycosuria, refer case no. 73.

CASE STUDY 75

A 35-year old man, with no family history of diabetes was found to excrete glucose in urine following a meal. His fasting glucose was found to be within normal limits-95 mg/dl. The following are the results of oral glucose tolerance test (OGTT).

	Fasting	1hr	2 hr
Blood glucose (mg/dl)	95	190	135
Urine glucose	–ve	+	–ve

Diagnosis and discussion

The subject had a transitory rise in blood glucose, following a meal. And this concentration exceeds renal threshold for glucose, it gets excreted in urine. This is a case of ***alimentary (lag storage) glycosuria***. This may occur in some normal people due to increased rate of intestinal absorption of glucose into blood stream.

CASE STUDY 76

A 35-year old man, being an alcoholic, regularly consumed whisky in the evenings. One day around 11 AM, this man had palpitations, severe sweating, confusion and fainted. His wife, fearing that her husband suffered heart attack, immediately rushed him to the nearby cardiac centre. All the tests indicative of myocardial infarction were negative in the subject. On enquiry, his wife informed the physician that he had not taken breakfast. The following are the laboratory data of the man.

Parameter	Subject	Reference range
Blood glucose	45 mg/dl	70-100 mg/dl
Serum creatine phosphokinase	120 U/l	50-200 U/l
Serum alanine transaminase	70 U/l	10-45 U/l
Serum aspartate transaminase	65 U/l	10-50 U/l
Serum lactate	15 mg/dl	0.3-7 mg/dl

The man was given intravenous glucose and responded immediately, and gained consciousness.

Diagnosis and discussion

This is a case of ***hypoglycemia due to alcohol consumption***. Since the man had not taken breakfast, his blood glucose concentration fell. It may be noted that the gluconeogenesis is reduced due to alcohol intake. This is so since more of NADH is produced in alcohol metabolism which converts pyruvate to lactate. This limits pyruvate conversion to glucose, a major pathway of gluconeogenesis.

CASE STUDY 77

A 30-year old man had fever, diarrhea, weight loss (despite normal diet intake). He had multiple opportunistic infections, generalized lymphadeopathy, besides neurological complications. He had 150 CD_4T-lymphocytes/μl blood (normal >800/ μl). Enzyme-linked immunosorbant assay (ELISA) was positive for the antibodies of human immuno-deficiency virus (HIV). The presence of HIV antibodies was confirmed by Western blot technique. The existence of HIV genome in the peripheral blood lymphocytes was established by polymerase chain reaction (PCR).

Diagnosis and discussion

The presence of HIV antibodies and the reduced concentration of CD_4T-lymphocytes in the circulation clearly indicate that this is a case of ***AIDS (acquired immuodeficiency syndrome)***. This is fully supported by the clinical manifestations. Due to deficiency of the immune system, the body of AIDS patient is highly susceptible to all sorts of infections (viral, bacterial, fungal). This is in contrast to healthy individuals.

Index

A

B

C

D

E

F

G

H

I

M

N

O

P

Q

R

S

T

U

V

W

X

Z